In particular, the square of the length of a vector $\mathbf{u}$ is $\|\mathbf{u}\|^2 = \mathbf{u} \cdot \mathbf{u} = \mathbf{u}^T\mathbf{u}$. Also, if is a *unit vector* (that is, $\|\mathbf{u}\| = 1$), then the $n \times n$ matrix $Y = \mathbf{u}\mathbf{u}^T$ satisfies $Y^2 = Y$ because

$$Y^2 = (\mathbf{u}\mathbf{u}^T)(\mathbf{u}\mathbf{u}^T) = \mathbf{u}(\mathbf{u}^T\mathbf{u})\mathbf{u}^T = \mathbf{u}\mathbf{u}^T = Y$$

since $\mathbf{u}^T\mathbf{u} = \mathbf{u} \cdot \mathbf{u} = \|\mathbf{u}\|^2 = 1$. **p. 75**

4. When faced with the product AB of matrices, it is often useful to view B as a series of columns. If $B = \begin{bmatrix} \mathbf{b}_1 & \mathbf{b}_2 & \cdots & \mathbf{b}_p \\ \downarrow & \downarrow & & \downarrow \end{bmatrix}$ has columns $\mathbf{b}_1, \mathbf{b}_2, \ldots, \mathbf{b}_p$, the product AB is the matrix

$$AB = A\begin{bmatrix} \mathbf{b}_1 & \mathbf{b}_2 & \cdots & \mathbf{b}_p \\ \downarrow & \downarrow & & \downarrow \end{bmatrix} = \begin{bmatrix} A\mathbf{b}_1 & A\mathbf{b}_2 & \cdots & A\mathbf{b}_p \\ \downarrow & \downarrow & & \downarrow \end{bmatrix}$$

whose columns are $A\mathbf{b}_1, A\mathbf{b}_2, \ldots, A\mathbf{b}_p$. Suppose $A = \begin{bmatrix} 1 & -2 & 0 \\ -2 & 0 & 2 \\ 0 & 2 & -1 \end{bmatrix}$, $\mathbf{x} = \begin{bmatrix} 2 \\ 1 \\ 2 \end{bmatrix}$, $\mathbf{y} = \begin{bmatrix} 1 \\ 1 \\ 1 \end{bmatrix}$, and $\mathbf{z} = \begin{bmatrix} -2 \\ 2 \\ 1 \end{bmatrix}$. You should check that $A\mathbf{x} = \mathbf{0}$, $A\mathbf{y} = \begin{bmatrix} -1 \\ 0 \\ 1 \end{bmatrix}$, and $A\mathbf{z} = 3\mathbf{z}$. This means, that if $B = \begin{bmatrix} \mathbf{x} & \mathbf{y} & \mathbf{z} \\ \downarrow & \downarrow & \downarrow \end{bmatrix} = \begin{bmatrix} 2 & -1 & -2 \\ 1 & 1 & 2 \\ 2 & 1 & 1 \end{bmatrix}$, then $AB = $

$$\begin{bmatrix} A\mathbf{x} & A\mathbf{y} & A\mathbf{z} \\ \downarrow & \downarrow & \downarrow \end{bmatrix} = \begin{bmatrix} 0 & -1 & -6 \\ 0 & 0 & 6 \\ 0 & 1 & 3 \end{bmatrix}.$$ **p. 77**

5. The product $A\mathbf{x}$ of an $m \times n$ matrix A and a vector $\mathbf{x}$ in $\mathbf{R}^n$ is a linear combination of the columns of A with coefficients the components of the vector. For example,

$$\begin{bmatrix} 1 & 4 & 7 \\ 2 & 5 & 8 \\ 3 & 6 & 9 \end{bmatrix}\begin{bmatrix} x_1 \\ x_2 \\ x_3 \end{bmatrix} = x_1\begin{bmatrix} 1 \\ 2 \\ 3 \end{bmatrix} + x_2\begin{bmatrix} 4 \\ 5 \\ 6 \end{bmatrix} + x_3\begin{bmatrix} 7 \\ 8 \\ 9 \end{bmatrix}.$$

So, if $\mathbf{b}$ is a vector in $\mathbf{R}^m$, the linear system $A\mathbf{x} = \mathbf{b}$ has a solution if and only if $\mathbf{b}$ is a linear combination of the columns of A—that is, if and only if $\mathbf{b}$ is in the column space of A. **p. 82**

6. The inverse of the product of invertible matrices A and B is the product of the inverses **order reversed**: $(AB)^{-1} = B^{-1}A^{-1}$. Similarly, the transpose of the product of A and B is the product of the transposes **order reversed**: $(AB)^T = B^TA^T$. **pp. 91 *and* 93**

Linear Algebra

Linear Algebra

A Pure & Applied First Course

Edgar G. Goodaire

Memorial University of Newfoundland

Pearson Education, Inc.
Upper Saddle River, NJ 07458

Library of Congress Cataloging-in-Publication Data

Goodaire, Edgar G.
    Linear algebra : a pure and applied first course / Edgar G. Goodaire.
      p. cm.
    Includes index.
    ISBN 0-13-047017-1
    1. Algebras, Linear. I. Title.

    QA184.2.G66 2002
    $512'.5$—dc21   2002044551

Acquisitions Editor: *George Lobell*
Editor in Chief: *Sally Yagan*
Vice-President/Director of Production and Manufacturing: *David W. Riccardi*
Executive Managing Editor: *Kathleen Schiaparelli*
Senior Managing Editor: *Linda Mihatov Behrens*
Production Editor: *Joan Wolk*
Manufacturing Buyer: *Lynda Castillo*
Manufacturing Manager: *Trudy Pisciotti*
Marketing Manager: *Halee Dinsey*
Marketing Assistant: *Rachel Beckman*
Director of Creative Services: *Paul Belfanti*
Art Editor: *Thomas Benfatti*
Creative Director: *Carole Anson*
Art Director: *Maureen Eide*
Interior/Cover Designer: *Joseph Sengotta*
Editorial Assistant: *Jennifer Brady*
Cover Image: A FINE ROMANCE @ 1996 by Michael James, $59'' \times 57.5''$w. Quilt made
of machine-pieced and machine-quilted cotton and silk. Photo credit: *David Caras*
Art Studio: *Laserwords Private Limited*

© 2003 Pearson Education, Inc.
Pearson Education, Inc.
Upper Saddle River, New Jersey 07458

Printed in the United States of America
10 9 8 7 6 5 4 3 2 1

ISBN 0-13-047017-1

Pearson Education LTD., *London*
Pearson Education Australia PTY, Limited, *Sydney*
Pearson Education Singapore, Pte. Ltd
Pearson Education North Asia Ltd., *Hong Kong*
Pearson Education Canada, Ltd., *Toronto*
Pearson Educación de Mexico, S.A. de C.V.
Pearson Education–Japan, *Tokyo*
Pearson Education Malaysia, Pte. Ltd.

To the girl who
cut and pasted figures
with me in the days
when ''cut and paste''
meant just that

To Linda, with heartfelt thanks

Contents

Preface

Why yet another book in linear algebra? At many universities, calculus, the usual prerequisite for linear algebra, has become a "course without proofs," with minimal reading requirements. In fact, students who are "successful" in calculus have developed the wonderful ability to match examples to homework with almost no reading. So, if one wishes to keep the algebra (vector spaces, linear transformations) in **linear algebra**, students need to be taught not only how to handle proofs, but more fundamentally they need to be taught how to read mathematics. My goal in this book is to use an intrinsically interesting subject to show students that there is more to mathematics than numbers.

The presentation in this text is frequently interrupted by "Reading Challenges," little questions a student can use to see if she really does understand the point just made. These Reading Challenges promote active reading. The goal is to get students reading actively by questioning the material as they read. Every section in this book concludes with answers to these Reading Challenges followed by a number of "True/False" questions that test definitions and concepts, not dexterity with numbers or buttons. Every exercise set concludes with "Critical Reading" exercises, which include harder problems, yes, but also problems that are easily solved if understood. Before the review exercises at the end of a chapter, there is a list of key words and concepts that have arisen in the chapter that I hope students will use as a mental check that they have understood the chapter.

There are several reasons for the "pure & applied" subtitle. While a pure mathematician myself, I have lost sympathy with the dry abstract vector space/linear transformation approach of many books that requires maturity which the typical sophomore does not have. My approach to linear algebra is through matrices. Essentially all vector spaces are (subspaces of) Euclidean n-space. Other standard examples, like polynomials and matrices, are merely mentioned. In my opinion, when students are thoroughly comfortable with $\mathbf{R}^n$, they have little trouble transferring concepts to general vector spaces. The emphasis on matrices and matrix factorizations, the recurring theme of projections starting in Chapter 1, and the idea of the "best (least squares) solution" to $A\mathbf{x} = \mathbf{b}$ around which Chapter 6 revolves, make this book more applied than most of its competition.

However, it does not take most readers long to discover the rigor and maturity that is required to move through the pages of this text. Whether a mathematical argument is always labeled "proof," virtually no statements are made without justification and a large number of exercises are of the "show or prove" variety. Since students tend to find such exercises difficult, I have included an appendix entitled "Show and Prove" designed to teach students how to write a mathematical proof and I tempt them actually to read this section by promising that they will find there the solutions to a number of exercises in the text itself! So despite its applied flavor, I believe that a linear algebra course that uses this book would serve well as the "introduction to proof" transition course that is now a part of many degree programs.

■ Organization, Philosophy, Style

It is somewhat atypical to begin with, even to include in a linear algebra course, a chapter on vector geometry of the plane and 3-space. My purpose is twofold. Such an introduction allows the immediate introduction of the terminology of linear algebra—linear combination, span (the noun and the verb), linear dependence and independence—in concrete settings that the student can readily understand. Furthermore, the student is quickly alerted to the fact that this course is not about manipulating numbers, that serious thinking and critical reading will be required to succeed. Many exercise sets contain writing exercises, marked as such by the symbol ✍, that ask for explanations, and other questions that can be answered very briefly, **if understood**.

The inclusion of vector geometry does not come at the expense of other more standard topics. In fact, some nonstandard topics like the pseudoinverse and singular value decomposition are even included. How has this efficiency been achieved?

While a great fan of technology myself, and after years of experimenting and delivering courses entirely via MAPLE worksheets, I now believe that technology can quickly become more distracting than useful for novice linear algebra students. The use of technology is certainly time-consuming and it gives students more to learn in addition to the subject itself. Of course, if you wish to integrate technology with this book, you can certainly do so. Prentice Hall can make available with this text various MATLAB and MAPLE technology manuals at relatively little cost. For more details, see the section on "Supplements".

Applications of linear algebra appear in the final chapter, leaving the instructor free to import a favorite application after earlier sections wherever appropriate or desired. (I think this is preferable to optional—and oft omitted—sections on applications interspersed throughout the text.) Again, it is my experience that there is not enough time in a short semester to do justice to the fundamental concepts and ideas of linear algebra, while at the same time exposing students to a wide variety of applications. Interestingly, linear algebra is the one course I teach nowadays in which I am not regularly asked "What is this used for?" As a matter of fact, I am often approached by students who want to tell me how linear algebra is making appearances in their other courses, from computer science to geology.

Students usually find the start of their linear algebra course easy with its typical heavy emphasis on matrix manipulation and linear systems, while comfort levels take a deep dive the day vector spaces are introduced. Sometimes, within a couple of lectures, terms like "linear combination," "span," "spanning set," "linear independence," "basis," and "dimension" are introduced and sophisticated theorems discussed, before the definitions have really settled in students' minds. Even the word "trivial," as in "trivial linear combination," is not so trivial for some!

To ease the transition from linear systems to vector spaces, I have moved a lot of the terminology of vector spaces to Chapter 1. Readers encounter and see in use the language of linear algebra throughout the first three chapters, before they meet the formal definition of "vector space." They see lots of examples that illustrate concepts usually not introduced until the typical "vector space" chapter. The idea of a plane being "spanned" by two vectors is not hard for beginning students; neither is the idea of "linear combination" or the fact that linear dependence of three or more vectors (in $\mathbf{R}^3$) means that the vectors all lie in a plane. That $A\mathbf{x}$ is a linear combination of the columns of A, surely one of the most useful ideas of linear algebra, is introduced early in Chapter 2 and used time and time again. Long before we introduce vector spaces, all the terminology and techniques of proof have been at work for a long time.

I endeavor to introduce new concepts only when they are needed, not just for their own sake. We solve linear systems by using Gaussian elimination in order to achieve row echelon form, postponing the definition of **reduced** row echelon form to where it is of benefit, in the calculation of the inverse of a matrix. The notion of the "transpose" of a matrix is introduced in the section on matrix multiplication because it is important for students to know that the **dot** product of two column vectors $\mathbf{u}$ and $\mathbf{v}$ is the **matrix** product $\mathbf{u}^T\mathbf{v}$. Symmetric matrices, whose LDU factorizations are so easy, appear for the first time in the section on LDU, and reappear as part of the characterization of a projection matrix.

There can be few aspects of linear algebra more useful, practical, and interesting than eigenvectors, eigenvalues, and diagonalizability. Moreover, these topics provide an excellent opportunity to discuss linear independence in a nonthreatening manner. Why do these ideas appear so late in so many books?

My writing style may be less formal than that of other authors, but my students quickly discover that informality is not synonymous with lack of rigor or demand for understanding. There is far less emphasis on computation and far more on mathematical reasoning in this book. I repeatedly ask students to explain "why." Already in Chapter 1, students are asked to show that two vectors are linearly dependent if and only if one is a scalar multiple of another. The concept of matrix inverse appears early, as well as its utility in solving matrix equations, long before we discuss how actually to find the inverse of a matrix (a skill arguably not as important as it once was). The fact that students find the early sections of this book quite difficult is evidence to my mind that I have succeeded in emphasizing the importance of asking "Why," discovering "Why," and then clearly communicating the reason "Why."

■ A Course Outline

A few comments about what I actually include in courses based on this book may be helpful. To achieve economy within the first three chapters, I omit Section 2.6 on LDU factorization (never the section on LU) and discuss only a few of the properties of determinants (Section 3.2), most of which are used, primarily, to assist in finding determinants, a task few people do by hand any more.

There is more material in Chapters 4 to 7 than I can ever manage. Thus I often discuss the matrix of a linear transformation only with respect to standard bases, omitting Sections 5.3 and 5.4. The material of Section 6.1, which is centered around the best least squares solution to overdetermined linear systems, is nontraditional, but try to resist the temptation to omit it. Many exercises on this topic have numerical answers (which students like!), there are lots of calculations, but lots of theoretical ideas are reinforced here too. For example, the fact that the formula $P = A(A^T A)^{-1} A^T$ works only for matrices A with linearly independent columns provides another opportunity to talk about linear independence.

If I insist on doing proper justice to Chapter 7—especially the unitary diagonalization of Hermitian matrices after some preliminary treatment of complex numbers and matrices—I have to cut Sections 4.4 and 6.4 on one-sided inverses and the pseudoinverse that are important, that students like, where a lot of ideas dealing with the "best solution" to $A\mathbf{x} = \mathbf{b}$ seem to converge, and where there are many opportunities to ask short simple theoretical questions. So, when I get to Chapter 7, I usually bypass the first three sections and head directly to the orthogonal diagonalization of real symmetric matrices in Section 7.4, after a brief review of the concepts of eigenvalues and eigenvectors.

■ Supplements

There is a solutions manual for this book that contains complete solutions to every exercise. There is no student's manual because we provide complete solutions, not just answers, to numerous exercises (marked BB) in the back of the book. In addition, there are a number of supplements that allow you to incorporate various kinds of technology with your course. All these are available very inexpensively if they are ordered shrink-wrapped with this textbook.

- *Visualizing Linear Algebra with MAPLE*, S. Keith (ISBN 0-13-041816-1)
  This manual offers syntax and projects. There are associated files that give strong geometric aspects to linear algebra.

- *Understanding Linear Algebra using MATLAB*, 1e, E. Kleinfeld and M. Kleinfeld (ISBN 0-13-060945-5)
  This manual contains syntax and projects. It is easy and somewhat computationally oriented.

- *Linear Algebra Labs with MATLAB*, 2e, D. Hill and D. Zitarelli (ISBN 0-13-505439-7)
  This manual gives extensive training in MATLAB, as it applied to linear algebra. Projects are included as well.

- *ATLAST Computer Exercises for Linear Algebra*, 2e, S. Leon, G. Herman, R. Faulkenberry (editors) (ISBN 0-13-101121-9)

This manual has syntax and projects that focus on the ideas of linear algebra made concrete.

Finally, **as of June 1, 2003**, there will be a website containing errata (we hope there are few), applications, and MAPLE worksheets.

```
http://www.prenhall.com/goodaire
```

■ Acknowledgments

Throughout the writing of this book, in some cases before drafts had even been tested in the classroom, I was helped and encouraged by a number of colleagues, including Peter Booth, Hermann Brunner, John Burry, George Miminis, Gerald O'Rielly, Michael Parmenter, Donald Rideout, Bruce Shawyer, and Ed Williams. My good friend Michael Parmenter, one of the best proofreaders I have ever come across, made numerous suggestions that improved this work immeasurably. The people whom Prentice Hall employed to review this book were of genuine assistance. Those whose suggestions have not yet been implemented should be assured that I continue to reflect on how best to incorporate them in the future. So I offer sincere thanks to

Anthony Bonato, Wilfrid Laurier University
Robert Boyer, Drexel University
Robert Craigen, University of Manitoba
Gather Isaak, Lehigh University
Mark Sepanski, Baylor University
Ronald Solomon, Ohio State University
Ilya Spitkovsky, College of William and Mary
Romuald Stanczak, University of Toronto
Cathleen M. Zucco Teveloff, Trinity College
Rebecca Wahl, Butler University.

This book would not have come to print without the help of many people at Prentice Hall and in other support agencies. Lynda Castillo coordinated the project, while Joan Wolk took charge of the editorial work and Thomas Benfatti the art. I acknowledge my profound respect for George Lobell, Acquisitions Editor, who, through the production of three books now, has become a friend. His enthusiasm and continual encouragement over the years have been extraordinary and so very welcome.

I hope you discover that this book provides a refreshing approach to an old familiar topic with lots of "neat ideas" that you perhaps have not noticed or fully appreciated previously. Thank you for giving this book a try. I genuinely welcome your comments.

Edgar G. Goodaire
edgar@math.mun.ca
St. John's, Newfoundland
November, 2002

To the Student

From aviation, to the design of cellular phone networks, to oil and gas exploration, to computer graphics, linear algebra is indispensable. With relatively little emphasis on sets and functions, linear algebra is "different"; most students find it enjoyable and a welcome change from calculus. Be careful though. The answers to most calculus problems are numerical and easily confirmed over the phone with a friend. The answers to many problems in this book are **not** numerical, but require explanations as to **why** things happen as they do. So let me begin this note to you with a word of caution: this is a book you **must read**.

For many students, linear algebra is the first course where many exercises ask students to explain, to answer why or how. It can be a shock to discover that there are mathematics courses (in fact, most of those above first year) where words are more important than numbers! More important, you cannot always match examples to homework. Gone are the days when the solution to a homework problem lies in finding an identical worked example in the text. Instead, homework problems are going to require some critical thinking.

When this book was still in draft form, a student came into my office one day to ask for help with a homework question. When this happens with a book of which I am an author, I am always eager to discover whether I have laid the proper groundwork in the section so that the average student could be expected to make a reasonable attempt at the exercises. From your point of view, in the ideal situation, the exercise would be very similar to a worked example. Right? In the instance I am recalling, I went through the section with my student page by page until we found such an example. In fact, we found precisely the question I had assigned, worked out as an example that I had forgotten to delete when I transferred it to the exercises! The student felt a little sheepish while I was completely shocked to be reminded, once again, that some students don't read their textbooks.

It is always tempting to start a homework problem right away, without preparing yourself first, but this approach isn't going to work very well here. You will find it imperative to read a section from start to finish before attempting the exercises at the end of a section. And please do more than just glance at the list of "key words" that appears before every exercise set. Contemplate each word there. Are you sure you know what it means? If you are not, you will not be able to answer a question where that word is used. At the back of the book, there is a glossary where every technical term is defined and where you will often find examples of things that fit the definition and things that don't. If you are not sure what is required when asked to prove something or to show that something is true, read the appendix "Show and Prove" that is also at the back. (You will find there the solutions to several exercises from the text itself!)

Have you noticed the endpapers of this text, those pages that line the front and back covers? There, under the heading of "Things I Must Remember," I have included many important ideas that my students have helped me to collect over the years. You will also find there some ideas that are often just what you need to solve homework problems. I hope you also appreciate the inclusion at the back of **complete solutions**, not simply answers, to many exercises in this text.

Many students have helped me improve this book and to make the subject easier for those that follow. In particular, I want to acknowledge the enthusiasm and assistance of Gerrard Barrington, Shauna Gammon, Ian Gillespie, Philip Johnson, and Melanie Ryan (whose most repeated advice to students is to **read Chapter 4 over and over again**).

I hope that you like my writing style, that you discover you like linear algebra, and that you soon surprise yourself with your ability to write a good clear mathematical proof. I hope that you do well in your linear algebra courses and all those other courses where linear algebra plays an important role. Let me know what you think of this book. I like receiving comments—good, bad, and ugly—from anyone.

Edgar G. Goodaire
edgar@math.mun.ca
St. John's, Newfoundland
November, 2002

Suggested Lecture Schedule

Linear Algebra

1

The Geometry of the Plane and 3-Space

1.1 Vectors

A *two-dimensional vector* is a pair of numbers, written in a column and surrounded by brackets. For example,

$$\begin{bmatrix} 1 \\ 3 \end{bmatrix}, \quad \begin{bmatrix} 2 \\ 4 \end{bmatrix}, \quad \begin{bmatrix} -2 \\ 3 \end{bmatrix}, \quad \begin{bmatrix} 0 \\ 0 \end{bmatrix}$$

are two-dimensional vectors. Different people use different notation for vectors. Some people underline, others use boldface type, and others use arrows. Thus, in various contexts, you may well see

$$\underline{v}, \quad \mathbf{v}, \quad \text{and} \quad \vec{v}$$

as notation for a vector. We prefer boldface type, the second of the above possibilities.

The *components* of the vector $\mathbf{v} = \begin{bmatrix} a \\ b \end{bmatrix}$ are the numbers a and b. By general agreement, vectors are *equal* if and only if they have the same corresponding components. Thus, if

$$\begin{bmatrix} a - 3 \\ 2b \end{bmatrix} = \begin{bmatrix} -1 \\ 6 \end{bmatrix},$$

then $a - 3 = -1$ and $2b = 6$, so that $a = 2$ and $b = 3$.

The vector $\begin{bmatrix} a \\ b \end{bmatrix}$ can be pictured by an arrow in the plane.

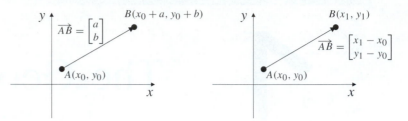

Figure 1.1 The arrow from $A(x_0, y_0)$ to $B(x_1, y_1)$ represents the vector $\begin{bmatrix} x_1 - x_0 \\ y_1 - y_0 \end{bmatrix}$.

1.1.1 Take any point $A(x_0, y_0)$ as a starting point and $B(x_0 + a, y_0 + b)$ as the terminal point (the tip of the arrowhead).

The arrow from A to B is a picture of the vector $\begin{bmatrix} a \\ b \end{bmatrix}$.

The notation $\overrightarrow{AB}$ means the vector pictured by the arrow from A to B, thus $\overrightarrow{AB} = \begin{bmatrix} a \\ b \end{bmatrix}$. The arrow is shown on the left in Figure 1.1.

Letting $x_1 = x_0 + a$ and $y_1 = y_0 + b$, so that the coordinates of B become (x_1, y_1), then $a = x_1 - x_0$ is the difference in the x-coordinates of A and B, and $b = y_1 - y_0$ is the difference in the y-coordinates. This observation makes it easy to determine the vector pictured or represented by any particular arrow.

1.1.2 The arrow from $A(x_0, y_0)$ to $B(x_1, y_1)$ represents the vector $\overrightarrow{AB} = \begin{bmatrix} x_1 - x_0 \\ y_1 - y_0 \end{bmatrix}$.

Figure 1.2 Four arrows, each one a picture of the vector $\begin{bmatrix} 1 \\ 2 \end{bmatrix}$.

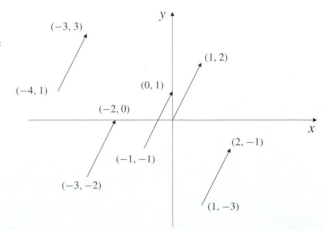

Reading Challenge 1 *If $A = (-2, 3)$ and $B = (1, 5)$, what is the vector $\overrightarrow{AB}$?*

Since an arrow can start anywhere, a given vector can be pictured by infinitely many different arrows. Each of the arrows in Figure 1.2 is an illustration of the vector $\begin{bmatrix} 1 \\ 2 \end{bmatrix}$.

Reading Challenge 2 *Suppose A and B are points in the plane and the arrow from A to B represents the vector $\overrightarrow{AB} = \begin{bmatrix} 3 \\ -1 \end{bmatrix}$. If $A = (-4, 2)$, what is B?*

In Chapter 7, scalars will be complex numbers. In general, scalars can come from any "field," just 0s and 1s, for instance.

■ Scalar Multiplication

We can multiply vectors by numbers, an operation called *scalar multiplication*. For now, "scalar" means "real number," so "scalar multiplication" means multiplication by real numbers. For example,

$$4 \begin{bmatrix} -1 \\ 3 \end{bmatrix} = \begin{bmatrix} -4 \\ 12 \end{bmatrix}; \quad -2 \begin{bmatrix} 2 \\ -3 \end{bmatrix} = \begin{bmatrix} -4 \\ 6 \end{bmatrix}; \quad 0 \begin{bmatrix} 4 \\ \sqrt{3} \end{bmatrix} = \begin{bmatrix} 0 \\ 0 \end{bmatrix};$$

$$-\begin{bmatrix} -3 \\ 8 \end{bmatrix} = (-1) \begin{bmatrix} -3 \\ 8 \end{bmatrix} = \begin{bmatrix} 3 \\ -8 \end{bmatrix}.$$

The last of these examples illustrates that $-\mathbf{v}$ means $(-1)\mathbf{v}$. Can you see the connection between an arrow for $\mathbf{v}$ and an arrow for $2\mathbf{v}$ or $-\mathbf{v}$? As shown in Figure 1.3, the vector $2\mathbf{v}$ has the same direction as $\mathbf{v}$ but is twice as long; $-\mathbf{v}$ has direction opposite to $\mathbf{v}$ but the same length; $0\mathbf{v} = \begin{bmatrix} 0 \\ 0 \end{bmatrix}$ is a vector of zero length, which we picture as a single point. This is called the *zero vector* and is denoted by a boldface $\mathbf{0}$ in books or by an underlined $\underline{0}$ in handwritten work.

As we have suggested, the vector $c\mathbf{v}$ has the same direction as $\mathbf{v}$ if $c > 0$ and the direction opposite to $\mathbf{v}$ if $c < 0$. This observation is the basis for the following definition:

1.1.3 DEFINITION Vectors $\mathbf{u}$ and $\mathbf{v}$ are *parallel* if one is a scalar multiple of the other, that is, if $\mathbf{u} = c\mathbf{v}$ or $\mathbf{v} = c\mathbf{u}$ for some scalar c.

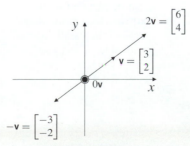

Figure 1.3

Thus parallel vectors **v** and c**v** have the *same direction* if $c > 0$ and the *opposite direction* if $c < 0$. Also, the length of c**v** is $|c|$ times the length of **v**.

■ Vector Addition

We add vectors in the obvious way, *componentwise*:

$$\begin{bmatrix} 0 \\ -3 \end{bmatrix} + \begin{bmatrix} 2 \\ 5 \end{bmatrix} = \begin{bmatrix} 2 \\ 2 \end{bmatrix}, \qquad \begin{bmatrix} 1 \\ 5 \end{bmatrix} + \begin{bmatrix} 3 \\ -2 \end{bmatrix} = \begin{bmatrix} 4 \\ 3 \end{bmatrix}, \qquad \begin{bmatrix} 2 \\ 3 \end{bmatrix} + \begin{bmatrix} 0 \\ 0 \end{bmatrix} = \begin{bmatrix} 2 \\ 3 \end{bmatrix}.$$

There is a nice connection between arrows for vectors **u** and **v** and an arrow for the sum **u** + **v**.

> **1.1.4 Parallelogram Rule:** If the arrows for **u** and **v** are drawn with the same starting point, **u** + **v** is the diagonal of the parallelogram whose sides are **u** and **v**.

We illustrate this principle in Figure 1.4 for the vectors $\mathbf{u} = \begin{bmatrix} 1 \\ 2 \end{bmatrix}$ and $\mathbf{v} = \begin{bmatrix} 3 \\ 1 \end{bmatrix}$. The sum of these vectors is

$$\mathbf{u} + \mathbf{v} = \begin{bmatrix} 1 \\ 2 \end{bmatrix} + \begin{bmatrix} 3 \\ 1 \end{bmatrix} = \begin{bmatrix} 4 \\ 3 \end{bmatrix},$$

a vector that can be pictured by the arrow $\overrightarrow{OC}$ from the origin to $C(4, 3)$. The parallelogram rule described in 1.1.4 says that $\overrightarrow{OC}$ is the diagonal of parallelogram $OACB$.

To see why this is the case, note that to move from $B(3, 1)$ to $C(4, 3)$, you move one unit to the right and two units up, so triangle BCQ is congruent to triangle OAP. In particular, angles AOP and CBQ are equal, so $\overrightarrow{BC}$ and $\overrightarrow{OA}$ are parallel and of the same length.

The vector **u** + **v** can also be pictured by the arrow $\overrightarrow{AC}$ from A to C, which is the third side of a triangle with sides **u** and **v**. So we have a second way to picture the sum of two vectors.

Figure 1.4 Vector **u** + **v** can be pictured by the arrow which is the diagonal of the parallelogram with sides **u** and **v**.

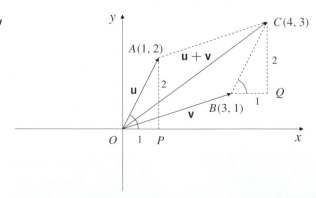

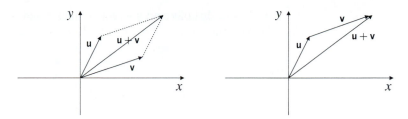

Figure 1.5 On the left, the parallelogram rule for vector addition; on the right, the triangle rule.

> **1.1.5 Triangle Rule:** If the arrow for **v** is drawn with its starting point being the terminal point of **u**, then **u** + **v** is the third side of the triangle with sides **u** and **v**.

Reading Challenge 3   *Suppose A, B, and C are the vertices of a triangle. What is* $\vec{AB} + \vec{BC}$?

Before continuing, let's record some properties of the addition and scalar multiplication of vectors. As noted, most of these properties have names that make them easy to reference.

Theorem 1.1.6   **(Properties of Vector Addition and Scalar Multiplication).**   *Let* **u**, **v**, *and* **w** *be vectors, and let c and d denote scalars.*

  1. *(Closure under addition)* **u** + **v** *is a vector.*

  2. *(Commutativity of addition)* **u** + **v** = **v** + **u**.

  3. *(Associativity of addition)* (**u** + **v**) + **w** = **u** + (**v** + **w**).

  4. *(Zero)* **u** + **0** = **0** + **u** = **u**.

  5. *(Negatives) There is a vector called the negative of* **u** *and denoted* −**u** *with the property that* **u** + (−**u**) = (−**u**) + **u** = **0**.

  6. *(Closure under scalar multiplication) c***u** *is a vector.*

  7. *(Scalar associativity) c*(*d***u**) = (*cd*)**u**.

  8. *(One)* 1**u** = **u**.

  9. *(Distributivity) c*(**u** + **v**) = *c***u** + *c***v** *and* (*c* + *d*)**u** = *c***u** + *d***u**.

These properties are so natural that you will find yourself using them without knowing it. We leave proofs to the exercises.

1.1.7 EXAMPLES
  - $4(\mathbf{u} - 3\mathbf{v}) + 6(-2\mathbf{u} + 3\mathbf{v}) = -8\mathbf{u} + 6\mathbf{v}$.
  - If $2\mathbf{u} + 3\mathbf{v} = 6\mathbf{x} + 4\mathbf{u}$, then $6\mathbf{x} = -2\mathbf{u} + 3\mathbf{v}$, so $\mathbf{x} = -\frac{1}{3}\mathbf{u} + \frac{1}{2}\mathbf{v}$.

- If $\mathbf{x} = 3\mathbf{u} - 2\mathbf{v}$ and $\mathbf{y} = \mathbf{u} + \mathbf{v}$, then $\mathbf{u}$ and $\mathbf{v}$ can be expressed in terms of $\mathbf{x}$ and $\mathbf{y}$ like this:

$$
\begin{array}{rl}
\mathbf{x} & = 3\mathbf{u} - 2\mathbf{v} \\
2\mathbf{y} & = 2\mathbf{u} + 2\mathbf{v} \\
\hline
\mathbf{x} + 2\mathbf{y} = & 5\mathbf{u},
\end{array}
$$

so $\mathbf{u} = \frac{1}{5}(\mathbf{x} + 2\mathbf{y})$ and $\mathbf{v} = \mathbf{y} - \mathbf{u} = \frac{4}{5}\mathbf{x} + \frac{3}{5}\mathbf{y}$. ∎

■ Subtracting Vectors

We subtract vectors using the rule

1.1.8 $\mathbf{u} - \mathbf{v} = \mathbf{u} + (-\mathbf{v})$.

For instance,

$$
\begin{bmatrix} 7 \\ 2 \end{bmatrix} - \begin{bmatrix} 5 \\ 3 \end{bmatrix} = \begin{bmatrix} 7 \\ 2 \end{bmatrix} + \left(-\begin{bmatrix} 5 \\ 3 \end{bmatrix} \right) = \begin{bmatrix} 7 \\ 2 \end{bmatrix} + \begin{bmatrix} -5 \\ -3 \end{bmatrix} = \begin{bmatrix} 2 \\ -1 \end{bmatrix}.
$$

Similarly,

$$
\begin{bmatrix} -1 \\ 3 \end{bmatrix} - \begin{bmatrix} -2 \\ 2 \end{bmatrix} = \begin{bmatrix} 1 \\ 1 \end{bmatrix} \quad \text{and} \quad \begin{bmatrix} 2 \\ 0 \end{bmatrix} - \begin{bmatrix} 0 \\ -3 \end{bmatrix} = \begin{bmatrix} 2 \\ 3 \end{bmatrix}.
$$

Again, we look for a geometrical interpretation of this operation. How does the arrow for $\mathbf{u} - \mathbf{v}$ correspond to arrows for $\mathbf{u}$ and $\mathbf{v}$? From Theorem 1.1.6, it follows that

$$
\mathbf{v} + (\mathbf{u} - \mathbf{v}) = \mathbf{u}.
$$

Reading Challenge 4   *How does* $\mathbf{v} + (\mathbf{u} - \mathbf{v}) = \mathbf{u}$ *follow from Theorem 1.1.6? Show and justify all required steps.*

From the triangle rule for vector addition, if we represent $\mathbf{u} - \mathbf{v}$ by an arrow that starts at the end of $\mathbf{v}$, then $\mathbf{u}$ is the third side of the triangle.

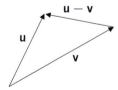

1.1.9 If vectors $\mathbf{u}$ and $\mathbf{v}$ are pictured by arrows with the same starting point, then $\mathbf{u} - \mathbf{v}$ can be pictured by the arrow from the end of $\mathbf{v}$ to the end of $\mathbf{u}$.

Figure 1.6 illustrates the vector equation $\begin{bmatrix} -1 \\ 3 \end{bmatrix} - \begin{bmatrix} 4 \\ 2 \end{bmatrix} = \begin{bmatrix} -5 \\ 1 \end{bmatrix}$.

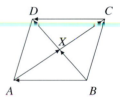

Figure 1.6 The vector $\mathbf{u} - \mathbf{v}$ can be pictured by the arrow from the end of $\mathbf{v}$ to the end of $\mathbf{u}$.

We have seen that vectors can be pictured by arrows and any two arrows with the same length and direction describe the same vector. These facts can be used to prove some familiar propositions from Euclidean geometry.

1.1.10 PROBLEM

Prove that the diagonals of a parallelogram bisect each other.

Solution. Label the vertices of the parallelogram A, B, C, D as shown and draw the diagonal AC. Let X be the midpoint of AC so that $\overrightarrow{AX} = \overrightarrow{XC}$. We wish to show that $\overrightarrow{BX} = \overrightarrow{XD}$. Now

$$\overrightarrow{BX} = \overrightarrow{BA} + \overrightarrow{AX} \quad \text{and} \quad \overrightarrow{XD} = \overrightarrow{XC} + \overrightarrow{CD}.$$

Since the arrows $\overrightarrow{BA}$ and $\overrightarrow{CD}$ have the same length and direction, we have $\overrightarrow{BA} = \overrightarrow{CD}$. Since $\overrightarrow{AX} = \overrightarrow{XC}$, we have the desired result.

■ **Linear Combinations**

A sum of scalar multiples of vectors, such as $-8\mathbf{u} + 6\mathbf{v}$, has a special name.

1.1.11 DEFINITION

A *linear combination* of vectors $\mathbf{u}$ and $\mathbf{v}$ is a vector of the form $a\mathbf{u} + b\mathbf{v}$, where a and b are real numbers. More generally, a *linear combination* of k vectors $\mathbf{u}_1, \mathbf{u}_2, \ldots, \mathbf{u}_k$ is a vector of the form $c_1\mathbf{u}_1 + c_2\mathbf{u}_2 + \cdots + c_k\mathbf{u}_k$, where $c_1, \ldots, c_k$ are real numbers.

1.1.12 EXAMPLES

- $\begin{bmatrix} -5 \\ 9 \end{bmatrix}$ is a linear combination of $\mathbf{u} = \begin{bmatrix} -2 \\ 3 \end{bmatrix}$ and $\mathbf{v} = \begin{bmatrix} -1 \\ 1 \end{bmatrix}$ since

$$\begin{bmatrix} -5 \\ 9 \end{bmatrix} = 4\begin{bmatrix} -2 \\ 3 \end{bmatrix} - 3\begin{bmatrix} -1 \\ 1 \end{bmatrix} = 4\mathbf{u} - 3\mathbf{v}.$$

- $\begin{bmatrix} 2 \\ -6 \end{bmatrix}$ is a linear combination of $\mathbf{u}_1 = \begin{bmatrix} -2 \\ 3 \end{bmatrix}$, $\mathbf{u}_2 = \begin{bmatrix} 6 \\ -5 \end{bmatrix}$ and $\mathbf{u}_3 = \begin{bmatrix} 4 \\ 5 \end{bmatrix}$ since

$$\begin{bmatrix} 2 \\ -6 \end{bmatrix} = 3\begin{bmatrix} -2 \\ 3 \end{bmatrix} + 2\begin{bmatrix} 6 \\ -5 \end{bmatrix} - \begin{bmatrix} 4 \\ 5 \end{bmatrix} = 3\mathbf{u}_1 + 2\mathbf{u}_2 + (-1)\mathbf{u}_3.$$ ■

1.1.13 PROBLEM Is $\begin{bmatrix} 3 \\ 0 \end{bmatrix}$ a linear combination of $\begin{bmatrix} -1 \\ 2 \end{bmatrix}$ and $\begin{bmatrix} -3 \\ 3 \end{bmatrix}$?

Solution. The question asks if there are scalars a and b such that $\begin{bmatrix} 3 \\ 0 \end{bmatrix} = a \begin{bmatrix} -1 \\ 2 \end{bmatrix} + b \begin{bmatrix} -3 \\ 3 \end{bmatrix}$. Equating corresponding components, we must have $3 = -a - 3b$ and $0 = 2a + 3b$. Adding these equations gives $a = 3$, so $-3b = 3 + a = 6$ and $b = -2$. Thus $\begin{bmatrix} 3 \\ 0 \end{bmatrix} = 3 \begin{bmatrix} -1 \\ 2 \end{bmatrix} - 2 \begin{bmatrix} -3 \\ 3 \end{bmatrix}$ is a linear combination of $\begin{bmatrix} -1 \\ 2 \end{bmatrix}$ and $\begin{bmatrix} -3 \\ 3 \end{bmatrix}$.

Reading Challenge 5 *Is $\begin{bmatrix} 0 \\ 0 \end{bmatrix}$ a linear combination of $\begin{bmatrix} -1 \\ 2 \end{bmatrix}$ and $\begin{bmatrix} -3 \\ 3 \end{bmatrix}$?*

Actually, the answer to Problem 1.1.13 does not depend on the particular vector $\begin{bmatrix} 3 \\ 0 \end{bmatrix}$: any two-dimensional vector is a linear combination of $\begin{bmatrix} -1 \\ 2 \end{bmatrix}$ and $\begin{bmatrix} -3 \\ 3 \end{bmatrix}$. For any vectors **u** and **v** that are not parallel, the set of all linear combinations of **u** and **v** is the entire xy-plane! Why in the world should this be true?

In Figure 1.7, we show two vectors **u** and **v**, which are not parallel, and an arbitrary vector **w**. The figure is intended to show that **w** is the diagonal of a parallelogram with sides that are multiples of **u** and **v**. To repeat, we have the following:

> **1.1.14** If two-dimensional vectors **u** and **v** are not parallel, then the set of linear combinations of **u** and **v** is the entire xy-plane.

This statement is easy to check if the vectors **u** and **v** are "nice."

1.1.15 DEFINITION The *standard basis vectors* in the plane are $\mathbf{i} = \begin{bmatrix} 1 \\ 0 \end{bmatrix}$ and $\mathbf{j} = \begin{bmatrix} 0 \\ 1 \end{bmatrix}$.

These vectors are pictured in Figure 1.8. Clearly, they are not parallel and notice how easy it is to verify 1.1.14: given any (two-dimensional) vector $\begin{bmatrix} a \\ b \end{bmatrix}$, we have

$$\begin{bmatrix} a \\ b \end{bmatrix} = \begin{bmatrix} a \\ 0 \end{bmatrix} + \begin{bmatrix} 0 \\ b \end{bmatrix} = a \begin{bmatrix} 1 \\ 0 \end{bmatrix} + b \begin{bmatrix} 0 \\ 1 \end{bmatrix} = a\mathbf{i} + b\mathbf{j}.$$

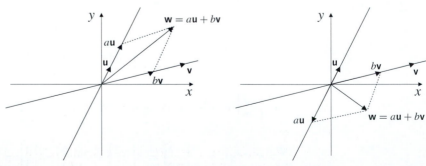

Figure 1.7 Any vector **w** is a linear combination of **u** and **v**.

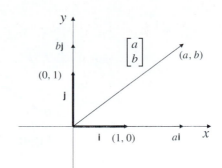

Figure 1.8 $\begin{bmatrix} a \\ b \end{bmatrix} =$ $a\mathbf{i} + b\mathbf{j}$.

For example, $\begin{bmatrix} 3 \\ 4 \end{bmatrix} = 3\begin{bmatrix} 1 \\ 0 \end{bmatrix} + 4\begin{bmatrix} 0 \\ 1 \end{bmatrix} = 3\mathbf{i} + 4\mathbf{j}$. We have demonstrated an important point that should be emphasized.

> **1.1.16** Every two-dimensional vector is a linear combination of $\mathbf{i}$ and $\mathbf{j}$:
> $$\begin{bmatrix} a \\ b \end{bmatrix} = a\mathbf{i} + b\mathbf{j}.$$

We conclude this section with the observation that most of what we have said about two-dimensional vectors applies equally to three-dimensional vectors.

1.1.17 DEFINITION A *three-dimensional vector* is a triple of numbers, written in a column and surrounded by brackets. The *components* of $\begin{bmatrix} a \\ b \\ c \end{bmatrix}$ are the three numbers a, b, and c.

1.1.18 EXAMPLES

- $\begin{bmatrix} 1 \\ 2 \\ 3 \end{bmatrix}$ is a three-dimensional vector with components 1, 2, and 3;

- $\begin{bmatrix} -1 \\ 1 \\ 0 \end{bmatrix}$ is a three-dimensional vector with components -1, 1, and 0;

- The three-dimensional *zero vector* is the vector $\begin{bmatrix} 0 \\ 0 \\ 0 \end{bmatrix}$ all of whose components are 0. ∎

Equality of Vectors. Two vectors $\mathbf{u} = \begin{bmatrix} a \\ b \\ c \end{bmatrix}$ and $\mathbf{v} = \begin{bmatrix} x \\ y \\ z \end{bmatrix}$ are equal if and only if $a = x$, $b = y$, and $c = z$.

■ **Vector Algebra**

We add three-dimensional vectors, multiply them by scalars, and form linear combinations all in the obvious way.

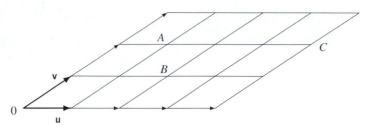

Figure 1.9   The set of all linear combinations of **u** and **v** is a plane.

1.1.19 EXAMPLES

- $\begin{bmatrix} -2 \\ 4 \\ 7 \end{bmatrix} + \begin{bmatrix} 1 \\ -1 \\ 6 \end{bmatrix} = \begin{bmatrix} -1 \\ 3 \\ 13 \end{bmatrix};$

- $-2 \begin{bmatrix} 4 \\ -1 \\ 5 \end{bmatrix} = \begin{bmatrix} -8 \\ 2 \\ -10 \end{bmatrix};$

- $4 \begin{bmatrix} -2 \\ 0 \\ 1 \end{bmatrix} - 3 \begin{bmatrix} 2 \\ 1 \\ -5 \end{bmatrix} = \begin{bmatrix} -14 \\ -3 \\ 19 \end{bmatrix}.$ ∎

We can also use arrows to picture three-dimensional vectors but, for obvious reasons, more often than not we use our imagination and the knowledge of what happens in the plane rather than attempting to draw pictures in three dimensions. If we think of **u** as an arrow, then we think of 2**u** as an arrow with the same direction but twice as long, and −**u** as an arrow the same length as **u** but with the opposite direction. The sum of two three-dimensional vectors **u** and **v**, drawn as arrows with the same starting point, is the diagonal of the parallelogram with sides **u** and **v**. With reference to Figure 1.9, the arrow from 0 to A represents **u** + 2**v** and the arrow from 0 to B represents 2**u** + **v**.

Reading Challenge 6   *What vector is represented by the arrow from 0 to C?*

Figure 1.9 suggests a fact about three-dimensional space analogous to 1.1.14.

1.1.20   If vectors **u** and **v** are not parallel, then the set of all linear combinations of **u** and **v** is a plane.

To imagine this plane, think of the corresponding arrows as the index and middle fingers of a hand held up in the air. Imagine a stiff sheet of cardboard resting against these fingers. This cardboard lies in the plane determined by your fingers. See Figure 1.10.

Figure 1.10

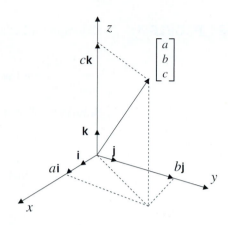

Figure 1.11 $\begin{bmatrix} a \\ b \\ c \end{bmatrix} =$

$ai + bj + ck.$

1.1.21 DEFINITION   The *standard basis vectors* in three-dimensional space are

$$\mathbf{i} = \begin{bmatrix} 1 \\ 0 \\ 0 \end{bmatrix}, \quad \mathbf{j} = \begin{bmatrix} 0 \\ 1 \\ 0 \end{bmatrix}, \quad \text{and} \quad \mathbf{k} = \begin{bmatrix} 0 \\ 0 \\ 1 \end{bmatrix}.$$

These are pictured in Figure 1.11, which also illustrates the fact that

> **1.1.22** Every vector in three-dimensional space is a linear combination of
> $\mathbf{i}, \mathbf{j}, \text{ and } \mathbf{k}:$ $\begin{bmatrix} a \\ b \\ c \end{bmatrix} = a\mathbf{i} + b\mathbf{j} + c\mathbf{k}.$

The fact expressed in 1.1.22 reminds the author of questions such as "Who wrote Brahms' *Lullaby*?" and "Who is buried in Grant's tomb?" To ask how to write $\begin{bmatrix} a \\ b \\ c \end{bmatrix}$ as a linear combination of $\mathbf{i}, \mathbf{j}, \mathbf{k}$ is to give away the answer!

Reading Challenge 7   *Who did write Brahms'* Lullaby?

1.1.23 EXAMPLE $\begin{bmatrix} 3 \\ -4 \\ 5 \end{bmatrix} = 3\begin{bmatrix} 1 \\ 0 \\ 0 \end{bmatrix} - 4\begin{bmatrix} 0 \\ 1 \\ 0 \end{bmatrix} + 5\begin{bmatrix} 0 \\ 0 \\ 1 \end{bmatrix} = 3\mathbf{i} - 4\mathbf{j} + 5\mathbf{k}.$ ∎

Reading Challenge 8   *Who wrote the* William Tell *Overture?*

1.1.24 PROBLEM

Determine whether the vector $\mathbf{x} = \begin{bmatrix} -1 \\ -2 \\ 2 \end{bmatrix}$ is a linear combination of $\mathbf{u} = \begin{bmatrix} 0 \\ 1 \\ 4 \end{bmatrix}$, $\mathbf{v} = \begin{bmatrix} -1 \\ 1 \\ 2 \end{bmatrix}$, and $\mathbf{w} = \begin{bmatrix} 3 \\ 1 \\ -2 \end{bmatrix}$.

Solution.   The question asks whether there exist scalars a, b, c so that

$$\mathbf{x} = a\mathbf{u} + b\mathbf{v} + c\mathbf{w};$$

that is, such that

$$\begin{bmatrix} -1 \\ -2 \\ 2 \end{bmatrix} = a\begin{bmatrix} 0 \\ 1 \\ 4 \end{bmatrix} + b\begin{bmatrix} -1 \\ 1 \\ 2 \end{bmatrix} + c\begin{bmatrix} 3 \\ 1 \\ -2 \end{bmatrix}.$$

The vector on the right is $\begin{bmatrix} -b + 3c \\ a + b + c \\ 4a + 2b - 2c \end{bmatrix}$, so the question is, are there numbers a, b, c such that

$$\begin{array}{rcrcrcr} & & -\,b & + & 3c & = & -1 \\ a & + & b & + & c & = & -2 \\ 4a & + & 2b & - & 2c & = & 2 \end{array} \quad ?$$

We find that $a = 1$, $b = -2$, $c = -1$ is a solution. Thus $\mathbf{x} = \mathbf{u} - 2\mathbf{v} - \mathbf{w}$ is a linear combination of $\mathbf{u}$, $\mathbf{v}$, and $\mathbf{w}$.

Whether you can find a, b, and c yourself at this point isn't important. Solving systems of equations such as the one here is the subject of Section 2.3.

Answers to Reading Challenges

1. $\overrightarrow{AB} = \begin{bmatrix} 1 - (-2) \\ 5 - 3 \end{bmatrix} = \begin{bmatrix} 3 \\ 2 \end{bmatrix}.$

2.   Let B have coordinates (x, y). Then $\overrightarrow{AB} = \begin{bmatrix} 3 \\ -1 \end{bmatrix} = \begin{bmatrix} x - (-4) \\ y - 2 \end{bmatrix}.$ Thus $x + 4 = 3$ and $y - 2 = -1$, so $x = -1$, $y = 1$. The point B has coordinates $(-1, 1)$.

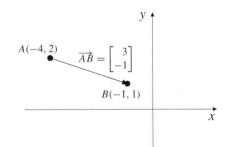

3. $\overrightarrow{AB} + \overrightarrow{BC} = \overrightarrow{AC}$, the third side of triangle ABC.

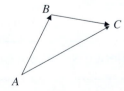

4. $\mathbf{v} + (\mathbf{u} - \mathbf{v}) = \mathbf{v} + (-\mathbf{v} + \mathbf{u})$ commutativity

$\qquad\qquad\quad = (\mathbf{v} + (-\mathbf{v})) + \mathbf{u}$ associativity

$\qquad\qquad\quad = \mathbf{0} + \mathbf{u}$ negatives

$\qquad\qquad\quad = \mathbf{u}$ zero

5. This question can be solved by the approach used in Problem 1.1.13, but it is easier simply to notice that $\begin{bmatrix} 0 \\ 0 \end{bmatrix} = 0 \begin{bmatrix} -1 \\ 2 \end{bmatrix} + 0 \begin{bmatrix} -3 \\ 3 \end{bmatrix}$. Indeed, the zero vector is a linear combination of any set of vectors, taking all scalars to be 0.

6. The arrow from 0 to C presents $4\mathbf{u} + 2\mathbf{v}$.

7. Johannes Brahms!

8. Gioacchino Rossini. (Don't press your luck.)

True/False Questions

Decide, with as little calculation as possible, whether each of the following statements is true or false and explain your answer whenever you say "false." (Answers can be found in the back of the book.)

1. If $A = (1, 2)$ and $B = (-3, 5)$, the vector $\overrightarrow{AB} = \begin{bmatrix} 4 \\ -3 \end{bmatrix}$.

2. Nonzero parallel vectors have the same direction.

3. If $\mathbf{u}$ and $\mathbf{v}$ are different nonzero vectors in the plane, the set of all linear combinations of $\mathbf{u}$ and $\mathbf{v}$ is all two-dimensional vectors.

4. If $\mathbf{u}$ and $\mathbf{v}$ are nonzero vectors in three-space that are not parallel, the set of all linear combinations of $\mathbf{u}$ and $\mathbf{v}$ is a plane.

5. $\begin{bmatrix} 1 \\ 2 \\ 3 \end{bmatrix}$ is a linear combination of $\begin{bmatrix} 1 \\ 0 \\ 1 \end{bmatrix}$ and $\begin{bmatrix} 2 \\ 0 \\ -1 \end{bmatrix}$.

6. The vectors $\begin{bmatrix} 1 \\ 2 \end{bmatrix}$ and $\begin{bmatrix} 0 \\ 0 \end{bmatrix}$ are parallel.

7. $4\mathbf{u} - 9\mathbf{v}$ is a linear combination of $2\mathbf{u}$ and $7\mathbf{v}$.

8. If vectors $\mathbf{u}$ and $\mathbf{v}$ are pictured by arrows with the same initial point, then $\mathbf{u} - \mathbf{v}$ can be pictured by the arrow from the end of $\mathbf{u}$ to the end of $\mathbf{v}$.

9. $\begin{bmatrix} 2 \\ -3 \\ 4 \end{bmatrix} = 2\mathbf{i} - 3\mathbf{j} + 4\mathbf{k}$.

10. The zero vector is a linear combination of any three vectors $\mathbf{u}, \mathbf{v}, \mathbf{w}$.

11. If a vector $\mathbf{u}$ is the sum of vectors $\mathbf{v}$ and $\mathbf{w}$, then $\mathbf{w}$ is a linear combination of $\mathbf{u}$ and $\mathbf{v}$.

12. If a vector $\mathbf{u}$ is a linear combination of vectors $\mathbf{v}$ and $\mathbf{w}$, then $\mathbf{w}$ is a linear combination of $\mathbf{u}$ and $\mathbf{v}$.

13. The set of all linear combinations of the vectors $\begin{bmatrix} 1 \\ 2 \end{bmatrix}$ and $\begin{bmatrix} 3 \\ 1 \end{bmatrix}$ is a plane.

Exercises

*Solutions to exercises marked [BB] can be found in the **B**ack of the **B**ook.*

1. Find the vector $\overrightarrow{AB}$ and illustrate with a picture if
   (a) [BB] $A = (-2, 1)$ and $B = (1, 4)$

   (b) $A = (1, -3)$ and $B = (-3, 2)$.

2. (a) [BB] If $A = (1, 4)$ and $\overrightarrow{AB} = \begin{bmatrix} -1 \\ 2 \end{bmatrix}$, find B.

   (b) If $B = (1, 4)$ and $\overrightarrow{AB} = \begin{bmatrix} -1 \\ 2 \end{bmatrix}$, find A.

3. [BB] Express $\mathbf{x} = \begin{bmatrix} -2 \\ 7 \\ 4 \end{bmatrix}$ as a scalar multiple of

   $\mathbf{u} = \begin{bmatrix} 8 \\ -28 \\ -16 \end{bmatrix}$, of $\mathbf{v} = \begin{bmatrix} 0 \\ 0 \\ 0 \end{bmatrix}$, and of $\mathbf{w} = \begin{bmatrix} -\frac{1}{2} \\ \frac{7}{4} \\ 1 \end{bmatrix}$.

4. Which of the following vectors are scalar multiples of other vectors in the list?

   $\mathbf{u}_1 = \begin{bmatrix} -4 \\ 2 \\ 2 \end{bmatrix}$, $\mathbf{u}_2 = \begin{bmatrix} 3 \\ -6 \\ 12 \end{bmatrix}$, $\mathbf{u}_3 = \begin{bmatrix} 6 \\ -3 \\ -3 \end{bmatrix}$,

   $\mathbf{u}_4 = \begin{bmatrix} 0 \\ 0 \\ 0 \end{bmatrix}$, $\mathbf{u}_5 = \begin{bmatrix} -5 \\ 10 \\ -20 \end{bmatrix}$.

5. Find the indicated vectors.
   (a) [BB] $4\begin{bmatrix} 2 \\ -3 \end{bmatrix} + 2\begin{bmatrix} 3 \\ 1 \end{bmatrix}$

   (b) $a\begin{bmatrix} -1 \\ 5 \end{bmatrix} - 3\begin{bmatrix} -a \\ 2 \end{bmatrix}$

   (c) [BB] $3\begin{bmatrix} 2 \\ 1 \\ 3 \end{bmatrix} - 2\begin{bmatrix} 1 \\ 0 \\ -5 \end{bmatrix} - 4\begin{bmatrix} 0 \\ -1 \\ 2 \end{bmatrix}$

   (d) $3\begin{bmatrix} 1 \\ 0 \\ -2 \end{bmatrix} - 4\begin{bmatrix} 6 \\ 1 \\ 5 \end{bmatrix} + 2\begin{bmatrix} -1 \\ 1 \\ 2 \end{bmatrix}$

   (e) $7\begin{bmatrix} 2 \\ 1 \\ 3 \end{bmatrix} + 4\begin{bmatrix} 0 \\ -4 \\ 2 \end{bmatrix} - 10\begin{bmatrix} 1 \\ -\frac{1}{2} \\ 3 \end{bmatrix}$

   (f) $\frac{1}{3}\begin{bmatrix} 9 \\ -4 \\ 7 \end{bmatrix} - \frac{1}{5}\begin{bmatrix} 25 \\ 6 \\ -2 \end{bmatrix} + 3\begin{bmatrix} 2 \\ 1 \\ -3 \end{bmatrix}$

   (g) $\frac{2}{5}\begin{bmatrix} 5 \\ 6 \\ -3 \end{bmatrix} - \frac{17}{5}\begin{bmatrix} 0 \\ 1 \\ 2 \end{bmatrix}$

   (h) $a\begin{bmatrix} -1 \\ 1 \\ 7 \end{bmatrix} + 2\begin{bmatrix} a \\ a \\ 0 \end{bmatrix} - 4\begin{bmatrix} 0 \\ -1 \\ a \end{bmatrix}$

   (i) $x\begin{bmatrix} -1 \\ 0 \\ 1 \end{bmatrix} + y\begin{bmatrix} -3 \\ 2 \\ -1 \end{bmatrix} + z\begin{bmatrix} 3 \\ -1 \\ 3 \end{bmatrix}$

6. Suppose $\mathbf{u} = \begin{bmatrix} a \\ -5 \end{bmatrix}$, $\mathbf{v} = \begin{bmatrix} 1 \\ 6 - b \end{bmatrix}$ and $\mathbf{w} = \begin{bmatrix} 3 \\ 1 \end{bmatrix}$. Find a and b if
   (a) [BB] $2\mathbf{u} - \mathbf{v} = \mathbf{w}$;   (b) $4\mathbf{u} - \mathbf{v} + 3\mathbf{w} = \mathbf{0}$;
   (c) $2\mathbf{u} - 3\mathbf{v} + 5\mathbf{w} = \mathbf{0}$;   (d) $6(\mathbf{u} - \mathbf{w}) = 12\mathbf{v}$.

7. [BB] Suppose $\mathbf{x}, \mathbf{y}, \mathbf{u}$, and $\mathbf{v}$ are vectors such that $\mathbf{x} - \mathbf{y} = \mathbf{u}$ and $2\mathbf{x} + 3\mathbf{y} = \mathbf{v}$. Express $\mathbf{x}$ and $\mathbf{y}$ in terms of $\mathbf{u}$ and $\mathbf{v}$.

8. In each of the following cases, find $\mathbf{u} - \mathbf{v}$ and illustrate with a picture like that in Figure 1.6.

   (a) [BB] $\mathbf{u} = \begin{bmatrix} -2 \\ -2 \end{bmatrix}$, $\mathbf{v} = \begin{bmatrix} 4 \\ 1 \end{bmatrix}$

   (b) $\mathbf{u} = \begin{bmatrix} -4 \\ 2 \end{bmatrix}$, $\mathbf{v} = \begin{bmatrix} 1 \\ -4 \end{bmatrix}$

9. Shown at the top of the next page are two nonparallel vectors $\mathbf{u}$ and $\mathbf{v}$ and four other vectors $\mathbf{w}_1, \mathbf{w}_2, \mathbf{w}_3, \mathbf{w}_4$. Reproduce $\mathbf{u}, \mathbf{v}$, and $\mathbf{w}_1$ in a picture by themselves and exhibit $\mathbf{w}_1$ as the diagonal of a parallelogram with sides parallel to $\mathbf{u}$ and $\mathbf{v}$. Guess values of a and b so that $\mathbf{w}_1 = a\mathbf{u} + b\mathbf{v}$ [BB]. Repeat using each of $\mathbf{w}_2, \mathbf{w}_3$, and $\mathbf{w}_4$ instead of $\mathbf{w}_1$.

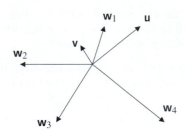

10. Express $\begin{bmatrix} 7 \\ 7 \end{bmatrix}$ as a linear combination of each of the following pairs of vectors:

    (a) [BB] $\begin{bmatrix} 2 \\ 3 \end{bmatrix}$ and $\begin{bmatrix} -1 \\ 2 \end{bmatrix}$ (b) $\begin{bmatrix} -1 \\ 1 \end{bmatrix}$ and $\begin{bmatrix} 5 \\ 2 \end{bmatrix}$

    (c) $\begin{bmatrix} 11 \\ 5 \end{bmatrix}$ and $\begin{bmatrix} 3 \\ 2 \end{bmatrix}$ (d) $\begin{bmatrix} 5 \\ 5 \end{bmatrix}$ and $\begin{bmatrix} 20 \\ 2 \end{bmatrix}$

    (e) $\begin{bmatrix} 2 \\ 6 \end{bmatrix}$ and $\begin{bmatrix} 14 \\ -8 \end{bmatrix}$.

11. (a) [BB] Is it possible to express $\begin{bmatrix} 1 \\ 2 \end{bmatrix}$ as a linear combination of $\begin{bmatrix} -2 \\ -2 \end{bmatrix}$ and $\begin{bmatrix} 3 \\ 3 \end{bmatrix}$? Explain.

    (b) Does your answer to part (a) contradict 1.1.14? Explain.

12. (a) [BB] Is it possible to express the zero vector as a linear combination of $\begin{bmatrix} 1 \\ 4 \end{bmatrix}$ and $\begin{bmatrix} -2 \\ -8 \end{bmatrix}$ in more than one way? Justify your answer.

    (b) Is it possible to express the zero vector as a linear combination of $\begin{bmatrix} 1 \\ 4 \end{bmatrix}$ and $\begin{bmatrix} 2 \\ -3 \end{bmatrix}$ in more than one way? Justify your answer.

13. Suppose **u** and **v** are vectors.

    (a) [BB] Show that any linear combination of $2\mathbf{u}$ and $-3\mathbf{v}$ is a linear combination of **u** and **v**.

    (b) Show that any linear combination of **u** and **v** is a linear combination of $2\mathbf{u}$ and $-3\mathbf{v}$.

14. Let $O(0, 0)$, $A(2, 2)$, $B(4, 4)$, $C(2, -1)$, $D(4, 1)$, $E(6, 3)$, $F(2, -4)$, $G(6, 0)$, $H(8, 2)$ be the indicated points in the xy-plane. Let $\mathbf{u} = \overrightarrow{OA}$ be the vector pictured by the arrow from O to A and $\mathbf{v} = \overrightarrow{OC}$ the vector pictured by the arrow from O to C.

    (a) [BB] Find **u** and **v**.

    (b) Express each of the following vectors as a linear combination of **u** and **v**.

      (a) [BB] the vector $\overrightarrow{OD}$ from 0 to D

      (b) the vector $\overrightarrow{OG}$ from 0 to G

      (c) $\overrightarrow{OH}$

      (d) $\overrightarrow{CA}$

      (e) $\overrightarrow{AF}$

      (f) $\overrightarrow{GE}$

      (g) $\overrightarrow{HC}$

    For example, $\overrightarrow{OE} = \begin{bmatrix} 6 \\ 3 \end{bmatrix} = 2\mathbf{u} + \mathbf{v}$.

15. [BB] Suppose ABC is a triangle. Let D be the midpoint of AB and E be the midpoint of AC. Use vectors to show that DE is parallel to BC and half its length.

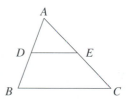

16. Let $ABCD$ be an arbitrary quadrilateral in the plane. Let P, Q, R, and S be, respectively, the midpoints of AB, BC, DC, and AD. Use vectors to show that $PQRS$ is a parallelogram.

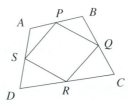

17. [BB] Let $A = (1, 2)$, $B = (4, -3)$, and $C = (-1, -2)$ be three points in the Euclidean plane. Find a fourth point D such that the A, B, C, and D are the vertices of a parallelogram and justify your answer. Is the answer unique?

18. Given three points $A(-1, 0)$, $B(2, 3)$, $C(4, -1)$ in the plane, find all points D such that the four points A, B, C, and D are the vertices of a parallelogram.

19. The vectors $\mathbf{u} = \begin{bmatrix} 1 \\ 2 \end{bmatrix}$ and $\mathbf{v} = \begin{bmatrix} -1 \\ 1 \end{bmatrix}$ are not parallel. According to 1.1.14, any vector $\mathbf{x}$ in the plane is a linear combination of $\mathbf{u}$ and $\mathbf{v}$.
Verify this statement with $\mathbf{x} = \begin{bmatrix} 1 \\ 0 \end{bmatrix}$, $\mathbf{x} = \begin{bmatrix} 0 \\ 1 \end{bmatrix}$ and $\mathbf{x} = \begin{bmatrix} 1 \\ 1 \end{bmatrix}$.

20. Use vectors to show that the midpoint of the line joining $A(x_1, y_1, z_1)$ to $B(x_2, y_2, z_2)$ is the point $C(\frac{x_1+x_2}{2}, \frac{y_1+y_2}{2}, \frac{z_1+z_2}{2})$.

21. [BB; 1, 2, 5] Prove all parts of Theorem 1.1.6.

■ Critical Reading

22. [BB] Show that any linear combination of $\begin{bmatrix} 1 \\ \frac{3}{2} \\ 0 \end{bmatrix}$ and $\begin{bmatrix} 0 \\ 3 \\ 6 \end{bmatrix}$ is also a linear combination of $\begin{bmatrix} 2 \\ 3 \\ 0 \end{bmatrix}$ and $\begin{bmatrix} 0 \\ 1 \\ 2 \end{bmatrix}$.

23. Let $\mathbf{u}$, $\mathbf{v}$, and $\mathbf{w}$ be vectors. Show that any linear combination of $\mathbf{u}$ and $\mathbf{v}$ is also a linear combination of $\mathbf{u}$, $\mathbf{v}$, and $\mathbf{w}$.

24. (a) [BB] Explain how one could use vectors to show that three points X, Y, and Z are *collinear*. (Points are collinear if they lie on a line.)

(b) Of the five points $A(-1, 1, 1)$, $B(2, 2, 1)$ $C(0, 3, 4)$, $D(2, 7, 10)$, $E(6, 0, -5)$, which triples, if any, are collinear?

25. Show that the vector $\overrightarrow{AX}$ is a linear combination of $\overrightarrow{AB}$ and $\overrightarrow{AC}$, the coefficients summing to 1.

26. Given a triangle $A_0B_0C_0$, the first *Gunther* triangle is constructed as follows: A_1 lies on A_0B_0 and is such that $\overrightarrow{A_0A_1} = \frac{1}{3}\overrightarrow{A_0B_0}$; B_1 lies on B_0C_0 and is such that $\overrightarrow{B_0B_1} = \frac{1}{3}\overrightarrow{B_0C_0}$; C_1 lies on C_0A_0 and is such that $\overrightarrow{C_0C_1} = \frac{1}{3}\overrightarrow{C_0A_0}$. Iterate this construction once more; that is, triangle $A_2B_2C_2$ is the first Gunther triangle of $A_1B_1C_1$. (This is called the second Gunther triangle of $A_0B_0C_0$.) Use vectors to show that the lengths of corresponding sides of $\triangle A_2B_2C_2$ and $\triangle A_0B_0C_0$ are in the same proportions, hence that these two triangles are similar.

27. (a) [BB] If $\mathbf{u}$ and $\mathbf{v}$ are parallel, show that some *nontrivial* linear combination of $\mathbf{u}$ and $\mathbf{v}$ is the zero vector. (*Nontrivial* means that $0\mathbf{u} + 0\mathbf{v} = \mathbf{0}$ doesn't count!)

(b) Show that if some nontrivial linear combination of vectors $\mathbf{u}$ and $\mathbf{v}$ is $\mathbf{0}$, then $\mathbf{u}$ and $\mathbf{v}$ are parallel.

1.2 Length and Direction

How long is an arrow representing the two-dimensional vector $\mathbf{v} = \begin{bmatrix} a \\ b \end{bmatrix}$? Remembering the Theorem of Pythagoras, it seems that the length of such an arrow is $\sqrt{a^2 + b^2}$. We call this number the *length* or *norm* of $\mathbf{v}$ and use the notation $\|\mathbf{v}\|$.

1.2.1 DEFINITION The *norm* of the vector $\mathbf{v} = \begin{bmatrix} a \\ b \end{bmatrix}$ is the number $\|\mathbf{v}\| = \sqrt{a^2 + b^2}$.

1.2.2 EXAMPLE If $\mathbf{u} = \begin{bmatrix} 3 \\ -4 \end{bmatrix}$, then $\|\mathbf{u}\| = \sqrt{3^2 + (-4)^2} = 5$. ■

If $\overrightarrow{OP} = \begin{bmatrix} a \\ b \\ c \end{bmatrix}$ is a three-dimensional vector, its norm is $\sqrt{a^2 + b^2 + c^2}$, as shown in Figure 1.12. Letting A be the point of the xy-plane directly below $P(a, b, c)$, the

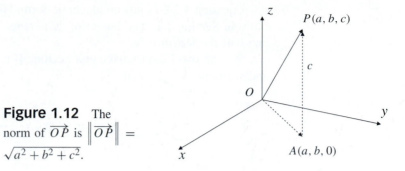

Figure 1.12 The norm of $\overrightarrow{OP}$ is $\left\|\overrightarrow{OP}\right\| = \sqrt{a^2 + b^2 + c^2}.$

triangle OAP is right-angled. Thus

$$\left\|\overrightarrow{OP}\right\|^2 = \left\|\overrightarrow{OA}\right\|^2 + \left\|\overrightarrow{AP}\right\|^2 = (a^2 + b^2) + c^2$$

so $\left\|\begin{bmatrix} a \\ b \\ c \end{bmatrix}\right\| = \sqrt{a^2 + b^2 + c^2}.$

1.2.3 EXAMPLE   If $\mathbf{u} = \begin{bmatrix} -1 \\ 2 \\ 1 \end{bmatrix}$ and $\mathbf{v} = \begin{bmatrix} 1 \\ 0 \\ -1 \end{bmatrix}$, then $\|\mathbf{u}\| = \sqrt{6}$ and $\|\mathbf{v}\| = \sqrt{2}$.   ∎

1.2.4 DEFINITION   A *unit vector* is a vector of norm 1.

1.2.5 EXAMPLES
- Each of the standard basis vectors has norm 1. For example, in the plane, $\mathbf{i} = \begin{bmatrix} 1 \\ 0 \end{bmatrix}$, so $\|\mathbf{i}\| = \sqrt{1^2 + 0^2} = \sqrt{1} = 1$.

- The vector $\begin{bmatrix} \frac{1}{\sqrt{2}} \\ \frac{1}{\sqrt{2}} \end{bmatrix}$ is a unit two-dimensional vector and $\begin{bmatrix} \frac{1}{\sqrt{6}} \\ -\frac{2}{\sqrt{6}} \\ \frac{1}{\sqrt{6}} \end{bmatrix}$ is a unit three-

dimensional vector.   ∎

Suppose $\mathbf{v} = \begin{bmatrix} a \\ b \end{bmatrix}$ is a two-dimensional vector. Then $c\mathbf{v} = \begin{bmatrix} ca \\ cb \end{bmatrix}$, so

$$\|c\mathbf{v}\| = \sqrt{(ca)^2 + (cb)^2} = \sqrt{c^2(a^2 + b^2)} = \sqrt{c^2}\sqrt{a^2 + b^2} = |c|\,\|\mathbf{v}\|.$$

1.2.6 $\|c\mathbf{v}\| = |c|\,\|\mathbf{v}\|.$

Reading Challenge 1   *If c is a real number, what is $\sqrt{c^2}$?*

Equation 1.2.6 is just an algebraic formulation of something we observed pictorially in Section 1.1. The length of $2\mathbf{v}$ is twice the length of $\mathbf{v}$; the length of $-\frac{1}{2}\mathbf{v}$ is one half the length of $\mathbf{v}$.

We can use 1.2.6 to make unit vectors. If $\mathbf{v}$ is any vector (except the zero vector) and we put $c = \frac{1}{\|\mathbf{v}\|}$, then

$$\|c\mathbf{v}\| = |c|\,\|\mathbf{v}\| = \frac{1}{\|\mathbf{v}\|}\,\|\mathbf{v}\| = 1.$$

Thus $\frac{1}{\|\mathbf{v}\|}\mathbf{v}$ is a unit vector, and it has the same direction as $\mathbf{v}$ since it is a positive scalar multiple of $\mathbf{v}$.

> **1.2.7** For any $\mathbf{v}$, the vector $\frac{1}{\|\mathbf{v}\|}\mathbf{v}$ is a unit vector in the direction of $\mathbf{v}$.

1.2.8 EXAMPLE   To find a unit vector in the same direction as $\mathbf{v} = \begin{bmatrix} 2 \\ -2 \\ 1 \end{bmatrix}$, we compute $c = \|\mathbf{v}\| = $

$\sqrt{2^2 + (-2)^2 + 1^2} = \sqrt{9} = 3$. The desired vector is then $\frac{1}{3}\mathbf{v} = \begin{bmatrix} \frac{2}{3} \\ -\frac{2}{3} \\ \frac{1}{3} \end{bmatrix}$. ■

■ Dot Product and the Angle Between Vectors

Any two-dimensional vector $\mathbf{v}$ is of the form $\begin{bmatrix} a \\ b \end{bmatrix}$, so if $\|\mathbf{v}\| = 1$, then $\sqrt{a^2 + b^2} = 1$. This means that if $\mathbf{v}$ is pictured by an arrow with initial point at the origin $(0, 0)$, then the end of $\mathbf{v}$ lies on the unit circle. Since any point on the unit circle has coordinates of the form $(\cos\theta, \sin\theta)$ (see Figure 1.13), it follows that

> **1.2.9** Any two-dimensional unit vector is of the form $\begin{bmatrix} \cos\theta \\ \sin\theta \end{bmatrix}$ for some
> angle θ.

Figure 1.13 Any vector of norm 1 is of the form $\begin{bmatrix} \cos\theta \\ \sin\theta \end{bmatrix}$ for some angle θ.

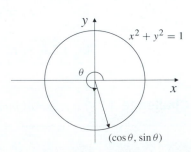

y

$x^2 + y^2 = 1$

θ

x

$(\cos\theta, \sin\theta)$

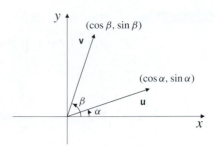

Figure 1.14 The angle between **u** and **v** is $\beta - \alpha$.

Suppose we have two vectors of norm 1, say $\mathbf{u} = \begin{bmatrix} \cos \alpha \\ \sin \alpha \end{bmatrix}$ and $\mathbf{v} = \begin{bmatrix} \cos \beta \\ \sin \beta \end{bmatrix}$. For convenience, let's suppose that $0 < \alpha < \beta < \frac{\pi}{2}$. See Figure 1.14. The angle between **u** and **v** is $\beta - \alpha$, and we hope you know that

$$\cos(\beta - \alpha) = \cos \beta \cos \alpha + \sin \beta \sin \alpha.$$

This is the product of the first coordinates of **u** and **v** added to the product of the second coordinates of **u** and **v**. This shows that if θ is the angle between unit vectors $\mathbf{u} = \begin{bmatrix} a_1 \\ b_1 \end{bmatrix}$ and $\mathbf{v} = \begin{bmatrix} a_2 \\ b_2 \end{bmatrix}$, then

$$\cos \theta = a_1 a_2 + b_1 b_2. \tag{1}$$

The vectors $\mathbf{u} = \begin{bmatrix} -\frac{3}{5} \\ \frac{4}{5} \end{bmatrix}$ and $\mathbf{v} = \begin{bmatrix} -\frac{4}{5} \\ -\frac{3}{5} \end{bmatrix}$ shown in Figure 1.15 are unit vectors, so the cosine of the angle θ between them is given by

$$\cos \theta = \frac{-3}{5}\frac{-4}{5} + \frac{4}{5}\frac{-3}{5} = 0.$$

So $\theta = \arccos(0) = \frac{\pi}{2}$ radians $= 90°$: **u** and **v** are perpendicular. The expression on the right of (1) has a special name: it is the *dot product* of **u** and **v**.

Figure 1.15 **u** and **v** are perpendicular.

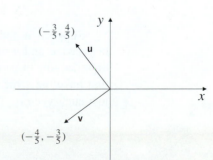

1.2.10 DEFINITION

The *dot product* of two-dimensional vectors $\mathbf{u} = \begin{bmatrix} a_1 \\ b_1 \end{bmatrix}$ and $\mathbf{v} = \begin{bmatrix} a_2 \\ b_2 \end{bmatrix}$ is the number $a_1 a_2 + b_1 b_2$. It is denoted $\mathbf{u} \cdot \mathbf{v}$.

$$\boxed{\mathbf{u} \cdot \mathbf{v} = a_1 a_2 + b_1 b_2.}$$

The *dot product* of three-dimensional vectors $\mathbf{u} = \begin{bmatrix} a_1 \\ b_1 \\ c_1 \end{bmatrix}$ and $\mathbf{v} = \begin{bmatrix} a_2 \\ b_2 \\ c_2 \end{bmatrix}$ is the number

$$\boxed{\mathbf{u} \cdot \mathbf{v} = a_1 a_2 + b_1 b_2 + c_1 c_2.}$$

1.2.11 EXAMPLES

- If $\mathbf{u} = \begin{bmatrix} -1 \\ 2 \end{bmatrix}$ and $\mathbf{v} = \begin{bmatrix} 4 \\ -3 \end{bmatrix}$, then $\mathbf{u} \cdot \mathbf{v} = -1(4) + 2(-3) = -10$.

- If $\mathbf{u} = \begin{bmatrix} -1 \\ 0 \\ 2 \end{bmatrix}$ and $\mathbf{v} = \begin{bmatrix} 1 \\ 2 \\ 3 \end{bmatrix}$, then $\mathbf{u} \cdot \mathbf{v} = -1(1) + 0(2) + 2(3) = 5$.   ∎

The dot product is useful in a number of ways. If $\mathbf{u} = \begin{bmatrix} a \\ b \\ c \end{bmatrix}$, then

$$\mathbf{u} \cdot \mathbf{u} = \begin{bmatrix} a \\ b \\ c \end{bmatrix} \cdot \begin{bmatrix} a \\ b \\ c \end{bmatrix} = a^2 + b^2 + c^2.$$

This is the square of the length of $\mathbf{u}$.

1.2.12 $\|\mathbf{u}\|^2 = \mathbf{u} \cdot \mathbf{u}$; equivalently, $\|\mathbf{u}\| = \sqrt{\mathbf{u} \cdot \mathbf{u}}$.

We have used several times the expression "angle between vectors." This is somewhat ambiguous, as shown in the figure to the right. To avoid this difficulty, by general agreement, *the angle between vectors is that angle θ in the range $0 \le \theta \le \pi$.*

If $\mathbf{u}$ and $\mathbf{v}$ are unit vectors and θ is the angle between them, we have seen that

$$\cos \theta = \mathbf{u} \cdot \mathbf{v}.$$

Suppose $\mathbf{u}$ and $\mathbf{v}$ are any, not necessarily unit, vectors. What is the angle θ between $\mathbf{u}$ and $\mathbf{v}$ in this case? The angle between $\mathbf{u}$ and $\mathbf{v}$ has nothing to do with their length. In particular, this angle is the same as the angle between unit vectors that have the same direction as $\mathbf{u}$ and $\mathbf{v}$. In 1.2.7, we showed that $\frac{1}{\|\mathbf{u}\|}\mathbf{u}$ and $\frac{1}{\|\mathbf{v}\|}\mathbf{v}$ are just such vectors. Thus

$$\cos \theta = \frac{1}{\|\mathbf{u}\|}\mathbf{u} \cdot \frac{1}{\|\mathbf{v}\|}\mathbf{v} = \frac{1}{\|\mathbf{u}\| \, \|\mathbf{v}\|}\mathbf{u} \cdot \mathbf{v}. \tag{2}$$

Otherwise put,

> **1.2.13** $\mathbf{u} \cdot \mathbf{v} = \|\mathbf{u}\| \, \|\mathbf{v}\| \cos \theta.$

This is a very important formula!

1.2.14 Remark The author confesses that he is getting a bit ahead of himself here. Proper justification for the move from $\frac{1}{\|\mathbf{u}\|}\mathbf{u} \cdot \frac{1}{\|\mathbf{v}\|}\mathbf{v}$ at the second equals sign in equation (2) is actually a consequence of part 3 of Theorem 1.2.19. On the other hand, the author doesn't like making obvious things seem hard and hopes the reader feels the same way.

1.2.15 EXAMPLE If $\mathbf{u} = \begin{bmatrix} \sqrt{3} \\ 1 \end{bmatrix}$ and $\mathbf{v} = \begin{bmatrix} 1 \\ \sqrt{3} \end{bmatrix}$, then $\mathbf{u} \cdot \mathbf{v} = \sqrt{3} + \sqrt{3} = 2\sqrt{3}$, $\|\mathbf{u}\| = \sqrt{(\sqrt{3})^2 + 1^2} = \sqrt{4} = 2 = \|\mathbf{v}\|$, so $2\sqrt{3} = 2(2)\cos\theta$, $\cos\theta = \frac{\sqrt{3}}{2}$ and $\theta = \frac{\pi}{6}$ radians $= 30°$. ∎

1.2.16 EXAMPLE Suppose $\mathbf{u} = \begin{bmatrix} 1 \\ 2 \\ 3 \end{bmatrix}$ and $\mathbf{v} = \begin{bmatrix} -2 \\ 1 \\ -1 \end{bmatrix}$. Then

$$\|\mathbf{u}\| = \sqrt{1^2 + 2^2 + 3^2} = \sqrt{14}, \qquad \|\mathbf{v}\| = \sqrt{(-2)^2 + 1^2 + (-1)^2} = \sqrt{6}$$

and

$$\mathbf{u} \cdot \mathbf{v} = 1(-2) + 2(1) + 3(-1) = -3.$$

So $-3 = \sqrt{14}\sqrt{6}\cos\theta = \sqrt{84}\cos\theta$ and hence $\cos\theta = \frac{-3}{\sqrt{84}} \approx -.327$. Therefore $\theta \approx \arccos(-.327) = 1.904$ radians $\approx 109°$. ∎

Vectors $\mathbf{u}$ and $\mathbf{v}$ are perpendicular if the angle between them is $\frac{\pi}{2}$ radians. Since $\cos\frac{\pi}{2} = 0$, we obtain from 1.2.13 yet another use of the dot product.

The word "orthogonal" is synonymous with "perpendicular," but is more commonly used in linear algebra.

> **1.2.17** Vectors $\mathbf{u}$ and $\mathbf{v}$ are orthogonal (or perpendicular) if and only if $\mathbf{u} \cdot \mathbf{v} = 0$. In particular, the zero vector is orthogonal to all other vectors.

1.2.18 EXAMPLES

- The vectors $\mathbf{i} = \begin{bmatrix} 1 \\ 0 \end{bmatrix}$ and $\mathbf{j} = \begin{bmatrix} 0 \\ 1 \end{bmatrix}$ are orthogonal because $\mathbf{i} \cdot \mathbf{j} = 1(0) + 0(1) = 0$.

- The vectors $\mathbf{u} = \begin{bmatrix} 3 \\ 4 \end{bmatrix}$ and $\mathbf{v} = \begin{bmatrix} -4 \\ 3 \end{bmatrix}$ are orthogonal because $\mathbf{u} \cdot \mathbf{v} = 3(-4) + 4(3) = 0$.

- The three-dimensional vectors $\mathbf{u} = \begin{bmatrix} -1 \\ 3 \\ 5 \end{bmatrix}$ and $\mathbf{v} = \begin{bmatrix} 2 \\ -1 \\ 1 \end{bmatrix}$ are orthogonal because $\mathbf{u} \cdot \mathbf{v} = (-1)2 + 3(-1) + 5(1) = 0$. ∎

We summarize the basic properties of the dot product in a theorem, leaving proofs for the exercises.

Theorem 1.2.19   **(Properties of the Dot Product).** *Let* **u**, **v**, *and* **w** *be vectors. Let c be a scalar.*

1. *(Commutativity)* $\mathbf{u} \cdot \mathbf{v} = \mathbf{v} \cdot \mathbf{u}$.

2. *(Distributivity)* $\mathbf{u} \cdot (\mathbf{v} + \mathbf{w}) = \mathbf{u} \cdot \mathbf{v} + \mathbf{u} \cdot \mathbf{w}$ *and* $(\mathbf{u} + \mathbf{v}) \cdot \mathbf{w} = \mathbf{u} \cdot \mathbf{w} + \mathbf{v} \cdot \mathbf{w}$.

3. *(Scalar associativity)* $(c\mathbf{u}) \cdot \mathbf{v} = c(\mathbf{u} \cdot \mathbf{v}) = \mathbf{u} \cdot (c\mathbf{v})$.

4. $\mathbf{v} \cdot \mathbf{v} = \|\mathbf{v}\|^2$.

5. $\mathbf{v} \cdot \mathbf{v} \geq 0$ *and* $\mathbf{v} \cdot \mathbf{v} = 0$ *if and only if* $\mathbf{v} = \mathbf{0}$.

As with Theorem 1.1.6, you will find yourself using Theorem 1.2.19 without realizing it.

1.2.20 EXAMPLE

$$(2\mathbf{u} - 3\mathbf{v}) \cdot (\mathbf{u} + 4\mathbf{v}) = (2\mathbf{u} - 3\mathbf{v}) \cdot \mathbf{u} + (2\mathbf{u} - 3\mathbf{v}) \cdot (4\mathbf{v})$$
$$= 2\mathbf{u} \cdot \mathbf{u} - 3\mathbf{v} \cdot \mathbf{u} + 8\mathbf{u} \cdot \mathbf{v} - 12\mathbf{v} \cdot \mathbf{v}$$
$$= 2\mathbf{u} \cdot \mathbf{u} + 5\mathbf{u} \cdot \mathbf{v} - 12\mathbf{v} \cdot \mathbf{v}$$

With both a lunar crater and a Paris street named after him, Augustin-Louis Cauchy (1789–1857) contributed to almost every area of mathematics. A student of Weierstrass, Hermann Amandus Schwarz (1843–1921) is best known for his work in conformal mappings.

Notice how this computation resembles the calculation of the product $(2x - 3y)(x + 4y)$. ∎

■ The Cauchy–Schwarz Inequality

We conclude this section with an important consequence of formula 1.2.13. Since $-1 \leq \cos\theta \leq 1$ and hence $|\cos\theta| \leq 1$ for any angle θ, it follows immediately that for any vectors **u** and **v**,

1.2.21 $\boxed{|\mathbf{u} \cdot \mathbf{v}| \leq \|\mathbf{u}\|\,\|\mathbf{v}\|.}$ **Cauchy–Schwarz Inequality**

This perhaps innocent-looking connection between the dot product and the lengths of two vectors has numerous applications.

1.2.22 PROBLEM

Show that $|ab + cd| \leq \sqrt{a^2 + c^2}\sqrt{b^2 + d^2}$ for any real numbers a, b, c, and d.

Solution.   Does the left side resemble a dot product? Does the right side resemble the product of the lengths of vectors? Let $\mathbf{u} = \begin{bmatrix} a \\ c \end{bmatrix}$ and $\mathbf{v} = \begin{bmatrix} b \\ d \end{bmatrix}$. Then $\mathbf{u} \cdot \mathbf{v} = ab + cd$, $\|\mathbf{u}\| = \sqrt{a^2 + c^2}$ and $\|\mathbf{v}\| = \sqrt{b^2 + d^2}$. The given inequality is exactly the Cauchy–Schwarz inequality for these two vectors.

Suppose a and b are nonnegative real numbers and we take $\mathbf{u} = \begin{bmatrix} \sqrt{a} \\ \sqrt{b} \end{bmatrix}$ and $\mathbf{v} = \begin{bmatrix} \sqrt{b} \\ \sqrt{a} \end{bmatrix}$. Then $\mathbf{u} \cdot \mathbf{v} = \sqrt{a}\sqrt{b} + \sqrt{a}\sqrt{b} = 2\sqrt{ab}$ while $\|\mathbf{u}\| = \|\mathbf{v}\| = \sqrt{a + b}$. The

Cauchy–Schwarz inequality says that

$$|2\sqrt{ab}| \le \sqrt{a+b}\sqrt{a+b} = a+b;$$

that is,

> **1.2.23 Arithmetic/Geometric Mean Inequality:**
> $$\sqrt{ab} \le \frac{a+b}{2} \text{ for any } a, b \ge 0.$$

The geometric mean appears in the work of Euclid. It is the number that "squares the rectangle," in the sense that if you want to build a square with area equal to that of a rectangle with sides of lengths a and b, then you'd better be able to find $\sqrt{ab}$.

The expression $\frac{a+b}{2}$ on the right-hand side is, of course, just the average, or *arithmetic mean*, of a and b. The expression on the left, $\sqrt{ab}$, is their *geometric mean*. So, as indicated, inequality 1.2.23 is known as the *arithmetic mean/geometric mean inequality*. For example, $6 = \sqrt{4(9)} \le \frac{4+9}{2} = 6\frac{1}{2}$.

Answers to Reading Challenges

1. Some people think $\sqrt{c^2} = c$, which would implry that $\sqrt{(-2)^2} = -2$. Any calculator, however, should tell you that $\sqrt{(-2)^2} = \sqrt{4} = +2 = |-2|$. For a given real number c, $\sqrt{c^2} = |c|$.

True/False Questions

Decide, with as little calculation as possible, whether each of the following statements is true or false and explain your answer whenever you say "false." (Answers can be found in the back of the book.)

1. The length of $\begin{bmatrix} -1 \\ 2 \\ 2 \end{bmatrix}$ is 3.

2. If c is a scalar and $\mathbf{v}$ is a vector, the length of $c\mathbf{v}$ is c times the length of $\mathbf{v}$.

3. The vectors $\begin{bmatrix} -1 \\ 2 \\ 1 \end{bmatrix}$ and $\begin{bmatrix} 1 \\ 2 \\ -3 \end{bmatrix}$ are orthogonal.

4. If $\mathbf{v}$ is a nonzero vector, a unit vector with the same direction as $\mathbf{v}$ is $\frac{1}{\|\mathbf{v}\|} \mathbf{v}$.

5. The cosine of the angle between vectors $\mathbf{u}$ and $\mathbf{v}$ is $\mathbf{u} \cdot \mathbf{v}$.

6. $\pi^3 \le \frac{\pi + \pi^2}{2}$. (Calculators not allowed! True/False questions are intended to be answered mentally.)

7. For any vectors $\mathbf{u}$ and $\mathbf{v}$, $\mathbf{u} \cdot \mathbf{v} \le \|\mathbf{u}\| \, \|\mathbf{v}\|$.

8. If $\mathbf{u}$ is a vector and $\mathbf{u} \cdot \mathbf{u} = 0$, then $\mathbf{u} = \mathbf{0}$.

9. There exist unit vectors $\mathbf{u}$ and $\mathbf{v}$ with $\mathbf{u} \cdot \mathbf{v} = 2$.

Exercises

*Solutions to exercises marked [BB] can be found in the **B**ack of the **B**ook.*

1. [BB] Let $\mathbf{u} = \begin{bmatrix} 0 \\ 3 \\ 4 \end{bmatrix}$ and $\mathbf{v} = \begin{bmatrix} 1 \\ 2 \\ 2 \end{bmatrix}$. Find a unit vector in the direction of $\mathbf{u}$ and a vector of length 2 in the direction opposite to $\mathbf{v}$.

2. Let $\mathbf{u} = \begin{bmatrix} 4 \\ 7 \\ -4 \end{bmatrix}$ and $\mathbf{v} = \begin{bmatrix} 2 \\ -6 \\ -1 \end{bmatrix}$. Find the following:

   (a) a unit vector in the direction of $\mathbf{v}$

   (b) a vector of length 5 in the direction of $\mathbf{v}$

   (c) a vector of length 3 in the direction opposite to $\mathbf{u}$.

3. Find the angle between each of the following pairs of vectors. Give your answers in radians and in degrees. If it is necessary to approximate, give radians to two decimal places of accuracy and degrees to the nearest degree.

   (a) [BB] $\mathbf{u} = \begin{bmatrix} 3 \\ 4 \end{bmatrix}$, $\mathbf{v} = \begin{bmatrix} 4 \\ -3 \end{bmatrix}$

   (b) [BB] $\mathbf{u} = \begin{bmatrix} 3 \\ 4 \end{bmatrix}$, $\mathbf{v} = \begin{bmatrix} -1 \\ 1 \end{bmatrix}$

   (c) $\mathbf{u} = \begin{bmatrix} 6 \\ 8 \end{bmatrix}$, $\mathbf{v} = \begin{bmatrix} -3 \\ 3 \end{bmatrix}$

   (d) $\mathbf{u} = \begin{bmatrix} 2 \\ -1 \\ 1 \end{bmatrix}$ and $\mathbf{v} = \begin{bmatrix} 1 \\ 1 \\ 2 \end{bmatrix}$

   (e) $\mathbf{u} = \begin{bmatrix} 1 \\ 3 \\ 2 \end{bmatrix}$ and $\mathbf{v} = \begin{bmatrix} -4 \\ -1 \\ 1 \end{bmatrix}$

4. Find $\mathbf{u} \cdot \mathbf{v}$ in each of the following cases.
   (a) [BB] $\|\mathbf{u}\| = 1$, $\|\mathbf{v}\| = 3$ and the angle between $\mathbf{u}$ and $\mathbf{v}$ is $45°$.

   (b) $\|\mathbf{u}\| = 4$, $\|\mathbf{v}\| = 7$ and the angle between $\mathbf{u}$ and $\mathbf{v}$ is 2 radians.

5. (a) [BB] Can $\mathbf{u} \cdot \mathbf{v} = -7$ if $\|\mathbf{u}\| = 3$ and $\|\mathbf{v}\| = 2$?

   (b) Can $\mathbf{u} \cdot \mathbf{v} = 9$ if $\|\mathbf{u}\| = 5$ and $\|\mathbf{v}\| = 2$? Justify your answers.

6. Let $\mathbf{v} = \begin{bmatrix} 1 \\ 0 \\ -3 \end{bmatrix}$. Find a vector of length 2 with the same direction as $\mathbf{v}$ and a vector of length 10 with direction opposite to $\mathbf{v}$.

7. [BB] Let $\mathbf{u} = \begin{bmatrix} 3 \\ 4 \\ 0 \end{bmatrix}$, $\mathbf{v} = \begin{bmatrix} 2 \\ 1 \\ 2 \end{bmatrix}$, and $\mathbf{w} = \begin{bmatrix} 1 \\ 1 \\ 1 \end{bmatrix}$.

   (a) Find $\|\mathbf{u}\|$, $\|\mathbf{u} + \mathbf{v}\|$, and $\left\| \dfrac{\mathbf{w}}{\|\mathbf{w}\|} \right\|$.

   (b) Find a unit vector in the same direction as $\mathbf{u}$.

   (c) Find a vector of norm 4 in the direction opposite to $\mathbf{v}$.

8. Let $\mathbf{u} = \begin{bmatrix} 2 \\ 5 \\ 3 \end{bmatrix}$, $\mathbf{v} = \begin{bmatrix} -4 \\ 1 \\ 0 \end{bmatrix}$, and $\mathbf{w} = \begin{bmatrix} -3 \\ 3 \\ 5 \end{bmatrix}$.

   (a) Find $\|\mathbf{u}\|$, $\|\mathbf{v}\|$, and $\|\mathbf{w}\|$.

   (b) Find $\|\mathbf{u} + \mathbf{v} - \mathbf{w}\|$.

   (c) Find a vector of norm 2 in the direction of $\mathbf{u} - 2\mathbf{v}$.

9. Let $\mathbf{u}$ and $\mathbf{v}$ be vectors of lengths 3 and 5, respectively, and suppose that $\mathbf{u} \cdot \mathbf{v} = 8$. Find the following:

   (a) [BB] $(\mathbf{u} - \mathbf{v}) \cdot (2\mathbf{u} - 3\mathbf{v})$

   (b) $(-3\mathbf{u} + 4\mathbf{v}) \cdot (2\mathbf{u} + 5\mathbf{v})$

   (c) $(2\mathbf{u} + 3\mathbf{v}) \cdot (8\mathbf{v} - \mathbf{u})$

   (d) $(-4\mathbf{v} + 5\mathbf{u}) \cdot (-3\mathbf{v} - 6\mathbf{u})$

   (e) $(2\mathbf{u} - 5\mathbf{v}) \cdot (3\mathbf{u} - 7\mathbf{v})$

   (f) [BB] $\|\mathbf{u} + \mathbf{v}\|^2$.

10. Let $\mathbf{u}$, $\mathbf{v}$, and $\mathbf{w}$ be vectors of lengths 3, 2, and 6, respectively. Suppose $\mathbf{u} \cdot \mathbf{v} = 3$, $\mathbf{u} \cdot \mathbf{w} = 5$, and $\mathbf{v} \cdot \mathbf{w} = 2$. Find the following:

    (a) $(\mathbf{u} - 2\mathbf{v}) \cdot (3\mathbf{w} + 2\mathbf{u})$

    (b) $(3\mathbf{w} - 2\mathbf{v}) \cdot (5\mathbf{v} - \mathbf{w})$

    (c) $(-3\mathbf{u} + 4\mathbf{w}) \cdot (6\mathbf{u} - 2\mathbf{v})$

    (d) $(8\mathbf{w} - 12\mathbf{v}) \cdot (37\mathbf{v} + 24\mathbf{w})$

    (e) $\|\mathbf{u} - \mathbf{v} - \mathbf{w}\|^2$.

11. Let $\mathbf{u} = \begin{bmatrix} 1 \\ 2 \end{bmatrix}$ and $\mathbf{v} = \begin{bmatrix} 1 \\ -1 \end{bmatrix}$.

    (a) [BB] Find all numbers k such that $\mathbf{u} + k\mathbf{v}$ has norm 3.

    (b) Find all numbers k such that $k\mathbf{u} - 3\mathbf{v}$ is orthogonal to $\begin{bmatrix} -4 \\ 5 \end{bmatrix}$.

12. Let $\mathbf{u} = \begin{bmatrix} 1 \\ 0 \\ -1 \end{bmatrix}$ and $\mathbf{v} = \begin{bmatrix} 2 \\ 1 \\ 0 \end{bmatrix}$.

    (a) Find a real number k so that $\mathbf{u} + k\mathbf{v}$ is orthogonal to $\mathbf{u}$, if such k exists.

    (b) Find a real number ℓ so that $\ell\mathbf{u} + \mathbf{v}$ is a unit vector, if such ℓ exists.

13. [BB] If a nonzero vector $\mathbf{u}$ is orthogonal to another vector $\mathbf{v}$ and k is a scalar, show that $\mathbf{u} + k\mathbf{v}$ is not orthogonal to $\mathbf{u}$.

14. Suppose $\mathbf{u}$ and $\mathbf{v}$ are vectors that are not orthogonal, k is a scalar, and $\mathbf{u} + k\mathbf{v}$ is orthogonal to $\mathbf{u}$. Find k.

15. [BB] Given the points $A(-1, 0)$ and $B(2, 3)$ in the plane, find all points C such that A, B, and C are the vertices of a right angled triangle with right angle at A, and AC of length 2.

16. Given the points $A(0, -2)$ and $B(1, 1)$, find all points C so that A, B, and C are the vertices of a right angled triangle with right angle at B and hypotenuse of length 10.

17. [BB] Find a point D such that $(1, 2)$, $(-3, -1)$, $(4, -2)$, and D are the vertices of a square and justify your answer.

18. Let $\mathbf{u} = \begin{bmatrix} 1 \\ 1 \\ 0 \end{bmatrix}$ and $\mathbf{v} = \begin{bmatrix} x \\ 1 \\ 1 \end{bmatrix}$. Find all numbers x (if any) such that the angle between $\mathbf{u}$ and $\mathbf{v}$ is $60°$.

19. Let $\mathbf{u} = \begin{bmatrix} 3 \\ 4 \\ -2 \end{bmatrix}$, $\mathbf{w} = \begin{bmatrix} -2 \\ 5 \\ 7 \end{bmatrix}$, and $\mathbf{v} = \begin{bmatrix} 4 \\ -2 \\ 2 \end{bmatrix}$.

    (a) [BB] Show that $\mathbf{u}$ is orthogonal to $\mathbf{v}$ and to $\mathbf{w}$.

    (b) Show that $\mathbf{u}$ is orthogonal to $\mathbf{v} - \mathbf{w}$.

    (c) Show that $\mathbf{u}$ is orthogonal to $a\mathbf{v} + b\mathbf{w}$ for any scalars a and b.

20. [BB; 1, 5] Prove all parts of Theorem 1.2.19 for two-dimensional vectors.

21. Use the important formula $\mathbf{u} \cdot \mathbf{v} = \|\mathbf{u}\| \, \|\mathbf{v}\| \cos\theta$ to derive the *Law of Cosines*, which goes as follows: If $\triangle ABC$ has sides of lengths a, b, c as shown, and $\theta = \angle BCA$, then $c^2 = a^2 + b^2 - 2ab\cos\theta$.

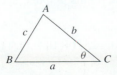

22. Verify the Cauchy–Schwarz inequality for each of the following pairs of vectors.

    (a) [BB] $\mathbf{u} = \begin{bmatrix} -2 \\ 1 \end{bmatrix}$, $\mathbf{v} = \begin{bmatrix} 3 \\ 5 \end{bmatrix}$

    (b) $\mathbf{u} = \begin{bmatrix} 1 \\ -2 \\ 3 \end{bmatrix}$, $\mathbf{v} = \begin{bmatrix} 2 \\ 0 \\ 1 \end{bmatrix}$

    (c) $\mathbf{u} = \begin{bmatrix} 1 \\ 1 \\ -2 \end{bmatrix}$, $\mathbf{v} = \begin{bmatrix} 1 \\ -3 \\ -2 \end{bmatrix}$

23. (a) Use the Cauchy–Schwarz inequality to prove the generalized Theorem of Pythagoras, also known as the triangle inequality: For vectors $\mathbf{u}$ and $\mathbf{v}$, $\|\mathbf{u} + \mathbf{v}\| \leq \|\mathbf{u}\| + \|\mathbf{v}\|$.

    (b) Verify the triangle inequality for each pair of vectors in Exercise 22.

24. [BB] Use the Cauchy–Schwarz inequality to prove that $(a\cos\theta + b\sin\theta)^2 \leq a^2 + b^2$ for all $a, b \in \mathbf{R}$.

25. Let a, b, c, d be real numbers.

    (a) [BB] Use the Cauchy–Schwarz inequality to prove that $(3ac + bd)^2 \leq (3a^2 + b^2)(3c^2 + d^2)$.

    (b) Use part (a) to prove that $abcd \leq \frac{1}{2}(a^2d^2 + b^2c^2)$.

26. Use the Arithmetic/Geometric Mean inequality to show that $\sqrt{\sin 2\theta} \leq \frac{\sqrt{2}}{2}(\sin\theta + \cos\theta)$ for any θ, $0 \leq \theta \leq \frac{\pi}{2}$.

■ **Critical Reading**

27. [BB] Suppose vectors $\mathbf{u}$ and $\mathbf{v}$ are represented by arrows OA and OB in the plane, as shown. Find a formula for a vector whose arrow bisects angle AOB.

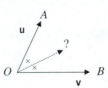

28. Can $\mathbf{u} \cdot \mathbf{v} = 9$ if $\|\mathbf{u}\| = 5$ and $\|\mathbf{v}\| = 1$? Justify your answer.

29. A rectangular box has sides of lengths in the ratio $1 : 2 : 3$. Find the angle between a diagonal to the box and one of its shortest sides.

30. Let $\mathcal{P}$ be a parallelogram with sides representing the vectors **u** and **v** as shown.

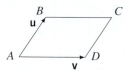

(a) [BB] Express the diagonals $\overrightarrow{AC}$ and $\overrightarrow{DB}$ of $\mathcal{P}$ in terms of **u** and **v**.

(b) Prove that the sum of the squares of the lengths of the diagonals equals the sum of the squares of the lengths of the four sides.

31. (a) Let $\mathbf{w} = \begin{bmatrix} 8 \\ 5 \\ 4 \end{bmatrix}$, $\mathbf{u} = \begin{bmatrix} 3 \\ -4 \\ -1 \end{bmatrix}$, and $\mathbf{v} = \begin{bmatrix} 1 \\ 0 \\ -2 \end{bmatrix}$.

  i. Show that **w** is orthogonal to **u** and to **v**.

  ii. Show that **w** is orthogonal to $3\mathbf{u} - 2\mathbf{v}$.

  iii. Show that **w** is orthogonal to $a\mathbf{u} + b\mathbf{v}$ for any scalars a and b.

(b) Generalize the results of part (a). Suppose a vector **w** is orthogonal to two vectors **u** and **v**. Show that **w** is orthogonal to every linear combination of **u** and **v**.

(c) What might be a good use for the principle discovered in part (b)?

32. Given vectors **u** and **v** that are not orthogonal, find a scalar k such that $\mathbf{u} + k\mathbf{v}$ is orthogonal to **u**.

33. Given six real numbers $x_1, y_1, x_2, y_2, x_3, y_3$, prove that

$$|x_1 y_1 + x_2 y_2 + x_3 y_3| \le \sqrt{x_1^2 + x_2^2 + x_3^2}\sqrt{y_1^2 + y_2^2 + y_3^2}.$$

[Hint: This is another application of the Cauchy–Schwarz inequality. See Problem 1.2.22.]

1.3 Lines, Planes, and the Cross Product of Vectors

To university students who have taken some mathematics courses, "$x^2 + y^2 = 1$ is a circle" probably doesn't seem a strange thing to say, but it would certainly confuse a seven-year-old who thinks she knows what a circle is but has no idea about equations. Of course, $x^2 + y^2 = 1$ is not a circle. It's an *equation* whose *graph* in the xy-plane happens to be a circle. What's the connection between an equation and its graph?

> The graph of an equation is the set of all points whose coordinates satisfy that equation.

1.3.1 EXAMPLE   A point (x, y) in the plane is on the line of slope 2 through the origin if and only if its y-coordinate is twice its x-coordinate. Thus $y = 2x$ is the equation of the line. ■

1.3.2 EXAMPLE   A point (x, y) is on the circle with center $(0, 0)$ and radius 1 if and only if its coordinates x and y satisfy $x^2 + y^2 = 1$. Thus the graph of $x^2 + y^2 = 1$ is a circle. ■

Implicit here, since the center of the circle has been identified with just two coordinates, is the fact that we are working in the xy-plane. What is the graph of $x^2 + y^2 = 1$ if we are working in 3-space? What points 3-*space* satisfy this equation? Here are some examples:

$$(1, 0, 0), (1, 0, 1), (1, 0, 2), (1, 0, 3), (1, 0, 4), \ldots, (1, 0, -1), (1, 0, -2), \ldots \quad (1)$$

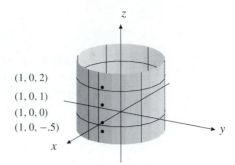

Figure 1.16 In 3-space, the graph of $x^2 + y^2 = 1$ is a cylinder.

$(1, 0, 2)$
$(1, 0, 1)$
$(1, 0, 0)$
$(1, 0, -.5)$

In 3-space, all of these points lie on the graph of $x^2 + y^2 = 1$. In every case, the sum of the squares of the x- and y-coordinates is 1. Think of the floor as the xy-plane, identify some point as the origin, and imagine the circle in the floor with its center at that origin and radius 1. In 3-space, not only is this circle on the graph of $x^2 + y^2 = 1$, but so also is every point directly above or below this circle. For example, as illustrated in (1), every point directly above or below $(1, 0, 0)$ is on the graph of $x^2 + y^2 = 1$, since $1^2 + 0^2 = 1$, $(1, 0, z)$ satisfies the equation for any z. In 3-space, the graph of $x^2 + y^2 = 1$ is a cylinder, specifically, that cylinder which slices through the xy-plane in the familiar unit circle with its center being the origin. (See Figure 1.16.)

What is the graph of the equation $y = 2x$? A line, perhaps? True, if we are working in the plane. The set of points (x, y) with $y = 2x$ is a line of slope 2 through the origin. In 3-space, however, not only do the points on this line satisfy $y = 2x$, but so does every point directly above and below this line. In 3-space, the graph of $y = 2x$ is a *plane*, specifically that plane which is perpendicular to the xy-plane and which cuts it in the familiar "line $y = 2x$."

In the plane, lines are precisely the graphs of *linear equations*, meaning equations of the form $ax + by = c$. The graph of such an equation in 3-space, however, is a plane perpendicular to the xy-plane. The graphs of linear equations in 3-space are *not* lines; in fact, they are planes.

A *normal* to a plane is a nonzero vector perpendicular to the plane. For example, $\mathbf{k}$ is a normal to the xy-plane. So are $-2\mathbf{k}$ and $7\mathbf{k}$. A plane has many normals, but all of them are parallel; each is a scalar multiple of any other. We say that a normal is unique only "up to scalar multiples."

■ The Equation of a Plane

We would like to obtain the equation of a plane. Remember that this means finding an equation in variables x, y, and z with the property that $P(x, y, z)$ is on the plane if and only if its coordinates x, y, z satisfy the equation. It is helpful to realize that a plane is precisely the set of all arrows perpendicular to a normal. [See Figure 1.17.] Let $P_0(x_0, y_0, z_0)$ be one particular point in a plane that has normal vector $\mathbf{n}$. Then $P(x, y, z)$ will lie in the plane precisely if the arrow from P_0 to P is perpendicular to $\mathbf{n}$. In Figure 1.18, we show the plane through $P_0(x_0, y_0, z_0)$ with normal $\mathbf{n} = \begin{bmatrix} a \\ b \\ c \end{bmatrix}$.

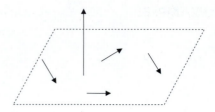

Figure 1.17 A plane consists of all arrows perpendicular to a given normal vector.

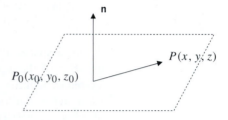

Figure 1.18 $P(x, y, z)$ is on the plane through $P_0(x_0, y_0, z_0)$ with normal **n** if and only if $\overrightarrow{P_0P}$ is perpendicular to **n**.

The point (x, y, z) is in this plane if and only if the arrow from (x_0, y_0, z_0) to (x, y, z) is perpendicular to **n**; that is, if and only if

$$\begin{bmatrix} x - x_0 \\ y - y_0 \\ z - z_0 \end{bmatrix} \cdot \begin{bmatrix} a \\ b \\ c \end{bmatrix} = 0.$$

This occurs if and only if

$$a(x - x_0) + b(y - y_0) + c(z - z_0) = 0,$$

an equation that can be rewritten

$$ax + by + cz = ax_0 + by_0 + cz_0.$$

The right-hand side is a definite number d, since a, b, c are the components of the given normal **n** and x_0, y_0, z_0 are the coordinates of the given point P_0. Thus the plane has equation

$$ax + by + cz = d. \tag{2}$$

Conversely, just as we showed that the equation $y = 1$ defines a plane, so it can be shown that any equation of the form (2) defines a plane, too. We summarize as follows:

1.3.3 Planes in 3-space are the graphs of equations of the form

$ax + by + cz = d$. The vector $\begin{bmatrix} a \\ b \\ c \end{bmatrix}$ is a normal to the plane with

equation $ax + by + cz = d$.

1.3.4 EXAMPLES

- The equation $2x - 3y + 4z = 0$ is the equation of a plane through the origin $(0, 0, 0)$ because $x = 0$, $y = 0$, $z = 0$ satisfies the equation. A normal to the plane is $\begin{bmatrix} 2 \\ -3 \\ 4 \end{bmatrix}$.

- The equation $2x - 3y + 4z = 10$ defines the plane containing $(5, 0, 0)$ and with normal $\begin{bmatrix} 2 \\ -3 \\ 4 \end{bmatrix}$. This plane is parallel to the previous one since the two planes have the same normal.

- The equation $y = 1$ defines the plane containing the point $(0, 1, 0)$ and which has normal $\mathbf{j} = \begin{bmatrix} 0 \\ 1 \\ 0 \end{bmatrix}$ (the coefficients of x, y, z in the equation $y = 1$).

- In 3-space, the xy-plane has the equation $z = 0$ since $\mathbf{k} = \begin{bmatrix} 0 \\ 0 \\ 1 \end{bmatrix}$ is a normal and $(0, 0, 0)$ is a point on the plane. ∎

1.3.5 PROBLEM

Find the equation of the plane perpendicular to $\mathbf{n} = \begin{bmatrix} 1 \\ 2 \\ 3 \end{bmatrix}$ and containing the point $(4, 0, -1)$.

Solution. The equation has the form $1x + 2y + 3z = d$, and $x = 4$, $y = 0$, $z = -1$ is a solution. Thus $1(4) + 2(0) + 3(-1) = d$, so $d = 1$ and the equation is $x + 2y + 3z = 1$.

1.3.6 PROBLEM

Find the equation of the plane through $(4, -1, 5)$ that is parallel to the plane with equation $2x - 3y = 6$.

Solution. The thing to realize here is that parallel planes have the same normals. So $\mathbf{n} = \begin{bmatrix} 2 \\ -3 \\ 0 \end{bmatrix}$ is a normal for the plane we seek and the desired equation has the form $2x - 3y = d$. Since $(4, -1, 5)$ is on the plane, $2(4) - 3(-1) = d$. So $d = 11$, and our plane has equation $2x - 3y = 11$.

1.3.7 EXAMPLE

In Section 1.1, we noted that a plane is the set of all linear combinations of any two nonparallel vectors it contains. Consider the plane with equation $-x + 4y + 5z = 0$. Since $(0, 0, 0)$ is in this plane, an arrow starting at the origin will represent the vector $\begin{bmatrix} x \\ y \\ z \end{bmatrix}$ if and only if the point (x, y, z) is in the plane too; that is, if and only if $x = 4y + 5z$.

So the plane consists of all vectors of the form $\begin{bmatrix} 4y + 5z \\ y \\ z \end{bmatrix} = y \begin{bmatrix} 4 \\ 1 \\ 0 \end{bmatrix} + z \begin{bmatrix} 5 \\ 0 \\ 1 \end{bmatrix}$, that

is, all linear combinations of $\begin{bmatrix} 4 \\ 1 \\ 0 \end{bmatrix}$ and $\begin{bmatrix} 5 \\ 0 \\ 1 \end{bmatrix}$. We say that these vectors *span* the plane

or, equivalently, that the plane is *spanned* by these vectors. Of course, there are many other pairs of vectors that span the same plane.   ■

Reading Challenge 1   *Find two vectors that span the plane with equation $-x + 4y + 5z = 0$ other than the two given in **1.3.7**.*

■ The Cross Product

The vectors $\mathbf{u} = \begin{bmatrix} 1 \\ 3 \\ 2 \end{bmatrix}$ and $\mathbf{v} = \begin{bmatrix} 0 \\ -2 \\ 1 \end{bmatrix}$ are not parallel, so they span a plane, namely,

that plane which consists of vectors of the form $a\mathbf{u} + b\mathbf{v} = a \begin{bmatrix} 1 \\ 3 \\ 2 \end{bmatrix} + b \begin{bmatrix} 0 \\ -2 \\ 1 \end{bmatrix} =$

$\begin{bmatrix} a \\ 3a - 2b \\ 2a + b \end{bmatrix}$. This is a perfectly good description of the plane in question, but what
is its equation? Since the plane in question must pass through the origin, we seek an equation of the form $ax + by + cz = 0$. As we have seen, the coefficients a, b, and c are the components of a vector normal to the plane. It would be convenient if there were an easy way to find these numbers, and there is. To describe what we have in mind, we first introduce the notion of 2×2 (read "two by two") determinant.

1.3.8 DEFINITION   If a, b, c, and d are four numbers, the 2×2 *determinant* $\begin{vmatrix} a & b \\ c & d \end{vmatrix}$ is the number $ad - bc$:

$$\begin{vmatrix} a & b \\ c & d \end{vmatrix} = ad - bc.$$

1.3.9 EXAMPLES

- $\begin{vmatrix} 2 & 1 \\ -4 & 5 \end{vmatrix} = 2(5) - 1(-4) = 14$

- $\begin{vmatrix} 0 & 5 \\ 1 & 1 \end{vmatrix} = 0(1) - 5(1) = -5$

- $\begin{vmatrix} 3 & -1 \\ -6 & 2 \end{vmatrix} = 3(2) - (-1)(-6) = 0.$ ■

1.3.10 DEFINITION The *cross product* of vectors $\mathbf{u} = \begin{bmatrix} u_1 \\ u_2 \\ u_3 \end{bmatrix}$ and $\mathbf{v} = \begin{bmatrix} v_1 \\ v_2 \\ v_3 \end{bmatrix}$ is the vector

$$\mathbf{u} \times \mathbf{v} = \begin{vmatrix} u_2 & u_3 \\ v_2 & v_3 \end{vmatrix} \mathbf{i} - \begin{vmatrix} u_1 & u_3 \\ v_1 & v_3 \end{vmatrix} \mathbf{j} + \begin{vmatrix} u_1 & u_2 \\ v_1 & v_2 \end{vmatrix} \mathbf{k}.$$

As indicated, the cross product of $\mathbf{u}$ and $\mathbf{v}$ is denoted $\mathbf{u} \times \mathbf{v}$, read "$\mathbf{u}$ cross $\mathbf{v}$." Expanding the three determinants, we obtain

$$\mathbf{u} \times \mathbf{v} = (u_2 v_3 - u_3 v_2)\mathbf{i} - (u_1 v_3 - u_3 v_1)\mathbf{j} + (u_1 v_2 - u_2 v_1)\mathbf{k},$$

although most people prefer to remember the coefficients of $\mathbf{i}$, $\mathbf{j}$, and $\mathbf{k}$ as given in terms of determinants, by thinking about the following scheme:

$$\textbf{1.3.11} \quad \mathbf{u} \times \mathbf{v} = \begin{vmatrix} \mathbf{i} & \mathbf{j} & \mathbf{k} \\ u_1 & u_2 & u_3 \\ v_1 & v_2 & v_3 \end{vmatrix}.$$

Here's how this works.

To find the coefficient of $\mathbf{i}$ in Definition 1.3.10, look at the array to the right of 1.3.11, mentally remove the row and column in which $\mathbf{i}$ appears (the first row and the leftmost column), and compute the determinant of the four numbers that remain,

$$\begin{vmatrix} u_2 & u_3 \\ v_2 & v_3 \end{vmatrix}.$$

The coefficient of $\mathbf{k}$ is obtained by removing the row and column in which $\mathbf{k}$ appears (the first row and rightmost column) and computing the determinant of the four numbers that remain,

$$\begin{vmatrix} u_1 & u_2 \\ v_1 & v_2 \end{vmatrix}.$$

The coefficient of $\mathbf{j}$ is obtained in a similar way, except for a sign change. Notice the minus sign highlighted by the arrow in the coefficient of $\mathbf{j}$ in the formula of 1.3.10.

1.3.12 EXAMPLE Let $\mathbf{u} = \begin{bmatrix} 1 \\ 3 \\ 2 \end{bmatrix}$ and $\mathbf{v} = \begin{bmatrix} 0 \\ 2 \\ -1 \end{bmatrix}$ be the two vectors considered earlier. Then

$$\mathbf{u} \times \mathbf{v} = \begin{vmatrix} \mathbf{i} & \mathbf{j} & \mathbf{k} \\ 1 & 3 & 2 \\ 0 & -2 & 1 \end{vmatrix} = \begin{vmatrix} 3 & 2 \\ -2 & 1 \end{vmatrix} \mathbf{i} - \begin{vmatrix} 1 & 2 \\ 0 & 1 \end{vmatrix} \mathbf{j} + \begin{vmatrix} 1 & 3 \\ 0 & -2 \end{vmatrix} \mathbf{k}$$

$$= 7\mathbf{i} - 1\mathbf{j} + (-2)\mathbf{k} = \begin{bmatrix} 7 \\ -1 \\ -2 \end{bmatrix}.$$

Note that this vector is orthogonal to the two given vectors because

$$(\mathbf{u} \times \mathbf{v}) \cdot \mathbf{u} = \begin{bmatrix} 7 \\ -1 \\ -2 \end{bmatrix} \cdot \begin{bmatrix} 1 \\ 3 \\ 2 \end{bmatrix} = 7 - 3 - 4 = 0$$

and

$$(\mathbf{u} \times \mathbf{v}) \cdot \mathbf{v} = \begin{bmatrix} 7 \\ -1 \\ -2 \end{bmatrix} \cdot \begin{bmatrix} 0 \\ 2 \\ -1 \end{bmatrix} = -2 + 2 = 0.$$

Thus $\begin{bmatrix} 7 \\ -1 \\ -2 \end{bmatrix}$ is a normal vector to the plane spanned by $\mathbf{u}$ and $\mathbf{v}$, so the equation is $7x - y - 2z = 0$. ∎

As we just did, it is very good policy always to check that what you think is the cross product of two vectors actually has dot product 0 with each of the two given vectors. This is an excellent way of making sure that your vector is probably correct.

1.3.13 PROBLEM   Find the equation of the plane through the points $A(1, 3, -2)$, $B(1, 1, 5)$, and $C(2, -2, 3)$.

Solution.   A normal vector to the plane is orthogonal to both $\overrightarrow{AB} = \begin{bmatrix} 0 \\ -2 \\ 7 \end{bmatrix}$ and $\overrightarrow{AC} = \begin{bmatrix} 1 \\ -5 \\ 5 \end{bmatrix}$. The cross product of $\overrightarrow{AB}$ and $\overrightarrow{AC}$ is such a vector, so we compute

$$\overrightarrow{AB} \times \overrightarrow{AC} = \begin{vmatrix} \mathbf{i} & \mathbf{j} & \mathbf{k} \\ 0 & -2 & 7 \\ 1 & -5 & 5 \end{vmatrix} = 25\mathbf{i} - (-7)\mathbf{j} + (-(-2))\mathbf{k} = 25\mathbf{i} + 7\mathbf{j} + 2\mathbf{k} = \begin{bmatrix} 25 \\ 7 \\ 2 \end{bmatrix}.$$

Thus the plane has equation $25x + 7y + 2z = d$, where d is determined by the fact that $A(1, 3, -2)$ (or B or C) is on the plane. Thus $25(1) + 7(3) + 2(-2) = d$. So $d = 42$ and the equation of the plane is $25x + 7y + 2z = 42$.

Reading Challenge 2   *Check that the same value for d is obtained by substituting the coordinates of $B(1, 1, 5)$ or of $C(2, -2, 3)$ into $25x + 7y + 2z = d$.*

Before continuing, let us record formally some basic properties of this new concept, the cross product.

Theorem 1.3.14 *If* **u**, **v**, **w** *are vectors in 3-space and c is a scalar,*

1. *(Anticommutativity)* $\mathbf{u} \times \mathbf{v} = -(\mathbf{v} \times \mathbf{u})$.

2. $\mathbf{u} \times (\mathbf{v} + \mathbf{w}) = (\mathbf{u} \times \mathbf{v}) + (\mathbf{u} \times \mathbf{w})$.

3. $(c\mathbf{u}) \times \mathbf{v} = c(\mathbf{u} \times \mathbf{v}) = \mathbf{u} \times (c\mathbf{v})$.

4. $\mathbf{u} \times \mathbf{0} = \mathbf{0} = \mathbf{0} \times \mathbf{u}$.

5. $\mathbf{u} \times \mathbf{v}$ *is orthogonal to both* **u** *and* **v**.

6. $\mathbf{u} \times \mathbf{u} = \mathbf{0}$.

7. $\mathbf{u} \times \mathbf{v} = \mathbf{0}$ *if and only if* **u** *is a scalar multiple of* **v** *or* **v** *is a scalar multiple of* **u***, that is, if and only if* **u** *and* **v** *are parallel.*

8. $\|\mathbf{u} \times \mathbf{v}\| = \|\mathbf{u}\| \, \|\mathbf{v}\| \sin\theta$, *where θ is the angle between* **u** *and* **v***. (Remember that we always assume $0 \leq \theta \leq \pi$.)*

9. $\|\mathbf{u} \times \mathbf{v}\|$ *is the area of the parallelogram having* **u** *and* **v** *as adjacent sides.*

That really is a minus sign in part 1. An operation $\star$ is *commutative* if $a \star b = b \star a$ for any a and b. Many operations are commutative, such as ordinary addition and multiplication of real numbers. The cross product is not commutative, but it comes close. It is *anticommutative*: $\mathbf{u} \times \mathbf{v} = -(\mathbf{v} \times \mathbf{u})$.

1.3.15 EXAMPLE To illustrate, take **u** and **v** as in **1.3.12**, and compute

$$\mathbf{v} \times \mathbf{u} = \begin{vmatrix} \mathbf{i} & \mathbf{j} & \mathbf{k} \\ 0 & -2 & 1 \\ 1 & 3 & 2 \end{vmatrix} = -7\mathbf{i} - (-1)\mathbf{j} + 2\mathbf{k} = \begin{bmatrix} -7 \\ 1 \\ 2 \end{bmatrix} = -(\mathbf{u} \times \mathbf{v}). \qquad \blacksquare$$

If we put $\mathbf{v} = \mathbf{u}$ in part 1 of Theorem 1.3.14, we obtain $\mathbf{u} \times \mathbf{u} = -(\mathbf{u} \times \mathbf{u})$, so $\mathbf{u} \times \mathbf{u} = \mathbf{0}$, which is part 6. Half of part 7 is then straightforward. If $\mathbf{v} = c\mathbf{u}$ is a scalar multiple of **u**, then $\mathbf{u} \times \mathbf{v} = \mathbf{u} \times (c\mathbf{v}) = c(\mathbf{u} \times \mathbf{v}) = c\mathbf{0} = \mathbf{0}$. The other direction, that $\mathbf{u} \times \mathbf{v} = \mathbf{0}$ implies **v** is a scalar multiple of **u**, we leave to the exercises. (See Exercise 22.)

The remarkable formula given in part 8 of Theorem 1.3.14 follows from

$$\mathbf{u} \cdot \mathbf{v} = \|\mathbf{u}\| \, \|\mathbf{v}\| \cos\theta \qquad (3)$$

and

$$\|\mathbf{u} \times \mathbf{v}\|^2 + (\mathbf{u} \cdot \mathbf{v})^2 = \|\mathbf{u}\|^2 \, \|\mathbf{v}\|^2 ,$$

which we ask the reader to verify in Exercise 10.

1.3.16 EXAMPLE Let $\mathbf{u} = \begin{bmatrix} -1 \\ 1 \\ 0 \end{bmatrix}$ and $\mathbf{v} = \begin{bmatrix} 1 \\ 1 \\ 1 \end{bmatrix}$. Then

$$\mathbf{u} \times \mathbf{v} = \begin{vmatrix} \mathbf{i} & \mathbf{j} & \mathbf{k} \\ -1 & 1 & 0 \\ 1 & 1 & 1 \end{vmatrix} = 1\mathbf{i} - (-1)\mathbf{j} - 2\mathbf{k} = \begin{bmatrix} 1 \\ 1 \\ -2 \end{bmatrix},$$

so $\|\mathbf{u} \times \mathbf{v}\| = \sqrt{6}$. Since $\|\mathbf{u}\| = \sqrt{2}$ and $\|\mathbf{v}\| = \sqrt{3}$, part 8 of Theorem 1.3.14 says

$$\sqrt{6} = \sqrt{2}\sqrt{3} \sin\theta,$$

and so $\sin\theta = 1$. Thus $\theta = \frac{\pi}{2}$, which is indeed the case: the given vectors are orthogonal since $\mathbf{u} \cdot \mathbf{v} = 0$. ■

What follows is another illustration of the formula $\|\mathbf{u} \times \mathbf{v}\| = \|\mathbf{u}\|\,\|\mathbf{v}\| \sin\theta$.

1.3.17 EXAMPLE Let $\mathbf{u} = \begin{bmatrix} 1 \\ 2 \\ 3 \end{bmatrix}$ and $\mathbf{v} = \begin{bmatrix} -2 \\ 1 \\ 4 \end{bmatrix}$. Then

$$\mathbf{u} \times \mathbf{v} = \begin{vmatrix} \mathbf{i} & \mathbf{j} & \mathbf{k} \\ 1 & 2 & 3 \\ -2 & 1 & 4 \end{vmatrix} = 5\mathbf{i} - 10\mathbf{j} + 5\mathbf{k} = \begin{bmatrix} 5 \\ -10 \\ 5 \end{bmatrix} = 5 \begin{bmatrix} 1 \\ -2 \\ 1 \end{bmatrix},$$

so that $\|\mathbf{u} \times \mathbf{v}\| = 5 \left\| \begin{bmatrix} 1 \\ -2 \\ 1 \end{bmatrix} \right\| = 5\sqrt{6}$. Since $\|\mathbf{u}\| = \sqrt{14}$, $\|\mathbf{v}\| = \sqrt{21}$ and $\|\mathbf{u} \times \mathbf{v}\| = \|\mathbf{u}\|\,\|\mathbf{v}\| \sin\theta$, we should have $5\sqrt{6} = \sqrt{14}\sqrt{21} \sin\theta$. Since $\sqrt{14}\sqrt{21} = \sqrt{2}\sqrt{7}\sqrt{3}\sqrt{7} = 7\sqrt{6}$, this is $5\sqrt{6} = 7\sqrt{6} \sin\theta$, which says $\sin\theta = \frac{5}{7}$.

Is this correct? We can check using the more familiar equation (3), which gives

$$\cos\theta = \frac{\mathbf{u} \cdot \mathbf{v}}{\|\mathbf{u}\|\,\|\mathbf{v}\|} = \frac{12}{7\sqrt{6}}$$

so that

$$\sin^2\theta = 1 - \cos^2\theta = 1 - \frac{144}{49(6)} = 1 - \frac{24}{49} = \frac{25}{49}.$$

For $0 \le \theta \le \pi$, $\sin\theta \ge 0$, so $\sin\theta = \frac{5}{7}$, in agreement with what we found previously. ■

Reading Challenge 3 *What is the angle between the vectors $\mathbf{u}$ and $\mathbf{v}$ in **1.3.17**? Give your answer in radians to three decimal place accuracy and in degrees to the nearest degree.*

Figure 1.19 explains part 9 of Theorem 1.3.14. The area of a parallelogram is the length of the base times the height. A parallelogram with sides $\mathbf{u}$ and $\mathbf{v}$ has base of length $\|\mathbf{u}\|$ and height $\|\mathbf{v}\| \sin\theta$. So the area is $\|\mathbf{u}\|\,\|\mathbf{v}\| \sin\theta = \|\mathbf{u} \times \mathbf{v}\|$.

Figure 1.19 The area of the parallelogram is base times height: $\|\mathbf{u}\|\,(\|\mathbf{v}\| \sin\theta)$, and this is precisely $\|\mathbf{u} \times \mathbf{v}\|$.

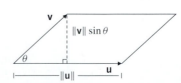

1.3.18 PROBLEM

What is the area of the parallelogram with adjacent sides $\mathbf{u} = \begin{bmatrix} -3 \\ 4 \\ 1 \end{bmatrix}$ and $\mathbf{v} = \begin{bmatrix} 1 \\ 0 \\ -2 \end{bmatrix}$?

Solution. We compute $\mathbf{u} \times \mathbf{v} = \begin{vmatrix} \mathbf{i} & \mathbf{j} & \mathbf{k} \\ -3 & 4 & 1 \\ 1 & 0 & -2 \end{vmatrix} = -8\mathbf{i} - 5\mathbf{j} - 4\mathbf{k} = \begin{bmatrix} -8 \\ -5 \\ -4 \end{bmatrix}$ and conclude that the area of the parallelogram is $\|\mathbf{u} \times \mathbf{v}\| = \sqrt{64 + 25 + 16} = \sqrt{105}$.

Reading Challenge 4

What is the area of the triangle two of whose sides are the vectors $\mathbf{u}$ and $\mathbf{v}$ of Problem 1.3.18?

One property that does not appear in Theorem 1.3.14 is *associativity*. An operation $\star$ is associative if $(a \star b) \star c = a \star (b \star c)$ for all a, b, and c. Ordinary addition and multiplication of real numbers are examples of associative operations. The cross product of vectors, however, is *not* associative. In general,

1.3.19 $(\mathbf{u} \times \mathbf{v}) \times \mathbf{w} \neq \mathbf{u} \times (\mathbf{v} \times \mathbf{w})$.

1.3.20 EXAMPLE

Let $\mathbf{u}$, $\mathbf{v}$, and $\mathbf{w}$ be the standard three-dimensional basis vectors $\mathbf{i}$, $\mathbf{j}$, and $\mathbf{k}$. Then

$$\mathbf{i} \times \mathbf{i} = \begin{vmatrix} \mathbf{i} & \mathbf{j} & \mathbf{k} \\ 1 & 0 & 0 \\ 1 & 0 & 0 \end{vmatrix} = 0\mathbf{i} - 0\mathbf{j} + 0\mathbf{k} = \begin{bmatrix} 0 \\ 0 \\ 0 \end{bmatrix}$$

(this is another illustration of the fact that $\mathbf{u} \times \mathbf{u} = \mathbf{0}$ for any $\mathbf{u}$), and hence

$$(\mathbf{i} \times \mathbf{i}) \times \mathbf{j} = \mathbf{0} \times \mathbf{j} = \mathbf{0}.$$

On the other hand,

$$\mathbf{i} \times \mathbf{j} = \begin{vmatrix} \mathbf{i} & \mathbf{j} & \mathbf{k} \\ 1 & 0 & 0 \\ 0 & 1 & 0 \end{vmatrix} = 0\mathbf{i} - 0\mathbf{j} + 1\mathbf{k} = \begin{bmatrix} 0 \\ 0 \\ 1 \end{bmatrix} = \mathbf{k} \tag{4}$$

so that

$$\mathbf{i} \times (\mathbf{i} \times \mathbf{j}) = \mathbf{i} \times \mathbf{k} = \begin{vmatrix} \mathbf{i} & \mathbf{j} & \mathbf{k} \\ 1 & 0 & 0 \\ 0 & 0 & 1 \end{vmatrix} = 0\mathbf{i} - 1\mathbf{j} + 0\mathbf{k} = -\mathbf{j}.$$

Most certainly, $(\mathbf{i} \times \mathbf{i}) \times \mathbf{j} \neq \mathbf{i} \times (\mathbf{i} \times \mathbf{j})$. ∎

In (4), we showed that $i \times j = k$. The cross products of any two of i, j, k are quite easy to remember. First of all, the cross product of any of these with itself is 0; remember that $u \times u = 0$ for any u. Think of i, j, k arranged clockwise in a circle, as at the right. The cross product of any two consecutive of these vectors is the next; that is,

1.3.21 $i \times j = k$, $\quad j \times k = i$, $\quad k \times i = j$.

Cross products in the reverse order are the negatives of these, by Theorem 1.3.14, part 1.

$$j \times i = -k, \quad k \times j = -i, \quad i \times k = -j.$$

Reading Challenge 5 *Verify these three statements using the definition of cross product—1.3.11.*

The cross product was introduced with the promise that it would give us a vector perpendicular to two given vectors. This is precisely what part 5 of Theorem 1.3.14 asserts.

1.3.22 PROBLEM Find a unit vector orthogonal to both $u = \begin{bmatrix} 1 \\ -4 \\ 1 \end{bmatrix}$ and $v = \begin{bmatrix} 2 \\ 3 \\ 0 \end{bmatrix}$.

Solution. $u \times v = \begin{vmatrix} i & j & k \\ 1 & -4 & 1 \\ 2 & 3 & 0 \end{vmatrix} = -3i - (-2)j + 11k = \begin{bmatrix} -3 \\ 2 \\ 11 \end{bmatrix}$. Notice that $\begin{bmatrix} -3 \\ 2 \\ 11 \end{bmatrix} \cdot u =$

$-3(1) + 2(-4) + 11(1) = 0$ and $\begin{bmatrix} -3 \\ 2 \\ 11 \end{bmatrix} \cdot v = -3(2) + 2(3) + 11(0) = 0$ so that, indeed,

$u \times v$ is orthogonal to both u and v. This vector has length $\sqrt{9 + 4 + 121} = \sqrt{134}$,

thus a unit vector orthogonal to the given two vectors is $\frac{1}{\sqrt{134}} \begin{bmatrix} -3 \\ 2 \\ 11 \end{bmatrix}$.

1.3.23 **The orientation of the cross product.** Thinking of u and v as vectors represented by arrows in the xy-plane, there are two possibilities for a perpendicular direction: up and down. Which is correct depends on the relative positions of the vectors and is illustrated by the fact that $i \times j = k$. If you are familiar with a screwdriver, you know that a screw goes into the wood if turned clockwise ⟳ and comes out of the wood if turned counterclockwise: ⟲ . The orientation of the cross product is determined in the same way depending on whether the turn from u to v (through that angle θ, $0 \leq \theta < \pi$) is counterclockwise or clockwise. Place the thumb, index finger, and middle finger of your right hand at right angles to each other. If your

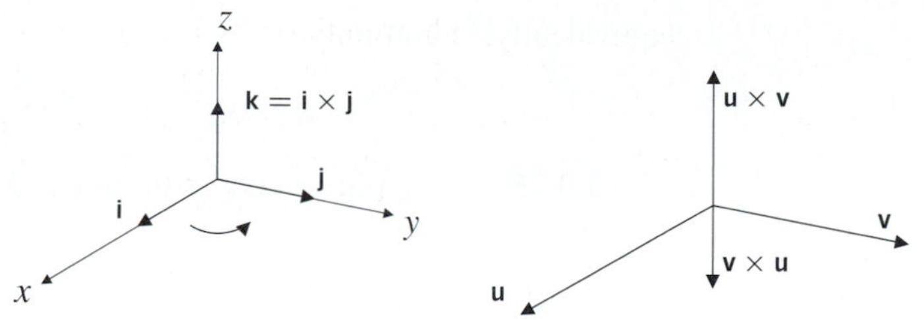

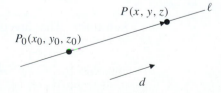

Figure 1.20 $\mathbf{u} \times \mathbf{v}$ points in the direction a screw would move if turned in the direction $\mathbf{u}$ to $\mathbf{v}$.

Figure 1.21 The right hand rule: If the thumb and first finger of your right hand represent $\mathbf{u}$ and $\mathbf{v}$, respectively, then your middle finger points in the direction of $\mathbf{u} \times \mathbf{v}$.

thumb corresponds to $\mathbf{u}$ and your index finger to $\mathbf{v}$, then your middle finger will point in the direction of $\mathbf{u} \times \mathbf{v}$. (See Figures 1.20 and 1.21.)

■ The Equation of a Line

We now turn our attention to the equation of a line in 3-space. What will the form of such an equation be? We know it is not linear; equations like $ax + by + cz = d$ represent planes.

With reference to Figure 1.22, observe that a line ℓ is determined by a direction $\mathbf{d} = \begin{bmatrix} a \\ b \\ c \end{bmatrix}$ and a point $P_0(x_0, y_0, z_0)$ on it. In fact, $P(x, y, z)$ is on ℓ if and only if the vector $\overrightarrow{P_0 P}$ is a multiple of $\mathbf{d}$; that is, if and only if

$$\begin{bmatrix} x - x_0 \\ y - y_0 \\ z - z_0 \end{bmatrix} = t\mathbf{d} = \begin{bmatrix} ta \\ tb \\ tc \end{bmatrix}$$

for some scalar t, and this occurs if and only if we have the following:

1.3.24	Vector equation of a line:	$\begin{bmatrix} x \\ y \\ z \end{bmatrix} = \begin{bmatrix} x_0 \\ y_0 \\ z_0 \end{bmatrix} + t \begin{bmatrix} a \\ b \\ c \end{bmatrix},$

Figure 1.22 A line is determined by a direction $\mathbf{d}$ and a point P_0 on the line.

equivalently, if and only if

1.3.25	**parametric equations of a line:**	$x = x_0 + ta$ $y = y_0 + tb$ $z = z_0 + tc$

for some scalar t.

1.3.26 PROBLEM

Find the vector and parametric equations of the line through $P(-1, 2, 5)$ in the direction of $\mathbf{d} = \begin{bmatrix} 2 \\ -3 \\ 4 \end{bmatrix}$.

Solution.   The vector equation is $\begin{bmatrix} x \\ y \\ z \end{bmatrix} = \begin{bmatrix} -1 \\ 2 \\ 5 \end{bmatrix} + t \begin{bmatrix} 2 \\ -3 \\ 4 \end{bmatrix}$ and the parametric equations

are $\begin{aligned} x &= -1 + 2t \\ y &= 2 - 3t \\ z &= 5 + 4t. \end{aligned}$

1.3.27 PROBLEM

Does the line in Problem 1.3.26 pass through the point $P(-7, 11, -7)$? Does it pass through $Q(0, -1, 2)$?

Solution.   The line passes through P if, for some number t, $\begin{bmatrix} -7 \\ 11 \\ -7 \end{bmatrix} = \begin{bmatrix} -1 \\ 2 \\ 5 \end{bmatrix} + t \begin{bmatrix} 2 \\ -3 \\ 4 \end{bmatrix}$.

We find that $t = -3$: the line does pass through $(-7, 11, -7)$.

It passes through Q if there is a number t such that $\begin{bmatrix} 0 \\ -1 \\ 2 \end{bmatrix} = \begin{bmatrix} -1 \\ 2 \\ 5 \end{bmatrix} + t \begin{bmatrix} 2 \\ -3 \\ 4 \end{bmatrix}$.

We would need

$$0 = -1 + 2t$$
$$-1 = 2 - 3t$$
$$2 = 5 + 4t.$$

There is no solution; the line does not pass through $(0, -1, 2)$.

1.3.28 PROBLEM

Find the equation of the line through $P(2, 0, 1)$ and $Q(4, -1, 1)$.

Solution.   The direction of the line is $\mathbf{d} = \overrightarrow{PQ} = \begin{bmatrix} 2 \\ -1 \\ 0 \end{bmatrix}$. So the line has vector

equation $\begin{bmatrix} x \\ y \\ z \end{bmatrix} = \begin{bmatrix} 2 \\ 0 \\ 1 \end{bmatrix} + t \begin{bmatrix} 2 \\ -1 \\ 0 \end{bmatrix}$.

Another correct answer is $\begin{bmatrix} x \\ y \\ z \end{bmatrix} = \begin{bmatrix} 4 \\ -1 \\ 1 \end{bmatrix} + t \begin{bmatrix} 2 \\ -1 \\ 0 \end{bmatrix}$.

Reading Challenge 6 *There are many possible answers to Problem 1.3.28. Find two (vector) equations for the line defined in Problem 1.3.28 other than the ones presented.*

1.3.29 PROBLEM

Find the equation of the line through $P(3, -1, 2)$ that is parallel to the line with equation $\begin{bmatrix} x \\ y \\ z \end{bmatrix} = \begin{bmatrix} -3 \\ 1 \\ 6 \end{bmatrix} + t \begin{bmatrix} 4 \\ -1 \\ -7 \end{bmatrix}$.

Solution. The two lines must have the same direction. So the direction of the sought-after line is any multiple of the given direction vector: The most obvious $\mathbf{d} = \begin{bmatrix} 4 \\ -1 \\ -7 \end{bmatrix}$ giving vector equation $\begin{bmatrix} x \\ y \\ z \end{bmatrix} = \begin{bmatrix} 3 \\ -1 \\ 2 \end{bmatrix} + t \begin{bmatrix} 4 \\ -1 \\ -7 \end{bmatrix}$.

1.3.30 PROBLEM

Do the lines with equations

$$\begin{bmatrix} x \\ y \\ z \end{bmatrix} = \begin{bmatrix} 1 \\ 2 \\ 1 \end{bmatrix} + t \begin{bmatrix} -3 \\ 5 \\ 1 \end{bmatrix} \quad \text{and} \quad \begin{bmatrix} x \\ y \\ z \end{bmatrix} = \begin{bmatrix} -1 \\ 3 \\ 1 \end{bmatrix} + t \begin{bmatrix} 1 \\ -4 \\ -1 \end{bmatrix}$$

intersect? If so, where?

Solution. If the lines intersect, say at the point (x, y, z), then (x, y, z) is on both lines. So there is a t such that

$$\begin{bmatrix} x \\ y \\ z \end{bmatrix} = \begin{bmatrix} 1 \\ 2 \\ 1 \end{bmatrix} + t \begin{bmatrix} -3 \\ 5 \\ 1 \end{bmatrix}$$

and an s *(not necessarily the same as t)* such that

$$\begin{bmatrix} x \\ y \\ z \end{bmatrix} = \begin{bmatrix} -1 \\ 3 \\ 1 \end{bmatrix} + s \begin{bmatrix} 1 \\ -4 \\ -1 \end{bmatrix}.$$

The question then is, do there exist parameters t and s so that

$$\begin{bmatrix} 1 \\ 2 \\ 1 \end{bmatrix} + t \begin{bmatrix} -3 \\ 5 \\ 1 \end{bmatrix} = \begin{bmatrix} -1 \\ 3 \\ 1 \end{bmatrix} + s \begin{bmatrix} 1 \\ -4 \\ -1 \end{bmatrix}?$$

We try to solve

$$1 - 3t = -1 + s$$
$$2 + 5t = 3 - 4s$$
$$1 + t = 1 - s;$$

that is,

$$s + 3t = 2$$
$$4s + 5t = 1$$
$$s + t = 0,$$

and find that $s = -1$, $t = 1$ is a solution. The lines intersect where

$$\begin{bmatrix} x \\ y \\ z \end{bmatrix} = \begin{bmatrix} 1 \\ 2 \\ 1 \end{bmatrix} + t \begin{bmatrix} -3 \\ 5 \\ 1 \end{bmatrix} = \begin{bmatrix} 1 \\ 2 \\ 1 \end{bmatrix} + 1 \begin{bmatrix} -3 \\ 5 \\ 1 \end{bmatrix} = \begin{bmatrix} -2 \\ 7 \\ 2 \end{bmatrix};$$

that is, at the point $(-2, 7, 2)$.

1.3.31 PROBLEM   Does the line with equation $\begin{bmatrix} x \\ y \\ z \end{bmatrix} = \begin{bmatrix} -1 \\ 7 \\ 0 \end{bmatrix} + t \begin{bmatrix} 1 \\ 2 \\ -1 \end{bmatrix}$ intersect the plane with equation $2x - 3y + 5z = 4$? If yes, find the point of intersection.

Solution.   The direction of the line is $\mathbf{d} = \begin{bmatrix} 1 \\ 2 \\ -1 \end{bmatrix}$. A normal to the plane is $\mathbf{n} = \begin{bmatrix} 2 \\ -3 \\ 5 \end{bmatrix}$.
Since $\mathbf{d} \cdot \mathbf{n} \neq 0$, $\mathbf{d}$ and $\mathbf{n}$ are not orthogonal, so the line is not parallel to the plane. The plane and the line must intersect. To find the point of intersection, substitute $x = -1 + t$, $y = 7 + 2t$, $z = -t$ in the equation for the plane as follows:

$$2(-1 + t) - 3(7 + 2t) + 5(-t) = 4$$
$$-23 - 9t = 4$$
$$-9t = 27$$
$$t = -3.$$

So the point of intersection occurs where $x = -1 + t = -4$, $y = 7 + 2t = 1$, $z = -t = 3$; that is, $(-4, 1, 3)$.

Answers to Reading Challenges

1. Any two nonparallel vectors in the plane will do. By inspection, $\begin{bmatrix} 9 \\ 1 \\ 1 \end{bmatrix}$ and $\begin{bmatrix} 13 \\ 2 \\ 1 \end{bmatrix}$ work.

2. Substituting the coordinates of B, we obtain $25(1) + 7(1) + 2(5) = 42$. Substituting the coordinates of C, we obtain $25(2) + 7(-2) + 2(3) = 42$, as before.

3. Since $\sin\theta > 0$ and $\cos\theta > 0$, we know $0 < \theta < \frac{\pi}{2}$. Since $\sin\theta = \frac{5}{7}$, $\theta = \arcsin\frac{5}{7} \approx .796$ rads $\approx 46°$.

4. The triangle has an area of one-half the area of the parallelogram, $\frac{1}{2}\sqrt{105}$.

5. $\mathbf{j} \times \mathbf{i} = \begin{vmatrix} \mathbf{i} & \mathbf{j} & \mathbf{k} \\ 0 & 1 & 0 \\ 1 & 0 & 0 \end{vmatrix} = \begin{vmatrix} 1 & 0 \\ 0 & 0 \end{vmatrix} \mathbf{i} - \begin{vmatrix} 0 & 0 \\ 1 & 0 \end{vmatrix} \mathbf{j} + \begin{vmatrix} 0 & 1 \\ 1 & 0 \end{vmatrix} \mathbf{k} = 0\mathbf{i} - 0\mathbf{j} + (-1)\mathbf{k} = -\mathbf{k}.$

$\mathbf{k} \times \mathbf{j} = \begin{vmatrix} \mathbf{i} & \mathbf{j} & \mathbf{k} \\ 0 & 0 & 1 \\ 0 & 1 & 0 \end{vmatrix} = \begin{vmatrix} 0 & 1 \\ 1 & 0 \end{vmatrix} \mathbf{i} - \begin{vmatrix} 0 & 1 \\ 0 & 0 \end{vmatrix} \mathbf{j} + \begin{vmatrix} 0 & 0 \\ 0 & 1 \end{vmatrix} \mathbf{k} = (-1)\mathbf{i} - 0\mathbf{j} + 0\mathbf{k} = -\mathbf{i}.$

$\mathbf{i} \times \mathbf{k} = \begin{vmatrix} \mathbf{i} & \mathbf{j} & \mathbf{k} \\ 1 & 0 & 0 \\ 0 & 0 & 1 \end{vmatrix} = \begin{vmatrix} 0 & 0 \\ 0 & 1 \end{vmatrix} \mathbf{i} - \begin{vmatrix} 1 & 0 \\ 0 & 1 \end{vmatrix} \mathbf{j} + \begin{vmatrix} 1 & 0 \\ 0 & 0 \end{vmatrix} \mathbf{k} = 0\mathbf{i} - 1\mathbf{j} + 0\mathbf{k} = -\mathbf{j}.$

6. We could replace the direction vector by any multiple of $\begin{bmatrix} 2 \\ -1 \\ 0 \end{bmatrix}$, so, for example,

$\begin{bmatrix} x \\ y \\ z \end{bmatrix} = \begin{bmatrix} 2 \\ 0 \\ 1 \end{bmatrix} + t \begin{bmatrix} 4 \\ -2 \\ 0 \end{bmatrix}$ is also an equation of the line. We could also replace

the point by any other point on the line. Given $\begin{bmatrix} x \\ y \\ z \end{bmatrix} = \begin{bmatrix} 2 \\ 0 \\ 1 \end{bmatrix} + t \begin{bmatrix} 2 \\ -1 \\ 0 \end{bmatrix}$, with

$t = 2$ we see that $(6, -2, 1)$ is also on the line, so another equation is $\begin{bmatrix} x \\ y \\ z \end{bmatrix} =$

$\begin{bmatrix} 6 \\ -2 \\ 1 \end{bmatrix} + t \begin{bmatrix} 2 \\ -1 \\ 0 \end{bmatrix}.$

True/False Questions

Decide, with as little calculation as possible, whether each of the following statements is true or false and explain your answer whenever you say "false." (Answers can be found in the back of the book.)

1. A line is an equation of the form $y = mx + b$.

2. For vectors $\mathbf{u}$ and $\mathbf{v}$ with θ the angle between, $|\mathbf{u} \times \mathbf{v}| = \|\mathbf{u}\| \|\mathbf{v}\| \sin \theta$.

3. For vectors $\mathbf{u}$ and $\mathbf{v}$ with θ the angle between, $\|\mathbf{u} \cdot \mathbf{v}\| = \|\mathbf{u}\| \|\mathbf{v}\| \cos \theta$.

4. For vectors $\mathbf{u}$ and $\mathbf{v}$ with θ the angle between, $|\mathbf{u} \cdot \mathbf{v}| = \|\mathbf{u}\| \|\mathbf{v}\| \cos \theta$.

5. If $\|\mathbf{u} \times \mathbf{v}\| = \|\mathbf{u}\| \|\mathbf{v}\|$, then $\mathbf{u}$ and $\mathbf{v}$ are orthogonal.

6. There exist vectors $\mathbf{u}$ and $\mathbf{v}$ such that $\|\mathbf{u}\| = 4$, $\|\mathbf{v}\| = 5$ and $\|\mathbf{u} \times \mathbf{v}\| = -10$.

7. The point $(1, 1, 1)$ lies on the plane with equation $2x - 3y + 5z = 4$.

8. The line with vector equation $\begin{bmatrix} x \\ y \\ z \end{bmatrix} = \begin{bmatrix} -1 \\ 0 \\ 2 \end{bmatrix} + t \begin{bmatrix} 1 \\ 2 \\ 3 \end{bmatrix}$ is parallel to the plane with

equation $x + 2y + 3z = 0$.

9. The planes with equations $-2x + 3y - z = 2$ and $4x - 6y + 2z = 0$ are parallel.

10. The line with vector equation $\begin{bmatrix} x \\ y \\ z \end{bmatrix} = \begin{bmatrix} -1 \\ 1 \\ 2 \end{bmatrix} + t \begin{bmatrix} 1 \\ 2 \\ 3 \end{bmatrix}$ is in the plane with equation

$x - y + 2z = 0$.

11. $\begin{vmatrix} 2 & -3 \\ 5 & 4 \end{vmatrix} = -7.$

12. If $\mathbf{u}$ and $\mathbf{v}$ are vectors and $\mathbf{u} \times \mathbf{v} = \mathbf{0}$, then $\mathbf{u} = \mathbf{v}$.

13. $\mathbf{i} \times (\mathbf{i} \times \mathbf{j}) = -\mathbf{j}$.

14. A line is parallel to a plane if the dot product of a direction vector of the line and a normal to the plane is 0.

Exercises

*Solutions to exercises marked [BB] can be found in the **Back** of the **Book**.*

1. Express the planes with each of the given equations as the set of all linear combinations of two nonparallel vectors. (In each case, there are many possibilities!)

   (a) [BB] $3x - 2y + z = 0$

   (b) $x - 3y = 0$

   (c) $2x - 3y + 4z = 0$

   (d) $2x + 5y - z = 0$

   (e) $3x - 2y + 4z = 0$

   (f) $4x - y + 3z = 0$

2. Find the equation of the plane that consists of all linear combinations of $\mathbf{u}$ and $\mathbf{v}$.

   (a) [BB] $\mathbf{u} = \begin{bmatrix} 1 \\ 0 \\ 4 \end{bmatrix}$, $\mathbf{v} = \begin{bmatrix} -1 \\ 2 \\ 0 \end{bmatrix}$

   (b) $\mathbf{u} = \begin{bmatrix} 1 \\ -3 \\ 0 \end{bmatrix}$, $\mathbf{v} = \begin{bmatrix} 1 \\ 2 \\ 6 \end{bmatrix}$

   (c) $\mathbf{u} = \begin{bmatrix} 2 \\ 5 \\ 1 \end{bmatrix}$, $\mathbf{v} = \begin{bmatrix} -1 \\ 1 \\ 1 \end{bmatrix}$

   (d) $\mathbf{u} = \begin{bmatrix} -3 \\ -1 \\ 1 \end{bmatrix}$, $\mathbf{v} = \begin{bmatrix} 2 \\ -6 \\ -2 \end{bmatrix}$

3. Find the equation of each of the following planes:

   (a) [BB] parallel to the plane with equation $4x + 2y - z = 7$ and passing through the point $(1, 2, 3)$

   (b) parallel to the plane with equation $18x + 6y - 5z = 0$ and passing through the point $(-1, 1, 7)$

   (c) parallel to the plane with equation $x - 3y + 7z = 2$ and passing through the point $(0, 4, -1)$.

4. Verify that $\mathbf{u} \times \mathbf{v} = -(\mathbf{v} \times \mathbf{u})$ in each of the following situations:

   (a) [BB] $\mathbf{u} = \begin{bmatrix} 3 \\ 0 \\ 2 \end{bmatrix}$ and $\mathbf{v} = \begin{bmatrix} -2 \\ 4 \\ 1 \end{bmatrix}$

   (b) $\mathbf{u} = \begin{bmatrix} 4 \\ -2 \\ -1 \end{bmatrix}$ and $\mathbf{v} = \begin{bmatrix} 5 \\ 6 \\ -3 \end{bmatrix}$

   (c) $\mathbf{u} = \begin{bmatrix} -3 \\ 1 \\ 4 \end{bmatrix}$ and $\mathbf{v} = \begin{bmatrix} 6 \\ -2 \\ -8 \end{bmatrix}$.

5. In each of the following cases, find the cross product $\mathbf{u} \times \mathbf{v}$, verify that this vector is perpendicular to $\mathbf{u}$ and to $\mathbf{v}$, and verify that $\mathbf{v} \times \mathbf{u} = -(\mathbf{u} \times \mathbf{v})$.

   (a) [BB] $\mathbf{u} = \begin{bmatrix} 1 \\ -2 \\ 1 \end{bmatrix}$, $\mathbf{v} = \begin{bmatrix} 3 \\ 1 \\ -2 \end{bmatrix}$

   (b) $\mathbf{u} = \begin{bmatrix} -3 \\ 1 \\ 0 \end{bmatrix}$, $\mathbf{v} = \begin{bmatrix} 4 \\ -1 \\ 5 \end{bmatrix}$

   (c) $\mathbf{u} = \begin{bmatrix} 1 \\ -3 \\ 0 \end{bmatrix}$ and $\mathbf{v} = \begin{bmatrix} 2 \\ 1 \\ -5 \end{bmatrix}$

6. Are there vectors $\mathbf{u}$ and $\mathbf{v}$ such that $\|\mathbf{u} \times \mathbf{v}\| = 15$, $\|\mathbf{u}\| = 3$, and $\|\mathbf{v}\| = 4$? Explain.

7. [BB] Find two vectors perpendicular to both $\mathbf{u} = \begin{bmatrix} 1 \\ 2 \\ 0 \end{bmatrix}$ and $\mathbf{v} = \begin{bmatrix} 0 \\ 0 \\ 3 \end{bmatrix}$.

8. Find a unit vector that is perpendicular to both $\begin{bmatrix} 4 \\ 3 \\ -1 \end{bmatrix}$

and $\begin{bmatrix} 2 \\ 3 \\ 0 \end{bmatrix}$.

9. [BB] Find a vector of length 5 that is perpendicular to both $\begin{bmatrix} 1 \\ 3 \\ -1 \end{bmatrix}$ and $\begin{bmatrix} 5 \\ 0 \\ 1 \end{bmatrix}$.

10. (a) As in **1.3.17**, verify that $\|\mathbf{u} \times \mathbf{v}\| = \|\mathbf{u}\| \, \|\mathbf{v}\| \sin \theta$ in each of the following cases:

    i. [BB] $\mathbf{u} = \begin{bmatrix} 1 \\ 2 \\ 3 \end{bmatrix}$, $\mathbf{v} = \begin{bmatrix} 0 \\ -1 \\ 1 \end{bmatrix}$

    ii. $\mathbf{u} = \begin{bmatrix} -2 \\ -4 \\ 1 \end{bmatrix}$, $\mathbf{v} = \begin{bmatrix} 1 \\ 5 \\ -2 \end{bmatrix}$

    iii. $\mathbf{u} = \begin{bmatrix} -1 \\ 2 \\ 1 \end{bmatrix}$, $\mathbf{v} = \begin{bmatrix} -3 \\ 0 \\ 2 \end{bmatrix}$

    iv. $\mathbf{u} = \begin{bmatrix} 0 \\ 1 \\ -1 \end{bmatrix}$ and $\mathbf{v} = \begin{bmatrix} 2 \\ 1 \\ 2 \end{bmatrix}$

    v. $\mathbf{u} = \begin{bmatrix} 3 \\ -4 \\ 1 \end{bmatrix}$ and $\mathbf{v} = \begin{bmatrix} -2 \\ 1 \\ 3 \end{bmatrix}$.

  (b) Verify the formula for general $\mathbf{u} = \begin{bmatrix} u_1 \\ u_2 \\ u_3 \end{bmatrix}$ and

$\mathbf{v} = \begin{bmatrix} v_1 \\ v_2 \\ v_3 \end{bmatrix}$ by the method suggested in the text;

that is, first show that $\|\mathbf{u} \times \mathbf{v}\|^2 + (\mathbf{u} \cdot \mathbf{v})^2 = \|\mathbf{u}\|^2 \, \|\mathbf{v}\|^2$.

11. Find the equation of each of the following planes:

  (a) [BB] passing through $A(2, 1, 3)$, $B(3, -1, 5)$, and $C(0, 2, -4)$

  (b) passing through $A(-1, 2, 1)$, $B(0, 1, 1)$, and $C(7, -3, 0)$

  (c) passing through $A(1, 2, 3)$, $B(-1, 5, 2)$, and $C(3, -1, 1)$

  (d) passing through $A(3, 2, 0)$, $B(-1, 0, 4)$, and $C(0, -3, 2)$.

12. Find the equation of each of the following planes:

  (a) [BB] containing $P(3, 1, -1)$ and the line with equation $\begin{bmatrix} x \\ y \\ z \end{bmatrix} = \begin{bmatrix} 1 \\ 0 \\ 2 \end{bmatrix} + t \begin{bmatrix} -4 \\ 3 \\ 1 \end{bmatrix}$

  (b) containing $P(-1, 2, 5)$ and the line with equation $\begin{bmatrix} x \\ y \\ z \end{bmatrix} = \begin{bmatrix} 3 \\ -1 \\ 1 \end{bmatrix} + t \begin{bmatrix} -5 \\ 7 \\ 7 \end{bmatrix}$

  (c) containing $P(2, -3, 1)$ and the line with equation $\begin{bmatrix} x \\ y \\ z \end{bmatrix} = \begin{bmatrix} 0 \\ -3 \\ 4 \end{bmatrix} + t \begin{bmatrix} 1 \\ -3 \\ 2 \end{bmatrix}$.

13. Find equations of the form $ax + by = c$ for the lines in the Euclidean plane with the following vector equations:

  (a) [BB] $\begin{bmatrix} x \\ y \end{bmatrix} = \begin{bmatrix} 1 \\ 2 \end{bmatrix} + t \begin{bmatrix} -1 \\ 1 \end{bmatrix}$

  (b) $\begin{bmatrix} x \\ y \end{bmatrix} = \begin{bmatrix} -3 \\ 5 \end{bmatrix} + t \begin{bmatrix} 1 \\ 2 \end{bmatrix}$

  (c) $\begin{bmatrix} x \\ y \end{bmatrix} = \begin{bmatrix} 2 \\ -7 \end{bmatrix} + t \begin{bmatrix} 4 \\ -3 \end{bmatrix}$

  (d) $\begin{bmatrix} x \\ y \end{bmatrix} = \begin{bmatrix} 4 \\ -1 \end{bmatrix} + t \begin{bmatrix} 2 \\ -2 \end{bmatrix}$

  (e) $\begin{bmatrix} x \\ y \end{bmatrix} = \begin{bmatrix} -7 \\ 3 \end{bmatrix} + t \begin{bmatrix} 6 \\ -5 \end{bmatrix}$.

14. Find vector equations for the lines in the Euclidean plane with each of the following equations:

  (a) [BB] $y = 2x - 3$ (b) $4x - 5y = 13$

  (c) $y = 9$ (d) $6y + 4x = 9$

  (e) $y = 5x + 7$.

15. [BB] The intersection of two planes is, in general, a line.

  (a) Find three points on the line that is the intersection of the planes with equations $2x + y + z = 5$ and $x - y + z = 1$.

  (b) Find a vector equation for the line in 15(a).

16. Find the equation of the line of intersection of
    (a) the planes with equations $x + 2y + 6z = 5$ and $x - y - 3z = -1$;
    (b) the planes with equations $x + y + z = 1$ and $2x - z = 0$;
    (c) the planes with equations $-2x + 3y + 7z = 4$ and $3x - y + 5z = 1$.

17. [BB] Find the point of intersection of the line whose equation is $\begin{bmatrix} x \\ y \\ z \end{bmatrix} = \begin{bmatrix} 1 \\ -2 \\ 3 \end{bmatrix} + t \begin{bmatrix} 2 \\ 5 \\ -1 \end{bmatrix}$ with the plane whose equation is $x - 3y + 2z = 4$.

18. Let ℓ_1 be the line with vector equation
    $\begin{bmatrix} x \\ y \\ z \end{bmatrix} = \begin{bmatrix} 2 \\ -3 \\ -4 \end{bmatrix} + t \begin{bmatrix} 1 \\ 2 \\ 3 \end{bmatrix}$ and let ℓ_2 be the line through $P(2, -2, 5)$ in the direction $\mathbf{d} = \begin{bmatrix} 2 \\ 3 \\ -3 \end{bmatrix}$.

    (a) [BB] Show that ℓ_1 and ℓ_2 intersect by finding the point of intersection.
    (b) Give a simple reason why ℓ_1 and the plane with equation $3x - 4y + z = 18$ must intersect. (No calculation is required.)
    (c) Find the point of intersection of ℓ_1 and the plane in 18(b)

19. Let π be the plane with equation $2x - 3y + 5z = 1$, let $\mathbf{u} = \begin{bmatrix} 1 \\ -2 \\ 3 \end{bmatrix}$ and $\mathbf{v} = \begin{bmatrix} 2 \\ 0 \\ 1 \end{bmatrix}$.

    (a) [BB] Find a vector perpendicular to π.
    (b) Find a vector perpendicular to both $\mathbf{u}$ and $\mathbf{v}$.
    (c) Find a point in the plane π.
    (d) Find an equation of the line through the point in (c) and parallel to $\mathbf{u}$.

20. Let $A(1, -1, -3)$, $B(2, 0, -3)$, $C(2, -5, 1)$, $D(3, -4, 1)$ be the indicated points of 3-space.

    (a) [BB] Show that A, B, C, D are the vertices of a parallelogram and find the area of this figure.
    (b) [BB] Find the area of triangle ABC.
    (c) Find the area of triangle ACD.

21. [BB] Find the area of the triangle in the plane with vertices $A(-2, 3)$, $B(4, 4)$, and $C(5, -3)$.

22. (a) [BB] Suppose a 2×2 determinant $\begin{vmatrix} a & b \\ c & d \end{vmatrix} = 0$.
    Show that one of the vectors $\begin{bmatrix} c \\ d \end{bmatrix}$, $\begin{bmatrix} a \\ b \end{bmatrix}$ is a scalar multiple of the other. There are four cases to consider, depending on whether $a = 0$ or $b = 0$.

    (b) Suppose $\mathbf{u} = \begin{bmatrix} u_1 \\ u_2 \\ u_3 \end{bmatrix}$, $\mathbf{v} = \begin{bmatrix} v_1 \\ v_2 \\ v_3 \end{bmatrix}$, and $\mathbf{u} \times \mathbf{v} = \mathbf{0}$.
    Show that for some scalar k, either $\mathbf{v} = k\mathbf{u}$ or $\mathbf{u} = k\mathbf{v}$, thus completing the proof of Theorem 1.3.14(7). [Hint: If $\mathbf{u} = \mathbf{0}$, then $\mathbf{u} = 0\mathbf{v}$, the desired result. So assume $\mathbf{u} \neq \mathbf{0}$ and use the result of 22(a) to show that $\mathbf{v}$ is a multiple of $\mathbf{u}$.]

■ Critical Reading

23. [BB] Let ℓ_1 and ℓ_2 be the lines with the equations
    $$\ell_1: \begin{bmatrix} x \\ y \\ z \end{bmatrix} = \begin{bmatrix} 2 \\ -1 \\ 3 \end{bmatrix} + t \begin{bmatrix} -1 \\ 2 \\ 1 \end{bmatrix};$$
    $$\ell_2: \begin{bmatrix} x \\ y \\ z \end{bmatrix} = \begin{bmatrix} 3 \\ 1 \\ a \end{bmatrix} + t \begin{bmatrix} 1 \\ -1 \\ b \end{bmatrix}.$$
    Given that these lines intersect at right angles, find the values of a and b.

24. Let π_1 and π_2 be the planes with equations $x - y + z = 3$ and $2x - 3y + z = 5$, respectively.
    (a) Find the equation of the line of intersection of π_1 and π_2.
    (b) For any scalars α and β, the equation $\alpha(x - y + z - 3) + \beta(2x - 3y + z - 5) = 0$ is that of a plane. Why?
    (c) Show that the plane in (b) contains the line of intersection of π_1 and π_2.

25. [BB] Find the equation of the plane each point of which is equidistant from the points $P(2, -1, 3)$ and $Q(1, 1, -1)$.

26. Show that the lines with vector equations
    $$\begin{bmatrix} x \\ y \\ z \end{bmatrix} = \begin{bmatrix} -1 \\ 4 \\ 4 \end{bmatrix} + t \begin{bmatrix} -1 \\ 5 \\ 2 \end{bmatrix} \text{ and } \begin{bmatrix} x \\ y \\ z \end{bmatrix} = \begin{bmatrix} 1 \\ -6 \\ 0 \end{bmatrix} + t \begin{bmatrix} 2 \\ -10 \\ -4 \end{bmatrix}$$
    are the same.

27. (a) Find the equation of the plane that contains the points $A(1, 0, -1)$, $B(0, 2, 3)$, and $C(-2, 1, 1)$.

(b) Find the area of triangle ABC, with A, B, C as in part (a).

(c) Find a point that is one unit away from the plane in part (a).

28. Let π be the plane through $P(5, 4, 8)$ that contains the line with equation $\begin{bmatrix} x \\ y \\ z \end{bmatrix} = \begin{bmatrix} 9 \\ 3 \\ 5 \end{bmatrix} + t \begin{bmatrix} 4 \\ 5 \\ -3 \end{bmatrix}$.

(a) What is the equation of π?

(b) Find the centers of two spheres that touch π at P and also touch the unit sphere in $\mathbf{R}^3$.

1.4 Projections

■ Projections onto Lines

What we call "projection," others call "orthogonal projection." Not all projections need be orthogonal, but they are in this book.

In this section, we discuss the notion of the *projection* of a vector onto a line or a plane. In each case, a picture really is worth a thousand words. In Figure 1.23, we show the projection of **u** on **v** in two different situations according to whether the angle between **u** and **v** is acute or obtuse.

Reading Challenge 1 *When does the projection* $\mathbf{p} = \text{proj}_\mathbf{v}\, \mathbf{u} = c\mathbf{v}$ *of* **u** *on* **v** *have the same direction as* **v**?

1.4.1 DEFINITION

The *projection* of a vector **u** on a (nonzero) vector **v** is the vector $\mathbf{p} = c\mathbf{v}$ with the property that $\mathbf{u} - \mathbf{p}$ is orthogonal to **v**. The projection of a vector **u** on a line is the projection of that vector onto any direction vector **v** for that line. We use the notation $\text{proj}_\mathbf{v}\, \mathbf{u}$ for the projection of **u** on **v**.

It is not very difficult to find the projection of one vector on another.

1.4.2 EXAMPLE

Suppose $\mathbf{u} = \begin{bmatrix} 1 \\ 1 \end{bmatrix}$ and $\mathbf{v} = \begin{bmatrix} 3 \\ 1 \end{bmatrix}$. The projection of **u** on **v** is a vector $\mathbf{p} = c\mathbf{v} = \begin{bmatrix} 3c \\ c \end{bmatrix}$ with the property that $\mathbf{u} - \mathbf{p} = \begin{bmatrix} 1 - 3c \\ 1 - c \end{bmatrix}$ is orthogonal to **v**. This leads to the equation

$$0 = (\mathbf{u} - \mathbf{p}) \cdot \mathbf{v} = 3(1 - 3c) + (1 - c) = -10c + 4$$

so, $c = \frac{2}{5}$ and $\mathbf{p} = \frac{2}{5}\begin{bmatrix} 3 \\ 1 \end{bmatrix} = \begin{bmatrix} \frac{6}{5} \\ \frac{2}{5} \end{bmatrix}$. ■

Figure 1.23 In both cases, $\mathbf{p} = \text{proj}_\mathbf{v}\, \mathbf{u}$ is the projection of **u** on **v**.

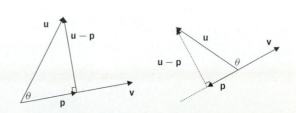

The method used in this example generalizes. We want $\mathbf{p} = c\mathbf{v}$ with $\mathbf{u} - \mathbf{p}$ orthogonal to $\mathbf{v}$, so

$$0 = (\mathbf{u} - \mathbf{p}) \cdot \mathbf{v} = \mathbf{u} \cdot \mathbf{v} - \mathbf{p} \cdot \mathbf{v} = \mathbf{u} \cdot \mathbf{v} - c(\mathbf{v} \cdot \mathbf{v})$$

and $c(\mathbf{v} \cdot \mathbf{v}) = \mathbf{u} \cdot \mathbf{v}$. Assuming $\mathbf{v} \neq \mathbf{0}$, we know $\|\mathbf{v}\| \neq 0$, so $\mathbf{v} \cdot \mathbf{v} = \|\mathbf{v}\|^2 \neq 0$ and $c = \frac{\mathbf{u} \cdot \mathbf{v}}{\mathbf{v} \cdot \mathbf{v}}$.

Theorem 1.4.3 *The projection of a vector $\mathbf{u}$ on a (nonzero) vector $\mathbf{v}$ is the vector*

$$\boxed{\operatorname{proj}_\mathbf{v} \mathbf{u} = \frac{\mathbf{u} \cdot \mathbf{v}}{\mathbf{v} \cdot \mathbf{v}} \mathbf{v}.}$$

Although our pictures and examples so far have been in the plane, everything we have done carries over to 3-space.

1.4.4 PROBLEM Find the projection of $\mathbf{u} = \begin{bmatrix} 1 \\ -3 \\ 4 \end{bmatrix}$ on $\mathbf{v} = \begin{bmatrix} -3 \\ 1 \\ 2 \end{bmatrix}$.

Solution. We have

$$\frac{\mathbf{u} \cdot \mathbf{v}}{\mathbf{v} \cdot \mathbf{v}} = \frac{-3 - 3 + 8}{9 + 1 + 4} = \frac{2}{14} = \frac{1}{7},$$

so the projection is $\operatorname{proj}_\mathbf{v} \mathbf{u} = \frac{1}{7}\mathbf{v} = \begin{bmatrix} -\frac{3}{7} \\ \frac{1}{7} \\ \frac{2}{7} \end{bmatrix}$. Notice that $\mathbf{u} - \operatorname{proj}_\mathbf{v} \mathbf{u} = \begin{bmatrix} \frac{10}{7} \\ -\frac{22}{7} \\ \frac{26}{7} \end{bmatrix}$ is indeed orthogonal to $\mathbf{v}$ since, checking the dot product, $-3\left(\frac{10}{7}\right) + \frac{-22}{7} + 2\left(\frac{26}{7}\right) = 0$.

1.4.5 PROBLEM Find two orthogonal vectors in the plane with equation $3x - 2y + z = 0$.

Solution. We find two nonparallel vectors $\mathbf{u}$ and $\mathbf{v}$ in the plane, find the projection $\mathbf{p}$ of $\mathbf{u}$ on $\mathbf{v}$, and take $\mathbf{v}$ and $\mathbf{u} - \mathbf{p}$ as our vectors. See Figure 1.23. For example, we might choose $\mathbf{u} = \begin{bmatrix} 1 \\ 0 \\ -3 \end{bmatrix}$ and $\mathbf{v} = \begin{bmatrix} 0 \\ 1 \\ 2 \end{bmatrix}$. (See Example 1.3.7 to recall how these are found.) Then

$$\mathbf{p} = \operatorname{proj}_\mathbf{v} \mathbf{u} = \frac{\mathbf{u} \cdot \mathbf{v}}{\mathbf{v} \cdot \mathbf{v}} \mathbf{v} = -\frac{6}{5} \begin{bmatrix} 0 \\ 1 \\ 2 \end{bmatrix}$$

and

$$\mathbf{u} - \mathbf{p} = \begin{bmatrix} 1 \\ 0 \\ -3 \end{bmatrix} + \frac{6}{5}\begin{bmatrix} 0 \\ 1 \\ 2 \end{bmatrix} = \begin{bmatrix} 1 \\ \frac{6}{5} \\ -\frac{3}{5} \end{bmatrix},$$

so we can answer the problem with the vectors $\begin{bmatrix} 0 \\ 1 \\ 2 \end{bmatrix}$ and $\begin{bmatrix} 1 \\ \frac{6}{5} \\ -\frac{3}{5} \end{bmatrix}$. In applications, we would probably multiply the second vector by 5 (this does not affect direction) and use $\begin{bmatrix} 0 \\ 1 \\ 2 \end{bmatrix}$ and $\begin{bmatrix} 5 \\ 6 \\ -3 \end{bmatrix}$.

■ The Distance from a Point to a Line

1.4.6 PROBLEM

Find the distance from the point $P(1, 8, -5)$ to the line with (vector) equation $\begin{bmatrix} x \\ y \\ z \end{bmatrix} = \begin{bmatrix} -1 \\ 1 \\ 6 \end{bmatrix} + t\begin{bmatrix} 3 \\ 0 \\ -4 \end{bmatrix}$.

Solution. Let Q be any point on the line, say $Q(-1, 1, 6)$, and let $\mathbf{w} = \overrightarrow{QP} = \begin{bmatrix} 2 \\ 7 \\ -11 \end{bmatrix}$.

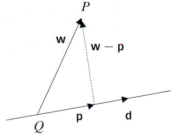

The line has direction $\mathbf{d} = \begin{bmatrix} 3 \\ 0 \\ -4 \end{bmatrix}$ and the distance we want is the length of $\mathbf{w} - \mathbf{p}$, where $\mathbf{p} = \text{proj}_{\mathbf{d}}\,\mathbf{w}$ is the projection of $\mathbf{w}$ on $\mathbf{d}$. The sketch to the right should make all this clear. We have

$$\mathbf{p} = \text{proj}_{\mathbf{d}}\,\mathbf{w} = \frac{\mathbf{w} \cdot \mathbf{d}}{\mathbf{d} \cdot \mathbf{d}}\,\mathbf{d} = \frac{50}{25}\begin{bmatrix} 3 \\ 0 \\ -4 \end{bmatrix} = \begin{bmatrix} 6 \\ 0 \\ -8 \end{bmatrix},$$

so $\mathbf{w} - \mathbf{p} = \begin{bmatrix} 2 \\ 7 \\ -11 \end{bmatrix} - \begin{bmatrix} 6 \\ 0 \\ -8 \end{bmatrix} = \begin{bmatrix} -4 \\ 7 \\ -3 \end{bmatrix}$. The required distance is $\|\mathbf{w} - \mathbf{p}\| = \sqrt{74}$.

■ The Distance from a Point to a Plane

1.4.7 PROBLEM

Find the distance from the point $P(-1, 2, 1)$ to the plane with equation $5x + y + 3z = 7$.

Solution.   Let Q be any point in the plane; for instance, $Q(1, -1, 1)$.

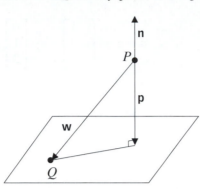

The required distance is the length of the projection $\mathbf{p}$ of $\mathbf{w} = \overrightarrow{PQ}$ on $\mathbf{n}$, a normal to the plane. We have $\mathbf{w} = \begin{bmatrix} 2 \\ -3 \\ 0 \end{bmatrix}$ and a normal $\mathbf{n} = \begin{bmatrix} 5 \\ 1 \\ 3 \end{bmatrix}$. Then $\operatorname{proj}_{\mathbf{n}} \mathbf{w} = \frac{\mathbf{w} \cdot \mathbf{n}}{\mathbf{n} \cdot \mathbf{n}}\, \mathbf{n} = \frac{7}{35}\, \mathbf{n} = \frac{1}{5}\mathbf{n}$. The required distance is

$$\left\| \frac{1}{5}\,\mathbf{n} \right\| = \frac{1}{5}\|\mathbf{n}\| = \frac{1}{5}\sqrt{35}.$$

■ Projections onto Planes through the Origin

As we turn our attention to the projection of a vector on a plane, it will be convenient to introduce some new terminology. In Section 1.1, we mentioned that the set of all linear combinations of two vectors determines a plane. (Remember the parallelogram rule for vector addition?) We call this the plane *spanned* by the vectors.

1.4.8 DEFINITION

The set of all linear combinations of vectors $\mathbf{u}$ and $\mathbf{v}$ is called the plane *spanned* by $\mathbf{u}$ and $\mathbf{v}$. We also say that $\mathbf{u}$ and $\mathbf{v}$ *span* the plane.

For example, $\mathbf{i} = \begin{bmatrix} 1 \\ 0 \\ 0 \end{bmatrix}$ and $\mathbf{j} = \begin{bmatrix} 0 \\ 1 \\ 0 \end{bmatrix}$ span the xy-plane in 3-space since the set of linear combinations of $\mathbf{i}$ and $\mathbf{j}$ is the set of vectors of the form

$$a\begin{bmatrix} 1 \\ 0 \\ 0 \end{bmatrix} + b\begin{bmatrix} 0 \\ 1 \\ 0 \end{bmatrix} = \begin{bmatrix} a \\ b \\ 0 \end{bmatrix}.$$

1.4.9 Remark The plane spanned by vectors $\mathbf{u}$ and $\mathbf{v}$ always passes through the origin since it contains the zero vector; $\mathbf{0} = 0\mathbf{u} + 0\mathbf{v}$ is a linear combination of $\mathbf{u}$ and $\mathbf{v}$.

1.4.10 PROBLEM

Let $\mathbf{u} = \begin{bmatrix} 1 \\ 3 \\ 2 \end{bmatrix}$ and $\mathbf{v} = \begin{bmatrix} 0 \\ -2 \\ 1 \end{bmatrix}$. Find an equation of the plane spanned by these vectors.

Solution.   This plane was investigated in Example 1.3.12. The answer is $7x - y - 2z = 0$.

1.4.11 DEFINITION

The *projection* of a vector $\mathbf{w}$ on a plane π is a vector $\mathbf{p}$ in π with the property that $\mathbf{w} - \mathbf{p}$ is orthogonal to every vector in π. We denote this vector $\operatorname{proj}_{\pi} \mathbf{w}$.

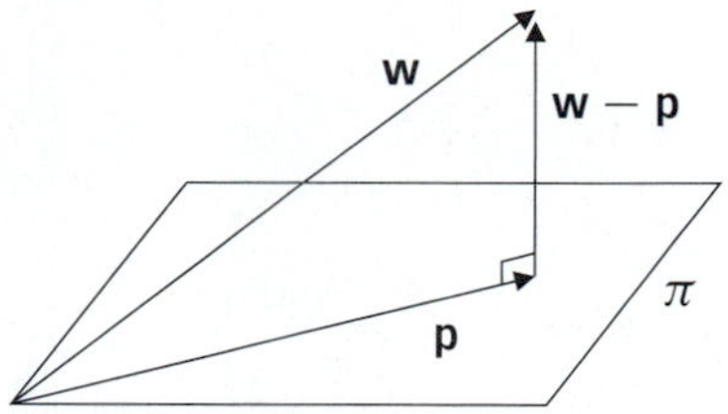

Figure 1.24 Vector $\mathbf{p} = \text{proj}_\pi \mathbf{w}$ is the projection of $\mathbf{w}$ on the plane π.

This concept is illustrated in Figure 1.24. The projection of $\mathbf{w} = \begin{bmatrix} a \\ b \\ c \end{bmatrix}$ on the

xy-plane is just $\mathbf{p} = \begin{bmatrix} a \\ b \\ 0 \end{bmatrix}$ since $\mathbf{p}$ is in the xy-plane and $\mathbf{w} - \mathbf{p} = \begin{bmatrix} 0 \\ 0 \\ c \end{bmatrix}$ is orthogonal

to every vector in the plane, so it is easy to find the projection of vectors in any of the three coordinate planes. But what about the projection of a vector onto some other plane? In the discussion that follows, we make use of a very important idea. (See Exercise 31 of Section 1.2.)

Theorem 1.4.12 *Suppose a plane π is spanned by vectors $\mathbf{u}$ and $\mathbf{v}$. If a vector is orthogonal just to $\mathbf{u}$ and $\mathbf{v}$, then it is orthogonal to π, that is, orthogonal to every vector in π.*

By the way, the symbol ■ is commonly used to mark the end of a proof.

Proof If $\mathbf{w}$ is in π, then $\mathbf{w}$ is a linear combination of $\mathbf{u}$ and $\mathbf{v}$, so $\mathbf{w} = a\mathbf{u} + b\mathbf{v}$ for some scalars a and b. Now suppose that $\mathbf{x}$ is orthogonal to both $\mathbf{u}$ and $\mathbf{v}$. Then $\mathbf{x} \cdot \mathbf{u} = \mathbf{x} \cdot \mathbf{v} = 0$, so using Theorem 1.2.19, $\mathbf{x} \cdot \mathbf{w} = a(\mathbf{x} \cdot \mathbf{u}) + b(\mathbf{x} \cdot \mathbf{v}) = 0 + 0 = 0$.■

One straightforward method for finding the projection of a vector on a plane is described in Exercise 18. Here, however, we adopt a different approach, which is a special case of a general procedure called the "Gram–Schmidt algorithm" that we discuss later. (See Section 6.2.)

Suppose our plane π is spanned by *orthogonal* vectors $\mathbf{e}$ and $\mathbf{f}$. (Remember that we know how to find such $\mathbf{e}$ and $\mathbf{f}$—see Problem 1.4.5.) This projection $\mathbf{p}$ is in π, so it is a linear combination of $\mathbf{e}$ and $\mathbf{f}$,

$$\mathbf{p} = a\mathbf{e} + b\mathbf{f},$$

for scalars a and b. Moreover, $\mathbf{p}$ must have the property that $\mathbf{w} - \mathbf{p}$ is perpendicular to every vector in π. By Theorem 1.4.12, it is sufficient to have $\mathbf{w} - \mathbf{p}$ perpendicular to just $\mathbf{e}$ and $\mathbf{f}$, so we require $(\mathbf{w} - \mathbf{p}) \cdot \mathbf{e} = 0$ and $(\mathbf{w} - \mathbf{p}) \cdot \mathbf{f} = 0$. Equivalently, we want $\mathbf{w} \cdot \mathbf{e} = \mathbf{p} \cdot \mathbf{e}$ and $\mathbf{w} \cdot \mathbf{f} = \mathbf{p} \cdot \mathbf{f}$. Here is where the orthogonality of $\mathbf{e}$ and $\mathbf{f}$ becomes useful:

$$\mathbf{w} \cdot \mathbf{e} = (a\mathbf{e} + b\mathbf{f}) \cdot \mathbf{e} = a(\mathbf{e} \cdot \mathbf{e}) \qquad (*)$$

because $\mathbf{f} \cdot \mathbf{e} = 0$. Assuming $\mathbf{e} \neq \mathbf{0}$ and hence $\mathbf{e} \cdot \mathbf{e} = \|\mathbf{e}\|^2 \neq 0$, the equation labeled (*) gives

$$a = \frac{\mathbf{w} \cdot \mathbf{e}}{\mathbf{e} \cdot \mathbf{e}}.$$

Similarly,

$$b = \frac{\mathbf{w} \cdot \mathbf{f}}{\mathbf{f} \cdot \mathbf{f}}$$

and so

$$\mathbf{p} = a\mathbf{e} + b\mathbf{f} = \frac{\mathbf{w} \cdot \mathbf{e}}{\mathbf{e} \cdot \mathbf{e}}\, \mathbf{e} + \frac{\mathbf{w} \cdot \mathbf{f}}{\mathbf{f} \cdot \mathbf{f}}\, \mathbf{f}.$$

We summarize as follows.

Theorem 1.4.13   *The projection of a vector* **w** *on the plane spanned by* orthogonal *vectors* **e** *and* **f** *is*

$$\mathrm{proj}_\pi\, \mathbf{w} = \frac{\mathbf{w} \cdot \mathbf{e}}{\mathbf{e} \cdot \mathbf{e}}\, \mathbf{e} + \frac{\mathbf{w} \cdot \mathbf{f}}{\mathbf{f} \cdot \mathbf{f}}\, \mathbf{f}.$$

Note the remarkable similarity of the formula here to that given in Theorem 1.4.3 for the projection of **w** on a single vector.

1.4.14 PROBLEM   Find the projection of $\mathbf{w} = \begin{bmatrix} 5 \\ -3 \\ -7 \end{bmatrix}$ on the plane π with equation $3x - 2y + z = 0$.

Solution.   We note that $\mathbf{e} = \begin{bmatrix} 5 \\ 6 \\ -3 \end{bmatrix}$ and $\mathbf{f} = \begin{bmatrix} 0 \\ 1 \\ 2 \end{bmatrix}$ are orthogonal vectors spanning π (see Problem 1.4.5) and simply write the answer as follows:

$$
\begin{aligned}
\mathrm{proj}_\pi\, \mathbf{w} &= \frac{\mathbf{w} \cdot \mathbf{e}}{\mathbf{e} \cdot \mathbf{e}}\, \mathbf{e} + \frac{\mathbf{w} \cdot \mathbf{f}}{\mathbf{f} \cdot \mathbf{f}}\, \mathbf{f} \\[2mm]
&= \frac{28}{70} \begin{bmatrix} 5 \\ 6 \\ -3 \end{bmatrix} - \frac{17}{5} \begin{bmatrix} 0 \\ 1 \\ 2 \end{bmatrix} = \frac{2}{5} \begin{bmatrix} 5 \\ 6 \\ -3 \end{bmatrix} - \frac{17}{5} \begin{bmatrix} 0 \\ 1 \\ 2 \end{bmatrix} = \begin{bmatrix} 2 \\ -1 \\ -8 \end{bmatrix}.
\end{aligned}
$$

Answers to Reading Challenges

1. This occurs when the angle θ between **u** and **v** is acute: $0 \le \theta < \frac{\pi}{2}$.

True/False Questions

Decide, with as little calculation as possible, whether each of the following statements is true or false and explain your answer whenever you say "false." (Answers can be found in the back of the book.)

1. The projection of a vector **w** on the plane spanned by vectors **e** and **f** is given by the formula $\mathrm{proj}_\pi\, \mathbf{w} = \frac{\mathbf{w} \cdot \mathbf{e}}{\mathbf{e} \cdot \mathbf{e}}\, \mathbf{e} + \frac{\mathbf{w} \cdot \mathbf{f}}{\mathbf{f} \cdot \mathbf{f}}\, \mathbf{f}$.

2. The projection of a vector $\mathbf{u}$ on a nonzero vector $\mathbf{v}$ is $\frac{\mathbf{u} \cdot \mathbf{v}}{\mathbf{v} \cdot \mathbf{v}} \mathbf{u}$.

3. If nonzero vectors $\mathbf{u}$ and $\mathbf{v}$ are orthogonal, you cannot find the projection of $\mathbf{u}$ on $\mathbf{v}$.

4. The projection of $\begin{bmatrix} 1 \\ 1 \\ 1 \end{bmatrix}$ on the xy-plane is $\begin{bmatrix} 1 \\ 1 \\ 0 \end{bmatrix}$.

5. The distance from $(-2, 3, 1)$ to the plane with equation $x - y + 5z = 0$ is 0.

6. The line whose vector equation is $\begin{bmatrix} x \\ y \\ z \end{bmatrix} = \begin{bmatrix} 0 \\ -3 \\ 5 \end{bmatrix} + t \begin{bmatrix} 2 \\ -4 \\ 1 \end{bmatrix}$ is parallel to the plane whose equation is $2x - 4y + z = 0$.

Exercises

*Solutions to exercises marked [BB] can be found in the **B**ack of the **B**ook.*

1. [BB] When does the projection of $\mathbf{u}$ on a nonzero vector $\mathbf{v}$ have a direction opposite that of $\mathbf{v}$?

2. Let $\mathbf{u} = \begin{bmatrix} -1 \\ 0 \\ 3 \end{bmatrix}$ and $\mathbf{v} = \begin{bmatrix} 3 \\ -1 \\ 0 \end{bmatrix}$. Determine whether $\mathbf{w}$ is in the plane spanned by $\mathbf{u}$ and $\mathbf{v}$ in each of the following cases:

   (a) [BB] $\mathbf{w} = \begin{bmatrix} 0 \\ 0 \\ 0 \end{bmatrix}$ (b) $\mathbf{w} = \begin{bmatrix} -7 \\ 3 \\ -6 \end{bmatrix}$

   (c) $\mathbf{w} = \begin{bmatrix} 1 \\ 1 \\ 1 \end{bmatrix}$ (d) $\mathbf{w} = \begin{bmatrix} 8 \\ -4 \\ 5 \end{bmatrix}$

   (e) $\mathbf{w} = \begin{bmatrix} -18 \\ 5 \\ 9 \end{bmatrix}$.

3. In each case below, find the projection of $\mathbf{u}$ on $\mathbf{v}$ and the projection of $\mathbf{v}$ on $\mathbf{u}$.

   (a) [BB] $\mathbf{u} = \begin{bmatrix} 1 \\ 2 \\ -1 \end{bmatrix}$ and $\mathbf{v} = \begin{bmatrix} 3 \\ 1 \\ 1 \end{bmatrix}$

   (b) $\mathbf{u} = \begin{bmatrix} 1 \\ 2 \\ 3 \end{bmatrix}$ and $\mathbf{v} = \begin{bmatrix} 4 \\ 5 \\ 6 \end{bmatrix}$

   (c) [BB] $\mathbf{u} = \begin{bmatrix} -1 \\ 0 \\ 1 \end{bmatrix}$ and $\mathbf{v} = \begin{bmatrix} 3 \\ 4 \\ -2 \end{bmatrix}$

   (d) $\mathbf{u} = \begin{bmatrix} 3 \\ -4 \\ -2 \end{bmatrix}$ and $\mathbf{v} = \begin{bmatrix} 7 \\ 2 \\ 0 \end{bmatrix}$

   (e) $\mathbf{u} = \begin{bmatrix} 6 \\ -2 \\ 3 \end{bmatrix}$ and $\mathbf{v} = \begin{bmatrix} -5 \\ 2 \\ 4 \end{bmatrix}$

   (f) $\mathbf{u} = \begin{bmatrix} 7 \\ 0 \\ 5 \end{bmatrix}$ and $\mathbf{v} = \begin{bmatrix} 3 \\ 1 \\ -4 \end{bmatrix}$

4. Find the distance from point P to line ℓ in each of the following cases:

   (a) [BB] $P(-1, 2, 1)$; ℓ has equation

   $$\begin{bmatrix} x \\ y \\ z \end{bmatrix} = \begin{bmatrix} 1 \\ 2 \\ 3 \end{bmatrix} + t \begin{bmatrix} 1 \\ 1 \\ 1 \end{bmatrix}$$

   (b) $P(1, 2, 3)$; ℓ has equation

   $$\begin{bmatrix} x \\ y \\ z \end{bmatrix} = \begin{bmatrix} -1 \\ 2 \\ 0 \end{bmatrix} + t \begin{bmatrix} 4 \\ -1 \\ 8 \end{bmatrix}$$

   (c) $P(4, 7, 3)$; ℓ has equation

   $$\begin{bmatrix} x \\ y \\ z \end{bmatrix} = \begin{bmatrix} 3 \\ 1 \\ 1 \end{bmatrix} + t \begin{bmatrix} 3 \\ -2 \\ 3 \end{bmatrix}.$$

5. In each of the given situations, find the distance from point P to the plane π. Also, find the point of π closest to P.

(a) [BB] P is $(2, 3, 0)$, π has equation $5x - y + z = 1$

(b) P is $(1, 1, 1)$, π has equation $x - 3y + 4z = 10$

(c) P is $(4, -1, 2)$, π has equation $3x - 2y + 4z = -2$

(d) P is $(1, 2, 1)$, π has equation $x - y + z = 0$.

6. (a) Find the equation of a plane π that is perpendicular to the line with equation

$$\begin{bmatrix} x \\ y \\ z \end{bmatrix} = \begin{bmatrix} 1 \\ 0 \\ -1 \end{bmatrix} + t \begin{bmatrix} 2 \\ -1 \\ 2 \end{bmatrix}$$

and 5 units away from the point $P(1, 4, 0)$. Is π unique? Explain.

(b) Find a point in π that is 5 units away from P.

7. Find two orthogonal vectors in each of the following planes:

(a) [BB] the plane spanned by $u = \begin{bmatrix} -1 \\ 1 \\ 4 \end{bmatrix}$ and

$v = \begin{bmatrix} 2 \\ 0 \\ 1 \end{bmatrix}$

(b) the plane spanned by $u = \begin{bmatrix} 1 \\ -2 \\ 3 \end{bmatrix}$ and $v = \begin{bmatrix} 1 \\ 1 \\ 1 \end{bmatrix}$

(c) the plane spanned by $u = \begin{bmatrix} -2 \\ 4 \\ 1 \end{bmatrix}$ and $v = \begin{bmatrix} 3 \\ 0 \\ -1 \end{bmatrix}$

(d) [BB] the plane with equation $2x - 3y + 4z = 0$

(e) the plane with equation $x - y + 5z = 0$

(f) the plane with equation $3x + 2y - z = 0$.

8. (a) [BB] Find two orthogonal vectors in the plane π with equation $2x - y + z = 0$.

(b) [BB] Use your answer to part (a) to find the projection of $w = \begin{bmatrix} 1 \\ 1 \\ 1 \end{bmatrix}$ on π.

9. (a) [BB] What is the projection of $w = \begin{bmatrix} 3 \\ 1 \\ 1 \end{bmatrix}$ onto the plane π with equation $2x + y - z = 0$?

(b) [BB] What is the projection of $w = \begin{bmatrix} x \\ y \\ z \end{bmatrix}$ onto π?

10. Find the projection of w on π in each of the following cases.

(a) [BB] $w = \begin{bmatrix} -2 \\ 3 \\ 0 \end{bmatrix}$, π has equation $x + y - 2z = 0$

(b) $w = \begin{bmatrix} 3 \\ 1 \\ -4 \end{bmatrix}$, π has equation $3x - y + 2z = 0$

(c) $w = \begin{bmatrix} x \\ y \\ z \end{bmatrix}$, π has equation $3x - 4y + 2z = 0$

11. Wendy has a theory that the projection of a vector w on the plane spanned by e and f should be the projection of w on the vector $e + f$. Let's test this theory with the vectors $w = \begin{bmatrix} 1 \\ 3 \\ 2 \end{bmatrix}$, $e = \begin{bmatrix} 2 \\ -1 \\ 1 \end{bmatrix}$, and

$f = \begin{bmatrix} 1 \\ 1 \\ -1 \end{bmatrix}$. Note that e and f are orthogonal.

(a) Find the projection p of w on the plane π spanned by e and f.

(b) Verify that $w - p$ is orthogonal to e and to f.

(c) Find the projection p' of w on the vector $e + f$.

(d) Is Wendy's theory correct?

(e) Is Wendy's vector in the right plane?

12. (a) [BB] Suppose a vector w is a linear combination of nonzero orthogonal vectors e and f. Prove that

$$w = \frac{w \cdot e}{e \cdot e} e + \frac{w \cdot f}{f \cdot f} f.$$

(b) The situation described in part (a) is pictured below. What happens if $e = i$, $f = j$, and $w = \begin{bmatrix} a \\ b \end{bmatrix}$?

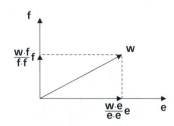

(c) Given that $\mathbf{w} = \begin{bmatrix} 13 \\ -20 \\ 15 \end{bmatrix}$ is in the span of the

orthogonal vectors $\mathbf{e} = \begin{bmatrix} 1 \\ -2 \\ 3 \end{bmatrix}$ and $\mathbf{f} = \begin{bmatrix} -1 \\ 1 \\ 1 \end{bmatrix}$,

write $\mathbf{w}$ as a linear combination of $\mathbf{e}$ and $\mathbf{f}$.

13. (a) Show that the distance of the point $P(x_0, y_0, z_0)$ from the plane with equation $ax + by + cz = d$ is $\frac{|ax_0 + by_0 + cz_0 - d|}{\sqrt{a^2 + b^2 + c^2}}$.

    [Yes, dear reader, there is a formula for the distance from a point to a plane. It was not mentioned in the text because the author wishes to emphasize the method, not the result.]

    (b) Use the formula in part (a) to determine the distance of $P(1, 2, 3)$ from the plane with equation $x - 3y + 5z = 4$.

    (c) Find the equation of the plane that contains the points $(1, 1, 0)$ and $(0, 1, 2)$ and is $\sqrt{6}$ units from the point $(3, -3, -2)$.

■ Critical Reading

14. (a) [BB] The lines with vector equations

    $\begin{bmatrix} x \\ y \\ z \end{bmatrix} = \begin{bmatrix} 0 \\ 1 \\ 4 \end{bmatrix} + t \begin{bmatrix} -2 \\ 1 \\ 2 \end{bmatrix}$ and

    $\begin{bmatrix} x \\ y \\ z \end{bmatrix} = \begin{bmatrix} 4 \\ -2 \\ 2 \end{bmatrix} + t \begin{bmatrix} 6 \\ -3 \\ -6 \end{bmatrix}$ are parallel. Why?

    (b) [BB] Find the distance between the lines in (a).

15. Let ℓ be the line with equation

    $\begin{bmatrix} x \\ y \\ z \end{bmatrix} = \begin{bmatrix} 3 \\ 5 \\ -4 \end{bmatrix} + t \begin{bmatrix} 2 \\ 1 \\ 3 \end{bmatrix}$ and let π be the plane with

    equation $-2x + y + z = 3$.
    (a) Show that ℓ is parallel to π.

    (b) Find a point P on ℓ and a point Q on π.

    (c) Find the (shortest) distance from ℓ to π.

16. Let ℓ_1 and ℓ_2 be the lines with equations

    $\ell_1 : \begin{bmatrix} x \\ y \\ z \end{bmatrix} = \begin{bmatrix} 3 \\ 0 \\ -1 \end{bmatrix} + t \begin{bmatrix} 4 \\ 3 \\ 1 \end{bmatrix}$,

    $\ell_2 : \begin{bmatrix} x \\ y \\ z \end{bmatrix} = \begin{bmatrix} 2 \\ -1 \\ -3 \end{bmatrix} + t \begin{bmatrix} 2 \\ 6 \\ 7 \end{bmatrix}$.

    (a) Find the equation of the plane that contains ℓ_1 and is parallel to ℓ_2.

    (b) Use the result of (a) to find the (shortest) distance between ℓ_1 and ℓ_2.

17. (a) Let π be the plane spanned by nonparallel vectors $\mathbf{u}$ and $\mathbf{v}$. Show that $\mathbf{u}$ and $\mathbf{v}$ are each in π.

    (b) Most theorems in mathematics can be stated in the form "If $\mathcal{P}$, then $\mathcal{Q}$," symbolically $\mathcal{P} \implies \mathcal{Q}$. The *converse* of this statement is $\mathcal{Q} \implies \mathcal{P}$: "If $\mathcal{Q}$, then $\mathcal{P}$."

        i. [BB] What is the converse of Theorem 1.4.12?

        ii. Is the converse true? Justify your answer.

18. [BB] Don has a theory that to find the projection of a vector $\mathbf{w}$ on a plane π (through the origin), one could project $\mathbf{w}$ on a normal $\mathbf{n}$ to the plane and take $\operatorname{proj}_\pi \mathbf{w} = \mathbf{w} - \operatorname{proj}_n \mathbf{w}$.

    (a) Is Don's theory correct? Explain by applying the definition of "projection onto a plane."

    (b) Illustrate Don's theory using the plane with

        equation $3x - 2y + z = 0$ and $\mathbf{w} = \begin{bmatrix} 1 \\ 1 \\ 1 \end{bmatrix}$.

19. Let ℓ_1 and ℓ_2 be the lines with the following equations:

    $\ell_1 : \begin{bmatrix} x \\ y \\ z \end{bmatrix} = \begin{bmatrix} 3 \\ 0 \\ 5 \end{bmatrix} + t \begin{bmatrix} 0 \\ -2 \\ 1 \end{bmatrix}$;

    $\ell_2 : \begin{bmatrix} x \\ y \\ z \end{bmatrix} = \begin{bmatrix} -2 \\ -1 \\ -1 \end{bmatrix} + t \begin{bmatrix} 3 \\ 2 \\ -2 \end{bmatrix}$.

    (a) Show that ℓ_1 and ℓ_2 are not parallel and that they do not intersect. (Such lines are called *skew*.)

    (b) Find the shortest distance between ℓ_1 and ℓ_2.

    (c) Let ℓ_3 be the line of shortest distance between ℓ_1 and ℓ_2; that is, ℓ_3 should intersect ℓ_1 and ℓ_2 at points A, B, respectively, the length of $\overrightarrow{AB}$ being the shortest distance between ℓ_1 and ℓ_2. Find A and B.

    (d) Find an equation of ℓ_3.

1.5 Euclidean *n*-Space

With the exception of our ability to picture vectors as arrows, everything we said in Section 1.1 applies not only to columns of two or three numbers, but also to columns of any (finite) length.

1.5.1 DEFINITION

Let $n \geq 1$ be an integer. An *n-dimensional vector* is a column of n numbers enclosed in brackets. The numbers are called the *components* of the vector.

Thus $\begin{bmatrix} 1 \\ 2 \\ 3 \\ 4 \end{bmatrix}$ is a four-dimensional vector with components 1, 2, 3, 4, and $\begin{bmatrix} -1 \\ 0 \\ 0 \\ 1 \\ 2 \end{bmatrix}$ is a

five-dimensional vector with components $-1, 0, 0, 1, 2$. In general, an n-dimensional

vector has the form $\begin{bmatrix} x_1 \\ x_2 \\ \vdots \\ x_n \end{bmatrix}$, the numbers $x_1, x_2, \ldots, x_n$ being the components.

1.5.2 DEFINITION

Euclidean n-space is the set of all *n*-dimensional vectors. It is denoted $\mathbf{R}^n$.

$$\mathbf{R}^n = \left\{ \begin{bmatrix} x_1 \\ x_2 \\ \vdots \\ x_n \end{bmatrix} \mid x_1, x_2, \ldots, x_n \in \mathbf{R} \right\}.$$

Euclidean 2-space,

$$\mathbf{R}^2 = \left\{ \begin{bmatrix} x \\ y \end{bmatrix} \mid x, y \in \mathbf{R} \right\}$$

is more commonly called the *Euclidean plane* and Euclidean 3-space

$$\mathbf{R}^3 = \left\{ \begin{bmatrix} x \\ y \\ z \end{bmatrix} \mid x, y, z \in \mathbf{R} \right\}$$

is often called simply 3-*space*.

The equality of n-dimensional vectors is the obvious extension of the concept for two- and three-dimensional vectors.

1.5.3 Vectors $\mathbf{x} = \begin{bmatrix} x_1 \\ x_2 \\ \vdots \\ x_n \end{bmatrix}$ and $\mathbf{y} = \begin{bmatrix} y_1 \\ y_2 \\ \vdots \\ y_m \end{bmatrix}$ are *equal* if and only if $n = m$ and

corresponding components are equal, that is,

$$x_1 = y_1, x_2 = y_2, \ldots, x_n = y_n.$$

If $\mathbf{x} = \begin{bmatrix} x_1 \\ x_2 \\ \vdots \\ x_n \end{bmatrix}$ and $\mathbf{y} = \begin{bmatrix} y_1 \\ y_2 \\ \vdots \\ y_n \end{bmatrix}$ are *n*-dimensional vectors, we define the *sum* of $\mathbf{x}$ and $\mathbf{y}$ in the following obvious way.

$$\mathbf{x} + \mathbf{y} = \begin{bmatrix} x_1 \\ x_2 \\ \vdots \\ x_n \end{bmatrix} + \begin{bmatrix} y_1 \\ y_2 \\ \vdots \\ y_n \end{bmatrix} = \begin{bmatrix} x_1 + y_1 \\ x_2 + y_2 \\ \vdots \\ x_n + y_n \end{bmatrix}.$$

For a scalar c, we define *scalar multiplication* by

$$c\mathbf{x} = c \begin{bmatrix} x_1 \\ x_2 \\ \vdots \\ x_n \end{bmatrix} = \begin{bmatrix} cx_1 \\ cx_2 \\ \vdots \\ cx_n \end{bmatrix}.$$

For obvious reasons, we say that addition and scalar multiplication in $\mathbf{R}^n$ are *componentwise* operations. Using our capabilities to add vectors and to multiply them by scalars, we again have the concept of linear combination.

1.5.4 DEFINITION

A *linear combination* of vectors $\mathbf{v}_1, \mathbf{v}_2, \ldots, \mathbf{v}_n$ is a vector of the form

$$c_1\mathbf{v}_1 + c_2\mathbf{v}_2 + \cdots + c_n\mathbf{v}_n$$

where $c_1, c_2, \ldots, c_n$ are scalars.

1.5.5 EXAMPLE

The equation

$$\begin{bmatrix} -11 \\ 7 \\ -4 \\ 2 \end{bmatrix} = 2 \begin{bmatrix} -2 \\ 0 \\ 5 \\ 1 \end{bmatrix} - 7 \begin{bmatrix} 1 \\ -1 \\ 2 \\ 0 \end{bmatrix}$$

says that $\begin{bmatrix} -11 \\ 7 \\ -4 \\ 2 \end{bmatrix}$ is a linear combination of $\begin{bmatrix} -2 \\ 0 \\ 5 \\ 1 \end{bmatrix}$ and $\begin{bmatrix} 1 \\ -1 \\ 2 \\ 0 \end{bmatrix}$. ∎

1.5.6 EXAMPLE

The vector $\begin{bmatrix} 1 \\ 5 \\ 2 \\ 0 \\ -1 \end{bmatrix}$ is not a linear combination of $\begin{bmatrix} 1 \\ 2 \\ -1 \\ 3 \\ 4 \end{bmatrix}$ and $\begin{bmatrix} 3 \\ 5 \\ 0 \\ 1 \\ 2 \end{bmatrix}$ because $\begin{bmatrix} 1 \\ 5 \\ 2 \\ 0 \\ -1 \end{bmatrix} =$

$a \begin{bmatrix} 1 \\ 2 \\ -1 \\ 3 \\ 4 \end{bmatrix} + b \begin{bmatrix} 3 \\ 5 \\ 0 \\ 1 \\ 2 \end{bmatrix}$ leads to the system of equations

$$\begin{aligned}
a + 3b &= 1 \\
2a + 5b &= 5 \\
-a &= 2 \\
3a + b &= 0 \\
4a + 2b &= -1,
\end{aligned}$$

which has no solution. The third equation says $a = -2$ and comparing with the first gives $b = 1$, but the pair $a = -2, b = 1$ does not satisfy the second equation. ∎

The n-dimensional zero vector is $\mathbf{0} = \begin{bmatrix} 0 \\ 0 \\ \vdots \\ 0 \end{bmatrix}$. As with the two-dimensional and three-dimensional zero vectors, it is always the case that $\mathbf{x} + \mathbf{0} = \mathbf{x}$ for any n-dimensional vector $\mathbf{x}$. In fact, all parts of Theorem 1.1.6 hold for n-dimensional vectors, with proofs requiring only the obvious change in the appearance of vectors (columns of n instead of two numbers).

Remember the standard basis vectors $\mathbf{i}, \mathbf{j},$ and $\mathbf{k}$ in 3-space? Among the properties of these special vectors is the fact that every vector in $\mathbf{R}^3$ is a linear combination of $\mathbf{i}, \mathbf{j},$ and $\mathbf{k}$. These vectors have their analogues in Euclidean n-space.

1.5.7 DEFINITION The *standard basis vectors* in $\mathbf{R}^n$ are the n-dimensional vectors $\mathbf{e}_1, \mathbf{e}_2, \ldots, \mathbf{e}_n$ where $\mathbf{e}_i$ has ith component 1 and all other components 0. Thus

$$\mathbf{e}_1 = \begin{bmatrix} 1 \\ 0 \\ 0 \\ 0 \\ \vdots \\ 0 \end{bmatrix}, \quad \mathbf{e}_2 = \begin{bmatrix} 0 \\ 1 \\ 0 \\ 0 \\ \vdots \\ 0 \end{bmatrix}, \quad \mathbf{e}_3 = \begin{bmatrix} 0 \\ 0 \\ 1 \\ 0 \\ \vdots \\ 0 \end{bmatrix}, \quad \ldots, \quad \mathbf{e}_n = \begin{bmatrix} 0 \\ 0 \\ 0 \\ \vdots \\ 0 \\ 1 \end{bmatrix}.$$

In $\mathbf{R}^2$, $\mathbf{e}_1 = \begin{bmatrix} 1 \\ 0 \end{bmatrix} = \mathbf{i}$ and $\mathbf{e}_2 = \begin{bmatrix} 0 \\ 1 \end{bmatrix} = \mathbf{j}$ are the standard basis vectors in the plane we introduced in 1.1.15; in $\mathbf{R}^3$, $\mathbf{e}_1 = \begin{bmatrix} 1 \\ 0 \\ 0 \end{bmatrix} = \mathbf{i}, \mathbf{e}_2 = \begin{bmatrix} 0 \\ 1 \\ 0 \end{bmatrix} = \mathbf{j},$ and $\mathbf{e}_3 = \begin{bmatrix} 0 \\ 0 \\ 1 \end{bmatrix} = \mathbf{k}$ are the standard basis vectors in 3-space introduced in Definition 1.1.21. Note that for scalars $x_1, x_2, \ldots, x_n,$

$$x_1\mathbf{e}_1 = \begin{bmatrix} x_1 \\ 0 \\ 0 \\ \vdots \\ 0 \end{bmatrix}, \quad x_2\mathbf{e}_2 = \begin{bmatrix} 0 \\ x_2 \\ 0 \\ \vdots \\ 0 \end{bmatrix}, \quad \ldots, \quad x_n\mathbf{e}_n = \begin{bmatrix} 0 \\ 0 \\ \vdots \\ 0 \\ x_n \end{bmatrix},$$

so a phenomenon already noted in the plane and in 3-space is true generally.

1.5.8 Every vector in $\mathbf{R}^n$ is a linear combination of the standard basis vectors:

$$\begin{bmatrix} x_1 \\ x_2 \\ \vdots \\ x_n \end{bmatrix} = x_1\mathbf{e}_1 + x_2\mathbf{e}_2 + \cdots + x_n\mathbf{e}_n.$$

■ Length and Angles in $\mathbf{R}^n$

Now we turn our attention to length and direction in $\mathbf{R}^n$. First of all, based on our experience with the Euclidean plane and Euclidean 3-space, we define the dot product of two *n*-dimensional vectors.

1.5.9 DEFINITION For any $n \geq 1$, the *dot product* of *n*-dimensional vectors $\mathbf{x} = \begin{bmatrix} x_1 \\ x_2 \\ \vdots \\ x_n \end{bmatrix}$ and $\mathbf{y} = \begin{bmatrix} y_1 \\ y_2 \\ \vdots \\ y_n \end{bmatrix}$ is

the number $\mathbf{x} \cdot \mathbf{y} = x_1 y_1 + x_2 y_2 + \cdots + x_n y_n$.

All the properties of Theorem 1.2.19 hold in this more general setting. In Section 1.2, we noted that the length of a three-dimensional vector is the square root of its dot product with itself. (See 1.2.12.) We use this idea to define the length of a vector in $\mathbf{R}^n$.

1.5.10 DEFINITION The *length* or *norm* of the *n*-dimensional vector $\mathbf{x} = \begin{bmatrix} x_1 \\ x_2 \\ \vdots \\ x_n \end{bmatrix}$ is the number

$$\|\mathbf{x}\| = \sqrt{\mathbf{x} \cdot \mathbf{x}} = \sqrt{x_1^2 + x_2^2 + \cdots + x_n^2}.$$

Reading Challenge 1 *Only the zero vector has norm* 0. *Why?*

■ The Cauchy–Schwarz and Triangle Inequalities

In Section 1.1, we saw that the angle between vectors could be expressed in terms of the dot product. There we showed

$$\mathbf{u} \cdot \mathbf{v} = \|\mathbf{u}\| \, \|\mathbf{v}\| \cos\theta. \tag{1}$$

See 1.2.13. We would like the same equation to hold in $\mathbf{R}^n$ for any n. There is, however, a difficulty here. Just what does θ mean when $n > 3$? Here we make a subtle, but important shift in viewpoint. Instead of proving (1), we use it to define the angle between vectors; that is, we are going to define the cosine of the angle θ between $\mathbf{u}$ and $\mathbf{v}$ to be the number

$$\cos\theta = \frac{\mathbf{u}\cdot\mathbf{v}}{\|\mathbf{u}\|\,\|\mathbf{v}\|}. \tag{2}$$

But we have a problem. The cosine of an angle is a number between -1 and $+1$. Our attempt to define $\cos\theta$ by (2) will be useless unless we know somehow that the quantity on the right is between -1 and $+1$. In other words, we must know that

$$-1 \le \frac{\mathbf{u}\cdot\mathbf{v}}{\|\mathbf{u}\|\,\|\mathbf{v}\|} \le 1$$

or, equivalently, that

$$\left|\frac{\mathbf{u}\cdot\mathbf{v}}{\|\mathbf{u}\|\,\|\mathbf{v}\|}\right| \le 1.$$

Remarkably, this turns out to be true because of an n-dimensional version of the Cauchy–Schwarz inequality.

Theorem 1.5.11   **(Cauchy–Schwarz Inequality).**   *If $\mathbf{u}$ and $\mathbf{v}$ are n-dimensional vectors, then*

$$\boxed{|\mathbf{u}\cdot\mathbf{v}| \le \|\mathbf{u}\|\,\|\mathbf{v}\|.}$$

In Section 1.1, we simply noted this inequality, which followed from a property of the cosine function, namely, that $|\cos\theta| \le 1$ for any angle θ (in $\mathbf{R}^2$ or $\mathbf{R}^3$). Here, such reasoning is invalid because we do not (yet) know what θ means when $n > 3$. So how can we prove Theorem 1.5.11 without using the idea of angle?

We begin with the observation that the theorem is certainly true if $\mathbf{u} = \mathbf{0}$, for then $|\mathbf{u}\cdot\mathbf{v}| = 0$ and $\|\mathbf{u}\|\,\|\mathbf{v}\| = 0$. This leaves the case $\mathbf{u} \ne \mathbf{0}$, and hence $\mathbf{u}\cdot\mathbf{u} \ne 0$. At this point, we use something that students usually call a "trick." A better term is "insight," an idea that somebody once had and which led to a useful discovery.

The "insight" is the fact $(x\mathbf{u} + \mathbf{v})\cdot(x\mathbf{u} + \mathbf{v}) \ge 0$ for any real number x because the dot product of any vector with itself, being a sum of squares, cannot be negative. Remembering that $(x\mathbf{u} + \mathbf{v})\cdot(x\mathbf{u} + \mathbf{v})$ expands just like $(xa + b)^2$—see Example 1.2.20—we obtain

$$(x\mathbf{u} + \mathbf{v})\cdot(x\mathbf{u} + \mathbf{v}) = x^2(\mathbf{u}\cdot\mathbf{u}) + 2x(\mathbf{u}\cdot\mathbf{v}) + \mathbf{v}\cdot\mathbf{v},$$

and so

$$x^2(\mathbf{u}\cdot\mathbf{u}) + 2x(\mathbf{u}\cdot\mathbf{v}) + \mathbf{v}\cdot\mathbf{v} \ge 0. \tag{3}$$

Letting $a = \mathbf{u}\cdot\mathbf{u}$, $b = 2(\mathbf{u}\cdot\mathbf{v})$, and $c = \mathbf{v}\cdot\mathbf{v}$, equation (3) says that $ax^2 + bx + c \ge 0$ (for any x). Also $a \ne 0$ because we are assuming $\mathbf{u} \ne \mathbf{0}$. The graph of $y = ax^2 + bx + c$ is a parabola and the fact that $y \ge 0$ for all x says that this parabola intersects

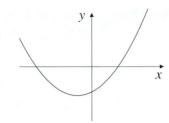

Figure 1.25 A parabola with equation $y = ax^2 + bx + c$ intersects the x-axis twice if some $y > 0$ and some $y < 0$.

the x-axis at most once. (See Figure 1.25.) The parabola intersects the x-axis where $ax^2 + bx + c = 0$; thus we have shown that the quadratic $ax^2 + bx + c$ has at most one root. Since the roots of $ax^2 + bx + c = 0$ are given by the well-known formula

$$x = \frac{-b \pm \sqrt{b^2 - 4ac}}{2a},$$

the term under the square root (the discriminant) cannot be positive (otherwise, we would have two roots). So

$$b^2 - 4ac \leq 0$$

$$b^2 \leq 4ac$$

$$4(\mathbf{u} \cdot \mathbf{v})^2 \leq 4(\mathbf{u} \cdot \mathbf{u})(\mathbf{v} \cdot \mathbf{v})$$

$$(\mathbf{u} \cdot \mathbf{v})^2 \leq (\mathbf{u} \cdot \mathbf{u})(\mathbf{v} \cdot \mathbf{v}) = \|\mathbf{u}\|^2 \|\mathbf{v}\|^2.$$

Taking the square root of each side and remembering that $\sqrt{t^2} = |t|$, we obtain the desired result:

$$|\mathbf{u} \cdot \mathbf{v}| \leq \|\mathbf{u}\| \|\mathbf{v}\|.$$

1.5.12 EXAMPLE   Suppose $\mathbf{u} = \begin{bmatrix} -1 \\ 0 \\ 2 \\ 1 \end{bmatrix}$ and $\mathbf{v} = \begin{bmatrix} 2 \\ -1 \\ 1 \\ -4 \end{bmatrix}$. Then

$$\|\mathbf{u}\| = \sqrt{1 + 0 + 4 + 1} = \sqrt{6},$$

$$\|\mathbf{v}\| = \sqrt{4 + 1 + 1 + 16} = \sqrt{22},$$

$$\mathbf{u} \cdot \mathbf{v} = -2 + 0 + 2 - 4 = -4.$$

The Cauchy–Schwarz inequality says that $|\mathbf{u} \cdot \mathbf{v}| \leq \|\mathbf{u}\| \|\mathbf{v}\|$; that is, $|-4| \leq \sqrt{6}\sqrt{22}$, which is certainly true: $4 \leq \sqrt{132}$. ∎

Where were we? Remember that we needed the Cauchy–Schwarz inequality in order to define the notion of *angle* in Euclidean *n*-space.

1.5.13 DEFINITION If $\mathbf{u}$ and $\mathbf{v}$ are nonzero vectors in $\mathbf{R}^n$, the *angle between* $\mathbf{u}$ and $\mathbf{v}$ is that angle θ, $0 \leq \theta \leq \pi$, with

$$\cos \theta = \frac{\mathbf{u} \cdot \mathbf{v}}{\|\mathbf{u}\| \, \|\mathbf{v}\|}.$$

This definition is only possible because $-1 \leq \frac{\mathbf{u} \cdot \mathbf{v}}{\|\mathbf{u}\|\|\mathbf{v}\|} \leq 1$ (by Cauchy–Schwarz).

1.5.14 EXAMPLE If $\mathbf{u} = \begin{bmatrix} -4 \\ 0 \\ 2 \\ -2 \end{bmatrix}$ and $\mathbf{v} = \begin{bmatrix} 2 \\ 0 \\ -1 \\ 1 \end{bmatrix}$,

$$\cos \theta = \frac{\mathbf{u} \cdot \mathbf{v}}{\|\mathbf{u}\| \, \|\mathbf{v}\|} = \frac{-8 - 2 - 2}{\sqrt{24}\sqrt{6}} = \frac{-12}{\sqrt{144}} = -1$$

so $\theta = \pi$, a fact consistent with the observation that $\mathbf{u} = -2\mathbf{v}$. ■

1.5.15 EXAMPLE If $\mathbf{u} = \begin{bmatrix} 1 \\ 2 \\ 1 \\ 0 \\ 3 \end{bmatrix}$ and $\mathbf{v} = \begin{bmatrix} 1 \\ -1 \\ 1 \\ 1 \\ 1 \end{bmatrix}$, $\cos \theta = \frac{\mathbf{u} \cdot \mathbf{v}}{\|\mathbf{u}\|\|\mathbf{v}\|} = \frac{3}{\sqrt{15}\sqrt{5}} = \frac{\sqrt{3}}{5} \approx .346$, so $\theta \approx$ 1.217 rads $\approx 70°$. ■

In $\mathbf{R}^2$ or $\mathbf{R}^3$, vectors are orthogonal if the angle θ between them is $\frac{\pi}{2}$; that is, if $\cos \theta = 0$. By Definition 1.5.13, $\cos \theta = 0$ if and only if $\mathbf{u} \cdot \mathbf{v} = 0$. Thus we extend the concept of orthogonality to any Euclidean space with the following definition.

1.5.16 DEFINITION Vectors $\mathbf{u}$ and $\mathbf{v}$ in $\mathbf{R}^n$ are *orthogonal* if $\mathbf{u} \cdot \mathbf{v} = 0$.

1.5.17 EXAMPLE The vectors $\mathbf{u} = \begin{bmatrix} 3 \\ 2 \\ -1 \\ 4 \end{bmatrix}$ and $\mathbf{v} = \begin{bmatrix} 1 \\ -1 \\ 1 \\ 0 \end{bmatrix}$ are orthogonal vectors in $\mathbf{R}^4$ since $\mathbf{u} \cdot \mathbf{v} =$ $3 - 2 - 1 + 0 = 0$. ■

A glance at Figure 1.26 shows why the next theorem is so-named.

Theorem 1.5.18 **(The Triangle Inequality).** *For any vectors $\mathbf{u}$ and $\mathbf{v}$ in $\mathbf{R}^n$, $\|\mathbf{u} + \mathbf{v}\| \leq \|\mathbf{u}\| + \|\mathbf{v}\|$.*

Figure 1.26 The triangle inequality: $\|\mathbf{u} + \mathbf{v}\| \leq \|\mathbf{u}\| + \|\mathbf{v}\|$.

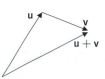

Proof The proof is direct.

$$\|\mathbf{u} + \mathbf{v}\|^2 = (\mathbf{u} + \mathbf{v}) \cdot (\mathbf{u} + \mathbf{v})$$

$$= \mathbf{u} \cdot \mathbf{u} + 2\mathbf{u} \cdot \mathbf{v} + \mathbf{v} \cdot \mathbf{v}$$

$$= \|\mathbf{u}\|^2 + 2\mathbf{u} \cdot \mathbf{v} + \|\mathbf{v}\|^2 \qquad (4)$$

$$\leq \|\mathbf{u}\|^2 + 2|\mathbf{u} \cdot \mathbf{v}| + \|\mathbf{v}\|^2$$

Applying the Cauchy–Schwarz inequality to the middle term in the last expression here, we get

$$\|\mathbf{u}\|^2 + 2|\mathbf{u} \cdot \mathbf{v}| + \|\mathbf{v}\|^2 \leq \|\mathbf{u}\|^2 + 2\,\|\mathbf{u}\|\,\|\mathbf{v}\| + \|\mathbf{v}\|^2 \,.$$

Since the expression on the right is $(\|\mathbf{u}\| + \|\mathbf{v}\|)^2$, we get

$$\|\mathbf{u} + \mathbf{v}\|^2 \leq (\|\mathbf{u}\| + \|\mathbf{v}\|)^2,$$

and the result follows by taking square roots. ■

At step (4), we had $\|\mathbf{u} + \mathbf{v}\|^2 = \|\mathbf{u}\|^2 + 2\mathbf{u} \cdot \mathbf{v} + \|\mathbf{v}\|^2$. Since vectors $\mathbf{u}$ and $\mathbf{v}$ are orthogonal if and only if $\mathbf{u} \cdot \mathbf{v} = 0$, the Theorem of Pythagoras is true in Euclidean n-space, for any n.

Theorem 1.5.19   (**Pythagoras**). *Vectors* $\mathbf{u}$ *and* $\mathbf{v}$ *in* $\mathbf{R}^n$ *are orthogonal if and only if* $\|\mathbf{u} + \mathbf{v}\|^2 = \|\mathbf{u}\|^2 + \|\mathbf{v}\|^2$.

■ Linear Dependence and Independence

One very important concept in linear algebra is that of *linear independence* (and its opposite, *linear dependence*).

Suppose we have some vectors $\mathbf{x}_1, \mathbf{x}_2, \ldots, \mathbf{x}_n$. It is easy to find a linear combination of these that equals the zero vector:

$$0\mathbf{x}_1 + 0\mathbf{x}_2 + \cdots + 0\mathbf{x}_n = \mathbf{0}.$$

This linear combination, with all coefficients 0, is called the *trivial* one. Vectors are linearly independent if the only linear combination of them that equals $\mathbf{0}$ is the trivial one.

1.5.20 DEFINITION   A set of vectors $\mathbf{x}_1, \mathbf{x}_2, \ldots, \mathbf{x}_n$ is *linearly independent* if $c_1\mathbf{x}_1 + c_2\mathbf{x}_2 + \cdots + c_n\mathbf{x}_n = \mathbf{0}$ implies $c_1 = c_2 = \cdots = c_n = 0$. Vectors that are not linearly independent are *linearly dependent*; that is, there is a linear combination $c_1\mathbf{x}_1 + c_2\mathbf{x}_2 + \cdots + c_n\mathbf{x}_n = \mathbf{0}$ with *some* coefficient not 0.

1.5.21 EXAMPLE   Are the vectors $\mathbf{x}_1 = \begin{bmatrix} 1 \\ 2 \end{bmatrix}$ and $\mathbf{x}_2 = \begin{bmatrix} -1 \\ 0 \end{bmatrix}$ linearly independent? Certainly

$$0\begin{bmatrix} 1 \\ 2 \end{bmatrix} + 0\begin{bmatrix} -1 \\ 0 \end{bmatrix} = \begin{bmatrix} 0 \\ 0 \end{bmatrix}, \tag{5}$$

but this is not the point! Here's the question: Is (5) the *only* way for a linear combination of $\mathbf{x}_1$ and $\mathbf{x}_2$ to be $\mathbf{0}$? Suppose $c_1\mathbf{x}_1 + c_2\mathbf{x}_2 = \mathbf{0}$. This is

$$c_1\begin{bmatrix} 1 \\ 2 \end{bmatrix} + c_2\begin{bmatrix} -1 \\ 0 \end{bmatrix} = \begin{bmatrix} 0 \\ 0 \end{bmatrix}, \quad \text{which is} \quad \begin{bmatrix} c_1 - c_2 \\ 2c_1 \end{bmatrix} = \begin{bmatrix} 0 \\ 0 \end{bmatrix}.$$

So we must solve the two equations

$$c_1 - c_2 = 0$$
$$2c_1 \qquad = 0$$

for the unknowns c_1 and c_2. The second equation gives $c_1 = 0$. Then putting $c_1 = 0$ in the first equation gives $c_2 = 0$. Thus the vectors x_1 and x_2 are linearly independent. ■

1.5.22 PROBLEM

Are the vectors $x_1 = \begin{bmatrix} 2 \\ 4 \end{bmatrix}$ and $x_2 = \begin{bmatrix} -3 \\ -6 \end{bmatrix}$ linearly dependent or linearly independent?

Solution. We must determine whether $c_1 = c_2 = 0$ is the only solution to the equation $c_1 x_1 + c_2 x_2 = 0$. This is

$$c_1 \begin{bmatrix} 2 \\ 4 \end{bmatrix} + c_2 \begin{bmatrix} -3 \\ -6 \end{bmatrix} = \begin{bmatrix} 0 \\ 0 \end{bmatrix}, \qquad \text{which is} \quad \begin{bmatrix} 2c_1 - 3c_2 \\ 4c_1 - 6c_2 \end{bmatrix} = \begin{bmatrix} 0 \\ 0 \end{bmatrix}.$$

We might also say that the zero vector is a *nontrivial* linear combination of x_1 and x_2.

There are many *nontrivial* solutions, such as $c_1 = 3$, $c_2 = 2$. Therefore, the vectors x_1 and x_2 are linearly dependent.

In Problem 1.5.22, the linearly dependent vectors x_1 and x_2 were parallel: $x_2 = -\frac{3}{2}x_1$. This illustrates an important point. Two vectors are linearly dependent if they are parallel (one is a scalar multiple of the other); otherwise, they are linearly independent.

> **1.5.23** Two vectors are linearly dependent if and only if they are parallel; that is, one is a scalar multiple of the other.

This is not hard to prove. If $c_1 x_1 + c_2 x_2 = 0$ is a nontrivial linear combination equal to 0, then at least one of the scalars c_1, c_2 is not 0. Suppose that scalar is c_1. Then we can divide by c_1 and obtain

$$x_1 + \frac{c_2}{c_1}x_2 = 0; \quad \text{that is,} \quad x_1 = -\frac{c_2}{c_1}x_2,$$

which says that x_1 is a scalar multiple of x_2. Similarly, if $c_2 \neq 0$, we could prove that x_2 is a scalar multiple of x_1.

Suppose three vectors x_1, x_2, and x_3 are linearly dependent. Then some *nontrivial* linear combination of these vectors is the zero vector:

$$c_1 x_1 + c_2 x_2 + c_3 x_3 = 0$$

for scalars c_1, c_2, c_3, not all of which are 0. Suppose $c_1 \neq 0$. As before, we can divide by c_1, giving

$$x_1 + \frac{c_2}{c_1}x_2 + \frac{c_3}{c_1}x_3 = 0, \quad \text{so that} \quad x_1 = -\frac{c_2}{c_1}x_2 - \frac{c_3}{c_1}x_3.$$

This says that x_1 is a linear combination of x_2 and x_3; hence, assuming we are in $\mathbf{R}^3$, it is in the plane spanned by x_2 and x_3. If three vectors in $\mathbf{R}^3$ are linearly dependent, they all lie in a plane. If three vectors are linearly independent, none lies in the plane determined by the other two.

> **1.5.24** Three vectors in $\mathbf{R}^3$ are linearly dependent if and only if they all lie in a plane; one is a linear combination of the other two.

Answers to Reading Challenges

1. If $\mathbf{x} = \begin{bmatrix} x_1 \\ x_2 \\ \vdots \\ x_n \end{bmatrix}$, then $\|\mathbf{x}\| = \sqrt{x_1^2 + x_2^2 + \cdots + x_n^2}$ is the square root of a sum of squares. A sum of squares is 0 if and only if each number being squared is 0.

So $\|\mathbf{x}\| = 0$ implies $x_1 = 0$, $x_2 = 0$, and so forth, $x_n = 0$. Thus $\mathbf{x} = \begin{bmatrix} 0 \\ 0 \\ \vdots \\ 0 \end{bmatrix}$ is the zero vector.

True/False Questions

Decide, with as little calculation as possible, whether each of the following statements is true or false and explain your answer whenever you say "false." (Answers can be found in the back of the book.)

1. Every vector in $\mathbf{R}^3$ is a linear combination of $\mathbf{i}$, $\mathbf{j}$, and $\mathbf{k}$.

2. Every vector in $\mathbf{R}^3$ is a linear combination of $\mathbf{i}$, $\mathbf{j}$, $\mathbf{k}$, and $\mathbf{u} = \begin{bmatrix} 1 \\ 2 \\ 3 \end{bmatrix}$.

3. The vectors $\begin{bmatrix} 1 \\ 1 \\ 1 \\ 1 \\ 1 \end{bmatrix}$ and $\begin{bmatrix} 1 \\ 0 \\ -1 \\ -1 \\ 1 \end{bmatrix}$ are orthogonal.

4. The vector $\frac{1}{15}\begin{bmatrix} 1 \\ 2 \\ 0 \\ 1 \\ 3 \end{bmatrix}$ is a unit vector.

5. The vector $\begin{bmatrix} 1 \\ -2 \\ -4 \\ 1 \end{bmatrix}$ is a linear combination of $\mathbf{u} = \begin{bmatrix} 1 \\ -2 \\ 0 \\ 1 \end{bmatrix}$ and $\mathbf{v} = \begin{bmatrix} 0 \\ 0 \\ 0 \\ 1 \end{bmatrix}$.

6. Linear independence of vectors $\mathbf{u}$ and $\mathbf{v}$ means that $0\mathbf{u} + 0\mathbf{v} = \mathbf{0}$.

7. Linear independence of vectors **u** and **v** means that $a\mathbf{u} + b\mathbf{v} = \mathbf{0}$ where a and b are 0.

8. For any **u** and **v**, the vectors **u**, **v**, **0** are linearly dependent.

Exercises

*Solutions to exercises marked [BB] can be found in the **B**ack of the **B**ook.*

1. Let $\mathbf{u} = \begin{bmatrix} 1 \\ 2 \\ 3 \\ -1 \end{bmatrix}$, $\mathbf{v} = \begin{bmatrix} 0 \\ -1 \\ 1 \end{bmatrix}$, $\mathbf{x} = \begin{bmatrix} 1 \\ 2 \\ 3 \end{bmatrix}$, and

$\mathbf{y} = \begin{bmatrix} 0 \\ -1 \\ 1 \\ 2 \end{bmatrix}$.

Find each of the following, if possible.

(a) [BB] $2\mathbf{x} - 3\mathbf{y} + \mathbf{u}$ (b) [BB] $\|\mathbf{y}\|$

(c) $3\mathbf{v} + 4\mathbf{x}$ (d) $\mathbf{x} \cdot \mathbf{v}$

(e) the cosine of the angle between **u** and **y**

2. Find a and b given the following equations:

(a) [BB] $\begin{bmatrix} a \\ -1 \\ 1 \\ b \\ 2 \end{bmatrix} + \begin{bmatrix} b \\ 0 \\ 6 \\ -2 \end{bmatrix} = \begin{bmatrix} 0 \\ -1 \\ 7 \\ 0 \\ 3 \end{bmatrix}$

(b) $a\begin{bmatrix} 1 \\ -1 \\ 1 \\ 2 \end{bmatrix} + b\begin{bmatrix} -1 \\ 1 \\ 2 \\ 3 \end{bmatrix} = \begin{bmatrix} -3 \\ a+1 \\ 12 \\ 4b-1 \end{bmatrix}$.

3. Let $\mathbf{u} = \begin{bmatrix} -1 \\ a \\ b \\ 2 \\ 0 \end{bmatrix}$, $\mathbf{v} = \begin{bmatrix} c \\ 1 \\ 1 \\ a \\ -4 \end{bmatrix}$, $\mathbf{x} = \begin{bmatrix} 1 \\ 2 \\ 3 \\ 4 \\ 5 \end{bmatrix}$, and

$\mathbf{y} = \begin{bmatrix} -1 \\ 1 \\ -2 \\ 4 \\ 8 \end{bmatrix}$.

Find a, b, and c, if possible, in each of the following situations.

(a) [BB] $3\mathbf{u} - 2\mathbf{v} = \mathbf{y}$ (b) $2\mathbf{u} + \mathbf{x} - \mathbf{y} = \mathbf{0}$

(c) **u** and **x** are orthogonal   (d) $\|\mathbf{v}\| = \sqrt{c^2 + 27}$

4. (a) [BB] Find a unit vector that has the same

direction as $\mathbf{u} = \begin{bmatrix} 1 \\ 0 \\ -1 \\ 1 \end{bmatrix}$.

(b) Find a vector of length 3 that has the same

direction as $\mathbf{u} = \begin{bmatrix} 1 \\ 2 \\ 1 \\ 1 \\ 1 \end{bmatrix}$.

(c) Find a vector of length 7 that has direction

opposite to $\mathbf{u} = \begin{bmatrix} 0 \\ -3 \\ 1 \\ 1 \\ 2 \end{bmatrix}$.

5. Let $\mathbf{u} = \begin{bmatrix} 1 \\ 1 \\ 0 \\ -2 \end{bmatrix}$ and $\mathbf{v} = \begin{bmatrix} -3 \\ 0 \\ 0 \\ 1 \end{bmatrix}$.

(a) [BB] Find all scalars c such that $\|c\mathbf{u}\| = 8$.

(b) Find all scalars c such that **u** is orthogonal to $\mathbf{u} + c\mathbf{v}$.

(c) Find all scalars c such that the angle between $c\mathbf{u} + \mathbf{v}$ and **v** is 60°.

6. Find the angle (to the nearest tenth of a radian and to the nearest degree) between each of the following pairs of vectors:

(a) [BB] $\mathbf{u} = \begin{bmatrix} 1 \\ -1 \\ 2 \end{bmatrix}$ and $\mathbf{v} = \begin{bmatrix} -1 \\ 1 \\ 0 \end{bmatrix}$

(b) $\mathbf{u} = \begin{bmatrix} 3 \\ 0 \\ 1 \\ -2 \end{bmatrix}$ and $\mathbf{v} = \begin{bmatrix} 1 \\ 1 \\ 1 \\ 5 \end{bmatrix}$

(c) $\mathbf{u} = \begin{bmatrix} 0 \\ 3 \\ 1 \\ -2 \end{bmatrix}$ and $\mathbf{v} = \begin{bmatrix} -3 \\ 6 \\ 1 \\ 1 \end{bmatrix}$

(d) $\mathbf{u} = \begin{bmatrix} 0 \\ 1 \\ 1 \\ 2 \\ 1 \end{bmatrix}$ and $\mathbf{v} = \begin{bmatrix} 2 \\ -2 \\ 3 \\ 0 \\ 1 \end{bmatrix}$.

7. [BB] Suppose **u** is a vector of length $\sqrt{6}$ and **v** is a unit vector orthogonal to $\mathbf{u} + \mathbf{v}$. Find $\|2\mathbf{u} - \mathbf{v}\|$.

8. Suppose **u**, **v**, **x**, and **y** are unit vectors such that $\mathbf{v} \cdot \mathbf{x} = -\frac{1}{3}$, **u** and **y** are orthogonal, the angle between **v** and **y** is $45°$, and the angle between **u** and **x** is $60°$. Find $(2\mathbf{u} - 3\mathbf{v}) \cdot (\mathbf{x} + 4\mathbf{y})$.

9. Determine, with justification, whether **v** is a linear combination of $\mathbf{v}_1, \dots, \mathbf{v}_5$, given $\mathbf{v} = \begin{bmatrix} 1 \\ 2 \\ 3 \\ 4 \\ 5 \end{bmatrix}$, $\mathbf{v}_1 = \begin{bmatrix} 3 \\ 4 \\ 5 \\ 1 \\ 2 \end{bmatrix}$,

$\mathbf{v}_2 = \begin{bmatrix} 2 \\ 3 \\ 4 \\ 5 \\ 1 \end{bmatrix}$, $\mathbf{v}_3 = \begin{bmatrix} -1 \\ -2 \\ -3 \\ -4 \\ -5 \end{bmatrix}$, $\mathbf{v}_4 = \begin{bmatrix} 4 \\ 5 \\ 1 \\ 2 \\ 3 \end{bmatrix}$, and $\mathbf{v}_5 = \begin{bmatrix} 5 \\ 1 \\ 2 \\ 3 \\ 4 \end{bmatrix}$.

10. Verify the Cauchy–Schwarz and triangle inequalities for each of the following pairs of vectors:

(a) [BB] $\mathbf{u} \begin{bmatrix} 1 \\ 0 \\ -2 \end{bmatrix}$, $\mathbf{v} = \begin{bmatrix} -3 \\ 1 \\ 0 \end{bmatrix}$

(b) $\mathbf{u} = \begin{bmatrix} -1 \\ 0 \\ 1 \\ 2 \end{bmatrix}$, $\mathbf{v} = \begin{bmatrix} 3 \\ -1 \\ 1 \\ 0 \end{bmatrix}$

(c) $\mathbf{u} = \begin{bmatrix} -1 \\ -1 \\ 0 \\ 0 \\ 1 \end{bmatrix}$, $\mathbf{v} = \begin{bmatrix} 2 \\ 1 \\ 1 \\ 3 \end{bmatrix}$

(d) $\mathbf{u} = \begin{bmatrix} 1 \\ 2 \\ 3 \\ 4 \\ 1 \end{bmatrix}$, $\mathbf{v} = \begin{bmatrix} 0 \\ 1 \\ 3 \\ -4 \\ 0 \end{bmatrix}$.

11. Determine whether each of the given sets of vectors is linearly independent or linearly dependent.

(a) [BB] $\mathbf{v}_1 = \begin{bmatrix} -1 \\ 1 \\ 2 \end{bmatrix}$, $\mathbf{v}_2 = \begin{bmatrix} 0 \\ 1 \\ 4 \end{bmatrix}$, $\mathbf{v}_3 = \begin{bmatrix} 3 \\ 1 \\ 2 \end{bmatrix}$,

$\mathbf{v}_4 = \begin{bmatrix} 2 \\ -2 \\ -4 \end{bmatrix}$

(b) $\mathbf{v}_1 = \begin{bmatrix} -1 \\ 1 \end{bmatrix}$, $\mathbf{v}_2 = \begin{bmatrix} 3 \\ 0 \end{bmatrix}$

(c) $\mathbf{v}_1 = \begin{bmatrix} 1 \\ 0 \\ 0 \end{bmatrix}$, $\mathbf{v}_2 = \begin{bmatrix} -2 \\ 1 \\ 2 \end{bmatrix}$, $\mathbf{v}_3 = \begin{bmatrix} 3 \\ 2 \\ 4 \end{bmatrix}$

(d) $\mathbf{v}_1 = \begin{bmatrix} 1 \\ 0 \\ 0 \end{bmatrix}$, $\mathbf{v}_2 = \begin{bmatrix} -2 \\ 1 \\ 2 \end{bmatrix}$, $\mathbf{v}_3 = \begin{bmatrix} 3 \\ 2 \\ 5 \end{bmatrix}$

(e) $\mathbf{v}_1 = \begin{bmatrix} 1 \\ 1 \\ 2 \\ 1 \end{bmatrix}$, $\mathbf{v}_2 = \begin{bmatrix} 3 \\ 3 \\ 6 \\ 3 \end{bmatrix}$, $\mathbf{v}_3 = \begin{bmatrix} 2 \\ 3 \\ 4 \\ 1 \end{bmatrix}$, $\mathbf{v}_4 = \begin{bmatrix} 1 \\ 2 \\ 3 \\ 1 \end{bmatrix}$

■ **Critical Reading**

12. If **u** is a vector in $\mathbf{R}^n$ that is orthogonal to every vector in $\mathbf{R}^n$, then **u** must be the zero vector. Why?

13. Suppose nonzero vectors **e** and **f** are orthogonal. Show that **e** and **f** are linearly independent.

14. Suppose **u** and **v** are vectors and **v** is a scalar multiple of **u**. Prove that **u** and **v** are linearly dependent.

15. Let **u**, **v**, **w** be vectors and suppose **w** is a linear combination of **u** and **w**. Show that **u**, **v**, and **w** are linearly dependent.

16. **(a)** Suppose **u**, **v**, and **w** are three vectors and $\mathbf{w} = \mathbf{u} + \mathbf{v}$. Show that every linear combination of **u**, **v**, **w** is actually just a linear combination of **u** and **v**.

(b) Let $\mathbf{u} = \begin{bmatrix} 1 \\ 0 \\ -1 \end{bmatrix}$, $\mathbf{v} = \begin{bmatrix} 2 \\ 1 \\ 0 \end{bmatrix}$, and $\mathbf{w} = \begin{bmatrix} 3 \\ 1 \\ -1 \end{bmatrix}$.

Show that the set of all linear combinations of these vectors is a plane and find the equation of this plane. [Hint: $\mathbf{w} = \mathbf{u} + \mathbf{v}$.]

CHAPTER KEY WORDS AND IDEAS: Here are some technical words and phrases that were used in this chapter. Do you know the meaning of each? If you're not sure, check the glossary or index at the back of the book.

anticommutative

associative

Cauchy–Schwarz inequality

commutative

component (of a vector)

componentwise addition

dot product

componentwise addition

dot product

Euclidean n-space

Euclidean plane

length of a vector

linear combination of vectors

linearly dependent vectors

linearly independent vectors

n-dimensional vector

negative (of a vector)

nontrivial linear combination

normal vector

norm of a vector

orthogonal vectors

parallelogram rule

parallel vectors

plane spanned by vectors

projection of a vector on a plane

projection of a vector onto a line

scalar multiple (of a vector)

standard basis (vectors)

Theorem of Pythagoras

three-dimensional vector

2×2 determinant

two-dimensional vector

triangle inequality

trivial linear combination

unit vector

vectors are equal

vectors have opposite direction

vectors have same direction

vectors span a plane

zero vector

Review Exercises for Chapter 1

1. Let $\mathbf{u} = \begin{bmatrix} 1 \\ -4 \\ 2 \end{bmatrix}$. Find a unit vector in the direction opposite $\mathbf{u}$ and a vector of length 3 with the same direction as $\mathbf{u}$.

2. Let $\mathbf{u} = \begin{bmatrix} 3 \\ 2 \\ 0 \end{bmatrix}$, $\mathbf{v} = \begin{bmatrix} 1 \\ -2 \\ 3 \end{bmatrix}$, and $\mathbf{w} = \begin{bmatrix} -1 \\ -2 \\ 3 \end{bmatrix}$.

(a) Find the cosine of the angle between $\mathbf{u}$ and $\mathbf{w}$.

(b) Find a vector of length 3 that is perpendicular to both $\mathbf{u}$ and $\mathbf{v}$.

(c) Is it possible to express $\mathbf{u}$ as a linear combination of $\mathbf{v}$ and $\mathbf{w}$? Explain.

(d) Find the area of the triangle two of whose sides are $\mathbf{u}$ and $\mathbf{v}$.

3. Let $\mathbf{u} = \begin{bmatrix} -1 \\ 1 \\ 4 \end{bmatrix}$ and $\mathbf{v} = \begin{bmatrix} 2 \\ 0 \\ 1 \end{bmatrix}$.

(a) Find a unit vector in the direction opposite that of $\mathbf{v}$.

(b) Find two orthogonal vectors in the plane spanned by $\mathbf{u}$ and $\mathbf{v}$.

(c) Find a real number k (if possible) so that $\mathbf{u} + k\mathbf{v}$ is orthogonal to $\mathbf{u}$.

4. Let $\mathbf{u} = \begin{bmatrix} -1 \\ 2 \\ 1 \end{bmatrix}$ and $\mathbf{v} = \begin{bmatrix} 2 \\ -3 \\ -5 \end{bmatrix}$.

(a) Find all numbers k, if any, so that $\|k\mathbf{u} - \mathbf{v}\| = 4$. Give k accurate to two decimal places.

(b) Find all numbers k, if any, so that $k\mathbf{u} + 2\mathbf{v}$ is orthogonal to $\mathbf{u}$.

(c) Find (approximately) the angle between the vectors. (Give your answer in radians to two decimal places and in degrees to the nearest degree.)

5. Suppose $\mathbf{u}$ and $\mathbf{v}$ are vectors in 3-space with $\|\mathbf{u}\| = 2$, $\|\mathbf{v}\| = 3$, and a $30°$ angle between them. Find the following:

(a) $\mathbf{u} \cdot \mathbf{v}$

(b) the area of the parallelogram with sides $\mathbf{u}$ and $\mathbf{v}$

(c) $\|\mathbf{u} + \mathbf{v}\|^2$.

6. Find the point on the line through $P(0, 0, 0)$ and $Q(1, -1, 7)$ that is closest to $\mathbf{w} = \begin{bmatrix} 2 \\ 4 \\ 5 \end{bmatrix}$.

7. Find an equation of the plane spanned by $\mathbf{u}$ and $\mathbf{v}$ in each of the following cases:

(a) $\mathbf{u} = \begin{bmatrix} -2 \\ 3 \\ 1 \end{bmatrix}$ and $\mathbf{v} = \begin{bmatrix} 1 \\ 0 \\ 1 \end{bmatrix}$

(b) $\mathbf{u} = \begin{bmatrix} 2 \\ 1 \\ 3 \end{bmatrix}$ and $\mathbf{v} = \begin{bmatrix} -1 \\ 0 \\ 4 \end{bmatrix}$

(c) $\mathbf{u} = \begin{bmatrix} -1 \\ 2 \\ 3 \end{bmatrix}$ and $\mathbf{v} = \begin{bmatrix} 2 \\ 0 \\ 5 \end{bmatrix}$

(d) $\mathbf{u} = \begin{bmatrix} 1 \\ -1 \\ 4 \end{bmatrix}$ and $\mathbf{v} = \begin{bmatrix} 1 \\ 1 \\ 2 \end{bmatrix}$.

8. (a) Find the equation of the plane containing the points $P(1, 0, 1)$, $Q(-1, 5, 3)$, and $R(3, 3, 0)$.

(b) Write the vector equation of the line through the points P and Q of part (a).

9. (a) Find the equation of the plane containing the points $A(1, -1, 2)$, $B(2, 1, 1)$, and $C(-1, -1, 2)$.

(b) Does this plane contain the line with equation $\begin{bmatrix} x \\ y \\ z \end{bmatrix} = \begin{bmatrix} \frac{3}{2} \\ \frac{3}{2} \\ 0 \end{bmatrix} + t \begin{bmatrix} 3 \\ 2 \\ -1 \end{bmatrix}$? Explain.

10. Let $\mathbf{u} = \begin{bmatrix} 3 \\ 4 \\ 0 \end{bmatrix}$ and $\mathbf{v} = \begin{bmatrix} 0 \\ 0 \\ 3 \end{bmatrix}$.

(a) Find the angle between $\mathbf{u}$ and $\mathbf{v}$.

(b) Find a vector of length 7 that is perpendicular to both $\mathbf{u}$ and $\mathbf{v}$.

(c) Find the equation of the plane passing through $P(1, 2, 3)$ and containing vectors parallel to $\mathbf{u}$ and $\mathbf{v}$.

11. Explain why the planes with equations $x - y + 3z = 2$ and $3x + y + z = 2$ intersect, and find the equation of the line of intersection.

12. Let P and Q be the points $P(2, 1, -1)$ and $Q(1, -3, 4)$ of Euclidean 3-space. Let π be the plane with equation $x + 4y - 5z = 53$.

(a) Find the vector equation of the line passing through the points P and Q.

(b) Is the line in (a) perpendicular to π? Explain.

(c) Where does the line through P and Q intersect the plane?

13. Find two orthogonal unit vectors in $\mathbf{R}^3$ each orthogonal to $\mathbf{u} = \begin{bmatrix} -1 \\ 6 \\ 1 \end{bmatrix}$.

14. Find two orthogonal vectors that span the planes with the given equations.

(a) $x + y + z = 0$

(b) $2x - 3y + z = 0$

(c) $x - 3y + 7z = 0$

15. (a) Determine whether $\begin{bmatrix} -1 \\ 3 \\ 2 \end{bmatrix}$ is a linear combination of $\begin{bmatrix} 2 \\ 0 \\ 1 \end{bmatrix}$ and $\begin{bmatrix} 0 \\ 2 \\ 4 \end{bmatrix}$.

(b) Do $\begin{bmatrix} 2 \\ 0 \\ 1 \end{bmatrix}$ and $\begin{bmatrix} 0 \\ 2 \\ 4 \end{bmatrix}$ span $\mathbf{R}^3$? Justify your answer.

16. In this exercise, π is the plane with equation $3x - 2y + 2z = 0$.

(a) Find two orthogonal vectors in plane π.

(b) What is the projection of $\mathbf{w} = \begin{bmatrix} -1 \\ 1 \\ 2 \end{bmatrix}$ onto π?

(c) What is the projection of $\mathbf{w} = \begin{bmatrix} x \\ y \\ z \end{bmatrix}$ onto π?

17. Find the projection of $\mathbf{w} = \begin{bmatrix} 1 \\ 1 \\ 1 \end{bmatrix}$ on the plane π with equation $x - 7y + 4z = 0$ by the following two methods: 1) projecting $\mathbf{w}$ on a normal to π, and 2) the method described in Problem 1.4.14.

18. Let ℓ be the line with equation
$$\begin{bmatrix} x \\ y \\ z \end{bmatrix} = \begin{bmatrix} -1 \\ 2 \\ 3 \end{bmatrix} + t \begin{bmatrix} 4 \\ 0 \\ -2 \end{bmatrix}$$ and let π be the plane with equation $2x - 3y + 4z = 2$.

    (a) Show that ℓ is parallel to π.

    (b) Find a point P on ℓ and a point Q on π.

    (c) Find the shortest distance from ℓ to π.

19. Find the distance from $P(1, 1, 1)$ to the plane through $(2, -1, 3)$ with normal $\begin{bmatrix} 3 \\ 0 \\ 1 \end{bmatrix}$.

20. Find the distance from the point $P(1, 1, 1)$ to the plane with equation $2x - y - 4z = 8$.

21. Let P be the point $(-1, 2, 1)$ and π the plane with equation $4x - 3y = 7$.

    (a) Find the distance from P to π.

    (b) Write the equation of any line that is contained in π.

22. (a) Find the equation of the line passing through the point $P(3, -4, 5)$ and perpendicular to the plane with equation $2x - 3y + z = 12$.

    (b) Find the distance from $P(3, -4, 5)$ to the plane in (a).

23. Find a point Q on the line with equation
$$\begin{bmatrix} x \\ y \\ z \end{bmatrix} = \begin{bmatrix} 4 \\ 1 \\ 0 \end{bmatrix} + t \begin{bmatrix} 2 \\ 1 \\ -1 \end{bmatrix}$$ that is $\sqrt{19}$ units distant from $P(3, -, 3, -2)$. Is Q unique?

24. Give an example of three nonzero vectors in $\mathbf{R}^3$ that span a plane and explain how you found these vectors.

25. If $\mathbf{u}$ and $\mathbf{v}$ are vectors, show that any linear combination of $\mathbf{u}$ and $\mathbf{v}$ is also a linear combination of $2\mathbf{u}$ and $-3\mathbf{v}$.

26. Suppose vectors $\mathbf{u}$ and $\mathbf{v}$ are each scalar multiples of a third vector $\mathbf{w}$. Show that any linear combination of $\mathbf{u}$ and $\mathbf{v}$ is a scalar multiple of $\mathbf{w}$.

2

Matrices and Linear Equations

2.1 The Algebra of Matrices

■ What Is a Matrix?

The heart of linear algebra is systems of linear equations. Here is such a system:

$$
\begin{aligned}
2x_1 + x_2 - x_3 &= 2 \\
3x_1 - 2x_2 + 6x_3 &= -1 \\
x_1 \qquad\quad - 5x_3 &= 1.
\end{aligned}
\tag{1}
$$

The idea is to find numbers x_1, x_2, and x_3 that make each equation true. Notice that each row is a statement about the dot product of a certain vector with the vector $\mathbf{x} = \begin{bmatrix} x_1 \\ x_2 \\ x_3 \end{bmatrix}$ of unknowns. Think of the coefficients in each row as the components of a vector. Specifically, let

$$
\mathbf{r}_1 = \begin{bmatrix} 2 \\ 1 \\ -1 \end{bmatrix}, \quad
\mathbf{r}_2 = \begin{bmatrix} 3 \\ -2 \\ 6 \end{bmatrix}, \quad
\mathbf{r}_3 = \begin{bmatrix} 1 \\ 0 \\ -5 \end{bmatrix}.
$$

Then our equations are just the assertions that

$$
\begin{aligned}
\mathbf{r}_1 \cdot \mathbf{x} &= 2 \\
\mathbf{r}_2 \cdot \mathbf{x} &= -1 \\
\mathbf{r}_3 \cdot \mathbf{x} &= 1.
\end{aligned}
$$

As we shall see in this chapter, these equations can also be written

$$\begin{bmatrix} 2 & 1 & -1 \end{bmatrix} \begin{bmatrix} x_1 \\ x_2 \\ x_3 \end{bmatrix} = 2$$

$$\begin{bmatrix} 3 & -2 & 6 \end{bmatrix} \begin{bmatrix} x_1 \\ x_2 \\ x_3 \end{bmatrix} = -1$$

$$\begin{bmatrix} 1 & 0 & -5 \end{bmatrix} \begin{bmatrix} x_1 \\ x_2 \\ x_3 \end{bmatrix} = 1$$

and summarized in the single *matrix equation*

$$\begin{bmatrix} 2 & 1 & -1 \\ 3 & -2 & 6 \\ 1 & 0 & -5 \end{bmatrix} \begin{bmatrix} x_1 \\ x_2 \\ x_3 \end{bmatrix} = \begin{bmatrix} 2 \\ -1 \\ 1 \end{bmatrix},$$

which is of the simple form

$$A\mathbf{x} = \mathbf{b},$$

with

$$A = \begin{bmatrix} 2 & 1 & -1 \\ 3 & -2 & 6 \\ 1 & 0 & -5 \end{bmatrix}, \quad \mathbf{x} = \begin{bmatrix} x_1 \\ x_2 \\ x_3 \end{bmatrix}, \quad \text{and} \quad \mathbf{b} = \begin{bmatrix} 2 \\ -1 \\ 1 \end{bmatrix}.$$

Look at the numbers appearing in A. Do you see that they are the coefficients of x_1, x_2, and x_3 in the equations of system (1)? In the form $A\mathbf{x} = \mathbf{b}$, the system doesn't look much different from the single equation $ax = b$ with a and b given real numbers and x the unknown. We begin this chapter with a discussion of *matrices*.

2.1.1 DEFINITION   A *matrix* is a rectangular array of numbers enclosed in square brackets. If there are m rows and n columns in the array, the matrix is called $m \times n$ (read "m by n") and said to have *size* $m \times n$.

For example, $\begin{bmatrix} 2 & 1 & -1 \\ 3 & -2 & 6 \\ 1 & 0 & -5 \end{bmatrix}$ is a 3 × 3 ("three by three") matrix, $\begin{bmatrix} 1 & 2 & 3 \\ 4 & 5 & 6 \end{bmatrix}$ is a

2 × 3 matrix, and $\begin{bmatrix} -4 \\ 0 \\ 5 \end{bmatrix}$ is a 3 × 1 matrix. An *n*-dimensional vector $\begin{bmatrix} x_1 \\ x_2 \\ \vdots \\ x_n \end{bmatrix}$ is just an

$n \times 1$ matrix (and conversely). Such a matrix is called a *column matrix*. A 1 × n matrix is called a *row matrix* because such a matrix looks like this: $\begin{bmatrix} x_1 & x_2 & \dots & x_n \end{bmatrix}$.

Ignoring the technicality that a vector should properly be enclosed in brackets, it is often convenient to think of an $m \times n$ matrix as n column vectors written side by side and enclosed in a single pair of brackets. We write

$$A = \begin{bmatrix} \mathbf{a}_1 & \mathbf{a}_2 & \cdots & \mathbf{a}_n \\ \downarrow & \downarrow & & \downarrow \end{bmatrix}$$

to indicate that the columns of A are the vectors $\mathbf{a}_1, \mathbf{a}_2, \ldots, \mathbf{a}_n$. Again ignoring the problem of brackets, each row of a matrix is the *transpose* of a vector in the following sense.

2.1.2 DEFINITION

The *transpose* of an $m \times n$ matrix A is the $n \times m$ matrix whose rows are the columns of A in the same order. The transpose of A is denoted A^T.

For example, if $A = \begin{bmatrix} 1 & 2 & 3 \\ 4 & 5 & 6 \\ 7 & 8 & 9 \end{bmatrix}$, then $A^T = \begin{bmatrix} 1 & 4 & 7 \\ 2 & 5 & 8 \\ 3 & 6 & 9 \end{bmatrix}$ has the three rows of A for its columns. If $A = \begin{bmatrix} -1 & 2 \\ 3 & 4 \\ 0 & -5 \end{bmatrix}$, then $A^T = \begin{bmatrix} 1 & 3 & 0 \\ 2 & 4 & -5 \end{bmatrix}$. More examples follow.

2.1.3 EXAMPLES

- $\begin{bmatrix} 1 & -3 \\ 0 & 2 \end{bmatrix}^T = \begin{bmatrix} 1 & 0 \\ -3 & 2 \end{bmatrix}$

- $\begin{bmatrix} -1 & 2 & 3 & 0 & 0 \\ 2 & 4 & 6 & -3 & 4 \\ 0 & 1 & 2 & 5 & 9 \end{bmatrix}^T = \begin{bmatrix} -1 & 2 & 0 \\ 2 & 4 & 1 \\ 3 & 6 & 2 \\ 0 & -3 & 5 \\ 0 & 4 & 9 \end{bmatrix}$

- $\begin{bmatrix} -1 & 0 & 2 \end{bmatrix}^T = \begin{bmatrix} -1 \\ 0 \\ 2 \end{bmatrix}$

- $\begin{bmatrix} 0 \\ 2 \\ 3 \\ 4 \end{bmatrix}^T = \begin{bmatrix} 0 & 2 & 3 & 4 \end{bmatrix}$ ◼

The last two examples here show that the transpose of a row matrix is a vector and the transpose of a vector is a row matrix. Sometimes, we wish to think of a matrix in terms of its rows and write

$$A = \begin{bmatrix} \mathbf{a}_1^T & \rightarrow \\ \mathbf{a}_2^T & \rightarrow \\ \vdots & \\ \mathbf{a}_m^T & \rightarrow \end{bmatrix}$$

to indicate that the rows of A are the transposes of the vectors $\mathbf{a}_1, \mathbf{a}_2, \ldots, \mathbf{a}_m$. For example, if

$$A = \begin{bmatrix} 1 & 2 & 3 & 4 \\ 8 & 7 & 6 & 5 \\ 0 & 1 & 1 & 0 \end{bmatrix}, \tag{2}$$

then the columns of A are the vectors

$$\begin{bmatrix} 1 \\ 8 \\ 0 \end{bmatrix}, \quad \begin{bmatrix} 2 \\ 7 \\ 1 \end{bmatrix}, \quad \begin{bmatrix} 3 \\ 6 \\ 1 \end{bmatrix}, \quad \begin{bmatrix} 4 \\ 5 \\ 0 \end{bmatrix}$$

and the rows of A are

$$\begin{bmatrix} 1 \\ 2 \\ 3 \\ 4 \end{bmatrix}^T, \quad \begin{bmatrix} 8 \\ 7 \\ 6 \\ 5 \end{bmatrix}^T, \quad \begin{bmatrix} 0 \\ 1 \\ 1 \\ 0 \end{bmatrix}^T.$$

Occasionally, we may want to refer to the individual entries of a matrix. With reference to the matrix A in (2), we call the numbers 1, 2, 3, and 4 in the first row the $(1, 1)$, $(1, 2)$, $(1, 3)$, and $(1, 4)$ *entries* of A and write

$$a_{11} = 1, \quad a_{12} = 2, \quad a_{13} = 3, \quad a_{14} = 4.$$

In general, the entry in row i and column j of a matrix called A is denoted a_{ij}. If the matrix were B, the (i, j) entry would be denoted b_{ij}.

2.1.4 EXAMPLE Let $A = \begin{bmatrix} 1 & 2 \\ 3 & 4 \\ 5 & 6 \end{bmatrix}$. Then $a_{31} = 5$, $a_{12} = 2$, and $a_{22} = 4$. ■

2.1.5 EXAMPLE Let A be the 2×3 matrix with $a_{ij} = i - j$. The $(1, 1)$ entry of A is $a_{11} = 1 - 1 = 0$, the $(1, 2)$ entry is $a_{12} = 1 - 2 = -1$, and so on. We obtain $A = \begin{bmatrix} 0 & -1 & -2 \\ 1 & 0 & -1 \end{bmatrix}$. ■

2.1.6 EXAMPLE If $A = [a_{ij}]$ and $B = A^T$, then $b_{ij} = a_{ji}$. ■

Equality of matrices is defined in terms of equality of vectors—see 1.5.3.

> **2.1.7** Matrices A and B are *equal* if and only if they have the same number of rows, the same number of columns and corresponding columns are equal.

2.1.8 EXAMPLE Let $A = \begin{bmatrix} -1 & x \\ 2y & -3 \end{bmatrix}$ and $B = \begin{bmatrix} a & -4 \\ 4 & a-b \end{bmatrix}$. If $A = B$, then $-1 = a$, $x = -4$, $2y = 4$, and $a - b = -3$. Thus $a = -1$, $b = a + 3 = 2$, $x = -4$, and $y = 2$. ■

■ Addition of Matrices

Matrices can also be added entry by entry, or row by row.

If matrices $A = \begin{bmatrix} \mathbf{a}_1 & \mathbf{a}_2 & \cdots & \mathbf{a}_n \\ \downarrow & \downarrow & & \downarrow \end{bmatrix}$ and $B = \begin{bmatrix} \mathbf{b}_1 & \mathbf{b}_2 & \cdots & \mathbf{b}_n \\ \downarrow & \downarrow & & \downarrow \end{bmatrix}$ *have the same size, they*

can be added column by column, like this:

$$A + B = \begin{bmatrix} \mathbf{a}_1 + \mathbf{b}_1 & \mathbf{a}_2 + \mathbf{b}_2 & \cdots & \mathbf{a}_n + \mathbf{b}_n \\ \downarrow & \downarrow & & \downarrow \end{bmatrix}.$$

2.1.9 EXAMPLE

Let $A = \begin{bmatrix} 1 & 2 & 3 \\ 4 & 5 & 6 \end{bmatrix}$, $B = \begin{bmatrix} -2 & 3 & 0 \\ 2 & 1 & -1 \end{bmatrix}$, and $C = \begin{bmatrix} 1 & 0 \\ 0 & -1 \end{bmatrix}$. Then $A + B = \begin{bmatrix} -1 & 5 & 3 \\ 6 & 6 & 5 \end{bmatrix}$ while $A + C$ and $C + B$ are not defined. *Only matrices of the same size can be added.* ■

For any positive integers m and n, there is an $m \times n$ *zero matrix*, which we denote with boldface type **0** (and underlining in handwritten work). This is the matrix each of whose columns is the zero vector. It has the property that $A + \mathbf{0} = \mathbf{0} + A = A$ for any $m \times n$ matrix A.

2.1.10 EXAMPLE

The 2×2 zero matrix is $\begin{bmatrix} 0 & 0 \\ 0 & 0 \end{bmatrix}$. This matrix works just like zero should, with respect to 2×2 matrices. For example,

$$\begin{bmatrix} -1 & 2 \\ 3 & 5 \end{bmatrix} + \begin{bmatrix} 0 & 0 \\ 0 & 0 \end{bmatrix} = \begin{bmatrix} -1 & 2 \\ 3 & 5 \end{bmatrix} = \begin{bmatrix} 0 & 0 \\ 0 & 0 \end{bmatrix} + \begin{bmatrix} -1 & 2 \\ 3 & 5 \end{bmatrix}.$$ ■

2.1.11 EXAMPLE

The 2×3 zero matrix is the matrix $\mathbf{0} = \begin{bmatrix} 0 & 0 & 0 \\ 0 & 0 & 0 \end{bmatrix}$. For any 2×3 matrix $A = \begin{bmatrix} a & c & e \\ b & d & f \end{bmatrix}$, we have

$$A + \mathbf{0} = \begin{bmatrix} a & c & e \\ b & d & f \end{bmatrix} + \begin{bmatrix} 0 & 0 & 0 \\ 0 & 0 & 0 \end{bmatrix} = \begin{bmatrix} a & c & e \\ b & d & f \end{bmatrix} = A$$

and, similarly, $\mathbf{0} + A = \mathbf{A}$. ■

■ Scalar Multiplication of Matrices

"Complex" here doesn't mean hard! See Section 7.1.

We have already said that the term "scalar" in linear algebra means "number," usually real, but sometimes complex. Just as there is a notion of multiplication of vectors by scalars, so also is there a *scalar multiplication* of matrices.

Of course, cA can also be formed by multiplying each entry of A by c.

If $A = \begin{bmatrix} \mathbf{a}_1 & \mathbf{a}_2 & \cdots & \mathbf{a}_n \\ \downarrow & \downarrow & & \downarrow \end{bmatrix}$ is an $m \times n$ matrix and c is a scalar, then cA is the matrix whose columns are c times the columns of A, as follows:

$$cA = \begin{bmatrix} c\mathbf{a}_1 & c\mathbf{a}_2 & \cdots & c\mathbf{a}_n \\ \downarrow & \downarrow & & \downarrow \end{bmatrix}.$$

2.1.12 EXAMPLE For $A = \begin{bmatrix} 1 & -1 \\ 2 & 0 \\ -3 & 5 \end{bmatrix}$ and $c = -4$, we have $(-4)A = \begin{bmatrix} -4 & 4 \\ -8 & 0 \\ 12 & -20 \end{bmatrix}$. ■

2.1.13 DEFINITION The *negative* of a matrix A is the matrix $(-1)A$, which we denote $-A$ and call "minus A": $-A = (-1)A$.

2.1.14 EXAMPLE If $A = \begin{bmatrix} -1 & 3 \\ -2 & 4 \end{bmatrix}$, then $-A = (-1)\begin{bmatrix} -1 & 3 \\ -2 & 4 \end{bmatrix} = \begin{bmatrix} 1 & -3 \\ 2 & -4 \end{bmatrix}$. ■

The terms "negative" and "minus A" are well chosen: for any matrix A, $A + (-A) = \mathbf{0}$ is the zero matrix. This is easily checked, as are all the other properties of addition and scalar multiplication of matrices, which are so reminiscent of the properties of *vector* addition and scalar multiplication recorded in Theorem 1.1.6.

Theorem 2.1.15 **(Properties of Matrix Addition and Scalar Multiplication).** *Let A, B, and C be matrices of the same size and let c and d be scalars.*

1. *(Closure under addition) $A + B$ is a matrix.*

2. *(Commutativity of addition) $A + B = B + A$.*

3. *(Addition is associative) $(A + B) + C = A + (B + C)$.*

4. *(Zero) There is a zero matrix $\mathbf{0}$ with the property that $A + \mathbf{0} = \mathbf{0} + A = A$.*

5. *(Negatives) For every matrix A, there is a matrix $-A$ called the negative of A with the property that $A + (-A) = \mathbf{0}$ is the zero matrix.*

6. *(Closure under scalar multiplication) cA is a matrix.*

7. *(Scalar associativity) $c(dA) = (cd)A$.*

8. *(One) $1A = A$.*

9. *(Distributivity) $c(A + B) = cA + cB$ and $(c + d)A = cA + dA$.*

10. *If $cA = \mathbf{0}$, then either $c = 0$ or $A = \mathbf{0}$.*

2.1.16 PROBLEM Let $A = \begin{bmatrix} 1 & 2 \\ 3 & 5 \\ -6 & -7 \end{bmatrix}$ and $B = \begin{bmatrix} -2 & 4 \\ 6 & 2 \\ -8 & 2 \end{bmatrix}$. Solve the equation $2A + 4X = 3B$.

Solution. Just as in ordinary algebra, $4X = 3B - 2A$, so $X = \frac{1}{4}(3B - 2A)$. We have

$$3B = \begin{bmatrix} -6 & 12 \\ 18 & 6 \\ -24 & 6 \end{bmatrix}, \quad 2A = \begin{bmatrix} 2 & 4 \\ 6 & 10 \\ -12 & -14 \end{bmatrix}, \quad 3B - 2A = \begin{bmatrix} -8 & 8 \\ 12 & -4 \\ -12 & 20 \end{bmatrix},$$

and $X = \frac{1}{4}\begin{bmatrix} -8 & 8 \\ 12 & -4 \\ -12 & 20 \end{bmatrix} = \begin{bmatrix} -2 & 2 \\ 3 & -1 \\ -3 & 5 \end{bmatrix}$.

■ Multiplication of Matrices

The definition of matrix multiplication here is going to seem pretty strange, but it comes about very naturally. See Section 5.2.

We explain how to multiply matrices in three stages. First we show how to multiply a row matrix by a column matrix, then how to multiply any matrix by a column matrix, and finally how to multiply any two matrices.

STAGE 1: Row times column. Remember that a row matrix is the transpose $\mathbf{a}^T$ of a vector $\mathbf{a}$, and a column matrix is a (column) vector $\mathbf{b}$. We define

$$\mathbf{a}^T \mathbf{b} = [\mathbf{a} \cdot \mathbf{b}]$$

as the 1×1 matrix whose single entry is the dot product of the vectors $\mathbf{a}$ and $\mathbf{b}$ (enclosed in square brackets).

2.1.17 EXAMPLES

- $\begin{bmatrix} 2 & 4 & -3 \end{bmatrix} \begin{bmatrix} 3 \\ -1 \\ 1 \end{bmatrix} = \begin{bmatrix} 2(3) + 4(-1) - 3(1) \end{bmatrix} = \begin{bmatrix} -1 \end{bmatrix}$

- $\begin{bmatrix} 3 & -1 & 2 & 5 \end{bmatrix} \begin{bmatrix} 4 \\ -1 \\ -1 \\ 2 \end{bmatrix} = \begin{bmatrix} 12 + 1 - 2 + 10 \end{bmatrix} = \begin{bmatrix} 21 \end{bmatrix}$ ■

2.1.18 The product of a row and a column is a dot product: $\mathbf{a}^T \mathbf{b} = [\mathbf{a} \cdot \mathbf{b}]$.

STAGE 2: Matrix times column. Knowing how to multiply a row by a column enables us to multiply a matrix with any number of rows by a column. The product is computed row by row.

Suppose the rows of A are $\mathbf{a}_1^T, \mathbf{a}_2^T, \ldots, \mathbf{a}_m^T$ and $\mathbf{b}$ is a column. The product

$$A\mathbf{b} = \begin{bmatrix} \mathbf{a}_1^T & \to \\ \mathbf{a}_2^T & \to \\ & \vdots \\ \mathbf{a}_m^T & \to \end{bmatrix} \begin{bmatrix} \mathbf{b} \\ \downarrow \end{bmatrix} = \begin{bmatrix} \mathbf{a}_1^T \mathbf{b} \\ \mathbf{a}_2^T \mathbf{b} \\ \vdots \\ \mathbf{a}_m^T \mathbf{b} \end{bmatrix} = \begin{bmatrix} \mathbf{a}_1 \cdot \mathbf{b} \\ \mathbf{a}_2 \cdot \mathbf{b} \\ \vdots \\ \mathbf{a}_m \cdot \mathbf{b} \end{bmatrix} \qquad (3)$$

is just another column, the column whose entries are the products of the rows of A with $\mathbf{b}$.

2.1.19 EXAMPLE

If $A = \begin{bmatrix} 0 & 1 & 3 \\ -1 & 1 & 2 \\ 2 & 0 & -1 \\ 4 & 5 & -1 \end{bmatrix}$ and $B = \begin{bmatrix} -1 \\ 2 \\ 3 \end{bmatrix}$, then AB is a column whose first entry is the product of the first row of A with B,

$$\begin{bmatrix} 0 & 1 & 3 \end{bmatrix} \begin{bmatrix} -1 \\ 2 \\ 3 \end{bmatrix} = 11,$$

whose second entry is the product of the second row of A with B,

$$\begin{bmatrix} -1 & 1 & 2 \end{bmatrix} \begin{bmatrix} -1 \\ 2 \\ 3 \end{bmatrix} = 9,$$

whose third entry is the product of the third row of A with B,

$$\begin{bmatrix} 2 & 0 & -1 \end{bmatrix} \begin{bmatrix} -1 \\ 2 \\ 3 \end{bmatrix} = -5$$

and whose fourth entry is the product of the fourth row of A with B,

$$\begin{bmatrix} 4 & 5 & -1 \end{bmatrix} \begin{bmatrix} -1 \\ 2 \\ 3 \end{bmatrix} = 3.$$

Thus

$$AB = \begin{bmatrix} 0 & 1 & 3 \\ -1 & 1 & 2 \\ 2 & 0 & -1 \\ 4 & 5 & -1 \end{bmatrix} \begin{bmatrix} -1 \\ 2 \\ 3 \end{bmatrix} = \begin{bmatrix} 11 \\ 9 \\ -5 \\ 3 \end{bmatrix}. \qquad \blacksquare$$

2.1.20 EXAMPLE   As noted at the start of this section, the system

$$\begin{aligned} 2x_1 + x_2 - x_3 &= 2 \\ 3x_1 - 2x_2 + 6x_3 &= -1 \\ x_1 \qquad - 5x_3 &= 1. \end{aligned}$$

can be written $A\mathbf{x} = \mathbf{b}$, with

$$A = \begin{bmatrix} 2 & 1 & -1 \\ 3 & -2 & 6 \\ 1 & 0 & -5 \end{bmatrix}, \quad \mathbf{x} = \begin{bmatrix} x_1 \\ x_2 \\ x_3 \end{bmatrix}, \quad \text{and} \quad \mathbf{b} = \begin{bmatrix} 2 \\ -1 \\ 1 \end{bmatrix}. \qquad \blacksquare$$

Recall that the standard basis vectors of $\mathbf{R}^n$ are labeled $\mathbf{e}_1, \ldots, \mathbf{e}_n$. Thus

$$\mathbf{e}_1 = \begin{bmatrix} 1 \\ 0 \\ 0 \\ \vdots \\ 0 \end{bmatrix}, \quad \mathbf{e}_2 = \begin{bmatrix} 0 \\ 1 \\ 0 \\ \vdots \\ 0 \end{bmatrix}, \quad \ldots, \quad \mathbf{e}_n = \begin{bmatrix} 0 \\ 0 \\ \vdots \\ 0 \\ 1 \end{bmatrix}.$$

Suppose $A = \begin{bmatrix} \mathbf{a}_1^T & \rightarrow \\ \mathbf{a}_2^T & \rightarrow \\ \vdots \\ \mathbf{a}_m^T & \rightarrow \end{bmatrix}$ is an $m \times n$ matrix with rows as indicated. As in (3), we have

$$Ae_1 = \begin{bmatrix} \mathbf{a}_1^T & \rightarrow \\ \mathbf{a}_2^T & \rightarrow \\ \vdots \\ \mathbf{a}_m^T & \rightarrow \end{bmatrix} \begin{bmatrix} 1 \\ 0 \\ \vdots \\ 0 \end{bmatrix} = \begin{bmatrix} \mathbf{a}_1 \cdot \mathbf{e}_1 \\ \mathbf{a}_2 \cdot \mathbf{e}_1 \\ \vdots \\ \mathbf{a}_m \cdot \mathbf{e}_1 \end{bmatrix},$$

which is just the first column of A, since $\mathbf{a}_1 \cdot \mathbf{e}_1$ is the first component of $\mathbf{a}_1$, $\mathbf{a}_2 \cdot \mathbf{e}_1$ is the first component of $\mathbf{a}_2$, and so on. In general, for any i,

2.1.21 $A\mathbf{e}_i =$ column i of A.

For example, if $n = 2$ and $A = \begin{bmatrix} 1 & 4 \\ 2 & 5 \\ 3 & 6 \end{bmatrix}$, then

$$Ae_1 = \begin{bmatrix} 1 & 4 \\ 2 & 5 \\ 3 & 6 \end{bmatrix} \begin{bmatrix} 1 \\ 0 \end{bmatrix} = \begin{bmatrix} 1 \\ 2 \\ 3 \end{bmatrix}$$

is the first column of A and

$$Ae_2 = \begin{bmatrix} 1 & 4 \\ 2 & 5 \\ 3 & 6 \end{bmatrix} \begin{bmatrix} 0 \\ 1 \end{bmatrix} = \begin{bmatrix} 4 \\ 5 \\ 6 \end{bmatrix}$$

is the second column of A.

2.1.22 PROBLEM   Suppose A and B are $m \times n$ matrices such that $A\mathbf{e}_i = B\mathbf{e}_i$ for each $\mathbf{e}_i \in \mathbf{R}^n$. Prove that $A = B$.

Solution.   By 2.1.21, the given condition says that each column of A equals the corresponding column of B, so $A = B$.

STAGE 3 (End of Story) Matrix by matrix. Now that we know how to multiply a matrix by a column, it is straightforward to explain how to find the product of any two matrices of *compatible* sizes. Suppose that A is an $m \times n$ matrix and B is an $n \times p$ matrix. (Notice the repeated n here.) Let B have columns $\mathbf{b}_1, \dots, \mathbf{b}_p$. Then

$$AB = A \begin{bmatrix} \mathbf{b}_1 & \mathbf{b}_2 & \cdots & \mathbf{b}_p \\ \downarrow & \downarrow & & \downarrow \end{bmatrix} = \begin{bmatrix} A\mathbf{b}_1 & A\mathbf{b}_2 & \cdots & A\mathbf{b}_p \\ \downarrow & \downarrow & & \downarrow \end{bmatrix} \qquad (4)$$

is the matrix whose columns are the products $A\mathbf{b}_1, A\mathbf{b}_2, \ldots, A\mathbf{b}_p$ of A with the columns of B.

2.1.23 EXAMPLE   Let $A = \begin{bmatrix} 2 & 4 & -1 \\ 3 & 5 & 1 \\ 0 & 4 & -2 \end{bmatrix}$ and $B = \begin{bmatrix} 2 & 4 & 0 & 1 & -1 \\ 3 & -2 & 0 & 2 & 4 \\ 1 & 2 & -6 & 2 & 1 \end{bmatrix}$. Then

$$AB = \begin{bmatrix} 2 & 4 & -1 \\ 3 & 5 & 1 \\ 0 & 4 & -2 \end{bmatrix}\begin{bmatrix} 2 & 4 & 0 & 1 & -1 \\ 3 & -2 & 0 & 2 & 4 \\ 1 & 2 & -6 & 2 & 1 \end{bmatrix} = \begin{bmatrix} 15 & -2 & 6 & 8 & 13 \\ 22 & 4 & -6 & 15 & 18 \\ 10 & -12 & 12 & 4 & 14 \end{bmatrix}.$$

The first column of AB is the product of A and the first column of B,

$$A\begin{bmatrix} 2 \\ 3 \\ 1 \end{bmatrix} = \begin{bmatrix} 2 & 4 & -1 \\ 3 & 5 & 1 \\ 0 & 4 & -2 \end{bmatrix}\begin{bmatrix} 2 \\ 3 \\ 1 \end{bmatrix} = \begin{bmatrix} 15 \\ 22 \\ 10 \end{bmatrix},$$

the second column of AB is the product of A and the second column of B,

$$A\begin{bmatrix} 4 \\ -2 \\ 2 \end{bmatrix} = \begin{bmatrix} 2 & 4 & -1 \\ 3 & 5 & 1 \\ 0 & 4 & -2 \end{bmatrix}\begin{bmatrix} 4 \\ -2 \\ 2 \end{bmatrix} = \begin{bmatrix} -2 \\ 4 \\ -12 \end{bmatrix},$$

and so on.   ∎

It is useful to note here that if

$$A = \begin{bmatrix} \mathbf{a}_1^T & \rightarrow \\ \mathbf{a}_2^T & \rightarrow \\ \vdots & \\ \mathbf{a}_m^T & \rightarrow \end{bmatrix} \quad \text{and} \quad B = \begin{bmatrix} \mathbf{b}_1 & \mathbf{b}_2 & \ldots & \mathbf{b}_p \\ \downarrow & \downarrow & & \downarrow \end{bmatrix}$$

then

$$AB = \begin{bmatrix} \mathbf{a}_1^T\mathbf{b}_1 & \mathbf{a}_1^T\mathbf{b}_2 & \ldots & \mathbf{a}_1^T\mathbf{b}_p \\ \mathbf{a}_2^T\mathbf{b}_1 & \mathbf{a}_2^T\mathbf{b}_2 & \ldots & \mathbf{a}_2^T\mathbf{b}_p \\ \vdots & \vdots & & \vdots \\ \mathbf{a}_m^T\mathbf{b}_1 & \mathbf{a}_m^T\mathbf{b}_2 & \ldots & \mathbf{a}_m^T\mathbf{b}_p \end{bmatrix} = \begin{bmatrix} \mathbf{a}_1 \cdot \mathbf{b}_1 & \mathbf{a}_1 \cdot \mathbf{b}_2 & \ldots & \mathbf{a}_1 \cdot \mathbf{b}_p \\ \mathbf{a}_2 \cdot \mathbf{b}_1 & \mathbf{a}_2 \cdot \mathbf{b}_2 & \ldots & \mathbf{a}_2 \cdot \mathbf{b}_p \\ \vdots & \vdots & & \vdots \\ \mathbf{a}_m \cdot \mathbf{b}_1 & \mathbf{a}_m \cdot \mathbf{b}_2 & \ldots & \mathbf{a}_m \cdot \mathbf{b}_p \end{bmatrix}$$

is a matrix of dot products.

2.1.24 The (i, j) entry of AB is the product of row i of A and column j of B.

In **2.1.23**, for example, the $(3, 2)$ entry is the product of the third row of A and the second column of B:

$$-12 = \begin{bmatrix} 0 & 4 & -2 \end{bmatrix} \begin{bmatrix} 4 \\ -2 \\ 2 \end{bmatrix}.$$

We have used the term *compatible sizes* a couple of times to emphasize that only certain pairs of matrices can be multiplied. The fact expressed in 2.1.24 makes it implicit that the number of entries in the ith row of A must match the number of entries in the jth column of B; that is, the number of columns of A must equal the number of rows of B.

Thus, if A is $m \times n$ and we wish to form the product AB for some other matrix B, then B must have n rows—that is, B must be $n \times p$ for some p—and then AB will be $m \times p$. Students sometimes use the equation

$$\frac{m \; \not{n}}{\not{n} \; p} = \frac{m}{p}$$

to help remember the sizes of matrices that can be multiplied and also the size of the product.

2.1.25 EXAMPLE   Suppose $A = \begin{bmatrix} 1 & -2 \\ 3 & -4 \end{bmatrix}$ and $B = \begin{bmatrix} -1 & 0 & 1 \\ 3 & 4 & 2 \end{bmatrix}$.

Then $AB = \begin{bmatrix} -7 & -8 & -3 \\ -15 & -16 & -5 \end{bmatrix}$ while BA is not defined: B is $2 \times ③$ but A is ② $\times 2$. (The circled numbers are different.)   ∎

2.1.26 EXAMPLE   If $A = \begin{bmatrix} 1 \\ 2 \\ 3 \\ 4 \end{bmatrix}$ and $B = \begin{bmatrix} -2 & 0 & 4 & 1 \end{bmatrix}$, then A is 4×1 and B is 1×4, so we can compute both AB and BA: AB will be 4×4; BA will be 1×1.

$$AB = \begin{bmatrix} -2 & 0 & 4 & 1 \\ -4 & 0 & 8 & 2 \\ -6 & 0 & 12 & 3 \\ -8 & 0 & 16 & 4 \end{bmatrix}, \quad BA = \begin{bmatrix} 14 \end{bmatrix}.$$ ∎

While a technical proof of the next statement is possible here, we much prefer an elegant argument that will explain why matrix multiplication is defined as it is. (See Section 5.2.)

2.1.27 Matrix multiplication is associative: $(AB)C = A(BC)$ whenever all products are defined.

On the other hand,

> **2.1.28** Matrix multiplication is not commutative. In general, $AB \neq BA$.

Examples of this failure of commutativity are easy to find. If A is a 3×4 matrix, for example, and B is 4×7, then AB is defined while BA is not, so there is no chance for $AB = BA$. Other times AB and BA are both defined but still $AB \neq BA$; we saw this in **2.1.26**. Even square matrices of the same size usually fail to commute. For example,

$$\begin{bmatrix} 1 & 2 \\ 3 & 4 \end{bmatrix} \begin{bmatrix} 0 & 1 \\ 1 & 0 \end{bmatrix} = \begin{bmatrix} 2 & 1 \\ 4 & 3 \end{bmatrix},$$

whereas

$$\begin{bmatrix} 0 & 1 \\ 1 & 0 \end{bmatrix} \begin{bmatrix} 1 & 2 \\ 3 & 4 \end{bmatrix} = \begin{bmatrix} 3 & 4 \\ 1 & 2 \end{bmatrix}.$$

The fact that matrices do not commute means special care must be exercised with algebraic expressions involving matrices. For instance, the product $ABAB$, in general, cannot be simplified. Note also that

$$(A + B)^2 = (A + B)(A + B)$$
$$= A(A + B) + B(A + B) = AA + AB + BA + BB$$
$$= A^2 + AB + BA + B^2$$

which, in general, is *not* the same as $A^2 + 2AB + B^2$. On the other hand, there are certain special matrices that commute with all matrices of an appropriate size. For example, if $\mathbf{0} = \begin{bmatrix} 0 & 0 \\ 0 & 0 \end{bmatrix}$ denotes the 2×2 zero matrix, then $A\mathbf{0} = \mathbf{0}A = \mathbf{0}$ for any 2×2 matrix A. For example, if $A = \begin{bmatrix} -1 & 2 \\ 3 & -4 \end{bmatrix}$, then

$$\begin{bmatrix} -1 & 2 \\ 3 & -4 \end{bmatrix} \begin{bmatrix} 0 & 0 \\ 0 & 0 \end{bmatrix} = \begin{bmatrix} 0 & 0 \\ 0 & 0 \end{bmatrix} = \begin{bmatrix} 0 & 0 \\ 0 & 0 \end{bmatrix} \begin{bmatrix} -1 & 2 \\ 3 & -4 \end{bmatrix}.$$

An *algebraic system* is a pair $(A, \star)$ where A is a set and $\star$ is a binary operation on A; that is, whenever a and b are in A, there is defined an element $a \star b$. An element e in such a system is called an *identity* if $a \star e = e \star a = a$ for all a. The number 1, for example, is an identity for the real numbers (when the operation is multiplication) and 0 is an identity when the operation is addition.

The matrix $I = \begin{bmatrix} 1 & 0 \\ 0 & 1 \end{bmatrix}$ is called the 2×2 *identity matrix* since it is an identity element for 2×2 matrices under multiplication. Just like $a1 = 1a = a$ for any real number a, so $AI = IA = A$ for any 2×2 matrix A. For example,

$$\begin{bmatrix} 1 & 2 \\ 3 & -4 \end{bmatrix} \begin{bmatrix} 1 & 0 \\ 0 & 1 \end{bmatrix} = \begin{bmatrix} 1 & 2 \\ 3 & -4 \end{bmatrix} = \begin{bmatrix} 1 & 0 \\ 0 & 1 \end{bmatrix} \begin{bmatrix} 1 & 2 \\ 3 & -4 \end{bmatrix}.$$

In general, the $n \times n$ *identity matrix* is the matrix whose columns are the standard basis vectors $\mathbf{e}_1, \mathbf{e}_2, \ldots, \mathbf{e}_n$ in order. It is denoted I_n, or just I when there is no chance of confusion.

2.1.29 EXAMPLES

$$I_2 = \begin{bmatrix} 1 & 0 \\ 0 & 1 \end{bmatrix}, \quad I_3 = \begin{bmatrix} 1 & 0 & 0 \\ 0 & 1 & 0 \\ 0 & 0 & 1 \end{bmatrix}, \quad I_4 = \begin{bmatrix} 1 & 0 & 0 & 0 \\ 0 & 1 & 0 & 0 \\ 0 & 0 & 1 & 0 \\ 0 & 0 & 0 & 1 \end{bmatrix}.$$

It is not hard to see that $AI_n = A$ for any $n \times n$ matrix A. The columns of I_n are $\mathbf{e}_1, \mathbf{e}_2, \ldots, \mathbf{e}_n$, so the columns of AI_n are $A\mathbf{e}_1, A\mathbf{e}_2, \ldots, A\mathbf{e}_n$, and these are just the columns of A, in order, by 2.1.21. It is almost as easy to see that $I_n A = A$, but we leave our reasoning for later. (See Problem 2.1.36.)

The next theorem summarizes some important properties associated with matrix multiplication. Most parts are quite straightforward with the exception of associativity, a proof of which is deferred to Section 5.2.

Theorem 2.1.30 (**Properties of Matrix Multiplication**). *Let A, B, and C be matrices.*

1. *Matrix multiplication is associative: $(AB)C = A(BC)$ whenever all products are defined.*

2. *Matrix multiplication is not commutative, in general: there exist matrices A and B for which $AB \neq BA$.*

3. *Matrix multiplication distributes over addition: $(A + B)C = AC + BC$ and $A(B + C) = AB + AC$ whenever all products are defined.*

4. *$A(cB) = (cA)B = c(AB)$ for any scalar c.*

5. *For any positive integer n, there is an $n \times n$ identity matrix denoted I_n with the property that $AI_n = A$ and $I_n A = A$ whenever these products are defined.*

2.1.31 PROBLEM

Show that $A(bB + cC) = b(AB) + c(AC)$ for any matrices A, B, C and scalars c, d whenever all products here are defined.

Solution.

$$\begin{aligned} A(bB + cC) &= A(bB) + A(cC) \quad \text{by Theorem 2.1.30, part 3} \\ &= b(AB) + c(AC) \quad \text{by Theorem 2.1.30, part 4.} \end{aligned}$$

We can combine the operations of addition and scalar multiplication and form *linear combinations* of matrices just as we can with vectors.

2.1.32 DEFINITION

A *linear combination* of matrices $A_1, A_2, \ldots, A_k$ (all of the same size) is a matrix of the form $c_1 A_1 + c_2 A_2 + \cdots + c_k A_k$, where $c_1, c_2, \ldots, c_k$ are scalars.

The matrix $bB + cC$ that appeared in Problem 2.1.31, for example, is a linear combination of the matrices B and C. This problem was designed to illustrate a general property of matrix multiplication. Suppose that A is an $m \times n$

matrix, that $B_1, B_2, \ldots, B_k$ are $n \times p$ matrices, and that $c_1, c_2, \ldots, c_k$ are scalars. Then

$$A(c_1 B_1 + c_2 B_2 + \cdots + c_k B_k) = c_1(AB_1) + c_2(AB_2) + \cdots + c_k(AB_k). \qquad (5)$$

The matrix $c_1 B_1 + c_2 B_2 + \cdots + c_k B_k$ is a linear combination of $B_1, \ldots, B_k$ so the left side of (5) is the product of A with a linear combination of matrices. The right side of (5) is a linear combination of the matrices $AB_1, \ldots, AB_k$. We summarize (5) by saying "matrix multiplication preserves linear combinations."

We have an important use in mind for this observation.

Let A be an $m \times n$ matrix and let $\mathbf{x} = \begin{bmatrix} x_1 \\ x_2 \\ \vdots \\ x_n \end{bmatrix}$ be a vector in $\mathbf{R}^n$. Remember—see 1.5.8—that $\mathbf{x} = x_1\mathbf{e}_1 + x_2\mathbf{e}_2 + \cdots + x_n\mathbf{e}_n$. Since matrix multiplication preserves linear combinations, we know that

$$A\mathbf{x} = A(x_1\mathbf{e}_1 + x_2\mathbf{e}_2 + \cdots + x_n\mathbf{e}_n) = x_1 A\mathbf{e}_1 + x_2 A\mathbf{e}_2 + \cdots + x_n A\mathbf{e}_n.$$

In view of 2.1.21, the expression on the right is x_1 times the first column of A, plus x_2 times the second column of A, and, finally, x_n times the last column of A.

The equation in 2.1.33 is the most important one in this book!

2.1.33 The product $A\mathbf{x}$ of a matrix A and a vector $\mathbf{x}$ is a linear combination of the columns of A, the coefficients being the components of $\mathbf{x}$:

$$\begin{bmatrix} \mathbf{a}_1 & \mathbf{a}_2 & \cdots & \mathbf{a}_n \\ \downarrow & \downarrow & & \downarrow \end{bmatrix} \begin{bmatrix} x_1 \\ x_2 \\ \vdots \\ x_n \end{bmatrix}$$

$$= x_1 \begin{bmatrix} \mathbf{a}_1 \\ \downarrow \end{bmatrix} + x_2 \begin{bmatrix} \mathbf{a}_2 \\ \downarrow \end{bmatrix} + \cdots + x_n \begin{bmatrix} \mathbf{a}_n \\ \downarrow \end{bmatrix}.$$

Conversely, any linear combination of vectors is $A\mathbf{x}$ for some matrix A and some vector $\mathbf{x}$.

Reading Challenge 1 *The last statement says any linear combination of vectors is of the form $A\mathbf{x}$. What is A? What is $\mathbf{x}$?*

2.1.34 EXAMPLE Let $A = \begin{bmatrix} 1 & 2 & 3 \\ 4 & 5 & 6 \\ 7 & 8 & 9 \end{bmatrix}$ and $\mathbf{x} = \begin{bmatrix} 2 \\ -1 \\ 4 \end{bmatrix}$. Then

$$
A\mathbf{x} = \begin{bmatrix} 1 & 2 & 3 \\ 4 & 5 & 6 \\ 7 & 8 & 9 \end{bmatrix} \begin{bmatrix} 2 \\ -1 \\ 4 \end{bmatrix}
$$

$$
= \begin{bmatrix} 12 \\ 27 \\ 42 \end{bmatrix} = 2 \begin{bmatrix} 1 \\ 4 \\ 7 \end{bmatrix} - \begin{bmatrix} 2 \\ 5 \\ 8 \end{bmatrix} + 4 \begin{bmatrix} 3 \\ 6 \\ 9 \end{bmatrix}
$$

$$
= \boxed{2} \times \text{ column } 1 + \boxed{-1} \times \text{ column } 2 + \boxed{4} \times \text{ column } 3.
$$

This is a linear combination of the columns of A, the coefficients, $2, -1, 4$, being the components of $\mathbf{x}$.

More generally,

$$
A \begin{bmatrix} x_1 \\ x_2 \\ x_3 \end{bmatrix} = x_1 \begin{bmatrix} 1 \\ 4 \\ 7 \end{bmatrix} + x_2 \begin{bmatrix} 2 \\ 5 \\ 8 \end{bmatrix} + x_3 \begin{bmatrix} 3 \\ 6 \\ 9 \end{bmatrix}. \qquad\blacksquare
$$

2.1.35 EXAMPLE $\begin{bmatrix} 2 & -3 \\ 4 & 5 \\ 6 & 8 \end{bmatrix} \begin{bmatrix} 5 \\ 2 \end{bmatrix} = \begin{bmatrix} 4 \\ 30 \\ 46 \end{bmatrix} = 5 \begin{bmatrix} 2 \\ 4 \\ 6 \end{bmatrix} + 2 \begin{bmatrix} -3 \\ 5 \\ 8 \end{bmatrix}. \qquad\blacksquare$

2.1.36 PROBLEM Let A be an $m \times n$ matrix and $I = I_m$ the $m \times m$ identity matrix. Explain why $IA = A$.

Solution. Let the columns of A be $\mathbf{a}_1, \mathbf{a}_2, \ldots, \mathbf{a}_n$. The first column of IA is $I\mathbf{a}_1$, which is a linear combination of the columns of I with coefficients the components of $\mathbf{a}_1$. The columns of I are the standard basis vectors $\mathbf{e}_1, \mathbf{e}_2, \ldots, \mathbf{e}_m$, so if $\mathbf{a}_1 = \begin{bmatrix} a_{11} \\ a_{21} \\ \vdots \\ a_{m1} \end{bmatrix}$, then the first column of IA is

$$
a_{11}\mathbf{e}_1 + a_{21}\mathbf{e}_2 + \cdots + a_{m1}\mathbf{e}_m = \begin{bmatrix} a_{11} \\ a_{21} \\ \vdots \\ a_{m1} \end{bmatrix} = \mathbf{a}_1.
$$

Similarly, one can show that column k of IA is $\mathbf{a}_k$, so $IA = A$.

2.1.37 PROBLEM

Suppose A is an $m \times n$ matrix with columns $a_1, a_2, \ldots, a_n$ and D is the $n \times n$ *diagonal* matrix whose *diagonal entries* are $d_1, \ldots, d_n$; that is,

$$A = \begin{bmatrix} a_1 & a_2 & \cdots & a_n \\ \downarrow & \downarrow & & \downarrow \end{bmatrix} \quad \text{and} \quad D = \begin{bmatrix} d_1 & 0 & 0 & \cdots & 0 \\ 0 & d_2 & 0 & \cdots & 0 \\ 0 & 0 & d_3 & & 0 \\ \vdots & \vdots & & \ddots & \\ 0 & 0 & \cdots & 0 & d_n \end{bmatrix}.$$

Describe the columns of AD.

Solution. The first column of AD is the product $A \begin{bmatrix} d_1 \\ 0 \\ \vdots \\ 0 \end{bmatrix}$ of A and the first column of D. This product is the linear combination of the columns of A whose coefficients are $d_1, 0, \ldots, 0$. So the product is just $d_1 a_1$, that is, the product of d_1 and the first column of A.

The second column of AD is the product $A \begin{bmatrix} 0 \\ d_2 \\ 0 \\ \vdots \\ 0 \end{bmatrix}$ of A and the second column of

D. This product is the linear combination of the columns of A whose coefficients are $0, d_2, 0, \ldots, 0$; that is, $d_2 a_2$. Similarly, the third column of AD is $d_3 a_3$. In general, column k of AD is $d_k a_k$; that is, d_k times the kth column of A.

$$AD = \begin{bmatrix} d_1 a_1 & d_2 a_2 & \cdots & d_n a_n \\ \downarrow & \downarrow & & \downarrow \end{bmatrix}.$$

Reading Challenge 2   *Write the product* $\begin{bmatrix} 1 & 1 & 1 & 1 \\ 1 & 1 & 1 & 1 \\ 1 & 1 & 1 & 1 \\ 1 & 1 & 1 & 1 \end{bmatrix} \begin{bmatrix} 2 & 0 & 0 & 0 \\ 0 & 3 & 0 & 0 \\ 0 & 0 & 4 & 0 \\ 0 & 0 & 0 & 5 \end{bmatrix}$ *without calculation.*

Answers to Reading Challenges

1. A is the matrix whose columns are the given vectors and x is the vector of coefficients.

2. Using the result of Problem 2.1.37,

$$\begin{bmatrix} 1 & 1 & 1 & 1 \\ 1 & 1 & 1 & 1 \\ 1 & 1 & 1 & 1 \\ 1 & 1 & 1 & 1 \end{bmatrix} \begin{bmatrix} 2 & 0 & 0 & 0 \\ 0 & 3 & 0 & 0 \\ 0 & 0 & 4 & 0 \\ 0 & 0 & 0 & 5 \end{bmatrix} = \begin{bmatrix} 2 & 3 & 4 & 5 \\ 2 & 3 & 4 & 5 \\ 2 & 3 & 4 & 5 \\ 2 & 3 & 4 & 5 \end{bmatrix}.$$

True/False Questions

Decide, with as little calculation as possible, whether each of the following statements is true or false and explain your answer whenever you say "false." (Answers can be found in the back of the book.)

1. If A and B are matrices such that both AB and BA are defined, then A and B are square.

2. If $\mathbf{u}$ and $\mathbf{v}$ are vectors, then $\mathbf{u} \cdot \mathbf{v}$ is the matrix product $\mathbf{u}^T\mathbf{v}$.

3. If A and B are matrices, $A \neq \mathbf{0}$, and $AB = \mathbf{0}$, then $B = \mathbf{0}$.

4. If A and B are matrices and $AB = \mathbf{0}$, then $AXB = \mathbf{0}$ for any matrix X (of appropriate size).

5. The $(2, 1)$ entry of $\begin{bmatrix} 1 & -1 & 0 \\ 2 & 3 & 4 \end{bmatrix}$ is -1.

6. If $A = \begin{bmatrix} 1 & 2 & 3 & 4 \\ 5 & 6 & 7 & 8 \\ 9 & 10 & 11 & 12 \end{bmatrix}$ and $\mathbf{b} = \begin{bmatrix} 0 \\ 0 \\ 1 \\ 0 \end{bmatrix}$, then $A\mathbf{b} = \begin{bmatrix} 3 \\ 7 \\ 11 \end{bmatrix}$.

7. If $A = \begin{bmatrix} 1 & 2 & 3 & 4 \\ 5 & 6 & 7 & 8 \\ 9 & 10 & 11 & 12 \end{bmatrix}$ and $\mathbf{x} = \begin{bmatrix} -2 \\ 3 \\ 5 \\ -8 \end{bmatrix}$, then $A\mathbf{x} = -2\begin{bmatrix} 1 \\ 5 \\ 9 \end{bmatrix} + 3\begin{bmatrix} 2 \\ 6 \\ 10 \end{bmatrix} + 5\begin{bmatrix} 3 \\ 7 \\ 11 \end{bmatrix}$

$- 8\begin{bmatrix} 4 \\ 8 \\ 12 \end{bmatrix}$.

8. The statement $(AB)C = A(BC)$ says that matrix multiplication is commutative.

9. If A is a matrix and $A\mathbf{x} = \mathbf{0}$ for every (compatible) vector $\mathbf{x}$, then $A = \mathbf{0}$ is the zero matrix.

10. If $\mathbf{x}_0$ is a solution to the system $A\mathbf{x} = \mathbf{b}$ with $\mathbf{b} \neq \mathbf{0}$, then $2\mathbf{x}_0$ is also a solution.

Exercises

*Solutions to exercises marked [BB] can be found in the **B**ack of the **B**ook.*

1. Find x and y, if possible, given $A = B$.

   (a) [BB] $A = \begin{bmatrix} 2x - 3y & -y \\ x - y & x + y \\ -x + y & x + 2y \end{bmatrix}$,

   $B = \begin{bmatrix} 8 & 2 \\ 3 & -1 \\ -3 & -3 \end{bmatrix}$

   (b) $A = \begin{bmatrix} 5x - 4y & -x \\ 3x + 3y & -3x - 3y \\ \frac{1}{2}x + y & x + 2y \\ -x + y & x - 2y \end{bmatrix}$,

   $B = \begin{bmatrix} 8 & 4 \\ -33 & 33 \\ -9 & -18 \\ -3 & 10 \end{bmatrix}$

   (c) $A = \begin{bmatrix} 6x - y & x + y \\ -x - y & 3x - y \\ x + y & x - y \end{bmatrix}$, $B = \begin{bmatrix} 10 & 4 \\ -4 & 4 \\ 6 & 0 \end{bmatrix}$

2. Find x, y, a, and b if possible, given $A = B$.

   (a) [BB] $A = \begin{bmatrix} 2a + 3x & 2 \\ -3 & y + b \end{bmatrix}$,

   $B = \begin{bmatrix} 0 & -2a - 2x \\ a + x & 2 \end{bmatrix}$

   (b) $A = \begin{bmatrix} a - b & x + a \\ y & 2 \\ -x & x + y \end{bmatrix}$, $B = \begin{bmatrix} 4 & b \\ x - 2 & x - y \\ a - b & b - a \end{bmatrix}$

(c) $A = \begin{bmatrix} 2a - b & x + y & -7 \\ a - b + x & y - a & 3y - b \end{bmatrix}$,

$B = \begin{bmatrix} x + 4 & 2a - x & x + 3y \\ -y - a & b - 7x & -x \end{bmatrix}$

3. Let $A = \begin{bmatrix} 2 & x \\ y & 1 \end{bmatrix}$, $B = \begin{bmatrix} z & -4 \\ -1 & 5 \end{bmatrix}$,

$C = \begin{bmatrix} 0 & 7 \\ 2 & -3 \end{bmatrix}$, and $D = \begin{bmatrix} 0 & 7 \\ 1 & -1 \end{bmatrix}$.

Find x, y, and z, if possible, so that

(a) [BB] $2A - B = C$

(b) [BB] $AB = C$

(c) $AB = D$

(d) $A + B + CD - DC = 13I$.

4. Let $A = \begin{bmatrix} 3 & x \\ y & 4 \end{bmatrix}$, $B = \begin{bmatrix} z & 3 \\ 4 & -1 \end{bmatrix}$,

$C = \begin{bmatrix} -1 & 6 \\ 4 & 14 \end{bmatrix}$, and $D = \begin{bmatrix} -3 & 9 \\ 15 & -1 \end{bmatrix}$.

Find x, y, and z, if possible, so that

(a) $3A - 2B = C$ **(b)** $AB = C$

(c) $AB = D$ **(d)** $BA - D + C = \begin{bmatrix} 2 & 7 \\ 2 & 3 \end{bmatrix}$.

5. Write the 2×3 matrix A for which

(a) [BB] $a_{ij} = j$ **(b)** $a_{ij} = 3i - 2j - 1$

(c) $a_{ij} = ij$ **(d)** a_{ij} is the larger of i and j.

6. Write the 4×4 matrix A whose (i, j) entry is

(a) $i - 3$ **(b)** $2^i - 3j$

(c) i^{j-1} **(d)** $i^2 j - ij^2$.

7. Find $A + B$, $A - B$, and $2A - B$ in each of the following cases.

(a) [BB] $A = \begin{bmatrix} 2 & 1 & 1 \\ -1 & -1 & 4 \end{bmatrix}$;

$B = \begin{bmatrix} 2 & -3 & 4 \\ -3 & 1 & 2 \end{bmatrix}$

(b) $A = \begin{bmatrix} 8 & 2 \\ -6 & -3 \end{bmatrix}$; $B = \begin{bmatrix} 7 & -4 \\ -3 & -1 \end{bmatrix}$

(c) $A = \begin{bmatrix} 3 & -4 & 8 \\ -7 & 1 & 0 \\ 2 & 2 & 3 \end{bmatrix}$; $B = \begin{bmatrix} 0 & -2 & 0 \\ 4 & 3 & -6 \\ 5 & 7 & 1 \end{bmatrix}$

(d) $A = \begin{bmatrix} -2 & 4 & -3 \\ 4 & -6 & 8 \end{bmatrix}$; $B = \begin{bmatrix} -2 & -4 & 3 \\ 5 & 6 & 8 \end{bmatrix}$.

8. Find AB and BA (if defined) in each of the following cases:

(a) [BB] $A = \begin{bmatrix} 1 & 2 & -4 \\ -6 & 8 & 1 \end{bmatrix}$; $B = \begin{bmatrix} -3 & 4 \\ 2 & 6 \end{bmatrix}$

(b) $A = \begin{bmatrix} 1 & 2 & -1 \\ -1 & 1 & 1 \\ 4 & 3 & -1 \end{bmatrix}$; $B = \begin{bmatrix} 2 & 1 \\ -1 & 4 \\ 2 & 3 \end{bmatrix}$

(c) $A = \begin{bmatrix} 2 \\ -2 \\ 0 \\ 3 \end{bmatrix}$; $B = \begin{bmatrix} 1 & 2 & 3 & -1 \end{bmatrix}$

(d) $A = \begin{bmatrix} -2 & 3 \\ 5 & 8 \end{bmatrix}$; $B = \begin{bmatrix} 3 & 0 \\ -5 & 3 \\ 7 & 4 \end{bmatrix}$.

9. Let $A = \begin{bmatrix} a & b & c \\ d & e & f \end{bmatrix}$ and $B = \begin{bmatrix} x & u \\ y & v \\ z & w \end{bmatrix}$.

(a) [BB] What are the $(1, 1)$ and $(2, 2)$ entries of AB?

(b) What are the diagonal entries of BA?

(c) Verify that the sum of the diagonal entries of AB and BA are equal.

10. If A is any $n \times n$ matrix and $\mathbf{x}$ is a vector in $\mathbf{R}^n$, what is the size of $\mathbf{x}^T A\mathbf{x}$ and why?

11. Write each of the following vectors in the form $A\mathbf{x}$, for a suitable matrix A and vector $\mathbf{x}$. [This question tests understanding of the very important fact expressed in 2.1.33.]

(a) [BB] $3\begin{bmatrix} -1 \\ 0 \end{bmatrix} + 4\begin{bmatrix} 2 \\ 1 \end{bmatrix} - \begin{bmatrix} 1 \\ 8 \end{bmatrix}$

(b) $5\begin{bmatrix} 1 \\ 3 \\ -5 \\ 7 \\ 9 \end{bmatrix} + \begin{bmatrix} -2 \\ 1 \\ 1 \\ 0 \\ 7 \end{bmatrix} - 3\begin{bmatrix} 2 \\ 4 \\ 6 \\ 0 \\ -1 \end{bmatrix}$

(c) $-2\begin{bmatrix} 1 \\ 1 \\ -2 \end{bmatrix} + 0\begin{bmatrix} 2 \\ 0 \\ 1 \end{bmatrix} + \begin{bmatrix} -3 \\ 1 \\ 1 \end{bmatrix} - \begin{bmatrix} 4 \\ 2 \\ 7 \end{bmatrix} + 3\begin{bmatrix} 6 \\ 5 \\ 4 \end{bmatrix}$

(d) $4\begin{bmatrix} 5 \\ 0 \\ 7 \\ 6 \end{bmatrix} - 2\begin{bmatrix} 2 \\ 3 \\ -1 \\ 7 \end{bmatrix} + \begin{bmatrix} 0 \\ 0 \\ 0 \\ 1 \end{bmatrix} - \begin{bmatrix} 5 \\ 1 \\ 4 \\ 5 \end{bmatrix}$

12. Express each of the following systems of equations in the form $A\mathbf{x} = \mathbf{b}$.

(a) [BB] $2x - y = 10$
$3x + 5y = 7$
$-x + y = -3$
$2x - 5y = 1$

(b) $y - z = 1$
$3x + 5y - 2z = -9$
$x \qquad + z = 0$

(c) $x - y + z \quad - \quad 2w = 0$
$3x + 5y - 2z + 2w = 1$
$3y + 4z \quad - \quad 7w = -1$

(d) $x_1 + x_2 + x_3 + x_4 - x_5 = 3$
$3x_1 - x_2 + 5x_3 \qquad - x_5 = -2$

13. [BB] The parabola with equation $y = ax^2 + bx + c$ passes through the points $(1, 4)$ and $(2, 8)$. Write a matrix equation whose solution is the vector $\begin{bmatrix} a \\ b \\ c \end{bmatrix}$.

14. It is conjectured that the points $(\frac{\pi}{3}, 2)$ and $(-\frac{\pi}{4}, 1)$ lie on a curve with equation of the form $y = a \sin x + b \cos x$. Assuming this is the case, write a matrix equation whose solution is $\mathbf{x} = \begin{bmatrix} a \\ b \end{bmatrix}$. [You are not being asked to find a and b.]

15. [BB] Complete the following sentence: "A linear combination of the columns of a matrix A is _____."

16. [BB] Let $A = \begin{bmatrix} 1 & 2 & 3 \\ 4 & 5 & 6 \\ 7 & 8 & 9 \end{bmatrix}$, $\mathbf{x} = \begin{bmatrix} -4 \\ 1 \\ 7 \end{bmatrix}$, and $\mathbf{b} = \begin{bmatrix} 19 \\ 31 \\ 43 \end{bmatrix}$. Given that $A\mathbf{x} = \mathbf{b}$, write $\mathbf{b}$ as a linear combination of the columns of A.

17. Repeat Exercise 16 with $A = \begin{bmatrix} -3 & 5 \\ 4 & -5 \end{bmatrix}$, $\mathbf{x} = \begin{bmatrix} 4 \\ 3 \end{bmatrix}$, and $\mathbf{b} = \begin{bmatrix} 3 \\ 1 \end{bmatrix}$.

18. [BB] Express $\mathbf{x} = \begin{bmatrix} 2 \\ 3 \end{bmatrix}$ as a linear combination of the columns of $A = \begin{bmatrix} 1 & 1 \\ 0 & 1 \end{bmatrix}$.

19. Express $\begin{bmatrix} 7 \\ -5 \\ 6 \end{bmatrix}$ as a linear combination of the columns of $A = \begin{bmatrix} 3 & 3 & -7 & -3 \\ 6 & 1 & 5 & 0 \\ -5 & 2 & -6 & 9 \end{bmatrix}$.

20. Express $\begin{bmatrix} -1 \\ 14 \\ 2 \end{bmatrix}$ as a linear combination of the columns of $A = \begin{bmatrix} 1 & 2 & 0 \\ 0 & 3 & 1 \\ 0 & 0 & 1 \end{bmatrix}$.

21. Let $I = \begin{bmatrix} 1 & 0 & 0 \\ 0 & 1 & 0 \\ 0 & 0 & 1 \end{bmatrix}$ be the 3×3 identity matrix. Express each of the following vectors as a linear combination of the columns of I.

(a) [BB] $\mathbf{x} = \begin{bmatrix} 20 \\ 35 \\ 47 \end{bmatrix}$ (b) $\mathbf{x} = \begin{bmatrix} -1 \\ 2 \\ 5 \end{bmatrix}$

(c) $\mathbf{x} = \mathbf{0}$ (d) $\mathbf{x} = \begin{bmatrix} x \\ y \\ z \end{bmatrix}$

22. [BB] Suppose $\mathbf{x}_1$, $\mathbf{x}_2$, and $\mathbf{x}_3$ are three-dimensional vectors and A is a 3×3 matrix such that $A\mathbf{x}_1 = \begin{bmatrix} 1 \\ 0 \\ 0 \end{bmatrix}$, $A\mathbf{x}_2 = \begin{bmatrix} 0 \\ 1 \\ 0 \end{bmatrix}$, $A\mathbf{x}_3 = \begin{bmatrix} 0 \\ 0 \\ 1 \end{bmatrix}$. Let $X = \begin{bmatrix} \mathbf{x}_1 & \mathbf{x}_2 & \mathbf{x}_3 \\ \downarrow & \downarrow & \downarrow \end{bmatrix}$ be the matrix with columns $\mathbf{x}_1$, $\mathbf{x}_2$, $\mathbf{x}_3$. What is AX?

23. Suppose $\mathbf{x}_1$, $\mathbf{x}_2$, and $\mathbf{x}_3$ are four-dimensional vectors and A is a 2×4 matrix such that $A\mathbf{x}_1 = \begin{bmatrix} -1 \\ 2 \end{bmatrix}$ and $A\mathbf{x}_2 = \begin{bmatrix} 10 \\ 12 \end{bmatrix}$. Find AX, given $X = \begin{bmatrix} \mathbf{x}_1 & \mathbf{x}_2 \\ \downarrow & \downarrow \end{bmatrix}$.

24. [BB] Let $A = \begin{bmatrix} 1 & 2 & 3 \\ 1 & 2 & 3 \\ 1 & 2 & 3 \end{bmatrix}$ and $D = \begin{bmatrix} -7 & 0 & 0 \\ 0 & 8 & 0 \\ 0 & 0 & 5 \end{bmatrix}$. Find AD without calculation and explain your reasoning.

25. Let $A = \begin{bmatrix} 1 & 2 & 3 \\ 4 & 5 & 6 \\ 7 & 8 & 9 \end{bmatrix}$ and $D = \begin{bmatrix} 2 & 0 & 0 \\ 0 & 3 & 0 \\ 0 & 0 & -1 \end{bmatrix}$. Find AD and DA without explicitly multiplying matrices. Explain your reasoning.

26. Find A^{10} with $A = \begin{bmatrix} 1 & 1 \\ 0 & 1 \end{bmatrix}$.

27. Let $\mathbf{v} = \begin{bmatrix} 1 \\ 2 \\ -3 \end{bmatrix}$ and $\mathbf{w} = \begin{bmatrix} 0 \\ 4 \\ 5 \end{bmatrix}$. Let $A = \begin{bmatrix} \mathbf{v} & \mathbf{w} \\ \downarrow & \downarrow \end{bmatrix}$ be the 3×2 matrix whose columns are $\mathbf{v}$ and $\mathbf{w}$ and let $B = \begin{bmatrix} \mathbf{v}^T & \rightarrow \\ \mathbf{w}^T & \rightarrow \end{bmatrix}$ be the 2×3 matrix whose rows are $\mathbf{v}^T$ and $\mathbf{w}^T$. Find $a_{11}, a_{13}, a_{21}, b_{32}, b_{12},$ and b_{22} if possible.

28. [BB] Let $A = \begin{bmatrix} -1 & 2 \\ 3 & 4 \end{bmatrix}$, $B = \begin{bmatrix} 0 & -3 & 4 \\ 4 & 1 & 2 \end{bmatrix}$, $C = \begin{bmatrix} 1 & 2 & 0 \\ -1 & 0 & 5 \end{bmatrix}$.
Compute AB, AC, $AB + AC$, $B + C$, and $A(B + C)$.

29. Let $A = \begin{bmatrix} 1 & 2 \\ 3 & 4 \end{bmatrix}$ and $B = \begin{bmatrix} 0 & -1 \\ 5 & -2 \end{bmatrix}$. Compute $(2A + B)(A - 3B)$ and $2A^2 - 5AB - 3B^2$. Are these equal? What is the correct expansion of $(2A + B)(A - 3B)$?

30. Suppose a 2×2 matrix A *commutes* with $B = \begin{bmatrix} 0 & -1 \\ 1 & 0 \end{bmatrix}$; that is, $AB = BA$. What can you say about the entries of A?

31. (a) [BB] Suppose A and B are matrices such that $AB = \mathbf{0}$. Does this imply $A = \mathbf{0}$ or $B = \mathbf{0}$? If you say "yes," give a proof. If you say "no," give an example of two nonzero matrices A and B for which $AB = \mathbf{0}$.

    (b) If A is a 2×2 matrix, $B = \begin{bmatrix} 1 & 2 \\ 0 & -1 \end{bmatrix}$, and $AB = \mathbf{0}$, show that $A = \mathbf{0}$. Does this result contradict your answer to part (a)?

    (c) If X and Y are any 2×2 matrices and B is the matrix of part (b), and if $XB = YB$, show that $X = Y$.

32. Let $A = \begin{bmatrix} 0 & -1 \\ 2 & 1 \end{bmatrix}$, $B = \begin{bmatrix} 1 & 0 \\ 3 & -1 \end{bmatrix}$, and $C = \begin{bmatrix} -2 & -3 \\ 0 & 5 \end{bmatrix}$.

    (a) Is C a linear combination of A and B?

    (b) Find a nontrivial linear combination of A, B, and C that equals $\begin{bmatrix} 0 & 0 \\ 0 & 0 \end{bmatrix}$. (*Nontrivial* means that not all coefficients can be 0.)

■ Critical Reading

33. [BB] If A is a matrix and $\mathbf{x}$ is a vector such that $\mathbf{x} + A\mathbf{x}$ is a vector, what can you say about the size of A (and why)?

34. Suppose $A = \begin{bmatrix} \mathbf{a}_1 & \mathbf{a}_2 & \cdots & \mathbf{a}_n \\ \downarrow & \downarrow & & \downarrow \end{bmatrix}$ is a matrix and suppose that the entries in each row of A sum to 0. Let $\mathbf{v} = \begin{bmatrix} 1 \\ 1 \\ \vdots \\ 1 \end{bmatrix}$ and show that $A\mathbf{v} = \mathbf{0}$.

35. In part 7 of Theorem 2.1.15, *scalar associativity* is defined as the property $c(dA) = (cd)A$ where A is a matrix and c, d are scalars. Show that you understand what this means by illustrating with $A = \begin{bmatrix} -7 & 2 \\ 1 & 5 \end{bmatrix}$, $c = -2$, and $d = 9$.

36. [BB] If A and B are $m \times n$ matrices and $A\mathbf{x} = B\mathbf{x}$ for each vector $\mathbf{x}$ in $\mathbf{R}^n$, show that $A = B$.

37. Let A be a square matrix. Does the equation $(A - 3I)(A + 2I) = \mathbf{0}$ imply $A = 3I$ or $A = -2I$? Answer by considering the matrix $A = \begin{bmatrix} 3 & -1 \\ 0 & -2 \end{bmatrix}$.

38. Let A be an $m \times n$ matrix such that $A^T A = \mathbf{0}$ is the zero matrix. Why must A be the zero matrix?

2.2 The Inverse and Transpose of a Matrix

We have observed that the $n \times n$ identity matrix I behaves with respect to matrices the way the number 1 does with numbers: $IA = A$ for any matrix A (when IA is defined) just as $1a = a$ for any real number a. How far does this analogy extend? If a is a nonzero real number, there is another number b with $ab = 1$. The number $b = \frac{1}{a} = a^{-1}$ is called the (multiplicative) inverse of a. This idea motivates the idea of *inverse of a matrix*.

2.2.1 DEFINITION   A matrix A is *invertible* or *has an inverse* if there is another matrix B such that $AB = I$ and $BA = I$. The matrix B is called the *inverse* of A and we write $B = A^{-1}$.

2.2.2 Remark Since matrix multiplication is not commutative, we *never* write something like $\frac{1}{A}$ or $\frac{AB}{B}$, which is ambiguous.

2.2.3 EXAMPLE   The matrix $A = \begin{bmatrix} 1 & -2 \\ 2 & -3 \end{bmatrix}$ is invertible. Its inverse is $B = \begin{bmatrix} -3 & 2 \\ -2 & 1 \end{bmatrix}$, since

$$AB = \begin{bmatrix} 1 & -2 \\ 2 & -3 \end{bmatrix} \begin{bmatrix} -3 & 2 \\ -2 & 1 \end{bmatrix} = \begin{bmatrix} 1 & 0 \\ 0 & 1 \end{bmatrix} = I$$

and

$$BA = \begin{bmatrix} -3 & 2 \\ -2 & 1 \end{bmatrix} \begin{bmatrix} 1 & -2 \\ 2 & -3 \end{bmatrix} = \begin{bmatrix} 1 & 0 \\ 0 & 1 \end{bmatrix} = I$$

too. Thus $B = A^{-1}$. ∎

The equations $AB = I$, $BA = I$, which say that A is invertible with inverse B, say also that B is invertible with inverse A. We say that A and B are *inverses*: each is the inverse of the other, $A^{-1} = B$, $B^{-1} = A$. Note that this implies $A = B^{-1} = (A^{-1})^{-1}$.

2.2.4 If A is invertible, so is A^{-1}, and $(A^{-1})^{-1} = A$.

2.2.5 EXAMPLE   Whereas "$a \neq 0$" in the real numbers implies that a is invertible, this is *not true* for matrices. Here's a nonzero matrix that is not invertible. Let $A = \begin{bmatrix} 1 & 2 \\ 2 & 4 \end{bmatrix}$ and $B = \begin{bmatrix} x & z \\ y & w \end{bmatrix}$. The second row of

$$AB = \begin{bmatrix} 1 & 2 \\ 2 & 4 \end{bmatrix} \begin{bmatrix} x & y \\ z & w \end{bmatrix} = \begin{bmatrix} x + 2z & y + 2w \\ 2x + 4z & 2y + 4w \end{bmatrix}$$

is twice the first, so AB can never be I. The matrix A does not have an inverse. ∎

2.2.6 Remark (Inverses are unique). We have called B *the* inverse of a matrix A if $AB = I$ and $BA = I$. This language suggests that a matrix can have at most one inverse. This is indeed the case, and here's why.
   If A has two inverses, B and C, then $AB = BA = I$ and also $AC = CA = I$. Now evaluate the product BAC, which, because matrix multiplication is associative,

does not require the insertion of parentheses. On the one hand, $BAC = (BA)C = IC = C$, while, on the other hand, $BAC = B(AC) = BI = B$. So $B = C$.

As we shall later see—Corollary 4.4.12—only a square matrix can have an inverse. On the other hand, not all square matrices have inverses. For example, just as the number 0 has no (multiplicative) inverse, so the $n \times n$ zero matrix does not have an inverse: $A\mathbf{0} = \mathbf{0} \neq I$ for any A. Unlike the situation with ordinary numbers, however, nonzero matrices may also fail to have inverses, as already noted.

According to Definition 2.2.1, to show that B is an inverse of A one must show that each product AB and BA is the identity matrix. Since matrix multiplication is not commutative, it seems logical that each of these products should be checked. Later (see Corollary 2.7.17), we shall show the following:

> **2.2.7** If A is a square matrix and $AB = I$, then $BA = I$; thus A and B are inverses.

This is a very useful fact that you are encouraged to start using right away.

2.2.8 PROBLEM

Show that the matrices $A = \begin{bmatrix} 1 & 2 & 3 \\ -1 & 0 & 1 \\ 4 & 1 & 1 \end{bmatrix}$ and

$B = \frac{1}{6} \begin{bmatrix} -1 & 1 & 2 \\ 5 & -11 & -4 \\ -1 & 7 & 2 \end{bmatrix}$ are inverses.

Solution. Since A is square, it suffices by 2.2.7 to show that $AB = I$. We have

$$AB = \frac{1}{6} \begin{bmatrix} 1 & 2 & 3 \\ -1 & 0 & 1 \\ 4 & 1 & 1 \end{bmatrix} \begin{bmatrix} -1 & 1 & 2 \\ 5 & -11 & -4 \\ -1 & 7 & 2 \end{bmatrix} = \frac{1}{6} \begin{bmatrix} 6 & 0 & 0 \\ 0 & 6 & 0 \\ 0 & 0 & 6 \end{bmatrix} = I,$$

so A and B are inverses.

2.2.9 Remark Notice how part 4 of Theorem 2.1.30 was used in the solution of this problem to move the $\frac{1}{6}$ to the front in the calculation of AB. For matrices X, Y and a scalar a, $X(aY) = a(XY)$.

2.2.10 EXAMPLE

If $A = \begin{bmatrix} a & b \\ c & d \end{bmatrix}$ and $ad - bc \neq 0$, then A is invertible and $A^{-1} = \frac{1}{ad-bc} \begin{bmatrix} d & -b \\ -c & a \end{bmatrix}$. To see why, we simply compute

$$\begin{bmatrix} a & b \\ c & d \end{bmatrix} \frac{1}{ad - bc} \begin{bmatrix} d & -b \\ -c & a \end{bmatrix}$$

$$= \frac{1}{ad - bc} \begin{bmatrix} ad - bc & 0 \\ 0 & ad - bc \end{bmatrix} = \begin{bmatrix} 1 & 0 \\ 0 & 1 \end{bmatrix}. \blacksquare$$

The solution to the next problem is motivated by the fact that if a and b are nonzero real numbers, then ab is invertible and $(ab)^{-1} = a^{-1}b^{-1}$.

2.2.11 PROBLEM

Suppose A and B are $n \times n$ matrices, each of which is invertible. Show that AB is invertible by finding its inverse.

Solution. We need a matrix X such that $(AB)X = I$. What might X be? The brief comment before this problem suggests we try $A^{-1}B^{-1}$. This does not seem to work, however, since in general the expression $ABA^{-1}B^{-1}$ cannot be simplified (matrix multiplication is not commutative). On the other hand, if we let $X = B^{-1}A^{-1}$, we get

$$(AB)X = (AB)(B^{-1}A^{-1}) = A(BB^{-1})A^{-1} = AIA^{-1} = AA^{-1} = I.$$

Since AB is square, the equation $(AB)X = I$ implies $X(AB) = I$, so X is the inverse of A.

We remember the result described in Problem 2.2.11 like this:

2.2.12 A product of invertible matrices is invertible. Its inverse is the product of the individual inverses, order reversed:
$$(AB)^{-1} = B^{-1}A^{-1}.$$

2.2.13 EXAMPLE

Let $A = \begin{bmatrix} 1 & 3 \\ -2 & 5 \end{bmatrix}$ and $B = \begin{bmatrix} 2 & -1 \\ 1 & -1 \end{bmatrix}$. Then $AB = \begin{bmatrix} 5 & -4 \\ 1 & -3 \end{bmatrix}$ and $(AB)^{-1} = \frac{1}{11}\begin{bmatrix} 3 & -4 \\ 1 & -5 \end{bmatrix}$ because the product of AB with this matrix, in either order, gives I;

$$\begin{bmatrix} 5 & -4 \\ 1 & -3 \end{bmatrix} \frac{1}{11}\begin{bmatrix} 3 & -4 \\ 1 & -5 \end{bmatrix} = \frac{1}{11}\begin{bmatrix} 11 & 0 \\ 0 & 11 \end{bmatrix} = I$$

and

$$\frac{1}{11}\begin{bmatrix} 3 & -4 \\ 1 & -5 \end{bmatrix}\begin{bmatrix} 5 & -4 \\ 1 & -3 \end{bmatrix} = \frac{1}{11}\begin{bmatrix} 11 & 0 \\ 0 & 11 \end{bmatrix} = I.$$

Now

$$A^{-1} = \frac{1}{11}\begin{bmatrix} 5 & -3 \\ 2 & 1 \end{bmatrix} \quad \text{and} \quad B^{-1} = \begin{bmatrix} 1 & -1 \\ 1 & -2 \end{bmatrix},$$

and so

$$B^{-1}A^{-1} = \frac{1}{11}\begin{bmatrix} 1 & -1 \\ 1 & -2 \end{bmatrix}\begin{bmatrix} 5 & -3 \\ 2 & 1 \end{bmatrix} = \frac{1}{11}\begin{bmatrix} 3 & -4 \\ 1 & -5 \end{bmatrix} = (AB)^{-1}. \quad \blacksquare$$

Reading Challenge 1   *With A and B as in **2.2.13**, verify that $A^{-1}B^{-1} \neq (AB)^{-1}$.*

2.2.14 PROBLEM

If A is an invertible matrix $n \times n$ matrix and B is an $n \times t$ matrix, show that the equation $AX = B$ has a solution and find it.

Solution. Multiplying $AX = B$ on the left by A^{-1} gives $A^{-1}AX = A^{-1}B$. Since $A^{-1}A = I$, this is $IX = A^{-1}B$, so $X = A^{-1}B$. This reasoning suggests that *if* there is an X with $AX = B$, then $X = A^{-1}B$. In fact, this is a solution since $A(A^{-1}B) = IB = B$.

The solution of the equation $ax = b$, with a, x, and b real numbers, is $x = \frac{b}{a}$, provided $a \neq 0$. Using exponents, this solution is just $x = a^{-1}b$. The condition "$a \neq 0$" is exactly the condition "a has a multiplicative inverse." As we just showed, the solution to $AX = B$ with A, X, and B matrices is similar.

If A has an inverse, $X = A^{-1}B$, which we discover simply by multiplying each side of $AX = B$ on the left by A^{-1} like this:

If $AX = B$, then $A^{-1}AX = A^{-1}B$, so $IX = A^{-1}B$ and hence $X = A^{-1}B$.

2.2.15 Remark Since matrix multiplication is not commutative, the solution $X = A^{-1}B$ is not (generally) the same as $X = BA^{-1}$ as, in general, $A^{-1}B \neq BA^{-1}$. To turn AX into X, we must multiply by A^{-1} *on the left*; thus we must multiply B also *on the left*. This is quite different from ordinary algebra, where, assuming $a \neq 0$, the solution to $ax = b$ can be written $x = a^{-1}b$ or $x = ba^{-1}$.

2.2.16 PROBLEM

Solve the system of equations $\begin{array}{l} x - 2y = 5 \\ 2x - 3y = 7. \end{array}$

Solution. The given system is $A\mathbf{x} = \mathbf{b}$, where $A = \begin{bmatrix} 1 & -2 \\ 2 & -3 \end{bmatrix}$ is the matrix of coefficients, $\mathbf{x} = \begin{bmatrix} x \\ y \end{bmatrix}$ is unknown, and $\mathbf{b} = \begin{bmatrix} 5 \\ 7 \end{bmatrix}$. As noted in **2.2.3**, the matrix A has an inverse, so

$$\mathbf{x} = A^{-1}\mathbf{b} = \begin{bmatrix} -3 & 2 \\ -2 & 1 \end{bmatrix} \begin{bmatrix} 5 \\ 7 \end{bmatrix} = \begin{bmatrix} -1 \\ -3 \end{bmatrix}.$$

The solution is $x = -1$, $y = -3$.

Reading Challenge 2

If A is an invertible $n \times n$ matrix and B is a $t \times n$ matrix, show that the equation $XA = B$ has a solution and find it.

Algebra with matrices is not terribly different from ordinary algebra with numbers, but one must always be careful not to assume commutativity. For example, a matrix of the form $A^{-1}XA$ cannot be simplified.

2.2.17 PROBLEM

Suppose A, B, and X are invertible matrices and $AX^{-1}A^{-1} = B^{-1}$. What is X?

Solution. First, we multiply the given equation on the right by A, obtaining $AX^{-1}A^{-1}A = B^{-1}A$. Since $A^{-1}A = I$, we obtain $AX^{-1} = B^{-1}A$. Now multiply on the left by A^{-1}. This gives $X^{-1} = A^{-1}B^{-1}A$. Thus $X = (X^{-1})^{-1} = (A^{-1}B^{-1}A)^{-1} = A^{-1}(B^{-1})^{-1}(A^{-1})^{-1} = A^{-1}BA$.

Some properties of the matrix inverse are summarized in the next theorem.

Theorem 2.2.18

If A is an invertible matrix, then

1. *so is A^k for any integer k, and $(A^k)^{-1} = (A^{-1})^k$;*

2. *so is cA for any nonzero scalar c, and $(cA)^{-1} = \frac{1}{c}A^{-1}$.*

Some Final Words about the Inverse. Our discussion of the inverse of a matrix has intentionally avoided mention of how to test a matrix for invertibility and how to find an inverse, when there is one. There are various ways to determine whether a matrix has an inverse, one of which is described in Section 2.7, and another of which involves the concept of *determinant* and is the subject of Chapter 3. We will show how to compute inverses in Section 2.7.

■ More on Transposes

We conclude this section with a summary of the most basic properties of the transpose of a matrix.

Theorem 2.2.19

Let A and B be matrices and let c be a scalar.

1. $(A + B)^T = A^T + B^T$ *(assuming A and B have the same size);*

2. $(cA)^T = cA^T$;

3. $(A^T)^T = A$;

4. *if A is invertible, so is A^T and $(A^T)^{-1} = (A^{-1})^T$.*

5. $(AB)^T = B^T A^T$ *(assuming AB is defined): the transpose of a product is the product of the transposes, order reversed.*

The last of these properties is the most interesting, and it should remind us of the formula for the inverse of the product of two invertible matrices. Just as $(AB)^{-1} = B^{-1}A^{-1}$, so also $(AB)^T = B^T A^T$.

The "obvious" rule for the transpose of a product, $(AB)^T = A^T B^T$, cannot possibly work. If A is 2×3 and B is 3×4, for example, then A^T is 3×2 and B^T is 4×3, so the product $A^T B^T$ is not even defined. On the other hand, the product $B^T A^T$ is defined and produces a 4×2 matrix, precisely the size of $(AB)^T$ because AB is 2×4. This argument does not *prove* that $(AB)^T = B^T A^T$, of course, but it gives us reason to hope. In a similar spirit, examples also give us hope.

2.2.20 EXAMPLE

Suppose $A = \begin{bmatrix} 1 & 2 & 3 \\ -4 & 0 & -1 \end{bmatrix}$ and $B = \begin{bmatrix} 0 & 1 \\ 1 & -3 \\ 4 & 2 \end{bmatrix}$. Then

$$AB = \begin{bmatrix} 1 & 2 & 3 \\ -4 & 0 & -1 \end{bmatrix} \begin{bmatrix} 0 & 1 \\ 1 & -3 \\ 4 & 2 \end{bmatrix} = \begin{bmatrix} 14 & 1 \\ -4 & -6 \end{bmatrix}.$$

Now $A^T = \begin{bmatrix} 1 & -4 \\ 2 & 0 \\ 3 & -1 \end{bmatrix}$ and $B^T = \begin{bmatrix} 0 & 1 & 4 \\ 1 & -3 & 2 \end{bmatrix}$, so

$$B^T A^T = \begin{bmatrix} 0 & 1 & 4 \\ 1 & -3 & 2 \end{bmatrix} \begin{bmatrix} 1 & -4 \\ 2 & 0 \\ 3 & -1 \end{bmatrix} = \begin{bmatrix} 14 & -4 \\ 1 & -6 \end{bmatrix} = (AB)^T. \qquad \blacksquare$$

One example does not prove $(AB)^T = B^T A^T$ for all matrices A and B, but here is a general argument that does. Let $\mathbf{a}_1^T, \mathbf{a}_2^T, \ldots, \mathbf{a}_m^T$ be the rows of A and let $\mathbf{b}_1, \mathbf{b}_2, \ldots, \mathbf{b}_p$ be the columns of B, so that we picture A and B like this:

$$A = \begin{bmatrix} \mathbf{a}_1^T & \rightarrow \\ \mathbf{a}_2^T & \rightarrow \\ \vdots \\ \mathbf{a}_m^T & \rightarrow \end{bmatrix}, \qquad B = \begin{bmatrix} \mathbf{b}_1 & \mathbf{b}_2 & \ldots & \mathbf{b}_p \\ \downarrow & \downarrow & & \downarrow \end{bmatrix}.$$

Then

$$AB = \begin{bmatrix} \mathbf{a}_1 \cdot \mathbf{b}_1 & \mathbf{a}_1 \cdot \mathbf{b}_2 & \ldots & \mathbf{a}_1 \cdot \mathbf{b}_p \\ \mathbf{a}_2 \cdot \mathbf{b}_1 & \mathbf{a}_2 \cdot \mathbf{b}_2 & \ldots & \mathbf{a}_2 \cdot \mathbf{b}_p \\ \vdots & \vdots & & \vdots \\ \mathbf{a}_m \cdot \mathbf{b}_1 & \mathbf{a}_m \cdot \mathbf{b}_2 & \ldots & \mathbf{a}_m \cdot \mathbf{b}_p \end{bmatrix},$$

so

$$(AB)^T = \begin{bmatrix} \mathbf{a}_1 \cdot \mathbf{b}_1 & \mathbf{a}_2 \cdot \mathbf{b}_1 & \ldots & \mathbf{a}_m \cdot \mathbf{b}_1 \\ \mathbf{a}_1 \cdot \mathbf{b}_2 & \mathbf{a}_2 \cdot \mathbf{b}_2 & \ldots & \mathbf{a}_m \cdot \mathbf{b}_2 \\ \vdots & \vdots & & \vdots \\ \mathbf{a}_1 \cdot \mathbf{b}_p & \mathbf{a}_2 \cdot \mathbf{b}_p & \ldots & \mathbf{a}_m \cdot \mathbf{b}_p \end{bmatrix}. \tag{1}$$

Now

$$A^T = \begin{bmatrix} \mathbf{a}_1 & \mathbf{a}_2 & \ldots & \mathbf{a}_m \\ \downarrow & \downarrow & & \downarrow \end{bmatrix} \text{ and } B^T = \begin{bmatrix} \mathbf{b}_1^T & \rightarrow \\ \mathbf{b}_2^T & \rightarrow \\ \vdots \\ \mathbf{b}_p^T & \rightarrow \end{bmatrix}$$

so that

$$
B^T A^T =
\begin{bmatrix}
\mathbf{b}_1^T & \rightarrow \\
\mathbf{b}_2^T & \rightarrow \\
\vdots \\
\mathbf{b}_p^T & \rightarrow
\end{bmatrix}
\begin{bmatrix}
\mathbf{a}_1 & \mathbf{a}_2 & \cdots & \mathbf{a}_m \\
\downarrow & \downarrow & & \downarrow
\end{bmatrix}
$$

$$
=
\begin{bmatrix}
\mathbf{b}_1 \cdot \mathbf{a}_1 & \mathbf{b}_1 \cdot \mathbf{a}_2 & \cdots & \mathbf{b}_1 \cdot \mathbf{a}_m \\
\mathbf{b}_2 \cdot \mathbf{a}_1 & \mathbf{b}_2 \cdot \mathbf{a}_2 & \cdots & \mathbf{b}_2 \cdot \mathbf{a}_m \\
\vdots & \vdots & & \vdots \\
\mathbf{b}_p \cdot \mathbf{a}_1 & \mathbf{b}_p \cdot \mathbf{a}_2 & \cdots & \mathbf{b}_p \cdot \mathbf{a}_m
\end{bmatrix}.
\tag{2}
$$

It remains only to observe that the matrices in (2) and (1) are the same because $\mathbf{a} \cdot \mathbf{b} = \mathbf{b} \cdot \mathbf{a}$ for any vectors $\mathbf{a}$ and $\mathbf{b}$.

Answers to Reading Challenges

1. $A^{-1}B^{-1} = \frac{1}{11} \begin{bmatrix} 5 & -3 \\ 2 & 1 \end{bmatrix} \begin{bmatrix} 1 & -1 \\ 1 & -2 \end{bmatrix} = \frac{1}{11} \begin{bmatrix} 2 & 1 \\ 3 & -4 \end{bmatrix}$, whereas, as shown in **2.2.13**, $(AB)^{-1} = \frac{1}{11} \begin{bmatrix} 3 & -4 \\ 1 & -5 \end{bmatrix}$.

2. Multiplying $XA = B$ on the right by A^{-1} gives $X = BA^{-1}$. In fact, this is the solution since $(BA^{-1})A = BI = B$.

True/False Questions

Decide, with as little calculation as possible, whether each of the following statements is true or false and explain your answer whenever you say "false." (Answers can be found in the back of the book.)

1. If A, B, C are matrices such that $AC = BC$, then $A = B$.

2. Only square matrices are invertible.

3. Only square matrices have transposes.

4. If A and B are matrices and A is invertible, then $A^{-1}BA = B$.

5. If A, B, and C are invertible matrices (of the same size), then $(ABC)^{-1} = C^{-1}B^{-1}A^{-1}$.

6. If A and B are invertible matrices and $XA = B$, then $X = A^{-1}B$.

7. If A is a 2×3 matrix and B is 3×5, then $(AB)^T$ is 5×2.

8. If A and B are matrices and both products $A^T B^T$ and $B^T A^T$ are defined, then A and B are square.

Exercises

*Solutions to exercises marked [BB] can be found in the **B**ack of the **B**ook.*

1. Determine whether each of the following pairs of matrices are inverses.

   (a) [BB] $A = \begin{bmatrix} 1 & 0 & 1 \\ 2 & 1 & 0 \\ 3 & -1 & 0 \end{bmatrix}$,

   $B = \dfrac{1}{5} \begin{bmatrix} 0 & 1 & 1 \\ 0 & 3 & -2 \\ 5 & -1 & -1 \end{bmatrix}$

   (b) $A = \begin{bmatrix} 0 & 1 \\ 1 & 0 \end{bmatrix}$, $B = A$

   (c) $A = \begin{bmatrix} 1 & 2 & 3 \\ 4 & 5 & 6 \end{bmatrix}$, $B = \begin{bmatrix} -1 & 1 \\ 0 & -1 \\ \frac{2}{3} & \frac{1}{3} \end{bmatrix}$

   (d) $A = \begin{bmatrix} 1 & 0 & 0 \\ 2 & 1 & 0 \\ -1 & 2 & 1 \end{bmatrix}$,

   $B = \begin{bmatrix} 1 & 0 & 0 \\ -2 & 1 & 1 \\ 5 & -1 & -1 \end{bmatrix}$

   (e) $A = \begin{bmatrix} 2 & 0 & -\frac{1}{2} \\ -1 & 0 & \frac{1}{2} \end{bmatrix}$, $B = \begin{bmatrix} 1 & 1 \\ -1 & -2 \\ 2 & 4 \end{bmatrix}$

   (f) $A = \begin{bmatrix} 1 & 0 & 0 & 0 \\ 0 & 2 & 0 & 0 \\ 0 & 0 & 3 & 0 \\ 0 & 0 & 0 & 4 \end{bmatrix}$,

   $B = \dfrac{1}{24} \begin{bmatrix} 24 & 0 & 0 & 0 \\ 0 & 12 & 0 & 0 \\ 0 & 0 & 8 & 0 \\ 0 & 0 & 0 & 6 \end{bmatrix}$.

2. Let $A = \begin{bmatrix} -1 & 2 & 2 \\ 3 & 0 & 5 \\ 2 & -1 & 0 \end{bmatrix}$. Use the fact that

   $A^{-1} = \frac{1}{9} \begin{bmatrix} 5 & -2 & 10 \\ 10 & -4 & 11 \\ -3 & 3 & -6 \end{bmatrix}$ to solve the system

   $$\begin{array}{rcrcrcr} -x & + & 2y & + & 2z & = & 12 \\ 3x & & & + & 5z & = & -1 \\ 2x & - & y & & & = & -8. \end{array}$$

3. Suppose A and B are matrices with B invertible.
   (a) If $A = B^{-1}AB$, show that A and B commute.

   (b) The *converse* of an implication "If $\mathcal{X}$, then $\mathcal{Y}$" is the implication "If $\mathcal{Y}$, then $\mathcal{X}$." Write the converse of the implication in part (a). Is the converse true? Explain.

4. [BB] Let $A = \begin{bmatrix} 1 & -7 & 1 \\ 2 & -9 & 1 \end{bmatrix}$ and $B = \begin{bmatrix} -1 & 3 \\ 0 & 1 \\ 2 & 4 \end{bmatrix}$.

   (a) Compute AB and BA.

   (b) Is A invertible? Explain.

5. Show that the matrix $A = \begin{bmatrix} 2 & 6 \\ 1 & 3 \end{bmatrix}$ is not invertible.

6. A matrix A is *diagonal* if its only nonzero entries occupy the diagonal positions $a_{11}, a_{22}, a_{33}, \ldots$.

   (a) Show that the diagonal matrix

   $A = \begin{bmatrix} 2 & 0 & 0 \\ 0 & -3 & 0 \\ 0 & 0 & 5 \end{bmatrix}$ is invertible.

   (b) Show that the diagonal matrix

   $A = \begin{bmatrix} 2 & 0 & 0 \\ 0 & 0 & 0 \\ 0 & 0 & 5 \end{bmatrix}$ is not invertible.

   (c) Complete the following sentence:
   "A square diagonal matrix with diagonal entries $d_1, d_2, \ldots, d_n$ is invertible if and only if _____."

7. [BB] Let $A = \begin{bmatrix} 3 & 1 \\ -1 & 2 \end{bmatrix}$ and $B = \begin{bmatrix} 2 & 0 \\ 4 & -3 \end{bmatrix}$.
   Compute A^T, B^T, AB, BA, $(AB)^T$, $(BA)^T$, $A^T B^T$, and $B^T A^T$.

8. Verify that $(AB)^T = B^T A^T \neq A^T B^T$ with
   $A = \begin{bmatrix} 1 & 3 \\ 2 & 1 \end{bmatrix}$ and $B = \begin{bmatrix} 1 & 3 \\ 0 & 1 \end{bmatrix}$.

9. [BB] Verify part 4 of Theorem 2.2.19 by showing that if A is an invertible matrix, then so is A^T, and $(A^T)^{-1} = (A^{-1})^T$.

10. Verify part 2 of Theorem 2.2.18: if A is an invertible matrix, so is cA for any nonzero scalar c, and $(cA)^{-1} = \frac{1}{c} A^{-1}$.

11. If a matrix A is invertible, show that A^2 is also invertible. (Do *not* use the result of Problem 2.2.11.)

12. Suppose A and B are $n \times n$ matrices and $I - AB$ is invertible. Show that $I - BA$ is invertible by showing that $(I - BA)^{-1} = I + B(I - AB)^{-1}A$.

13. [BB] Suppose A and B are square matrices that commute. Prove that $(AB)^T = A^T B^T$.

14. Let A, B, and C be matrices with C invertible.

(a) [BB] If $AC = BC$, then $A = B$.

(b) If $CA = CB$, then $A = B$.

15. Let A be any $m \times n$ matrix. Let $\mathbf{x}$ be a vector in $\mathbf{R}^n$ and let $\mathbf{y}$ be a vector in $\mathbf{R}^m$.

(a) [BB] Explain the equation $\mathbf{y} \cdot (A\mathbf{x}) = \mathbf{y}^T A\mathbf{x}$.

(b) Prove that $\|A\mathbf{x}\|^2 = \mathbf{x}^T A^T A\mathbf{x}$.

16. Let $A = \begin{bmatrix} 1 & 0 & 1 & 1 \\ 2 & 1 & 0 & 1 \\ 1 & 2 & 1 & 0 \end{bmatrix}$, $\mathbf{x} = \begin{bmatrix} 1 \\ 2 \\ 1 \\ 1 \end{bmatrix}$, and $\mathbf{y} = \begin{bmatrix} 1 \\ 0 \\ 1 \end{bmatrix}$.

(a) Find $A\mathbf{x}$.

(b) Find $\mathbf{x}^T A^T$ by two different methods.

(c) Find $\mathbf{y}^T A\mathbf{x}$.

17. Find A, given that A is a 2×2 matrix and

$$\begin{bmatrix} 1 & 2 \\ 3 & 0 \end{bmatrix}^{-1} A \begin{bmatrix} 5 & 1 \\ -1 & 1 \end{bmatrix}^{-1} = \begin{bmatrix} -3 & 4 \\ 0 & 2 \end{bmatrix}.$$

18. (a) [BB] Suppose A and B are invertible matrices and X is a matrix such that $AXB = A + B$. What is X?

(b) Suppose A, B, C, and X are matrices, X and C invertible, such that $X^{-1}A = C - X^{-1}B$. What is X?

(c) Suppose A, B, and X are invertible matrices such that $BAX = XABX$. What is X?

19. (a) Let $A = \begin{bmatrix} 1 & 0 & 0 & 0 \\ 2 & 1 & 0 & 0 \\ 4 & 4 & 1 & 0 \\ 8 & 12 & 6 & 1 \end{bmatrix}$ and

$B = \begin{bmatrix} 1 & 0 & 0 & 0 \\ 3 & 1 & 0 & 0 \\ 9 & 6 & 1 & 0 \\ 27 & 27 & 9 & 1 \end{bmatrix}$. Find AB.

(b) Let $P(a) = \begin{bmatrix} 1 & 0 & 0 & 0 \\ a & 1 & 0 & 0 \\ a^2 & 2a & 1 & 0 \\ a^3 & 3a^2 & 3a & 1 \end{bmatrix}$ and

$P(b) = \begin{bmatrix} 1 & 0 & 0 & 0 \\ b & 1 & 0 & 0 \\ b^2 & 2b & 1 & 0 \\ b^3 & 3b^2 & 3b & 1 \end{bmatrix}$.

Show that $P(a)P(b)$ is $P(x)$ for some x.

(c) The identity matrix has the form $P(a)$ for some a. Explain.

(d) Without any calculation, show that the matrix A in part (a) has an inverse by exhibiting the inverse.

[This exercise was suggested by an article in the *American Mathematical Monthly*, "Inverting the Pascal Matrix Plus One," by Rita Aggarwala and Michael P. Lamoureux, **109** (April 2002), no. 9.]

20. Let $\mathbf{x} = \begin{bmatrix} x_1 \\ x_2 \\ \vdots \\ x_n \end{bmatrix}$ be a vector in $\mathbf{R}^n$ and let

$A = \begin{bmatrix} \mathbf{a}_1^T & \rightarrow \\ \mathbf{a}_2^T & \rightarrow \\ \vdots \\ \mathbf{a}_n^T & \rightarrow \end{bmatrix}$ be an $n \times m$ matrix. Show that $\mathbf{x}^T A$

is a linear combination of the rows of A with coefficients the components of $\mathbf{x}$. [Hint: 2.1.33 and Theorem 2.2.19.]

21. Suppose $A = \begin{bmatrix} \mathbf{a}_1^T & \rightarrow \\ \mathbf{a}_2^T & \rightarrow \\ \vdots \\ \mathbf{a}_m^T & \rightarrow \end{bmatrix}$ is an $m \times n$ matrix with

rows as indicated and suppose B is an $n \times t$ matrix.

Show that $AB = \begin{bmatrix} \mathbf{a}_1^T B & \rightarrow \\ \mathbf{a}_2^T B & \rightarrow \\ \vdots \\ \mathbf{a}_m^T B & \rightarrow \end{bmatrix}$.

[Hint: Theorem 2.2.19 and equation (4), p. 77.]

■ Critical Reading

22. Show that the matrix $A = \begin{bmatrix} 1 & 3 & 3 \\ 3 & 1 & 3 \\ -3 & -3 & -5 \end{bmatrix}$ has an inverse that is a linear polynomial in A; that is, of the form $aI + bA$ for scalars a and b.

23. Suppose A, B, and C are invertible matrices and X is a matrix such that $A = BXC$. Prove that X is invertible.

24. Suppose A, B, and C are matrices with $A = BC$ and $B^2 = C^2 = I$. Explain why A is invertible.

25. If A is a matrix and $A^T A = \mathbf{0}$ is the zero matrix, show that $A = \mathbf{0}$. [Hint: What is the $(1, 1)$ entry of $A^T A$?]

26. Given that A is a 3×3 matrix and

$$\begin{bmatrix} 4 & 0 & 6 \\ 0 & -3 & 1 \\ 0 & 5 & -4 \end{bmatrix}^{-1} A \begin{bmatrix} 2 & 0 & 1 \\ 1 & 4 & -1 \\ 1 & 3 & 0 \end{bmatrix}^{-1} = $$

$$\begin{bmatrix} 1 & 2 & -1 \\ -4 & 0 & 6 \\ -2 & 1 & 1 \end{bmatrix}, \text{ find } A.$$

27. Let A be a 3×3 matrix such that $A \begin{bmatrix} 1 \\ 2 \\ 3 \end{bmatrix} = \begin{bmatrix} 0 \\ 0 \\ 0 \end{bmatrix}$. Can A be invertible? Explain.

28. Suppose A is an $n \times n$ matrix such that $A + A^2 = I$. Show that A is invertible.

29. Find a formula for $((AB)^T)^{-1}$ in terms of $(A^T)^{-1}$ and $(B^T)^{-1}$.

30. Is the product of invertible matrices invertible? What about the sum?

31. If A and B are matrices with both AB and B invertible, prove that A is invertible.

2.3   Systems of Linear Equations

A *linear equation* in variables $x_1, x_2, \ldots$ is an equation such as $2x_1 - 3x_2 + x_3 - \sqrt{2}x_4 = 17$, where each variable appears by itself (and to the first power) and the coefficients are scalars. A set of one or more linear equations is called a *linear system. To solve* a linear system means to find values of the variables that make each equation true.

In this section, we describe a procedure for solving systems of linear equations. Many people could find numbers x and y so that

$$\begin{aligned} x + y &= 7 \\ x - 2y &= -2 \end{aligned}$$

via some ad hoc process and, in fact, we have assumed you have this ability several times already in previous sections of this book. Suppose you were faced with a system of eight equations in 17 unknowns, however. How would you go about finding a solution in this case? The procedure we have in mind involves three simple principles.

First, in any system of linear equations, we can always rearrange the equations without changing a solution. For example, the numbers x, y, and z that satisfy

$$\begin{aligned} -2x + y + z &= -9 \\ x + y + z &= 0 \\ x + 3y \phantom{{}+ z} &= -3 \end{aligned} \tag{1}$$

are the same as the numbers that satisfy

$$\begin{aligned} x + y + z &= 0 \\ -2x + y + z &= -9 \\ x + 3y \phantom{{}+ z} &= -3, \end{aligned}$$

which is the same as system (1), except that the first and second equations have been interchanged.

Second, in any system of equations, we may always multiply an equation by a number (different from 0) without changing any solution. For example,

$$x + 3y = -3 \quad \text{if and only if (multiplying by } \tfrac{1}{3}) \quad \tfrac{1}{3}x + y = -1.$$

Third, and this is less obvious, in any system of equations, we may always subtract a multiple of one equation from another without changing a solution. Here's an instance of what we mean:

Suppose a system of equations contains the equations

$$\begin{aligned} x + 3y - z &= 2 \\ 2x - y + 4z &= 7. \end{aligned} \tag{2}$$

Any values of x, y, and z that satisfy this system must also satisfy the system in which equations (2) are replaced by equations (3), where the second equation of (2) has been replaced by itself minus three times the first equation:

$$\begin{aligned} x + 3y - z &= 2 \\ (2x - y + 4z) - 3(x + 3y - z) &= 7 - 3(2) \end{aligned} \tag{3}$$

Conversely, any values of x, y, and z that satisfy a system containing the two equations in (3) must also satisfy the system with these replaced by the two equations in (2). To see why, just note that (3) implies

$$\begin{aligned} 2x - y + 4z &= [(2x - y + 4z) - 3(x + 3y - z)] + 3(x + 3y - z) \\ &= [7 - 3(2)] + 3(2) = 7. \end{aligned}$$

Let's summarize. When solving a system of linear equations, you may alter the system in one of three ways without changing any solution:

1. interchange two equations;

2. multiply an equation by any number other than 0; and

3. replace an equation by that equation minus any multiple of another.

By the way, it makes no difference whether we say "minus" or "plus" in Statement 3 since, for example, adding $2E$ is the same as subtracting $-2E$, but we prefer the language of subtraction for reasons that will become clear later.

We illustrate how these principles can be applied to solve system (1).

First, interchange the first two equations to get

$$\begin{aligned} x + y + z &= 0 \\ -2x + y + z &= -9 \\ x + 3y \phantom{{}+ z} &= -3. \end{aligned}$$

Now replace the second equation by that equation plus twice the first—we denote this as $E2 \to E2 + 2(E1)$—to obtain

$$\begin{aligned} x + y + z &= 0 \\ 3y + 3z &= -9 \\ x + 3y &= -3. \end{aligned} \qquad (4)$$

We describe the transformation from system (1) to system (4) like this:

$$\begin{aligned} -2x + y + z &= -9 \\ x + y + z &= 0 \\ x + 3y &= -3 \end{aligned} \quad \xrightarrow{E1 \leftrightarrow E2} \quad \begin{aligned} x + y + z &= 0 \\ -2x + y + z &= -9 \\ x + 3y &= -3 \end{aligned}$$

$$\xrightarrow{E2 \to E2 + 2(E1)} \quad \begin{aligned} x + y + z &= 0 \\ 3y + 3z &= -9 \\ x + 3y &= -3. \end{aligned}$$

We continue with our solution, replacing equation 3 with equation 3 minus equation 1:

$$\begin{aligned} x + y + z &= 0 \\ 3y + 3z &= -9 \\ x + 3y &= -3 \end{aligned} \quad \xrightarrow{E3 \to E3 - E1} \quad \begin{aligned} x + y + z &= 0 \\ 3y + 3z &= -9 \\ 2y - z &= -3. \end{aligned}$$

Now we multiply the equation 2 by $\frac{1}{3}$ and replace equation 3 with equation 3 minus twice equation 2:

$$\begin{aligned} x + y + z &= 0 \\ 3y + 3z &= -9 \\ 2y - z &= -3 \end{aligned} \quad \xrightarrow{E2 \to \frac{1}{3} E2} \quad \begin{aligned} x + y + z &= 0 \\ y + z &= -3 \\ 2y - z &= -3 \end{aligned}$$

$$\xrightarrow{E3 \leftrightarrow E3 - 2(E2)} \quad \begin{aligned} x + y + z &= 0 \\ y + z &= -3 \\ -3z &= 3. \end{aligned}$$

We have reached our goal, which was to transform the original system to one that is *upper triangular*, like this: $\diagdown$. At this point, we can obtain a solution in a straightforward manner by using a technique called *back substitution*, which means solving the equations from the bottom up. The third equation gives $z = -1$. The second says $y + z = -3$, so $y = -3 - z = -2$. The first says $x + y + z = 0$, so $x = -y - z = 3$.

It is easy to check that $x = 3$, $y = -2$, $z = -1$ satisfies the original system (1), so we have our solution. **We strongly advise you always to check solutions!**

Did you get tired of writing the x, the y, the z, and the $=$ in the solution just illustrated? Only the coefficients change from one stage to another, so it is only the coefficients we really have to write down. Let's solve system (1) again, by different means.

First, we write the system in matrix form, like this:

$$\begin{bmatrix} -2 & 1 & 1 \\ 1 & 1 & 1 \\ 1 & 3 & 0 \end{bmatrix} \begin{bmatrix} x \\ y \\ z \end{bmatrix} = \begin{bmatrix} -9 \\ 0 \\ 3 \end{bmatrix}.$$

This is $A\mathbf{x} = \mathbf{b}$ with $A = \begin{bmatrix} -2 & 1 & 1 \\ 1 & 1 & 1 \\ 1 & 3 & 0 \end{bmatrix}$, the *matrix of coefficients*, $\mathbf{x} = \begin{bmatrix} x \\ y \\ z \end{bmatrix}$ and $\mathbf{b} = \begin{bmatrix} -9 \\ 0 \\ -3 \end{bmatrix}$. Next, we write the *augmented matrix*

$$[A|\mathbf{b}] = \begin{bmatrix} -2 & 1 & 1 & | & -9 \\ 1 & 1 & 1 & | & 0 \\ 1 & 3 & 0 & | & -3 \end{bmatrix},$$

which consists of A followed by a vertical line and then $\mathbf{b}$.

Just keeping track of coefficients, with augmented matrices, the solution to system (1) now looks like this:

$$\begin{bmatrix} -2 & 1 & 1 & | & -9 \\ 1 & 1 & 1 & | & 0 \\ 1 & 3 & 0 & | & -3 \end{bmatrix} \xrightarrow{R1 \leftrightarrow R2} \begin{bmatrix} 1 & 1 & 1 & | & 0 \\ -2 & 1 & 1 & | & -9 \\ 1 & 3 & 0 & | & -3 \end{bmatrix}$$

$$\xrightarrow[\begin{subarray}{l} R2 \to R2 + 2(R1) \\ R3 \to R3 - R1 \end{subarray}]{} \begin{bmatrix} 1 & 1 & 1 & | & 0 \\ 0 & 3 & 3 & | & -9 \\ 0 & 2 & -1 & | & -3 \end{bmatrix} \xrightarrow{R2 \to \frac{1}{3}R2} \begin{bmatrix} 1 & 1 & 1 & | & 0 \\ 0 & 1 & 1 & | & -3 \\ 0 & 2 & -1 & | & -3 \end{bmatrix} \quad (5)$$

$$\xrightarrow{R3 \to R3 - 2(R2)} \begin{bmatrix} 1 & 1 & 1 & | & 0 \\ 0 & 1 & 1 & | & -3 \\ 0 & 0 & -3 & | & 3 \end{bmatrix} \xrightarrow{R3 \to -\frac{1}{3}R3} \begin{bmatrix} 1 & 1 & 1 & | & 0 \\ 0 & 1 & 1 & | & -3 \\ 0 & 0 & 1 & | & -1 \end{bmatrix}.$$

Here we have used R for *row* rather than E for *equation* to describe the changes to each system, writing, for example, $R3 \to R3 - 2(R2)$ at the last step, rather than $E3 \to E3 - 2(E2)$ as we did before. The three basic operations that we have been performing on equations now become operations on rows, called the *elementary row operations*.

2.3.1 The Elementary Row Operations.

1. Interchange two rows.
2. Multiply a row by any scalar except 0.
3. Replace a row by that row minus a multiple of another.

We could equally well have stated the third elementary row operation as

3. replace a row by that row *plus* a multiple of another,

since $R1 + c(R2) = R1 - (-c)(R2)$, but, as we said before, there are advantages to using the language of subtraction. In particular, when doing so, the multiple has a special name. It is called a *multiplier*, $-c$ in our example, a concept that will be important to us later.

The author is a ninth-generation student of Gauss, through Gudermann, Weierstrass, Frobenius, Schur, Brauer, Bruck, Kleinfeld, and Anderson.

The procedure illustrated in (5), whereby a matrix is transformed to an upper triangular matrix via elementary row operations, is called *Gaussian elimination* after the German mathematician Karl Friedrich Gauss (1777–1855), who first used this process to solve systems of 17 linear equations in order to determine the orbit of a new planet. (See 2.5.9 for a precise definition of *upper triangular matrix*.)

2.3.2 Gaussian Elimination. To move a matrix to one that is upper triangular,

1. If the first row is all 0s, move it to the bottom; otherwise, use the first nonzero entry of the first row and the third elementary row operation to get 0s below it;

2. Move to row two and repeat step 1;

3. Continue, sweeping across the matrix from left to right, turning into a 0 each entry below the first nonzero entry in a row.

When doing calculations by hand, it is often easiest to form 0s after making the first nonzero entry of a row a 1 (with the second elementary operation).

When the Gaussian elimination process is complete, interpret the augmented matrix as a system of equations, in our case, the system

$$x + y + z = 0$$
$$y + z = -3$$
$$z = -1,$$

and then solve by back substitution, as before, obtaining $z = -1$, $y = -2$, and $x = 3$. This solution is *unique* because it is the only one; 3, -2, and -1 are the only values of x, y, and z that satisfy the original system (1). Equivalently, $\begin{bmatrix} x \\ y \\ z \end{bmatrix}$ is the only vector that satisfies

$$\begin{bmatrix} -2 & 1 & 1 \\ 1 & 1 & 1 \\ 1 & 3 & 0 \end{bmatrix} \begin{bmatrix} x \\ y \\ z \end{bmatrix} = \begin{bmatrix} -9 \\ 0 \\ -3 \end{bmatrix}.$$

2.3.3 To Solve a System of Linear Equations

1. Write the system in the form $A\mathbf{x} = \mathbf{b}$, where A is the matrix of coefficients, $\mathbf{x}$ is the vector of unknowns, and $\mathbf{b}$ is the vector whose components are the constants to the right of the equals signs.

2. Add a vertical line and $\mathbf{b}$ to the right of A to form the *augmented matrix* $[A|\mathbf{b}]$.

3. Reduce the augmented matrix to an upper triangular matrix by Gaussian elimination.

4. Write the equations that correspond to the triangular matrix and solve by back substitution.

We illustrate our method of solving equations with a number of examples. As you will see, solutions are not always unique.

2.3.4 PROBLEM

Solve the system
$$\begin{aligned} x + 2y &= 1 \\ 4x + 8y &= 3. \end{aligned}$$

Solution. This is $A\mathbf{x} = \mathbf{b}$ with $A = \begin{bmatrix} 1 & 2 \\ 4 & 8 \end{bmatrix}$, $\mathbf{x} = \begin{bmatrix} x \\ y \end{bmatrix}$, and $\mathbf{b} = \begin{bmatrix} 1 \\ 3 \end{bmatrix}$. Gaussian elimination on the augmented matrix requires only one step.

$$[A|\mathbf{b}] = \begin{bmatrix} 1 & 2 & | & 1 \\ 4 & 8 & | & 3 \end{bmatrix} \xrightarrow{R2 \to R2 - 4(R1)} \begin{bmatrix} 1 & 2 & | & 1 \\ 0 & 0 & | & -1 \end{bmatrix}.$$

The equations that correspond to this triangular matrix are

$$\begin{aligned} x + 2y &= 1 \\ 0 &= -1. \end{aligned}$$

The last equation is not true, so there is no solution, a fact we might have noted earlier. If $x + 2y = 1$, then certainly $4x + 8y = 4(x + 2y) = 4$, not 3. So if the first equation is satisfied, the second one is not. A system like this, with no solution, is called *inconsistent*.

2.3.5 PROBLEM

Solve the system
$$\begin{aligned} 2x - 3y + 6z &= 14 \\ x + y - 2z &= -3. \end{aligned}$$

Solution. This is $A\mathbf{x} = \mathbf{b}$, with

$$A = \begin{bmatrix} 2 & -3 & 6 \\ 1 & 1 & -2 \end{bmatrix}, \quad \mathbf{x} = \begin{bmatrix} x \\ y \\ z \end{bmatrix}, \quad \text{and} \quad \mathbf{b} = \begin{bmatrix} 14 \\ -3 \end{bmatrix}. \tag{6}$$

We form the augmented matrix

$$[A|\mathbf{b}] = \begin{bmatrix} 2 & -3 & 6 & | & 14 \\ 1 & 1 & -2 & | & -3 \end{bmatrix}$$

and reduce it to upper triangular form by Gaussian elimination.

$$\begin{bmatrix} 2 & -3 & 6 & | & 14 \\ 1 & 1 & -2 & | & -3 \end{bmatrix} \xrightarrow{R1 \leftrightarrow R2} \begin{bmatrix} 1 & 1 & -2 & | & -3 \\ 2 & -3 & 6 & | & 14 \end{bmatrix}$$

$$\xrightarrow{R2 \to R2-2(R1)} \begin{bmatrix} 1 & 1 & -2 & | & -3 \\ 0 & -5 & 10 & | & 20 \end{bmatrix} \xrightarrow{R2 \to -\frac{1}{5}(R2)} \begin{bmatrix} 1 & 1 & -2 & | & -3 \\ 0 & 1 & -2 & | & -4 \end{bmatrix}.$$

The last matrix is upper triangular, so Gaussian elimination is complete. The equations corresponding to the final matrix are

$$\begin{aligned} x + y - 2z &= -3 \\ y - 2z &= -4. \end{aligned}$$

Now we apply back substitution. The last equation says $y = 2z - 4$. The first says $x + y - 2z = -3$, so $x = -y + 2z - 3 = -(2z - 4) + 2z - 3 = 1$. The solution is

$$\begin{aligned} x &= \quad 1 \\ y &= 2z - 4 \\ z &= \quad z. \end{aligned}$$

Our solution shows that $x = 1$ and that z can be chosen arbitrarily, but that y is then determined in terms of z. To emphasize the idea that z can be any number, we introduce the *parameter* t and let $z = t$. The solution then takes the form

$$\begin{aligned} x &= \quad 1 \\ y &= 2t - 4 \\ z &= \quad t. \end{aligned}$$

The variable z is called *free* because, as the equation $z = t$ tries to show, z can be any number. The system of equations in this problem has infinitely many solutions, one for each value of t. We write the solution as a vector, like this:

$$\begin{bmatrix} x \\ y \\ z \end{bmatrix} = \begin{bmatrix} 1 \\ 2t - 4 \\ t \end{bmatrix} = \begin{bmatrix} 1 \\ -4 \\ 0 \end{bmatrix} + t \begin{bmatrix} 0 \\ 2 \\ 1 \end{bmatrix}.$$

Such an equation should look familiar. It is the equation of a line, the one through $(1, -4, 0)$ with direction $\mathbf{d} = \begin{bmatrix} 0 \\ 2 \\ 1 \end{bmatrix}$. Geometrically, this is what we expect. Each of the two given equations describes a plane in 3-space. When we solve the system, we are finding the points that lie on both planes. Since the planes are not parallel, the points that lie on both form a line.

Reading Challenge 1   *How do you know that the planes described by the equations in Problem 2.3.5 are not parallel?*

Each solution to the linear system in Problem 2.3.5 is the sum of the vector $\begin{bmatrix} 1 \\ -4 \\ 0 \end{bmatrix}$ and a scalar multiple of $\begin{bmatrix} 0 \\ 2 \\ 1 \end{bmatrix}$. The solution $\begin{bmatrix} 1 \\ -4 \\ 0 \end{bmatrix}$ is called a *particular solution*.

The vector $\begin{bmatrix} 0 \\ 2 \\ 1 \end{bmatrix}$ is a solution to the system $A\mathbf{x} = \mathbf{0}$. It is characteristic of linear systems that whenever there is more than one solution, each solution is the sum of a particular solution and the solution to $A\mathbf{x} = \mathbf{0}$. We shall have more to say about this in Section 2.4.

Reading Challenge 2   *Verify that* $\begin{bmatrix} 1 \\ -4 \\ 0 \end{bmatrix}$ *is a solution to the system* $A\mathbf{x} = \mathbf{b}$ *in Problem 2.3.5, and that* $\begin{bmatrix} 0 \\ 2 \\ 1 \end{bmatrix}$ *is a solution to* $A\mathbf{x} = \begin{bmatrix} 0 \\ 0 \end{bmatrix}$.

In general, there are three possible outcomes when we attempt to solve a system of linear equations.

> **2.3.6 A system of linear equations may have**
>
> **1.** a unique solution,
>
> **2.** no solution, or
>
> **3.** infinitely many solutions.

The solution to system (1) was unique. In Problem 2.3.4, you met an inconsistent system (no solution) and in Problem 2.3.5 a system with infinitely many solutions. Interestingly, a linear system cannot have precisely five solutions, for instance, because if a solution is not unique, then there is a *free* variable that can be chosen in infinitely many ways. But we are getting a bit ahead of ourselves!

Reading Challenge 3   *How many solutions does the system* $\begin{aligned} x^2 + y^2 &= 8 \\ x - y &= 0 \end{aligned}$ *have? (Note that this system is not* linear!*)*

A row echelon matrix takes its name from the French word "échelon" meaning "step." When a matrix is in row echelon form, the path formed by the leading nonzero entries resembles a staircase.

At this point we confess that the goal of Gaussian elimination is to transform a matrix not simply into any upper triangular matrix, but into a special kind of upper triangular matrix called a *row echelon matrix*.

In the following definition, it is convenient to refer to the first nonzero entry in a nonzero row as its *leading nonzero entry*.

2.3.7 DEFINITION

A *row echelon matrix* is an upper triangular matrix with the following properties:

1. All rows consisting entirely of 0s are at the bottom;
2. The leading nonzero entries of the nonzero rows step from left to right as you read down the matrix;
3. All the entries in a column below a leading nonzero entry are 0.

A *row echelon form* of a matrix A is a row echelon matrix U to which A can be moved via elementary row operations.

It is convenient to have a name for the leading nonzero entries in nonzero rows of a row echelon matrix, and for the columns in which these numbers sit.

2.3.8 DEFINITION

A *pivot* in a row echelon matrix U is a leading nonzero entry in a nonzero row. A column containing a pivot is called a *pivot column of U*. If U is a row echelon form of a matrix A, then a *pivot column of A* is a column that corresponds to a pivot column of U. If U was obtained without any multiplication of rows by nonzero scalars, then a *pivot of A* is defined to be any pivot of U.

2.3.9 EXAMPLES

- The matrix $U = \begin{bmatrix} 1 & 1 & 3 & 4 & 5 \\ 0 & 0 & 1 & 3 & 1 \\ 0 & 0 & 0 & 0 & 1 \end{bmatrix}$ is a row echelon matrix. The pivots are circled.

  Since the pivots are in columns one, three, and five, the pivot columns are columns one, three, and five.

- The matrix $U = \begin{bmatrix} 0 & 0 & 2 & 1 & 5 \\ 0 & 0 & 0 & 0 & 1 \\ 0 & 0 & 0 & 0 & 0 \end{bmatrix}$ is a row echelon matrix. The pivots are the circled entries. The pivot columns are columns three and five.

- Let $A = \begin{bmatrix} -2 & 3 & 4 \\ 1 & 0 & 1 \\ 4 & -5 & 6 \end{bmatrix}$. Then

$$
A \xrightarrow{R1 \leftrightarrow R2} \begin{bmatrix} 1 & 0 & 1 \\ -2 & 3 & 4 \\ 4 & -5 & 6 \end{bmatrix} \xrightarrow[\substack{R3 \to R3 - 4(R1)}]{R2 \to R2 + 2(R1)} \begin{bmatrix} 1 & 0 & 1 \\ 0 & 3 & 6 \\ 0 & -5 & 2 \end{bmatrix}
$$

$$
\xrightarrow{R3 \to R3 + \frac{5}{3}(R2)} \begin{bmatrix} 1 & 0 & 1 \\ 0 & 3 & 6 \\ 0 & 0 & 12 \end{bmatrix} = U,
$$

so the pivots of A are 1, 3, and 12, and the pivot columns of A are columns one, two, and three.

- Let $A = \begin{bmatrix} 3 & -2 & 1 & -4 \\ 6 & -4 & 8 & -7 \\ 9 & -6 & 9 & -12 \end{bmatrix}$. Then

$$
A \to \begin{bmatrix} 3 & -2 & 1 & -4 \\ 0 & 0 & 6 & 1 \\ 0 & 0 & 6 & 0 \end{bmatrix} \to \begin{bmatrix} 3 & -2 & 1 & -4 \\ 0 & 0 & 6 & 1 \\ 0 & 0 & 0 & -1 \end{bmatrix} = U
$$

so the pivots of A are 3, 6, and -1, and the pivot columns of A are one, three, and four.   ■

Reading Challenge 4   *Find the pivots and identify the pivot columns of $A = \begin{bmatrix} -3 & 6 & 4 \\ 2 & -4 & 0 \end{bmatrix}$.*

2.3.10 EXAMPLES   Here are some matrices that are not in row echelon form:

$$\begin{bmatrix} 2 & 0 & 7 \\ 0 & 0 & 1 \\ 0 & 0 & 1 \end{bmatrix}, \quad \begin{bmatrix} 6 & 0 & 3 & 2 \\ 0 & 0 & 5 & 4 \\ 0 & 1 & 0 & 2 \end{bmatrix}, \quad \begin{bmatrix} 3 & 0 & -1 & 5 \\ 0 & 0 & 0 & 0 \\ 0 & -4 & 2 & -3 \end{bmatrix}. \qquad ■$$

Reading Challenge 5   *Explain why each of the matrices in **2.3.10** is not in row echelon form.*

2.3.11 Remark   As the phrase "a row echelon form" suggests, row echelon form is not unique, since different sequences of elementary row operations can transform a matrix A into different row echelon matrices. Thus we say *a* row echelon form, not *the* row echelon form. On the other hand, the columns in which the pivots eventually appear—that is, the pivot columns—are unique. This follows from the uniqueness of the *reduced* row echelon matrix. See Definition 2.7.4 and Appendix A of *Linear Algebra and Its Applications* by David C. Lay, Addison-Wesley (2000).

When a system of equations involves more than two or three variables, we usually label the variables $x_1, x_2, x_3, \ldots$ (so that we don't run out of letters). In the rest of this section, we regularly divide each row containing a pivot by that pivot, thus turning the leading entry in that row into a 1. Putting 0s below a 1 by hand is much easier (and less susceptible to errors) than putting 0s below $\frac{2}{3}$, for instance.

The definition of row echelon form varies from author to author. Many authors still require that the leading nonzero entry in a nonzero row of a matrix in row echelon form be a 1. This makes a lot of sense when doing calculations by hand, but it is much less important in today's highly technological society.

2.3.12 EXAMPLE   Suppose we wish to solve the system

$$\begin{aligned} 2x - y + z &= -7 \\ x + y + z &= -2 \\ 3x + y + z &= 0 \end{aligned}$$

without the use of a computer. The steps of Gaussian elimination we would probably employ are these:

$$[A|b] = \begin{bmatrix} 2 & -1 & 1 & -7 \\ 1 & 1 & 1 & -2 \\ 3 & 1 & 1 & 0 \end{bmatrix} \xrightarrow{R1 \leftrightarrow R2} \begin{bmatrix} 1 & 1 & 1 & -2 \\ 2 & -1 & 1 & -7 \\ 3 & 1 & 1 & 0 \end{bmatrix}$$

$$\begin{matrix} R2 \leftarrow R2 - 2(R1) \\ R3 \leftarrow R3 - 3(R1) \\ \longrightarrow \end{matrix} \left[\begin{array}{ccc|c} 1 & 1 & 1 & -2 \\ 0 & -3 & -1 & -3 \\ 0 & -2 & -2 & 6 \end{array}\right] \xrightarrow{R2 \leftrightarrow R3} \left[\begin{array}{ccc|c} 1 & 1 & 1 & -2 \\ 0 & -2 & -2 & 6 \\ 0 & -3 & -1 & -3 \end{array}\right]$$

$$\xrightarrow{R2 \leftarrow (-\frac{1}{2})R2} \left[\begin{array}{ccc|c} 1 & 1 & 1 & -2 \\ 0 & 1 & 1 & -3 \\ 0 & -3 & -1 & -3 \end{array}\right] \xrightarrow{R3 \rightarrow R3+3(R2)} \left[\begin{array}{ccc|c} 1 & 1 & 1 & -2 \\ 0 & 1 & 1 & -3 \\ 0 & 0 & 2 & -12 \end{array}\right]$$

$$\xrightarrow{R3 \rightarrow \frac{1}{2}(R3)} \left[\begin{array}{ccc|c} 1 & 1 & 1 & -2 \\ 0 & 1 & 1 & -3 \\ 0 & 0 & 1 & -6 \end{array}\right].$$

The final matrix is in row echelon form and, by back substitution, we quickly read off the solution: $z = -6$, $y = -3 - z = 3$, $x = -2 - y - z = 1$.

We went to a lot of effort to make all leading nonzero entries 1. Look, however, at how much more efficiently the Gaussian elimination process would proceed if we didn't so insist:

$$[A|b] = \left[\begin{array}{ccc|c} 2 & -1 & 1 & -7 \\ 1 & 1 & 1 & -2 \\ 3 & 1 & 1 & 0 \end{array}\right] \begin{matrix} R2 \rightarrow R2 - \frac{1}{2}(R1) \\ R3 \rightarrow R3 - \frac{3}{2}(R1) \\ \longrightarrow \end{matrix} \left[\begin{array}{ccc|c} 2 & -1 & 1 & -7 \\ 0 & \frac{3}{2} & \frac{1}{2} & \frac{3}{2} \\ 0 & \frac{5}{2} & -\frac{1}{2} & \frac{21}{2} \end{array}\right]$$

$$\xrightarrow{R3 \rightarrow R3 - \frac{5}{3}(R2)} \left[\begin{array}{ccc|c} 2 & -1 & 1 & -7 \\ 0 & \frac{3}{2} & \frac{1}{2} & \frac{3}{2} \\ 0 & 0 & -\frac{4}{3} & 8 \end{array}\right]$$

(two steps instead of six). The first elementary row operation—multiply a row by a nonzero scalar—is by far the least important. It is useful when doing calculations by hand, but not otherwise. ■

The pivot columns of a matrix are extremely important for many reasons. The variable that corresponds to a pivot column is never free. For example, look at the pivot in column three of

$$\left[\begin{array}{cccc|c} 1 & 1 & 3 & 4 & 5 \\ 0 & 0 & 1 & 3 & 1 \\ 0 & 0 & 0 & 1 & -3 \end{array}\right]. \tag{7}$$

The second equation reads $x_3 + 3x_4 = 1$. Thus $x_3 = 1 - x_4$ is not free; it is determined by x_4. On the other hand, a variable that corresponds to a column that is not a pivot column is always free. The equations that correspond to the rows of the matrix in (7) are

$$\begin{aligned} x_1 + x_2 + 3x_3 + 4x_4 &= 5 \\ x_3 + 3x_4 &= 1 \\ x_4 &= -3. \end{aligned}$$

The variables x_1, x_3, and x_4 are not free since

$$x_4 = -3$$

$$x_3 = 1 - 3x_4$$

$$x_1 = 5 - x_2 - 3x_3 - 4x_4,$$

but there is no equation of the form $x_2 = *$, so x_2 is free. To summarize,

> **2.3.13** Free variables are those that correspond to columns which are not pivot columns.

Reading Challenge 6 *Suppose a system of equations in the variables x_1, x_2, x_3, x_4 leads to the row echelon matrix* $\begin{bmatrix} 0 & 1 & 2 & 7 & 5 \\ 0 & 0 & 0 & 1 & 1 \\ 0 & 0 & 0 & 0 & 0 \end{bmatrix}$. *What are the free variables?*

2.3.14 PROBLEM Solve $\begin{array}{rcl} x_1 - x_2 - x_3 &=& 2 \\ 2x_1 - x_2 - 3x_3 &=& 6 \\ x_1 \phantom{{}- x_2} - 2x_3 &=& 4. \end{array}$

Solution. The given system is $A\mathbf{x} = \mathbf{b}$ where

$$A = \begin{bmatrix} 1 & -1 & -1 \\ 2 & -1 & -3 \\ 1 & 0 & -2 \end{bmatrix}, \quad \mathbf{x} = \begin{bmatrix} x_1 \\ x_2 \\ x_3 \end{bmatrix}, \quad \text{and} \quad \mathbf{b} = \begin{bmatrix} 2 \\ 6 \\ 4 \end{bmatrix}.$$

The augmented matrix is

$$[A|\mathbf{b}] = \begin{bmatrix} 1 & -1 & -1 & 2 \\ 2 & -1 & -3 & 6 \\ 1 & 0 & -2 & 4 \end{bmatrix}.$$

Gaussian elimination begins

$$\begin{bmatrix} 1 & -1 & -1 & 2 \\ 2 & -1 & -3 & 6 \\ 1 & 0 & -2 & 4 \end{bmatrix} \xrightarrow[\begin{subarray}{l} E2 \to E2 - 2(E1) \\ E3 \to E3 - E1 \end{subarray}]{} \begin{bmatrix} 1 & -1 & -1 & 2 \\ 0 & 1 & -1 & 2 \\ 0 & 1 & -1 & 2 \end{bmatrix}$$

(the first pivot, the first 1 in row one, was used to get 0s below it)

$$\xrightarrow{E3 \to E3 - E2} \begin{bmatrix} 1 & -1 & -1 & 2 \\ 0 & 1 & -1 & 2 \\ 0 & 0 & 0 & 0 \end{bmatrix}$$

(the second pivot, the first 1 in row two, was used to get 0s below it). This is row echelon form. There are two pivots, in columns one and two. There is one column that does not contain a pivot: column three. The corresponding variable, x_3, is free, and we emphasize this fact by setting $x_3 = t$, a parameter. The equations corresponding to the rows of the row echelon matrix are

$$\begin{aligned} x_1 - x_2 - x_3 &= 2 \\ x_2 - x_3 &= 2 \\ 0 &= 0. \end{aligned}$$

We solve these by back substitution (that is, from the bottom up).

The second equation reads $x_2 - x_3 = 2$, so $x_2 = x_3 + 2 = t + 2$.
The first equation reads $x_1 - x_2 - x_3 = 2$,
so $x_1 = x_2 + x_3 + 2 = (t + 2) + t + 2 = 2t + 4$.

The solution is

$$\begin{aligned} x_1 &= 2t + 4 \\ x_2 &= t + 2 \\ x_3 &= t \end{aligned}$$

or, in vector form,

$$\mathbf{x} = \begin{bmatrix} x_1 \\ x_2 \\ x_3 \end{bmatrix} = \begin{bmatrix} 2t + 4 \\ t + 2 \\ t \end{bmatrix} = \begin{bmatrix} 4 \\ 2 \\ 0 \end{bmatrix} + t \begin{bmatrix} 2 \\ 1 \\ 1 \end{bmatrix}.$$

Again, thinking geometrically, our solution is a line. There are infinitely many solutions, each of which is the sum of the particular solution $\begin{bmatrix} 4 \\ 2 \\ 0 \end{bmatrix}$ and a multiple of $\begin{bmatrix} 2 \\ 1 \\ 1 \end{bmatrix}$, which is a solution of $A\mathbf{x} = \mathbf{0}$.

2.3.15 PROBLEM Express the vector $\begin{bmatrix} 2 \\ 6 \\ 4 \end{bmatrix}$ as a linear combination of the columns of $A = \begin{bmatrix} 1 & -1 & -1 \\ 2 & -1 & -3 \\ 1 & 0 & -2 \end{bmatrix}$.

Solution. We seek a, b, and c so that

$$\begin{bmatrix} 2 \\ 6 \\ 4 \end{bmatrix} = a \begin{bmatrix} 1 \\ 2 \\ 1 \end{bmatrix} + b \begin{bmatrix} -1 \\ -1 \\ 0 \end{bmatrix} + c \begin{bmatrix} -1 \\ -3 \\ -2 \end{bmatrix}.$$

By 2.1.33, this is just $A \begin{bmatrix} a \\ b \\ c \end{bmatrix} = \begin{bmatrix} 2 \\ 6 \\ 4 \end{bmatrix}$, the system considered in Problem 2.3.14. There are many solutions, such as $\begin{bmatrix} a \\ b \\ c \end{bmatrix} = \begin{bmatrix} 4 \\ 2 \\ 0 \end{bmatrix}$.

So, for instance, $\begin{bmatrix} 2 \\ 6 \\ 4 \end{bmatrix} = 4\begin{bmatrix} 1 \\ 2 \\ 1 \end{bmatrix} + 2\begin{bmatrix} -1 \\ -1 \\ 0 \end{bmatrix} + 0\begin{bmatrix} -1 \\ -3 \\ -2 \end{bmatrix}$, the coefficients being the

components of the vector $\begin{bmatrix} 4 \\ 2 \\ 0 \end{bmatrix}$.

2.3.16 PROBLEM Is $\begin{bmatrix} 1 \\ -6 \end{bmatrix}$ a linear combination of $\begin{bmatrix} 1 \\ -4 \end{bmatrix}$, $\begin{bmatrix} 2 \\ 0 \end{bmatrix}$, and $\begin{bmatrix} 3 \\ -1 \end{bmatrix}$?

Solution. The question asks if there exist scalars a, b, c so that

$$\begin{bmatrix} 1 \\ -6 \end{bmatrix} = a\begin{bmatrix} 1 \\ -4 \end{bmatrix} + b\begin{bmatrix} 2 \\ 0 \end{bmatrix} + c\begin{bmatrix} 3 \\ -1 \end{bmatrix} = \begin{bmatrix} 1 & 2 & 3 \\ -4 & 0 & -1 \end{bmatrix}\begin{bmatrix} a \\ b \\ c \end{bmatrix}.$$

So we want to know whether there exists a solution to $A\mathbf{x} = \mathbf{b}$, with $\mathbf{x} = \begin{bmatrix} a \\ b \\ c \end{bmatrix}$ and

$\mathbf{b} = \begin{bmatrix} 1 \\ -6 \end{bmatrix}$. We apply Gaussian elimination to the augmented matrix as follows:

$$[A|\mathbf{b}] = \begin{bmatrix} 1 & 2 & 3 & | & 1 \\ -4 & 0 & -1 & | & -6 \end{bmatrix} \rightarrow \begin{bmatrix} 1 & 2 & 3 & | & 1 \\ 0 & 8 & 11 & | & -2 \end{bmatrix}.$$

This is row echelon form. There is one column that is not a pivot column: column three. Thus the third variable, c, is free. There are infinitely many solutions. Yes, the vector $\begin{bmatrix} 1 \\ -6 \end{bmatrix}$ is a linear combination of the other three.

Reading Challenge 7 *Find specific numbers a, b, and c so that* $\begin{bmatrix} 1 \\ -6 \end{bmatrix} = a\begin{bmatrix} 1 \\ -4 \end{bmatrix} + b\begin{bmatrix} 2 \\ 0 \end{bmatrix} + c\begin{bmatrix} 3 \\ -1 \end{bmatrix}$.

Answers to Reading Challenges

1. The two normals, $\begin{bmatrix} 2 \\ -3 \\ 6 \end{bmatrix}$ and $\begin{bmatrix} 1 \\ 1 \\ -2 \end{bmatrix}$, are not parallel.

2. $A\begin{bmatrix} 1 \\ -4 \\ 0 \end{bmatrix} = \begin{bmatrix} 2 & -3 & 6 \\ 1 & 1 & -2 \end{bmatrix}\begin{bmatrix} 1 \\ -4 \\ 0 \end{bmatrix} = \begin{bmatrix} 14 \\ -3 \end{bmatrix} = \mathbf{b}$;

$A\begin{bmatrix} 0 \\ 2 \\ 1 \end{bmatrix} = \begin{bmatrix} 2 & -3 & 6 \\ 1 & 1 & -2 \end{bmatrix}\begin{bmatrix} 0 \\ 2 \\ 1 \end{bmatrix} = \begin{bmatrix} 0 \\ 0 \end{bmatrix}.$

3. Setting $x = y$ in the first equation, we obtain $2x^2 = 8$, so $x^2 = 4$. The system has *two* solutions, $x = 2$, $y = 2$ and $x = -2$, $y = -2$.

4. Since $A \to \begin{bmatrix} -3 & 6 & 4 \\ 0 & 0 & \frac{8}{3} \end{bmatrix} = U$, the pivots are -3 and $\frac{8}{3}$, and the pivot columns are columns one and three.

5. The 1 in the $(2, 3)$ position of $\begin{bmatrix} 2 & 0 & 7 \\ 0 & 0 & 1 \\ 0 & 0 & 1 \end{bmatrix}$ is a pivot, so the number below it should be a 0.

The leading nonzero entries in $\begin{bmatrix} 6 & 0 & 3 & 2 \\ 0 & 0 & 5 & 4 \\ 0 & 1 & 0 & 2 \end{bmatrix}$ are 6, 5, and 1, but these do not step from left to right as you read down the matrix.

The zero row of $\begin{bmatrix} 3 & 0 & -1 & 5 \\ 0 & 0 & 0 & 0 \\ 0 & -4 & 2 & -3 \end{bmatrix}$ is not at the bottom of the matrix.

6. There are pivots in columns two and four. There are no pivots in columns one and three, so the corresponding variables x_1 and x_3 are free.

7. We must solve the system whose augmented matrix has row echelon form $\begin{bmatrix} 1 & 2 & 3 & 1 \\ 0 & 8 & 11 & -2 \end{bmatrix}$. Thus $c = t$ is free, $8b = -2 - 11c$, so $b = -\frac{1}{4} - \frac{11}{8}t$ and $a + 2b + 3c = 1$, so that $a = 1 - 2b - 3c = 1 - 2(-\frac{1}{4} - \frac{11}{8}t) - 3t = \frac{3}{2} - \frac{1}{4}t$.
One particular solution is obtained with $t = 0$, giving $a = \frac{3}{2}$, $b = -\frac{1}{4}$, $c = 0$.
Checking our answer, we have $\begin{bmatrix} 1 \\ -6 \end{bmatrix} = \frac{3}{2}\begin{bmatrix} 1 \\ -4 \end{bmatrix} - \frac{1}{4}\begin{bmatrix} 2 \\ 0 \end{bmatrix} + 0\begin{bmatrix} 3 \\ -1 \end{bmatrix}$.

True/False Questions

Decide, with as little calculation as possible, whether each of the following statements is true or false and explain your answer whenever you say "false." (Answers can be found in the back of the book.)

1. $\pi x_1 + \sqrt{3}x_2 + 5\sqrt{x_3} = \ln 7$ is a linear equation.

2. The systems

$$\begin{array}{rcr} x - y - z &=& 0 \\ -4x + y + 2z &=& -2 \\ 2x + y - z &=& 3 \end{array} \quad \text{and} \quad \begin{array}{rcr} x - y - z &=& 0 \\ -4x + y + 2z &=& -2 \\ 3y + z &=& 3 \end{array}$$

have the same solutions.

3. It is conceivable that a system of linear equations could have exactly two (different) solutions.

4. The matrix $\begin{bmatrix} 0 & 4 & 3 & -4 & 1 \\ 0 & 0 & 0 & 5 & 1 \\ 0 & 0 & 0 & 0 & -1 \end{bmatrix}$ is in row echelon form.

5. If a matrix A has row echelon form U, the pivot columns of A are the pivot columns of U.

6. If Gaussian elimination on the augmented matrix of a system of linear equations leads to the matrix $\begin{bmatrix} 0 & 1 & 3 & -4 & 1 \\ 0 & 0 & 0 & 1 & 1 \\ 0 & 0 & 0 & 0 & 1 \end{bmatrix}$, the system has a unique solution.

7. In the solution to a system of linear equations, free variables are those that correspond to columns of row echelon form that contain the pivots.

8. If a linear system $A\mathbf{x} = \mathbf{b}$ has a solution, then $\mathbf{b}$ is a linear combination of the columns of A.

9. The pivots of the matrix $\begin{bmatrix} 4 & 0 & 2 \\ 0 & 7 & 3 \\ 0 & 0 & 1 \end{bmatrix}$ are 4, 7, and 1.

10. The pivots of the matrix $\begin{bmatrix} 4 & 0 & 2 \\ 0 & 7 & 1 \\ 0 & 0 & 0 \end{bmatrix}$ are 4, 7, and 0.

Exercises

Solutions to exercises marked [BB] can be found in the Back of the Book.

1. Which of the following matrices are *not* in row echelon form? Give a reason in each case.

   (a) [BB] $\begin{bmatrix} 0 & 1 \\ 1 & 0 \end{bmatrix}$ (b) $\begin{bmatrix} 1 & 2 & 3 \\ 0 & 1 & 0 \\ 0 & 0 & 1 \end{bmatrix}$

   (c) $\begin{bmatrix} 1 & 5 \\ 0 & 0 \\ 0 & 1 \\ 0 & 0 \\ 0 & 0 \end{bmatrix}$ (d) $\begin{bmatrix} 1 & 2 & 3 & 4 \\ 0 & 0 & -1 & 10 \\ 0 & 0 & 0 & 0 \end{bmatrix}$.

2. Reduce each of the following matrices to row echelon form. In each case, identify the pivots and the pivot columns.

   (a) [BB] $\begin{bmatrix} 1 & -1 & -2 \\ 2 & -3 & -5 \\ -1 & 4 & 5 \end{bmatrix}$

   (b) $\begin{bmatrix} 2 & 1 & 4 & 2 \\ 3 & 0 & 2 & 1 \\ 5 & 2 & 3 & 5 \end{bmatrix}$

   (c) [BB] $\begin{bmatrix} 1 & 4 & 5 & 2 \\ 3 & 13 & 20 & 8 \\ -2 & -10 & -16 & -4 \\ 1 & 10 & 38 & 28 \end{bmatrix}$

   (d) $\begin{bmatrix} 1 & 2 & 1 & 3 & 1 \\ -1 & -1 & 2 & -1 & 4 \\ 2 & 8 & 15 & 11 & 26 \\ 0 & 2 & 8 & -2 & 21 \end{bmatrix}$

   (e) $\begin{bmatrix} 0 & -1 & 2 & 1 & 2 & 1 & -1 \\ 0 & 1 & -2 & 2 & 7 & 2 & 4 \\ 0 & -2 & 4 & 3 & 7 & 1 & 0 \\ 0 & 3 & -6 & 1 & 6 & 4 & 1 \end{bmatrix}$.

3. Repeat Exercise 2 for the matrices given below. Identify the pivot columns, but this time it is not necessary to identify the pivots.

   (a) [BB] $\begin{bmatrix} 3 & -1 & 5 & 3 & 6 & -3 \\ 7 & -7 & 1 & 5 & 4 & 2 \\ -4 & 6 & 4 & -2 & 4 & 10 \\ 16 & -10 & 16 & 14 & 22 & -10 \\ -13 & 9 & -11 & -11 & -16 & 12 \end{bmatrix}$

   (b) $\begin{bmatrix} 6 & -4 & 0 & -5 \\ 3 & -7 & -2 & 2 \\ -9 & 5 & 83 & 12 \\ -3 & 5 & 0 & 2 \\ -1 & 1 & 4 & -2 \\ 5 & 7 & 0 & 0 \end{bmatrix}$.

4. In each case, solve $A\mathbf{x} = \mathbf{b}$, expressing your solution as a vector or as a linear combination of vectors as appropriate.

   (a) [BB] $A = \begin{bmatrix} 2 & 2 & 2 \\ 4 & 6 & 6 \\ 6 & 6 & 10 \end{bmatrix}$, $\mathbf{x} = \begin{bmatrix} x_1 \\ x_2 \\ x_3 \end{bmatrix}$, $\mathbf{b} = \begin{bmatrix} 2 \\ 4 \\ 2 \end{bmatrix}$

   (b) $A = \begin{bmatrix} 3 & 0 & 6 \\ 3 & 1 & 9 \\ -2 & 4 & 10 \end{bmatrix}$, $\mathbf{x} = \begin{bmatrix} x_1 \\ x_2 \\ x_3 \end{bmatrix}$, $\mathbf{b} = \begin{bmatrix} 3 \\ 5 \\ 10 \end{bmatrix}$

   (c) $A = \begin{bmatrix} 1 & 0 & 1 & 2 & 0 \\ 2 & 2 & 4 & 6 & -4 \\ 0 & 1 & 1 & 5 & 6 \end{bmatrix}$, $\mathbf{x} = \begin{bmatrix} x_1 \\ x_2 \\ x_3 \\ x_4 \\ x_5 \end{bmatrix}$, $\mathbf{b} = \begin{bmatrix} 1 \\ 8 \\ 19 \end{bmatrix}$

   (d) $A = \begin{bmatrix} 2 & 2 & 2 & -8 \\ 4 & 6 & 6 & 0 \\ 6 & 6 & 10 & -4 \end{bmatrix}$, $\mathbf{x} = \begin{bmatrix} x_1 \\ x_2 \\ x_3 \\ x_4 \end{bmatrix}$, $\mathbf{b} = \begin{bmatrix} 2 \\ 4 \\ 2 \end{bmatrix}$.

5. [BB] In solving the system $Ax = b$, Gaussian elimination on the augmented matrix $[A|b]$ led to the row echelon matrix $\begin{bmatrix} 1 & -2 & 3 & -1 & | & 5 \end{bmatrix}$. The variables were x_1, x_2, x_3, x_4.

   (a) Circle the pivots. Identify the free variables.

   (b) Write the solution to the given system as a vector or as a linear combination of vectors, as appropriate.

6. In solving the system $Ax = b$, Gaussian elimination on the augmented matrix $[A|b]$ led to the row echelon matrix $\begin{bmatrix} 1 & 2 & 3 & | & 0 \\ 0 & 1 & -1 & | & 0 \end{bmatrix}$. The variables were x_1, x_2, x_3.

   (a) Circle the pivots. Identify the free variables.

   (b) Write the solution to the given system as a vector or as a linear combination of vectors, as appropriate.

7. In solving the system $Ax = b$, Gaussian elimination on the augmented matrix $[A|b]$ led to the row echelon matrix $\begin{bmatrix} 1 & -1 & 0 & 1 & | & 1 \\ 0 & 0 & 1 & 2 & | & 3 \\ 0 & 0 & 0 & 0 & | & 0 \end{bmatrix}$. The variables were x_1, x_2, x_3, x_4.

   (a) Circle the pivots. Identify the free variables.

   (b) Write the solution to the given system as a vector or as a linear combination of vectors, as appropriate.

8. Shown below are some matrices in row echelon form. Each represents the final row echelon matrix after Gaussian elimination was applied to the matrix $[A|b]$ in an attempt to solve $Ax = b$. Find the solution (in vector form) of $Ax = b$. In each case, if there is a solution, state whether this is unique or whether there are infinitely many solutions. In each case, $x = \begin{bmatrix} x_1 \\ x_2 \\ \vdots \end{bmatrix}$.

   (a) [BB] $\begin{bmatrix} 1 & 0 & 0 & 3 & | & 2 \\ 0 & 1 & 1 & 2 & | & 3 \\ 0 & 0 & 0 & 1 & | & \frac{1}{3} \\ 0 & 0 & 0 & 0 & | & 0 \end{bmatrix}$

   (b) $\begin{bmatrix} 1 & 6 & 0 & 3 & 0 & | & 0 \\ 0 & 0 & 1 & -4 & 0 & | & 5 \\ 0 & 0 & 0 & 0 & 1 & | & 7 \end{bmatrix}$

   (c) $\begin{bmatrix} 0 & 3 & 1 & 2 & | & 1 \\ 0 & 0 & 0 & 0 & | & 4 \\ 0 & 0 & 0 & 0 & | & 0 \end{bmatrix}$

   (d) $\begin{bmatrix} 1 & 4 & 0 & 2 & 3 & 0 & | & 2 \\ 0 & 0 & 1 & 1 & 0 & 5 & | & -3 \\ 0 & 0 & 0 & 0 & 1 & -4 & | & 0 \\ 0 & 0 & 0 & 0 & 0 & 1 & | & -7 \end{bmatrix}$.

9. Solve each of the following systems of linear equations by Gaussian elimination and back substitution. Write your answers as vectors or as linear combinations of vectors if appropriate.

   (a) [BB] $\begin{aligned} 2x - y + 2z &= -4 \\ 3x + 2y &= 1 \\ x + 3y - 6z &= 5 \end{aligned}$

   (b) $\begin{aligned} x + 3z &= 6 \\ 3x + 4y - z &= 4 \\ 2x + 5y - 4z &= -3 \end{aligned}$

   (c) [BB] $x - y + 2z = 4$

   (d) $\begin{aligned} -x_1 + x_2 + x_3 + 2x_4 &= 4 \\ 2x_3 - x_4 &= -7 \\ 3x_1 - 3x_2 - 7x_3 - 4x_4 &= 2 \end{aligned}$

   (e) [BB] $\begin{aligned} x + y + 7z &= 2 \\ 2x - 4y + 14z &= -1 \\ 5x + 11y - 7z &= 8 \\ 2x + 5y - 4z &= -3 \end{aligned}$

   (f) $\begin{aligned} 2x - y + z &= 3 \\ 4x - y - 2z &= 7 \end{aligned}$

   (g) $\begin{aligned} 2x - y + z &= 2 \\ 3x + y - 6z &= -9 \\ -x + 2y - 5z &= -4 \end{aligned}$

   (h) $\begin{aligned} 2x_1 + 4x_2 - 2x_3 &= 2 \\ 4x_1 + 9x_2 - 3x_3 &= 8 \\ -2x_1 - 3x_2 + 7x_3 &= 10 \end{aligned}$

   (i) $\begin{aligned} 3x_1 + x_2 &= -10 \\ x_1 + 3x_2 + x_3 &= 0 \\ x_2 + 3x_3 + x_4 &= 0 \\ x_3 + 3x_4 &= 10 \end{aligned}$

   (j) $\begin{aligned} x_1 + x_2 - x_4 &= 3 \\ x_1 - x_2 + 4x_3 + x_4 &= 5 \\ x_2 - 2x_3 - 2x_4 &= 3 \\ x_3 - 2x_4 &= 9 \end{aligned}$

   (k) $\begin{aligned} x - y + z - w &= 0 \\ 2x - 2z + 3w &= 11 \\ 5x - 2y + z - w &= 6 \\ -x + y + w &= 0 \end{aligned}$

(l) $\begin{aligned} 2x_1 - 3x_2 + 4x_3 - x_4 &= 5 \\ -x_1 + x_2 + x_4 &= -1 \\ 2x_2 - x_3 - 3x_4 &= 1 \\ 3x_1 + x_3 + 4x_4 &= 7 \end{aligned}$

(m) [BB] $\begin{aligned} -6x_1 + 8x_2 - 5x_3 - 5x_4 &= -11 \\ -6x_1 + 7x_2 - 10x_3 - 8x_4 &= -9 \\ -8x_1 + 10x_2 - 10x_3 - 9x_4 &= -13 \end{aligned}$

(n) $\begin{aligned} -x_1 + 2x_2 + 3x_3 + 5x_4 - x_5 &= 0 \\ -2x_1 + 5x_2 + 10x_3 + 13x_4 - 4x_5 &= -5 \\ -3x_1 + 7x_2 + 13x_3 + 19x_4 - 11x_5 &= 1 \\ -x_1 + 4x_2 + 11x_3 + 11x_4 - 5x_5 &= -10 \end{aligned}$

(o) $\begin{aligned} x_1 - 2x_2 + x_3 + x_4 + 4x_5 + 3x_6 &= 3 \\ -x_1 + x_2 + x_3 + 4x_6 &= 7 \\ -3x_1 + 5x_2 - x_3 - 2x_4 - 9x_5 &= 3 \\ -2x_1 + 4x_2 - 2x_3 - 2x_4 - 9x_5 - 4x_6 &= -4 \\ -5x_1 + 9x_2 - 3x_3 - 4x_4 - 16x_5 - 8x_6 &= -5. \end{aligned}$

10. [BB] Determine whether $\begin{bmatrix} 2 \\ -11 \\ -3 \end{bmatrix}$ is a linear

combination of $\begin{bmatrix} 0 \\ -1 \\ 5 \end{bmatrix}$ and $\begin{bmatrix} -1 \\ 4 \\ 9 \end{bmatrix}$.

11. Determine whether $\begin{bmatrix} 3 \\ -2 \\ 4 \end{bmatrix}$ is a linear combination of

$\begin{bmatrix} -1 \\ 2 \\ 3 \end{bmatrix}$, $\begin{bmatrix} 0 \\ 1 \\ 1 \end{bmatrix}$, and $\begin{bmatrix} 5 \\ -3 \\ 1 \end{bmatrix}$.

12. [BB] Determine whether $\begin{bmatrix} 1 \\ 6 \\ -4 \end{bmatrix}$ is a linear

combination of the columns of

$A = \begin{bmatrix} 2 & 3 & 4 \\ 4 & 7 & 5 \\ 6 & -1 & 9 \end{bmatrix}$.

13. Determine whether each of the given vectors is a linear combination of the columns of

$A = \begin{bmatrix} 2 & -1 \\ -1 & 4 \\ 5 & 9 \end{bmatrix}$.

(a) $\begin{bmatrix} 8 \\ -11 \\ -3 \end{bmatrix}$ (b) $\begin{bmatrix} 0 \\ -1 \\ 3 \end{bmatrix}$

(c) $\begin{bmatrix} 0 \\ 0 \\ 0 \end{bmatrix}$ (d) $\begin{bmatrix} -17 \\ 33 \\ 38 \end{bmatrix}$

14. Let $A = \begin{bmatrix} 2 & 5 \\ 1 & 3 \end{bmatrix}$ and $b = \begin{bmatrix} b_1 \\ b_2 \end{bmatrix}$.

(a) Solve the system $Ax = b$ for $x = \begin{bmatrix} x_1 \\ x_2 \end{bmatrix}$.

(b) Write b as a linear combination of the columns of A.

15. [BB] Let $v_1 = \begin{bmatrix} 2 \\ 0 \\ 2 \end{bmatrix}$, $v_2 = \begin{bmatrix} 1 \\ 1 \\ 1 \end{bmatrix}$, $v_3 = -\begin{bmatrix} -1 \\ 1 \\ -1 \end{bmatrix}$, and

$v_4 = \begin{bmatrix} 0 \\ 2 \\ 0 \end{bmatrix}$.

Determine whether every vector in $\mathbf{R}^3$ is a linear combination of v_1, v_2, v_3, and v_4.

16. Determine whether every vector in $\mathbf{R}^4$ is a linear combination of the given vectors.

(a) $v_1 = \begin{bmatrix} 1 \\ 1 \\ 1 \\ 1 \end{bmatrix}$, $v_2 = \begin{bmatrix} 1 \\ 1 \\ 1 \\ 0 \end{bmatrix}$, $v_3 = \begin{bmatrix} 1 \\ 1 \\ 0 \\ 0 \end{bmatrix}$, $v_4 = \begin{bmatrix} 1 \\ 0 \\ 0 \\ 0 \end{bmatrix}$

(b) $v_1 = \begin{bmatrix} 1 \\ 2 \\ 3 \\ 4 \end{bmatrix}$, $v_2 = \begin{bmatrix} 5 \\ 6 \\ 7 \\ 8 \end{bmatrix}$, $v_3 = \begin{bmatrix} 9 \\ 10 \\ 11 \\ 12 \end{bmatrix}$, $v_4 = \begin{bmatrix} 13 \\ 14 \\ 15 \\ 16 \end{bmatrix}$

17. [BB] The planes with equations $x + 2z = 5$ and $2x + y = 2$ intersect in a line.

(a) Find two points on that line.

(b) Find a vector in the direction of that line.

(c) Find an equation of the line.

18. Answer Exercise 17 for the planes with equations $x + 2y + 6z = 5$ and $x - y - 3z = -1$.

19. [BB] Suppose we want to find a quadratic polynomial $p(x) = a + bx + cx^2$ that passes through the points $(-2, 3)$, $(0, -11)$, $(5, 24)$.

(a) Write the system of equations that must be solved in order to determine a, b, and c.

(b) Find the desired polynomial.

20. Repeat Exercise 19 using the points $(3, 0)$, $(-1, 4)$, $(0, 6)$.

21. Find a polynomial of degree two that passes through the points $(-1, 1)$, $(0, 1)$, $(1, 4)$. Graph your polynomial showing clearly the given points.

22. Find a polynomial of degree three that passes through the points $(-2, 1)$, $(0, 3)$, $(3, -2)$, $(4, 5)$. Graph this polynomial showing clearly the given points.

23. Find values of a, b, and c such that the function $f(x) = a2^x + b2^{2x} + c2^{3x}$ passes through the points $(-1, 2)$, $(0, 0)$, and $(1, -2)$.

24. Find all matrices A and B such that
$$2A + 3B^T = \begin{bmatrix} 3 & -1 \\ 1 & 2 \end{bmatrix}.$$

25. [BB] Consider the system $\begin{aligned} 5x + 2y &= a \\ -15x - 6y &= b. \end{aligned}$

    (a) Under what conditions (on a and b), if any, does this system fail to have a solution?

    (b) Under what conditions, if any, does the system have a unique solution? Find any unique solution that may exist.

    (c) Under what conditions are there infinitely many solutions? Find these solutions if and when they exist.

26. Find conditions on a, b, and c (if any) such that the system
$$\begin{aligned} x \quad\quad + z &= -1 \\ 2x - y \quad\quad &= 2 \\ y + 2z &= -4 \\ ax + by + cz &= 3 \end{aligned}$$

    - has a unique solution;
    - has no solution;
    - has infinitely many solutions.

27. Find a condition or conditions on k so that the system
$$\begin{aligned} x + y + 2z &= 1 \\ 3x - y - 2z &= 3 \\ -x + 3y + 6z &= k \end{aligned}$$

    - has no solution;
    - has infinitely many solutions.

28. Find necessary and sufficient conditions on a, b, and c in order that the system
$$\begin{aligned} x + y - z &= a \\ 2x - 3y + 5z &= b \\ 5x \quad\quad + 2z &= c \end{aligned}$$

should have a solution. When there is a solution, is this unique? Explain.

29. **(a)** [BB] Find a condition on a, b, and c that is necessary and sufficient for the system
$$\begin{aligned} x - 2y &= a \\ -5x + 3y &= b \\ 3x + y &= c \end{aligned}$$
to have a solution.

    (b) [BB] Could the system in part (a) have infinitely many solutions? Explain.

30. Answer Exercise 29 for the system
$$\begin{aligned} x - 5y &= a \\ -2x + 6y &= b \\ 3x + 1y &= c. \end{aligned}$$

31. Find numbers a, b, and c so that the system
$$\begin{aligned} 7x_1 + x_2 - 4x_3 + 5x_4 &= a \\ x_1 + x_2 + 2x_3 - x_4 &= b \\ 3x_1 + x_2 \quad\quad + x_4 &= c \end{aligned}$$
does *not* have a solution.

■ Critical Reading

32. Find the equation of the line of intersection of the planes with equations $x - y + z = 3$ and $2x - 3y + z = 5$.

33. In the solution of $A\mathbf{x} = \mathbf{b}$, elementary row operations on the augmented matrix $[A|\mathbf{b}]$ produce the matrix
$$\left[\begin{array}{cccc|c} 1 & 2 & 3 & 4 & 5 \\ 0 & -3 & 4 & 0 & -1 \\ 0 & 0 & 2 & 4 & 0 \\ 0 & 0 & -1 & -2 & 0 \end{array}\right].$$
Does $A\mathbf{x} = \mathbf{b}$ have a unique solution, infinitely many solutions, or no solution?

34. If A is an invertible matrix, explain why the system $A\mathbf{x} = \mathbf{0}$ has only the trivial solution, $\mathbf{x} = \mathbf{0}$.

2.4 Homogeneous Systems and Linear Independence

In this section, we consider systems of linear equations of the form $A\mathbf{x} = \mathbf{0}$, in which the vector to the right of the $=$ sign is the zero vector. As astute readers might quickly observe, solving $A\mathbf{x} = \mathbf{0}$ is equivalent to determining whether the columns of the matrix A are linearly independent.

Reading Challenge 1   *Why?*

A system $A\mathbf{x} = \mathbf{0}$ of linear equations, the vector on the right being the zero vector, is called *homogeneous*. For example, the linear system

$$
\begin{aligned}
x_1 + \ x_2 + x_3 &= 0 \\
x_1 + 3x_2 \quad\quad &= 0 \\
2x_1 - \ x_2 - x_3 &= 0
\end{aligned}
$$

is homogeneous. When we speak of the homogeneous system *corresponding to* the system $A\mathbf{x} = \mathbf{b}$, we mean the system $A\mathbf{x} = \mathbf{0}$, with the same A.

For example, the homogeneous system that corresponds to the linear system

$$
\begin{aligned}
x_1 - \ x_2 + 2x_3 \quad\quad + 3x_5 - \ 5x_6 &= -1 \\
-3x_1 + 4x_2 - 2x_3 + \ x_4 - 7x_5 + 15x_6 &= \ \ 5 \\
2x_1 - 2x_2 + 4x_3 + \ x_4 + 5x_5 - \ 9x_6 &= -2 \\
x_1 - 2x_2 - 2x_3 - 2x_4 + 2x_5 - \ 5x_6 &= -2
\end{aligned}
\tag{1}
$$

is

$$
\begin{aligned}
x_1 - \ x_2 + 2x_3 \quad\quad + 3x_5 - \ 5x_6 &= 0 \\
-3x_1 + 4x_2 - 2x_3 + \ x_4 - 7x_5 + 15x_6 &= 0 \\
2x_1 - 2x_2 + 4x_3 + \ x_4 + 5x_5 - \ 9x_6 &= 0 \\
x_1 - 2x_2 - 2x_3 - 2x_4 + 2x_5 - \ 5x_6 &= 0.
\end{aligned}
\tag{2}
$$

When solving a homogeneous system of equations by Gaussian elimination, it is not necessary to augment the matrix of coefficients with a column of 0s, since the elementary row operations will not change these 0s. Thus, to solve the homogeneous system (2), we apply Gaussian elimination directly to the matrix of coefficients,

$$
\begin{bmatrix}
1 & -1 & 2 & 0 & 3 & -5 \\
-3 & 4 & -2 & 1 & -7 & 15 \\
2 & -2 & 4 & 1 & 5 & -9 \\
1 & -2 & -2 & -2 & 2 & -5
\end{bmatrix},
$$

proceeding like this:

$$
\begin{bmatrix}
1 & -1 & 2 & 0 & 3 & -5 \\
-3 & 4 & -2 & 1 & -7 & 15 \\
2 & -2 & 4 & 1 & 5 & -9 \\
1 & -2 & -2 & -2 & 2 & -5
\end{bmatrix}
\rightarrow
\begin{bmatrix}
1 & -1 & 2 & 0 & 3 & -5 \\
0 & 1 & 4 & 1 & 2 & 0 \\
0 & 0 & 0 & 1 & -1 & 1 \\
0 & -1 & -4 & -2 & -1 & 0
\end{bmatrix}
$$

$$
\rightarrow
\begin{bmatrix}
1 & -1 & 2 & 0 & 3 & -5 \\
0 & 1 & 4 & 1 & 2 & 0 \\
0 & 0 & 0 & 1 & -1 & 1 \\
0 & 0 & 0 & -1 & 1 & 0
\end{bmatrix}
\rightarrow
\begin{bmatrix}
1 & -1 & 2 & 0 & 3 & -5 \\
0 & 1 & 4 & 1 & 2 & 0 \\
0 & 0 & 0 & 1 & -1 & 1 \\
0 & 0 & 0 & 0 & 0 & 1
\end{bmatrix}.
$$

This is row echelon form. There are four pivot columns and two columns without pivots (columns three and five). The variables that correspond to the columns without pivots, x_3 and x_5, are free, and we so indicate by setting $x_3 = t$, $x_5 = s$ for parameters t and s. We interpret each row of the row echelon matrix as one side of an equation, the other side of which is 0. The corresponding equations are

$$
\begin{aligned}
x_1 - x_2 + 2x_3 \qquad + 3x_5 - x_6 &= 0 \\
x_2 + 4x_3 + x_4 + 2x_5 &= 0 \\
x_4 - x_5 + x_6 &= 0 \\
x_6 &= 0.
\end{aligned}
$$

Using back substitution, we obtain

$x_4 - x_5 + x_6 = 0$, so $x_4 = x_5 - x_6 = s$,

$x_2 + 4x_3 + x_4 + 2x_5 = 0$, so

$\qquad x_2 = -4x_3 - x_4 - 2x_5 = -4t - s - 2s = -4t - 3s$, and

$x_1 - x_2 + 2x_3 + 3x_5 - 5x_6 = 0$, so

$\qquad x_1 = x_2 - 2x_3 - 3x_5 + 5x_6 = -4t - 3s - 2t - 3s = -6t - 6s.$

The solution to the homogeneous system (2) is

$$
\mathbf{x} =
\begin{bmatrix}
-6t - 6s \\
-4t - 3s \\
t \\
s \\
s \\
0
\end{bmatrix}
= t
\begin{bmatrix}
-6 \\
-4 \\
1 \\
0 \\
0 \\
0
\end{bmatrix}
+ s
\begin{bmatrix}
-6 \\
-3 \\
0 \\
1 \\
1 \\
0
\end{bmatrix}. \tag{3}
$$

Interestingly, this solution is closely related to the solution of the *nonhomogeneous system* (1). To solve (1), we can employ the same sequence of steps in Gaussian elimination as we did before, except this time we start with the augmented matrix

$$
[A|\mathbf{b}] =
\left[
\begin{array}{cccccc|c}
1 & -1 & 2 & 0 & 3 & -5 & -1 \\
-3 & 4 & -2 & 1 & -7 & 15 & 5 \\
2 & -2 & 4 & 1 & 5 & -9 & -2 \\
1 & -2 & -2 & -2 & 2 & -5 & -2
\end{array}
\right].
$$

We proceed

$$[A|\mathbf{b}] \rightarrow \left[\begin{array}{cccccc|c} 1 & -1 & 2 & 0 & 3 & -5 & -1 \\ 0 & 1 & 4 & 1 & 2 & 0 & 2 \\ 0 & 0 & 0 & 1 & -1 & 1 & 0 \\ 0 & -1 & -4 & -2 & -1 & 0 & -1 \end{array}\right]$$

$$\rightarrow \left[\begin{array}{cccccc|c} 1 & -1 & 2 & 0 & 3 & -5 & -1 \\ 0 & 1 & 4 & 1 & 2 & 0 & 2 \\ 0 & 0 & 0 & 1 & -1 & 1 & 0 \\ 0 & 0 & 0 & -1 & 1 & 0 & 1 \end{array}\right] \rightarrow \left[\begin{array}{cccccc|c} 1 & -1 & 2 & 0 & 3 & -5 & -1 \\ 0 & 1 & 4 & 1 & 2 & 0 & 2 \\ 0 & 0 & 0 & 1 & -1 & 1 & 0 \\ 0 & 0 & 0 & 0 & 0 & 1 & 1 \end{array}\right].$$

This is row echelon form. As before, the free variables are $x_3 = t$ and $x_5 = s$. This time, by back substitution, we obtain

$x_6 = 1,$

$x_4 - x_5 + x_6 = 0$, so $x_4 = x_5 - x_6 = s - 1,$

$x_2 + 4x_3 + x_4 + 2x_5 = 2$, so
$\quad x_2 = 2 - 4x_3 - x_4 - 2x_5 = 2 - 4t - (s - 1) - 2s = 3 - 4t - 3s,$ and

$x_1 - x_2 + 2x_3 + 3x_5 - 5x_6 = -1$, so
$\quad\begin{aligned} x_1 &= -1 + x_2 - 2x_3 - 3x_5 + 5x_6 \\ &= -1 + (3 - 4t - 3s) - 2t - 3s + 5 = 7 - 6t - 6s. \end{aligned}$

In vector form, the solution is

$$\mathbf{x} = \begin{bmatrix} x_1 \\ x_2 \\ x_3 \\ x_4 \\ x_5 \\ x_6 \end{bmatrix} = \begin{bmatrix} 7 - 6t - 6s \\ 3 - 4t - 3s \\ t \\ -1 + s \\ s \\ 1 \end{bmatrix} = \begin{bmatrix} 7 \\ 3 \\ 0 \\ -1 \\ 0 \\ 1 \end{bmatrix} + t \begin{bmatrix} -6 \\ -4 \\ 1 \\ 0 \\ 0 \\ 0 \end{bmatrix} + s \begin{bmatrix} -6 \\ -3 \\ 0 \\ 1 \\ 1 \\ 0 \end{bmatrix}. \tag{4}$$

Part of this solution should seem familiar. The vector

$$\mathbf{x}_h = t \begin{bmatrix} -6 \\ -4 \\ 1 \\ 0 \\ 0 \\ 0 \end{bmatrix} + s \begin{bmatrix} -6 \\ -3 \\ 0 \\ 1 \\ 1 \\ 0 \end{bmatrix}$$

is the solution to the homogeneous system (2). The first vector in (4), namely,

$$\mathbf{x}_p = \begin{bmatrix} 7 \\ 3 \\ 0 \\ -1 \\ 0 \\ 1 \end{bmatrix},$$

is a *particular solution* to system (1); when $t = 0$ and $s = 0$, we obtain one specific solution to (1). Any other solution is $x_p + x_h$, the sum of a particular solution and a solution to the corresponding homogeneous system. This is one instance of a general pattern.

Theorem 2.4.1   *Suppose x_p is a particular solution to the system $Ax = b$. Then any solution to $Ax = b$ is the sum $x = x_p + x_h$ of x_p and a solution x_h to the homogeneous system $Ax = 0$.*

Proof   Suppose that x is a solution to $Ax = b$. We are given that x_p is another, so $Ax_p = b$ and $Ax = b$. It follows that $A(x - x_p) = 0$, so $x - x_p = x_h$ is a solution to $Ax = 0$. So $x = x_p + x_h$.   ∎

In Problem 2.3.5, we considered the linear system

$$\begin{aligned} 2x - 3y + 6z &= 14 \\ x + y - 2z &= -3, \end{aligned}$$

and showed that the solutions were all of the form

$$\begin{bmatrix} x \\ y \\ z \end{bmatrix} = \begin{bmatrix} 1 \\ 2z - 4 \\ z \end{bmatrix} = \begin{bmatrix} 1 \\ -4 \\ 0 \end{bmatrix} + t \begin{bmatrix} 0 \\ 2 \\ 1 \end{bmatrix}.$$

We noted that the vector $\begin{bmatrix} 1 \\ -4 \\ 0 \end{bmatrix}$ is one particular solution to the given system and

that the other vector, $\begin{bmatrix} 0 \\ 2 \\ 1 \end{bmatrix}$, is a solution to the corresponding homogeneous system.

Here $x_p = \begin{bmatrix} 1 \\ -4 \\ 0 \end{bmatrix}$ and $x_h = t \begin{bmatrix} 0 \\ 2 \\ 1 \end{bmatrix}$. (See also Problem 2.3.14.)

2.4.2 EXAMPLE   The system

$$\begin{aligned} x_1 + 2x_2 + 3x_3 + 4x_4 &= 1 \\ 2x_1 + 4x_2 + 8x_3 + 10x_4 &= 6 \\ 3x_1 + 7x_2 + 11x_3 + 14x_4 &= 7 \end{aligned}$$

is $Ax = b$ with

$$A = \begin{bmatrix} 1 & 2 & 3 & 4 \\ 2 & 4 & 8 & 10 \\ 3 & 7 & 11 & 14 \end{bmatrix}, \quad x = \begin{bmatrix} x_1 \\ x_2 \\ x_3 \\ x_4 \end{bmatrix}, \quad b = \begin{bmatrix} 1 \\ 6 \\ 7 \end{bmatrix}.$$

Row reduction of the augmented matrix to row echelon form proceeds

$$\begin{bmatrix} 1 & 2 & 3 & 4 & | & 1 \\ 2 & 4 & 8 & 10 & | & 6 \\ 3 & 7 & 11 & 14 & | & 7 \end{bmatrix} \rightarrow \begin{bmatrix} 1 & 2 & 3 & 4 & | & 1 \\ 0 & 0 & 2 & 2 & | & 4 \\ 0 & 1 & 2 & 2 & | & 4 \end{bmatrix} \rightarrow \begin{bmatrix} 1 & 2 & 3 & 4 & | & 1 \\ 0 & 1 & 2 & 2 & | & 4 \\ 0 & 0 & 1 & 1 & | & 2 \end{bmatrix},$$

so $x_4 = t$ is free,

$$x_3 = 2 - x_4 = 2 - t,$$
$$x_2 = 4 - 2x_3 - 2x_4 = 0, \text{ and}$$
$$x_1 = 1 - 2x_2 - 3x_3 - 4x_4 = -5 - t.$$

The solution to $A\mathbf{x} = \mathbf{b}$ is

$$\mathbf{x} = \begin{bmatrix} x_1 \\ x_2 \\ x_3 \\ x_4 \end{bmatrix} = \begin{bmatrix} -5 - t \\ 0 \\ 2 - t \\ t \end{bmatrix} = \begin{bmatrix} -5 \\ 0 \\ 2 \\ 0 \end{bmatrix} + t \begin{bmatrix} -1 \\ 0 \\ -1 \\ 1 \end{bmatrix}.$$

Here, $\mathbf{x}_p = \begin{bmatrix} -5 \\ 0 \\ 2 \\ 0 \end{bmatrix}$ and $\mathbf{x}_h = t \begin{bmatrix} -1 \\ 0 \\ -1 \\ 1 \end{bmatrix}.$ ■

Reading Challenge 2 *Verify that* $\mathbf{x}_p = \begin{bmatrix} -5 \\ 0 \\ 2 \\ 0 \end{bmatrix}$ *is a solution to* $A\mathbf{x} = \mathbf{b}$ *in* **2.4.2** *and that* $\mathbf{x}_h = t \begin{bmatrix} 1 \\ 0 \\ -1 \\ 1 \end{bmatrix}$ *is the*

solution to $A\mathbf{x} = \mathbf{0}.$

■ More on Linear Independence

Recall that vectors $\mathbf{x}_1, \mathbf{x}_2, \dots, \mathbf{x}_n$ are linearly independent if the condition

$$c_1 \mathbf{x}_1 + c_2 \mathbf{x}_2 + \cdots + c_n \mathbf{x}_n = \mathbf{0} \tag{5}$$

implies that $c_1 = 0$, $c_2 = 0$, $\dots$, $c_n = 0$. Vectors that are not linearly independent are linearly dependent, that is, there exists an equation of the form (5) where not all the coefficients $c_1, c_2, \dots, c_n$ are 0. Using 2.1.33, we see that the left side of (5) is just $A\mathbf{c}$, where A is the matrix whose columns are $\mathbf{x}_1, \mathbf{x}_2, \dots, \mathbf{x}_n$ and $\mathbf{c} = \begin{bmatrix} c_1 \\ c_2 \\ \vdots \\ c_n \end{bmatrix}.$

Thus $\mathbf{x}_1, \dots, \mathbf{x}_n$ are linearly independent if and only if $A\mathbf{c} = \mathbf{0}$ implies $\mathbf{c} = \mathbf{0}$. If $A\mathbf{c} = \mathbf{0}$ with $\mathbf{c} \neq \mathbf{0}$, the vectors are linearly dependent.

2.4.3 EXAMPLE The standard basis vectors of $\mathbf{R}^n$ are linearly independent. To see why, we suppose that some linear combination of these is the zero vector, say

$$c_1 \mathbf{e}_1 + c_2 \mathbf{e}_2 + \cdots + c_n \mathbf{e}_n = \mathbf{0}.$$

This is $A\mathbf{c} = \mathbf{0}$ where $\mathbf{c} = \begin{bmatrix} c_1 \\ c_2 \\ \vdots \\ c_n \end{bmatrix}$ and $A = \begin{bmatrix} \mathbf{e}_1 & \mathbf{e}_2 & \cdots & \mathbf{e}_n \\ \downarrow & \downarrow & & \downarrow \end{bmatrix}$ is the $n \times n$ identity matrix. Thus $A\mathbf{c} = I\mathbf{c} = \mathbf{0}$ certainly means $\mathbf{c} = \mathbf{0}$, so the vectors are linearly independent. ∎

2.4.4 PROBLEM   Decide whether the vectors $\mathbf{x}_1 = \begin{bmatrix} 2 \\ 3 \\ 1 \\ 4 \end{bmatrix}$, $\mathbf{x}_2 = \begin{bmatrix} -1 \\ 1 \\ 2 \\ 3 \end{bmatrix}$, and $\mathbf{x}_3 = \begin{bmatrix} 4 \\ 0 \\ -1 \\ 1 \end{bmatrix}$ are linearly independent or linearly dependent.

Solution.   Suppose $c_1\mathbf{x}_1 + c_2\mathbf{x}_2 + c_3\mathbf{x}_3 = \mathbf{0}$. This is the homogeneous system $A\mathbf{c} = \mathbf{0}$ with $A = \begin{bmatrix} 2 & -1 & 4 \\ 3 & 1 & 0 \\ 1 & 2 & -1 \\ 4 & 3 & 1 \end{bmatrix}$ and $\mathbf{c} = \begin{bmatrix} c_1 \\ c_2 \\ c_3 \end{bmatrix}$. Using Gaussian elimination, we obtain

$$\begin{bmatrix} 2 & -1 & 4 \\ 3 & 1 & 0 \\ 1 & 2 & -1 \\ 4 & 3 & 1 \end{bmatrix} \rightarrow \begin{bmatrix} 1 & 2 & -1 \\ 0 & -5 & 6 \\ 0 & -5 & 3 \\ 0 & -5 & 5 \end{bmatrix} \rightarrow \begin{bmatrix} 1 & 2 & -1 \\ 0 & 1 & -1 \\ 0 & 0 & 1 \\ 0 & 0 & -2 \end{bmatrix} \rightarrow \begin{bmatrix} 1 & 2 & -1 \\ 0 & 1 & -1 \\ 0 & 0 & 1 \\ 0 & 0 & 0 \end{bmatrix}.$$

Back substitution gives $c_3 = 0$, then $c_2 - c_3 = 0$, so $c_2 = 0$ and, finally, $c_1 + 2c_2 - c_3 = 0$, so $c_1 = 0$. Thus $\mathbf{c} = \mathbf{0}$, so the vectors are linearly independent.

Answers to Reading Challenges

1. $A\mathbf{x}$ is a linear combination of the columns of A.

2. $A\mathbf{x}_p = \begin{bmatrix} 1 & 2 & 3 & 4 \\ 2 & 4 & 8 & 10 \\ 3 & 7 & 11 & 14 \end{bmatrix} \begin{bmatrix} -5 \\ 0 \\ 2 \\ 0 \end{bmatrix} = \begin{bmatrix} 1 \\ 6 \\ 7 \end{bmatrix}$;

$A\mathbf{x}_h = t \begin{bmatrix} 1 & 2 & 3 & 4 \\ 2 & 4 & 8 & 10 \\ 3 & 7 & 11 & 14 \end{bmatrix} \begin{bmatrix} -1 \\ 0 \\ -1 \\ 1 \end{bmatrix} = \begin{bmatrix} 0 \\ 0 \\ 0 \end{bmatrix}$.

True/False Questions

Decide, with as little calculation as possible, whether each of the following statements is true or false and explain your answer whenever you say "false." (Answers can be found in the back of the book.)

1. The system $\begin{aligned} -2x_1 \quad\quad\;\; - \;\; x_3 &= 0 \\ 5x_2 + 6x_3 &= 0 \\ 7x_1 + 5x_2 - 4x_3 &= 0 \end{aligned}$ is homogeneous.

2. If a 4×4 matrix A has row echelon form $U = \begin{bmatrix} 1 & 2 & 3 & 4 \\ 0 & 1 & 5 & -1 \\ 0 & 0 & 1 & 7 \\ 0 & 0 & 0 & 0 \end{bmatrix}$, the system $A\mathbf{x} = \mathbf{0}$ has only the trivial solution.

3. If $\mathbf{u}$ and $\mathbf{v}$ are solutions to $A\mathbf{x} = \mathbf{b}$, then $\mathbf{u} - \mathbf{v}$ is a solution to the corresponding homogeneous system.

4. If vectors $\mathbf{u}$, $\mathbf{v}$, and $\mathbf{w}$ are linearly dependent and A is the matrix whose columns are these vectors, then the homogeneous system $A\mathbf{x} = \mathbf{0}$ has a nontrivial solution.

Exercises

*Solutions to exercises marked [BB] can be found in the **B**ack of the **B**ook.*

1. Solve each of the following systems of linear equations by Gaussian elimination and back substitution. Write your answers as vectors or as linear combinations of vectors if appropriate.

   (a) [BB]
   $$\begin{aligned} x_1 + x_2 + x_3 &= 0 \\ x_1 + 3x_2 \quad\quad &= 0 \\ 2x_1 - x_2 - x_3 &= 0 \end{aligned}$$

   (b)
   $$\begin{aligned} x - 2y + z &= 0 \\ 3x - 7y + 2z &= 0 \end{aligned}$$

   (c)
   $$\begin{aligned} 2x_1 - 7x_2 + x_3 + x_4 &= 0 \\ x_1 - 2x_2 + x_3 \quad\quad &= 0 \\ 3x_1 + 6x_2 + 7x_3 - 4x_4 &= 0 \end{aligned}$$

   (d)
   $$\begin{aligned} x_1 - 2x_2 + x_3 + 3x_4 - x_5 &= 0 \\ 4x_1 - 7x_2 + 5x_3 + 16x_4 - 2x_5 &= 0 \\ -3x_1 + 4x_2 - 4x_3 - 18x_4 + 6x_5 &= 0 \end{aligned}$$

2. (a) [BB] Express the zero vector $\mathbf{0} = \begin{bmatrix} 0 \\ 0 \\ 0 \end{bmatrix}$ as a nontrivial linear combination of the columns of
   $$A = \begin{bmatrix} 1 & -1 & 1 \\ 0 & 1 & 1 \\ 1 & 0 & 2 \end{bmatrix}.$$

   (b) [BB] Use part (a) to show that A is not invertible.

3. Determine whether each of the given sets of vectors is linearly independent or linearly dependent.

   (a) [BB] $\mathbf{v}_1 = \begin{bmatrix} 1 \\ 1 \\ 2 \\ 1 \end{bmatrix}$, $\mathbf{v}_2 = \begin{bmatrix} 3 \\ 3 \\ 3 \\ 1 \end{bmatrix}$, $\mathbf{v}_3 = \begin{bmatrix} 2 \\ 5 \\ 10 \\ 4 \end{bmatrix}$, $\mathbf{v}_4 = \begin{bmatrix} 1 \\ 2 \\ 3 \\ 1 \end{bmatrix}$

   (b) $\mathbf{v}_1 = \begin{bmatrix} -1 \\ 5 \\ -1 \\ 3 \\ 1 \end{bmatrix}$, $\mathbf{v}_2 = \begin{bmatrix} -2 \\ 9 \\ 1 \\ 6 \\ 1 \end{bmatrix}$, $\mathbf{v}_3 = \begin{bmatrix} 3 \\ -11 \\ -1 \\ -9 \\ -3 \end{bmatrix}$

   (c) $\mathbf{v}_1 = \begin{bmatrix} 1 \\ -1 \\ -1 \\ -4 \end{bmatrix}$, $\mathbf{v}_2 = \begin{bmatrix} -1 \\ 2 \\ -1 \\ 1 \end{bmatrix}$, $\mathbf{v}_3 = \begin{bmatrix} 2 \\ 4 \\ -2 \\ 6 \end{bmatrix}$, $\mathbf{v}_4 = \begin{bmatrix} 3 \\ 5 \\ -1 \\ 7 \end{bmatrix}$

   (d) $\mathbf{v}_1 = \begin{bmatrix} 1 \\ 2 \\ 3 \end{bmatrix}$, $\mathbf{v}_2 = \begin{bmatrix} -3 \\ 0 \\ 1 \end{bmatrix}$, $\mathbf{v}_3 = \begin{bmatrix} 4 \\ 2 \\ 2 \end{bmatrix}$, $\mathbf{v}_4 = \begin{bmatrix} 1 \\ 1 \\ 13 \end{bmatrix}$

   (e) $\mathbf{v}_1 = \begin{bmatrix} 4 \\ 5 \\ 1 \\ 7 \end{bmatrix}$, $\mathbf{v}_2 = \begin{bmatrix} 8 \\ 6 \\ 10 \\ 8 \end{bmatrix}$, $\mathbf{v}_3 = \begin{bmatrix} 4 \\ 2 \\ -3 \\ 4 \end{bmatrix}$

   (f) $\mathbf{v}_1 = \begin{bmatrix} -2 \\ 1 \\ 6 \\ 1 \\ 1 \end{bmatrix}$, $\mathbf{v}_2 = \begin{bmatrix} 0 \\ -1 \\ 2 \\ 9 \\ 7 \end{bmatrix}$, $\mathbf{v}_3 = \begin{bmatrix} 1 \\ 3 \\ 4 \\ 0 \\ -1 \end{bmatrix}$, $\mathbf{v}_4 = \begin{bmatrix} 1 \\ 1 \\ 1 \\ 2 \\ 1 \end{bmatrix}$.

4. [BB] Find conditions on a, b, and c (if any) such that the system
   $$\begin{aligned} x + y \quad\quad &= 0 \\ y + z &= 0 \\ x \quad\quad - z &= 0 \\ ax + by + cz &= 0 \end{aligned}$$
   has

   - a unique solution;

   - no solution;

   - an infinite number of solutions.

5. A homogeneous system of four linear equations in the six unknowns $x_1, x_2, x_3, x_4, x_5, x_6$ has coefficient matrix A whose row echelon form is

$$\begin{bmatrix} 1 & 2 & 0 & 3 & 1 & 1 \\ 0 & 0 & 0 & 1 & -2 & 0 \\ 0 & 0 & 0 & 0 & 0 & 1 \\ 0 & 0 & 0 & 0 & 0 & 0 \end{bmatrix}.$$

Solve the system.

6. Write the solution to each of the following linear systems in the form $\mathbf{x} = \mathbf{x}_p + \mathbf{x}_h$ where $\mathbf{x}_p$ is a particular solution and $\mathbf{x}_h$ is a solution of the corresponding homogeneous system.

(a) [BB] $\begin{aligned} x_1 + 5x_2 + 7x_3 &= -2 \\ 9x_3 &= 3 \end{aligned}$

(b) $\begin{aligned} 3x_1 + 6x_2 \quad\; - 2x_4 + 7x_5 &= -1 \\ -2x_1 - 4x_2 + x_3 + 4x_4 - 9x_5 &= 0 \\ x_1 + 2x_2 \quad\; - x_4 + 2x_5 &= 4 \end{aligned}$

(c) $\begin{aligned} -x_1 \qquad\qquad - x_4 + 2x_5 &= -2 \\ 3x_1 + x_2 + 2x_3 + 4x_4 - x_5 &= 1 \\ 4x_1 - x_2 + x_3 \qquad + 2x_5 &= 1 \end{aligned}$

(d)
$$\begin{aligned} x_1 + 2x_2 - x_3 \qquad + 3x_5 + x_6 &= 0 \\ -2x_1 - 4x_2 + 3x_3 - 3x_4 - 6x_5 - 2x_6 &= 2 \\ 3x_1 + 6x_2 - x_3 - 6x_4 + 11x_5 + 7x_6 &= 2 \\ x_1 + 2x_2 - 2x_3 + 3x_4 \qquad + 2x_6 &= -1 \end{aligned}$$

7. Consider the following system of linear equations.

$$\begin{aligned} x_1 - 2x_2 + 3x_3 + x_4 + 3x_5 + 4x_6 &= -1 \\ -3x_1 + 6x_2 - 8x_3 + 2x_4 - 11x_5 - 15x_6 &= 2 \\ x_1 - 2x_2 + 2x_3 - 4x_4 + 6x_5 + 9x_6 &= 3 \\ -2x_1 + 4x_2 - 6x_3 - 2x_4 - 6x_5 - 7x_6 &= 1 \end{aligned}$$

(a) Write this system in the form $A\mathbf{x} = \mathbf{b}$.

(b) Solve the system expressing your answer as a vector $\mathbf{x} = \mathbf{x}_p + \mathbf{x}_h$, where $\mathbf{x}_p$ is a particular solution and $\mathbf{x}_h$ is a solution to $A\mathbf{x} = \mathbf{0}$.

(c) Express $\begin{bmatrix} -1 \\ 2 \\ 3 \\ 1 \end{bmatrix}$ as a linear combination of the columns of A.

8. Let $\mathbf{v}_1 = \begin{bmatrix} 2 \\ 0 \\ 2 \end{bmatrix}$, $\mathbf{v}_2 = \begin{bmatrix} 1 \\ 1 \\ 1 \end{bmatrix}$, $\mathbf{v}_3 = \begin{bmatrix} -1 \\ 1 \\ -1 \end{bmatrix}$, and $\mathbf{v}_4 = \begin{bmatrix} 0 \\ 2 \\ 0 \end{bmatrix}$. Determine whether the vectors $\mathbf{v}_1, \mathbf{v}_2, \mathbf{v}_3, \mathbf{v}_4$ are linearly independent.

9. Find a condition on k which is *necessary and sufficient* for the vectors $\begin{bmatrix} 1 \\ -1 \\ 0 \end{bmatrix}$, $\begin{bmatrix} k \\ 1 \\ 0 \end{bmatrix}$, and $\begin{bmatrix} 0 \\ 2 \\ 3 \end{bmatrix}$ to be linearly independent; that is, complete the following sentence:

"The vectors $\begin{bmatrix} 1 \\ -1 \\ 0 \end{bmatrix}$, $\begin{bmatrix} k \\ 1 \\ 0 \end{bmatrix}$ and $\begin{bmatrix} 0 \\ 2 \\ 3 \end{bmatrix}$ are linearly independent if and only if _____ ."

Gaussian elimination proceeds

$$\begin{bmatrix} 1 & k & 0 \\ -1 & 1 & 2 \\ 0 & 0 & 3 \end{bmatrix} \rightarrow \begin{bmatrix} 1 & k & 0 \\ 0 & k+1 & 2 \\ 0 & 0 & 3 \end{bmatrix}.$$

If $k \neq -1$, we may continue, $\rightarrow \begin{bmatrix} 1 & k & 0 \\ 0 & 1 & \frac{2}{k+1} \\ 0 & 0 & 1 \end{bmatrix}$

and obtain the unique solution $c_1 = c_2 = c_3 = 0$. So, if $k \neq -1$, the vectors are linearly independent. On the other hand, if $k = -1$, then

$$1\begin{bmatrix} 1 \\ -1 \\ 0 \end{bmatrix} + 1\begin{bmatrix} k \\ 1 \\ 0 \end{bmatrix} + 0\begin{bmatrix} 0 \\ 2 \\ 3 \end{bmatrix} = \mathbf{0},$$ so the vectors are not linearly independent. In summary, the vectors are linearly independent if and only if $k \neq -1$.

■ Critical Reading

10. Suppose the homogeneous system $A\mathbf{x} = \mathbf{0}$ has a nontrivial solution $\mathbf{x}_0$. Show that the system has infinitely many nontrivial solutions.

11. Suppose the quadratic polynomial $ax^2 + bx + c$ is 0 when evaluated at three different values x_1, x_2, x_3 of x. Show that $a = b = c = 0$.

12. Suppose A and B are matrices with linearly independent columns. Prove that AB also has linearly independent columns (assuming this matrix product is defined).

2.5 The LU Factorization of a Matrix

The three elementary row operations on a matrix are

1. the interchange of two rows;
2. multiplication of a row by a nonzero scalar; and
3. replacing a row by that row less a multiple of another.

2.5.1 DEFINITION

An *elementary matrix* is a square matrix obtained from the identity matrix by a single elementary row operation.

2.5.2 EXAMPLES

- $E_1 = \begin{bmatrix} 0 & 1 & 0 \\ 1 & 0 & 0 \\ 0 & 0 & 1 \end{bmatrix}$ is an elementary matrix, obtained from the 3×3 identity matrix I by interchanging rows one and two.

- The matrix $E_2 = \begin{bmatrix} 1 & 0 & 0 \\ 0 & 3 & 0 \\ 0 & 0 & 1 \end{bmatrix}$ is elementary; it is obtained from I by multiplying row two by 3.

- The matrix $E_3 = \begin{bmatrix} 1 & 0 & 0 \\ 0 & 1 & 0 \\ -4 & 0 & 1 \end{bmatrix}$ is elementary; it is obtained from I by replacing row three by row three minus 4 times row one. ∎

The next observation is very useful.

2.5.3 If A is a matrix and E is an elementary matrix with EA defined, then EA is that matrix obtained by applying to A the elementary operation that defined E.

For some reason, this always reminds the author of the golden rule. An elementary matrix E does unto others what was done unto the identity when E was created.

To illustrate, let

$$A = \begin{bmatrix} a & d \\ b & e \\ c & f \end{bmatrix}$$

and leave E_1, E_2, and E_3 as above. The matrix $E_1 A$ is

$$E_1 A = \begin{bmatrix} 0 & 1 & 0 \\ 1 & 0 & 0 \\ 0 & 0 & 1 \end{bmatrix} \begin{bmatrix} a & d \\ b & e \\ c & f \end{bmatrix} = \begin{bmatrix} b & e \\ a & d \\ c & f \end{bmatrix},$$

which is just A with rows one and two interchanged. The matrix $E_2 A$ is

$$E_2 A = \begin{bmatrix} 1 & 0 & 0 \\ 0 & 3 & 0 \\ 0 & 0 & 1 \end{bmatrix} \begin{bmatrix} a & d \\ b & e \\ c & f \end{bmatrix} = \begin{bmatrix} a & d \\ 3b & 3e \\ c & f \end{bmatrix},$$

which is A with its second row multiplied by 3, and

$$E_3 A = \begin{bmatrix} 1 & 0 & 0 \\ 0 & 1 & 0 \\ -4 & 0 & 1 \end{bmatrix} \begin{bmatrix} a & d \\ b & e \\ c & f \end{bmatrix} = \begin{bmatrix} a & d \\ b & e \\ -4a + c & -4d + f \end{bmatrix},$$

which is A, but with row three replaced by row three minus 4 times row one. The inverse of an elementary matrix is another elementary matrix.

> **2.5.4** If E is an elementary matrix, then E has an inverse that is also an elementary matrix, namely, that elementary matrix that reverses (or undoes) what E does.

We illustrate with a few examples.

2.5.5 EXAMPLE $E_1 = \begin{bmatrix} 0 & 1 & 0 \\ 1 & 0 & 0 \\ 0 & 0 & 1 \end{bmatrix}$ is the elementary matrix that interchanges rows one and two of a matrix. To undo such an interchange, we interchange rows one and two again. According to 2.5.4, we should have $E_1^{-1} = E_1$. Since E_1 is square, to prove that E_1 is the inverse of E_1 it is enough to show that $E_1 E_1 = I$. This follows from 2.5.3 since $E_1 E_1$ is E_1 with its first two rows interchanged. Alternatively, we can compute

$$E_1 E_1 = \begin{bmatrix} 0 & 1 & 0 \\ 1 & 0 & 0 \\ 0 & 0 & 1 \end{bmatrix} \begin{bmatrix} 0 & 1 & 0 \\ 1 & 0 & 0 \\ 0 & 0 & 1 \end{bmatrix} = \begin{bmatrix} 1 & 0 & 0 \\ 0 & 1 & 0 \\ 0 & 0 & 1 \end{bmatrix} = I. \qquad \blacksquare$$

2.5.6 EXAMPLE   To multiply row two of a matrix by 3, multiply (on the left) by the elementary matrix $E_2 = \begin{bmatrix} 1 & 0 & 0 \\ 0 & 3 & 0 \\ 0 & 0 & 1 \end{bmatrix}$. To undo the effect of this operation, we should multiply row two by $\frac{1}{3}$. According to 2.5.4, E_2^{-1} should be $B = \begin{bmatrix} 1 & 0 & 0 \\ 0 & \frac{1}{3} & 0 \\ 0 & 0 & 1 \end{bmatrix}$, the elementary matrix that multiplies row two of a matrix by $\frac{1}{3}$. Again, 2.5.3 helps, for it says that $E_2 B$ is B with row two multiplied by 3. Alternatively, we can compute

$$E_2 B = \begin{bmatrix} 1 & 0 & 0 \\ 0 & 3 & 0 \\ 0 & 0 & 1 \end{bmatrix} \begin{bmatrix} 1 & 0 & 0 \\ 0 & \frac{1}{3} & 0 \\ 0 & 0 & 1 \end{bmatrix} = \begin{bmatrix} 1 & 0 & 0 \\ 0 & 1 & 0 \\ 0 & 0 & 1 \end{bmatrix} = I. \qquad \blacksquare$$

2.5.7 EXAMPLE   Multiplying a matrix by the elementary matrix $E_3 = \begin{bmatrix} 1 & 0 & 0 \\ 0 & 1 & 0 \\ -4 & 0 & 1 \end{bmatrix}$ replaces row three of that matrix by row three minus 4 times row one. To undo this operation, we should replace row three by row three plus 4 times row one. Apparently then, the inverse of E_3 is $B = \begin{bmatrix} 1 & 0 & 0 \\ 0 & 1 & 0 \\ 4 & 0 & 1 \end{bmatrix}$. Since $E_3 B$ is B, but with row three replaced by that row minus 4 times row one, $E_3 B = I$. Alternatively, we can compute directly

$$E_3 B = \begin{bmatrix} 1 & 0 & 0 \\ 0 & 1 & 0 \\ -4 & 0 & 1 \end{bmatrix} \begin{bmatrix} 1 & 0 & 0 \\ 0 & 1 & 0 \\ 4 & 0 & 1 \end{bmatrix} = \begin{bmatrix} 1 & 0 & 0 \\ 0 & 1 & 0 \\ 0 & 0 & 1 \end{bmatrix} = I. \qquad \blacksquare$$

What can one do with these ideas about elementary matrices?

2.5.8 EXAMPLE

We find a row echelon form of the matrix $A = \begin{bmatrix} -1 & 1 & -2 \\ 2 & 1 & 7 \end{bmatrix}$, making note of the elementary matrices that effect the row operations we use in the Gaussian elimination process.

To begin, multiply the first row of A by -1. This can be achieved by multiplying (on the left) by a certain elementary matrix, which must be 2×2 since A is 2×3. The elementary matrix we need is $E_1 = \begin{bmatrix} -1 & 0 \\ 0 & 1 \end{bmatrix}$:

$$E_1 A = \begin{bmatrix} -1 & 0 \\ 0 & 1 \end{bmatrix} \begin{bmatrix} -1 & 1 & -2 \\ 2 & 1 & 7 \end{bmatrix} = \begin{bmatrix} 1 & -1 & 2 \\ 2 & 1 & 7 \end{bmatrix}.$$

Now we replace the second row of $E_1 A$ by row two minus twice the first. This is achieved by multiplying by $E_2 = \begin{bmatrix} 1 & 0 \\ -2 & 1 \end{bmatrix}$:

$$E_2(E_1 A) = \begin{bmatrix} 1 & 0 \\ -2 & 1 \end{bmatrix} \begin{bmatrix} 1 & -1 & 2 \\ 2 & 1 & 7 \end{bmatrix} = \begin{bmatrix} 1 & -1 & 2 \\ 0 & 3 & 3 \end{bmatrix}.$$

Finally, we multiply the second row by $\frac{1}{3}$. This corresponds to multiplication by $E_3 = \begin{bmatrix} 1 & 0 \\ 0 & \frac{1}{3} \end{bmatrix}$:

$$E_3(E_2 E_1 A) = \begin{bmatrix} 1 & 0 \\ 0 & \frac{1}{3} \end{bmatrix} \begin{bmatrix} 1 & -1 & 2 \\ 0 & 3 & 3 \end{bmatrix} = \begin{bmatrix} 1 & -1 & 2 \\ 0 & 1 & 1 \end{bmatrix} = U.$$

We have shown that $(E_3 E_2 E_1)A = U$. This is $BA = U$ with $B = E_3 E_2 E_1$. Since each of E_3, E_2, and E_1 are invertible, so is their product B, and $B^{-1} = E_1^{-1} E_2^{-1} E_3^{-1}$ because the inverse of the product of invertible matrices is the product of the inverses, *order reversed*—see 2.2.12. Therefore,

$$B^{-1} = E_1^{-1} E_2^{-1} E_3^{-1} = \begin{bmatrix} -1 & 0 \\ 0 & 1 \end{bmatrix} \begin{bmatrix} 1 & 0 \\ 2 & 1 \end{bmatrix} \begin{bmatrix} 1 & 0 \\ 0 & 3 \end{bmatrix} = \begin{bmatrix} -1 & 0 \\ 2 & 3 \end{bmatrix},$$

and $BA = U$ implies

$$A = B^{-1}U = \begin{bmatrix} -1 & 0 \\ 2 & 3 \end{bmatrix} \begin{bmatrix} 1 & -1 & 2 \\ 0 & 1 & 1 \end{bmatrix}, \tag{1}$$

an equation the student should verify. $\qquad \blacksquare$

Equation (1) expresses the matrix A as the product of a *lower triangular* and an *upper triangular* matrix. This idea is the subject of this section—L for lower and U for upper.

2.5.9 DEFINITIONS

The *main diagonal* of an $m \times n$ matrix A is the list of elements $a_{11}, a_{22}, \ldots$ A matrix is *diagonal* if its only nonzero entries lie on the main diagonal, *upper triangular* if all entries below (and to the left of) the main diagonal are 0, and *lower triangular* if all entries above (and to the right of) the main diagonal are 0.

Figure 2.1 Schematic representations of a diagonal, upper triangular, and lower triangular matrix.

Diagonal　　　　Upper triangular　　　　Lower triangular

2.5.10 EXAMPLES

The main diagonal of $\begin{bmatrix} 1 & 2 & 3 \\ 4 & 5 & 6 \\ 7 & 8 & 9 \end{bmatrix}$ is $1, 5, 9$, and the main diagonal of $\begin{bmatrix} 7 & 9 & 3 \\ 6 & 5 & 4 \end{bmatrix}$ is $7, 5$.

The matrix $\begin{bmatrix} 1 & -1 & 2 \\ 0 & 1 & 1 \end{bmatrix}$ that appeared in (1) is upper triangular, while the matrix

$\begin{bmatrix} -1 & 0 \\ 2 & 3 \end{bmatrix}$ is lower triangular. The matrix $\begin{bmatrix} 1 & 0 & 0 \\ 0 & -7 & 0 \\ 0 & 0 & 2 \end{bmatrix}$ is diagonal. ∎

2.5.11 EXAMPLE

Let D, L, and U be the matrices

$$D = \begin{bmatrix} 2 & 0 \\ 0 & -4 \\ 0 & 0 \\ 0 & 0 \end{bmatrix}, \quad L = \begin{bmatrix} 6 & 0 & 0 \\ 1 & 0 & 0 \\ 3 & 2 & 1 \\ 0 & 3 & 1 \end{bmatrix}, \quad U = \begin{bmatrix} -1 & 2 & 1 & 6 \\ 0 & 5 & -2 & 4 \\ 0 & 0 & 0 & 8 \end{bmatrix}.$$

Then D is a diagonal 4×2 matrix, L is a lower triangular 4×3 matrix, and U is an upper triangular 3×4 matrix. ∎

2.5.12 DEFINITION

An *LU factorization* of a matrix A is the representation of A as the product $A = LU$ of a (necessarily square) lower triangular matrix L and an upper triangular matrix U, which is the same size as A.

2.5.13 EXAMPLE

$\begin{bmatrix} 1 & 2 \\ 4 & 6 \end{bmatrix} = \begin{bmatrix} 1 & 0 \\ 4 & 1 \end{bmatrix}\begin{bmatrix} 1 & 2 \\ 0 & -2 \end{bmatrix}$ is an LU factorization of $A = \begin{bmatrix} 1 & 2 \\ 4 & 6 \end{bmatrix}$. ∎

An LU factorization of $A = \begin{bmatrix} -1 & 1 & -2 \\ 2 & 1 & 7 \end{bmatrix}$ is given in equation (1). Note that the U here is a row echelon form of A. Carrying a matrix to row echelon form U always yields an LU factorization.

2.5.14 PROBLEM

Find an LU factorization of $A = \begin{bmatrix} 1 & 2 & 3 \\ 4 & -6 & -6 \\ -7 & -7 & 9 \end{bmatrix}$.

Solution. We reduce A to row echelon form with elementary matrices. First, we replace row two by row two minus 4 times row one. The elementary matrix required is

$$E_1 = \begin{bmatrix} 1 & 0 & 0 \\ -4 & 1 & 0 \\ 0 & 0 & 1 \end{bmatrix}, \quad \text{thus} \quad E_1 A = \begin{bmatrix} 1 & 2 & 3 \\ 0 & -14 & -18 \\ -7 & -7 & 9 \end{bmatrix}.$$

Then we replace row three by row three plus 7 times row one. Thus

$$E_2 = \begin{bmatrix} 1 & 0 & 0 \\ 0 & 1 & 0 \\ 7 & 0 & 1 \end{bmatrix} \quad \text{and} \quad E_2(E_1 A) = \begin{bmatrix} 1 & 2 & 3 \\ 0 & -14 & -18 \\ 0 & 7 & 30 \end{bmatrix}.$$

Next, we replace row three by row three minus $-\frac{1}{2}$ times row two. (We say "minus," not "plus.") We need

$$E_3 = \begin{bmatrix} 1 & 0 & 0 \\ 0 & 1 & 0 \\ 0 & \frac{1}{2} & 1 \end{bmatrix}, \quad \text{thus} \quad E_3(E_2 E_1 A) = \begin{bmatrix} 1 & 2 & 3 \\ 0 & -14 & -18 \\ 0 & 0 & 21 \end{bmatrix} = U.$$

Since $(E_3 E_2 E_1)A = U$, $A = (E_3 E_2 E_1)^{-1}U = E_1^{-1}E_2^{-1}E_3^{-1}U$, so

$$L = E_1^{-1}E_2^{-1}E_3^{-1} = \begin{bmatrix} 1 & 0 & 0 \\ 4 & 1 & 0 \\ 0 & 0 & 1 \end{bmatrix} \begin{bmatrix} 1 & 0 & 0 \\ 0 & 1 & 0 \\ -7 & 0 & 1 \end{bmatrix} \begin{bmatrix} 1 & 0 & 0 \\ 0 & 1 & 0 \\ 0 & -\frac{1}{2} & 1 \end{bmatrix}$$

$$= \begin{bmatrix} 1 & 0 & 0 \\ 4 & 1 & 0 \\ -7 & -\frac{1}{2} & 1 \end{bmatrix}.$$

Thus $A = LU$;

$$\begin{bmatrix} 1 & 2 & 3 \\ 4 & -6 & -6 \\ -7 & -7 & 9 \end{bmatrix} = \begin{bmatrix} 1 & 0 & 0 \\ 4 & 1 & 0 \\ -7 & -\frac{1}{2} & 1 \end{bmatrix} \begin{bmatrix} 1 & 2 & 3 \\ 0 & -14 & -18 \\ 0 & 0 & 21 \end{bmatrix},$$

an equation that should be checked (of course).

Preceding examples suggest the following theorem.

Theorem 2.5.15 *If a matrix A can be reduced to row echelon form without row interchanges, then A has an LU factorization.*

Any row echelon form of A is a suitable upper triangular matrix U, and L is the product of the inverses of elementary matrices corresponding to elementary operations of the last two types (no row exchanges). The inverse of an elementary matrix is an elementary matrix of the same type, hence lower triangular. The product of lower triangular matrices is lower triangular. This explains why L is lower triangular and why no row interchanges are allowed: the elementary matrix corresponding to an interchange of rows is not lower triangular. Consider, for example, $\begin{bmatrix} 0 & 1 \\ 1 & 0 \end{bmatrix}$.

■ *L* without Effort

Our primary purpose in introducing elementary matrices in this section was theoretical. They show why Theorem 2.5.15 is true: the lower triangular L in an LU factorization is the product of lower triangular elementary matrices. In practice, this is *not* how L is determined, however. If you use only the third elementary row operation in the Gaussian elimination process—that is, only an operation of the type $R_1 \rightarrow R_1 - cR_2$—then L can be found with ease.

Consider again Problem 2.5.14. There we found

$$\begin{bmatrix} 1 & 2 & 3 \\ 4 & -6 & -6 \\ -7 & -7 & 9 \end{bmatrix} = \begin{bmatrix} 1 & 0 & 0 \\ 4 & 1 & 0 \\ -7 & -\frac{1}{2} & 1 \end{bmatrix} \begin{bmatrix} 1 & 2 & 3 \\ 0 & -14 & -18 \\ 0 & 0 & 21 \end{bmatrix}$$

using only the third elementary row operation, like this:

$$A = \begin{bmatrix} 1 & 2 & 3 \\ 4 & -6 & -6 \\ -7 & -7 & 9 \end{bmatrix} \xrightarrow{R2 \rightarrow R2-4(R1)} \begin{bmatrix} 1 & 2 & 3 \\ 0 & -14 & -18 \\ -7 & -7 & 9 \end{bmatrix}$$

$$\xrightarrow{R3 \rightarrow R3-(-7)(R1)} \begin{bmatrix} 1 & 2 & 3 \\ 0 & -14 & -18 \\ 0 & 7 & 30 \end{bmatrix} \xrightarrow{R3 \rightarrow R3-(-\frac{1}{2})(R2)} \begin{bmatrix} 1 & 2 & 3 \\ 0 & -14 & -18 \\ 0 & 0 & 21 \end{bmatrix}.$$

So $U = \begin{bmatrix} 1 & 2 & 3 \\ 0 & -14 & -18 \\ 0 & 0 & 21 \end{bmatrix}$, and we found $L = \begin{bmatrix} 1 & 0 & 0 \\ 4 & 1 & 0 \\ -7 & -\frac{1}{2} & 1 \end{bmatrix}$ by multiplying together certain elementary matrices. In this particular instance, however, we could have simply written L without any calculation whatsoever! Remember our insistence on using the language of subtraction to describe the third elementary row operation.

Have a close look at L. It's square and lower triangular with diagonal entries all 1. That's easy to remember. What about the entries below the diagonal, the 4, the -7, and the $-\frac{1}{2}$? These are just the *multipliers* that were needed in the row reduction, *since we used the language of subtraction*. Our operations were

$$R2 \rightarrow R2-\mathbf{4}(R1)$$
$$R3 \rightarrow R3-(\mathbf{-7})(R1)$$
$$R3 \rightarrow R3-(-\tfrac{1}{2})R2,$$

hence L has the multipliers

$$4 \quad \text{in row two, column one,}$$
$$-7 \quad \text{in row three, column one, and}$$
$$-\tfrac{1}{2} \quad \text{in row three, column two.}$$

In general, if A can be reduced to an upper triangular matrix using only the third elementary row operation (no row interchanges, no multiplications of a row by a scalar), then the lower triangular matrix L has 1s on the diagonal, while the multipliers required in the row reduction are recorded below the diagonal.

2.5.16 To find an LU factorization of a matrix A, try to reduce A to an upper triangular matrix U from the top down using only the third elementary row operation.

2.5.17 EXAMPLE We bring $A = \begin{bmatrix} 1 & 2 & 3 & 4 \\ 5 & 6 & 7 & 8 \\ 9 & 10 & 0 & 1 \\ 0 & -2 & 3 & 4 \\ -1 & 6 & 4 & 9 \end{bmatrix}$ to upper triangular form using only the third row operation.

$$A \xrightarrow{R2 \to R2 - 5(R1)} \begin{bmatrix} 1 & 2 & 3 & 4 \\ 0 & -4 & -8 & -12 \\ 9 & 10 & 0 & 1 \\ 0 & -2 & 3 & 4 \\ -1 & 6 & 4 & 9 \end{bmatrix}$$

$$\xrightarrow{R3 \to R3 - 9(R1)} \begin{bmatrix} 1 & 2 & 3 & 4 \\ 0 & -4 & -8 & -12 \\ 0 & -8 & -27 & -35 \\ 0 & -2 & 3 & 4 \\ -1 & 6 & 4 & 9 \end{bmatrix} \xrightarrow{R5 \to R5 - (-1)(R1)} \begin{bmatrix} 1 & 2 & 3 & 4 \\ 0 & -4 & -8 & -12 \\ 0 & -8 & -27 & -35 \\ 0 & -2 & 3 & 4 \\ 0 & 8 & 7 & 13 \end{bmatrix}$$

$$\xrightarrow{R3 \to R3 - 2(R2)} \begin{bmatrix} 1 & 2 & 3 & 4 \\ 0 & -4 & -8 & -12 \\ 0 & 0 & -11 & -11 \\ 0 & -2 & 3 & 4 \\ 0 & 8 & 7 & 13 \end{bmatrix} \xrightarrow{R4 \to R4 - \frac{1}{2} R2} \begin{bmatrix} 1 & 2 & 3 & 4 \\ 0 & -4 & -8 & -12 \\ 0 & 0 & -11 & -11 \\ 0 & 0 & 7 & 10 \\ 0 & 8 & 7 & 13 \end{bmatrix}$$

$$\xrightarrow{R5 \to R5 - (-2)(R2)} \begin{bmatrix} 1 & 2 & 3 & 4 \\ 0 & -4 & -8 & -12 \\ 0 & 0 & -11 & -11 \\ 0 & 0 & 7 & 10 \\ 0 & 0 & -9 & -11 \end{bmatrix} \xrightarrow{R4 \to R4 - (-\frac{7}{11}) R3} \begin{bmatrix} 1 & 2 & 3 & 4 \\ 0 & -4 & -8 & -12 \\ 0 & 0 & -11 & -11 \\ 0 & 0 & 0 & 3 \\ 0 & 0 & -9 & -11 \end{bmatrix}$$

$$R5 \rightarrow R5 - \tfrac{9}{11}R3 \xrightarrow{\hspace{1cm}} \begin{bmatrix} 1 & 2 & 3 & 4 \\ 0 & -4 & -8 & -12 \\ 0 & 0 & -11 & -11 \\ 0 & 0 & 0 & 3 \\ 0 & 0 & 0 & -2 \end{bmatrix} \quad R5 \rightarrow R5 - (-\tfrac{2}{3})R4 \xrightarrow{\hspace{1cm}} \begin{bmatrix} 1 & 2 & 3 & 4 \\ 0 & -4 & -8 & -12 \\ 0 & 0 & -11 & -11 \\ 0 & 0 & 0 & 3 \\ 0 & 0 & 0 & 0 \end{bmatrix} = U.$$

We conclude that $A = LU$ with a lower triangular L that we can just write down: put 1s on the main diagonal and the multipliers below.

$$L = \begin{bmatrix} 1 & 0 & 0 & 0 & 0 \\ 5 & 1 & 0 & 0 & 0 \\ 9 & 2 & 1 & 0 & 0 \\ 0 & \tfrac{1}{2} & -\tfrac{7}{11} & 1 & 0 \\ -1 & -2 & \tfrac{9}{11} & -\tfrac{2}{3} & 1 \end{bmatrix}. \qquad \blacksquare$$

Reading Challenge 1   *If A is an $m \times n$ matrix and we factor $A = LU$ by the method described, the matrix L is always square. Why?*

2.5.18 Remark Why did we say "from the top down" in Definition 2.5.16? Suppose $A = \begin{bmatrix} 1 & 2 & 3 \\ 1 & 1 & 1 \\ 2 & 2 & 3 \end{bmatrix}$ and we apply the following sequence of third elementary row operations:

$$A \xrightarrow{R3 \rightarrow R3 - 2(R2)} \begin{bmatrix} 1 & 2 & 3 \\ 1 & 1 & 1 \\ 0 & 0 & 1 \end{bmatrix} \xrightarrow{R2 \rightarrow R2 - R1} \begin{bmatrix} 1 & 2 & 3 \\ 1 & 1 & 1 \\ 0 & 0 & 1 \end{bmatrix} = U$$

and the matrix recording the multipliers is $L = \begin{bmatrix} 1 & 0 & 0 \\ 1 & 1 & 0 \\ 0 & 2 & 1 \end{bmatrix}$, but $A \neq LU$. For this observation and example, the author is indebted to one of the reviewers of this book.

Forcing ourselves to use only the third row operation often leads to a messy row reduction when working by hand, but remember that such a calculation is easy by machine and it leads to enormous savings in the calculation of L. Henceforth,

2.5.19 When trying to find an LU factorization, use only the third elementary row operation.

■ **Why LU?**

A major use for LU factorization is the simplification of the solution of systems of linear equations. Suppose we want to solve the system $A\mathbf{x} = \mathbf{b}$ and we know that $A = LU$. What we want to solve is $LU\mathbf{x} = \mathbf{b}$. Think of this as $L(U\mathbf{x}) = \mathbf{b}$ and let $\mathbf{y} = U\mathbf{x}$. We have $L\mathbf{y} = \mathbf{b}$, which is easily solved since L is lower triangular. Having obtained $\mathbf{y}$, we then solve $U\mathbf{x} = \mathbf{y}$ for $\mathbf{x}$. This is also easy since U is upper triangular.

2.5.20 EXAMPLE Suppose we want to solve $A\mathbf{x} = \mathbf{b}$ for $\mathbf{x} = \begin{bmatrix} x_1 \\ x_2 \\ x_3 \end{bmatrix}$, where $\mathbf{b} = \begin{bmatrix} 0 \\ 1 \\ 0 \end{bmatrix}$, and $A = $

$\begin{bmatrix} 1 & 2 & 3 \\ 4 & -6 & -6 \\ -7 & -7 & 9 \end{bmatrix}$ is the matrix of Problem 2.5.14. We found $A = LU$, with

$$L = \begin{bmatrix} 1 & 0 & 0 \\ 4 & 1 & 0 \\ -7 & -\frac{1}{2} & 1 \end{bmatrix} \text{ and } U = \begin{bmatrix} 1 & 2 & 3 \\ 0 & -14 & -18 \\ 0 & 0 & 21 \end{bmatrix}.$$

To solve $A\mathbf{x} = \mathbf{b}$, which is $L(U\mathbf{x}) = \mathbf{b}$, we first solve $L\mathbf{y} = \mathbf{b}$ for $\mathbf{y} = \begin{bmatrix} y_1 \\ y_2 \\ y_3 \end{bmatrix}$. This is the system

$$\begin{aligned} y_1 \qquad\qquad &= 0 \\ 4y_1 + y_2 \qquad &= 1 \\ -7y_1 - \tfrac{1}{2}y_2 + y_3 &= 0. \end{aligned}$$

By *forward substitution* we have $y_1 = 0$, $y_2 = 1 - 4y_1 = 1$, and $y_3 = 7y_1 + \frac{1}{2}y_2 = \frac{1}{2}$. Thus $\mathbf{y} = \begin{bmatrix} 0 \\ 1 \\ \frac{1}{2} \end{bmatrix}$. Now we solve $U\mathbf{x} = \mathbf{y}$ by back substitution. We have

$$\begin{aligned} x_1 + 2x_2 + 3x_3 &= 0 \\ -14x_2 - 18x_3 &= 1 \\ 21x_3 &= \tfrac{1}{2}, \end{aligned}$$

so, by back substitution,

$x_3 = \frac{1}{42}$,

$-14x_2 = 1 + 18x_3 = \frac{10}{7}$, so $x_2 = -\frac{5}{49}$, and

$x_1 = -2x_2 - 3x_3 = \frac{10}{49} - \frac{1}{14} = \frac{13}{98}$.

Our solution is $\mathbf{x} = \begin{bmatrix} \frac{13}{98} \\ -\frac{5}{49} \\ \frac{1}{42} \end{bmatrix}$. We check that this is correct by verifying that $A\mathbf{x} = \mathbf{b}$:

$$\begin{bmatrix} 1 & 2 & 3 \\ 4 & -6 & -6 \\ -7 & -7 & 9 \end{bmatrix} \begin{bmatrix} \frac{13}{98} \\ -\frac{5}{49} \\ \frac{1}{42} \end{bmatrix} = \begin{bmatrix} 0 \\ 1 \\ 0 \end{bmatrix}.$$

∎

> **2.5.21** Given $A = LU$, an LU factorization of A, solve $A\mathbf{x} = \mathbf{b}$ by first solving $L\mathbf{y} = \mathbf{b}$ for $\mathbf{y}$ and then $U\mathbf{x} = \mathbf{y}$ for $\mathbf{x}$.

To make more precise the assertion that solving a triangular system is "faster" than solving an arbitrary system, we count the number of elementary operations required to solve $A\mathbf{x} = \mathbf{b}$, where A is $n \times n$. In the first step, we conceivably would have to put a 0 in the first position of every row below the first. This requires $n - 1$ row operations of the sort $R \to R - aR_1$. Since there are $n - 1$ elements to be changed in each row, we require a total of $2(n - 1)(n - 1) = 2(n - 1)^2$ operations to put 0s below the $(1, 1)$ entry.

Similarly, to put 0s below the $(2, 2)$ position requires $2(n - 2)^2$ operations (the previous number with $n - 1$ replacing n), and to put our matrix in row echelon form requires

$$[2(n - 1)^2] + [2(n - 2)^2] + \cdots + [2(1)^2] + 0$$

$$= 2[1^2 + 2^2 + \cdots + (n - 1)^2] = 2\frac{(n - 1)(n)(2n - 1)}{6}$$

$$= \frac{(n - 1)n(2n - 1)}{3}.$$

Roughly n^3 operations are involved.

2.5.22 Remark Here we assume that the reader is familiar with the formula

$$1^2 + 2^2 + 3^2 + \cdots + k^2 = \frac{k(k + 1)(2k + 1)}{6}$$

for the sum of the squares of the first k integers. In what follows, we assume that the reader is also aware of the fact that the sum of the squares of the first k odd integers is k^2, that is,

$$1 + 3 + 5 + \cdots + [2k - 1] = k^2.$$

Now we count the number of operations involved in back substitution. The last equation, of the form $bx_n = a$, gives x_n with at most one division. The previous equation, of the form $ax_{n-1} + bx_n = c$, gives x_{n-1} with at most one multiplication, one subtraction, and one division—three operations. The third equation from the bottom, of the form $ax_{n-2} + bx_{n-1} + cx_n = d$, gives x_{n-2} with at most two multiplications, two subtractions, and one division—five operations in all. The equation fourth from the bottom, of the form $ax_{n-3} + bx_{n-2} + cx_{n-1} + dx_n = e$, gives x_{n-3} with at most three multiplications, three subtractions, and one division—seven operations total. The first equation, of the form $a_1x_1 + a_2x_2 + \cdots + a_nx_n = b$, gives x_1 after $n - 1$ multiplications, $n - 1$ subtractions, and one division, or $2(n - 1) + 1 = 2n - 1$ operations altogether. In total, the number of operations is

$$1 + 3 + 5 + \cdots + [2n - 1] = n^2.$$

We conclude that solving $A\mathbf{x} = \mathbf{b}$ in general requires roughly $n^3 + n^2$ operations.

On the other hand, the solution of an upper triangular system, $U\mathbf{x} = \mathbf{y}$, involves just n^2 operations (only back substitution is required), and the same goes for the solution of a lower triangular system $L\mathbf{y} = \mathbf{b}$. Factoring $A = LU$ reduces the "work" involved in solving $A\mathbf{x} = \mathbf{b}$ from n^3 to $2n^2$. This difference is substantial for large n. For instance, if $n = 1000$, then $n^2 = 10^6$ and $n^3 = 10^9$. On a personal computer performing 10^8 operations per second (this is very slow!), the difference is 0.02 seconds versus 10 seconds.

■ The *PLU* Factorization; Row Interchanges

We conclude this section by discussing a possibility hitherto avoided. To find an LU factorization, we have said that we should *try* to move A to an upper triangular matrix without interchanging rows. This suggests that such a goal may not always be attainable. For example, if we apply Gaussian elimination to $A = \begin{bmatrix} 2 & 4 & 2 \\ 1 & 2 & 3 \\ 3 & 2 & 1 \end{bmatrix}$, trying to avoid row interchanges, we begin

$$A \longrightarrow \begin{bmatrix} 2 & 4 & 2 \\ 0 & 0 & 2 \\ 0 & -4 & -2 \end{bmatrix},$$

but now are forced to interchange rows two and three. To avoid this problem, we change A at the outset by interchanging rows two and three. This gives a matrix A', which we can reduce to an upper triangular matrix without row interchanges as follows:

$$A' = \begin{bmatrix} 2 & 4 & 2 \\ 3 & 2 & 1 \\ 1 & 2 & 3 \end{bmatrix} \longrightarrow \begin{bmatrix} 2 & 4 & 2 \\ 0 & -4 & -2 \\ 0 & 0 & 2 \end{bmatrix} = U.$$

So

$$A' = LU = \begin{bmatrix} 1 & 0 & 0 \\ \frac{3}{2} & 1 & 0 \\ \frac{1}{2} & 0 & 1 \end{bmatrix} \begin{bmatrix} 2 & 4 & 2 \\ 0 & -4 & -2 \\ 0 & 0 & 2 \end{bmatrix}. \tag{2}$$

Of course, this is a factorization of A', not of A, but fortunately there is a nice connection between A' and A, and a connection we have seen before. The matrix A' is obtained from A by interchanging rows two and three, so $A' = P'A$ where $P' = \begin{bmatrix} 1 & 0 & 0 \\ 0 & 0 & 1 \\ 0 & 1 & 0 \end{bmatrix}$ is the elementary matrix obtained from the 3×3 identity matrix by interchanging rows two and three. Thus (2) is an equation of the form $P'A = LU$. The inverse of the elementary matrix P' is another elementary matrix P—here $P = (P')^{-1} = P'$—so we obtain the factorization $A = PLU$.

In general, row reduction may involve several interchanges of rows so that the initial matrix A' is A with its rows scrambled, or *permuted*. In this case, $A' = P'A$

where P' is the matrix obtained from the identity matrix by reordering rows just as the rows of A were reordered to give A'. Such a matrix is called a *permutation matrix*, that is, a matrix whose rows are the rows of the identity matrix in some order (equivalently, whose columns are the columns of the identity matrix, in some order).

2.5.23 EXAMPLE $\begin{bmatrix} 0 & 1 \\ 1 & 0 \end{bmatrix}$, $\begin{bmatrix} 0 & 1 & 0 \\ 0 & 0 & 1 \\ 1 & 0 & 0 \end{bmatrix}$ and $\begin{bmatrix} 0 & 0 & 1 & 0 \\ 0 & 1 & 0 & 0 \\ 0 & 0 & 0 & 1 \\ 1 & 0 & 0 & 0 \end{bmatrix}$ are all permutation matrices.   ■

Reading Challenge 2   *For each of the permutation matrices P' in 2.5.23, describe the relationship between A and $P'A$, where A is a matrix of appropriate size.*

It is easy to find the inverse of a permutation matrix.

2.5.24 If P is a permutation matrix, then $P^{-1} = P^T$.

To see why, we compute PP^T. The (i, j) entry of PP^T is the dot product of row i of P and column j of P^T. This is the dot product of row i of P and row j of P, which is the dot product of two of the standard basis vectors of $\mathbf{R}^n$,

$$\mathbf{e}_1 = \begin{bmatrix} 1 \\ 0 \\ 0 \\ \vdots \\ 0 \end{bmatrix}, \quad \mathbf{e}_2 = \begin{bmatrix} 0 \\ 1 \\ 0 \\ \vdots \\ 0 \end{bmatrix}, \quad \dots, \quad \mathbf{e}_n = \begin{bmatrix} 0 \\ 0 \\ \vdots \\ 0 \\ 1 \end{bmatrix}.$$

The dot product $\mathbf{e}_1 \cdot \mathbf{e}_1 = 1(1) + 0(1) + 0(0) + \cdots + 0(0) = 1$ and, similarly, the dot product of any $\mathbf{e}_i$ with itself is 1. Thus each diagonal entry of PP^T is 1. On the other hand, $\mathbf{e}_1 \cdot \mathbf{e}_2 = 1(0) + 0(1) + 0(0) + \cdots + 0(0) = 0$ and, similarly, the dot product of any two different vectors $\mathbf{e}_i$ and $\mathbf{e}_j$ is 0. So each off-diagonal entry of PP^T is 0. This shows that $PP^T = I$, the $n \times n$ identity matrix. Since permutation matrices are square, $PP^T = I$ is sufficient to guarantee that P^T is the inverse of P; $P^T = P^{-1}$.

Reading Challenge 3   *Find the inverse of the matrix $\begin{bmatrix} 0 & 1 & 0 \\ 0 & 0 & 1 \\ 1 & 0 & 0 \end{bmatrix}$ without any calculation. Explain your reasoning.*

In general, we cannot expect to factor a given matrix A in the form $A = LU$, but we can always achieve $P'A = LU$, so that $A = PLU$, with $P = (P')^{-1} = (P')^T$.

2.5.25 PROBLEM

Find an LU factorization of $A = \begin{bmatrix} 0 & 1 & 1 & 0 \\ 0 & -1 & -1 & 4 \\ -2 & 3 & 2 & 1 \\ -6 & 4 & 2 & -8 \end{bmatrix}$ if possible; otherwise find a

PLU factorization.

Solution. The first step in row reduction to an upper triangular matrix requires a row interchange. Thus there is no LU factorization. Rearranging the rows of A in the order $3, 4, 1, 2$ leads to the matrix

$$P'A = \begin{bmatrix} -2 & 3 & 2 & 1 \\ -6 & 4 & 2 & -8 \\ 0 & 1 & 1 & 0 \\ 0 & -1 & -1 & 4 \end{bmatrix},$$

with P' the permutation matrix formed from the identity by reordering its rows as we have reordered those of A, like this:

$$P' = \begin{bmatrix} 0 & 0 & 1 & 0 \\ 0 & 0 & 0 & 1 \\ 1 & 0 & 0 & 0 \\ 0 & 1 & 0 & 0 \end{bmatrix}.$$

Row reduction, using only the third elementary row operation, yields

$$P'A = \begin{bmatrix} -2 & 3 & 2 & 1 \\ -6 & 4 & 2 & -8 \\ 0 & 1 & 1 & 0 \\ 0 & -1 & -1 & 4 \end{bmatrix} \longrightarrow \begin{bmatrix} -2 & 3 & 2 & 1 \\ 0 & -5 & -4 & -11 \\ 0 & 1 & 1 & 0 \\ 0 & -1 & -1 & 4 \end{bmatrix}$$

$$\longrightarrow \begin{bmatrix} -2 & 3 & 2 & 1 \\ 0 & -5 & -4 & -11 \\ 0 & 0 & \frac{1}{5} & -\frac{11}{5} \\ 0 & 0 & -\frac{1}{5} & \frac{31}{5} \end{bmatrix} \longrightarrow \begin{bmatrix} -2 & 3 & 2 & 1 \\ 0 & -5 & -4 & -11 \\ 0 & 0 & \frac{1}{5} & -\frac{11}{5} \\ 0 & 0 & 0 & 4 \end{bmatrix} = U',$$

with $L = \begin{bmatrix} 1 & 0 & 0 & 0 \\ 3 & 1 & 0 & 0 \\ 0 & -\frac{1}{5} & 1 & 0 \\ 0 & \frac{1}{5} & -1 & 1 \end{bmatrix}$. Thus $P'A = LU'$ and hence $A = PLU'$, with $P = $

$(P')^{-1} = (P')^T = \begin{bmatrix} 0 & 0 & 1 & 0 \\ 0 & 0 & 0 & 1 \\ 1 & 0 & 0 & 0 \\ 0 & 1 & 0 & 0 \end{bmatrix}.$

Answers to Reading Challenges

1. U is also $m \times n$ since U is just A to which some row operations have been applied. Thus L is $? \times m$ and, since LU is $m \times n$, L must be $m \times m$.

2. With $P' = \begin{bmatrix} 0 & 1 \\ 1 & 0 \end{bmatrix}$ and A a $2 \times n$ matrix, $P'A$ is A with its two rows interchanged.

With $P' = \begin{bmatrix} 0 & 1 & 0 \\ 0 & 0 & 1 \\ 1 & 0 & 0 \end{bmatrix}$ and A a $3 \times n$ matrix, $P'A$ is the $3 \times n$ matrix whose rows, in order, are the second, third, and first rows of A.

With $P' = \begin{bmatrix} 0 & 0 & 1 & 0 \\ 0 & 1 & 0 & 0 \\ 0 & 0 & 0 & 1 \\ 1 & 0 & 0 & 0 \end{bmatrix}$ and A a $4 \times n$ matrix, $P'A$ is the matrix whose rows are the rows of A, but in the order three, two, four, one.

3. The given matrix is a permutation matrix, so its inverse is its transpose:

$$\begin{bmatrix} 0 & 1 & 0 \\ 0 & 0 & 1 \\ 1 & 0 & 0 \end{bmatrix}^{-1} = \begin{bmatrix} 0 & 0 & 1 \\ 1 & 0 & 0 \\ 0 & 1 & 0 \end{bmatrix}.$$

True/False Questions

Decide, with as little calculation as possible, whether each of the following statements is true or false and explain your answer whenever you say "false." (Answers can be found in the back of the book.)

1. The matrix $\begin{bmatrix} 1 & 0 & 0 \\ 0 & 2 & 1 \\ 0 & 0 & 1 \end{bmatrix}$ is elementary.

2. If $E = \begin{bmatrix} 1 & 0 & 0 \\ 0 & 1 & -2 \\ 0 & 0 & 1 \end{bmatrix}$, then $E^{-1} = \begin{bmatrix} 1 & 0 & 0 \\ 0 & 1 & 2 \\ 0 & 0 & 1 \end{bmatrix}$.

3. If $E = \begin{bmatrix} 1 & 0 & -3 \\ 0 & 1 & 0 \\ 0 & 0 & 1 \end{bmatrix}$ and A is a $3 \times n$ matrix, rows two and three of EA are the same as rows two and three of A.

4. The elements of the main diagonal of $U = \begin{bmatrix} 0 & 4 & 0 & -2 \\ 0 & 0 & -1 & 1 \\ 0 & 0 & 0 & 7 \end{bmatrix}$ are 4, -1, and 7.

5. The product of elementary matrices is necessarily lower triangular.

6. Any matrix that can be reduced to row echelon form without row interchanges has an LU factorization.

7. If $A = LU$ is an LU factorization of A, we can solve $A\mathbf{x} = \mathbf{b}$ by first solving $U\mathbf{y} = \mathbf{b}$ for $\mathbf{y}$ and then $L\mathbf{x} = \mathbf{y}$ for $\mathbf{x}$.

8. If $P = \begin{bmatrix} 0 & 1 & 0 \\ 1 & 0 & 0 \\ 0 & 0 & 1 \end{bmatrix}$, then $P^{-1} = P$.

Exercises

*Solutions to exercises marked [BB] can be found in the **B**ack of the **B**ook.*

1. [BB] Let $E = \begin{bmatrix} 1 & 0 & 0 \\ 0 & 1 & 4 \\ 0 & 0 & 1 \end{bmatrix}$ and let A be an arbitrary $3 \times n$ matrix, $n \geq 1$.

   (a) Explain the connection between A and EA.

   (b) What is the name for a matrix such as E?

   (c) Is E invertible? If so, what is its inverse?

2. Suppose A is a $3 \times n$ matrix and we would like to modify A in the following ways:

   (a) [BB] subtracting the second row from the third, or

   (b) adding the third row to the first, or

   (c) interchanging the first and third rows, or

   (d) dividing the second row by -2.

   Write the elementary matrices E for which EA is A modified as specified in each of these ways.

3. Let $A = \begin{bmatrix} 1 & 2 & 3 \\ 4 & 5 & 6 \\ 6 & 8 & 9 \end{bmatrix}$. Write an elementary matrix E so that

   (a) [BB] $EA = \begin{bmatrix} 1 & 2 & 3 \\ 4 & 5 & 6 \\ 0 & -4 & -9 \end{bmatrix}$

   (b) $EA = \begin{bmatrix} 1 & 2 & 3 \\ 0 & -3 & -6 \\ 6 & 8 & 9 \end{bmatrix}$

   (c) $EA = \begin{bmatrix} 1 & 2 & 3 \\ 4 & 5 & 6 \\ 0 & \frac{1}{2} & 0 \end{bmatrix}$

   (d) $EA = \begin{bmatrix} 1 & 2 & 3 \\ 6 & 8 & 9 \\ 4 & 5 & 6 \end{bmatrix}$.

4. Let $A = \begin{bmatrix} 2 & 1 & 0 \\ 4 & -2 & 1 \\ 5 & 0 & 2 \\ -3 & 4 & 1 \end{bmatrix}$. Find a matrix E so that EA has a 0 in the (2, 1) position and a matrix F so that FA has a 0 in the (4, 1) position.

5. Let $E = \begin{bmatrix} -2 & 0 \\ 0 & 1 \end{bmatrix}$ and $F = \begin{bmatrix} 1 & 0 \\ 3 & 0 \end{bmatrix}$. Suppose A is a 2×5 matrix. Describe EA and FA.

6. [BB] Let E be the 3×3 elementary matrix that adds 4 times row one to row two and F the 3×3 elementary matrix that subtracts 3 times row two from row three.

   (a) Find E, F, EF, and FE without calculation.

   (b) Write the inverses of the matrices in part (a) without calculation.

7. Let E be the 4×4 elementary matrix that subtracts 5 times row two from row four and F the 4×4 elementary matrix that adds $\frac{1}{2}$ row three to row one.

   (a) Find E, F, EF, and FE without calculation.

   (b) Write the inverses of the matrices in part (a) without calculation.

8. [BB] What number x will force a row exchange when Gaussian elimination is applied to $\begin{bmatrix} 2 & 5 & 1 \\ 4 & x & 1 \\ 0 & 1 & -1 \end{bmatrix}$?

9. Suppose Gaussian elimination is applied to the matrix $\begin{bmatrix} -1 & 0 & 1 & 1 \\ 3 & 0 & x & -4 \\ 4 & 1 & 7 & 8 \end{bmatrix}$. Explain why a row exchange will be necessary no matter what the value of x might be.

10. Write the inverse of each of the following matrices without calculation. Explain your reasoning.

    (a) [BB] $A = \begin{bmatrix} 1 & 3 \\ 0 & 1 \end{bmatrix}$

    (b) $A = \begin{bmatrix} 1 & 0 \\ -5 & 1 \end{bmatrix}$

    (c) [BB] $A = \begin{bmatrix} 1 & 0 & 0 \\ 0 & 1 & -2 \\ 0 & 0 & 1 \end{bmatrix}$

    (d) $A = \begin{bmatrix} 1 & 4 & 0 \\ 0 & 1 & 0 \\ 0 & 0 & 1 \end{bmatrix}$

(e) $A = \begin{bmatrix} 0 & 0 & 0 & 1 \\ 0 & 1 & 0 & 0 \\ 0 & 0 & 1 & 0 \\ 1 & 0 & 0 & 0 \end{bmatrix}$

(f) $A = \begin{bmatrix} 1 & 0 & 0 \\ 0 & \frac{1}{2} & 0 \\ 0 & 0 & 1 \end{bmatrix}$.

11. Find the inverse of each of the following matrices simply by writing it down. No calculation is necessary, but explain your reasoning.

(a) [BB] $\begin{bmatrix} 0 & 1 \\ 1 & 0 \end{bmatrix}$

(b) $\begin{bmatrix} 0 & 0 & 1 \\ 1 & 0 & 0 \\ 0 & 1 & 0 \end{bmatrix}$

(c) $\begin{bmatrix} 0 & 0 & 1 \\ 0 & 1 & 0 \\ 1 & 0 & 0 \end{bmatrix}$

(d) $\begin{bmatrix} 0 & 1 & 0 & 0 \\ 0 & 0 & 1 & 0 \\ 0 & 0 & 0 & 1 \\ 1 & 0 & 0 & 0 \end{bmatrix}$.

12. Find the inverse of each of the following matrices without using 2.5.4, and explain your answers.

(a) $E = \begin{bmatrix} a & 0 \\ 0 & 1 \end{bmatrix}$ (assume $a \neq 0$)

(b) $E = \begin{bmatrix} 1 & 0 & 0 \\ 0 & 1 & a \\ 0 & 0 & 1 \end{bmatrix}$.

13. (a) [BB] Show that the lower triangular matrix $L = \begin{bmatrix} 1 & 0 \\ a & 1 \end{bmatrix}$ is invertible by finding a matrix M with $LM = I$, the 2×2 identity matrix.

(b) Show that the lower triangular matrix $L = \begin{bmatrix} 1 & 0 & 0 \\ a & 1 & 0 \\ b & c & 1 \end{bmatrix}$ is invertible.

14. [BB] Let A be a $3 \times n$ matrix, $E = \begin{bmatrix} 1 & 0 & 0 \\ -1 & 1 & 0 \\ 2 & 0 & 1 \end{bmatrix}$, $D = \begin{bmatrix} 4 & 0 & 0 \\ 0 & 1 & 0 \\ 0 & 0 & -1 \end{bmatrix}$ and $P = \begin{bmatrix} 0 & 0 & 1 \\ 1 & 0 & 0 \\ 0 & 1 & 0 \end{bmatrix}$.

(a) How are the rows of EA, of DA, and of PA related to the rows of A?

(b) Find P^{-1}.

15. Answer Exercise 14 with $E = \begin{bmatrix} 1 & 0 & 3 \\ 0 & 1 & 0 \\ 0 & -2 & 1 \end{bmatrix}$, $D = \begin{bmatrix} 1 & 0 & 0 \\ 0 & -1 & 0 \\ 0 & 0 & 2 \end{bmatrix}$, and $P = \begin{bmatrix} 1 & 0 & 0 \\ 0 & 0 & 1 \\ 0 & 1 & 0 \end{bmatrix}$.

16. Show that $\begin{bmatrix} 0 & 1 \\ 1 & 0 \end{bmatrix}$ does not have an LU factorization.

17. Suppose A is an $m \times n$ matrix that can be reduced by Gaussian elimination to an upper triangular matrix U without the use of row interchanges. How does this lead to an LU factorization of A? Explain clearly how the lower triangular matrix L arises.

18. [BB] Ian says $\begin{bmatrix} 0 & 1 \\ 1 & 2 \end{bmatrix}$ has no LU factorization but Lynn disagrees. Lynn adds the second row to the first, then subtracts the first from the second, like this

$$\begin{bmatrix} 0 & 1 \\ 1 & 2 \end{bmatrix} \to \begin{bmatrix} 1 & 3 \\ 1 & 2 \end{bmatrix} \to \begin{bmatrix} 1 & 3 \\ 0 & -1 \end{bmatrix} = U.$$

She gets L as the product of certain elementary matrices. Who is right? Explain.

19. In each case, write L as the product of elementary matrices (without calculation) and use this factorization to find L^{-1}.

(a) [BB] $L = \begin{bmatrix} 1 & 0 & 0 \\ -2 & 1 & 0 \\ 3 & 5 & 1 \end{bmatrix}$

(b) $L = \begin{bmatrix} 1 & 0 & 0 & 0 \\ -3 & 1 & 0 & 0 \\ 4 & 2 & 1 & 0 \\ -1 & -5 & 7 & 1 \end{bmatrix}$.

20. Write $U = \begin{bmatrix} 1 & -2 & -4 & 3 \\ 0 & 1 & -7 & -5 \\ 0 & 0 & 1 & 4 \\ 0 & 0 & 0 & 1 \end{bmatrix}$ as the product of elementary matrices and use this factorization to find U^{-1}.

21. If possible, find an LU factorization of $A = \begin{bmatrix} 2 & -6 & -2 & 4 \\ -1 & 3 & 3 & 2 \\ -1 & 3 & 7 & 10 \end{bmatrix}$ and express L as the product of elementary matrices.

22. [BB] Let $A = \begin{bmatrix} 2 & 1 & 0 \\ 0 & 4 & 2 \\ 6 & 3 & 5 \end{bmatrix}$ and $\mathbf{v} = \begin{bmatrix} -2 \\ 3 \\ 5 \end{bmatrix}$.

(a) Find an elementary matrix E and an upper triangular matrix U such that $EA = U$.

(b) Find a lower triangular matrix L such that $A = LU$.

(c) Express $A\mathbf{v}$ as a linear combination of the columns of A.

23. Let $A = \begin{bmatrix} 3 & 4 & 8 \\ 6 & 2 & 9 \\ 0 & 0 & -3 \end{bmatrix}$ and $\mathbf{v} = \begin{bmatrix} 4 \\ -7 \\ 2 \end{bmatrix}$.

(a) Find an elementary matrix E and an upper triangular matrix U such that $EA = U$.

(b) Find a lower triangular matrix L such that $A = LU$.

(c) Express $A\mathbf{v}$ as a linear combination of the columns of A.

24. Let E be a 4×4 matrix such that for any $4 \times n$ matrix $A = \begin{bmatrix} \mathbf{a}_1^T & \to \\ \mathbf{a}_2^T & \to \\ \mathbf{a}_3^T & \to \\ \mathbf{a}_4^T & \to \end{bmatrix}$ with rows $\mathbf{a}_1^T, \mathbf{a}_2^T, \mathbf{a}_3^T, \mathbf{a}_4^T$ as shown, $EA = \begin{bmatrix} \mathbf{a}_1^T + 2\mathbf{a}_3^T & \to \\ \mathbf{a}_2^T & \to \\ \mathbf{a}_3^T & \to \\ \mathbf{a}_4^T & \to \end{bmatrix}$. What is E? What is E^{-1}?

25. [BB] Let $A = \begin{bmatrix} -3 & 3 & 6 \\ 2 & 5 & 10 \\ 0 & 1 & 4 \end{bmatrix}$.

(a) Find an LU factorization of A by row reduction to an upper triangular matrix, keeping track of the elementary matrices. Express L as the product of elementary matrices.

(b) Find an LU factorization A by using only the third elementary row operation and simply writing L down.

26. Repeat Exercise 25 for the matrix

$A = \begin{bmatrix} 2 & 4 & 6 & 18 \\ 4 & 5 & 6 & 24 \\ 3 & 1 & -2 & 4 \end{bmatrix}$.

27. Find, if possible, an LU factorization of each of the following matrices:

(a) [BB] $A = \begin{bmatrix} 2 & 4 \\ -1 & 6 \end{bmatrix}$

(b) $A = \begin{bmatrix} 2 & -4 & 1 \\ 1 & 3 & -1 \end{bmatrix}$

(c) [BB] $A = \begin{bmatrix} 5 & 0 & 15 \\ 0 & 1 & 2 \\ 4 & -3 & 8 \\ -2 & 2 & -1 \end{bmatrix}$

(d) $A = \begin{bmatrix} 1 & -1 & 1 \\ 1 & -1 & 1 \\ 1 & -1 & 1 \end{bmatrix}$

(e) $A = \begin{bmatrix} 1 & 3 & 8 \\ 2 & 5 & 21 \\ 1 & 7 & -5 \end{bmatrix}$

(f) $A = \begin{bmatrix} 1 & -1 & -2 & 1 & 1 \\ 2 & -4 & -5 & 6 & 0 \\ -1 & -5 & 2 & 12 & -2 \end{bmatrix}$

(g) $A = \begin{bmatrix} -2 & 4 & 6 & -8 \\ 1 & 2 & -8 & 2 \\ -4 & 0 & -5 & 3 \\ 2 & 1 & -4 & 1 \end{bmatrix}$

(h) $A = \begin{bmatrix} -1 & 2 & 2 & 3 & 4 \\ 5 & -10 & -6 & -18 & -12 \\ 2 & -2 & -3 & -5 & -7 \end{bmatrix}$.

28. [BB] Let $A = \begin{bmatrix} 2 & 1 \\ 6 & 8 \end{bmatrix} = \begin{bmatrix} 2 & 0 \\ 6 & 5 \end{bmatrix}\begin{bmatrix} 1 & \frac{1}{2} \\ 0 & 1 \end{bmatrix}$ $= LU$.

(a) Use this factorization of A to solve the system

$$A\begin{bmatrix} x \\ y \end{bmatrix} = \begin{bmatrix} -2 \\ 9 \end{bmatrix}.$$

(b) Express $\begin{bmatrix} -2 \\ 9 \end{bmatrix}$ as a linear combination of the columns of A.

29. Let $A = \begin{bmatrix} 1 & 6 & 7 \\ 5 & 25 & 36 \\ -2 & -27 & -4 \end{bmatrix}$

$= \begin{bmatrix} 1 & 0 & 0 \\ 5 & 1 & 0 \\ -2 & 3 & 1 \end{bmatrix}\begin{bmatrix} 1 & 6 & 7 \\ 0 & -5 & 1 \\ 0 & 0 & 7 \end{bmatrix} = LU$.

(a) Use the given factorization of A to solve the linear system $A\mathbf{x} = \mathbf{b}$, where $\mathbf{b} = \begin{bmatrix} 1 \\ 2 \\ 3 \end{bmatrix}$.

(b) Express $\mathbf{b}$ as a linear combination of the columns of A.

30. (a) Use the factorization

$$A = \begin{bmatrix} 1 & 4 & 3 & 4 \\ 3 & 10 & 10 & 17 \\ -4 & -20 & -14 & -1 \\ 7 & 38 & 28 & -6 \end{bmatrix}$$

$$= \begin{bmatrix} 1 & 0 & 0 & 0 \\ 3 & 1 & 0 & 0 \\ -4 & 2 & 1 & 0 \\ 7 & -5 & -3 & 1 \end{bmatrix} \begin{bmatrix} 1 & 4 & 3 & 4 \\ 0 & -2 & 1 & 5 \\ 0 & 0 & -4 & 5 \\ 0 & 0 & 0 & 6 \end{bmatrix}$$

$$= LU$$

to solve the linear system $A\mathbf{x} = \mathbf{b}$, where

$$\mathbf{b} = \begin{bmatrix} 3 \\ 4 \\ 2 \\ -2 \end{bmatrix}.$$

(b) Express $\mathbf{b}$ as a linear combination of the columns of A.

31. [BB] Use the factorization

$$\begin{bmatrix} -2 & -6 \\ 1 & 15 \end{bmatrix} = \begin{bmatrix} -2 & 0 \\ 1 & 1 \end{bmatrix} \begin{bmatrix} 1 & 3 \\ 0 & 12 \end{bmatrix} \text{ to solve}$$

$A\mathbf{x} = \mathbf{b}$ for $\mathbf{x} = \begin{bmatrix} x_1 \\ x_2 \end{bmatrix}$, with $\mathbf{b} = \begin{bmatrix} -3 \\ 4 \end{bmatrix}$, where

$$A = \begin{bmatrix} -2 & -6 \\ 1 & 15 \end{bmatrix}.$$

32. Use the factorization

$$\begin{bmatrix} 3 & 6 \\ 2 & -5 \end{bmatrix} = \begin{bmatrix} 3 & 0 \\ 2 & 1 \end{bmatrix} \begin{bmatrix} 1 & 2 \\ 0 & -9 \end{bmatrix} \text{ to solve}$$

$A\mathbf{x} = \mathbf{b}$ for $\mathbf{x} = \begin{bmatrix} x_1 \\ x_2 \end{bmatrix}$, with $\mathbf{b} = \begin{bmatrix} 6 \\ -11 \end{bmatrix}$, where

$$A = \begin{bmatrix} 3 & 6 \\ 2 & -5 \end{bmatrix}.$$

33. Use the factorization

$$\begin{bmatrix} 2 & 2 & 2 \\ 4 & 6 & 6 \\ 4 & 8 & 10 \end{bmatrix} = \begin{bmatrix} 1 & 0 & 0 \\ 2 & 1 & 0 \\ 2 & 2 & 1 \end{bmatrix} \begin{bmatrix} 2 & 2 & 2 \\ 0 & 2 & 2 \\ 0 & 0 & 2 \end{bmatrix} \text{ to}$$

solve $A\mathbf{x} = \mathbf{b}$ for $\mathbf{x} = \begin{bmatrix} x_1 \\ x_2 \\ x_3 \end{bmatrix}$, where $\mathbf{b} = \begin{bmatrix} 2 \\ 4 \\ 2 \end{bmatrix}$ and

$$A = \begin{bmatrix} 2 & 2 & 2 \\ 4 & 6 & 6 \\ 4 & 8 & 10 \end{bmatrix}.$$

34. [BB] Let $A = \begin{bmatrix} 1 & -1 & 0 \\ 3 & 0 & 2 \\ -1 & 0 & -1 \end{bmatrix}$ and $\mathbf{b} = \begin{bmatrix} 1 \\ 0 \\ 0 \end{bmatrix}$.

(a) Solve the system $A\mathbf{x} = \mathbf{b}$ by applying the method of Gaussian elimination to the augmented matrix.

(b) Factor $A = LU$.

(c) Why is A invertible?

(d) Find the first column of A^{-1}.

35. Answer Exercise 34 with

$$A = \begin{bmatrix} 1 & 0 & 1 & -1 \\ -1 & 1 & 0 & 4 \\ -5 & 4 & 0 & 1 \\ 2 & -2 & 3 & 0 \end{bmatrix} \text{ and } \mathbf{b} = \begin{bmatrix} 0 \\ 0 \\ 1 \\ 0 \end{bmatrix}.$$

For (d), find the third column of A^{-1}.

36. Let $L = \begin{bmatrix} 1 & 0 & 0 & 0 \\ -1 & 1 & 0 & 0 \\ 0 & 2 & 1 & 0 \\ 1 & 0 & 0 & 1 \end{bmatrix}$,

$$U = \begin{bmatrix} 3 & 1 & 0 & 1 \\ 0 & 1 & 1 & 0 \\ 0 & 0 & 2 & 0 \\ 0 & 0 & 0 & 1 \end{bmatrix}, \mathbf{b} = \begin{bmatrix} 3 \\ -4 \\ -2 \\ 2 \end{bmatrix}, \text{ and } A = LU.$$

Use the given factorization of A to solve the equation $A\mathbf{x} = \mathbf{b}$.

37. Solve each of the following systems as follows.

i. Write the system in the form $A\mathbf{x} = \mathbf{b}$.

ii. Factor $A = LU$.

iii. Solve $L\mathbf{y} = \mathbf{b}$ for $\mathbf{y}$ by forward substitution and $U\mathbf{x} = \mathbf{y}$ by back substitution. Write the solution to $A\mathbf{x} = \mathbf{b}$ in vector form.

(a) [BB] $\begin{aligned} 2x_1 + 2x_2 &= 8 \\ 4x_1 + 9x_2 &= 21 \end{aligned}$

(b) $\begin{aligned} 2x_1 + 4x_2 - 2x_3 &= 2 \\ 4x_1 + 9x_2 - 3x_3 &= 8 \\ -2x_1 - 3x_2 + 7x_3 &= 10 \end{aligned}$

(c) $\begin{aligned} 3x_1 + 4x_2 + x_3 &= 1 \\ 2x_1 + 3x_2 &= 0 \\ 4x_1 + 3x_2 - x_3 &= -2 \end{aligned}$

(d) $\begin{aligned} 2x_1 + 2x_2 + 2x_3 &= 2 \\ 2x_1 + 6x_2 + 6x_3 &= 4 \\ 4x_1 + 8x_2 + 10x_3 &= 2 \end{aligned}$

(e) $\begin{aligned} x_1 \quad - x_3 + 5x_4 &= 1 \\ -x_1 + x_2 \quad - 3x_4 &= 2 \\ 2x_2 - 3x_3 + 7x_4 &= 3 \\ 2x_1 - x_2 - 2x_3 + 12x_4 &= 4 \end{aligned}$

(f) $\begin{aligned} -2x_1 + 4x_2 - 8x_3 \quad - 8x_5 &= -14 \\ -6x_1 + 15x_2 - 24x_3 - 9x_4 - 21x_5 &= -36 \\ x_1 - x_2 + 4x_3 + x_4 - 2x_5 &= -8. \end{aligned}$

38. In each of the following cases,

    i. find an LU factorization of A, and

    ii. use this to solve $A\mathbf{x} = \mathbf{b}$.

(a) [BB] $A = \begin{bmatrix} 1 & 1 & 1 \\ 1 & 2 & 3 \\ 0 & 2 & 1 \end{bmatrix}$ and $\mathbf{b} = \begin{bmatrix} -1 \\ 0 \\ 1 \end{bmatrix}$

(b) $A = \begin{bmatrix} 2 & -3 & 0 \\ 4 & -5 & 1 \\ 2 & 0 & 4 \end{bmatrix}$ and $\mathbf{b} = \begin{bmatrix} 8 \\ 15 \\ 1 \end{bmatrix}$

(c) $A = \begin{bmatrix} 3 & -1 & 1 \\ 12 & -1 & 8 \\ -3 & 10 & 16 \end{bmatrix}$ and $\mathbf{b} = \begin{bmatrix} 0 \\ 2 \\ 1 \end{bmatrix}$

(d) $A = \begin{bmatrix} 1 & 0 & 1 & 1 \\ 1 & 1 & 2 & 0 \\ 0 & 1 & 1 & -4 \end{bmatrix}$ and $\mathbf{b} = \begin{bmatrix} 2 \\ 5 \\ 3 \end{bmatrix}$

(e) $A = \begin{bmatrix} 1 & 8 & 2 \\ 6 & 50 & 8 \\ 4 & 40 & -11 \\ -1 & 2 & 9 \end{bmatrix}$ and $\mathbf{b} = \begin{bmatrix} 10 \\ 44 \\ -39 \\ -45 \end{bmatrix}$

(f) $A = \begin{bmatrix} 2 & 1 & 0 \\ -4 & -3 & 1 \\ 6 & 4 & 7 \\ 0 & -2 & 6 \\ 2 & 1 & 8 \end{bmatrix}$ and $\mathbf{b} = \begin{bmatrix} -2 \\ 5 \\ 1 \\ 6 \\ 6 \end{bmatrix}$.

39. Let $A = \begin{bmatrix} 0 & 1 & -2 \\ 0 & 0 & 7 \\ 2 & 4 & 9 \end{bmatrix}$. Write down a permutation matrix P such that PA is upper triangular.

40. Find, if possible, an LU factorization of each of the following matrices:

(a) [BB] $\begin{bmatrix} 2 & 1 \\ 6 & 8 \end{bmatrix}$

(b) $\begin{bmatrix} 1 & 2 & 3 \\ 2 & 3 & 4 \\ 3 & 4 & 5 \end{bmatrix}$

(c) $\begin{bmatrix} 2 & -6 & 5 \\ -4 & 12 & -9 \\ 2 & -9 & 8 \end{bmatrix}$

(d) $\begin{bmatrix} -3 & 8 & 10 & 1 \\ 3 & -2 & -1 & 0 \\ -3 & 4 & 4 & 0 \\ 6 & -4 & -2 & 1 \\ x9 & 0 & 1 & 0 \end{bmatrix}$.

41. Find an equation of the form $A = PLU$ in each of the following cases:

(a) [BB] $A = \begin{bmatrix} 0 & 3 & 1 \\ 1 & 2 & 1 \end{bmatrix}$

(b) $A = \begin{bmatrix} 0 & 2 & 4 \\ 1 & 1 & 0 \\ 3 & 4 & 5 \end{bmatrix}$

(c) $A = \begin{bmatrix} 0 & 3 & 0 & 2 \\ 1 & 2 & 1 & -1 \\ 1 & 2 & 1 & 1 \\ 0 & -4 & 2 & 0 \end{bmatrix}$

(d) $A = \begin{bmatrix} 2 & -1 & 2 \\ 6 & -3 & 1 \\ 4 & -2 & 4 \\ -2 & 1 & 8 \end{bmatrix}$.

■ Critical Reading

42. Is the product of elementary matrices invertible? Explain.

43. If the rows on an $n \times n$ matrix P are the (transposes of the) standard basis vectors $\mathbf{e}_1, \dots, \mathbf{e}_n$ in some order, is the same true for the columns? If yes, is the order the same? [Hint: What is PP^T?]

44. Prove that the product of two $n \times n$ permutation matrices is a permutation matrix. [Hint: Let P_1 and P_2 be permutation matrices and examine the columns of $P_1 P_2$.]

45. The system $A\mathbf{x} = \mathbf{b}$ has been solved by applying Gaussian elimination to the augmented matrix $[A|\mathbf{b}]$, thereby reducing A to an upper triangular matrix U and the augmented matrix $[A|\mathbf{b}]$ to $[U|\mathbf{u}]$ for some vector $\mathbf{u}$. In the elimination, only the third elementary operation was used. Now suppose we wish to solve $A\mathbf{x} = \mathbf{b}$ by factoring $A = LU$, solving $L\mathbf{y} = \mathbf{b}$ for $\mathbf{y}$ and then $U\mathbf{x} = \mathbf{y}$ for $\mathbf{x}$. Gerard claims $\mathbf{y} = \mathbf{u}$. Is he right?

Guess an answer by examining $A\mathbf{x} = \mathbf{b}$, with

$A = \begin{bmatrix} 1 & 0 & 1 \\ 2 & -2 & 3 \\ -45 & 4 & -2 \end{bmatrix}$, $\mathbf{x} = \begin{bmatrix} x_1 \\ x_2 \\ x_3 \end{bmatrix}$, and

$\mathbf{b} = \begin{bmatrix} 2 \\ 8 \\ -16 \end{bmatrix}$. Then explain what is happening in general.

46. Explain how an LU factorization of a matrix A can be used to obtain an LU factorization of A^T.

47. Suppose P is the 5×5 permutation matrix whose columns are the standard basis vectors of $\mathbf{R}^5$ in the order $\mathbf{e}_1, \mathbf{e}_5, \mathbf{e}_3, \mathbf{e}_4, \mathbf{e}_2$. Let A be a 5×5 matrix with columns $\mathbf{c}_1, \mathbf{c}_2, \mathbf{c}_3, \mathbf{c}_4, \mathbf{c}_5$ and rows $\mathbf{r}_1, \mathbf{r}_2, \mathbf{r}_3, \mathbf{r}_4, \mathbf{r}_5$.

(a) Describe the matrix AP.

(b) Describe the matrix $P^T A$.

48. Let P be an $n \times n$ permutation matrix. Let $\mathbf{x}, \mathbf{y}$ be n-dimensional vectors. Show that $(P\mathbf{x}) \cdot (P\mathbf{y}) = \mathbf{x} \cdot \mathbf{y}$. [Hint: By 2.1.18, the dot product $\mathbf{u} \cdot \mathbf{v}$ of vectors $\mathbf{u}$ and $\mathbf{v}$ is the matrix product $\mathbf{u}^T \mathbf{v}$.]

49. Suppose A is an invertible matrix and its first two rows are exchanged to give a matrix B. Is B invertible? Explain.

2.6 LDU Factorizations

In Section 2.5, we found an LU factorization of $A = \begin{bmatrix} 1 & 2 & 3 & 4 \\ 5 & 6 & 7 & 8 \\ 9 & 10 & 0 & 1 \\ 0 & -2 & 3 & 4 \\ -1 & 6 & 4 & 9 \end{bmatrix}$. (See **2.5.17**.)

We found $A = LU'$ with

$$L = \begin{bmatrix} 1 & 0 & 0 & 0 & 0 \\ 5 & 1 & 0 & 0 & 0 \\ 9 & 2 & 1 & 0 & 0 \\ 0 & \frac{1}{2} & -\frac{7}{11} & 1 & 0 \\ -1 & -2 & \frac{9}{11} & -\frac{2}{3} & 1 \end{bmatrix} \quad \text{and} \quad U' = \begin{bmatrix} 1 & 2 & 3 & 4 \\ 0 & -4 & -8 & -12 \\ 0 & 0 & -11 & -11 \\ 0 & 0 & 0 & 3 \\ 0 & 0 & 0 & 0 \end{bmatrix}.$$

Since the diagonal entries of U' are not zero, we can factor $U' = DU$ with D diagonal and U an upper triangular matrix with 1s on the diagonal by factoring the diagonal entries from the rows of U' as follows:

$$\begin{bmatrix} 1 & 2 & 3 & 4 \\ 0 & -4 & -8 & -12 \\ 0 & 0 & -11 & -11 \\ 0 & 0 & 0 & 3 \\ 0 & 0 & 0 & 0 \end{bmatrix} = \begin{bmatrix} 1 & 0 & 0 & 0 & 0 \\ 0 & -4 & 0 & 0 & 0 \\ 0 & 0 & -11 & 0 & 0 \\ 0 & 0 & 0 & 3 & 0 \\ 0 & 0 & 0 & 0 & 1 \end{bmatrix} \begin{bmatrix} 1 & 2 & 3 & 4 \\ 0 & 1 & 2 & 3 \\ 0 & 0 & 1 & 1 \\ 0 & 0 & 0 & 1 \\ 0 & 0 & 0 & 0 \end{bmatrix}.$$

The LU factorization of $A = \begin{bmatrix} 1 & 2 & 3 & 4 \\ 5 & 6 & 7 & 8 \\ 9 & 10 & 0 & 1 \\ 0 & -2 & 3 & 4 \\ -1 & 6 & 4 & 9 \end{bmatrix}$ becomes an *LDU factorization*, like this:

$$\begin{bmatrix} 1 & 2 & 3 & 4 \\ 5 & 6 & 7 & 8 \\ 9 & 10 & 0 & 1 \\ 0 & -2 & 3 & 4 \\ -1 & 6 & 4 & 9 \end{bmatrix} = \begin{bmatrix} 1 & 0 & 0 & 0 & 0 \\ 5 & 1 & 0 & 0 & 0 \\ 9 & 2 & 1 & 0 & 0 \\ 0 & \frac{1}{2} & -\frac{7}{11} & 1 & 0 \\ 0 & 0 & 0 & 0 & 1 \end{bmatrix} \begin{bmatrix} 1 & 0 & 0 & 0 & 0 \\ 0 & -4 & 0 & 0 & 0 \\ 0 & 0 & -11 & 0 & 0 \\ 0 & 0 & 0 & 3 & 0 \\ 0 & 0 & 0 & 0 & 1 \end{bmatrix} \begin{bmatrix} 1 & 2 & 3 & 4 \\ 0 & 1 & 2 & 3 \\ 0 & 0 & 1 & 1 \\ 0 & 0 & 0 & 1 \\ 0 & 0 & 0 & 0 \end{bmatrix}.$$

2.6.1 DEFINITION

An *LDU factorization* of a matrix A is a representation of $A = LDU$ as the product of a (necessarily square) lower triangular matrix L with 1s on the diagonal, a (square) diagonal matrix D, and an upper triangular matrix U (the same size as A) with 1s on the diagonal.

2.6.2 To find an LDU factorization of an $m \times n$ matrix A, find an LU factorization $A = LU'$, if possible, using only the third elementary row operation. If none of the diagonal entries of U' are 0, factor these from the rows of U' to obtain $U' = DU$ with U upper triangular and 1s on the diagonal.

2.6.3 PROBLEM

Find an LDU factorization of $A = \begin{bmatrix} -1 & 2 & 0 & -2 \\ 2 & 4 & 6 & 8 \\ -3 & 2 & 4 & 1 \end{bmatrix}$.

Solution. We attempt to bring A to upper triangular form *using only the third elementary row operation.*

$$A \xrightarrow[\begin{subarray}{c} R2 \to R2 - (-2)(R1) \\ R3 \to R3 - 3R1 \end{subarray}]{} \begin{bmatrix} -1 & 2 & 0 & -2 \\ 0 & 8 & 6 & 4 \\ 0 & -4 & 4 & 7 \end{bmatrix}$$

$$\xrightarrow[R3 \to R3 - (-\frac{1}{2})R2]{} \begin{bmatrix} -1 & 2 & 0 & -2 \\ 0 & 8 & 6 & 4 \\ 0 & 0 & 7 & 9 \end{bmatrix} = U'.$$

At this point, we have $A = LU'$ with

$$L = \begin{bmatrix} 1 & 0 & 0 \\ -2 & 1 & 0 \\ 3 & -\frac{1}{2} & 1 \end{bmatrix},$$

the lower triangular matrix with 1s on the diagonal and multipliers below. The final LDU factorization of A is obtained by factoring the diagonal entries of U' from its rows.

$$\begin{bmatrix} -1 & 2 & 0 & -2 \\ 2 & 4 & 6 & 8 \\ -3 & 2 & 4 & 1 \end{bmatrix}$$

$$= \begin{bmatrix} 1 & 0 & 0 \\ -2 & 1 & 0 \\ 3 & -\frac{1}{2} & 1 \end{bmatrix} \begin{bmatrix} -1 & 0 & 0 \\ 0 & 8 & 0 \\ 0 & 0 & 7 \end{bmatrix} \begin{bmatrix} 1 & -2 & 0 & 2 \\ 0 & 1 & \frac{3}{4} & \frac{1}{2} \\ 0 & 0 & 1 & \frac{9}{7} \end{bmatrix}.$$

■ **Uniqueness Questions**

In general, neither LU nor LDU factorizations are unique. Here are two LU factorizations of $A = \begin{bmatrix} -4 & 2 \\ 3 & 1 \end{bmatrix}$:

$$\begin{bmatrix} -4 & 2 \\ 3 & 1 \end{bmatrix} = \begin{bmatrix} 1 & 0 \\ -\frac{3}{4} & 1 \end{bmatrix} \begin{bmatrix} -4 & 2 \\ 0 & \frac{5}{2} \end{bmatrix} = \begin{bmatrix} -4 & 0 \\ 3 & 1 \end{bmatrix} \begin{bmatrix} 1 & -\frac{1}{2} \\ 0 & \frac{5}{2} \end{bmatrix}.$$

Furthermore, for any real numbers a and b, the fact

$$\begin{bmatrix} 0 & 0 \\ 0 & 1 \end{bmatrix} = \begin{bmatrix} 1 & 0 \\ a & 1 \end{bmatrix} \begin{bmatrix} 0 & 0 \\ 0 & 1 \end{bmatrix} \begin{bmatrix} 1 & b \\ 0 & 1 \end{bmatrix}$$

shows that $A = \begin{bmatrix} 0 & 0 \\ 0 & 1 \end{bmatrix}$ can be written in the form LDU in many ways.

On the other hand, if we require that A be invertible and that L have 1s on the diagonal, then there is just one way to factor A in the form $A = LU$. To explain why, we first list a few facts about triangular matrices that we invite the reader to examine with examples. (The proofs are not hard, but they are too technical for the author's taste to write out here.)

1. A square triangular matrix is invertible if it has no 0s on the diagonal. [See 3.1.14 in Section 3.1.]

2. The product of lower triangular matrices is lower triangular and the product of upper triangular matrices is upper triangular. In each case, the diagonal entries of the product are the products of the diagonal entries of the two matrices. For example,

$$\begin{bmatrix} 2 & 4 & 5 \\ 0 & -1 & 6 \\ 0 & 0 & 3 \end{bmatrix} \begin{bmatrix} -4 & 1 & 2 \\ 0 & 7 & 8 \\ 0 & 0 & -3 \end{bmatrix} = \begin{bmatrix} -8 & \star & \star \\ 0 & -7 & \star \\ 0 & 0 & -9 \end{bmatrix}.$$

In particular, the product of two lower triangular matrices with 1s on the diagonal is a lower triangular matrix with 1s on the diagonal, and the same holds true for upper triangular matrices.

3. The inverse of a lower triangular matrix is lower triangular and the inverse of an upper triangular matrix is upper triangular.

Now suppose we have two LU factorizations of an invertible matrix A,

$$A = LU = L_1 U_1 \tag{1}$$

where L and L_1 are (necessarily square) lower triangular matrices with 1s on the diagonal, and U and U_1 are upper triangular matrices. The matrices L and L_1 are invertible, hence so are U and U_1 because each is the product of two invertible matrices, $U = L^{-1}A$ and $U_1 = L_1^{-1}A$ (see 2.2.12). From equation (1), we obtain

$$L^{-1}L_1 = UU_1^{-1},$$

which is interesting because the matrix on the left is lower triangular while the matrix on the right is upper triangular. This can happen only if each matrix is diagonal.

Reading Challenge 1   *Why?*

Since the diagonal entries of $L^{-1}L_1$ are 1s, the diagonal matrix in question must be the identity. So $L^{-1}L_1 = UU_1^{-1} = I$, hence $L_1 = L$ and $U_1 = U$, and the factorization $A = LU$ is unique.

An LDU factorization is often unique as well. We leave the proof of the following theorem to the exercises.

Theorem 2.6.4   *If an invertible matrix A can be factored $A = LDU$ and also $A = L_1 D_1 U_1$ with L and L_1 lower triangular matrices with 1s on the diagonal, U and U_1 upper triangular matrices with 1s on the diagonal, and D and D_1 diagonal matrices, then $L = L_1$, $U = U_1$, and $D = D_1$.*

■ **Symmetric Matrices**

We describe a neat application of Theorem 2.6.4.

2.6.5 DEFINITION   A matrix A is *symmetric* if it equals its transpose: $A^T = A$.

The name "symmetric" is derived from the fact that the entries of a symmetric matrix are symmetric with respect to the main diagonal. The $(2, 3)$ entry equals the $(3, 2)$ entry, the $(1, 4)$ entry equals the $(4, 1)$ entry, the (i, j) entry equals the (j, i) entry for any i and j.

For example, $\begin{bmatrix} 1 & 2 \\ 2 & 4 \end{bmatrix}$ is a symmetric matrix. Its first column equals its first row and its second column equals its second row. The matrix $\begin{bmatrix} 1 & 2 & 3 \\ 2 & 7 & -8 \\ 3 & -8 & 5 \end{bmatrix}$ is also symmetric.

Reading Challenge 2   *A symmetric matrix must be square. Why?*

It is easy to make symmetric matrices. Take any matrix A and form $S = AA^T$. This is symmetric because it equals its transpose: $S^T = (AA^T)^T = (A^T)^T A^T = AA^T = S$. (Remember Theorem 2.2.19: $(AB)^T = B^T A^T$.) For example, if

$$A = \begin{bmatrix} -1 & 0 \\ 1 & 3 \\ 2 & -2 \end{bmatrix},$$

then

$$A^T = \begin{bmatrix} -1 & 1 & 2 \\ 0 & 3 & -2 \end{bmatrix},$$

and

$$AA^T = \begin{bmatrix} 1 & -1 & -2 \\ -1 & 10 & -4 \\ -2 & -4 & 8 \end{bmatrix}$$

is symmetric.

Now suppose we have a factorization $A = LDU$ of a symmetric matrix A. Then $LDU = A = A^T = (LDU)^T = U^T D^T L^T$. The transpose of an upper triangular matrix is lower triangular and the transpose of a lower triangular matrix is upper triangular, so $LDU = U^T D^T L^T$ gives two LDU factorizations of A. Provided A is invertible, we have seen that the LDU factorization is unique. So $L = U^T$, $U = L^T$, and $LDU = U^T DU$. It will never be easier to find L!

2.6.6 Remark Our argument that $L = U^T$ in the LDU factorization of a symmetric matrix A assumed that A was invertible. In fact, this restriction is not necessary. In the exercises, we ask the reader to show that if a symmetric matrix can be factored $A = LDU$, then it can also be factored $A = U^T DU$. (See Exercise 10.)

2.6.7 PROBLEM   Find an LDU factorization of the symmetric matrix

$$A = \begin{bmatrix} 1 & 2 & -1 \\ 2 & 3 & 0 \\ -1 & 0 & 5 \end{bmatrix}.$$

Solution.   We reduce A to an upper triangular matrix using only the third elementary row operation,

$$A \to \begin{bmatrix} 1 & 2 & -1 \\ 0 & -1 & 2 \\ 0 & 2 & 4 \end{bmatrix} \to \begin{bmatrix} 1 & 2 & -1 \\ 0 & -1 & 2 \\ 0 & 0 & 8 \end{bmatrix} = U',$$

and factor $U' = DU$

$$\begin{bmatrix} 1 & 2 & -1 \\ 0 & -1 & 2 \\ 0 & 0 & 8 \end{bmatrix} = \begin{bmatrix} 1 & 0 & 0 \\ 0 & -1 & 0 \\ 0 & 0 & 8 \end{bmatrix} \begin{bmatrix} 1 & 2 & -1 \\ 0 & 1 & -2 \\ 0 & 0 & 1 \end{bmatrix},$$

obtaining $D = \begin{bmatrix} 1 & 0 & 0 \\ 0 & -1 & 0 \\ 0 & 0 & 8 \end{bmatrix}$ and $U = \begin{bmatrix} 1 & 2 & -1 \\ 0 & 1 & -2 \\ 0 & 0 & 1 \end{bmatrix}$.

We conclude that $L = U^T = \begin{bmatrix} 1 & 0 & 0 \\ 2 & 1 & 0 \\ -1 & -2 & 1 \end{bmatrix}$ and leave it to the reader to confirm that $A = LDU$;

$$\begin{bmatrix} 1 & 2 & -1 \\ 2 & 3 & 0 \\ -1 & 0 & 5 \end{bmatrix} = \begin{bmatrix} 1 & 0 & 0 \\ 2 & 1 & 0 \\ -1 & -2 & 1 \end{bmatrix} \begin{bmatrix} 1 & 0 & 0 \\ 0 & -1 & 0 \\ 0 & 0 & 8 \end{bmatrix} \begin{bmatrix} 1 & 2 & -1 \\ 0 & 1 & -2 \\ 0 & 0 & 1 \end{bmatrix}.$$

■ The *PLDU* Factorization

Suppose one or more row interchanges are needed in the Gaussian elimination process when a matrix A is carried to an upper triangular matrix U'. In Section 2.5, we observed that while A will not have an LU factorization in this case, it can be factored $A = PLU'$, with P a permutation matrix. If U' can be factored $U' = DU$, then we can also factor $A = PLDU$.

2.6.8 EXAMPLE

Let $A = \begin{bmatrix} 2 & 4 & 2 \\ 1 & 2 & 3 \\ 3 & 2 & 1 \end{bmatrix}$. As in Section 2.5, let $P' = \begin{bmatrix} 1 & 0 & 0 \\ 0 & 0 & 1 \\ 0 & 1 & 0 \end{bmatrix}$. Then $P'A = A' = \begin{bmatrix} 2 & 4 & 2 \\ 3 & 2 & 1 \\ 1 & 2 & 3 \end{bmatrix}$ and

$$A' = LU' = \begin{bmatrix} 1 & 0 & 0 \\ \frac{3}{2} & 1 & 0 \\ \frac{1}{2} & 0 & 1 \end{bmatrix} \begin{bmatrix} 2 & 4 & 2 \\ 0 & -4 & -2 \\ 0 & 0 & 2 \end{bmatrix}.$$

Factoring $U' = DU$ gives

$$A' = LDU = \begin{bmatrix} 1 & 0 & 0 \\ \frac{3}{2} & 1 & 0 \\ \frac{1}{2} & 0 & 1 \end{bmatrix} \begin{bmatrix} 2 & 0 & 0 \\ 0 & -4 & 0 \\ 0 & 0 & 2 \end{bmatrix} \begin{bmatrix} 1 & 2 & 1 \\ 0 & 1 & \frac{1}{2} \\ 0 & 0 & 1 \end{bmatrix}.$$

The inverse of the elementary matrix P' is $(P')^{-1} = P'$, so we have $A = PLDU$ with $P = P'$. ■

2.6.9 PROBLEM

Find an LDU factorization of $A = \begin{bmatrix} 0 & 1 & 1 & 0 \\ 0 & -1 & -1 & 4 \\ -2 & 3 & 2 & 1 \\ -6 & 4 & 2 & -8 \end{bmatrix}$ if possible. Otherwise find a *PLDU* factorization.

Solution. In Problem 2.5.25 of Section 2.5, we noted that A does not have an LU factorization, but $A = PLU'$ with

$$P = \begin{bmatrix} 0 & 0 & 1 & 0 \\ 0 & 0 & 0 & 1 \\ 1 & 0 & 0 & 0 \\ 0 & 1 & 0 & 0 \end{bmatrix},$$

$$L = \begin{bmatrix} 1 & 0 & 0 & 0 \\ 3 & 1 & 0 & 0 \\ 0 & -\frac{1}{5} & 1 & 0 \\ 0 & \frac{1}{5} & -1 & 1 \end{bmatrix}, \quad \text{and} \quad U' = \begin{bmatrix} -2 & 3 & 2 & 1 \\ 0 & -5 & -4 & -11 \\ 0 & 0 & \frac{1}{5} & -\frac{11}{5} \\ 0 & 0 & 0 & 4 \end{bmatrix}.$$

Since the diagonal entries of U' are not 0, we can factor $U' = DU$ with

$$D = \begin{bmatrix} -2 & 0 & 0 & 0 \\ 0 & -5 & 0 & 0 \\ 0 & 0 & \frac{1}{5} & 0 \\ 0 & 0 & 0 & 4 \end{bmatrix} \quad \text{and} \quad U = \begin{bmatrix} 1 & -\frac{3}{2} & -1 & -\frac{1}{2} \\ 0 & 1 & \frac{4}{5} & \frac{11}{5} \\ 0 & 0 & 1 & -11 \\ 0 & 0 & 0 & 1 \end{bmatrix},$$

hence obtaining a factorization $A = PLDU$.

Answers to Reading Challenges

1. A lower triangular matrix has 0s above the diagonal and an upper triangular matrix has 0s below the diagonal. Therefore, if a matrix is both lower and upper triangular, all entries off the diagonal (above and below) are 0. This is what we mean by a *diagonal* matrix.

2. If A is $m \times n$, then A^T is $n \times m$. If A is to equal A^T, these two matrices must have the same size, so $m = n$.

True/False Questions

Decide, with as little calculation as possible, whether each of the following statements is true or false and explain your answer whenever you say "false." (Answers can be found in the back of the book.)

1. If $A = LU$ is an LU factorization of a matrix, it is understood that L and U have 1s on the diagonal.

2. If $A = LDU$ is an LDU factorization of a matrix, it is understood that L and U have 1s on the diagonal.

3. The matrix $\begin{bmatrix} 0 & 0 \\ 0 & 0 \end{bmatrix}$ has an LDU factorization.

4. An LDU factorization of a matrix is unique.

5. The product of symmetric matrices is symmetric.

6. If A is invertible and symmetric, then A^{-1} is symmetric.

Exercises

*Solutions to exercises marked [BB] can be found in the **B**ack of the **B**ook.*

1. Find an LDU factorization of each of the following matrices. In each case, find L without calculation, just by keeping track of multipliers.

   (a) [BB] $\begin{bmatrix} -2 & 1 & 6 \\ 4 & 0 & 8 \\ 6 & 3 & -10 \end{bmatrix}$

   (b) $\begin{bmatrix} 1 & -1 & -2 \\ 2 & -3 & -5 \\ -1 & 3 & 5 \end{bmatrix}$

   (c) $\begin{bmatrix} 2 & 8 & -5 \\ 2 & 12 & -8 \\ 8 & 40 & -29 \\ 6 & 20 & -6 \end{bmatrix}$

   (d) $\begin{bmatrix} 1 & 0 & -1 & 5 & -1 \\ -1 & 4 & 0 & -3 & 2 \\ 0 & 2 & -3 & 7 & 6 \\ 2 & -1 & -2 & 12 & 3 \end{bmatrix}$.

2. Find an LDU factorization of each of the following:

   (a) [BB] the matrix $A = \begin{bmatrix} -3 & 3 & 6 \\ 2 & 5 & 10 \\ 0 & 1 & 4 \end{bmatrix}$ in Exercise 25, Section 2.5

   (b) the matrix $A = \begin{bmatrix} 2 & 4 & 6 & 18 \\ 4 & 5 & 6 & 24 \\ 3 & 1 & -2 & 4 \end{bmatrix}$ in Exercise 26, Section 2.5.

3. [BB; (a), (c)] Find an LDU factorization of each of the matrices of Exercise 27, Section 2.5.

4. Show that $\begin{bmatrix} 1 & 3 & -1 \\ 2 & 6 & 1 \end{bmatrix}$ does not have an LDU factorization. Why does this fact not contradict Theorem 2.6.4?

5. Suppose A and B are symmetric matrices.

   (a) [BB] Is AB symmetric (in general)?

   (b) What about A^2, ABA, and $A - B$?

6. Suppose that A and P are $n \times n$ matrices and A is symmetric. Prove that $P^T A P$ is symmetric.

7. [BB] A matrix A is *skew-symmetric* if $A^T = -A$. Show that if A and B are $n \times n$ skew-symmetric matrices, then so is $A + B$.

8. Let A be an $n \times n$ matrix.

   (a) Show that $A - A^T$ is skew-symmetric. (The concept of skew-symmetric matrix is defined in Exercise 7.)

   (b) Show that $A + A^T$ is symmetric.

   (c) Show that $A = S + K$ can be written as the sum of a symmetric matrix S and a skew-symmetric matrix K.

9. Find an LDU factorization of each of the following symmetric matrices:

   (a) [BB] $A = \begin{bmatrix} 2 & 5 \\ 5 & 9 \end{bmatrix}$

   (b) $A = \begin{bmatrix} 1 & 2 & 0 \\ 2 & 6 & 4 \\ 0 & 4 & 11 \end{bmatrix}$

   (c) $A = \begin{bmatrix} 2 & -1 & 0 \\ -1 & 2 & -1 \\ 0 & -1 & 2 \end{bmatrix}$

   (d) $A = \begin{bmatrix} -2 & 1 & 4 \\ 1 & 0 & 1 \\ 4 & 1 & 3 \end{bmatrix}$

   (e) $A = \begin{bmatrix} 1 & -4 & 2 \\ -4 & 1 & -2 \\ 2 & -2 & 2 \end{bmatrix}$

   (f) $A = \begin{bmatrix} 5 & 4 & -4 \\ 4 & 5 & 4 \\ -4 & 4 & 5 \end{bmatrix}$

   (g) $A = \begin{bmatrix} 1 & 2 & 2 & 1 \\ 2 & 7 & -2 & -1 \\ 2 & -2 & 3 & 8 \\ 1 & -1 & 8 & 4 \end{bmatrix}$

   (h) $\begin{bmatrix} 7 & 5 & 0 & 3 & 0 \\ 5 & 4 & -2 & 1 & -2 \\ 0 & -2 & 1 & 0 & 1 \\ 3 & 1 & 0 & 1 & 0 \\ 0 & -2 & 1 & 0 & 1 \end{bmatrix}$.

10. (a) If A is a symmetric matrix and $A = LDU$ is an LDU factorization, show that $A = U^T DU$.

    (b) If A is an invertible symmetric matrix and $A = LDU$, then $L = U^T$.

11. Find an LU or LDU factorization of

$$A = \begin{bmatrix} 0 & 1 & 1 & -1 \\ 1 & 2 & 3 & 1 \\ 2 & 7 & 9 & 3 \\ -3 & 1 & 1 & 2 \end{bmatrix}, \text{ if possible, and otherwise}$$

find a factorization $A = PLU$. Is there a factorization $A = PLDU$?

12. Prove Theorem 2.6.4.

■ **Critical Reading**

13. Suppose a matrix A has two (different) row echelon forms U_1 and U_2. Find two (different) LU factorizations of U_2.

14. If A is a symmetric $n \times n$ matrix and $\mathbf{x}$ is a vector in $\mathbf{R}^n$, then $\mathbf{x}^T A^2 \mathbf{x} = \|A\mathbf{x}\|^2$.

15. **(a)** Let A be any $m \times n$ matrix. Show that AA^T and $A^T A$ are both symmetric matrices. What is the size of each?

(b) If A is a matrix such that $A^T = A^T A$, show that A is symmetric and that $A = A^2$.

16. If A is a symmetric $n \times n$ matrix and $\mathbf{x}, \mathbf{y}$ are vectors in $\mathbf{R}^n$, prove that $A\mathbf{x} \cdot \mathbf{y} = \mathbf{x} \cdot A\mathbf{y}$. [Hint: 2.1.18.]

2.7 Finding the Inverse of a Matrix

In Section 2.1, we introduced the concept of the inverse of a matrix. A matrix A is said to have an inverse or to be invertible if there is another matrix B, called the inverse of A, such that $AB = I$ and $BA = I$. We write $B = A^{-1}$ for the inverse of A just as 5^{-1} denotes the inverse of the number 5. An invertible matrix is necessarily square and, as we shall see, if A is square, it is only necessary to verify that $AB = I$ (or $BA = I$) to establish that $B = A^{-1}$ (Corollary 2.7.17). This idea has been used previously, and it is the very best way to determine whether two square matrices A and B are inverses; compute AB or BA and see if you get the identity.

2.7.1 PROBLEM

Determine whether $A = \begin{bmatrix} -3 & 2 \\ -2 & 1 \end{bmatrix}$ and $B = \begin{bmatrix} 1 & -2 \\ 2 & -3 \end{bmatrix}$ are inverses.

Solution. We compute $AB = \begin{bmatrix} -3 & 2 \\ -2 & 1 \end{bmatrix} \begin{bmatrix} 1 & -2 \\ 2 & -3 \end{bmatrix} = \begin{bmatrix} 1 & 0 \\ 0 & 1 \end{bmatrix} = I$ and, since the matrices are square, conclude that A and B are inverses.

2.7.2 PROBLEM

Let $A = \begin{bmatrix} 1 & 2 & 3 \\ 4 & 5 & 6 \end{bmatrix}$ and $B = \begin{bmatrix} -1 & 1 \\ 0 & -1 \\ \frac{2}{3} & \frac{1}{3} \end{bmatrix}$. Then

$$AB = \begin{bmatrix} 1 & 2 & 3 \\ 4 & 5 & 6 \end{bmatrix} \begin{bmatrix} -1 & 1 \\ 0 & -1 \\ \frac{2}{3} & \frac{1}{3} \end{bmatrix} = \begin{bmatrix} 1 & 0 \\ 0 & 1 \end{bmatrix}.$$

Is A invertible?

Solution. No, A is not invertible because it is not square. Alternatively, you may check that $BA \neq I$.

Reading Challenge 1   *Are* $A = \begin{bmatrix} 2 & 7 & 1 \\ 0 & -3 & 1 \\ 1 & 1 & 0 \end{bmatrix}$ *and* $B = \begin{bmatrix} 3 & -5 & 0 \\ -1 & 2 & 1 \\ 2 & -4 & -7 \end{bmatrix}$ *inverses?*

■ A Method for Finding the Inverse

Given two matrices A and B, it is easy to determine if they are inverses: they must be square and AB must be the identity matrix. Suppose we are given just one matrix and we wish to know whether it has an inverse. If it does, we would also like to find this inverse.

Let A be an $n \times n$ matrix. If A is invertible, there is a matrix B with

$$AB = I = \begin{bmatrix} 1 & 0 & \cdots & 0 \\ 0 & 1 & \cdots & 0 \\ \vdots & \vdots & \ddots & \vdots \\ 0 & 0 & \cdots & 1 \end{bmatrix} = \begin{bmatrix} \mathbf{e}_1 & \mathbf{e}_2 & \cdots & \mathbf{e}_n \\ \downarrow & \downarrow & & \downarrow \end{bmatrix}.$$

Let the columns of B be the vectors $\mathbf{x}_1, \mathbf{x}_2, \ldots, \mathbf{x}_n$. So

$$B = \begin{bmatrix} \mathbf{x}_1 & \mathbf{x}_2 & \cdots & \mathbf{x}_n \\ \downarrow & \downarrow & & \downarrow \end{bmatrix} \quad \text{and} \quad AB = \begin{bmatrix} A\mathbf{x}_1 & A\mathbf{x}_2 & \cdots & A\mathbf{x}_n \\ \downarrow & \downarrow & & \downarrow \end{bmatrix}.$$

We want $AB = I$, so we wish to find vectors $\mathbf{x}_1, \ldots, \mathbf{x}_n$ such that

$$A\mathbf{x}_1 = \begin{bmatrix} 1 \\ 0 \\ 0 \\ \vdots \\ 0 \end{bmatrix} = \mathbf{e}_1, \quad A\mathbf{x}_2 = \begin{bmatrix} 0 \\ 1 \\ 0 \\ \vdots \\ 0 \end{bmatrix} = \mathbf{e}_2, \quad \ldots, \quad A\mathbf{x}_n = \begin{bmatrix} 0 \\ 0 \\ \vdots \\ 0 \\ 1 \end{bmatrix} = \mathbf{e}_n.$$

2.7.3 EXAMPLE   Suppose $A = \begin{bmatrix} 1 & 4 \\ 2 & 7 \end{bmatrix}$. Let $B = \begin{bmatrix} x & z \\ y & w \end{bmatrix}$. We want $AB = I$, that is,

$$\begin{bmatrix} x + 4y & z + 4w \\ 2x + 7y & 2z + 7w \end{bmatrix} = \begin{bmatrix} 1 & 0 \\ 0 & 1 \end{bmatrix}.$$

Comparing columns, this leads to these two systems of equations:

$$\begin{array}{ll} x + 4y = 1 \\ 2x + 7y = 0 \end{array} \quad \text{and} \quad \begin{array}{ll} z + 4w = 0 \\ 2z + 7w = 1. \end{array}$$

This is just

$$A\begin{bmatrix} x \\ y \end{bmatrix} = \begin{bmatrix} 1 \\ 0 \end{bmatrix} = \mathbf{e}_1, \quad A\begin{bmatrix} z \\ w \end{bmatrix} = \begin{bmatrix} 0 \\ 1 \end{bmatrix} = \mathbf{e}_2.$$

Solving the first system,

$$\left[\begin{array}{cc|c} 1 & 4 & 1 \\ 2 & 7 & 0 \end{array}\right], \rightarrow \left[\begin{array}{cc|c} 1 & 4 & 1 \\ 0 & -1 & -2 \end{array}\right] \rightarrow \left[\begin{array}{cc|c} 1 & 4 & 1 \\ 0 & ① & 2 \end{array}\right],$$

so $y = 2$, $x + 4y = 1$, so $x = 1 - 4y = -7$. This solution can also be obtained by continuing the Gaussian elimination, using the circled pivot as indicated to get a zero *above* it:

$$\left[\begin{array}{cc|c} 1 & 4 & 1 \\ 0 & ① & 2 \end{array}\right] \rightarrow \left[\begin{array}{cc|c} 1 & 0 & -7 \\ 0 & 1 & 2 \end{array}\right],$$

thus $x = -7$, $y = 2$ comes directly. Solving the second system this way,

$$\left[\begin{array}{cc|c} 1 & 4 & 0 \\ 2 & 7 & 1 \end{array}\right] \rightarrow \left[\begin{array}{cc|c} 1 & 4 & 0 \\ 0 & -1 & 1 \end{array}\right] \rightarrow \left[\begin{array}{cc|c} 1 & 4 & 0 \\ 0 & 1 & -1 \end{array}\right] \rightarrow \left[\begin{array}{cc|c} 1 & 0 & 4 \\ 0 & 1 & -1 \end{array}\right],$$

so $z = 4$, $w = -1$, and $B = A^{-1} = \left[\begin{array}{cc} -7 & 4 \\ 2 & -1 \end{array}\right]$.

In the two systems of linear equations just solved, the row operations required to bring each matrix to row echelon form were precisely the same because the coefficient matrix was the same in each case. It follows that both systems can be solved at the same time, like this:

$$\left[\begin{array}{cc|cc} 1 & 4 & 1 & 0 \\ 2 & 7 & 0 & 1 \end{array}\right] \rightarrow \left[\begin{array}{cc|cc} 1 & 4 & 1 & 0 \\ 0 & -1 & -2 & 1 \end{array}\right] \rightarrow \left[\begin{array}{cc|cc} 1 & 4 & 1 & 0 \\ 0 & 1 & 2 & -1 \end{array}\right]$$

$$\rightarrow \left[\begin{array}{cc|cc} 1 & 0 & -7 & 4 \\ 0 & 1 & 2 & -1 \end{array}\right].$$

The final matrix is in *reduced* row echelon form with I on the left and A^{-1} to the right. ■

2.7.4 DEFINITION

A matrix is in *reduced row echelon form* if it is in row echelon form, each pivot (that is, each leading nonzero entry in a nonzero row) is a 1, and each such pivot is the only nonzero entry in its column.

Thus a matrix in reduced row echelon has 0s *above* as well as below each leading 1.

2.7.5 EXAMPLES

The following matrices are all in reduced row echelon form:

$$\left[\begin{array}{cc} 1 & 0 \\ 0 & 1 \end{array}\right], \left[\begin{array}{ccc} 1 & 0 & -2 \\ 0 & 1 & 1 \end{array}\right], \left[\begin{array}{cccc} 1 & 0 & 0 & 0 \\ 0 & 0 & 1 & 0 \\ 0 & 0 & 0 & 1 \end{array}\right], \text{ and } \left[\begin{array}{cccccc} 1 & 0 & 0 & 0 & 0 & 4 \\ 0 & 1 & 0 & 0 & 0 & -2 \\ 0 & 0 & 0 & 0 & 1 & 3 \end{array}\right].$$ ■

2.7.6 Remark Unlike row echelon form, the *reduced* row echelon form of a matrix is unique, so we refer to "the reduced row echelon form." See Appendix A of *Linear Algebra and Its Applications* by David C. Lay, Addison-Wesley (2000) for a proof.

To bring a matrix to reduced row echelon form, we use Gaussian elimination in the usual way, sweeping across the columns of the matrix from left to right making sure all nonzero leading entries are 1, and then, each time we get a leading 1, using this to get 0s in the rest of the corresponding column, below and *above* the 1.

2.7.7 EXAMPLE

$$\begin{bmatrix} ①&2&-5 \\ 3&1&5 \\ -2&3&4 \end{bmatrix} \to \begin{bmatrix} 1&2&-5 \\ 0&-5&20 \\ 0&7&-6 \end{bmatrix} \to \begin{bmatrix} 1&2&-5 \\ 0&①&-4 \\ 0&7&-6 \end{bmatrix}$$

$$\to \begin{bmatrix} 1&0&3 \\ 0&1&-4 \\ 0&0&22 \end{bmatrix} \to \begin{bmatrix} 1&0&3 \\ 0&1&-4 \\ 0&0&① \end{bmatrix} \to \begin{bmatrix} 1&0&0 \\ 0&1&0 \\ 0&0&1 \end{bmatrix}. \qquad \blacksquare$$

Our interest in reduced row echelon form derives from a method for finding the inverse of a matrix, illustrated in Example 2.7.3.

> **2.7.8** To find the inverse of A, apply Gaussian elimination to $[A|I]$, attempting to move A to reduced row echelon form. If this is $[I|B]$ for some B, then $B = A^{-1}$. If you cannot move A to I, then A is not invertible.

2.7.9 EXAMPLE

Suppose $A = \begin{bmatrix} 1&2&0 \\ 3&-1&2 \\ -2&3&-2 \end{bmatrix}$. Gaussian elimination applied to $[A|I]$ proceeds

$$\left[\begin{array}{ccc|ccc} 1&2&0&1&0&0 \\ 3&-1&2&0&1&0 \\ -2&3&-2&0&0&1 \end{array} \right] \to \left[\begin{array}{ccc|ccc} 1&2&0&1&0&0 \\ 0&-7&2&-3&1&0 \\ 0&7&-2&2&0&1 \end{array} \right]$$

$$\to \left[\begin{array}{ccc|ccc} 1&2&0&1&0&0 \\ 0&-7&2&-3&1&0 \\ 0&0&0&-1&1&1 \end{array} \right].$$

There is no point in continuing. The three initial 0s in the third row cannot be changed. We cannot make the left 3×3 block the identity matrix. The given matrix A has no inverse. $\qquad \blacksquare$

2.7.10 EXAMPLE

Suppose $A = \begin{bmatrix} 2&7&0 \\ 1&4&-1 \\ 1&3&0 \end{bmatrix}$.

Gaussian elimination proceeds $[A|I] = \left[\begin{array}{ccc|ccc} 2&7&0&1&0&0 \\ 1&4&-1&0&1&0 \\ 1&3&0&0&0&1 \end{array} \right]$

$$\to \left[\begin{array}{ccc|ccc} 1&4&-1&0&1&0 \\ 2&7&0&1&0&0 \\ 1&3&0&0&0&1 \end{array} \right] \to \left[\begin{array}{ccc|ccc} 1&4&-1&0&1&0 \\ 0&-1&2&1&-2&0 \\ 0&-1&1&0&-1&1 \end{array} \right]$$

$$\rightarrow \left[\begin{array}{ccc|ccc} 1 & 4 & -1 & 0 & 1 & 0 \\ 0 & 1 & -2 & -1 & 2 & 0 \\ 0 & -1 & 1 & 0 & -1 & 1 \end{array}\right] \rightarrow \left[\begin{array}{ccc|ccc} 1 & 0 & 7 & 4 & -7 & 0 \\ 0 & 1 & -2 & -1 & 2 & 0 \\ 0 & 0 & -1 & -1 & 1 & 1 \end{array}\right]$$

$$\rightarrow \left[\begin{array}{ccc|ccc} 1 & 0 & 7 & 4 & -7 & 0 \\ 0 & 1 & -2 & -1 & 2 & 0 \\ 0 & 0 & 1 & 1 & -1 & -1 \end{array}\right] \rightarrow \left[\begin{array}{ccc|ccc} 1 & 0 & 0 & -3 & 0 & 7 \\ 0 & 1 & 0 & 1 & 0 & -2 \\ 0 & 0 & 1 & 1 & -1 & -1 \end{array}\right].$$

If we haven't made a mistake, A^{-1} should be the matrix to the right of the vertical line: $B = \left[\begin{array}{ccc} -3 & 0 & 7 \\ 1 & 0 & -2 \\ 1 & -1 & -1 \end{array}\right]$. We check that

$$AB = \left[\begin{array}{ccc} 2 & 7 & 0 \\ 1 & 4 & -1 \\ 1 & 3 & 0 \end{array}\right]\left[\begin{array}{ccc} -3 & 0 & 7 \\ 1 & 0 & -2 \\ 1 & -1 & -1 \end{array}\right] = \left[\begin{array}{ccc} 1 & 0 & 0 \\ 0 & 1 & 0 \\ 0 & 0 & 1 \end{array}\right] = I.$$

As noted at the start of this section, $AB = I$ with A square ensures that $BA = I$ too, so, indeed, $B = A^{-1}$. ∎

Suppose a and b are real numbers and we want to solve the equation $ax = b$. If $a \neq 0$, then $x = \frac{b}{a}$, a solution that can also be written $x = a^{-1}b$. Similarly, suppose we want to solve the system of linear equations $A\mathbf{x} = \mathbf{b}$. If A is invertible, then $\mathbf{x} = A^{-1}\mathbf{b}$ because $A\mathbf{x} = A(A^{-1}\mathbf{b}) = I\mathbf{b} = \mathbf{b}$. Moreover, this is the only solution to $A\mathbf{x} = \mathbf{b}$—the solution is unique—because if $A\mathbf{x}_1 = A\mathbf{x}_2 = \mathbf{b}$, multiplying on the left by A^{-1} would give $A^{-1}A\mathbf{x}_1 = A^{-1}A\mathbf{x}_2$, so $I\mathbf{x}_1 = I\mathbf{x}_2$ and $\mathbf{x}_1 = \mathbf{x}_2$.

2.7.11 EXAMPLE   The system

$$\begin{array}{rcl} 2x_1 + 7x_2 & = & -2 \\ x_1 + 4x_2 - x_3 & = & 4 \\ x_1 + 3x_2 & = & 5 \end{array}$$

is $A\mathbf{x} = \mathbf{b}$ with $A = \left[\begin{array}{ccc} 2 & 7 & 0 \\ 1 & 4 & -1 \\ 1 & 3 & 0 \end{array}\right]$, the matrix of **2.7.10**, $\mathbf{x} = \left[\begin{array}{c} x_1 \\ x_2 \\ x_3 \end{array}\right]$, and $\mathbf{b} = \left[\begin{array}{c} -2 \\ 4 \\ 5 \end{array}\right]$.

Since A is invertible, the unique solution is

$$\mathbf{x} = A^{-1}\mathbf{b} = \left[\begin{array}{ccc} -3 & 0 & 7 \\ 1 & 0 & -2 \\ 1 & -1 & -1 \end{array}\right]\left[\begin{array}{c} -2 \\ 4 \\ 5 \end{array}\right] = \left[\begin{array}{c} 41 \\ -12 \\ -11 \end{array}\right].$$

We check that this is the right answer by computing $A\mathbf{x}$. We have

$$A\mathbf{x} = \left[\begin{array}{ccc} 2 & 7 & 0 \\ 1 & 4 & -1 \\ 1 & 3 & 0 \end{array}\right]\left[\begin{array}{c} 41 \\ -12 \\ -11 \end{array}\right] = \left[\begin{array}{c} -2 \\ 4 \\ 5 \end{array}\right] = \mathbf{b},$$

so our answer is correct. ∎

2.7.12 PROBLEM Express the vector $\mathbf{b} = \begin{bmatrix} -2 \\ 4 \\ 5 \end{bmatrix}$ as a linear combination of the columns of the matrix

A in **2.7.11**.

Solution. We just saw that $A\mathbf{x} = \mathbf{b}$ with $\mathbf{x} = \begin{bmatrix} 41 \\ -12 \\ -11 \end{bmatrix}$. Since $A\mathbf{x}$ is a linear combina-

tion of the columns of A with coefficients the components of $\mathbf{x}$—please recall and never forget 2.1.33—we have

$$\mathbf{b} = 41 \begin{bmatrix} 2 \\ 1 \\ 1 \end{bmatrix} - 12 \begin{bmatrix} 7 \\ 4 \\ 3 \end{bmatrix} - 11 \begin{bmatrix} 0 \\ -1 \\ 0 \end{bmatrix}.$$

2.7.13 Remark Whether working by hand or with a computer, we virtually never solve $A\mathbf{x} = \mathbf{b}$ by the method employed in **2.7.11**. By hand, we much prefer Gaussian elimination and back substitution on the augmented matrix $[A|\mathbf{b}]$. In the first place, this is easier. Additionally, at the outset, there is no reason at all to suppose that A^{-1} actually exists, so trying to find A^{-1} in order to solve $A\mathbf{x} = \mathbf{b}$ is potentially a waste of time. The real value to the approach illustrated in **2.7.11** is its contribution to the theory of linear equations, which we discuss in more detail in Section 4.1. For now, we simply remark that

> If A has an inverse, then $A\mathbf{x} = \mathbf{b}$ has the unique solution $\mathbf{x} = A^{-1}\mathbf{b}$.

We conclude this section with a discussion of two important theorems about invertible matrices.

Theorem 2.7.14 *A matrix is invertible if and only if it can be written as the product of elementary matrices.*

Proof Suppose A is the product of elementary matrices. An elementary matrix is invertible. Also, the product of invertible matrices is invertible. Thus A is invertible.

On the other hand, suppose A is invertible. Then we can transform A to the identity matrix by a sequence of elementary row operations. The row reduction looks like this:

$$A \to E_1 A \to E_2(E_1 A) \to \cdots \to (E_k E_{k-1} \cdots E_2 E_1)A = I,$$

where $E_1, E_2, \ldots, E_k$ are elementary matrices.

So $A = (E_k E_{k-1} \cdots E_2 E_1)^{-1} = E_1^{-1} E_2^{-1} \cdots E_{k-1}^{-1} E_k^{-1}$. Since the inverse of an elementary matrix is invertible, $A = E_1^{-1} E_2^{-1} \cdots E_k^{-1}$ is an expression of A as the product of elementary matrices. ∎

2.7.15 PROBLEM Express $A = \begin{bmatrix} 2 & 3 \\ 1 & 1 \end{bmatrix}$ as the product of elementary matrices.

Solution. We move A to the identity matrix by a sequence of elementary row operations.

$$A = \begin{bmatrix} 2 & 3 \\ 1 & 1 \end{bmatrix} \rightarrow \begin{bmatrix} 1 & 1 \\ 2 & 3 \end{bmatrix} = E_1 A$$

$$\rightarrow \begin{bmatrix} 1 & 1 \\ 0 & 1 \end{bmatrix} = E_2 E_1 A \rightarrow \begin{bmatrix} 1 & 0 \\ 0 & 1 \end{bmatrix} = E_3 E_2 E_1 A = I$$

with

$$E_1 = \begin{bmatrix} 0 & 1 \\ 1 & 0 \end{bmatrix}, \quad E_2 = \begin{bmatrix} 1 & 0 \\ -2 & 1 \end{bmatrix}, \quad \text{and} \quad E_3 = \begin{bmatrix} 1 & -1 \\ 0 & 1 \end{bmatrix}.$$

So $(E_3 E_2 E_1) A = I$ and $A = (E_3 E_2 E_1)^{-1} = E_1^{-1} E_2^{-1} E_3^{-1}$;

$$\begin{bmatrix} 2 & 3 \\ 1 & 1 \end{bmatrix} = \begin{bmatrix} 0 & 1 \\ 1 & 0 \end{bmatrix} \begin{bmatrix} 1 & 0 \\ 2 & 1 \end{bmatrix} \begin{bmatrix} 1 & 1 \\ 0 & 1 \end{bmatrix}.$$

2.7.16 PROBLEM   Express $A = \begin{bmatrix} -1 & 1 & 3 \\ -2 & 2 & 7 \\ 4 & -3 & -12 \end{bmatrix}$ as the product of elementary matrices.

Solution. Row reduction to I is

$$A = \begin{bmatrix} -1 & 1 & 3 \\ -2 & 2 & 7 \\ 4 & -3 & -12 \end{bmatrix} \rightarrow \begin{bmatrix} 1 & -1 & -3 \\ -2 & 2 & 7 \\ 4 & -3 & -12 \end{bmatrix} = E_1 A$$

$$\rightarrow \begin{bmatrix} 1 & -1 & -3 \\ 0 & 0 & 1 \\ 4 & -3 & -12 \end{bmatrix} = E_2 E_1 A \rightarrow \begin{bmatrix} 1 & -1 & -3 \\ 0 & 0 & 1 \\ 0 & 1 & 0 \end{bmatrix} = E_3 E_2 E_1 A$$

$$\rightarrow \begin{bmatrix} 1 & -1 & -3 \\ 0 & 1 & 0 \\ 0 & 0 & 1 \end{bmatrix} = E_4 E_3 E_2 E_1 A \rightarrow \begin{bmatrix} 1 & 0 & -3 \\ 0 & 1 & 0 \\ 0 & 0 & 1 \end{bmatrix} = E_5 E_4 E_3 E_2 E_1 A$$

$$\rightarrow \begin{bmatrix} 1 & 0 & 0 \\ 0 & 1 & 0 \\ 0 & 0 & 1 \end{bmatrix} = E_6 E_5 E_4 E_3 E_2 E_1 A = I,$$

where

$$E_1 = \begin{bmatrix} -1 & 0 & 0 \\ 0 & 1 & 0 \\ 0 & 0 & 1 \end{bmatrix}, \quad E_2 = \begin{bmatrix} 1 & 0 & 0 \\ 2 & 1 & 0 \\ 0 & 0 & 1 \end{bmatrix}, \quad E_3 = \begin{bmatrix} 1 & 0 & 0 \\ 0 & 1 & 0 \\ -4 & 0 & 1 \end{bmatrix},$$

$$E_4 = \begin{bmatrix} 1 & 0 & 0 \\ 0 & 0 & 1 \\ 0 & 1 & 0 \end{bmatrix}, \quad E_5 = \begin{bmatrix} 1 & 1 & 0 \\ 0 & 1 & 0 \\ 0 & 0 & 1 \end{bmatrix}, \quad E_6 = \begin{bmatrix} 1 & 0 & 3 \\ 0 & 1 & 0 \\ 0 & 0 & 1 \end{bmatrix}.$$

Since

$$(E_6 E_5 E_4 E_3 E_2 E_1) A = I, \quad A = (E_6 E_5 E_4 E_3 E_2 E_1)^{-1} = E_1^{-1} E_2^{-1} E_3^{-1} E_4^{-1} E_5^{-1} E_6^{-1},$$

which is

$$\begin{bmatrix} -1 & 1 & 3 \\ -2 & 2 & 7 \\ 4 & -3 & -12 \end{bmatrix} = \begin{bmatrix} -1 & 0 & 0 \\ 0 & 1 & 0 \\ 0 & 0 & 1 \end{bmatrix} \begin{bmatrix} 1 & 0 & 0 \\ -2 & 1 & 0 \\ 0 & 0 & 1 \end{bmatrix} \begin{bmatrix} 1 & 0 & 0 \\ 0 & 1 & 0 \\ 4 & 0 & 1 \end{bmatrix}$$

$$\cdot \begin{bmatrix} 1 & 0 & 0 \\ 0 & 0 & 1 \\ 0 & 1 & 0 \end{bmatrix} \begin{bmatrix} 1 & -1 & 0 \\ 0 & 1 & 0 \\ 0 & 0 & 1 \end{bmatrix} \begin{bmatrix} 1 & 0 & -3 \\ 0 & 1 & 0 \\ 0 & 0 & 1 \end{bmatrix}.$$

We now justify a property of square matrices we have used on a number of occasions.

Corollary 2.7.17 *If A and B are square $n \times n$ matrices and $AB = I$ (the $n \times n$ identity matrix), then $BA = I$ too, so A and B are invertible.*

Proof If $AB = I$, then the system $B\mathbf{x} = \mathbf{0}$ has only the solution $\mathbf{x} = \mathbf{0}$: from $B\mathbf{x} = \mathbf{0}$, we have $AB\mathbf{x} = A\mathbf{0}$, so $I\mathbf{x} = \mathbf{0}$ and $\mathbf{x} = \mathbf{0}$. Thus, when we try to solve $B\mathbf{x} = \mathbf{0}$ by reducing B to row echelon form, there are no free variables (free variables imply infinitely many solutions), every column contains a pivot, and the reduced row echelon form of B is I. As we have seen, this means that $EB = I$ for some product E of elementary matrices. Since E is invertible, $B = E^{-1}$ is the product of elementary matrices. Thus B is invertible by Theorem 2.7.14.

It follows readily that A is also invertible, for multiplying $AB = I$ on the right by B^{-1} gives $A = B^{-1}$. (See 2.2.4.) ∎

Reading Challenge 2 *In the statement of Corollary 2.7.17, the hypothesis that both matrices be square is redundant. It is sufficient, for example, that only B be square. Why?*

Reading Challenge 3 *Where was the fact that B is square used in the proof of Corollary 2.7.17?*

Recall that vectors $\mathbf{x}_1, \mathbf{x}_2, \ldots, \mathbf{x}_n$ are *linearly independent* if and only if the only linear combination of them that is the zero vector is the trivial one—all coefficients 0;

$$c_1 \mathbf{x}_1 + c_2 \mathbf{x}_2 + \cdots + c_n \mathbf{x}_n = \mathbf{0} \quad \text{implies} \quad c_1 = 0, c_2 = 0, \ldots, c_n = 0.$$

Since $c_1\mathbf{x}_1 + c_2\mathbf{x}_2 + \cdots + c_n\mathbf{x}_n = A\mathbf{x}$ where $A = \begin{bmatrix} \mathbf{x}_1 & \mathbf{x}_2 & \cdots & \mathbf{x}_n \\ \downarrow & \downarrow & & \downarrow \end{bmatrix}$, the condition

for linear independence can be summarized by stating that $A\mathbf{x} = \mathbf{0}$ implies $\mathbf{x} = \mathbf{0}$; that is, the linear system $A\mathbf{x} = \mathbf{0}$ has only the trivial solution. It follows that a row echelon form of A has no free variables (it is free variables that give infinitely many solutions and, in particular, nontrivial solutions). Thus a row echelon form of A must have a pivot in every column.

Now suppose A is square. In this case, a pivot in every column of a row echelon form guarantees that A can be moved to the identity matrix with elementary row operations. So A is the product of elementary matrices and hence invertible. We have shown that if the columns of a square matrix A are linearly independent, then A is invertible.

On other hand, if A is invertible, its columns are surely linearly independent since if $A\mathbf{x} = \mathbf{0}$, then $A^{-1}A\mathbf{x} = \mathbf{0}$, so $\mathbf{x} = \mathbf{0}$.

Theorem 2.7.18 *A square matrix is invertible if and only if its columns are linearly independent.*

Answers to Reading Challenges

1. $AB = \begin{bmatrix} 1 & 0 & 0 \\ 5 & -10 & -10 \\ 2 & -3 & 1 \end{bmatrix} \neq I$, so A and B are not inverses.

2. Suppose A is $r \times s$. Since we can form the product AB with B a square $n \times n$ matrix, we must have $s = n$. Thus AB is $r \times n$. Since AB is the $n \times n$ identity matrix, $r = n$ too.

3. We discovered that B can be reduced to a row echelon matrix U with a pivot in every column. Since B is square, so is U; therefore U has a pivot in every row. It follows that the *reduced* row echelon form of B is I.

True/False Questions

Decide, with as little calculation as possible, whether each of the following statements is true or false and explain your answer whenever you say "false." (Answers can be found in the back of the book.)

1. The matrix $\begin{bmatrix} 3 & 0 & 4 \\ 0 & 0 & 9 \\ 0 & 0 & 0 \end{bmatrix}$ is in row echelon form.

2. A row echelon form for a matrix is unique.

3. If A and B are invertible matrices, so is AB.

4. If A and B are invertible matrices, so is $A + B$.

5. If E and F are elementary matrices, EF is invertible.

6. If ABC is an invertible matrix, then its inverse is $A^{-1}B^{-1}C^{-1}$.

7. If A and B are matrices with $AB = I$, then $BA = I$.

Exercises

Solutions to exercises marked [BB] can be found in the Back of the Book.

1. Determine whether each of the following pairs of matrices are inverses of each other:

   (a) [BB] $A = \begin{bmatrix} -1 & 2 \\ -7 & 10 \end{bmatrix}$, $B = \frac{1}{4}\begin{bmatrix} 10 & -2 \\ 7 & -1 \end{bmatrix}$

   (b) $A = \begin{bmatrix} 1 & 2 \\ 3 & 4 \end{bmatrix}$, $B = \frac{1}{2}\begin{bmatrix} -4 & 2 & 1 \\ 3 & -1 & 0 \end{bmatrix}$

   (c) $A = \begin{bmatrix} 0 & 2 & 3 \\ 4 & 5 & 6 \\ 7 & 8 & 9 \end{bmatrix}$,

   $B = \frac{1}{3}\begin{bmatrix} -3 & 6 & -3 \\ 6 & -21 & 12 \\ -3 & 14 & -8 \end{bmatrix}$

   (d) $A = \begin{bmatrix} -1 & 0 & 1 \\ 2 & 4 & -1 \\ 0 & 1 & 1 \end{bmatrix}$,

   $B = \begin{bmatrix} -5 & -1 & 4 \\ 2 & 1 & -1 \\ -2 & -1 & 4 \end{bmatrix}$

   (e) $A = \begin{bmatrix} 0 & 4 & 3 & 5 \\ 1 & 7 & 2 & -1 \\ 5 & 3 & -1 & 0 \\ -2 & 1 & 7 & 1 \end{bmatrix}$,

   $B = \begin{bmatrix} 1 & -3 & 4 & 2 \\ 5 & -1 & 6 & 1 \\ -7 & 5 & 2 & 3 \\ 0 & -2 & 5 & 1 \end{bmatrix}$

   (f) $A = \frac{1}{2}\begin{bmatrix} 1 & -2 & 2 & 1 \\ 2 & 1 & 1 & -2 \\ 1 & 0 & 1 & 3 \\ 1 & -1 & 1 & 1 \end{bmatrix}$,

   $B = \frac{1}{4}\begin{bmatrix} -7 & 1 & -2 & 15 \\ 2 & 2 & 4 & -10 \\ 10 & 2 & 4 & -18 \\ -1 & -1 & 2 & 1 \end{bmatrix}$.

2. Find the reduced row echelon form of each of the following matrices:

   (a) [BB] $\begin{bmatrix} 1 & -1 & 2 \\ -2 & 0 & 1 \\ -3 & 3 & -5 \\ 2 & -1 & 4 \end{bmatrix}$

   (b) $\begin{bmatrix} 1 & 1 & 5 & 17 \\ -1 & 0 & -2 & -8 \\ 1 & 5 & 18 & 56 \\ 0 & 3 & 8 & 24 \end{bmatrix}$

   (c) $\begin{bmatrix} 1 & 0 & 2 & 2 & 0 \\ 4 & 1 & 12 & 8 & 4 \\ 0 & -1 & -4 & 1 & -2 \\ 1 & 2 & 10 & 6 & 14 \end{bmatrix}$

   (d) $\begin{bmatrix} 0 & -1 & 2 & 1 & 2 & 1 & -1 \\ 0 & 1 & -2 & 2 & 7 & 2 & 4 \\ 0 & -2 & 4 & 3 & 7 & 1 & 0 \\ 0 & 3 & -6 & 1 & 6 & 4 & 1 \end{bmatrix}$.

3. Determine whether each of the following matrices has an inverse. Find the inverse whenever this exists.

   (a) [BB] $\begin{bmatrix} 3 & 1 \\ -6 & -3 \end{bmatrix}$ (b) $\begin{bmatrix} 1 & 2 \\ -3 & 4 \end{bmatrix}$

   (c) $\begin{bmatrix} -2 & 3 \\ 4 & -6 \end{bmatrix}$ (d) $\begin{bmatrix} -2 & 2 \\ 1 & 1 \end{bmatrix}$

   (e) [BB] $\begin{bmatrix} 2 & 4 & 2 \\ 1 & 2 & 3 \\ 3 & 2 & 1 \end{bmatrix}$ (f) $\begin{bmatrix} 1 & 1 & 2 \\ 3 & 2 & 3 \\ 4 & 2 & 1 \end{bmatrix}$

   (g) $\begin{bmatrix} -2 & 3 & 4 \\ 1 & 0 & 1 \\ -1 & 5 & 8 \end{bmatrix}$ (h) $\begin{bmatrix} 1 & 1 & 1 \\ 0 & 2 & 3 \\ 5 & 5 & 1 \end{bmatrix}$

   (i) $\begin{bmatrix} 0 & -1 & 2 \\ 2 & 1 & 4 \\ 1 & -1 & 5 \end{bmatrix}$ (j) $\begin{bmatrix} 0 & 1 & 0 & 0 & 0 \\ 0 & 0 & 0 & 1 & 0 \\ 1 & 0 & 0 & 0 & 0 \\ 0 & 0 & 0 & 0 & 1 \\ 0 & 0 & 1 & 0 & 0 \end{bmatrix}$.

4. Determine whether $A = \begin{bmatrix} -\frac{1}{4} & \frac{3}{4} & 1 \\ \frac{3}{4} & -\frac{5}{4} & -1 \\ \frac{1}{4} & \frac{1}{4} & 0 \end{bmatrix}$ has an inverse. Find A^{-1} if this exists.

5. [BB] Consider the system of equations

   $$\begin{aligned} x_1 - 2x_2 + 2x_3 &= 3 \\ 2x_1 + x_2 + x_3 &= 0 \\ x_1 \qquad\;\; + x_3 &= -2. \end{aligned}$$

   i. Write this in the form $A\mathbf{x} = \mathbf{b}$.

   ii. Solve the system by finding A^{-1}.

iii. Express $\begin{bmatrix} 3 \\ 0 \\ -2 \end{bmatrix}$ as a linear combination of the

vectors $\begin{bmatrix} 1 \\ 2 \\ 1 \end{bmatrix}, \begin{bmatrix} -2 \\ 1 \\ 0 \end{bmatrix}, \begin{bmatrix} 2 \\ 1 \\ 1 \end{bmatrix}$.

6. Consider the system of equations

$$\begin{aligned} x_2 - x_3 &= 8 \\ x_1 + 2x_2 + x_3 &= 5 \\ x_1 \qquad + x_3 &= -7. \end{aligned}$$

   i. Write this in the form $A\mathbf{x} = \mathbf{b}$.

   ii. Solve the system by finding A^{-1}.

   iii. Express $\begin{bmatrix} 8 \\ 5 \\ -7 \end{bmatrix}$ as a linear combination of the

   vectors $\begin{bmatrix} 0 \\ 1 \\ 1 \end{bmatrix}, \begin{bmatrix} 1 \\ 2 \\ 0 \end{bmatrix}, \begin{bmatrix} -1 \\ 1 \\ 1 \end{bmatrix}$.

7. Consider the system of equations

$$\begin{aligned} x_1 + 2x_2 - 4x_3 &= 4 \\ -x_1 + 3x_2 + 2x_3 &= -2 \\ 5x_1 - 2x_2 - x_3 &= 0. \end{aligned}$$

   i. Write this in the form $A\mathbf{x} = \mathbf{b}$.

   ii. Solve the system by finding A^{-1}.

   iii. Express $\begin{bmatrix} 4 \\ -2 \\ 0 \end{bmatrix}$ as a linear combination of the

   vectors $\begin{bmatrix} 1 \\ -1 \\ 5 \end{bmatrix}, \begin{bmatrix} 2 \\ 3 \\ -2 \end{bmatrix}, \begin{bmatrix} -4 \\ 2 \\ -1 \end{bmatrix}$.

8. [BB] Given that A is a 2×2 matrix and

$$\begin{bmatrix} 1 & 2 \\ 3 & 0 \end{bmatrix}^{-1} A \begin{bmatrix} 5 & 1 \\ -1 & 1 \end{bmatrix} = \begin{bmatrix} 1 & 0 \\ 0 & 1 \end{bmatrix}, \text{ find } A.$$

9. Find a matrix A satisfying

$$\begin{bmatrix} 4 & 7 & 6 \\ 2 & -3 & 1 \\ 0 & 5 & -4 \end{bmatrix}^{-1} A = \begin{bmatrix} 1 & 2 & 3 \\ -4 & 0 & 6 \\ -2 & 1 & 1 \end{bmatrix} \begin{bmatrix} 2 & 0 & 0 \\ 1 & 4 & -1 \\ 1 & 3 & 0 \end{bmatrix}^{-1}.$$

10. Let $A = \begin{bmatrix} 1 & 1 \\ 2 & 4 \end{bmatrix}$ and $C = \begin{bmatrix} 5 & 3 \\ 2 & 2 \end{bmatrix}$.

    Given that B is a 2×2 matrix and that $ABC^{-1} = I$, the identity matrix, find B.

11. Express each of the following invertible matrices as the product of elementary matrices:

(a) [BB] $\begin{bmatrix} 0 & -1 \\ 2 & 1 \end{bmatrix}$

(b) $\begin{bmatrix} 1 & 4 \\ 1 & -2 \end{bmatrix}$

(c) $\begin{bmatrix} -1 & 2 \\ 3 & 1 \end{bmatrix}$

(d) $\begin{bmatrix} 0 & -2 & 1 \\ 0 & 1 & 0 \\ 1 & -5 & 2 \end{bmatrix}$

(e) $\begin{bmatrix} 0 & -1 & -2 \\ 0 & 2 & 6 \\ 1 & -1 & 3 \end{bmatrix}$

(f) $\begin{bmatrix} 1 & -1 & 1 \\ 0 & 2 & -1 \\ 2 & 1 & 0 \end{bmatrix}$

(g) $\begin{bmatrix} 3 & 0 & 3 & -11 \\ 5 & 1 & 5 & -20 \\ 1 & 0 & 1 & -4 \\ -9 & -2 & -10 & 40 \end{bmatrix}$

(h) $\begin{bmatrix} 0 & -2 & 0 & 0 & 14 \\ -4 & 0 & 0 & 12 & 1 \\ 0 & 0 & 1 & 0 & 0 \\ 0 & 0 & 2 & 1 & 0 \\ 1 & 0 & 0 & -3 & 0 \end{bmatrix}$.

■ Critical Reading

12. (a) [BB] Given two $n \times n$ matrices X and Y, how do you determine whether $Y = X^{-1}$?
    Let A be an $n \times n$ matrix and let I denote the $n \times n$ identity matrix.

    (b) If $A^2 = \mathbf{0}$, verify that $(I - A)^{-1} = I + A$.

    (c) If $A^3 = \mathbf{0}$, verify that $(I - A)^{-1} = I + A + A^2$.

    (d) Use part (c) to find the inverse of
    $$\begin{bmatrix} 1 & 2 & -1 \\ 0 & 1 & 3 \\ 0 & 0 & 1 \end{bmatrix}.$$

13. Suppose A is a square matrix with linearly independent columns and $BA = \mathbf{0}$. What can you say about B and why?

14. [BB] Is the factorization of an invertible matrix as the product of elementary matrices unique? Explain.

15. Find the inverse of $A = \begin{bmatrix} 1 & a & b & c \\ 0 & 1 & x & y \\ 0 & 0 & 1 & z \\ 0 & 0 & 0 & 1 \end{bmatrix}$.

CHAPTER KEY WORDS AND IDEAS: Here are some technical words and phrases that were used in this chapter. Do you know the meaning of each? If you're not sure, check the glossary or index at the back of the book.

associative
augmented matrix
back substitution
commutative
commute (matrices do not)
elementary matrix
elementary row operations
equal matrices
forward substitution
Gaussian elimination
has an inverse
homogeneous system
identity matrix
inconsistent system
inverse of a matrix
invertible matrix
linear combination of matrices
linear equation
linearly dependent vectors
linearly independent vectors
lower triangular matrix

LDU factorization
LU factorization
main diagonal of a matrix
matrix
matrix of coefficients
multiplier
negative of a matrix
parameter
particular solution
permutation matrix
pivot
reduced row echelon form
row echelon form
standard basis vectors
symmetric matrix
system of linear equations
transpose
unique solution
upper triangular matrix
zero matrix

Review Exercises for Chapter 2

1. Find each of the following products:

   (a) $\begin{bmatrix} 1 & 3 & 0 & 4 \\ 2 & 1 & 1 & 0 \end{bmatrix} \begin{bmatrix} 1 & 0 \\ 0 & 1 \\ 1 & 0 \\ 0 & 1 \end{bmatrix}$

   (b) $\begin{bmatrix} 1 & 3 & 2 \\ 0 & 2 & 1 \\ 1 & 3 & 1 \end{bmatrix} \begin{bmatrix} 1 & 2 \\ 0 & 1 \\ 1 & 1 \end{bmatrix}$

   (c) $\begin{bmatrix} 1 \\ 2 \\ 3 \\ 4 \end{bmatrix} \begin{bmatrix} 1 \\ 2 \\ 3 \\ 4 \end{bmatrix}^T$

   (d) $\begin{bmatrix} 1 \\ 2 \\ 3 \\ 4 \end{bmatrix}^T \begin{bmatrix} 1 \\ 2 \\ 3 \\ 4 \end{bmatrix}$.

2. Evaluate the product $\begin{bmatrix} x & y & z \end{bmatrix} \begin{bmatrix} a & d & e \\ d & b & f \\ e & f & c \end{bmatrix} \begin{bmatrix} x \\ y \\ z \end{bmatrix}$.

3. Compute A^2 and A^3, given $A = \begin{bmatrix} 1 & 2 & 3 \\ 2 & 1 & 0 \\ 1 & 0 & 1 \end{bmatrix}$.

4. Let $A = \begin{bmatrix} 1 & 2 \\ -1 & 11 \end{bmatrix}$ and $B = \begin{bmatrix} -2 & 3 \\ -2 & 1 \end{bmatrix}$.

   (a) Compute AB, BA, A^2, B^2, $A - B$, $(A - B)^2$, $A^2 - 2AB + B^2$, and $A^2 - AB - BA + B^2$.

   (b) In general, for arbitrary matrices A and B, is it the case that $(A - B)^2 = A^2 - 2AB + B^2$. If not, find a correct expansion of $(A - B)^2$.

5. Let $A = \begin{bmatrix} 3 & 1 & -2 \\ 2 & -2 & 0 \\ -1 & 1 & 2 \end{bmatrix}$ and

$B = \begin{bmatrix} 1 & 1 & 1 \\ 1 & -1 & 1 \\ 0 & 1 & 2 \end{bmatrix}$. Use the fact that $AB = 4I$ to

solve the system

$$\begin{aligned} x + y + z &= 2 \\ x - y + z &= 0 \\ y + 2z &= -1. \end{aligned}$$

6. Solve each of the following systems of equations, in each case expressing any solution as a vector or a linear combination of vectors.

(a) $\begin{aligned} x + 2y + z &= 1 \\ 2x + y + 3z &= 3 \\ 5x - y + 2z &= 2 \end{aligned}$

(b) $\begin{aligned} -2x + y + z &= 0 \\ y - 3z &= 0 \\ -4x + 3y - z &= 0 \end{aligned}$

(c) $\begin{aligned} x_1 + x_2 + x_3 + x_4 &= 4 \\ 2x_1 - x_2 + 5x_3 + x_4 &= -1 \\ 4x_1 + 3x_2 + x_3 - x_4 &= 3 \\ 5x_1 + 6x_2 - 4x_3 - 4x_4 &= 3 \end{aligned}$

(d) $\begin{aligned} x_1 + x_2 + x_3 + 2x_4 &= 10 \\ 3x_1 + 2x_2 - x_3 + 4x_4 &= 16 \\ x_1 - x_2 + 2x_3 - 5x_4 &= -6 \end{aligned}$

(e) $\begin{aligned} x_1 + x_2 + x_3 + x_4 &= 10 \\ x_1 - x_2 - x_3 + 2x_4 &= 4 \\ 2x_1 + x_2 + 3x_3 - x_4 &= 9 \\ 5x_2 + 7x_3 - 6x_4 &= 6 \end{aligned}$

(f) $\begin{aligned} x_1 + x_2 + x_3 + x_4 &= 10 \\ x_1 - x_2 - x_3 + 2x_4 &= 4 \\ 2x_1 + x_2 + 3x_3 - x_4 &= 9 \\ 5x_2 + 7x_3 - 6x_4 &= 7. \end{aligned}$

7. (a) Determine whether $\begin{bmatrix} -1 \\ 0 \\ 6 \\ 3 \end{bmatrix}$ is a linear combination

of $\begin{bmatrix} -1 \\ 0 \\ 2 \\ 1 \end{bmatrix}, \begin{bmatrix} 0 \\ 1 \\ 3 \\ 2 \end{bmatrix}$, and $\begin{bmatrix} -1 \\ 1 \\ 0 \\ 1 \end{bmatrix}$.

(b) Do $\begin{bmatrix} -1 \\ 0 \\ 2 \\ 1 \end{bmatrix}, \begin{bmatrix} 0 \\ 1 \\ 3 \\ 2 \end{bmatrix}$, and $\begin{bmatrix} -1 \\ 1 \\ 0 \\ 1 \end{bmatrix}$ span $\mathbf{R}^4$? Justify

your answer.

8. Determine whether each of the following sets of vectors is linearly independent or linearly dependent.

(a) $\begin{bmatrix} 1 \\ 1 \\ 0 \end{bmatrix}, \begin{bmatrix} 2 \\ 3 \\ 1 \end{bmatrix}, \begin{bmatrix} 1 \\ 2 \\ 1 \end{bmatrix}$

(b) $\begin{bmatrix} 1 \\ 1 \\ 0 \end{bmatrix}, \begin{bmatrix} 2 \\ 3 \\ 1 \end{bmatrix}, \begin{bmatrix} 1 \\ 2 \\ 0 \end{bmatrix}$

(c) $\begin{bmatrix} 1 \\ 3 \\ -1 \\ -1 \end{bmatrix}, \begin{bmatrix} 1 \\ 2 \\ 0 \\ 2 \end{bmatrix}, \begin{bmatrix} 3 \\ 3 \\ 1 \\ 5 \end{bmatrix}, \begin{bmatrix} 0 \\ -3 \\ 1 \\ 3 \end{bmatrix}, \begin{bmatrix} 5 \\ 17 \\ -5 \\ -5 \end{bmatrix}$

(d) $\begin{bmatrix} -2 \\ 0 \\ -4 \\ 0 \end{bmatrix}, \begin{bmatrix} 1 \\ 1 \\ 3 \\ 5 \end{bmatrix}, \begin{bmatrix} 1 \\ -3 \\ -1 \\ 1 \end{bmatrix}$.

9. Find necessary and sufficient conditions on k so that

the vectors $\begin{bmatrix} 1 \\ -1 \\ 0 \end{bmatrix}, \begin{bmatrix} k \\ -4 \\ 0 \end{bmatrix}$, and $\begin{bmatrix} 0 \\ 2 \\ 3 \end{bmatrix}$ are linearly

independent.

10. If A is an $n \times n$ matrix and $\mathbf{x}$ is a vector in $\mathbf{R}^n$, what is the size of the matrix $\mathbf{x}^T A \mathbf{x}$? Explain.

11. Suppose A is a matrix such that $A^2 + A = I$, the identity matrix.

(a) A must be square. Why?

(b) A must be invertible. Why?

12. Let $A = \begin{bmatrix} 4 & -1 \\ 1 & 5 \end{bmatrix}$. Does there exist a *nonzero*

vector $\mathbf{x} = \begin{bmatrix} x \\ y \end{bmatrix}$ such that $A\mathbf{x} = 2\mathbf{x}$? Explain your

answer.

13. Let $\mathbf{u} = \begin{bmatrix} 1 \\ 2 \\ 1 \end{bmatrix}, \mathbf{v} = \begin{bmatrix} 2 \\ 1 \\ -4 \end{bmatrix}$, and $\mathbf{w} = \begin{bmatrix} 10 \\ 6 \\ -4 \end{bmatrix}$. Determine

whether $\mathbf{w}$ is in the plane spanned by $\mathbf{u}$ and $\mathbf{v}$.

14. Consider the following system of linear equations.

$$\begin{aligned} -3x_1 + 8x_2 - 5x_3 + 19x_4 - 9x_5 - 7x_6 &= 5 \\ x_1 - 3x_2 + 2x_3 - 9x_4 + 5x_5 - 2x_6 &= -3 \\ -x_1 + 2x_2 - x_3 + 2x_4 - 8x_6 &= 0 \\ -5x_1 + 13x_2 - 8x_3 + 29x_4 - 13x_5 - 15x_6 &= 3. \end{aligned}$$

(a) Write this system in the form $A\mathbf{x} = \mathbf{b}$.

(b) Solve the system by reducing the augmented matrix of coefficients to a matrix in row echelon form. Express your answer in the form $\mathbf{x} = \mathbf{x}_p + \mathbf{x}_n$, where $\mathbf{x}_p$ is a particular solution vector and $\mathbf{x}_n$ is the general solution to the homogeneous system $A\mathbf{x} = \mathbf{0}$.

(c) Express $\begin{bmatrix} 5 \\ -3 \\ 0 \\ 3 \end{bmatrix}$ as a linear combination of the columns of A.

15. Consider the system
$$\begin{aligned} y + z &= a \\ -x - y + 4z &= b \\ 2x \quad\quad - 3z &= c. \end{aligned}$$

(a) Write this in the form $A\mathbf{x} = \mathbf{b}$.

(b) Under what condition or conditions on a, b, and c (if any) does the system have a solution?

(c) If and when the system has a solution, is it unique?

16. Answer all parts of Exercise 15 for the system
$$\begin{aligned} y + 2z &= a \\ -x - y + 4z &= b \\ 2x \quad\quad - 12z &= c. \end{aligned}$$

17. Consider the system
$$\begin{aligned} x + 2y + kz + \quad\quad\quad w &= 0 \\ 2x + 3y - 2z + (1-k)w &= k \\ x + 2y - z + (2k+1)w &= k+1 \\ kx + y + z + (k^2 - 2)w &= 3. \end{aligned}$$

(a) Find a value or values of k so that this system has no solution.

(b) Find a value or values of k so that this system has infinitely many solutions.

(c) For each value of k found in (b), find all solutions (in vector form).

18. Solve the following system of equations, expressing each answer as a linear combination of vectors:
$$\begin{aligned} 2x - y + 7z + 10w &= -4 \\ 3x + 2y \quad\quad + w &= 1 \\ x + 3y - 7z - 9w &= 5. \end{aligned}$$

19. For which a, if any, is the following system inconsistent?
$$\begin{aligned} 2x_2 + 3x_3 + x_4 &= 11 \\ x_1 + 4x_2 + 6x_3 + 2x_4 &= 8 \\ 2x_1 + 3x_2 + 4x_3 + 2x_4 &= 3 \\ -5x_1 + 3x_2 + 5x_3 + x_4 &= a. \end{aligned}$$

20. (a) Let $P = \begin{bmatrix} 0 & 0 & 1 & 0 \\ 1 & 0 & 0 & 0 \\ 0 & 0 & 0 & 1 \\ 0 & 1 & 0 & 0 \end{bmatrix}$ and let A be a $4 \times n$ matrix. What kind of a matrix is P? Explain the connection between PA and A.

(b) Let P be an $n \times n$ permutation matrix and let J be the $n \times n$ matrix consisting entirely of 1s. Find PJ, JP^T, and PJP^T.

(c) Suppose A and B are matrices such that $AB = B$. Does this imply $A = I$? Explain.

21. Is the inverse of a symmetric matrix symmetric? Explain.

22. Suppose $A = \begin{bmatrix} \mathbf{x}_1 & \mathbf{x}_2 & \mathbf{x}_3 & \mathbf{x}_4 \\ \downarrow & \downarrow & \downarrow & \downarrow \end{bmatrix}$ is a matrix with four columns and $B = \begin{bmatrix} -2 & 0 & -1 & 0 \\ 0 & 0 & 0 & 0 \\ 0 & 1 & 0 & 0 \\ 0 & 0 & 3 & 4 \end{bmatrix}$. Find AB and explain.

23. (a) What is meant by an *elementary matrix*?

(b) Given that E is an elementary matrix such that $E \begin{bmatrix} 1 & 4 \\ 5 & 3 \end{bmatrix} = \begin{bmatrix} 1 & 4 \\ 0 & -17 \end{bmatrix}$, write down E and E^{-1} without any calculation.

24. Write a short note, in good clear English, explaining why every invertible matrix is the product of elementary matrices. Illustrate your answer with reference to the matrix $A = \begin{bmatrix} 1 & -2 \\ 3 & 1 \end{bmatrix}$.

25. Let A be the matrix $\begin{bmatrix} 1 & 1 \\ 0 & 3 \\ -2 & 1 \end{bmatrix}$.

(a) Reduce A to row echelon form.

(b) Show that the equation $A\mathbf{x} = \mathbf{b}$ has a unique solution for any $\mathbf{b} = \begin{bmatrix} b_1 \\ b_2 \\ b_3 \end{bmatrix}$ provided
$$2b_1 - b_2 + b_3 = 0.$$

(c) Show that there is a unique matrix B such that

$$AB = \begin{bmatrix} 1 & -1 & -2 \\ 1 & 2 & -3 \\ -1 & 4 & 1 \end{bmatrix}.$$

26. Find an LU factorization of

$$A = \begin{bmatrix} 1 & 2 & 3 \\ -7 & -7 & 9 \\ 4 & -6 & -6 \end{bmatrix}.$$

27. Find an LU factorization of

$$A = \begin{bmatrix} 1 & -1 & 1 & -1 \\ 2 & -1 & 3 & 0 \\ 3 & -2 & 2 & 1 \\ 1 & 0 & 1 & 2 \end{bmatrix} \quad \text{and use this to solve the}$$

system

$$\begin{array}{rcl}
x_1 - x_2 + x_3 - x_4 &=& 1 \\
2x_1 - x_2 + 3x_3 &=& 2 \\
3x_1 - 2x_2 + 2x_3 + x_4 &=& 1 \\
x_1 + x_3 + 2x_4 &=& 0.
\end{array}$$

Express your answer in the form $\mathbf{x} = \mathbf{x}_p + \mathbf{x}_H$ where $\mathbf{x}_p$ is a particular solution and $\mathbf{x}_h$ is the solution of the corresponding homogeneous system.

28. Find an LDU factorization of

$$A = \begin{bmatrix} -2 & 4 \\ 0 & 5 \\ 3 & 1 \end{bmatrix} \quad \text{or explain why there is no such}$$

factorization.

29. Find an LDU factorization of

$$A = \begin{bmatrix} -1 & -1 & 0 & -4 \\ 2 & -1 & -6 & 8 \\ -1 & -1 & 2 & 6 \\ -3 & -6 & -6 & -8 \\ 0 & 6 & 14 & 10 \end{bmatrix}.$$

30. Find an LDU factorization of the symmetric matrix

$$A = \begin{bmatrix} 5 & 2 & 2 \\ 2 & 2 & -4 \\ 2 & -4 & 2 \end{bmatrix}.$$

31. Factor $A = PLDU$ with

$$A = \begin{bmatrix} 0 & 2 & 1 & 3 & 1 \\ 3 & 1 & 2 & 1 & 1 \\ 2 & 0 & -1 & 1 & 1 \\ 4 & 1 & 2 & 2 & 1 \end{bmatrix}.$$

32. (a) Find A^{-1}, given $A = \begin{bmatrix} 4 & 1 & -1 \\ 1 & 0 & -1 \\ 2 & 2 & 3 \end{bmatrix}.$

(b) Using your answer to (a), write down the solution to the system of equations $A\mathbf{x} = \mathbf{b}$ where

$$\mathbf{b} = \begin{bmatrix} 4 \\ 1 \\ 1 \end{bmatrix}.$$

33. Given $A = \begin{bmatrix} 2 & 3 & 1 \\ 1 & 1 & 1 \\ -2 & 3 & 0 \end{bmatrix}$, find A^{-1} and use this to solve the system

$$\begin{array}{rcl}
2x + 3y + z &=& 2 \\
x + y + z &=& 5 \\
-2x + 3y &=& 3.
\end{array}$$

3

Determinants, Eigenvalues, Eigenvectors

3.1 The Determinant of a Matrix

In Section 1.3, we defined the determinant of a 2×2 matrix and used this to evaluate the cross product of two three-dimensional vectors. The determinant of a matrix A is a number, denoted $\det A$ or $|A|$. The determinant of a 1×1 matrix $[a]$ is just the number a: $\det[a] = a$. The determinant of the 2×2 matrix $A = \begin{bmatrix} a & b \\ c & d \end{bmatrix}$ is

$$\det A = \begin{vmatrix} a & b \\ c & d \end{vmatrix} = ad - bc.$$

This number, $ad - bc$, is closely connected to the invertibility of A. Let's see why. Let $B = \begin{bmatrix} d & -b \\ -c & a \end{bmatrix}$ and observe that

$$AB = \begin{bmatrix} a & b \\ c & d \end{bmatrix} \begin{bmatrix} d & -b \\ -c & a \end{bmatrix}$$

$$= \begin{bmatrix} ad - bc & 0 \\ 0 & ad - bc \end{bmatrix} = (ad - bc) \begin{bmatrix} 1 & 0 \\ 0 & 1 \end{bmatrix} = (\det A)I. \qquad (1)$$

Thus, if $\det A \neq 0$, then $\frac{1}{\det A} AB = I$. So, $A\left(\frac{1}{\det A} B\right) = I$ and A is invertible with $A^{-1} = \frac{1}{\det A} B$.

Only **square** matrices have a determinant.

In this section, we define the determinant of a square matrix A of any size and show that if $\det A \neq 0$, then A is always invertible with an inverse of the form $\frac{1}{\det A} B$.

3.1.1 DEFINITION   If A is a square $n \times n$ matrix, the (i, j) *minor* is the determinant of the $(n-1) \times (n-1)$ matrix obtained from A by deleting row i and column j. The (i, j) minor is denoted m_{ij}. The (i, j) *cofactor* of A is $(-1)^{i+j}m_{ij}$ and is denoted c_{ij}.

This may seem like a mouthful, but as with many new ideas, the concepts of "minor" and "cofactor" will soon seem second nature. Let's look at some examples.

3.1.2 EXAMPLE   Let $A = \begin{bmatrix} 1 & 2 & 3 \\ 0 & -2 & 2 \\ 3 & 7 & -4 \end{bmatrix}$.

The $(1, 1)$ minor is the determinant of the matrix

$$\begin{bmatrix} -2 & 2 \\ 7 & -4 \end{bmatrix},$$

which is what remains when we remove row one and column one of A. So $m_{11} = -2(-4) - 2(7) = -6$. The $(1, 1)$ cofactor is $c_{11} = (-1)^{1+1}m_{11} = (-1)^2(-6) = -6$.

The $(2, 3)$ minor of A is the determinant of

$$\begin{bmatrix} 1 & 2 \\ 3 & 7 \end{bmatrix},$$

which is what remains after removing row two and column three of A. Thus $m_{23} = 1$ and $c_{23} = (-1)^{2+3}m_{23} = (-1)^5(1) = -1$.

The $(3, 1)$ minor of A is the determinant of

$$\begin{bmatrix} 2 & 3 \\ -2 & 2 \end{bmatrix},$$

which is the matrix obtained from A by removing row three and column one. Thus $m_{31} = 10$ and $c_{31} = (-1)^{3+1}m_{31} = (-1)^4(10) = +10$.

Continuing in this way, we can find a minor and a cofactor corresponding to every position (i, j) of the matrix A. The *matrix of minors* of A is

$$M = \begin{bmatrix} -6 & -6 & 6 \\ -29 & -13 & 1 \\ 10 & 2 & -2 \end{bmatrix}$$

and the *matrix of cofactors* is

$$C = \begin{bmatrix} -6 & 6 & 6 \\ 29 & -13 & -1 \\ 10 & -2 & -2 \end{bmatrix}. \qquad \blacksquare$$

Since the only powers of -1 are $+1$ and -1, the factor $(-1)^{i+j}$ that appears in the definition of the (i, j) cofactor is ± 1. Thus any cofactor is either plus or minus the corresponding minor. Whether it is plus or minus is most easily remembered by the following pattern, which shows the values of $(-1)^{i+j}$ for all positive integers i and j.

$$
\begin{array}{ccccc}
+ & - & + & - & \cdots \\
- & + & - & + & \cdots \\
+ & - & + & - & \cdots \\
- & + & - & + & \cdots \\
\vdots & \vdots & & &
\end{array}
\tag{2}
$$

Each "+" means "multiply by +1," that is, leave the minor unchanged. Each "−" means "multiply by −1," that is, change the sign of the minor.

3.1.3 EXAMPLE

Let $A = \begin{bmatrix} -1 & 0 & 2 \\ 3 & -7 & 8 \\ -4 & 4 & -5 \end{bmatrix}$. The matrix of minors of A is

$$
M = \begin{bmatrix} 3 & 17 & -16 \\ -8 & 13 & -4 \\ 14 & -14 & 7 \end{bmatrix}.
$$

For example, $m_{22} = \begin{vmatrix} -1 & 2 \\ -4 & -5 \end{vmatrix} = 5 - (-8) = 13$ and $m_{32} = \begin{vmatrix} -1 & 2 \\ 3 & 8 \end{vmatrix} = -8 - 6 = -14$. To obtain the matrix of cofactors of A, we modify the entries of M according to the pattern given in (2). Thus, we leave the $(1, 1)$ entry of M alone, but multiply the $(1, 2)$ entry by -1. We leave the $(1, 3)$ entry alone, but multiply the $(2, 3)$ entry by -1, and so forth. The matrix of cofactors is

$$
C = \begin{bmatrix} 3 & -17 & -16 \\ 8 & 13 & 4 \\ 14 & 14 & 7 \end{bmatrix}.
$$ ∎

You can be excused from believing what comes next, but the **transpose of the matrix of cofactors** is an extraordinary object.

Let

$$
A = \begin{bmatrix} a & b \\ c & d \end{bmatrix}.
$$

The matrix of minors is

$$
M = \begin{bmatrix} d & c \\ b & a \end{bmatrix},
$$

the matrix of cofactors is

$$
C = \begin{bmatrix} d & -c \\ -b & a \end{bmatrix},
$$

and the transpose of the matrix of cofactors is

$$
C^T = \begin{bmatrix} d & -b \\ -c & a \end{bmatrix}.
$$

The product of A and the transpose of its matrix of cofactors is $AC^T = (ad - bc)I = (\det A)I$.

3.1.4 EXAMPLE

Suppose $A = \begin{bmatrix} 1 & 2 & 3 \\ 0 & -2 & 2 \\ 3 & 7 & -4 \end{bmatrix}$ is the matrix of **3.1.2**. In that example, we found that the

matrix of cofactors is $C = \begin{bmatrix} -6 & 6 & 6 \\ 29 & -13 & -1 \\ 10 & -2 & -2 \end{bmatrix}$, so $C^T = \begin{bmatrix} -6 & 29 & 10 \\ 6 & -13 & -2 \\ 6 & -1 & -2 \end{bmatrix}$. The product

of A and the transpose of its matrix of cofactors is

$$AC^T = \begin{bmatrix} 1 & 2 & 3 \\ 0 & -2 & 2 \\ 3 & 7 & -4 \end{bmatrix} \begin{bmatrix} -6 & 29 & 10 \\ 6 & -13 & -2 \\ 6 & -1 & -2 \end{bmatrix} = 24 \begin{bmatrix} 1 & 0 & 0 \\ 0 & 1 & 0 \\ 0 & 0 & 1 \end{bmatrix} = 24I,$$

a scalar multiple of the identity matrix, just as in the 2×2 case.   ■

3.1.5 EXAMPLE

Let $A = \begin{bmatrix} -1 & 0 & 2 \\ 3 & -7 & 8 \\ -4 & 4 & -5 \end{bmatrix}$ be the matrix we considered in **3.1.3**. There we found that

the matrix of cofactors is $C = \begin{bmatrix} 3 & -17 & -16 \\ 8 & 13 & 4 \\ 14 & 14 & 7 \end{bmatrix}$, so $C^T = \begin{bmatrix} 3 & 8 & 14 \\ -17 & 13 & 14 \\ -16 & 4 & 7 \end{bmatrix}$. The

product of A and the transpose of its matrix of cofactors is

$$AC^T = \begin{bmatrix} -1 & 0 & 2 \\ 3 & -7 & 8 \\ -4 & 4 & -5 \end{bmatrix} \begin{bmatrix} 3 & 8 & 14 \\ -17 & 13 & 14 \\ -16 & 4 & 7 \end{bmatrix} = \begin{bmatrix} -35 & 0 & 0 \\ 0 & -35 & 0 \\ 0 & 0 & -35 \end{bmatrix} = -35I,$$

again a scalar multiple of the identity matrix.   ■

It's a fact, which we shall not justify, that

> **3.1.6** A square matrix A commutes with the transpose C^T of its matrix of cofactors, and the product $AC^T = C^T A = kI$ is a scalar multiple k of the identity matrix.

See W. Keith Nicholson, *Linear Algebra with Applications*, Fourth edition, McGraw-Hill Ryerson, Toronto, 2003, for a proof.

The number k is called the *determinant* of A, so the determinant of a matrix is defined by the equation

> **3.1.7** $AC^T = C^T A = (\det A)I$.

So, the determinant of the matrix A in **3.1.4** is 24—since $AC^T = 24I$—and the determinant of the matrix A in **3.1.5** is -35.

Actually, it is not necessary to compute the entire matrix of cofactors if we are only interested in finding the determinant of a matrix. Suppose $A = [a_{ij}]$ is the general 3×3 matrix,

$$A = \begin{bmatrix} a_{11} & a_{12} & a_{13} \\ a_{21} & a_{22} & a_{23} \\ a_{31} & a_{32} & a_{33} \end{bmatrix}.$$

The matrix of cofactors is

$$C = \begin{bmatrix} c_{11} & c_{12} & c_{13} \\ c_{21} & c_{22} & c_{23} \\ c_{31} & c_{32} & c_{33} \end{bmatrix}.$$

Thus

$$AC^T = \begin{bmatrix} a_{11} & a_{12} & a_{13} \\ a_{21} & a_{22} & a_{23} \\ a_{31} & a_{32} & a_{33} \end{bmatrix} \begin{bmatrix} c_{11} & c_{21} & c_{31} \\ c_{12} & c_{22} & c_{32} \\ c_{13} & c_{23} & c_{33} \end{bmatrix} = (\det A)I.$$

The $(1, 1)$, the $(2, 2)$, and the $(3, 3)$ entries of AC^T all equal $\det A$:

$$a_{11}c_{11} + a_{12}c_{12} + a_{13}c_{13} = \det A$$
$$a_{21}c_{21} + a_{22}c_{22} + a_{23}c_{23} = \det A$$
$$a_{31}c_{31} + a_{32}c_{32} + a_{33}c_{33} = \det A.$$

Entries not on the diagonal are 0. For example, the $(2, 3)$ and $(1, 2)$ entries are

$$a_{21}c_{31} + a_{22}c_{32} + a_{23}c_{33} = 0$$

and

$$a_{11}c_{21} + a_{12}c_{22} + a_{13}c_{23} = 0.$$

These calculations suggest that

3.1.8 The determinant of A is the dot product of any row (or column) of A with the corresponding row (or column) of cofactors.

Reading Challenge 1   *Why? Show that the dot product of any **column** of a 3×3 matrix A with the corresponding column of cofactors is again $\det A$. [Hint: Use $C^T A = (\det A)I$.]*

Looking again at **3.1.5**, the dot product of the first column of A and the first column of C is -35:

$$\begin{bmatrix} -1 \\ 3 \\ -4 \end{bmatrix} \cdot \begin{bmatrix} 3 \\ 8 \\ 14 \end{bmatrix} = -35 = \det A.$$

Strictly speaking, we shouldn't be talking about the dot product of rows since we have only defined the dot product of vectors and, in this book, vectors are columns. On the other hand, it is cumbersome to refer to "the vector whose transpose is the second row of A" and "the vector whose transpose is the second row of C."

The dot product of the second row of A and the second row of C is -35:

$$\begin{bmatrix} 3 \\ -7 \\ 8 \end{bmatrix} \cdot \begin{bmatrix} 8 \\ 13 \\ 4 \end{bmatrix} = -35 = \det A.$$

Suppose we had to compute the determinant of $A = \begin{bmatrix} -1 & 0 & 2 \\ 3 & -7 & 8 \\ -4 & 4 & -5 \end{bmatrix}$, which we met in **3.1.5**. To find the determinant of A, we take any row or column and compute the dot product with the corresponding row of cofactors. Given this degree of choice, we should take advantage of the 0 in the $(1, 2)$ position and choose either the first row of A or second column of A. If we choose to *expand by cofactors of the first row*, we get

$$-1 \begin{vmatrix} -7 & 8 \\ 4 & -5 \end{vmatrix} + 2 \begin{vmatrix} 3 & -7 \\ -4 & 4 \end{vmatrix} = -1(3) + 2(-16) = -35,$$

whereas, if we *expand by cofactors of the second column*, we get

$$-7 \begin{vmatrix} -1 & 2 \\ -4 & -5 \end{vmatrix} \ominus 4 \begin{vmatrix} -1 & 2 \\ 3 & 8 \end{vmatrix} = -7(13) - 4(-14) = -35.$$

Did you notice the minus sign in the last expansion, the one circled? We must change the sign of the 4 in the $(3, 2)$ position of A because the corresponding cofactor requires multiplication of the minor by -1.

3.1.9 PROBLEM   Find the determinant of the matrix $A = \begin{bmatrix} -1 & -2 & 0 \\ 0 & 1 & 5 \\ 2 & 3 & -4 \end{bmatrix}$.

Solution. Expanding by cofactors of the first row gives

$$\det A = -1 \begin{vmatrix} 1 & 5 \\ 3 & -4 \end{vmatrix} - (-2) \begin{vmatrix} 0 & 5 \\ 2 & -4 \end{vmatrix} = -(-19) + 2(-10) = -1.$$

Reading Challenge 2   *Verify that $\det A = -1$ by expanding by cofactors of the third column.*

3.1.10 PROBLEM   Find the determinant of $A = \begin{bmatrix} -1 & 2 & 3 & 4 \\ 0 & -2 & 3 & 0 \\ 1 & 2 & 3 & 5 \\ 8 & 9 & -1 & 2 \end{bmatrix}$.

Solution. The easiest thing to do here is to expand by cofactors of the second row. We get

$$\det A = -2 \begin{vmatrix} -1 & 3 & 4 \\ 1 & 3 & 5 \\ 8 & -1 & 2 \end{vmatrix} - 3 \begin{vmatrix} -1 & 2 & 4 \\ 1 & 2 & 5 \\ 8 & 9 & 2 \end{vmatrix}.$$

We require the determinants of two 3×3 matrices. Expanding by cofactors of the first row, we have

$$\begin{vmatrix} -1 & 3 & 4 \\ 1 & 3 & 5 \\ 8 & -1 & 2 \end{vmatrix} = -1 \begin{vmatrix} 3 & 5 \\ -1 & 2 \end{vmatrix} - 3 \begin{vmatrix} 1 & 5 \\ 8 & 2 \end{vmatrix} + 4 \begin{vmatrix} 1 & 3 \\ 8 & -1 \end{vmatrix}$$

$$= -1(11) - 3(-38) + 4(-25) = 3$$

and, expanding by cofactors of the first row,

$$\begin{vmatrix} -1 & 2 & 4 \\ 1 & 2 & 5 \\ 8 & 9 & 2 \end{vmatrix} = -1 \begin{vmatrix} 2 & 5 \\ 9 & 2 \end{vmatrix} - 2 \begin{vmatrix} 1 & 5 \\ 8 & 2 \end{vmatrix} + 4 \begin{vmatrix} 1 & 2 \\ 8 & 9 \end{vmatrix}$$

$$= -1(-41) - 2(-38) + 4(-7) = 89.$$

Thus $\det A = -2(3) - 3(89) = -273$.

We began this section with the observation that if the determinant of a 2×2 matrix is not zero, then that matrix is invertible. We conclude by pointing out that the same is true for a square matrix of any size. In **3.1.4**, for instance, we noted that the cofactor matrix of $A = \begin{bmatrix} 1 & 2 & 3 \\ 0 & -2 & 2 \\ 3 & 7 & -4 \end{bmatrix}$ is $C = \begin{bmatrix} -6 & 6 & 6 \\ 29 & -13 & -1 \\ 10 & -2 & -2 \end{bmatrix}$ and found that $AC^T = 24I = (\det A)I$. Since $\det A \neq 0$, we can divide by this number, obtaining $A \left(\frac{1}{24} C^T \right) = I$. This equation says that A is invertible with inverse

$$A^{-1} = \frac{1}{24} C^T = \frac{1}{24} \begin{bmatrix} -6 & 29 & 10 \\ 6 & -13 & -2 \\ 6 & -1 & -2 \end{bmatrix}.$$

Theorem 3.1.11 *If A is a square matrix and $\det A \neq 0$, then A is invertible and $A^{-1} = \frac{1}{\det A}$ times the transpose of the matrix of cofactors of A.*

Proof Let C^T denote the transpose of the matrix of cofactors of A. Since $AC^T = (\det A)I$ and $\det A \neq 0$, we have $A\left(\frac{1}{\det A} C^T\right) = I$, so $\frac{1}{\det A} C^T$ is the inverse of A. ∎

3.1.12 EXAMPLE In **3.1.5**, with $A = \begin{bmatrix} -1 & 0 & 2 \\ 3 & -7 & 8 \\ -4 & 4 & -5 \end{bmatrix}$, we had $AC^T = -35I$. So A is invertible and

$$A^{-1} = -\frac{1}{35} C^T = -\frac{1}{35} \begin{bmatrix} 3 & 8 & 14 \\ -17 & 13 & 14 \\ -16 & 4 & 7 \end{bmatrix}.$$ ∎

The converse of Theorem 3.1.11 is also true; that is, if A is invertible, then $\det A \neq 0$. There is an easy way to see this, which we describe in Section 3.2. For now, we content ourselves with an example.

3.1.13 EXAMPLE

Let $A = \begin{bmatrix} -1 & 2 & 2 \\ 3 & 0 & -1 \\ 1 & 4 & 3 \end{bmatrix}$.

The matrix of cofactors is $C = \begin{bmatrix} 4 & -10 & 12 \\ 2 & -5 & 6 \\ -2 & 5 & -6 \end{bmatrix}$, so $C^T = \begin{bmatrix} 4 & 2 & -2 \\ -10 & -5 & 5 \\ 12 & 6 & -6 \end{bmatrix}$ and

$$AC^T = \begin{bmatrix} -1 & 2 & 2 \\ 3 & 0 & -1 \\ 1 & 4 & 3 \end{bmatrix} \begin{bmatrix} 4 & 2 & -2 \\ -10 & -5 & 5 \\ 12 & 6 & -6 \end{bmatrix} = \begin{bmatrix} 0 & 0 & 0 \\ 0 & 0 & 0 \\ 0 & 0 & 0 \end{bmatrix}$$

is the zero matrix. Thus $\det A = 0$. If A were invertible, from $AC^T = \mathbf{0}$, we could derive $A^{-1}(AC^T) = A^{-1}\mathbf{0}$, hence $C^T = \mathbf{0}$. Since this is not true, A is not invertible. ∎

■ Triangular Matrices

Consider the general 4×4 lower triangular matrix

$$A = \begin{bmatrix} a & 0 & 0 & 0 \\ b & c & 0 & 0 \\ d & e & f & 0 \\ g & h & i & k \end{bmatrix}.$$

The easiest way to find the determinant of A is to expand by cofactors of the first row:

$$\det A = a \begin{vmatrix} c & 0 & 0 \\ e & f & 0 \\ h & i & k \end{vmatrix}.$$

Expanding the 3×3 determinant here also by cofactors of the first row, we obtain

$$\begin{vmatrix} c & 0 & 0 \\ e & f & 0 \\ h & i & k \end{vmatrix} = c \begin{vmatrix} f & 0 \\ i & k \end{vmatrix} = cfk.$$

Thus $\det A = acfk$, the product of the diagonal entries of A. It is not hard to see that this is true for square triangular matrices (lower or upper) of any size.

> **3.1.14** The determinant of a (square) triangular matrix is the product of the entries on its main diagonal. In particular, the determinant of a diagonal matrix is the product of its main diagonal entries.

3.1.15 EXAMPLES

• $\begin{vmatrix} -3 & 0 & 0 \\ 4 & 5 & 0 \\ -1 & 2 & 2 \end{vmatrix} = (-3)(5)(2) = -30$

• $\begin{vmatrix} 1 & 2 & 3 & 4 \\ 0 & 5 & 6 & 7 \\ 0 & 0 & 8 & 9 \\ 0 & 0 & 0 & 10 \end{vmatrix} = 1(5)(8)(10) = 400$

• $\begin{vmatrix} 8 & 0 \\ 0 & -4 \end{vmatrix} = (8)(-4) = -32$

• $\begin{vmatrix} -2 & 0 & 0 & 0 & 0 \\ 0 & 3 & 0 & 0 & 0 \\ 0 & 0 & 5 & 0 & 0 \\ 0 & 0 & 0 & -1 & 0 \\ 0 & 0 & 0 & 0 & -7 \end{vmatrix} = (-2)(3)(5)(-1)(-7) = -210.$ ∎

Reading Challenge 3   *What is the determinant of the 100×100 identity matrix? What is the determinant of the 101×101 matrix that is **minus** the identity matrix?*

Answers to Reading Challenges

1. Since $C^T A = \begin{bmatrix} c_{11} & c_{21} & c_{31} \\ c_{12} & c_{22} & c_{32} \\ c_{13} & c_{23} & c_{33} \end{bmatrix} \begin{bmatrix} a_{11} & a_{12} & a_{13} \\ a_{21} & a_{22} & a_{23} \\ a_{31} & a_{32} & a_{33} \end{bmatrix} = (\det A)I$, the $(1, 1)$, the $(2, 2)$, and the $(3, 3)$ entries of this product are all equal to $\det A$. These numbers are $c_{11}a_{11} + c_{21}a_{21} + c_{31}a_{31}$, $c_{12}a_{12} + c_{22}a_{22} + c_{32}a_{32}$, and $c_{13}a_{13} + c_{23}a_{23} + c_{33}a_{33}$, these being the dot products of the columns of A with the corresponding cofactors.

2. Expanding by cofactors of the third column,

$$\begin{vmatrix} -1 & -2 & 0 \\ 0 & 1 & 5 \\ 2 & 3 & -4 \end{vmatrix} = 0 \begin{vmatrix} \star & \star \\ \star & \star \end{vmatrix} - 5 \begin{vmatrix} -1 & -2 \\ 2 & 3 \end{vmatrix} - 4 \begin{vmatrix} -1 & -2 \\ 0 & 1 \end{vmatrix}$$

$$= -5(1) - 4(-1) = -1.$$

3. Any identity matrix is diagonal with 1s on the diagonal. Its determinant is $+1$. Minus the identity has -1s on the diagonal, so the 101×101 matrix which is $-I$ has determinant $(-1)^{101} = -1$.

True/False Questions

Decide, with as little calculation as possible, whether each of the following statements is true or false and explain your answer whenever you say "false." (Answers can be found in the back of the book.)

1. The $(1, 2)$ cofactor of $\begin{bmatrix} -6 & 7 \\ 2 & 1 \end{bmatrix}$ is -2.

2. The product AC^T of a matrix and the transpose of its matrix of cofactors is a scalar multiple of the identity matrix.

3. If I is the 2×2 identity matrix, then $\det(-2)I = -2$.

4. A triangular matrix with no 0s on the diagonal is invertible.

5. If A and B are square matrices of the same size, $\det(A + B) = \det A + \det B$.

6. If the matrix of cofactors of a matrix A is the zero matrix, then $A = \mathbf{0}$ too.

7. If $A = \begin{bmatrix} -2 & 0 & 0 \\ 1 & 3 & 0 \\ 1 & 5 & -4 \end{bmatrix}$, then $\det A = 24$.

8. If an $n \times n$ matrix A has nonzero determinant, every vector $\mathbf{b}$ in $\mathbf{R}^n$ is a linear combination of the columns of A.

Exercises

*Solutions to exercises marked [BB] can be found in the **B**ack of the **B**ook.*

1. [BB] Can a 3×2 matrix have determinant -1? Explain.

2. Explain why $\begin{vmatrix} \cos\theta & -\sin\theta \\ \sin\theta & \cos\theta \end{vmatrix}$ is independent of θ.

3. For each of the following matrices A,

   i.   find the matrix M of minors, find the matrix C of cofactors, and compute the product AC^T;

   ii. find $\det A$;

   iii. if A is invertible, find A^{-1}.

   (a) [BB] $A = \begin{bmatrix} 2 & -4 \\ -7 & 9 \end{bmatrix}$

   (b) $A = \begin{bmatrix} 5 & 3 \\ 15 & 9 \end{bmatrix}$

   (c) [BB] $A = \begin{bmatrix} 2 & 3 & 4 \\ 0 & -1 & 3 \\ 4 & 7 & 5 \end{bmatrix}$

   (d) $A = \begin{bmatrix} 3 & -1 & 4 \\ 6 & 2 & 0 \\ -4 & 5 & 1 \end{bmatrix}$

   (e) $A = \begin{bmatrix} 3 & 5 & 2 \\ 4 & 8 & 9 \\ -1 & 2 & 5 \end{bmatrix}$

   (f) $A = \begin{bmatrix} -6 & 8 & 9 \\ 7 & -4 & 0 \\ 11 & 2 & -5 \end{bmatrix}$.

4. For what values of x are the given matrices not invertible? (Such matrices are called *singular*.)

   (a) [BB] $\begin{bmatrix} x & 1 \\ -2 & x+3 \end{bmatrix}$ **(b)** $\begin{bmatrix} -1 & x \\ 3 & x \end{bmatrix}$

   (c) [BB] $\begin{bmatrix} 1 & 2 & x \\ 0 & -3 & 0 \\ 4 & x & 7 \end{bmatrix}$ **(d)** $\begin{bmatrix} -1 & x & x \\ 2 & 0 & x \\ 4 & 5 & -2 \end{bmatrix}$.

5. Let $A = \begin{bmatrix} 1 & -1 & 2 \\ 3 & 1 & 1 \\ 2 & -1 & 3 \end{bmatrix}$.

   (a) [BB] Find the determinant of A expanding by cofactors of the third row.

   (b) Find the determinant of A expanding by cofactors of the second column.

6. Let $A = \begin{bmatrix} -4 & 2 & 1 \\ 0 & -3 & 5 \\ 6 & 4 & -1 \end{bmatrix}$.

   (a) Find the determinant of A expanding by cofactors of the first column.

   (b) Find the determinant of A expanding by cofactors of the third row.

7. [BB] Let $A = \begin{bmatrix} 1 & 0 & 1 \\ 2 & 1 & 1 \\ 1 & 2 & 1 \end{bmatrix}$ and suppose

   $C = \begin{bmatrix} -1 & -1 & c_{13} \\ c_{21} & 0 & -2 \\ -1 & c_{32} & 1 \end{bmatrix}$ is the matrix of cofactors of A.

   (a) Find the values of c_{21}, c_{13}, and c_{32}.

   (b) Find $\det A$.

   (c) Is A invertible? Explain without further calculation.

   (d) Find A^{-1}, if this exists.

8. The matrix of cofactors of

   $A = \begin{bmatrix} -1 & 2 & 3 \\ 0 & 4 & 4 \\ -2 & 8 & 10 \end{bmatrix}$ is $C = \begin{bmatrix} 8 & c_{12} & 8 \\ 4 & -4 & c_{23} \\ c_{31} & 4 & -4 \end{bmatrix}$.

   (a) Find the values of c_{12}, c_{23}, and c_{31}.

   (b) Find $\det A$.

   (c) Is A invertible? Explain without further calculation.

   (d) Find A^{-1}, if this exists.

9. [BB] Professor G. was making up a homework problem for his class when he spilled a cup of coffee over his work. (This happens to Prof. G. all too frequently.) All that remained legible was this:

   $A = \begin{bmatrix} -1 & 2 \\ 0 & 3 \\ 2 & -2 \end{bmatrix}$; matrix of cofactors $= C =$

   $\begin{bmatrix} 19 & 10 \\ -14 & -11 \\ -2 & 5 \end{bmatrix}$.

   (a) Find the missing entries.

10. (b) Find A^{-1}.

10. Answer Exercise 9 with $A = \begin{bmatrix} 2 & -3 \\ -4 & 2 \\ 1 & 0 \end{bmatrix}$ and

    $C = \begin{bmatrix} -10 & -21 \\ -15 & -6 \\ 11 & 18 \end{bmatrix}$.

■ Critical Reading

11. [BB] If the matrix of cofactors of a matrix A is $\begin{bmatrix} -1 & 7 \\ 2 & 4 \end{bmatrix}$, find A.

12. The 2×2 matrix A has cofactor matrix $C = \begin{bmatrix} 3 & -4 \\ 2 & 1 \end{bmatrix}$. Find A.

13. Given $\det A$ and cofactor matrix C, find the matrix A in each of the following cases:

    (a) [BB] $\det A = 5$, $C = \begin{bmatrix} 3 & 1 \\ -2 & 1 \end{bmatrix}$

    (b) $\det A = -24$, $C = \begin{bmatrix} -1 & 2 \\ 6 & 3 \end{bmatrix}$

    (c) $\det A = 11$, $C = \begin{bmatrix} 1 & 5 & 4 \\ 2 & -1 & -3 \\ 10 & -5 & -4 \end{bmatrix}$

    (d) $\det A = 100$, $C = \begin{bmatrix} 2 & 34 & 6 \\ 8 & 36 & 24 \\ 13 & -29 & -11 \end{bmatrix}$.

14. Suppose the matrix $T = \begin{bmatrix} a & b & c \\ p & q & r \\ x & y & z \end{bmatrix}$ has cofactor matrix $\begin{bmatrix} A & B & C \\ P & Q & R \\ X & Y & Z \end{bmatrix}$.

    (a) Find P, Q, and R.

    (b) What is $aA + pP + xX$? Explain.

    (c) What is $aB + pQ + xY$? Explain.

15. Explain why a square triangular matrix with no zero entries on its diagonal must be invertible.

3.2 Properties of Determinants

Readers of this book should be familiar with the word "function." For our purposes, a function is just a method of transforming elements of one set to elements of another.

"Addition" is a function that turns a pair (a, b) of numbers into another number, $a+b$. The determinant is a function that transforms $n \times n$ matrices into real numbers. In this section, we show that the determinant function has many fascinating and unexpected properties, which we discuss under separate headings. First, we summarize.

Theorem 3.2.1

1. *The determinant is linear in each row.*

2. *The determinant of any (square) matrix that has a row of zeros is* 0.

3. *The determinant of a matrix changes sign if two rows are interchanged.*

4. *The determinant of a matrix with two equal rows is* 0.

5. *The determinant of a matrix in which one row is a scalar multiple of another is* 0.

6. *The determinant is* multiplicative: $\det AB = \det A \det B$.

7. *If A is an invertible matrix, then* $\det A \neq 0$ *and* $\det A^{-1} = \frac{1}{\det A}$.

8. *For any (square) matrix A,* $\det A^T = \det A$; *hence, any property of a matrix that is stated in terms of rows holds equally when stated in terms of columns.*

Property 1: The determinant is linear in each row. In a calculus or college algebra course, the student has probably encountered the term "linear function" applied to functions described by equations of the form $f(x) = ax + b$ (whose graphs are straight lines). In linear algebra, the concept of linearity is more restrictive.

3.2.2 DEFINITION

A *linear function* from $\mathbf{R}^n$ to $\mathbf{R}^m$ is a function f satisfying

1. $f(\mathbf{x} + \mathbf{y}) = f(\mathbf{x}) + f(\mathbf{y})$ for all vectors $\mathbf{x}$ and $\mathbf{y}$, and
2. $f(c\mathbf{x}) = cf(\mathbf{x})$ for all scalars c and all vectors $\mathbf{x}$.

3.2.3 EXAMPLES

- The function $f: \mathbf{R} \to \mathbf{R}$ defined by $f(x) = 3x$ is linear since, for any real numbers x and y and any scalar c,

  1. $f(x + y) = 3(x + y) = 3x + 3y = f(x) + f(y)$ and
  2. $f(cx) = 3(cx) = c(3x) = cf(x)$.

[In this example, our vector space is $\mathbf{R}^1$, which is the same as $\mathbf{R}$, except that we should, but do not, put square brackets around the "vectors."]

- The function $f: \mathbf{R}^2 \to \mathbf{R}$ defined by $f\left(\begin{bmatrix} x_1 \\ x_2 \end{bmatrix}\right) = 2x_1 - 3x_2$ is linear since, for any vectors $\mathbf{x} = \begin{bmatrix} x_1 \\ x_2 \end{bmatrix}$ and $\mathbf{y} = \begin{bmatrix} y_1 \\ y_2 \end{bmatrix}$, and for any scalar c,

  1. $f(\mathbf{x}+\mathbf{y}) = f\left(\begin{bmatrix} x_1 + y_1 \\ x_2 + y_2 \end{bmatrix}\right) = 2(x_1 + y_1) - 3(x_2 + y_2) = (2x_1 - 3x_2) + (2y_1 - 3y_2) = f(\mathbf{x}) + f(\mathbf{y})$, and
  2. $f(c\mathbf{x}) = f\left(\begin{bmatrix} cx_1 \\ cx_2 \end{bmatrix}\right) = 2(cx_1) - 3(cx_2) = c(2x_1 - 3x_2) = cf(\mathbf{x})$.

- The function $f: \mathbf{R} \to \mathbf{R}^3$ defined by $f(x) = \begin{bmatrix} x \\ 2x \\ 3x \end{bmatrix}$ is linear since, for any real numbers x and y, and for any scalar c,

$$
\textbf{1. } f(x + y) = \begin{bmatrix} x + y \\ 2(x + y) \\ 3(x + y) \end{bmatrix} = \begin{bmatrix} x \\ 2x \\ 3x \end{bmatrix} + \begin{bmatrix} y \\ 2y \\ 3y \end{bmatrix} = f(x) + f(y), \text{ and}
$$

$$
\textbf{2. } f(cx) = \begin{bmatrix} cx \\ 2(cx) \\ 3(cx) \end{bmatrix} = c \begin{bmatrix} x \\ 2x \\ 3x \end{bmatrix} = cf(x).
$$

3.2.4 EXAMPLES

- The function $g: \mathbf{R} \to \mathbf{R}$ defined by $g(x) = 3x + 1$ is not linear (in the sense of Definition 3.2.2) since, for example, $g(x+y) = 3(x+y)+1$, while $g(x)+g(y) = (3x + 1) + (3y + 1)$. [One could also point out that $g(cx) \neq cg(x)$.]

- The sine function $\mathbf{R} \to \mathbf{R}$ is not linear: $\sin(\alpha + \beta) \neq \sin \alpha + \sin \beta$. (Also, $\sin c\alpha \neq c \sin \alpha$.)

- The determinant is a function from matrices to numbers that is certainly not linear: $\det(A + B) \neq \det A + \det B$. For example, if

$$
A = \begin{bmatrix} 1 & 2 \\ 3 & 4 \end{bmatrix} \quad \text{and} \quad B = \begin{bmatrix} -1 & -2 \\ -3 & -4 \end{bmatrix},
$$

then $A+B = \begin{bmatrix} 0 & 0 \\ 0 & 0 \end{bmatrix}$, so $\det(A+B) = 0$, while $\det A + \det B = -2 + (-2) = -4$.

Our last example notwithstanding, all is not lost! The determinant is linear **in each row**. Suppose $A = \begin{bmatrix} \mathbf{a}_1 & \to \\ \mathbf{a}_2 & \to \\ & \vdots \\ \mathbf{a}_n & \to \end{bmatrix}$ is an $n \times n$ matrix with rows $\mathbf{a}_1, \dots, \mathbf{a}_n$. It will be helpful to use the notation $\det(\mathbf{a}_1, \dots, \mathbf{a}_n)$ for the determinant of A. The statement "The determinant is a linear function of each row" means

1. $\det(\mathbf{a}_1, \dots, \mathbf{x} + \mathbf{y}, \dots, \mathbf{a}_n) = \det(\mathbf{a}_1, \dots, \mathbf{x}, \dots, \mathbf{a}_n) + \det(\mathbf{a}_1, \dots, \mathbf{y}, \dots, \mathbf{a}_n),$ and

2. $\det(\mathbf{a}_1, \dots, c\mathbf{a}_i, \dots, \mathbf{a}_n) = c \det(\mathbf{a}_1, \dots \mathbf{a}_i, \dots, \mathbf{a}_n).$

The first point says that if one row of a matrix is the sum of two rows $\mathbf{x}$ and $\mathbf{y}$, then the determinant of the new matrix is the sum of the determinants of the matrices with rows $\mathbf{x}$ and $\mathbf{y}$, respectively. For instance,

$$
\det \begin{bmatrix} a + a' & b + b' \\ c & d \end{bmatrix} = (a + a')d - (b + b')c \tag{1}
$$

$$
= ad + a'd - bc - b'c = (ad - bc) + (a'd - b'c)
$$

$$
= \det \begin{bmatrix} a & b \\ c & d \end{bmatrix} + \det \begin{bmatrix} a' & b' \\ c & d \end{bmatrix}.
$$

The second point says that if one particular row of A is multiplied by a scalar c, then so is the determinant. We illustrate for a 2×2 matrix $A = \begin{bmatrix} x & y \\ z & w \end{bmatrix}$. If row two is multiplied by c, the determinant of the new matrix,

$$\det \begin{bmatrix} x & y \\ cz & cw \end{bmatrix} = x(cw) - y(cz) = c(xw - yz) = c \det A, \qquad (2)$$

is c times the determinant of A. The second point does **not** say that $\det(cA) = c \det A$, which is not true. With A as above, $cA = \begin{bmatrix} cx & cy \\ cz & cw \end{bmatrix}$ and

$$\det cA = (cx)(cw) - (cy)(cz) = c^2(xw - yz) = c^2 \det A \neq c \det A.$$

This observation does, however, illustrate a nice formula for $\det cA$.

Suppose A is an $n \times n$ matrix with rows $\mathbf{a}_1, \mathbf{a}_2, \dots, \mathbf{a}_n$. Then cA has rows $c\mathbf{a}_1, c\mathbf{a}_2, \dots, c\mathbf{a}_n$. Factoring c from each row,

$$\det cA = \det(c\mathbf{a}_1, c\mathbf{a}_2, \dots, c\mathbf{a}_n) = \underbrace{(c \cdot c \cdot \cdots \cdot c)}_{n \text{ times}} \det(\mathbf{a}_1, \mathbf{a}_2, \dots, \mathbf{a}_n) = c^n \det A.$$

3.2.5 If A is an $n \times n$ matrix, $\det cA = c^n \det A$.

3.2.6 EXAMPLE

If A is a 5×5 matrix and $\det A = 3$, then $\det(-2A) = (-2)^5 \det A = (-2)^5 3 = -32(3) = -96$. ∎

Property 2: The determinant of a matrix with a row of 0s is 0. This is quite easy to see—if one row of a matrix A consists entirely of 0s, expanding by cofactors of that row shows quickly that the determinant is 0.

Property 3: The determinant of a matrix changes sign if two rows are interchanged.

We illustrate the main idea behind this important result in the specific case of 2×2 matrices (and, in the exercises, ask the reader to investigate the 3×3 case). If $A = \begin{bmatrix} a & b \\ c & d \end{bmatrix}$ and we interchange the two rows of A,

See W. Keith Nicholson, *Linear Algebra with Applications*, Fourth edition, McGraw-Hill Ryerson, Toronto, 2003, for a proof in general.

$$\det \begin{bmatrix} c & d \\ a & b \end{bmatrix} = cb - da = -(ad - bc) = - \det A.$$

Recall that a permutation matrix is a matrix whose rows (or columns) are those of the identity matrix, in some order. Since a rearrangement of the rows of a matrix can be achieved by successively interchanging pairs of rows,

3.2.7 The determinant of any permutation matrix is ± 1.

3.2.8 EXAMPLE

The matrix $P = \begin{bmatrix} 0 & 1 & 0 & 0 \\ 0 & 0 & 1 & 0 \\ 0 & 0 & 0 & 1 \\ 1 & 0 & 0 & 0 \end{bmatrix}$ is a permutation matrix whose rows are the rows of I_4,

the 4×4 identity matrix, in order 2, 3, 4, 1. We can compute $\det P$ by interchanging rows until we get I_4, as follows:

$$\det P = \begin{vmatrix} 0 & 1 & 0 & 0 \\ 0 & 0 & 1 & 0 \\ 0 & 0 & 0 & 1 \\ 1 & 0 & 0 & 0 \end{vmatrix}$$

$$= - \begin{vmatrix} 1 & 0 & 0 & 0 \\ 0 & 0 & 1 & 0 \\ 0 & 0 & 0 & 1 \\ 0 & 1 & 0 & 0 \end{vmatrix} \qquad \text{interchanging rows one and four}$$

$$= + \begin{vmatrix} 1 & 0 & 0 & 0 \\ 0 & 1 & 0 & 0 \\ 0 & 0 & 0 & 1 \\ 0 & 0 & 1 & 0 \end{vmatrix} \qquad \text{interchanging rows two and four}$$

$$= - \begin{vmatrix} 1 & 0 & 0 & 0 \\ 0 & 1 & 0 & 0 \\ 0 & 0 & 1 & 0 \\ 0 & 0 & 0 & 1 \end{vmatrix} \qquad \text{interchanging rows three and four}$$

$$= -1. \qquad \blacksquare$$

Property 4: The determinant of a matrix with two equal rows is 0. Suppose we interchange two rows of a matrix A. Property 3 says the determinant changes sign, but if the rows are equal, the matrix hasn't changed.

$$\det(\mathbf{a}_1, \dots, \mathbf{b}, \dots, \mathbf{b}, \dots, \mathbf{a}_n) = -\det(\mathbf{a}_1, \dots, \mathbf{b}, \dots, \mathbf{b}, \dots, \mathbf{a}_n)$$

(did you see the row interchange?!), so $\det A = -\det A$, $2 \det A = 0$, and $\det A = 0$.

Property 5: The determinant of a matrix in which one row is a scalar multiple of another is 0. Suppose A is a matrix with two rows of the form $\mathbf{b}$ and $c\mathbf{b}$. Using linearity to factor c from row $c\mathbf{b}$ gives a matrix with two equal rows,

$$\det(\mathbf{a}_1, \dots, \mathbf{b}, \dots, c\mathbf{b}, \dots, \mathbf{a}_n) = c \det(\mathbf{a}_1, \dots, \mathbf{b}, \dots, \mathbf{b}, \dots, \mathbf{a}_n),$$

which is c times the determinant of a matrix with two equal rows. So we get 0.

Property 6: The determinant is multiplicative: $\det AB = \det A \det B$.

A proof in the general case can be found in W. Keith Nicholson, *Linear Algebra with Applications*, Fourth edition, McGraw-Hill Ryerson, Toronto, 2003.

We give a proof for 2×2 matrices. So let $A = \begin{bmatrix} a & b \\ c & d \end{bmatrix}$ and $B = \begin{bmatrix} p & q \\ r & s \end{bmatrix}$. Then

$$\det AB = \begin{vmatrix} ap + br & aq + bs \\ cp + dr & cq + ds \end{vmatrix}$$

$$= (ap + br)(cq + ds) - (aq + bs)(cp + dr)$$

$$= (apcq + apds + brcq + brds) - (aqcp + aqdr + bscp + bsdr)$$

$$= adps - adqr - bcps + bcqr$$

$$= (ad - bc)(ps - qr) = \det A \det B.$$

For example, if $A = \begin{bmatrix} -1 & 2 \\ 4 & 5 \end{bmatrix}$ and $B = \begin{bmatrix} 0 & 3 \\ 4 & -2 \end{bmatrix}$, then $AB = \begin{bmatrix} 8 & -7 \\ 20 & 2 \end{bmatrix}$ and $\det AB = 156 = (-13)(-12) = \det A \det B$.

Property 7: If A is invertible, then $\det A \neq 0$ and $\det A^{-1} = \frac{1}{\det A}$. This follows from the fact that the determinant is multiplicative. We know that $AA^{-1} = I$, the identity matrix, so $1 = \det I = \det AA^{-1} = \det A \det A^{-1}$. If the product of two numbers is not 0, neither number can be 0. Here, this means $\det A \neq 0$ and, moreover, $\det A^{-1} = \frac{1}{\det A}$.

What follows is a nice application of this fact.

Proposition 3.2.9   *If P is a permutation matrix,* $\det P^{-1} = \det P$.

Proof   As noted in 3.2.7, $\det P = \pm 1$. Thus

$$\det P^{-1} = \frac{1}{\det P} = \begin{cases} +1 & \text{if } \det P = 1 \\ -1 & \text{if } \det P = -1 \end{cases}$$

and we have $\det P^{-1} = \det P$ in either case.   ∎

In Theorem 3.1.11, we noted that if a matrix has a nonzero determinant, then it is invertible. Property 7 says that the "converse" is also true: if A is invertible, then $\det A \neq 0$. Thus the determinant gives a remarkable test for invertibility.

Theorem 3.2.10   *A square matrix A is invertible if and only if* $\det A \neq 0$.

Under what conditions is a real number a invertible? Surely, a is invertible if and only if $a \neq 0$. For matrices, A is invertible if and only if the number $\det A \neq 0$. The analogy is beautiful!

3.2.11 EXAMPLE $A = \begin{bmatrix} -1 & 2 \\ 5 & 9 \end{bmatrix}$ is invertible because $\det A = -19 \neq 0$.   ∎

3.2.12 PROBLEM   Show that the matrix $A = \begin{bmatrix} 1 & 0 & 3 \\ -4 & 1 & 6 \\ -2 & 1 & 4 \end{bmatrix}$ is invertible.

Solution.   Expanding by cofactors of the second column, we have

$$\det A = \begin{vmatrix} 1 & 3 \\ -2 & 4 \end{vmatrix} - \begin{vmatrix} 1 & 3 \\ -4 & 6 \end{vmatrix} = 10 - 18 = -8 \neq 0,$$

so A is invertible by Theorem 3.2.10.

There is a commonly used term for "not invertible."

3.2.13 DEFINITION   A square matrix is *singular* if it is not invertible.

By Theorem 3.2.10, a matrix is singular if and only if its determinant is 0.

3.2.14 PROBLEM   For what values of c is the matrix $A = \begin{bmatrix} 1 & 0 & -c \\ -1 & 3 & 1 \\ 0 & 2c & -4 \end{bmatrix}$ singular?

Solution.   Expanding by cofactors of the first row, we have

$$\det A = \begin{vmatrix} 3 & 1 \\ 2c & -4 \end{vmatrix} - c \begin{vmatrix} -1 & 3 \\ 0 & 2c \end{vmatrix} = (-12 - 2c) - c(-2c)$$

$$= 2c^2 - 2c - 12 = 2(c^2 - c - 6) = 2(c - 3)(c + 2).$$

The matrix is singular if and only if $\det A = 0$, and this is the case if and only if $c = 3$ or $c = -2$.

Property 8: $\det A^T = \det A$; hence, any property of the determinant involving rows holds equally for columns. Why should this be the case? We return to equation 3.1.7 of Section 3.1, which said that

$$AC^T = C^T A = (\det A)I, \tag{3}$$

where C is the matrix of cofactors of A. We can do two things with this equation. We can take the transpose of each piece and use the fact that the diagonal matrix $(\det A)I$ is its own transpose. This gives

$$CA^T = A^T C = (\det A)I. \tag{4}$$

Alternatively, we can apply (3) to the matrix A^T (whose cofactor matrix is C^T) and obtain

$$A^T C = CA^T = (\det A^T)I. \tag{5}$$

Comparing equations (4) and (5) shows that, indeed, $\det A^T = \det A$.

3.2.15 EXAMPLE   Suppose A and B are square $n \times n$ matrices with $\det A = 2$ and $\det B = 5$, and one would like $\det A^3 B^{-1} A^T$. By Property 6, $\det A^3 = \det AAA = (\det A)(\det A)(\det A) = (\det A)^3 = 2^3 = 8$. By Property 7, $\det B^{-1} = \frac{1}{\det B} = \frac{1}{5}$. By Property 8, $\det A^T = \det A = 2$. A final application of Property 6 gives $\det A^3 B^{-1} A^T = 8(\frac{1}{5})(2) = \frac{16}{5}$. ∎

Reading Challenge 1   *Compute* $\det(-A^3 B^{-1} A^T)$ *with A and B as in **3.2.15**, or can this be done? Explain.*

We illustrate how the fact that $\det A^T = \det A$ implies that any property of the determinant function concerning rows holds equally for columns. For example, interchanging two columns of a matrix changes the sign of the determinant. Suppose we interchange columns one and two of the matrix $\begin{bmatrix} a & d & g \\ b & e & h \\ c & f & i \end{bmatrix}$.

$$\begin{vmatrix} d & a & g \\ e & b & h \\ f & c & i \end{vmatrix} = \begin{vmatrix} d & e & f \\ a & b & c \\ g & h & i \end{vmatrix} \qquad \text{since } \det A = \det A^T$$

$$= - \begin{vmatrix} a & b & c \\ d & e & f \\ g & h & i \end{vmatrix} \qquad \text{interchanging rows one and two}$$

$$= - \begin{vmatrix} a & d & g \\ b & e & h \\ c & f & i \end{vmatrix} \qquad \text{since } \det A = \det A^T.$$

The determinant has changed sign.

Consider the effect of multiplying a column of a matrix by a scalar. For example,

$$\begin{vmatrix} ka & d & g \\ kb & e & h \\ kc & f & i \end{vmatrix} = \begin{vmatrix} ka & kb & kc \\ d & e & f \\ g & h & i \end{vmatrix} \qquad \text{since } \det A = \det A^T$$

$$= k \begin{vmatrix} a & b & c \\ d & e & f \\ g & h & i \end{vmatrix} \qquad \text{factoring } k \text{ from the first row}$$

$$= k \begin{vmatrix} a & d & g \\ b & e & h \\ c & f & i \end{vmatrix} \qquad \text{since } \det A = \det A^T,$$

and thus, if B is obtained from A by multiplying column one by k, $\det B = k \det A$. In a similar manner, using $\det A = \det A^T$ it is possible to show that all the properties of determinants stated about rows are also true for columns.

■ Evaluating a Determinant with Elementary Row Operations

There is another and an easier way to calculate a determinant, a way related to Gaussian elimination. Remember the three elementary row operations:

1. Interchange two rows.
2. Multiply a row by any nonzero scalar.
3. Replace a row by that row less a multiple of another.

Earlier in this section, we showed the effect of the first two of these operations on the determinant. The determinant changes sign if two rows are interchanged and the determinant is linear in each row, so

$$\det(\mathbf{a}_1, \ldots, c\mathbf{x}, \ldots, \mathbf{a}_n) = c \det(\mathbf{a}_1, \ldots, \mathbf{x}, \ldots, \mathbf{a}_n).$$

What happens to the determinant when we apply the third elementary row operation to a matrix? Suppose A has rows $\mathbf{a}_1, \mathbf{a}_2, \ldots, \mathbf{a}_n$ and we replace row j by row j less c times row i to get a new matrix B. Since the determinant is linear in row j,

$$\det B = \det(\mathbf{a}_1, \ldots, \mathbf{a}_i, \ldots, \mathbf{a}_j - c\mathbf{a}_i, \ldots, \mathbf{a}_n)$$
$$= \det(\mathbf{a}_1, \ldots, \mathbf{a}_i, \ldots, \mathbf{a}_j, \ldots, \mathbf{a}_n) + \det(\mathbf{a}_1, \ldots, \mathbf{a}_i, \ldots, -c\mathbf{a}_i, \ldots, \mathbf{a}_n).$$

Now the second determinant here is 0 because one row is a scalar multiple of another. So

$$\det B = \det(\mathbf{a}_1, \ldots, \mathbf{a}_i, \ldots, \mathbf{a}_j, \ldots, \mathbf{a}_n) = \det A.$$

The determinant has not changed! We summarize.

Matrix operation	Effect on determinant
interchange two rows	multiply by -1
factor $c \neq 0$ from a row	factor c from the determinant
replace a row by that row less a multiple of another	no change

This suggests the following method of computing the determinant of a matrix.

3.2.16 To find the determinant of a matrix A, reduce to upper triangular with elementary row operations, making note of the effect of each operation on the determinant. Then use the fact that the determinant of a triangular matrix is the product of its diagonal entries.

3.2.17 EXAMPLE

$$\begin{vmatrix} 1 & 2 & 3 \\ 4 & 5 & 6 \\ 7 & 8 & 9 \end{vmatrix} = \begin{vmatrix} 1 & 2 & 3 \\ 0 & -3 & -6 \\ 7 & 8 & 9 \end{vmatrix} \quad \text{replacing row two by row two less 4 times row one}$$

$$= \begin{vmatrix} 1 & 2 & 3 \\ 0 & -3 & -6 \\ 0 & -6 & -12 \end{vmatrix} \quad \text{replacing row three by row three less 7 times row one}$$

$$= 0 \quad \text{since row three is a multiple of row two.} \quad \blacksquare$$

3.2.18 EXAMPLE

$$
\begin{vmatrix} 1 & -1 & 4 \\ 2 & 1 & -1 \\ -2 & 0 & 3 \end{vmatrix} = \begin{vmatrix} 1 & -1 & 4 \\ 0 & 3 & -9 \\ -2 & 0 & 3 \end{vmatrix} \quad
\begin{array}{l} \text{replacing row two} \\ \text{by row two minus} \\ \text{2 times row one} \end{array}
$$

$$
= \begin{vmatrix} 1 & -1 & 4 \\ 0 & 3 & -9 \\ 0 & -2 & 11 \end{vmatrix} \quad
\begin{array}{l} \text{replacing row three by row three plus 2 times row} \\ \text{one} \end{array}
$$

$$
= 3 \begin{vmatrix} 1 & -1 & 4 \\ 0 & 1 & -3 \\ 0 & -2 & 11 \end{vmatrix} \quad \text{factoring a 3 from row two}
$$

$$
= 3 \begin{vmatrix} 1 & -1 & 4 \\ 0 & 1 & -3 \\ 0 & 0 & 5 \end{vmatrix} \quad
\begin{array}{l} \text{replacing row three by row three plus 2 times row} \\ \text{two} \end{array}
$$

$$
= 3(5) = 15 \quad
\begin{array}{l} \text{the determinant of a triangular matrix is the} \\ \text{product of its diagonal entries.} \end{array} \quad \blacksquare
$$

3.2.19 EXAMPLE   Let $A = \begin{bmatrix} 1 & 0 & 3 & 1 \\ 2 & 2 & 6 & 0 \\ 4 & 0 & -3 & 1 \\ 4 & 1 & 12 & 1 \end{bmatrix}$. Using the 1 in the $(1, 1)$ position to put 0s below it (via the third elementary row operation), we have

$$
\det A = \begin{vmatrix} 1 & 0 & 3 & 1 \\ 0 & 2 & 0 & -2 \\ 0 & 0 & -15 & -3 \\ 0 & 1 & 0 & -3 \end{vmatrix}.
$$

Factoring 2 from two and -3 from row three, we obtain

$$
2(-3) \begin{vmatrix} 1 & 0 & 3 & 1 \\ 0 & 1 & 0 & -1 \\ 0 & 0 & 5 & 1 \\ 0 & 1 & 0 & -3 \end{vmatrix} = -6 \begin{vmatrix} 1 & 0 & 3 & 1 \\ 0 & 1 & 0 & -1 \\ 0 & 0 & 5 & 1 \\ 0 & 0 & 0 & -2 \end{vmatrix},
$$

replacing row four by row four less row two. Since the determinant of a triangular matrix is the product of its diagonal entries, we obtain finally $\det A = -6(5)(-2) = 60$. $\blacksquare$

3.2.20 PROBLEM   Suppose $A = \begin{bmatrix} a & b & c \\ d & e & f \\ g & h & i \end{bmatrix}$ is a 3×3 matrix with $\det A = 2$. Find the determinant of

$$
\begin{bmatrix} -g & -h & -i \\ 3d + 5a & 3e + 5b & 3f + 5c \\ 7d & 7e & 7f \end{bmatrix}.
$$

Solution.   We use elementary row operations to move the new matrix towards A. Factoring 7 from the third row,

$$\begin{vmatrix} -g & -h & -i \\ 3d+5a & 3e+5b & 3f+5c \\ 7d & 7e & 7f \end{vmatrix} = 7\begin{vmatrix} -g & -h & -i \\ 3d+5a & 3e+5b & 3f+5c \\ d & e & f \end{vmatrix}$$

$$= 7\begin{vmatrix} -g & -h & -i \\ 5a & 5b & 5c \\ d & e & f \end{vmatrix} \qquad \text{subtracting 3 times row three from row two}$$

$$= 35\begin{vmatrix} -g & -h & -i \\ a & b & c \\ d & e & f \end{vmatrix} \qquad \text{factoring 5 from the second row}$$

$$= -35\begin{vmatrix} g & h & i \\ a & b & c \\ d & e & f \end{vmatrix} \qquad \text{factoring } -1 \text{ from the first row}$$

$$= 35\begin{vmatrix} a & b & c \\ g & h & i \\ d & e & f \end{vmatrix} \qquad \text{interchanging rows one and two}$$

$$= -35\begin{vmatrix} a & b & c \\ d & e & f \\ g & h & i \end{vmatrix} \qquad \text{interchanging rows two and three.}$$

The desired determinant is $-35 \det A$. The answer is -70.

As we have said, the fact that $\det A^T = \det A$ means all properties of the determinant that are true for rows are true for columns, and the converse. While the author's strong preference is to use row operations as far as possible, it is nonetheless true that it sometimes helps a lot to take advantage of column as well as row properties.

3.2.21 EXAMPLE

$$\begin{vmatrix} 3 & -1 & 2 \\ 2 & 4 & -9 \\ -3 & 5 & 6 \end{vmatrix} = -\begin{vmatrix} -1 & 3 & 2 \\ 4 & 2 & -9 \\ 5 & -3 & 6 \end{vmatrix} \qquad \begin{array}{l}\text{interchanging} \\ \text{columns one and} \\ \text{two}\end{array}$$

$$= -\begin{vmatrix} -1 & 3 & 2 \\ 0 & 14 & -1 \\ 0 & 12 & 16 \end{vmatrix} \qquad \begin{array}{l}\text{using the } -1 \text{ in the } (1,1) \text{ position to put} \\ \text{0s below it, via the third elementary row} \\ \text{operation}\end{array}$$

$$= -(-1)\begin{vmatrix} -1 & 2 & 3 \\ 0 & -1 & 14 \\ 0 & 16 & 12 \end{vmatrix} \qquad \text{interchanging columns two and three}$$

$$= \begin{vmatrix} -1 & 2 & 3 \\ 0 & -1 & 14 \\ 0 & 0 & 236 \end{vmatrix} \qquad \begin{array}{l}\text{using the } -1 \text{ in the } (2,2) \text{ position to get} \\ \text{a 0 below, with the third elementary row} \\ \text{operation}\end{array}$$

$$= 236 \qquad \begin{array}{l}\text{since the determinant of a triangular} \\ \text{matrix is the product of the diagonal} \\ \text{entries.}\end{array}$$

■ Concluding Remarks

This section and the previous one have introduced a number of ways to compute determinants. Which is best depends on the matrix and who is doing the calculation. The author's preference is often to use a computer algebra package like MAPLE. Figure 3.1 shows how to enter a matrix, which MAPLE echos, and then how easy it is to find the determinant.

```
>   A := matrix(3,3,[-1,2,0,4,5,-3,-2,1,6]);
```

$$A := \begin{bmatrix} -1 & 2 & 0 \\ 4 & 5 & -3 \\ -2 & 1 & 6 \end{bmatrix}$$

Figure 3.1   How to enter a matrix and find its determinant with MAPLE.

```
>   det(A);
```

$$-69$$

If no computing tools are available, the author finds 2×2 determinants in his head, 3×3 determinants by the method of cofactors (because 2×2 determinants are mental exercises), and higher order determinants by reducing to triangular form, usually exclusively by row operations. There are always exceptions to these general rules, of course, depending on the nature of the matrix.

3.2.22 EXAMPLES

- $\begin{vmatrix} 2 & 3 \\ -1 & 7 \end{vmatrix} = 17$

- $\begin{vmatrix} 1 & -2 & 3 \\ 4 & 5 & 6 \\ 9 & 8 & 7 \end{vmatrix}$

$$= 1(-13) - (-2)(-26) + 3(-13) = -13 - 52 - 39 = -104,$$

 expanding by cofactors along the first row.

- $\begin{vmatrix} 6 & -4 & 0 & -5 \\ 3 & -7 & -2 & 2 \\ -9 & 5 & -5 & 12 \\ -3 & 5 & 1 & 0 \end{vmatrix} = - \begin{vmatrix} 3 & -7 & -2 & 2 \\ 6 & -4 & 0 & -5 \\ -9 & 5 & -5 & 12 \\ -3 & 5 & 1 & 0 \end{vmatrix}$ interchanging rows one and two

$$= - \begin{vmatrix} 3 & -7 & -2 & 2 \\ 0 & 10 & 4 & -9 \\ 0 & -16 & -11 & 18 \\ 0 & -2 & -1 & 2 \end{vmatrix}$$ using the 3 in the $(1, 1)$ position to put 0s in the rest of column one

$$= \begin{vmatrix} 3 & -7 & -2 & 2 \\ 0 & -2 & -1 & 2 \\ 0 & -16 & -11 & 18 \\ 0 & 10 & 4 & -9 \end{vmatrix}$$ interchanging rows two and four

$$= \begin{vmatrix} 3 & -7 & -2 & 2 \\ 0 & -2 & -1 & 2 \\ 0 & 0 & -3 & 2 \\ 0 & 0 & -1 & 1 \end{vmatrix}$$ using the -2 in the $(2, 2)$ position to put 0s below it

$$= - \begin{vmatrix} 3 & -7 & -2 & 2 \\ 0 & -2 & -1 & 2 \\ 0 & 0 & -1 & 1 \\ 0 & 0 & -3 & 2 \end{vmatrix}$$ interchanging rows three and four

$$= - \begin{vmatrix} 3 & -7 & -2 & 2 \\ 0 & -2 & -1 & 2 \\ 0 & 0 & -1 & 1 \\ 0 & 0 & 0 & -1 \end{vmatrix}$$ using the -1 in the $(3, 3)$ position to put a 0 below it

$$= -3(-2)(-1)(-1) = 6$$ because the determinant of a triangular matrix is the product of its diagonal entries.

- $$\begin{vmatrix} 1 & 2 & 3 & 4 & 5 \\ 6 & 7 & 8 & 9 & 10 \\ 1 & 2 & 3 & 4 & 5 \\ 10 & 9 & 8 & 7 & 6 \\ 5 & 4 & 3 & 2 & 1 \end{vmatrix} = 0$$ because there are two equal rows. No calculation is required here!

- The three 0s in column three quickly turn the next 4×4 into a 3×3 determinant.

$$\begin{vmatrix} 2 & -3 & 0 & 7 \\ 9 & 2 & 0 & -3 \\ -5 & 6 & -4 & 8 \\ -7 & 0 & 0 & 2 \end{vmatrix}$$

$$= -4 \begin{vmatrix} 2 & -3 & 7 \\ 9 & 2 & -3 \\ -7 & 0 & 2 \end{vmatrix}$$

$$= -4 \left[-(-3) \begin{vmatrix} 9 & -3 \\ -7 & 2 \end{vmatrix} + 2 \begin{vmatrix} 2 & 7 \\ -7 & 2 \end{vmatrix} \right]$$ expanding by cofactors down the second column

$$= -4[3(-3) + 2(53)]$$

$$= -4[-9 + 106] = -4(97) = -388. \qquad \blacksquare$$

Answers to Reading Challenges

1. Using **3.2.5**,

$$\det(-A^3 B^{-1} A^T) = \det((-1)A^3 B^{-1} A^T)$$

$$= (-1)^n \det(A^3 B^{-1} A^T) = (-1)^n \frac{16}{5}$$

(using **3.2.15**) and we cannot say more unless we know whether n is even or odd. If n is even, the answer is $+\frac{16}{5}$; if n is odd, it is $-\frac{16}{5}$.

True/False Questions

Decide, with as little calculation as possible, whether each of the following statements is true or false and explain your answer whenever you say "false." (Answers can be found in the back of the book.)

1. If A is a square matrix, $\det A = -\det A$.

2. A matrix is singular if its determinant is 0.

3. If A is not singular, then A^{-1} is not singular.

4. If $\det A \neq 0$, the matrix A is invertible.

5. If A is 3×3, $\det A = 5$ and C is the matrix of cofactors of A, then $\det C^T = 25$.

6. An elementary row operation does not change the determinant.

7. If A is a square matrix and $\det A = 0$, then A must have a row of 0s.

8. If A and B are square matrices of the same size, then $\det(A+B) = \det A + \det B$.

9. If A and B are square matrices of the same size, then $\det AB = \det BA$.

Exercises

*Solutions to exercises marked [BB] can be found in the **B**ack of the **B**ook.*

1. [BB] Let A and B be 5×5 matrices with $\det(-3A) = 4$ and $\det B^{-1} = 2$. Find $\det A$, $\det B$, and $\det AB$.

2. Suppose A and B are 3×3 matrices with $\det A = 2$ and $\det B = -5$. Find $\det A^T B^{-1} A^3 (-B)$ and $\det(-A^T B^{-1} A^3 (-B))$.

3. A 6×6 matrix A has determinant 5. Find $\det 2A$, $\det(-A)$, $\det A^{-1}$, and $\det A^2$.

4. Let A and B be 3×3 matrices such that $\det A = 10$ and $\det B = 12$. Find $\det AB$, $\det A^4$, $\det 2B$, $\det(AB)^T$, $\det A^{-1}$, and $\det A^{-1} B^{-1} AB$.

5. If A and B are 5×5 matrices, $\det A = 4$, and $\det B = \frac{1}{2}$, find $\det B^{-1} A^2 (-B)^T$.

6. [BB] Let $A = \begin{bmatrix} 1 & 7 & 0 & 17 & 9 \\ 0 & 2 & -35 & 10 & 15 \\ 0 & 0 & 3 & -5 & 11 \\ 0 & 0 & 0 & 2 & 77 \\ 0 & 0 & 0 & 0 & 5 \end{bmatrix}$.

Find $\det A$, $\det A^{-1}$, and $\det A^2$.

7. [BB] Let $A = \begin{bmatrix} -1 & -1 & 1 & 0 \\ 2 & 1 & 1 & 3 \\ 0 & 1 & 1 & 2 \\ 1 & 3 & -1 & 2 \end{bmatrix}$.

(a) Find the determinant of A expanding by cofactors of the first column.

(b) Find the determinant of A by reducing it to an upper triangular matrix.

8. Let $A = \begin{bmatrix} 5 & 15 & 10 & 0 \\ 4 & 11 & 12 & 2 \\ 7 & 25 & 0 & -2 \\ -3 & -14 & 17 & 23 \end{bmatrix}$.

(a) Find the determinant of A expanding by cofactors of the fourth column.

(b) Find the determinant of A expanding by cofactors of the third row.

(c) Find the determinant of A by reducing it to an upper triangular matrix.

(d) Is A invertible? If yes, what is $\det A^{-1}$? If no, why not?

9. Find the determinants of each of the following matrices by reducing to upper triangular matrices.

(a) [BB] $\begin{bmatrix} -2 & 1 & 2 \\ 1 & 3 & 6 \\ -4 & 5 & 9 \end{bmatrix}$

(b) $\begin{bmatrix} -3 & 1 & 1 \\ 0 & -7 & 1 \\ 3 & 2 & 1 \end{bmatrix}$

(c) $\begin{bmatrix} 8 & 1 & -2 \\ -6 & 1 & 1 \\ 4 & -2 & 3 \end{bmatrix}$

(d) [BB] $\begin{bmatrix} -3 & 0 & 1 & 1 \\ 3 & 1 & 2 & 2 \\ -6 & -2 & -4 & 2 \\ 1 & -1 & 0 & -1 \end{bmatrix}$

(e) $\begin{bmatrix} 4 & -1 & 3 & -1 \\ 0 & 1 & 2 & 2 \\ 3 & 1 & 0 & 2 \\ 1 & 2 & -1 & 1 \end{bmatrix}$

(f) $\begin{bmatrix} -2 & 1 & 1 & 2 \\ 1 & -2 & 1 & 2 \\ 4 & -3 & 5 & 1 \\ 0 & -2 & 2 & 3 \end{bmatrix}$

(g) $\begin{bmatrix} 1 & 0 & 0 & 1 & 0 \\ 0 & 2 & 4 & 0 & -8 \\ -3 & 0 & 1 & 0 & 5 \\ 4 & 1 & 0 & 1 & -2 \\ -1 & 0 & 1 & 0 & 1 \end{bmatrix}$

(h) $\begin{bmatrix} 1 & 0 & 0 & 0 & 5 \\ 0 & 0 & 1 & 0 & 1 \\ 2 & 5 & 0 & 0 & 0 \\ 0 & 0 & 1 & -4 & 0 \\ 0 & 1 & 0 & 1 & 0 \end{bmatrix}$

(i) $\begin{bmatrix} -3 & 0 & 1 & 1 & 0 \\ 3 & 1 & 2 & -2 & 2 \\ -6 & -2 & -4 & 1 & 2 \\ 1 & 0 & -4 & 4 & 5 \\ -1 & 0 & 0 & 0 & -1 \end{bmatrix}$

(j) $\begin{bmatrix} 0 & 6 & 42 & -18 & 12 \\ 1 & 3 & 4 & 5 & 1 \\ 2 & 11 & 40 & -20 & 6 \\ -3 & -12 & -27 & 25 & 7 \\ 0 & 2 & 11 & -14 & 29 \end{bmatrix}$.

10. Let $A = \begin{bmatrix} 1 & 2 & 3 \\ 1 & 1 & 1 \\ 1 & 0 & 1 \end{bmatrix}$ and $B = \begin{bmatrix} 2 & 0 & 1 \\ 1 & 1 & 0 \\ 0 & 1 & 1 \end{bmatrix}$.

(a) Find each of the following matrices and their determinants: AB, $A^T B$, AB^T, $A^T B^T$.

(b) Could any of the results of part (a) have been anticipated? Explain.

11. [BB] If P is a permutation matrix and

$$PA = \begin{bmatrix} 1 & 6 & 7 \\ 5 & 25 & 36 \\ -2 & -27 & -4 \end{bmatrix}$$

$$= \begin{bmatrix} 1 & 0 & 0 \\ 5 & 1 & 0 \\ -2 & 3 & 1 \end{bmatrix} \begin{bmatrix} 1 & 6 & 7 \\ 0 & -5 & 1 \\ 0 & 0 & 7 \end{bmatrix} = LU,$$

what are the possible values for det A?

12. Let $A = \begin{bmatrix} a & b & c \\ d & e & f \\ g & h & i \end{bmatrix}$. Show that the determinant of the matrix obtained from A by interchanging rows two and three is $-$ det A. (You may use the fact, established in the text, that the corresponding property holds for 2×2 matrices.)

13. Write at least three conditions on the rows of a matrix A that guarantee that A is singular.

14. [BB] Suppose each column of a 3×3 matrix A is a linear combination of two vectors $\mathbf{u}$ and $\mathbf{v}$. Find an argument involving the determinant that shows that A is singular.

15. Find the following determinants by methods of your choice.

(a) [BB] $\begin{vmatrix} 6 & -3 \\ 1 & -4 \end{vmatrix}$

(b) $\begin{vmatrix} -1 & 2 & 0 \\ 2 & -7 & -8 \\ 3 & 1 & 5 \end{vmatrix}$

(c) [BB] $\begin{vmatrix} 5 & 3 & 8 \\ -4 & 1 & 4 \\ -2 & 3 & 6 \end{vmatrix}$

(d) $\begin{vmatrix} 4 & 28 & 16 & 12 \\ 12 & 81 & 27 & 18 \\ 4 & 31 & 36 & 28 \\ 1 & 15 & 56 & 41 \end{vmatrix}$

(e) [BB] $\begin{vmatrix} -1 & 2 & 3 & 4 \\ 3 & -9 & 2 & 1 \\ 0 & -5 & 7 & 6 \\ 2 & -4 & -6 & -8 \end{vmatrix}$

(f) $\begin{vmatrix} 1 & 0 & 0 & 0 & 1 \\ 0 & 0 & 1 & 0 & 1 \\ 1 & 1 & 0 & 0 & 0 \\ 0 & 0 & 1 & 1 & 0 \\ 0 & 1 & 0 & 1 & 0 \end{vmatrix}$

(g) $\begin{vmatrix} 1 & -1 & 2 & 0 & -2 \\ 0 & 1 & 0 & 4 & 1 \\ 1 & 1 & 3 & 0 & 0 \\ 0 & 0 & 0 & 3 & -1 \\ 0 & 0 & 0 & 1 & 1 \end{vmatrix}$

(h) $\begin{vmatrix} 12 & 48 & 72 & 36 \\ 3 & 17 & 28 & -26 \\ 2 & 11 & 11 & 6 \\ -2 & -10 & -21 & 31 \end{vmatrix}$.

16. [BB] Suppose $\begin{vmatrix} a & b & c \\ d & e & f \\ g & h & i \end{vmatrix} = -3$. Find

$\begin{vmatrix} a & -g & 2d \\ b & -h & 2e \\ c & -i & 2f \end{vmatrix}$.

17. Let $A = \begin{bmatrix} a & b & c \\ p & q & r \\ u & v & w \end{bmatrix}$ and suppose $\det A = 5$. Find

$\begin{vmatrix} 2p & -a+u & 3u \\ 2q & -b+v & 3v \\ 2r & -c+w & 3w \end{vmatrix}$.

18. Let $A =$
$\begin{bmatrix} a+h+g & h+b+f & g+f+c \\ 2a+3h+4g & 2h+3b+4f & 2g+3f+4c \\ 3a+2h+5g & 3h+2b+5f & 3g+2f+5c \end{bmatrix}$,

$B = 4\begin{vmatrix} a & h & g \\ h & b & f \\ g & f & c \end{vmatrix}$.

Show that $\det A = 4\det B$.

■ Critical Reading

19. [BB] Suppose A is a square matrix. Explain how an LDU factorization of A could prove helpful in finding the determinant of A and illustrate with

$A = \begin{bmatrix} 2 & -1 & 4 & 1 \\ 1 & 1 & -10 & -2 \\ 4 & 0 & -7 & 6 \\ 6 & -3 & 0 & 1 \end{bmatrix}$.

20. If A is $n \times n$, n is odd, and $A^T = -A$, then A is singular (that is, not invertible). Why?

21. [BB] Suppose A is an $n \times n$ invertible matrix and $A^{-1} = \frac{1}{2}(I - A)$, where I is the $n \times n$ identity matrix. Show that $\det A$ is a power of 2.

22. Let $A = \begin{bmatrix} 1 & 2 & 3 & 4 \\ 0 & -1 & 1 & 2 \\ 1 & 2 & 3 & 4 \\ 8 & -2 & 5 & 5 \end{bmatrix}$. Find $\det A$ by inspection and explain.

23. Let $B = \begin{bmatrix} 1 & 2 & 1 \\ 3 & -2 & 0 \\ -1 & 4 & 1 \end{bmatrix}$.

(a) Find $\det B$, $\det \frac{1}{3}B$, and $\det B^{-1}$.

(b) Suppose A is a matrix whose inverse is B. Find the cofactor matrix of A.

24. Suppose A is a 4×4 matrix and P is an invertible 4×4 matrix such that

$$PAP^{-1} = D = \begin{bmatrix} 1 & 0 & 0 & 0 \\ 0 & -1 & 0 & 0 \\ 0 & 0 & -1 & 0 \\ 0 & 0 & 0 & -1 \end{bmatrix}.$$

(a) Find $\det A$.

(b) Find A^2.

(c) Find A^{101}.

25. (a) Let a, b, c be three real numbers and let

$A = \begin{bmatrix} 1 & a & a^2 \\ 1 & b & b^2 \\ 1 & c & c^2 \end{bmatrix}$. Show that

$\det A = (a-b)(a-c)(b-c)$.

(b) Suppose a, b, c, d are four real numbers. Make a

guess as to the value of $\det \begin{bmatrix} 1 & a & a^2 & a^3 \\ 1 & b & b^2 & b^3 \\ 1 & c & c^2 & c^3 \\ 1 & d & d^2 & d^3 \end{bmatrix}$.

[No proof required.]

26. Let A be the $n \times n$ matrix

$$A = \begin{bmatrix} 3 & 1 & 1 & \cdots & 1 \\ 1 & 3 & 1 & \cdots & 1 \\ 1 & 1 & 3 & & 1 \\ \vdots & \vdots & & \ddots & \\ 1 & 1 & \cdots & 1 & 3 \end{bmatrix}.$$

(a) Find $\det A$. [Hint: Investigate small values of n and look for a way to apply elementary row operations.]

(b) Does there exist a matrix B, with integer entries, such that $A = BB^T$ when $n = 40$? Explain.

27. Let A be a 3×3 matrix with columns $\mathbf{a}_1, \mathbf{a}_2, \mathbf{a}_3$, respectively. Let B be the matrix with columns $\mathbf{a}_1 + \alpha\mathbf{a}_2, \mathbf{a}_2 + \beta\mathbf{a}_1, \mathbf{a}_3$, respectively, for scalars α and β. Show that $\det B = (1 - \alpha\beta)\det A$. [Hint: Use the notation $\det(\mathbf{a}_1, \mathbf{a}_2, \mathbf{a}_3)$ for $\det A$.]

28. Let

$$A = \begin{bmatrix} 4x & -x-y & 2y+z \\ 4x+2y-z & -x-3y+z & 2x+4y+z \\ x+y & -x-y & 3x+y+z \end{bmatrix}$$

and

$$B = \begin{bmatrix} 1 & 1 & 1 \\ 2 & 3 & 1 \\ 1 & 1 & 0 \end{bmatrix}.$$

(a) Find AB.

(b) Express $\det A$ as the product of factors that are linear in x, y, and z.

29. Let $A = \begin{bmatrix} a_1 & a_2 & a_3 \\ b_1 & b_2 & b_3 \\ c_1 & c_2 & c_3 \end{bmatrix}$ and

$$B = \begin{bmatrix} a_1 & a_2 & a_3 & 0 \\ a_1b_1+a_2c_1 & a_1b_2+a_2c_2 & a_1b_3+a_2c_3 & a_3 \\ b_1b_1+b_2c_1 & b_1b_2+b_2c_2 & b_1b_3+b_2c_3 & b_3 \\ c_1b_1+c_2c_1 & c_1b_2+c_2c_2 & c_1b_3+c_2c_3 & c_3 \end{bmatrix}.$$

Prove that $\det B = (\det A)^2$. [Hint: $\det XY = \det X \det Y$.]

30. Let B and C be square matrices and D be a matrix such that $A = \begin{bmatrix} B & D \\ 0 & C \end{bmatrix}$ is a matrix. Prove that $\det A = \det B \det C$.

3.3 The Eigenvalues and Eigenvectors of a Matrix

Perhaps the single most important concept in linear algebra is that of *eigenvalues* and *eigenvectors*. From statisticians to people who work in dynamical systems to those who make their living in computer graphics, the author has been urged repeatedly to define and illustrate these terms as early as possible, and so we do.

This section revolves around the equation

$$A\mathbf{x} = \lambda\mathbf{x},$$

where A is a (necessarily square) matrix and $\mathbf{x}$ is a vector. By tradition, the Greek symbol λ, pronounced "lambda," is used to denote a scalar in this special equation, which says that multiplication by A transforms $\mathbf{x}$ into a multiple of itself.

Reading Challenge 1   *Why does $A\mathbf{x} = \lambda\mathbf{x}$ imply that A is square?*

3.3.1 DEFINITIONS   If A is an $n \times n$ matrix, a real number λ is called an *eigenvalue* of A if

$$A\mathbf{x} = \lambda\mathbf{x}$$

for some **nonzero** vector $\mathbf{x}$. The nonzero vector $\mathbf{x}$ is called an *eigenvector of A corresponding to λ*. The set of all solutions to $A\mathbf{x} = \lambda\mathbf{x}$ is called the *eigenspace of A corresponding to λ*.

3.3.2 Remark The equation $A\mathbf{x} = \lambda\mathbf{x}$ is not very interesting if $\mathbf{x} = \mathbf{0}$ since, in this case, it holds for any scalar λ. This is why eigenvectors are required to be **nonzero** vectors. On the other hand, the eigenspace of A does include the zero vector. The eigenspace of A corresponding to λ consists of all eigenvectors corresponding to λ together with the zero vector.

3.3.3 EXAMPLE

Let $A = \begin{bmatrix} 1 & 4 \\ 2 & 3 \end{bmatrix}$ and $\mathbf{x} = \begin{bmatrix} 1 \\ 1 \end{bmatrix}$. Then $\mathbf{x}$ is an eigenvector of A corresponding to $\lambda = 5$ because

$$A\mathbf{x} = \begin{bmatrix} 1 & 4 \\ 2 & 3 \end{bmatrix} \begin{bmatrix} 1 \\ 1 \end{bmatrix} = \begin{bmatrix} 5 \\ 5 \end{bmatrix} = 5\mathbf{x}.$$

The eigenspace of A corresponding to $\lambda = 5$ is the set of solutions to $A\mathbf{x} = 5\mathbf{x}$. Rewriting this equation in the form $(A - 5I)\mathbf{x} = \mathbf{0}$, we must solve the homogeneous system with coefficient matrix $A - 5I$. Gaussian elimination proceeds $A - 5I = \begin{bmatrix} -4 & 4 \\ 2 & -2 \end{bmatrix} \rightarrow \begin{bmatrix} 1 & -1 \\ 0 & 0 \end{bmatrix}$. If $\mathbf{x} = \begin{bmatrix} x_1 \\ x_2 \end{bmatrix}$ is an eigenvector, then $x_2 = t$ is free and $x_1 - x_2 = 0$, so $x_1 = x_2 = t$ and $\mathbf{x} = \begin{bmatrix} t \\ t \end{bmatrix} = t \begin{bmatrix} 1 \\ 1 \end{bmatrix}$. The eigenspace corresponding to $\lambda = 5$ is the set of all multiples of $\begin{bmatrix} 1 \\ 1 \end{bmatrix}$. ∎

Reading Challenge 2

In 3.3.3, we rewrote $A\mathbf{x} = 5\mathbf{x}$ as $(A - 5I)\mathbf{x} = \mathbf{0}$. Why didn't we write $(A - 5)\mathbf{x} = \mathbf{0}$? Why $5I$?

Suppose $\mathbf{x}$ is an eigenvector of A corresponding to λ; thus $A\mathbf{x} = \lambda\mathbf{x}$. Any vector on the line through $(0, 0)$ with direction $\mathbf{x}$ is of the form $c\mathbf{x}$ and

$$A(c\mathbf{x}) = c(A\mathbf{x}) = c(\lambda\mathbf{x}) = (c\lambda)\mathbf{x}.$$

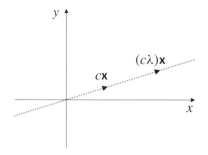

Thus multiplication (on the left) by A moves $c\mathbf{x}$ to $(c\lambda)\mathbf{x}$, which is another multiple of $\mathbf{x}$.

Vectors on the line determined by $\mathbf{x}$ are moved to other vectors on this same line. The line itself is fixed under multiplication by A.

> **3.3.4** Eigenvectors correspond to lines left invariant under multiplication by A.

We are not saying here that every vector on an invariant line is left fixed, just that every vector on the line is moved upon multiplication by A to another vector on the line. The line, as a whole, is fixed. In **3.3.3**, for instance, the line consisting of scalar multiples of $\begin{bmatrix} 1 \\ 1 \end{bmatrix}$ (this is the line with equation $y = x$) is invariant under multiplication by A. If $\mathbf{x}$ is on this line, however, $\mathbf{x}$ is not fixed: $A\mathbf{x} = 5\mathbf{x}$.

3.3.5 EXAMPLE

The scalar $\lambda = -3$ is an eigenvalue of $A = \begin{bmatrix} 5 & 8 & 16 \\ 4 & 1 & 8 \\ -4 & -4 & -11 \end{bmatrix}$ and $\mathbf{x} = \begin{bmatrix} 3 \\ -1 \\ -1 \end{bmatrix}$ is a corresponding eigenvector because

$$Ax = \begin{bmatrix} 5 & 8 & 16 \\ 4 & 1 & 8 \\ -4 & -4 & -11 \end{bmatrix}\begin{bmatrix} 3 \\ -1 \\ -1 \end{bmatrix} = \begin{bmatrix} -9 \\ 3 \\ 3 \end{bmatrix} = -3\mathbf{x}.$$

To find the eigenspace of $\lambda = -3$, we must find all the vectors $\mathbf{x}$ that satisfy $A\mathbf{x} = -3\mathbf{x}$. Rewriting this equation as $(A + 3I)\mathbf{x} = \mathbf{0}$, where I is the 3×3 identity matrix, the desired eigenspace is the set of solutions to the homogeneous system with coefficient matrix $A + 3I$. Gaussian elimination proceeds

$$A + 3I = \begin{bmatrix} 8 & 8 & 16 \\ 4 & 4 & 8 \\ -4 & -4 & -8 \end{bmatrix} \to \begin{bmatrix} 1 & 1 & 2 \\ 4 & 4 & 8 \\ -4 & -4 & -8 \end{bmatrix} \to \begin{bmatrix} 1 & 1 & 2 \\ 0 & 0 & 0 \\ 0 & 0 & 0 \end{bmatrix}.$$

With $\mathbf{x} = \begin{bmatrix} x_1 \\ x_2 \\ x_3 \end{bmatrix}$, we have free variables $x_2 = t$ and $x_3 = s$, and $x_1 = -x_2 - 2x_3 = -t - 2s$. So $\mathbf{x} = \begin{bmatrix} -t - 2s \\ t \\ s \end{bmatrix} = t\begin{bmatrix} -1 \\ 1 \\ 0 \end{bmatrix} + s\begin{bmatrix} -2 \\ 0 \\ 1 \end{bmatrix}$. The eigenspace of A corresponding to $\lambda = -3$ consists of all linear combinations of $\begin{bmatrix} -1 \\ 1 \\ 0 \end{bmatrix}$ and $\begin{bmatrix} -2 \\ 0 \\ 1 \end{bmatrix}$, a plane in Euclidean 3-space—see 1.1.20. ∎

Reading Challenge 3 *What is the equation of this plane?*

3.3.6 EXAMPLE

Multiplication by the matrix $A = \begin{bmatrix} 1 & 0 \\ 0 & -1 \end{bmatrix}$ transforms $\begin{bmatrix} x \\ y \end{bmatrix}$ to $\begin{bmatrix} x \\ -y \end{bmatrix}$:

$$A\begin{bmatrix} x \\ y \end{bmatrix} = \begin{bmatrix} 1 & 0 \\ 0 & -1 \end{bmatrix}\begin{bmatrix} x \\ y \end{bmatrix} = \begin{bmatrix} x \\ -y \end{bmatrix}.$$

So, multiplication by A reflects vectors in the x-axis. Reflection in a line fixes the line. In this particular case, it fixes every vector on the line because

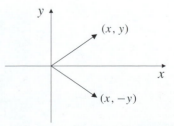

$$A\begin{bmatrix} x \\ 0 \end{bmatrix} = \begin{bmatrix} 1 & 0 \\ 0 & -1 \end{bmatrix}\begin{bmatrix} x \\ 0 \end{bmatrix} = \begin{bmatrix} x \\ 0 \end{bmatrix}.$$

Thus every vector on the x-axis is an eigenvector corresponding to $\lambda = 1$. The reflection in question also fixes the y-axis because every vector on this line is moved to its negative (which is still on the line). Here's why:

$$A\begin{bmatrix} 0 \\ y \end{bmatrix} = \begin{bmatrix} 1 & 0 \\ 0 & -1 \end{bmatrix}\begin{bmatrix} 0 \\ y \end{bmatrix} = \begin{bmatrix} 0 \\ -y \end{bmatrix} = -\begin{bmatrix} 0 \\ y \end{bmatrix}.$$

Every vector on the y-axis is an eigenvector corresponding to $\lambda = -1$. The matrix A has two eigenspaces, the two axes. Both these lines are fixed under multiplication by A. ■

3.3.7 EXAMPLE   The matrix $A = \begin{bmatrix} 0 & 1 \\ -1 & 0 \end{bmatrix}$ moves $\begin{bmatrix} x \\ y \end{bmatrix}$ to $\begin{bmatrix} y \\ -x \end{bmatrix}$:

$$A\begin{bmatrix} x \\ y \end{bmatrix} = \begin{bmatrix} 0 & 1 \\ -1 & 0 \end{bmatrix}\begin{bmatrix} x \\ y \end{bmatrix} = \begin{bmatrix} y \\ -x \end{bmatrix}.$$

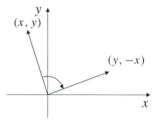

Multiplication by A rotates vectors clockwise through the angle $\pi/2$. There are no fixed lines, so there are no eigenspaces. Verification is straightforward if we first notice that $A^2 = -I$, for if $A\mathbf{x} = \lambda\mathbf{x}$, then

$$A(A\mathbf{x}) = A(\lambda\mathbf{x}) = \lambda(A\mathbf{x}) = \lambda(\lambda\mathbf{x}) = \lambda^2\mathbf{x},$$

so $A^2\mathbf{x} = \lambda^2\mathbf{x}$, or $-\mathbf{x} = \lambda^2\mathbf{x}$. Since $\mathbf{x} \neq \mathbf{0}$, this last equation says $\lambda^2 = -1$. This cannot be the case for a real number λ. This matrix A has no **real** eigenvalues. ■

■ How to Find Eigenvalues and Eigenvectors

Examples **3.3.3** and **3.3.5** illustrate the general method of finding the eigenvalues and eigenspaces of a matrix. The key idea is that if $A\mathbf{x} = \lambda\mathbf{x}$ with $\mathbf{x} \neq \mathbf{0}$, then $A\mathbf{x} - \lambda\mathbf{x} = \mathbf{0}$, so $(A - \lambda I)\mathbf{x} = \mathbf{0}$. Since $\mathbf{x} \neq \mathbf{0}$, the matrix $A - \lambda I$ cannot be invertible. Hence,

3.3.8 $\det(A - \lambda I) = 0.$

The expression $\det(A - \lambda I)$ is a polynomial in the variable λ called the *characteristic polynomial* of A. Its roots are the eigenvalues of A.

3.3.9 To Find Eigenvalues and Eigenvectors of a (Square) Matrix A:

1. Compute the characteristic polynomial $\det(A - \lambda I)$.
2. Set $\det(A - \lambda I) = 0$. The roots of this polynomial are the eigenvalues.
3. For each eigenvalue λ, solve the homogeneous system $(A - \lambda I)\mathbf{x} = \mathbf{0}$. The solutions to this system are the vectors of the eigenspace corresponding to λ.

3.3.10 EXAMPLE

The characteristic polynomial of the matrix $A = \begin{bmatrix} 1 & 4 \\ 2 & 3 \end{bmatrix}$ is

$$\det(A - \lambda I) = \begin{vmatrix} 1 - \lambda & 4 \\ 2 & 3 - \lambda \end{vmatrix} = (1 - \lambda)(3 - \lambda) - 8 = \lambda^2 - 4\lambda - 5. \quad (1)$$

This is a polynomial of degree 2 in λ. It factors, $\lambda^2 - 4\lambda - 5 = (\lambda - 5)(\lambda + 1)$, and its roots, $\lambda = 5$, $\lambda = -1$, are the eigenvalues of A. We found the eigenspace corresponding to $\lambda = 5$ in **3.3.3**. To find the eigenspace corresponding to $\lambda = -1$, we solve the homogeneous system $(A + 1I)\mathbf{x} = \mathbf{0}$. The coefficient matrix is $A - \lambda I$ with $\lambda = -1$; that is, $\begin{bmatrix} 2 & 4 \\ 2 & 4 \end{bmatrix}$—see (1). Gaussian elimination proceeds

$$\begin{bmatrix} 2 & 4 \\ 2 & 4 \end{bmatrix} \to \begin{bmatrix} 1 & 2 \\ 2 & 4 \end{bmatrix} \to \begin{bmatrix} 1 & 2 \\ 0 & 0 \end{bmatrix}.$$

Variable $x_2 = t$ is free and $x_1 = -2x_2 = -2t$. The eigenspace corresponding to $\lambda = -1$ is the set of vectors of the form

$$\mathbf{x} = \begin{bmatrix} x_1 \\ x_2 \end{bmatrix} = \begin{bmatrix} -2t \\ t \end{bmatrix} = t \begin{bmatrix} -2 \\ 1 \end{bmatrix},$$

which is, geometrically, another line through $(0, 0)$.

To check that our calculations are correct, we compute

$$A\mathbf{x} = \begin{bmatrix} 1 & 4 \\ 2 & 3 \end{bmatrix} \begin{bmatrix} -2 \\ 1 \end{bmatrix} = \begin{bmatrix} 2 \\ -1 \end{bmatrix},$$

and since this is $(-1) \begin{bmatrix} -2 \\ 1 \end{bmatrix}$, $\mathbf{x} = \begin{bmatrix} -2 \\ 1 \end{bmatrix}$ is indeed an eigenvector corresponding to $\lambda = -1$. ◼

3.3.11 Remark (Understanding matrix multiplication via eigenspaces). When we multiply a vector $\mathbf{x}$ by a matrix A, the vector $\mathbf{x}$ moves to the vector $A\mathbf{x}$. Knowing the eigenvalues and eigenvectors of a matrix A can help us understand the nature of the transformation from $\mathbf{x}$ to $A\mathbf{x}$. In **3.3.10**, for instance, the matrix A has eigenvalues -1 and 5 corresponding to eigenspaces $t \begin{bmatrix} -2 \\ 1 \end{bmatrix}$ and $t \begin{bmatrix} 1 \\ 1 \end{bmatrix}$, respectively. If $\mathbf{x}$ is a multiple of $\begin{bmatrix} -2 \\ 1 \end{bmatrix}$, for example, then $A\mathbf{x} = -\mathbf{x}$, while if $\mathbf{x}$ is a multiple of $\begin{bmatrix} 1 \\ 1 \end{bmatrix}$, then $A\mathbf{x} = 5\mathbf{x}$. What does multiplication by A do to a vector $\mathbf{v}$ that is not a multiple of $\mathbf{x}_1 = \begin{bmatrix} -2 \\ 1 \end{bmatrix}$ or $\mathbf{x}_2 = \begin{bmatrix} 1 \\ 1 \end{bmatrix}$? We construct a parallelogram with diagonal $\mathbf{v}$ and whose sides are on the lines with directions $\mathbf{x}_1$ and $\mathbf{x}_2$. See Figure 3.2. Algebraically, we have expressed $\mathbf{v}$ as a linear combination of $\mathbf{x}_1$ and $\mathbf{x}_2$: $\mathbf{v} = a\mathbf{x}_1 + b\mathbf{x}_2$. Since matrix multiplication preserves linear combinations,

$$A\mathbf{v} = A(a\mathbf{x}_1 + b\mathbf{x}_2) = a(A\mathbf{x}_1) + b(A\mathbf{x}_2) = a(-\mathbf{x}_1) + b(5\mathbf{x}_2) = -a\mathbf{x}_1 + 5b\mathbf{x}_2,$$

so $A\mathbf{v}$ is the diagonal of the parallelogram with sides $-a\mathbf{x}_1$ and $5b\mathbf{x}_2$.

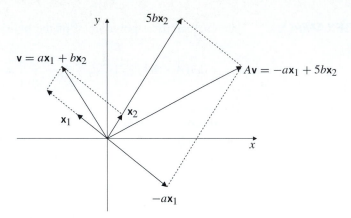

Figure 3.2

3.3.12 PROBLEM

Find all eigenvalues and the corresponding eigenspaces of

$$A = \begin{bmatrix} 5 & 8 & 16 \\ 4 & 1 & 8 \\ -4 & -4 & -11 \end{bmatrix}.$$

Solution. The characteristic polynomial of A is the determinant of

$$A - \lambda I = \begin{bmatrix} 5 & 8 & 16 \\ 4 & 1 & 8 \\ -4 & -4 & -11 \end{bmatrix} - \begin{bmatrix} \lambda & 0 & 0 \\ 0 & \lambda & 0 \\ 0 & 0 & \lambda \end{bmatrix}$$

$$= \begin{bmatrix} 5-\lambda & 8 & 16 \\ 4 & 1-\lambda & 8 \\ -4 & -4 & -11-\lambda \end{bmatrix}. \tag{2}$$

Expanding by cofactors of the first row,

$$\det(A - \lambda I) = (5-\lambda) \begin{vmatrix} 1-\lambda & 8 \\ -4 & -11-\lambda \end{vmatrix} - 8 \begin{vmatrix} 4 & 8 \\ -4 & -11-\lambda \end{vmatrix}$$

$$+ 16 \begin{vmatrix} 4 & 1-\lambda \\ -4 & -4 \end{vmatrix}$$

$$= (5-\lambda)[(1-\lambda)(-11-\lambda) + 32] - 8[4(-11-\lambda) + 32]$$

$$+ 16[-16 + 4(1-\lambda)]$$

$$= (5-\lambda)(21 + 10\lambda + \lambda^2) - 8(-12 - 4\lambda) + 16(-12 - 4\lambda)$$

$$= -\lambda^3 - 5\lambda^2 - 3\lambda + 9 = -(\lambda^3 + 5\lambda^2 + 3\lambda - 9).$$

In **3.3.5**, we noted that $\lambda = -3$ is an eigenvalue. Thus $\lambda = -3$ is one root of this polynomial and we can find the others by long division—see Figure 3.3. We obtain

$$\lambda^3 + 5\lambda^2 + 3\lambda - 9 = (\lambda + 3)(\lambda^2 + 2\lambda - 3)$$

$$= (\lambda + 3)(\lambda + 3)(\lambda - 1) = (\lambda + 3)^2(\lambda - 1).$$

$$
\begin{array}{r}
\lambda^2 + 2\lambda - 3 \\
\hline
\lambda + 3\, \big|\, \lambda^3 + 5\lambda^2 + 3\lambda - 9 \\
\lambda^3 + 3\lambda^2 \\
\hline
2\lambda^2 + 3\lambda - 9 \\
2\lambda^2 + 6\lambda \\
\hline
-3\lambda - 9 \\
-3\lambda - 9 \\
\hline
0
\end{array}
$$

Figure 3.3

So the eigenvalues are $\lambda = 1$ and $\lambda = -3$. We found the eigenspace corresponding to $\lambda = -3$ in **3.3.5**. The eigenspace corresponding to $\lambda = 1$ is the set of solutions to $(A - \lambda I)\mathbf{x} = \mathbf{0}$, with $\lambda = 1$. The coefficient matrix of this homogeneous system is just the matrix in (2) with $\lambda = 1$, namely,

$$
\begin{bmatrix} 4 & 8 & 16 \\ 4 & 0 & 8 \\ -4 & -4 & -12 \end{bmatrix}.
$$

Gaussian elimination proceeds

$$
\begin{bmatrix} 4 & 8 & 16 \\ 4 & 0 & 8 \\ -4 & -4 & -12 \end{bmatrix} \rightarrow
\begin{bmatrix} 1 & 2 & 4 \\ 4 & 0 & 8 \\ -4 & -4 & -12 \end{bmatrix} \rightarrow
\begin{bmatrix} 1 & 2 & 4 \\ 0 & -8 & -8 \\ 0 & 4 & 4 \end{bmatrix}
$$

$$
\rightarrow \begin{bmatrix} 1 & 2 & 4 \\ 0 & 1 & 1 \\ 0 & 4 & 4 \end{bmatrix} \rightarrow
\begin{bmatrix} 1 & 2 & 4 \\ 0 & 1 & 1 \\ 0 & 0 & 0 \end{bmatrix}.
$$

With $\mathbf{x} = \begin{bmatrix} x_1 \\ x_2 \\ x_3 \end{bmatrix}$, $x_3 = t$ is a free variable. Then $x_2 = -x_3 = -t$ and $x_1 = -2x_2 - 4x_3 = 2t - 4t = -2t$. The eigenspace corresponding to $\lambda = 1$ is the set of vectors of the form

$$
\mathbf{x} = \begin{bmatrix} x_1 \\ x_2 \\ x_3 \end{bmatrix} = \begin{bmatrix} -2t \\ -t \\ t \end{bmatrix} = t \begin{bmatrix} -2 \\ -1 \\ 1 \end{bmatrix},
$$

the line in 3-space through $(0, 0, 0)$ with direction $\begin{bmatrix} -2 \\ -1 \\ 1 \end{bmatrix}$. To check that $\mathbf{x} = \begin{bmatrix} -2 \\ -1 \\ 1 \end{bmatrix}$ really is an eigenvector corresponding to $\lambda = 1$, we compute

$$
A\mathbf{x} = \begin{bmatrix} 5 & 8 & 16 \\ 4 & 1 & 8 \\ -4 & -4 & -11 \end{bmatrix} \begin{bmatrix} -2 \\ -1 \\ 1 \end{bmatrix} = \begin{bmatrix} -2 \\ -1 \\ 1 \end{bmatrix} = +1 \begin{bmatrix} -2 \\ -1 \\ 1 \end{bmatrix}.
$$

3.3.13 Remark Finding the eigenvalues and eigenvectors of a matrix can involve a lot of calculation. Without the assistance of a computer or sophisticated calculator, errors can creep into the work of even the most careful person. Fortunately, there are ways to check our work as it progresses. In Problem 3.3.12, for instance, we found that the characteristic polynomial was $(\lambda + 3)^2(\lambda - 1)$ and concluded that $\lambda = 1$ was an eigenvalue of A. If this really is the case, then the homogeneous system $(A - \lambda I)\mathbf{x} = \mathbf{0}$ must have nonzero solutions when $\lambda = 1$. It was reassuring to find that $A - I$ had row echelon form $\begin{bmatrix} 1 & 2 & 4 \\ 0 & 1 & 1 \\ 0 & 0 & 0 \end{bmatrix}$, guaranteeing nonzero solutions.

■ A Final Thought

We have found the eigenvalues of a lot of matrices in this section but, in closing, we must caution the reader that not all matrices have (real) eigenvalues. Consider again the matrix $A = \begin{bmatrix} 0 & 1 \\ -1 & 0 \end{bmatrix}$ of **3.3.7**. This matrix has characteristic polynomial

$$\det(A - \lambda I) = \begin{vmatrix} -\lambda & 1 \\ -1 & -\lambda \end{vmatrix} = \lambda^2 + 1.$$

Since this polynomial has no **real** roots, A has no **real** eigenvalues. In Chapter 7, we extend our discussion of eigenvalues and eigenvectors to the complex numbers.

Answers to Reading Challenges

1. Let A be $m \times n$. Since $A\mathbf{x}$ is defined, $\mathbf{x}$ must be $n \times 1$, and then $A\mathbf{x}$ is $m \times 1$. But $A\mathbf{x} = \lambda\mathbf{x}$ is the size of $\mathbf{x}$, $n \times 1$. So $m = n$.

2. If A is a 2×2 matrix, $A - 5$ has no meaning! What is a matrix less a number?

3. We seek an equation for the plane that consists of all vectors of the form $\begin{bmatrix} -t - 2s \\ t \\ s \end{bmatrix}$. Let $x = -t - 2s$, $y = t$, $z = s$. Then $x = -y - 2z$, so an equation is $x + y + 2z = 0$.

True/False Questions

Decide, with as little calculation as possible, whether each of the following statements is true or false and explain your answer whenever you say "false." (Answers can be found in the back of the book.)

1. The eigenvectors of a matrix A that correspond to a scalar λ are the solutions to the homogeneous system $(A - \lambda I)\mathbf{x} = \mathbf{0}$.

2. If $\mathbf{x}$ is an eigenvector of a matrix A, then so is every vector on the line through $\mathbf{x}$.

3. The characteristic polynomial of an 4×4 matrix is a polynomial of degree 4.

4. Every 2×2 matrix has at least one eigenvector.

5. If all the eigenvalues of a matrix are 0, then A is the zero matrix.

6. If 1 is the only eigenvalue of a 2×2 matrix A, then $A = I$.

Exercises

*Solutions to exercises marked [BB] can be found in the **B**ack of the **B**ook.*

1. [BB] Let $A = \begin{bmatrix} 2 & 1 & 1 \\ -2 & 5 & 0 \\ 2 & -1 & 4 \end{bmatrix}$ and $\mathbf{v} = \begin{bmatrix} 2 \\ 4 \\ 0 \end{bmatrix}$. Is $\mathbf{v}$ an eigenvector for A? Explain and, if your answer is yes, give the corresponding eigenvalue.

2. In each of the following cases, determine whether the given vectors are eigenvectors of the given matrix. Justify your answers and state the eigenvalues that correspond to any eigenvectors.

   (a) [BB] $A = \begin{bmatrix} 1 & 2 \\ 2 & 4 \end{bmatrix}$; $\mathbf{v}_1 = \begin{bmatrix} 1 \\ 2 \end{bmatrix}$, $\mathbf{v}_2 = \begin{bmatrix} -2 \\ 1 \end{bmatrix}$,

   $\mathbf{v}_3 = \begin{bmatrix} 1 \\ 1 \end{bmatrix}$, $\mathbf{v}_4 = \begin{bmatrix} 1 \\ -2 \end{bmatrix}$, $\mathbf{v}_5 = \begin{bmatrix} 0 \\ 0 \end{bmatrix}$.

   (b) $A = \begin{bmatrix} 5 & -7 & 7 \\ 4 & -3 & 4 \\ 4 & -1 & 2 \end{bmatrix}$; $\mathbf{v}_1 = \begin{bmatrix} 0 \\ 0 \\ 0 \end{bmatrix}$, $\mathbf{v}_2 = \begin{bmatrix} 1 \\ 0 \\ 0 \end{bmatrix}$,

   $\mathbf{v}_3 = \begin{bmatrix} 0 \\ 2 \\ 2 \end{bmatrix}$, $\mathbf{v}_4 = \begin{bmatrix} 1 \\ 0 \\ -1 \end{bmatrix}$, $\mathbf{v}_5 = \begin{bmatrix} 3 \\ 3 \\ 3 \end{bmatrix}$, $\mathbf{v}_6 = \begin{bmatrix} 1 \\ 0 \\ 1 \end{bmatrix}$.

   (c) $A = \begin{bmatrix} 5 & 8 & 16 \\ 4 & 1 & 8 \\ -4 & -4 & -11 \end{bmatrix}$; $\mathbf{v}_1 = \begin{bmatrix} 3 \\ 4 \\ -1 \end{bmatrix}$,

   $\mathbf{v}_2 = \begin{bmatrix} -1 \\ 1 \\ 0 \end{bmatrix}$, $\mathbf{v}_3 = \begin{bmatrix} -1 \\ 1 \\ -1 \end{bmatrix}$, $\mathbf{v}_4 = \begin{bmatrix} -2 \\ 0 \\ 1 \end{bmatrix}$,

   $\mathbf{v}_5 = \begin{bmatrix} 2 \\ 1 \\ -1 \end{bmatrix}$.

3. In each of the following situations, determine whether each of the given scalars is an eigenvalue of the given matrix. Justify your answers.

   (a) [BB] $A = \begin{bmatrix} 1 & 2 \\ 2 & 4 \end{bmatrix}$; $-1, 0, 1, 3, 5$

   (b) $A = \begin{bmatrix} 5 & -2 \\ 4 & -1 \end{bmatrix}$; $-2, -1, 1, 3, 4$

   (c) $A = \begin{bmatrix} -4 & -2 \\ 3 & 1 \end{bmatrix}$; $0, \sqrt{3}, \frac{-3+\sqrt{5}}{2}, -1, 4$

   (d) [BB] $A = \begin{bmatrix} 5 & -7 & 7 \\ 4 & -3 & 4 \\ 4 & -1 & 2 \end{bmatrix}$; $1, 2, 4, 5, 6$

   (e) $A = \begin{bmatrix} 3 & 4 & 2 \\ 2 & -2 & 6 \\ 0 & -7 & 5 \end{bmatrix}$; $2, 1, 3, 6, -2$

   (f) $A = \begin{bmatrix} 2 & 1 & -1 \\ -2 & -1 & 2 \\ 2 & 1 & 0 \end{bmatrix}$; $2, -1, 3, 0, 4$.

4. [BB] Let $A = \begin{bmatrix} 1 & 2 \\ 3 & 2 \end{bmatrix}$.

   (a) What is the characteristic polynomial of A?

   (b) What are the eigenvalues of A?

   (c) Show that $\begin{bmatrix} 2 \\ 3 \end{bmatrix}$ is an eigenvector of A and state the corresponding eigenvalue.

5. Let $A = \begin{bmatrix} 5 & 1 \\ 1 & -3 \end{bmatrix}$.

   (a) What is the characteristic polynomial of A?

   (b) What are the eigenvalues of A?

   (c) Show that $\begin{bmatrix} -3 \\ 3 \end{bmatrix}$ is an eigenvector of A and state the corresponding eigenvalue.

6. Find the characteristic polynomial, the (real) eigenvalues, and the corresponding eigenspaces of each of the following matrices.

   (a) [BB] $\begin{bmatrix} 5 & 8 \\ 4 & 1 \end{bmatrix}$

   (b) $\begin{bmatrix} -1 & 3 \\ 2 & 0 \end{bmatrix}$

   (c) $\begin{bmatrix} 1 & 2 \\ 3 & 2 \end{bmatrix}$

   (d) [BB] $\begin{bmatrix} 1 & -2 & 3 \\ 2 & 6 & -6 \\ 1 & 2 & -1 \end{bmatrix}$

   (e) $\begin{bmatrix} 0 & 1 & 1 \\ -2 & 3 & 0 \\ 2 & -1 & 2 \end{bmatrix}$

   (f) $\begin{bmatrix} 4 & 2 & 2 \\ 4 & 2 & -4 \\ -2 & 0 & 4 \end{bmatrix}$

(g) $\begin{bmatrix} 3 & 4 & -4 & -4 \\ 4 & 3 & -4 & -4 \\ 0 & 4 & -1 & -4 \\ 4 & 0 & -4 & -1 \end{bmatrix}$

(h) $\begin{bmatrix} 1 & 6 & 0 & 0 \\ 5 & 2 & 0 & 0 \\ 0 & 0 & 3 & -2 \\ 0 & 0 & 1 & 0 \end{bmatrix}$.

7. [BB] Let $A = \begin{bmatrix} 0 & 1 \\ 1 & 0 \end{bmatrix}$, let P be the point (x, y) in the Cartesian plane, and let $\mathbf{v}$ be the vector $\overrightarrow{OP}$. Let Q be the point such that $A\mathbf{v} = \overrightarrow{OQ}$.

   (a) Find the coordinates of Q.

   (b) Show that the line ℓ with equation $y = x$ is the right bisector of the line PQ joining P to Q; that is, PQ is perpendicular to ℓ and the lines intersect at the midpoint of PQ.

   (c) Describe the action of multiplication of A in geometrical terms.

   (d) Find the eigenvalues and corresponding eigenspaces of A without calculation.

8. Multiplication by each of the following matrices is reflection in a line. Find the equation of this line. Use the fact that the matrix is a reflection to find all eigenvectors and eigenvalues.

   (a) [BB] $A = \begin{bmatrix} -1 & 0 \\ 0 & 1 \end{bmatrix}$ (b) $A = \begin{bmatrix} -\frac{3}{5} & \frac{4}{5} \\ \frac{4}{5} & \frac{3}{5} \end{bmatrix}$

   (c) $A = \begin{bmatrix} \frac{7}{25} & -\frac{24}{25} \\ -\frac{24}{25} & -\frac{7}{25} \end{bmatrix}$ (d) $A = \begin{bmatrix} \frac{4}{5} & \frac{3}{5} \\ \frac{3}{5} & -\frac{4}{5} \end{bmatrix}$

9. Let A be an $n \times n$ matrix and λ be a scalar.

   (a) If $A\mathbf{u} = \lambda\mathbf{u}$, then $A(a\mathbf{u}) = \lambda(a\mathbf{u})$ for any scalar a. Thus, if $a \neq 0$ is a scalar and $\mathbf{u}$ is an eigenvector of A, then $a\mathbf{u}$ is also an eigenvector of A.

   (b) If $A\mathbf{u} = \lambda\mathbf{u}$ and $A\mathbf{v} = \lambda\mathbf{v}$, then $A(\mathbf{u} + \mathbf{v}) = \lambda(\mathbf{u} + \mathbf{v})$.

   [Together, these two properties show that the eigenspace of A corresponding to λ is a *subspace* of $\mathbf{R}^n$. See Section 4.2.]

10. [BB] Let λ be an eigenvalue of an $n \times n$ matrix A and let $\mathbf{u}, \mathbf{v}$ be eigenvectors corresponding to λ. If a and b are scalars, show that $A(a\mathbf{u} + b\mathbf{v}) = \lambda(a\mathbf{u} + b\mathbf{v})$. Thus, if $a\mathbf{u} + b\mathbf{v} \neq \mathbf{0}$, it is also an eigenvector of A corresponding to λ.

11. If $\begin{bmatrix} 2 \\ 2 \end{bmatrix}$ is an eigenvector of a matrix A and the corresponding eigenvalue is $\lambda = 6$, find the sum of the columns of A.

12. [BB] Suppose the components of each row of an $n \times n$ matrix A add to 17. Prove that 17 is an eigenvalue of A.

13. Let A be a 3×3 matrix, the entries in each row adding to 0.

    (a) Show that $\begin{bmatrix} 1 \\ 1 \\ 1 \end{bmatrix}$ is an eigenvector of A and find the corresponding eigenvalue.

    (b) Explain why $\det A = 0$.

    (c) **Without calculation**, explain why
    $$\begin{vmatrix} 2 & -1 & -1 \\ -3 & 1 & 2 \\ 0 & -2 & 2 \end{vmatrix} = 0.$$

14. [BB] Suppose A, B, and P are $n \times n$ matrices with P invertible and $PA = BP$. Show that every eigenvalue of A is an eigenvalue of B.

15. Suppose $\mathbf{x}$ is an eigenvector of an invertible matrix A corresponding to λ.

    (a) Show that $\lambda \neq 0$.

    (b) Show that $\mathbf{x}$ is also an eigenvector of A^{-1}. What is the corresponding eigenvalue?

■ Critical Reading

16. [BB] Suppose A is a square matrix. Explain why the characteristic polynomial of A^T is the same as the characteristic polynomial of A.

17. Suppose $\begin{bmatrix} 1 \\ 2 \\ 3 \end{bmatrix}$ is an eigenvector of a matrix A with corresponding eigenvalue -3 and $\begin{bmatrix} -1 \\ 0 \\ 5 \end{bmatrix}$ is an eigenvector with corresponding eigenvalue 5. What is
    $$A \begin{bmatrix} -1 & 1 \\ 0 & 2 \\ 5 & 3 \end{bmatrix}?$$

18. Let $\mathbf{v}$ be an eigenvector of a matrix A with corresponding eigenvalue λ.

    (a) [BB] Is $\mathbf{v}$ an eigenvector of $5A$? If so, what is the corresponding eigenvalue?

(b) Is $5\mathbf{v}$ an eigenvector of $5A$? If so, what is the corresponding eigenvalue?

(c) Is $5\mathbf{v}$ an eigenvector of A? If so, what is the corresponding eigenvalue?

Explain your answers.

19. If $\mathbf{v}$ is an eigenvector of the matrix A and an eigenvector of the matrix B, show that $\mathbf{v}$ is also an eigenvector of AB. What is the corresponding eigenvalue?

20. Given a matrix A, let λ be an eigenvalue of the matrix $A^T A$. Show that $\lambda \geq 0$.

21. If 0 is an eigenvalue of a matrix A, then A cannot be invertible. Why not?

22. Suppose A is a 3×3 matrix and $\mathbf{x} = \begin{bmatrix} -2 \\ -2 \\ -2 \end{bmatrix}$ is an eigenvector corresponding to $\lambda = 5$. Find the sum of the columns of A.

23. Let A be a square matrix with characteristic polynomial $f(\lambda)$ which factors $f(\lambda) = (\lambda_1 - \lambda)(\lambda_2 - \lambda) \cdots (\lambda_n - \lambda)$ into the product of linear polynomials over $\mathbf{R}$. Explain why $\det A$ is the product of the eigenvalues of A.

3.4 Similarity and Diagonalization

In this section, we focus attention on diagonal matrices. Remember that a diagonal matrix is a matrix like $\begin{bmatrix} 2 & 0 \\ 0 & 7 \end{bmatrix}$ whose only nonzero entries lie on the (main) diagonal. Calculations with diagonal matrices are much easier than with arbitrary matrices. For example, compare

$$\begin{bmatrix} 2 & 0 \\ 0 & -5 \end{bmatrix} \begin{bmatrix} -17 & 0 \\ 0 & 8 \end{bmatrix} = \begin{bmatrix} -34 & 0 \\ 0 & -40 \end{bmatrix}$$

with

$$\begin{bmatrix} 9 & -7 \\ 6 & 25 \end{bmatrix} \begin{bmatrix} 12 & 11 \\ -18 & 8 \end{bmatrix} = \begin{bmatrix} ? & ? \\ ? & ? \end{bmatrix}.$$

Finding a power of an arbitrary matrix can seem impossible compared with the ease of finding a power of a diagonal matrix. Compare

$$\begin{bmatrix} 2 & 0 \\ 0 & -5 \end{bmatrix}^{100} = \begin{bmatrix} 2^{100} & 0 \\ 0 & 5^{100} \end{bmatrix}$$

with

$$\begin{bmatrix} 3 & -4 \\ 5 & 2 \end{bmatrix}^{100} = ??.$$

There are certainly obvious advantages to working with diagonal matrices. In this section, we focus attention on matrices that, even if not diagonal, can at least be made diagonal, in a sense we shall make clear. Such matrices are called *diagonalizable*.

3.4.1 Suppose A is an $n \times n$ matrix and we discover that A has n eigenvalues $\lambda_1, \dots, \lambda_n$ with corresponding eigenvectors $\mathbf{x}_1, \dots, \mathbf{x}_n$. Put these eigenvectors down the columns of a matrix P. The rules for matrix multiplication give

$$AP = A \begin{bmatrix} \mathbf{x}_1 & \mathbf{x}_2 & \cdots & \mathbf{x}_n \\ \downarrow & \downarrow & & \downarrow \end{bmatrix} = \begin{bmatrix} A\mathbf{x}_1 & A\mathbf{x}_2 & \cdots & A\mathbf{x}_n \\ \downarrow & \downarrow & & \downarrow \end{bmatrix}.$$

Since $A\mathbf{x}_1 = \lambda_1\mathbf{x}_1$, $A\mathbf{x}_2 = \lambda_2\mathbf{x}_2$, and so on, we obtain

$$AP = \begin{bmatrix} \lambda_1\mathbf{x}_1 & \lambda_2\mathbf{x}_2 & \cdots & \lambda_n\mathbf{x}_n \\ \downarrow & \downarrow & & \downarrow \end{bmatrix}.$$

As we saw in Problem 2.1.37, this matrix is the same as

$$\begin{bmatrix} \mathbf{x}_1 & \mathbf{x}_2 & \cdots & \mathbf{x}_n \\ \downarrow & \downarrow & & \downarrow \end{bmatrix} \begin{bmatrix} \lambda_1 & 0 & \cdots & 0 \\ 0 & \lambda_2 & \cdots & 0 \\ \vdots & & \ddots & \\ 0 & \cdots & & \lambda_n \end{bmatrix},$$

which is the product PD of P and the diagonal matrix

$$D = \begin{bmatrix} \lambda_1 & 0 & \cdots & 0 \\ 0 & \lambda_2 & \cdots & 0 \\ \vdots & & \ddots & \\ 0 & \cdots & & \lambda_n \end{bmatrix}.$$

We obtain the equation $AP = PD$. If P happens to be invertible, we can multiply on the left by P^{-1} and obtain $P^{-1}AP = D$. We shall meet this kind of equation many times in this section and in later chapters of this book. When $P^{-1}AP = D$, we say that A is *similar* to the diagonal matrix D.

3.4.2 DEFINITION   Matrices A and B are *similar* if $P^{-1}AP = B$ for some invertible matrix P.

As the wording implies, this definition is symmetric: You don't have to remember whether to write $P^{-1}AP = B$ or $Q^{-1}BQ = A$. Either equation implies the other. Either statement "A is similar to B" or "B is similar to A" means "A and B are similar."

Reading Challenge 1   *Show that $B = P^{-1}AP$ implies $A = Q^{-1}BQ$ for some Q.*

3.4.3 EXAMPLE

Matrices $A = \begin{bmatrix} 1 & 2 \\ 3 & 4 \end{bmatrix}$ and $B = \begin{bmatrix} -55 & -97 \\ 34 & 60 \end{bmatrix}$ are similar. To see why, let $P = \begin{bmatrix} 3 & 5 \\ 1 & 2 \end{bmatrix}$, note that $P^{-1} = \begin{bmatrix} 2 & -5 \\ -1 & 3 \end{bmatrix}$, and simply compute:

$$P^{-1}AP = \begin{bmatrix} 2 & -5 \\ -1 & 3 \end{bmatrix} \begin{bmatrix} 1 & 2 \\ 3 & 4 \end{bmatrix} \begin{bmatrix} 3 & 5 \\ 1 & 2 \end{bmatrix} = \begin{bmatrix} -55 & -97 \\ 34 & 60 \end{bmatrix} = B. \qquad \blacksquare$$

As the name suggests, similar matrices share many properties. They have the same determinant since

$$\det(P^{-1}AP) = \det(P^{-1}) \det A \det P = \frac{1}{\det P} \det A \det P = \det A.$$

A calculation not unlike this shows that they have the same characteristic polynomial and hence the same eigenvalues. It is not a coincidence that the sum of the diagonal entries of the matrices A and B in Example **3.4.3** is 5 in each case. Similar matrices even have the same trace. The *trace* of a matrix is the sum of its diagonal entries.

3.4.4 DEFINITION

A matrix A is *diagonalizable* if it is similar to a diagonal matrix—that is, if there exists an invertible matrix P and a diagonal matrix D such that $P^{-1}AP = D$.

If A is diagonalizable, say $P^{-1}AP = D$, then $AP = PD$ and it follows, as above, that the columns of P are eigenvectors of A corresponding to eigenvalues that are precisely the diagonal entries of D, in order.

> **3.4.5** $P^{-1}AP = D$ if and only if the columns of P are eigenvectors of A and the diagonal entries of D are eigenvalues that correspond in order to the columns of P.

3.4.6 EXAMPLE

Let $A = \begin{bmatrix} 1 & 4 \\ 2 & 3 \end{bmatrix}$ be the matrix considered first in Example 3.3.3 and again in 3.3.10. There are two eigenvalues, $\lambda = -1$ and $\lambda = 5$, with corresponding eigenvectors $\begin{bmatrix} -2 \\ 1 \end{bmatrix}$ and $\begin{bmatrix} 1 \\ 1 \end{bmatrix}$, respectively. If we put these into the columns of a matrix $P = \begin{bmatrix} -2 & 1 \\ 1 & 1 \end{bmatrix}$, we have

$$AP = \begin{bmatrix} 1 & 4 \\ 2 & 3 \end{bmatrix} \begin{bmatrix} -2 & 1 \\ 1 & 1 \end{bmatrix} = \begin{bmatrix} 2 & 5 \\ -1 & 5 \end{bmatrix} = \begin{bmatrix} -2 & 1 \\ 1 & 1 \end{bmatrix} \begin{bmatrix} -1 & 0 \\ 0 & 5 \end{bmatrix} = PD$$

with $D = \begin{bmatrix} -1 & 0 \\ 0 & 5 \end{bmatrix}$. Since P is invertible, $P^{-1}AP = D$. Notice that the first column of P is an eigenvector of A corresponding to $\lambda = -1$ and the second column of P is an eigenvector corresponding to $\lambda = 5$. If we change the order of the columns of P and use $Q = \begin{bmatrix} 1 & -2 \\ 1 & 1 \end{bmatrix}$, we find $Q^{-1}AQ = \begin{bmatrix} 5 & 0 \\ 0 & -1 \end{bmatrix}$. The order of the columns of Q and the diagonal entries of D must match. $\qquad \blacksquare$

3.4.7 Remarks

1. As just noted, the *diagonalizing matrix* P is far from unique. You can get a variety of Ds by rearranging the columns of P. And even for a fixed D, there are many Ps that work. For example, you can always replace a column of P by a nonzero multiple of this column. Throughout Section 3.3, we noted if $\mathbf{x}$ is an eigenvector corresponding to λ, then so is $c\mathbf{x}$ for any $c \neq 0$. See also Exercise 9 of Section 3.3.

2. Not all matrices are diagonalizable. Suppose we try to find the eigenvalues of $A = \begin{bmatrix} 0 & 1 \\ 0 & 0 \end{bmatrix}$. We have

$$\det(A - \lambda I) = \begin{vmatrix} -\lambda & 1 \\ 0 & -\lambda \end{vmatrix} = \lambda^2,$$

so $\lambda = 0$ is the only eigenvalue. If A were diagonalizable, the diagonal matrix D, having eigenvalues of A on its diagonal, would be the zero matrix. Since the equation $P^{-1}AP = 0$ implies $A = 0$, A cannot be diagonalizable.

Reading Challenge 2   *Alternatively, show that the only eigenvectors of $A = \begin{bmatrix} 0 & 1 \\ 0 & 0 \end{bmatrix}$ are multiples of $\begin{bmatrix} 1 \\ 0 \end{bmatrix}$. Conclude again that A is not diagonalizable.*

3.4.8 Remark   The diagonalizability of a matrix is very much connected to the *field of scalars* over which we are working. Over the real numbers, for example, $A = \begin{bmatrix} 0 & -1 \\ 1 & 0 \end{bmatrix}$ is not diagonalizable because its characteristic polynomial,

$$\det(A - \lambda I) = \begin{vmatrix} -\lambda & -1 \\ 1 & -\lambda \end{vmatrix} = \lambda^2 + 1,$$

has no real roots. Over the complex numbers, however, this polynomial has the roots $\pm i$, so A has two eigenvalues and, in fact, A is diagonalizable (via a matrix P with complex, not all real, entries). To this point in our studies, all scalars have been assumed real, so we do not push this issue further at the moment. The subject of complex matrices will be discussed in earnest in Chapter 7.

We have seen that there is an equation of the type $P^{-1}AP = D$ if and only if the columns of P are eigenvectors of A. We also know that a matrix is invertible if and only if its columns are linearly independent—this was Theorem 2.7.18. These are the two key ingredients in the proof of a fundamental theorem characterizing diagonalizable matrices.

Theorem 3.4.9   *A square $n \times n$ matrix is diagonalizable if and only if it has n linearly independent eigenvectors.*

Proof First assume that A is a diagonalizable matrix. Then there is an invertible matrix P such that $P^{-1}AP = D$ is a diagonal matrix. Since P is invertible, its columns are linearly independent (by Theorem 2.7.18). By 3.4.5, the columns of P are eigenvectors of A, so A has n linearly independent eigenvectors.

Conversely, assume that A has n linearly independent eigenvectors. In this case, Theorem 2.7.18 implies that the matrix P whose columns are these eigenvectors is invertible. As shown in paragraph 3.4.1, we have an equation $AP = PD$ where D is the diagonal matrix whose entries are the eigenvalues of A. Since P is invertible, we obtain $P^{-1}AP = D$, which says that A is diagonalizable. ■

To *diagonalize* a matrix means to find an invertible matrix P such that $P^{-1}AP = D$ is diagonal. Theorem 3.4.9 (and its proof) imply a procedure for diagonalization.

3.4.10 Diagonalization Algorithm: Given an $n \times n$ matrix A,

1. find the eigenvalues of A;
2. find (if possible) n linearly independent eigenvectors $\mathbf{x}_1, \ldots, \mathbf{x}_n$;
3. let P be the matrix whose columns are $\mathbf{x}_1, \ldots, \mathbf{x}_n$;
4. then P is invertible and $P^{-1}AP = D$ is diagonal, the diagonal entries being the eigenvalues of A in the order corresponding to the columns of P.

3.4.11 EXAMPLE

Let $A = \begin{bmatrix} 1 & 4 \\ 2 & 3 \end{bmatrix}$. In Example 3.3.10, we saw that the characteristic polynomial of A is $\lambda^2 - 4\lambda - 5 = (\lambda - 5)(\lambda + 1)$. There are two eigenvalues, $\lambda = 5$ and $\lambda = -1$. An eigenvector for $\lambda = 5$ is $\mathbf{x}_1 = \begin{bmatrix} 1 \\ 1 \end{bmatrix}$, while $\mathbf{x}_2 = \begin{bmatrix} -2 \\ 1 \end{bmatrix}$ is an eigenvector for $\lambda = -1$.

With $P = \begin{bmatrix} \mathbf{x}_1 & \mathbf{x}_2 \\ \downarrow & \downarrow \end{bmatrix}$, we have

$$AP = \begin{bmatrix} A\mathbf{x}_1 & A\mathbf{x}_2 \\ \downarrow & \downarrow \end{bmatrix}$$

$$= \begin{bmatrix} 5\mathbf{x}_1 & -\mathbf{x}_2 \\ \downarrow & \downarrow \end{bmatrix} = \begin{bmatrix} 5 & 2 \\ 5 & -1 \end{bmatrix} = \begin{bmatrix} \mathbf{x}_1 & \mathbf{x}_2 \\ \downarrow & \downarrow \end{bmatrix} \begin{bmatrix} 5 & 0 \\ 0 & -1 \end{bmatrix} = PD,$$

with $D = \begin{bmatrix} 5 & 0 \\ 0 & -1 \end{bmatrix}$. The columns of P are linearly independent since our eigenvectors are not parallel. So the matrix $P = \begin{bmatrix} 1 & -2 \\ 1 & 1 \end{bmatrix}$ is invertible, and $P^{-1}AP = D = \begin{bmatrix} 5 & 0 \\ 0 & -1 \end{bmatrix}$, the diagonal matrix whose entries are the eigenvalues of A, in the order determined by the columns of P. ■

3.4.12 EXAMPLE Let $A = \begin{bmatrix} 5 & 8 & 16 \\ 4 & 1 & 8 \\ -4 & -4 & -11 \end{bmatrix}$. In Problem 3.3.12, we recorded the fact that A has eigen-

values $\lambda = -3$ and $\lambda = 1$. For $\lambda = 1$, we found that $\mathbf{x}_1 = \begin{bmatrix} -2 \\ -1 \\ 1 \end{bmatrix}$. In Example 3.3.5,

we discovered that $\mathbf{x}_2 = \begin{bmatrix} -1 \\ 1 \\ 0 \end{bmatrix}$ and $\mathbf{x}_3 = \begin{bmatrix} -2 \\ 0 \\ 1 \end{bmatrix}$ are eigenvectors for $\lambda = -3$. We
leave it to the reader to verify that $\mathbf{x}_1, \mathbf{x}_2, \mathbf{x}_3$ are linearly independent.

Thus the matrix $P = \begin{bmatrix} -2 & -1 & -2 \\ -1 & 1 & 0 \\ 1 & 0 & 1 \end{bmatrix}$ whose columns are $\mathbf{x}_1, \mathbf{x}_2, \mathbf{x}_3$ is invertible,

and $P^{-1}AP = D = \begin{bmatrix} 1 & 0 & 0 \\ 0 & -3 & 0 \\ 0 & 0 & -3 \end{bmatrix}$, the diagonal matrix whose diagonal entries are
the eigenvalues of A, in the order determined by the columns of P. Since the first
column of P is an eigenvector corresponding to $\lambda = 1$, the first diagonal entry of D
is 1; since the last two columns of P are eigenvectors corresponding to $\lambda = -3$, the
last two diagonal entries of D are -3. ∎

Reading Challenge 3 *With A as in **3.4.12**, but $P = \begin{bmatrix} -2 & -2 & -1 \\ 0 & -1 & 1 \\ 1 & 1 & 0 \end{bmatrix}$, what is $P^{-1}AP$?*

3.4.13 EXAMPLE Let $A = \begin{bmatrix} 2 & 1 & -12 \\ 0 & 1 & 11 \\ 1 & 1 & 4 \end{bmatrix}$. Expanding the determinant by cofactors down the first column, the characteristic polynomial of A is

$$\begin{vmatrix} 2-\lambda & 1 & -12 \\ 0 & 1-\lambda & 11 \\ 1 & 1 & 4-\lambda \end{vmatrix} = (2-\lambda)[(1-\lambda)(4-\lambda) - 11] + [11 + 12(1-\lambda)]$$

$$= -\lambda^3 + 7\lambda^2 - 15\lambda + 9 = -(\lambda - 1)(\lambda - 3)^2.$$

The eigenvalues of A are $\lambda = 1$ and $\lambda = 3$. To find the eigenvectors for $\lambda = 1$, we
must solve $(A - \lambda I)\mathbf{x} = \mathbf{0}$. Gaussian elimination proceeds

$$\begin{bmatrix} 2-\lambda & 1 & -12 \\ 0 & 1-\lambda & 11 \\ 1 & 1 & 4-\lambda \end{bmatrix} = \begin{bmatrix} 1 & 1 & -12 \\ 0 & 0 & 11 \\ 1 & 1 & 3 \end{bmatrix}$$

$$\rightarrow \begin{bmatrix} 1 & 1 & -12 \\ 0 & 0 & 1 \\ 0 & 0 & 15 \end{bmatrix} \rightarrow \begin{bmatrix} 1 & 1 & -12 \\ 0 & 0 & 1 \\ 0 & 0 & 0 \end{bmatrix},$$

so $x_2 = t$ is free, $x_3 = 0$, and $x_1 + x_2 - 12x_3 = 0$, so $x_1 = -x_2 + 12x_3 = -t$. Eigenvectors are of the form

$$\begin{bmatrix} x_1 \\ x_2 \\ x_3 \end{bmatrix} = \begin{bmatrix} -t \\ t \\ 0 \end{bmatrix} = t \begin{bmatrix} -1 \\ 1 \\ 0 \end{bmatrix}.$$

To find the eigenvectors for $\lambda = 3$,

$$\begin{bmatrix} 2-\lambda & 1 & -12 \\ 0 & 1-\lambda & 11 \\ 1 & 1 & 4-\lambda \end{bmatrix} = \begin{bmatrix} -1 & 1 & -12 \\ 0 & -2 & 11 \\ 1 & 1 & 1 \end{bmatrix}$$

$$\to \begin{bmatrix} 1 & -1 & 12 \\ 1 & 1 & 1 \\ 0 & -2 & 11 \end{bmatrix} \to \begin{bmatrix} 1 & -1 & 12 \\ 0 & 2 & -11 \\ 0 & -2 & 11 \end{bmatrix} \to \begin{bmatrix} 1 & -1 & 12 \\ 0 & 1 & -\frac{11}{2} \\ 0 & 0 & 0 \end{bmatrix},$$

so $x_3 = t$ is free, $x_2 = \frac{11}{2}x_3 = \frac{11}{2}t$, and $x_1 - x_2 + 12x_3 = 0$, so $x_1 = x_2 - 12x_3 = \frac{11}{2}t - 12t = -\frac{13t}{2}$. Eigenvectors are of the form

$$\begin{bmatrix} x_1 \\ x_2 \\ x_3 \end{bmatrix} = \begin{bmatrix} -\frac{13t}{2} \\ \frac{11t}{2} \\ t \end{bmatrix} = t \begin{bmatrix} -\frac{13}{2} \\ \frac{11}{2} \\ 1 \end{bmatrix}.$$

This time we can find only two linearly independent eigenvectors, so A is not diagonalizable. ∎

There is one instance when we are assured that the condition of Theorem 3.4.9 holds and hence that a matrix is diagonalizable.

Theorem 3.4.14 *Eigenvectors corresponding to different eigenvalues are linearly independent. Hence, an $n \times n$ matrix with n different eigenvalues is diagonalizable.*

Proof Let A be a matrix with eigenvectors $x_1, x_2, \ldots, x_k$ corresponding to different eigenvalues $\lambda_1, \lambda_2, \ldots, \lambda_k$. We claim that $x_1, \ldots, x_k$ are linearly independent. Suppose then that we have an equation of the form

$$c_1 x_1 + c_2 x_2 + \cdots + c_k x_k = 0. \tag{1}$$

To make life simpler (a general argument is given in Exercise 16), we suppose that $k = 2$, so that we have

$$c_1 x_1 + c_2 x_2 = 0. \tag{2}$$

Multiplying both sides by A, and remembering that $A x_1 = \lambda_1 x_1$ and $A x_2 = \lambda_2 x_2$, we obtain

$$c_1 \lambda_1 x_1 + c_2 \lambda_2 x_2 = 0.$$

Multiplying (2) by λ_1 gives

$$c_1\lambda_1\mathbf{x}_1 + c_2\lambda_1\mathbf{x}_2 = \mathbf{0},$$

then subtracting these equations gives

$$(c_2\lambda_2 - c_2\lambda_1)\mathbf{x}_2 = \mathbf{0},$$

so that $c_2(\lambda_2 - \lambda_1)\mathbf{x}_2 = \mathbf{0}$. Here's where the fact that the eigenvalues are different becomes important. Since $\lambda_2 - \lambda_1 \neq 0$, we may divide, obtaining $c_2\mathbf{x}_2 = \mathbf{0}$. Eigenvectors are not $\mathbf{0}$, so $c_2 = 0$. Returning to (2), we get $c_1\mathbf{x}_1 = \mathbf{0}$ and so $c_1 = 0$ too. In general, this sort of argument proves that all the coefficients in (1) are 0; that is, the eigenvectors $\mathbf{x}_1, \ldots, \mathbf{x}_k$ are linearly independent. ∎

As an illustration of this theorem, in **3.4.11**, the minute we discovered that A had different eigenvalues 5 and -1, we could have concluded that A was diagonalizable (without computing eigenvectors). Here is a similar example.

3.4.15 PROBLEM   Show that $A = \begin{bmatrix} 1 & 2 \\ 1 & 1 \end{bmatrix}$ is diagonalizable.

Solution.   The characteristic polynomial of A is

$$\begin{vmatrix} 1 - \lambda & 2 \\ 1 & 1 - \lambda \end{vmatrix} = (1 - \lambda)^2 - 2.$$

The roots of this polynomial satisfy $(1 - \lambda)^2 = 2$, so $|1 - \lambda| = \sqrt{2}$ and $\lambda = 1 \pm \sqrt{2}$. These are distinct, so A is diagonalizable by Theorem 3.4.14; in fact, A is similar to $\begin{bmatrix} 1 + \sqrt{2} & 0 \\ 0 & 1 - \sqrt{2} \end{bmatrix}$.

3.4.16 Remark   Theorem 3.4.14 doesn't quite explain why the matrix A in **3.4.12** is diagonalizable. This 3×3 matrix has just two eigenvalues, $\lambda = -3$ and $\lambda = 1$. For $\lambda = 1$, we found an eigenvector $\mathbf{x}_1$. For $\lambda = -3$, we found two eigenvectors $\mathbf{x}_2$ and $\mathbf{x}_3$. Theorem 3.4.14 guarantees that $\mathbf{x}_1$ and $\mathbf{x}_2$ are linearly independent since they correspond to different eigenvalues, and, for the same reason, it says that $\mathbf{x}_1$ and $\mathbf{x}_3$ are linearly independent. It doesn't say, however, that the **three** vectors $\mathbf{x}_1, \mathbf{x}_2, \mathbf{x}_3$ are linearly independent. A slight modification of the proof of Theorem 3.4.14 handles this case as well, however. See Exercise 26.

Answers to Reading Challenges

1. Solving $B = P^{-1}AP$ for A gives $A = PBP^{-1}$, which is $Q^{-1}BQ$, with $Q = P^{-1}$.

2. As shown in the text, the only eigenvalue of A is 0. To find the corresponding eigenvalues, we reduce A to row echelon form. In this case, row echelon form is the original matrix $A = \begin{bmatrix} 0 & 1 \\ 0 & 0 \end{bmatrix}$. We have $x_1 = t$ free and $x_2 = 0$, giving

$\mathbf{x} = \begin{bmatrix} x_1 \\ x_2 \end{bmatrix} = \begin{bmatrix} t \\ 0 \end{bmatrix} = t \begin{bmatrix} 1 \\ 0 \end{bmatrix}$ as desired. If A is similar to a diagonal matrix, the (invertible) matrix P has columns that are eigenvectors of A. Since all eigenvectors of A are multiples of $\begin{bmatrix} 1 \\ 0 \end{bmatrix}$, A does not have two linearly independent eigenvectors, so no invertible P exists.

3. The columns of P are eigenvectors of A with eigenvalues $-3, 1, -3$, respectively, so $P^{-1}AP = \begin{bmatrix} -3 & 0 & 0 \\ 0 & 1 & 0 \\ 0 & 0 & -3 \end{bmatrix}$.

True/False Questions

Decide, with as little calculation as possible, whether each of the following statements is true or false and explain your answer whenever you say "false." (Answers can be found in the back of the book.)

1. The characteristic polynomial of $\begin{bmatrix} 1 & 2 & 3 \\ 0 & 4 & 5 \\ 0 & 0 & 6 \end{bmatrix}$ is $(1 - \lambda)(4 - \lambda)(6 - \lambda)$.

2. If $A = \begin{bmatrix} -1 & 0 \\ 0 & 2 \end{bmatrix}$, then $\lambda = +1$ is an eigenvalue of A^{10}.

3. The matrices $\begin{bmatrix} 2 & 0 \\ 0 & -1 \end{bmatrix}$ and $\begin{bmatrix} 2 & 1 \\ 0 & 1 \end{bmatrix}$ are similar.

4. The matrix $A = \begin{bmatrix} -1 & 0 \\ 0 & 5 \end{bmatrix}$ is diagonalizable.

5. If a 2×2 matrix has just one eigenvalue, it is not diagonalizable.

6. If an $n \times n$ matrix has n different eigenvalues, then it is diagonalizable.

7. If an $n \times n$ matrix is diagonalizable, then it must have n different eigenvalues.

8. If A is a symmetric matrix, then the eigenvalues of A are its diagonal entries.

Exercises

*Solutions to exercises marked [BB] can be found in the **B**ack of the **B**ook.*

1. (a) [BB] Suppose a matrix A is similar to the identity matrix. What can you say about A? Explain.

   (b) [BB] Can $\begin{bmatrix} 1 & 2 \\ 0 & 1 \end{bmatrix}$ be similar to the 2×2 identity matrix? Explain.

   (c) [BB] Is $A = \begin{bmatrix} 3 & 2 \\ -2 & -1 \end{bmatrix}$ diagonalizable? Explain.

2. (a) Show that similar matrices have the same characteristic polynomial.

   (b) Are the matrices $A = \begin{bmatrix} 2 & 1 \\ 1 & -1 \end{bmatrix}$ and $B = \begin{bmatrix} 3 & 0 \\ 1 & -1 \end{bmatrix}$ similar? Explain.

   (c) State the "converse" of Exercise 2(a). Show that this converse is false by considering the matrices $A = \begin{bmatrix} 1 & 0 \\ 0 & 1 \end{bmatrix}$ and $B = \begin{bmatrix} 1 & 1 \\ 0 & 1 \end{bmatrix}$.

3. [BB] Suppose A, B, and C are $n \times n$ matrices. If A is similar to B and B is similar to C, show that A is similar to C.

4. [BB] Explain why the matrix $A = \begin{bmatrix} 5 & 3 \\ 2 & 4 \end{bmatrix}$ is diagonalizable. To how many matrices is A similar? Explain. (You are not being asked to find eigenvectors.)

5. Answer Exercise 4 for the matrix
$$A = \begin{bmatrix} 5 & -2 & 6 \\ 0 & -1 & 9 \\ 0 & 0 & 3 \end{bmatrix}.$$

6. (a) [BB] Find all eigenvalues and eigenvectors of
$$A = \begin{bmatrix} 0 & 1 \\ 0 & 0 \end{bmatrix}.$$

(b) [BB] Explain why the results of part (a) show that A is not diagonalizable.

7. Given that the eigenvalues of
$$A = \begin{bmatrix} -2 & 3 & 1 & 0 \\ -1 & 2 & 1 & 0 \\ -4 & 3 & 3 & 1 \\ 0 & 0 & 0 & 2 \end{bmatrix} \text{ are } -1 \text{ and } 2, \text{ determine}$$
(with reasons) whether A is diagonalizable.

8. [BB] Let $A = \begin{bmatrix} 1 & 2 \\ 2 & -2 \end{bmatrix}.$

(a) Show that $\lambda = 2$ is an eigenvalue of A and find the corresponding eigenspace.

(b) Given that the eigenvalues of A are 2 and -3, what is the characteristic polynomial of A?

(c) Why is A diagonalizable?

(d) Given that $\begin{bmatrix} -1 \\ 2 \end{bmatrix}$ is an eigenvector of A corresponding to $\lambda = -3$, find a matrix P such that $P^{-1}AP = \begin{bmatrix} -3 & 0 \\ 0 & 2 \end{bmatrix}.$

(e) Let $Q = \begin{bmatrix} -2 & -6 \\ 4 & -3 \end{bmatrix}.$ Find $Q^{-1}AQ$ without calculation.

9. [BB] Let $A = \begin{bmatrix} 4 & 2 & 2 \\ -5 & -3 & -2 \\ 5 & 5 & 4 \end{bmatrix}.$

(a) What is the characteristic polynomial of A?

(b) Why is A diagonalizable?

(c) Find a matrix P such that
$$P^{-1}AP = \begin{bmatrix} 4 & 0 & 0 \\ 0 & -1 & 0 \\ 0 & 0 & 2 \end{bmatrix}.$$

10. Let $A = \begin{bmatrix} 3 & 1 & 1 \\ -4 & -2 & -5 \\ 2 & 2 & 5 \end{bmatrix}.$

(a) What is the characteristic polynomial of A?

(b) Why is A diagonalizable?

(c) Find a matrix P such that
$$P^{-1}AP = \begin{bmatrix} 1 & 0 & 0 \\ 0 & 2 & 0 \\ 0 & 0 & 3 \end{bmatrix}.$$

11. Let A be the matrix $A = \begin{bmatrix} 5 & -7 & 7 \\ 4 & -3 & 4 \\ 4 & -1 & 2 \end{bmatrix}.$

(a) Show that $\lambda = -2$ is an eigenvalue of A and find the corresponding eigenspace.

(b) Given that the eigenvalues of A are 1, -2, 5, what is the characteristic polynomial of A?

(c) Given that $\begin{bmatrix} 1 \\ 1 \\ 1 \end{bmatrix}$ is an eigenvector of A corresponding to $\lambda = 5$ and $\begin{bmatrix} 0 \\ 1 \\ 1 \end{bmatrix}$ is an eigenvector of A corresponding to $\lambda = 1$, find a matrix P such that $P^{-1}AP = \begin{bmatrix} 1 & 0 & 0 \\ 0 & -2 & 0 \\ 0 & 0 & 5 \end{bmatrix}.$

(d) Let $Q = \begin{bmatrix} -1 & 1 & 0 \\ 0 & 1 & 1 \\ 1 & 1 & 1 \end{bmatrix}.$ Find $Q^{-1}AQ$ without calculation.

12. [BB] What is an easy way to see that the matrix
$$A = \begin{bmatrix} 2 & -16 & -2 \\ 0 & 5 & 0 \\ 2 & -8 & -3 \end{bmatrix}$$ is diagonalizable? Find a diagonal matrix to which A is similar.

13. Answer Exercise 12 with $A = \begin{bmatrix} -1 & 2 & 2 \\ 2 & 2 & 2 \\ -3 & -6 & -6 \end{bmatrix}.$

14. Determine whether each of the matrices A given below is diagonalizable. If it is, find an invertible matrix P and a diagonal matrix D such that $P^{-1}AP = D$. If it is not, explain why.

(a) [BB] $A = \begin{bmatrix} 1 & 0 \\ 2 & 3 \end{bmatrix}$

(b) $A = \begin{bmatrix} 2 & 1 \\ -1 & 0 \end{bmatrix}$

(c) $A = \begin{bmatrix} 5 & -2 \\ 1 & 2 \end{bmatrix}$

(d) [BB] $A = \begin{bmatrix} -1 & 3 & 0 \\ 0 & 2 & 0 \\ 2 & 1 & -1 \end{bmatrix}$

(e) $A = \begin{bmatrix} 8 & 9 & -9 \\ 0 & 2 & 0 \\ 4 & 6 & -4 \end{bmatrix}$

(f) $A = \begin{bmatrix} 4 & -3 & 3 \\ 0 & 1 & 0 \\ -6 & 6 & -5 \end{bmatrix}$

(g) $A = \begin{bmatrix} 22 & 0 & 9 \\ -9 & -5 & -3 \\ 9 & 0 & -2 \end{bmatrix}$

(h) [BB] $A = \begin{bmatrix} 2 & 1 & 1 \\ 0 & 1 & 0 \\ 1 & -1 & 2 \end{bmatrix}$

(i) $A = \begin{bmatrix} 3 & 0 & 6 \\ 0 & -3 & 0 \\ 5 & 0 & 2 \end{bmatrix}$

(j) $A = \begin{bmatrix} 1 & 6 & 0 & 0 \\ 5 & 2 & 0 & 0 \\ 0 & 0 & 3 & -2 \\ 0 & 0 & 1 & 0 \end{bmatrix}$.

15. [BB] Let $A = \begin{bmatrix} 3 & -2 \\ 1 & 0 \end{bmatrix}$.

   (a) Find an invertible matrix P and a diagonal matrix D such that $P^{-1}AP = D$.

   (b) Find a matrix B such that $B^2 = A$. [Hint: First, find a matrix D_1 such that $D = D_1^2$.]

16. Let $A = \begin{bmatrix} 10 & -5 & 7 \\ -5 & 22 & -5 \\ 7 & -5 & 10 \end{bmatrix}$.

   (a) Find an invertible matrix P and a diagonal matrix D such that $P^{-1}AP = D$.

   (b) Find a matrix B such that $B^2 = A$. [Hint: Exercise 15.]

17. [BB] One of the most important and useful theorems in linear algebra says that a real symmetric matrix A is always diagonalizable—in fact, *orthogonally diagonalizable*; that is, there is a matrix P **with orthogonal columns** such that $P^{-1}AP$ is a diagonal

matrix. (See Corollary 7.3.12.) Illustrate this fact with respect to the matrix $A = \begin{bmatrix} 103 & -96 \\ -96 & 47 \end{bmatrix}$.

18. Repeat Exercise 17 using each of the following symmetric matrices.

   (a) [BB] $A = \begin{bmatrix} 3 & -5 \\ -5 & 3 \end{bmatrix}$

   (b) $A = \begin{bmatrix} 2 & 1 & 0 \\ 1 & 2 & 0 \\ 0 & 0 & 1 \end{bmatrix}$

   (c) $A = \begin{bmatrix} 2 & -2 & -1 \\ -2 & 2 & 1 \\ -1 & 1 & 5 \end{bmatrix}$

   (d) $A = \begin{bmatrix} 1 & -2 & -1 \\ -2 & 4 & 2 \\ -1 & 2 & 1 \end{bmatrix}$.

■ Critical Reading

19. [BB] Given that a matrix A is similar to the matrix $\begin{bmatrix} -1 & 2 & 3 & 4 \\ 0 & 1 & 1 & 2 \\ 0 & 0 & -3 & 5 \\ 0 & 0 & 0 & -3 \end{bmatrix}$, what is the characteristic polynomial of A? Explain.

20. Given that the eigenvalues of a 3×3 matrix A are -1 and 2, what are the possibilities for the characteristic polynomial of A?

21. Find the eigenvalues and the corresponding eigenspaces of the matrix $A = \begin{bmatrix} 2 & 0 & 0 & 0 \\ 0 & 2 & 0 & 0 \\ 0 & 0 & 2 & 0 \\ 0 & 0 & 0 & 2 \end{bmatrix}$.

22. Let $A = \begin{bmatrix} a & b \\ c & d \end{bmatrix}$.

   (a) Find the characteristic polynomial of A.

   (b) Identify the constant term of the characteristic polynomial.

   (c) If $\det A = 1$ and the characteristic polynomial of A factors as the product of polynomials of degree 1, show that each eigenvalue of A is the reciprocal of the other.

23. [BB] If A is a diagonalizable matrix with all eigenvalues nonnegative, show that A has a square root; that is, $A = B^2$ for some matrix B. [Hint: Exercise 15.]

24. Suppose A and B are matrices, one of which is invertible, and $AB = BA$.

   (a) Show that AB and BA are similar.

   (b) Explain why AB and BA have the same eigenvalues. (This is true without the invertibility assumption, but the author is unaware of an elementary reason.)

25. The *converse* of an implication "if $\mathcal{A}$, then $\mathcal{B}$" is the implication "if $\mathcal{B}$, then $\mathcal{A}$." Write the converse of Theorem 3.4.14 and describe, with a proof or a counterexample, whether the converse is true.

26. Suppose λ and μ are different eigenvectors of a matrix A. Suppose that $\mathbf{u}$ and $\mathbf{v}$ are linearly independent eigenvectors corresponding to λ and $\mathbf{w}$ is an eigenvector corresponding to μ. Show that the **three** vectors $\mathbf{u}, \mathbf{v}, \mathbf{w}$ are linearly independent. (See 3.4.16.) [Hint: Use the fact that $a\mathbf{u} + b\mathbf{v}$ is either $\mathbf{0}$ or an eigenvector of A corresponding to λ. See Exercises 9 and 10 of Section 3.3.]

27. Let $A = \begin{bmatrix} -11 & 18 \\ -6 & 10 \end{bmatrix}$. Show that there does not exist a matrix B such that $B^2 = A$. [Hint: Show that A is diagonalizable and then examine the solution to Exercise 15.]

28. The instructor of a linear algebra course preparing a homework assignment for her class wants to produce a 3×3 matrix with eigenvalues $1, 2, 3$ with corresponding eigenvectors $\mathbf{x}_1, \mathbf{x}_2, \mathbf{x}_3$, respectively. How should she proceed?

29. Find an invertible matrix P such that

$$P^{-1}\begin{bmatrix} 0 & 1 & 0 \\ 0 & 0 & 1 \\ 2 & 1 & -2 \end{bmatrix} P = \begin{bmatrix} -1 & 0 & 0 \\ 0 & -2 & 0 \\ 0 & 0 & 1 \end{bmatrix}.$$

30. **(a)** Explain how the equation $P^{-1}AP = D$ makes it easy to find powers of A.

   (b) Find A^{10} if $A = \begin{bmatrix} 1 & 2 \\ 0 & 3 \end{bmatrix}$.

31. Prove that an $n \times n$ real matrix with n distinct nonzero (real) eigenvalues is invertible.

CHAPTER KEY WORDS AND IDEAS: Here are some technical words and phrases that were used in this chapter. Do you know the meaning of each? If you're not sure, check the glossary or index at the back of the book.

characteristic polynomial
cofactor
commute (matrices do not)
determinant of a 2×2 matrix
diagonalizable
eigenvalue

eigenvector
linear function
minor (of a matrix)
similar matrices
singular matrix

Review Exercises for Chapter 3

1. An $n \times n$ matrix A has the property that $A^T = -A$. Suppose $\det A \neq 0$. Show that n is even.

2. An invertible matrix A satisfies $A^{-1} = A$. Show that $\det A = \pm 1$.

3. Let $A = \begin{bmatrix} 1 & 3 & 5 \\ 2 & 1 & 1 \\ 3 & 4 & 2 \end{bmatrix}$.

   (a) Find the matrix C of cofactors of A.

   (b) Compute the product AC^T.

   (c) What is $\det A$?

   (d) What is A^{-1}?

   (e) Use your answer to (d) to solve the system

$$A\mathbf{x} = \mathbf{b}, \text{ where } \mathbf{b} = \begin{bmatrix} 1 \\ 0 \\ 0 \end{bmatrix}.$$

4. Let A be an $n \times n$ matrix and let $\mathbf{x}$, $\mathbf{y}$, $\mathbf{z}$ be vectors in $\mathbf{R}^n$. Suppose

$$A\mathbf{x} = 3\mathbf{y}, \quad A\mathbf{y} = \mathbf{x} - 2\mathbf{z}, \quad \text{and} \quad A\mathbf{z} = \mathbf{x} + \mathbf{y} + \mathbf{z}.$$

Let $B = \begin{bmatrix} \mathbf{x} & \mathbf{y} & \mathbf{z} \\ \downarrow & \downarrow & \downarrow \end{bmatrix}$ be the $n \times 3$ matrix as shown.

(a) Compute AB.

(b) Find a 3×3 matrix X such that $AB = BX$.

5. (a) Let $A = \begin{bmatrix} 1 & 3 & 5 \\ 2 & 4 & 6 \\ 3 & 5 & 9 \end{bmatrix}$. Find $\det A$ by expanding by cofactors of the first row and the third column.

(b) Let $B = \begin{bmatrix} 6 & -4 & -3 \\ 2 & 8 & 5 \\ 8 & 4 & 0 \end{bmatrix}$. Find $\det B$ by expanding by cofactors of the second row and the second column.

(c) Are either of the matrices A and B singular? Explain.

6. Find the determinant of each of the following matrices by reducing each matrix to a triangular one.

(a) $\begin{bmatrix} 2 & 1 & 1 & 1 \\ 0 & 3 & 0 & 1 \\ 1 & 0 & 2 & 1 \\ 0 & 0 & 0 & 1 \end{bmatrix}$

(b) $\begin{bmatrix} -3 & 0 & 1 & 1 \\ 3 & 1 & 2 & 2 \\ -6 & -2 & -4 & 2 \\ 1 & -1 & 0 & -1 \end{bmatrix}$

(c) $\begin{bmatrix} -2 & 1 & 2 & 1 \\ 0 & 2 & 0 & 3 \\ 1 & 3 & -1 & 0 \\ 1 & 2 & 1 & 1 \end{bmatrix}$.

7. Show that $\begin{bmatrix} 1 & 8 & 2 & 4 \\ 3 & 0 & 1 & -2 \\ 1 & 7 & 4 & 1 \\ -3 & 2 & 1 & 3 \end{bmatrix}$ and

$\begin{bmatrix} 2 & 8 & 1 & 4 \\ 4 & 7 & 1 & 1 \\ 1 & 0 & 3 & -2 \\ 1 & 2 & -3 & 3 \end{bmatrix}$ have the same determinant, without finding the determinant of either matrix.

8. Find $\begin{vmatrix} a & b & c & 0 \\ b & a & d & 0 \\ c & d & 0 & a \\ d & c & 0 & b \end{vmatrix}$.

9. For which values of x is the matrix

$A = \begin{bmatrix} 1 & -1 & x \\ 2 & 1 & x^2 \\ 4 & -1 & x^3 \end{bmatrix}$ singular?

10. Suppose A is a 4×4 with row echelon form

$U = \begin{bmatrix} 1 & 2 & 3 & 4 \\ 0 & 1 & 5 & -1 \\ 0 & 0 & 1 & 7 \\ 0 & 0 & 0 & 1 \end{bmatrix}$. For each of the following

statements, either prove that it is true or provide a counterexample to show that it is false.

(a) $\det A = \det U$.

(b) $\det A \neq 0$.

11. Suppose $A = \begin{bmatrix} 50 & -1 & \cdots & -1 \\ -1 & 50 & \cdots & -1 \\ -1 & -1 & \cdots & -1 \\ \vdots & \vdots & \ddots & \vdots \\ -1 & -1 & \cdots & -1 \\ -1 & -1 & \cdots & 50 \end{bmatrix}$ is a 50×50

matrix. Find $\det A$.

12. Let

$A = \begin{bmatrix} 1 + x_1 & x_2 & x_3 & \cdots & x_n \\ x_1 & 1 + x_2 & x_3 & \cdots & x_n \\ x_1 & x_2 & 1 + x_3 & \cdots x_n \\ \vdots & \vdots & \vdots & & \vdots \\ x_1 & x_2 & x_3 & \cdots & 1 + x_n \end{bmatrix}$.

(a) Find $\det A$.

(b) Find $\begin{vmatrix} 0 & 1 & 1 & \cdots & 1 \\ 1 & 0 & 1 & \cdots & 1 \\ 1 & 1 & 0 & \cdots & 1 \\ \vdots & \vdots & \vdots & & \vdots \\ 1 & 1 & 1 & \cdots & 0 \end{vmatrix}$.

13. Let A and B be 6×6 matrices with $\det A = \frac{1}{4}$ and $\det B = -2$. Let $C = -ABA^T B^T$.

(a) Find $\det C$.

(b) Matrix C is invertible. Why?

(c) Find $\det C^{-1}$.

14. Show that $\mathbf{v} = \begin{bmatrix} 2 \\ 3 \end{bmatrix}$ is an eigenvector of

$A = \begin{bmatrix} 1 & 2 \\ 3 & 2 \end{bmatrix}$. Find a corresponding eigenvalue.

15. Let $A = \begin{bmatrix} 2 & 1 & 1 \\ -2 & 5 & 0 \\ 2 & -1 & 4 \end{bmatrix}$ and $\mathbf{v} = \begin{bmatrix} 2 \\ 4 \\ 0 \end{bmatrix}$. Is $\mathbf{v}$ an eigenvector of A? If not, explain; if so, what is the corresponding eigenvalue?

16. Let $A = \begin{bmatrix} 4 & 8 \\ 6 & 2 \end{bmatrix}$.

(a) What is the characteristic polynomial of A?

(b) Find the eigenvalues and corresponding eigenvectors of A.

(c) Exhibit an invertible matrix P such that $P^{-1}AP$ is a diagonal matrix D whose smallest diagonal entry is in position $(1, 1)$.

17. Let $A = \begin{bmatrix} 6 & -1 & -3 \\ 8 & 0 & -6 \\ 8 & -1 & -5 \end{bmatrix}$.

(a) Find all eigenvalues of A and the corresponding eigenspaces.

(b) Is A diagonalizable? Explain and, if the answer is "yes," find an invertible matrix P such that $P^{-1}AP$ is diagonal.

18. (a) If λ is an eigenvalue of an invertible matrix A and $\mathbf{x}$ is a corresponding eigenvector, then $\lambda \neq 0$. Why not?

(b) Show that $\mathbf{x}$ is also an eigenvector for A^{-1}. What is the corresponding eigenvalue?

19. Let $A = \begin{bmatrix} -1 & -14 & 7 \\ -7 & 6 & -7 \\ 7 & -14 & -1 \end{bmatrix}$. Given that

$\mathbf{x}_1 = \begin{bmatrix} -1 \\ 0 \\ 1 \end{bmatrix}$ and $\mathbf{x}_2 = \begin{bmatrix} 2 \\ 1 \\ 0 \end{bmatrix}$ are eigenvectors of A and that 20 is an eigenvalue, find an invertible matrix P and a diagonal matrix D such that $P^{-1}AP = D$.

20. Let $A = \begin{bmatrix} 3 & -2 & 0 \\ 0 & 3 & 0 \\ 0 & 0 & 5 \end{bmatrix}$.

(a) Find all (real) eigenvalues of A and corresponding eigenspaces.

(b) Is A similar to a diagonal matrix?

21. Without calculation, explain why $\begin{bmatrix} 3 & -2 & 0 \\ -2 & 3 & 0 \\ 0 & 0 & 5 \end{bmatrix}$ is a diagonalizable matrix.

22. (a) What does it mean to say that matrices A and B are *similar*?

(b) Prove that similar matrices have the same determinant.

23. If A and B are similar matrices, show that A^2 and B^2 are also similar.

24. Suppose A is a matrix that satisfies $A^9 = \mathbf{0}$. Find all eigenvalues of A.

4

Vector Spaces

The Theory of Linear Equations

As we know, any linear system of equations can be written in the form $A\mathbf{x} = \mathbf{b}$, where A is the matrix that records the coefficients of the variables, $\mathbf{x}$ is the vector of variables whose values we want to find, and $\mathbf{b}$ is the vector whose components are the constants appearing to the right of the $=$ signs. For example, the system

$$\begin{array}{rcr} -2x_1 + x_2 + x_3 &=& -9 \\ x_1 + x_2 + x_3 &=& 0 \\ x_1 + 3x_2 &=& -3 \end{array}$$

is $A\mathbf{x} = \mathbf{b}$ with

$$A = \begin{bmatrix} -2 & 1 & 1 \\ 1 & 1 & 1 \\ 1 & 3 & 0 \end{bmatrix}, \quad \mathbf{x} = \begin{bmatrix} x_1 \\ x_2 \\ x_3 \end{bmatrix}, \quad \text{and } \mathbf{b} = \begin{bmatrix} -9 \\ 0 \\ -3 \end{bmatrix}.$$

In Section 2.3, we gave examples of linear systems for which there is no solution, others for which the solution is unique, and still others with infinitely many solutions.

Thinking geometrically for a moment, it is not hard to see how each of these possibilities might come about. Suppose we are solving two equations in three unknowns, x, y, and z. In 3-space, we know that the set of points (x, y, z) that satisfy an equation of the form $ax + by + cz = d$ is a plane. So, solving two such equations means we are trying to find the points that lie in two planes. If the planes are parallel and not identical—think of the floor and the ceiling of a room—there is no intersection; the equations do not have a solution. If the planes are not parallel, they intersect in a line—think of a wall and the ceiling of a room—and the equations have infinitely many solutions. Except in special cases, the solution of three equations in three unknowns is unique; the first two equations determine a line, and the unique solution is the point where this line punctures the plane determined by the third equation.

One of the goals of this section is to make it possible to determine in advance what sort of solution, if any, to expect from a given system. In so doing, we get a lot of mileage from the fact that $A\mathbf{x}$ is a linear combination of the columns of the matrix A. Remember the important fact highlighted in 2.1.33.

■ The Null Space of a Matrix

Suppose A is an $m \times n$ matrix, $\mathbf{x}$ and $\mathbf{b}$ are vectors, and $A\mathbf{x} = \mathbf{b}$. In what Euclidean spaces do $\mathbf{x}$ and $\mathbf{b}$ live? The product $A\mathbf{x}$ is the product of an $m \times n$ matrix by what size vector $\mathbf{x}$? The vector $\mathbf{x}$ must be $n \times 1$, so $\mathbf{b}$, which is the product $A\mathbf{x}$, is $m \times 1$. Remember the mnemonic we presented in our discussion of matrix multiplication in Section 2.1:

$$\frac{m}{n} \times \frac{n}{1} = \frac{m}{1}.$$

> **4.1.1** Multiplication by an $m \times n$ matrix is a function that sends vectors in $\mathbf{R}^n$ to vectors in $\mathbf{R}^m$. We write $A \colon \mathbf{R}^n \to \mathbf{R}^m$.

This observation is illustrated in Figure 4.1.

4.1.2 DEFINITION   The *null space* of a matrix A is the set of all vectors $\mathbf{x}$ that satisfy $A\mathbf{x} = \mathbf{0}$, that is, the set of all solutions to the homogeneous system $A\mathbf{x} = \mathbf{0}$. We use the notation null sp A for the null space of A.

If A is $m \times n$, the null space is a set of vectors in $\mathbf{R}^n$. This set always contains the zero vector; sometimes, this is all there is.

4.1.3 EXAMPLE   Let $A = \begin{bmatrix} 1 & 0 \\ 0 & 1 \end{bmatrix}$ be the 2×2 identity matrix. Since $A\mathbf{x} = \mathbf{x}$ for any $\mathbf{x}$, if $A\mathbf{x} = \mathbf{0}$, then $\mathbf{x} = \mathbf{0}$. The null space of A consists of just the zero vector.   ■

4.1.4 EXAMPLE   Let $A = \begin{bmatrix} 1 & 0 & 2 \\ 0 & 1 & 3 \end{bmatrix}$. Then $A \colon \mathbf{R}^3 \to \mathbf{R}^2$ and the null space of A is the set of those vectors $\mathbf{x} = \begin{bmatrix} x \\ y \\ z \end{bmatrix}$ in $\mathbf{R}^3$ that satisfy $A\mathbf{x} = \mathbf{0}$. The matrix is already in row echelon

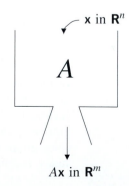

Figure 4.1   Multiplication by an $m \times n$ matrix A is a function $\mathbf{R}^n \to \mathbf{R}^m$.

x in $\mathbf{R}^n$

A

$A\mathbf{x}$ in $\mathbf{R}^m$

form. The variable $z = t$ is free, $y = -3t$, and $x = -2t$. The null space is the set of vectors of the form $\begin{bmatrix} -2t \\ -3t \\ t \end{bmatrix} = t \begin{bmatrix} -2 \\ -3 \\ 1 \end{bmatrix}$, geometrically, the line in $\mathbf{R}^3$ through the origin and $(-2, -3, 1)$. ∎

4.1.5 EXAMPLE Suppose $A = \begin{bmatrix} 1 & 5 \\ 0 & 1 \end{bmatrix}$. Notice that A is invertible, so if $A\mathbf{x} = \mathbf{0}$, then $\mathbf{x} = \mathbf{0}$. The null space of A consists of the zero vector alone. ∎

4.1.6 EXAMPLE Let $A = \begin{bmatrix} 1 & -3 & -2 & 4 \\ 2 & 0 & 2 & 2 \\ 0 & 4 & 4 & -4 \end{bmatrix}$. The null space of A is the set of all $\mathbf{x} = \begin{bmatrix} x_1 \\ x_2 \\ x_3 \\ x_4 \end{bmatrix}$ such that $A\mathbf{x} = \mathbf{0}$. Gaussian elimination proceeds

$$A \to \begin{bmatrix} 1 & -3 & -2 & 4 \\ 0 & 6 & 6 & -6 \\ 0 & 1 & 1 & -1 \end{bmatrix} \to \begin{bmatrix} 1 & -3 & -2 & 4 \\ 0 & 1 & 1 & -1 \\ 0 & 0 & 0 & 0 \end{bmatrix},$$

so $x_3 = t$ and $x_4 = s$ are free variables, $x_2 = -x_3 + x_4 = -t + s$, and $x_1 = 3x_2 + 2x_3 - 4x_4 = -t - s$. The null space consists of all vectors of the form

$$\begin{bmatrix} x_1 \\ x_2 \\ x_3 \\ x_4 \end{bmatrix} = \begin{bmatrix} -t - s \\ -t + s \\ t \\ s \end{bmatrix} = t \begin{bmatrix} -1 \\ -1 \\ 1 \\ 0 \end{bmatrix} + s \begin{bmatrix} -1 \\ 1 \\ 0 \\ 1 \end{bmatrix}.$$ ∎

4.1.7 PROBLEM Find and identify geometrically the null space of $A = \begin{bmatrix} 1 & 3 \\ 2 & 6 \end{bmatrix}$.

Solution. Gaussian elimination proceeds $A \to \begin{bmatrix} 1 & 3 \\ 0 & 0 \end{bmatrix}$, so $A\mathbf{x} = \mathbf{0}$ with $\mathbf{x} = \begin{bmatrix} x \\ y \end{bmatrix}$ gives $y = t$ free and $x + 3y = 0$, so $x = -3y = -3t$. Thus $\mathbf{x} = \begin{bmatrix} -3t \\ t \end{bmatrix} = t \begin{bmatrix} -3 \\ 1 \end{bmatrix}$.

The null space of A is the set of all scalar multiples of $\begin{bmatrix} -3 \\ 1 \end{bmatrix}$. Geometrically, this is the line through the origin and the point $(-3, 1)$.

As we have remarked, the null space of a matrix always contains the zero vector. The homogeneous system $A\mathbf{x} = \mathbf{0}$ always has $\mathbf{x} = \mathbf{0}$ as a solution. This is known as the *trivial solution*, for obvious reasons. The interesting question about a homogeneous system is whether the solution $\mathbf{x} = \mathbf{0}$ is unique. Are there other solutions to $A\mathbf{x} = \mathbf{0}$?

Suppose A is a 5×7 matrix. A row echelon form of A might look like this:

$$U = \begin{bmatrix} 1 & \star & \star & \star & \star & \star & \star \\ 0 & 1 & \star & \star & \star & \star & \star \\ 0 & 0 & 0 & 0 & 1 & \star & \star \\ 0 & 0 & 0 & 0 & 0 & 0 & 1 \\ 0 & 0 & 0 & 0 & 0 & 0 & 0 \end{bmatrix}, \tag{1}$$

where we use the symbol $\star$ to denote a number of no interest to us. On the other hand, we care very much about the pivots. The columns that are not pivot columns correspond to free variables; variables associated with pivot columns can be expressed in terms of free variables. The matrix U shown in (1) has three free variables, so there are many nontrivial solutions to $A\mathbf{x} = \mathbf{0}$.

Can you see why any 5×7 matrix would have a row echelon form with at least **two** free variables? There are only five rows, hence at most five pivots, but there are seven columns. So at least two columns cannot be pivot columns and the corresponding variables will be free. Free variables imply many solutions, in particular many nonzero solutions.

This little example generalizes.

Theorem 4.1.8 *If A is an $m \times n$ matrix and $m < n$, then A has a nontrivial null space; there exists a vector $\mathbf{x} \neq \mathbf{0}$ such that $A\mathbf{x} = \mathbf{0}$.*

[Remark. In some books, this theorem sometimes appears like this: A homogeneous system of linear equations with more unknowns than equations has a nontrivial solution.]

Proof The matrix A has at most m pivot columns, one per row. Since $m < n$, there are columns that are not pivot columns. So there is at least one free variable in the solution of $A\mathbf{x} = \mathbf{0}$, hence there are an infinite number of solutions; in particular, there are nontrivial solutions. ∎

We have observed Theorem 4.1.8 on many occasions, most recently in **4.1.6**. Here is another example.

4.1.9 EXAMPLE Let A be the 2×5 matrix $\begin{bmatrix} 1 & 1 & 1 & -1 & 1 \\ 2 & 2 & -1 & 4 & -1 \end{bmatrix}$. With Gaussian elimination, we obtain

$$\begin{bmatrix} 1 & 1 & 1 & -1 & 1 \\ 2 & 2 & -1 & 4 & -1 \end{bmatrix} \to \begin{bmatrix} 1 & 1 & 1 & -1 & 1 \\ 0 & 0 & -3 & 6 & -3 \end{bmatrix} = U,$$

a row echelon form of A. The columns without pivots are two, four, and five; variables $x_2 = t$, $x_4 = s$, and $x_5 = r$ are free. Then

$$-3x_3 + 6x_4 - 3x_5 = 0, \text{ so } 3x_3 = 6x_4 - 3x_5 = 6s - 3r, \ x_3 = 2s - r, \text{ and}$$
$$x_1 + x_2 + x_3 - x_4 + x_5 = 0, \text{ so } x_1 = -x_2 - x_3 + x_4 - x_5 = -t - s.$$

As promised by Theorem 4.1.8, the null space of A is certainly different from **0**. It consists of all vectors of the form

$$\mathbf{x} = \begin{bmatrix} x_1 \\ x_2 \\ x_3 \\ x_4 \\ x_5 \end{bmatrix} = \begin{bmatrix} -t - s \\ t \\ 2s - r \\ s \\ r \end{bmatrix} = t \begin{bmatrix} -1 \\ 1 \\ 0 \\ 0 \\ 0 \end{bmatrix} + s \begin{bmatrix} -1 \\ 0 \\ 2 \\ 1 \\ 0 \end{bmatrix} + r \begin{bmatrix} 0 \\ 0 \\ -1 \\ 0 \\ 1 \end{bmatrix}.$$

Equivalently, the homogeneous system

$$\begin{aligned} x_1 + x_2 + x_3 - x_4 + x_5 &= 0 \\ 2x_1 + 2x_2 - x_3 + 4x_4 - x_5 &= 0 \end{aligned}$$

has infinitely many solutions. ∎

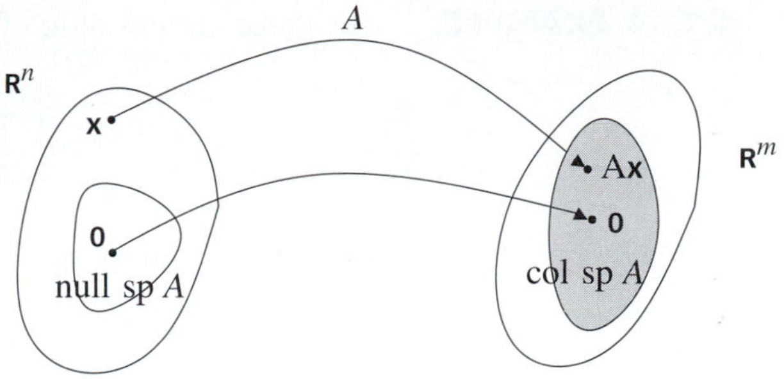

Figure 4.2 If $A: \mathbf{R}^n \to \mathbf{R}^m$ is an $m \times n$ matrix, the null space is in $\mathbf{R}^n$ and the column space is in $\mathbf{R}^m$.

■ The Column Space of a Matrix

4.1.10 DEFINITION The *column space* of a matrix A is the set of all linear combinations of the columns of A. It is denoted col sp A.

Since A is $m \times n$, the columns of A are vectors in $\mathbf{R}^m$, so col sp A is a collection of vectors in $\mathbf{R}^m$, whereas the null space consists of vectors in $\mathbf{R}^n$. These concepts are illustrated in Figure 4.2.

4.1.11 EXAMPLE Let $A = \begin{bmatrix} 1 & 0 \\ 0 & 1 \end{bmatrix}$. Then A is 2×2, so it maps $\mathbf{R}^2$ to $\mathbf{R}^2$. Its column space is some set of vectors in $\mathbf{R}^2$, the Euclidean plane. In fact, the column space is all of $\mathbf{R}^2$ since any $\begin{bmatrix} x \\ y \end{bmatrix}$ in $\mathbf{R}^2$ is a linear combination of the columns of A:

$$\begin{bmatrix} x \\ y \end{bmatrix} = x \begin{bmatrix} 1 \\ 0 \end{bmatrix} + y \begin{bmatrix} 0 \\ 1 \end{bmatrix}. \qquad ■$$

4.1.12 EXAMPLE Let $A = \begin{bmatrix} 1 & 0 & 2 \\ 0 & 1 & 3 \end{bmatrix}$ be the matrix whose null space we found in **4.1.4**. Then $A: \mathbf{R}^3 \to \mathbf{R}^2$ and the column space of A is a collection of vectors in $\mathbf{R}^2$. As we have just seen, every vector in $\mathbf{R}^2$ is a linear combination of the first two columns of A. We cannot get fewer vectors by taking linear combinations of the first two as well as the third, so again col sp $A = \mathbf{R}^2$. $\qquad ■$

Reading Challenge 1 *Express the vector $\begin{bmatrix} x \\ y \end{bmatrix}$ as a linear combination of the columns of the matrix A in 4.1.12.*

4.1.13 EXAMPLE Suppose $A = \begin{bmatrix} 1 & 5 \\ 0 & 1 \end{bmatrix}$ is the matrix of **4.1.5**. The columns of this matrix are nonparallel vectors in $\mathbf{R}^2$, so they span the xy-plane. (Recall 1.1.14.) The column space of this matrix is also $\mathbf{R}^2$. $\qquad ■$

At this point, the student may well be thinking that the column space of any matrix with two rows is $\mathbf{R}^2$. Such is clearly **not** the case, however. The column space of $\begin{bmatrix} 0 & 0 & 0 & 0 \\ 0 & 0 & 0 & 0 \end{bmatrix}$ is just the zero vector. Look at the next example.

4.1.14 EXAMPLE A linear combination of the columns of $A = \begin{bmatrix} 1 & 3 \\ 2 & 6 \end{bmatrix}$ is a vector of the form

$$a\begin{bmatrix} 1 \\ 2 \end{bmatrix} + b\begin{bmatrix} 3 \\ 6 \end{bmatrix} = a\begin{bmatrix} 1 \\ 2 \end{bmatrix} + 3b\begin{bmatrix} 1 \\ 2 \end{bmatrix} = (a + 3b)\begin{bmatrix} 1 \\ 2 \end{bmatrix}.$$

Here, the column space is not all of $\mathbf{R}^2$, but merely those vectors in the plane that are scalar multiples of $\begin{bmatrix} 1 \\ 2 \end{bmatrix}$. Geometrically, this is the line through the origin and the point $(1, 2)$. ∎

4.1.15 EXAMPLE The 3×4 matrix $A = \begin{bmatrix} 1 & -3 & -2 & 4 \\ 2 & 0 & 2 & 2 \\ 0 & 4 & 4 & -4 \end{bmatrix}$ encountered in **4.1.6** maps $\mathbf{R}^4$ to $\mathbf{R}^3$, so its column space is a collection of vectors in $\mathbf{R}^3$. Call the columns $\mathbf{a}_1, \mathbf{a}_2, \mathbf{a}_3, \mathbf{a}_4$ and notice that $\mathbf{a}_3$ is the sum of the first two columns while $\mathbf{a}_4$ is their difference. A linear combination of the columns of A is a vector of the form

$$c_1\mathbf{a}_1 + c_2\mathbf{a}_2 + c_3\mathbf{a}_3 + c_4\mathbf{a}_4 = c_1\mathbf{a}_1 + c_2\mathbf{a}_2 + c_3(\mathbf{a}_1 + \mathbf{a}_2) + c_4(\mathbf{a}_1 - \mathbf{a}_2)$$
$$= (c_1 + c_3 + c_4)\mathbf{a}_1 + (c_2 + c_3 - c_4)\mathbf{a}_2.$$

In this case, the column space of A is not all of $\mathbf{R}^3$, but merely the set of linear combinations of $\mathbf{a}_1$ and $\mathbf{a}_2$. Geometrically, this is a plane. ∎

Reading Challenge 2 *What is the equation of this plane?*

Our interest in the column space of a matrix in a section concerned with linear equations is the fact that

4.1.16 A vector $\mathbf{b}$ is in the column space of A if and only if $\mathbf{b} = A\mathbf{x}$ for some vector $\mathbf{x}$—that is, if and only if the linear system $A\mathbf{x} = \mathbf{b}$ has a solution.

Reading Challenge 3 *Why?*

Thus we have another description of the column space of a matrix A. Remember, the column space of A was defined as the set of all linear combinations of the columns A. Now we see also that

4.1.17 The column space of an $m \times n$ matrix A is the set of all $A\mathbf{x}$ as $\mathbf{x}$ ranges over all vectors in $\mathbf{R}^n$.

So the column space is precisely the range of the function described in 4.1.1.

4.1.18 EXAMPLE

Is $\mathbf{b} = \begin{bmatrix} 1 \\ 2 \\ 2 \end{bmatrix}$ in the column space of $A = \begin{bmatrix} 1 & 0 & -2 & -1 \\ -1 & 3 & 5 & 4 \\ 2 & 1 & -3 & -1 \end{bmatrix}$?

We are just asking if the system $A\mathbf{x} = \mathbf{b}$ has a solution $\mathbf{x} = \begin{bmatrix} x_1 \\ x_2 \\ x_3 \\ x_4 \end{bmatrix}$.

The augmented matrix is $\begin{bmatrix} 1 & 0 & -2 & -1 & | & 1 \\ -1 & 3 & 5 & 4 & | & 2 \\ 2 & 1 & -3 & -1 & | & 2 \end{bmatrix}$ and Gaussian elimination proceeds

$$\begin{bmatrix} 1 & 0 & -2 & -1 & | & 1 \\ -1 & 3 & 5 & 4 & | & 2 \\ 2 & 1 & -3 & -1 & | & 2 \end{bmatrix} \rightarrow \begin{bmatrix} 1 & 0 & -2 & -1 & | & 1 \\ 0 & 3 & 3 & 3 & | & 3 \\ 0 & 1 & 1 & 1 & | & 0 \end{bmatrix}$$

$$\rightarrow \begin{bmatrix} 1 & 0 & -2 & -1 & | & 1 \\ 0 & 1 & 1 & 1 & | & 1 \\ 0 & 1 & 1 & 1 & | & 0 \end{bmatrix} \rightarrow \begin{bmatrix} 1 & 0 & -2 & -1 & | & 1 \\ 0 & 1 & 1 & 1 & | & 1 \\ 0 & 0 & 0 & 0 & | & -1 \end{bmatrix}.$$

The equation corresponding to the last row is $0 = -1$. This absurdity tells us there are no solutions. The system is *inconsistent*; the vector $\mathbf{b} = \begin{bmatrix} 1 \\ 2 \\ 2 \end{bmatrix}$ is **not** in the column space of A.

What is the column space of A? Well, $\mathbf{b} = \begin{bmatrix} b_1 \\ b_2 \\ b_3 \end{bmatrix}$ is in the column space if and only if $A\mathbf{x} = \mathbf{b}$ has a solution. The augmented matrix in this more general situation is

$$\begin{bmatrix} 1 & 0 & -2 & -1 & | & b_1 \\ -1 & 3 & 5 & 4 & | & b_2 \\ 2 & 1 & -3 & -1 & | & b_3 \end{bmatrix}, \tag{2}$$

and Gaussian elimination proceeds

$$\begin{bmatrix} 1 & 0 & -2 & -1 & | & b_1 \\ -1 & 3 & 5 & 4 & | & b_2 \\ 2 & 1 & -3 & -1 & | & b_3 \end{bmatrix} \rightarrow \begin{bmatrix} 1 & 0 & -2 & -1 & | & b_1 \\ 0 & 3 & 3 & 3 & | & b_2 + b_1 \\ 0 & 1 & 1 & 1 & | & b_3 - 2b_1 \end{bmatrix}$$

$$\rightarrow \begin{bmatrix} 1 & 0 & -2 & -1 & | & b_1 \\ 0 & 1 & 1 & 1 & | & \frac{1}{3}(b_2 + b_1) \\ 0 & 1 & 1 & 1 & | & b_3 - 2b_1 \end{bmatrix}$$

$$\rightarrow \begin{bmatrix} 1 & 0 & -2 & -1 & | & b_1 \\ 0 & 1 & 1 & 1 & | & \frac{1}{3}(b_2 + b_1) \\ 0 & 0 & 0 & 0 & | & b_3 - 2b_1 - \frac{1}{3}(b_2 + b_1) \end{bmatrix}.$$

Just as a system of equations leads to an augmented matrix, any augmented matrix can be interpreted as a system of equations. In particular, each line in the augmented matrix we have just obtained corresponds to an equation. The first line, for example, corresponds to the equation $x_1 - 2x_3 - x_4 = b_1$, but it is the last line that is most helpful right now. The last line corresponds to the equation $0 = b_3 - 2b_1 - \frac{1}{3}(b_2 + b_1)$. This says that if the system has a solution, then it is necessary for $3b_3 - 6b_1 - b_1 - b_2 = 0$ or, equivalently, for $7b_1 + b_2 - 3b_3 = 0$.

On the other hand, if b_1, b_2, and b_3 satisfy this condition, the system with augmented matrix (2) has a solution (in fact, infinitely many solutions), for in this case the final augmented matrix is

$$\left[\begin{array}{cccc|c} 1 & 0 & -2 & -1 & b_1 \\ 0 & 1 & 1 & 1 & \frac{1}{3}(b_2 + b_1) \\ 0 & 0 & 0 & 0 & 0 \end{array}\right],$$

so the variables x_3 and x_4 are free and there are infinitely many solutions.

We conclude that the given system has a solution if and only if $7b_1 + b_2 - 3b_3 = 0$. Another way of expressing this fact is to say that the vector $\mathbf{b} = \begin{bmatrix} b_1 \\ b_2 \\ b_3 \end{bmatrix}$ is in the column space of A if and only if $7b_1 + b_2 - 3b_3 = 0$. So we see that the column space of A is the plane with equation $7x + y - 3z = 0$. In particular, this gives us another way to see why $\mathbf{b} = \begin{bmatrix} 1 \\ 2 \\ 2 \end{bmatrix}$ is not in the column space: Its components do not satisfy the equation $7x + y - 3z = 0$.   ■

■ The Rank of a Matrix

4.1.19 DEFINITION   The *rank* of a matrix is the number of pivot columns.

4.1.20 EXAMPLE   The matrix $A = \begin{bmatrix} 4 & 0 & 2 \\ 0 & 1 & 3 \end{bmatrix}$ has rank 2. It is already in row echelon form and there are two pivot columns.   ■

4.1.21 EXAMPLE   The matrix $A = \begin{bmatrix} 1 & 2 & 3 \\ 2 & 4 & 6 \end{bmatrix}$ has rank 1 since a row echelon form of A is $\begin{bmatrix} 1 & 2 & 3 \\ 0 & 0 & 0 \end{bmatrix}$, which has just one pivot column.   ■

Suppose A is an $m \times n$ matrix. As you might guess, the particular cases rank $A = m$ and rank $A = n$ are special and, as we shall see, are related to the nature of the solutions to the linear system $A\mathbf{x} = \mathbf{b}$. In what follows, it is important to remember that a linear system is *inconsistent* (meaning no solutions) if and only if, when the augmented matrix is reduced to row echelon form, there is a row at the bottom of the form

$$0 \quad 0 \quad 0 \quad \cdots \quad 0 \mid \star \tag{3}$$

where $\star$ is not zero, since such a row corresponds to the equation $0 = \star$, which is not true. Note that the row in (3) is a row of all zeros in a row echelon form of A, followed by a nonzero scalar (from the last component of **b**).

Suppose rank $A = m$. In this case, the number of pivot columns of A is the number of rows. There are no zero rows; there is no possibility that the system $A\mathbf{x} = \mathbf{b}$ is inconsistent, no matter what **b** is. For any **b**, the system $A\mathbf{x} = \mathbf{b}$ has at least one solution. Here's an example.

4.1.22 EXAMPLE   The 3×4 matrix $A = \begin{bmatrix} 1 & 2 & 3 & 4 \\ 5 & 6 & 7 & 8 \\ 0 & 12 & 11 & 10 \end{bmatrix}$ has row echelon form $U = \begin{bmatrix} 1 & 2 & 3 & 4 \\ 0 & 1 & 2 & 3 \\ 0 & 0 & 1 & 2 \end{bmatrix}$, so

rank $A = m = 3$, the number of rows. Suppose we wish to solve $A\mathbf{x} = \mathbf{b} = \begin{bmatrix} b_1 \\ b_2 \\ b_3 \end{bmatrix}$.

Using Gaussian elimination,

$$\begin{bmatrix} 1 & 2 & 3 & 4 & \bigm| & b_1 \\ 5 & 6 & 7 & 8 & \bigm| & b_2 \\ 0 & 12 & 11 & 10 & \bigm| & b_3 \end{bmatrix} \to \begin{bmatrix} 1 & 2 & 3 & 4 & \bigm| & b_1 \\ 0 & -4 & -8 & -12 & \bigm| & b_2 - 5b_1 \\ 0 & 12 & 11 & 10 & \bigm| & b_3 \end{bmatrix}$$

$$\to \begin{bmatrix} 1 & 2 & 3 & 4 & \bigm| & b_1 \\ 0 & 1 & 2 & 3 & \bigm| & -\frac{1}{4}(b_2 - 5b_1) \\ 0 & 12 & 11 & 10 & \bigm| & b_3 \end{bmatrix}$$

$$\to \begin{bmatrix} 1 & 2 & 3 & 4 & \bigm| & b_1 \\ 0 & 1 & 2 & 3 & \bigm| & -\frac{1}{4}(b_2 - 5b_1) \\ 0 & 0 & -13 & -26 & \bigm| & b_3 + 3(b_2 - 5b_1) \end{bmatrix}$$

$$\to \begin{bmatrix} 1 & 2 & 3 & 4 & \bigm| & b_1 \\ 0 & 1 & 2 & 3 & \bigm| & -\frac{1}{4}(b_2 - 5b_1) \\ 0 & 0 & 1 & 2 & \bigm| & -\frac{1}{13}[b_3 + 3(b_2 - 5b_1)] \end{bmatrix},$$

so $\mathbf{x}_4 = t$ is free and there are infinitely many solutions.   ■

Suppose rank $A = n$. This time, the number of pivot columns of A is the number of columns of A. There are no free variables, and the system $A\mathbf{x} = \mathbf{b}$ has at most one solution (depending on whether **b** is in the column space of A). Here's an example.

4.1.23 EXAMPLE   A row echelon form of $A = \begin{bmatrix} 1 & 2 \\ 3 & 4 \\ 5 & 6 \end{bmatrix}$ is $U = \begin{bmatrix} 1 & 2 \\ 0 & 1 \\ 0 & 0 \end{bmatrix}$, so rank $A = n = 2$, the

number of columns of A. What happens when we try to solve $A\mathbf{x} = \mathbf{b} = \begin{bmatrix} b_1 \\ b_2 \\ b_3 \end{bmatrix}$?

Row reduction is

$$
\begin{bmatrix}
1 & 2 & b_1 \\
3 & 4 & b_2 \\
5 & 6 & b_3
\end{bmatrix}
\rightarrow
\begin{bmatrix}
1 & 2 & b_1 \\
0 & -2 & b_2 - 3b_1 \\
0 & -4 & b_3 - 5b_1
\end{bmatrix}
$$

$$
\rightarrow
\begin{bmatrix}
1 & 2 & b_1 \\
0 & 1 & -\frac{1}{2}(b_2 - 3b_1) \\
0 & -4 & b_3 - 5b_1
\end{bmatrix}
\rightarrow
\begin{bmatrix}
1 & 2 & b_1 \\
0 & 1 & -\frac{1}{2}(b_2 - 3b_1) \\
0 & 0 & b_3 - 5b_1 - 2(b_2 - 3b_1)
\end{bmatrix}.
$$

The equation corresponding to the last row reads $0 = b_3 - 5b_1 - 2(b_2 - 3b_1) = b_1 - 2b_2 + b_3$. If the vector $\mathbf{b}$ does not satisfy this equation, there is no solution. If it does, there is a unique solution, since back substitution would give

$$
x_2 = -\tfrac{1}{2}(b_2 - 3b_1)
$$
$$
x_1 = b_1 - 2b_2 = -2b_1 + b_2.
$$

Our discussion has suggested proofs of the first two parts of the next theorem. The third part begins "The following statements are equivalent," and this is followed by five statements numbered i, ii, iii, iv, and v. The wording means that, for a given matrix A, all five statements are true or all five are false. A typical method of proof is to establish the truth of these five statements:

> If i is true, then ii is true. (This is denoted i $\implies$ ii and reads "i *implies* ii.")
> If ii is true, then iii is true. (This is denoted ii $\implies$ iii.)
> If iii, then iv. (iii $\implies$ iv.)
> If iv, then v. (iv $\implies$ v.)
> If v, then i. (v $\implies$ i.)

If we can prove these five statements, we will indeed have established the equivalence of all five. For example, to determine if the truth of iv implies the truth of ii, use the facts that the truth of iv implies the truth of v, which, in turn, implies the truth of i, which, in turn, implies the truth of ii.

Theorem 4.1.24   *Let A be an $m \times n$ matrix.*

1. *Let* rank $A = m$. *Then, for any vector $\mathbf{b}$ in $\mathbf{R}^m$, $A\mathbf{x} = \mathbf{b}$ has at least one solution.*

2. *Let* rank $A = n$. *Then, for any vector $\mathbf{b}$ in $\mathbf{R}^m$, $A\mathbf{x} = \mathbf{b}$ has at most one solution.*

3. *If $m = n$, the following statements are equivalent:*

   i. rank $A = n$.

   ii. *For any vector $\mathbf{b}$ in $\mathbf{R}^m$, $A\mathbf{x} = \mathbf{b}$ has a unique solution.*

   iii. *For any vector $\mathbf{b}$ in $\mathbf{R}^m$, $A\mathbf{x} = \mathbf{b}$ has a solution.*

   iv. *A is invertible.*

   v. $\det A \neq 0$.

Proof We have already given proofs of parts 1 and 2 and illustrated these with examples. For part 3, we first note that the equivalence of iv and v was Theorem 3.2.10, so we have only to establish the equivalence of the first four statements, which we do by showing that

$$i \implies ii \implies iii \implies iv \implies i.$$

We begin by assuming that i is true and proving that ii is true. So let $\mathbf{b}$ be a vector in $\mathbf{R}^m$. We have to show that $A\mathbf{x} = \mathbf{b}$ has a unique solution. Since rank $A = n = m$, part 1 says that $A\mathbf{x} = \mathbf{b}$ has at least one solution. Since rank $A = n$, part 2 says $A\mathbf{x} = \mathbf{b}$ has at most one solution. Putting these facts together, we deduce that $A\mathbf{x} = \mathbf{b}$ has exactly one solution, which is just what we wanted to know.

Next we assume that ii holds and prove iii. This is immediate, since if a system has just one solution, it certainly has a solution.

Now assume iii. We prove iv. So we assume that $A\mathbf{x} = \mathbf{b}$ has a solution for any $\mathbf{b}$ and exploit this fact by choosing some very special vectors $\mathbf{b}$. Let $\mathbf{e}_1, \mathbf{e}_2, \dots, \mathbf{e}_n$ be the standard basis vectors of $\mathbf{R}^n$. Our system says we can solve

$$\begin{aligned} A\mathbf{x} &= \mathbf{e}_1 \quad \text{for } \mathbf{x}_1, \\ A\mathbf{x} &= \mathbf{e}_2 \quad \text{for } \mathbf{x}_2, \\ &\;\;\vdots \\ A\mathbf{x} &= \mathbf{e}_n \quad \text{for } \mathbf{x}_n. \end{aligned}$$

So $\mathbf{x}_1, \mathbf{x}_2, \dots, \mathbf{x}_n$ are vectors satisfying $A\mathbf{x}_1 = \mathbf{e}_1, A\mathbf{x}_2 = \mathbf{e}_2, \dots, A\mathbf{x}_n = \mathbf{e}_n$.

Let $B = \begin{bmatrix} \mathbf{x}_1 & \mathbf{x}_2 & \cdots & \mathbf{x}_n \\ \downarrow & \downarrow & & \downarrow \end{bmatrix}$ be the matrix whose columns are $\mathbf{x}_1, \mathbf{x}_2, \dots, \mathbf{x}_n$.

Then

$$AB = A \begin{bmatrix} \mathbf{x}_1 & \mathbf{x}_2 & \cdots & \mathbf{x}_n \\ \downarrow & \downarrow & & \downarrow \end{bmatrix}$$

$$= \begin{bmatrix} A\mathbf{x}_1 & A\mathbf{x}_2 & \cdots & A\mathbf{x}_n \\ \downarrow & \downarrow & & \downarrow \end{bmatrix} = \begin{bmatrix} \mathbf{e}_1 & \mathbf{e}_2 & \cdots & \mathbf{e}_n \\ \downarrow & \downarrow & & \downarrow \end{bmatrix} = I.$$

Thus A has a *right inverse* B. Since A is square, Corollary 2.7.17 implies A is invertible, which is iv.

Finally, we assume iv and establish i. Assume that A is invertible. This means we can find an inverse by the technique described in Section 2.7, that is, by reducing the matrix $[A|I]$ to $[I|A^{-1}]$. Since A can be reduced to I via elementary row operations, a row echelon form has a pivot in every row. Thus rank $A = m = n$, which is i. ∎

Reading Challenge 4 *Part 3 of Theorem 4.1.24 claims that four statements are equivalent and hence that 12 implications are true:*

$$i \iff ii, \quad i \iff iii, \quad i \iff iv,$$

$$ii \iff iii, \quad ii \iff iv, \quad iii \iff iv,$$

yet we proved only four implications: i $\implies$ ii; ii $\implies$ iii; iii $\implies$ iv; and iv $\implies$ i. Is this sufficient?

Theorem 4.1.24 has a corollary that is quite incredible at first sight: it relates the possibility of solving $A\mathbf{x} = \mathbf{b}$ for any $\mathbf{b}$ to the single system $A\mathbf{x} = \mathbf{0}$.

Corollary 4.1.25   *If A is a square $n \times n$ matrix, then the following statements are equivalent:*

   i. *For any vector $\mathbf{b}$ in $\mathbf{R}^n$, $A\mathbf{x} = \mathbf{b}$ has a unique solution.*

   ii. *The homogeneous system $A\mathbf{x} = \mathbf{0}$ has just the trivial solution $\mathbf{x} = \mathbf{0}$.*

   iii. rank $A = n$.

Proof i $\implies$ ii: If $A\mathbf{x} = \mathbf{b}$ has a unique solution for any $\mathbf{b}$, then, in particular, it has a unique solution for $\mathbf{b} = \mathbf{0}$. Since $\mathbf{0}$ is one solution to $A\mathbf{x} = \mathbf{0}$, this is the only one.

ii $\implies$ iii: Suppose that $A\mathbf{x} = \mathbf{0}$ has just the solution $\mathbf{x} = \mathbf{0}$. This means that there are no free variables in the solution to $A\mathbf{x} = \mathbf{0}$, so every column of A is a pivot column. There are n columns, so rank $A = n$.

iii $\implies$ i: This is part 3 of Theorem 4.1.24.   ∎

Answers to Reading Challenges

1. $\begin{bmatrix} x \\ y \end{bmatrix} = x \begin{bmatrix} 1 \\ 0 \end{bmatrix} + y \begin{bmatrix} 0 \\ 1 \end{bmatrix} + 0 \begin{bmatrix} 2 \\ 3 \end{bmatrix}.$

2. The plane is the set of all vectors of the form $a\mathbf{a_1} + b\mathbf{a_2} = a \begin{bmatrix} 1 \\ 2 \\ 0 \end{bmatrix} + b \begin{bmatrix} -3 \\ 0 \\ 4 \end{bmatrix} = \begin{bmatrix} a-3b \\ 2a \\ 4b \end{bmatrix}.$ Let $x = a - 3b$, $y = 2a$, and $z = 4b$. Then $a = \frac{1}{2}y$ and $b = \frac{1}{4}z$, so $x = \frac{1}{2}y - \frac{3}{4}z$. The equation is $4x - 2y + 3z = 0$.

3. Remember 2.1.33: A linear combination of the columns of A is a vector of the form $A\mathbf{x}$.

4. The sequence of four implications

   i $\implies$ ii,   ii $\implies$ iii,   iii $\implies$ iv,   iv $\implies$ i

indeed implies all 12 statements. For example, i $\implies$ iii follows because i $\implies$ ii and ii $\implies$ iii, and iv $\implies$ ii holds because iv $\implies$ i, which in turn implies ii.

True/False Questions

Decide, with as little calculation as possible, whether each of the following statements is true or false and explain your answer whenever you say "false." (Answers can be found in the back of the book.)

1. Left multiplication by a $k \times \ell$ matrix defines a function from $\mathbf{R}^k \to \mathbf{R}^\ell$.

2. The column space of $\begin{bmatrix} -1 & 0 & 4 & 1 \\ 0 & 0 & 0 & 0 \\ 0 & 2 & 5 & 1 \end{bmatrix}$ is the xz-plane.

3. The matrix $A = \begin{bmatrix} 1 & 2 & 3 & 4 \\ -7 & 15 & 9 & 18 \\ 2 & -6 & 8 & -1 \end{bmatrix}$ has a nontrivial null space.

4. If an $m \times n$ matrix A has rank m, then $m \le n$.

5. If the columns of a matrix A are linearly independent, the null space of A is **0**.

6. If A and B are matrices of appropriate sizes, rank $AB = (\text{rank } A)(\text{rank } B)$.

7. If A and B are $n \times n$ matrices and $AB = cI$ for some $c \ne 0$, then rank $A = $ rank B.

8. The rank of a matrix is the number of leading 1s in its reduced row echelon form.

9. If $A\mathbf{x} = 2\mathbf{x}$ for some matrix A and vector $\mathbf{x}$, then $\mathbf{x}$ is in the column space of A.

10. If an $m \times n$ matrix A has rank n, then $m \le n$.

11. If A is a 7×5 matrix of rank 5, the linear system $A\mathbf{x} = \mathbf{b}$ has at least one solution for any **b**.

12. The columns of a matrix A are linearly independent if and only if the null space of A consists only of the zero vector.

13. Let $\mathbf{x}_p$ be a particular solution to the system $A\mathbf{x} = \mathbf{b}$. Then any solution can be written in the form $\mathbf{x} = \mathbf{x}_p + \mathbf{x}_h$, where $\mathbf{x}_h$ is a vector in the null space of A.

14. If U is a row echelon form of the matrix A, null sp $A = $ null sp U.

15. If U is a row echelon form of the matrix A, col sp $A = $ col sp U.

16. An $n \times n$ singular matrix has rank less than n.

17. Corollary 4.1.25 says that if A is a square matrix, then the homogeneous system $A\mathbf{x} = \mathbf{0}$ has only the trivial solution.

Exercises

Solutions to exercises marked [BB] can be found in the Back of the Book.

1. [BB] Suppose A is a 7×9 matrix with null space in $\mathbf{R}^n$ and column space in $\mathbf{R}^m$. What are m and n?

2. Is $\begin{bmatrix} 3 \\ 2 \\ 1 \end{bmatrix}$ in the null space of $A = \begin{bmatrix} -1 & 1 & 1 \\ 1 & 1 & -5 \\ 1 & 2 & 3 \end{bmatrix}$?

   What about $\begin{bmatrix} -1 \\ 0 \\ 1 \end{bmatrix}$?

3. Express the null space of each of the following matrices as the set linear combinations of certain vectors. Also give the rank of each matrix.

   (a) [BB] $\begin{bmatrix} 1 & -2 & 1 & 1 \\ 3 & 0 & 2 & -2 \\ 0 & 4 & -1 & -1 \\ 5 & 0 & 3 & -1 \end{bmatrix}$

   (b) $\begin{bmatrix} 0 & 0 & 2 & 3 & 1 \\ 1 & -2 & 2 & 1 & -2 \\ 1 & -2 & 6 & 7 & 0 \end{bmatrix}$

   (c) $\begin{bmatrix} 1 & 1 & 1 & -1 \\ -1 & 1 & -1 & 1 \\ 2 & -1 & 1 & 3 \end{bmatrix}$

   (d) $\begin{bmatrix} 1 & 2 & 3 & 4 \\ 2 & 4 & 8 & 10 \\ 3 & 6 & 11 & 14 \end{bmatrix}$.

4. (a) [BB] Is $\begin{bmatrix} 1 \\ 3 \end{bmatrix}$ in the column space of $\begin{bmatrix} 2 & 0 \\ 2 & 3 \end{bmatrix}$?

(b) Is $\begin{bmatrix} -6 \\ 7 \\ 3 \end{bmatrix}$ in the column space of $\begin{bmatrix} 1 & 1 & 1 \\ 0 & 1 & 1 \\ 0 & 0 & 1 \end{bmatrix}$?

(c) [BB] Is $\begin{bmatrix} 1 \\ 0 \\ 0 \end{bmatrix}$ in the column space of

$$\begin{bmatrix} -1 & 4 \\ 2 & 5 \\ 3 & 6 \end{bmatrix}?$$

(d) Is $\begin{bmatrix} 5 \\ 3 \\ 4 \end{bmatrix}$ in the column space of $\begin{bmatrix} 2 & 2 & -4 \\ 0 & 3 & 5 \\ 1 & 2 & -1 \end{bmatrix}$?

(e) Is $\begin{bmatrix} -4 \\ -6 \\ 1 \\ 0 \end{bmatrix}$ in the column space of $\begin{bmatrix} 1 & 2 \\ -5 & 3 \\ -4 & 9 \\ 2 & 4 \end{bmatrix}$?

Explain your answers.

5. Let $A = \begin{bmatrix} 2 & -1 & 4 \\ 0 & 2 & 3 \\ -1 & -5 & 1 \\ 1 & -3 & 2 \end{bmatrix}$, $b_1 = \begin{bmatrix} 1 \\ 1 \\ 1 \\ 1 \end{bmatrix}$, and

$b_2 = \begin{bmatrix} 1 \\ 1 \\ -1 \\ 0 \end{bmatrix}$.

(a) [BB] Is b_1 in the column space of A? If it is, express this vector as a linear combination of the columns of A. If it is not, explain why.

(b) Same as (a) for b_2.

(c) Find conditions on b_1, b_2, b_3, b_4 that are necessary and sufficient for $b = \begin{bmatrix} b_1 \\ b_2 \\ b_3 \\ b_4 \end{bmatrix}$ to belong to the column space of A.

6. Let $A = \begin{bmatrix} 3 & 1 & 7 & 2 & 1 \\ 2 & -4 & 14 & -2 & 0 \\ 5 & 11 & -7 & 10 & 3 \\ 2 & 5 & -4 & -3 & 1 \end{bmatrix}$, $u = \begin{bmatrix} 1 \\ 2 \\ 5 \\ 9 \end{bmatrix}$,

$v = \begin{bmatrix} 13 \\ -10 \\ 59 \\ 39 \end{bmatrix}$, and $b = \begin{bmatrix} b_1 \\ b_2 \\ b_3 \\ b_4 \end{bmatrix}$.

(a) Find the general solution of $Ax = u$ and of $Ax = v$ [BB], if a solution exists.

(b) Find a condition on b_1, b_2, b_3, b_4 that is necessary and sufficient for $Ax = b$ to have a solution; that is, complete the sentence

$$Ax = b \text{ has a solution if and only if}$$

_____.

(c) The column space of A is a subset of $\mathbf{R}^n$ for what n?

(d) Complete the following sentence: $\begin{bmatrix} x_1 \\ x_2 \\ \vdots \\ x_n \end{bmatrix}$ is in the

column space of A if and only if _____. (Here, n is the n of part (c).)

7. The column space of each of the following matrices is a plane in $\mathbf{R}^3$. Find the equation of this plane in each case. Also find the null space of the given matrix.

(a) [BB] $A = \begin{bmatrix} 2 & 3 & 4 \\ 4 & 7 & 5 \\ 0 & -1 & 3 \end{bmatrix}$

(b) $A = \begin{bmatrix} 1 & 6 & 5 & 7 & -8 \\ -5 & -4 & 1 & -9 & 14 \\ 0 & 2 & 2 & 2 & -2 \end{bmatrix}$

(c) $A = \begin{bmatrix} 1 & 7 & 4 & 3 & 8 \\ 8 & 59 & 35 & 21 & 69 \\ -3 & -24 & -15 & -6 & -29 \end{bmatrix}$

(d) $A = \begin{bmatrix} 2 & -5 & 4 & 3 \\ 6 & -14 & 15 & 7 \\ 4 & -14 & -4 & 14 \end{bmatrix}$.

8. Compute the rank of each of the following matrices.

(a) [BB] $\begin{bmatrix} 1 & 2 \\ 4 & 8 \end{bmatrix}$ **(b)** $\begin{bmatrix} 1 & 2 & 3 \\ 4 & 5 & 6 \\ 7 & 8 & 9 \end{bmatrix}$

(c) $\begin{bmatrix} 1 & 2 & 3 & 5 \\ 1 & 0 & 1 & 3 \\ 2 & 3 & 5 & 9 \end{bmatrix}$

(d) $\begin{bmatrix} -1 & 2 & 3 & 4 & 0 \\ 0 & 1 & 1 & 2 & 1 \\ -3 & 7 & 10 & 14 & 1 \end{bmatrix}$.

9. [BB] Let $A = \begin{bmatrix} 2 & 3 & 4 \\ 4 & 7 & 5 \\ 0 & -1 & 3 \end{bmatrix}$.

(a) Find the null space of A.

(b) What is the rank of A?

(c) Solve $A \begin{bmatrix} x \\ y \\ z \end{bmatrix} = \begin{bmatrix} 1 \\ 6 \\ -4 \end{bmatrix}$.

(d) Express $\begin{bmatrix} 1 \\ 6 \\ -4 \end{bmatrix}$ as a linear combination of the columns of A.

(e) Find the column space of A and describe this in geometrical terms.

10. Let $A = \begin{bmatrix} 1 & 0 & 3 \\ 3 & 1 & 16 \\ 5 & 2 & 29 \end{bmatrix}$.

(a) Find the null space of A.

(b) What is the rank of A?

(c) Solve $A \begin{bmatrix} x \\ y \\ z \end{bmatrix} = \begin{bmatrix} 2 \\ 3 \\ 4 \end{bmatrix}$.

(d) Express $\begin{bmatrix} 2 \\ 3 \\ 4 \end{bmatrix}$ as a linear combination of the columns of A.

(e) Find the column space of A and describe this in geometrical terms.

11. (a) [BB] If $\mathbf{x}$ is in the null space of A, show that $\mathbf{x}$ is in the null space of BA for any (compatible) matrix B.

(b) If $A = BC$ and $\mathbf{x}$ is in the column space of A, show that $\mathbf{x}$ is in the column space of B.

12. [BB] Find the column space and the null space of
$$A = \begin{bmatrix} 1 & 0 & 0 \\ 0 & 1 & 2 \\ 0 & 0 & 1 \end{bmatrix} \text{ without calculations.}$$

13. [BB] Fill in the missing entries of
$$A = \begin{bmatrix} 1 & 2 & 3 & 4 \\ 2 & \star & \star & \star \\ 3 & \star & \star & \star \end{bmatrix} \text{ so that } A \text{ has rank 1.}$$

14. How does the rank of $\begin{bmatrix} 1 & 2 & 3 \\ a & 4 & 6 \end{bmatrix}$ depend on a?

15. [BB] Let $A = \begin{bmatrix} 1 & 2 & 3 & 2 \\ -1 & -2 & -2 & 1 \\ 2 & 4 & 8 & 12 \end{bmatrix}$ and
$$\mathbf{b} = \begin{bmatrix} -1 \\ 7 \\ 4 \end{bmatrix}.$$

(a) Find the null space of A.

(b) Solve $A\mathbf{x} = \mathbf{b}$.

(c) What is the rank of A and why?

(d) Find the column space of A and describe this in geometrical terms.

16. Let $A = \begin{bmatrix} 6 & -12 & -5 & 16 & -2 \\ -3 & 6 & 3 & -9 & 1 \\ -4 & 8 & 3 & -10 & 1 \end{bmatrix}$,
$$\mathbf{u} = \begin{bmatrix} 3 \\ 2 \\ -1 \\ 0 \\ 1 \end{bmatrix}, \text{ and } \mathbf{b} = \begin{bmatrix} -53 \\ 29 \\ 33 \end{bmatrix}.$$

(a) Is $\mathbf{u}$ in the null space of A? Explain.

(b) Is $\mathbf{b}$ in the column space of A? If it is, express $\mathbf{b}$ as a linear combination of the columns of A. If it is not, explain why.

(c) Find the null space of A.

(d) Find the column space of A.

(e) Find rank A.

17. Let $A = \begin{bmatrix} 2 & -8 & 7 & -4 & 3 \\ 6 & -24 & 24 & -7 & 10 \\ 10 & -40 & 41 & -8 & 24 \end{bmatrix}, \mathbf{u} = \begin{bmatrix} 1 \\ -2 \\ -2 \\ 1 \\ 0 \end{bmatrix}$,
and $\mathbf{b} = \begin{bmatrix} -4 \\ 5 \\ 7 \end{bmatrix}$.

(a) Is $\mathbf{u}$ in the null space of A? Explain.

(b) Is $\mathbf{b}$ in the column space of A? If it is, express $\mathbf{b}$ as a linear combination of the columns of A. If it is not, explain why.

(c) Find the null space of A.

(d) Find the column space of A.

(e) Find rank A.

18. [BB] Give an example of a 3×3 matrix whose column space contains $\mathbf{i} = \begin{bmatrix} 1 \\ 0 \\ 0 \end{bmatrix}$ and $\mathbf{j} = \begin{bmatrix} 0 \\ 1 \\ 0 \end{bmatrix}$, but not
$$\mathbf{k} = \begin{bmatrix} 0 \\ 0 \\ 1 \end{bmatrix}.$$

The problems on the left also include:

(a) Find the null space of A.

(b) Solve $A\mathbf{x} = \mathbf{b}$.

19. (a) [BB] If **x** is in the null space of AB, is $B\mathbf{x}$ in the null space of A? Why?

(b) If A is invertible and **x** is in the null space of AB, is **x** in the null space of B? Why?

20. (a) Complete the following sentence:

"An $n \times n$ matrix A does not have an inverse if and only if the matrix equation $A\mathbf{x} = \mathbf{0}$ has _____ solution."

(b) Use your answer to (a) to explain whether the following sentence is true or false.

"If A is a 3×3 matrix such that $A\mathbf{x} = \mathbf{x}$ for some nonzero column vector
$$\mathbf{x} = \begin{bmatrix} x_1 \\ x_2 \\ x_3 \end{bmatrix},$$
then $I - A$ must have an inverse." (Here I denotes the 3×3 identity matrix.)

21. [BB] The column space of $\begin{bmatrix} 2 & -1 & 1 \\ 1 & 2 & 5 \end{bmatrix}$ is $\mathbf{R}^2$. Why?

22. Let $A = \begin{bmatrix} 2 & -1 & -1 & 1 & 2 \\ -2 & 1 & 1 & 0 & -\frac{3}{2} \\ -2 & 2 & 5 & 4 & 0 \\ -6 & 3 & 3 & 1 & -4 \end{bmatrix}$.

(a) Find an equation of the form $PA = LU$, where L is lower triangular with 1s on the diagonal, U is upper triangular, and P is a permutation matrix.

(b) What is the rank of A and why?

(c) Find the null space of A.

(d) Find a condition on b_1, b_2, b_3, b_4 that is necessary and sufficient for $A\mathbf{x} = \begin{bmatrix} b_1 \\ b_2 \\ b_3 \\ b_4 \end{bmatrix}$ to have a solution.

(e) What is the column space of A and why?

23. [BB] Let $A = \begin{bmatrix} 2 & 1 & 5 & 4 \\ 1 & -1 & 1 & -1 \\ 1 & 0 & 2 & 1 \end{bmatrix}$.

(a) Find the rank of A.

(b) For which values of c does the system
$$2x_1 + x_2 + 5x_3 + 4x_4 = c^2 + 2$$
$$x_1 - x_2 + x_3 - x_4 = 4c + 5$$
$$x_1 \quad\quad + 2x_3 + x_4 = c + 3$$

have a solution? For each such c, find all solutions as linear combinations of certain vectors.

■ Critical Reading

24. [BB] Suppose A and B are matrices and $A = ABA$. Show that A and AB have the same column space.

25. What can you say about a matrix of rank 0?

26. (a) Let x be a real number and
$$A = \begin{bmatrix} 1 & x & x^2 \\ x & x^2 & 1 \\ x^2 & 1 & x \end{bmatrix}.$$
Explain how the rank of A depends on x.

(b) Find a necessary and sufficient condition on c which guarantees that the system
$$x + cy + c^2z = b_1$$
$$cx + c^2y + z = b_2$$
$$c^2x + y + cz = b_3$$
has a solution for all scalars b_1, b_2, b_3. Explain your answer.

(c) Determine whether the system
$$x + cy + c^2z = 1$$
$$cx + c^2y + z = 1$$
$$c^2x + y + cz = 1$$
has a solution. Does your answer depend on c? Explain.

27. [BB] The eigenspace of a matrix A corresponding to the eigenvalue λ is the null space of $A - \lambda I$. Explain.

28. Can a 1×3 matrix A have scalar multiples of $\begin{bmatrix} 1 \\ 0 \\ 1 \end{bmatrix}$ as its null space? Explain.

29. Suppose A is an $n \times n$ invertible matrix. What is the column space of A. Why?

30. If A is a square matrix whose column space is contained in its null space, what can you say about A^2? Why?

31. Suppose A is an $m \times n$ matrix and its null space is $\mathbf{R}^n$. What can you conclude about A, and why?

32. Prove that an $n \times n$ matrix A is invertible if and only if the homogeneous system $A\mathbf{x} = \mathbf{0}$ has only the trivial solution.

33. **(a)** Suppose the system $A\mathbf{x} = \mathbf{0}$ has only the trivial solution. Show that **if** $A\mathbf{x} = \mathbf{b}$ has a solution, then this solution is unique.

(b) Suppose A is an $m \times n$ matrix such that $A\mathbf{x} = \mathbf{0}$ has only the trivial solution. Suppose also that $A\mathbf{x} = \mathbf{b}$ has no solution for some $\mathbf{b}$. What can you conclude about the relative sizes of m and n, and why?

(c) Give an example of a matrix A and a vector $\mathbf{b}$ such that $A\mathbf{x} = \mathbf{0}$ has only the trivial solution and $A\mathbf{x} = \mathbf{b}$ has **no** solution.

34. **(a)** Suppose A is an $n \times n$ matrix and $\mathbf{x}^T A \mathbf{x} > 0$ for every nonzero $\mathbf{x} \in \mathbf{R}^n$. Show that all eigenvalues of A are positive.

(b) Let B be an $m \times n$ matrix of rank n. Show that the eigenvalues of $B^T B$ are positive.

4.2 Basic Terminology and Concepts

Much of this chapter should be easy reading since we review ideas and concepts discussed at length in previous sections that should have firmly settled in your head by now.

In Section 1.5, we introduced *Euclidean n-space*—$\mathbf{R}^n$—which is the set of all n-dimensional vectors:

$$\mathbf{R}^n = \{ \begin{bmatrix} x_1 \\ x_2 \\ \vdots \\ x_n \end{bmatrix} \mid x_1, x_2, \dots, x_n \in \mathbf{R}\}.$$

Euclidean n-space is an example, and in a very real sense the only example, of a *finite dimensional vector space*.

4.2.1 DEFINITION

A *vector space* is a nonempty set V of objects called vectors that can be added and multiplied by scalars in special ways. For vectors $\mathbf{u}$, $\mathbf{v}$, and $\mathbf{w}$ and scalars c and d,

"Nonempty" means that there is at least one vector in the set.

1. (Closure under addition) $\mathbf{u} + \mathbf{v}$ is in V.
2. (Commutativity of addition) $\mathbf{u} + \mathbf{v} = \mathbf{v} + \mathbf{u}$.
3. (Associativity of addition) $(\mathbf{u} + \mathbf{v}) + \mathbf{w} = \mathbf{u} + (\mathbf{v} + \mathbf{w})$.
4. (Zero) There is an element denoted $\mathbf{0}$ in V and called the *zero vector* with the property that $\mathbf{0} + \mathbf{v} = \mathbf{v}$ for any $\mathbf{v} \in V$.
5. (Negatives) There exists a vector called the *negative* of $\mathbf{v}$ and denoted $-\mathbf{v}$ with the property that $\mathbf{v} + (-\mathbf{v}) = \mathbf{0}$.
6. (Closure under scalar multiplication) $c\mathbf{v}$ is in V.
7. (Scalar associativity) $c(d\mathbf{v}) = (cd)\mathbf{v}$.
8. (One) $1\mathbf{v} = \mathbf{v}$.
9. (Distributivity) $c(\mathbf{v} + \mathbf{w}) = c\mathbf{v} + c\mathbf{w}$ and $(c + d)\mathbf{v} = c\mathbf{v} + d\mathbf{v}$.

These are exactly the properties of two-dimensional vectors listed in Theorem 1.1.6 and observed also to hold in $\mathbf{R}^3$ and, more generally, in $\mathbf{R}^n$. They are exactly the properties of matrix addition and scalar multiplication recorded in Theorem 2.1.15 and are not so very difficult to remember.

The first five properties deal just with addition and the next three just with scalar multiplication; the last (distributivity) is the only one that involves both addition and scalar multiplication.

4.2.2 Remark As in previous sections of this book, the word "scalar" means "real number." In Chapter 7, scalars can be complex numbers. In fact, scalars can be restricted to any "field," a special kind of algebraic system where you can add, subtract, multiply, and divide (by nonzero elements). Scalars can be just 0s and 1s, for instance, with addition and multiplication "mod 2." There are very few results in this book that require scalars to be arbitrary real numbers.

4.2.3 EXAMPLES

1. Euclidean n-space, $\mathbf{R}^n$.

2. The set of $m \times n$ matrices over $\mathbf{R}$, where m and n are positive integers. In this vector space, "vectors" are matrices and the zero "vector" is the $m \times n$ matrix, all of whose entries are 0. If $A = [a_{ij}]$, then $-A = [-a_{ij}]$ is the matrix whose entries are the negatives of those of A.

3. As a special case of the previous example, take $m = 1$ so that our matrices are row matrices. If $\mathbf{a}$ and $\mathbf{b}$ are row matrices, say

$$\mathbf{a} = [a_1 \ a_2 \ \cdots \ a_n] \quad \text{and} \quad \mathbf{b} = [b_1 \ b_2 \ \cdots \ b_n],$$

then

$$\mathbf{a} + \mathbf{b} = [a_1 + b_1 \ a_2 + b_2 \ \cdots \ a_n + b_n]$$

and, for any scalar c,

$$c\mathbf{a} = [ca_1 \ ca_2 \ \cdots \ ca_n].$$

In this case, the zero vector is

$$\mathbf{0} = [0 \ 0 \ \cdots \ 0]$$

and the negative of $\mathbf{a}$ is

$$-\mathbf{a} = [-a_1 \ -a_2 \ \cdots \ -a_n].$$

4. The set of polynomials over $\mathbf{R}$ in a variable x with usual addition and scalar multiplication; that is, if

$$f(x) = a_0 + a_1 x + a_2 x^2 + \cdots + a_n x^n$$

and

$$g(x) = b_0 + b_1 x + b_2 x^2 + \cdots + b_m x^m,$$

then addition and scalar multiplication by c are defined by

$$f(x) + g(x) = (a_0 + b_0) + (a_1 + b_1)x + (a_2 + b_2)x^2 + \cdots$$
$$cf(x) = ca_0 + ca_1 x + ca_2 x^2 + \cdots + ca_n x^n.$$

In this vector space, "vectors" are polynomials and the zero vector is the polynomial $0 + 0x + 0x^2 + \cdots$, all of whose coefficients are zero. Also, if $f(x) = a_0 + a_1x + a_2x^2 + \cdots + a_nx^n$, then

$$-f(x) = -a_0 - a_1x - a_2x^2 - \cdots - a_nx^n.$$

5. The set of all functions $\mathbf{R} \to \mathbf{R}$ with "usual" addition and scalar multiplication; that is, $f + g: \mathbf{R} \to \mathbf{R}$ is defined by

$$(f + g)(x) = f(x) + g(x), \quad \text{for any real number } x$$

and, for a scalar c, $cf: \mathbf{R} \to \mathbf{R}$ is defined by

$$(cf)(x) = c[f(x)], \quad \text{for any real number } x.$$

6. The set of continuous functions $\mathbf{R} \to \mathbf{R}$ with usual addition and scalar multiplication.

7. The set of differentiable functions $\mathbf{R} \to \mathbf{R}$ with usual addition and scalar multiplication.

See the discussion of coordinate vectors in Chapter 5 and especially 5.3.6.

There are many other examples, but those that are "finite dimensional" are, as mentioned before, essentially $\mathbf{R}^n$. For this reason, in this book, the word "vector" almost always means "column vector," so that "vector space" means "Euclidean n-space." Occasionally, "vector" will mean "row vector," that is, a $1 \times n$ matrix as described in Example **4.2.3**(3).

4.2.4 DEFINITION

A *subspace* of a vector space V is a subset U of V that is itself a vector space, under the operations of addition and scalar multiplication it inherits from V.

Lemma 4.2.5

A subset U of a vector space V is a subspace if and only if it is

- *nonempty,*
- *closed under addition: if* **u** *and* **v** *are in U, so is* **u** + **v**, *and*
- *closed under scalar multiplication: if* **u** *is in U and c is a scalar, then c**u** is also in U.*

Proof If U is a subspace, then U is nonempty because it contains at least the zero vector, and it is closed under addition and scalar multiplication since these are two of the nine axioms for a vector space (see Definition 4.2.1).

Conversely, suppose U is a nonempty subset of a vector space V that is closed under addition and scalar multiplication. We must show that U is a subspace, hence that U satisfies the nine axioms of Definition 4.2.1. By hypothesis, U satisfies axioms 1 and 6. Axioms 2, 3, 7, 8, and 9 hold in U because they hold in V. This leaves axioms 4 and 5 to establish.

First, we show that U contains the zero vector. Since U is not empty, there is some **u** in U. Thus $\mathbf{0} = 0\mathbf{u}$ is in U because U is closed under scalar multiplication. Finally, we must show that U contains the negative of each of its vectors. This is equally straightforward since if $\mathbf{u}$ is in U, so is $-\mathbf{u} = (-1)\mathbf{u}$, because U is closed under scalar multiplication. ∎

Lemma 4.2.5 is both important and useful. In practice, to show that a subset of a vector space is a subspace, we have only three, not nine, axioms to check.

4.2.6 EXAMPLE

Every vector space V contains at least two subspaces, namely, V itself and (at the other extreme) the one element set $\{\mathbf{0}\}$ consisting of just the zero vector. Neither V nor $\{\mathbf{0}\}$ is empty; both are closed under addition and scalar multiplication. ∎

Reading Challenge 1 *Prove that $\{\mathbf{0}\}$ is closed under addition and scalar multiplication.*

4.2.7 EXAMPLE

The set $U = \left\{ \begin{bmatrix} x \\ x - y \\ y \end{bmatrix} \mid x, y \in \mathbf{R} \right\}$ is a subspace of $\mathbf{R}^3$. Note that U consists precisely of those vectors whose middle component is the first minus the third. This set is nonempty because it contains $\mathbf{0} = \begin{bmatrix} 0 \\ 0 - 0 \\ 0 \end{bmatrix}$. It is closed under addition because if $\mathbf{u}_1 = \begin{bmatrix} x_1 \\ x_1 - y_1 \\ y_1 \end{bmatrix}$ and $\mathbf{u}_2 = \begin{bmatrix} x_2 \\ x_2 - y_2 \\ y_2 \end{bmatrix}$ are in U, so is their sum, because $\mathbf{u}_1 + \mathbf{u}_2 = \begin{bmatrix} x_1 + x_2 \\ (x_1 + x_2) - (y_1 + y_2) \\ y_1 + y_2 \end{bmatrix}$. It is closed under scalar multiplication because if $\mathbf{u} = \begin{bmatrix} x \\ x - y \\ y \end{bmatrix}$ is in U and c is a scalar, then $c\mathbf{u}$ is in U because $c\mathbf{u} = \begin{bmatrix} cx \\ cx - cy \\ cy \end{bmatrix}$. ∎

4.2.8 EXAMPLE

The set U of all symmetric $n \times n$ matrices is a subspace of the vector space V of all $n \times n$ matrices. This set is not empty because, for example, the $n \times n$ identity matrix is symmetric. It is closed under addition since, if A and B are symmetric, then $(A + B)^T = A^T + B^T = A + B$, so $A + B$ is also symmetric. It is closed under scalar multiplication since, if A is symmetric and c is a scalar, $(cA)^T = cA^T = cA$ is symmetric as well. ∎

4.2.9 EXAMPLE

The set $U = \left\{ \begin{bmatrix} x_1 \\ x_2 \end{bmatrix} \in \mathbf{R}^2 \mid x_1 = 0 \text{ or } x_2 = 0 \right\}$ is not a subspace of $\mathbf{R}^2$. While it is closed under scalar multiplication, it is **not** closed under addition. For example, $\mathbf{u} = \begin{bmatrix} 1 \\ 0 \end{bmatrix}$ is in U and $\mathbf{v} = \begin{bmatrix} 0 \\ 1 \end{bmatrix}$ is in U, but $\mathbf{u} + \mathbf{v} = \begin{bmatrix} 1 \\ 1 \end{bmatrix}$ is not in U. ∎

Reading Challenge 2 *Why is the set U in **4.2.9** closed under scalar multiplication?*

4.2.10 EXAMPLE

The set $U = \{\begin{bmatrix} x \\ x-y+1 \\ y \end{bmatrix} \mid x, y \in \mathbf{R}\}$ is not a subspace of $\mathbf{R}^3$ since it is not closed under addition. For example, $\mathbf{u}_1 = \begin{bmatrix} 1 \\ 1 \\ 1 \end{bmatrix}$ and $\mathbf{u}_2 = \begin{bmatrix} 2 \\ 2 \\ 1 \end{bmatrix}$ are in U, but their sum $\mathbf{u}_1 + \mathbf{u}_2 = \begin{bmatrix} 3 \\ 3 \\ 2 \end{bmatrix}$ is not in U because its middle component is not $3 - 2 + 1$. ∎

In this last example, we could equally have observed that U is not a subspace because it is not closed under scalar multiplication. For example, the vector $\mathbf{u} = \begin{bmatrix} 1 \\ 1 \\ 1 \end{bmatrix}$ is in U but $5\mathbf{u} = \begin{bmatrix} 5 \\ 5 \\ 5 \end{bmatrix}$ is not, because the middle component of this vector is not $5 - 5 + 1$. Alternatively, U is not a subspace because it does not contain the zero vector. So there are (at least) three possible ways to see that U is not a subspace of $\mathbf{R}^3$, any of which provides a perfectly acceptable explanation.

4.2.11 EXAMPLE

The set $U = \{\begin{bmatrix} x \\ x+1 \end{bmatrix} \mid x \in \mathbf{R}\}$ of vectors on the line with equation $y = x + 1$ is not a subspace of $\mathbf{R}^2$ since it does not contain the zero vector. ∎

4.2.12 PROBLEM

Matthew says $\mathbf{R}^2$ is not a subspace of $\mathbf{R}^3$ because vectors in $\mathbf{R}^2$ have two components while vectors in $\mathbf{R}^3$ have three. Jayne, on the other hand, says this is just a minor irritation: "Everybody knows the xy-plane is part of 3-space." Who's right?

Solution. Technically, Matthew is correct, but Jayne has given the mature answer. Writing the vector $\begin{bmatrix} x \\ y \end{bmatrix}$ as $\begin{bmatrix} x \\ y \\ 0 \end{bmatrix}$ allows us to imagine $\mathbf{R}^2$ as part of $\mathbf{R}^3$. With this understanding, $\mathbf{R}^2$ is indeed a subspace of $\mathbf{R}^3$, and indeed $\mathbf{R}^m$ is a subspace of $\mathbf{R}^n$ as long as $m \leq n$.

4.2.13 Remark It is useful and important to be rather flexible about what you think of as "Euclidean n-space." Suppose two people are asked to draw a picture that illustrates the points in the plane that satisfy $x \geq 0$, $y \geq 0$, $x \leq 3 - 3y$. Mary goes to the front blackboard and produces a picture like the one shown in Figure 4.3, while Johnny draws more or less the same picture on a blackboard on

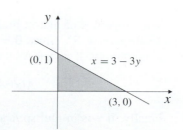

Figure 4.3

the side wall. Has each student drawn a picture in $\mathbf{R}^2$? Most of us would say yes, while acknowledging that the planes in which the pictures were drawn were not the same. Are both planes "essentially" $\mathbf{R}^2$? Surely, yes. Does each picture make the point that $\mathbf{R}^2$ is a subspace of $\mathbf{R}^3$? Surely, yes.

4.2.14 PROBLEM Show that $U = \{\begin{bmatrix} x \\ y \\ z \end{bmatrix} \mid 2x - y + 3z = 0\}$ is a subspace of $\mathbf{R}^3$.

Solution. The set U is a plane containing the origin, so it is a subspace, as already observed. Here is another argument.

The set U is nonempty because it contains $\mathbf{0} = \begin{bmatrix} 0 \\ 0 \\ 0 \end{bmatrix}$; indeed, $2(0)-0+3(0)=0$.

If $\mathbf{v}_1 = \begin{bmatrix} x_1 \\ y_1 \\ z_1 \end{bmatrix}$ and $\mathbf{v}_2 = \begin{bmatrix} x_2 \\ y_2 \\ z_2 \end{bmatrix}$ are in U, then

$$2x_1 - y_1 + 3z_1 = 0 \text{ and } 2x_2 - y_2 + 3z_2 = 0.$$

Adding these equations gives

$$2(x_1 + y_1) - (x_2 + y_2) + 3(z_1 + z_2) = 0,$$

which says that $\mathbf{v}_1 + \mathbf{v}_2$ is in the plane. Finally, if $\mathbf{v} = \begin{bmatrix} x \\ y \\ z \end{bmatrix}$ is in the plane and c is a scalar, then $2x - y + 3z = 0$, so $2(cx) - (cy) + 3(cz) = c(2x - y + 3z) = 0$, and $c\mathbf{v}$ is in the plane too.

Reading Challenge 3 *The set* $U = \{\begin{bmatrix} x \\ y \\ z \end{bmatrix} \mid 2x - y + 3z = 1\}$ *is not a subspace of* $\mathbf{R}^3$. *Why not?*

4.2.15 EXAMPLE **(The null space of a matrix).** In Section 4.1, we defined the *null space* of a matrix. If A is an $m \times n$ matrix,

$$\text{null sp } A = \{\mathbf{x} \in \mathbf{R}^n \mid A\mathbf{x} = \mathbf{0}\}.$$

The properties of matrix multiplication allow us to show that the null space of A is a subspace of $\mathbf{R}^n$. Since $\mathbf{0} = A\mathbf{0}$, the zero vector of $\mathbf{R}^n$ is in null sp A, so null sp A is not empty. If $\mathbf{x}_1$ and $\mathbf{x}_2$ are in null sp A, so is their sum since

$$A(\mathbf{x}_1 + \mathbf{x}_2) = A\mathbf{x}_1 + A\mathbf{x}_2 = \mathbf{0} + \mathbf{0} = \mathbf{0}.$$

Finally, if $\mathbf{x}$ is in null sp A and c is a scalar,

$$A(c\mathbf{x}) = cA\mathbf{x} = c\mathbf{0} = \mathbf{0},$$

which shows that $c\mathbf{x}$ is in null sp A too. Since null sp A is not empty and closed under both addition and scalar multiplication, it is a subspace of $\mathbf{R}^n$. ■

4.2.16 EXAMPLE

(Lines through 0 are subspaces). In any vector space, V, "line" means the set

$$\mathbf{Rv} = \{a\mathbf{v} \mid a \in \mathbf{R}\} \qquad (1)$$

of all scalar multiples of some vector $\mathbf{v}$. In $\mathbf{R}^2$ or $\mathbf{R}^3$, a "line" is a line (in the usual sense of the word) that contains the origin, since $\mathbf{0} = 0\mathbf{v}$. The set defined in (1) is not empty because it contains (for example) the zero vector; $\mathbf{0} = 0\mathbf{v}$ is a scalar multiple of $\mathbf{v}$. It is closed under addition because if $a_1\mathbf{v}$ and $a_2\mathbf{v}$ are in $\mathbf{Rv}$, so is their sum, because $a_1\mathbf{v} + a_2\mathbf{v} = (a_1 + a_2)\mathbf{v}$ is again a scalar multiple of $\mathbf{v}$. It is closed under scalar multiplication because if $a\mathbf{v}$ is in $\mathbf{Rv}$ and c is a scalar, then $c(a\mathbf{v})$ is in $\mathbf{Rv}$ because $c(a\mathbf{v}) = (ca)\mathbf{v}$ is also a scalar multiple of $\mathbf{v}$. ■

4.2.17 EXAMPLE

(Planes through 0 are subspaces). Let $\mathbf{u}$ and $\mathbf{v}$ be nonparallel vectors in a vector space V. The *plane spanned by* $\mathbf{u}$ *and* $\mathbf{v}$ is defined to be the set

$$U = \{a\mathbf{u} + b\mathbf{v} \mid a, b \in \mathbf{R}\}$$

of all linear combinations of $\mathbf{u}$ and $\mathbf{v}$. We first met this concept in Section 1.4. If $V = \mathbf{R}^3$, then U is a real plane—see 1.1.20—so it seems sensible to use this term even in a general vector space. The plane U spanned by two vectors is always a subspace. It is nonempty since $\mathbf{0} = 0\mathbf{u} + 0\mathbf{v}$ is in U. If $\mathbf{x} = a\mathbf{u} + b\mathbf{v}$ and $\mathbf{y} = c\mathbf{u} + d\mathbf{v}$ are in U, so is their sum since

$$\mathbf{x} + \mathbf{y} = (a + c)\mathbf{u} + (b + d)\mathbf{v}$$

is also a linear combination of $\mathbf{u}$ and $\mathbf{v}$, so U is closed under addition. Finally, U is closed under scalar multiplication since if $\mathbf{x} = a\mathbf{u} + b\mathbf{v}$ is in U and c is a scalar, then

$$c\mathbf{x} = (ca)\mathbf{u} + (cb)\mathbf{v}$$

is a linear combination of $\mathbf{u}$ and $\mathbf{v}$, hence also in U. ■

■ The Span of Vectors

Lines and planes (through $\mathbf{0}$) are special cases of a more general kind of subspace that we now describe.

Recall that a *linear combination* of vectors $\mathbf{u}_1, \mathbf{u}_2, \ldots, \mathbf{u}_n$ is a sum of scalar multiples of these vectors, that is, a vector of the form

$$c_1\mathbf{u}_1 + c_2\mathbf{u}_2 + \cdots + c_n\mathbf{u}_n,$$

where $c_1, c_2, \ldots, c_n$ are scalars. (See Section 1.1.)

4.2.18 DEFINITION

The *span* of vectors $u_1, u_2, \ldots, u_n$ is the set of all linear combinations of them. It is denoted $\text{sp}\{u_1, \ldots, u_n\}$. Thus

$$\text{sp}\{u_1, \ldots, u_n\} = \{c_1 u_1 + c_2 u_2 + \cdots + c_n u_n \mid c_1, \ldots, c_n \in \mathbf{R}\}.$$

The vectors $u_1, u_2, \ldots, u_n$ are called a *spanning set* for U.

4.2.19 EXAMPLE

The vector space $\mathbf{R}^n$ is the span of the standard basis vectors

$$e_1 = \begin{bmatrix} 1 \\ 0 \\ 0 \\ \vdots \\ 0 \end{bmatrix}, \quad e_2 = \begin{bmatrix} 0 \\ 1 \\ 0 \\ \vdots \\ 0 \end{bmatrix}, \ldots, e_n = \begin{bmatrix} 0 \\ 0 \\ \vdots \\ 0 \\ 1 \end{bmatrix} \tag{2}$$

because the most general vector $x = \begin{bmatrix} x_1 \\ x_2 \\ \vdots \\ x_n \end{bmatrix}$ of $\mathbf{R}^n$ is a linear combination of $e_1, \ldots, e_n$:

$$\begin{bmatrix} x_1 \\ x_2 \\ x_3 \\ \vdots \\ x_n \end{bmatrix} = \begin{bmatrix} x_1 \\ 0 \\ 0 \\ \vdots \\ 0 \end{bmatrix} + \begin{bmatrix} 0 \\ x_2 \\ 0 \\ \vdots \\ 0 \end{bmatrix} + \cdots + \begin{bmatrix} 0 \\ 0 \\ \vdots \\ 0 \\ x_n \end{bmatrix}$$

$$= x_1 \begin{bmatrix} 1 \\ 0 \\ 0 \\ \vdots \\ 0 \end{bmatrix} + x_2 \begin{bmatrix} 0 \\ 1 \\ 0 \\ \vdots \\ 0 \end{bmatrix} + \cdots + x_n \begin{bmatrix} 0 \\ 0 \\ \vdots \\ 0 \\ 1 \end{bmatrix} = x_1 e_1 + x_2 e_2 + \cdots + x_n e_n. \quad \blacksquare$$

4.2.20 EXAMPLE

In $\mathbf{R}^3$, the span of $e_1 = \begin{bmatrix} 1 \\ 0 \\ 0 \end{bmatrix}$ and $e_2 = \begin{bmatrix} 0 \\ 1 \\ 0 \end{bmatrix}$ is the set of all vectors of the form

$x e_1 + y e_2 = \begin{bmatrix} x \\ y \\ 0 \end{bmatrix}$, that is, the xy-plane. The set $\{e_1, e_2\}$ is a spanning set for the xy-plane. $\blacksquare$

4.2.21 EXAMPLE

The span of a single vector v is $\{av \mid a \in \mathbf{R}\} = \mathbf{R}v$ is just the set of all scalar multiples of v, a set we have called the "line through 0." $\blacksquare$

4.2.22 EXAMPLE

The span of two nonparallel vectors u and v is the set of all linear combinations of u and v, a set we have called a "plane through 0." $\blacksquare$

The word "span" is used both as a noun and as a verb. Suppose $U = \text{sp}\{u_1, \ldots, u_n\}$ is the span of vectors $u_1, \ldots, u_n$. Here we have used "span" as a noun, but we also say that $\mathbf{u}_1, \ldots, \mathbf{u}_n$ *span* U or that U is *spanned by* $\mathbf{u}_1, \ldots, \mathbf{u}_n$. Here "span" has been used as a verb.

4.2.23 EXAMPLE The plane with equation $2x - y + 3z = 0$ is spanned by $\mathbf{u} = \begin{bmatrix} 1 \\ 2 \\ 0 \end{bmatrix}$ and $\mathbf{v} = \begin{bmatrix} 0 \\ 3 \\ 1 \end{bmatrix}$. In fact, any two nonparallel vectors in this plane span it. ∎

4.2.24 Remark There is nothing unique about a spanning set. For example, as noted in 4.2.19, the three standard basis vectors $\mathbf{i} = \begin{bmatrix} 1 \\ 0 \\ 0 \end{bmatrix}$, $\mathbf{j} = \begin{bmatrix} 0 \\ 1 \\ 0 \end{bmatrix}$, and $\mathbf{k} = \begin{bmatrix} 0 \\ 0 \\ 1 \end{bmatrix}$ give a spanning set for $\mathbf{R}^3$, but so do $\mathbf{i}, \mathbf{j}, \mathbf{k}, \begin{bmatrix} 4 \\ 5 \\ 8 \end{bmatrix}$ since any vector in $\mathbf{R}^3$ is a linear combination of these four vectors:

$$\begin{bmatrix} x \\ y \\ z \end{bmatrix} = x\begin{bmatrix} 1 \\ 0 \\ 0 \end{bmatrix} + y\begin{bmatrix} 0 \\ 1 \\ 0 \end{bmatrix} + z\begin{bmatrix} 0 \\ 0 \\ 1 \end{bmatrix} + 0\begin{bmatrix} 4 \\ 5 \\ 8 \end{bmatrix}.$$

Reading Challenge 4 *Suppose* $\mathbf{u}$ *and* $\mathbf{v}$ *are vectors. Show that the span of* $\mathbf{u}$, $\mathbf{v}$, *and* $\mathbf{0}$ *is the same as the span of* $\mathbf{u}$ *and* $\mathbf{v}$.

4.2.25 EXAMPLE A more interesting spanning set for $\mathbf{R}^3$ is $\{\begin{bmatrix} 1 \\ 2 \\ 3 \end{bmatrix}, \begin{bmatrix} -1 \\ 3 \\ 5 \end{bmatrix}, \begin{bmatrix} 7 \\ 10 \\ -1 \end{bmatrix}\}$. Why is this a spanning set? We have to show that given any $\begin{bmatrix} x \\ y \\ z \end{bmatrix}$, there exist scalars a, b, c so that

$$\begin{bmatrix} x \\ y \\ z \end{bmatrix} = a\begin{bmatrix} 1 \\ 2 \\ 3 \end{bmatrix} + b\begin{bmatrix} -1 \\ 3 \\ 5 \end{bmatrix} + c\begin{bmatrix} 7 \\ 10 \\ -1 \end{bmatrix}. \tag{3}$$

In matrix form, this is

$$\begin{bmatrix} x \\ y \\ z \end{bmatrix} = \begin{bmatrix} 1 & -1 & 7 \\ 2 & 3 & 10 \\ 3 & 5 & -1 \end{bmatrix}\begin{bmatrix} a \\ b \\ c \end{bmatrix}. \tag{4}$$

It is easy to verify that the coefficient matrix here has rank 3, so the system has a (unique) solution for any x, y, z, by part 3 of Theorem 4.1.24. ∎

Reading Challenge 5 *Explain equation* (4).

We have shown that the plane spanned by two vectors is a subspace. More generally, the span of any number of vectors is a subspace.

Proposition 4.2.26

The span of any set of vectors is a subspace.

Proof The proof is a straightforward generalization of that provided in the case of two vectors. (See the discussion of planes through $\mathbf{0}$ in **4.2.17**.) Let $U = \mathrm{sp}\{\mathbf{u}_1, \mathbf{u}_2, \ldots, \mathbf{u}_n\}$ be the span of the vectors $\mathbf{u}_1, \mathbf{u}_2, \ldots, \mathbf{u}_n$. Since $\mathbf{0} = 0\mathbf{u}_1 + 0\mathbf{u}_2 + \cdots + 0\mathbf{u}_n$ is a linear combination of $\mathbf{u}_1, \ldots, \mathbf{u}_n$, the set U contains the zero vector. To show that U is closed under addition, we take two general vectors from U, say

$$\mathbf{x} = x_1\mathbf{u}_1 + x_2\mathbf{u}_2 + \cdots + x_n\mathbf{u}_n$$

and

$$\mathbf{y} = y_1\mathbf{u}_1 + y_2\mathbf{u}_2 + \cdots + y_n\mathbf{u}_n,$$

and note that

$$\mathbf{x} + \mathbf{y} = (x_1 + y_1)\mathbf{u}_1 + (x_2 + y_2)\mathbf{u}_2 + \cdots + (x_n + y_n)\mathbf{u}_n$$

is again a linear combination of $\mathbf{u}_1, \ldots, \mathbf{u}_n$, and hence in U. To show that U is closed under scalar multiplication, we take a vector

$$\mathbf{x} = x_1\mathbf{u}_1 + x_2\mathbf{u}_2 + \cdots + x_n\mathbf{u}_n$$

in U and a scalar c and note that

$$c\mathbf{x} = c(x_1\mathbf{u}_1 + x_2\mathbf{u}_2 + \cdots + x_n\mathbf{u}_n) = (cx_1)\mathbf{u}_1 + (cx_2)\mathbf{u}_2 + \cdots + (cx_n)\mathbf{u}_n$$

is again a linear combination of $\mathbf{u}_1, \ldots, \mathbf{u}_n$, and hence in U. ∎

Almost always, to show that an alleged subspace is not empty, we show that it contains $\mathbf{0}$. If we can't do that, we are in big trouble!

■ The Column Space and the Row Space of a Matrix

We have met specific instances of spanning sets in previous sections. In Section 4.1, we defined the column space of a matrix A as the set of all linear combinations of the columns of A and denoted this col sp A. Thus col sp A is the span of the columns of A. If A is $m \times n$, the columns are vectors in $\mathbf{R}^m$, thus

4.2.27 The column space of an $m \times n$ matrix is a subspace of $\mathbf{R}^m$.

In a similar fashion, we define the *row space* of a matrix A to be the set of all linear combinations of its rows, that is, the span of the rows of A, thought of as row vectors, as in Example 3 of **4.2.3**. This is denoted row sp A. Since spanning sets are always subspaces,

4.2.28 The row space of an $m \times n$ matrix A is a subspace of a vector space just like $\mathbf{R}^n$, but where the vectors are row vectors.

Given a matrix A, there are three *elementary row operations* we can perform, namely,

1. the interchange of two rows,
2. the multiplication of a row by a nonzero scalar, and
3. the replacement of a row by that row less a multiple of another row.

We wish to show that these operations do not change the row space of a matrix and illustrate assuming that

$$A = \begin{bmatrix} \mathbf{u}_1 & \rightarrow \\ \mathbf{u}_2 & \rightarrow \\ \mathbf{u}_3 & \rightarrow \end{bmatrix}$$

has just three rows, $\mathbf{u}_1, \mathbf{u}_2, \mathbf{u}_3$. Thus the row space of A is the subspace spanned by $\mathbf{u}_1, \mathbf{u}_2, \mathbf{u}_3$:

$$\text{row sp } A = \text{sp}\{\mathbf{u}_1, \mathbf{u}_2, \mathbf{u}_3\}.$$

Suppose we interchange the first two rows of A to get a matrix B. The row space of B is $\text{sp}\{\mathbf{u}_2, \mathbf{u}_1, \mathbf{u}_3\}$, which is surely the same as row sp A—any linear combination of $\mathbf{u}_2, \mathbf{u}_1, \mathbf{u}_3$ is a linear combination of $\mathbf{u}_1, \mathbf{u}_2, \mathbf{u}_3$, and conversely. This argument shows that **interchanging rows does not change the row space.**

The notation $X \subseteq Y$ means "Set X is contained in set Y."

Suppose B is obtained from A by multiplying the third row by 2. The row space of B is $\text{sp}\{\mathbf{u}_1, \mathbf{u}_2, 2\mathbf{u}_3\}$, which we assert is also the same as row sp A. The fact that

$$a\mathbf{u}_1 + b\mathbf{u}_2 + c(2\mathbf{u}_3) = a\mathbf{u}_1 + b\mathbf{u}_2 + (2c)\mathbf{u}_3$$

shows that row sp $B \subseteq$ row sp A, and the fact that

$$a\mathbf{u}_1 + b\mathbf{u}_2 + c\mathbf{u}_3 = a\mathbf{u}_1 + b\mathbf{u}_2 + \frac{c}{2}(2\mathbf{u}_3)$$

To show sets X and Y are equal, we almost always show that $X \subseteq Y$ and $Y \subseteq X$.

shows that row sp $A \subseteq$ row sp B, so the row spaces of A and B are the same. **Multiplying a row by a nonzero scalar does not change the row space.**

Finally, suppose B is obtained from A by replacing the second row of A by the sum of this row and 4 times the first: $R2 \rightarrow R2 + 4(R1)$. The row space of B is $\text{sp}\{\mathbf{u}_1, \mathbf{u}_2 + 4\mathbf{u}_1, \mathbf{u}_3\}$ and, again, it is not hard to see that this is the row space of A. The fact that

$$a\mathbf{u}_1 + b(\mathbf{u}_2 + 4\mathbf{u}_1) + c\mathbf{u}_3 = (a + 4b)\mathbf{u}_1 + b\mathbf{u}_2 + c\mathbf{u}_3$$

shows that row sp $B \subseteq$ row sp A, and the equation

$$a\mathbf{u}_1 + b\mathbf{u}_2 + c\mathbf{u}_3 = (a - 4b)\mathbf{u}_1 + b(\mathbf{u}_2 + 4\mathbf{u}_1) + c\mathbf{u}_3$$

shows that row sp $A \subseteq$ row sp B. So the row spaces of A and B are the same. **Replacing a row by that row less a multiple of another row does not change the row space.**

In the author's view, generalizing the arguments just presented in the $m = 3$ case to the case of general m contributes little to the learning process, so we state with no further comment:

> **4.2.29** The elementary row operations do not change the row space of a matrix. Thus the row space of a matrix A is the row space of any matrix obtained from A by performing elementary row operations.

4.2.30 EXAMPLE Suppose $A = \begin{bmatrix} 1 & 3 & 1 & -1 & 1 \\ -1 & -1 & 3 & 1 & 3 \\ 1 & 4 & 3 & -1 & 3 \end{bmatrix}$. The row space of A is the span of

$$\begin{bmatrix} 1 & 3 & 1 & -1 & 1 \end{bmatrix}, \quad \begin{bmatrix} -1 & -1 & 3 & 1 & 3 \end{bmatrix}, \text{ and } \begin{bmatrix} 1 & 4 & 3 & -1 & 3 \end{bmatrix}.$$

Gaussian elimination gives

$$A \to \begin{bmatrix} 1 & 3 & 1 & -1 & 1 \\ 0 & 2 & 4 & 0 & 4 \\ 0 & 1 & 2 & 0 & 2 \end{bmatrix} \to \begin{bmatrix} 1 & 3 & 1 & -1 & 1 \\ 0 & 1 & 2 & 0 & 2 \\ 0 & 0 & 0 & 0 & 0 \end{bmatrix}.$$

Thus the row space of A is also the span of

$$\begin{bmatrix} 1 & 3 & 1 & -1 & 1 \end{bmatrix} \text{ and } \begin{bmatrix} 0 & 1 & 2 & 0 & 2 \end{bmatrix}$$

(see Reading Challenge 4). ∎

4.2.31 PROBLEM Find a set of two vectors that spans the column space of the matrix A in **4.2.30**.

Solution. We take the transpose of A, so that the subspace in question is the row space of the new matrix, and then use Gaussian elimination.

$$A^T = \begin{bmatrix} 1 & -1 & 1 \\ 3 & -1 & 4 \\ 1 & 3 & 3 \\ -1 & 1 & -1 \\ 1 & 3 & 3 \end{bmatrix} \to \begin{bmatrix} 1 & -1 & 1 \\ 0 & 2 & 1 \\ 0 & 4 & 2 \\ 0 & 0 & 0 \\ 0 & 4 & 2 \end{bmatrix} \to \begin{bmatrix} 1 & -1 & 1 \\ 0 & 2 & 1 \\ 0 & 0 & 0 \\ 0 & 0 & 0 \\ 0 & 0 & 0 \end{bmatrix}.$$

The row space of this last matrix is the same as the row space of A^T, so the column space of A is spanned by $\left\{ \begin{bmatrix} 1 \\ -1 \\ 1 \end{bmatrix}, \begin{bmatrix} 0 \\ 2 \\ 1 \end{bmatrix} \right\}$.

■ Linear Dependence and Independence

At the close of this section, we record some facts about linear independence and linear dependence that will be of use later. We had our first brush with these concepts in Section 1.5 and met them again in Section 2.1.

Given any vectors $\mathbf{v}_1, \mathbf{v}_2, \ldots, \mathbf{v}_n$, it is easy to write the zero vector as a linear combination of these:

$$\mathbf{0} = 0\mathbf{v}_1 + 0\mathbf{v}_2 + \cdots + 0\mathbf{v}_n.$$

We have termed this particular linear combination *trivial*. Any other linear combination is *nontrivial*.

4.2.32 DEFINITION   Vectors $\mathbf{v}_1, \mathbf{v}_2, \ldots, \mathbf{v}_n$ are *linearly independent* if and only if

$$c_1\mathbf{v}_1 + c_2\mathbf{v}_2 + \cdots + c_n\mathbf{v}_n = \mathbf{0} \text{ implies } c_1 = 0, c_2 = 0, \ldots, c_n = 0. \qquad (5)$$

Vectors are *linearly dependent* if they are not linearly independent; that is, there exists a linear combination $c_1\mathbf{v}_1 + c_2\mathbf{v}_2 + \cdots + c_n\mathbf{v}_n = \mathbf{0}$, without **all** coefficients equal to 0.

4.2.33 Remarks

1. Colloquially, equation (5) is stated like this:

   The only linear combination of vectors that equals the zero vector is the trivial one, in which all coefficients are 0.

2. Way back in Section 2.4, we noted that linear independence of $\mathbf{v}_1, \mathbf{v}_2, \ldots, \mathbf{v}_n$ is equivalent to the statement

   $A\mathbf{x} = \mathbf{0}$ implies $\mathbf{x} = \mathbf{0}$,

   where A is the matrix whose columns are $\mathbf{v}_1, \mathbf{v}_2, \ldots, \mathbf{v}_n$.

   The columns of a matrix A are linearly independent if and only if $A\mathbf{x} = \mathbf{0}$ implies $\mathbf{x} = \mathbf{0}$.

   This is an extremely useful idea.

4.2.34 PROBLEM   Are $\mathbf{v}_1 = \begin{bmatrix} 1 \\ 2 \\ 5 \end{bmatrix}$, $\mathbf{v}_2 = \begin{bmatrix} 1 \\ 0 \\ 0 \end{bmatrix}$, and $\mathbf{v}_3 = \begin{bmatrix} -2 \\ 1 \\ 3 \end{bmatrix}$ linearly independent or linearly dependent?

Solution.   We consider the possibility that

$$c_1\mathbf{v}_1 + c_2\mathbf{v}_2 + c_3\mathbf{v}_3 = \mathbf{0}.$$

This is $A\mathbf{c} = \mathbf{0}$ with

$$A = \begin{bmatrix} 1 & 1 & -2 \\ 2 & 0 & 1 \\ 5 & 0 & 3 \end{bmatrix} \quad \text{and} \quad \mathbf{c} = \begin{bmatrix} c_1 \\ c_2 \\ c_3 \end{bmatrix}.$$

It is easy to check that rank $A = 3$, so this system has only the trivial solution $\mathbf{c} = \mathbf{0}$, by Corollary 4.1.25. Thus $c_1 = c_2 = c_3 = 0$ and the given vectors are linearly independent.

4.2.35 PROBLEM

Suppose $\mathbf{v}_1, \mathbf{v}_2, \mathbf{v}_3$ are as in Problem 4.2.34 and $\mathbf{v}_4 = \begin{bmatrix} 1 \\ 1 \\ 1 \end{bmatrix}$. Are the four vectors $\mathbf{v}_1, \mathbf{v}_2, \mathbf{v}_3, \mathbf{v}_4$ linearly dependent or linearly independent?

Solution. The equation $c_1\mathbf{v}_1 + c_2\mathbf{v}_2 + c_3\mathbf{v}_3 + c_4\mathbf{v}_4 = \mathbf{0}$ is $A\mathbf{c} = \mathbf{0}$ with

$$A = \begin{bmatrix} 1 & 1 & -2 & 1 \\ 2 & 0 & 1 & 1 \\ 5 & 0 & 3 & 1 \end{bmatrix}, \quad \mathbf{c} = \begin{bmatrix} c_1 \\ c_2 \\ c_3 \\ c_4 \end{bmatrix}, \quad \mathbf{0} = \begin{bmatrix} 0 \\ 0 \\ 0 \end{bmatrix}.$$

Gaussian elimination is

$$\begin{bmatrix} 1 & 1 & -2 & 1 \\ 2 & 0 & 1 & 1 \\ 5 & 0 & 3 & 1 \end{bmatrix} \rightarrow \begin{bmatrix} 1 & 1 & -2 & 1 \\ 0 & -2 & 5 & -1 \\ 0 & -5 & 13 & -4 \end{bmatrix} \rightarrow \begin{bmatrix} 1 & 1 & -2 & 1 \\ 0 & 1 & -\frac{5}{2} & \frac{1}{2} \\ 0 & 0 & \frac{1}{2} & -\frac{3}{2} \end{bmatrix}.$$

Thus $c_4 = t$ is a free variable and we need go no further. There are certainly nontrivial solutions. So the vectors are linearly dependent.

Upon reflection, we see that Gaussian elimination wasn't needed here at all. We have a homogeneous system of three equations in four knowns. A homogeneous system with more unknowns than equations always has a nontrivial solution! (See Theorem 4.1.8.)

Theorem 4.2.36   *If $n > m$, any n vectors in $\mathbf{R}^m$ are linearly dependent.*

Proof   Put the vectors into the columns of an $m \times n$ matrix A. The homogeneous system $A\mathbf{x} = \mathbf{0}$ has more unknowns than equations, so there is a nontrivial solution. So the columns are linearly dependent. ∎

In Section 1.5, we showed that two vectors are linearly dependent if and only if one is a scalar multiple of the other, and that three vectors are linearly dependent if and only if one is a linear combination of the other two—see 1.5.23 and 1.5.24. These facts generalize.

Theorem 4.2.37   *Vectors are linearly dependent if and only if one of them is a linear combination of the others.*

Proof   Suppose vectors $\mathbf{v}_1, \mathbf{v}_2, \ldots, \mathbf{v}_n$ are linearly dependent. Then some nontrivial linear combination of them is the zero vectors; there are scalars $c_1, c_2, \ldots, c_n$, not all 0, such that

$$c_1\mathbf{v}_1 + c_2\mathbf{v}_2 + \cdots + c_n\mathbf{v}_n = \mathbf{0}.$$

Suppose $c_1 \neq 0$. Then we can take the equation

$$c_1\mathbf{v}_1 = -c_2\mathbf{v}_2 - \cdots - c_n\mathbf{v}_n,$$

divide by c_1, and obtain

$$\mathbf{v}_1 = -\frac{c_2}{c_1}\mathbf{v}_2 - \cdots - \frac{c_n}{c_1}\mathbf{v}_n,$$

which says that $\mathbf{v}_1$ is a linear combination of the other vectors. If $c_1 = 0$ but $c_2 \neq 0$, a similar argument would show that $\mathbf{v}_2$ is a linear combination of the rest. In general, since some $c_k \neq 0$, we can show that $\mathbf{v}_k$ is a linear combination of the others.

Conversely, suppose $\mathbf{v}_1, \mathbf{v}_2, \ldots, \mathbf{v}_n$ are vectors and one of them, say $\mathbf{v}_1$, is a linear combination of the others. Thus there are scalars $c_2, \ldots, c_n$ so that

$$\mathbf{v}_1 = c_2\mathbf{v}_2 + \cdots + c_n\mathbf{v}_n.$$

From this, we obtain

$$\underset{\Uparrow}{1}\mathbf{v}_1 - c_2\mathbf{v}_2 - \cdots - c_n\mathbf{v}_n = \mathbf{0},$$

which expresses $\mathbf{0}$ as a nontrivial linear combination of all the vectors (the arrow points to one coefficient which is certainly not 0). A similar argument would apply if any $\mathbf{v}_k$ is a linear combination of the other vectors. ∎

4.2.38 EXAMPLE   The three vectors $\begin{bmatrix} 1 \\ 1 \\ 2 \end{bmatrix}, \begin{bmatrix} 0 \\ 3 \\ 4 \end{bmatrix}, \begin{bmatrix} 3 \\ 0 \\ 2 \end{bmatrix}$ are linearly dependent because the second vector is a linear combination of the first and the third:

$$\begin{bmatrix} 0 \\ 3 \\ 4 \end{bmatrix} = 3\begin{bmatrix} 1 \\ 1 \\ 2 \end{bmatrix} - \begin{bmatrix} 3 \\ 0 \\ 2 \end{bmatrix}.$$ ∎

4.2.39 Remark   It is sometimes useful to have available a statement somewhat stronger than that of Theorem 4.2.37. In the exercises, we ask you to show that if $\mathbf{v}_1, \mathbf{v}_2, \ldots, \mathbf{v}_n$ are linearly dependent vectors, then one of these is a linear combination of those vectors that **precede** it in the list. See Exercise 21.

This remark makes it easy to show that the pivot columns and the nonzero rows of a row echelon matrix are linearly independent.

4.2.40 EXAMPLE The matrix $U = \begin{bmatrix} 3 & -1 & 4 & 1 & 2 & 8 \\ 0 & -2 & 0 & 5 & 7 & 5 \\ 0 & 0 & 0 & 8 & 3 & -2 \\ 0 & 0 & 0 & 0 & 0 & 9 \\ 0 & 0 & 0 & 0 & 0 & 0 \end{bmatrix}$ is in row echelon form. The pivots, 3, −2, 8,

and 9 are in columns one, two, four, and six. Using Remark 4.2.39, we know that if these columns are linearly dependent, then one of them is a linear combination of columns that precede it. But this is not possible because all the elements of a row before a pivot are 0. For example, column four is not a linear combination of columns one and two;

$$\begin{bmatrix} 1 \\ 5 \\ 8 \\ 0 \\ 0 \end{bmatrix} = a \begin{bmatrix} 3 \\ 0 \\ 0 \\ 0 \\ 0 \end{bmatrix} + b \begin{bmatrix} -1 \\ -2 \\ 0 \\ 0 \\ 0 \end{bmatrix}$$

says $8 = 0$, which is not true.

The nonzero rows of U (written from the last nonzero row up), namely,

$$\begin{bmatrix} 0 & 0 & 0 & 0 & 0 & 9 \end{bmatrix},$$
$$\begin{bmatrix} 0 & 0 & 0 & 8 & 3 & -2 \end{bmatrix},$$
$$\begin{bmatrix} 0 & -2 & 0 & 5 & 7 & 5 \end{bmatrix}, \text{ and}$$
$$\begin{bmatrix} 3 & -1 & 4 & 1 & 2 & 8 \end{bmatrix},$$

are also linearly independent. As before, it is easy to see that none of these is a linear combination of rows that precede it, so they are linearly dependent. ∎

A general proof of the next theorem would follow very closely the arguments just presented in a special case.

Theorem 4.2.41 *Let U be a matrix in row echelon form.*

1. *The pivot columns of U are linearly independent.*

2. *The nonzero rows of U are linearly independent.*

■ Uniqueness of Coefficients

Let V be the subspace of $\mathbf{R}^4$ spanned by

$$\mathbf{v}_1 = \begin{bmatrix} -1 \\ 2 \\ 0 \\ 1 \end{bmatrix}, \quad \mathbf{v}_2 = \begin{bmatrix} 2 \\ -4 \\ 1 \\ 3 \end{bmatrix}, \quad \mathbf{v}_3 = \begin{bmatrix} -1 \\ 2 \\ 1 \\ 6 \end{bmatrix}.$$

The vector $\mathbf{v} = \begin{bmatrix} 2 \\ -4 \\ 0 \\ -2 \end{bmatrix}$ is in V because $\mathbf{v} = -2\mathbf{v}_1 = -2\mathbf{v}_1 + 0\mathbf{v}_2 + 0\mathbf{v}_3$. There is

nothing particularly special about the coefficients $-2, 0, 0$ used here. For example, the reader can check that v can also be written $\mathbf{v} = \mathbf{v}_1 + \mathbf{v}_2 - \mathbf{v}_3$.

In general, there are many ways to express one vector as a linear combination of others, but not if the others are linearly independent. To see why, suppose $\mathbf{v}_1, \mathbf{v}_2, \ldots, \mathbf{v}_n$ are linearly independent vectors and a vector $\mathbf{v}$ is in the span of these. Thus

$$\mathbf{v} = c_1\mathbf{v}_1 + c_2\mathbf{v}_2 + \cdots + c_n\mathbf{v}_n$$

for certain scalars $c_1, c_2, \ldots, c_n$. If also

$$\mathbf{v} = d_1\mathbf{v}_1 + d_2\mathbf{v}_2 + \cdots + d_n\mathbf{v}_n,$$

then, subtracting, we would have

$$(c_1 - d_1)\mathbf{v}_1 + (c_2 - d_2)\mathbf{v}_2 + \cdots + (c_n - d_n)\mathbf{v}_n = \mathbf{0},$$

so all the coefficients would be 0. Thus $c_1 = d_1,\ c_2 = d_2,\ \ldots,\ c_n = d_n$.

Theorem 4.2.42 *If $\mathbf{v}_1, \mathbf{v}_2, \ldots, \mathbf{v}_n$ are linearly independent vectors in a vector space V, then any vector in their span can be expressed in exactly one way as a linear combination of these vectors: the coefficients are unique.*

Answers to Reading Challenges

1. The set $U = \{\mathbf{0}\}$ is closed under addition because $\mathbf{0} + \mathbf{0} = \mathbf{0}$ is in U and closed under scalar multiplication because, for any scalar c, $c\mathbf{0} = \mathbf{0}$ is in U.

2. Let $\mathbf{x} = \begin{bmatrix} x_1 \\ x_2 \end{bmatrix}$ belong to U and let c be a scalar. We know that $x_1 = 0$ or $x_2 = 0$.

   If $x_1 = 0$, then $c\mathbf{x} = \begin{bmatrix} 0 \\ cx_2 \end{bmatrix}$ is in U because the first component of this vector is 0. Similarly, if $x_2 = 0$, then $c\mathbf{x} = \begin{bmatrix} cx_1 \\ 0 \end{bmatrix}$ is in U.

3. The set U is closed under neither addition nor scalar multiplication. For example, $\mathbf{u} = \begin{bmatrix} -1 \\ 0 \\ 1 \end{bmatrix}$ is in U, but $2\mathbf{u} = \begin{bmatrix} -2 \\ 0 \\ 2 \end{bmatrix}$ is not, because $2(-2) - 0 + 3(2) \neq 1$.

4. We have to show that any linear combination of $\mathbf{u}$, $\mathbf{v}$, and $\mathbf{0}$ is a linear combination of $\mathbf{u}$ and $\mathbf{v}$, and conversely. This follows from $a\mathbf{u} + b\mathbf{v} + c\mathbf{0} = a\mathbf{u} + b\mathbf{v}$ and $a\mathbf{u} + b\mathbf{v} = a\mathbf{u} + b\mathbf{v} + c\mathbf{0}$.

5. $A\mathbf{x}$ is a linear combination of the columns of A with coefficients the components of $\mathbf{x}$. See 2.1.33. So equation (4) is simply a rewriting of (3).

True/False Questions

Decide, with as little calculation as possible, whether each of the following statements is true or false and explain your answer whenever you say "false." (Answers can be found in the back of the book.)

1. The line with equation $x + y = 1$ is a subspace of $\mathbf{R}^2$.

2. A plane is spanned by any two vectors it contains.

3. The set of vectors in $\mathbf{R}^3$ of the form $\begin{bmatrix} x \\ x \\ x \end{bmatrix}$ is closed under scalar multiplication.

4. If $\mathbf{u}$ and $\mathbf{v}$ are any vectors, the zero vector belongs to their span.

5. The vector $\begin{bmatrix} 0 \\ 1 \end{bmatrix}$ is in the column space of $\begin{bmatrix} -1 & 2 & 1 & 5 \\ 0 & 1 & 7 & -9 \end{bmatrix}$.

6. The matrices $\begin{bmatrix} 1 & 2 & 3 \\ -1 & -2 & -3 \\ 2 & 5 & 6 \end{bmatrix}$ and $\begin{bmatrix} 1 & 2 & 3 \\ 0 & 1 & 0 \\ 0 & 0 & 0 \end{bmatrix}$ have the same row space.

7. The matrices $\begin{bmatrix} 1 & 2 & 3 \\ -1 & -2 & -3 \\ 2 & 5 & 6 \end{bmatrix}$ and $\begin{bmatrix} 1 & 2 & 3 \\ 0 & 1 & 0 \\ 0 & 0 & 0 \end{bmatrix}$ have the same column space.

8. If $\mathbf{u}$, $\mathbf{v}$, and $\mathbf{w}$ are vectors, $\mathbf{v}$ is not a scalar multiple of $\mathbf{u}$, and $\mathbf{w}$ is not in the span of $\mathbf{u}$ and $\mathbf{v}$, then $\mathbf{u}$, $\mathbf{v}$, $\mathbf{w}$ are linearly independent.

9. If $\mathcal{S}$ is a set of linearly independent vectors that spans a subspace U, then the zero vector is not in $\mathcal{S}$.

10. If $\mathcal{S}$ is a set of linearly independent vectors that spans a subspace U, then the zero vector is not in U.

11. To say that vectors $\mathbf{v}_1, \mathbf{v}_2, \dots, \mathbf{v}_n$ are linearly independent means that $c_1\mathbf{v}_1 + c_2\mathbf{v}_2 + \cdots + c_n\mathbf{v}_n = \mathbf{0}$, where $c_1 = 0, c_2 = 0, \dots, c_n = 0$.

12. If a set $\{\mathbf{v}_1, \mathbf{v}_2, \dots, \mathbf{v}_n\}$ of vectors is linearly dependent, then each vector in this set is a linear combination of the others.

13. If the columns of a matrix A are linearly dependent, there is a nonzero vector $\mathbf{x}$ satisfying $A\mathbf{x} = \mathbf{0}$.

14. If $\mathbf{u}_1, \mathbf{u}_2, \dots, \mathbf{u}_n$ are linearly dependent vectors, then there is an equation of the form $c_1\mathbf{u}_1 + c_2\mathbf{u}_2 + \cdots + c_n\mathbf{u}_n = \mathbf{0}$ with none of the scalars c_i equal to 0.

Exercises

Solutions to exercises marked [BB] can be found in the Back of the Book.

1. Determine whether the given vector $\mathbf{v}$ is a linear combination of the other vectors $\mathbf{v}_1, \mathbf{v}_2, \dots$ in each of the following cases.

   (a) [BB] $\mathbf{v} = \begin{bmatrix} 1 \\ 3 \\ 2 \end{bmatrix}$; $\mathbf{v}_1 = \begin{bmatrix} 0 \\ 1 \\ 1 \end{bmatrix}$, $\mathbf{v}_2 = \begin{bmatrix} 1 \\ 1 \\ 0 \end{bmatrix}$,

   $\mathbf{v}_3 = \begin{bmatrix} 1 \\ 0 \\ 2 \end{bmatrix}$

   (b) $\mathbf{v} = \begin{bmatrix} 1 \\ 2 \\ 3 \\ 4 \\ 5 \end{bmatrix}$; $\mathbf{v}_1 = \begin{bmatrix} 5 \\ 1 \\ 2 \\ 3 \\ 4 \end{bmatrix}$, $\mathbf{v}_2 = \begin{bmatrix} 4 \\ 5 \\ 1 \\ 2 \\ 3 \end{bmatrix}$, $\mathbf{v}_3 = \begin{bmatrix} 2 \\ 3 \\ 4 \\ 5 \\ 1 \end{bmatrix}$,

   $\mathbf{v}_4 = \begin{bmatrix} -1 \\ -2 \\ -3 \\ -4 \\ -5 \end{bmatrix}$, $\mathbf{v}_5 = \begin{bmatrix} 0 \\ 1 \\ 2 \\ 3 \\ -1 \end{bmatrix}$

(c) $\mathbf{v} = \begin{bmatrix} 1 \\ 2 \\ 3 \\ 4 \end{bmatrix}$; $\mathbf{v}_1 = \begin{bmatrix} -2 \\ 1 \\ 0 \\ 1 \end{bmatrix}$, $\mathbf{v}_2 = \begin{bmatrix} 3 \\ 4 \\ 0 \\ -3 \end{bmatrix}$, $\mathbf{v}_3 = \begin{bmatrix} 6 \\ 5 \\ 0 \\ 4 \end{bmatrix}$

(d) $\mathbf{v} = \begin{bmatrix} 1 \\ -2 \\ 3 \\ 0 \end{bmatrix}$; $\mathbf{v}_1 = \begin{bmatrix} 2 \\ -3 \\ 1 \\ 0 \end{bmatrix}$, $\mathbf{v}_2 = \begin{bmatrix} 4 \\ -2 \\ 1 \\ 0 \end{bmatrix}$,

$\mathbf{v}_3 = \begin{bmatrix} 0 \\ 1 \\ 1 \\ 1 \end{bmatrix}$, $\mathbf{v}_4 = \begin{bmatrix} 1 \\ -1 \\ 1 \\ -1 \end{bmatrix}$.

2. In each of the following cases, determine whether the vector $\mathbf{v}$ is in the span of the other vectors $\mathbf{v}_1, \mathbf{v}_2, \ldots$.

(a) [BB] $\mathbf{v} = \begin{bmatrix} 1 \\ 3 \\ -1 \\ 1 \end{bmatrix}$; $\mathbf{v}_1 = \begin{bmatrix} 1 \\ -1 \\ 2 \\ 1 \end{bmatrix}$, $\mathbf{v}_2 = \begin{bmatrix} 2 \\ 1 \\ 1 \\ 0 \end{bmatrix}$

(b) $\mathbf{v} = \begin{bmatrix} 1 \\ -7 \\ 8 \\ 5 \end{bmatrix}$; $\mathbf{v}_1 = \begin{bmatrix} 1 \\ -1 \\ 2 \\ 1 \end{bmatrix}$, $\mathbf{v}_2 = \begin{bmatrix} 2 \\ 1 \\ 1 \\ 0 \end{bmatrix}$

(c) $\mathbf{v} = \begin{bmatrix} 1 \\ -1 \\ 1 \\ 9 \end{bmatrix}$; $\mathbf{v}_1 = \begin{bmatrix} 0 \\ -1 \\ 1 \\ 2 \end{bmatrix}$, $\mathbf{v}_2 = \begin{bmatrix} 3 \\ 0 \\ 0 \\ -1 \end{bmatrix}$,

$\mathbf{v}_3 = \begin{bmatrix} 2 \\ 1 \\ -1 \\ 1 \end{bmatrix}$

(d) $\mathbf{v} = \begin{bmatrix} 2 \\ 0 \\ 1 \\ -3 \end{bmatrix}$; $\mathbf{v}_1 = \begin{bmatrix} 0 \\ -1 \\ 1 \\ 2 \end{bmatrix}$, $\mathbf{v}_2 = \begin{bmatrix} 3 \\ 0 \\ 0 \\ -1 \end{bmatrix}$,

$\mathbf{v}_3 = \begin{bmatrix} 2 \\ 1 \\ -1 \\ 1 \end{bmatrix}$.

3. (a) [BB] Do the vectors $\mathbf{u}_1 = \begin{bmatrix} -2 \\ 4 \end{bmatrix}$ and $\mathbf{u}_2 = \begin{bmatrix} 3 \\ -6 \end{bmatrix}$ span $\mathbf{R}^2$? If the answer is yes, provide a proof; otherwise explain why the answer is no.

(b) Answer 3(a) with $\mathbf{u}_1 = \begin{bmatrix} 1 \\ 2 \end{bmatrix}$ and $\mathbf{u}_2 = \begin{bmatrix} 2 \\ 1 \end{bmatrix}$.

4. (a) Do the vectors $\mathbf{u}_1 = \begin{bmatrix} 2 \\ 1 \\ -3 \end{bmatrix}$ and $\mathbf{u}_2 = \begin{bmatrix} 6 \\ 0 \\ 1 \end{bmatrix}$ span $\mathbf{R}^3$? If the answer is yes, provide a proof; otherwise explain why the answer is no.

(b) Answer 4(a) for the vectors $\mathbf{u}_1, \mathbf{u}_2, \mathbf{u}_3$, with $\mathbf{u}_1$ and $\mathbf{u}_2$ as before and $\mathbf{u}_3 = \begin{bmatrix} 1 \\ 1 \\ 1 \end{bmatrix}$.

(c) Answer 4(a) for the vectors $\mathbf{u}_1, \mathbf{u}_2, \mathbf{u}_3$, with $\mathbf{u}_1$ and $\mathbf{u}_2$ as before and $\mathbf{u}_3 = \begin{bmatrix} 2 \\ -2 \\ 7 \end{bmatrix}$. [Hint: $\mathbf{u}_3 = \mathbf{u}_2 - 2\mathbf{u}_1$.]

5. [BB] Determine whether the vectors $\mathbf{v}_1 = \begin{bmatrix} 1 \\ 0 \\ 0 \\ 1 \end{bmatrix}$,

$\mathbf{v}_2 = \begin{bmatrix} 1 \\ 0 \\ 0 \\ -1 \end{bmatrix}$, $\mathbf{v}_3 = \begin{bmatrix} 0 \\ 1 \\ 1 \\ 0 \end{bmatrix}$, and $\mathbf{v}_4 = \begin{bmatrix} 0 \\ 1 \\ -1 \\ 0 \end{bmatrix}$ span $\mathbf{R}^4$. If the answer is yes, supply a proof; otherwise explain why the answer is no.

6. Show that each of the following sets is closed under the usual addition and scalar multiplication of vectors in $\mathbf{R}^3$.

(a) [BB] $U = \left\{ \begin{bmatrix} x \\ y \\ 2x - 3y \end{bmatrix} \mid x, y \in \mathbf{R} \right\}$

(b) $U = \left\{ \begin{bmatrix} x \\ y \\ z \end{bmatrix} \mid x + 4y - 5z = 0 \right\}$.

7. Determine whether each of the following sets U is a subspace of the indicated vector space V. Justify your answers.

(a) [BB] $V = \mathbf{R}^3$, $U = \left\{ \begin{bmatrix} a \\ a - 1 \\ c \end{bmatrix} \mid a, c \in \mathbf{R} \right\}$

(b) $V = \mathbf{R}^3$, $U = \left\{ \begin{bmatrix} a \\ a \\ c \end{bmatrix} \mid a, c \in \mathbf{R} \right\}$

(c) $V = \mathbf{R}^3$, $U = \left\{ \begin{bmatrix} x \\ y \\ x + y + 1 \end{bmatrix} \mid x, y \in \mathbf{R} \right\}$

(d) $V = \mathbf{R}^4$, $U = \left\{ \begin{bmatrix} x \\ y \\ z \\ x+y-z \end{bmatrix} \ \middle|\ x, y, z \in \mathbf{R} \right\}$

(e) $V = \mathbf{R}^3$, $U = \left\{ \begin{bmatrix} x \\ y \\ z \end{bmatrix} \ \middle|\ x, y, z \text{ are integers} \right\}$.

8. Let A be an $n \times n$ matrix and let λ be a scalar. In Section 3.3, we defined the eigenspace of A corresponding to λ as the set $U = \{\mathbf{x} \in \mathbf{R}^n \mid A\mathbf{x} = \lambda\mathbf{x}\}$. Show that U is a subspace of $\mathbf{R}^n$.

9. [BB] Let U be a subspace of a vector space V and let $\mathbf{u}_1$ and $\mathbf{u}_2$ be vectors in U. Explain clearly why the vector $\mathbf{u}_1 - \mathbf{u}_2$ is also in U. Refer to the definition of subspace.

10. Determine whether each of the following sets of vectors is linearly independent or linearly dependent.

(a) [BB] $\mathbf{v}_1 = \begin{bmatrix} 1 \\ 0 \\ 1 \end{bmatrix}$, $\mathbf{v}_2 = \begin{bmatrix} 1 \\ 2 \\ -1 \end{bmatrix}$, $\mathbf{v}_3 = \begin{bmatrix} 2 \\ 1 \\ -4 \end{bmatrix}$

(b) $\mathbf{v}_1 = \begin{bmatrix} 1 \\ 2 \\ 3 \end{bmatrix}$, $\mathbf{v}_2 = \begin{bmatrix} 2 \\ 1 \\ -1 \end{bmatrix}$, $\mathbf{v}_3 = \begin{bmatrix} -1 \\ 4 \\ 11 \end{bmatrix}$

(c) $\mathbf{v}_1 = \begin{bmatrix} -2 \\ 3 \\ 3 \\ 1 \end{bmatrix}$, $\mathbf{v}_2 = \begin{bmatrix} 3 \\ 4 \\ 0 \\ 7 \end{bmatrix}$, $\mathbf{v}_3 = \begin{bmatrix} 2 \\ 1 \\ 1 \\ 3 \end{bmatrix}$

(d) $\mathbf{v}_1 = \begin{bmatrix} -4 \\ 2 \\ 0 \\ -3 \end{bmatrix}$, $\mathbf{v}_2 = \begin{bmatrix} 1 \\ 3 \\ 4 \\ 1 \end{bmatrix}$, $\mathbf{v}_3 = \begin{bmatrix} -2 \\ 8 \\ 8 \\ 3 \end{bmatrix}$,

$\mathbf{v}_4 = \begin{bmatrix} 3 \\ -5 \\ -4 \\ -3 \end{bmatrix}$

(e) $\mathbf{u} = \begin{bmatrix} 1 \\ 1 \\ 2 \\ 7 \\ 8 \\ -2 \end{bmatrix}$, $\mathbf{v} = \begin{bmatrix} -2 \\ 3 \\ 1 \\ 6 \\ 4 \\ -11 \end{bmatrix}$, $\mathbf{w} = \begin{bmatrix} 1 \\ 2 \\ 3 \\ 11 \\ 12 \\ -5 \end{bmatrix}$.

11. [BB] Suppose $\mathbf{v}_1, \mathbf{v}_2, \ldots, \mathbf{v}_n$ are vectors and one of them is the zero vector. To be specific, suppose $\mathbf{v}_1 = \mathbf{0}$. Show that $\mathbf{v}_1, \mathbf{v}_2, \ldots, \mathbf{v}_n$ are linearly

dependent. (Thus the zero vector is never part of a linearly independent set.)

12. Let U be the row echelon matrix
$$\begin{bmatrix} 0 & 1 & -3 & 2 & 4 & 5 \\ 0 & 0 & 1 & 1 & 2 & 0 \\ 0 & 0 & 0 & 0 & 1 & 7 \\ 0 & 0 & 0 & 0 & 0 & 1 \\ 0 & 0 & 0 & 0 & 0 & 0 \end{bmatrix}.$$

(a) Prove that the nonzero rows of U are linearly independent.

(b) Prove that the pivot columns of U are linearly independent.

(c) Find a set of linearly independent vectors that spans the null space of U and explain fully.

13. Repeat Exercise 12 with
$$U = \begin{bmatrix} 2 & 0 & -1 & 4 & 1 \\ 0 & 0 & 1 & 3 & 2 \\ 0 & 0 & 0 & 5 & 4 \\ 0 & 0 & 0 & 0 & 0 \\ 0 & 0 & 0 & 0 & 0 \end{bmatrix}.$$

14. For each of the following matrices, find a linearly independent set of vectors that spans

i. the null space;

ii. the row space;

iii. the column space.

(a) [BB] $A = \begin{bmatrix} -1 & 1 & 0 & 2 \\ 3 & 1 & 1 & 10 \\ 2 & 4 & -1 & 0 \end{bmatrix}$

(b) $A = \begin{bmatrix} 0 & 1 & 3 \\ -1 & 3 & 8 \\ -13 & 2 & -7 \\ 1 & 1 & 4 \\ 4 & -2 & -2 \end{bmatrix}$

(c) $A = \begin{bmatrix} 1 & 2 & -1 & 0 \\ 1 & 3 & -2 & 0 \\ 1 & -1 & 1 & -1 \\ 5 & 4 & 0 & -1 \\ 1 & 8 & -3 & 4 \\ 0 & 7 & -2 & 5 \end{bmatrix}$

(d) $A = \begin{bmatrix} 1 & 2 & -1 & 0 & 1 & 3 \\ -3 & -10 & -3 & 3 & 8 & -18 \\ 3 & 6 & -3 & 2 & -7 & 11 \\ 1 & 4 & 2 & -2 & -2 & 7 \\ 2 & 2 & -5 & 3 & 0 & 3 \end{bmatrix}.$

15. Prove Theorem 4.2.36 without appealing to Theorem 4.1.8.

16. [BB] Prove Theorem 3.4.14 in general; that is, if $x_1, x_2, \ldots, x_k$ are eigenvectors of a matrix A whose corresponding eigenvalues $\lambda_1, \lambda_2, \ldots, \lambda_k$ are all different, then $x_1, x_2, \ldots, x_k$ are linearly independent. [Hint: If the result is false, then there are scalars $c_1, c_2, \ldots, c_k$, not all 0, such that $c_1 x_1 + c_2 x_2 + \cdots + c_k x_k = 0$. Omitting the terms with $c_i = 0$ and renaming the terms that remain, we may assume that

$$c_1 x_1 + c_2 x_2 + \cdots + c_\ell x_\ell = 0 \qquad (\dagger)$$

for some ℓ, where $c_1 \neq 0, c_2 \neq 0, \ldots, c_\ell \neq 0$. Among all such equations, suppose the one exhibited is shortest; that is, whenever we have an equation like (†) (and no coefficients 0), there are at least ℓ terms. Note that $\ell > 1$, since otherwise $c_1 x_1 = 0$ and $x_1 \neq 0$ gives $c_1 = 0$, a contradiction. Now imitate the proof of Theorem 3.4.14.]

■ Critical Reading

17. [BB] Let u, v, and w be vectors such that $u \neq 0$, v is not a scalar multiple of u, and w is not in the span of u and v. Show that u, v, and w are linearly independent.

18. [BB] Suppose U and W are subspaces of a vector space V. Show that $U \cup W = \{v \in V \mid v \in U \text{ or } v \in W\}$, the set of vectors that are in either U or W (possibly both) is a subspace of V if and only if one of the subspaces U, W contains the other. [Hint: In the part of the proof in which you assume $U \cup W$ is a subspace, suppose neither U nor W is contained in the other and derive a contradiction.]

19. Suppose A and U are $m \times n$ matrices and E is an invertible $m \times m$ matrix such that $U = EA$.

    (a) If the columns of U are linearly independent, show that the columns of A are linearly independent [BB], and conversely.

    (b) If the columns of U span $\mathbf{R}^m$, show that the columns of A span $\mathbf{R}^m$, and [BB] conversely.

20. Suppose A is an $m \times n$ matrix and B is an $n \times t$ matrix. Suppose the columns of A span $\mathbf{R}^m$ and the columns of B span $\mathbf{R}^n$. Show that the columns of AB span $\mathbf{R}^m$.

21. Suppose $v_1, v_2, \ldots, v_n$ are linearly dependent vectors. Show that one of these is a linear combination of vectors that precede it in the list. [Hint: Examine the proof of Theorem 4.2.37.]

4.3   Basis and Dimension; Rank and Nullity

In this section, we prove that if a vector space has a finite linearly independent set of spanning vectors, then the number of such vectors is an invariant of the vector space.

> Two finite linearly independent spanning sets of vectors for the same vector space must contain the same number of vectors.

This result, the most important in the theory of linear algebra, hinges on a theorem—Theorem 4.3.1—which the author likes to remember as $|\mathcal{L}| \leq |\mathcal{S}|$. The proof of this theorem will seem much easier if we first remind ourselves of a couple of things.

First, if A is a matrix and x is a vector, then Ax is a linear combination of the columns of A with coefficients the components of x. By now, the reader should be tired of being reminded of this central idea, which first appeared as 2.1.33. Second, if A is an $m \times n$ matrix and $m < n$, then A has a nontrivial null space; that is, there is a nonzero vector x satisfying $Ax = 0$—see Theorem 4.1.8.

Theorem 4.3.1 $(|\mathcal{L}| \leq |\mathcal{S}|)$. *Let V be a Euclidean space (or, more generally, a subspace of a Euclidean space). Let $\mathcal{L}$ be a finite set of linearly independent vectors in V and $\mathcal{S}$ a* *finite set of vectors that span V. Then the number of vectors in $\mathcal{L}$ is less than or equal to the number of vectors in $\mathcal{S}$, a fact that we remember symbolically by $|\mathcal{L}| \leq |\mathcal{S}|$.*

For any finite set A, the number of elements in A is often denoted $|A|$ (and called the *cardinality* of A). For example, if $A = \{\alpha, 2, x, \star\}$, then $|A| = 4$.

Proof   Suppose $\mathcal{L} = \{\mathbf{v}_1, \mathbf{v}_2, \dots, \mathbf{v}_n\}$ and $\mathcal{S} = \{\mathbf{u}_1, \mathbf{u}_2, \dots, \mathbf{u}_m\}$. We wish to prove that $n \leq m$.

Each of the vectors $\mathbf{v}_1, \dots, \mathbf{v}_n$ is a linear combination of $\mathbf{u}_1, \mathbf{u}_2, \dots, \mathbf{u}_m$, in particular $\mathbf{v}_1$. Thus there exist scalars $a_{11}, a_{21}, \dots, a_{m1}$ such that

$$\mathbf{v}_1 = a_{11}\mathbf{u}_1 + a_{21}\mathbf{u}_2 + \cdots + a_{m1}\mathbf{u}_m.$$

This is just the matrix equation

$$\mathbf{v}_1 = \begin{bmatrix} \mathbf{u}_1 & \mathbf{u}_2 & \cdots & \mathbf{u}_m \\ \downarrow & \downarrow & & \downarrow \end{bmatrix} \begin{bmatrix} a_{11} \\ a_{21} \\ \vdots \\ a_{m1} \end{bmatrix} = U \begin{bmatrix} a_{11} \\ a_{21} \\ \vdots \\ a_{m1} \end{bmatrix}$$

where $U = \begin{bmatrix} \mathbf{u}_1 & \mathbf{u}_2 & \cdots & \mathbf{u}_m \\ \downarrow & \downarrow & & \downarrow \end{bmatrix}$. Similarly, there exist scalars $a_{12}, a_{22}, \dots, a_{m2}$ such that

$$\mathbf{v}_2 = a_{12}\mathbf{u}_1 + a_{22}\mathbf{u}_2 + \cdots + a_{m2}\mathbf{u}_m.$$

This is $\mathbf{v}_2 = U \begin{bmatrix} a_{12} \\ a_{22} \\ \vdots \\ a_{m2} \end{bmatrix}$, and there are similar equations for $\mathbf{v}_3, \dots, \mathbf{v}_n$. Letting

$$A = \begin{bmatrix} a_{11} & a_{12} & \cdots & a_{1n} \\ a_{21} & a_{22} & \cdots & a_{2n} \\ \vdots & \vdots & & \vdots \\ a_{m1} & a_{n2} & \cdots & a_{mn} \end{bmatrix} \quad \text{and} \quad V = \begin{bmatrix} \mathbf{v}_1 & \mathbf{v}_2 & \cdots & \mathbf{v}_n \\ \downarrow & \downarrow & & \downarrow \end{bmatrix},$$

we have $V = UA$. Now suppose $m < n$. Since A is $m \times n$, the null space of A is nontrivial, so there is a nonzero vector $\mathbf{x}$ with $A\mathbf{x} = \mathbf{0}$. Thus $V\mathbf{x} = U(A\mathbf{x}) = \mathbf{0}$, and this says that the columns of V are linearly dependent. Since this is not true, $n \leq m$. ∎

4.3.2 EXAMPLE   Let π be a plane in $\mathbf{R}^3$. Then π is spanned by any two nonparallel vectors it contains—1.1.20—so π has a spanning set $\mathcal{S}$ of two vectors. It follows that any set of linearly independent set $\mathcal{L}$ can contain at most two vectors: $|\mathcal{L}| \leq |\mathcal{S}| = 2$. This implies that any three or more vectors in π are linearly dependent. ∎

4.3.3 EXAMPLE   In Problem 4.2.35, we showed that a certain set $\mathbf{v}_1, \mathbf{v}_2, \mathbf{v}_3, \mathbf{v}_4$ of four vectors in $\mathbf{R}^3$ is linearly dependent. This fact does not depend on the particular vectors. Since

R^3 has a spanning set of three vectors, namely, e_1, e_2, e_3—see 4.2.19—any linearly independent set contains at most three vectors, by Theorem 4.3.1.

More generally, since $e_1, e_2, \ldots, e_m$ are m vectors that span R^m, any set of more than m vectors in R^m must be linearly dependent, a fact first observed in Theorem 4.2.36.

By the same token, the standard basis vectors are linearly independent—see 2.4.3—so any spanning set of R^m must contain at least m vectors. ∎

> **4.3.4** Any linearly independent set of vectors in R^m contains at most m vectors. Any spanning set for R^m contains at least m vectors.

4.3.5 PROBLEM

Let $u_1 = \begin{bmatrix} 1 \\ 2 \\ 3 \\ 4 \end{bmatrix}$, $u_2 = \begin{bmatrix} 5 \\ 6 \\ 7 \\ 8 \end{bmatrix}$, $u_3 = \begin{bmatrix} 9 \\ 10 \\ 11 \\ 12 \end{bmatrix}$, and $u_4 = \begin{bmatrix} 13 \\ 14 \\ 15 \\ 16 \end{bmatrix}$.

Do u_1, u_2, u_3, u_4 span R^4?

Solution. Let $A = \begin{bmatrix} 1 & 2 & 3 & 4 \\ 5 & 6 & 7 & 8 \\ 9 & 10 & 11 & 12 \\ 13 & 14 & 15 & 16 \end{bmatrix}$ be the matrix whose rows are the (transposes

of) the given vectors. Row echelon form for A is $U = \begin{bmatrix} 1 & 2 & 3 & 4 \\ 0 & 1 & 2 & 3 \\ 0 & 0 & 0 & 0 \\ 0 & 0 & 0 & 0 \end{bmatrix}$, so the span of

the given vectors is the span of the two vectors $\begin{bmatrix} 1 \\ 2 \\ 3 \\ 4 \end{bmatrix}$ and $\begin{bmatrix} 0 \\ 1 \\ 2 \\ 3 \end{bmatrix}$. (Recall 4.2.29.) These

vectors do not span R^4 since a spanning set for R^4 requires at least four vectors. So the vectors u_1, u_2, u_3, u_4 do not span R^4 either.

■ Basis

4.3.6 DEFINITION

A *basis* of a vector space V is a linearly independent set of vectors that spans V.

> A basis is a linearly independent spanning set.

4.3.7 EXAMPLE

The standard basis vectors $e_1, e_2, \ldots, e_n$ in R^n—see 4.2.19—form a basis for R^n. ∎

4.3.8 EXAMPLE

If $\begin{bmatrix} x \\ y \\ z \end{bmatrix}$ is in the plane π with equation $2x - y + z = 0$, then $\begin{bmatrix} x \\ y \\ z \end{bmatrix} = \begin{bmatrix} x \\ y \\ -2x + y \end{bmatrix} =$

$x \begin{bmatrix} 1 \\ 0 \\ -2 \end{bmatrix} + y \begin{bmatrix} 0 \\ 1 \\ 1 \end{bmatrix}$. So the vectors $u_1 = \begin{bmatrix} 1 \\ 0 \\ -2 \end{bmatrix}$ and $u_2 = \begin{bmatrix} 0 \\ 1 \\ 1 \end{bmatrix}$ span π. Since they are

also linearly independent, they comprise a basis for π. ∎

4.3.9 DEFINITION

A vector space V is *finite dimensional* or has *finite dimension* if $V = \{\mathbf{0}\}$ consists only of the zero vector or if V has a basis of n vectors for some $n \geq 1$. A vector space that is not finite dimensional is *infinite dimensional*.

Although all the vector spaces of interest to us in this book have finite dimension, the reader should be aware that there are many important infinite dimensional vector spaces as well, such as the set of all polynomials over $\mathbf{R}$ in a variable x, with usual addition and scalar multiplication. (See Exercise 24.)

Suppose $\mathcal{V} = \{\mathbf{v}_1, \ldots, \mathbf{v}_n\}$ and $\mathcal{U} = \{\mathbf{u}_1, \ldots, \mathbf{u}_m\}$ are both bases for the same vector space, one set containing n vectors and the other m. According to Theorem 4.3.1, the number of vectors in any linearly independent set is less than or equal to the number in any spanning set. Now $\mathcal{V}$ is linearly independent and $\mathcal{U}$ spans, so $n \leq m$. On the other hand, $\mathcal{U}$ is linearly independent and $\mathcal{V}$ spans, so $m \leq n$. We conclude that $m = n$ and emphasize what we have just shown.

Theorem 4.3.10

If $V \neq \{\mathbf{0}\}$ is a finite dimensional vector space, then any two bases of V contain the same number of vectors.

This number of vectors, which according to the theorem does not depend on any particular basis, has a name.

4.3.11 DEFINITION

The *dimension* of a nonzero finite dimensional vector space V is the number of elements in any basis of V. The notation $\dim V$ is shorthand for the dimension of V. The dimension of the zero vector space, $V = \{\mathbf{0}\}$, is 0, by general agreement.

Thus, if V has a basis consisting of n vectors, any basis of V will contain n vectors, and $\dim V = n$.

4.3.12 EXAMPLE

Let U be a line through the origin in Euclidean 3-space, that is, $U = \mathbf{R}\mathbf{u} = \{c\mathbf{u} \mid c \in \mathbf{R}\}$ for some nonzero vector $\mathbf{u}$. Since $\mathbf{u} \neq \mathbf{0}$, $\{\mathbf{u}\}$ is a linearly independent set and since this set also spans the line, it is a basis for U. It follows that any basis for U will contain exactly **one** vector, so $\dim U = 1$. ∎

4.3.13 EXAMPLE

Let π be a plane in Euclidean 3-space. We know that π is spanned by any two nonparallel vectors. Since two nonparallel vectors are linearly independent, these vectors comprise a basis for π. Any basis for π will contain exactly **two** vectors, so $\dim \pi = 2$. ∎

4.3.14 EXAMPLE

The standard basis $\{\mathbf{e}_1, \mathbf{e}_2, \ldots, \mathbf{e}_n\}$ is linearly independent and spans $\mathbf{R}^n$, so it is a basis and $\dim \mathbf{R}^n = n$. ∎

Euclidean n-space has dimension n: $\dim \mathbf{R}^n = n$.

4.3.15 EXAMPLE

We find the dimensions of the row space and the null space of

$$A = \begin{bmatrix} 1 & 2 & -1 & 1 & 3 \\ -3 & -6 & 4 & -13 & -17 \\ 4 & 8 & -6 & 25 & 29 \\ -1 & -2 & 0 & 10 & 6 \end{bmatrix}.$$

The row space of A is the same as the row space of a row echelon form of A. (See 4.2.29.) Gaussian elimination proceeds

$$A \rightarrow \begin{bmatrix} 1 & 2 & -1 & 1 & 3 \\ 0 & 0 & 1 & -10 & -8 \\ 0 & 0 & -2 & 21 & 17 \\ 0 & 0 & -1 & 11 & 9 \end{bmatrix} \rightarrow \begin{bmatrix} 1 & 2 & -1 & 1 & 3 \\ 0 & 0 & 1 & -10 & -8 \\ 0 & 0 & 0 & 1 & 1 \\ 0 & 0 & 0 & 1 & 1 \end{bmatrix}$$

$$\rightarrow \begin{bmatrix} 1 & 2 & -1 & 1 & 3 \\ 0 & 0 & 1 & -10 & -8 \\ 0 & 0 & 0 & 1 & 1 \\ 0 & 0 & 0 & 0 & 0 \end{bmatrix} = U.$$

So the row space of A is spanned by

$$\begin{bmatrix} 1 & 2 & -1 & 1 & 3 \end{bmatrix}, \quad \begin{bmatrix} 0 & 0 & 1 & -10 & -8 \end{bmatrix}, \text{ and } \begin{bmatrix} 0 & 0 & 0 & 1 & 1 \end{bmatrix}.$$

As the nonzero rows of a row echelon matrix, these vectors are also linearly independent—see Theorem 4.2.41—so they are a basis for the row space. The row space of A has dimension 3. Notice that this is also the rank of A, the number of pivots.

The null space of A is the set of solutions to $A\mathbf{x} = \mathbf{0}$. With $\mathbf{x} = \begin{bmatrix} x_1 \\ x_2 \\ x_3 \\ x_4 \\ x_5 \end{bmatrix}$, we see

that $x_2 = t$ and $x_5 = s$ are free variables. Then

$$x_4 = -x_5 = -s$$

$$x_3 = 10x_4 + 8x_5 = -10s + 8s = -2s$$

$$x_1 = -2x_2 + x_3 - x_4 - 3x_5 = -2t - 2s + s - 3s = -2t - 4s.$$

Thus

$$\mathbf{x} = \begin{bmatrix} x_1 \\ x_2 \\ x_3 \\ x_4 \\ x_5 \end{bmatrix} = \begin{bmatrix} -2t - 4s \\ t \\ -2s \\ -s \\ s \end{bmatrix} = t \begin{bmatrix} -2 \\ 1 \\ 0 \\ 0 \\ 0 \end{bmatrix} + s \begin{bmatrix} -4 \\ 0 \\ -2 \\ -1 \\ 1 \end{bmatrix}.$$

This equation shows that every vector in the null space is a linear combination of $\begin{bmatrix} -2 \\ 1 \\ 0 \\ 0 \\ 0 \end{bmatrix}$ and $\begin{bmatrix} -4 \\ 0 \\ -2 \\ -1 \\ 1 \end{bmatrix}$, so these vectors span the null space. It also shows that these vectors are linearly independent since

$$t \begin{bmatrix} -2 \\ 1 \\ 0 \\ 0 \\ 0 \end{bmatrix} + s \begin{bmatrix} -4 \\ 0 \\ -2 \\ -1 \\ 1 \end{bmatrix} = \begin{bmatrix} \star \\ t \\ \star \\ \star \\ s \end{bmatrix} = \begin{bmatrix} 0 \\ 0 \\ 0 \\ 0 \\ 0 \end{bmatrix}$$

certainly means $t = s = 0$. (As on previous occasions, we use $\star$ for entries that don't matter.) So these vectors are a basis for the null space. The null space has dimension 2. ∎

■ Rank Plus Nullity $= n$

4.3.16 DEFINITION   The *nullity* of a matrix A is the dimension of its null space.

The matrix in Example **4.3.15** has nullity 2. This example also suggests that the nullity of an $m \times n$ matrix A is the number of free variables in the solution of $A\mathbf{x} = \mathbf{0}$. This is the number of columns of A that are not pivot columns.

The number of pivot columns is, by definition, the rank of the matrix. (See Section 4.1.) Since the number of columns without pivots plus the number of pivot columns is n, the total number of columns of A, we are led to one of the most important results in linear algebra.

Theorem 4.3.17   *For any $m \times n$ matrix A, rank A + nullity $A = n$, the number of columns of A.*

4.3.18 EXAMPLES
- Let $A = \begin{bmatrix} -3 & 6 \\ 2 & 5 \end{bmatrix}$. Gaussian elimination proceeds

$$A \to \begin{bmatrix} 1 & -2 \\ 2 & 5 \end{bmatrix} \to \begin{bmatrix} 1 & -2 \\ 0 & 9 \end{bmatrix}.$$

There are no free variables in the solution of $A\mathbf{x} = \mathbf{0}$, so nullity $A = 0$. There are two pivots, so rank $A = 2$ and rank A + nullity $A = 2 + 0 = 2 = n$, the number of columns of A.

- Let $A = \begin{bmatrix} 1 & 2 & 3 \\ 4 & 5 & 6 \\ 7 & 8 & 9 \end{bmatrix}$. Gaussian elimination proceeds

$$A \to \begin{bmatrix} 1 & 2 & 3 \\ 0 & -3 & -6 \\ 0 & -6 & -12 \end{bmatrix} \to \begin{bmatrix} 1 & 2 & 3 \\ 0 & 1 & 2 \\ 0 & 0 & 0 \end{bmatrix}.$$

There is one free variable in the solution of $A\mathbf{x} = \mathbf{0}$, so nullity $A = 1$. There are two pivots, so rank $A = 2$ and rank A + nullity $A = 2 + 1 = 3 = n$, the number of columns of A.

- Let $A = \begin{bmatrix} -3 & 4 & 5 & 8 & 1 \\ 1 & 0 & -2 & 1 & -1 \\ -2 & 4 & 3 & 9 & 0 \\ 4 & -4 & -7 & -7 & 3 \end{bmatrix}$. Gaussian elimination proceeds

$$A \to \begin{bmatrix} 1 & 0 & -2 & 1 & -1 \\ -3 & 4 & 5 & 8 & 1 \\ -2 & 4 & 3 & 9 & 0 \\ 4 & -4 & -7 & -7 & 3 \end{bmatrix} \to \begin{bmatrix} 1 & 0 & -2 & 1 & -1 \\ 0 & 4 & -1 & 11 & -2 \\ 0 & 4 & -1 & 11 & -2 \\ 0 & -4 & 1 & -11 & 7 \end{bmatrix}$$

$$\to \begin{bmatrix} 1 & 0 & -2 & 1 & -1 \\ 0 & 4 & -1 & 11 & -2 \\ 0 & 0 & 0 & 0 & 0 \\ 0 & 0 & 0 & 0 & 5 \end{bmatrix} \to \begin{bmatrix} 1 & 0 & -2 & 1 & -1 \\ 0 & 4 & -1 & 11 & -2 \\ 0 & 0 & 0 & 0 & 1 \\ 0 & 0 & 0 & 0 & 0 \end{bmatrix}.$$

There are two free variables in the solution of $A\mathbf{x} = \mathbf{0}$, so nullity $A = 2$. There are three pivot columns, so rank $A = 3$ and rank $A +$ nullity $A = 3 + 2 = 5 = n$, the number of columns of A. ∎

Finite Dimensional Vector Spaces: Basic Facts

Let V be a vector space of dimension n and let $\mathcal{V}$ be a basis for V. Then $\mathcal{V}$ contains n vectors. Let $\mathcal{L}$ be a linearly independent set of vectors. Since $\mathcal{V}$ spans V, Theorem 4.3.1 says that $|\mathcal{L}| \le |\mathcal{V}| = n$.

> **4.3.19** If dim $V = n$, any set of linearly independent vectors contains at most n vectors. Any set of more than n vectors is linearly dependent.

For example, every set of four or more vectors in $\mathbf{R}^3$ is linearly dependent.

Suppose $\mathcal{S}$ is a spanning set for V. Since $\mathcal{V}$ is a linearly independent set, Theorem 4.3.1 says that $|\mathcal{V}| \le |\mathcal{S}|$, so $|\mathcal{S}| \ge n$.

> **4.3.20** If dim $V = n$, any spanning set must contain at least n vectors. No set of fewer than n vectors spans V.

For example, $\mathbf{R}^3$ cannot be spanned by one vector, nor by two. Does this agree with intuition? One vector spans a line, two vectors span a plane (or a line, if they are parallel). To span $\mathbf{R}^3$, you need at least **three** vectors.

Our next few basic facts make use of a little lemma.

Lemma 4.3.21   *Suppose V is a finite dimensional vector space. Let $\mathbf{v}$ be a vector in V and let $\mathcal{L}$ be a set of linearly independent vectors in V. If $\mathbf{v}$ is not in the span of $\mathcal{L}$, then the set $\mathcal{L} \cup \{\mathbf{v}\}$ obtained by adding $\mathbf{v}$ to $\mathcal{L}$ is still linearly independent.*

The expression "$\mathcal{L} \cup \{\mathbf{v}\}$" is read "$\mathcal{L}$ union $\mathbf{v}$"; it means the set consisting of all the vectors in $\mathcal{L}$ together with $\mathbf{v}$.

Proof   Suppose $\mathcal{L} = \{\mathbf{v}_1, \mathbf{v}_2, \ldots, \mathbf{v}_k\}$. We must show that

$$\mathcal{L} \cup \{\mathbf{v}\} = \{\mathbf{v}, \mathbf{v}_1, \mathbf{v}_2, \ldots, \mathbf{v}_k\}$$

is linearly independent. So suppose

$$cv + c_1v_1 + c_2v_2 + \cdots + c_kv_k = 0 \tag{1}$$

for scalars $c, c_1, c_2, \ldots, c_k$. If $c \neq 0$, we can divide by c and rewrite (1) as

$$v = -\frac{c_1}{c}v_1 - \frac{c_2}{c}v_2 - \cdots - \frac{c_k}{c}v_k.$$

This expresses v as a linear combination of $v_1, \ldots, v_k$, which says that v is in the span of $\mathcal{L}$, which is not true. Thus $c = 0$ and (1) reads

$$c_1v_1 + c_2v_2 + \cdots + c_kv_k = 0.$$

Since the vectors in $\mathcal{L}$ are linearly independent, all the coefficients $c_i = 0$, so all the coefficients in (1) are 0. The vectors $v, v_1, v_2, \ldots, v_k$ are linearly independent, as claimed. ∎

Corollary 4.3.22 *If V is a finite dimensional vector space, any set of linearly independent vectors can be extended to a basis of V; that is, V has a basis that contains the given vectors.*

Proof Suppose V has dimension n and $\mathcal{L}$ is a set of linearly independent vectors in V. A basis is a set of vectors that is linearly independent and spans V. So, if $\mathcal{L}$ isn't already a basis, this is because $\mathcal{L}$ doesn't span V. If this is so, there exists a vector v_1 not in the span of $\mathcal{L}$. By Lemma 4.3.21, $\mathcal{L} \cup \{v_1\}$ is a linearly independent set. If this spans V, it forms a basis. Otherwise, there is a vector v_2 in V that is not in the span of $\mathcal{L} \cup \{v_1\}$. Thus $\mathcal{L} \cup \{v_1, v_2\}$ is linearly independent. We repeat this procedure, adding vectors to sets that don't span, if necessary, and getting increasingly larger linearly independent sets. At some point this process has to stop since, according to 4.3.19, no set of more than n vectors is linearly independent. Eventually, then, we obtain a basis for V of the form $\mathcal{L} \cup \{v_1, v_2, \ldots, v_k\}$. ∎

Corollary 4.3.23 *If V is a vector space of dimension n, then any set of n linearly independent vectors is a basis.*

Proof Suppose $v_1, v_2, \ldots, v_n$ are n linearly independent vectors. If these vectors do not form a basis, we can keep adding more vectors until a basis is obtained. This cannot occur since such a basis would have more than n vectors, so the given vectors must have already been a basis. ∎

4.3.24 EXAMPLE Let $v_1 = \begin{bmatrix} 1 \\ 2 \\ 3 \end{bmatrix}$ and $v_2 = \begin{bmatrix} -1 \\ 0 \\ 1 \end{bmatrix}$. The set $\mathcal{L} = \{v_1, v_2\}$ is linearly independent, so it can be extended to a basis of $\mathbf{R}^3$, consisting necessarily of three vectors. Thus we require a third vector v_3, which, together with v_1 and v_2, forms a linearly independent set (and hence a basis for $\mathbf{R}^3$). There are many possibilities, for example, $v_3 = \begin{bmatrix} 1 \\ 0 \\ 0 \end{bmatrix}$. We leave it to the reader to verify that $c_1v_1 + c_2v_2 + c_3v_3 = 0$ implies $c_1 = c_2 = c_3 = 0$. ∎

4.3.25 EXAMPLE   The vectors $\mathbf{v}_1 = \begin{bmatrix} 1 \\ 0 \\ 1 \\ 1 \end{bmatrix}$ and $\mathbf{v}_2 = \begin{bmatrix} 0 \\ -1 \\ 2 \\ 1 \end{bmatrix}$ are linearly independent in $\mathbf{R}^4$ (because neither is a scalar multiple of the other), so the set $\{\mathbf{v}_1, \mathbf{v}_2\}$ can be extended to a basis. There are numerous ways to do this. Since the span of these vectors has dimension 2, it can contain at most two of the four standard basis vectors $\mathbf{e}_1, \mathbf{e}_2, \mathbf{e}_3, \mathbf{e}_4$. In fact, $\{\mathbf{v}_1, \mathbf{v}_2, \mathbf{e}_1, \mathbf{e}_2\}$ is a basis of $\mathbf{R}^4$, and it extends the given set. ∎

Reading Challenge 1   *Verify that* $\{\mathbf{v}_1, \mathbf{v}_2, \mathbf{e}_1, \mathbf{e}_2\}$ *is a basis for* $\mathbf{R}^4$.

We wish now to make some comments about spanning sets, but first we require another lemma that gives a condition under which we can remove a vector from a given set of vectors, without changing the subspace which they span.

Lemma 4.3.26   *Let V be a vector space and let S be a finite spanning set for V. Let $\mathbf{v}$ be one of the vectors in S. If $\mathbf{v}$ is a linear combination of the vectors in $S \smallsetminus \{\mathbf{v}\}$, then $\mathbf{v}$ is redundant; that is, V is spanned by $S \smallsetminus \{\mathbf{v}\}$.*

The expression $S \smallsetminus \{\mathbf{v}\}$ is read "S minus $\mathbf{v}$". As you might guess, it means the set obtained by removing $\mathbf{v}$ from S.

Proof   We are given that S is a set of the form

$$S = \{\mathbf{v}, \mathbf{v}_1, \mathbf{v}_2, \dots, \mathbf{v}_k\}$$

and that

$$\mathbf{v} = a_1\mathbf{v}_1 + a_2\mathbf{v}_2 + \cdots + a_k\mathbf{v}_k \tag{2}$$

is a linear combination of $\mathbf{v}_1, \dots, \mathbf{v}_k$. Every vector in V is a linear combination of the vectors in S, and we are asked to show that $\mathbf{v}$ is redundant, that every vector in V is a linear combination of just $\mathbf{v}_1, \dots, \mathbf{v}_k$. A vector in $V = \mathrm{sp}\, S$ has the form

$$c\mathbf{v} + c_1\mathbf{v}_1 + c_2\mathbf{v}_2 + \cdots + c_k\mathbf{v}_k. \tag{3}$$

If we replace $\mathbf{v}$ with the expression on the right of (2), we get a linear combination of just $\mathbf{v}_1, \mathbf{v}_2, \dots, \mathbf{v}_k$, so the linear combination in (3) is really a linear combination of the vectors in $S \smallsetminus \{\mathbf{v}\}$. This is exactly what we wanted! ∎

Corollary 4.3.27   *If a vector space V is finite dimensional, any spanning set of vectors contains a basis.*

Proof   Suppose that $\{\mathbf{v}_1, \mathbf{v}_2, \dots, \mathbf{v}_k\}$ spans V. If these vectors are also linearly independent, they are already a basis. If they are not linearly independent, one of them is a linear combination of the others—see Theorem 4.2.37—and, by Lemma 4.3.26, this vector is redundant; that is, it can be removed, the remaining $k - 1$ vectors still spanning V. If these $k - 1$ vectors are linearly independent, they form a basis for V; otherwise, we can throw away another vector with the assurance that the remaining $k - 2$ vectors still span V. This process cannot be continued indefinitely, else we would be left with a set of zero vectors that spans V! The process must terminate, and it terminates because we eventually get a set of linearly independent vectors from within the original set that forms a basis for V. ∎

Corollary 4.3.28 *If V is a vector space of dimension n, any n vectors that span V form a basis.*

Proof Suppose $v_1, v_2, \ldots, v_n$ are n vectors that span V. If these vectors do not form a basis, some subset of fewer than n vectors does a form a basis. This contradicts $\dim V = n$. ∎

4.3.29 EXAMPLE

Let $u_1 = \begin{bmatrix} -1 \\ 2 \\ 1 \end{bmatrix}$, $u_2 = \begin{bmatrix} -1 \\ 8 \\ 2 \end{bmatrix}$, $u_3 = \begin{bmatrix} 3 \\ 0 \\ -2 \end{bmatrix}$, $u_4 = \begin{bmatrix} 2 \\ 1 \\ -3 \end{bmatrix}$, let $S = \{u_1, u_2, u_3, u_4\}$, and let V be the subspace of $\mathbf{R}^3$ spanned by S. Let's find a basis for V contained in S. Notice that u_1 and u_2 are linearly independent, so $\dim V \geq 2$. Are u_1, u_2, u_3 linearly independent? We solve $Ax = 0$ with $A = \begin{bmatrix} -1 & -1 & 3 \\ 2 & 8 & 0 \\ 1 & 2 & -2 \end{bmatrix}$, the matrix whose columns are these three vectors. The elementary row operations give

$$A \to \begin{bmatrix} 1 & 1 & -3 \\ 0 & 6 & 6 \\ 0 & 1 & 1 \end{bmatrix} \to \begin{bmatrix} 1 & 1 & -3 \\ 0 & 1 & 1 \\ 0 & 0 & 0 \end{bmatrix}$$

and show that there are nontrivial solutions. Thus the vectors u_1, u_2, u_3 are linearly dependent. It follows from Lemma 4.3.21 that u_3 is in the span of u_1, u_2 and hence redundant. ∎

Reading Challenge 2 *Why?*

What about u_1, u_2, u_4? Are these linearly independent?

Solving $Ax = 0$ with $A = \begin{bmatrix} -1 & -1 & 2 \\ 2 & 8 & 1 \\ 1 & 2 & -3 \end{bmatrix}$, we reduce A to row echelon form

$$A \to \begin{bmatrix} 1 & 1 & -2 \\ 0 & 6 & 5 \\ 0 & 1 & -1 \end{bmatrix} \to \begin{bmatrix} 1 & 1 & -2 \\ 0 & 1 & -1 \\ 0 & 0 & 11 \end{bmatrix} \to \begin{bmatrix} 1 & 1 & -2 \\ 0 & 1 & -1 \\ 0 & 0 & 1 \end{bmatrix}.$$

The only solution to $Ax = 0$ is $x = 0$. Vectors u_1, u_2, u_4 are linearly independent and hence a basis for $\mathbf{R}^3$. We have shown that $\operatorname{sp} S = V = \mathbf{R}^3$ and that S contains the basis $\{u_1, u_2, u_4\}$.

4.3.30 PROBLEM

Show that $\begin{bmatrix} 1 \\ 2 \\ 3 \\ 0 \end{bmatrix}, \begin{bmatrix} 4 \\ 6 \\ 2 \\ 1 \end{bmatrix}, \begin{bmatrix} 9 \\ 10 \\ 11 \\ 12 \end{bmatrix}, \begin{bmatrix} 8 \\ -4 \\ 23 \\ -2 \end{bmatrix}$ comprise a basis for $\mathbf{R}^4$.

Solution.   One Approach: Since $\dim \mathbf{R}^4 = 4$, by Corollary 4.3.28, it is sufficient to show that these vectors span $\mathbf{R}^4$. We apply Gaussian elimination to the matrix whose rows are the (transposes of) the given vectors; this does not change the row space.

$$\begin{bmatrix} 1 & 2 & 3 & 0 \\ 4 & 6 & 2 & 1 \\ 9 & 10 & 11 & 12 \\ 8 & -4 & 23 & -2 \end{bmatrix} \rightarrow \begin{bmatrix} 1 & 2 & 3 & 0 \\ 0 & -2 & -10 & 1 \\ 0 & -8 & -16 & 12 \\ 0 & -20 & -1 & -2 \end{bmatrix} \rightarrow \begin{bmatrix} 1 & 2 & 3 & 0 \\ 0 & -2 & -10 & 1 \\ 0 & 0 & 24 & 8 \\ 0 & 0 & 0 & -12 \end{bmatrix}.$$

The rows of the last matrix are independent and there are four of them, so they span $\mathbf{R}^4$. The elementary row operations do not change the span of the rows; thus the given vectors span $\mathbf{R}^4$.

A Second Approach: Since $\dim \mathbf{R}^4 = 4$, by Corollary 4.3.23, it is sufficient to show that these vectors are linearly independent. Let $A = \begin{bmatrix} 1 & 4 & 9 & 8 \\ 2 & 6 & 10 & -4 \\ 3 & 2 & 11 & 23 \\ 0 & 1 & 12 & -2 \end{bmatrix}$ be the matrix whose columns are the given vectors. The columns are linearly independent if and only if $A\mathbf{x} = \mathbf{0}$ has only the trivial solution $\mathbf{x} = \mathbf{0}$. Gaussian elimination proceeds

$$\begin{bmatrix} 1 & 4 & 9 & 8 \\ 2 & 6 & 10 & -4 \\ 3 & 2 & 11 & 23 \\ 0 & 1 & 12 & -2 \end{bmatrix} \rightarrow \begin{bmatrix} 1 & 4 & 9 & 8 \\ 0 & -2 & -8 & -20 \\ 0 & -10 & -16 & -1 \\ 0 & 1 & 12 & -2 \end{bmatrix}$$

$$\rightarrow \begin{bmatrix} 1 & 4 & 9 & 8 \\ 0 & -2 & -8 & -20 \\ 0 & 0 & 24 & 99 \\ 0 & 0 & 8 & -12 \end{bmatrix} \rightarrow \begin{bmatrix} 1 & 4 & 9 & 8 \\ 0 & -2 & -8 & -20 \\ 0 & 0 & 24 & 99 \\ 0 & 0 & 0 & -45 \end{bmatrix}.$$

Every column is a pivot column, so there are no free variables and the system $A\mathbf{x} = \mathbf{0}$ has only the trivial solution. The given vectors are linearly independent.

4.3.31 EXAMPLE   Let $\mathbf{v}_1 = \begin{bmatrix} 1 \\ 0 \\ 0 \end{bmatrix}$, $\mathbf{v}_2 = \begin{bmatrix} 1 \\ 1 \\ 0 \end{bmatrix}$, $\mathbf{v}_3 = \begin{bmatrix} 1 \\ 1 \\ 1 \end{bmatrix}$, and $\mathbf{v}_4 = \begin{bmatrix} 1 \\ 2 \\ 3 \end{bmatrix}$. The set $\mathcal{S} = \{\mathbf{v}_1, \mathbf{v}_2, \mathbf{v}_3, \mathbf{v}_4\}$ spans $\mathbf{R}^3$, so it contains a basis (of necessarily three vectors). We leave it to the reader to verify that the removal of any one of the given four vectors yields a spanning set, and hence a basis. For example, $\{\mathbf{v}_2, \mathbf{v}_3, \mathbf{v}_4\}$ is a basis. Perhaps the easiest way to establish this is to prove that $\mathbf{v}_2, \mathbf{v}_3, \mathbf{v}_4$ are linearly independent. ∎

4.3.32 EXAMPLE   Let $\mathbf{v}_1 = \begin{bmatrix} -1 \\ 0 \\ 1 \\ 1 \end{bmatrix}$, $\mathbf{v}_2 = \begin{bmatrix} 0 \\ 1 \\ 1 \\ 0 \end{bmatrix}$, $\mathbf{v}_3 = \begin{bmatrix} 2 \\ 1 \\ 1 \\ -1 \end{bmatrix}$, $\mathbf{v}_4 = \begin{bmatrix} -4 \\ -4 \\ -6 \\ 1 \end{bmatrix}$, and $\mathbf{v}_5 = \begin{bmatrix} 4 \\ 4 \\ -2 \\ -5 \end{bmatrix}$. Let V be the span of these vectors, which is the row space of $A = \begin{bmatrix} -1 & 0 & 1 & 1 \\ 0 & 1 & 1 & 0 \\ 2 & 1 & 1 & -1 \\ -4 & -4 & -6 & 1 \\ 4 & 4 & -2 & -5 \end{bmatrix}$. Gaussian elimination proceeds

$$
\begin{bmatrix} -1 & 0 & 1 & 1 \\ 0 & 1 & 1 & 0 \\ 2 & 1 & 1 & -1 \\ -4 & -4 & -6 & 1 \\ 4 & 4 & -2 & -5 \end{bmatrix}
\rightarrow
\begin{bmatrix} -1 & 0 & 1 & 1 \\ 0 & 1 & 1 & 0 \\ 0 & 1 & 3 & 1 \\ 0 & -4 & -10 & -3 \\ 0 & 4 & 2 & -1 \end{bmatrix}
$$

$$
\rightarrow
\begin{bmatrix} -1 & 0 & 1 & 1 \\ 0 & 1 & 1 & 0 \\ 0 & 0 & 2 & 1 \\ 0 & 0 & -6 & -3 \\ 0 & 0 & -2 & -1 \end{bmatrix}
\rightarrow
\begin{bmatrix} -1 & 0 & 1 & 1 \\ 0 & 1 & 1 & 0 \\ 0 & 0 & 2 & 1 \\ 0 & 0 & 0 & 0 \\ 0 & 0 & 0 & 0 \end{bmatrix}.
$$

The last matrix is upper triangular, so its nonzero rows are linearly independent. It follows that V has dimension 3 and that some subset of the given five vectors forms a basis for V. The fourth and fifth rows of zeros in the indicated row echelon form of A suggest that v_4 and v_5 are linear combinations of v_1, v_2, and v_3. Thus $\{v_1, v_2, v_3\}$ is a basis for V. ∎

■ Row Rank = Column Rank

In Section 4.2, we noted that the elementary row operations do not change the row space of a matrix. We also noted that the (nonzero) rows of a row echelon matrix are linearly independent.

Suppose then that Gaussian elimination moves a matrix A to a row echelon matrix U. The row space of A is the row space of U. Thus the nonzero rows of U span the row space of A and they are also linearly independent, so they are a basis for the row space. Therefore, the number of such nonzero rows is the dimension of the row space, a number rather naturally called the *row rank* of A. Notice that this number is just the number of pivot columns of A. Thus the row rank of a matrix is the same as its rank. (See Section 4.1.)

4.3.33 EXAMPLE   Let $A = \begin{bmatrix} 1 & 2 & -1 & 1 & 3 \\ -3 & -6 & 4 & -13 & -17 \\ 4 & 8 & -6 & 25 & 29 \\ -1 & -2 & 0 & 10 & 6 \end{bmatrix}$. In Example **4.3.15**, we found that the row echelon form of A was

$$
U = \begin{bmatrix} 1 & 2 & -1 & 1 & 3 \\ 0 & 0 & 1 & -10 & -8 \\ 0 & 0 & 0 & 1 & 1 \\ 0 & 0 & 0 & 0 & 0 \end{bmatrix}.
$$

The nonzero rows of U, namely,

$$
\begin{bmatrix} 1 & 2 & -1 & 1 & 3 \end{bmatrix}, \quad \begin{bmatrix} 0 & 0 & 1 & -10 & -8 \end{bmatrix}, \text{ and } \begin{bmatrix} 0 & 0 & 0 & 1 & 1 \end{bmatrix},
$$

form a basis for the row space of A, so dim row sp $A = 3 =$ row rank $A =$ rank A. ∎

Now let's think about the column space of a matrix. We wish to show that the dimension of the column space, which is called the *column rank*, is also the number of pivot columns; that is, we want to show that for any matrix A

row rank A = column rank A = rank A.

It will be useful first to consider the special case where the matrix in question is an $m \times n$ matrix U in row echelon form. For example, U might be the matrix of **4.3.33**.

There, there are three pivot columns—columns one, three, and four—and these are linearly independent (by Theorem 4.2.41). Thus the column space has dimension at least 3.

But think about the columns of U. Even though they are vectors in $\mathbf{R}^4$, since the last components are all 0, we can think of them as belonging to $\mathbf{R}^3$ (in the same way that we often think of vectors of the form $\begin{bmatrix} x \\ y \\ 0 \end{bmatrix}$ as belonging to the plane $\mathbf{R}^2$).

So columns one, three, and four of U are linearly independent vectors in a vector space that is essentially $\mathbf{R}^3$. Corollary 4.3.23 says that these columns form a basis for the column space of U.

We argue similarly for a general $m \times n$ matrix U in row echelon form. Suppose there are r pivot columns. These are linearly independent. Since U has m rows and r pivots, U has $m - r$ zero rows, so the columns of U are vectors in $\mathbf{R}^m$ whose last $m - r$ components are 0.

$$U = \begin{bmatrix} \star & \cdots & \star \\ & \star & & \star \\ \vdots & & \vdots \\ \star & & \star \\ 0 & \cdots & 0 \\ \vdots & & \vdots \\ 0 & \cdots & 0 \end{bmatrix} \begin{array}{l} \left.\vphantom{\begin{matrix}a\\a\\a\\a\end{matrix}}\right\} r \text{ rows containing the pivots} \\[2em] \left.\vphantom{\begin{matrix}a\\a\\a\end{matrix}}\right\} m - r \text{ zero rows} \end{array}$$

Technically, U is "isomorphic" to $\mathbf{R}^r$; the only difference is that the vectors in U have an additional $m - r$ rows of 0s at the bottom.

Effectively, the columns of U are vectors in $\mathbf{R}^r$. Thus the pivot columns of U are r linearly independent vectors in a vector space of dimension r. It follows that these columns form a basis for the column space, so the dimension of the column space is r.

4.3.34 If U is in row echelon form, the pivot columns of U comprise a basis for the column space of U.

Now let's think about the general situation. Suppose that A is any $m \times n$ matrix and let U be a row echelon form of A. The column spaces of A and U are usually quite different. For example, suppose $A = \begin{bmatrix} 1 & 2 \\ 2 & 4 \end{bmatrix}$. One step of the Gaussian elimination process carries A to row echelon form:

$$A = \begin{bmatrix} 1 & 2 \\ 2 & 4 \end{bmatrix} \rightarrow \begin{bmatrix} 1 & 2 \\ 0 & 0 \end{bmatrix} = U.$$

The column space of A is the set of scalar multiples of $\begin{bmatrix} 1 \\ 2 \end{bmatrix}$, but the column space of U is the set of scalar multiples of $\begin{bmatrix} 1 \\ 0 \end{bmatrix}$, a totally different vector space. Notice, however, that the column spaces of A and U do have the same dimension and the first column of each matrix is a basis for its column space.

In general, suppose that the pivot columns of U are columns $j_1, j_2, \ldots, j_r$.

$$
\begin{array}{ccccc}
 & \cdots & j_1 & \cdots & j_2 & \cdots \\
U = & \begin{bmatrix} & & 1 & & & \\ & & & & 1 & \\ & & & & & \end{bmatrix} \\
 & & \uparrow & & \uparrow & \\
 & & \text{first} & & \text{second} & \\
 & & \text{pivot} & & \text{pivot} &
\end{array}
$$

We are going to show that the pivot columns of A are a basis for the column space of A. Our argument depends on work in Section 2.5, where we saw that U can be written $U = EA$ with E invertible (E is the product of elementary matrices). Also, letting $a_1, \ldots, a_{j_1}, \ldots, a_{j_2}, \ldots, a_n$ denote the columns of A, it will be important to remember that the columns of EA are $Ea_1, \ldots, Ea_{j_1}, \ldots, Ea_{j_2}, \ldots, Ea_n$:

Do you regularly review the end-papers at the front and back of this book? See especially point 4.

$$
U = EA = E \begin{bmatrix} a_1 & \cdots & a_{j_1} & \cdots & a_{j_2} & \cdots & a_{j_r} & \cdots & a_n \\ \downarrow & & \downarrow & & \downarrow & & \downarrow & & \downarrow \end{bmatrix}
$$

$$
= \begin{bmatrix} Ea_1 & \cdots & Ea_{j_1} & \cdots & Ea_{j_2} & \cdots & Ea_{j_r} & \cdots & Ea_n \\ \downarrow & & \downarrow & & \downarrow & & \downarrow & & \downarrow \end{bmatrix}.
$$

The author has always been puzzled as to why some students don't want to learn the definition of linear independence **exactly** as formulated. We often prove vectors are linearly independent by appealing directly to the definition. It's a good one to know.

The pivots of U are in columns $j_1, j_2, \ldots, j_r$, so the pivot columns of A are $a_{j_1}, a_{j_2}, \ldots, a_{j_r}$, and it is these vectors we claim form a basis for the column space of A. Therefore, we have to show that they are linearly independent and that they span the column space of A. First, we show they are linearly independent and offer the "classic" proof of linear independence, appealing directly to the definition.

Thus we suppose

$$c_1 a_{j_1} + c_2 a_{j_2} + \cdots + c_r a_{j_r} = 0$$

and multiply this equation by E. This gives

$$E(c_1 a_{j_1} + c_2 a_{j_2} + \cdots + c_r a_{j_r}) = 0,$$

so

$$c_1(E\mathbf{a}_{j_1}) + c_2(E\mathbf{a}_{j_2}) + \cdots + c_r(E\mathbf{a}_{j_r}) = \mathbf{0}. \tag{4}$$

Now $E\mathbf{a}_{j_1}$ is column j_1 of U, the first pivot column of U. Similarly, $E\mathbf{a}_{j_2}$ is column j_2 of U, which is the second pivot column of U. Columns $E\mathbf{a}_1, \ldots, E\mathbf{a}_{j_r}$ are the pivot columns of U, which we know form a basis for the column space of U (see 4.3.34); in particular, they are linearly independent.

Thus the left side of (4) is a linear combination of linearly independent vectors, so $c_1 = 0, c_2 = 0, \ldots, c_r = 0$. This proves that $\mathbf{a}_{j_1}, \ldots, \mathbf{a}_{j_r}$ are linearly independent. Now we must prove that these vectors also span the column space of A.

To do this, we let $\mathbf{y}$ be a vector in the column space of A and show that $\mathbf{y}$ is a linear combination of $\mathbf{a}_{j_1}, \ldots, \mathbf{a}_{j_r}$. Since $\mathbf{y}$ is in the column space of A, we know that $\mathbf{y} = A\mathbf{x}$ for some vector $\mathbf{x}$. Multiplying by E, we get $E\mathbf{y} = EA\mathbf{x} = U\mathbf{x}$, which is in the column space of U. Now columns $j_1, \ldots, j_r$ of U—these are the pivot columns $E\mathbf{a}_{j_1}, E\mathbf{a}_{j_2}, \ldots, E\mathbf{a}_{j_r}$ of U—are a basis for the column space of U (4.3.34 again); in particular, they span the column space of U. Since $E\mathbf{y}$ is in the column space of U,

$$E\mathbf{y} = c_1 E\mathbf{a}_{j_1} + c_2 E\mathbf{a}_{j_2} + \cdots + c_r E\mathbf{a}_{j_r} \tag{5}$$

for certain scalars $c_1, c_2, \ldots, c_r$. The right side of (5) is $E(c_1\mathbf{a}_{j_1} + c_2\mathbf{a}_{j_2} + \cdots + c_r\mathbf{a}_{j_r})$, so

$$E\mathbf{y} = E(c_1\mathbf{a}_{j_1} + c_2\mathbf{a}_{j_2} + \cdots + c_r\mathbf{a}_{j_r}).$$

Since E is invertible, $\mathbf{y} = c_1\mathbf{a}_{j_1} + c_2\mathbf{a}_{j_2} + \cdots + c_r\mathbf{a}_{j_r}$, which says that $\mathbf{y}$ is in the span of $\mathbf{a}_{j_1}, \ldots, \mathbf{a}_{j_r}$. Since $\mathbf{y}$ was arbitrary, we have proven that these vectors indeed span the column space. Since they are linearly independent, they comprise a basis. We summarize.

4.3.35 The pivot columns of a matrix A form a basis for the column space of A.

4.3.36 EXAMPLE The matrix $A = \begin{bmatrix} 1 & 2 & -1 & 1 & 3 \\ -3 & -6 & 4 & -13 & -17 \\ 4 & 8 & -6 & 25 & 29 \end{bmatrix}$ of **4.3.15** has row echelon form $U = \begin{bmatrix} 1 & 2 & -1 & 1 & 3 \\ 0 & 0 & 1 & -10 & -8 \\ 0 & 0 & 0 & 1 & 1 \end{bmatrix}$. The pivot columns are columns one, three, and four. Thus columns one, three, and four of A, that is,

$$\begin{bmatrix} 1 \\ -3 \\ 4 \end{bmatrix}, \quad \begin{bmatrix} -1 \\ 4 \\ -6 \end{bmatrix}, \quad \text{and} \quad \begin{bmatrix} 1 \\ -13 \\ 25 \end{bmatrix},$$

comprise a basis for the column space of A. ∎

The dimension of the column space of a matrix A is its column rank. We have just seen that this number is r, the number of pivot columns of A. This number r, of course, is the rank of A, and we saw earlier that this was also the row rank. This establishes what we have been after.

Theorem 4.3.37   *For any matrix A, row rank $A = $ column rank $A = $ rank A.*

That the row rank and the column rank of a matrix are equal is quite an astonishing fact. If A is 15×28, for example, there seems to be no particular reason why the row space, a subspace of $\mathbf{R}^{28}$, and the column space, a subspace of $\mathbf{R}^{15}$, should have the same dimension, but they do!

4.3.38 EXAMPLE   Let $A = \begin{bmatrix} 1 & 2 & 3 & -9 & -3 & 6 \\ -4 & -8 & 0 & 0 & -12 & 12 \\ 3 & 6 & 5 & -15 & -1 & 18 \\ -2 & -4 & 1 & -3 & -8 & 4 \end{bmatrix}$. We leave it to the reader to verify that a row

echelon form of A is $U = \begin{bmatrix} 1 & 2 & 3 & -9 & -3 & 6 \\ 0 & 0 & 1 & -3 & -2 & 3 \\ 0 & 0 & 0 & 0 & 0 & 1 \\ 0 & 0 & 0 & 0 & 0 & 0 \end{bmatrix}$. There are three pivots, so the row

rank of A is 3 and the nonzero rows of U comprise a basis for the row space of A, that is, the rows

$$\begin{bmatrix} 1 & 2 & 3 & -9 & -3 & 6 \end{bmatrix}, \quad \begin{bmatrix} 0 & 0 & 1 & -3 & -2 & -3 \end{bmatrix}, \text{ and } \begin{bmatrix} 0 & 0 & 0 & 0 & 0 & 1 \end{bmatrix}.$$

The pivot columns of A are columns one, three, and six. Thus the first, third, and sixth columns of A are a basis for the column space of A. The column rank is 3 and

the column space of A has basis $\begin{bmatrix} 1 \\ -4 \\ 3 \\ -2 \end{bmatrix}, \begin{bmatrix} 3 \\ 0 \\ 5 \\ 1 \end{bmatrix}$, and $\begin{bmatrix} 6 \\ 12 \\ 18 \\ 4 \end{bmatrix}$. ∎

Because of Theorem 4.3.37, when we talk about the rank of a matrix A, we can think in terms of column rank or row rank, whichever we please, because these numbers are the same.

Suppose we have a matrix A. The rank of the transpose A^T is its column rank, the dimension of its column space. This column space is the row space of A, whose dimension is the row rank of A.

$$\text{rank } A^T = \text{column rank } A^T = \text{row rank } A = \text{rank } A.$$

We have obtained another remarkable property of matrices.

Theorem 4.3.39   *For any matrix A, rank $A = $ rank A^T.*

We close this section with a second surprising consequence of "row rank equals column rank." If A is a 5×20 matrix, say, then $A^T A$ is 20×20, and one might expect that this matrix could have a rank as big as 20. This cannot occur, however.

Theorem 4.3.40   *For any matrix A, rank A = rank $A^T A$ = rank AA^T.*

Proof   Suppose A is $m \times n$. First, we show that rank A = rank $A^T A$. Since each of these matrices has n columns, rank+nullity = n for each matrix (by Theorem 4.3.17), so we can show A and $A^T A$ have the same rank by showing they have the same nullity. Thus we examine the null space of each matrix.

Suppose $\mathbf{x}$ is in the null space of A. Thus $A\mathbf{x} = \mathbf{0}$, so certainly $A^T A \mathbf{x} = \mathbf{0}$. This observation shows that null sp A is contained in null sp $A^T A$. On the other hand, if $\mathbf{x}$ is in the null space of $A^T A$, then $A^T A \mathbf{x} = \mathbf{0}$, so $\mathbf{x}^T A^T A \mathbf{x} = \mathbf{x}^T \mathbf{0} = [0]$; that is, $(A\mathbf{x})^T (A\mathbf{x}) = [0]$. Now remember that $\mathbf{u}^T \mathbf{u} = \mathbf{u} \cdot \mathbf{u} = \|\mathbf{u}\|^2$ for any vector $\mathbf{u}$. Here, we have $\|A\mathbf{x}\| = 0$, so $A\mathbf{x} = 0$. This shows that null sp $A^T A$ is contained in null sp A (null sp $A^T A \subseteq$ null sp A), so null sp A = null sp $A^T A$. In particular, the dimensions of these two spaces are the same: nullity A = nullity $A^T A$, as desired. This shows that rank A = rank $A^T A$.

The second equality of the theorem—rank A = rank AA^T—follows by applying what we have just shown to the matrix $B = A^T$. We conclude that rank B = rank $B^T B$. The theorem then follows because $B^T B = AA^T$ and rank B = rank A, by Theorem 4.3.39.   ∎

Symbolically, "null sp A is contained in null sp $A^T A$" is written null sp $A \subseteq$ null sp $A^T A$, the symbol "$\subseteq$" denoting containment, or "subset of."

Reading Challenge 3   *If A is 5×20, what is the biggest possible rank of $A^T A$?*

Answers to Reading Challenges

1. We form the matrix $A = \begin{bmatrix} \mathbf{v}_1 & \mathbf{v}_2 & \mathbf{e}_1 & \mathbf{e}_2 \\ \downarrow & \downarrow & \downarrow & \downarrow \\ & & & \end{bmatrix} = \begin{bmatrix} 1 & 0 & 1 & 0 \\ 0 & -1 & 0 & 1 \\ 1 & 2 & 0 & 0 \\ 1 & 1 & 0 & 0 \end{bmatrix}$.

Since det $A = \begin{vmatrix} 1 & 0 & 1 & 0 \\ 0 & -1 & 0 & 1 \\ 1 & 2 & 0 & 0 \\ 1 & 1 & 0 & 0 \end{vmatrix} = \begin{vmatrix} 1 & 0 & 1 & 0 \\ 0 & -1 & 0 & 1 \\ 0 & 2 & -1 & 0 \\ 0 & 1 & -1 & 0 \end{vmatrix}$

$= \begin{vmatrix} 1 & 0 & 1 & 0 \\ 0 & -1 & 0 & 1 \\ 0 & 0 & -1 & 2 \\ 0 & 0 & -1 & 1 \end{vmatrix} = \begin{vmatrix} 1 & 0 & 1 & 0 \\ 0 & -1 & 0 & 1 \\ 0 & 0 & -1 & 2 \\ 0 & 0 & 0 & -1 \end{vmatrix} \neq 0,$

the system $A\mathbf{x} = \mathbf{0}$ has only the trivial solution. The vectors are linearly independent. Since there are four of them, they give a basis for $\mathbf{R}^4$.

2. If $\mathbf{u}_3$ is not in the span of $\mathcal{L} = \{\mathbf{u}_1, \mathbf{u}_2\}$ (that is, if $\mathbf{u}_3$ is not a linear combination of the vectors in $\mathcal{L}$), then $\mathcal{L} \cup \{\mathbf{u}_3\}$ is a linearly independent set, by Lemma 4.3.21, and this is not true.

3. Theorem 4.3.40 says that rank $A^T A$ = rank $A \leq 5$.

True/False Questions

Decide, with as little calculation as possible, whether each of the following statements is true or false and explain your answer whenever you say "false." (Answers can be found in the back of the book.)

1. The vectors $\mathbf{v}_1 = \begin{bmatrix} -1 \\ 2 \\ 3 \\ 4 \end{bmatrix}$, $\mathbf{v}_2 = \begin{bmatrix} 0 \\ 0 \\ -1 \\ 2 \end{bmatrix}$, $\mathbf{v}_3 = \begin{bmatrix} 2 \\ 1 \\ 3 \\ 5 \end{bmatrix}$ span $\mathbf{R}^4$.

2. The vectors $\mathbf{u} = \begin{bmatrix} 1 \\ 2 \end{bmatrix}$ and $\mathbf{v} = \begin{bmatrix} 3 \\ 1 \end{bmatrix}$ span $\mathbf{R}^2$.

3. The vectors $\mathbf{v}_1 = \begin{bmatrix} 1 \\ 2 \\ 3 \end{bmatrix}$, $\mathbf{v}_2 = \begin{bmatrix} -1 \\ -3 \\ 7 \end{bmatrix}$, $\mathbf{v}_3 = \begin{bmatrix} 0 \\ -1 \\ 1 \end{bmatrix}$, $\mathbf{v}_4 = \begin{bmatrix} 2 \\ 3 \\ -7 \end{bmatrix}$ are linearly independent.

4. Any set of linearly independent vectors in $\mathbf{R}^n$ contains at least n vectors.

5. Any set of vectors of $\mathbf{R}^n$ that spans $\mathbf{R}^n$ contains at least n vectors.

6. A finite dimensional vector space is a vector space that has only a finite number of vectors.

7. The vectors $\begin{bmatrix} 1 \\ 0 \\ 0 \end{bmatrix}$, $\begin{bmatrix} 0 \\ 1 \\ 0 \end{bmatrix}$, and $\begin{bmatrix} 0 \\ 0 \\ 1 \end{bmatrix}$ constitute a basis for the column space of
$\begin{bmatrix} 1 & 3 & 2 & 0 & 0 \\ 0 & 0 & 1 & 0 & 2 \\ 0 & 0 & 0 & 1 & 4 \end{bmatrix}$.

8. A basis for the row space of a matrix can be determined by reducing it to row echelon form.

9. A basis for the column space of a matrix can be determined by reducing it to row echelon form.

10. Given that $U = \begin{bmatrix} 1 & 2 & 3 & 4 \\ 0 & 1 & 2 & 3 \\ 0 & 0 & 0 & 0 \end{bmatrix}$ is a row echelon form of $A = \begin{bmatrix} 1 & 2 & 3 & 4 \\ 5 & 6 & 7 & 8 \\ 9 & 10 & 11 & 12 \end{bmatrix}$, the vectors $\begin{bmatrix} 1 \\ 0 \\ 0 \end{bmatrix}$ and $\begin{bmatrix} 2 \\ 1 \\ 0 \end{bmatrix}$ constitute a basis for A.

11. If A is an $m \times n$ matrix and $n > m$, the columns of A linearly dependent.

12. If A is a 4×3 matrix, the maximum value for the rank of AA^T is 3.

13. Given that the rows of $A = \begin{bmatrix} 1 & 1 & 1 \\ 1 & 2 & 3 \\ 1 & 3 & 6 \end{bmatrix}$ are linearly independent, the columns of A must be linearly independent too.

14. If the columns of an $n \times n$ matrix span $\mathbf{R}^n$, then the columns must be linearly independent.

Exercises

*Solutions to exercises marked [BB] can be found in the **B**ack of the **B**ook.*

1. Decide whether each of the following sets of vectors is a basis for the identified vector space.

   (a) [BB] $\left\{ \begin{bmatrix} 3 \\ 6 \\ 3 \\ -6 \end{bmatrix}, \begin{bmatrix} 0 \\ -1 \\ -1 \\ 0 \end{bmatrix}, \begin{bmatrix} 0 \\ -8 \\ -12 \\ -4 \end{bmatrix}, \begin{bmatrix} 1 \\ 0 \\ -1 \\ 2 \end{bmatrix} \right\}$ for $\mathbf{R}^4$

   (b) $\left\{ \begin{bmatrix} 1 \\ 4 \\ 7 \end{bmatrix}, \begin{bmatrix} 2 \\ 5 \\ 8 \end{bmatrix}, \begin{bmatrix} 3 \\ 6 \\ 9 \end{bmatrix} \right\}$ for $\mathbf{R}^3$

   (c) [BB] $\left\{ \begin{bmatrix} 1 \\ 4 \\ 7 \end{bmatrix}, \begin{bmatrix} 2 \\ 5 \\ 8 \end{bmatrix} \right\}$ for $\mathbf{R}^3$

   (d) $\left\{ \begin{bmatrix} 1 \\ 8 \\ -9 \\ 16 \end{bmatrix}, \begin{bmatrix} -1 \\ -6 \\ 10 \\ -15 \end{bmatrix}, \begin{bmatrix} -3 \\ 6 \\ 11 \\ -14 \end{bmatrix}, \begin{bmatrix} 4 \\ 5 \\ -12 \\ 13 \end{bmatrix} \right\}$ for $\mathbf{R}^4$

   (e) $\left\{ \begin{bmatrix} 1 \\ 8 \\ 9 \\ 16 \end{bmatrix}, \begin{bmatrix} 2 \\ 7 \\ 10 \\ 15 \end{bmatrix}, \begin{bmatrix} 3 \\ 6 \\ 11 \\ 14 \end{bmatrix}, \begin{bmatrix} 4 \\ 5 \\ 12 \\ 13 \end{bmatrix} \right\}$ for $\mathbf{R}^4$

   (f) $\left\{ \begin{bmatrix} -1 \\ 2 \\ 0 \\ 1 \end{bmatrix}, \begin{bmatrix} 0 \\ 0 \\ 1 \\ 1 \end{bmatrix}, \begin{bmatrix} -3 \\ 4 \\ 0 \\ 2 \end{bmatrix}, \begin{bmatrix} 0 \\ -5 \\ 6 \\ 14 \end{bmatrix}, \begin{bmatrix} -3 \\ 2 \\ 4 \\ 1 \end{bmatrix} \right\}$ for $\mathbf{R}^4$.

2. [BB] Let $\mathbf{v}_1$, $\mathbf{v}_2$, $\mathbf{v}_3$, $\mathbf{v}_4$, and $\mathbf{v}_5$ be the vectors defined in Example 4.3.32. Let A be the matrix with these vectors as columns. Use the principle stated in 4.3.35 to find a basis for the span of these vectors.

3. Repeat Exercise 2 with $\mathbf{v}_1 = \begin{bmatrix} 2 \\ 8 \\ 9 \\ -1 \end{bmatrix}$, $\mathbf{v}_2 = \begin{bmatrix} 4 \\ 24 \\ 27 \\ -3 \end{bmatrix}$,

   $\mathbf{v}_3 = \begin{bmatrix} 1 \\ 0 \\ 0 \\ 0 \end{bmatrix}$, $\mathbf{v}_4 = \begin{bmatrix} 3 \\ -6 \\ 11 \\ -2 \end{bmatrix}$, and $\mathbf{v}_5 = \begin{bmatrix} 4 \\ 5 \\ 2 \\ 6 \end{bmatrix}$.

4. (a) [BB] Let $\mathbf{v}_1 = \begin{bmatrix} 1 \\ 2 \end{bmatrix}$ and $\mathcal{L} = \{\mathbf{v}_1\}$. Extend $\mathcal{L}$ to a basis of $\mathbf{R}^2$.

   (b) Let $\mathbf{v}_1 = \begin{bmatrix} 1 \\ 0 \\ -1 \end{bmatrix}$, $\mathbf{v}_2 = \begin{bmatrix} 0 \\ 1 \\ 2 \end{bmatrix}$, and $\mathcal{L} = \{\mathbf{v}_1, \mathbf{v}_2\}$. Extend $\mathcal{L}$ to a basis of $\mathbf{R}^3$.

   (c) Let $\mathbf{v}_1 = \begin{bmatrix} 0 \\ 0 \\ 1 \\ -1 \end{bmatrix}$, $\mathbf{v}_2 = \begin{bmatrix} 1 \\ 2 \\ -1 \\ -3 \end{bmatrix}$, $\mathbf{v}_3 = \begin{bmatrix} -1 \\ -2 \\ 0 \\ 1 \end{bmatrix}$, and $\mathcal{L} = \{\mathbf{v}_1, \mathbf{v}_2, \mathbf{v}_3\}$. Extend $\mathcal{L}$ to a basis of $\mathbf{R}^4$. In every case, explain your reasoning (which is more important than your answer).

5. In each of the following cases, you are given a set $\mathcal{S}$ of vectors. Let $V = \mathrm{sp}\,\mathcal{S}$ be the span of $\mathcal{S}$. Find a basis for V contained in $\mathcal{S}$.

   (a) [BB] $\mathcal{S} = \left\{ \begin{bmatrix} 1 \\ 2 \\ 3 \end{bmatrix}, \begin{bmatrix} 0 \\ 1 \\ 0 \end{bmatrix}, \begin{bmatrix} 4 \\ 5 \\ 6 \end{bmatrix}, \begin{bmatrix} 1 \\ 0 \\ 0 \end{bmatrix}, \begin{bmatrix} 0 \\ 0 \\ 1 \end{bmatrix} \right\}$

   (b) $\mathcal{S} = \left\{ \begin{bmatrix} 1 \\ 1 \\ -1 \end{bmatrix}, \begin{bmatrix} 1 \\ 0 \\ -3 \end{bmatrix}, \begin{bmatrix} 0 \\ 1 \\ 2 \end{bmatrix}, \begin{bmatrix} 2 \\ 4 \\ 2 \end{bmatrix} \right\}$

   (c) $\mathcal{S} = \left\{ \begin{bmatrix} 1 \\ 1 \\ -1 \end{bmatrix}, \begin{bmatrix} 1 \\ 0 \\ -3 \end{bmatrix}, \begin{bmatrix} 1 \\ 1 \\ 2 \end{bmatrix}, \begin{bmatrix} 2 \\ 4 \\ 1 \end{bmatrix} \right\}$

   (d) $\mathcal{S} = \left\{ \begin{bmatrix} 1 \\ 1 \\ 1 \\ 1 \end{bmatrix}, \begin{bmatrix} 2 \\ -1 \\ -1 \\ 2 \end{bmatrix}, \begin{bmatrix} 3 \\ 2 \\ 1 \\ 2 \end{bmatrix}, \begin{bmatrix} 0 \\ -4 \\ -3 \\ 1 \end{bmatrix} \right\}$

6. [BB] Let $\mathcal{S} = \left\{ \begin{bmatrix} 1 \\ -1 \\ 3 \\ 2 \end{bmatrix}, \begin{bmatrix} 2 \\ 1 \\ 1 \\ 3 \end{bmatrix}, \begin{bmatrix} 1 \\ 5 \\ -7 \\ 0 \end{bmatrix}, \begin{bmatrix} 4 \\ -1 \\ 7 \\ 7 \end{bmatrix} \right\}$ and $V = \mathrm{sp}\,\mathcal{S}$.

   (a) Find a basis and the dimension of V.

   (b) Find all subsets of $\mathcal{S}$ that are bases for V. Explain.

7. Repeat Exercise 6 with $\mathcal{S} = \left\{ \begin{bmatrix} -1 \\ 0 \\ 2 \\ 1 \end{bmatrix}, \begin{bmatrix} 0 \\ 0 \\ 1 \\ 1 \end{bmatrix}, \begin{bmatrix} 1 \\ 1 \\ 5 \\ 2 \end{bmatrix}, \begin{bmatrix} 3 \\ 1 \\ 1 \\ 0 \end{bmatrix} \right\}$.

8. Suppose $\mathbf{u}$, $\mathbf{v}$, and $\mathbf{w}$ are vectors with $\{\mathbf{u}, \mathbf{v}\}$ linearly independent but $\{\mathbf{u}, \mathbf{v}, \mathbf{w}\}$ linearly dependent. Show that $\mathbf{w}$ is in the span of $\mathbf{u}$ and $\mathbf{v}$.

9. [BB] Let W be the subspace of $\mathbf{R}^4$ consisting of all

vectors $\mathbf{x} = \begin{bmatrix} x_1 \\ x_2 \\ x_3 \\ x_4 \end{bmatrix}$ that satisfy

$$\begin{aligned} x_1 + 2x_2 + 3x_3 + 5x_4 &= 0 \\ x_1 \quad\;\; + \;\; x_3 + 3x_4 &= 0 \\ 2x_1 + 3x_2 + 5x_3 + 9x_4 &= 0. \end{aligned}$$

Find a basis for and the dimension of W.

10. Let V be a finite dimensional vector space. A set $S = \{\mathbf{v}_1, \mathbf{v}_2, \dots, \mathbf{v}_k\}$ of vectors that spans V is said to be a *minimal spanning set* if no subset formed by removing one or more vectors of S spans V. Prove that a minimal spanning set of vectors in V is a basis for V.

11. [BB] Let V be a finite dimensional vector space. A set $S = \{\mathbf{v}_1, \mathbf{v}_2, \dots, \mathbf{v}_k\}$ of linearly independent vectors is said to be *maximal* if $S \cup \{\mathbf{v}\}$ is linearly dependent for any $\mathbf{v} \in V$. Prove that a maximal linearly independent set of vectors in V is a basis for V.

12. Suppose $\mathbf{v}_1, \dots, \mathbf{v}_r$ span a vector space V. Write a short note explaining why some subset of these vectors forms a basis. [Hint: Review the proof of Corollary 4.3.27.]

13. Let V be a vector space of dimension n and let U be a subspace of V.

   (a) [BB] Explain why any basis of U is part of a basis for V.

   (b) Why is $\dim U \le \dim V$?

14. Let V be a finite dimensional vector space. Suppose U is a subspace of V and $\dim U = \dim V$. Prove that $U = V$.

15. [BB] Let $A = \begin{bmatrix} 1 & -1 & 0 & 2 & 3 & 1 \\ 3 & -2 & 4 & 4 & 11 & 6 \\ 5 & -4 & 4 & 9 & 12 & 9 \\ 2 & -1 & 4 & 2 & 8 & 5 \end{bmatrix}.$

   (a) Find a row echelon form U for A.

   (b) Find a basis for the row space of U and a basis for the row space of A. Explain your answer.

   (c) What is the row rank of U? What is the row rank of A? Explain.

   (d) What are the pivot columns of U? Show that these columns are linearly independent.

   (e) Let the columns of A be $\mathbf{a}_1, \mathbf{a}_2, \mathbf{a}_3, \dots$ in order. Show that the pivot columns of A are also linearly independent.

   (f) Show that each of the "other" columns of A (those that are not pivot columns) is in the span of the ones identified in part (e). (Be explicit.)

   (g) It follows from part (f) that the columns identified in (e) span the column space of A and hence form a basis for the column space. What is the dimension of this space? (This shows that row rank = column rank for this matrix.)

   (h) Why is rank A^T = rank A?

   (i) What is the nullity of A? Why?

16. Repeat Exercise 15 using the matrix

$$A = \begin{bmatrix} 1 & -3 & 2 & 4 & 5 \\ 0 & 1 & 1 & 4 & 3 \\ 2 & -9 & 1 & -4 & 2 \\ -1 & 6 & 1 & 8 & 4 \end{bmatrix}.$$

17. [BB] Let $A = \begin{bmatrix} -1 & 0 & -1 & 1 \\ 0 & 1 & -1 & 2 \\ 1 & 1 & 0 & 1 \\ 2 & -1 & 3 & 1 \\ 0 & -1 & 1 & 0 \end{bmatrix}.$

   (a) Find a basis and the dimension of the row space of A.

   (b) Exhibit some columns of A that form a basis for the column space. What is the dimension of the column space?

   (c) Find a basis of the null space of A. What is the nullity of A?

18. Repeat Exercise 17 with $A = \begin{bmatrix} 11 & 8 & -2 & 3 \\ 2 & 3 & -1 & 2 \\ 7 & -15 & 7 & -17 \\ 4 & -11 & 5 & -12 \end{bmatrix}.$

19. Let $A = CR$ where C is a nonzero $m \times 1$ (column) matrix and R is a nonzero $1 \times n$ (row) matrix.

   (a) Show that C is a basis for col sp A.

   (b) Find nullity A (and explain).

   (c) Show that null sp A = null sp R.

20. Let the vectors $\mathbf{v}_1, \mathbf{v}_2, \dots, \mathbf{v}_n$ be a basis of $\mathbf{R}^n$ and let A be an $n \times n$ matrix.

   (a) [BB] Prove that the $n \times n$ matrix B whose columns are $\mathbf{v}_1, \mathbf{v}_2, \dots, \mathbf{v}_n$ is invertible.

   (b) Suppose $\{A\mathbf{v}_1, A\mathbf{v}_2, \dots, A\mathbf{v}_n\}$ is a basis of $\mathbf{R}^n$. Prove that A is invertible.

(c) Establish the converse of (b). If A is invertible, then the set $\{A\mathbf{v}_1, \ldots, A\mathbf{v}_n\}$ is a basis for $\mathbf{R}^n$.

21. For each of the following matrices A,

i. find a basis for the null space and the nullity;

ii. find a basis for the row space and the row rank;

iii. find a basis for the column space and the column rank;

iv. verify that rank A + nullity A is the number of columns of A.

(a) [BB] $A = \begin{bmatrix} 1 & 1 \\ 1 & -1 \\ 2 & 0 \\ 1 & 2 \end{bmatrix}$

(b) $A = \begin{bmatrix} 1 & 2 & 3 \\ 3 & 5 & 7 \\ 1 & 1 & 1 \end{bmatrix}$

(c) $A = \begin{bmatrix} 1 & 2 & 3 & -1 \\ 2 & 4 & 6 & 4 \end{bmatrix}$

(d) [BB] $A = \begin{bmatrix} 1 & 0 & 1 & 1 \\ 1 & 1 & 2 & 0 \\ 0 & 1 & 1 & -1 \end{bmatrix}$

(e) $A = \begin{bmatrix} 1 & 1 & -2 & 0 \\ 3 & -1 & 2 & -4 \\ -1 & 2 & -4 & 3 \end{bmatrix}$

(f) $A = \begin{bmatrix} 2 & 3 & 4 \\ 0 & 7 & 10 \\ -1 & 2 & 3 \\ 4 & -1 & -2 \end{bmatrix}$

(g) $A = \begin{bmatrix} 1 & 0 & 3 & 0 & 1 \\ 2 & 5 & -1 & 5 & -4 \\ -3 & 2 & 5 & 2 & 3 \\ 2 & -5 & 13 & -5 & 8 \end{bmatrix}$

(h) $A = \begin{bmatrix} 1 & -3 & 1 & -1 & 0 & -1 & 2 \\ -1 & 3 & 0 & 3 & 1 & 3 & -7 \\ 2 & -6 & 3 & 0 & -1 & 2 & 7 \\ -1 & 3 & 1 & 5 & 2 & 6 & -3 \\ 2 & -6 & 5 & 4 & 7 & 0 & -27 \end{bmatrix}$.

22. Suppose A is a 7×12 matrix. What are the sizes of $A^T A$ and AA^T? What are the possible values for the rank of each matrix? Explain your answer.

23. Let A be an $m \times n$ matrix with linearly independent columns.

(a) Why are the columns of $A^T A$ linearly independent?

(b) Is $A^T A$ invertible? Explain.

(c) Is AA^T invertible? Explain.

24. Let V be the vector space of all polynomials in the variable x with real coefficients. Prove that V does not have finite dimension. [Hint: It is sufficient to show that $1, x, x^2, \ldots, x^t$ are linearly independent for any $t \geq 1$. Why?]

■ Critical Reading

25. [BB] Given that two vectors $\mathbf{u}$ and $\mathbf{v}$ span $\mathbf{R}^n$, what can you say about n, and why?

26. Suppose vectors $\mathbf{v}_1, \mathbf{v}_2, \ldots, \mathbf{v}_n$ form a basis for a vector space V and $\mathbf{x}$ is a vector in V. What do you know?

27. Let $\mathbf{w}, \mathbf{v}_1, \mathbf{v}_2, \mathbf{v}_3$ be vectors. State a condition under which

$$\mathrm{sp}\{\mathbf{v}_1, \mathbf{v}_2, \mathbf{v}_3\} = \mathrm{sp}\{\mathbf{w}, \mathbf{v}_1, \mathbf{v}_2, \mathbf{v}_3\}.$$

28. (a) [BB] Suppose you wanted to exhibit three nonzero vectors in $\mathbf{R}^3$ that span a plane. What would you do?

(b) [BB] Suppose you were offered \$100 to produce three linearly independent vectors in $\mathbf{R}^3$ that span a plane. What would you do?

29. If A and B are matrices, B has rank 1, and AB is defined, then rank $AB \leq 1$.

30. [BB] If every row of a nonzero 7×7 matrix A is a multiple of the first, explain why every column of A is a multiple of one of its columns.

31. Show that the system $A\mathbf{x} = \mathbf{b}$ has a solution if and only if the rank of A is the rank of the augmented matrix $[A \mid \mathbf{b}]$.

32. (a) Let $\mathbf{x}$ and $\mathbf{y}$ be nonzero vectors in $\mathbf{R}^n$, thought of as column matrices. What is the size of the matrix $\mathbf{x}\mathbf{y}^T$? Show that this matrix has rank 1.

(b) Suppose A is an $m \times n$ matrix of rank 1. Show that there are column matrices $\mathbf{x}$ and $\mathbf{y}$ such that $A = \mathbf{x}\mathbf{y}^T$. [Hint: Think about the columns of A.]

Remark. Note the difference between $\mathbf{x}^T\mathbf{y}$ and $\mathbf{x}\mathbf{y}^T$. The 1×1 matrix $\mathbf{x}^T\mathbf{y}$ is a scalar, the dot product of $\mathbf{x}$ and $\mathbf{y}$. The dot product is a special case of a more general concept known as the *inner product*. The matrix $\mathbf{x}\mathbf{y}^T$ (which is not a scalar unless $n = 1$) is called the *outer product* of $\mathbf{x}$ and $\mathbf{y}$.

33. Given four nonzero vectors $\mathbf{v}_1$, $\mathbf{v}_2$, $\mathbf{v}_3$, and $\mathbf{v}_4$ in a vector space V, suppose that $\{\mathbf{v}_1, \mathbf{v}_2, \mathbf{v}_3\}$ and $\{\mathbf{v}_2, \mathbf{v}_3, \mathbf{v}_4\}$ are linearly dependent sets of vectors, but $\{\mathbf{v}_1, \mathbf{v}_2, \mathbf{v}_4\}$ is a linearly independent set. Prove that $\mathbf{v}_2$ is a multiple of $\mathbf{v}_3$.

34. If matrices B and AB have the same rank, prove that they have the same null spaces. [Hint: First observe that null sp B is contained in null sp AB. Why?]

35. If A is an $n \times n$ matrix of rank r and $A^2 = 0$, explain why $r \leq \frac{n}{2}$.

36. Suppose an $n \times n$ matrix A satisfies $A^2 = A$. Show that rank$(I - A) = n -$ rank A, where I denotes the $n \times n$ identity matrix.

37. Suppose $\{\mathbf{u}_1, \mathbf{u}_2, \dots, \mathbf{u}_r\}$ and $\{\mathbf{w}_1, \mathbf{w}_2, \dots, \mathbf{w}_s\}$ are linearly independent sets of vectors in a vector space V, each of which can be extended to a basis of V with the same set of vectors $\{\mathbf{x}_1, \mathbf{x}_2, \dots, \mathbf{x}_t\}$. Explain why $r = s$.

38. Suppose $\mathbf{u}$ and $\mathbf{w}$ are nonzero vectors in a (finite dimensional) vector space V, dim $V \geq 2$, and neither $\mathbf{u}$ nor $\mathbf{w}$ is a scalar multiple of the other. Show that there exists a vector $\mathbf{v}$ that is not a scalar multiple of $\mathbf{u}$ and not a scalar multiple of $\mathbf{w}$.

39. Suppose $\{\mathbf{u}_1, \mathbf{u}_2, \dots, \mathbf{u}_r\}$ and $\{\mathbf{w}_1, \mathbf{w}_2, \dots, \mathbf{w}_s\}$ are bases for vector spaces U and W, respectively. Let V be the set of all ordered pairs $(\mathbf{u}, \mathbf{w})$ of vectors, with $\mathbf{u}$ in U and $\mathbf{w}$ in W, and the agreement that $(\mathbf{u}_1, \mathbf{w}_1) = (\mathbf{u}_2, \mathbf{w}_2)$ if and only if $\mathbf{u}_1 = \mathbf{u}_2$ and $\mathbf{w}_1 = \mathbf{w}_2$. Then V is a vector space with addition and scalar multiplication defined by

$$(\mathbf{u}_1, \mathbf{w}_1) + (\mathbf{u}_2, \mathbf{w}_2) = (\mathbf{u}_1 + \mathbf{u}_2, \mathbf{w}_1 + \mathbf{w}_2)$$

and

$$a(\mathbf{u}, \mathbf{w}) = (a\mathbf{u}, a\mathbf{w})$$

for vectors $\mathbf{u}, \mathbf{u}_1, \mathbf{u}_2$ in U, $\mathbf{w}, \mathbf{w}_1, \mathbf{w}_2$ in W, and a scalar a (the proof is routine), but what is the dimension of V, and why?

4.4 One-Sided Inverses

Let V be a vector space of dimension n and let U be a subspace of V. If $\mathbf{u}_1, \mathbf{u}_2, \dots, \mathbf{u}_k$ are vectors in U that are linearly independent in U, then they are certainly linearly independent when we think of them as belonging to V. Thus $k \leq n$. It follows that U contains a finite maximal linearly independent set of at most n vectors (see Exercise 11 of Section 4.3). So U is finite dimensional of dimension at most n.

> **4.4.1** If V is a finite dimensional vector space and U is a subspace of V, then U is finite dimensional and dim $U \leq$ dim V.

For instance, since dim $\mathbf{R}^3 = 3$, the only subspaces of $\mathbf{R}^3$ have dimension 0, 1, 2, and 3. The only subspace of dimension 0 is the one consisting only of $\mathbf{0}$, the zero vector, and the only subspace of dimension 3 is $\mathbf{R}^3$ itself. The subspaces of dimension 1 are the lines through $\mathbf{0}$ and those of dimension 2 are the planes that contain $\mathbf{0}$.

Here's another application of 4.4.1.

Theorem 4.4.2 *If A is an $m \times n$ matrix, then* rank $A \leq m$ *and* rank $A \leq n$.

Proof The rank of A is the dimension of the row space of A. The row space of A is a subspace of $\mathbf{R}^n$ that has dimension n. By 4.4.1, rank $A = \dim$ row sp $A \leq n$. The rank of A is also the dimension of the column space of A. The column space of A is a subspace of $\mathbf{R}^m$ so, by 4.4.1, the dimension of the column space of A is at most m; that is, rank $A = \dim$ col sp $A \leq m$. ∎

Theorem 4.4.3 (**Matrix multiplication cannot increase rank**). *Suppose A is an $m \times n$ matrix, B is an $n \times p$ matrix, and C is an $r \times m$ matrix. Then*

(**a**) rank $AB \leq$ rank A, *and*

(**b**) rank $CA \leq$ rank A.

Proof (a) If $\mathbf{y}$ is in the column space of AB, then $\mathbf{y} = (AB)\mathbf{x}$ for some vector $\mathbf{x}$. Since $(AB)\mathbf{x} = A(B\mathbf{x})$ is in the column space of A, we have col sp $AB \subseteq$ col sp A. Thus rank $AB = \dim$ col sp $AB \leq \dim$ col sp $A =$ rank A.

  (b) We have rank $CA = \text{rank}(CA)^T = \text{rank } A^T C^T \leq \text{rank } A^T$ by the first part. Since rank $A^T =$ rank A, we have the second as well. ∎

The following result was first stated in Theorem 4.1.24.

Corollary 4.4.4 *An $n \times n$ matrix A is invertible if and only if* rank $A = n$.

Proof If A is invertible, there is a matrix B such that $AB = I$, the $n \times n$ identity matrix. Thus $n = \text{rank } I = \text{rank } AB \leq \text{rank } A$, using Theorem 4.4.3. On the other hand, since A is $n \times n$, rank $A \leq n$, so rank $A = n$.

  Conversely, suppose rank $A = n$. In particular, the column rank of A (which is the dimension of the column space) is n, so the column space of A is a subspace of $\mathbf{R}^n$ of dimension n. Therefore the column space of A **is** $\mathbf{R}^n$: Every vector in $\mathbf{R}^n$ is a linear combination of the columns of A; equivalently, the equation $A\mathbf{x} = \mathbf{b}$ has a solution for any $\mathbf{b}$ in $\mathbf{R}^n$. Exactly as in the proof of part (c) of Theorem 4.1.24 (iii $\implies$ iv), we construct a *right inverse* for A, that is, a matrix B with $AB = I$. Since A is square, B is also a left inverse, so A is invertible. (See Corollary 2.7.17.) ∎

As we have already used the terms "right inverse" and "left inverse," it is appropriate finally to give formal definitions.

4.4.5 DEFINITION Let A be an $m \times n$ matrix. A *right inverse* of A is an $n \times m$ matrix B such that $AB = I_m$, the $m \times m$ identity matrix and a *left inverse* of A is an $n \times m$ matrix C such that $CA = I_n$, the $n \times n$ identity matrix.

4.4.6 EXAMPLE Let $A = \begin{bmatrix} 1 & 2 & 3 \\ 4 & 5 & 6 \end{bmatrix}$ and $B = \begin{bmatrix} -\frac{2}{3} & -\frac{1}{3} \\ -\frac{2}{3} & \frac{5}{3} \\ 1 & -1 \end{bmatrix}$. The reader may check that $AB = I$. Thus B is a right inverse of A and A is a left inverse of B. The reader may check that $C = \begin{bmatrix} -\frac{5}{3} & \frac{8}{3} \\ \frac{4}{3} & -\frac{13}{3} \\ 0 & 2 \end{bmatrix}$ is also a right inverse for A, so we see that the right inverse of a matrix is not necessarily unique. (In fact, the right or left inverse of a matrix that is not square is never unique. See Exercise 17.) ∎

Reading Challenge 1   *If a matrix A has both a right inverse X and a left inverse Y, prove that X = Y.*
[Hint: Consider Y AX.]

Theorem 4.4.7   *Let A be an m × n matrix. Then the following statements are equivalent.*

1. *A has a right inverse.*

2. *For any* **b** *in* $\mathbf{R}^m$, *the system A***x** = **b** *has at least one solution.*

3. *The columns of A span* $\mathbf{R}^m$.

4. rank $A = m$.

5. *The rows of A are linearly independent.*

Proof   1 $\implies$ 2: If there is an $n \times m$ matrix B with $AB = I$, then $\mathbf{x} = B\mathbf{b}$ is a solution to $A\mathbf{x} = \mathbf{b}$ since $A\mathbf{x} = AB\mathbf{b} = I\mathbf{b} = \mathbf{b}$.

2 $\implies$ 3: Remember that $A\mathbf{x}$ is just a linear combination of the columns of A. Thus statement 2 says that every **b** in $\mathbf{R}^m$ is a linear combination of the columns of A, so the columns span $\mathbf{R}^m$.

3 $\implies$ 4: If the columns of A span $\mathbf{R}^m$, then the column space of A is $\mathbf{R}^m$, so the column rank is m. Therefore rank $A = m$.

4 $\implies$ 5: The row space of A has dimension m and is spanned by the m rows. These are a basis by Corollary 4.3.28 (and hence linearly independent).

5 $\implies$ 1: The row space of A is spanned by m linearly independent rows, so the rows are a basis for the row space, the dimension of the row space is m, and rank $A = m$. By Theorem 4.3.40, rank $AA^T = m$. Since AA^T is $m \times m$, AA^T is invertible, so $AA^T B = I_m$ for some matrix B. Thus $A^T B$ is a right inverse for A.   ∎

4.4.8 EXAMPLE   Let's look again at the matrix $A = \begin{bmatrix} 1 & 2 & 3 \\ 4 & 5 & 6 \end{bmatrix}$ that appeared in **4.4.6**. Since A has rank 2, the number of rows, A has a right inverse by Theorem 4.4.7. To find one, we use an idea from the proof of this theorem. We have $AA^T = \begin{bmatrix} 14 & 32 \\ 32 & 77 \end{bmatrix}$. This is invertible (as it must be) with inverse $B = \frac{1}{54}\begin{bmatrix} 77 & -32 \\ -32 & 14 \end{bmatrix}$. Since $AA^T B = I$, a right inverse for A is

$$A^T B = \frac{1}{54}\begin{bmatrix} 1 & 4 \\ 2 & 5 \\ 3 & 6 \end{bmatrix}\begin{bmatrix} 77 & -32 \\ -32 & 14 \end{bmatrix} = \frac{1}{18}\begin{bmatrix} -17 & 8 \\ -2 & 2 \\ 13 & -4 \end{bmatrix}.$$

So we have one right inverse for A (which the reader should confirm), but there are, in fact, many others, as we proceed to show.

To find all of these, we ask the reader first to reread Section 2.7 and recall **why** the method given in 2.7.8 for finding the inverse of a square matrix works.

Let $B = \begin{bmatrix} x_1 & x_2 \\ y_1 & y_2 \\ z_1 & z_2 \end{bmatrix}$ denote a right inverse of A. The matrix equation $AB = \begin{bmatrix} 1 & 0 \\ 0 & 1 \end{bmatrix}$ leads to the two systems

$$A \begin{bmatrix} x_1 \\ y_1 \\ z_1 \end{bmatrix} = \begin{bmatrix} 1 \\ 0 \end{bmatrix} \quad \text{and} \quad A \begin{bmatrix} x_2 \\ y_2 \\ z_2 \end{bmatrix} = \begin{bmatrix} 0 \\ 1 \end{bmatrix},$$

which we can solve simultaneously:

$$\begin{bmatrix} 1 & 2 & 3 & | & 1 & 0 \\ 4 & 5 & 6 & | & 0 & 1 \end{bmatrix}$$

$$\rightarrow \begin{bmatrix} 1 & 2 & 3 & | & 1 & 0 \\ 0 & -3 & -6 & | & -4 & 1 \end{bmatrix} \rightarrow \begin{bmatrix} 1 & 2 & 3 & | & 1 & 0 \\ 0 & 1 & 2 & | & \frac{4}{3} & -\frac{1}{3} \end{bmatrix}.$$

In the first system, $z_1 = t$ is free,

$$y_1 = \tfrac{4}{3} - 2z_1 = \tfrac{4}{3} - 2t,$$

and

$$x_1 = 1 - 2y_1 - 3z_1 = 1 - 2(\tfrac{4}{3} - 2t) - 3t = -\tfrac{5}{3} + t$$

so that $\begin{bmatrix} x_1 \\ y_1 \\ z_1 \end{bmatrix} = \begin{bmatrix} -\frac{5}{3} + 2t \\ \frac{4}{3} - 2t \\ t \end{bmatrix}$. In the second system, $z_2 = s$ is free,

$$y_2 = -\tfrac{1}{3} - 2z_2 = -\tfrac{1}{3} - 2s$$

$$x_2 = -2y_2 - 3z_2 = -2(-\tfrac{1}{3} - 2s) - 3s = \tfrac{2}{3} + s$$

so that $\begin{bmatrix} x_2 \\ y_2 \\ z_2 \end{bmatrix} = \begin{bmatrix} \frac{2}{3} + s \\ -\frac{1}{3} - 2s \\ s \end{bmatrix}$. The most general right inverse for A is

$$\begin{bmatrix} t - \frac{5}{3} & s + \frac{2}{3} \\ \frac{4}{3} - 2t & -\frac{1}{3} - 2s \\ t & s \end{bmatrix}.$$

Reading Challenge 2 *Earlier in **4.4.8**, we exhibited $\frac{1}{18} \begin{bmatrix} -17 & 8 \\ -2 & 2 \\ 13 & -4 \end{bmatrix}$ as a right inverse of A. Show that this matrix is a particular case of the "general" right inverse.*

Theorem 4.4.9 *Let A be an $m \times n$ matrix. Then the following statements are equivalent.*

1. *A has a left inverse.*

2. *The system $A\mathbf{x} = \mathbf{b}$ has at most one solution for any $\mathbf{b} \in \mathbf{R}^m$.*

3. *The columns of A are linearly independent.*

4. *rank $A = n$.*

5. *The rows of A span $\mathbf{R}^n$.*

Proof 1 $\implies$ 2: By 1, there is a matrix C with $CA = I_n$. Assume $\mathbf{x}_1$ and $\mathbf{x}_2$ are both solutions to $A\mathbf{x} = \mathbf{b}$. Then $A\mathbf{x}_1 = \mathbf{b} = A\mathbf{x}_2$. Multiplying on the left by C gives $CA\mathbf{x}_1 = CA\mathbf{x}_2$, so $\mathbf{x}_1 = \mathbf{x}_2$. Thus $A\mathbf{x} = \mathbf{b}$ has at most one solution.

2 $\implies$ 3: Suppose some linear combination of the columns of A is $\mathbf{0}$. Then $A\mathbf{x} = \mathbf{0}$ for some $\mathbf{x}$. Since $A\mathbf{0} = \mathbf{0}$ too, we must have $\mathbf{x} = \mathbf{0}$ by 2. Thus the columns are linearly independent.

3 $\implies$ 4: The n columns of A span the column space. By 3, they are also linearly independent, so they are basis for the column space. So $n =$ column rank $A =$ rank A.

4 $\implies$ 5: The row space of A has dimension n and is a subspace of $\mathbf{R}^n$. Thus the row space is $\mathbf{R}^n$, so the rows span $\mathbf{R}^n$.

5 $\implies$ 1: Statement 5 says that the row space of A is $\mathbf{R}^n$, thus rank $A = n$. By Theorem 4.3.40, rank $A^T A = n$. Since $A^T A$ is $n \times n$, $A^T A$ is invertible, so $(A^T A)^{-1}(A^T A) = I_n$. Thus $(A^T A)^{-1} A^T$ is a left inverse for A. ∎

4.4.10 EXAMPLE Let $A = \begin{bmatrix} 1 & 4 \\ 2 & 5 \\ 3 & 6 \end{bmatrix}$. Since this matrix has rank 2, which is the number of columns, it has a left inverse by Theorem 4.4.9. To find the most general left inverse, we note that if A has left inverse C (so $CA = I$), then A^T has right inverse C^T (since $A^T C^T = I$). In **4.4.8**, we found

$$\begin{bmatrix} 1 & 2 & 3 \\ 4 & 5 & 6 \end{bmatrix} \begin{bmatrix} -\frac{5}{3}+t & \frac{2}{3}+s \\ \frac{4}{3}-2t & -\frac{1}{3}-2s \\ t & s \end{bmatrix} = \begin{bmatrix} 1 & 0 \\ 0 & 1 \end{bmatrix},$$

so, taking transposes,

$$\begin{bmatrix} -\frac{5}{3}+t & \frac{4}{3}-2t & t \\ \frac{2}{3}+s & -\frac{1}{3}-2s & s \end{bmatrix} \begin{bmatrix} 1 & 4 \\ 2 & 5 \\ 3 & 6 \end{bmatrix} = \begin{bmatrix} 1 & 0 \\ 0 & 1 \end{bmatrix}.$$

The general left inverse of A is $\begin{bmatrix} -\frac{5}{3}+t & \frac{4}{3}-2t & t \\ \frac{2}{3}+s & -\frac{1}{3}-2s & s \end{bmatrix}$. ∎

Theorems 4.4.7 and 4.4.9 have several interesting consequences.

Corollary 4.4.11 *If A is $m \times n$ and the rows of A are linearly independent, then AA^T is invertible. If A is $m \times n$ and the columns of A are linearly independent, then $A^T A$ is invertible.*

Proof We use the fact that rank $A =$ rank $AA^T =$ rank $A^T A$—see Theorem 4.3.40. If the rows of A are linearly independent, rank $A = m =$ rank AA^T. Since AA^T is $m \times m$, AA^T is invertible. If the columns of A are linearly independent, rank $A = n =$ rank $A^T A$. Since $A^T A$ is $n \times n$, $A^T A$ is invertible. ∎

Corollary 4.4.12 *If a matrix A has both a right and a left inverse, then A is square and hence invertible.*

Proof Suppose A is $m \times n$. Since A has a right inverse, rank $A = m$ by Theorem 4.4.7. Since A has a left inverse, rank $A = n$ by Theorem 4.4.9. Thus $m = n$. Now apply Corollary 4.4.4. ∎

Corollary 4.4.13 *If A is a square matrix, then A has a right inverse if and only if A has a left inverse (and these are necessarily equal—see Reading Challenge 1).*

Proof Suppose A is $n \times n$. Then A has a right inverse if and only if rank $A = n$ by Theorem 4.4.7 and, by Theorem 4.4.9, this occurs if and only if A has a left inverse. ∎

Corollary 4.4.14 *Let A be an $n \times n$ matrix. Then the following statements are equivalent.*

1. *The rows of A are linearly independent.*
2. *The rows of A span $\mathbf{R}^n$.*
3. *The columns of A are linearly independent.*
4. *The columns of A span $\mathbf{R}^n$.*
5. *A is invertible.*
6. rank $A = n$.

Proof Statements 1, 4, and 6 are equivalent by Theorem 4.4.7. Statements 2, 3, and 6 are equivalent by Theorem 4.4.9. Corollary 4.4.4 gives the equivalence of 5 and 6, hence the theorem. ∎

Answers to Reading Challenges

1. Following the hint, we have $(YA)X = IX = X$ and $Y(AX) = YI = Y$. Since $(YA)X = Y(AX)$ (matrix multiplication is associative), $X = Y$.

2. Setting $t = \frac{13}{18}$ and $s = -\frac{4}{18}$ in the general right inverse gives the specific matrix $\frac{1}{18} \begin{bmatrix} -17 & 8 \\ -2 & 2 \\ 13 & -4 \end{bmatrix}$.

True/False Questions

Decide, with as little calculation as possible, whether each of the following statements is true or false and explain your answer whenever you say "false." (Answers can be found in the back of the book.)

1. A vector space of dimension 5 has precisely five subspaces.

2. There exists a 5×8 matrix of rank 6.

3. If $A = \begin{bmatrix} 1 & 2 & 3 \\ 2 & 4 & 6 \\ 17 & 8 & 92 \end{bmatrix}$ and $B = \begin{bmatrix} 1 & 2 & 3 \\ 0 & 7 & 5 \\ 0 & 0 & 4 \end{bmatrix}$, then rank $AB <$ rank B.

4. If A and B are matrices and B is not invertible, then rank $AB <$ rank A.

5. If A and B are matrices and rank $BA =$ rank A, then B is invertible.

6. If the equation $A\mathbf{x} = \mathbf{b}$ has a solution for any vector $\mathbf{b}$, the matrix A has a right inverse.

7. If the rows of an $m \times n$ matrix A are linearly independent, the columns of A span $\mathbf{R}^m$.

8. If the rows of an $m \times n$ matrix A span $\mathbf{R}^n$, the columns of A are linearly independent.

9. If a matrix A has a left inverse, the system $A\mathbf{x} = \mathbf{b}$ has a unique solution.

10. The matrix $A = \begin{bmatrix} 1 & -1 & 2 \\ -1 & 2 & 0 \end{bmatrix}$ has a right inverse.

11. The matrix $A = \begin{bmatrix} 1 & 3 & 7 \\ 9 & -10 & 8 \end{bmatrix}$ has a left inverse.

12. If A is $m \times n$, B is $n \times t$, and $AB = cI$ for some $c \neq 0$, then rank $A =$ rank B.

Exercises

Solutions to exercises marked [BB] can be found in the Back of the Book.

1. [BB] The matrix $A = \begin{bmatrix} 1 & 1 & -2 & 0 \\ 3 & -1 & 2 & -4 \\ -1 & 2 & -4 & 3 \end{bmatrix}$ has neither a right nor a left one-sided inverse. Why?

2. The matrix $A = \begin{bmatrix} 1 & 2 & 1 \\ 3 & 7 & 4 \\ 2 & 5 & 3 \\ 1 & 3 & 2 \end{bmatrix}$ has neither a right nor a left one-sided inverse. Why?

3. [BB] Let $A = \begin{bmatrix} -1 & 2 & 1 \\ 3 & 1 & 0 \end{bmatrix}$.

   (a) This matrix has a one-sided inverse. Which side and why?

   (b) Does A have a left inverse and a right inverse? Explain.

   (c) Find the most general one-sided inverse of A.

4. Answer Exercise 3 with $A = \begin{bmatrix} 1 & 1 \\ -1 & -2 \\ 2 & 4 \end{bmatrix}$.

5. Answer Exercise 3 with $A = \begin{bmatrix} 1 & 3 \\ 2 & 4 \\ 6 & 22 \end{bmatrix}$.

6. Answer Exercise 3 with $A = \begin{bmatrix} 1 & 2 & 3 \\ -2 & 1 & 4 \end{bmatrix}$.

7. [BB] Let $A = \begin{bmatrix} 1 & 0 & 3 & 4 \\ -1 & 1 & 4 & 0 \end{bmatrix}$.

   (a) This matrix has a one-sided inverse. Which side and why?

   (b) Find the most general one-sided inverse for A.

   (c) Give a specific example of a one-sided inverse of A.

8. Answer Exercise 7 with $A = \begin{bmatrix} 1 & 3 \\ 3 & 1 \\ -2 & 2 \end{bmatrix}$.

9. Answer Exercise 7 with
$$A = \begin{bmatrix} 3 & 12 & -6 & 0 \\ 3 & 13 & -3 & -1 \\ -5 & -18 & 16 & -6 \end{bmatrix}.$$

10. Let A be an $m \times n$ matrix, B an invertible $n \times n$ matrix, and C an invertible $m \times m$ matrix.

   (a) [BB] Show that rank $AB =$ rank A.

   (b) Show that rank $CA =$ rank A.

11. Let A be an $m \times n$ and B an $n \times t$ matrix.

   (a) [BB] If A and B each have a left inverse, show that AB also has a left inverse.

   (b) True or false: If A and B have linearly independent columns, so does AB. Justify your answer.

12. Suppose A is an $m \times n$ matrix and B is an $n \times t$ matrix.

   (a) If A and B each have a right inverse, show that AB also has a right inverse.

   (b) True or false: If the columns of A span $\mathbf{R}^m$ and the columns of B span $\mathbf{R}^n$, then the columns of AB span $\mathbf{R}^m$. Justify your answer.

13. Is the converse of each statement of Corollary 4.4.11 true? To be specific:

   (a) [BB] If A is $m \times n$ and $A^T A$ is invertible, need the columns of A be linearly independent?

   (b) If A is $m \times n$ and AA^T is invertible, need the rows of A be linearly independent?

   Justify your answers.

14. (a) [BB] Suppose an $m \times n$ matrix A has a unique right inverse. Prove that $m = n$ and A is invertible. [Hint: If $A\mathbf{x} = \mathbf{0}$, then $\mathbf{x} = \mathbf{0}$.]

   (b) Suppose an $m \times n$ matrix A has a unique left inverse. Prove that $m = n$ and A is invertible. (Use Exercise 14(a) if you wish.)

■ Critical Reading

15. [BB] Let U be a row echelon matrix without zero rows. Does U have a one-sided inverse? Explain.

16. Suppose U is a row echelon matrix that has a pivot in every column. Does U have a one-sided inverse? Explain.

17. Suppose A is an $m \times n$ matrix.

   (a) If $m < n$ and A has one right inverse, show that A has infinitely many right inverses. [Hint: First argue that null sp $A \neq \{\mathbf{0}\}$.]

   (b) If $n < m$ and A has one left inverse, show that A has infinitely many left inverses. [Hint: There is an easy way to see this using the result of part (a).]

18. Give an example of a matrix A with the property that $A\mathbf{x} = \mathbf{b}$ does not have a solution for all $\mathbf{b}$, but it does have a unique solution for some $\mathbf{b}$. Explain your answer.

19. Give an example of a matrix A for which the system $A\mathbf{x} = \mathbf{b}$ can be solved for any $\mathbf{b}$, yet not always uniquely.

20. Does there exist a matrix A with the property that $A\mathbf{x} = \mathbf{b}$ has a solution for any $\mathbf{b}$, sometimes a unique solution, sometimes not? Explain.

21. Let x_1, x_2, x_3, x_4 be real numbers and let

$$B = \begin{bmatrix} 4 & x_1 + x_2 + x_3 + x_4 \\ x_1 + x_2 + x_3 + x_4 & x_1^2 + x_2^2 + x_3^2 + x_4^2 \end{bmatrix}.$$

Show that det $B = 0$ if and only if $x_1 = x_2 = x_3 = x_4$.

[Hint: Let $A = \begin{bmatrix} 1 & x_1 \\ 1 & x_2 \\ 1 & x_3 \\ 1 & x_4 \end{bmatrix}$ and compute $A^T A$.]

CHAPTER KEY WORDS AND IDEAS: Here are some technical words and phrases that were used in this chapter. Do you know the meaning of each? If you're not sure, check the glossary or index at the back of the book.

associative (vector addition is)
basis
closed under addition
closed under scalar multiplication

nontrivial linear combination
null space
nullity
subspace spanned by vectors

(continued)

column rank

column space

commutative (vector addition is)

dimension

extend vectors to a basis

finite dimensional vector space

homogeneous linear system

inconsistent linear system

infinite dimensional vector space

left inverse of a matrix

linear combination of vectors

linearly dependent vectors

linearly independent vectors

rank

right inverse of a matrix

row rank

row space

span a subspace

span of vectors

spanning set

standard basis vectors

subspace

trivial linear combination

trivial solution

vector space

Review Exercises for Chapter 4

1. Complete the following sentence:

"An $n \times n$ matrix A has an inverse if and only if the homogeneous system $A\mathbf{x} = 0$ has _____ ."

2. Let $A = \begin{bmatrix} -2 & -1 & 3 & 1 \\ 1 & 1 & -2 & 1 \\ 1 & 0 & -2 & 0 \\ 0 & 1 & -2 & 1 \\ 1 & 1 & 3 & -1 \end{bmatrix}$, $\mathbf{b}_1 = \begin{bmatrix} 0 \\ 1 \\ 0 \\ -1 \\ 4 \end{bmatrix}$, and

$\mathbf{b}_2 = \begin{bmatrix} 1 \\ 0 \\ 1 \\ 0 \\ 1 \end{bmatrix}$.

(a) Is $\mathbf{b}_1$ in the column space of A? If it is, express this vector as a linear combination of the columns of A. If it is not, explain why not.

(b) Same as (a) for $\mathbf{b}_2$.

(c) Find a condition on the scalars $b_1, \ldots, b_5$ that is necessary and sufficient for $\mathbf{b} = \begin{bmatrix} b_1 \\ b_2 \\ b_3 \\ b_4 \\ b_5 \end{bmatrix}$ to belong to the column space of A; that is, complete the sentence

$\mathbf{b} = \begin{bmatrix} b_1 \\ b_2 \\ b_3 \\ b_4 \\ b_5 \end{bmatrix}$ is in the column space of A

if and only if _____.

3. Find the rank, the null space, and the nullity of each of the following matrices. In each case, verify that the sum of the rank and the nullity is what it should be.

(a) $A = \begin{bmatrix} 2 & 3 & 1 & 6 \\ 4 & 5 & 2 & 12 \\ 3 & 4 & 2 & 9 \end{bmatrix}$

(b) $A = \begin{bmatrix} 1 & 2 & 3 & 0 & 1 \\ 2 & 1 & 0 & 1 & 0 \\ -1 & 2 & 2 & 2 & 1 \\ 2 & 2 & -2 & 0 & 0 \\ 4 & 1 & 2 & 3 & 0 \end{bmatrix}$.

4. Suppose A and B are matrices and $AB = \mathbf{0}$ is the zero matrix.

(a) Explain why the column space of B is contained in the null space of A.

(b) Prove that rank A + rank $B \leq n$, the number of columns of A.

5. Suppose the $n \times n$ matrix A has rank $n - r$. Let $\mathbf{x}_1, \mathbf{x}_2, \ldots, \mathbf{x}_r$ be linearly independent vectors satisfying $A\mathbf{x} = \mathbf{0}$. Suppose $\mathbf{x}$ is a vector and $A\mathbf{x} = \mathbf{0}$. Explain why $\mathbf{x}$ must be a linear combination of $\mathbf{x}_1, \mathbf{x}_2, \ldots, \mathbf{x}_r$.

6. Suppose A is an $m \times n$ matrix and $\mathbf{x}$ is a vector in $\mathbf{R}^n$. Suppose $\mathbf{y}^T A\mathbf{x} = \mathbf{0}$ for all vectors $\mathbf{y}$ in $\mathbf{R}^m$. Explain why $\mathbf{x}$ must be in the null space of A.

7. Let $\mathbf{v}_1 = \begin{bmatrix} 7 \\ -8 \\ -4 \\ 3 \end{bmatrix}$, $\mathbf{v}_2 = \begin{bmatrix} -1 \\ 2 \\ 2 \\ 1 \end{bmatrix}$, $\mathbf{v}_3 = \begin{bmatrix} 2 \\ -1 \\ 1 \\ 3 \end{bmatrix}$,

$\mathbf{v}_4 = \begin{bmatrix} 3 \\ -3 \\ -1 \\ 2 \end{bmatrix}$, $\mathbf{v}_5 = \begin{bmatrix} 5 \\ -4 \\ 0 \\ 5 \end{bmatrix}$.

(a) Are these vectors linearly independent or linearly dependent?

(b) Do $\mathbf{v}_1$, $\mathbf{v}_4$, and $\mathbf{v}_5$ span $\mathbf{R}^4$?

(c) Do $\mathbf{v}_1$ and $\mathbf{v}_3$ span a two-dimensional subspace of $\mathbf{R}^4$?

(d) What is the dimension of the subspace of $\mathbf{R}^4$ spanned by the five given vectors?

Most of the answers here are very short.

8. Let $U = \{ \begin{bmatrix} x+y \\ x-2y \\ 3x+y \end{bmatrix} \mid x, y \in \mathbf{R} \}$.

Show that U is a subspace of $\mathbf{R}^3$ and find a basis for U.

9. Let U be the subspace of $\mathbf{R}^4$ spanned by

$\mathbf{u}_1 = \begin{bmatrix} 1 \\ 0 \\ 1 \\ 0 \end{bmatrix}$, $\mathbf{u}_2 = \begin{bmatrix} 0 \\ 1 \\ 2 \\ -1 \end{bmatrix}$, $\mathbf{u}_3 = \begin{bmatrix} 1 \\ -1 \\ 2 \\ -2 \end{bmatrix}$.

(a) Are $\mathbf{u}_1$, $\mathbf{u}_2$, $\mathbf{u}_3$ linearly independent or linearly dependent? Explain.

(b) Is $\begin{bmatrix} 1 \\ 0 \\ 0 \\ 0 \end{bmatrix}$ a linear combination of $\mathbf{u}_1$, $\mathbf{u}_2$, and $\mathbf{u}_3$?

(c) Do $\mathbf{u}_1$, $\mathbf{u}_2$, $\mathbf{u}_3$ span $\mathbf{R}^4$?

(d) What is $\dim U$?

(e) Illustrate the Cauchy–Schwarz inequality with respect to the vectors $\mathbf{u}_2$ and $\mathbf{u}_3$.

10. Let $A = \begin{bmatrix} 2 & 1 & -1 & 0 & 1 \\ -1 & -1 & 2 & -3 & 1 \\ 1 & 1 & -2 & 0 & -1 \\ 0 & 0 & 1 & 1 & 1 \end{bmatrix}$.

(a) Find a basis for the null space of A.

(b) Find a basis for the column space of A.

(c) Verify that rank A + nullity $A = n$. What is n?

(d) Are the columns of A linearly independent?

(e) Find the dimension of the subspace of $\mathbf{R}^5$ spanned by the rows of A.

11. Show that $U = \{ \begin{bmatrix} a \\ b \\ c \end{bmatrix} \mid a + 2b - 3c = 0 \}$ is a subspace of $\mathbf{R}^3$ and find a basis for U.

12. Let $U = \{ \begin{bmatrix} a \\ b \\ c \\ d \end{bmatrix} \mid a + b + c + d = 0 \}$.

(a) Show that U is a subspace of $\mathbf{R}^4$.

(b) Find a basis for U. What is the dimension of U?

13. Let $\mathbf{v}_1 = \begin{bmatrix} 1 \\ 1 \\ 1 \\ 1 \end{bmatrix}$ and $\mathbf{v}_2 = \begin{bmatrix} 1 \\ 0 \\ 1 \\ 0 \end{bmatrix}$. Extend $\mathbf{v}_1$, $\mathbf{v}_2$ to a basis of $\mathbf{R}^4$.

14. Let $\mathbf{v}_1 = \begin{bmatrix} 1 \\ 2 \\ -1 \\ 1 \end{bmatrix}$, $\mathbf{v}_2 = \begin{bmatrix} -2 \\ -4 \\ 2 \\ -2 \end{bmatrix}$, $\mathbf{v}_3 = \begin{bmatrix} 1 \\ 2 \\ -4 \\ 0 \end{bmatrix}$, $\mathbf{v}_4 = \begin{bmatrix} 2 \\ 4 \\ 1 \\ 3 \end{bmatrix}$.

(a) Find a basis $\mathcal{B}_0$ for the subspace of $\mathbf{R}^4$ spanned by $\mathbf{v}_1, \ldots, \mathbf{v}_4$.

(b) Extend $\mathcal{B}_0$ to a basis $\mathcal{B}$ of $\mathbf{R}^4$.

(c) Express $\begin{bmatrix} 1 \\ 2 \\ 3 \\ 5 \end{bmatrix}$ as linear combination of the vectors of $\mathcal{B}$.

15. Let $\mathbf{v}_1 = \begin{bmatrix} 1 \\ 2 \\ 3 \end{bmatrix}$, $\mathbf{v}_2 = \begin{bmatrix} 2 \\ 4 \\ 6 \end{bmatrix}$, $\mathbf{v}_3 = \begin{bmatrix} -1 \\ 0 \\ 1 \end{bmatrix}$, $\mathbf{v}_4 = \begin{bmatrix} 1 \\ 4 \\ 7 \end{bmatrix}$,

$\mathbf{v}_5 = \begin{bmatrix} 1 \\ 1 \\ 1 \end{bmatrix}$, and $\mathbf{v}_6 = \begin{bmatrix} 0 \\ 1 \\ 0 \end{bmatrix}$. Find all possible subsets of

$\{\mathbf{v}_1, \mathbf{v}_2, \mathbf{v}_3, \mathbf{v}_4, \mathbf{v}_5, \mathbf{v}_6\}$ that form bases of $\mathbf{R}^3$.

16. Let A be an $m \times n$ matrix of rank n. Prove that $A^T A$ is invertible.

17. Use the fact that $\begin{bmatrix} 4 & 0 & -2 \\ 3 & 1 & -1 \\ 1 & 3 & 1 \end{bmatrix} \begin{bmatrix} 1 \\ -1 \\ 2 \end{bmatrix} = \begin{bmatrix} 0 \\ 0 \\ 0 \end{bmatrix}$ to

show that there exist scalars a, b, c for which the system

$$\begin{array}{rcl} 4x & - 2z & = a \\ 3x + y - & z & = b \\ x + 3y + & z & = c \end{array}$$

has no solution.

18. Let $S = \{v_1, v_2, \ldots, v_m\}$ be a set of m vectors in $\mathbf{R}^n$. Determine, if possible and with reasons, whether S is linearly independent or spans $\mathbf{R}^n$ in each of the following cases:

    (a) $m = n$,   **(b)** $m > n$,   **(c)** $m < n$.

19. Suppose A is a 13×21 matrix. What are the possible values for nullity A, and why?

20. Suppose A is an $m \times n$ matrix and B is an $n \times p$ matrix.

    (a) Show that col sp $AB \subseteq$ col sp A.

    (b) Why is rank $AB \le$ rank A?

    (c) If A is 7×3 and B is 3×7, is it possible that $AB = I$? Explain.

21. Show that 0 is an eigenvalue of a matrix A if and only if A is not invertible.

22. (a) The matrix $A = \begin{bmatrix} 1 & 2 & 3 \\ 0 & 1 & -1 \end{bmatrix}$ has a right inverse. Why? Find the most general right inverse.

    (b) The matrix $A = \begin{bmatrix} 1 & 0 \\ 2 & 1 \\ 3 & -1 \end{bmatrix}$ has a left inverse. Why? Find the most general left inverse.

23. The matrix $A = \begin{bmatrix} -2 & 1 & 3 & 0 \\ 1 & -\frac{1}{2} & -1 & 1 \end{bmatrix}$ has a right inverse. Why? Find the most general right inverse of A.

24. The matrix $A = \begin{bmatrix} 1 & 3 \\ 3 & 11 \\ -2 & 2 \end{bmatrix}$ has a left inverse.

    (a) Why?

    (b) Does it also have a right inverse?

    (c) Find the most general left inverse.

25. How many statements can you think of that are equivalent to "matrix A is invertible?"

26. Let A be an $m \times n$ matrix. The linear system $A\mathbf{x} = \mathbf{b}$ has a solution for any $\mathbf{b}$ if and only if the column space of A is $\mathbf{R}^m$. Explain why.

27. Explain why a homogeneous system of equations in which the number of unknowns exceeds the number of equations always has a nontrivial solution.

28. Suppose A is a square matrix and the homogeneous system $A\mathbf{x} = \mathbf{0}$ has only the solution $\mathbf{x} = \mathbf{0}$. Explain why $A\mathbf{x} = \mathbf{b}$ has a solution for any $\mathbf{b}$.

29. Let A be an $m \times n$ matrix whose columns span $\mathbf{R}^n$. Explain how you know that $A\mathbf{x} = \mathbf{b}$ can be solved for any $\mathbf{b}$ in $\mathbf{R}^n$.

30. Let A be an $m \times n$ matrix. Suppose the system $A\mathbf{x} = \mathbf{b}$ has at least one solution for any $\mathbf{b} \in \mathbf{R}^m$. Explain why the columns of A must span $\mathbf{R}^m$.

5

Linear Transformations

5.1   Linear Transformations of $\mathbf{R}^n$

5.1.1 DEFINITION

A *linear transformation* is a function $T : V \to W$ from a vector space V to another vector space W that satisfies these two conditions:

1. $T(\mathbf{u} + \mathbf{v}) = T(\mathbf{u}) + T(\mathbf{v})$ and
2. $T(c\mathbf{v}) = cT(\mathbf{v})$

for all vectors $\mathbf{u}, \mathbf{v}$ in V and all scalars c.

5.1.2 EXAMPLES

- The function $T : \mathbf{R}^1 \to \mathbf{R}^1$ defined by $T(x) = 2x$ is a linear transformation because $T(x + y) = 2(x + y) = 2x + 2y = T(x) + T(y)$ and $T(cx) = 2(cx) = c(2x) = cT(x)$ for all "vectors" x, y in $\mathbf{R}^1 = \mathbf{R}$ and all scalars c.

- Let $\mathbf{x}$ be any vector in $\mathbf{R}^2$. Then the function $T : \mathbf{R}^2 \to \mathbf{R}^2$ defined by $T(\mathbf{v}) = \mathbf{v} \cdot \mathbf{x}$ is linear since for any vectors $\mathbf{u}, \mathbf{v}$ and any scalar c, $T(\mathbf{u}+\mathbf{v}) = (\mathbf{u}+\mathbf{v}) \cdot \mathbf{x} = \mathbf{u} \cdot \mathbf{x} + \mathbf{v} \cdot \mathbf{x}$ and $T(c\mathbf{v}) = (c\mathbf{v}) \cdot \mathbf{x} = c(\mathbf{v} \cdot \mathbf{x})$. (See Theorem 1.2.19.)

- The function $\sin : \mathbf{R} \to \mathbf{R}$ does **not** define a linear transformation; this function satisfies neither of the required properties. Readers should know that $\sin(x+y) \neq \sin x + \sin y$ and $\sin cx \neq c \sin x$ (in general).

- The function $T : \mathbf{R}^3 \to \mathbf{R}$ defined by

$$T(\mathbf{x}) = \det \begin{bmatrix} x_1 & x_2 & x_3 \\ 1 & 2 & 3 \\ 4 & 5 & 6 \end{bmatrix}$$

for $\mathbf{x} = \begin{bmatrix} x_1 \\ x_2 \\ x_3 \end{bmatrix}$ is a linear transformation; this is exactly what is meant by the assertion, "The determinant is **linear in its first row**." See Section 3.2.

285

- The "differentiation operator," D, which maps a polynomial to its derivative by the familiar rule

$$D(a_0 + a_1x + a_2x^2 + a_3x^3 + \cdots + a_nx^n)$$

$$= a_1 + 2a_2x + 3a_3x^2 + \cdots + na_nx^{n-1}.$$

Any student of calculus should know that $D(f+g) = D(f)+D(g)$ and $D(cf) = cf$ for any scalar c. (These rules may have been learned as $(f + g)' = f' + g'$ and $(cf)' = cf'$.) ∎

Reading Challenge 1   *As noted, the sine function is not a linear transformation* $\mathbf{R} \to \mathbf{R}$. *There are, however, nice expressions for* $\sin(x + y)$ *and* $\sin 2x$. *What are they?*

5.1.3 PROBLEM   Show that the function $T: \mathbf{R}^3 \to \mathbf{R}^2$ defined by $T\left(\begin{bmatrix} x_1 \\ x_2 \\ x_3 \end{bmatrix}\right) = \begin{bmatrix} x_1 - x_3 \\ x_1 + 2x_2 \end{bmatrix}$ is a linear transformation.

Solution.   Let $\mathbf{x} = \begin{bmatrix} x_1 \\ x_2 \\ x_3 \end{bmatrix}$ and $\mathbf{y} = \begin{bmatrix} y_1 \\ y_2 \\ y_3 \end{bmatrix}$ be vectors in $\mathbf{R}^3$. Then $\mathbf{x}+\mathbf{y} = \begin{bmatrix} x_1 + y_1 \\ x_2 + y_2 \\ x_3 + y_3 \end{bmatrix}$, so

$$T(\mathbf{x} + \mathbf{y}) = \begin{bmatrix} (x_1 + y_1) - (x_3 + y_3) \\ (x_1 + y_1) + 2(x_2 + y_2) \end{bmatrix}$$

$$= \begin{bmatrix} x_1 - x_3 + y_1 - y_3 \\ x_1 + 2x_2 + y_1 + 2y_2 \end{bmatrix}$$

$$= \begin{bmatrix} x_1 - x_3 \\ x_1 + 2x_2 \end{bmatrix} + \begin{bmatrix} y_1 - y_3 \\ y_1 + 2y_2 \end{bmatrix} = T(\mathbf{x}) + T(\mathbf{y}).$$

Also, for any scalar c, $c\mathbf{x} = c\begin{bmatrix} x_1 \\ x_2 \\ x_3 \end{bmatrix} = \begin{bmatrix} cx_1 \\ cx_2 \\ cx_3 \end{bmatrix}$, so

$$T(c\mathbf{x}) = \begin{bmatrix} cx_1 - cx_3 \\ cx_1 + 2cx_2 \end{bmatrix} = c\begin{bmatrix} x_1 - x_3 \\ x_1 + x_2 \end{bmatrix} = cT(\mathbf{x}).$$

5.1.4 EXAMPLES   Other examples of linear transformations include the following:

- The projection onto the xy-plane, that is, the function $T: \mathbf{R}^3 \to \mathbf{R}^3$ defined by $T\left(\begin{bmatrix} x \\ y \\ z \end{bmatrix}\right) = \begin{bmatrix} x \\ y \\ 0 \end{bmatrix}$ (see the leftmost picture in Figure 5.1) and

- Rotation in the plane through the angle θ, that is, the function $T: \mathbf{R}^2 \to \mathbf{R}^2$ where $T(\mathbf{x})$ is $\mathbf{x}$ rotated through θ (see the rightmost picture in Figure 5.1).

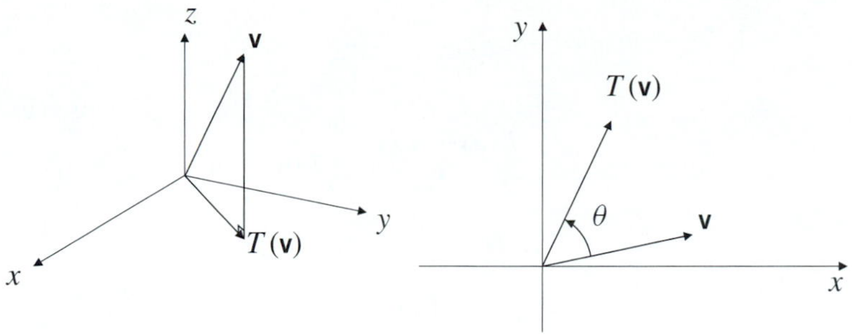

Figure 5.1   A projection (on the left) and a rotation

To see that the projection described here is a linear transformation, let $\mathbf{x} = \begin{bmatrix} x_1 \\ x_2 \\ x_3 \end{bmatrix}$

and $\mathbf{y} = \begin{bmatrix} y_1 \\ y_2 \\ y_3 \end{bmatrix}$ be vectors in $\mathbf{R}^3$ and let c be a scalar. Then $\mathbf{x} + \mathbf{y} = \begin{bmatrix} x_1 + y_1 \\ x_2 + y_2 \\ x_3 + y_3 \end{bmatrix}$ and

$c\mathbf{x} = \begin{bmatrix} cx_1 \\ cx_2 \\ cx_3 \end{bmatrix}$. So

$$T(\mathbf{x} + \mathbf{y}) = \begin{bmatrix} x_1 + y_1 \\ x_2 + y_2 \\ 0 \end{bmatrix} = \begin{bmatrix} x_1 \\ x_2 \\ 0 \end{bmatrix} + \begin{bmatrix} y_1 \\ y_2 \\ 0 \end{bmatrix} = T(\mathbf{x}) + T(\mathbf{y})$$

and

$$T(c\mathbf{x}) = \begin{bmatrix} cx_1 \\ cx_2 \\ 0 \end{bmatrix} = c \begin{bmatrix} x_1 \\ x_2 \\ 0 \end{bmatrix} = cT(\mathbf{x}).$$

To see that rotation in the plane through the angle θ is a linear transformation, it is probably easiest to draw pictures such as those shown in Figure 5.2. Geometrically, the sum $\mathbf{u} + \mathbf{v}$ of vectors is the diagonal of the parallelogram with sides the vectors $\mathbf{u}$ and $\mathbf{v}$. The statement $T(\mathbf{u} + \mathbf{v}) = T(\mathbf{u}) + T(\mathbf{v})$ expresses the fact that the rotated vector $\mathbf{u}+\mathbf{v}$ is the diagonal of the parallelogram with sides $T(\mathbf{u})$ and $T(\mathbf{v})$. Similarly, the statement $T(c\mathbf{u}) = cT(\mathbf{u})$ expresses the fact that the rotated vector $c\mathbf{u}$ is c times the vector $T(\mathbf{u})$. ■

This is not true for linear transformations on infinite dimensional vector spaces such as the differentiation operator on polynomials.

The most important and, as we shall see, essentially the only example of a linear transformation (on a finite dimensional vector space) is "left multiplication by a matrix."

5.1.5 EXAMPLE

This example is very important!

(Left Multiplication by a Matrix). Any $m \times n$ matrix A defines a linear transformation $T_A \colon \mathbf{R}^n \to \mathbf{R}^m$ by $T_A(\mathbf{x}) = A\mathbf{x}$. This function is linear because

$$T_A(\mathbf{x} + \mathbf{y}) = A(\mathbf{x} + \mathbf{y}) = A\mathbf{x} + A\mathbf{y}$$

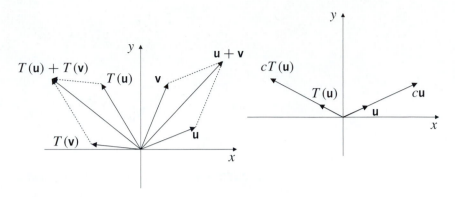

Figure 5.2 A rotation is a linear transformation. On the left, $T(\mathbf{u}+\mathbf{v}) = T(\mathbf{u}) + T(\mathbf{v})$; on the right, $T(c\mathbf{u}) = cT(\mathbf{u})$.

and

$$T_A(c\mathbf{x}) = A(c\mathbf{x}) = c(A\mathbf{x})$$

for all vectors $\mathbf{x}$, $\mathbf{y}$ in $\mathbf{R}^n$ and all scalars c. ∎

■ Properties of Linear Transformations

Recall that a vector space is a set on which there is defined an operation called addition and another called scalar multiplication. A linear transformation T is just a function that "preserves" these two operations. Applying T to the sum of two vectors $\mathbf{u}$, $\mathbf{v}$ is the same as applying T first, then adding. Applying T to a scalar multiple $c\mathbf{v}$ is the same as applying T to $\mathbf{v}$ and then multiplying by c.

In fact, a linear transformation preserves more than just addition and scalar multiplication.

5.1.6 A linear transformation preserves the zero vector: $T(\mathbf{0}) = \mathbf{0}$.

Let's see why this should be so. Suppose $T: \mathbf{R}^n \to \mathbf{R}^m$ is a linear transformation. Take any vector $\mathbf{v}$ in $\mathbf{R}^n$. Then $0\mathbf{v} = \begin{bmatrix} 0 \\ \vdots \\ 0 \end{bmatrix} = \mathbf{0}$ is the zero vector, so $T(\mathbf{0}) = T(0\mathbf{v}) = 0T(\mathbf{v})$, using Property 2 in Definition 5.1.1 to factor out the scalar 0. Since $0T(\mathbf{v})$ is the zero vector (of $\mathbf{R}^m$) no matter what $T(\mathbf{v})$ happens to be, we see that $T(\mathbf{0}) = \mathbf{0}$.

5.1.7 PROBLEM Explain why the function $T: \mathbf{R}^4 \to \mathbf{R}^3$ defined by $T\left(\begin{bmatrix} x_1 \\ x_2 \\ x_3 \\ x_4 \end{bmatrix}\right) = \begin{bmatrix} x_1 + 2x_4 \\ 3x_1 + x_3 - 5x_4 \\ 2 \end{bmatrix}$ is not linear.

Solution. In $\mathbf{R}^4$, $\mathbf{0} = \begin{bmatrix} 0 \\ 0 \\ 0 \\ 0 \\ 0 \end{bmatrix}$. Here, however, $T(\mathbf{0}) = \begin{bmatrix} 0 \\ 0 \\ 2 \end{bmatrix}$ is not the zero vector of $\mathbf{R}^3$. So T is not linear.

5.1.8 A linear transformation preserves negatives: $T(-\mathbf{x}) = -T(\mathbf{x})$.

The proof of this fact is very similar to the proof that a linear transformation preserves $\mathbf{0}$. If $\mathbf{x} = \begin{bmatrix} x_1 \\ x_2 \\ \vdots \\ x_n \end{bmatrix}$ is a vector in $\mathbf{R}^n$, the negative of $\mathbf{x}$ is the vector

$$-\mathbf{x} = \begin{bmatrix} -x_1 \\ \vdots \\ -x_n \end{bmatrix} = (-1)\begin{bmatrix} x_1 \\ \vdots \\ x_n \end{bmatrix} = (-1)\mathbf{x}.$$

Since T preserves scalar multiplication, $T(-\mathbf{x}) = T((-1)\mathbf{x}) = (-1)T(\mathbf{x}) = -T(\mathbf{x})$.

5.1.9 A linear transformation preserves linear combinations:
$T(c_1\mathbf{v}_1 + c_2\mathbf{v}_2 + \cdots + c_k\mathbf{v}_k) = c_1T(\mathbf{v}_1) + c_2T(\mathbf{v}_2) + \cdots + c_kT(\mathbf{v}_k)$
for any vectors $\mathbf{v}_1, \ldots, \mathbf{v}_k$ and scalars $c_1, \ldots, c_k$.

Let's see why this is so in the case of a linear combination of two vectors $\mathbf{v}_1$ and $\mathbf{v}_2$. Since T preserves addition,

$$T(c_1\mathbf{v}_1 + c_2\mathbf{v}_2) = T(c_1\mathbf{v}_1) + T(c_2\mathbf{v}_2)$$

for any scalars c_1 and c_2. Since T preserves scalar multiplication, we can continue, obtaining

$$T(c_1\mathbf{v}_1) + T(c_2\mathbf{v}_2) = c_1T(\mathbf{v}_1) + c_2T(\mathbf{v}_2).$$

Altogether, we have

$$T(c_1\mathbf{v}_1 + c_2\mathbf{v}_2) = c_1T(\mathbf{v}_1) + c_2T(\mathbf{v}_2).$$

Applying T to a linear combination of $\mathbf{v}_1$ and $\mathbf{v}_2$ gives a linear combination of $T(\mathbf{v}_1)$ and $T(\mathbf{v}_2)$ with the same coefficients.

That a linear transformation T preserves linear combinations is extremely important. It says, for instance, that if $\mathcal{S} = \{\mathbf{v}_1, \mathbf{v}_2, \ldots, \mathbf{v}_k\}$ is a spanning set for a vector space V, and if you know the values $T(\mathbf{v}_1), T(\mathbf{v}_2), \ldots, T(\mathbf{v}_k)$ of T on each vector in $\mathcal{S}$, then you know $T(\mathbf{v})$ for any $\mathbf{v}$. Why should this be the case? Any $\mathbf{v}$ is a linear combination of $\mathbf{v}_1, \mathbf{v}_2, \ldots, \mathbf{v}_k$ and T preserves linear combinations!

5.1.10 A linear transformation $T: V \to W$ is determined by its effect on any spanning set for V.

5.1.11 PROBLEM

If $f : \mathbf{R} \to \mathbf{R}$ is a function such that $f(1) = 2$, what can you say about f? Does it matter if f is a linear transformation?

Solution. In general, you can say nothing about f. Think, for example, of the functions defined by $f(x) = x^2 + 1$, $f(x) = 2\cos(x - 1)$, and $f(x) = 2^x$. On the other hand, suppose f is a linear transformation. Then, for any real number x, $f(x) = f(x \cdot 1) = xf(1)$ because x is a scalar and f preserves scalar multiplication. Thus $f(x) = 2x$.

5.1.12 EXAMPLE

Suppose $T : \mathbf{R}^2 \to \mathbf{R}^2$ is a linear transformation such that $T(\mathbf{i}) = \begin{bmatrix} 2 \\ -1 \end{bmatrix}$ and $T(\mathbf{j}) = \begin{bmatrix} 4 \\ 7 \end{bmatrix}$. Any $\mathbf{v} = \begin{bmatrix} a \\ b \end{bmatrix}$ in $\mathbf{R}^2$ can be written $\mathbf{v} = a\mathbf{i} + b\mathbf{j}$, so $T(\mathbf{v}) = aT(\mathbf{i}) + bT(\mathbf{j})$ because T preserves linear combinations. Thus $T(\mathbf{v}) = a\begin{bmatrix} 2 \\ -1 \end{bmatrix} + b\begin{bmatrix} 4 \\ 7 \end{bmatrix} = \begin{bmatrix} 2a + 4b \\ -a + 7b \end{bmatrix}$. Knowing only $T(\mathbf{i})$ and $T(\mathbf{j})$ tells us $T(\mathbf{v})$ for any $\mathbf{v}$. ∎

We **said** Example 5.1.5 was important!

We generalize the idea of this example to show that any linear transformation $\mathbf{R}^n \to \mathbf{R}^m$ is left multiplication by a matrix.

Theorem 5.1.13

Let $T : \mathbf{R}^n \to \mathbf{R}^m$ be a linear transformation. Let A be the $m \times n$ matrix whose columns are $T(\mathbf{e}_1), T(\mathbf{e}_2), \dots , T(\mathbf{e}_n)$. Then $T = T_A$ is left multiplication by A, that is, $T(\mathbf{x}) = A\mathbf{x}$ for any $\mathbf{x}$ in $\mathbf{R}^n$.

Proof Let

$$e_1 = \begin{bmatrix} 1 \\ 0 \\ 0 \\ \vdots \\ 0 \end{bmatrix}, \quad e_2 = \begin{bmatrix} 0 \\ 1 \\ 0 \\ \vdots \\ 0 \end{bmatrix}, \quad \dots, \quad e_n = \begin{bmatrix} 0 \\ 0 \\ \vdots \\ 0 \\ 1 \end{bmatrix}$$

denote the standard basis vectors of $\mathbf{R}^n$. If $\mathbf{x} = \begin{bmatrix} x_1 \\ x_2 \\ \vdots \\ x_n \end{bmatrix}$ is any vector in $\mathbf{R}^n$, then we can write

$$\mathbf{x} = x_1\mathbf{e}_1 + x_2\mathbf{e}_2 + \cdots + x_n\mathbf{e}_n$$

as a linear combination of $\mathbf{e}_1, \dots , \mathbf{e}_n$. Since T preserves linear combinations,

$$T(\mathbf{x}) = x_1 T(\mathbf{e}_1) + x_2 T(\mathbf{e}_2) + \cdots + x_n T(\mathbf{e}_n)$$

$$= \begin{bmatrix} T(\mathbf{e}_1) & T(\mathbf{e}_2) & \cdots & T(\mathbf{e}_n) \\ \downarrow & \downarrow & & \downarrow \end{bmatrix} \begin{bmatrix} x_1 \\ x_2 \\ \vdots \\ x_n \end{bmatrix} = A\mathbf{x}$$

where $A = \begin{bmatrix} T(\mathbf{e}_1) & T(\mathbf{e}_2) & \cdots & T(\mathbf{e}_n) \\ \downarrow & \downarrow & & \downarrow \end{bmatrix}$. So $T = T_A$ is left multiplication by A. ∎

Theorem 5.1.13 says there is a one-to-one correspondence between linear transformations $\mathbf{R}^n \to \mathbf{R}^m$ and $m \times n$ matrices.

$$\text{Linear transformations} \quad \longleftrightarrow \quad \text{Matrices}$$

$$T \quad \longrightarrow \quad \begin{bmatrix} T(e_1) & T(e_2) & \cdots & T(e_n) \\ \downarrow & \downarrow & & \downarrow \end{bmatrix}.$$

$$T_A \quad \longleftarrow \quad A$$

5.1.14 EXAMPLE

Let $T: \mathbf{R}^2 \to \mathbf{R}^2$ be the linear transformation defined by $T\left(\begin{bmatrix} x \\ y \end{bmatrix}\right) = \begin{bmatrix} 3x - y \\ 4x + 5y \end{bmatrix}$. Theorem 5.1.13 says that T is left multiplication by the matrix whose columns are $T(e_1)$ and $T(e_2)$. Since $e_1 = \begin{bmatrix} 1 \\ 0 \end{bmatrix}$ and $e_2 = \begin{bmatrix} 0 \\ 1 \end{bmatrix}$, we have $T(e_1) = \begin{bmatrix} 3 \\ 4 \end{bmatrix}$ and $T(e_2) = \begin{bmatrix} -1 \\ 5 \end{bmatrix}$, so $A = \begin{bmatrix} T(e_1) & T(e_2) \\ \downarrow & \downarrow \end{bmatrix} = \begin{bmatrix} 3 & -1 \\ 4 & 5 \end{bmatrix}$. For $x = \begin{bmatrix} x \\ y \end{bmatrix}$, notice that

$$Ax = \begin{bmatrix} 3 & -1 \\ 4 & 5 \end{bmatrix}\begin{bmatrix} x \\ y \end{bmatrix} = \begin{bmatrix} 3x - y \\ 4x + 5y \end{bmatrix} = T(x),$$

so T is indeed left multiplication by A. ∎

5.1.15 PROBLEM

Define $T: \mathbf{R}^3 \to \mathbf{R}^4$ by $T\left(\begin{bmatrix} x_1 \\ x_2 \\ x_3 \end{bmatrix}\right) = \begin{bmatrix} x_1 + 2x_2 + x_3 \\ 2x_1 - 4x_3 \\ -x_1 + x_2 - 2x_3 \\ 3x_1 - 2x_2 + 5x_3 \end{bmatrix}.$

Find a matrix A such that T is left multiplication by A.

Solution. Let $e_1 = \begin{bmatrix} 1 \\ 0 \\ 0 \end{bmatrix}$, $e_2 = \begin{bmatrix} 0 \\ 1 \\ 0 \end{bmatrix}$, $e_3 = \begin{bmatrix} 0 \\ 0 \\ 1 \end{bmatrix}$ be the standard basis vectors for $\mathbf{R}^3$. We have

$$T(e_1) = \begin{bmatrix} 1 \\ 2 \\ -1 \\ 3 \end{bmatrix}, \quad T(e_2) = \begin{bmatrix} 2 \\ 0 \\ 1 \\ -2 \end{bmatrix}, \quad T(e_3) = \begin{bmatrix} 1 \\ -4 \\ -2 \\ 5 \end{bmatrix}.$$

Let

$$A = \begin{bmatrix} T(e_1) & T(e_2) & T(e_3) \\ \downarrow & \downarrow & \downarrow \end{bmatrix} = \begin{bmatrix} 1 & 2 & 1 \\ 2 & 0 & -4 \\ -1 & 1 & -2 \\ 3 & -2 & 5 \end{bmatrix}.$$

Then

$$Ax = \begin{bmatrix} 1 & 2 & 1 \\ 2 & 0 & -4 \\ -1 & 1 & -2 \\ 3 & -2 & 5 \end{bmatrix} \begin{bmatrix} x_1 \\ x_2 \\ x_3 \end{bmatrix} = \begin{bmatrix} x_1 + 2x_2 + x_3 \\ 2x_1 - 4x_3 \\ -x_1 + x_2 - 2x_3 \\ 3x_1 - 2x_2 + 5x_3 \end{bmatrix} = T(\mathbf{x}),$$

so T is left multiplication by A.

5.1.16 EXAMPLE

(Projections). In 5.1.4, we defined the projection $T: \mathbf{R}^3 \to \mathbf{R}^3$ by $T\left(\begin{bmatrix} x \\ y \\ z \end{bmatrix}\right) = \begin{bmatrix} x \\ y \\ 0 \end{bmatrix}$.

Since $T(\mathbf{e}_1) = T\left(\begin{bmatrix} 1 \\ 0 \\ 0 \end{bmatrix}\right) = \begin{bmatrix} 1 \\ 0 \\ 0 \end{bmatrix}$, $T(\mathbf{e}_2) = T\left(\begin{bmatrix} 0 \\ 1 \\ 0 \end{bmatrix}\right) = \begin{bmatrix} 0 \\ 1 \\ 0 \end{bmatrix}$ and $T(\mathbf{e}_3) = T\left(\begin{bmatrix} 0 \\ 0 \\ 1 \end{bmatrix}\right) =$

$\begin{bmatrix} 0 \\ 0 \\ 0 \end{bmatrix}$, T is left multiplication by the matrix $A = \begin{bmatrix} 1 & 0 & 0 \\ 0 & 1 & 0 \\ 0 & 0 & 0 \end{bmatrix}$:

$$T\left(\begin{bmatrix} x \\ y \\ z \end{bmatrix}\right) = A \begin{bmatrix} x \\ y \\ z \end{bmatrix} = \begin{bmatrix} 1 & 0 & 0 \\ 0 & 1 & 0 \\ 0 & 0 & 0 \end{bmatrix} \begin{bmatrix} x \\ y \\ z \end{bmatrix} = \begin{bmatrix} x \\ y \\ 0 \end{bmatrix}. \qquad \blacksquare$$

5.1.17 PROBLEM

Let $T: \mathbf{R}^3 \to \mathbf{R}^3$ be the linear transformation that projects vectors onto the plane with equation $x + y - z = 0$. Find a matrix A such that T is left multiplication by A.

Solution. The given plane—call it π—is spanned by the orthogonal vectors $\mathbf{f}_1 = \begin{bmatrix} 1 \\ 0 \\ 1 \end{bmatrix}$

and $\mathbf{f}_2 = \begin{bmatrix} -1 \\ 2 \\ 1 \end{bmatrix}$. Using results from Section 1.4—see especially Theorem 1.4.13—the

projection of $\mathbf{e}_1 = \begin{bmatrix} 1 \\ 0 \\ 0 \end{bmatrix}$ on π is

$$T(\mathbf{e}_1) = \frac{\mathbf{e}_1 \cdot \mathbf{f}_1}{\mathbf{f}_1 \cdot \mathbf{f}_1} \mathbf{f}_1 + \frac{\mathbf{e}_1 \cdot \mathbf{f}_2}{\mathbf{f}_2 \cdot \mathbf{f}_2} \mathbf{f}_2 = \frac{1}{2} \begin{bmatrix} 1 \\ 0 \\ 1 \end{bmatrix} + \frac{-1}{6} \begin{bmatrix} -1 \\ 2 \\ 1 \end{bmatrix} = \begin{bmatrix} \frac{2}{3} \\ -\frac{1}{3} \\ \frac{1}{3} \end{bmatrix},$$

the projection of $\mathbf{e}_2 = \begin{bmatrix} 0 \\ 1 \\ 0 \end{bmatrix}$ on π is

$$T(\mathbf{e}_2) = \frac{\mathbf{e}_2 \cdot \mathbf{f}_1}{\mathbf{f}_1 \cdot \mathbf{f}_1} \mathbf{f}_1 + \frac{\mathbf{e}_2 \cdot \mathbf{f}_2}{\mathbf{f}_2 \cdot \mathbf{f}_2} \mathbf{f}_2 = \mathbf{0} + \frac{2}{6} \begin{bmatrix} -1 \\ 2 \\ 1 \end{bmatrix} = \begin{bmatrix} -\frac{1}{3} \\ \frac{2}{3} \\ \frac{1}{3} \end{bmatrix},$$

and the projection of $\mathbf{e}_3 = \begin{bmatrix} 0 \\ 0 \\ 1 \end{bmatrix}$ on π is

$$T(\mathbf{e}_3) = \frac{\mathbf{e}_3 \cdot \mathbf{f}_1}{\mathbf{f}_1 \cdot \mathbf{f}_1} \mathbf{f}_1 + \frac{\mathbf{e}_3 \cdot \mathbf{f}_2}{\mathbf{f}_2 \cdot \mathbf{f}_2} \mathbf{f}_2 = \frac{1}{2} \begin{bmatrix} 1 \\ 0 \\ 1 \end{bmatrix} + \frac{1}{6} \begin{bmatrix} -1 \\ 2 \\ 1 \end{bmatrix} = \begin{bmatrix} \frac{1}{3} \\ \frac{1}{3} \\ \frac{2}{3} \end{bmatrix}.$$

Thus T is left multiplication by $A = \begin{bmatrix} \frac{2}{3} & -\frac{1}{3} & \frac{1}{3} \\ -\frac{1}{3} & \frac{2}{3} & \frac{1}{3} \\ \frac{1}{3} & \frac{1}{3} & \frac{2}{3} \end{bmatrix}.$

5.1.18 EXAMPLE

(Rotations). In 5.1.4, we exhibited rotation in the plane as an example of a linear transformation and later gave a geometrical argument making this statement plausible. With reference to Figure 5.3, we see that $T(\mathbf{e}_1) = T\left(\begin{bmatrix} 1 \\ 0 \end{bmatrix}\right) = \begin{bmatrix} \cos\theta \\ \sin\theta \end{bmatrix}$ and $T(\mathbf{e}_2) = T\left(\begin{bmatrix} 0 \\ 1 \end{bmatrix}\right) = \begin{bmatrix} -\sin\theta \\ \cos\theta \end{bmatrix}$. Thus T is left multiplication by the matrix

$$A = \begin{bmatrix} \cos\theta & -\sin\theta \\ \sin\theta & \cos\theta \end{bmatrix}.$$

For any $\begin{bmatrix} x \\ y \end{bmatrix}$ in the plane, the rotated vector is

$$T\left(\begin{bmatrix} x \\ y \end{bmatrix}\right) = A\begin{bmatrix} x \\ y \end{bmatrix} = \begin{bmatrix} \cos\theta & -\sin\theta \\ \sin\theta & \cos\theta \end{bmatrix}\begin{bmatrix} x \\ y \end{bmatrix} = \begin{bmatrix} x\cos\theta - y\sin\theta \\ x\sin\theta + y\cos\theta \end{bmatrix},$$

a fact perhaps not so easily seen directly. For example, rotation through $\theta = 60° = \frac{\pi}{3}$ moves the vector $\begin{bmatrix} x \\ y \end{bmatrix}$ to $\frac{1}{2}\begin{bmatrix} x - \sqrt{3}y \\ \sqrt{3}x + y \end{bmatrix}.$ ∎

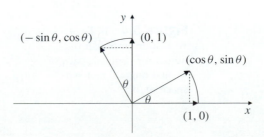

Figure 5.3

■ **Kernel and Range**

5.1.19 DEFINITION

The *kernel* of a linear transformation $T : V \to W$ is the set of vectors in V that T sends to the zero vector of W:

$$\ker T = \{\mathbf{v} \in V \mid T(\mathbf{v}) = \mathbf{0}\}.$$

The *range* of T is the set of vectors in W that are T of some vector in V:

$$\text{range } T = \{\mathbf{w} \in W \mid \mathbf{w} = T(\mathbf{v}) \text{ for some } \mathbf{v} \in V\}.$$

(See Figure 5.4.)

Proposition 5.1.20 *If $T : V \to W$ is a linear transformation, the kernel of T is a subspace of V and the range of T is a subspace of W.*

Proof Suppose $\mathbf{u}$ and $\mathbf{v}$ are in the kernel of T. Then $\mathbf{u} + \mathbf{v}$ is also in the kernel because $T(\mathbf{u} + \mathbf{v}) = T(\mathbf{u}) + T(\mathbf{v}) = \mathbf{0} + \mathbf{0} = \mathbf{0}$. Also, if $\mathbf{v}$ is in the kernel and c is any scalar, then $c\mathbf{v}$ is in the kernel because $T(c\mathbf{v}) = cT(\mathbf{v}) = c\mathbf{0} = \mathbf{0}$.

We leave the proof that range T is a subspace of W to the exercises (Exercise 10). ■

5.1.21 EXAMPLE

Let $T : \mathbf{R}^2 \to \mathbf{R}^3$ be the linear transformation defined by $T\left(\begin{bmatrix} x \\ y \end{bmatrix}\right) = \begin{bmatrix} x - y \\ 2x + y \\ -x + 4y \end{bmatrix}$.

Then $\mathbf{x} = \begin{bmatrix} x \\ y \end{bmatrix}$ is in $\ker T$ if and only if

$$\begin{array}{rcr} x - & y & = 0 \\ 2x + & y & = 0 \\ -x + & 4y & = 0. \end{array}$$

We obtain $x = y = 0$, so $\mathbf{x} = \mathbf{0}$: $\ker T = \{\mathbf{0}\}$. A vector $\mathbf{b} = \begin{bmatrix} b_1 \\ b_2 \\ b_3 \end{bmatrix}$ is in range T if and only if $\mathbf{b} = T(\mathbf{x})$ for some $\mathbf{x} = \begin{bmatrix} x \\ y \end{bmatrix}$. This is the case if and only if there exist x and y so that

$$\begin{array}{rcr} x - & y & = b_1 \\ 2x + & y & = b_2 \\ -x + & 4y & = b_3. \end{array}$$

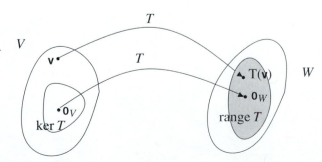

Figure 5.4 If $T : V \to W$ is a linear transformation, $\ker T$ is a subspace of V, and range T is a subspace of W.

Applying Gaussian elimination to the augmented matrix of coefficients, we get

$$\begin{bmatrix} 1 & -1 & b_1 \\ 2 & 1 & b_2 \\ -1 & 4 & b_3 \end{bmatrix} \rightarrow \begin{bmatrix} 1 & -1 & b_1 \\ 0 & 3 & b_2 - 2b_1 \\ 0 & 3 & b_1 + b_3 \end{bmatrix}$$

$$\rightarrow \begin{bmatrix} 1 & -1 & b_1 \\ 0 & 3 & b_2 - 2b_1 \\ 0 & 0 & (b_1 + b_3) - (b_2 - 2b_1) \end{bmatrix}$$

and so $\mathbf{b} \in \text{range } T$ if and only if $(b_1 + b_3) - (b_2 - 2b_1) = 0$, that is, if and only if

$3b_1 - b_2 + b_3 = 0$: $\text{range } T = \{\begin{bmatrix} x \\ y \\ z \end{bmatrix} \in \mathbf{R}^3 \mid 3x - y + z = 0\}$. ■

Suppose A is an $m \times n$ matrix and $T: \mathbf{R}^n \rightarrow \mathbf{R}^m$ is left multiplication by A: $T(\mathbf{x}) = A\mathbf{x}$ for $\mathbf{x}$ in $\mathbf{R}^n$. A vector $\mathbf{x}$ is in the kernel of T if and only if $A\mathbf{x} = \mathbf{0}$, that is, if and only if $\mathbf{x}$ is in the null space of the matrix A. A vector $\mathbf{y}$ is in the range of T if and only if $\mathbf{y} = A\mathbf{x}$ for some $\mathbf{x}$ in $\mathbf{R}^n$, that is, if and only if $\mathbf{y}$ is in the column space of A. Thus the concepts of "kernel" and "range" generalize familiar notions.

> **5.1.22** If $T: \mathbf{R}^n \rightarrow \mathbf{R}^m$ is left multiplication by the matrix A, then $\ker T = \text{null sp } A$ and $\text{range } T = \text{col sp } A$.

5.1.23 EXAMPLE   Let $T: \mathbf{R}^4 \rightarrow \mathbf{R}^3$ be left multiplication by $A = \begin{bmatrix} 1 & 1 & -3 & 5 \\ 3 & -2 & 1 & 0 \\ 4 & -2 & 0 & 2 \end{bmatrix}$.

Then $\mathbf{x} = \begin{bmatrix} x_1 \\ x_2 \\ x_3 \\ x_4 \end{bmatrix} \in \ker T$ if and only if $A\mathbf{x} = \mathbf{0}$. Using Gaussian elimination,

$$A \rightarrow \begin{bmatrix} 1 & 1 & -3 & 5 \\ 0 & -5 & 10 & -15 \\ 0 & -6 & 12 & -18 \end{bmatrix} \rightarrow \begin{bmatrix} 1 & 1 & -3 & 5 \\ 0 & 1 & -2 & 3 \\ 0 & 1 & -2 & 3 \end{bmatrix} \rightarrow \begin{bmatrix} 1 & 1 & -3 & 5 \\ 0 & 1 & -2 & 3 \\ 0 & 0 & 0 & 0 \end{bmatrix},$$

so $x_3 = t$ and $x_4 = s$ are free,

$$x_2 = 2x_3 - 3x_4 = 2t - 3s$$

$$x_1 = -x_2 + 3x_3 - 5x_4 = -(2t - 3s) + 3t - 5s = t - 2s.$$

Thus $\ker T = \{\begin{bmatrix} t - 2s \\ 2t - 3s \\ t \\ s \end{bmatrix} \mid t, s \in \mathbf{R}\}$ is the two-dimensional subspace of $\mathbf{R}^4$ with basis

$\begin{bmatrix} 1 \\ 2 \\ 1 \\ 0 \end{bmatrix}$ and $\begin{bmatrix} -2 \\ -3 \\ 0 \\ 1 \end{bmatrix}$. Now $\mathbf{y} = \begin{bmatrix} y_1 \\ y_2 \\ y_3 \end{bmatrix} \in \text{range } T$ if and only if $\mathbf{y} = A\mathbf{x}$ for some $\mathbf{x}$, that is, if and only if $\mathbf{y}$ is in the column space of A. Recall that the pivot columns of A are a basis for the column space—see 4.3.35. Here, the pivot columns are columns one and two, so range T is the two-dimensional subspace of $\mathbf{R}^3$ with basis $\begin{bmatrix} 1 \\ 3 \\ 4 \end{bmatrix}, \begin{bmatrix} 1 \\ -2 \\ -2 \end{bmatrix}$. ■

The following important theorem is a straightforward restatement of the rank + nullity theorem, Theorem 4.3.17.

Theorem 5.1.24   *If $T : \mathbf{R}^n \to \mathbf{R}^m$ is a linear transformation, then* $\dim \text{range } T + \dim \ker T = n$.

Proof   We know that $T : \mathbf{R}^n \to \mathbf{R}^m$ is left multiplication by a matrix A. By 5.1.22,

$$\ker T = \text{null sp } A, \quad \text{so } \dim \ker T = \dim \text{null sp } A = \text{nullity } A,$$

and

$$\text{range } T = \text{col sp } A, \quad \text{so } \dim \text{range } T = \dim \text{col sp } A = \text{rank } A.$$

Since $\text{rank } A + \text{nullity } A = n$, the number of columns of A, the result follows.   ■

(See Exercise 24 for another proof, one that does not use Theorem 4.3.17.)

■ **One-to-One and Onto**

5.1.25 DEFINITION   A linear transformation $T : V \to W$ is *one-to-one* if and only if $T(\mathbf{v}_1) = T(\mathbf{v}_2)$ implies $\mathbf{v}_1 = \mathbf{v}_2$.

The student may have encountered the notion of a one-to-one function before. The "doubling function" $\mathbf{R} \to \mathbf{R}$, which sends x to $2x$, is one-to-one since $2x_1 = 2x_2$ implies $x_1 = x_2$, whereas the "squaring function," which sends x to x^2, is not one-to-one. For example, $2^2 = (-2)^2$, but $2 \neq -2$.

5.1.26 EXAMPLE   The linear transformation $T : \mathbf{R}^2 \to \mathbf{R}^3$ defined by $T\left(\begin{bmatrix} x \\ y \end{bmatrix}\right) = \begin{bmatrix} x + y \\ x - y \\ 2x \end{bmatrix}$ is one-to-one

since, if $T\left(\begin{bmatrix} x_1 \\ y_1 \end{bmatrix}\right) = T\left(\begin{bmatrix} x_2 \\ y_2 \end{bmatrix}\right)$, then $\begin{bmatrix} x_1 + y_1 \\ x_1 - y_1 \\ 2x_1 \end{bmatrix} = \begin{bmatrix} x_2 + y_2 \\ x_2 - y_2 \\ 2x_2 \end{bmatrix}$ and it is easy to see that

$x_1 = x_2$, $y_1 = y_2$; that is, $\begin{bmatrix} x_1 \\ y_1 \end{bmatrix} = \begin{bmatrix} x_2 \\ y_2 \end{bmatrix}$. This, however, is not the best way to show that a linear transformation is one-to-one.   ■

Proposition 5.1.27   *A linear transformation T is one-to-one if and only if* $\ker T = \{\mathbf{0}\}$.

Proof   Suppose $T : V \to W$ is one-to-one and $\mathbf{v} \in \ker T$. Then $T(\mathbf{v}) = \mathbf{0}$. But $T(\mathbf{0}) = \mathbf{0}$ too, so $\mathbf{v} = \mathbf{0}$. On the other hand, if $\ker T = \mathbf{0}$ and $T(\mathbf{v}_1) = T(\mathbf{v}_2)$, then $T(\mathbf{v}_1 - \mathbf{v}_2) = T(\mathbf{v}_1) - T(\mathbf{v}_2) = \mathbf{0}$, so $\mathbf{v}_1 - \mathbf{v}_2 \in \ker T = \{\mathbf{0}\}$, so $\mathbf{v}_1 = \mathbf{v}_2$.   ■

Reading Challenge 2 *In this proof, we used $T(\mathbf{v}_1 - \mathbf{v}_2) = T(\mathbf{v}_1) - T(\mathbf{v}_2)$. Why is this the case?*

5.1.28 EXAMPLE To prove that the linear transformation T of Example 5.1.26 is one-to-one, observe merely that $\ker T = \{0\}$, for if $\mathbf{v} = \begin{bmatrix} x \\ y \end{bmatrix}$ and $T(\mathbf{v}) = \begin{bmatrix} 0 \\ 0 \end{bmatrix}$, then $x + y = 0$, $x - y = 0$, $2x = 0$, so $x = y = 0$. Thus $\mathbf{v} = \mathbf{0}$. ∎

5.1.29 DEFINITION A linear transformation $T: V \to W$ is *onto* if and only if range $T = W$, that is, if and only if every $\mathbf{w}$ in W is of the form $\mathbf{w} = T(\mathbf{v})$ for some $\mathbf{v}$ in V.

5.1.30 EXAMPLES
- The linear transformation $T: \mathbf{R}^2 \to \mathbf{R}^2$ defined by $T\left(\begin{bmatrix} x \\ y \end{bmatrix}\right) = \begin{bmatrix} x + y \\ x - y \end{bmatrix}$ is onto because any $\mathbf{w} = \begin{bmatrix} w_1 \\ w_2 \end{bmatrix}$ is $T\mathbf{v}$ with $\mathbf{v} = \begin{bmatrix} \frac{1}{2}(w_1 + w_2) \\ \frac{1}{2}(w_1 - w_2) \end{bmatrix}$.

- The linear transformation $T: \mathbf{R}^2 \to \mathbf{R}^2$ defined by $T\left(\begin{bmatrix} x \\ y \end{bmatrix}\right) = \begin{bmatrix} x \\ 2x \end{bmatrix}$ is not onto because, for instance, $\begin{bmatrix} 1 \\ 3 \end{bmatrix}$ is not $T(\mathbf{v})$ for any $\mathbf{v}$. Only vectors with second coordinate double the first are of the form $\mathbf{v}$. ∎

  If V is any vector space, the function id$: V \to V$ defined by id$(\mathbf{v}) = \mathbf{v}$ for any $\mathbf{v}$ is a linear transformation:
  1. id$(\mathbf{u} + \mathbf{v}) = \mathbf{u} + \mathbf{v} = $ id$(\mathbf{u}) + $ id$(\mathbf{v})$ for any $\mathbf{u}, \mathbf{v}$ in V, and
  2. id$(c\mathbf{v}) = c\mathbf{v} = c$ id$(\mathbf{v})$ for any scalar c and any $\mathbf{v}$ in V.

  This special linear transformation is called the *identity* on V.

Reading Challenge 3 *Prove that* id$: V \to V$ *is one-to-one and onto.*

Reading Challenge 4 *Suppose $T: V \to V$ is a linear transformation and* id *is the identity on V. Prove that* id $T = T$ id $= T$. *(The identity linear transformation behaves like the number 1, which satisfies $1a = a1 = a$ for any real number a.)*

Answers to Reading Challenges

1. $\sin(x + y) = \sin x \cos y + \cos x \sin y$; $\sin 2x = 2 \sin x \cos x$.
2. A linear transformation preserves linear combinations.
3. If id$(\mathbf{v}) = \mathbf{0}$, then $\mathbf{v} = \mathbf{0}$, so $\ker$ id $= \{0\}$ and id is one-to-one by Proposition 5.1.27. To see that id is onto, let $\mathbf{v}$ be a vector in V. Then $\mathbf{v} = $ id$(\mathbf{v})$ is in the range of id. So range id $= V$ and id is onto.
4. For any $\mathbf{v}$, we have id $T(\mathbf{v}) = $ id$(T(\mathbf{v})) = T(\mathbf{v})$, so id $T = T$. Also, T id$(\mathbf{v}) = T($id$(\mathbf{v})) = T(\mathbf{v})$ so T id $= T$.

True/False Questions

Decide, with as little calculation as possible, whether each of the following statements is true or false and explain your answer whenever you say "false." (Answers can be found in the back of the book.)

1. The square root function $\mathbf{R} \to \mathbf{R}$ is a linear transformation.

2. The function $T: \mathbf{R}^3 \to \mathbf{R}^4$ defined by $T\left(\begin{bmatrix} x \\ y \\ z \end{bmatrix}\right) = \begin{bmatrix} x - y \\ y - z \\ z - x + 2 \\ x + y + z \end{bmatrix}$ is a linear transformation.

3. If $T: \mathbf{R}^2 \to \mathbf{R}^2$ is a linear transformation such that $T\left(\begin{bmatrix} 1 \\ 2 \end{bmatrix}\right) = \begin{bmatrix} 0 \\ 1 \end{bmatrix}$ and $T\left(\begin{bmatrix} 2 \\ 1 \end{bmatrix}\right) = \begin{bmatrix} 1 \\ 0 \end{bmatrix}$, then $T\left(\begin{bmatrix} 3 \\ 3 \end{bmatrix}\right) = \begin{bmatrix} 1 \\ 1 \end{bmatrix}$.

4. A linear transformation is a function that is closed under addition and scalar multiplication.

5. If $T: \mathbf{R}^3 \to \mathbf{R}^2$ is a linear transformation such that $T\left(\begin{bmatrix} -3 \\ 1 \end{bmatrix}\right) = \begin{bmatrix} 0 \\ 0 \end{bmatrix}$, then $\begin{bmatrix} -3 \\ 1 \end{bmatrix}$ is in the kernel of T.

6. If $T: \mathbf{R}^3 \to \mathbf{R}^2$ is a linear transformation such that $T\left(\begin{bmatrix} -3 \\ 1 \end{bmatrix}\right) = \begin{bmatrix} 0 \\ 0 \end{bmatrix}$, then T is not one-to-one.

7. A linear transformation $T: \mathbf{R}^3 \to \mathbf{R}^4$ cannot be onto.

8. If V is a finite dimensional vector space and $T: V \to V$ is a linear transformation, then T is one-to-one if and only if T is onto.

Exercises

*Solutions to exercises marked [BB] can be found in the **B**ack of the **B**ook.*

1. Determine whether each of the following functions is a linear transformation.

   (a) [BB] $T: \mathbf{R}^3 \to \mathbf{R}^3$ is defined by
   $$T\left(\begin{bmatrix} x \\ y \\ z \end{bmatrix}\right) = \begin{bmatrix} x + 2y + 3z \\ 1 \\ y - x - z \end{bmatrix}$$

   (b) [BB] $T: \mathbf{R}^3 \to \mathbf{R}^2$ is defined by
   $$T\left(\begin{bmatrix} x_1 \\ x_2 \\ x_3 \end{bmatrix}\right) = \begin{bmatrix} x_1 + x_2 \\ x_2 + x_3 \end{bmatrix}$$

   (c) $T: \mathbf{R}^2 \to \mathbf{R}^2$ rotates vectors $45°$ about the point $(1, 0)$

   (d) $T: \mathbf{R}^2 \to \mathbf{R}^3$ is defined by $T\left(\begin{bmatrix} x \\ y \end{bmatrix}\right) = \begin{bmatrix} x + y \\ xy \\ y \end{bmatrix}$

   (e) $T: \mathbf{R}^2 \to \mathbf{R}^2$ is defined by $T\left(\begin{bmatrix} x \\ y \end{bmatrix}\right) = \begin{bmatrix} 2x + y \\ y \end{bmatrix}$.

2. Let $T: \mathbf{R}^2 \to \mathbf{R}^3$ be a linear transformation such that $T\left(\begin{bmatrix} 1 \\ 0 \end{bmatrix}\right) = \begin{bmatrix} 3 \\ -1 \\ 1 \end{bmatrix}$ and $T\left(\begin{bmatrix} 0 \\ 1 \end{bmatrix}\right) = \begin{bmatrix} 4 \\ 5 \\ -2 \end{bmatrix}$. Find $T\left(\begin{bmatrix} x \\ y \end{bmatrix}\right)$.

3. [BB] Suppose $T: \mathbf{R}^3 \to \mathbf{R}^2$ is a linear transformation such that $T\left(\begin{bmatrix} 1 \\ 0 \\ 0 \end{bmatrix}\right) = \begin{bmatrix} 3 \\ 1 \end{bmatrix}$, $T\left(\begin{bmatrix} 0 \\ 1 \\ 0 \end{bmatrix}\right) = \begin{bmatrix} 1 \\ -2 \end{bmatrix}$, and $T\left(\begin{bmatrix} 0 \\ 0 \\ 1 \end{bmatrix}\right) = \begin{bmatrix} -1 \\ 0 \end{bmatrix}$. Find $T\left(\begin{bmatrix} x_1 \\ x_2 \\ x_3 \end{bmatrix}\right)$.

4. Suppose $T: \mathbf{R}^2 \to \mathbf{R}^3$ is a linear transformation such

that $T\left(\begin{bmatrix} 1 \\ -1 \end{bmatrix}\right) = \begin{bmatrix} 1 \\ 2 \\ 3 \end{bmatrix}$ and $T\left(\begin{bmatrix} -2 \\ 3 \end{bmatrix}\right) = \begin{bmatrix} -2 \\ -4 \\ 7 \end{bmatrix}$. Find

$T\left(\begin{bmatrix} x \\ y \end{bmatrix}\right)$.

5. [BB] Suppose $T: \mathbf{R}^3 \to \mathbf{R}^2$ is a linear transformation

such that $T\left(\begin{bmatrix} 1 \\ 0 \\ 1 \end{bmatrix}\right) = \begin{bmatrix} -2 \\ 3 \end{bmatrix}$, $T\left(\begin{bmatrix} -1 \\ 2 \\ 1 \end{bmatrix}\right) = \begin{bmatrix} 1 \\ 1 \end{bmatrix}$, and

$T\left(\begin{bmatrix} 0 \\ 2 \\ 3 \end{bmatrix}\right) = \begin{bmatrix} -2 \\ 5 \end{bmatrix}$. Find $T\left(\begin{bmatrix} x \\ y \\ z \end{bmatrix}\right)$.

6. Suppose $T: \mathbf{R}^3 \to \mathbf{R}^4$ is a linear transformation such

that $T\left(\begin{bmatrix} -1 \\ 1 \\ 1 \end{bmatrix}\right) = \begin{bmatrix} 0 \\ 1 \\ 2 \\ 3 \end{bmatrix}$, $T\left(\begin{bmatrix} -2 \\ 1 \\ 0 \end{bmatrix}\right) = \begin{bmatrix} -4 \\ -5 \\ 1 \\ 1 \end{bmatrix}$, and

$T\left(\begin{bmatrix} 1 \\ 0 \\ -1 \end{bmatrix}\right) = \begin{bmatrix} 2 \\ 2 \\ -3 \\ 1 \end{bmatrix}$. Find $T\left(\begin{bmatrix} x \\ y \\ z \end{bmatrix}\right)$.

7. Does there exist a linear transformation

(a) [BB] $T: \mathbf{R}^3 \to \mathbf{R}^2$ with $T\left(\begin{bmatrix} 1 \\ 0 \\ -2 \end{bmatrix}\right) = \begin{bmatrix} 1 \\ 2 \\ 3 \end{bmatrix}$ and

$T\left(\begin{bmatrix} 2 \\ 0 \\ -4 \end{bmatrix}\right) = \begin{bmatrix} 2 \\ 3 \\ 9 \end{bmatrix}$?

(b) $T: \mathbf{R}^3 \to \mathbf{R}^3$ with $T\left(\begin{bmatrix} 2 \\ 3 \\ 1 \end{bmatrix}\right) = \begin{bmatrix} 1 \\ 1 \\ 1 \end{bmatrix}$,

$T\left(\begin{bmatrix} 1 \\ 0 \\ 9 \end{bmatrix}\right) = \begin{bmatrix} 1 \\ 2 \\ 3 \end{bmatrix}$, and $T\left(\begin{bmatrix} 3 \\ 3 \\ 10 \end{bmatrix}\right) = \begin{bmatrix} 0 \\ 1 \\ -1 \end{bmatrix}$?

(c) $T: \mathbf{R}^2 \to \mathbf{R}^2$ with $T\left(\begin{bmatrix} 1 \\ 0 \end{bmatrix}\right) = \begin{bmatrix} 1 \\ 2 \end{bmatrix}$ and

$T\left(\begin{bmatrix} 0 \\ 1 \end{bmatrix}\right) = \begin{bmatrix} -1 \\ 3 \end{bmatrix}$?

(d) $T: \mathbf{R}^2 \to \mathbf{R}^2$ with $T\left(\begin{bmatrix} 1 \\ 1 \end{bmatrix}\right) = \begin{bmatrix} 1 \\ 2 \end{bmatrix}$ and

$T\left(\begin{bmatrix} -1 \\ 1 \end{bmatrix}\right) = \begin{bmatrix} -1 \\ 4 \end{bmatrix}$?

Explain your answers.

8. For each of the following linear transformations T, find a matrix A such that $T(\mathbf{x}) = A\mathbf{x}$.

(a) [BB] $T: \mathbf{R}^2 \to \mathbf{R}^3$, $T(\mathbf{x}) = \begin{bmatrix} 3x_1 \\ x_1 - x_2 \\ -x_1 + 5x_2 \end{bmatrix}$ for

$\mathbf{x} = \begin{bmatrix} x_1 \\ x_2 \end{bmatrix}$

(b) $T: \mathbf{R}^3 \to \mathbf{R}^2$, $T(\mathbf{x}) = \begin{bmatrix} x_1 - 2x_2 + x_3 \\ x_1 + 5x_2 - x_3 \end{bmatrix}$ for

$\mathbf{x} = \begin{bmatrix} x_1 \\ x_2 \\ x_3 \end{bmatrix}$.

9. Find a matrix A such that the linear transformation T is left multiplication by A in each of the following cases.

(a) [BB] $T: \mathbf{R}^2 \to \mathbf{R}^2$ has the property that

$T\left(\begin{bmatrix} -3 \\ -2 \end{bmatrix}\right) = \begin{bmatrix} 3 \\ 4 \end{bmatrix}$ and $T\left(\begin{bmatrix} 2 \\ 1 \end{bmatrix}\right) = \begin{bmatrix} 2 \\ 5 \end{bmatrix}$.

(b) $T: \mathbf{R}^3 \to \mathbf{R}^3$ has the property that

$T\left(\begin{bmatrix} 2 \\ 1 \\ 1 \end{bmatrix}\right) = \begin{bmatrix} 1 \\ -1 \\ 0 \end{bmatrix}$, $T\left(\begin{bmatrix} 7 \\ 4 \\ 3 \end{bmatrix}\right) = \begin{bmatrix} 2 \\ -1 \\ 3 \end{bmatrix}$, and

$T\left(\begin{bmatrix} 0 \\ -1 \\ 0 \end{bmatrix}\right) = \begin{bmatrix} -1 \\ -1 \\ 2 \end{bmatrix}$

(c) $T: \mathbf{R}^2 \to \mathbf{R}^3$ has the property that

$T\left(\begin{bmatrix} 1 \\ -3 \end{bmatrix}\right) = \begin{bmatrix} -1 \\ 0 \\ -1 \end{bmatrix}$ and $T\left(\begin{bmatrix} 2 \\ -5 \end{bmatrix}\right) = \begin{bmatrix} -1 \\ -1 \\ 0 \end{bmatrix}$.

10. Let $T: V \to W$ be a linear transformation. Show that the range of T is a subspace of W.

11. [BB] Define $T: \mathbf{R}^4 \to \mathbf{R}^3$ by

$T\left(\begin{bmatrix} x_1 \\ x_2 \\ x_3 \\ x_4 \end{bmatrix}\right) = \begin{bmatrix} -x_1 + x_2 + x_3 - 4x_4 \\ x_1 + 2x_2 + 5x_3 + x_4 \\ x_1 + x_3 + 3x_4 \end{bmatrix}$.

(a) Find the matrix of T relative to the standard bases of $\mathbf{R}^4$ and $\mathbf{R}^3$.

(b) Is $\begin{bmatrix} 1 \\ -1 \\ 0 \\ 0 \end{bmatrix}$ in ker T? (c) Is $\begin{bmatrix} -1 \\ 7 \\ 3 \end{bmatrix}$ in range T?

12. For each of the linear transformations defined below,

   i. find a matrix A such that T is left multiplication by A.

   ii. find a basis for the kernel of T and a basis for the range of T.

   iii. verify that $\dim \ker T + \dim \operatorname{range} T = \dim V$ and identify V.

(a) [BB] $T: \mathbf{R}^3 \to \mathbf{R}^3$ is

$$T\left(\begin{bmatrix} x \\ y \\ z \end{bmatrix}\right) = \begin{bmatrix} x - 2y \\ y + 3z \\ 2x - 3y + 3z \end{bmatrix}$$

(b) $T: \mathbf{R}^4 \to \mathbf{R}^3$ is

$$T\left(\begin{bmatrix} x_1 \\ x_2 \\ x_3 \\ x_4 \end{bmatrix}\right) = \begin{bmatrix} x_1 - x_4 \\ x_2 - x_3 \\ x_1 + x_2 - x_3 - x_4 \end{bmatrix}$$

(c) $T: \mathbf{R}^3 \to \mathbf{R}^3$ is defined by

$$T\left(\begin{bmatrix} x_1 \\ x_2 \\ x_3 \end{bmatrix}\right) = \begin{bmatrix} 0 \\ x_1 - x_2 \\ x_2 + x_3 \end{bmatrix}$$

(d) $T: \mathbf{R}^4 \to \mathbf{R}^3$ is defined by

$$T\left(\begin{bmatrix} x_1 \\ x_2 \\ x_3 \\ x_4 \end{bmatrix}\right) = \begin{bmatrix} -x_1 - x_2 + 2x_3 + 3x_4 \\ 2x_1 + 2x_2 - 3x_3 \\ -4x_1 - 4x_2 + 4x_3 + 4x_4 \end{bmatrix}.$$

13. Let $T: V \to W$ be a linear transformation from a vector space V to a vector space W.

(a) [BB] If T is one-to-one and $\mathbf{v}_1, \mathbf{v}_2, \ldots, \mathbf{v}_n$ are linearly independent vectors of V, show that $T(\mathbf{v}_1), T(\mathbf{v}_2), \ldots, T(\mathbf{v}_n)$ are linearly independent vectors of W.

(b) If T is onto and $\mathbf{v}_1, \mathbf{v}_2, \ldots, \mathbf{v}_n$ span V, show that $T(\mathbf{v}_1), T(\mathbf{v}_2), \ldots, T(\mathbf{v}_n)$ span W.

(c) If T is both one-to-one and onto and if $\{\mathbf{v}_1, \ldots, \mathbf{v}_n\}$ is a basis for V, then $\{T(\mathbf{v}_1), \ldots, T(\mathbf{v}_n)\}$ is a basis for W.

(d) [The converse of 13(c)] If $\{\mathbf{v}_1, \mathbf{v}_2, \ldots, \mathbf{v}_n\}$ is a basis for V with the property that $\{T(\mathbf{v}_1), T(\mathbf{v}_2), \ldots, T(\mathbf{v}_n)\}$ is a basis of W, prove that T is one-to-one and onto.

14. Let $U = \mathbf{R}^n$, $V = \mathbf{R}^m$, $W = \mathbf{R}^p$, and let $T: U \to V$ and $S: V \to W$ be linear transformations. Answer true or false and justify your answers.

(a) [BB] $\operatorname{range}(ST) \subseteq \operatorname{range}(S)$.

(b) $\ker T \subseteq \ker ST$.

(c) [BB] $\dim \operatorname{range} ST \leq \dim \operatorname{range} S$.

(d) $\dim \operatorname{range} ST \leq \dim \operatorname{range} T$.

■ Critical Reading

15. [BB] Suppose $T: \mathbf{R}^n \to \mathbf{R}^m$ is a linear transformation that is left multiplication by the matrix A. It is the case that T is one-to-one if and only if A has a left inverse. Why?

16. Suppose $T: \mathbf{R}^n \to \mathbf{R}^m$ is a linear transformation that is left multiplication by the matrix A. It is the case that T is onto if and only if A has a right inverse. Why?

17. [BB] Suppose $T: \mathbf{R}^2$ rotates vectors through an angle θ and $S: \mathbf{R}^2$ rotates vectors through $-\theta$. If T is left multiplication by the matrix A and S is left multiplication by B, what is the connection between A and B?

18. Show that any linear transformation $T: \mathbf{R}^2 \to \mathbf{R}^2$ has the form $T\left(\begin{bmatrix} x \\ y \end{bmatrix}\right) = \begin{bmatrix} ax + by \\ cx + dy \end{bmatrix}$ for certain constants a, b, c, d.

19. Let $T: \mathbf{R}^2 \to \mathbf{R}^2$ be the linear transformation that reflects vectors in the line with equation $y = mx$. Find the matrix of T.

20. [BB] Suppose V is the span of vectors $\mathbf{v}_1, \mathbf{v}_2, \ldots, \mathbf{v}_k$ and $T: V \to W$ is a linear transformation from V to a vector space W. Let $\mathbf{w}_1 = T(\mathbf{v}_1)$, $\mathbf{w}_2 = T(\mathbf{v}_2)$, and so on, $\mathbf{w}_k = T(\mathbf{v}_k)$. Let $\mathbf{v}$ be a vector in V. Show that $T(\mathbf{v})$ is a linear combination of $\mathbf{w}_1, \ldots, \mathbf{w}_k$.

21. Let $T: V \to W$ be a linear transformation with $\dim \ker T = \dim \operatorname{range} T$. Show that $\dim V$ is even.

22. If $T: V \to W$ is a linear transformation and $\dim V > \dim W$, prove that there is some nonzero vector $\mathbf{v}$ in V with $T(\mathbf{v}) = \mathbf{0}$.

23. (a) A linear transformation $\mathbf{R}^2 \to \mathbf{R}^2$ has kernel a line through the origin. Explain why its range is also a line through the origin.

(b) Give an example of a linear transformation $\mathbf{R}^2 \to \mathbf{R}^2$ whose kernel is the line with equation $2x + 5y = 0$. What is the range of your linear transformation?

24. Theorem 5.1.24 says that if $T: V \to W$ is a linear transformation and $\dim V = n$, then $\dim \operatorname{range} T + \dim \ker T = n$. Give a proof different from the one in the text that follows these lines. Let $\mathbf{v}_1, \mathbf{v}_2, \ldots, \mathbf{v}_k$ be a basis for $\ker T$ and extend this to a basis $\mathbf{v}_1, \ldots, \mathbf{v}_k, \mathbf{u}_1, \ldots, \mathbf{u}_\ell$ of V.

(a) It suffices to show that $\dim \operatorname{range} T = \ell$. Why?

(b) It suffices to show that $T(\mathbf{u}_1), \ldots, T(\mathbf{u}_\ell)$ form a basis for $\operatorname{range} T$. Why?

(c) Show that $T(\mathbf{u}_1), \ldots, T(\mathbf{u}_\ell)$ are linearly independent vectors.

(d) Show that $T(\mathbf{u}_1), \ldots, T(\mathbf{u}_\ell)$ span $\operatorname{range} T$.

(e) The result follows. Why?

5.2 Matrix Multiplication Revisited

Given matrices $A = \begin{bmatrix} 1 & 2 \\ 3 & 4 \end{bmatrix}$ and $B = \begin{bmatrix} 2 & 3 \\ 4 & 5 \end{bmatrix}$, why not define AB to be the matrix $\begin{bmatrix} 2 & 6 \\ 12 & 20 \end{bmatrix}$, whose entries are the products of the corresponding entries of A and B? Where did the definition of matrix multiplication originate? Why should it be that defining the product of matrices in terms of dot products should be "correct?" We answer these questions in this short section.

In a calculus course, the student will have met the concept of *composition* of functions. If f and g are functions from **R** to **R**, the *composition* $f \circ g$ of f and g is defined by

$$(f \circ g)(x) = f(g(x)).$$

For example, if $f(x) = x^2$ and $g(x) = 3x - 1$, then

$$(f \circ g)(x) = f(g(x)) = f(3x - 1) = (3x - 1)^2$$

while

$$(g \circ f)(x) = g(f(x)) = f(x^2) = 3x^2 - 1.$$

Such examples illustrate the point that composition is not a commutative operation: in general, $f \circ g \neq g \circ f$. On the other hand, composition of functions is associative.

Theorem 5.2.1 *Composition of functions is an associative operation: $(f \circ g) \circ h = f \circ (g \circ h)$ whenever both sides of this equation make sense.*

Proof Suppose $h: V \to W$, $g: W \to U$, and $f: U \to X$ are functions. Then, for any u in U,

$$[(f \circ g) \circ h](u) = (f \circ g)(h(u)) = f(g(h(u))),$$

which is precisely the same as

$$[f \circ (g \circ h)](u) = f\big(g \circ h\big)(u) = f(g(h(u))). \qquad \blacksquare$$

Let V, W, and U be vector spaces and suppose $T: V \to W$ and $S: W \to U$ are linear transformations. The *composition* of S and T is defined as always, but usually denoted ST rather than $S \circ T$ (suppress the $\circ$). Thus ST is the function $V \to U$ defined by $(ST)(\mathbf{v}) = S(T(\mathbf{v}))$ for $\mathbf{v}$ in V. It is convenient to picture the situation as shown to the right.

Of importance is the fact that the composition of S and T is another linear transformation. To see why, let $\mathbf{v}_1$ and $\mathbf{v}_2$ be vectors in V. Then

$$
\begin{aligned}
(ST)(\mathbf{v}_1 + \mathbf{v}_2) &= S(T(\mathbf{v}_1 + \mathbf{v}_2)) && \text{definition of } ST \\
&= S(T(\mathbf{v}_1) + T(\mathbf{v}_2)) && \text{since } T \text{ is linear} \\
&= S(T(\mathbf{v}_1)) + S(T(\mathbf{v}_2)) && \text{since } S \text{ is linear} \\
&= (ST)(\mathbf{v}_1) + (ST)(\mathbf{v}_2) && \text{definition of } ST
\end{aligned}
$$

and, for any vector $\mathbf{v}$ in V and any scalar c,

$$
\begin{aligned}
(ST)(c\mathbf{v}) &= S(T(c\mathbf{v})) && \text{definition of } ST \\
&= S(c(T(\mathbf{v})) && \text{since } T \text{ is linear} \\
&= cS(T(\mathbf{v})) && \text{since } S \text{ is linear} \\
&= c(ST)(\mathbf{v}) && \text{definition of } ST.
\end{aligned}
$$

In Section 5.1, we saw that every linear transformation is multiplication by a matrix. How is the matrix for ST related to the matrices for S and for T?

Suppose $V = \mathbf{R}^n$, $W = \mathbf{R}^m$, and $U = \mathbf{R}^p$. Then $S\colon \mathbf{R}^m \to \mathbf{R}^p$ is left multiplication by a $p \times m$ matrix A and $T\colon \mathbf{R}^n \to \mathbf{R}^m$ is left multiplication by an $m \times n$ matrix B.

$$\mathbf{R}^n \xrightarrow[B]{T} \mathbf{R}^m \xrightarrow[A]{S} \mathbf{R}^p$$

Let $\mathbf{e}_1, \ldots, \mathbf{e}_n$ be the standard basis vectors for $\mathbf{R}^n$. By Theorem 5.1.13, the first column of the matrix for ST is $(ST)(\mathbf{e}_1) = S(T(\mathbf{e}_1))$. Now $T(\mathbf{e}_1) = \mathbf{b}_1$ is the first column of B, the matrix for T. So the first column of the matrix for ST is

$$(ST)(\mathbf{e}_1) = S(T(\mathbf{e}_1)) = S(\mathbf{b}_1) = A\mathbf{b}_1$$

because S is multiplication by A. The second column of ST is $(ST)(\mathbf{e}_2) = S(T(\mathbf{e}_2)) = S(\mathbf{b}_2) = A\mathbf{b}_2$, and so on. In general, column k of the matrix for ST is $A\mathbf{b}_k$. The matrix for ST is

$$
\begin{bmatrix} A\mathbf{b}_1 & A\mathbf{b}_2 & \cdots & A\mathbf{b}_n \\ \downarrow & \downarrow & & \downarrow \end{bmatrix} = A \begin{bmatrix} \mathbf{b}_1 & \mathbf{b}_2 & \cdots & \mathbf{b}_n \\ \downarrow & \downarrow & & \downarrow \end{bmatrix} = AB,
$$

which is the product of the matrices for S and for T.

Theorem 5.2.2 *Multiplication of matrices corresponds to composition of linear transformations. If S has matrix A and T has matrix B, then ST has matrix AB.*

(See Theorem 5.3.12 for a more general version of this theorem.)

5.2.3 EXAMPLE

Suppose $T : \mathbf{R}^3 \to \mathbf{R}^2$ and $S : \mathbf{R}^2 \to \mathbf{R}^3$ are defined by

$$T\left(\begin{bmatrix} x \\ y \\ z \end{bmatrix}\right) = \begin{bmatrix} 2x - y \\ y + z \end{bmatrix} \quad \text{and} \quad S\left(\begin{bmatrix} x \\ y \end{bmatrix}\right) = \begin{bmatrix} x + y \\ x - y \\ 3x + 2y \end{bmatrix}.$$

Then $ST : \mathbf{R}^3 \to \mathbf{R}^3$ is defined by

$$ST\left(\begin{bmatrix} x \\ y \\ z \end{bmatrix}\right) = S\left(T\left(\begin{bmatrix} x \\ y \\ z \end{bmatrix}\right)\right) = S\left(\begin{bmatrix} 2x - y \\ y + z \end{bmatrix}\right)$$

$$= \begin{bmatrix} (2x - y) + (y + z) \\ (2x - y) - (y + z) \\ 3(2x - y) + 2(y + z) \end{bmatrix} = \begin{bmatrix} 2x + z \\ 2x - 2y - z \\ 6x - y + 2z \end{bmatrix}.$$

By Theorem 5.1.13, the matrices A, B, and C for S, T, and ST, respectively, are

$$A = \begin{bmatrix} 1 & 1 \\ 1 & -1 \\ 3 & 2 \end{bmatrix}, \quad B = \begin{bmatrix} 2 & -1 & 0 \\ 0 & 1 & 1 \end{bmatrix}, \quad \text{and} \quad C = \begin{bmatrix} 2 & 0 & 1 \\ 2 & -2 & -1 \\ 6 & -1 & 2 \end{bmatrix}.$$

Observe that $C = AB$:

$$\begin{bmatrix} 1 & 1 \\ 1 & -1 \\ 3 & 2 \end{bmatrix} \begin{bmatrix} 2 & -1 & 0 \\ 0 & 1 & 1 \end{bmatrix} = \begin{bmatrix} 2 & 0 & 1 \\ 2 & -2 & -1 \\ 6 & -1 & 2 \end{bmatrix}.$$

The matrix for ST is the product of the matrices for S and T. ∎

5.2.4 EXAMPLE

Let S and T be the linear transformations $\mathbf{R}^2 \to \mathbf{R}^2$ that rotate vectors through angles α and β, respectively. In Example 5.1.18, we saw that $S = S_A$ is left multiplication by $A = \begin{bmatrix} \cos\alpha & -\sin\alpha \\ \sin\alpha & \cos\alpha \end{bmatrix}$ and $T = T_B$ is left multiplication by $B = \begin{bmatrix} \cos\beta & -\sin\beta \\ \sin\beta & \cos\beta \end{bmatrix}$. The composition ST is rotation through $\alpha + \beta$, which has matrix $C = \begin{bmatrix} \cos(\alpha + \beta) & -\sin(\alpha + \beta) \\ \sin(\alpha + \beta) & \cos(\alpha + \beta) \end{bmatrix}$. On the other hand, the matrix of ST is AB; thus $AB = C$. This implies

$$\begin{bmatrix} \cos\alpha & -\sin\alpha \\ \sin\alpha & \cos\alpha \end{bmatrix} \begin{bmatrix} \cos\beta & -\sin\beta \\ \sin\beta & \cos\beta \end{bmatrix} = \begin{bmatrix} \cos(\alpha + \beta) & -\sin(\alpha + \beta) \\ \sin(\alpha + \beta) & \cos(\alpha + \beta) \end{bmatrix}$$

and hence

$$\begin{bmatrix} \cos\alpha\cos\beta - \sin\alpha\sin\beta & -\cos\alpha\sin\beta - \sin\alpha\cos\beta \\ \sin\alpha\cos\beta + \cos\alpha\sin\beta & -\sin\alpha\sin\beta + \cos\alpha\cos\beta \end{bmatrix}$$

$$= \begin{bmatrix} \cos(\alpha + \beta) & -\sin(\alpha + \beta) \\ \sin(\alpha + \beta) & \cos(\alpha + \beta) \end{bmatrix}.$$

Equating corresponding entries, we obtain the familiar *addition rules* of trigonometry:

$$\cos(\alpha + \beta) = \cos\alpha\cos\beta - \sin\alpha\sin\beta$$
$$\sin(\alpha + \beta) = \sin\alpha\cos\beta + \cos\alpha\sin\beta.$$ ∎

■ Matrix Multiplication is Associative

When you first learned how to multiply matrices, the operation may have seemed strange. Why all these dot products? Why not multiply entry by entry the way matrices are added, for example,

$$\begin{bmatrix} 1 & 2 \\ 3 & 4 \end{bmatrix}\begin{bmatrix} a & b \\ c & d \end{bmatrix} = \begin{bmatrix} a & 2b \\ 3c & 4d \end{bmatrix}?$$

In this section, however, we see that matrix multiplication is perfectly natural. It is defined precisely so that the product of matrices corresponds to the composition of the corresponding linear transformations. Since $f \circ g$ is rarely the same as $g \circ f$, we gain a deeper understanding as to why matrices rarely commute. Since composition of functions is an associative operation, we get an especially pleasing explanation as to why matrix multiplication is associative.

If A, B, and C are matrices corresponding to the linear transformations S, T, and U, respectively (that is, S is left multiplication by A, T is left multiplication by B, and U is left multiplication by C), and if the composition $(ST)U = S(TU)$ is defined, then so are each of the products $(AB)C$ and $A(BC)$. Moreover, since $(AB)C$ is the matrix for $(ST)U$ and $A(BC)$ is the matrix for $S(TU)$, we have $(AB)C = A(BC)$. This proof of the associativity of matrix multiplication is much preferable to that which uses the explicit definition of matrix multiplication.

■ Invertible Linear Transformations

5.2.5 DEFINITION

A linear transformation $T: V \to W$ is *invertible* if and only if there exists a linear transformation $S: W \to V$ such that $TS = \mathrm{id}_W$ and $ST = \mathrm{id}_V$. The linear transformation S is called the *inverse* of T.

5.2.6 EXAMPLE

Define $T, S: \mathbf{R}^2 \to \mathbf{R}^2$ by

$$T\left(\begin{bmatrix} x \\ y \end{bmatrix}\right) = \begin{bmatrix} x + y \\ x - y \end{bmatrix}, \quad S\left(\begin{bmatrix} x \\ y \end{bmatrix}\right) = \begin{bmatrix} \frac{1}{2}(x + y) \\ \frac{1}{2}(x - y) \end{bmatrix}.$$

Then T is invertible with inverse S (and S is invertible with inverse T) because for any $\begin{bmatrix} x \\ y \end{bmatrix} \in \mathbf{R}^2$,

$$TS\left(\begin{bmatrix} x \\ y \end{bmatrix}\right) = T\left(\begin{bmatrix} \frac{1}{2}(x + y) \\ \frac{1}{2}(x - y) \end{bmatrix}\right) = \begin{bmatrix} \frac{1}{2}(x + y) + \frac{1}{2}(x - y) \\ \frac{1}{2}(x + y) - \frac{1}{2}(x - y) \end{bmatrix} = \begin{bmatrix} x \\ y \end{bmatrix}$$

and

$$ST\left(\begin{bmatrix}x\\y\end{bmatrix}\right) = S\left(\begin{bmatrix}x+y\\x-y\end{bmatrix}\right) = \begin{bmatrix}\frac{1}{2}[(x+y)+(x-y)]\\\frac{1}{2}[(x+y)-(x-y)]\end{bmatrix} = \begin{bmatrix}x\\y\end{bmatrix}$$

so that both TS and ST are the identity on $\mathbf{R}^2$. ■

Theorem 5.2.7 *A linear transformation $T: V \to W$ is invertible if and only if it is one-to-one and onto.*

Proof Suppose T is invertible with inverse $S: W \to V$. Then $ST = \mathrm{id}_V$ and $TS = \mathrm{id}_W$. To see that T is one-to-one, suppose $\mathbf{v}$ is in $\ker T$. Then $T(\mathbf{v}) = \mathbf{0}$, so $S(T(\mathbf{v})) = \mathbf{0}$, $\mathrm{id}_V(\mathbf{v}) = \mathbf{0}$, and $\mathbf{v} = \mathbf{0}$. To see that T is onto, let $\mathbf{w}$ be any vector in W. Then $\mathbf{w} = \mathrm{id}_W(\mathbf{w}) = T(S(\mathbf{w}))$.

Conversely, suppose T is one-to-one and onto. If $\mathbf{w} \in W$, then $\mathbf{w} = T(\mathbf{v})$ for some $\mathbf{v} \in V$ (because T is onto) and $\mathbf{v}$ is unique because T is one-to-one: $\mathbf{w} = T(\mathbf{v}_1) = T(\mathbf{v}_2) \implies \mathbf{v}_1 = \mathbf{v}_2$. Now define $S(\mathbf{w}) = \mathbf{v}$. We leave it to the reader to check that S is the inverse of T. (See Exercise 12.) ■

If T is left multiplication by a matrix, the concepts of one-to-one and onto have familiar interpretations.

Theorem 5.2.8 *Suppose A is an $m \times n$ matrix and $T = T_A: \mathbf{R}^n \to \mathbf{R}^m$ is left multiplication by A.*

1. *T is one-to-one if and only if rank $A = n$, that is, A has linearly independent columns.*

2. *T is onto if and only if rank $A = m$, that is, A has linearly independent rows.*

Proof 1. In Theorem 5.1.24, we noted that $\dim \mathrm{range}\, T + \dim \ker T = n$. Thus $\ker T = \{\mathbf{0}\}$ if and only if $\dim \mathrm{range}\, T = n$. Since $\mathrm{range}\, T = \mathrm{col\ sp}\, A$, this occurs if and only if rank $A = n$, equivalently, if and only if the columns of A are linearly independent. (See also Theorem 4.4.9.)

2. The linear transformation T is onto if and only if every $\mathbf{y}$ in $\mathbf{R}^m$ is of the form $A\mathbf{x}$. This occurs if and only if the columns span $\mathbf{R}^m$, equivalently, by Theorem 4.4.7, if rank $A = m$ or the rows of A are linearly independent. ■

Corollary 5.2.9 *Let $T = T_A: \mathbf{R}^n \to \mathbf{R}^n$ be a linear transformation which is left multiplication by the matrix A. Then the following statements are equivalent:*

i. *T is one-to-one;*

ii. *T is onto;*

iii. *T is invertible;*

iv. *The matrix A is invertible.*

Proof The first two statements are equivalent since each is equivalent to the assertion that the rank of the matrix of T is n by Theorem 5.2.8. Equivalence with iii follows from Theorem 5.2.7 and with iv from Theorem 4.1.24 (an $n \times n$ matrix is invertible if and only if it has rank n). ∎

True/False Questions

Decide, with as little calculation as possible, whether each of the following statements is true or false and explain your answer whenever you say "false." (Answers can be found in the back of the book.)

1. If $f, g: \mathbf{R} \to \mathbf{R}$ are defined by $f(x) = 1$ and $g(x) = x$ for all real numbers x, then $(f \circ g)(x) = x$ for all x.

2. If $f, g: \mathbf{R} \to \mathbf{R}$ are defined by $f(x) = 1$ and $g(x) = x$ for all real numbers x, then $(g \circ f)(x) = x$ for all x.

3. Composition of functions is a commutative operation.

4. Composition of functions is an associative operation.

5. If $T: \mathbf{R}^2: \mathbf{R}^3$ is a linear transformation, the matrix of T is 2×3.

6. If $T: \mathbf{R}^n \to \mathbf{R}^m$ is an invertible linear transformation, then T is one-to-one.

7. A linear transformation $T: \mathbf{R}^n \to \mathbf{R}^m$ is invertible if and only if $n = m$.

Exercises

Solutions to exercises marked [BB] can be found in the Back of the Book.

1. Let A and B be the matrices of linear transformations S and T, respectively. In each of the following situations,
   i. find A and B;
   ii. find the vector that is $ST(\mathbf{x})$;
   iii. use your answer to part ii to find the matrix for ST and verify that this is AB.

   (a) [BB] $S: \mathbf{R}^3 \to \mathbf{R}^4$ is defined by
   $$S\left(\begin{bmatrix} x \\ y \\ z \end{bmatrix}\right) = \begin{bmatrix} x+y-z \\ y-z \\ x+z \\ y \end{bmatrix};$$
   $T: \mathbf{R}^2 \to \mathbf{R}^3$ is defined by $T\left(\begin{bmatrix} x \\ y \end{bmatrix}\right) = \begin{bmatrix} 2x+y \\ x-3y \\ 4x \end{bmatrix}$; $\mathbf{x} = \begin{bmatrix} x \\ y \end{bmatrix}$.

   (b) $S: \mathbf{R}^3 \to \mathbf{R}^2$ is defined by
   $$S\left(\begin{bmatrix} x \\ y \\ z \end{bmatrix}\right) = \begin{bmatrix} -x+y \\ 2x+y-z \end{bmatrix};$$
   $T: \mathbf{R}^2 \to \mathbf{R}^3$ is defined by $T\left(\begin{bmatrix} x \\ y \end{bmatrix}\right) = \begin{bmatrix} 4x+y \\ 2x \\ x-3y \end{bmatrix}$; $\mathbf{x} = \begin{bmatrix} x \\ y \end{bmatrix}$.

   (c) $S: \mathbf{R}^2 \to \mathbf{R}^4$ is defined by $S\left(\begin{bmatrix} x \\ y \end{bmatrix}\right) = \begin{bmatrix} 3x-2y \\ x+y \\ -2x+7y \\ -x+3y \end{bmatrix}$;
   $T: \mathbf{R}^3 \to \mathbf{R}^2$ is defined by $T\left(\begin{bmatrix} x \\ y \\ z \end{bmatrix}\right) = \begin{bmatrix} x+2y-z \\ 2x-3y+z \end{bmatrix}$; $\mathbf{x} = \begin{bmatrix} x \\ y \\ z \end{bmatrix}$.

2. Define $S: \mathbf{R}^2 \to \mathbf{R}^3$, and $T: \mathbf{R}^3 \to \mathbf{R}^4$ by
   $$S\left(\begin{bmatrix} x \\ y \end{bmatrix}\right) = \begin{bmatrix} 2x+3y \\ y \\ -x \end{bmatrix} \text{ and } T\left(\begin{bmatrix} x \\ y \\ z \end{bmatrix}\right) = \begin{bmatrix} x-y \\ y-z \\ z-x \\ x+y+z \end{bmatrix}.$$
   (a) One of ST, TS is defined. Which is it?
   (b) Find n, $\mathbf{x} \in \mathbf{R}^n$ and the vector $ST(\mathbf{x})$ or $TS(\mathbf{x})$, as the case may be.

3. [BB] Define $S, T: \mathbf{R}^2 \to \mathbf{R}^2$ by $S\left(\begin{bmatrix} x \\ y \end{bmatrix}\right) = \begin{bmatrix} x - y \\ -x + 2y \end{bmatrix}$, $T\left(\begin{bmatrix} x \\ y \end{bmatrix}\right) = \begin{bmatrix} y \\ x \end{bmatrix}$. Suppose S is multiplication by A, T is multiplication by B, ST is multiplication by C, and TS is multiplication by D.

(a) Find $ST\left(\begin{bmatrix} x \\ y \end{bmatrix}\right)$ and $TS\left(\begin{bmatrix} x \\ y \end{bmatrix}\right)$.

(b) Find A, B, C, and D.

(c) Verify that $AB = C$ and $BA = D$.

4. Define $S, T: \mathbf{R}^2 \to \mathbf{R}^2$ by $S\left(\begin{bmatrix} x \\ y \end{bmatrix}\right) = \begin{bmatrix} 2x - 3y \\ x + 5y \end{bmatrix}$, $T\left(\begin{bmatrix} x \\ y \end{bmatrix}\right) = \begin{bmatrix} -y \\ x + 7y \end{bmatrix}$. Suppose S is multiplication by A, T is multiplication by B, ST is multiplication by C, and TS is multiplication by D.

(a) Find $ST\left(\begin{bmatrix} x \\ y \end{bmatrix}\right)$ and $TS\left(\begin{bmatrix} x \\ y \end{bmatrix}\right)$.

(b) Find A, B, C, and D.

(c) Verify that $AB = C$ and $BA = D$.

5. [BB] Given that $S: \mathbf{R}^3 \to \mathbf{R}^2$ and $T: \mathbf{R}^2 \to \mathbf{R}^2$ are linear transformations such that $T\left(\begin{bmatrix} x \\ y \end{bmatrix}\right) = \begin{bmatrix} x - 2y \\ 4x \end{bmatrix}$ and $TS\left(\begin{bmatrix} x \\ y \\ z \end{bmatrix}\right) = \begin{bmatrix} -5x - 4y - z \\ 12x - 4z \end{bmatrix}$, find the matrix of S. Also find $S\left(\begin{bmatrix} x \\ y \\ z \end{bmatrix}\right)$.

6. Given that $S: \mathbf{R}^3 \to \mathbf{R}^2$ and $T: \mathbf{R}^2 \to \mathbf{R}^2$ are linear transformations such that $T\left(\begin{bmatrix} x \\ y \end{bmatrix}\right) = \begin{bmatrix} x - 2y \\ 3x - 5y \end{bmatrix}$ and $TS\left(\begin{bmatrix} x \\ y \\ z \end{bmatrix}\right) = \begin{bmatrix} x - 2y - z \\ 3x - z \end{bmatrix}$, find the matrix of S. Also find $S\left(\begin{bmatrix} x \\ y \\ z \end{bmatrix}\right)$.

7. [BB] Define $S, T, U: \mathbf{R}^2 \to \mathbf{R}^2$ by $S\left(\begin{bmatrix} x \\ y \end{bmatrix}\right) = \begin{bmatrix} y \\ x \end{bmatrix}$, $T\left(\begin{bmatrix} x \\ y \end{bmatrix}\right) = \begin{bmatrix} x + y \\ x - y \end{bmatrix}$, and $U\left(\begin{bmatrix} x \\ y \end{bmatrix}\right) = \begin{bmatrix} 2x \\ 3x \end{bmatrix}$.

(a) Find $ST\left(\begin{bmatrix} x \\ y \end{bmatrix}\right)$ and $TU\left(\begin{bmatrix} x \\ y \end{bmatrix}\right)$.

(b) Verify that $[(ST)U]\left(\begin{bmatrix} x \\ y \end{bmatrix}\right) = [S(TU)]\left(\begin{bmatrix} x \\ y \end{bmatrix}\right)$.

8. Define $S, T, U: \mathbf{R}^3 \to \mathbf{R}^3$ by
$$S\left(\begin{bmatrix} x \\ y \\ z \end{bmatrix}\right) = \begin{bmatrix} y \\ z \\ x \end{bmatrix}, \quad T\left(\begin{bmatrix} x \\ y \\ z \end{bmatrix}\right) = \begin{bmatrix} x - y \\ y - z \\ z - x \end{bmatrix},$$
$$\text{and} \quad U\left(\begin{bmatrix} x \\ y \\ z \end{bmatrix}\right) = \begin{bmatrix} -x \\ y \\ -z \end{bmatrix}.$$

(a) Find $ST\left(\begin{bmatrix} x \\ y \end{bmatrix}\right)$ and $TU\left(\begin{bmatrix} x \\ y \end{bmatrix}\right)$.

(b) Verify that $[(ST)U]\left(\begin{bmatrix} x \\ y \end{bmatrix}\right) = [S(TU)]\left(\begin{bmatrix} x \\ y \end{bmatrix}\right)$.

9. Define $S: \mathbf{R}^4 \to \mathbf{R}^2$, $T: \mathbf{R}^3 \to \mathbf{R}^4$, and $U: \mathbf{R}^2 \to \mathbf{R}^3$ by
$$S\left(\begin{bmatrix} x_1 \\ x_2 \\ x_3 \\ x_4 \end{bmatrix}\right) = \begin{bmatrix} x_1 + x_2 - x_3 - x_4 \\ -x_1 + 2x_2 + x_3 - 4x_4 \end{bmatrix},$$
$$T\left(\begin{bmatrix} x_1 \\ x_2 \\ x_3 \end{bmatrix}\right) = \begin{bmatrix} 2x_1 - x_2 + x_3 \\ x_2 - 4x_3 \\ x_1 - x_2 \\ x_1 + x_2 + x_3 \end{bmatrix},$$
$$\text{and} \quad U\left(\begin{bmatrix} x_1 \\ x_2 \end{bmatrix}\right) = \begin{bmatrix} 2x_2 \\ 3x_2 \\ 4x_1 \end{bmatrix}.$$

(a) Find $ST\left(\begin{bmatrix} x_1 \\ x_2 \\ x_3 \end{bmatrix}\right)$ and $TU\left(\begin{bmatrix} x_1 \\ x_2 \end{bmatrix}\right)$.

(b) Find $[(ST)U]\left(\begin{bmatrix} x_1 \\ x_2 \end{bmatrix}\right)$ and $[S(TU)]\left(\begin{bmatrix} x_1 \\ x_2 \end{bmatrix}\right)$.

10. Use Theorem 5.2.2 to prove that the linear transformation $T = T_A: \mathbf{R}^n \to \mathbf{R}^n$ that is left multiplication by A is invertible if and only if A is invertible.

11. [BB] Use the approach of Example 5.2.4 to derive the trigonometric identities
$$\cos(\alpha - \beta) = \cos\alpha\cos\beta + \sin\alpha\sin\beta$$
$$\sin(\alpha - \beta) = \sin\alpha\cos\beta - \cos\alpha\sin\beta.$$

12. Let $T: V \to W$ be a linear transformation that is one-to-one and onto. For any $\mathbf{w}$ in W, we have $\mathbf{w} = T(\mathbf{v})$ because T is onto. Also, as shown in the proof of Theorem 5.2.7, the vector $\mathbf{v}$ is unique. Thus we can define a function $S: W \to V$ by $S(\mathbf{w}) = \mathbf{v}$.

(a) [BB; half] Show that S is a linear transformation.

(b) Show that S is the inverse of T.

■ **Critical Reading**

13. Suppose V is a vector space and $T: V \to V$ is a linear transformation. Suppose $\mathbf{v}$ is a vector in V such that $T(\mathbf{v}) \neq \mathbf{0}$, but $T^2(\mathbf{v}) = \mathbf{0}$ [T^2 means TT, that is $T \circ T$]. Prove that $\mathbf{v}$ and $T(\mathbf{v})$ are linearly independent.

14. Suppose $n > m$ and $T: \mathbf{R}^n \to \mathbf{R}^m$ is a linear transformation. Explain why T cannot be one-to-one.

5.3 The Matrices of a Linear Transformation

In Section 5.1, we defined the term "linear transformation," discussed some elementary properties of linear transformations, and showed that any linear transformation $\mathbf{R}^n \to \mathbf{R}^m$ is multiplication by an $m \times n$ matrix. We propose here to investigate linear transformations more deeply and to show that the matrix of a linear transformation is far from unique—a linear transformation has many matrices associated with it.

Suppose $\mathcal{V} = \{\mathbf{v}_1, \mathbf{v}_2, \ldots, \mathbf{v}_n\}$ is a basis for a vector space V. If $\mathbf{x}$ is a vector in V, then $\mathbf{x}$ can be written uniquely as a linear combination of the basis vectors:

$$\mathbf{x} = x_1\mathbf{v}_1 + x_2\mathbf{v}_2 + \cdots + x_n\mathbf{v}_n.$$

The scalars $x_1, x_2, \ldots, x_n$ are called the *coordinates of* $\mathbf{x}$ *relative to* $\mathcal{V}$ and the n-dimensional vector $\begin{bmatrix} x_1 \\ x_2 \\ \vdots \\ x_n \end{bmatrix}$ is called the *coordinate vector of* $\mathbf{x}$ *relative to* $\mathcal{V}$. We use the notation $\mathbf{x}_\mathcal{V}$ for this vector. Thus

5.3.1 $\quad \mathbf{x}_\mathcal{V} = \begin{bmatrix} x_1 \\ x_2 \\ \vdots \\ x_n \end{bmatrix}$ if and only if $\quad \mathbf{x} = x_1\mathbf{v}_1 + x_2\mathbf{v}_2 + \cdots + x_n\mathbf{v}_n.$

5.3.2 EXAMPLE

If $\mathbf{x} = \begin{bmatrix} x_1 \\ x_2 \end{bmatrix}$ is a vector in $\mathbf{R}^2$, the coordinates of $\mathbf{x}$ relative to the standard basis $\mathcal{E} = \{\mathbf{e}_1, \mathbf{e}_2\}$ are x_1 and x_2 because $\mathbf{x} = x_1\mathbf{e}_1 + x_2\mathbf{e}_2$. Thus $\mathbf{x}_\mathcal{E} = \begin{bmatrix} x_1 \\ x_2 \end{bmatrix}$. More generally, if $\mathcal{E}$ is the standard basis of $\mathbf{R}^n$ and $\mathbf{x} = \begin{bmatrix} x_1 \\ x_2 \\ \vdots \\ x_n \end{bmatrix}$, then $\mathbf{x}_\mathcal{E} = \begin{bmatrix} x_1 \\ x_2 \\ \vdots \\ x_n \end{bmatrix}$. ■

5.3.3 EXAMPLE Let $\mathbf{v}_1 = \begin{bmatrix} -1 \\ 2 \end{bmatrix}$ and $\mathbf{v}_2 = \begin{bmatrix} 2 \\ -3 \end{bmatrix}$. Then $\mathcal{V} = \{\mathbf{v}_1, \mathbf{v}_2\}$ is a basis for $\mathbf{R}^2$. We ask you to check for yourself that

$$\begin{bmatrix} x_1 \\ x_2 \end{bmatrix} = (3x_1 + 2x_2)\mathbf{v}_1 + (2x_1 + x_2)\mathbf{v}_2, \tag{1}$$

so the coordinates of $\mathbf{x} = \begin{bmatrix} x_1 \\ x_2 \end{bmatrix}$ relative to $\mathcal{V}$ are $3x_1 + 2x_2$ and $2x_1 + x_2$: $\mathbf{x}_\mathcal{V} = \begin{bmatrix} 3x_1 + 2x_2 \\ 2x_1 + x_2 \end{bmatrix}$. Note the difference between the vector $\mathbf{x} = \begin{bmatrix} x_1 \\ x_2 \end{bmatrix}$ and its **coordinate vector** $\mathbf{x}_\mathcal{V} = \begin{bmatrix} 3x_1 + 2x_2 \\ 2x_1 + x_2 \end{bmatrix}$ relative to the basis $\mathcal{V}$. ∎

Reading Challenge 1 *How would you find the coefficients $3x_1 + 2x_2$ and $2x_1 + x_2$ in equation (1) if they weren't handed to you on a platter?*

5.3.4 EXAMPLE Let $\mathbf{v}_1 = \begin{bmatrix} 1 \\ 0 \\ 0 \end{bmatrix}$, $\mathbf{v}_2 = \begin{bmatrix} 1 \\ 1 \\ 0 \end{bmatrix}$, and $\mathbf{v}_3 = \begin{bmatrix} 1 \\ 1 \\ 1 \end{bmatrix}$. The set $\mathcal{V} = \{\mathbf{v}_1, \mathbf{v}_2, \mathbf{v}_3\}$ is a basis for $\mathbf{R}^3$. The vector $\mathbf{x} = \begin{bmatrix} x_1 \\ x_2 \\ x_3 \end{bmatrix}$ can be written uniquely

$$\begin{bmatrix} x_1 \\ x_2 \\ x_3 \end{bmatrix} = (x_1 - x_2)\mathbf{v}_1 + (x_2 - x_3)\mathbf{v}_2 + x_3\mathbf{v}_3, \tag{2}$$

so the coordinates of $\mathbf{x}$ relative to $\mathcal{V}$ are $x_1 - x_2$, $x_2 - x_3$ and x_3: $\mathbf{x}_\mathcal{V} = \begin{bmatrix} x_1 - x_2 \\ x_2 - x_3 \\ x_3 \end{bmatrix}$. ∎

Before continuing, we record a fact that will soon prove useful.

Lemma 5.3.5 *If V is a vector space of dimension n with basis $\mathcal{V}$, then the function $V \to \mathbf{R}^n$ defined by $\mathbf{x} \mapsto \mathbf{x}_\mathcal{V}$, which assigns to each vector in V its coordinate vector relative to $\mathcal{V}$, is a linear transformation.*

Proof Call the given function T, so that $T(\mathbf{x}) = \mathbf{x}_\mathcal{V}$ for $\mathbf{x} \in V$. Let

$$\mathbf{x} = x_1\mathbf{v}_1 + x_2\mathbf{v}_2 + \cdots + x_n\mathbf{v}_n$$

and

$$\mathbf{y} = y_1\mathbf{v}_1 + y_2\mathbf{v}_2 + \cdots + y_n\mathbf{v}_n$$

be vectors in V. The associated coordinate vectors are

$$\mathbf{x}_\mathcal{V} = \begin{bmatrix} x_1 \\ x_2 \\ \vdots \\ x_n \end{bmatrix} \quad \text{and} \quad \mathbf{y}_\mathcal{V} = \begin{bmatrix} y_1 \\ y_2 \\ \vdots \\ y_n \end{bmatrix}.$$

Since

$$\mathbf{x} + \mathbf{y} = (x_1 + y_1)\mathbf{v}_1 + (x_2 + y_2)\mathbf{v}_2 + \cdots + (x_n + y_n)\mathbf{v}_n,$$

the coordinate vector of $\mathbf{x} + \mathbf{y}$ relative to $\mathcal{V}$ is

$$T(\mathbf{x} + \mathbf{y}) = \begin{bmatrix} x_1 + y_1 \\ x_2 + y_2 \\ \vdots \\ x_n + y_n \end{bmatrix},$$

and this is

$$\begin{bmatrix} x_1 \\ x_2 \\ \vdots \\ x_n \end{bmatrix} + \begin{bmatrix} y_1 \\ y_2 \\ \vdots \\ y_n \end{bmatrix} = T(\mathbf{x}) + T(\mathbf{y}).$$

Thus T preserves vector addition. Let c be a scalar. Since

$$c\mathbf{x} = (cx_1)\mathbf{v}_1 + (cx_2)\mathbf{v}_2 + \cdots + (cx_n)\mathbf{v}_n,$$

the coordinate vector of $c\mathbf{x}$ is $\begin{bmatrix} cx_1 \\ cx_2 \\ \vdots \\ cx_n \end{bmatrix}$, and this is $c\begin{bmatrix} x_1 \\ x_2 \\ \vdots \\ x_n \end{bmatrix} = cT(\mathbf{x})$. Thus T also

preserves scalar multiples. ∎

5.3.6 Remark It is not hard to prove that the map T defined in Lemma 5.3.5 is one-to-one—see Exercise 11—so the lemma says that any finite dimensional vector space V is "essentially" $\mathbf{R}^n$, where vectors $\mathbf{x}$ in V are replaced by their coordinate vectors $\mathbf{x}_\mathcal{V}$ relative to a basis $\mathcal{V}$. This is the reason we have chosen so frequently in this work to make "vector space" synonymous with "Euclidean n-space."

To illustrate this point further, the set $V = \{a + bx + cx^2 \mid a, b, c \in \mathbf{R}\}$ of polynomials with real coefficients of degree at most 2 is a three-dimensional vector space with basis $\{1, x, x^2\}$ relative to which $f = a + bx + cx^2$ has coordinate vector $\begin{bmatrix} a \\ b \\ c \end{bmatrix}$. The linear transformation T of Lemma 5.3.5 is $f \mapsto \begin{bmatrix} a \\ b \\ c \end{bmatrix}$. Thus working with V is essentially working in $\mathbf{R}^3$.

The set $V = \{\begin{bmatrix} a & b \\ c & d \end{bmatrix} \mid a, b, c, d \in \mathbf{R}\}$ of 2×2 matrices over $\mathbf{R}$ is a vector space of dimension 4 with basis $\{E_{11}, E_{12}, E_{21}, E_{22}\}$, with

$$E_{11} = \begin{bmatrix} 1 & 0 \\ 0 & 0 \end{bmatrix}, \quad E_{12} = \begin{bmatrix} 0 & 1 \\ 0 & 0 \end{bmatrix}, \quad E_{21} = \begin{bmatrix} 0 & 0 \\ 1 & 0 \end{bmatrix}, \quad E_{22} = \begin{bmatrix} 0 & 0 \\ 0 & 1 \end{bmatrix},$$

relative to which $A = \begin{bmatrix} a & b \\ c & d \end{bmatrix}$ has coordinate vector $\begin{bmatrix} a \\ b \\ c \\ d \end{bmatrix}$. The linear transformation

T of Lemma 5.3.5 is $\begin{bmatrix} a & b \\ c & d \end{bmatrix} \mapsto \begin{bmatrix} a \\ b \\ c \\ d \end{bmatrix}$. Working with V is essentially working in $\mathbf{R}^4$.

■ The Matrix of T Relative to $\mathcal{V}$ and $\mathcal{W}$

Suppose we have a linear transformation $T : V \to W$, where

- V is a vector space with basis $\mathcal{V} = \{\mathbf{v}_1, \mathbf{v}_2, \dots, \mathbf{v}_n\}$, and
- W is a (finite dimensional) vector space with basis $\mathcal{W}$.

Let $\mathbf{x}$ be a vector in V and set $\mathbf{y} = T(\mathbf{x})$. We are going to explore the connection between the coordinate vector $\mathbf{x}_\mathcal{v}$ of $\mathbf{x}$ relative to $\mathcal{V}$ and the coordinate vector $\mathbf{y}_\mathcal{w}$ of $\mathbf{y}$ relative to $\mathcal{W}$.

To begin, let the coordinate vector of $\mathbf{x}$ relative to $\mathcal{V}$ be $\mathbf{x}_\mathcal{v} = \begin{bmatrix} x_1 \\ x_2 \\ \vdots \\ x_n \end{bmatrix}$. By the

definition of "coordinate vector," we know that $\mathbf{x} = x_1\mathbf{v}_1 + x_2\mathbf{v}_2 + \cdots + x_n\mathbf{v}$. Since T preserves linear combinations,

$$\mathbf{y} = T(\mathbf{x}) = x_1 T(\mathbf{v}_1) + x_2 T(\mathbf{v}_2) + \cdots + x_n T(\mathbf{v}_n).$$

By Lemma 5.3.5, the function $\mathbf{w} \mapsto \mathbf{w}_\mathcal{w}$ is a linear transformation, so it too preserves linear combinations. Thus the coordinate vector of $\mathbf{y}$ relative to $\mathcal{W}$ is

$$\mathbf{y}_\mathcal{w} = [T(\mathbf{x})]_\mathcal{w} = x_1[T(\mathbf{v}_1)]_\mathcal{w} + x_2[T(\mathbf{v}_2)]_\mathcal{w} + \cdots + x_n[T(\mathbf{v}_n)]_\mathcal{w}$$

$$= \begin{bmatrix} [T(\mathbf{v}_1)]_\mathcal{w} & [T(\mathbf{v}_2)]_\mathcal{w} & \cdots & [T(\mathbf{v}_n)]_\mathcal{w} \\ \downarrow & \downarrow & & \downarrow \end{bmatrix} \begin{bmatrix} x_1 \\ x_2 \\ \vdots \\ x_n \end{bmatrix}$$

$$= A \begin{bmatrix} x_1 \\ x_2 \\ \vdots \\ x_n \end{bmatrix} = A\mathbf{x}_\mathcal{v}, \tag{3}$$

where

$$A = \begin{bmatrix} [T(\mathbf{v}_1)]_W & [T(\mathbf{v}_2)]_W & \cdots & [T(\mathbf{v}_n)]_W \\ \downarrow & \downarrow & & \downarrow \end{bmatrix}$$

is the matrix whose columns are the coordinate vectors of $T(\mathbf{v}_1)$, $T(\mathbf{v}_2)$, ..., $T(\mathbf{v}_n)$ relative to W. Again, as in Section 5.1, T is "multiplication by a matrix":

$$T(\mathbf{x})_W = A\mathbf{x}_V. \tag{4}$$

We use the notation $M_{W \leftarrow V}(T)$ for the matrix A in equation (4) and call this matrix the "matrix of T relative to V and W."

5.3.7 DEFINITION   Let $V = \{\mathbf{v}_1, \mathbf{v}_2, \ldots, \mathbf{v}_n\}$ and W be bases for vector spaces V and W, of dimensions n and m, respectively. Let $T: V \to W$ be a linear transformation. Then the *matrix of T relative to V and W* is the matrix

$$M_{W \leftarrow V}(T) = \begin{bmatrix} [T(\mathbf{v}_1)]_W & [T(\mathbf{v}_2)]_W & \cdots & [T(\mathbf{v}_n)]_W \\ \downarrow & \downarrow & & \downarrow \end{bmatrix}$$

whose columns are the coordinate vectors of $T(\mathbf{v}_1)$, $T(\mathbf{v}_2)$, ..., $T(\mathbf{v}_n)$ relative to W. This matrix has the property that

$$\boxed{[T(\mathbf{x})]_W = M_{W \leftarrow V}(T)\mathbf{x}_V}$$

for any vector $\mathbf{x}$ in V.

In plain words, the boxed equation says that if you want the **coordinates** of $T(\mathbf{x})$ relative to W, you take the coordinate vector of $\mathbf{x}$ relative to V and multiply by the matrix $M_{W \leftarrow V}(T)$.

5.3.8 EXAMPLE   Let $V = W = \mathbf{R}^2$ and let $T: V \to V$ be the linear transformation defined by $T\left(\begin{bmatrix} x_1 \\ x_2 \end{bmatrix}\right) = \begin{bmatrix} x_2 \\ x_1 \end{bmatrix}$. The set $V = \{\mathbf{v}_1, \mathbf{v}_2\}$ where $\mathbf{v}_1 = \begin{bmatrix} -1 \\ 2 \end{bmatrix}$ and $\mathbf{v}_2 = \begin{bmatrix} 2 \\ -3 \end{bmatrix}$ is a basis for V. Let $\mathcal{E} = \{\mathbf{e}_1, \mathbf{e}_2\}$ be the standard basis for V. We find $M_{\mathcal{E} \leftarrow \mathcal{E}}(T)$, the matrix of T relative to $\mathcal{E}$ and $\mathcal{E}$. The columns of this matrix are the coordinate vectors of $T(\mathbf{e}_1)$ and $T(\mathbf{e}_2)$ relative to $\mathcal{E}$. We have

$$T(\mathbf{e}_1) = T\left(\begin{bmatrix} 1 \\ 0 \end{bmatrix}\right) = \begin{bmatrix} 0 \\ 1 \end{bmatrix} \quad \text{and} \quad T(\mathbf{e}_2) = T\left(\begin{bmatrix} 0 \\ 1 \end{bmatrix}\right) = \begin{bmatrix} 1 \\ 0 \end{bmatrix},$$

and we have observed that the coordinate vector of a vector relative to the standard basis $\mathcal{E}$ is the vector itself—see **5.3.2**. Thus

$$M_{\mathcal{E} \leftarrow \mathcal{E}}(T) = \begin{bmatrix} 0 & 1 \\ 1 & 0 \end{bmatrix}$$

just as in Section 5.1 where, implicitly, all matrices of linear transformations were formed with respect to standard bases. The boxed equation in Definition 5.3.7 states the obvious fact:

$$[T(\mathbf{x})]_{\mathcal{E}} = \begin{bmatrix} x_2 \\ x_1 \end{bmatrix} = \begin{bmatrix} 0 & 1 \\ 1 & 0 \end{bmatrix} \begin{bmatrix} x_1 \\ x_2 \end{bmatrix} = M_{\mathcal{E} \leftarrow \mathcal{E}}(T)\mathbf{x}_{\mathcal{E}}.$$

Now let us find $M_{\mathcal{E} \leftarrow \mathcal{V}}(T)$, the matrix of T relative to $\mathcal{V}$ and $\mathcal{E}$. The columns of this matrix are the coordinate vectors of $T(\mathbf{v}_1)$ and $T(\mathbf{v}_2)$ relative to $\mathcal{E}$. We find

$$T\left(\begin{bmatrix} -1 \\ 2 \end{bmatrix}\right) = \begin{bmatrix} 2 \\ -1 \end{bmatrix} = 2\mathbf{e}_1 + (-1)\mathbf{e}_2, \quad \text{so that} \quad [T(\mathbf{v}_1)]_{\mathcal{E}} = \begin{bmatrix} 2 \\ -1 \end{bmatrix};$$

$$T\left(\begin{bmatrix} 2 \\ -3 \end{bmatrix}\right) = \begin{bmatrix} -3 \\ 2 \end{bmatrix} = -3\mathbf{e}_1 + 2\mathbf{e}_2, \quad \text{so that} \quad [T(\mathbf{v}_2)]_{\mathcal{E}} = \begin{bmatrix} -3 \\ 2 \end{bmatrix}.$$

Thus $M_{\mathcal{E} \leftarrow \mathcal{V}}(T) = \begin{bmatrix} 2 & -3 \\ -1 & 2 \end{bmatrix}$. Let's be sure we understand the sense in which T is multiplication by this matrix.

Using **5.3.3**, we know that the coordinate vector of $\mathbf{x} = \begin{bmatrix} x_1 \\ x_2 \end{bmatrix}$ relative to $\mathcal{V}$ is $\mathbf{x}_{\mathcal{V}} = \begin{bmatrix} 3x_1 + 2x_2 \\ 2x_1 + x_2 \end{bmatrix}$. Thus

$$M_{\mathcal{E} \leftarrow \mathcal{V}}(T)\mathbf{x}_{\mathcal{V}} = \begin{bmatrix} 2 & -3 \\ -1 & 2 \end{bmatrix} \begin{bmatrix} 3x_1 + 2x_2 \\ 2x_1 + x_2 \end{bmatrix}$$

$$= \begin{bmatrix} 2(3x_1 + 2x_2) - 3(2x_1 + x_2) \\ -(3x_1 + 2x_2) + 2(2x_1 + x_2) \end{bmatrix} = \begin{bmatrix} x_2 \\ x_1 \end{bmatrix} = [T(\mathbf{x})]_{\mathcal{E}},$$

the coordinate vector $T(\mathbf{x})$ relative to $\mathcal{E}$. ∎

5.3.9 EXAMPLE   Let $V = W = \mathbf{R}^2$ and let $\mathcal{V}$, $\mathcal{E}$, and T be as in **5.3.8**. Let us find $M_{\mathcal{V} \leftarrow \mathcal{E}}(T)$, the matrix of T relative to $\mathcal{E}$ and $\mathcal{V}$. The columns of T are the coordinate vectors of $T(\mathbf{e}_1)$ and $T(\mathbf{e}_2)$ relative to $\mathcal{V}$. Again, using what we discovered in **5.3.3**, the coordinates of $\begin{bmatrix} 1 \\ 0 \end{bmatrix}$ relative to $\mathcal{V}$ are 3 and 2, and the coordinates of $\begin{bmatrix} 0 \\ 1 \end{bmatrix}$ relative to $\mathcal{V}$ are 2 and 1. We have

$$T(\mathbf{e}_1) = \begin{bmatrix} 0 \\ 1 \end{bmatrix} \quad \text{so that} \quad [T(\mathbf{e}_1)]_{\mathcal{V}} = \begin{bmatrix} 2 \\ 1 \end{bmatrix};$$

$$T(\mathbf{e}_2) = \begin{bmatrix} 1 \\ 0 \end{bmatrix} \quad \text{so that} \quad [T(\mathbf{e}_2)]_{\mathcal{V}} = \begin{bmatrix} 3 \\ 2 \end{bmatrix}.$$

Thus $M_{\mathcal{V} \leftarrow \mathcal{E}}(T) = \begin{bmatrix} 2 & 3 \\ 1 & 2 \end{bmatrix}$. Again, we emphasize the sense in which T is multiplication by this matrix:

$$M_{\mathcal{V} \leftarrow \mathcal{E}}(T)\mathbf{x}_{\mathcal{E}} = \begin{bmatrix} 2 & 3 \\ 1 & 2 \end{bmatrix} \begin{bmatrix} x_1 \\ x_2 \end{bmatrix} = \begin{bmatrix} 2x_1 + 3x_2 \\ x_1 + 2x_2 \end{bmatrix},$$

which (by **5.3.3**) is the coordinate vector $[T(\mathbf{x})]_{\mathcal{V}}$ of $T(\mathbf{x})$ relative to $\mathcal{V}$.

Finally, let us find the matrix $M_{\mathcal{V}\leftarrow\mathcal{V}}(T)$, the matrix of T relative to $\mathcal{V}$ and $\mathcal{V}$. The columns of T are the coordinate vectors of $T(\mathbf{v}_1)$ and $T(\mathbf{v}_2)$ relative to $\mathcal{V}$. Again **5.3.3** proves useful. We have

$$T(\mathbf{v}_1) = \begin{bmatrix} 2 \\ -1 \end{bmatrix} = 4\mathbf{v}_1 + 3\mathbf{v}_2 \quad \text{so that} \quad [T(\mathbf{v}_1)]_{\mathcal{V}} = \begin{bmatrix} 4 \\ 3 \end{bmatrix}$$

and

$$T(\mathbf{v}_2) = \begin{bmatrix} -3 \\ 2 \end{bmatrix} = -5\mathbf{v}_1 - 4\mathbf{v}_2 \quad \text{so that} \quad [T(\mathbf{v}_2)]_{\mathcal{V}} = \begin{bmatrix} -5 \\ -4 \end{bmatrix}.$$

Thus $M_{\mathcal{V}\leftarrow\mathcal{V}}(T) = \begin{bmatrix} 4 & -5 \\ 3 & -4 \end{bmatrix}$. Since the coordinate vector of $\mathbf{x} = \begin{bmatrix} x_1 \\ x_2 \end{bmatrix}$ relative to $\mathcal{V}$ is $\mathbf{x}_{\mathcal{V}} = \begin{bmatrix} 3x_1 + 2x_2 \\ 2x_1 + x_2 \end{bmatrix}$, the linear transformation T is left multiplication by $M_{\mathcal{V}\leftarrow\mathcal{V}}(T)$ in the sense that

$$M_{\mathcal{V}\leftarrow\mathcal{V}}(T)\mathbf{x}_{\mathcal{V}} = \begin{bmatrix} 4 & -5 \\ 3 & -4 \end{bmatrix} \begin{bmatrix} 3x_1 + 2x_2 \\ 2x_1 + x_2 \end{bmatrix}$$

$$= \begin{bmatrix} 4(3x_1 + 2x_2) - 5(2x_1 + x_2) \\ 3(3x_1 + 2x_2) - 4(2x_1 + x_2) \end{bmatrix} = \begin{bmatrix} 2x_1 + 3x_2 \\ x_1 + 2x_2 \end{bmatrix} = [T(\mathbf{x})]_{\mathcal{V}},$$

the coordinate vector $T(\mathbf{x}) = \begin{bmatrix} x_2 \\ x_1 \end{bmatrix}$ relative to $\mathcal{V}$. $\blacksquare$

■ Matrices of the Identity

Suppose $\mathcal{V} = \{\mathbf{v}_1, \mathbf{v}_2, \dots, \mathbf{v}_n\}$ is a basis for a vector space V. What is the matrix $P = M_{\mathcal{V}\leftarrow\mathcal{V}}(\text{id})$ of the identity linear transformation $\text{id}: V \to V$ relative to $\mathcal{V}$ and $\mathcal{V}$? By Definition 5.3.7, the first column of P is the coordinate vector of $\text{id}(\mathbf{v}_1)$ relative to $\mathcal{V}$. Since $\text{id}(\mathbf{v}_1) = \mathbf{v}_1 = 1\mathbf{v}_1 + 0\mathbf{v}_2 + \cdots + 0\mathbf{v}_n$, this vector is $\begin{bmatrix} 1 \\ 0 \\ \vdots \\ 0 \end{bmatrix}$.

The second column of P is the coordinate vector of $\text{id}(\mathbf{v}_2)$ relative to $\mathcal{V}$. Since $\text{id}(\mathbf{v}_2) = \mathbf{v}_2 = 0\mathbf{v}_1 + 1\mathbf{v}_2 + 0\mathbf{v}_3 + \cdots + 0\mathbf{v}_n$, this vector is $\begin{bmatrix} 0 \\ 1 \\ 0 \\ \vdots \\ 0 \end{bmatrix}$.

We obtain $P = \begin{bmatrix} 1 & 0 & \cdots & 0 \\ 0 & 1 & & 0 \\ \vdots & & \ddots & \\ 0 & 0 & \cdots & 1 \end{bmatrix} = I_n$, the $n \times n$ identity matrix.

5.3.10 The matrix of the identity linear transformation $V \to V$ relative to one fixed basis for V is the identity matrix.

Reading Challenge 2 *Let $V = \mathbf{R}^2$. Let $\mathcal{E}$ denote the standard basis of V and let $\mathcal{V} = \{\mathbf{v}_1, \mathbf{v}_2\}$ be the basis with $\mathbf{v}_1 = \begin{bmatrix} -1 \\ 2 \end{bmatrix}$ and $\mathbf{v}_2 = \begin{bmatrix} 2 \\ -3 \end{bmatrix}$. Find the matrix $M_{\mathcal{E} \leftarrow \mathcal{E}}(\mathrm{id})$ of the identity $V \to V$ relative to $\mathcal{E}$ and $\mathcal{E}$ and the matrix $M_{\mathcal{V} \leftarrow \mathcal{V}}(\mathrm{id})$ of the identity $V \to V$ relative to $\mathcal{V}$ and $\mathcal{V}$.*

One of the main goals of this section is to have the reader realize that a given linear transformation has many different matrices associated with it. The identity matrix I is just one of many matrices associated with the identity linear transformation id.

Proposition 5.3.11 *Any invertible matrix is a matrix of the identity linear transformation.*

Proof Let $P = \begin{bmatrix} \mathbf{v}_1 & \mathbf{v}_2 & \cdots & \mathbf{v}_n \\ \downarrow & \downarrow & & \downarrow \end{bmatrix}$ be an invertible $n \times n$ matrix. The columns

of P form a basis $\mathcal{V} = \{\mathbf{v}_1, \mathbf{v}_2, \ldots, \mathbf{v}_n\}$ of $\mathbf{R}^n$. (See Corollary 4.4.14, for example.) We claim that $P = M_{\mathcal{E} \leftarrow \mathcal{V}}(\mathrm{id})$ is the matrix of the identity linear transformation relative to $\mathcal{V}$ and $\mathcal{E}$, the standard basis of $\mathbf{R}^n$. To see this, recall that the first column of $M_{\mathcal{E} \leftarrow \mathcal{V}}(\mathrm{id})$ is the coordinate vector of $\mathrm{id}(\mathbf{v}_1)$ relative to $\mathcal{E}$. But $\mathrm{id}(\mathbf{v}_1) = \mathbf{v}_1$, whose coordinate vector relative to $\mathcal{E}$ is $\mathbf{v}_1$ itself (see **5.3.2**). Similarly, column two of the matrix is $\mathbf{v}_2$, and so on. The matrix of the identity in this situation is the given matrix. ∎

■ Matrix Multiplication Is Composition of Linear Transformations

In Section 5.2, we showed that matrix multiplication corresponds to composition of linear transformations—see Theorem 5.2.2. At that point, all matrices were with respect to standard bases. Here is the more general statement, whose proof is left to the exercises.

Theorem 5.3.12 *Let $T: V \to W$ and $S: W \to U$ be linear transformations. Let $\mathcal{V}$, $\mathcal{W}$, and $\mathcal{U}$ be bases for V, W, and U, respectively. Then*

$$\boxed{M_{\mathcal{U} \leftarrow \mathcal{V}}(ST) = M_{\mathcal{U} \leftarrow \mathcal{W}}(S) M_{\mathcal{W} \leftarrow \mathcal{V}}(T).}$$

It is no accident that the matrices $M_{\mathcal{E} \leftarrow \mathcal{V}} = \begin{bmatrix} 2 & -3 \\ -1 & 2 \end{bmatrix}$ and $M_{\mathcal{V} \leftarrow \mathcal{E}} = \begin{bmatrix} 2 & 3 \\ 1 & 2 \end{bmatrix}$ in Examples 5.3.8 and 5.3.9, respectively, are inverses:

$$\begin{bmatrix} 2 & -3 \\ -1 & 2 \end{bmatrix} \begin{bmatrix} 2 & 3 \\ 1 & 2 \end{bmatrix} = \begin{bmatrix} 1 & 0 \\ 0 & 1 \end{bmatrix}.$$

Since $T\left(T\left(\begin{bmatrix} x_1 \\ x_2 \end{bmatrix} \right) \right) = T\left(\begin{bmatrix} x_2 \\ x_1 \end{bmatrix} \right) = \begin{bmatrix} x_1 \\ x_2 \end{bmatrix}$, we have $TT = \mathrm{id}$, thus

$$M_{\mathcal{E} \leftarrow \mathcal{V}}(T) M_{\mathcal{V} \leftarrow \mathcal{E}}(T) = M_{\mathcal{E} \leftarrow \mathcal{E}}(TT) = M_{\mathcal{E} \leftarrow \mathcal{E}}(\mathrm{id}) = I.$$

Here is another illustration of Theorem 5.3.12.

5.3.13 EXAMPLE   Let $V = \mathbf{R}^2$, $W = \mathbf{R}^3$, and $U = \mathbf{R}^4$. Let $T\colon V \to W$ and $S\colon W \to U$ be the linear transformations defined by

$$T\left(\begin{bmatrix} x_1 \\ x_2 \end{bmatrix}\right) = \begin{bmatrix} 2x_1 - 3x_2 \\ x_2 \\ x_1 + x_2 \end{bmatrix} \quad \text{and} \quad S\left(\begin{bmatrix} x_1 \\ x_2 \\ x_3 \end{bmatrix}\right) = \begin{bmatrix} x_1 - x_2 - x_3 \\ 2x_2 + x_3 \\ x_1 - x_3 \\ 3x_1 + x_2 - 2x_3 \end{bmatrix}.$$

Let V, W, U have, respectively, the bases

$$V = \{\mathbf{v}_1, \mathbf{v}_2\}, \quad \text{where } \mathbf{v}_1 = \begin{bmatrix} 1 \\ 1 \end{bmatrix}, \mathbf{v}_2 = \begin{bmatrix} -1 \\ 1 \end{bmatrix};$$

$$W = \{\mathbf{w}_1, \mathbf{w}_2, \mathbf{w}_3\}, \quad \text{where } \mathbf{w}_1 = \begin{bmatrix} 1 \\ 0 \\ 0 \end{bmatrix}, \mathbf{w}_2 = \begin{bmatrix} 1 \\ 1 \\ 0 \end{bmatrix}, \mathbf{w}_3 = \begin{bmatrix} 1 \\ 1 \\ 1 \end{bmatrix}; \quad \text{and}$$

$$U = \{\mathbf{e}_1, \mathbf{e}_2, \mathbf{e}_3, \mathbf{e}_4\} = \mathcal{E}, \quad \text{the standard basis of } \mathbf{R}^4.$$

First, we find $M_{W \leftarrow V}(T)$, the matrix of T relative to V and W. In view of Definition 5.3.7, to do so we must find the coordinate vectors of $T(\mathbf{v}_1)$ and $T(\mathbf{v}_2)$ relative to W. (These vectors are the columns of the desired matrix.) In **5.3.4**, we showed that the coordinate vector of $\mathbf{x} = \begin{bmatrix} x_1 \\ x_2 \\ x_3 \end{bmatrix}$ relative to W is $\begin{bmatrix} x_1 - x_2 \\ x_2 - x_3 \\ x_3 \end{bmatrix}$. Thus

$$T(\mathbf{v}_1) = \begin{bmatrix} -1 \\ 1 \\ 2 \end{bmatrix} \quad \text{has coordinate vector} \quad [T(\mathbf{v}_1)]_W = \begin{bmatrix} -2 \\ -1 \\ 2 \end{bmatrix},$$

$$T(\mathbf{v}_2) = \begin{bmatrix} -5 \\ 1 \\ 0 \end{bmatrix} \quad \text{has coordinate vector} \quad [T(\mathbf{v}_2)]_W = \begin{bmatrix} -6 \\ 1 \\ 0 \end{bmatrix},$$

and $M_{W \leftarrow V}(T) = \begin{bmatrix} -2 & -6 \\ -1 & 1 \\ 2 & 0 \end{bmatrix}$. Next, we find $M_{\mathcal{E} \leftarrow W}(S)$, the matrix of S relative to W and $\mathcal{E}$. We have

$$S(\mathbf{w}_1) = \begin{bmatrix} 1 \\ 0 \\ 1 \\ 3 \end{bmatrix} = [S(\mathbf{w}_1)]_{\mathcal{E}}, \quad S(\mathbf{w}_2) = \begin{bmatrix} 0 \\ 2 \\ 1 \\ 4 \end{bmatrix} = [S(\mathbf{w}_2)]_{\mathcal{E}},$$

$$\text{and} \quad S(\mathbf{w}_3) = \begin{bmatrix} -1 \\ 3 \\ 0 \\ 2 \end{bmatrix} = [S(\mathbf{w}_3)]_{\mathcal{E}},$$

so $M_{\mathcal{E}\leftarrow\mathcal{W}}(S) = \begin{bmatrix} 1 & 0 & -1 \\ 0 & 2 & 3 \\ 1 & 1 & 0 \\ 3 & 4 & 2 \end{bmatrix}$. The linear transformation $ST: V \to U$ is defined by $ST(\mathbf{v}) = S(T(\mathbf{v}))$. We have

$$ST(\mathbf{v}_1) = S\left(\begin{bmatrix} -1 \\ 1 \\ 2 \end{bmatrix}\right) = \begin{bmatrix} -4 \\ 4 \\ -3 \\ -6 \end{bmatrix} = [ST(\mathbf{v}_1)]_{\mathcal{E}} \text{ and}$$

$$ST(\mathbf{v}_2) = S\left(\begin{bmatrix} -5 \\ 1 \\ 0 \end{bmatrix}\right) = \begin{bmatrix} -6 \\ 2 \\ -5 \\ -14 \end{bmatrix} = [ST(\mathbf{v}_2)]_{\mathcal{E}},$$

so that the matrix of ST relative to V and $\mathcal{E}$ is $M_{\mathcal{E}\leftarrow V}(ST) = \begin{bmatrix} -4 & -6 \\ 4 & 2 \\ -3 & -5 \\ -6 & -14 \end{bmatrix}$. We ask the reader to check that $M_{\mathcal{E}\leftarrow\mathcal{W}}(S)M_{\mathcal{W}\leftarrow V}(T) = M_{\mathcal{E}\leftarrow V}(ST)$:

$$\begin{bmatrix} 1 & 0 & -1 \\ 0 & 2 & 3 \\ 1 & 1 & 0 \\ 3 & 4 & 2 \end{bmatrix} \begin{bmatrix} -2 & -6 \\ -1 & 1 \\ 2 & 0 \end{bmatrix} = \begin{bmatrix} -4 & -6 \\ 4 & 2 \\ -3 & -5 \\ -6 & -14 \end{bmatrix},$$

as asserted in Theorem 5.3.12. ∎

Now consider the special case of Theorem 5.3.12 where $U = V$ and the linear transformation $T: V \to W$ is invertible. Since $TT^{-1} = \mathrm{id}_W$, Theorem 5.3.12 says that

$$M_{W\leftarrow V}(T)M_{V\leftarrow W}(T^{-1}) = M_{W\leftarrow W}(TT^{-1}) = M_{W\leftarrow W}(\mathrm{id}) = I, \qquad (5)$$

the matrix of the identity transformation $W \to W$ relative to the single basis $\mathcal{W}$ (see 5.3.10). Thus each matrix $M_{W\leftarrow V}(T)$ and $M_{V\leftarrow W}(T^{-1})$ is invertible and each is the inverse of the other: $M_{V\leftarrow W}(T^{-1}) = [M_{W\leftarrow V}(T)]^{-1}$.

Theorem 5.3.14 *Let V be a vector space and $T: V \to V$ a linear transformation with matrix A relative to certain bases. If T is invertible, then A is invertible and the matrix of T^{-1} is A^{-1}:*

$$\boxed{M_{V\leftarrow W}(T^{-1}) = [M_{W\leftarrow V}(T)]^{-1}.}$$

The converse of Theorem 5.3.14 is true as well; that is, if the matrix $A = M_{W\leftarrow V}(T)$ of a linear transformation T is invertible, then so is T. The argument, briefly, is this: Let $V = \{\mathbf{v}_1, \mathbf{v}_2, \dots, \mathbf{v}_n\}$. If A is invertible, its columns, which are the coordinate vectors of $T(\mathbf{v}_1), T(\mathbf{v}_2), \dots, T(\mathbf{v}_n)$ relative to $\mathcal{W}$, are linearly independent and hence a basis for $\mathbf{R}^n$. It follows that the vectors $T(\mathbf{v}_1), T(\mathbf{v}_2), \dots, T(\mathbf{v}_n)$ themselves comprise a basis for W. Thus T is one-to-one and onto (see Exercise 13(d) of

Section 5.1) and hence invertible. Again, (5) shows that $M_{W \leftarrow V}(T)M_{V \leftarrow W}(T^{-1}) = M_{W \leftarrow W}(\text{id}) = I_n$ so that the matrix for T^{-1} is A^{-1}.

5.3.15 EXAMPLE   Suppose $T \colon \mathbf{R}^3 \to \mathbf{R}^3$ is defined by $T\left(\begin{bmatrix} x_1 \\ x_2 \\ x_3 \end{bmatrix}\right) = \begin{bmatrix} x_1 + x_2 \\ x_1 - x_2 \\ x_1 + x_2 + x_3 \end{bmatrix}$. The matrix of T

relative to $\mathcal{E}$ and $\mathcal{E}$, the standard basis of $\mathbf{R}^3$, is $A = \begin{bmatrix} 1 & 1 & 0 \\ 1 & -1 & 0 \\ 1 & 1 & 1 \end{bmatrix}$. This is invertible,

with $A^{-1} = \frac{1}{2}\begin{bmatrix} 1 & 1 & 0 \\ 1 & -1 & 0 \\ -2 & 0 & 2 \end{bmatrix}$. Thus T is invertible, with corresponding matrix A^{-1}:

$$T^{-1}\left(\begin{bmatrix} x_1 \\ x_2 \\ x_3 \end{bmatrix}\right) = A^{-1}\begin{bmatrix} x_1 \\ x_2 \\ x_3 \end{bmatrix} = \frac{1}{2}\begin{bmatrix} x_1 + x_2 - 2x_3 \\ x_1 - x_2 \\ 2x_3 \end{bmatrix}. \qquad \blacksquare$$

Proposition 5.3.11 said that any invertible matrix is the matrix of the identity transformation. On the other hand, the identity linear transformation is invertible, so Theorem 5.3.14 says any of its matrices is also invertible.

Corollary 5.3.16   *A matrix is invertible if and only if it is the matrix of the identity transformation relative to appropriate bases.*

Corollary 5.3.17   *Let A and B be $n \times n$ matrices. Then $AB = I$ if and only if $BA = I$.*

Proof   We have seen this result before—see Corollary 2.7.17 and Corollary 4.4.13—but here we present a very different proof.

Suppose $AB = I$. Let $T \colon \mathbf{R}^n \to \mathbf{R}^n$ be left multiplication by A. Then T is onto since, for any $\mathbf{w}$ in $\mathbf{R}^n$, $\mathbf{w} = I\mathbf{w} = (AB)\mathbf{w} = A(B\mathbf{w}) = T(B\mathbf{w})$. By Corollary 5.2.9, T is invertible, so A is invertible by Theorem 5.3.14 and $BA = A^{-1}(AB)A = A^{-1}IA = A^{-1}A = I$. Similarly, if $BA = I$, then $AB = I$. $\blacksquare$

Answers to Reading Challenges

1. We wish to find the scalars a and b such that $\mathbf{x} = a\mathbf{v}_1 + b\mathbf{v}_2 = a\begin{bmatrix} -1 \\ 2 \end{bmatrix} + b\begin{bmatrix} 2 \\ -3 \end{bmatrix}$.

   This is the matrix equation $\begin{bmatrix} x_1 \\ x_2 \end{bmatrix} = \begin{bmatrix} -1 & 2 \\ 2 & -3 \end{bmatrix}\begin{bmatrix} a \\ b \end{bmatrix}$ whose solution is $a = 3x_1 + 2x_2$, $b = 2x_1 + x_2$.

2. As described in 5.3.10, the answer in each case is the 2×2 identity matrix $\begin{bmatrix} 1 & 0 \\ 0 & 1 \end{bmatrix}$. The reader is urged, however, to carry out the reasoning, without simply quoting the result of a worked example.

True/False Questions

Decide, with as little calculation as possible, whether each of the following statements is true or false and explain your answer whenever you say "false." (Answers can be found in the back of the book.)

1. The coordinate vector of $\begin{bmatrix} -2 \\ 3 \end{bmatrix}$ relative to the standard basis of $\mathbf{R}^2$ is $\begin{bmatrix} -2 \\ 3 \end{bmatrix}$.

2. The coordinate vector of $\begin{bmatrix} -2 \\ 3 \end{bmatrix}$ relative to the basis $\{\begin{bmatrix} 0 \\ 1 \end{bmatrix}, \begin{bmatrix} 1 \\ 0 \end{bmatrix}\}$ is $\begin{bmatrix} -2 \\ 3 \end{bmatrix}$.

3. Let $\mathbf{v}_1 = \begin{bmatrix} 2 \\ 5 \end{bmatrix}$ and $\mathbf{v}_2 = \begin{bmatrix} -5 \\ 1 \end{bmatrix}$ and let $\mathcal{V} = \{\mathbf{v}_1, \mathbf{v}_2\}$. The coordinate vector of $\mathbf{v}_1$ relative to $\mathcal{V}$ is $\begin{bmatrix} 1 \\ 0 \end{bmatrix}$.

4. Let $\mathbf{v}_1 = \begin{bmatrix} 2 \\ 5 \end{bmatrix}$ and $\mathbf{v}_2 = \begin{bmatrix} 5 \\ 1 \end{bmatrix}$ and let $\mathcal{V} = \{\mathbf{v}_1, \mathbf{v}_2\}$. The coordinate vector of $\begin{bmatrix} 3 \\ -4 \end{bmatrix}$ relative to the is $\begin{bmatrix} -1 \\ 1 \end{bmatrix}$.

5. The matrix of the identity transformation $\mathbf{R}^n \to \mathbf{R}^n$ is the identity matrix.

Exercises

*Solutions to exercises marked [BB] can be found in the **B**ack of the **B**ook.*

1. [BB] Let $\mathcal{V} = \{\mathbf{v}_1, \mathbf{v}_2\}$ be the basis of $\mathbf{R}^2$ where $\mathbf{v}_1 = \begin{bmatrix} 2 \\ 5 \end{bmatrix}$ and $\mathbf{v}_2 = \begin{bmatrix} -3 \\ 1 \end{bmatrix}$. Find the coordinates of $\mathbf{x} = \begin{bmatrix} x_1 \\ x_2 \end{bmatrix}$ relative to $\mathcal{V}$.

2. Let $\mathcal{V} = \{\mathbf{v}_1, \mathbf{v}_2, \mathbf{v}_3\}$ be the basis of $\mathbf{R}^3$ where $\mathbf{v}_1 = \begin{bmatrix} -1 \\ 1 \\ 1 \end{bmatrix}$, $\mathbf{v}_2 = \begin{bmatrix} 1 \\ 0 \\ 1 \end{bmatrix}$, and $\mathbf{v}_3 = \begin{bmatrix} 0 \\ 2 \\ -1 \end{bmatrix}$. Find the coordinates of $\mathbf{x} = \begin{bmatrix} x_1 \\ x_2 \\ x_3 \end{bmatrix}$ relative to $\mathcal{V}$.

3. [BB] Let $\mathcal{V} = \{\mathbf{v}_1, \mathbf{v}_2, \mathbf{v}_3\}$ be the basis of $\mathbf{R}^3$ where $\mathbf{v}_1 = \begin{bmatrix} 2 \\ 3 \\ -1 \end{bmatrix}$, $\mathbf{v}_2 = \begin{bmatrix} -1 \\ -2 \\ 4 \end{bmatrix}$, and $\mathbf{v}_3 = \begin{bmatrix} 0 \\ 1 \\ -1 \end{bmatrix}$. Find the coordinate vector $\mathbf{x}_\mathcal{V}$ of $\mathbf{x} = \begin{bmatrix} x_1 \\ x_2 \\ x_3 \end{bmatrix}$ relative to $\mathcal{V}$.

4. [BB] Let $\mathbf{v}_1 = \begin{bmatrix} 1 \\ 0 \end{bmatrix}$, $\mathbf{v}_2 = \begin{bmatrix} 1 \\ 1 \end{bmatrix}$, and $\mathcal{V} = \{\mathbf{v}_1, \mathbf{v}_2\}$. Let $\mathcal{E}$ denote the standard basis of $\mathbf{R}^2$. Let $A = \begin{bmatrix} 1 & 2 \\ 3 & 4 \end{bmatrix}$

be the matrix of a linear transformation $T: \mathbf{R}^2 \to \mathbf{R}^2$. Find $T\left(\begin{bmatrix} x_1 \\ x_2 \end{bmatrix}\right)$ in each of the following cases.

(a) $A = M_{\mathcal{E} \leftarrow \mathcal{E}}(T)$ (b) $A = M_{\mathcal{E} \leftarrow \mathcal{V}}(T)$
(c) $A = M_{\mathcal{V} \leftarrow \mathcal{E}}(T)$ (d) $A = M_{\mathcal{V} \leftarrow \mathcal{V}}(T)$.

5. Answer all parts of Exercise 4 with $\mathbf{v}_1 = \begin{bmatrix} -1 \\ 2 \end{bmatrix}$, $\mathbf{v}_2 = \begin{bmatrix} 2 \\ 1 \end{bmatrix}$, and $A = \begin{bmatrix} -2 & 0 \\ 1 & -3 \end{bmatrix}$.

6. Let $T: \mathbf{R}^3 \to \mathbf{R}^3$ be the linear transformation defined by the rule $T\left(\begin{bmatrix} x \\ y \\ z \end{bmatrix}\right) = \begin{bmatrix} 2x + y \\ x + y - z \\ -z \end{bmatrix}$. Find $M_{\mathcal{V} \leftarrow \mathcal{V}}(T)$, the matrix of T relative to the basis $\mathcal{V}$ (in each instance), where $\mathcal{V} = \{\mathbf{v}_1, \mathbf{v}_2, \mathbf{v}_3\}$, $\mathbf{v}_1 = \begin{bmatrix} 1 \\ 0 \\ 0 \end{bmatrix}$, $\mathbf{v}_2 = \begin{bmatrix} 1 \\ 1 \\ 0 \end{bmatrix}$, $\mathbf{v}_3 = \begin{bmatrix} 1 \\ 1 \\ 1 \end{bmatrix}$.

7. [BB] Let $\mathbf{v}_1 = \begin{bmatrix} 1 \\ 0 \end{bmatrix}$, $\mathbf{v}_2 = \begin{bmatrix} 1 \\ 1 \end{bmatrix}$, and $\mathcal{V} = \{\mathbf{v}_1, \mathbf{v}_2\}$. Let $\mathbf{w}_1 = \begin{bmatrix} 1 \\ 0 \\ 1 \end{bmatrix}$, $\mathbf{w}_2 = \begin{bmatrix} 0 \\ 1 \\ -1 \end{bmatrix}$, $\mathbf{w}_3 = \begin{bmatrix} 1 \\ 1 \\ 0 \end{bmatrix}$, and

$W = \{w_1, w_2, w_3\}$. Let $\mathcal{E} = \{e_1, e_2\}$ and $\mathcal{F} = \{f_1, f_2, f_3\}$ denote the standard bases of $\mathbf{R}^2$ and $\mathbf{R}^3$, respectively. Let $A = \begin{bmatrix} 0 & -1 \\ 2 & 1 \\ -4 & 1 \end{bmatrix}$ be the

matrix of a linear transformation $T: \mathbf{R}^2 \to \mathbf{R}^3$. Find $T\left(\begin{bmatrix} x_1 \\ x_2 \end{bmatrix}\right)$ in each of the following cases.

(a) $A = M_{\mathcal{F} \leftarrow \mathcal{E}}(T)$ (b) $A = M_{W \leftarrow \mathcal{E}}(T)$

(c) $A = M_{\mathcal{F} \leftarrow \mathcal{V}}(T)$ (d) $A = M_{W \leftarrow \mathcal{V}}(T)$.

8. Let $v_1 = \begin{bmatrix} -1 \\ 1 \end{bmatrix}$, $v_2 = \begin{bmatrix} 2 \\ 0 \end{bmatrix}$, and $\mathcal{V} = \{v_1, v_2\}$. Let $\mathcal{E}$ denote the standard basis of $\mathbf{R}^2$ and id: $\mathbf{R}^2 \to \mathbf{R}^2$ the identity map. Find each of the following matrices of id.

(a) $M_{\mathcal{E} \leftarrow \mathcal{E}}(\mathrm{id})$ (b) $M_{\mathcal{V} \leftarrow \mathcal{E}}(\mathrm{id})$

(c) $M_{\mathcal{E} \leftarrow \mathcal{V}}(\mathrm{id})$ (d) $M_{\mathcal{V} \leftarrow \mathcal{V}}(\mathrm{id})$.

9. [BB] Let $T: \mathbf{R}^2 \to \mathbf{R}^2$ be the linear transformation defined by $T\left(\begin{bmatrix} x \\ y \end{bmatrix}\right) = \begin{bmatrix} x + 4y \\ 2x + 7y \end{bmatrix}$. Let $\mathcal{V} = \{v_1, v_2\}$ where $v_1 = \begin{bmatrix} 1 \\ -1 \end{bmatrix}$ and $v_2 = \begin{bmatrix} -2 \\ 1 \end{bmatrix}$ and let $\mathcal{E} = \{e_1, e_2\}$ be the standard basis for V.

(a) Find $M_{\mathcal{E} \leftarrow \mathcal{E}}(T)$, the matrix of T relative to $\mathcal{E}$ and $\mathcal{E}$.

(b) Find $M_{\mathcal{E} \leftarrow \mathcal{V}}(T)$, the matrix of T relative to $\mathcal{V}$ and $\mathcal{E}$.

(c) Find $M_{\mathcal{V} \leftarrow \mathcal{E}}(T)$, the matrix of T relative to $\mathcal{E}$ and $\mathcal{V}$.

(d) Find $M_{\mathcal{V} \leftarrow \mathcal{V}}(T)$, the matrix of T relative to $\mathcal{V}$ and $\mathcal{V}$.

10. Let $v_1 = \begin{bmatrix} -1 \\ 1 \\ 1 \end{bmatrix}$, $v_2 = \begin{bmatrix} 2 \\ 0 \\ 1 \end{bmatrix}$, $v_3 = \begin{bmatrix} 1 \\ 1 \\ 3 \end{bmatrix}$, and

$\mathcal{V} = \{v_1, v_2, v_3\}$. Let $w_1 = \begin{bmatrix} 1 \\ 1 \end{bmatrix}$, $w_2 = \begin{bmatrix} -2 \\ 4 \end{bmatrix}$, and $W = \{w_1, w_2\}$. Let $\mathcal{E} = \{e_1, e_2, e_3\}$ and $\mathcal{F} = \{f_1, f_2\}$ denote the standard bases of $\mathbf{R}^3$ and $\mathbf{R}^2$, respectively. Let $T: \mathbf{R}^3 \to \mathbf{R}^2$ be the linear transformation defined by $T\left(\begin{bmatrix} x \\ y \\ z \end{bmatrix}\right) = \begin{bmatrix} 2x - y + z \\ -x + y + 4z \end{bmatrix}$.

(a) Find $M_{\mathcal{F} \leftarrow \mathcal{E}}(T)$, the matrix of T relative to $\mathcal{E}$ and $\mathcal{F}$.

(b) Find $M_{\mathcal{F} \leftarrow \mathcal{V}}(T)$, the matrix of T relative to $\mathcal{V}$ and $\mathcal{F}$.

(c) Find $M_{W \leftarrow \mathcal{E}}(T)$, the matrix of T relative to $\mathcal{E}$ and W.

(d) Find $M_{W \leftarrow \mathcal{V}}(T)$, the matrix of T relative to $\mathcal{V}$ and W.

11. [BB] Prove that the linear transformation defined in Lemma 5.3.5 is one-to-one.

12. Let $T: \mathbf{R}^3 \to \mathbf{R}^2$ and $S: \mathbf{R}^2 \to \mathbf{R}^4$ be the linear transformations defined by

$$T\left(\begin{bmatrix} x \\ y \\ z \end{bmatrix}\right) = \begin{bmatrix} x - y \\ y + 2z \end{bmatrix} \quad \text{and} \quad S\left(\begin{bmatrix} x \\ y \end{bmatrix}\right) = \begin{bmatrix} 2x \\ 3y \\ -x \\ x + y \end{bmatrix}.$$

Let $\mathbf{R}^3$, $\mathbf{R}^2$, and $\mathbf{R}^4$ have, respectively, the bases

$$\mathcal{V} = \{v_1, v_2, v_3\}, \quad \text{where } v_1 = \begin{bmatrix} 1 \\ 0 \\ 0 \end{bmatrix}, v_2 = \begin{bmatrix} 1 \\ 1 \\ 0 \end{bmatrix},$$

$$v_3 = \begin{bmatrix} 1 \\ 1 \\ 1 \end{bmatrix};$$

$$W = \{w_1, w_2\}, \quad \text{where } w_1 = \begin{bmatrix} -1 \\ 1 \end{bmatrix}, w_2 = \begin{bmatrix} 0 \\ 2 \end{bmatrix}; \text{ and}$$

$\mathcal{E} = \{e_1, e_2, e_3, e_4\}$, the standard basis of $\mathbf{R}^4$.

(a) Find $M_{W \leftarrow \mathcal{V}}(T)$, the matrix of T relative to $\mathcal{V}$ and W.

(b) Find $M_{\mathcal{E} \leftarrow W}(S)$, the matrix of S relative to W and $\mathcal{E}$.

(c) Find $ST\left(\begin{bmatrix} x \\ y \\ z \end{bmatrix}\right)$.

(d) Find $M_{\mathcal{E} \leftarrow \mathcal{V}}(ST)$, the matrix of ST relative to $\mathcal{V}$ and $\mathcal{E}$.

(e) Verify that $M_{\mathcal{E} \leftarrow \mathcal{V}}(ST) = M_{\mathcal{E} \leftarrow W}(S) M_{W \leftarrow \mathcal{V}}(T)$.

13. [BB] Let $T: \mathbf{R}^3 \to \mathbf{R}^4$ be the linear transformation defined by $T\left(\begin{bmatrix} x \\ y \\ z \end{bmatrix}\right) = \begin{bmatrix} x - y + 2z \\ x \\ 5x - 2y + 4z \\ 3x - y + 2z \end{bmatrix}$. Let $\mathcal{E}$ and $\mathcal{F}$ denote the standard bases for $\mathbf{R}^3$ and $\mathbf{R}^4$, respectively.

(a) Find the matrix $M_{\mathcal{F} \leftarrow \mathcal{E}}(T)$ of T with respect to $\mathcal{E}$ and $\mathcal{F}$.

(b) Find the kernel and range of T and verify that $\dim \ker T + \dim \mathrm{range}\, T$ agrees with that predicted by Theorem 5.1.24.

14. Prove Theorem 5.3.12. [Hint: Let $A = M_{U \leftarrow W}(S)$, $B = M_{W \leftarrow V}(T)$, and $C = M_{U \leftarrow V}(ST)$. Use the definition of "matrix of a linear transformation" (5.3.7) repeatedly to show that $(AB)\mathbf{x} = C\mathbf{x}$ for any column vector $\mathbf{x}$.]

15. [BB] Define $T : \mathbf{R}^2 \to \mathbf{R}^2$ by $T\left(\begin{bmatrix} x \\ y \end{bmatrix}\right) = \begin{bmatrix} x + 3y \\ x + 4y \end{bmatrix}$.

    Let $\mathcal{E}$ denote the standard basis of $\mathbf{R}^2$.

    (a) Find $A = M_{\mathcal{E} \leftarrow \mathcal{E}}(T)$, the matrix of T relative to $\mathcal{E}$ and $\mathcal{E}$.

    (b) What is A^{-1}?

    (c) Use your answer to (b) to find $T^{-1}\left(\begin{bmatrix} x \\ y \end{bmatrix}\right)$.

16. Let $\mathbf{v}_1 = \begin{bmatrix} 2 \\ -1 \end{bmatrix}$, $\mathbf{v}_2 = \begin{bmatrix} 7 \\ 4 \end{bmatrix}$. Answer Exercise 15 for each of the following matrices.

    (a) $A = M_{\mathcal{E} \leftarrow V}(T)$ (b) $A = M_{V \leftarrow \mathcal{E}}(T)$

    (c) $A = M_{V \leftarrow V}(T)$.

17. State in your words the meaning of the equation $M_{V \leftarrow W}(T^{-1}) = [M_{W \leftarrow V}(T)]^{-1}$.

■ Critical Reading

18. [BB] Let $\mathcal{V} = \{\mathbf{v}_1, \mathbf{v}_2, \dots, \mathbf{v}_n\}$ be a basis for a vector space V and let $\mathbf{x}$ be a vector in V. Explain clearly why $\mathbf{x}$ can be written in just one way as a linear combination of the vectors in $\mathcal{V}$. (Thus the notion of "coordinate vector" is well-defined.)

19. Let $\mathcal{V}$ and $\mathcal{V}'$ be bases for a vector space V and let $T : V \to W$ be a linear transformation. Is there any connection between the matrices $M_{V' \leftarrow V}(T)$ and $M_{V \leftarrow V'}(T)$?

5.4 Changing Coordinates

In Section 5.3, we introduced the concept of "coordinates" of a vector relative to a basis and saw the need, on several occasions, to be able to convert the coordinates of a vector from one basis to another. In this section, we describe a straightforward way to accomplish this task, with a formula!

Suppose $T : V \to W$ is a linear transformation from a vector space V to a vector space W. As expressed in Definition 5.3.7, once bases $\mathcal{V}$ and $\mathcal{W}$ have been specified for V and W, respectively, T is multiplication by a matrix we have denoted $M_{W \leftarrow V}(T)$ and which works like this:

$$[T(\mathbf{x})]_W = M_{W \leftarrow V}(T)\mathbf{x}_V.$$

Now suppose that $W = V$ and $T = \mathrm{id}$ is the identity linear transformation $V \to V$. If $\mathcal{V}$ and $\mathcal{V}'$ are bases for V, then for any vector $\mathbf{x}$,

$$\mathbf{x}_{V'} = M_{V' \leftarrow V}(\mathrm{id})\mathbf{x}_V.$$

This says that $\mathbf{x}_{V'}$, the coordinate vector of $\mathbf{x}$ relative to $\mathcal{V}'$, can be obtained from its coordinate vector $\mathbf{x}_V$ relative to $\mathcal{V}$ simply by multiplying $\mathbf{x}_V$ by the matrix $M_{V' \leftarrow V}(\mathrm{id})$. This matrix is called the *change of coordinates matrix* $\mathcal{V}$ to $\mathcal{V}'$ and denoted $P_{V' \leftarrow V}$.

5.4.1 DEFINITION   Given bases $\mathcal{V}$ and $\mathcal{V}'$ for a vector space V, the *change of coordinates matrix* from $\mathcal{V}$ to $\mathcal{V}'$ is denoted $P_{V' \leftarrow V}$ and defined by

$$P_{V' \leftarrow V} = M_{V' \leftarrow V}(\mathrm{id}).$$

Recalling the definition of $M_{W \leftarrow V}$ (5.3.7), the change of coordinates matrix is the matrix whose columns are the coordinate vectors of $v_1, v_2, \ldots, v_n$ relative to V', where $V = \{v_1, v_2, \ldots, v_n\}$.

To change coordinates from V to V', multiply by the change of coordinates matrix:

5.4.2 $\mathbf{x}_{V'} = P_{V' \leftarrow V} \mathbf{x}_V.$

Sometimes, the matrix $P_{V' \leftarrow V}$ seems hard to compute, whereas the matrix $P_{V \leftarrow V'}$ needed to change coordinates in the other direction $V' \to V$ is easy. In such a case, since the identity linear transformation is invertible and $\text{id}^{-1} = \text{id}$, we are helped by the following immediate consequence of Theorem 5.3.14:

5.4.3 $P_{V' \leftarrow V} = [P_{V \leftarrow V'}]^{-1}.$

5.4.4 EXAMPLE

Let $V = \mathbf{R}^2$. Let $\mathcal{E} = \{e_1, e_2\}$ be the standard basis of V and let $V = \{v_1, v_2\}$ be the basis with

$$v_1 = \begin{bmatrix} -1 \\ 2 \end{bmatrix} \quad \text{and} \quad v_2 = \begin{bmatrix} 2 \\ -3 \end{bmatrix}.$$

The change of coordinates matrix V to $\mathcal{E}$ is

$$P = P_{\mathcal{E} \leftarrow V} = M_{\mathcal{E} \leftarrow V}(\text{id}).$$

Since $\text{id}(v_1) = v_1 = \begin{bmatrix} -1 \\ 2 \end{bmatrix}$ has coordinates -1 and 2 relative to $\mathcal{E}$, the first column of P is $\begin{bmatrix} -1 \\ 2 \end{bmatrix}$. Since $\text{id}(v_2) = v_2 = \begin{bmatrix} 2 \\ -3 \end{bmatrix}$ has coordinates 2 and -3 relative to $\mathcal{E}$, the second column of P is $\begin{bmatrix} 2 \\ -3 \end{bmatrix}$. Thus $P = \begin{bmatrix} -1 & 2 \\ 2 & -3 \end{bmatrix}$. It is usually the matrix of base change $\mathcal{E}$ to V in which we are interested, however, and this can be computed with the help of 5.4.3:

$$P_{V \leftarrow \mathcal{E}} = [P_{\mathcal{E} \leftarrow V}]^{-1} = \begin{bmatrix} -1 & 2 \\ 2 & -3 \end{bmatrix}^{-1} = \begin{bmatrix} 3 & 2 \\ 2 & 1 \end{bmatrix}.$$

The import of this matrix is this: the coordinate vector of a vector $\mathbf{x} = \begin{bmatrix} x_1 \\ x_2 \end{bmatrix}$ relative to V is

$$P_{V \leftarrow \mathcal{E}} \mathbf{x}_{\mathcal{E}} = \begin{bmatrix} 3 & 2 \\ 2 & 1 \end{bmatrix} \begin{bmatrix} x_1 \\ x_2 \end{bmatrix} = \begin{bmatrix} 3x_1 + 2x_2 \\ 2x_1 + x_2 \end{bmatrix}$$

as noted much earlier, in **5.3.3**.

5.4.5 EXAMPLE   Let $\mathbf{v}_1 = \begin{bmatrix} 1 \\ 0 \\ 0 \\ 0 \end{bmatrix}$, $\mathbf{v}_2 = \begin{bmatrix} 1 \\ 0 \\ 0 \\ 1 \end{bmatrix}$, $\mathbf{v}_3 = \begin{bmatrix} 0 \\ 1 \\ 1 \\ 0 \end{bmatrix}$, and $\mathbf{v}_4 = \begin{bmatrix} 1 \\ 1 \\ 0 \\ 1 \end{bmatrix}$. These four vectors form a basis $\mathcal{V}$

of $\mathbf{R}^4$, so it is reasonable to ask what the coordinates of a vector $\mathbf{x} = \begin{bmatrix} x_1 \\ x_2 \\ x_3 \\ x_4 \end{bmatrix}$ in $\mathbf{R}^4$ are

relative to $\mathcal{V}$. Let $\mathcal{E} = \{\mathbf{e}_1, \mathbf{e}_2, \mathbf{e}_3, \mathbf{e}_4\}$ be the standard basis for $\mathbf{R}^4$. We first compute

$$P_{\mathcal{E} \leftarrow \mathcal{V}} = \begin{bmatrix} 1 & 1 & 0 & 1 \\ 0 & 0 & 1 & 1 \\ 0 & 0 & 1 & 0 \\ 0 & 1 & 0 & 1 \end{bmatrix},$$

whose columns are the coordinate vectors of $\mathbf{v}_1, \mathbf{v}_2, \mathbf{v}_3, \mathbf{v}_4$ relative to $\mathcal{E}$. By 5.4.3, this is the inverse of the matrix we want:

$$P_{\mathcal{V} \leftarrow \mathcal{E}} = [P_{\mathcal{E} \leftarrow \mathcal{V}}]^{-1} = \begin{bmatrix} 1 & 1 & 0 & 1 \\ 0 & 0 & 1 & 1 \\ 0 & 0 & 1 & 0 \\ 0 & 1 & 0 & 1 \end{bmatrix}^{-1} = \begin{bmatrix} 1 & 0 & 0 & -1 \\ 0 & -1 & 1 & 1 \\ 0 & 0 & 1 & 0 \\ 0 & 1 & -1 & 0 \end{bmatrix}.$$

Thus the coordinates of $\mathbf{x}$ relative to $\mathcal{V}$ are the components of the vector

$$P_{\mathcal{V} \leftarrow \mathcal{E}} \begin{bmatrix} x_1 \\ x_2 \\ x_3 \\ x_4 \end{bmatrix} = \begin{bmatrix} 1 & 0 & 0 & -1 \\ 0 & -1 & 1 & 1 \\ 0 & 0 & 1 & 0 \\ 0 & 1 & -1 & 0 \end{bmatrix} \begin{bmatrix} x_1 \\ x_2 \\ x_3 \\ x_4 \end{bmatrix} = \begin{bmatrix} x_1 - x_4 \\ -x_2 + x_3 + x_4 \\ x_3 \\ x_2 - x_3 \end{bmatrix},$$

namely, the numbers $x_1 - x_4$, $-x_2 + x_3 + x_4$, x_3, and $x_2 - x_3$:

$$\begin{bmatrix} x_1 \\ x_2 \\ x_3 \\ x_4 \end{bmatrix} = (x_1 - x_4)\mathbf{v}_1 + (-x_2 + x_3 + x_4)\mathbf{v}_2 + x_3\mathbf{v}_3 + (x_2 - x_3)\mathbf{v}_4 \qquad (1)$$

$$= (x_1 - x_4)\begin{bmatrix} 1 \\ 0 \\ 0 \\ 0 \end{bmatrix} + (-x_2 + x_3 + x_4)\begin{bmatrix} 1 \\ 0 \\ 0 \\ 1 \end{bmatrix} + x_3\begin{bmatrix} 0 \\ 1 \\ 1 \\ 0 \end{bmatrix} + (x_2 - x_3)\begin{bmatrix} 1 \\ 1 \\ 0 \\ 1 \end{bmatrix}. \quad \blacksquare$$

5.4.6 Remark Equation (1) shows how to write the vector $\begin{bmatrix} x_1 \\ x_2 \\ x_3 \\ x_4 \end{bmatrix}$ as a linear combination of

$$\mathbf{v}_1 = \begin{bmatrix} 1 \\ 0 \\ 0 \\ 0 \end{bmatrix}, \quad \mathbf{v}_2 = \begin{bmatrix} 1 \\ 0 \\ 0 \\ 1 \end{bmatrix}, \quad \mathbf{v}_3 = \begin{bmatrix} 0 \\ 1 \\ 1 \\ 0 \end{bmatrix}, \quad \text{and} \quad \mathbf{v}_4 = \begin{bmatrix} 1 \\ 1 \\ 0 \\ 1 \end{bmatrix}.$$

This result can also be obtained straight from the definition of "coordinates." We seek the numbers a, b, c, d so that

$$\begin{bmatrix} x_1 \\ x_2 \\ x_3 \\ x_4 \end{bmatrix} = a\mathbf{v}_1 + b\mathbf{v}_2 + c\mathbf{v}_3 + d\mathbf{v}_4$$

$$= \begin{bmatrix} \mathbf{v}_1 & \mathbf{v}_2 & \mathbf{v}_3 & \mathbf{v}_4 \\ \downarrow & \downarrow & \downarrow & \downarrow \\ & & & \end{bmatrix} \begin{bmatrix} a \\ b \\ c \\ d \end{bmatrix} = \begin{bmatrix} 1 & 1 & 0 & 1 \\ 0 & 0 & 1 & 1 \\ 0 & 0 & 1 & 0 \\ 0 & 1 & 0 & 1 \end{bmatrix} \begin{bmatrix} a \\ b \\ c \\ d \end{bmatrix}.$$

Thus

$$\begin{bmatrix} a \\ b \\ c \\ d \end{bmatrix} = A^{-1} \begin{bmatrix} x_1 \\ x_2 \\ x_3 \\ x_4 \end{bmatrix} = \begin{bmatrix} 1 & 0 & 0 & -1 \\ 0 & -1 & 1 & 1 \\ 0 & 0 & 1 & 0 \\ 0 & 1 & -1 & 0 \end{bmatrix} \begin{bmatrix} x_1 \\ x_2 \\ x_3 \\ x_4 \end{bmatrix} = \begin{bmatrix} x_1 - x_4 \\ -x_2 + x_3 + x_4 \\ x_3 \\ x_2 - x_3 \end{bmatrix},$$

as before.

5.4.7 EXAMPLE

Let $V = \mathbf{R}^2$ and suppose $T: V \to V$ is the linear transformation defined by $T\left(\begin{bmatrix} x_1 \\ x_2 \end{bmatrix}\right) = \begin{bmatrix} 2x_1 + 5x_2 \\ 4x_1 - 3x_2 \end{bmatrix}$. Let $\mathcal{V}$ and $\mathcal{W}$ be the bases of V defined by

$$\mathcal{V} = \{\mathbf{v}_1, \mathbf{v}_2\} \quad \text{where } \mathbf{v}_1 = \begin{bmatrix} -1 \\ 2 \end{bmatrix}, \mathbf{v}_2 = \begin{bmatrix} 2 \\ -3 \end{bmatrix}$$

and

$$\mathcal{W} = \{\mathbf{w}_1, \mathbf{w}_2\} \quad \text{where } \mathbf{w}_1 = \begin{bmatrix} 1 \\ 1 \end{bmatrix}, \mathbf{w}_2 = \begin{bmatrix} -1 \\ 1 \end{bmatrix}.$$

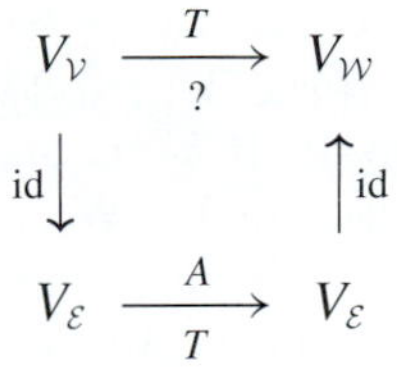

The matrix $A = M_{\mathcal{E}\leftarrow\mathcal{E}}(T)$ of T relative to the standard basis of V is easy to write down:

$$A = \begin{bmatrix} 2 & 5 \\ 4 & -3 \end{bmatrix}.$$

But what is the matrix $M_{\mathcal{W}\leftarrow\mathcal{V}}(T)$ of T relative to the bases $\mathcal{V}$ and $\mathcal{W}$? With the aid of Theorem 5.3.12, the answer is not hard to find.

As Figure 5.5 shows, the linear transformation T is also the composition $(\text{id})T(\text{id})$—follow the arrow representing the identity on the left down, then go to the right with T and up with the other identity. By Theorem 5.3.12, the desired matrix, denoted ? in the figure, is

$$M_{\mathcal{W}\leftarrow\mathcal{V}}(T) = M_{\mathcal{W}\leftarrow\mathcal{E}}(\text{id})M_{\mathcal{E}\leftarrow\mathcal{E}}(T)M_{\mathcal{E}\leftarrow\mathcal{V}}(\text{id}) = P_{\mathcal{W}\leftarrow\mathcal{E}}M_{\mathcal{E}\leftarrow\mathcal{E}}(T)P_{\mathcal{E}\leftarrow\mathcal{V}}.$$

Now $P_{\mathcal{E}\leftarrow\mathcal{V}} = \begin{bmatrix} -1 & 2 \\ 2 & -3 \end{bmatrix}$ and $P_{\mathcal{E}\leftarrow\mathcal{W}} = \begin{bmatrix} 1 & -1 \\ 1 & 1 \end{bmatrix}$, so $P_{\mathcal{W}\leftarrow\mathcal{E}} = [P_{\mathcal{E}\leftarrow\mathcal{W}}]^{-1} = \begin{bmatrix} \frac{1}{2} & \frac{1}{2} \\ -\frac{1}{2} & \frac{1}{2} \end{bmatrix}$

and $P_{\mathcal{V}\leftarrow\mathcal{E}} = [P_{\mathcal{E}\leftarrow\mathcal{V}}]^{-1} = \begin{bmatrix} 3 & 2 \\ 2 & 1 \end{bmatrix}$. Thus

$$M_{\mathcal{W}\leftarrow\mathcal{V}}(T) = \begin{bmatrix} \frac{1}{2} & \frac{1}{2} \\ -\frac{1}{2} & \frac{1}{2} \end{bmatrix} \begin{bmatrix} 2 & 5 \\ 4 & -3 \end{bmatrix} \begin{bmatrix} -1 & 2 \\ 2 & -3 \end{bmatrix} = \begin{bmatrix} -1 & 3 \\ -9 & 14 \end{bmatrix}.$$

In what sense is T "multiplication" by this matrix? Take a vector $\mathbf{x} = \begin{bmatrix} x_1 \\ x_2 \end{bmatrix}$ in V. Its coordinate vector relative to $\mathcal{V}$ is

$$\mathbf{x}_{\mathcal{V}} = P_{\mathcal{V}\leftarrow\mathcal{E}} \begin{bmatrix} x_1 \\ x_2 \end{bmatrix} = \begin{bmatrix} 3x_1 + 2x_2 \\ 2x_1 + x_2 \end{bmatrix}$$

and multiplication by $M_{\mathcal{W}\leftarrow\mathcal{V}}(T)$ gives

$$M_{\mathcal{W}\leftarrow\mathcal{V}}(T)\mathbf{x}_{\mathcal{V}} = \begin{bmatrix} 3x_1 + x_2 \\ x_1 - 4x_2 \end{bmatrix},$$

which is simply the coordinate vector of $T(\mathbf{x})$ relative to $\mathcal{W}$. Note

$$(3x_1 + x_2)\mathbf{w}_1 + (x_1 - 4x_2)\mathbf{w}_2 = (3x_1 + x_2)\begin{bmatrix} 1 \\ 1 \end{bmatrix} + (x_1 - 4x_2)\begin{bmatrix} -1 \\ 1 \end{bmatrix}$$

$$= \begin{bmatrix} 2x_1 + 5x_2 \\ 4x_1 - 3x_2 \end{bmatrix} = T(\mathbf{x}).$$

$$V_{\mathcal{V}} \xrightarrow[\ ?\ ]{T} V_{\mathcal{W}}$$

$$\text{id} \downarrow \qquad \uparrow \text{id}$$

$$V_{\mathcal{E}} \xrightarrow[T]{A} V_{\mathcal{E}}$$

Figure 5.5

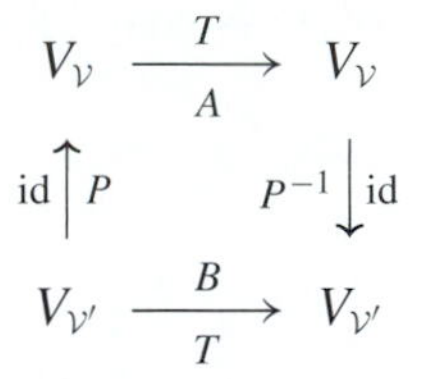

$$V_{\mathcal{V}} \xrightarrow[A]{T} V_{\mathcal{V}}$$

$$\text{id} \uparrow P \qquad P^{-1} \downarrow \text{id}$$

$$V_{\mathcal{V}'} \xrightarrow[T]{B} V_{\mathcal{V}'}$$

Figure 5.6 One linear transformation T and two of its matrices A and $B = P^{-1}AP$.

■ Similarity of Matrices

Suppose $T: V \to V$ is a linear transformation of a vector space V and $\mathcal{V}$ is a basis for V. Then T has a matrix $A = M_{\mathcal{V} \leftarrow \mathcal{V}}(T)$ relative to $\mathcal{V}$. If $\mathcal{V}'$ is another basis for V, then T also has a matrix $B = M_{\mathcal{V}' \leftarrow \mathcal{V}'}(T)$ relative to $\mathcal{V}'$. We might expect a connection between the matrices A and B. Figure 5.6 gives the clue. The figure displays a *commutative diagram*. The linear transformation $T: V \to V$ is the same as the composition $(\text{id})T(\text{id})$ of three linear transformations. Using Theorem 5.3.12,

$$B = M_{\mathcal{V}' \leftarrow \mathcal{V}'}(T) = M_{\mathcal{V}' \leftarrow \mathcal{V}}(\text{id})M_{\mathcal{V} \leftarrow \mathcal{V}}(T)M_{\mathcal{V} \leftarrow \mathcal{V}'}(\text{id}) = P_{\mathcal{V}' \leftarrow \mathcal{V}}A P_{\mathcal{V} \leftarrow \mathcal{V}'} = P^{-1}AP,$$

where $P = M_{\mathcal{V} \leftarrow \mathcal{V}'}(\text{id})$. (By 5.4.3, $P_{\mathcal{V}' \leftarrow \mathcal{V}} = P^{-1}$.) The matrices A and B are **similar**, a concept we first encountered in Section 3.4.

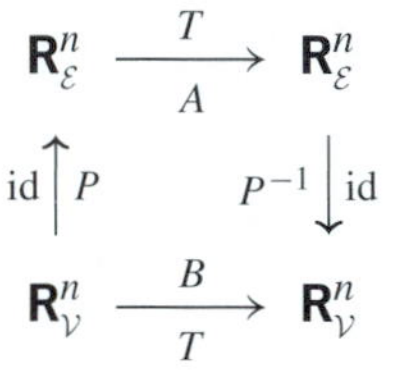

$$\mathbf{R}_{\mathcal{E}}^n \xrightarrow[A]{T} \mathbf{R}_{\mathcal{E}}^n$$

$$\text{id} \uparrow P \qquad P^{-1} \downarrow \text{id}$$

$$\mathbf{R}_{\mathcal{V}}^n \xrightarrow[T]{B} \mathbf{R}_{\mathcal{V}}^n$$

Figure 5.7 Similar matrices A and $B = P^{-1}AP$ are matrices for the same linear transformation T.

Conversely, suppose A and B are similar $n \times n$ matrices and $B = P^{-1}AP$. The columns $\mathbf{v}_1, \mathbf{v}_2, \ldots, \mathbf{v}_n$ of P form a basis $\mathcal{V}$ for $\mathbf{R}^n$ and $P = M_{\mathcal{E} \leftarrow \mathcal{V}}(\text{id})$ is the change of coordinates matrix from $\mathcal{V}$ to $\mathcal{E}$, the standard basis of $\mathbf{R}^n$. By 5.4.3, $P^{-1} = M_{\mathcal{V} \leftarrow \mathcal{E}}(\text{id})$ is the change of coordinates matrix from $\mathcal{E}$ to $\mathcal{V}$. Let $T: \mathbf{R}^n \to \mathbf{R}^n$ be the linear transformation that is left multiplication by A; thus $A = M_{\mathcal{E} \leftarrow \mathcal{E}}(T)$. By Theorem 5.3.12,

$$B = P^{-1}AP = M_{\mathcal{V} \leftarrow \mathcal{E}}(\text{id})M_{\mathcal{E} \leftarrow \mathcal{E}}(T)M_{\mathcal{E} \leftarrow \mathcal{V}}(\text{id}) = M_{\mathcal{V} \leftarrow \mathcal{V}}(T)$$

is the matrix of T relative to $\mathcal{V}$ and $\mathcal{V}$. (See Figure 5.7.) Thus A and B are each matrices for the same linear transformation.

Theorem 5.4.8 *Matrices are similar if and only if they are matrices of the same linear transformation $V \to V$, in each case with respect to a single basis for the vector space V.*

5.4.9 EXAMPLE

Let $T: \mathbf{R}^4 \to \mathbf{R}^4$ be the linear transformation whose matrix relative to $\mathcal{E}$ and $\mathcal{E}$, the standard basis of $\mathbf{R}^4$, is

$$A = M_{\mathcal{E} \leftarrow \mathcal{E}}(T) = \begin{bmatrix} -1 & 2 & 3 & 0 \\ 5 & 0 & 1 & 1 \\ 2 & -3 & 0 & -2 \\ 0 & 4 & 1 & 1 \end{bmatrix}.$$

Let $\mathcal{V} = \{\mathbf{v}_1, \mathbf{v}_2, \mathbf{v}_3, \mathbf{v}_4\}$ be the basis of $\mathbf{R}^4$ whose vectors are those of **5.4.5**. The matrix of T relative to $\mathcal{V}$ and $\mathcal{V}$ is

$$B = M_{\mathcal{V} \leftarrow \mathcal{V}}(T) = P_{\mathcal{V} \leftarrow \mathcal{E}}M_{\mathcal{E} \leftarrow \mathcal{E}}(T)P_{\mathcal{E} \leftarrow \mathcal{V}} = P^{-1}AP$$

where $P = P_{\mathcal{E} \leftarrow \mathcal{V}} = \begin{bmatrix} 1 & 1 & 0 & 1 \\ 0 & 0 & 1 & 1 \\ 0 & 0 & 1 & 0 \\ 0 & 1 & 0 & 1 \end{bmatrix}$. Since $P^{-1} = \begin{bmatrix} 1 & 0 & 0 & -1 \\ 0 & -1 & 1 & 1 \\ 0 & 0 & 1 & 0 \\ 0 & 1 & -1 & 0 \end{bmatrix}$, we have

$$
B = \begin{bmatrix} 1 & 0 & 0 & -1 \\ 0 & -1 & 1 & 1 \\ 0 & 0 & 1 & 0 \\ 0 & 1 & -1 & 0 \end{bmatrix} \begin{bmatrix} -1 & 2 & 3 & 0 \\ 5 & 0 & 1 & 1 \\ 2 & -3 & 0 & -2 \\ 0 & 4 & 1 & 1 \end{bmatrix} \begin{bmatrix} 1 & 1 & 0 & 1 \\ 0 & 0 & 1 & 1 \\ 0 & 0 & 1 & 0 \\ 0 & 1 & 0 & 1 \end{bmatrix}
$$

$$
= \begin{bmatrix} -1 & -2 & 0 & -4 \\ -3 & -5 & 1 & -4 \\ 2 & 0 & -3 & -3 \\ 3 & 6 & 4 & 9 \end{bmatrix}.
$$

According to Definition 5.3.7, T is multiplication by B in the sense of

$$[T(\mathbf{x})]_\mathcal{V} = B\mathbf{x}_\mathcal{V}. \tag{2}$$

Let's check that this is true. Let $\mathbf{x} = \begin{bmatrix} x_1 \\ x_2 \\ x_3 \\ x_4 \end{bmatrix}$. By **5.4.5**,

$$
\mathbf{x}_\mathcal{V} = P_{\mathcal{V} \leftarrow \mathcal{E}} \mathbf{x}_\mathcal{E} = \begin{bmatrix} 1 & 0 & 0 & -1 \\ 0 & -1 & 1 & 1 \\ 0 & 0 & 1 & 0 \\ 0 & 1 & -1 & 0 \end{bmatrix} \begin{bmatrix} x_1 \\ x_2 \\ x_3 \\ x_4 \end{bmatrix} = \begin{bmatrix} x_1 - x_4 \\ -x_2 + x_3 + x_4 \\ x_3 \\ x_2 - x_3 \end{bmatrix}.
$$

Now $T(\mathbf{x}) = A\mathbf{x} = \begin{bmatrix} -x_1 + 2x_2 + 3x_3 \\ 5x_1 + x_3 + x_4 \\ 2x_1 - 3x_2 - 2x_4 \\ 4x_2 + x_3 + x_4 \end{bmatrix}$, so

$$[(T(\mathbf{x})]_\mathcal{V} = P_{\mathcal{V} \leftarrow \mathcal{E}}[T(x)]_\mathcal{E}$$

$$
= \begin{bmatrix} 1 & 0 & 0 & -1 \\ 0 & -1 & 1 & 1 \\ 0 & 0 & 1 & 0 \\ 0 & 1 & -1 & 0 \end{bmatrix} \begin{bmatrix} -x_1 + 2x_2 + 3x_3 \\ 5x_1 + x_3 + x_4 \\ 2x_1 - 3x_2 - 2x_4 \\ 4x_2 + x_3 + x_4 \end{bmatrix} = \begin{bmatrix} -x_1 - 2x_2 + 2x_3 - x_4 \\ -3x_1 + x_2 - 2x_4 \\ 2x_1 - 3x_2 - 2x_4 \\ 3x_1 + 3x_2 + x_3 + 3x_4 \end{bmatrix}.
$$

We let the reader verify equation (2), which asserts that T is multiplication by B with respect to the basis $\mathcal{V}$. ∎

True/False Questions

Decide, with as little calculation as possible, whether each of the following statements is true or false and explain your answer whenever you say "false." (Answers can be found in the back of the book.)

1. If $V = \{v_1, v_2\}$ and $W = \{w_1, w_2\}$ are bases for $\mathbf{R}^2$ and $P_{W \leftarrow V} \begin{bmatrix} a \\ b \end{bmatrix} = \begin{bmatrix} c \\ d \end{bmatrix}$, then $av_1 + bv_2 = cw_1 + dw_2$.

2. Let V be a vector space with bases V and V'. The change of coordinates matrix from V to V' is the matrix of the identity linear transformation $V_V \to V_{V'}$.

3. Let $v_1 = \begin{bmatrix} 1 \\ 2 \end{bmatrix}$, $v_2 = \begin{bmatrix} -2 \\ 3 \end{bmatrix}$, $v_3 = \begin{bmatrix} 1 \\ 0 \end{bmatrix}$, and $v_4 = \begin{bmatrix} 0 \\ 1 \end{bmatrix}$. Let $V = \{v_1, v_2\}$ and $V' = \{v_3, v_4\}$. Then the change of coordinates matrix $V \to V'$ is $\begin{bmatrix} 1 & -2 \\ 2 & 3 \end{bmatrix}$.

4. Let $v_1 = \begin{bmatrix} 1 \\ 2 \end{bmatrix}$, $v_2 = \begin{bmatrix} -2 \\ 3 \end{bmatrix}$, $v_3 = \begin{bmatrix} 0 \\ 1 \end{bmatrix}$, and $v_4 = \begin{bmatrix} 1 \\ 0 \end{bmatrix}$. Let $V = \{v_3, v_4\}$ and $V' = \{v_1, v_2\}$. Then the change of coordinates matrix $V \to V'$ is $\begin{bmatrix} 2 & 3 \\ 1 & -2 \end{bmatrix}$.

5. Let V and V' be bases for $\mathbf{R}^2$ and let $T: \mathbf{R}^2 \to \mathbf{R}^2$ be a linear transformation. If $A = M_{V \leftarrow V}(T)$ and $B = M_{V' \leftarrow V'}(T)$, then A and B have the determinant.

Exercises

*Solutions to exercises marked [BB] can be found in the **B**ack of the **B**ook.*

1. [BB] Let $\mathcal{E} = \{e_1, e_2, e_3\}$ be the standard basis for $\mathbf{R}^3$ and let $V = \{v_1, v_2, v_3\}$ be a new basis, with

$$v_1 = \begin{bmatrix} 1 \\ 1 \\ 0 \end{bmatrix}, \quad v_2 = \begin{bmatrix} 0 \\ 0 \\ 1 \end{bmatrix}, \quad v_3 = \begin{bmatrix} 3 \\ 0 \\ 0 \end{bmatrix}.$$

   (a) Find the changes of coordinates matrix $P_{V \leftarrow \mathcal{E}}$ from $\mathcal{E}$ to V.

   (b) What are the coordinates of $v = \begin{bmatrix} 2 \\ 1 \\ -1 \end{bmatrix}$ relative to V?

2. Let $\mathcal{E} = \{e_1, e_2, e_3, e_4\}$ be the standard basis for $\mathbf{R}^4$ and let V be the basis $\{v_1, v_2, v_3, v_4\}$ where

$$v_1 = \begin{bmatrix} 1 \\ 1 \\ 0 \\ 0 \end{bmatrix}, \quad v_2 = \begin{bmatrix} 1 \\ 0 \\ 1 \\ 0 \end{bmatrix}, \quad v_3 = \begin{bmatrix} 1 \\ 0 \\ 0 \\ 1 \end{bmatrix}, \quad v_4 = \begin{bmatrix} 0 \\ 1 \\ 1 \\ 0 \end{bmatrix}.$$

   (a) Find the change of coordinates matrix $P_{V \leftarrow \mathcal{E}}$ from $\mathcal{E} \to V$.

   (b) What are the coordinates of $v = \begin{bmatrix} 1 \\ 3 \\ 2 \\ 4 \end{bmatrix}$ relative to V?

3. Let $v_1 = \begin{bmatrix} 2 \\ 2 \end{bmatrix}$, $v_2 = \begin{bmatrix} 4 \\ 0 \end{bmatrix}$, $w_1 = \begin{bmatrix} 1 \\ 3 \end{bmatrix}$, $w_2 = \begin{bmatrix} -1 \\ -1 \end{bmatrix}$, and $u = \begin{bmatrix} 3 \\ 6 \end{bmatrix}$. Let $V = \{v_1, v_2\}$ and $W = \{w_1, w_2\}$.

   (a) Find the change of coordinates matrices from $V \to W$ and $W \to V$.

   (b) Find the coordinate vectors of u with respect to V and to W.

4. [BB] Let B be the matrix found in **5.4.9** and consider the equation $B \begin{bmatrix} 2 \\ -2 \\ -3 \\ 1 \end{bmatrix} = \begin{bmatrix} -2 \\ -3 \\ 10 \\ -9 \end{bmatrix}$. This says that if

$$\mathbf{x}_\mathcal{V} = \begin{bmatrix} 2 \\ -2 \\ -3 \\ 1 \end{bmatrix} \text{ and } \mathbf{y}_\mathcal{V} = \begin{bmatrix} -2 \\ -3 \\ 109 \end{bmatrix}, \text{ then } T(\mathbf{x}) = \mathbf{y}. \text{ Verify}$$

that this is correct.

5. Let $\mathcal{E}$ denote the standard basis of $\mathbf{R}^2$ and let $\mathcal{V} = \{\mathbf{v}_1, \mathbf{v}_2\}$ and $\mathcal{V}' = \{\mathbf{v}_1', \mathbf{v}_2'\}$ be the bases with

$$\mathbf{v}_1 = \begin{bmatrix} -1 \\ 1 \end{bmatrix}, \mathbf{v}_2 = \begin{bmatrix} 4 \\ -5 \end{bmatrix}, \mathbf{v}_1' = \begin{bmatrix} 3 \\ 1 \end{bmatrix}, \mathbf{v}_2' = \begin{bmatrix} 1 \\ 2 \end{bmatrix}.$$

Let $T: \mathbf{R}^2 \to \mathbf{R}^2$ be a linear transformation and let
$$A = \begin{bmatrix} -109 & 494 \\ -24 & 109 \end{bmatrix}.$$

Let $\mathbf{u} = \begin{bmatrix} 1 \\ 0 \end{bmatrix}, \mathbf{v} = \begin{bmatrix} -2 \\ 1 \end{bmatrix}, \text{ and } \mathbf{w} = \begin{bmatrix} 3 \\ 3 \end{bmatrix}.$

(a) [BB] Suppose $A = M_{\mathcal{E} \leftarrow \mathcal{E}}(T)$ is the matrix of T relative to $\mathcal{E}$ and $\mathcal{E}$. Find $T(\mathbf{u})$, $T(\mathbf{v})$, and $T(\mathbf{w})$.

(b) [BB] Suppose $A = M_{\mathcal{E} \leftarrow \mathcal{V}}(T)$ is the matrix of T relative to $\mathcal{E}$ and $\mathcal{V}$. Find $T(\mathbf{u})$, $T(\mathbf{v})$, and $T(\mathbf{w})$.

(c) Suppose $A = M_{\mathcal{V} \leftarrow \mathcal{E}}(T)$ is the matrix of T relative to $\mathcal{V}$ and $\mathcal{E}$. Find $T(\mathbf{u})$, $T(\mathbf{v})$, and $T(\mathbf{w})$.

(d) Suppose $A = M_{\mathcal{V} \leftarrow \mathcal{V}}(T)$ is the matrix of T relative to $\mathcal{V}$ and $\mathcal{V}$. Find $T(\mathbf{u})$, $T(\mathbf{v})$, and $T(\mathbf{w})$.

(e) Suppose $A = M_{\mathcal{V} \leftarrow \mathcal{V}}(T)$. Find $B = M_{\mathcal{E} \leftarrow \mathcal{E}}(T)$.

(f) Suppose $A = M_{\mathcal{V} \leftarrow \mathcal{V}}(T)$. Find $B = M_{\mathcal{V}' \leftarrow \mathcal{V}'}(T)$.

6. Let $\mathcal{E} = \{\mathbf{e}_1, \mathbf{e}_2, \mathbf{e}_3, \mathbf{e}_4\}$ be the standard basis for $\mathbf{R}^4$ and let $\mathcal{V}$ be the basis $\{\mathbf{v}_1, \mathbf{v}_2, \mathbf{v}_3, \mathbf{v}_4\}$ where

$$\mathbf{v}_1 = \begin{bmatrix} 1 \\ 1 \\ 0 \\ 0 \end{bmatrix}, \quad \mathbf{v}_2 = \begin{bmatrix} 1 \\ 0 \\ 1 \\ 0 \end{bmatrix},$$

$$\mathbf{v}_3 = \begin{bmatrix} 0 \\ 0 \\ 1 \\ 1 \end{bmatrix}, \quad \mathbf{v}_4 = \begin{bmatrix} 1 \\ 0 \\ 0 \\ 1 \end{bmatrix}.$$

(a) Find the change of coordinates matrix $P_{\mathcal{V} \leftarrow \mathcal{E}}$ $\mathcal{E} \to \mathcal{V}$ and use this to find the coordinates of

$$\mathbf{v} = \begin{bmatrix} 1 \\ 2 \\ -1 \\ 0 \end{bmatrix} \text{ relative to } \mathcal{V}.$$

(b) Let $T: \mathbf{R}^4 \to \mathbf{R}^4$ be the linear transformation whose matrix relative to $\mathcal{E}$ and $\mathcal{E}$, the standard basis of $\mathbf{R}^4$, is

$$A = M_{\mathcal{E} \leftarrow \mathcal{E}} = \begin{bmatrix} -1 & 2 & 3 & 0 \\ 5 & 0 & 1 & 1 \\ 2 & -3 & 0 & -2 \\ 0 & 4 & 1 & 1 \end{bmatrix}.$$

Find the matrix $B = M_{\mathcal{V} \leftarrow \mathcal{V}}(T)$ of T relative to $\mathcal{V}$ and $\mathcal{V}$.

(c) Consider the equation $B \begin{bmatrix} 0 \\ 2 \\ 0 \\ -4 \end{bmatrix} = \begin{bmatrix} -12 \\ 13 \\ -9 \\ 7 \end{bmatrix}$. This

says that if $\mathbf{x}_\mathcal{V} = \begin{bmatrix} 0 \\ 2 \\ 0 \\ -4 \end{bmatrix}$ and $\mathbf{y}_\mathcal{V} = \begin{bmatrix} -12 \\ 13 \\ -9 \\ 7 \end{bmatrix}$, then

$T(\mathbf{x}) = \mathbf{y}$. Verify that this is correct.

7. [BB] Let $\mathbf{v}_1 = \begin{bmatrix} 0 \\ 1 \end{bmatrix}, \mathbf{v}_2 = \begin{bmatrix} 2 \\ 4 \end{bmatrix}, \mathbf{w}_1 = \begin{bmatrix} 2 \\ -3 \end{bmatrix},$ and

$\mathbf{w}_2 = \begin{bmatrix} -2 \\ 5 \end{bmatrix}.$ If the matrix of a linear transformation

$T: \mathbf{R}^2 \to \mathbf{R}^2$ is $A = \begin{bmatrix} 3 & 2 \\ 2 & 1 \end{bmatrix}$ relative to the bases
$\mathcal{V} = \{\mathbf{v}_1, \mathbf{v}_2\}$ and $\mathcal{V}$, find the matrix B of T relative to the bases $\mathcal{W} = \{\mathbf{w}_1, \mathbf{w}_2\}$ and $\mathcal{W}$.

8. Define $T: \mathbf{R}^3 \to \mathbf{R}^3$ by $T \begin{bmatrix} x \\ y \\ z \end{bmatrix} = \begin{bmatrix} 2x + 3y \\ -2x + y + z \\ x - y - z \end{bmatrix}$.

(a) [BB] Let $\mathcal{V} = \{\begin{bmatrix} 1 \\ 0 \\ 1 \end{bmatrix}, \begin{bmatrix} 0 \\ 1 \\ 1 \end{bmatrix}, \begin{bmatrix} 1 \\ 1 \\ 0 \end{bmatrix}\}$. Find

$M_{\mathcal{V} \leftarrow \mathcal{V}}(T)$, the matrix of T relative to $\mathcal{V}$ and $\mathcal{V}$.

(b) Let $\mathcal{V}' = \{\begin{bmatrix} 0 \\ 0 \\ 1 \end{bmatrix}, \begin{bmatrix} 0 \\ 1 \\ 1 \end{bmatrix}, \begin{bmatrix} 1 \\ 1 \\ 1 \end{bmatrix}\}$. Find $M_{\mathcal{V}' \leftarrow \mathcal{V}'}(T)$,

the matrix of T relative to $\mathcal{V}'$ and $\mathcal{V}'$.

(c) True or false: The matrices found in the previous two parts of this exercise are similar. Explain.

9. Let $\mathcal{V} = \{\mathbf{v}_1, \mathbf{v}_2, \mathbf{v}_3\}$ with $\mathbf{v}_1 = \begin{bmatrix} 1 \\ 2 \\ -1 \end{bmatrix}, \mathbf{v}_2 = \begin{bmatrix} 0 \\ 1 \\ 1 \end{bmatrix},$ and

$$\mathbf{v}_3 = \begin{bmatrix} -3 \\ 0 \\ 6 \end{bmatrix}.$$

(a) [BB] Suppose $\mathbf{u}$, $\mathbf{v}$, and $\mathbf{w}$ are vectors in $\mathbf{R}^3$ with

$$\mathbf{u}_\mathcal{V} = \begin{bmatrix} 1 \\ 2 \\ 3 \end{bmatrix}, \mathbf{v}_\mathcal{V} = \begin{bmatrix} 0 \\ 1 \\ 0 \end{bmatrix}, \text{ and } \mathbf{w}_\mathcal{V} = \begin{bmatrix} -1 \\ 0 \\ 1 \end{bmatrix}. \text{ What}$$

are $\mathbf{u}$, $\mathbf{v}$, and $\mathbf{w}$?

(b) If $\mathbf{u} = \begin{bmatrix} 1 \\ 2 \\ 3 \end{bmatrix}, \mathbf{v} = \begin{bmatrix} 0 \\ 1 \\ 0 \end{bmatrix},$ and $\mathbf{w} = \begin{bmatrix} -1 \\ 0 \\ 1 \end{bmatrix},$ find $\mathbf{u}_\mathcal{V}$,

$\mathbf{v}_\mathcal{V}$, and $\mathbf{w}_\mathcal{V}$.

10. Let $V = \mathbf{R}^3$, $W = \mathbf{R}^2$, and $U = \mathbf{R}^3$. Let $T: V \to W$ and $S: W \to U$ be the linear transformations defined by

$$T\left(\begin{bmatrix} x_1 \\ x_2 \\ x_3 \end{bmatrix}\right) = \begin{bmatrix} 2x_1 - x_2 + x_3 \\ x_2 + x_3 \end{bmatrix}$$

and $$S\left(\begin{bmatrix} x_1 \\ x_2 \end{bmatrix}\right) = \begin{bmatrix} x_1 + 2x_2 \\ x_1 - x_2 \\ 3x_1 \end{bmatrix}.$$

Let V, W, U have, respectively, the bases

$$\mathcal{V} = \{\mathbf{v}_1, \mathbf{v}_2, \mathbf{v}_3\},$$

where $\mathbf{v}_1 = \begin{bmatrix} -1 \\ 1 \\ 1 \end{bmatrix}, \mathbf{v}_2 = \begin{bmatrix} 1 \\ 0 \\ 1 \end{bmatrix}, \mathbf{v}_3 = \begin{bmatrix} 1 \\ 1 \\ 1 \end{bmatrix};$

$$\mathcal{W} = \{\mathbf{w}_1, \mathbf{w}_2\}, \quad \text{where } \mathbf{w}_1 = \begin{bmatrix} 1 \\ 2 \end{bmatrix}, \mathbf{w}_2 = \begin{bmatrix} 2 \\ 1 \end{bmatrix}; \text{ and}$$

$$\mathcal{U} = \{\mathbf{u}_1, \mathbf{u}_2, \mathbf{u}_3\},$$

where $\mathbf{u}_1 = \begin{bmatrix} 1 \\ 0 \\ -1 \end{bmatrix}, \mathbf{u}_2 = \begin{bmatrix} 0 \\ 0 \\ 1 \end{bmatrix}, \mathbf{u}_3 = \begin{bmatrix} -1 \\ 1 \\ 1 \end{bmatrix}.$

(a) Find $A = M_{\mathcal{U} \leftarrow \mathcal{W}}(S)$ and $B = M_{\mathcal{W} \leftarrow \mathcal{V}}(T)$.

(b) Find $ST\left(\begin{bmatrix} x_1 \\ x_2 \\ x_3 \end{bmatrix}\right)$.

(c) Find $C = M_{\mathcal{U} \leftarrow \mathcal{V}}(ST)$ and verify that $AB = C$.

11. Let $\mathcal{E} = \{\mathbf{e}_1, \mathbf{e}_2, \mathbf{e}_3\}$ be the standard basis of $\mathbf{R}^3$, and let $\mathcal{V} = \{\mathbf{v}_1, \mathbf{v}_2, \mathbf{v}_3\}$ and $\mathcal{W} = \{\mathbf{w}_1, \mathbf{w}_2, \mathbf{w}_3\}$ be other bases, where

$$\mathbf{v}_1 = \begin{bmatrix} 1 \\ 1 \\ 1 \end{bmatrix}, \mathbf{v}_2 = \begin{bmatrix} 1 \\ -1 \\ 0 \end{bmatrix}, \mathbf{v}_3 = \begin{bmatrix} -1 \\ 0 \\ 1 \end{bmatrix},$$

and $\mathbf{w}_1 = \mathbf{e}_3$, $\mathbf{w}_2 = \mathbf{e}_2$, $\mathbf{w}_3 = \mathbf{e}_1$.

(a) [BB] Find the change of coordinates matrix $\mathcal{V} \to \mathcal{E}$, the change of coordinates matrix $\mathcal{E} \to \mathcal{W}$, and the change of coordinates matrix $\mathcal{V} \to \mathcal{W}$.

(b) Verify that $P_{\mathcal{W} \leftarrow \mathcal{V}} = P_{\mathcal{W} \leftarrow \mathcal{E}} P_{\mathcal{E} \leftarrow \mathcal{V}}$.

(c) Express $\mathbf{v} = \begin{bmatrix} a \\ b \\ c \end{bmatrix}$ as a linear combination of $\mathbf{v}_1, \mathbf{v}_2, \mathbf{v}_3$.

12. Let A be a 2×2 matrix, $A \neq \mathbf{0}$, with $A^2 = \mathbf{0}$. Let $\mathbf{v}$ be a vector in $\mathbf{R}^2$ such that $A\mathbf{v} \neq \mathbf{0}$.

(a) Prove that $\{\mathbf{v}, A\mathbf{v}\}$ is a basis for $\mathbf{R}^2$.

(b) Prove that A is similar to $\begin{bmatrix} 0 & 0 \\ 1 & 0 \end{bmatrix}$.

(c) Prove that A is similar to $\begin{bmatrix} 0 & 1 \\ 0 & 0 \end{bmatrix}$.

✍ 13. In a sentence or two, describe the meaning of the formula $[T(\mathbf{x})]_\mathcal{W} = M_{\mathcal{W} \leftarrow \mathcal{V}}(T)\mathbf{x}_\mathcal{V}$.

✍ 14. In your own words, express the meaning of the formula $[P_{\mathcal{V} \leftarrow \mathcal{V}'}]^{-1} = P_{\mathcal{V}' \leftarrow \mathcal{V}}$.

■ Critical Reading

15. Let $\mathcal{V} = \{\mathbf{v}_1, \mathbf{v}_2, \mathbf{v}_3\}$ be a basis for $\mathbf{R}^3$ and let $A = M_{\mathcal{V} \leftarrow \mathcal{V}}(T)$ be the matrix of a linear transformation $T: \mathbf{R}^3 \to \mathbf{R}^3$ relative to $\mathcal{V}$ and $\mathcal{V}$. Suppose we rearrange the order of the vectors in $\mathcal{V}$ to obtain the basis $\mathcal{W} = \{\mathbf{v}_2, \mathbf{v}_3, \mathbf{v}_1\}$. Let $B = M_{\mathcal{W} \leftarrow \mathcal{W}}(T)$ be the matrix of T relative to $\mathcal{W}$ and $\mathcal{W}$. What is the relationship between A and B? For example, if

$$A = \begin{bmatrix} 1 & 2 & 3 \\ 4 & 5 & 6 \\ 7 & 8 & 9 \end{bmatrix}, \text{ what is } B? \text{ Explain.}$$

16. Your linear algebra professor wants to construct some matrices A that satisfy $A^2 = A$. She knows all about projection matrices and could find suitable A using a formula like $A = B(B^T B)^{-1} B^T$, but such matrices are also symmetric. Find a method of constructing suitable A that are **not** symmetric. Explain your answer with at least one example you consider interesting. [Hint: Consider that $A(A\mathbf{x}) = A\mathbf{x}$ for all vectors $\mathbf{x}$.]

CHAPTER KEY WORDS AND IDEAS: Here are some technical words and phrases that were used in this chapter. Do you know the meaning of each? If you're not sure, check the glossary or index at the back of the book.

associative (composition is)
commutative (composition is not)
composition
     of linear transformations
coordinate vector
coordinates of a vector
identity linear transformation
inverse (of a linear transformation)
invertible linear transformation
kernel
linear transformation

linear transformations preserve
• scalar multiplication
• the zero vector
• linear combinations
matrix
     of a linear transformation
matrix of basis change
one-to-one
onto
range
similar matrices

Review Exercises for Chapter 5

1. Let $T: \mathbf{R}^2 \to \mathbf{R}^2$ be a linear transformation such that
$$T\left(\begin{bmatrix} 1 \\ 0 \end{bmatrix}\right) = \begin{bmatrix} 2 \\ -1 \end{bmatrix} \text{ and } T\left(\begin{bmatrix} 1 \\ 1 \end{bmatrix}\right) = \begin{bmatrix} 4 \\ 4 \end{bmatrix}.$$

(a) Find $T\left(\begin{bmatrix} 0 \\ 1 \end{bmatrix}\right)$.

(b) Find a matrix A such that $T = T_A$ is multiplication by A.

2. Suppose $T: \mathbf{R}^2 \to \mathbf{R}^4$ is a linear transformation with the property that $T\left(\begin{bmatrix} 3 \\ -1 \end{bmatrix}\right) = \begin{bmatrix} 2 \\ -1 \\ 1 \\ 0 \end{bmatrix}$ and

$T\left(\begin{bmatrix} -1 \\ 2 \end{bmatrix}\right) = \begin{bmatrix} 4 \\ 0 \\ 1 \\ -3 \end{bmatrix}$. Find $T\left(\begin{bmatrix} x \\ y \end{bmatrix}\right)$.

3. Define $T: \mathbf{R}^2 \to \mathbf{R}^3$ by $T\begin{bmatrix} x \\ y \end{bmatrix} = \begin{bmatrix} x + 2y \\ 2x + 5y \\ x + y \end{bmatrix}$.

(a) Show that T is a linear transformation.

(b) Find a matrix A such that $T(\mathbf{v}) = A\begin{bmatrix} x \\ y \end{bmatrix}$ for any $\mathbf{v} = \begin{bmatrix} x \\ y \end{bmatrix}$ in $\mathbf{R}^2$.

(c) Find a basis for the kernel and a basis for the range of T.

(d) Verify that $\dim \ker T + \dim \operatorname{range} T = \dim V$. What is V?

4. Let $T: \mathbf{R}^4 \to \mathbf{R}^3$ be the linear transformation defined by
$$T\left(\begin{bmatrix} x_1 \\ x_2 \\ x_3 \\ x_4 \end{bmatrix}\right) = \begin{bmatrix} -2x_1 + 2x_2 + 6x_3 + 4x_4 \\ 2x_1 - x_3 - x_4 \\ 3x_1 + 4x_2 + 5x_3 + 8x_4 \end{bmatrix}.$$

Determine whether T is one-to-one. Determine whether T is onto.

5. Let $A = \begin{bmatrix} 2 & -6 & 1 & -16 & 9 \\ 1 & -3 & -2 & 2 & 1 \\ 3 & -9 & -4 & -2 & 7 \end{bmatrix}$, $\mathbf{b} = \begin{bmatrix} 6 \\ 1 \\ 7 \end{bmatrix}$, and

$\mathbf{x} = \begin{bmatrix} x_1 \\ x_2 \\ x_3 \\ x_4 \\ x_5 \end{bmatrix}$.

(a) Solve the system $A\mathbf{x} = \mathbf{b}$.

(b) Is $\mathbf{b}$ in the range of T? Explain.

(c) Show that every solution to the homogeneous system with matrix A is a linear combination of
$$\begin{bmatrix} 3 \\ 1 \\ 0 \\ 0 \\ 0 \end{bmatrix} \text{ and } \begin{bmatrix} 6 \\ 0 \\ 4 \\ 1 \\ 0 \end{bmatrix}.$$

Now let $T = T_A : \mathbf{R}^5 \to \mathbf{R}^3$ be "left multiplication by A."

(d) Find the kernel of T.

6. Define linear transformations $S : \mathbf{R}^2 \to \mathbf{R}^3$, $T : \mathbf{R}^4 \to \mathbf{R}^2$, and $U : \mathbf{R}^1 \to \mathbf{R}^4$ by

$$S\left(\begin{bmatrix} x_1 \\ x_2 \end{bmatrix}\right) = \begin{bmatrix} x_1 - x_2 \\ 2x_1 \\ 3x_1 + x_2 \end{bmatrix},$$

$$T\left(\begin{bmatrix} x_1 \\ x_2 \\ x_3 \\ x_4 \end{bmatrix}\right) = \begin{bmatrix} x_1 - x_2 - x_3 + x_4 \\ 3x_1 + x_3 \end{bmatrix},$$

$$U(x) = \begin{bmatrix} -x \\ 0 \\ 3x \\ x \end{bmatrix}.$$

(a) Compute $ST\left(\begin{bmatrix} x_1 \\ x_2 \\ x_3 \\ x_4 \end{bmatrix}\right)$ and $TU(x)$.

(b) Compute $[(ST)U](x)$ and $[S(TU)](x)$ and comment on your answers.

7. Let $T : \mathbf{R}^3 \to \mathbf{R}^3$ be the linear transformation defined by $T\left(\begin{bmatrix} x \\ y \\ z \end{bmatrix}\right) = \begin{bmatrix} x - y \\ x - z \\ x \end{bmatrix}$. Find the matrix of T relative to the basis $V = \{\mathbf{v}_1, \mathbf{v}_2, \mathbf{v}_3\}$ (in each instance), where

$$\mathbf{v}_1 = \begin{bmatrix} 1 \\ 1 \\ 1 \end{bmatrix}, \mathbf{v}_2 = \begin{bmatrix} 1 \\ 1 \\ 0 \end{bmatrix}, \mathbf{v}_3 = \begin{bmatrix} 1 \\ 0 \\ 0 \end{bmatrix}.$$

8. Let $\mathbf{v}_1 = \begin{bmatrix} 1 \\ 2 \\ 0 \\ 1 \end{bmatrix}, \mathbf{v}_2 = \begin{bmatrix} -1 \\ 1 \\ 3 \\ 0 \end{bmatrix}, \mathbf{v}_3 = \begin{bmatrix} 0 \\ 0 \\ 1 \\ 0 \end{bmatrix}, \mathbf{v}_4 = \begin{bmatrix} 0 \\ 1 \\ 2 \\ 3 \end{bmatrix},$ and

$V = \{\mathbf{v}_1, \mathbf{v}_2, \mathbf{v}_3, \mathbf{v}_4\}$. Let $\mathbf{w}_1 = \begin{bmatrix} 3 \\ 4 \end{bmatrix}, \mathbf{w}_2 = \begin{bmatrix} -1 \\ 5 \end{bmatrix},$ and

$W = \{\mathbf{w}_1, \mathbf{w}_2\}$. Define $T : \mathbf{R}^4 \to \mathbf{R}^2$ by $T(\mathbf{v}_1) = \mathbf{w}_1$, $T(\mathbf{v}_2) = \mathbf{w}_1$, $T(\mathbf{v}_3) = \mathbf{w}_2$, $T(\mathbf{v}_4) = \mathbf{w}_2$.

(a) Find the matrix $M_{W \leftarrow V}(T)$ of T relative to V and W.

(b) Find the matrix $M_{\mathcal{E} \leftarrow V}(T)$ of T relative to V and the standard basis of $\mathbf{R}^2$.

(c) Find the matrix $M_{\mathcal{E} \leftarrow \mathcal{E}}(T)$ of T relative to the standard bases of $\mathbf{R}^4$ and $\mathbf{R}^2$.

9. Let $\mathbf{v}_1 = \begin{bmatrix} 1 \\ 1 \\ 1 \end{bmatrix}, \mathbf{v}_2 = \begin{bmatrix} 4 \\ 2 \\ 1 \end{bmatrix}, \mathbf{v}_3 = \begin{bmatrix} 3 \\ 2 \\ 1 \end{bmatrix}$. Define

$$T : \mathbf{R}^3 \to \mathbf{R}^4 \text{ by } T(\mathbf{v}_1) = \begin{bmatrix} 5 \\ 7 \\ -3 \\ -1 \end{bmatrix}, T(\mathbf{v}_2) = \begin{bmatrix} -3 \\ 0 \\ 2 \\ -2 \end{bmatrix},$$

$$T(\mathbf{v}_3) = \begin{bmatrix} -1 \\ 7 \\ 1 \\ -5 \end{bmatrix}.$$

(a) Show that $V = \{\mathbf{v}_1, \mathbf{v}_2, \mathbf{v}_3\}$ is a basis for $\mathbf{R}^3$.

(b) Find $M_{\mathcal{E} \leftarrow V}(T)$, the matrix of T relative to V and the standard basis $\mathcal{E}$ of $\mathbf{R}^4$.

(c) Find the matrix of T relative to the standard bases of $\mathbf{R}^3$ and $\mathbf{R}^4$.

(d) Find a basis for the range of T. What's the dimension of the range of T?

10. Define the linear transformation $T : \mathbf{R}^3 \to \mathbf{R}^3$ by

$$T\left(\begin{bmatrix} x_1 \\ x_2 \\ x_3 \end{bmatrix}\right) = \begin{bmatrix} -x_1 + 2x_2 + 2x_3 \\ x_2 + 2x_3 \\ 2x_1 + 5x_2 \end{bmatrix}.$$

Show that T is invertible and find $T^{-1}\left(\begin{bmatrix} x_1 \\ x_2 \\ x_3 \end{bmatrix}\right)$.

11. "Any invertible matrix is the matrix of the identity linear transformation." Explain.

12. Let $V = \{\mathbf{v}_1, \mathbf{v}_2, \mathbf{v}_3\}$ be the basis of $\mathbf{R}^3$ where

$$\mathbf{v}_1 = \begin{bmatrix} 1 \\ 0 \\ 1 \end{bmatrix}, \quad \mathbf{v}_2 = \begin{bmatrix} -1 \\ 2 \\ 2 \end{bmatrix}, \quad \mathbf{v}_3 = \begin{bmatrix} 0 \\ 1 \\ 2 \end{bmatrix}$$

and $W = \{\mathbf{w}_1, \mathbf{w}_2\}$ be the basis of $\mathbf{R}^2$ with

$$\mathbf{w}_1 = \begin{bmatrix} 3 \\ 1 \end{bmatrix}, \quad \mathbf{w}_2 = \begin{bmatrix} 5 \\ 2 \end{bmatrix}.$$

(a) Find the matrices of basis change $\mathcal{E} \to V$ and $V \to \mathcal{E}$, where $\mathcal{E}$ denotes the standard basis of $\mathbf{R}^3$.

(b) Find the coordinate vectors of $\begin{bmatrix} 1 \\ 0 \\ 0 \end{bmatrix}$ and $\begin{bmatrix} 1 \\ 0 \\ 1 \end{bmatrix}$ relative to V.

Let $T : \mathbf{R}^2 \to \mathbf{R}^3$ be a linear transformation such that $T\left(\begin{bmatrix} 3 \\ 1 \end{bmatrix}\right) = \begin{bmatrix} 1 \\ 0 \\ 0 \end{bmatrix}$ and $T\left(\begin{bmatrix} 5 \\ 2 \end{bmatrix}\right) = \begin{bmatrix} 1 \\ 0 \\ 1 \end{bmatrix}$.

(c) Find $T\left(\begin{bmatrix} 0 \\ 1 \end{bmatrix}\right)$ and $T\left(\begin{bmatrix} 1 \\ 0 \end{bmatrix}\right)$.

(d) Find the matrix of T relative to the standard bases of $\mathbf{R}_2$ and $\mathbf{R}_3$.

(e) Find the matrix of T relative to $\mathcal{V}$ and $\mathcal{W}$.

13. Let $T: \mathbf{R}^4 \rightarrow \mathbf{R}^3$ be the linear transformation whose matrix relative to the standard bases of $\mathbf{R}^4$ and $\mathbf{R}^3$ is
$$A = \begin{bmatrix} 7 & 1 & -4 & 5 \\ 1 & 1 & 2 & -1 \\ 3 & 1 & 0 & 1 \end{bmatrix}.$$ Find bases for the kernel and range of T. Verify that $\dim \ker T + \dim \operatorname{range} T = \dim V$. What is V in this case?

6

Orthogonality

Projection Matrices and Least Squares Approximation

■ Orthogonality: A Quick Review

We have met the idea of "orthogonality" many times in this book and begin this chapter with a quick review of the basic terminology associated with this concept.

If $\mathbf{x} = \begin{bmatrix} x_1 \\ x_2 \\ \vdots \\ x_n \end{bmatrix}$ and $\mathbf{y} = \begin{bmatrix} y_1 \\ y_2 \\ \vdots \\ y_n \end{bmatrix}$ are vectors in $\mathbf{R}^n$, the dot product of $\mathbf{x}$ and $\mathbf{y}$ is

the number $x_1 y_1 + x_2 y_2 + \cdots + x_n y_n$, which is also the matrix product $\mathbf{x}^T \mathbf{y}$ (square brackets omitted):

> **6.1.1** $\mathbf{x} \cdot \mathbf{y} = x_1 y_1 + x_2 y_2 + \cdots + x_n y_n = \mathbf{x}^T \mathbf{y}.$

Vectors $\mathbf{x}$ and $\mathbf{y}$ are *orthogonal* if $\mathbf{x} \cdot \mathbf{y} = 0$. This coincides with reality when $n = 2$ or $n = 3$ and is a definition for $n > 3$. The *norm* or *length* of $\mathbf{x}$ is the number

> **6.1.2** $\|\mathbf{x}\| = \sqrt{\mathbf{x} \cdot \mathbf{x}} = \sqrt{x_1^2 + x_2^2 + \cdots + x_n^2}.$

Again, this really is the length of $\mathbf{x}$ in $\mathbf{R}^2$ or $\mathbf{R}^3$ where vectors can be pictured by arrows, but it is simply a definition in $\mathbf{R}^n$ when $n > 3$. Since a sum of squares of real numbers is 0 if and only if each number is 0, it follows that

> **6.1.3** $\|\mathbf{x}\| = 0$ if and only if $\mathbf{x} = \mathbf{0}.$

If a vector $\mathbf{x}$ in $\mathbf{R}^n$ is orthogonal to all other vectors, in particular, it is orthogonal to itself, so $\mathbf{x} \cdot \mathbf{x} = 0$. As we have just seen, this means $\mathbf{x} = \mathbf{0}$:

6.1.4 $\mathbf{x} \cdot \mathbf{y} = 0$ for all $\mathbf{y}$ if and only if $\mathbf{x} = \mathbf{0}$.

Since $\|\mathbf{x} + \mathbf{y}\|^2 = (\mathbf{x} + \mathbf{y}) \cdot (\mathbf{x} + \mathbf{y}) = \mathbf{x} \cdot \mathbf{x} + 2\mathbf{x} \cdot \mathbf{y} + \mathbf{y} \cdot \mathbf{y} = \|\mathbf{x}\|^2 + 2\mathbf{x} \cdot \mathbf{y} + \|\mathbf{y}\|^2$, we obtain the Theorem of Pythagoras:

Theorem 6.1.5   **(Pythagoras).** *Vectors $\mathbf{x}$ and $\mathbf{y}$ are orthogonal if and only if*

$$\|\mathbf{x} + \mathbf{y}\|^2 = \|\mathbf{x}\|^2 + \|\mathbf{y}\|^2 .$$

■ The "Best Solution" to $A\mathbf{x} = \mathbf{b}$

The primary focus of this section is the linear system $A\mathbf{x} = \mathbf{b}$, where A is an $m \times n$ matrix and $m > n$. In this situation, there are generally no solutions. There are more equations than unknowns. We are asking too much of a relatively few number of variables. Consider, for example, the following system:

$$\begin{aligned} x_1 + x_2 &= 2 \\ x_1 - x_2 &= 0 \\ 2x_1 + x_2 &= -4. \end{aligned} \tag{1}$$

The first two equations already give $x_1 = x_2 = 1$, so we would have to be awfully lucky for these values to satisfy more equations. Here, the third equation $2x_1 + x_2 = -4$ is not satisfied by $x_1 = x_2 = 1$, so the system (1) has no solution.

Suppose your boss doesn't find this answer acceptable, however. Is there a pair of numbers x_1, x_2 that might satisfy her, providing a "best solution" to (1) in some sense under the circumstances?

Our system is $A\mathbf{x} = \mathbf{b}$ with

$$A = \begin{bmatrix} 1 & 1 \\ 1 & -1 \\ 2 & 1 \end{bmatrix}, \quad \mathbf{x} = \begin{bmatrix} x_1 \\ x_2 \end{bmatrix}, \quad \text{and} \quad \mathbf{b} = \begin{bmatrix} 2 \\ 0 \\ -4 \end{bmatrix}.$$

For any $\mathbf{x}$, $A\mathbf{x}$ is in the column space of A, so $A\mathbf{x} = \mathbf{b}$ has no solution precisely when $\mathbf{b}$ is not in the column space of A. In this case, we can't make $\mathbf{b} - A\mathbf{x} = \mathbf{0}$, but we can at least try to minimize the length of this vector; symbolically, we wish to make $\|\mathbf{b} - A\mathbf{x}\|$ as small as possible. We call this number the *distance* between $\mathbf{b}$ and $A\mathbf{x}$. (See Figure 6.1.)

6.1.6 DEFINITION   The *distance* between vectors $\mathbf{u}$ and $\mathbf{v}$ in some Euclidean space V is $\|\mathbf{u} - \mathbf{v}\|$. If $\mathbf{v}$ is in V and U is a subspace of V, a vector $\mathbf{u}_0$ in U is *closest* to $\mathbf{v}$ if $\|\mathbf{u}_0 - \mathbf{v}\| \le \|\mathbf{u} - \mathbf{v}\|$ for all $\mathbf{u}$ in U.

So we shall take as our "best solution" to $A\mathbf{x} = \mathbf{b}$ any vector $\mathbf{x}^+$ with the property that the distance from $A\mathbf{x}^+$ to $\mathbf{b}$ is minimal; equivalently, any vector $\mathbf{x}^+$ with the property that $A\mathbf{x}^+$ is closest to $\mathbf{b}$. Such a vector, if it exists and is unique,

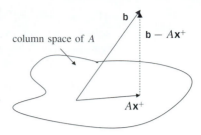

Figure 6.1 The best solution to $A\mathbf{x} = \mathbf{b}$ is the vector $\mathbf{x}^+$ with $A\mathbf{x}^+ = P\mathbf{b}$, the projection of $\mathbf{b}$ on the column space of A.

will be called the *best solution in the sense of least squares*, the term deriving from the fact that the length of a vector is the square root of a sum of squares:

$$\mathbf{x} = \begin{bmatrix} x_1 \\ x_2 \\ \vdots \\ x_n \end{bmatrix} \quad \text{implies} \quad \|\mathbf{x}\| = \sqrt{x_1^2 + x_2^2 + \cdots + x_n^2}.$$

So the length of $\mathbf{b} - A\mathbf{x}^+$ is minimal when a certain sum of squares is least. As Figure 6.1 suggests, this will occur when $A\mathbf{x}^+$ is the (orthogonal) *projection* of $\mathbf{b}$ onto the column space of A; that is, the vector $\mathbf{b} - A\mathbf{x}^+$ is orthogonal to the column space of A in the sense that it is orthogonal to every vector in the column space. (This is perhaps "obvious" geometrically—see Figure 6.1—but see Exercise 24 for a formal proof.)

We wish to find a formula for $\mathbf{x}^+$. Although the calculation is not difficult, it does require two important ideas, which we highlight before proceeding.

1. A dot product is just a matrix product, thinking of vectors as $n \times 1$ matrices (except for the minor point that there are no brackets around a dot product)— $[\mathbf{u} \cdot \mathbf{v}] = \mathbf{u}^T \mathbf{v}$—see 2.1.18.

2. If $\mathbf{u}$ is a vector that is orthogonal to all other vectors (in the same Euclidean space)—$\mathbf{u} \cdot \mathbf{v} = \mathbf{0}$ for all $\mathbf{v}$—then $\mathbf{u} = \mathbf{0}$. See 6.1.4.

We want $\mathbf{b} - A\mathbf{x}^+$ to be orthogonal to every vector in the column space of A. A vector in the column space of A is $A\mathbf{y}$. So the condition we require of $\mathbf{x}^+$ is that

$$(A\mathbf{y}) \cdot (\mathbf{b} - A\mathbf{x}^+) = 0$$

for all $\mathbf{y}$. We have

$$(A\mathbf{y}) \cdot (\mathbf{b} - A\mathbf{x}^+) = (A\mathbf{y})^T (\mathbf{b} - A\mathbf{x}^+)$$

$$= \mathbf{y}^T A^T (\mathbf{b} - A\mathbf{x}^+)$$

$$= \mathbf{y}^T (A^T \mathbf{b} - A^T A\mathbf{x}^+) = \mathbf{y} \cdot (A^T \mathbf{b} - A^T A\mathbf{x}^+).$$

Now this is to be 0 for all $\mathbf{y}$ and, as we have mentioned, the only vector orthogonal to every vector $\mathbf{y}$ is the zero vector. So $A^T \mathbf{b} - A^T A\mathbf{x}^+ = \mathbf{0}$; otherwise put,

$$A^T A\mathbf{x}^+ = A^T \mathbf{b}. \tag{2}$$

This is a system of equations with coefficient matrix $A^T A$. It will have a unique solution $\mathbf{x}^+$, for example, if the $n \times n$ matrix $A^T A$ is invertible, in which case the solution to (2) is

6.1.7 $\mathbf{x}^+ = (A^T A)^{-1} A^T \mathbf{b}.$

We know a condition under which $A^T A$ is invertible, namely, linear independence of the columns of A—see Corollary 4.4.11. So we obtain the following theorem.

Theorem 6.1.8 *If A has linearly independent columns, the best solution in the sense of least squares to the linear system $A\mathbf{x} = \mathbf{b}$ is $\mathbf{x}^+ = (A^T A)^{-1} A^T \mathbf{b}$.*

6.1.9 EXAMPLE

Let's return to the system of equations (1) that appeared at the start of this section. There we had $A = \begin{bmatrix} 1 & 1 \\ 1 & -1 \\ 2 & 1 \end{bmatrix}$ and $\mathbf{b} = \begin{bmatrix} 2 \\ 0 \\ -4 \end{bmatrix}$. The columns of A are linearly independent, so $A^T A$ is invertible. We have $A^T A = \begin{bmatrix} 1 & 1 & 2 \\ 1 & -1 & 1 \end{bmatrix} \begin{bmatrix} 1 & 1 \\ 1 & -1 \\ 2 & 1 \end{bmatrix} = \begin{bmatrix} 6 & 2 \\ 2 & 3 \end{bmatrix}$ and $(A^T A)^{-1} = \frac{1}{14} \begin{bmatrix} 3 & -2 \\ -2 & 6 \end{bmatrix}$. Thus

$$\mathbf{x}^+ = (A^T A)^{-1} A^T \mathbf{b} = \frac{1}{14} \begin{bmatrix} 3 & -2 \\ -2 & 6 \end{bmatrix} \begin{bmatrix} 1 & 1 & 2 \\ 1 & -1 & 1 \end{bmatrix} \begin{bmatrix} 2 \\ 0 \\ -4 \end{bmatrix} = \begin{bmatrix} -1 \\ 0 \end{bmatrix}.$$

This vector is the "best solution in the sense of least squares" to the system (1). Let's make sure we understand exactly what this means. The vector of errors,

The vector of errors $\mathbf{b} - A\mathbf{x}^+$ shows how far each equation in $A\mathbf{x} = \mathbf{b}$ is from equality.

$$\mathbf{b} - A\mathbf{x}^+ = \begin{bmatrix} 2 \\ 0 \\ -4 \end{bmatrix} - \begin{bmatrix} 1 & 1 \\ 1 & -1 \\ 2 & 1 \end{bmatrix} \begin{bmatrix} -1 \\ 0 \end{bmatrix} = \begin{bmatrix} 2 \\ 0 \\ -4 \end{bmatrix} - \begin{bmatrix} -1 \\ -1 \\ -2 \end{bmatrix} = \begin{bmatrix} 3 \\ 1 \\ -2 \end{bmatrix},$$

has length $d_{\min} = \sqrt{9 + 1 + 4} = \sqrt{14}$, and this is less than the length of any other $\mathbf{b} - A\mathbf{x}$. For any other vector $\mathbf{x} = \begin{bmatrix} x_1 \\ x_2 \end{bmatrix}$,

$$\mathbf{b} - A\mathbf{x} = \begin{bmatrix} 2 - x_1 - x_2 \\ -x_1 + x_2 \\ -4 - 2x_1 - x_2 \end{bmatrix}$$

has length

$$d = \sqrt{(2 - x_1 - x_2)^2 + (-x_1 + x_2)^2 + (-4 - 2x_1 - x_2)^2},$$

and we have proven that $d_{\min} \leq d$ for any x_1, x_2. For example, with $x_1 = x_2 = 1$—the exact solution to the first two equations in (1)—we have $d = \sqrt{0 + 0 + 49} = 7 \geq \sqrt{14} = d_{\min}$. ∎

6.1.10 PROBLEM Suppose we wish to solve the system $A\mathbf{x} = \mathbf{b}$ where A is an $m \times n$ matrix with linearly independent columns and $\mathbf{b}$ is a given vector in $\mathbf{R}^m$. Multiplying by A^T gives $A^T A\mathbf{x} = A^T \mathbf{b}$. Since A has linearly independent columns, $A^T A$ is invertible, so $\mathbf{x} = (A^T A)^{-1} A^T \mathbf{b}$. Apparently the system $A\mathbf{x} = \mathbf{b}$ can always be solved exactly. Is this right? Explain.

Solution. The answer is no! Certainly $A\mathbf{x} = \mathbf{b}$ cannot always be solved. The system

$$x_1 = 1$$
$$x_2 = 2$$
$$x_2 = 3$$

is $A\mathbf{x} = \mathbf{b}$ with $A = \begin{bmatrix} 1 & 0 \\ 0 & 1 \\ 0 & 1 \end{bmatrix}$, $\mathbf{x} = \begin{bmatrix} x_1 \\ x_2 \end{bmatrix}$, and $\mathbf{b} = \begin{bmatrix} 1 \\ 2 \\ 3 \end{bmatrix}$. This clearly has no solution. The problem with our reasoning is one of **logic**. In starting with the equation $A\mathbf{x} = \mathbf{b}$, we assumed that $\mathbf{x}$ was a solution and proved that $\mathbf{x} = (A^T A)^{-1} A^T \mathbf{b}$. This is correct **only if we know there is a solution in the first place.**

Reading Challenge 1 *Suppose the linear system $A\mathbf{x} = \mathbf{0}$ has only the trivial solution. What is the matter with the following argument, which purports to show that $A\mathbf{x} = \mathbf{b}$ has a unique solution for any $\mathbf{b}$?*

"If $A\mathbf{x}_1 = \mathbf{b}$ and $A\mathbf{x}_2 = \mathbf{b}$, then $A(\mathbf{x}_1 - \mathbf{x}_2) = \mathbf{0}$, so, by assumption, $\mathbf{x}_1 - \mathbf{x}_2 = \mathbf{0}$, so $\mathbf{x}_1 = \mathbf{x}_2$."

Now let's have another look at Figure 6.1 and the calculations leading to the formula for $\mathbf{x}^+$ given in Theorem 6.1.8. The vector $A\mathbf{x}^+$ has the property that $\mathbf{b} - A\mathbf{x}^+$ is perpendicular to the column space of A. We call $A\mathbf{x}^+$ the *projection* of $\mathbf{b}$ on the column space. Notice that $A\mathbf{x}^+ = A(A^T A)^{-1} A^T \mathbf{b} = P\mathbf{b}$ where

6.1.11 $P = A(A^T A)^{-1} A^T.$

The matrix P is called, quite naturally, a *projection matrix*. For example, with

$$A = \begin{bmatrix} 1 & 1 \\ 1 & -1 \\ 2 & 1 \end{bmatrix},$$

$$P = A(A^T A)^{-1} A^T = \frac{1}{14} \begin{bmatrix} 1 & 1 \\ 1 & -1 \\ 2 & 1 \end{bmatrix} \begin{bmatrix} 3 & -2 \\ -2 & 6 \end{bmatrix} \begin{bmatrix} 1 & 1 & 2 \\ 1 & -1 & 1 \end{bmatrix}$$

$$= \frac{1}{14} \begin{bmatrix} 5 & -3 & 6 \\ -3 & 13 & 2 \\ 6 & 2 & 10 \end{bmatrix}.$$

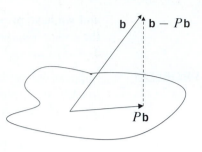

Figure 6.2 The action of a projection matrix: $P\mathbf{b}$ is the projection of $\mathbf{b}$ on the column space of P if $\mathbf{b} - P\mathbf{b}$ is orthogonal to the column space.

Reading Challenge 2 *For any $m \times n$ matrix A, the matrix P defined in 6.1.11 must be square. Why?*

6.1.12 DEFINITION A (necessarily square) matrix P is a *projection matrix* if and only if $\mathbf{b} - P\mathbf{b}$ is orthogonal to the column space of P for all vectors $\mathbf{b}$. (See Figure 6.2.)

Reading Challenge 3 *Why did we enclose "necessarily square" in parentheses in Definition 6.1.12? If $\mathbf{b} - P\mathbf{b}$ is defined, show that P must be square.*

In our previous discussion, where $P = A(A^T A)^{-1} A^T$, we noted that $\mathbf{b} - P\mathbf{b} = \mathbf{b} - A\mathbf{x}^+$ is orthogonal to the column space of A. Thus P is a projection matrix in the sense of Definition 6.1.12 if the column space of A and the column space of P are the same.

Since $P\mathbf{b} = A[(A^T A)^{-1} A^T \mathbf{b}]$ is of the form $A\mathbf{x}$, the column space of P is contained in the column space of A. On the other hand, since $A = PA$ (see Reading Challenge 4), any $A\mathbf{x} = P(A\mathbf{x})$ has the form $P\mathbf{b}$, hence is in the column space of P, so these spaces are indeed the same. We have established the following theorem.

Theorem 6.1.13 *For any matrix A with linearly independent columns, the matrix $P = A(A^T A)^{-1} A^T$ is a projection matrix. It projects vectors (orthogonally) onto the column space of A, which is the column space of P.*

Reading Challenge 4 *With $P = A(A^T A)^{-1} A^T$, why is $PA = A$?*

6.1.14 EXAMPLE Let $A = \begin{bmatrix} 1 & 1 \\ 1 & -1 \\ 2 & 1 \end{bmatrix}$ be the coefficient matrix of the system (1) considered earlier. We have seen that the projection matrix defined in Theorem 6.1.13 is $P = A(A^T A)^{-1} A^T = \frac{1}{14} \begin{bmatrix} 5 & -3 & 6 \\ -3 & 13 & 2 \\ 6 & 2 & 10 \end{bmatrix}$. Suppose we want to project $\mathbf{b} = \begin{bmatrix} 2 \\ 0 \\ -4 \end{bmatrix}$ onto the

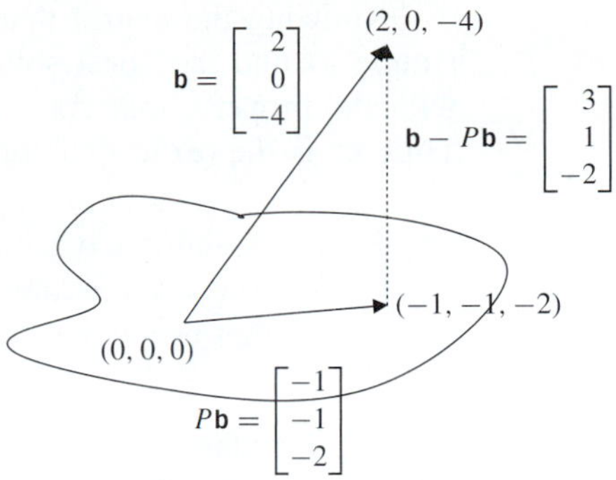

Figure 6.3

plane. The projection is

$$Pb = \frac{1}{14} \begin{bmatrix} 5 & -3 & 6 \\ -3 & 13 & 2 \\ 6 & 2 & 10 \end{bmatrix} \begin{bmatrix} 2 \\ 0 \\ -4 \end{bmatrix} = \begin{bmatrix} -1 \\ -1 \\ -2 \end{bmatrix}.$$

Make sure you understand the significance of what we have done (with an eye to Figure 6.3). The vector $Pb = \begin{bmatrix} -1 \\ -1 \\ -2 \end{bmatrix}$ is in the column space of P (which is the column space of A) and the vector

$$b - Pb = \begin{bmatrix} 2 \\ 0 \\ -4 \end{bmatrix} - \begin{bmatrix} -1 \\ -1 \\ -2 \end{bmatrix} = \begin{bmatrix} 3 \\ 1 \\ -2 \end{bmatrix}$$

is orthogonal to any vector in the column space. We can check by taking the dot product with each column of A:

$$\begin{bmatrix} 3 \\ 1 \\ -2 \end{bmatrix} \cdot \begin{bmatrix} 1 \\ 1 \\ 2 \end{bmatrix} = 3 + 1 - 4 = 0, \qquad \begin{bmatrix} 3 \\ 1 \\ -2 \end{bmatrix} \cdot \begin{bmatrix} 1 \\ -1 \\ 1 \end{bmatrix} = 3 - 1 - 2 = 0.$$

So Pb is the projection of b onto the plane, which is the column space of A. Note that $Pb = \begin{bmatrix} -1 \\ -1 \\ -2 \end{bmatrix}$ is the vector in this plane closest to b (and $(-1, -1, -2)$ is the point of the plane closest to $(2, 0, -4)$). ∎

Reading Challenge 5   *Let* $u = \begin{bmatrix} 1 \\ 1 \\ 2 \end{bmatrix}$ *and* $v = \begin{bmatrix} 1 \\ -1 \\ 1 \end{bmatrix}$ *be the columns of the matrix A in* **6.1.14** *and let*

$x = \begin{bmatrix} 3 \\ 1 \\ -2 \end{bmatrix}$. *We just showed that* $x \cdot u = 0$ *and* $x \cdot v = 0$ *and implied that this makes* x *orthogonal to every vector in the column space. Why is this the case?*

Knowing the matrix that projects vectors onto the column space of A makes it simple to find the "best solution" to $A\mathbf{x} = \mathbf{b}$. The "best solution" is the vector $\mathbf{x}^+$ with the property that $A\mathbf{x}^+ = P\mathbf{b}$ is the projection of $\mathbf{b}$ on the column space of A. Thus $\mathbf{x}^+$ is the (exact) solution to $A\mathbf{x} = P\mathbf{b}$.

> **6.1.15** Assuming A has linearly independent columns, the "best solution" to $A\mathbf{x} = \mathbf{b}$ is the exact solution to $A\mathbf{x} = P\mathbf{b}$, where P is the matrix that projects vectors onto the column space of A.

The language of 6.1.15 suggests that there is just one solution to $A\mathbf{x} = P\mathbf{b}$. We ask you to prove this in the exercises.

6.1.16 EXAMPLE With $A = \begin{bmatrix} 1 & 1 \\ 1 & -1 \\ 2 & 1 \end{bmatrix}$ and $\mathbf{b} = \begin{bmatrix} 2 \\ 0 \\ -4 \end{bmatrix}$, we saw in **6.1.14** that $P = \frac{1}{14}\begin{bmatrix} 5 & -3 & 6 \\ -3 & 3 & 2 \\ 6 & 2 & 10 \end{bmatrix}$ and $P\mathbf{b} = \begin{bmatrix} -1 \\ -1 \\ -2 \end{bmatrix}$. The best solution to $A\mathbf{x} = \mathbf{b}$ is the exact solution to $A\mathbf{x} = P\mathbf{b}$. Gaussian elimination on the augmented matrix $[A|P\mathbf{b}]$ proceeds

$$\begin{bmatrix} 1 & 1 & | & -1 \\ 1 & -1 & | & -1 \\ 2 & 1 & | & -2 \end{bmatrix} \to \begin{bmatrix} 1 & 1 & | & -1 \\ 0 & -2 & | & 0 \\ 0 & -1 & | & 0 \end{bmatrix} \begin{bmatrix} 1 & 1 & | & -1 \\ 0 & 1 & | & 0 \\ 0 & 0 & | & 0 \end{bmatrix},$$

so $x_2 = 0$, $x_1 = -1 - x_2 = -1$ and $\mathbf{x}^+ = \begin{bmatrix} -1 \\ 0 \end{bmatrix}$ as before. ∎

■ How to Recognize a Projection Matrix

Theorem 6.1.17 *A matrix P is a projection matrix if and only if P is symmetric (meaning $P^T = P$) and $P^2 = P$.*

Proof Suppose first that P is a matrix satisfying $P^T = P$ and $P^2 = P$. We must show that $\mathbf{b} - P\mathbf{b}$ is orthogonal to the column space of P for all $\mathbf{b}$, that is, orthogonal to any vector of the form $P\mathbf{x}$. This follows because

$$(P\mathbf{x}) \cdot (\mathbf{b} - P\mathbf{b}) = \mathbf{x}^T P^T (\mathbf{b} - P\mathbf{b})$$

$$= \mathbf{x}^T P(\mathbf{b} - P\mathbf{b}) = \mathbf{x}^T (P\mathbf{b} - P^2\mathbf{b}) = \mathbf{x}^T (P\mathbf{b} - P\mathbf{b}) = 0.$$

Conversely, suppose P is a projection matrix. Thus $\mathbf{b} - P\mathbf{b}$ is orthogonal to $P\mathbf{x}$ for any vectors $\mathbf{x}$ and $\mathbf{b}$. So, for any $\mathbf{x}$ and any $\mathbf{b}$,

$$0 = (P\mathbf{x}) \cdot (\mathbf{b} - P\mathbf{b})$$

$$= (P\mathbf{x})^T (\mathbf{b} - P\mathbf{b}) \tag{3}$$

$$= \mathbf{x}^T P^T (\mathbf{b} - P\mathbf{b})$$

$$= \mathbf{x}^T (P^T \mathbf{b} - P^T P\mathbf{b})$$

$$= \mathbf{x} \cdot (P^T \mathbf{b} - P^T P\mathbf{b}).$$

As noted before, the only vector orthogonal to all vectors $\mathbf{y}$ is the zero vector; hence

$$P^T \mathbf{b} - P^T P\mathbf{b} = \mathbf{0}; \quad \text{that is,} \quad P^T P\mathbf{b} = P^T \mathbf{b}$$

for all $\mathbf{b}$. So $P^T P = P^T$. (See Reading Challenge 7.) Thus $P = (P^T)^T = (P^T P)^T = P^T P = P^T$ is symmetric, and $P^2 = PP = P^T P = P^T = P$. ∎

Reading Challenge 6   *Explain the step marked (3) in the proof of Theorem 6.1.17. How did we move from* $(P\mathbf{x}) \cdot (\mathbf{b} - P\mathbf{b})$ *to* $(P\mathbf{x})^T (\mathbf{b} - P\mathbf{b})$?

Reading Challenge 7   *In the proof of Theorem 6.1.17, we asserted that* $P^T P\mathbf{b} = P^T\mathbf{b}$ *for all* $\mathbf{b}$ *implies* $P^T P = P^T$. *Prove that if A and B are matrices such that* $A\mathbf{x} = B\mathbf{x}$ *for all* $\mathbf{x}$, *then* $A = B$.

■ A "Nice" Form for a Projection Matrix

In Section 1.4, we showed that the projection of a vector $\mathbf{u}$ onto another vector $\mathbf{v}$ is

$$\text{proj}_\mathbf{v}\, \mathbf{u} = \frac{\mathbf{u} \cdot \mathbf{v}}{\mathbf{v} \cdot \mathbf{v}}\, \mathbf{v}. \tag{4}$$

This vector can also be obtained by multiplying $\mathbf{u}$ by a certain projection matrix. To find this matrix, we write

$$\frac{\mathbf{u} \cdot \mathbf{v}}{\mathbf{v} \cdot \mathbf{v}}\, \mathbf{v} = \frac{1}{\mathbf{v} \cdot \mathbf{v}}\, (\mathbf{u} \cdot \mathbf{v})\mathbf{v}$$

and manipulate $(\mathbf{u} \cdot \mathbf{v})\mathbf{v}$ as follows:

$$(\mathbf{u} \cdot \mathbf{v})\mathbf{v} = (\mathbf{v} \cdot \mathbf{u})\mathbf{v} \qquad \text{the dot product is commutative}$$

$$= \mathbf{v}[\mathbf{v} \cdot \mathbf{u}] \qquad \text{if } \mathbf{v} \text{ is a vector and } c \text{ a scalar, } c\mathbf{v} = \begin{bmatrix} \mathbf{v} \\ \downarrow \end{bmatrix} [c] \tag{5}$$

$$= \mathbf{v}(\mathbf{v}^T \mathbf{u}) \qquad \text{see the answer to Reading Challenge 6}$$

$$= (\mathbf{v}\mathbf{v}^T)\mathbf{u} \qquad \text{matrix multiplication is associative.}$$

Thus

$$\text{proj}_\mathbf{v}\,\mathbf{u} = \frac{\mathbf{u}\cdot\mathbf{v}}{\mathbf{v}\cdot\mathbf{v}}\,\mathbf{v} = \frac{1}{\mathbf{v}\cdot\mathbf{v}}\,(\mathbf{u}\cdot\mathbf{v})\mathbf{v} = \frac{1}{\mathbf{v}\cdot\mathbf{v}}\,(\mathbf{v}\mathbf{v}^T)\mathbf{u} = \frac{\mathbf{v}\mathbf{v}^T}{\mathbf{v}\cdot\mathbf{v}}\,\mathbf{u},$$

which is $P\mathbf{u}$ with $P = \frac{\mathbf{v}\mathbf{v}^T}{\mathbf{v}\cdot\mathbf{v}}$.

Reading Challenge 8 *Use Theorem 6.1.17 to show that the matrix $P = \frac{\mathbf{v}\mathbf{v}^T}{\mathbf{v}\cdot\mathbf{v}}$ really is a projection matrix.*

6.1.18 EXAMPLE Suppose we wish to project vectors in the Euclidean plane onto the vector $\mathbf{v} = \begin{bmatrix} 1 \\ 2 \end{bmatrix}$. Since

$$\mathbf{v}\mathbf{v}^T = \begin{bmatrix} 1 \\ 2 \end{bmatrix}\begin{bmatrix} 1 & 2 \end{bmatrix} = \begin{bmatrix} 1 & 2 \\ 2 & 4 \end{bmatrix},$$

the projection matrix we need is $P = \frac{\mathbf{v}\mathbf{v}^T}{\mathbf{v}\cdot\mathbf{v}} = \frac{1}{5}\begin{bmatrix} 1 & 2 \\ 2 & 4 \end{bmatrix}$. This matrix is symmetric—$P^T = P$—and the reader can check that $P^2 = P$. Thus, by Theorem 6.1.17, it is a projection matrix, and it projects vectors onto its column space, which is multiples of $\mathbf{v} = \begin{bmatrix} 1 \\ 2 \end{bmatrix}$, as desired. ∎

In Section 1.4 (see Theorem 1.4.13 especially), we showed that the projection of a vector $\mathbf{w}$ on the plane π spanned by orthogonal vectors $\mathbf{e}$ and $\mathbf{f}$ is

$$\text{proj}_\pi\,\mathbf{w} = \frac{\mathbf{w}\cdot\mathbf{e}}{\mathbf{e}\cdot\mathbf{e}}\,\mathbf{e} + \frac{\mathbf{w}\cdot\mathbf{f}}{\mathbf{f}\cdot\mathbf{f}}\,\mathbf{f}. \tag{6}$$

This vector can also be obtained by multiplying $\mathbf{w}$ by a certain projection matrix. With the same manipulations as before, we deduce that

$$(\mathbf{w}\cdot\mathbf{e})\mathbf{e} = (\mathbf{e}\cdot\mathbf{w})\mathbf{e} = \mathbf{e}(\mathbf{e}\cdot\mathbf{w}) = \mathbf{e}(\mathbf{e}^T\mathbf{w}) = (\mathbf{e}\mathbf{e}^T)\mathbf{w}$$

and, similarly, that $(\mathbf{w}\cdot\mathbf{f})\mathbf{f} = (\mathbf{f}\mathbf{f}^T)\mathbf{w}$. Thus

$$\text{proj}_\pi\,\mathbf{w} = \frac{\mathbf{e}\mathbf{e}^T}{\mathbf{e}\cdot\mathbf{e}}\,\mathbf{w} + \frac{\mathbf{f}\mathbf{f}^T}{\mathbf{f}\cdot\mathbf{f}}\,\mathbf{w},$$

which is $P\mathbf{w}$, with

$$P = \frac{\mathbf{e}\mathbf{e}^T}{\mathbf{e}\cdot\mathbf{e}} + \frac{\mathbf{f}\mathbf{f}^T}{\mathbf{f}\cdot\mathbf{f}}.$$

[In Exercise 12, we ask the reader to verify that this P satisfies the conditions of Theorem 6.1.17 and hence really is a projection matrix.]

6.1.19 EXAMPLE   Suppose we want to project vectors onto the plane π in $\mathbf{R}^3$ whose equation is $3x - 2y + z = 0$. We find two orthogonal vectors $\mathbf{e} = \begin{bmatrix} 5 \\ 6 \\ -3 \end{bmatrix}$ and $\mathbf{f} = \begin{bmatrix} 0 \\ 1 \\ 2 \end{bmatrix}$ in π and compute the matrix

$$P = \frac{\mathbf{e}\mathbf{e}^T}{\mathbf{e} \cdot \mathbf{e}} + \frac{\mathbf{f}\mathbf{f}^T}{\mathbf{f} \cdot \mathbf{f}} = \frac{1}{70}\mathbf{e}\mathbf{e}^T + \frac{1}{5}\mathbf{f}\mathbf{f}^T.$$

We compute

$$\mathbf{e}\mathbf{e}^T = \begin{bmatrix} 5 \\ 6 \\ -3 \end{bmatrix}\begin{bmatrix} 5 & 6 & -3 \end{bmatrix} = \begin{bmatrix} 25 & 30 & -15 \\ 30 & 36 & -18 \\ -15 & -18 & 9 \end{bmatrix}$$

and

$$\mathbf{f}\mathbf{f}^T = \begin{bmatrix} 0 \\ 1 \\ 2 \end{bmatrix}\begin{bmatrix} 0 & 1 & 2 \end{bmatrix} = \begin{bmatrix} 0 & 0 & 0 \\ 0 & 1 & 2 \\ 0 & 2 & 4 \end{bmatrix}$$

so that

$$P = \frac{1}{70}\begin{bmatrix} 25 & 30 & -15 \\ 30 & 36 & -18 \\ -15 & -18 & 9 \end{bmatrix} + \frac{1}{5}\begin{bmatrix} 0 & 0 & 0 \\ 0 & 1 & 2 \\ 0 & 2 & 4 \end{bmatrix} = \frac{1}{14}\begin{bmatrix} 5 & 6 & -3 \\ 6 & 10 & 2 \\ -3 & 2 & 13 \end{bmatrix}.$$

The projection of a vector $\mathbf{w}$ onto π is just $\text{proj}_\pi \mathbf{w} = P\mathbf{w}$. For example, if $\mathbf{w} = \begin{bmatrix} 5 \\ -3 \\ -7 \end{bmatrix}$, then $P\mathbf{w} = \begin{bmatrix} 2 \\ -1 \\ -8 \end{bmatrix}$. ∎

Reading Challenge 9   *Verify that the column space of the matrix*

$$P = \frac{1}{14}\begin{bmatrix} 5 & 6 & -3 \\ 6 & 10 & 2 \\ -3 & 2 & 13 \end{bmatrix}$$

*found in **6.1.19** is the plane π with equation $3x - 2y + z = 0$ so P really works as advertised.*

All this generalizes in a natural way.

6.1.20 DEFINITION

If U is a subspace of some Euclidean space V and $\mathbf{w}$ is in V, the *projection of $\mathbf{w}$ on* U is the vector $\mathbf{p} = \text{proj}_U \mathbf{w}$ in U with the property that $\mathbf{w} - \mathbf{p}$ is orthogonal to U, that is, orthogonal to every vector in U.

Pairwise orthogonal means that any two (different) vectors are orthogonal.

We have considered projections onto one-dimensional and two-dimensional subspaces of $\mathbf{R}^n$. In general, let U be any subspace of $\mathbf{R}^n$ and let $\mathbf{f}_1, \mathbf{f}_2, \ldots, \mathbf{f}_k$ be a basis of k pairwise orthogonal vectors. For any $\mathbf{x}$ in $\mathbf{R}^n$, we find a formula for $\text{proj}_U \mathbf{x}$ that is the natural extension of (4) and (6).

Let

$$A = \begin{bmatrix} \mathbf{f}_1 & \mathbf{f}_2 & \cdots & \mathbf{f}_k \\ \downarrow & \downarrow & & \downarrow \end{bmatrix}$$

be the matrix whose column space is U. Then

$$A^T A = \begin{bmatrix} \mathbf{f}_1^T & \rightarrow \\ \mathbf{f}_2^T & \rightarrow \\ \vdots \\ \mathbf{f}_k^T & \rightarrow \end{bmatrix} \begin{bmatrix} \mathbf{f}_1 & \mathbf{f}_2 & \cdots & \mathbf{f}_k \\ \downarrow & \downarrow & & \downarrow \end{bmatrix} = \begin{bmatrix} \mathbf{f}_1 \cdot \mathbf{f}_1 & 0 & \cdots & 0 \\ 0 & \mathbf{f}_2 \cdot \mathbf{f}_2 & \cdots & 0 \\ \vdots & & \ddots & \vdots \\ 0 & 0 & \cdots & \mathbf{f}_k \cdot \mathbf{f}_k \end{bmatrix}$$

and so

$$(A^T A)^{-1} = \begin{bmatrix} \frac{1}{\mathbf{f}_1 \cdot \mathbf{f}_1} & 0 & \cdots & 0 \\ 0 & \frac{1}{\mathbf{f}_2 \cdot \mathbf{f}_2} & \cdots & 0 \\ \vdots & & \ddots & \vdots \\ 0 & 0 & \cdots & \frac{1}{\mathbf{f}_k \cdot \mathbf{f}_k} \end{bmatrix}.$$

By Theorem 6.1.13, the matrix that projects onto U is $P = A(A^T A)^{-1} A^T$, so the projection of $\mathbf{w}$ onto the column space is

$$\text{proj}_U \mathbf{w} = P\mathbf{w} = A(A^T A)^{-1} A^T \mathbf{w} = A(A^T A)^{-1} \begin{bmatrix} \mathbf{f}_1^T & \rightarrow \\ \mathbf{f}_2^T & \rightarrow \\ \vdots \\ \mathbf{f}_k^T & \rightarrow \end{bmatrix} \begin{bmatrix} \mathbf{w} \\ \downarrow \end{bmatrix}$$

$$= A(A^T A)^{-1} \begin{bmatrix} \mathbf{f}_1 \cdot \mathbf{w} \\ \mathbf{f}_2 \cdot \mathbf{w} \\ \vdots \\ \mathbf{f}_k \cdot \mathbf{w} \end{bmatrix} = A \begin{bmatrix} \frac{\mathbf{f}_1 \cdot \mathbf{w}}{\mathbf{f}_1 \cdot \mathbf{f}_1} & 0 & \cdots & 0 \\ 0 & \frac{\mathbf{f}_2 \cdot \mathbf{w}}{\mathbf{f}_2 \cdot \mathbf{f}_2} & & \\ \vdots & & \ddots & \\ 0 & 0 & \cdots & \frac{\mathbf{f}_k \cdot \mathbf{w}}{\mathbf{f}_k \cdot \mathbf{f}_k} \end{bmatrix}.$$

Because $A\mathbf{x}$ is that linear combination of the columns of A in which the coefficients are the components of $\mathbf{x}$ (does the reader agree?), we obtain the following generalization of formula (6) for the projection of $\mathbf{w}$ on the span of $\mathbf{f}_1, \ldots, \mathbf{f}_k$.

$$\textbf{6.1.21} \quad \text{proj}_U \, \mathbf{w} = P\mathbf{w} = \frac{\mathbf{w} \cdot \mathbf{f}_1}{\mathbf{f}_1 \cdot \mathbf{f}_1} \, \mathbf{f}_1 + \frac{\mathbf{w} \cdot \mathbf{f}_2}{\mathbf{f}_2 \cdot \mathbf{f}_2} \, \mathbf{f}_2 + \cdots + \frac{\mathbf{w} \cdot \mathbf{f}_k}{\mathbf{f}_k \cdot \mathbf{f}_k} \, \mathbf{f}_k$$

Using manipulations we have seen before, this can be rewritten

$$\frac{\mathbf{f}_1 \mathbf{f}_1^T}{\mathbf{f}_1 \cdot \mathbf{f}_1} \, \mathbf{w} + \frac{\mathbf{f}_2 \mathbf{f}_2^T}{\mathbf{f}_2 \cdot \mathbf{f}_2} \, \mathbf{w} + \cdots + \frac{\mathbf{f}_k \mathbf{f}_k^T}{\mathbf{f}_k \cdot \mathbf{f}_k} \, \mathbf{w}$$

$$= \left(\frac{\mathbf{f}_1 \mathbf{f}_1^T}{\mathbf{f}_1 \cdot \mathbf{f}_1} + \frac{\mathbf{f}_2 \mathbf{f}_2^T}{\mathbf{f}_2 \cdot \mathbf{f}_2} + \cdots + \frac{\mathbf{f}_k \mathbf{f}_k^T}{\mathbf{f}_k \cdot \mathbf{f}_k} \right) \mathbf{w} = P\mathbf{w}$$

where the projection matrix P takes the beautiful form

$$\textbf{6.1.22} \quad P = \frac{\mathbf{f}_1 \mathbf{f}_1^T}{\mathbf{f}_1 \cdot \mathbf{f}_1} + \frac{\mathbf{f}_2 \mathbf{f}_2^T}{\mathbf{f}_2 \cdot \mathbf{f}_2} + \cdots + \frac{\mathbf{f}_k \mathbf{f}_k^T}{\mathbf{f}_k \cdot \mathbf{f}_k}.$$

Theorem 6.1.23 *Let U be the span of pairwise orthogonal vectors $\mathbf{f}_1, \mathbf{f}_2, \dots, \mathbf{f}_k$. Then the projection $\text{proj}_U \, \mathbf{w}$ of $\mathbf{w}$ on U is the vector $P\mathbf{w}$, where*

$$P = \frac{\mathbf{f}_1 \mathbf{f}_1^T}{\mathbf{f}_1 \cdot \mathbf{f}_1} + \frac{\mathbf{f}_2 \mathbf{f}_2^T}{\mathbf{f}_2 \cdot \mathbf{f}_2} + \cdots + \frac{\mathbf{f}_k \mathbf{f}_k^T}{\mathbf{f}_k \cdot \mathbf{f}_k}.$$

The vector $P\mathbf{w}$ has the property that $\mathbf{w} - P\mathbf{w}$ is orthogonal to every vector in U.

Reading Challenge 10 *The vectors $\mathbf{u} = \begin{bmatrix} 1 \\ 1 \\ 0 \end{bmatrix}$ and $\mathbf{v} = \begin{bmatrix} 1 \\ 0 \\ -1 \end{bmatrix}$ span the plane U with equation $x + y + z = 0$. So,*

to project $\mathbf{w}$ on U, we multiply by $P = \dfrac{\mathbf{u}\mathbf{u}^T}{\mathbf{u} \cdot \mathbf{u}} + \dfrac{\mathbf{v}\mathbf{v}^T}{\mathbf{v} \cdot \mathbf{v}} = \dfrac{1}{2} \begin{bmatrix} 1 & 1 & 0 \\ 1 & 1 & 0 \\ 0 & 0 & 0 \end{bmatrix} + \dfrac{1}{2} \begin{bmatrix} 1 & 0 & -1 \\ 0 & 0 & 0 \\ -1 & 0 & 1 \end{bmatrix} =$

$\dfrac{1}{2} \begin{bmatrix} 2 & 1 & -1 \\ 1 & 1 & 0 \\ -1 & 0 & 1 \end{bmatrix}$. Is this correct?

6.1.24 EXAMPLE The vectors $\mathbf{f}_1 = \begin{bmatrix} 1 \\ 0 \\ -1 \\ 1 \end{bmatrix}$, $\mathbf{f}_2 = \begin{bmatrix} 1 \\ 1 \\ 1 \\ 0 \end{bmatrix}$, and $\mathbf{f}_3 = \begin{bmatrix} -1 \\ 0 \\ 1 \\ 2 \end{bmatrix}$ are pairwise orthogonal in $\mathbf{R}^4$. If U is the span of these vectors and we wish to project onto U, we need the matrix

$$P = \frac{\mathbf{f}_1 \mathbf{f}_1^T}{\mathbf{f}_1 \cdot \mathbf{f}_1} + \frac{\mathbf{f}_2 \mathbf{f}_2^T}{\mathbf{f}_2 \cdot \mathbf{f}_2} + \frac{\mathbf{f}_3 \mathbf{f}_3^T}{\mathbf{f}_3 \cdot \mathbf{f}_3} = \tfrac{1}{3} \mathbf{f}_1 \mathbf{f}_1^T + \tfrac{1}{3} \mathbf{f}_2 \mathbf{f}_2^T + \tfrac{1}{6} \mathbf{f}_3 \mathbf{f}_3^T.$$

Since

$$\mathbf{f}_1\mathbf{f}_1^T = \begin{bmatrix} 1 \\ 0 \\ -1 \\ 1 \end{bmatrix} \begin{bmatrix} 1 & 0 & -1 & 1 \end{bmatrix} = \begin{bmatrix} 1 & 0 & -1 & 1 \\ 0 & 0 & 0 & 0 \\ -1 & 0 & 1 & -1 \\ 1 & 0 & -1 & 1 \end{bmatrix},$$

$$\mathbf{f}_2\mathbf{f}_2^T = \begin{bmatrix} 1 \\ 1 \\ 1 \\ 0 \end{bmatrix} \begin{bmatrix} 1 & 1 & 1 & 0 \end{bmatrix} = \begin{bmatrix} 1 & 1 & 1 & 0 \\ 1 & 1 & 1 & 0 \\ 1 & 1 & 1 & 0 \\ 0 & 0 & 0 & 0 \end{bmatrix}, \text{ and}$$

$$\mathbf{f}_3\mathbf{f}_3^T = \begin{bmatrix} -1 \\ 0 \\ 1 \\ 2 \end{bmatrix} \begin{bmatrix} -1 & 0 & 1 & 2 \end{bmatrix} = \begin{bmatrix} 1 & 0 & -1 & -2 \\ 0 & 0 & 0 & 0 \\ -1 & 0 & 1 & 2 \\ -2 & 0 & 2 & 4 \end{bmatrix},$$

the desired matrix P is

$$\frac{1}{3} \begin{bmatrix} 1 & 0 & -1 & 1 \\ 0 & 0 & 0 & 0 \\ -1 & 0 & 1 & -1 \\ 1 & 0 & -1 & 1 \end{bmatrix} + \frac{1}{3} \begin{bmatrix} 1 & 1 & 1 & 0 \\ 1 & 1 & 1 & 0 \\ 1 & 1 & 1 & 0 \\ 0 & 0 & 0 & 0 \end{bmatrix} + \frac{1}{6} \begin{bmatrix} 1 & 0 & -1 & -2 \\ 0 & 0 & 0 & 0 \\ -1 & 0 & 1 & 2 \\ -2 & 0 & 2 & 4 \end{bmatrix}$$

$$= \frac{1}{6} \begin{bmatrix} 5 & 2 & -1 & 0 \\ 2 & 2 & 2 & 0 \\ -1 & 2 & 5 & 0 \\ 0 & 0 & 0 & 6 \end{bmatrix}. \qquad \blacksquare$$

Answers to Reading Challenges

1. The given argument is correct **provided there is a solution**. The matrix A in Problem 6.1.10 has the property that $A\mathbf{x} = \mathbf{0}$ has only the trivial solution $\mathbf{x} = \mathbf{0}$, but it is not true that $A\mathbf{x} = \mathbf{b}$ has a unique solution for any $\mathbf{b}$.

2. Since A is $m \times n$, A^T is $n \times m$, so P is the product of an $m \times n$, an $n \times n$, and an $n \times m$ matrix. So P is $m \times m$.

3. If P is $m \times n$ and $P\mathbf{b}$ is defined, then $\mathbf{b}$ must be $n \times 1$ and then $P\mathbf{b}$ is $m \times 1$. Since $\mathbf{b} - P\mathbf{b}$ is the difference of a vector in $\mathbf{R}^n$ and another in $\mathbf{R}^m$, we must have $m = n$.

4. $PA = [A(A^TA)^{-1}A^T]A = A(A^TA)^{-1}(A^TA) = A$ since $(A^TA)^{-1}(A^TA) = I$ is the identity matrix.

5. A vector in the column space has the form $a\mathbf{u} + b\mathbf{v}$ for some scalars a and b, and $\mathbf{x}$ is orthogonal to such a vector because $\mathbf{x} \cdot (a\mathbf{u} + b\mathbf{v}) = a\mathbf{x} \cdot \mathbf{u} + b\mathbf{x} \cdot \mathbf{v} = 0$.

6. For any vectors $\mathbf{u}$ and $\mathbf{v}$, the dot product $\mathbf{u} \cdot \mathbf{v}$ is the **matrix product** $\mathbf{u}^T\mathbf{v}$, thinking of $\mathbf{u}$ and $\mathbf{v}$ as $n \times 1$ matrices.

7. Let $\mathbf{x} = \mathbf{e}_1 = \begin{bmatrix} 1 \\ 0 \\ \vdots \\ 0 \end{bmatrix}$. Then $A\mathbf{x} = A\mathbf{e}_1$ is a linear combination of the columns of A with coefficients $1, 0, \ldots, 0$, that is, the first column of A. Similarly, $B\mathbf{e}_1$ is the first column of B, so A and B have the same first column. Replacing $\mathbf{e}_1$ with $\mathbf{e}_2 = \begin{bmatrix} 0 \\ 1 \\ 0 \\ \vdots \\ 0 \end{bmatrix}$ shows that A and B also have the same second column. And so on.

8. To verify that the matrix is symmetric we use these three properties of transpose (see Theorem 2.2.19):

   i. $(cA)^T = cA^T$ for any constant c,

   ii. $(AB)^T = B^T A^T$ for any (compatible) matrices A and B, and

   iii. $(A^T)^T = A$.

   Here then (with $c = \frac{1}{\mathbf{v} \cdot \mathbf{v}}$), we have $P^T = (\frac{1}{\mathbf{v} \cdot \mathbf{v}} \mathbf{v}\mathbf{v}^T)^T = \frac{1}{\mathbf{v} \cdot \mathbf{v}}(\mathbf{v}^T)^T \mathbf{v}^T = \frac{1}{\mathbf{v} \cdot \mathbf{v}} \mathbf{v}\mathbf{v}^T = P$. Now $P^2 = \frac{\mathbf{v}\mathbf{v}^T \mathbf{v}\mathbf{v}^T}{(\mathbf{v} \cdot \mathbf{v})(\mathbf{v} \cdot \mathbf{v})}$ and $\mathbf{v}^T \mathbf{v} = \mathbf{v} \cdot \mathbf{v}$, so the numerator is $\mathbf{v}(\mathbf{v} \cdot \mathbf{v})\mathbf{v}^T$. Cancelling $\mathbf{v} \cdot \mathbf{v}$ from numerator and denominator gives $P^2 = \frac{\mathbf{v}\mathbf{v}^T}{\mathbf{v} \cdot \mathbf{v}} = P$.

9. The components of each column of P satisfy $3x - 2y + z = 0$, so col sp $P \subseteq \pi$. Since the columns of P are not all multiples of the same vector, dim col sp $P > 1$, so dim col sp $P = 2$ and col sp $P = \pi$.

10. This is not correct. The reader has misread Theorem 6.1.23. The vectors $\mathbf{u}$ and $\mathbf{v}$ are not orthogonal.

True/False Questions

Decide, with as little calculation as possible, whether each of the following statements is true or false and explain your answer whenever you say "false." (Answers can be found in the back of the book.)

1. The length of $\begin{bmatrix} 1 \\ 1 \\ 1 \\ 1 \end{bmatrix}$ is 4.

2. If $\mathbf{x}$ is a vector and $\mathbf{x} \cdot \mathbf{x} = 0$, then $\mathbf{x} = \mathbf{0}$.

3. If $\mathbf{u}$ and $\mathbf{v}$ are vectors, $\mathbf{u} \cdot \mathbf{v} = \mathbf{u}\mathbf{v}^T$.

4. For any matrix A, the "best solution" to $A\mathbf{x} = \mathbf{b}$ in the sense of least squares is $\mathbf{x}^+ = (A^T A)^{-1} A^T \mathbf{b}$.

5. The "best solution" to $A\mathbf{x} = \mathbf{b}$ is a solution to $A\mathbf{x} = \mathbf{b}$.

6. A projection matrix is a square matrix.

7. The only invertible projection matrix is the identity.

8. For any unit vector $\mathbf{u}$, the matrix $\mathbf{u}\mathbf{u}^T$ is a projection matrix.

9. If vectors **u** and **v** span a plane π, then the matrix that projects vectors onto π is $P = \dfrac{\mathbf{u}\mathbf{u}^T}{\mathbf{u}\cdot\mathbf{u}} + \dfrac{\mathbf{v}\mathbf{v}^T}{\mathbf{v}\cdot\mathbf{v}}$.

10. If **w** is a vector in a subspace U, the projection of **w** on U is **w**.

Exercises

*Solutions to exercises marked [BB] can be found in the **B**ack of the **B**ook.*

1. [BB] In this section, a lot of attention was paid to the matrix $A(A^T A)^{-1}A^T$. What's the problem with the calculation $A(A^T A)^{-1}A^T = A[A^{-1}(A^T)^{-1}]A^T = I$? Was this whole section about the identity matrix?

2. Let **x** be a vector in $\mathbf{R}^n$.
   (a) Show that $\mathbf{x}^T\mathbf{x} = 0$ implies $\mathbf{x} = \mathbf{0}$.
   (b) Show that $\mathbf{x}\mathbf{x}^T = 0$ implies $\mathbf{x} = \mathbf{0}$.

3. [BB] Let A be an $m \times n$ matrix and let **x** and **b** be vectors. Suppose $\mathbf{y} \cdot (A^T\mathbf{b} - A^T A\mathbf{x}) = 0$ for all vectors **y**. Prove that $\mathbf{b} - A\mathbf{x}$ is perpendicular to the column space of A.

4. (a) Find the best solution $\mathbf{x}^+$ in the sense of least squares to
$$\begin{aligned} x + y &= 1 \\ x - y &= 2 \\ 2x + y &= 5. \end{aligned}$$
   (b) What does "best solution in the sense of least squares" mean? Explain with reference to part (a).

5. Find the best solution in the sense of least squares to each of the following systems of linear equations.
   (a) [BB]
$$\begin{aligned} 2x_1 - x_2 &= 1 \\ x_1 - x_2 &= 0 \\ 2x_1 + x_2 &= -1 \\ x_1 + 3x_2 &= 3 \end{aligned}$$
   (b)
$$\begin{aligned} x_1 + 2x_2 &= 1 \\ x_1 - x_2 &= -1 \\ 3x_1 - x_2 &= 0 \\ -4x_1 + x_2 &= -3 \end{aligned}$$
   (c) [BB]
$$\begin{aligned} x_1 \quad\;\; - 2x_3 &= 0 \\ x_1 + x_2 - x_3 &= 2 \\ - x_2 - x_3 &= 3 \\ 2x_1 + 3x_2 + x_3 &= -4 \end{aligned}$$
   (d)
$$\begin{aligned} x_2 + 2x_3 &= 1 \\ x_1 \quad\;\; - x_3 &= -1 \\ x_1 + x_2 + x_3 &= 0 \\ -x_1 + x_2 + x_3 &= 1 \\ -x_1 + x_2 \quad\;\; &= 2. \end{aligned}$$

6. Which of the following matrices are projection matrices? Explain your answers.
   (a) [BB] $P = \dfrac{1}{10}\begin{bmatrix} 2 & 5 \\ 5 & 10 \end{bmatrix}$
   (b) $P = \dfrac{1}{3}\begin{bmatrix} 2 & 1 & 1 \\ 1 & 2 & -1 \\ 1 & -1 & 2 \end{bmatrix}$
   (c) $\dfrac{1}{25}\begin{bmatrix} 1 & 2 & -3 \\ 2 & 4 & 5 \\ -1 & 5 & 10 \end{bmatrix}$
   (d) [BB] $\dfrac{1}{6}\begin{bmatrix} 5 & -2 & -1 \\ -2 & 2 & -2 \\ -1 & -2 & 5 \end{bmatrix}$
   (e) $\dfrac{1}{3}\begin{bmatrix} 1 & 0 & 1 & 1 \\ 0 & 1 & 1 & -1 \\ 1 & 1 & 2 & 0 \\ 1 & -1 & 0 & 2 \end{bmatrix}$
   (f) $\begin{bmatrix} 2 & -1 & 1 & 1 \\ 1 & 0 & 1 & 1 \\ 1 & -1 & 2 & 1 \\ -2 & 2 & -2 & -1 \end{bmatrix}$.

7. Find the matrix P that projects vectors in $\mathbf{R}^4$ onto the column space of each of the given matrices.
   (a) [BB] $A = \begin{bmatrix} 1 & 1 \\ 2 & 1 \\ -2 & 1 \\ 0 & 0 \end{bmatrix}$
   (b) $A = \begin{bmatrix} -2 & 1 \\ -1 & 1 \\ 0 & 1 \\ 1 & -1 \end{bmatrix}$
   (c) $A = \begin{bmatrix} 1 & 1 & -1 \\ 1 & 2 & 0 \\ 1 & -3 & 1 \\ 2 & -1 & 1 \end{bmatrix}$
   (d) [BB] $A = \begin{bmatrix} 1 & 0 & 1 & 0 \\ 1 & 1 & 1 & 0 \\ -1 & 0 & 3 & 1 \\ -1 & 1 & -1 & 0 \end{bmatrix}$.

8. Let $A = \begin{bmatrix} 1 & -2 & 0 \\ -1 & 0 & 1 \\ 0 & 0 & 1 \\ 1 & -1 & 1 \end{bmatrix}$.

(a) Find the column space of A. [What is required is a linear equation in b_1, b_2, b_3, b_4 that is necessary and sufficient for $\mathbf{b} = \begin{bmatrix} b_1 \\ b_2 \\ b_3 \\ b_4 \end{bmatrix}$ to belong to the column space.]

(b) Prove that the columns of A are linearly independent.

(c) Without calculation, explain why the matrix $A^T A$ must be invertible.

For the rest of this exercise, let $\mathbf{b} = e_1 = \begin{bmatrix} 1 \\ 0 \\ 0 \\ 0 \end{bmatrix}$.

(d) Without calculation, explain why $A\mathbf{x} = \mathbf{b}$ has no solution.

(e) Find the best solution $\mathbf{x}^+$ in the sense of least squares to the system $A\mathbf{x} = \mathbf{b}$.

(f) What is the significance of your answer to 8(e)?

(g) Find the matrix P that projects vectors in $\mathbf{R}^4$ onto the column space of A.

(h) Find P^2 **without calculation**, and explain.

(i) Solve the system $A\mathbf{x} = P\mathbf{b}$ and comment on your answer.

9. Let A be any $m \times n$ matrix. Prove that $A^T A$ and A have the same null space. [Hint: $\|A\mathbf{x}\|^2 = \mathbf{x}^T A^T A\mathbf{x}$. See Exercise 15 of Section 2.1.]

10. [BB] Prove that any linear combination of projection matrices is symmetric.

11. Let P be a projection matrix. Show that $I - P$ is also a projection matrix.

12. Let $\mathbf{e}$ and $\mathbf{f}$ be orthogonal vectors. Show that the matrix $P = \frac{\mathbf{e}\mathbf{e}^T}{\mathbf{e}\cdot\mathbf{e}} + \frac{\mathbf{f}\mathbf{f}^T}{\mathbf{f}\cdot\mathbf{f}}$ is symmetric and satisfies $P^2 = P$, and hence is a projection matrix. [Hint: See the solution to Reading Challenge 8.]

13. [BB] If A is $m \times n$ and $A^T A = I$, show that $P = AA^T$ is a projection matrix.

14. For each of the following matrices A and vectors $\mathbf{b}$,

    i. use formula 6.1.7 to find the best (least squares) solution $\mathbf{x}^+$ of the system $A\mathbf{x} = \mathbf{b}$;

    ii. compute the vector $A\mathbf{x}^+$ and explain its significance;

    iii. find the matrix P that projects vectors onto the column space of A;

    iv. solve $A\mathbf{x} = P\mathbf{b}$ by Gaussian elimination and compare your answer with that found in part i.

(a) [BB] $A = \begin{bmatrix} 1 & -1 \\ 0 & 2 \\ 1 & 1 \\ 1 & -3 \end{bmatrix}$, $\mathbf{b} = \begin{bmatrix} 1 \\ 2 \\ -1 \\ 3 \end{bmatrix}$

(b) $A = \begin{bmatrix} 1 & -1 \\ 2 & 1 \\ 1 & -3 \end{bmatrix}$, $\mathbf{b} = \begin{bmatrix} 1 \\ -1 \\ 1 \end{bmatrix}$

(c) [BB] $A = \begin{bmatrix} 1 & 0 & 1 \\ 1 & 1 & 1 \\ 2 & 0 & 0 \\ 1 & 1 & -1 \end{bmatrix}$, $\mathbf{b} = \begin{bmatrix} -1 \\ 0 \\ 0 \\ 3 \end{bmatrix}$

(d) $A = \begin{bmatrix} 1 & 1 & 0 \\ 1 & 0 & 2 \\ -1 & 1 & -1 \\ 0 & 0 & -2 \end{bmatrix}$, $\mathbf{b} = \begin{bmatrix} 0 \\ 2 \\ -2 \\ 0 \end{bmatrix}$.

15. [BB] Find the matrix that projects vectors in $\mathbf{R}^3$ onto the plane spanned by $\begin{bmatrix} 1 \\ 0 \\ 1 \end{bmatrix}$ and $\begin{bmatrix} 1 \\ 1 \\ 0 \end{bmatrix}$.

16. (a) Find the matrix that projects vectors in $\mathbf{R}^3$ onto the plane with equation $x + 3y - z = 0$.

    (b) Find the projection of $\mathbf{b} = \begin{bmatrix} 1 \\ 1 \\ 1 \end{bmatrix}$ onto the plane in (a) and verify that $\mathbf{b} - P\mathbf{b}$ is orthogonal to every vector in the plane.

17. [BB] Let U be the set of vectors in $\mathbf{R}^4$ of the form $\begin{bmatrix} 2s - t \\ s \\ t \\ -5s + 2t \end{bmatrix}$. Find the matrix that projects vectors in $\mathbf{R}^4$ onto U.

18. Let $U = \{ \begin{bmatrix} x_1 \\ x_2 \\ x_3 \\ x_4 \end{bmatrix} \in \mathbf{R}^4 \mid x_1 - x_2 + x_4 = 0 \}$.

    (a) [BB] Find three vectors that span U.

    (b) Find a matrix that projects vectors in $\mathbf{R}^4$ onto U.

19. Find the matrix P that projects vectors onto each of the following spaces.

    (a) [BB] the subspace of $\mathbf{R}^3$ spanned by $\mathbf{v} = \begin{bmatrix} 1 \\ -3 \\ 5 \end{bmatrix}$

    (b) the subspace of $\mathbf{R}^3$ spanned by $\mathbf{v} = \begin{bmatrix} -1 \\ 2 \\ 4 \end{bmatrix}$

    (c) the subspace of $\mathbf{R}^3$ spanned by $\mathbf{v} = \begin{bmatrix} 2 \\ 0 \\ -3 \end{bmatrix}$

    (d) [BB] the plane spanned by the orthogonal vectors
    $\mathbf{e} = \begin{bmatrix} -1 \\ 2 \\ 1 \end{bmatrix}$ and $\mathbf{f} = \begin{bmatrix} 3 \\ 1 \\ 1 \end{bmatrix}$

    (e) the plane spanned by the orthogonal vectors
    $\mathbf{e} = \begin{bmatrix} 0 \\ 2 \\ 1 \end{bmatrix}$ and $\mathbf{f} = \begin{bmatrix} 5 \\ 3 \\ -6 \end{bmatrix}$

    (f) [BB] the plane in $\mathbf{R}^3$ with equation $2x - y + 5z = 0$

    (g) the plane in $\mathbf{R}^3$ with equation $x + 3y + z = 0$

    (h) the subspace U of $\mathbf{R}^4$ spanned by the orthogonal
    vectors $\mathbf{e} = \begin{bmatrix} 1 \\ -2 \\ 3 \\ 4 \end{bmatrix}$ and $\mathbf{f} = \begin{bmatrix} 0 \\ 1 \\ 2 \\ -1 \end{bmatrix}$

    (i) the subspace U of $\mathbf{R}^4$ spanned by the pairwise
    orthogonal vectors $\mathbf{f}_1 = \begin{bmatrix} 1 \\ 0 \\ 0 \\ 1 \end{bmatrix}$, $\mathbf{f}_2 = \begin{bmatrix} -1 \\ 0 \\ 2 \\ 1 \end{bmatrix}$,
    $\mathbf{f}_3 = \begin{bmatrix} -1 \\ 2 \\ -1 \\ 1 \end{bmatrix}$.

20. Let $A = \begin{bmatrix} 1 & 1 & -2 & 0 \\ 3 & -1 & 2 & -4 \\ -1 & 2 & -4 & 3 \end{bmatrix}$.

    (a) Find two vectors that span the null space of A.

    (b) Use the result of 20(a) to find the matrix that projects vectors onto the null space of A.

    (c) Find two orthogonal vectors that span the null space of A.

    (d) Use the result of 20(c) to find the matrix that projects vectors onto the null space of A. Compare this matrix with the one found in 20(b).

    (e) Find the vector $\mathbf{u}$ in the null space of A for which
    $\|\mathbf{u} - \mathbf{w}\|$ is a minimum, with $\mathbf{w} = \begin{bmatrix} 1 \\ 0 \\ -1 \\ 0 \end{bmatrix}$.

21. (a) [BB] Suppose one wants to project vectors in $\mathbf{R}^3$
    onto the column space of the matrix
    $A = \begin{bmatrix} 1 & -2 \\ 2 & -4 \\ 3 & -6 \end{bmatrix}$. Do the methods of this section
    apply? Explain, then find the matrix that projects vectors onto the column space of A.a

    (b) Find the matrix that projects vectors in $\mathbf{R}^2$ onto
    the column space of $A = \begin{bmatrix} -2 & -1 & 0 & 1 \\ 1 & 1 & 1 & -1 \end{bmatrix}$.

    (c) Find the matrix that projects vectors in $\mathbf{R}^3$ onto
    the column space of $A = \begin{bmatrix} 1 & 0 & -1 \\ 0 & -1 & -1 \\ 1 & 3 & 2 \end{bmatrix}$.

22. Suppose A is a matrix with linearly independent columns.
    Let $P = A(A^T A)^{-1} A^T$.

    (a) [BB] Prove that $[(A^T A)^T]^{-1} = [(A^T A)^{-1}]^T$.

    (b) Use Theorem 6.1.17 to prove that P is a projection matrix.

23. [BB] Let Q be a matrix whose columns are the
    pairwise orthogonal vectors $\mathbf{q}_1, \mathbf{q}_2, \ldots, \mathbf{q}_n$ of length 1.

    (a) Show that QQ^T is a projection matrix. [Hint:
    Show that $Q^T Q = I$ and use Exercise 13.]

    (b) Show that $Q^T Q = \mathbf{q}_1^T \mathbf{q}_1 + \mathbf{q}_2^T \mathbf{q}_2 + \cdots + \mathbf{q}_n^T \mathbf{q}_n$.
    [Hint: Apply the left side to a vector $\mathbf{x}$ and use
    the fact that $(\mathbf{v} \cdot \mathbf{u})\mathbf{v}$ is the matrix product $\mathbf{v}(\mathbf{v}^T \mathbf{u})$.]

24. Let U be a subspace of a vector space V. Let $\mathbf{b}$ and $\mathbf{u}$
    be vectors with $\mathbf{u}$ in U. Suppose that $\mathbf{b} - \mathbf{u}$ is
    orthogonal to every vector in U. Show that
    $\|\mathbf{b} - \mathbf{u}\| \le \|\mathbf{b} - \mathbf{u}_1\|$ for every vector $\mathbf{u}_1$ in U. [Hint:
    $\mathbf{b} - \mathbf{u}_1 = \mathbf{x} + \mathbf{y}$ with $\mathbf{x} = \mathbf{b} - \mathbf{u}$ and $\mathbf{y} = \mathbf{u} - \mathbf{u}_1$.
    Explain why the Theorem of Pythagoras applies and
    use it.]

■ Critical Reading

25. [BB] Let $P: \mathbf{R}^3 \to \mathbf{R}^3$ be the projection matrix that sends $\begin{bmatrix} x \\ y \\ z \end{bmatrix}$ to $\begin{bmatrix} x \\ y \\ 0 \end{bmatrix}$. Find the eigenvectors and eigenvalues of P without any calculation.

26. What are the possible values for the eigenvalues of a projection matrix P and why?

27. [BB] Let $A = \begin{bmatrix} 1 & 2 & 0 & 0 \\ 0 & 0 & 0 & 0 \\ 0 & 1 & 1 & 0 \end{bmatrix}$. **Without any calculation**, identify the column space of A and write down the matrix P that projects vectors in $\mathbf{R}^k$ onto the column space. A brief answer, in a sentence or two, should also reveal the value of k.

28. Suppose A has linearly independent columns and $A\mathbf{x} = \mathbf{b}$ has a genuine solution $\mathbf{x}_0$. Show that the best solution to $A\mathbf{x} = \mathbf{b}$ in the sense of least squares is this vector $\mathbf{x}_0$.

29. If a matrix A has linearly independent columns, show that the "best solution" to $A\mathbf{x} = \mathbf{b}$ is unique.

30. Let $\mathbf{u}$ and $\mathbf{v}$ be orthogonal vectors of length 1. Let $P = \mathbf{u}\mathbf{u}^T + \mathbf{v}\mathbf{v}^T$.

   (a) Prove that P is a projection matrix. [Use Theorem 6.1.17.]

   (b) Identify the space onto which P project vectors. Say something more precise than "the column space of P." [Hint: A vector is in the range of P if and only if it has the form $P\mathbf{x}$ for some vector $\mathbf{x}$. Look at the calculations preceding 6.1.18.]

   (c) Compute P with $\mathbf{u} = \begin{bmatrix} 1 \\ 0 \\ 0 \end{bmatrix}$ and $\mathbf{v} = \begin{bmatrix} 0 \\ 1 \\ 0 \end{bmatrix}$. Is this the matrix you would expect? Explain.

31. Prove that the invertible projection matrix is the identity.

32. Suppose A is an $m \times n$ matrix with linearly independent columns and $P = A(A^T A)^{-1} A^T$. Let $\mathbf{y}$ be any vector in $\mathbf{R}^m$.

   (a) Without any calculation, explain why the system $A\mathbf{x} = P\mathbf{y}$ must have a solution.

   (b) Show that the solution is unique.

33. If P is a projection matrix, its column space and row space are the same (ignoring the cosmetic difference between a row and a column). Why?

34. **(a)** Let P be a projection matrix and let $\mathbf{x}$ be a vector.

   i. If $\mathbf{x}$ is in the null space of P, prove that $\mathbf{x}$ is orthogonal to the column space of P.

   ii. If $\mathbf{x}$ is orthogonal to the column space of P, prove that $\mathbf{x}$ is in the null space of P.

   [Remark: These statements should be geometrically obvious. If you project a vector onto a plane and the vector is perpendicular to the plane, that vector goes to the zero vector.]

   (b) Find a 3×3 matrix whose null space is spanned by the vector $\mathbf{u} = \begin{bmatrix} 1 \\ 1 \\ 1 \end{bmatrix}$.

6.2 The Gram–Schmidt Algorithm and QR Factorization

6.2.1 DEFINITION

A set $\{\mathbf{f}_1, \mathbf{f}_2, \ldots, \mathbf{f}_k\}$ of vectors is said to be *orthogonal* if and only if the vectors are orthogonal *pairwise*, that is, if and only if $\mathbf{f}_i \cdot \mathbf{f}_j = 0$ when $i \neq j$. If, in addition, each $\mathbf{f}_i$ has norm (length) 1, then the set is called *orthonormal*.

6.2.2 EXAMPLES

• Any subset of the set of standard basis vectors $\mathbf{e}_1, \mathbf{e}_2, \ldots, \mathbf{e}_n$ is an orthonormal set in $\mathbf{R}^n$, for example, $\left\{ \begin{bmatrix} 0 \\ 1 \\ 0 \end{bmatrix}, \begin{bmatrix} 0 \\ 0 \\ 1 \end{bmatrix} \right\}$ in $\mathbf{R}^3$.

- The four vectors $\mathbf{f}_1 = \begin{bmatrix} 1 \\ 1 \\ 1 \\ -1 \end{bmatrix}$, $\mathbf{f}_2 = \begin{bmatrix} 1 \\ 0 \\ 1 \\ 2 \end{bmatrix}$, $\mathbf{f}_3 = \begin{bmatrix} -1 \\ 0 \\ 1 \\ 0 \end{bmatrix}$, $\mathbf{f}_4 = \begin{bmatrix} -1 \\ 3 \\ -1 \\ 1 \end{bmatrix}$ are orthogonal

in $\mathbf{R}^4$. A neat way to see this is to form the matrix

$$Q = \begin{bmatrix} 1 & 1 & -1 & -1 \\ 1 & 0 & 0 & 3 \\ 1 & 1 & 1 & -1 \\ -1 & 2 & 0 & 1 \end{bmatrix},$$

whose columns are $\mathbf{f}_1, \mathbf{f}_2, \mathbf{f}_3, \mathbf{f}_4$, and to compute

$$Q^T Q = \begin{bmatrix} 1 & 1 & 1 & -1 \\ 1 & 0 & 1 & 2 \\ -1 & 0 & 1 & 0 \\ -1 & 3 & -1 & 1 \end{bmatrix} \begin{bmatrix} 1 & 1 & -1 & -1 \\ 1 & 0 & 0 & 3 \\ 1 & 1 & 1 & -1 \\ -1 & 2 & 0 & 1 \end{bmatrix}$$

$$= \begin{bmatrix} 4 & 0 & 0 & 0 \\ 0 & 6 & 0 & 0 \\ 0 & 0 & 2 & 0 \\ 0 & 0 & 0 & 12 \end{bmatrix}.$$

Since $Q^T Q$ is a diagonal matrix, the set $\{\mathbf{f}_1, \mathbf{f}_2, \mathbf{f}_3, \mathbf{f}_4\}$ is orthogonal. ■

Reading Challenge 1 *How does knowing $Q^T Q$ is diagonal establish orthogonality of the vectors $\mathbf{f}_1$, $\mathbf{f}_2$, $\mathbf{f}_3$, and $\mathbf{f}_4$?*

Reading Challenge 2 *Suppose $\{\mathbf{f}_1, \mathbf{f}_2, \ldots, \mathbf{f}_k\}$ is an **orthonormal** set of vectors. Let Q be the matrix whose columns are these vectors. What is $Q^T Q$ and why?*

It is useful to be able to find orthogonal sets of vectors, in order to project vectors onto various subspaces, for example—see Section 6.1. We explain how to do so in this section, after recording a basic fact about orthogonal vectors. In $\mathbf{R}^2$ or $\mathbf{R}^3$, where vectors can be easily pictured, two orthogonal vectors are certainly linearly independent since they point in different directions. More generally, we have the following theorem.

Theorem 6.2.3 *An orthogonal set of nonzero vectors in $\mathbf{R}^n$ is a linearly independent set.*

Proof Let $\{\mathbf{f}_1, \mathbf{f}_2, \ldots, \mathbf{f}_k\}$ be an orthogonal set of vectors and suppose that

$$c_1 \mathbf{f}_1 + c_2 \mathbf{f}_2 + \cdots + c_k \mathbf{f}_k = \mathbf{0}. \tag{1}$$

We must prove that each of the scalars $c_1, c_2, \ldots, c_k$ is 0. Taking the dot product of each side of (1) with $\mathbf{f}_1$ gives

$$c_1(\mathbf{f}_1 \cdot \mathbf{f}_1) + c_2(\mathbf{f}_1 \cdot \mathbf{f}_2) + \cdots + c_k(\mathbf{f}_1 \cdot \mathbf{f}_k) = 0.$$

If $i \neq 1$, the vectors $\mathbf{f}_i$ and $\mathbf{f}_1$ are orthogonal, so $\mathbf{f}_1 \cdot \mathbf{f}_i = 0$. Thus $c_1 \|\mathbf{f}_1\|^2 = 0$ and, since $\mathbf{f}_1 \neq \mathbf{0}$, we must have $c_1 = 0$. In a similar way, it follows that all the other scalars c_i are 0 too. ∎

■ Linearly Independent to Orthogonal

Way back in Section 1.4, we showed how to turn two linearly independent vectors $\mathbf{v}$ and $\mathbf{u}$ into orthogonal vectors $\mathbf{e}$ and $\mathbf{f}$: take

$$\mathbf{e} = \mathbf{u}$$

$$\text{and} \quad \mathbf{f} = \mathbf{v} - \text{proj}_{\mathbf{u}} \mathbf{v}.$$

See Figure 1.23 and examine Problem 1.4.5 once again.

Reading Challenge 3   *Let $U = \text{sp}\{\mathbf{u}, \mathbf{v}\}$ be the span of vectors $\mathbf{u}$ and $\mathbf{v}$ and let $\mathbf{e}$ and $\mathbf{f}$ be the vectors just constructed.*

(a) $\mathbf{f} = \mathbf{v} + a\mathbf{u}$ *for some constant a. What is a?*

(b) *Show that $\text{sp}\{\mathbf{e}, \mathbf{f}\}$ is also U.*

In Section 1.4, we also discussed the projection of a vector $\mathbf{w}$ on the plane π spanned by two vectors $\mathbf{v}$ and $\mathbf{u}$. We can use this idea to turn $\mathbf{w}, \mathbf{v}$, and $\mathbf{u}$ into orthogonal vectors. Turn $\mathbf{v}$ and $\mathbf{u}$ into orthogonal $\mathbf{e}$ and $\mathbf{f}$ as before, then replace $\mathbf{w}$ by $\mathbf{g} = \mathbf{w} - \text{proj}_{\pi} \mathbf{w}$—see Figure 1.24:

$$\mathbf{e} = \mathbf{u},$$

$$\mathbf{f} = \mathbf{v} - \text{proj}_{\mathbf{u}} \mathbf{v},$$

$$\text{and} \quad \mathbf{g} = \mathbf{w} - \text{proj}_{\pi} \mathbf{w}.$$

Just as in Reading Challenge 3, the vectors $\mathbf{e}$ and $\mathbf{f}$ have the same span as $\mathbf{u}$ and $\mathbf{v}$ (the plane π), and one can also show that the three vectors $\mathbf{e}, \mathbf{f}, \mathbf{g}$ have the same span as $\mathbf{u}, \mathbf{v}, \mathbf{w}$. All this generalizes to a procedure for converting linearly independent sets of any size to orthogonal sets (of the same size). The procedure, called the "Gram–Schmidt algorithm," is described in Figure 6.4.

The key idea in the proof that the algorithm works as claimed is that if $\mathbf{f}_1, \mathbf{f}_2, \ldots,$ $\mathbf{f}_t$ are orthogonal vectors spanning a subspace U, then, for any $\mathbf{w}$, the projection of $\mathbf{w}$ on U,

$$\text{proj}_U \mathbf{w} = \frac{\mathbf{w} \cdot \mathbf{f}_1}{\mathbf{f}_1 \cdot \mathbf{f}_1} \mathbf{f}_1 + \frac{\mathbf{w} \cdot \mathbf{f}_2}{\mathbf{f}_2 \cdot \mathbf{f}_2} \mathbf{f}_2 + \cdots + \frac{\mathbf{w} \cdot \mathbf{f}_t}{\mathbf{f}_t \cdot \mathbf{f}_t} \mathbf{f}_t,$$

Gram–Schmidt Algorithm. The Gram–Schmidt algorithm defines a procedure for turning a set of k linearly independent vectors $\mathbf{u}_1, \ldots, \mathbf{u}_k$ into a set of k orthogonal vectors $\mathbf{f}_1, \ldots, \mathbf{f}_k$. The orthogonal vectors are constructed in a special way. Suppose $\{\mathbf{u}_1, \mathbf{u}_2, \ldots, \mathbf{u}_k\}$ is a set of linearly independent vectors in $\mathbf{R}^n$. Let

$U_1 = \mathrm{sp}\{\mathbf{u}_1\}$ be the subspace spanned by the vector $\mathbf{u}_1$,

$U_2 = \mathrm{sp}\{\mathbf{u}_1, \mathbf{u}_2\}$ be the subspace spanned by the vectors $\mathbf{u}_1$ and $\mathbf{u}_2$,

$U_3 = \mathrm{sp}\{\mathbf{u}_1, \mathbf{u}_2, \mathbf{u}_3\}$ be the subspace spanned by the vectors $\mathbf{u}_1$, $\mathbf{u}_2$, and $\mathbf{u}_3$,

$$\vdots$$

$U_k = \mathrm{sp}\{\mathbf{u}_1, \mathbf{u}_2, \ldots, \mathbf{u}_k\}$ be the subspace spanned by $\mathbf{u}_1, \mathbf{u}_2, \ldots, \mathbf{u}_k$.

Define vectors $\mathbf{f}_1, \mathbf{f}_2, \ldots, \mathbf{f}_k$ as follows:

$$\mathbf{f}_1 = \mathbf{u}_1$$

$$\mathbf{f}_2 = \mathbf{u}_2 - \frac{\mathbf{u}_2 \cdot \mathbf{f}_1}{\mathbf{f}_1 \cdot \mathbf{f}_1}\, \mathbf{f}_1$$

$$\mathbf{f}_3 = \mathbf{u}_3 - \frac{\mathbf{u}_3 \cdot \mathbf{f}_1}{\mathbf{f}_1 \cdot \mathbf{f}_1}\, \mathbf{f}_1 - \frac{\mathbf{u}_3 \cdot \mathbf{f}_2}{\mathbf{f}_2 \cdot \mathbf{f}_2}\, \mathbf{f}_2$$

$$\vdots$$

$$\mathbf{f}_k = \mathbf{u}_k - \frac{\mathbf{u}_k \cdot \mathbf{f}_1}{\mathbf{f}_1 \cdot \mathbf{f}_1}\, \mathbf{f}_1 - \frac{\mathbf{u}_k \cdot \mathbf{f}_2}{\mathbf{f}_2 \cdot \mathbf{f}_2}\, \mathbf{f}_2 - \cdots - \frac{\mathbf{u}_k \cdot \mathbf{f}_{k-1}}{\mathbf{f}_{k-1} \cdot \mathbf{f}_{k-1}}\, \mathbf{f}_{k-1}.$$

Then $\{\mathbf{f}_1, \mathbf{f}_2, \ldots, \mathbf{f}_k\}$ is an orthogonal set of vectors with the property that

$$\begin{aligned}
\mathrm{sp}\{\mathbf{f}_1\} &= U_1 \\
\mathrm{sp}\{\mathbf{f}_1, \mathbf{f}_2\} &= U_2 \\
\mathrm{sp}\{\mathbf{f}_1, \mathbf{f}_2, \mathbf{f}_3\} &= U_3 \\
&\vdots \\
\mathrm{sp}\{\mathbf{f}_1, \mathbf{f}_2, \ldots, \mathbf{f}_k\} &= U_k.
\end{aligned}$$

Figure 6.4

has the property that $\mathbf{w} - \mathrm{proj}_U\, \mathbf{w}$ is orthogonal to every vector in U—see Figure 6.5 and 6.1.21. Why does this process lead to an orthogonal set of vectors as described?

First, since $\mathbf{f}_1$ is a nonzero scalar multiple of $\mathbf{u}_1$, the first subspace U_1 is certainly spanned by $\mathbf{f}_1$. That $\mathrm{sp}\{\mathbf{f}_1, \mathbf{f}_2\} = U_2$ was exactly Reading Challenge 3, but here is another argument that has the advantage that it generalizes and enables us to prove that every U_i is spanned by $\mathbf{f}_1, \ldots, \mathbf{f}_i$ as claimed.

We know that $\mathbf{f}_2 = \mathbf{u}_2 - \mathrm{proj}_{U_1}\, \mathbf{u}_2$ is orthogonal to $\mathbf{f}_1$, by the definition of "projection," and it is nonzero because $\mathbf{u}_1$ and $\mathbf{u}_2$ are linearly independent.

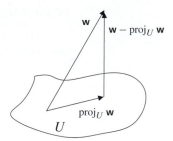

Figure 6.5 **w** is the projection on U if **w** $-$ proj_U **w** is orthogonal to U.

Reading Challenge 4   *Do you really understand this last statement?*

By Theorem 6.2.3, the orthogonal vectors $\mathbf{f}_1$ and $\mathbf{f}_2$ are linearly independent. Since they are contained in the two-dimensional space U_2, they must be a basis for this subspace. In particular, $U_2 = \text{sp}\{\mathbf{f}_1, \mathbf{f}_2\}$.

Similarly, $\mathbf{f}_3 = \mathbf{u}_3 - \text{proj}_{U_2} \mathbf{u}_3$ is orthogonal to all vectors in U_2 (definition of "projection") and also nonzero. In particular, it is orthogonal to both $\mathbf{f}_1$ and $\mathbf{f}_2$. The orthogonal vectors $\mathbf{f}_1, \mathbf{f}_2, \mathbf{f}_3$ are linearly independent in the three-dimensional space U_3, so they form a basis for U_3. In particular, $U_3 = \text{sp}\{\mathbf{f}_1, \mathbf{f}_2, \mathbf{f}_3\}$. Continuing in this way, we see that each U_i is spanned by the vectors $\mathbf{f}_1, \ldots, \mathbf{f}_i$ and the Gram–Schmidt Algorithm works as advertised.

6.2.4 EXAMPLE   Suppose $\mathbf{u}_1 = \begin{bmatrix} 1 \\ 0 \\ -1 \\ 1 \end{bmatrix}$, $\mathbf{u}_2 = \begin{bmatrix} 1 \\ -2 \\ 1 \\ -1 \end{bmatrix}$, and $\mathbf{u}_3 = \begin{bmatrix} 2 \\ 0 \\ -1 \\ 0 \end{bmatrix}$.

The Gram–Schmidt algorithm begins

$$\mathbf{f}_1 = \mathbf{u}_1 = \begin{bmatrix} 1 \\ 0 \\ -1 \\ 1 \end{bmatrix}$$

$$\mathbf{f}_2 = \mathbf{u}_2 - \frac{\mathbf{u}_2 \cdot \mathbf{f}_1}{\mathbf{f}_1 \cdot \mathbf{f}_1} \mathbf{f}_1 = \mathbf{u}_2 - \frac{-1}{3}\mathbf{f}_1 = \mathbf{u}_2 + \frac{1}{3}\mathbf{f}_1 = \begin{bmatrix} 1 \\ -2 \\ 1 \\ -1 \end{bmatrix} + \frac{1}{3}\begin{bmatrix} 1 \\ 0 \\ -1 \\ 1 \end{bmatrix}$$

$$= \begin{bmatrix} \frac{4}{3} \\ -2 \\ \frac{2}{3} \\ -\frac{2}{3} \end{bmatrix}.$$

When working by hand, the calculation of $\mathbf{f}_3$ will be easier if we multiply $\mathbf{f}_2$ by $\frac{3}{2}$,

obtaining a new $\mathbf{f}_2 = \begin{bmatrix} 2 \\ -3 \\ 1 \\ -1 \end{bmatrix}$ that is still orthogonal to $\mathbf{f}_1$. Continuing,

$$\mathbf{f}_3 = \mathbf{u}_3 - \frac{\mathbf{u}_3 \cdot \mathbf{f}_1}{\mathbf{f}_1 \cdot \mathbf{f}_1}\mathbf{f}_1 - \frac{\mathbf{u}_3 \cdot \mathbf{f}_2}{\mathbf{f}_2 \cdot \mathbf{f}_2}\mathbf{f}_2 = \mathbf{u}_3 - \frac{3}{3}\mathbf{f}_1 - \frac{3}{15}\mathbf{f}_2 = \mathbf{u}_3 - \mathbf{f}_1 - \frac{1}{5}\mathbf{f}_2$$

$$= \begin{bmatrix} 2 \\ 0 \\ -1 \\ 0 \end{bmatrix} - \begin{bmatrix} 1 \\ 0 \\ -1 \\ 1 \end{bmatrix} - \frac{1}{5}\begin{bmatrix} 2 \\ -3 \\ 1 \\ -1 \end{bmatrix} = \begin{bmatrix} \frac{3}{5} \\ \frac{3}{5} \\ -\frac{1}{5} \\ -\frac{4}{5} \end{bmatrix},$$

which we would probably replace by $5\mathbf{f}_3$, giving as our final set of orthogonal vectors

$$\mathbf{f}_1 = \begin{bmatrix} 1 \\ 0 \\ -1 \\ 1 \end{bmatrix}, \quad \mathbf{f}_2 = \begin{bmatrix} 2 \\ -3 \\ 1 \\ -1 \end{bmatrix}, \quad \mathbf{f}_3 = \begin{bmatrix} 3 \\ 3 \\ -1 \\ -4 \end{bmatrix}.$$ ∎

6.2.5 EXAMPLE   Suppose U is the subspace of $\mathbf{R}^5$ with basis

$$\mathbf{u}_1 = \begin{bmatrix} 0 \\ 1 \\ 1 \\ 0 \\ 0 \end{bmatrix}, \quad \mathbf{u}_2 = \begin{bmatrix} 1 \\ 0 \\ 1 \\ -1 \\ 1 \end{bmatrix}, \quad \mathbf{u}_3 = \begin{bmatrix} 1 \\ 4 \\ -1 \\ 1 \\ -1 \end{bmatrix}, \quad \mathbf{u}_4 = \begin{bmatrix} 2 \\ 0 \\ 2 \\ 3 \\ 1 \end{bmatrix},$$

and we wish to find an orthogonal basis for U. The Gram–Schmidt algorithm begins

$$\mathbf{f}_1 = \mathbf{u}_1 = \begin{bmatrix} 0 \\ 1 \\ 1 \\ 0 \\ 0 \end{bmatrix}$$

$$\mathbf{f}_2 = \mathbf{u}_2 - \frac{\mathbf{u}_2 \cdot \mathbf{f}_1}{\mathbf{f}_1 \cdot \mathbf{f}_1}\mathbf{f}_1 = \mathbf{u}_2 - \frac{1}{2}\mathbf{f}_1 = \begin{bmatrix} 1 \\ 0 \\ 1 \\ -1 \\ 1 \end{bmatrix} - \frac{1}{2}\begin{bmatrix} 0 \\ 1 \\ 1 \\ 0 \\ 0 \end{bmatrix} = \begin{bmatrix} 1 \\ -\frac{1}{2} \\ \frac{1}{2} \\ -1 \\ 1 \end{bmatrix},$$

and it would be natural at this point immediately to replace $\mathbf{f}_2$ by $2\mathbf{f}_2$, giving a new

$$\mathbf{f}_2 = \begin{bmatrix} 2 \\ -1 \\ 1 \\ -2 \\ 2 \end{bmatrix}.$$ Continuing with the algorithm,

$$\mathbf{f}_3 = \mathbf{u}_3 - \frac{\mathbf{u}_3 \cdot \mathbf{f}_1}{\mathbf{f}_1 \cdot \mathbf{f}_1} \mathbf{f}_1 - \frac{\mathbf{u}_3 \cdot \mathbf{f}_2}{\mathbf{f}_2 \cdot \mathbf{f}_2} \mathbf{f}_2$$

$$= \mathbf{u}_3 - \frac{3}{2}\mathbf{f}_1 - \frac{-7}{14}\mathbf{f}_2 = \mathbf{u}_3 - \frac{3}{2}\mathbf{f}_1 + \frac{1}{2}\mathbf{f}_2$$

$$= \begin{bmatrix} 1 \\ 4 \\ -1 \\ 1 \\ -1 \end{bmatrix} - \frac{3}{2}\begin{bmatrix} 0 \\ 1 \\ 1 \\ 0 \\ 0 \end{bmatrix} + \frac{1}{2}\begin{bmatrix} 2 \\ -1 \\ 1 \\ -2 \\ 2 \end{bmatrix} = \begin{bmatrix} 2 \\ 2 \\ -2 \\ 0 \\ 0 \end{bmatrix},$$

which, at least in applications, would be replaced by $\frac{1}{2}\mathbf{f}_3$, giving a new $\mathbf{f}_3 = \begin{bmatrix} 1 \\ 1 \\ -1 \\ 0 \\ 0 \end{bmatrix}$.

Finally,

$$\mathbf{f}_4 = \mathbf{u}_4 - \frac{\mathbf{u}_4 \cdot \mathbf{f}_1}{\mathbf{f}_1 \cdot \mathbf{f}_1} \mathbf{f}_1 - \frac{\mathbf{u}_4 \cdot \mathbf{f}_2}{\mathbf{f}_2 \cdot \mathbf{f}_2} \mathbf{f}_2 - \frac{\mathbf{u}_4 \cdot \mathbf{f}_3}{\mathbf{f}_3 \cdot \mathbf{f}_3} \mathbf{f}_3$$

$$= \mathbf{u}_4 - \frac{2}{2}\mathbf{f}_1 - \frac{1}{7}\mathbf{f}_2 - \frac{0}{3}\mathbf{f}_3 = \mathbf{u}_4 - \mathbf{f}_1 - \frac{1}{7}\mathbf{f}_2$$

$$= \begin{bmatrix} 2 \\ 0 \\ 2 \\ 3 \\ 1 \end{bmatrix} - \begin{bmatrix} 0 \\ 1 \\ 1 \\ 0 \\ 0 \end{bmatrix} - \frac{1}{7}\begin{bmatrix} 2 \\ -1 \\ 1 \\ -2 \\ 2 \end{bmatrix} = \begin{bmatrix} \frac{12}{7} \\ -\frac{6}{7} \\ \frac{6}{7} \\ \frac{23}{7} \\ \frac{5}{7} \end{bmatrix},$$

which might well be replaced by $\begin{bmatrix} 12 \\ -6 \\ 6 \\ 23 \\ 5 \end{bmatrix}.$ ■

■ QR Factorization

Let $A = \begin{bmatrix} 2 & 1 \\ 1 & 1 \\ -2 & 0 \end{bmatrix}$ and label the columns of this matrix $\mathbf{a}_1$ and $\mathbf{a}_2$, respectively. We apply the Gram–Schmidt algorithm to these vectors. Thus $\mathbf{f}_1 = \mathbf{a}_1$,

$$\mathbf{f}_2 = \mathbf{a}_2 - \frac{\mathbf{a}_2 \cdot \mathbf{f}_1}{\mathbf{f}_1 \cdot \mathbf{f}_1} \mathbf{f}_1 = \mathbf{a}_2 - \frac{3}{9}\mathbf{a}_1 = \mathbf{a}_2 - \frac{1}{3}\mathbf{a}_1 = \begin{bmatrix} 1 \\ 1 \\ 0 \end{bmatrix} - \frac{1}{3}\begin{bmatrix} 2 \\ 1 \\ -2 \end{bmatrix} = \begin{bmatrix} \frac{1}{3} \\ \frac{2}{3} \\ \frac{2}{3} \end{bmatrix},$$

and it is natural to replace $\mathbf{f}_2$ by $3\mathbf{f}_2 = \begin{bmatrix} 1 \\ 2 \\ 2 \end{bmatrix}$. So we obtain

$$\mathbf{f}_1 = \mathbf{a}_1 = \begin{bmatrix} 2 \\ 1 \\ -2 \end{bmatrix} \quad \text{and} \quad \mathbf{f}_2 = \begin{bmatrix} 1 \\ 2 \\ 2 \end{bmatrix}$$

as orthogonal vectors with $\mathrm{sp}\{\mathbf{f}_1\} = \mathrm{sp}\{\mathbf{a}_1\}$ and $\mathrm{sp}\{\mathbf{f}_1, \mathbf{f}_2\} = \mathrm{sp}\{\mathbf{a}_1, \mathbf{a}_2\}$. Dividing each of $\mathbf{f}_1$ and $\mathbf{f}_2$ by its length, we obtain orthonormal vectors

$$\mathbf{q}_1 = \frac{1}{\|\mathbf{f}_1\|}\mathbf{f}_1 = \tfrac{1}{3}\mathbf{f}_1 = \tfrac{1}{3}\begin{bmatrix} 2 \\ 1 \\ -2 \end{bmatrix} \quad \text{and} \quad \mathbf{q}_2 = \frac{1}{\|\mathbf{f}_2\|}\mathbf{f}_2 = \tfrac{1}{3}\mathbf{f}_2 = \tfrac{1}{3}\begin{bmatrix} 1 \\ 2 \\ 2 \end{bmatrix}.$$

Now we **solve for $\mathbf{a}_1$ and $\mathbf{a}_2$, first in terms of $\mathbf{f}_1$ and $\mathbf{f}_2$, and then in terms of $\mathbf{q}_1$ and $\mathbf{q}_2$.**

$$\mathbf{f}_1 = \mathbf{a}_1 \qquad\qquad \text{so} \qquad \mathbf{a}_1 = \mathbf{f}_1 = 3\mathbf{q}_1$$

$$\mathbf{f}_2 = 3(\mathbf{a}_2 - \tfrac{1}{3}\mathbf{a}_1) = 3\mathbf{a}_2 - \mathbf{f}_1 \qquad \text{so} \qquad \mathbf{a}_2 = \tfrac{1}{3}\mathbf{f}_2 + \tfrac{1}{3}\mathbf{f}_1 = \mathbf{q}_2 + \mathbf{q}_1.$$

We obtain

$$\begin{bmatrix} \mathbf{a}_1 & \mathbf{a}_2 \\ \downarrow & \downarrow \end{bmatrix} = \begin{bmatrix} \mathbf{q}_1 & \mathbf{q}_2 \\ \downarrow & \downarrow \end{bmatrix}\begin{bmatrix} 3 & 1 \\ 0 & 1 \end{bmatrix}$$

$$\begin{bmatrix} 2 & 1 \\ 1 & 1 \\ -2 & 0 \end{bmatrix} = \begin{bmatrix} \frac{2}{3} & \frac{1}{3} \\ \frac{1}{3} & \frac{2}{3} \\ -\frac{2}{3} & \frac{2}{3} \end{bmatrix}\begin{bmatrix} 3 & 1 \\ 0 & 1 \end{bmatrix}.$$

The matrix A has been factored $A = QR$ where $Q = \begin{bmatrix} \frac{2}{3} & \frac{1}{3} \\ \frac{1}{3} & \frac{2}{3} \\ -\frac{2}{3} & \frac{2}{3} \end{bmatrix}$ is a matrix the same size as A, but with orthonormal columns, and $R = \begin{bmatrix} 3 & 1 \\ 0 & 1 \end{bmatrix}$ is a square upper triangular matrix.

The following theorem, just illustrated, is a consequence of the Gram–Schmidt algorithm.

Theorem 6.2.6   *Let A be an m × n matrix with linearly independent columns. Then A can be factored A = QR where Q is an m × n matrix with orthonormal columns and R is a square n × n upper triangular matrix.*

Reading Challenge 5   *Suppose the m × n matrix A has linearly independent columns and we factor A = QR as in Theorem 6.2.6.*

(a) *R also has linearly independent columns. Why? [Hint: Use the definition of linear independence.]*

(b) *R is invertible. Why?*

6.2.7 PROBLEM   Let $A = \begin{bmatrix} 0 & 1 & 1 \\ 1 & 0 & 4 \\ 1 & 1 & -1 \\ 0 & -1 & 1 \\ 0 & 1 & -1 \end{bmatrix}$. Then use the results of **6.2.5** to find a QR factorization.

Solution.   The columns of A are the vectors $\mathbf{a}_1 = \begin{bmatrix} 0 \\ 1 \\ 1 \\ 0 \\ 0 \end{bmatrix}$, $\mathbf{a}_2 = \begin{bmatrix} 1 \\ 0 \\ 1 \\ -1 \\ 1 \end{bmatrix}$, and

$\mathbf{a}_3 = \begin{bmatrix} 1 \\ 4 \\ -1 \\ 1 \\ -1 \end{bmatrix}$. In **6.2.5**, we used the Gram–Schmidt algorithm to produce the orthogonal vectors

$$\mathbf{f}_1 = \begin{bmatrix} 0 \\ 1 \\ 1 \\ 0 \\ 0 \end{bmatrix}, \quad \mathbf{f}_2 = \begin{bmatrix} 2 \\ -1 \\ 1 \\ -2 \\ 2 \end{bmatrix}, \quad \mathbf{f}_3 = \begin{bmatrix} 1 \\ 1 \\ -1 \\ 0 \\ 0 \end{bmatrix}.$$

Dividing each of these by their length, we obtain the orthonormal vectors

$$\mathbf{q}_1 = \frac{1}{\|\mathbf{f}_1\|}\mathbf{f}_1 = \frac{1}{\sqrt{2}}\mathbf{f}_1 = \frac{1}{\sqrt{2}}\begin{bmatrix} 0 \\ 1 \\ 1 \\ 0 \\ 0 \end{bmatrix},$$

$$q_2 = \frac{1}{\|f_2\|} f_2 = \frac{1}{\sqrt{14}} f_2 = \frac{1}{\sqrt{14}} \begin{bmatrix} 2 \\ -1 \\ 1 \\ -2 \\ 2 \end{bmatrix},$$

$$q_3 = \frac{1}{\|f_3\|} f_3 = \frac{1}{\sqrt{3}} f_3 = \frac{1}{\sqrt{3}} \begin{bmatrix} 1 \\ 1 \\ -1 \\ 0 \\ 0 \end{bmatrix}.$$

As before, we **solve for the a_i in terms of f_i and then in terms of the q_i**. In **6.2.5**, we found

$$f_1 = a_1 \qquad \text{so that} \qquad a_1 = f_1 = \sqrt{2}\,q_1,$$

$$f_2 = 2(a_2 - \tfrac{1}{2}f_1) \qquad \text{so that} \qquad a_2 = \tfrac{1}{2}f_1 + \tfrac{1}{2}f_2 = \tfrac{\sqrt{2}}{2}\,q_1 + \tfrac{\sqrt{14}}{2}\,q_2,$$

$$f_3 = \tfrac{1}{2}(a_3 - \tfrac{3}{2}f_1 + \tfrac{1}{2}f_2) \qquad \text{so that} \qquad a_3 = 2f_3 + \tfrac{3}{2}f_1 - \tfrac{1}{2}f_2$$

$$= \tfrac{3\sqrt{2}}{2}\,q_1 - \tfrac{\sqrt{14}}{2}\,q_2 + 2\sqrt{3}\,q_3.$$

Thus

$$\begin{bmatrix} a_1 & a_2 & a_3 \\ \downarrow & \downarrow & \downarrow \end{bmatrix} = \begin{bmatrix} q_1 & q_2 & q_3 \\ \downarrow & \downarrow & \downarrow \end{bmatrix} \begin{bmatrix} \sqrt{2} & \frac{\sqrt{2}}{2} & \frac{3\sqrt{2}}{2} \\ 0 & \frac{\sqrt{14}}{2} & -\frac{\sqrt{14}}{2} \\ 0 & 0 & 2\sqrt{3} \end{bmatrix}$$

$$\begin{bmatrix} 0 & 1 & 1 \\ 1 & 0 & 4 \\ 1 & 1 & -1 \\ 0 & -1 & 1 \\ 0 & 1 & -1 \end{bmatrix} = \begin{bmatrix} 0 & \frac{2}{\sqrt{14}} & \frac{1}{\sqrt{3}} \\ \frac{1}{\sqrt{2}} & -\frac{1}{\sqrt{14}} & \frac{1}{\sqrt{3}} \\ \frac{1}{\sqrt{2}} & \frac{1}{\sqrt{14}} & -\frac{1}{\sqrt{3}} \\ 0 & -\frac{2}{\sqrt{14}} & 0 \\ 0 & \frac{2}{\sqrt{14}} & 0 \end{bmatrix} \begin{bmatrix} \sqrt{2} & \frac{\sqrt{2}}{2} & \frac{3\sqrt{2}}{2} \\ 0 & \frac{\sqrt{14}}{2} & -\frac{\sqrt{14}}{2} \\ 0 & 0 & 2\sqrt{3} \end{bmatrix}. \quad \blacksquare$$

■ Orthogonal Matrices

This section has focused attention on matrices with orthonormal columns. Such matrices have some nice properties. Since the (i, j) entry of $Q^T Q$ is the dot product of row i of Q^T and column j of Q, that is, the dot product of columns i and j of Q, and since these columns are orthonormal, the dot product is 0 if $i \neq j$ and 1 otherwise. In other words, $Q^T Q = I$. Conversely, if $Q^T Q = I$, its columns are orthonormal.

6.2.8 A matrix Q has orthonormal columns if and only if $Q^T Q = I$.

Recall that the dot product of vectors $\mathbf{x}$ and $\mathbf{y}$ is just the matrix product $\mathbf{x}^T\mathbf{y}$. If Q has orthonormal columns and $\mathbf{x}$ and $\mathbf{y}$ are vectors, the dot product of $Q\mathbf{x}$ and $Q\mathbf{y}$ is

$$Q\mathbf{x} \cdot Q\mathbf{y} = (Q\mathbf{x})^T (Q\mathbf{y}) = \mathbf{x}^T Q^T Q\mathbf{y} = \mathbf{x}^T\mathbf{y} = \mathbf{x} \cdot \mathbf{y}$$

since $Q^T Q = I$. This condition is usually summarized by saying that multiplication by an orthogonal matrix "preserves" the dot product. In the special case $\mathbf{y} = \mathbf{x}$, we get $Q\mathbf{x} \cdot Q\mathbf{x} = \mathbf{x} \cdot \mathbf{x}$. The right side is $\|\mathbf{x}\|^2$ and the left side is $\|Q\mathbf{x}\|^2$, so $\|Q\mathbf{x}\| = \|\mathbf{x}\|$. Multiplication by Q also "preserves" the norm (or length) of a vector.

6.2.9 Multiplication by a matrix with orthonormal columns preserves the dot product and the norms (lengths) of vectors.

Reading Challenge 6 *This fact implies that multiplication by an orthogonal matrix preserves the angle between nonzero vectors. Why?*

Now we turn our attention to matrices with orthonormal columns that are also **square**. Such matrices have a special name.

6.2.10 DEFINITION An *orthogonal matrix* is a square matrix with orthonormal columns.

Since an orthogonal matrix is square, the equation $Q^T Q = I$ says that Q is invertible (Corollary 4.4.14) with $Q^{-1} = Q^T$. In particular, $QQ^T = I$ too, and this says that the rows of Q are also orthonormal. (Is this obvious?)

6.2.11 EXAMPLES

- For any angle θ, the *rotation matrix* $Q = \begin{bmatrix} \cos\theta & -\sin\theta \\ \sin\theta & \cos\theta \end{bmatrix}$ is orthogonal. Its columns are orthogonal and have length 1 because $\cos^2\theta + \sin^2\theta = 1$. For example, with $\theta = \frac{\pi}{4}$, we see that $\begin{bmatrix} \frac{1}{\sqrt{2}} & -\frac{1}{\sqrt{2}} \\ \frac{1}{\sqrt{2}} & \frac{1}{\sqrt{2}} \end{bmatrix}$ is orthogonal.

The name "rotation matrix" derives from the fact that if $\mathbf{x}$ is pictured in $\mathbf{R}^2$ as an arrow based at the origin, then multiplication by Q rotates this arrow through the angle θ. See Figure 6.6 and also Section 5.1.

- Any permutation matrix is orthogonal because its columns are just the standard basis vectors $\mathbf{e}_1, \mathbf{e}_2, \dots, \mathbf{e}_n$ rearranged. For example, $P = \begin{bmatrix} 0 & 0 & 1 \\ 1 & 0 & 0 \\ 0 & 1 & 0 \end{bmatrix}$ is an orthogonal matrix.

Figure 6.6 Multiplication by $Q = \begin{bmatrix} \cos\theta & -\sin\theta \\ \sin\theta & \cos\theta \end{bmatrix}$ rotates $\mathbf{x}$ through the angle θ.

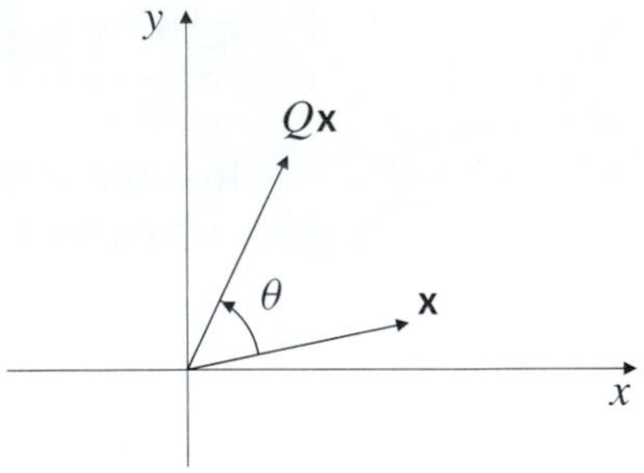

6.2.12 EXAMPLE Let $Q = \begin{bmatrix} \cos\theta & -\sin\theta \\ \sin\theta & \cos\theta \end{bmatrix}$, $\mathbf{x} = \begin{bmatrix} x \\ y \end{bmatrix}$, $\mathbf{x}' = \begin{bmatrix} x' \\ y' \end{bmatrix}$, and suppose $\begin{bmatrix} x' \\ y' \end{bmatrix} = Q\begin{bmatrix} x \\ y \end{bmatrix}$. Thus

$$x' = x\cos\theta - y\sin\theta$$

$$y' = x\sin\theta + y\cos\theta.$$

The norm (or length) of $Q\mathbf{x}$ is $x'^2 + y'^2 = (x\cos\theta - y\sin\theta)^2 + (x\sin\theta + y\cos\theta)^2$

$$= x^2\cos^2\theta - 2xy\cos\theta\sin\theta + y^2\sin^2\theta + x^2\sin^2\theta + 2xy\cos\theta\sin\theta + y^2\cos^2\theta$$

$$= x^2(\cos^2\theta + \sin^2\theta) + y^2(\cos^2\theta + \sin^2\theta) = x^2 + y^2 = \|\mathbf{x}\|^2.$$

Consistent with 6.2.9, $\|Q\mathbf{x}\| = \|\mathbf{x}\|$. ∎

6.2.13 PROBLEM Let $\mathbf{w}$ be a vector in $\mathbf{R}^n$ of length 1. Show that the *Householder matrix* $Q = I - 2\mathbf{w}\mathbf{w}^T$ is symmetric and orthogonal.

Solution. To show that Q is symmetric, we must show that $Q^T = Q$. This follows from $Q^T = I^T - 2(\mathbf{w}\mathbf{w}^T)^T = I - 2(\mathbf{w}^T)^T\mathbf{w}^T = I - 2\mathbf{w}\mathbf{w}^T = Q$.

Since Q is symmetric, it is square, so to establish orthogonality, it is sufficient to show that $Q^T Q = I$. We compute

$$Q^T Q = (I - 2\mathbf{w}\mathbf{w}^T)(I - 2\mathbf{w}\mathbf{w}^T) = I - 2\mathbf{w}\mathbf{w}^T - 2\mathbf{w}\mathbf{w}^T + 4\mathbf{w}\mathbf{w}^T\mathbf{w}\mathbf{w}^T.$$

Now $\mathbf{w}^T\mathbf{w} = \mathbf{w}\cdot\mathbf{w} = \|\mathbf{w}\|^2 = 1$, so $Q^T Q = I - 4\mathbf{w}\mathbf{w}^T + 4\mathbf{w}\mathbf{w}^T = I$ and Q is indeed orthogonal.

6.2.14 PROBLEM Let $\mathbf{w} = \frac{1}{\sqrt{3}}\begin{bmatrix} 1 \\ -1 \\ 1 \end{bmatrix}$. Compute the Householder matrix $Q = I - 2\mathbf{w}\mathbf{w}^T$ and verify that its rows are orthonormal.

Solution. The matrix $Q = I - \frac{2}{3} \begin{bmatrix} 1 \\ -1 \\ 1 \end{bmatrix} \begin{bmatrix} 1 & -1 & 1 \end{bmatrix}$

$$= \begin{bmatrix} 1 & 0 & 0 \\ 0 & 1 & 0 \\ 0 & 0 & 1 \end{bmatrix} - \frac{2}{3} \begin{bmatrix} 1 & -1 & 1 \\ -1 & 1 & -1 \\ 1 & -1 & 1 \end{bmatrix} = \frac{1}{3} \begin{bmatrix} 1 & 2 & -2 \\ 2 & 1 & 2 \\ -2 & 2 & 1 \end{bmatrix}.$$

The rows of Q are orthogonal since

$$QQ^T = \frac{1}{9} \begin{bmatrix} 1 & 2 & -2 \\ 2 & 1 & 2 \\ -2 & 2 & 1 \end{bmatrix} \begin{bmatrix} 1 & 2 & -2 \\ 2 & 1 & 2 \\ -2 & 2 & 1 \end{bmatrix} = \begin{bmatrix} 1 & 0 & 0 \\ 0 & 1 & 0 \\ 0 & 0 & 1 \end{bmatrix}.$$

■ Least Squares Revisited

In Section 6.1, we introduced the idea of *least squares approximation*. The idea was to find the "best solution" to a linear system $A\mathbf{x} = \mathbf{b}$ when, in fact, there weren't any solutions. For us, "best" meant the vector $\mathbf{x}^+$ which minimizes $\|\mathbf{b} - A\mathbf{x}\|$. Since this number is (the square root of) a sum of squares, we say that $\mathbf{x}^+$ is best "in the sense of least squares."

We had an explicit formula for $\mathbf{x}^+$ when the matrix A has linearly independent columns, namely,

$$\mathbf{x}^+ = (A^T A)^{-1} A^T \mathbf{b}.$$

(See Theorem 6.1.8.)

Of course, if A has linearly independent columns, we now know that $A = QR$ can be factored as the product of a matrix Q with orthonormal columns and an upper triangular matrix R. In this case, $A^T A = R^T Q^T Q R = R^T R$ because $Q^T Q = I$, and so $\mathbf{x}^+ = (R^T R)^{-1} R^T Q^T \mathbf{b}$.

There is a world of advantage in replacing $(Q^T Q)^{-1}$ by $(R^T R)^{-1}$. The matrix R is invertible—see Reading Challenge 5(*b*)—and so $(R^T R)^{-1} = R^{-1}(R^T)^{-1}$. We obtain

$$\mathbf{x}^+ = (R^T R)^{-1} R^T Q^T \mathbf{b} = R^{-1}(R^T)^{-1} R^T Q^T \mathbf{b} = R^{-1} Q^T \mathbf{b}.$$

Rewriting this as

$$R\mathbf{x}^+ = Q^T \mathbf{b},$$

we see that the desired vector $\mathbf{x}^+$ is nothing but the solution to a triangular system of equations (which is easily solved by back substitution).

6.2.15 EXAMPLE

In Section 6.1, in **6.1.9**, we showed that the least squares solution to

$$x_1 + x_2 = 2$$
$$x_1 - x_2 = 0$$
$$2x_1 + x_2 = -4$$

is $\mathbf{x}^+ = \begin{bmatrix} -1 \\ 0 \end{bmatrix}$. This is $A \begin{bmatrix} x_1 \\ x_2 \end{bmatrix} = \mathbf{b}$, with $A = \begin{bmatrix} 1 & 1 \\ 1 & -1 \\ 2 & 1 \end{bmatrix}$ and $\mathbf{b} = \begin{bmatrix} 2 \\ 0 \\ -4 \end{bmatrix}$. The QR factorization of A is

$$A = \begin{bmatrix} \frac{1}{\sqrt{6}} & \frac{2}{\sqrt{21}} \\ \frac{1}{\sqrt{6}} & -\frac{4}{\sqrt{21}} \\ \frac{2}{\sqrt{6}} & \frac{1}{\sqrt{21}} \end{bmatrix} \begin{bmatrix} \sqrt{6} & \frac{\sqrt{6}}{3} \\ 0 & \frac{\sqrt{21}}{3} \end{bmatrix},$$

so the desired vector $\mathbf{x}^+$ is the solution to $R\mathbf{x} = Q^T\mathbf{b}$. We have

$$R\mathbf{x} = \begin{bmatrix} \sqrt{6} & \frac{\sqrt{6}}{3} \\ 0 & \frac{\sqrt{21}}{3} \end{bmatrix} \begin{bmatrix} x_1 \\ x_2 \end{bmatrix}$$

and

$$Q^T\mathbf{b} = \begin{bmatrix} \frac{1}{\sqrt{6}} & \frac{1}{\sqrt{6}} & \frac{2}{\sqrt{6}} \\ \frac{2}{\sqrt{21}} & -\frac{4}{\sqrt{21}} & \frac{1}{\sqrt{21}} \end{bmatrix} \begin{bmatrix} 2 \\ 0 \\ -4 \end{bmatrix} = \begin{bmatrix} -\frac{6}{\sqrt{6}} \\ 0 \end{bmatrix}.$$

The augmented matrix for the triangular system $R\mathbf{x} = Q^T\mathbf{b}$ is

$$\begin{bmatrix} \sqrt{6} & \frac{\sqrt{6}}{3} & \bigg| & -\frac{6}{\sqrt{6}} \\ 0 & \frac{\sqrt{21}}{3} & \bigg| & 0 \end{bmatrix}.$$

Back substitution gives

$$\frac{\sqrt{21}}{3}x_2 = 0 \qquad \text{so} \qquad x_2 = 0$$

$$\sqrt{6}x_1 + \frac{\sqrt{6}}{3}x_2 = -\frac{6}{\sqrt{6}} \qquad \text{so} \qquad x_1 = -\frac{6}{\sqrt{6}\sqrt{6}} = -1.$$

We obtain $\mathbf{x}^+ = \begin{bmatrix} -1 \\ 0 \end{bmatrix}$ as before. ∎

Reading Challenge 7   *Verify the QR factorization of A in* **6.2.15**.

Answers to Reading Challenges

1. The (i, j) entry of $Q^T Q$ is the dot product of $\mathbf{f}_i$ and $\mathbf{f}_j$. If $Q^T Q$ is diagonal, this dot product is 0 for $i \neq j$. Thus $\mathbf{f}_i$ and $\mathbf{f}_j$ are orthogonal.

2. The matrix $Q^T Q = I$ is the identity matrix. The (i, j) entry, which is $\mathbf{f}_i \cdot \mathbf{f}_j$, is 0 when $i \neq j$ since $\mathbf{f}_i$ and $\mathbf{f}_j$ are orthogonal. It is 1 when $i = j$ since $\mathbf{f}_i \cdot \mathbf{f}_i = \|\mathbf{f}_i\|^2 = 1$.

3. (a) $\mathbf{f} = \mathbf{v} - \mathrm{proj}_\mathbf{u}\, \mathbf{v} = \mathbf{f} - \frac{\mathbf{v} \cdot \mathbf{u}}{\mathbf{u} \cdot \mathbf{u}}\, \mathbf{u}$, so $a = -\frac{\mathbf{v} \cdot \mathbf{u}}{\mathbf{u} \cdot \mathbf{u}}$.

   (b) Any vector in sp$\{\mathbf{e}, \mathbf{f}\}$ has the form $\mathbf{x} = c\mathbf{e} + d\mathbf{f}$. This is $c\mathbf{u} + d(\mathbf{v} + a\mathbf{u}) = (c + da)\mathbf{u} + e\mathbf{v}$, a linear combination of $\mathbf{u}$ and $\mathbf{v}$, hence a vector of U. Conversely, if $\mathbf{x}$ is in U, then $\mathbf{x} = c\mathbf{u} + d\mathbf{v} = c\mathbf{e} + d(\mathbf{f} - a\mathbf{u}) = c\mathbf{e} + d(\mathbf{f} - a\mathbf{e}) = (c - da)\mathbf{e} + d\mathbf{f}$ is a linear combination of $\mathbf{e}$ and $\mathbf{f}$. So $U = $ sp$\{\mathbf{e}, \mathbf{f}\}$.

4. $\mathrm{proj}_{U_1}\, \mathbf{u}_2 = c\mathbf{u}_1$ is a scalar multiple of $\mathbf{u}_1$. If $\mathbf{f}_2 = \mathbf{0}$, then $\mathbf{u}_2 - c\mathbf{u}_1 = \mathbf{0}$, which says that $\mathbf{u}_1$ and $\mathbf{u}_2$ are linearly dependent.

5. (a) If a linear combination of the columns of R is $\mathbf{0}$, then $R\mathbf{x} = \mathbf{0}$ for some vector $\mathbf{x}$. Thus $QR\mathbf{x} = \mathbf{0}$, so $A\mathbf{x} = \mathbf{0}$. Since the columns of A are linearly independent, $\mathbf{x} = \mathbf{0}$. Thus the columns of R are linearly independent.

   (b) The columns of R are linearly independent, so col rank $R = n = $ rank R.

6. $\cos \theta = \frac{\mathbf{u} \cdot \mathbf{v}}{\|\mathbf{u}\|\|\mathbf{v}\|}$, and multiplication by Q preserves $\mathbf{u} \cdot \mathbf{v}$, $\|\mathbf{u}\|$, and $\|\mathbf{v}\|$.

7. The columns of A are the vectors $\mathbf{a}_1 = \begin{bmatrix} 1 \\ 1 \\ 2 \end{bmatrix}$ and $\mathbf{a}_2 = \begin{bmatrix} 1 \\ -1 \\ 1 \end{bmatrix}$. We apply the Gram–Schmidt algorithm to these.

$$\mathbf{f}_1 = \mathbf{a}_1 = \begin{bmatrix} 1 \\ 1 \\ 2 \end{bmatrix}$$

$$\mathbf{f}_2 = \mathbf{a}_2 - \frac{\mathbf{a}_2 \cdot \mathbf{f}_1}{\mathbf{f}_1 \cdot \mathbf{f}_1}\, \mathbf{f}_1 = \mathbf{a}_2 - \tfrac{2}{6}\mathbf{f}_1 = \begin{bmatrix} 1 \\ -1 \\ 1 \end{bmatrix} - \tfrac{1}{3}\begin{bmatrix} 1 \\ 1 \\ 2 \end{bmatrix} = \begin{bmatrix} \frac{2}{3} \\ -\frac{4}{3} \\ \frac{1}{3} \end{bmatrix},$$

which we replace with $3\mathbf{f}_2$ to ease further calculations with this vector. Thus we take $\mathbf{f}_2 = \begin{bmatrix} 2 \\ -4 \\ 1 \end{bmatrix}$. Dividing $\mathbf{f}_1$ and $\mathbf{f}_2$ by their lengths, we obtain the orthonormal vectors

$$\mathbf{q}_1 = \frac{1}{\|\mathbf{f}_1\|}\mathbf{f}_1 = \frac{1}{\sqrt{6}}\mathbf{f}_1 = \begin{bmatrix} \frac{1}{\sqrt{6}} \\ \frac{1}{\sqrt{6}} \\ \frac{2}{\sqrt{6}} \end{bmatrix} \text{ and } \mathbf{q}_2 = \frac{1}{\|\mathbf{f}_2\|}\mathbf{f}_2 = \frac{1}{\sqrt{21}}\mathbf{f}_2 = \begin{bmatrix} \frac{2}{\sqrt{21}} \\ -\frac{41}{\sqrt{21}} \\ \frac{1}{\sqrt{21}} \end{bmatrix}.$$

Now

$$\mathbf{f}_1 = \mathbf{a}_1 \qquad \text{so that} \qquad \mathbf{a}_1 = \mathbf{f}_1 = \sqrt{6}\mathbf{q}_1$$

$$\mathbf{f}_2 = 3(\mathbf{a}_2 - \tfrac{1}{3}\mathbf{f}_1) \qquad \text{so that} \qquad \mathbf{a}_2 = \tfrac{1}{3}\mathbf{f}_1 + \tfrac{1}{3}\mathbf{f}_2 = \tfrac{\sqrt{6}}{3}\mathbf{q}_1 + \tfrac{\sqrt{21}}{3}\mathbf{q}_2.$$

$$\text{Thus } Q = \begin{bmatrix} \mathbf{q}_1 & \mathbf{q}_2 \\ \downarrow & \downarrow \end{bmatrix} = \begin{bmatrix} \frac{1}{\sqrt{6}} & \frac{2}{\sqrt{21}} \\ \frac{1}{\sqrt{6}} & -\frac{4}{\sqrt{21}} \\ \frac{2}{\sqrt{6}} & \frac{1}{\sqrt{21}} \end{bmatrix} \text{ and } R = \begin{bmatrix} \sqrt{6} & \frac{\sqrt{6}}{3} \\ 0 & \frac{\sqrt{21}}{3} \end{bmatrix}.$$

True/False Questions

Decide, with as little calculation as possible, whether each of the following statements is true or false and explain your answer whenever you say "false." (Answers can be found in the back of the book.)

1. An orthogonal set of nonzero vectors is a linearly independent set.

2. The Gram–Schmidt algorithm would convert the vectors $\mathbf{u}_1 = \begin{bmatrix} 1 \\ 2 \\ 3 \\ 4 \end{bmatrix}$, $\mathbf{u}_2 = \begin{bmatrix} 5 \\ 6 \\ 7 \\ 8 \end{bmatrix}$, $\mathbf{u}_3 = \begin{bmatrix} 2 \\ 4 \\ 6 \\ 8 \end{bmatrix}$ into three orthogonal vectors.

3. A vector space has an orthogonal basis if and only if it has an orthonormal basis.

4. A permutation matrix is an orthogonal matrix.

5. The matrix $A = \begin{bmatrix} 1 & -2 \\ 2 & 1 \end{bmatrix}$ is orthogonal.

6. $\begin{bmatrix} \frac{1}{\sqrt{2}} & \frac{1}{\sqrt{3}} \\ 0 & \frac{1}{\sqrt{3}} \\ -\frac{1}{\sqrt{2}} & \frac{1}{\sqrt{3}} \end{bmatrix}$ is an orthogonal matrix.

7. The inverse of $\begin{bmatrix} \cos\theta & -\sin\theta \\ \sin\theta & \cos\theta \end{bmatrix}$ is $\begin{bmatrix} \cos(-\theta) & -\sin(-\theta) \\ \sin(-\theta) & \cos(-\theta) \end{bmatrix}$.

8. If $\mathbf{q}_1, \ldots, \mathbf{q}_k$ are orthonormal vectors in $\mathbf{R}^n$, then these can be extended to a basis of $\mathbf{R}^n$.

9. If $A = QR$ is a QR factorization the matrix A, then Q is the same size as A and R is square.

10. If Q is an orthogonal matrix, then $\det Q = 1$.

11. If A is a matrix and $\det A = 1$, then A is orthogonal.

12. If AA^T is the identity matrix, then A has orthonormal rows.

13. The matrix $\begin{bmatrix} 1 & 0 \\ 0 & -1 \end{bmatrix}$ is a rotation matrix.

Exercises

*Solutions to exercises marked [BB] can be found in the **B**ack of the **B**ook.*

1. Apply the Gram–Schmidt algorithm to each of the following sets of vectors.

   (a) [BB] $\mathbf{u}_1 = \begin{bmatrix} 0 \\ 1 \\ 3 \end{bmatrix}$, $\mathbf{u}_2 = \begin{bmatrix} -1 \\ 0 \\ 1 \end{bmatrix}$

   (b) $\mathbf{u}_1 = \begin{bmatrix} 1 \\ -1 \\ 1 \end{bmatrix}$, $\mathbf{u}_2 = \begin{bmatrix} 1 \\ 0 \\ 1 \end{bmatrix}$, $\mathbf{u}_3 = \begin{bmatrix} 1 \\ 1 \\ 2 \end{bmatrix}$

   (c) $\mathbf{u}_1 = \begin{bmatrix} 1 \\ 0 \\ 1 \\ 0 \end{bmatrix}$, $\mathbf{u}_2 = \begin{bmatrix} 0 \\ 0 \\ 1 \\ 1 \end{bmatrix}$, $\mathbf{u}_3 = \begin{bmatrix} 1 \\ 0 \\ 1 \\ 1 \end{bmatrix}$

   (d) $\mathbf{u}_1 = \begin{bmatrix} 1 \\ 0 \\ 2 \\ 1 \end{bmatrix}$, $\mathbf{u}_2 = \begin{bmatrix} 0 \\ 1 \\ 0 \\ -3 \end{bmatrix}$, $\mathbf{u}_3 = \begin{bmatrix} 2 \\ 1 \\ -3 \\ 1 \end{bmatrix}$.

2. Let $U = \text{sp}\{\begin{bmatrix} 1 \\ 0 \\ 1 \\ 0 \end{bmatrix}, \begin{bmatrix} 1 \\ 1 \\ 1 \\ 0 \end{bmatrix}, \begin{bmatrix} 1 \\ 1 \\ 0 \\ 0 \end{bmatrix}\}$ and $\mathbf{w} = \begin{bmatrix} 2 \\ 0 \\ -1 \\ 3 \end{bmatrix}$.

   (a) Apply the Gram–Schmidt algorithm to exhibit an orthogonal basis for U.

   (b) Find the vector in U that is closest to $\mathbf{w}$.

3. Find a QR factorization of each of the following matrices.

   (a) [BB] $A = \begin{bmatrix} 1 & 4 \\ 1 & 0 \end{bmatrix}$ (b) $A = \begin{bmatrix} 1 & 2 \\ 3 & 4 \end{bmatrix}$

   (c) $A = \begin{bmatrix} 2 & -4 \\ 1 & 3 \end{bmatrix}$ (d) $A = \begin{bmatrix} -1 & 1 \\ 0 & 1 \\ 1 & -2 \end{bmatrix}$

   (e) [BB] $A = \begin{bmatrix} 1 & -1 & 2 \\ 0 & 1 & 1 \\ 1 & 1 & 1 \end{bmatrix}$

   (f) $A = \begin{bmatrix} 0 & 0 & 1 \\ 0 & 2 & -1 \\ 1 & 4 & 2 \end{bmatrix}$

   (g) $A = \begin{bmatrix} 1 & 3 & 1 \\ -1 & 1 & 0 \\ 0 & 2 & 1 \end{bmatrix}$

   (h) $A = \begin{bmatrix} 1 & 1 & 2 \\ 0 & -2 & 0 \\ -1 & 1 & -1 \\ 1 & -1 & 0 \end{bmatrix}$ (You may use the results of **6.2.4**.)

   (i) $A = \begin{bmatrix} 3 & -1 & 6 & -2 \\ 2 & -6 & 8 & 16 \\ 1 & 5 & -2 & -2 \\ 1 & -3 & 14 & 10 \\ -1 & 3 & 6 & 10 \end{bmatrix}$

   (j) $A = \begin{bmatrix} -7 & 3 & 22 & 25 \\ -2 & 6 & -4 & 2 \\ -2 & -12 & 5 & 2 \\ -4 & 12 & -8 & 13 \\ -2 & 6 & 23 & 2 \\ -2 & 6 & -4 & 2 \end{bmatrix}$.

4. In each case, find a QR factorization of A and use this to find A^{-1}.

   (a) [BB] $A = \begin{bmatrix} 1 & 4 \\ 1 & 0 \end{bmatrix}$ [See Exercise 3(a).]

   (b) $A = \begin{bmatrix} 1 & -1 & 2 \\ 0 & 1 & 1 \\ 1 & 1 & 1 \end{bmatrix}$ [See Exercise 3(e).]

   (c) $A = \begin{bmatrix} 0 & 3 & 2 \\ 3 & 5 & 5 \\ 4 & 0 & 5 \end{bmatrix}$

   (d) $A = \begin{bmatrix} 0 & 1 & 1 & 0 \\ 1 & 0 & 4 & 1 \\ 1 & 1 & -1 & 1 \\ 0 & -1 & 1 & 1 \end{bmatrix}$.

5. (a) [BB] Find a QR factorization of $A = \begin{bmatrix} 1 & 1 \\ 2 & 3 \\ 2 & 1 \end{bmatrix}$.

   (b) [BB] Use 5(a) to find the least squares solution to
   $$\begin{aligned} x + y &= 1 \\ 2x + 3y &= -1 \\ 2x + y &= 1. \end{aligned}$$

6. Use a QR factorization of an appropriate matrix to find the best least squares solution to each of the given systems.

$$\begin{aligned} x - y &= 1 \\ \text{(a) [BB]} \quad 2x + y &= -2 \\ x + 3y &= -3. \end{aligned}$$

$$\begin{aligned} x \quad\quad + z &= -1 \\ x + y + z &= 0 \\ \text{(b)} \quad 2x \quad\quad &= 0 \\ x + y - z &= 3. \end{aligned}$$

7. [BB] In this section, we seem to have been finding QR factorizations for $m \times n$ matrices only when $m \geq n$. Why?

8. Suppose P is a projection matrix (see Section 6.1). Show that $Q = I - 2P$ is an orthogonal, symmetric matrix.

9. [BB] A matrix K is *skew-symmetric* if $K^T = -K$. Suppose K is a skew-symmetric matrix such that $I - K$ is invertible. Show that $(I + K)(I - K)^{-1}$ is orthogonal.

10. If Q is an orthogonal matrix and $\mathbf{x}$ is a nonzero vector such that $Q\mathbf{x} = c\mathbf{x}$, show that $c = \pm 1$.

11. (a) [BB] If Q is orthogonal and symmetric, show that $Q^2 = I$.

    (b) If Q is orthogonal and symmetric, show that $P = \frac{1}{2}(I - Q)$ is a projection matrix.

12. Show that the determinant of an orthogonal matrix is ± 1.

13. (a) [BB] If Q_1 is $m \times n$ with orthonormal columns and Q_2 is $n \times t$ with orthonormal columns, show that $Q_1 Q_2$ has orthonormal columns.

    (b) Is the product of orthogonal matrices orthogonal? Explain.

14. If Q is a real $n \times n$ matrix that preserves the dot product in $\mathbf{R}^n$—that is, $Q\mathbf{x} \cdot Q\mathbf{y} = \mathbf{x} \cdot \mathbf{y}$ for any vectors $\mathbf{x}, \mathbf{y}$—is Q orthogonal? Explain.

■ Critical Reading

15. [BB] Suppose the Gram–Schmidt algorithm is applied to a set $\mathbf{u}_1, \mathbf{u}_2, \dots, \mathbf{u}_k$ of vectors that are **not** linearly independent.

    (a) The algorithm must break down. Why?

    (b) Explain **how** the algorithm will break.

16. Apply the Gram–Schmidt algorithm to the vectors
$$\mathbf{u}_1 = \begin{bmatrix} -1 \\ -2 \\ 1 \end{bmatrix}, \mathbf{u}_2 = \begin{bmatrix} 3 \\ 1 \\ 2 \end{bmatrix}, \mathbf{u}_3 = \begin{bmatrix} 0 \\ 1 \\ -1 \end{bmatrix}.$$ What goes wrong? Explain.

17. [BB] Any finite dimensional vector space has an orthogonal basis. Why?

18. If $\mathbf{q}_1, \mathbf{q}_2, \dots, \mathbf{q}_k$ are orthonormal vectors in $\mathbf{R}^n$, then these vectors can be extended to an orthonormal basis of $\mathbf{R}^n$. Why?

19. [BB] Suppose $\{\mathbf{v}_1, \mathbf{v}_2, \dots, \mathbf{v}_n\}$ is a basis for a vector space V and $\mathbf{x}$ is a vector with the property that $\mathbf{x} \cdot \mathbf{v}_i = 0$ for all i. Prove that $\mathbf{x} = \mathbf{0}$.

20. Suppose $\mathbf{f}_1, \dots, \mathbf{f}_n$ is an orthogonal basis for a vector space V. Let $\mathbf{v}$ be any vector in V. Then $\mathbf{v}$ can be written as a linear combination of $\mathbf{f}_1, \dots, \mathbf{f}_n$.

    (a) How does the orthogonality of these vectors make it especially easy to calculate coefficients?

    (b) Why are the coefficients unique?

21. Let $\mathbf{w}$ be a vector in $\mathbf{R}^n$ of length 1 and $H = I - 2\mathbf{w}\mathbf{w}^T$ be the corresponding Householder matrix. Let $\mathbf{x}$ be any vector in $\mathbf{R}^n$ and $\mathbf{y} = H\mathbf{x} - \mathbf{x}$.

    (a) [BB] Show that $\mathbf{y}$ is a scalar multiple of $\mathbf{w}$.

    (b) [BB] Show that $\mathbf{x} + \frac{1}{2}\mathbf{y}$ is orthogonal to $\mathbf{w}$.

    (c) Add the vectors $\mathbf{y}$ and $\mathbf{x} + \frac{1}{2}\mathbf{y}$ to the picture below. Your picture should make it appear that $H\mathbf{x}$ is the reflection of $\mathbf{x}$ in a plane perpendicular to $\mathbf{w}$.

22. Let V be the subspace of $\mathbf{R}^4$ spanned by the vectors
$$\mathbf{u}_1 = \begin{bmatrix} 1 \\ -1 \\ 0 \\ 0 \end{bmatrix}, \mathbf{u}_2 = \begin{bmatrix} 1 \\ 0 \\ -1 \\ 0 \end{bmatrix}, \text{ and } \mathbf{u}_3 = \begin{bmatrix} 1 \\ 0 \\ 0 \\ -1 \end{bmatrix}.$$ Apply the Gram–Schmidt algorithm to find an **orthonormal**

basis $\mathcal{B}$ of V. Is $\mathcal{B}$ the only orthonormal basis for V? Explain.

23. Roberta is having trouble finding the QR factorization of a 3×4 matrix. What would you say to her?

24. Suppose we want to solve the system $A\mathbf{x} = \mathbf{b}$ and we are able to factor $A = QR$. Then $QR\mathbf{x} = \mathbf{b}$. Multiplying each side by Q^T gives $Q^T QR\mathbf{x} = Q^T\mathbf{x}$ and, since $Q^T Q = I$, we have $R\mathbf{x} = Q^T\mathbf{b}$. Since R is invertible, $\mathbf{x} = R^{-1}Q^T\mathbf{b}$, so $A\mathbf{x} = \mathbf{b}$ has a solution that is even unique. Is there a flaw in our reasoning? [Hint: See Problem 6.1.10.]

25. If $Q^{-1}AQ$ is diagonal and Q is orthogonal, show that A is symmetric.

26. True or false: A Householder matrix is a projection matrix. (Justify your answer.)

6.3 Orthogonal Subspaces and Complements

We begin by introducing some definitions and notation that are very much a part of the language of vector spaces.

6.3.1 DEFINITION

The *sum* of subspaces U and W of a vector space V is the set

$$U + W = \{\mathbf{u} + \mathbf{w} \mid u \in U, w \in W\}$$

of vectors that can be written as the sum of a vector in U and a vector in W. The *intersection* of U and W is the set

$$U \cap W = \{\mathbf{x} \in V \mid \mathbf{x} \in U \text{ and } \mathbf{x} \in W\}$$

of vectors that are in both U and W.

It is straightforward to show that sums and intersections of subspaces of a vector space V are also subspaces of V—see Exercise 8.

6.3.2 EXAMPLE

If U and W are the subspaces

$$U = \{\begin{bmatrix} c \\ 2c \end{bmatrix} \mid c \in \mathbf{R}\}, \quad W = \{\begin{bmatrix} 0 \\ c \end{bmatrix} \mid c \in \mathbf{R}\},$$

then $U + W = \mathbf{R}^2$ since every $\mathbf{v} = \begin{bmatrix} x \\ y \end{bmatrix} \in \mathbf{R}^2$ can be written

$$\begin{bmatrix} x \\ y \end{bmatrix} = \begin{bmatrix} x \\ 2x \end{bmatrix} + \begin{bmatrix} 0 \\ y - 2x \end{bmatrix}$$

as the sum of a vector in U and a vector in W. If a vector $\mathbf{v} = \begin{bmatrix} x \\ y \end{bmatrix}$ is in the intersection of U and W, then

$$\mathbf{v} = \begin{bmatrix} x \\ y \end{bmatrix} = \begin{bmatrix} c \\ 2c \end{bmatrix} = \begin{bmatrix} 0 \\ d \end{bmatrix}$$

for certain scalars c and d. So $x = c = 0$ and $y = 2c = d$, so $\mathbf{v} = \mathbf{0}$: $U \cap W = \{\mathbf{0}\}$. ∎

6.3.3 EXAMPLE

Let U and W be the subspaces of $\mathbf{R}^3$ defined by

$$U = \left\{ \begin{bmatrix} x \\ y \\ z \end{bmatrix} \;\middle|\; x + y - z = 0 \right\} \quad \text{and} \quad W = \left\{ \begin{bmatrix} x \\ y \\ z \end{bmatrix} \;\middle|\; 2x - 3y + z = 0 \right\}.$$

Then $U + W = \mathbf{R}^3$ since any $\begin{bmatrix} x \\ y \\ z \end{bmatrix}$ in $\mathbf{R}^3$ can be written

$$\begin{bmatrix} x \\ y \\ z \end{bmatrix} = \begin{bmatrix} -y + \frac{1}{3}z \\ -x \\ -x - y + \frac{1}{3}z \end{bmatrix} + \begin{bmatrix} x + y - \frac{1}{3}z \\ x + y \\ x + y + \frac{2}{3}z \end{bmatrix}$$

as the sum of a vector in U and a vector in W.

What about $U \cap W$, the intersection of U and W? Each of these subspaces is a plane, the two planes are not parallel, so they intersect in a line. Solving the system

$$\begin{aligned} x + y - z &= 0 \\ 2x - 3y + z &= 0 \end{aligned}$$

gives $\begin{bmatrix} x \\ y \\ z \end{bmatrix} = t \begin{bmatrix} 2 \\ 3 \\ 5 \end{bmatrix}$, which is the equation of the line of intersection: $U \cap W = \left\{ t \begin{bmatrix} 2 \\ 3 \\ 5 \end{bmatrix} \;\middle|\; t \in \mathbf{R} \right\}$. ∎

Reading Challenge 1   *How were we able to tell so quickly that the planes in Example 6.3.3 were not parallel?*

6.3.4 EXAMPLE

If U is the subspace of $\mathbf{R}^4$ spanned by $\mathbf{u}_1 = \begin{bmatrix} -1 \\ 1 \\ 2 \\ 3 \end{bmatrix}$ and $\mathbf{u}_2 = \begin{bmatrix} 2 \\ 0 \\ -1 \\ 3 \end{bmatrix}$ and W is the

subspace of $\mathbf{R}^4$ spanned by $\mathbf{w}_1 = \begin{bmatrix} 0 \\ 1 \\ -3 \\ 1 \end{bmatrix}$ and $\mathbf{w}_2 = \begin{bmatrix} 1 \\ 0 \\ 4 \\ 5 \end{bmatrix}$, then $U + W$ is spanned by

$\{\mathbf{u}_1, \mathbf{u}_2, \mathbf{w}_1, \mathbf{w}_2\}$. To find a basis for $U + W$, we might put these vectors into the rows of a matrix and reduce to an upper triangular matrix whose nonzero rows provide the basis we are after. Alternatively, and this will pay dividends later, we can form the matrix

$$A = \begin{bmatrix} \mathbf{u}_1 & \mathbf{u}_2 & \mathbf{w}_1 & \mathbf{w}_2 \\ \downarrow & \downarrow & \downarrow & \downarrow \end{bmatrix} = \begin{bmatrix} -1 & 2 & 0 & 1 \\ 1 & 0 & 1 & 0 \\ 2 & -1 & -3 & 4 \\ 3 & 3 & 1 & 5 \end{bmatrix} \tag{1}$$

whose columns are $\mathbf{u}_1, \mathbf{u}_2, \mathbf{w}_1, \mathbf{w}_2$, reduce this to row echelon form, and use the fact that the pivot columns are a basis for the column space (which is $U + W$)—see 4.3.35 of Section 4.3.

Gaussian elimination on A proceeds

$$
\begin{bmatrix}
-1 & 2 & 0 & 1 \\
1 & 0 & 1 & 0 \\
2 & -1 & -3 & 4 \\
3 & 3 & 1 & 5
\end{bmatrix}
\rightarrow
\begin{bmatrix}
1 & -2 & 0 & -1 \\
0 & 2 & 1 & 1 \\
0 & 3 & -3 & 6 \\
0 & 9 & 1 & 8
\end{bmatrix}
\rightarrow
\begin{bmatrix}
1 & -2 & 0 & -1 \\
0 & 1 & -1 & 2 \\
0 & 0 & 3 & -3 \\
0 & 0 & 10 & -10
\end{bmatrix}
$$

$$
\rightarrow
\begin{bmatrix}
1 & -2 & 0 & -1 \\
0 & 1 & -1 & 2 \\
0 & 0 & 1 & -1 \\
0 & 0 & 0 & 0
\end{bmatrix}.
$$

The pivots are in columns one, two, and three, so the column space of A has basis columns one, two, and three. A basis for $U + W$ is $\mathbf{u}_1, \mathbf{u}_2, \mathbf{w}_1$. Now we turn our attention to $U \cap W$.

Suppose $\mathbf{y}$ is in $U \cap W$. Then $\mathbf{y} = a_1\mathbf{u}_1 + a_2\mathbf{u}_2 = b_1\mathbf{w}_1 + b_2\mathbf{w}_2$ for certain scalars a_1, a_2, b_1, b_2. Thus $a_1\mathbf{u}_1 + a_2\mathbf{u}_2 - b_1\mathbf{w}_1 - b_2\mathbf{w}_2 = \mathbf{0}$, so $\mathbf{x} = \begin{bmatrix} a_1 \\ a_2 \\ -b_1 \\ -b_2 \end{bmatrix}$ is in the null space of the matrix A defined in (1) (using, once again, the famous formula in 2.1.33).

Conversely, if $\mathbf{x} = \begin{bmatrix} x_1 \\ x_2 \\ y_1 \\ y_2 \end{bmatrix}$ is in the null space of A (it is useful to label the components of $\mathbf{x}$ this way), then $x_1\mathbf{u}_1 + x_2\mathbf{u}_2 + y_1\mathbf{w}_1 + y_2\mathbf{w}_2 = \mathbf{0}$, so $x_1\mathbf{u}_1 + x_2\mathbf{u}_2 = -y_1\mathbf{w}_1 - y_2\mathbf{w}_2$ is in $U \cap W$. So we can find $U \cap W$ by finding the null space of A.

From the Gaussian elimination process already undertaken, we see that $y_2 = t$ is free, $y_1 = y_2 = t$, $x_2 = y_1 - 2y_2 = -t$, and $x_1 = 2x_2 + y_2 = -t$, so $\mathbf{x} = \begin{bmatrix} -t \\ -t \\ t \\ t \end{bmatrix}$. Thus

$-t\mathbf{u}_1 - t\mathbf{u}_2 + t\mathbf{w}_1 + t\mathbf{w}_2 = \mathbf{0}$, so vectors in $U \cap W$ have the form $t(\mathbf{w}_1 + \mathbf{w}_2) = t(\mathbf{u}_1 + \mathbf{u}_2)$.

A basis for $U \cap W$ is $\mathbf{u}_1 + \mathbf{u}_2 = \mathbf{w}_1 + \mathbf{w}_2 = \begin{bmatrix} 1 \\ 1 \\ 1 \\ 6 \end{bmatrix}$. ∎

In this example, we had $\dim(U + W) = 3$ and $\dim(U \cap W) = 1$, so that $\dim(U + W) + \dim(U \cap W) = 4 = 2 + 2 = \dim U + \dim W$. The reader should verify that this formula also holds in Examples 6.3.2 and 6.3.3. In fact, it is true in general.

Theorem 6.3.5   *Let U and W be subspaces of $\mathbf{R}^n$. Then $\dim(U + W) + \dim(U \cap W) = \dim U + \dim W$.*

Proof Let $\{u_1, u_2, \dots, u_k\}$ be a basis for U and let $\{w_1, w_2, \dots, w_\ell\}$ be a basis for W. Let

$$A = \begin{bmatrix} u_1 & u_2 & \cdots & u_k & w_1 & w_2 & \cdots & w_\ell \\ \downarrow & \downarrow & & \downarrow & \downarrow & \downarrow & & \downarrow \end{bmatrix}$$

be the $n \times (k + \ell)$ matrix whose columns are as indicated. The column space of A is $U+W$, so $\dim(U+W) = \operatorname{rank} A$. If we could show that $\dim(U \cap W) = \operatorname{nullity} A$, then we would have $\dim(U+W)+\dim(U \cap W) = \operatorname{rank} A + \operatorname{nullity} A = k+\ell$, the number of columns of A. This would give the desired result because $k = \dim U$ and $\ell = \dim W$. To summarize, we can complete the proof by showing that $\dim(U \cap W) = \operatorname{nullity} A$. We do this by showing that the nice relationship between $U \cap W$ and the null space of A, that we discovered in Example 6.3.4, holds in general.

So let y be a vector in $U \cap W$. Then y is a linear combination of $u_1, \dots, u_k$ and also of $w_1, \dots, w_\ell$:

$$y = x_1 u_1 + \cdots + x_k u_k = y_1 w_1 + \cdots + y_\ell w_\ell,$$

so $x = \begin{bmatrix} x_1 \\ \vdots \\ x_k \\ -y_1 \\ \vdots \\ -y_\ell \end{bmatrix}$ is in the null space of A. On the other hand, if $x = \begin{bmatrix} x_1 \\ \vdots \\ x_k \\ y_1 \\ \vdots \\ y_\ell \end{bmatrix}$ is in the

null space of A, then $x_1 u_1 + \cdots + x_k u_k + y_1 w_1 + \cdots + y_\ell w_\ell = 0$, so

$$x_1 u_1 + \cdots + x_k u_k = -y_1 w_1 - \cdots - y_\ell w_\ell$$

is a vector in $U \cap W$. It follows that $U \cap W$ is "essentially" the null space of A; in particular, these vector spaces have the same dimension. This completes the proof. ∎

6.3.6 Remark Readers uncomfortable with our use of the word "essentially" in the proof of Theorem 6.3.5 may prefer the following more formal argument. Our proof showed that the null space of A is precisely the set of vectors

$$x = \begin{bmatrix} x_1 \\ \vdots \\ x_k \\ y_1 \\ \vdots \\ y_\ell \end{bmatrix}$$

that have the property that $x_1\mathbf{u}_1 + \cdots + x_k\mathbf{u}_k = -y_1\mathbf{w}_1 - \cdots - y_\ell\mathbf{w}_\ell$ is in $U \cap W$. So the function null sp $A \to U \cap W$ defined by

$$
\begin{bmatrix} x_1 \\ \vdots \\ x_k \\ y_1 \\ \vdots \\ y_\ell \end{bmatrix} \to x_1\mathbf{u}_1 + \cdots + x_k\mathbf{u}_k
$$

establishes a one-to-one correspondence between null sp A and $U \cap W$. Moreover, this function is a "linear transformation," that is, it preserves sums and scalar multiples. See Exercise 16 for details. The vector spaces null sp A and $U \cap W$ are the same. (The technical term is "isomorphic.") What is important to us here is that they have the same dimension.

6.3.7 Remark In Theorem 6.3.5, it was important that the given spanning sets $\mathcal{U} = \{\mathbf{u}_1, \mathbf{u}_2, \dots, \mathbf{u}_k\}$ and $\mathcal{W} = \{\mathbf{w}_1, \mathbf{w}_2, \dots, \mathbf{w}_\ell\}$ were bases for U and W, respectively, in order that the rank of A was $k + \ell = \dim U + \dim W$. If we are interested only in finding a basis for $U \cap W$, then it is sufficient that $\mathcal{U}$ and $\mathcal{W}$ be merely spanning sets for U and W, as illustrated in our next example, where the vectors $\mathbf{w}_1, \mathbf{w}_2, \mathbf{w}_3$ are not a basis for W.

Reading Challenge 2   *Why not?*

6.3.8 EXAMPLE   Let U be the subspace of $\mathbf{R}^3$ spanned by $\mathbf{u}_1 = \begin{bmatrix} 2 \\ 4 \\ -2 \end{bmatrix}$ and $\mathbf{u}_2 = \begin{bmatrix} 1 \\ 2 \\ 1 \end{bmatrix}$. Let W be the

subspace of $\mathbf{R}^3$ spanned by $\mathbf{w}_1 = \begin{bmatrix} 0 \\ 3 \\ -1 \end{bmatrix}$, $\mathbf{w}_2 = \begin{bmatrix} 1 \\ 0 \\ 1 \end{bmatrix}$, and $\mathbf{w}_3 = \begin{bmatrix} 3 \\ 6 \\ 1 \end{bmatrix}$. Let

$$
A = \begin{bmatrix} 2 & 1 & 0 & 1 & 3 \\ 4 & 2 & 3 & 0 & 6 \\ -2 & 1 & -1 & 1 & 1 \end{bmatrix}
$$

be the matrix whose columns are $\mathbf{u}_1, \mathbf{u}_2, \mathbf{w}_1, \mathbf{w}_2, \mathbf{w}_3$. Gaussian elimination proceeds

$$
A \to \begin{bmatrix} 2 & 1 & 0 & 1 & 3 \\ 0 & 0 & 3 & -2 & 0 \\ 0 & 2 & -1 & 2 & 4 \end{bmatrix} \to \begin{bmatrix} 2 & 1 & 0 & 1 & 3 \\ 0 & 2 & -1 & 2 & 4 \\ 0 & 0 & 3 & -2 & 0 \end{bmatrix}.
$$

If $\mathbf{x} = \begin{bmatrix} x_1 \\ x_2 \\ y_1 \\ y_2 \\ y_3 \end{bmatrix}$ is in the null space of A, then $y_2 = t$ and $y_3 = s$ are free, $y_1 = \frac{2}{3}y_2 = \frac{2}{3}t$,

$$2x_2 = y_1 - 2y_2 - 4y_3 = \frac{2}{3}t - 2t - 42 = -\frac{4}{3}t - 4s, \qquad \text{so that} \quad x_2 = -\frac{2}{3}t - 2s$$

$$2x_1 = -x_2 - y_2 - 3y_3 = \frac{2}{3}t + 2s - t - 3s = -\frac{1}{3}t - s, \quad \text{so that} \quad x_1 = -\frac{1}{6}t - \frac{1}{2}s$$

and $\mathbf{x} = \begin{bmatrix} -\frac{1}{6}t - \frac{1}{2}s \\ -\frac{2}{3}t - 2s \\ \frac{2}{3}t \\ t \\ s \end{bmatrix}$. It follows that $(-\frac{1}{6}t-\frac{1}{2}s)\mathbf{u}_1+(-\frac{2}{3}t-2s)\mathbf{u}_2+\frac{2}{3}t\mathbf{w}_1+t\mathbf{w}_2+s\mathbf{w}_3 =$

$\mathbf{0}$, so that vectors in $U \cap W$ have the form

$$(-\tfrac{1}{6}t - \tfrac{1}{2}s)\mathbf{u}_1 + (-\tfrac{2}{3}t - 2s)\mathbf{u}_2 \;=\; -\tfrac{2}{3}t\mathbf{w}_1 - t\mathbf{w}_2 - s\mathbf{w}_3$$

$$= \begin{bmatrix} -t - 3s \\ -2t - 6s \\ -\frac{1}{3}t - s \end{bmatrix} = (t + 3s)\begin{bmatrix} -1 \\ -2 \\ -\frac{1}{3} \end{bmatrix}.$$

Thus $U \cap W$ has basis the vector $\begin{bmatrix} -1 \\ -2 \\ -\frac{1}{3} \end{bmatrix}$, so $\dim(U \cap W) = 1$. (Note that this is **not** the nullity of A.)

Incidentally, $U+W$ is again the column space of A, so $\dim(U+W) = \operatorname{rank} A = 3$ and $\dim(U + W) + \dim(U \cap W) = 3 + 1 = 4 = \dim U + \dim W$. (This number is no longer the number of columns in A.) ◼

Reading Challenge 3   *With U and W as in **6.3.8**, find a basis for $U + W$.*

6.3.9 DEFINITION   Subspaces U and W of $\mathbf{R}^n$ are *orthogonal* if $\mathbf{u} \cdot \mathbf{w} = 0$ for any vector $\mathbf{u}$ in U and any vector $\mathbf{w}$ in W.

6.3.10 EXAMPLE   In $\mathbf{R}^3$, the x-axis and the z-axis are orthogonal subspaces since if $\mathbf{u} = \begin{bmatrix} x \\ 0 \\ 0 \end{bmatrix}$ is on the

x-axis and $\mathbf{w} = \begin{bmatrix} 0 \\ 0 \\ z \end{bmatrix}$ is on the z-axis, $\mathbf{u} \cdot \mathbf{w} = 0$. ◼

6.3.11 EXAMPLE

Let $U = \left\{ \begin{bmatrix} u_1 \\ u_2 \\ u_3 \\ u_4 \end{bmatrix} \in \mathbf{R}^4 \mid 2u_1 - u_2 + u_3 - u_4 = 0 \right\}$ and $W = \left\{ c \begin{bmatrix} 2 \\ -1 \\ 1 \\ -1 \end{bmatrix} \mid c \in \mathbf{R} \right\}.$

Then U and W are orthogonal since if $\mathbf{u} = \begin{bmatrix} u_1 \\ u_2 \\ u_3 \\ u_4 \end{bmatrix} \in U$ and $\mathbf{w} = c \begin{bmatrix} 2 \\ -1 \\ 1 \\ -1 \end{bmatrix} \in W$, then

$$\mathbf{u} \cdot \mathbf{w} = c(2u_1 - u_2 + u_3 - u_4) = 0.$$ ∎

6.3.12 EXAMPLE

Let U be the subspace of $\mathbf{R}^4$ spanned by $\mathbf{u}_1 = \begin{bmatrix} -1 \\ 1 \\ 2 \\ 1 \end{bmatrix}$ and $\mathbf{u}_2 = \begin{bmatrix} 2 \\ -1 \\ 0 \\ 1 \end{bmatrix}$. Let W be

the subspace of $\mathbf{R}^4$ spanned by $\mathbf{w} = \begin{bmatrix} 0 \\ -1 \\ 1 \\ -1 \end{bmatrix}$. Then U and W are orthogonal because

$\mathbf{w} \cdot \mathbf{u}_1 = 0$ and $\mathbf{w} \cdot \mathbf{u}_2 = 0$, and these equations imply that any multiple of $\mathbf{w}$ is orthogonal to any linear combination of $\mathbf{u}_1$ and $\mathbf{u}_2$. ∎

6.3.13 EXAMPLE

The row space and the null space of a matrix are orthogonal. To see why, let

$$A = \begin{bmatrix} \mathbf{r}_1 & \to \\ \mathbf{r}_2 & \to \\ \vdots & \\ \mathbf{r}_m & \to \end{bmatrix}$$

Here and elsewhere in this section, we say "row" when we really mean the (column) vector that is its transpose. For example, we think of the rows of a 2×3 matrix as vectors in $\mathbf{R}^3$.

be an $m \times n$ matrix with rows as indicated and take a vector $\mathbf{x}$ in the null space of A. Since $A\mathbf{x} = \begin{bmatrix} \mathbf{r}_1 \cdot \mathbf{x} \\ \mathbf{r}_2 \cdot \mathbf{x} \\ \vdots \\ \mathbf{r}_m \cdot \mathbf{x} \end{bmatrix}$, each $\mathbf{r}_i \cdot \mathbf{x} = \mathbf{0}$. Any $\mathbf{r}$ in the row space of A is a linear combination of the rows of A, so $\mathbf{r} \cdot \mathbf{x} = 0$ for any $\mathbf{r}$ in the row space of A—see Exercise 9—which is what we wanted to show. Here's a specific illustration. ∎

6.3.14 EXAMPLE

Let $A = \begin{bmatrix} 1 & 4 & 0 \\ 2 & 3 & 10 \end{bmatrix}$. First, we find the null space by reducing A to row echelon form in the usual way:

$$A \to \begin{bmatrix} 1 & 4 & 0 \\ 0 & -5 & 10 \end{bmatrix} \to \begin{bmatrix} 1 & 4 & 0 \\ 0 & 1 & -2 \end{bmatrix}.$$

If $\begin{bmatrix} x \\ y \\ z \end{bmatrix}$ is in the null space of A, $z = t$ is free, $y = 2z = 2t$, and $x = -4y = -8t$. The null space of A is the one-dimensional subspace

$$\text{null sp } A = \left\{ \mathbf{u} \in \mathbf{R}^3 \mid \mathbf{u} = t \begin{bmatrix} -8 \\ 2 \\ 1 \end{bmatrix} \right\}$$

and we observe that this is orthogonal to the row space of A since $\begin{bmatrix} -8 \\ 2 \\ 1 \end{bmatrix}$ is orthogonal to each row of A

$$\begin{bmatrix} -8 \\ 2 \\ 1 \end{bmatrix} \cdot \begin{bmatrix} 1 \\ 4 \\ 0 \end{bmatrix} = -8 + 8 = 0, \qquad \begin{bmatrix} -8 \\ 2 \\ 1 \end{bmatrix} \cdot \begin{bmatrix} 2 \\ 3 \\ 10 \end{bmatrix} = -16 + 6 + 10 = 0$$

and hence to any linear combination of these rows. ∎

6.3.15 DEFINITION

The *orthogonal complement* of a subspace U of a vector space V is the set of vectors in V that are orthogonal to every vector of U. The orthogonal complement of U is denoted $U^\perp$, read "U perp."

$$\boxed{U^\perp = \{\mathbf{v} \in V \mid \mathbf{v} \cdot \mathbf{u} = 0 \text{ for all } \mathbf{u} \in U\}}$$

6.3.16 Remarks

1. The orthogonal complement of a subspace of V is also a subspace of V. See Exercise 10.

2. The assertion that subspaces U and W are orthogonal is precisely the assertion that each is contained in the orthogonal complement of the other: U is contained in $W^\perp$; W is contained in $U^\perp$. See Exercise 14.

6.3.17 EXAMPLE

Let $V = \mathbf{R}^3$ and U let be the z-axis. Then $U^\perp$, the orthogonal complement of U, is the entire xy-plane: Those vectors of $\mathbf{R}^3$ that are orthogonal to the z-axis are precisely those in the xy-plane. ∎

The following theorem is illustrated in Figure 6.7.

Theorem 6.3.18

The orthogonal complement of the row space of a matrix is its null space. The orthogonal complement of the column space of a matrix is the null space of its transpose:

$$(\text{row sp } A)^\perp = \text{null sp } A; \qquad (\text{col sp } A)^\perp = \text{null sp } A^T.$$

Figure 6.7 An $m \times n$ matrix A induces decompositions of $\mathbf{R}^n$ and $\mathbf{R}^m$ as shown.

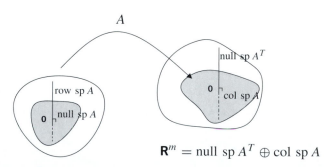

$$\mathbf{R}^m = \text{null sp } A^T \oplus \text{col sp } A$$

$$\mathbf{R}^n = \text{null sp } A \oplus \text{row sp } A$$

Proof Let A be an $m \times n$ matrix. The second statement follows by applying the first statement to the matrix A^T. (Remember that row sp A^T = col sp A.) For the first statement, think of A in terms of its rows, $A = \begin{bmatrix} r_1 & \to \\ r_2 & \to \\ \vdots \\ r_m & \to \end{bmatrix}$. In 6.3.13, we showed that null sp A is contained in (row sp $A)^\perp$. Conversely, if $\mathbf{x}$ is in (row sp $A)^\perp$, then $r_i \cdot \mathbf{x} = 0$ for each row r_i, so $A\mathbf{x} = \begin{bmatrix} r_1 \cdot \mathbf{x} \\ r_2 \cdot \mathbf{x} \\ \vdots \\ r_m \cdot \mathbf{x} \end{bmatrix} = \mathbf{0}$, which says $\mathbf{x}$ is in the null space of A. ∎

We can use Theorem 6.3.18 to find orthogonal complements.

6.3.19 PROBLEM   Let U be the subspace of $\mathbf{R}^3$ spanned by $\mathbf{u}_1 = \begin{bmatrix} 1 \\ 4 \\ 0 \end{bmatrix}$ and $\mathbf{u}_2 = \begin{bmatrix} 2 \\ 3 \\ 10 \end{bmatrix}$. Find $U^\perp$.

Solution.   An easy (if somewhat cheeky!) answer is

   "$U^\perp$ is the set of all vectors orthogonal to U,"

but this, of course, is not what we have in mind! Let $A = \begin{bmatrix} 1 & 4 & 0 \\ 2 & 3 & 10 \end{bmatrix}$ be the matrix whose row space is U. This was the matrix that appeared in Example 6.3.14. There we found that the null space of A is the one-dimensional subspace of $\mathbf{R}^3$ with basis $\begin{bmatrix} -8 \\ 2 \\ 1 \end{bmatrix}$. This is the orthogonal complement of the row space of A, that is, $U^\perp$.

Reading Challenge 4   *Can you think of another way to find this vector* $\begin{bmatrix} -8 \\ 2 \\ 1 \end{bmatrix}$ *?*

6.3.20 PROBLEM   Let U and W be as in **6.3.12**. In that example, we showed that U and W are orthogonal subspaces. Does this mean that $W = U^\perp$?

Solution.   The subspace U is the row space of $A = \begin{bmatrix} -1 & 1 & 2 & 1 \\ 2 & -1 & 0 & 1 \end{bmatrix}$, so the orthogonal complement of U is the null space of A. Gaussian elimination proceeds

$$\begin{bmatrix} -1 & 1 & 2 & 1 \\ 2 & -1 & 0 & 1 \end{bmatrix} \to \begin{bmatrix} 1 & -1 & -2 & -1 \\ 0 & 1 & 4 & 3 \end{bmatrix},$$

so, if $\mathbf{x} = \begin{bmatrix} x_1 \\ x_2 \\ x_3 \\ x_4 \end{bmatrix}$ is in null sp A, then $x_3 = s$ and $x_4 = t$ are free, $x_2 = -4x_3 - 3x_4 =$

$-4s - 3t$, $x_1 = x_2 + 2x_3 + x_4 = -2s - 2t$, and $\mathbf{x} = \begin{bmatrix} -2s - 2t \\ -4s - 3t \\ s \\ t \end{bmatrix}$. The orthogonal

complement of U is spanned by $\begin{bmatrix} -2 \\ -4 \\ 1 \\ 0 \end{bmatrix}$ and $\begin{bmatrix} -2 \\ -3 \\ 0 \\ 1 \end{bmatrix}$. We have $W \subseteq U^\perp$, but $W \neq U^\perp$.

We conclude with a theorem that summarizes most of the important properties of orthogonal complements. One part requires familiarity with the concept of *direct sum*.

6.3.21 DEFINITION

The sum $V = U + W$ of subspaces is *direct*, and we write $V = U \oplus W$, if every vector $\mathbf{v}$ in V can be written uniquely in the form $\mathbf{v} = \mathbf{u} + \mathbf{w}$, with $\mathbf{u} \in U$ and $\mathbf{w} \in W$; that is, if $\mathbf{v} = \mathbf{u}_1 + \mathbf{w}_1$ with $\mathbf{u}_1$ in U and $\mathbf{w}_1$ in W, and also $\mathbf{v} = \mathbf{u}_2 + \mathbf{w}_2$ with $\mathbf{u}_2$ in U and $\mathbf{w}_2$ in W, then $\mathbf{u}_1 = \mathbf{u}_2$ and $\mathbf{w}_1 = \mathbf{w}_2$.

Lemma 6.3.22

The sum $U + W$ of subspaces is direct if and only if the intersection $U \cap W = \mathbf{0}$, the zero vector.

Proof Suppose that $U \cap W = \{\mathbf{0}\}$. We have to show that if $\mathbf{u}_1 + \mathbf{w}_1 = \mathbf{u}_2 + \mathbf{w}_2$ with $\mathbf{u}_1, \mathbf{u}_2$ in U and $\mathbf{w}_1, \mathbf{w}_2$ in W, then $\mathbf{u}_1 = \mathbf{u}_2$ and $\mathbf{w}_1 = \mathbf{w}_2$. Now the equation $\mathbf{u}_1 + \mathbf{w}_1 = \mathbf{u}_2 + \mathbf{w}_2$ implies

$$\mathbf{u}_1 - \mathbf{u}_2 = \mathbf{w}_2 - \mathbf{w}_1.$$

The vector $\mathbf{u}_1 - \mathbf{u}_2$ is in U (because U is a subspace), and the vector $\mathbf{w}_2 - \mathbf{w}_1$ is in W (because W is a subspace). These vectors are equal, so we have a vector in U and in W. Since $U \cap W = \{\mathbf{0}\}$, it must be that $\mathbf{u}_1 - \mathbf{u}_2 = \mathbf{w}_1 - \mathbf{w}_2 = \mathbf{0}$, so $\mathbf{u}_1 = \mathbf{u}_2$ and $\mathbf{w}_1 = \mathbf{w}_2$.

Conversely, suppose $U + W$ is direct. We must show that $U \cap W = \mathbf{0}$. So we take a vector $\mathbf{x}$ that is in both U and W and show that $\mathbf{x} = \mathbf{0}$. Now $\mathbf{x}$ is in both U and W, so it's the zero vector. This gives two ways to write $\mathbf{x}$ as the sum of a vector in U and a vector in W, namely, $\mathbf{x} = \mathbf{x} + \mathbf{0}$ and $\mathbf{x} = \mathbf{0} + \mathbf{x}$. By uniqueness, $\mathbf{x} = \mathbf{0}$. ∎

6.3.23 EXAMPLE

If $U = \{\begin{bmatrix} c \\ 2c \end{bmatrix} \mid c \in \mathbf{R}\}$ and $W = \{\begin{bmatrix} 0 \\ c \end{bmatrix} \mid c \in \mathbf{R}\}$ are as in **6.3.2**, then $\mathbf{R}^2 = U \oplus W$ is a direct sum since we showed there that $\mathbf{R}^2 = U + W$ and $U \cap W = \{\mathbf{0}\}$. ∎

6.3.24 EXAMPLE

If U and W are the subspaces of $\mathbf{R}^3$ defined in **6.3.3**, then $U + W$ is not direct, since the intersection of U and W is a line, not just the zero vector. ∎

Theorem 6.3.25

Let U be a subspace of a (finite dimensional) vector space V.

1. *$U \cap U^\perp = \{\mathbf{0}\}$; that is, if $\mathbf{x}$ is in U and also in $U^\perp$, then $\mathbf{x} = \mathbf{0}$.*

2. *$V = U \oplus U^\perp$; that is, every $\mathbf{v} \in V$ can be written uniquely in the form $\mathbf{v} = \mathbf{u} + \mathbf{w}$ with $\mathbf{u} \in U$ and $\mathbf{w} \in U^\perp$.*

3. $\dim V = \dim U + \dim U^{\perp}$.

4. $(U^{\perp})^{\perp} = U$.

Proof 1. Let $\mathbf{u}$ be a vector that is in U and also in $U^{\perp}$. Since $\mathbf{u}$ is in $U^{\perp}$, it is orthogonal to every vector in U, in particular to itself. Thus $0 = \mathbf{u} \cdot \mathbf{u} = \|\mathbf{u}\|^2$, so $\mathbf{u} = \mathbf{0}$.

2. We have to show that any vector in V can be written uniquely as the sum $\mathbf{u} + \mathbf{w}$ of a vector $\mathbf{u}$ in U and a vector $\mathbf{w}$ in $U^{\perp}$. Uniqueness follows immediately from part 1 and Lemma 6.3.22, so we have only to show how to find $\mathbf{u}$ and $\mathbf{w}$.

Take $\mathbf{v}$ in V and let $\mathbf{u} = \text{proj}_U \mathbf{v}$ be the projection of $\mathbf{v}$ on U—see Theorem 6.1.23. By definition of projection, $\mathbf{v} - \mathbf{u} = \mathbf{v} - \text{proj}_U \mathbf{v}$ is orthogonal to every vector in U; that is, $\mathbf{w} = \mathbf{v} - \mathbf{u}$ is in $U^{\perp}$. The obvious equation $\mathbf{v} = \mathbf{u} + (\mathbf{v} - \mathbf{u})$ expresses $\mathbf{v}$ as the sum of a vector in U and a vector in $U^{\perp}$.

3. Since $V = U + U^{\perp}$, $\dim V = \dim(U + U^{\perp}) = \dim U + \dim U^{\perp} - \dim(U \cap U^{\perp})$ by Theorem 6.3.5. Since $U \cap U^{\perp} = \{0\}$, we obtain $\dim V = \dim U + \dim U^{\perp}$. (See Exercise 28 for a different proof.)

4. As always, to prove that two sets are equal, we must show that an element in either of the sets is also in the other. So our proof is in two parts.

Suppose $\mathbf{u} \in U$. Let $\mathbf{x} \in U^{\perp}$. Then $\mathbf{u} \cdot \mathbf{x} = 0$. Thus $\mathbf{u}$ is orthogonal to $\mathbf{x}$. Since $\mathbf{x}$ was arbitrary, $\mathbf{u}$ is orthogonal to $U^{\perp}$; that is, $\mathbf{u}$ is in $(U^{\perp})^{\perp}$. We conclude that U is contained in $(U^{\perp})^{\perp}$. On the other hand, suppose $\mathbf{v}$ is in $(U^{\perp})^{\perp}$. By part 2, we can write $\mathbf{v} = \mathbf{u} + \mathbf{w}$, $\mathbf{u} \in U$ and $\mathbf{w} \in U^{\perp}$. Thus $\mathbf{v} - \mathbf{u} = \mathbf{w}$. Since U is contained in $(U^{\perp})^{\perp}$, which is a subspace, the vector $\mathbf{w} = \mathbf{v} - \mathbf{u}$ is in $(U^{\perp})^{\perp}$ as well as in $U^{\perp}$. So it is $\mathbf{0}$, applying part 1 to the subspace $U^{\perp}$. Thus $\mathbf{v} = \mathbf{u}$ is in U. Since $\mathbf{v}$ was arbitrary, we have $(U^{\perp})^{\perp}$ contained in U. These subspaces must be the same. ∎

The most important part of Theorem 6.3.25 is part 2. Given a subspace U of a vector space V, then V is the direct sum of U and its orthogonal complement; that is, every $\mathbf{v}$ in V can be written uniquely as the sum of a vector in U and a vector in $\mathbf{U}^{\perp}$.

6.3.26 EXAMPLE If $V = \mathbf{R}^3$ and $U = \{ \begin{bmatrix} 0 \\ 0 \\ z \end{bmatrix} \mid z \in \mathbf{R} \}$ is the z-axis, then $U^{\perp} = \{ \begin{bmatrix} x \\ y \\ 0 \end{bmatrix} \mid x, y \in \mathbf{R} \}$ is the xy-plane and the assertion $V = U \oplus U^{\perp}$ simply expresses the fact that every vector $\begin{bmatrix} x \\ y \\ z \end{bmatrix}$ in V can be written uniquely $\begin{bmatrix} 0 \\ 0 \\ z \end{bmatrix} + \begin{bmatrix} x \\ y \\ 0 \end{bmatrix}$ as the sum of a vector on the z-axis and a vector in the xy-plane. ∎

6.3.27 PROBLEM Let $V = \mathbf{R}^3$ and let U be the subspace

$$U = \{ \mathbf{x} = \begin{bmatrix} x \\ y \\ z \end{bmatrix} \in \mathbf{R}^3 \mid 2x - y + z = 0 \}.$$

Write $\mathbf{v} = \begin{bmatrix} x \\ y \\ z \end{bmatrix} \in V$ in the form $\mathbf{v} = \mathbf{u} + \mathbf{w}$ with $\mathbf{u} \in U$ and $\mathbf{w} \in U^{\perp}$.

Solution. Exactly as in the proof of the theorem, we take $\mathbf{u} = \text{proj}_U \mathbf{v}$ and $\mathbf{w} = \mathbf{v} - \mathbf{u}$.

An orthogonal basis for U is $\{\mathbf{f}_1, \mathbf{f}_2\}$ with $\mathbf{f}_1 = \begin{bmatrix} 0 \\ 1 \\ 1 \end{bmatrix}$ and $\mathbf{f}_2 = \begin{bmatrix} 1 \\ 1 \\ -1 \end{bmatrix}$. So

$$\mathbf{u} = \text{proj}_U \mathbf{v} = \frac{\mathbf{v} \cdot \mathbf{f}_1}{\mathbf{f}_1 \cdot \mathbf{f}_1} \mathbf{f}_1 + \frac{\mathbf{v} \cdot \mathbf{f}_2}{\mathbf{f}_2 \cdot \mathbf{f}_2} \mathbf{f}_2$$

$$= \frac{y+z}{2} \begin{bmatrix} 0 \\ 1 \\ 1 \end{bmatrix} + \frac{x+y-z}{3} \begin{bmatrix} 1 \\ 1 \\ -1 \end{bmatrix} = \begin{bmatrix} \frac{x+y-z}{3} \\ \frac{2x+5y+z}{6} \\ \frac{-2x+y+5z}{6} \end{bmatrix}.$$

Note that $\mathbf{u}$ is in U since $2(\frac{x+y-z}{3}) - \frac{2x+5y+z}{6} + \frac{-2x+y+5z}{6} = 0$. Finally,

$$\mathbf{w} = \mathbf{v} - \mathbf{u} = \begin{bmatrix} \frac{2x-y+z}{3} \\ \frac{-2x+y-z}{6} \\ \frac{2x-y+z}{6} \end{bmatrix} = \frac{2x-y+z}{6} \begin{bmatrix} 2 \\ -1 \\ 1 \end{bmatrix}.$$

Reading Challenge 5 *The astute reader will not be surprised that the vector $\mathbf{w}$ in Problem 6.3.27 is a multiple of* $\begin{bmatrix} 2 \\ -1 \\ 1 \end{bmatrix}$. *Why?*

6.3.28 Remark Part 4 of Theorem 6.3.25 says that the term "orthogonal complement" is *symmetric* in the sense that if W is the orthogonal complement of U, then U is the orthogonal complement of W: if $W = U^\perp$, then $U = (U^\perp)^\perp = W^\perp$. We say that U and W are *orthogonal complements*. For example, we have shown that the orthogonal complement of the row space of a matrix is its null space; thus the orthogonal complement of the null space is the row space. The null space and row space of a matrix are orthogonal complements.

Answers to Reading Challenges

1. The two normals, $\begin{bmatrix} 1 \\ 1 \\ -1 \end{bmatrix}$ and $\begin{bmatrix} 2 \\ -3 \\ 1 \end{bmatrix}$, are not parallel.

2. Since $\mathbf{w}_3 = 2\mathbf{w}_1 + 3\mathbf{w}_2$, the vectors are not linearly independent.

3. We have seen that $\dim(U + W) = 3$. Since $U + W$ is a subspace of $\mathbf{R}^3$, we have $U + W = \mathbf{R}^3$. One basis is $\{\mathbf{i}, \mathbf{j}, \mathbf{k}\}$.

4. The subspace U is a plane in $\mathbf{R}^3$, so its orthogonal complement is the one-dimensional subspace with basis the cross product of $\mathbf{u}_1$ and $\mathbf{u}_2$.

$$\mathbf{u}_1 \times \mathbf{u}_2 = \begin{vmatrix} \mathbf{i} & \mathbf{j} & \mathbf{k} \\ 1 & 4 & 0 \\ 2 & 3 & 10 \end{vmatrix} = 40\mathbf{i} - 10\mathbf{j} - 5\mathbf{k} = -5 \begin{bmatrix} -8 \\ 2 \\ 1 \end{bmatrix}.$$

5. $\mathbf{w}$ is supposed to be orthogonal to U, which is a plane with normal $\begin{bmatrix} 2 \\ -1 \\ 1 \end{bmatrix}$.

True/False Questions

Decide, with as little calculation as possible, whether each of the following statements is true or false and explain your answer whenever you say "false." (Answers can be found in the back of the book.)

1. In $\mathbf{R}^2$, the sum of $U = \{\begin{bmatrix} x \\ x \end{bmatrix} \mid x \in \mathbf{R}\}$ and $W = \{\begin{bmatrix} x \\ 2x \end{bmatrix} \mid x \in \mathbf{R}$ is the entire xy-plane.

2. In $\mathbf{R}^3$, the intersection of two (different) two-dimensional subspaces is a one-dimensional subspace.

3. If U and W are three-dimensional subspaces of $\mathbf{R}^5$, then $U \cap W \neq \{\mathbf{0}\}$.

4. If U and W are subspaces of a vector space V and U is contained in $W^\perp$, then W is contained in $U^\perp$.

5. The sum of the xy- and yz-planes in $\mathbf{R}^3$ is direct.

6. If U and W are orthogonal subspaces of a vector space V, then $\dim U + \dim W = \dim V$.

7. In $\mathbf{R}^2$, the orthogonal complement of the x-axis is the y-axis.

8. In $\mathbf{R}^3$, the orthogonal complement of the x-axis is the y-axis.

9. The orthogonal complement of the null space of a matrix is its row space.

10. If U is a subspace of a finite dimensional vector space V, there exists a subspace W such that $V = U + W$ and $U \cap W = \{\mathbf{0}\}$.

11. If A is a symmetric matrix with column space U and $A\mathbf{x} = \mathbf{0}$, then $\mathbf{x}$ is in $U^\perp$.

▌ Exercises

*Solutions to exercises marked [BB] can be found in the **Back** of the **Book**.*

1. In each of the following cases,

   - find a basis and the dimension of $U + W$,

   - find a basis and the dimension of $U \cap W$,

   - verify that $\dim(U + W) + \dim(U \cap W) = \dim U + \dim W$.

   (a) [BB] $V = \mathbf{R}^4$, $U = \{\begin{bmatrix} x - y \\ x + y \\ 2x - y \\ y \end{bmatrix} \mid x, y \in \mathbf{R}\}$; W is the subspace of $\mathbf{R}^4$ with basis $\{\begin{bmatrix} -1 \\ 5 \\ 0 \\ 3 \end{bmatrix}, \begin{bmatrix} 0 \\ 0 \\ 1 \\ 0 \end{bmatrix}\}$.

(b) $V = \mathbf{R}^4$, $U = \left\{ \begin{bmatrix} -2x+4y-3z \\ -x+y-z \\ x-7y+4z \\ x-5y+3z \end{bmatrix} \mid x, y \in \mathbf{R} \right\}$; W

is the subspace of $\mathbf{R}^4$ spanned by $\left\{ \begin{bmatrix} 0 \\ 1 \\ 1 \\ 1 \end{bmatrix} \text{ and } \begin{bmatrix} 5 \\ 2 \\ 3 \\ 0 \end{bmatrix} \right\}$.

(c) [BB] $V = \mathbf{R}^5$, U has basis $\mathbf{u}_1 = \begin{bmatrix} -2 \\ 0 \\ 2 \\ 4 \\ 1 \end{bmatrix}$,

$\mathbf{u}_2 = \begin{bmatrix} 3 \\ 5 \\ 2 \\ 4 \\ 1 \end{bmatrix}$, $\mathbf{u}_3 = \begin{bmatrix} 6 \\ 3 \\ 3 \\ -6 \\ -1 \end{bmatrix}$, and W has basis

$\mathbf{w}_1 = \begin{bmatrix} 1 \\ 2 \\ -5 \\ 2 \\ 0 \end{bmatrix}$ and $\mathbf{w}_2 = \begin{bmatrix} 3 \\ 1 \\ 10 \\ -4 \\ 0 \end{bmatrix}$.

(d) $V = \mathbf{R}^4$, U has basis $\mathbf{u}_1 = \begin{bmatrix} 1 \\ 3 \\ 1 \\ 1 \end{bmatrix}$, $\mathbf{u}_2 = \begin{bmatrix} 1 \\ 5 \\ -1 \\ 2 \end{bmatrix}$ and

W is spanned by $\mathbf{w}_1 = \begin{bmatrix} 2 \\ 2 \\ 3 \\ 1 \end{bmatrix}$, $\mathbf{w}_2 = \begin{bmatrix} 1 \\ 1 \\ 0 \\ 1 \end{bmatrix}$,

$\mathbf{w}_3 = \begin{bmatrix} 0 \\ 6 \\ -3 \\ 2 \end{bmatrix}$, $\mathbf{w}_4 = \begin{bmatrix} 1 \\ -1 \\ 2 \\ 0 \end{bmatrix}$.

(e) $V = \mathbf{R}^5$, U is spanned by $\mathbf{u}_1 = \begin{bmatrix} 1 \\ 1 \\ 1 \\ 1 \\ 3 \end{bmatrix}$ and

$\mathbf{u}_2 = \begin{bmatrix} 0 \\ -1 \\ 2 \\ 1 \\ 2 \end{bmatrix}$ and W is spanned by $\mathbf{w}_1 = \begin{bmatrix} 2 \\ 5 \\ -4 \\ -1 \\ 0 \end{bmatrix}$,

$\mathbf{w}_2 = \begin{bmatrix} 0 \\ -2 \\ 6 \\ 3 \\ 6 \end{bmatrix}$, $\mathbf{w}_3 = \begin{bmatrix} -1 \\ 1 \\ 1 \\ 0 \\ -1 \end{bmatrix}$, $\mathbf{w}_4 = \begin{bmatrix} 1 \\ 0 \\ 3 \\ 2 \\ 5 \end{bmatrix}$.

2. Find $U^\perp$ in each of the following cases:

(a) [BB] $V = \mathbf{R}^3$ and $U = \left\{ c \begin{bmatrix} 2 \\ -3 \\ -7 \end{bmatrix} \mid c \in \mathbf{R} \right\}$

(b) $V = \mathbf{R}^3$ and $U = \left\{ c \begin{bmatrix} 2 \\ -6 \\ 7 \end{bmatrix} \mid c \in \mathbf{R} \right\}$

(c) [BB] $V = \mathbf{R}^3$, U is the plane with equation $x - 2y + z = 0$

(d) $V = \mathbf{R}^3$, U is the plane with equation $2x + y - 3z = 0$

(e) [BB] $V = \mathbf{R}^3$, $\mathbf{u}_1 = \begin{bmatrix} 1 \\ 1 \\ -3 \end{bmatrix}$, $\mathbf{u}_2 = \begin{bmatrix} 1 \\ 2 \\ -1 \end{bmatrix}$, and
$U = \mathrm{sp}\{\mathbf{u}_1, \mathbf{u}_2\}$

(f) $V = \mathbf{R}^3$, $\mathbf{u}_1 = \begin{bmatrix} -1 \\ 2 \\ 2 \end{bmatrix}$, $\mathbf{u}_2 = \begin{bmatrix} 4 \\ 0 \\ 1 \end{bmatrix}$, $U = \mathrm{sp}\{\mathbf{u}_1, \mathbf{u}_2\}$

(g) $V = \mathbf{R}^4$, $\mathbf{u}_1 = \begin{bmatrix} 2 \\ 0 \\ -5 \\ 6 \end{bmatrix}$, $\mathbf{u}_2 = \begin{bmatrix} 1 \\ 0 \\ -3 \\ 2 \end{bmatrix}$,
$U = \mathrm{sp}\{\mathbf{u}_1, \mathbf{u}_2\}$

(h) $V = \mathbf{R}^5$, $\mathbf{u}_1 = \begin{bmatrix} -2 \\ 0 \\ 1 \\ 0 \\ 2 \end{bmatrix}$, $\mathbf{u}_2 = \begin{bmatrix} 5 \\ 4 \\ -4 \\ 0 \\ 1 \end{bmatrix}$, $\mathbf{u}_3 = \begin{bmatrix} 1 \\ 4 \\ -2 \\ 0 \\ 3 \end{bmatrix}$,
$U = \mathrm{sp}\{\mathbf{u}_1, \mathbf{u}_2\}$

(i) $V = \mathbf{R}^4$, $U = \left\{ \begin{bmatrix} x-y \\ x+y \\ 2x-y \\ y \end{bmatrix} \mid x, y \in \mathbf{R} \right\}$.

3. In each case below, an orthogonal basis is given for a subspace U. Write the given vector $\mathbf{v} = \mathbf{u} + \mathbf{w}$ as the sum of a vector $\mathbf{u}$ in U and a vector $\mathbf{w}$ in $U^\perp$.

(a) [BB] $U = \mathrm{sp}\left\{ \begin{bmatrix} 1 \\ 1 \\ 1 \\ 1 \end{bmatrix}, \begin{bmatrix} 1 \\ 1 \\ -1 \\ -1 \end{bmatrix}, \begin{bmatrix} 1 \\ -1 \\ 1 \\ -1 \end{bmatrix} \right\}$; $\mathbf{v} = \begin{bmatrix} 2 \\ 0 \\ 1 \\ 6 \end{bmatrix}$

(b) $U = \mathrm{sp}\left\{ \begin{bmatrix} 1 \\ 0 \\ 0 \\ 1 \end{bmatrix}, \begin{bmatrix} -1 \\ 0 \\ 1 \\ 1 \end{bmatrix}, \begin{bmatrix} 1 \\ 0 \\ 2 \\ -1 \end{bmatrix} \right\}$; $\mathbf{v} = \begin{bmatrix} a \\ b \\ c \\ d \end{bmatrix}$.

4. Write $\mathbf{v} = \mathbf{u} + \mathbf{w}$ with $\mathbf{u}$ in U and $\mathbf{w}$ in $U^\perp$ in each of the following situations.

   (a) [BB] $V = \mathbf{R}^2$, $\mathbf{v} = \begin{bmatrix} x \\ y \end{bmatrix}$, $U = \{\begin{bmatrix} t \\ t \end{bmatrix} \mid t \in \mathbf{R}\}$

   (b) $V = \mathbf{R}^3$, $\mathbf{v} = \begin{bmatrix} x \\ y \\ z \end{bmatrix}$, U is the span of $\begin{bmatrix} -2 \\ 1 \\ 0 \end{bmatrix}$

   (c) [BB] $V = \mathbf{R}^3$, $\mathbf{v} = \begin{bmatrix} 3 \\ 1 \\ -1 \end{bmatrix}$,

   $U = \{\begin{bmatrix} x \\ y \\ z \end{bmatrix} \mid 2x - y + 3z = 0\}$

   (d) $V = \mathbf{R}^3$, $\mathbf{v} = \begin{bmatrix} x \\ y \\ z \end{bmatrix}$,

   $U = \{\begin{bmatrix} x \\ y \\ z \end{bmatrix} \mid 2x - y + 3z = 0\}$

   (e) $V = \mathbf{R}^4$, $\mathbf{v} = \begin{bmatrix} x_1 \\ x_2 \\ x_3 \\ x_4 \end{bmatrix}$, $U = \{\begin{bmatrix} 2t - s \\ s + t \\ s - t \\ t \end{bmatrix} \mid t, s \in \mathbf{R}\}$.

5. [BB] Let U be the line in $\mathbf{R}^3$ with equation
   $\begin{bmatrix} x \\ y \\ z \end{bmatrix} = t \begin{bmatrix} -2 \\ 5 \\ 1 \end{bmatrix}$.

   (a) Find the matrix that projects vectors onto U.

   (b) What is $U^\perp$?

   (c) Find the matrix that projects vectors onto $U^\perp$.

   (d) Find the vector in $U^\perp$ that is closest to $\begin{bmatrix} -1 \\ 2 \\ 3 \end{bmatrix}$.

6. (a) Find the matrix P that projects vectors in $\mathbf{R}^4$ onto the orthogonal complement of the subspace with equation $x_1 + x_2 - x_3 + x_4 = 0$.

   (b) Find the vector in the orthogonal complement of the subspace in 6(a) that is closest to the vector $\begin{bmatrix} 1 \\ 1 \\ 1 \\ 1 \end{bmatrix}$.

7. [BB] Let U be a subspace of $\mathbf{R}^n$ and $\mathbf{v}$ be a vector in $\mathbf{R}^n$. Show that $\text{proj}_U \mathbf{v}$ is unique in the sense that if $\mathbf{v} = \mathbf{u} + \mathbf{w}$ with $\mathbf{u} \in U$ and $\mathbf{w} \in U^\perp$, then $\mathbf{u} = \text{proj}_U \mathbf{v}$.

8. Suppose U and W are subspaces of $\mathbf{R}^n$. Prove that $U + W$ [BB] and $U \cap W$ are also subspaces.

9. [BB] Let $\mathbf{u}_1, \mathbf{u}_2, \ldots, \mathbf{u}_k$ be vectors in $\mathbf{R}^n$ and let $U = \text{sp}\{\mathbf{u}_1, \ldots, \mathbf{u}_k\}$. Show that $\mathbf{x} \in U^\perp$ if and only if $\mathbf{x} \cdot \mathbf{u}_i = 0$ for all i.

10. If V is a vector space and U is a subspace of V, show that $U^\perp$ is a subspace of V.

11. [BB] Let U be the subspace of $\mathbf{R}^4$ that consists of all vectors of the form $\begin{bmatrix} 2s - t \\ s \\ t \\ -5s + 2t \end{bmatrix}$.

    (a) Find a basis for U and a basis for $U^\perp$.

    (b) Find the vector in U that is closest to $\begin{bmatrix} -1 \\ 0 \\ 0 \\ 1 \end{bmatrix}$ and the vector in $U^\perp$ that is closest to this same vector.

12. Let U be the subspace of $\mathbf{R}^4$ that consists of all vectors of the form $\begin{bmatrix} 2s \\ -t \\ 4s - 2t \\ 2t - 2s \end{bmatrix}$.

    (a) Find a basis for U and a basis for $U^\perp$.

    (b) Find the vector in U that is closest to $\begin{bmatrix} 1 \\ -1 \\ 1 \\ -1 \end{bmatrix}$ and the vector in $U^\perp$ that is closest to this same vector.

13. [BB] Let U be the subspace of $\mathbf{R}^4$ that is the intersection of the planes with equations $x_1 - x_3 + 2x_4 = 0$ and $2x_1 + 3x_2 + x_3 + x_4 = 0$.

    (a) Find an orthogonal basis for U.

    (b) Find a spanning set for $U^\perp$.

    (c) Find the matrix P that projects vectors of $\mathbf{R}^4$ onto U.

    (d) Find the vector in U that is closest to $\begin{bmatrix} 1 \\ 2 \\ -1 \\ 1 \end{bmatrix}$.

14. Suppose U and W are subspaces of a finite dimensional vector space V.

    (a) If $U \subseteq W$, show that $W^\perp \subseteq U^\perp$.

(b) [BB] Prove that the following three statements are equivalent:

    i. U and W are orthogonal;

    ii. $U \subseteq W^{\perp}$;

    iii. $W \subseteq U^{\perp}$.

15. Explain why the orthogonal complement of the column space of a matrix A is the null space of A^T; symbolically, $(\text{col sp } A)^{\perp} = \text{null sp } A^T$.

16. Let U and W be subspaces of $\mathbf{R}^n$. Let $\{\mathbf{u}_1, \mathbf{u}_2, \dots, \mathbf{u}_k\}$ be a basis for U and let $\{\mathbf{w}_1, \mathbf{w}_2, \dots, \mathbf{w}_\ell\}$ be a basis for W. Let V be the set of vectors of the form

$$\begin{bmatrix} x_1 \\ \vdots \\ x_k \\ y_1 \\ \vdots \\ y_\ell \end{bmatrix}$$

that satisfy $x_1\mathbf{u}_1 + \cdots + x_k\mathbf{u}_k = -y_1\mathbf{w}_1 - \cdots - y_\ell\mathbf{w}_\ell$.

(a) [BB] Show that the function $T: V \to U \cap W$ defined by

$$T\left(\begin{bmatrix} x_1 \\ \vdots \\ x_k \\ y_1 \\ \vdots \\ y_\ell \end{bmatrix}\right) = x_1\mathbf{u}_1 + \cdots + x_k\mathbf{u}_k$$

is a linear transformation; that is, $T(\mathbf{x} + \mathbf{x}') = T(\mathbf{x}) + T(\mathbf{x}')$ for any vectors $\mathbf{x}, \mathbf{x}'$ in V and $T(c\mathbf{x}) = cT(\mathbf{x})$ for any $\mathbf{x}$ in V and any scalar c.

(b) Show that T is *one-to-one*; that is, if $T(\mathbf{x}) = T(\mathbf{x}')$, then $\mathbf{x} = \mathbf{x}'$.

(c) Show that T is *onto*; that is, given a vector $\mathbf{u}$ in $U \cap W$, there exists an $\mathbf{x}$ in V such that $T(\mathbf{x}) = \mathbf{u}$.

■ Critical Reading

17. [BB] Suppose U and W are subspaces of some Euclidean space V with $U \cap W = \mathbf{0}$. Under what condition(s) will $U + W = V$?

18. Suppose U, W, and S are subspaces of a Euclidean space V such that W is contained in S and $U \cap S = \{\mathbf{0}\}$.

(a) Prove that $(U + W) \cap S = W$.

(b) Find $\dim(U + W + S)$.

19. [BB] If A is an $m \times n$ matrix with linearly independent columns and U is the column space of $A^T A$, find $U^{\perp}$.

20. Suppose an $n \times n$ matrix A satisfies $A^2 = A$. Show that $\mathbf{R}^n = \text{null sp } A \oplus \text{range } A$. [Hint: $\mathbf{x} = (\mathbf{x} - A\mathbf{x}) + A\mathbf{x}$.]

21. [BB] Let P be a projection matrix. Given that the eigenvalues of P are $\lambda = 0$ and $\lambda = 1$, show that the corresponding eigenspaces are orthogonal complements.

22. Let A and B be matrices of the same size.

(a) Show that $\text{col sp}(A + B)$ is contained in $\text{col sp } A + \text{col sp } B$.

(b) Show that $\text{rank}(A + B) \leq \text{rank } A + \text{rank } B$.

23. Suppose $\{\mathbf{f}_1, \mathbf{f}_2, \mathbf{f}_3, \mathbf{f}_4\}$ is an orthogonal basis for $V = \mathbf{R}^4$. Let U be the subspace spanned by $\mathbf{f}_1$ and $\mathbf{f}_2$. Identify $U^{\perp}$ and explain your answer.

24. Is it possible for a 3×3 matrix to have row space spanned by $\mathbf{i}$ and $\mathbf{j}$ and $\mathbf{v} = \begin{bmatrix} 1 \\ 1 \\ 1 \end{bmatrix}$ in the null space? Explain.

25. Suppose U and W are subspaces of the vector space V. Prove that $(U + W)^{\perp} = U^{\perp} \cap W^{\perp}$.

26. Have another look at Problem 6.3.27 on p. 381.

The vector $\mathbf{w}$ in $U^{\perp}$ is $\frac{2x-y+z}{6}\begin{bmatrix} 2 \\ -1 \\ 1 \end{bmatrix}$. It's not surprising that this is a multiple of $\begin{bmatrix} 2 \\ -1 \\ 1 \end{bmatrix}$ since this vector is normal to the plane U, but the coefficient contains the expression $2x - y + z$, which comes from the equation of the plane. Could this have been predicted? Do Problem 6.3.27 again, but this time computing $\mathbf{w}$ first, and then obtaining $\mathbf{u}$ as $\mathbf{v} - \mathbf{w}$.

[Hint: Write $\mathbf{w} = \alpha \begin{bmatrix} 2 \\ -1 \\ 1 \end{bmatrix}$ and find α.]

27. (a) Let V be a vector space. Suppose the vectors $\mathbf{u}_1, \mathbf{u}_2, \dots, \mathbf{u}_k$ are a basis for a subspace U and the vectors $\mathbf{w}_1, \mathbf{w}_2, \dots, \mathbf{w}_\ell$ are a basis for $U^{\perp}$. Show that $\mathbf{u}_1, \dots, \mathbf{u}_k, \mathbf{w}_1, \dots, \mathbf{w}_\ell$ are a basis for V.

(b) Use part (a) to show how to write a vector $\mathbf{v}$ in V in the form $\mathbf{v} = \mathbf{u} + \mathbf{w}$ with $\mathbf{u}$ in U and $\mathbf{w}$ in $U^{\perp}$.

(c) Let $V = \mathbf{R}^4$, $U = \left\{ \begin{bmatrix} y \\ x + 2y \\ x \\ x + y \end{bmatrix} \mid x, y \in \mathbf{R} \right\}$, and

$\mathbf{v} = \begin{bmatrix} 1 \\ 1 \\ 1 \\ 1 \end{bmatrix}$. Apply the result of part (b) to write $\mathbf{v}$

as the sum of a vector in U and a vector in $U^\perp$.

28. According to Theorem 6.3.25, if V is a finite dimensional vector space, say $\dim V = n$, and U is a subspace of V, then $\dim U + \dim U^\perp = n$. Give a proof of this result different from the one presented, which follows these lines. Let $\mathcal{U} = \{\mathbf{u}_1, \dots, \mathbf{u}_k\}$ be a basis for U and extend $\mathcal{U}$ to a basis $\mathcal{V}$ of V. Now

apply the Gram–Schmidt process to $\mathcal{V}$, thus obtaining an orthogonal basis $\mathcal{F} = \{\mathbf{f}_1, \dots, \mathbf{f}_n\}$ of V.

(a) The vectors $\mathbf{f}_1, \dots, \mathbf{f}_k$ form a basis of U. Why?

(b) It suffices to show that $\mathbf{f}_{k+1}, \dots, \mathbf{f}_n$ form a basis for $U^\perp$. Why?

(c) The vectors $\mathbf{f}_{k+1}, \dots, \mathbf{f}_n$ are linearly independent. Why?

(d) The vectors $\mathbf{f}_{k+1}, \dots, \mathbf{f}_n$ span $U^\perp$. Why?

(e) The result follows. Why?

29. Let $V = \mathbf{R}^n$. Use the result $\dim U + \dim U^\perp = n$ (Theorem 6.3.25) to give another proof that the sum of the rank and nullity of an $m \times n$ matrix is n.

6.4 The Pseudoinverse of a Matrix

In this section, we continue the discussion of "How to solve a linear system that doesn't have a solution!" which we started in Section 6.1.

Suppose we are given the linear system $A\mathbf{x} = \mathbf{b}$ with A an $m \times n$ matrix. If A has linearly independent columns, the $n \times n$ matrix $A^T A$ is invertible, and we have seen that the vector

$$\mathbf{x}^+ = (A^T A)^{-1} A^T \mathbf{b}$$

is the "best solution" to $A\mathbf{x} = \mathbf{b}$ in the sense that $\|\mathbf{b} - A\mathbf{x}^+\| \leq \|\mathbf{b} - A\mathbf{x}\|$ for any $\mathbf{x} \in \mathbf{R}^n$.

Reading Challenge 1   *Why does A having linearly independent columns mean that $A^T A$ is invertible?*

Our concern in this section is what to do if the columns of A are not linearly independent. In this case, it is useful to remember that the vector we are looking for is a (genuine) solution to $A\mathbf{x} = P\mathbf{b}$, where P is the projection matrix onto the column space of A. (See Figure 6.1.) When the columns of A are linearly independent, the system $A\mathbf{x} = P\mathbf{b}$ has a unique solution.

Reading Challenge 2   *I hope the reader remembers that we had a formula for P: $P = A(A^T A)^{-1} A^T$. Why?*

The astute reader might well ask if there could be several such vectors of minimal length. We shall soon see that this is not the case. There really is a unique shortest vector satisfying $A\mathbf{x} = P\mathbf{b}$.

When the columns of A are not linearly independent, however, $A\mathbf{x} = P\mathbf{b}$ will still have a solution (since $P\mathbf{b}$ is in the column space of A), but no longer a unique one. In fact, when the columns of A are linearly dependent, there is a nonzero vector $\mathbf{x}_0$ satisfying $A\mathbf{x}_0 = \mathbf{0}$, so if $\mathbf{x}_1$ satisfies $A\mathbf{x} = P\mathbf{b}$, so does $\mathbf{x}_1 + \mathbf{x}_0$ because

$A(\mathbf{x}_1 + \mathbf{x}_0) = A\mathbf{x}_1 + A\mathbf{x}_0 = P\mathbf{b} + \mathbf{0} = P\mathbf{b}$. Thus, in general, there is more than one vector $\mathbf{x}$ with $A\mathbf{x} = P\mathbf{b}$, more than one $\mathbf{x}$ with $A\mathbf{x}$ of least distance from $\mathbf{b}$.

We need a convention: Among all the solutions to $A\mathbf{x} = P\mathbf{b}$, let us agree that $\mathbf{x}^+$ will be the shortest one.

6.4.1 EXAMPLE   Suppose $A = \begin{bmatrix} -1 & 0 & 0 \\ 0 & \frac{1}{2} & 0 \\ 0 & 0 & 0 \end{bmatrix}$. The column space of A is the set of all linear combi-

nations of $\begin{bmatrix} -1 \\ 0 \\ 0 \end{bmatrix}$ and $\begin{bmatrix} 0 \\ \frac{1}{2} \\ 0 \end{bmatrix}$, that is, the xy-plane, so the projection $\mathbf{b} = \begin{bmatrix} b_1 \\ b_2 \\ b_3 \end{bmatrix}$ on the

column space of A is $P\mathbf{b} = \begin{bmatrix} b_1 \\ b_2 \\ 0 \end{bmatrix}$. The equation $A\mathbf{x} = P\mathbf{b}$ gives

$$-x_1 = b_1$$
$$\tfrac{1}{2}x_2 = b_2$$
$$0 = 0,$$

so $x_1 = -b_1$, $x_2 = 2b_2$, and x_3 is arbitrary: $\mathbf{x} = \begin{bmatrix} -b_1 \\ 2b_2 \\ x_3 \end{bmatrix}$, a vector of length

$\sqrt{b_1^2 + 4b_2^2 + x_3^2}$. For a given $\mathbf{b}$, the $\mathbf{x}$ of shortest length is the one with $x_3 = 0$,

so $\mathbf{x}^+ = \begin{bmatrix} -b_1 \\ 2b_2 \\ 0 \end{bmatrix}$.   ∎

6.4.2 EXAMPLE   Suppose $A = \begin{bmatrix} \frac{1}{2} & 0 & 0 & 0 \\ 0 & 5 & 0 & 0 \\ 0 & 0 & 0 & 0 \end{bmatrix}$. Again, the column space of A is the xy-plane and $P\mathbf{b} = $

$\begin{bmatrix} b_1 \\ b_2 \\ 0 \end{bmatrix}$ for $\mathbf{b} = \begin{bmatrix} b_1 \\ b_2 \\ b_3 \end{bmatrix}$. With $\mathbf{x} = \begin{bmatrix} x_1 \\ x_2 \\ x_3 \\ x_4 \end{bmatrix}$, the equation $A\mathbf{x} = P\mathbf{b}$ says

$$\tfrac{1}{2}x_1 = b_1$$
$$5x_2 = b_2$$
$$0 = 0,$$

so $x_1 = 2b_1$, $x_2 = \frac{1}{5}b_2$, and x_3 and x_4 are arbitrary. So $\mathbf{x} = \begin{bmatrix} 2b_1 \\ \frac{1}{5}b_2 \\ x_3 \\ x_4 \end{bmatrix}$, a vector of

length $\sqrt{4b_1^2 + \frac{1}{25}b_2^2 + x_3^2 + x_4^2}$. For given $\mathbf{b}$, the shortest $\mathbf{x}$ is $\mathbf{x}^+ = \begin{bmatrix} 2b_1 \\ \frac{1}{5}b_2 \\ 0 \\ 0 \end{bmatrix}$.   ∎

6.4.3 EXAMPLE Suppose $A = \begin{bmatrix} 1 & 1 & 1 \\ 1 & 1 & 1 \end{bmatrix}$. The column space of A is the set of all scalar multiples of $\mathbf{v} = \begin{bmatrix} 1 \\ 1 \end{bmatrix}$. The matrix that projects $\mathbf{b} = \begin{bmatrix} b_1 \\ b_2 \end{bmatrix}$ on the column space of A is $P = \frac{1}{\mathbf{v}^T\mathbf{v}}\mathbf{v}\mathbf{v}^T = \frac{1}{2}\begin{bmatrix} 1 & 1 \\ 1 & 1 \end{bmatrix}$. (See the discussion of a "nice" form for a projection matrix in Section 6.1.) With $\mathbf{x} = \begin{bmatrix} x_1 \\ x_2 \\ x_3 \end{bmatrix}$, the equation $A\mathbf{x} = P\mathbf{b}$ gives $x_1 + x_2 + x_3 = \frac{1}{2}(b_1 + b_2)$,

so $\mathbf{x} = \begin{bmatrix} \frac{1}{2}(b_1 + b_2) - x_2 - x_3 \\ x_2 \\ x_3 \end{bmatrix}$ and

$$\|\mathbf{x}\|^2 = [\tfrac{1}{2}(b_1 + b_2) - x_2 - x_3]^2 + x_2^2 + x_3^2$$

$$= \tfrac{1}{4}(b_1 + b_2)^2 + 2x_2^2 + 2x_3^2 - (b_1 + b_2)(x_2 + x_3) + 2x_2x_3 = f(x_2, x_3)$$

is a function of the two variables x_2, x_3. We use calculus to find the values of x_2 and x_3 that minimize this function. The partial derivatives with respect to x_2 and x_3 are

> If your calculus background makes this next part difficult to follow, don't worry. A better method is just ahead!

$$\frac{\partial f}{\partial x_2} = 4x_2 - (b_1 + b_2) + 2x_3$$

$$\frac{\partial f}{\partial x_3} = 4x_3 - (b_1 + b_2) + 2x_2.$$

Setting these to 0 gives $x_2 = x_3 = \frac{1}{6}(b_1 + b_2)$. Since the Jacobian is positive,

$$\begin{vmatrix} \dfrac{\partial^2 f}{\partial x_2^2} & \dfrac{\partial^2 f}{\partial x_2 \partial x_3} \\ \dfrac{\partial^2 f}{\partial x_3 \partial x_2} & \dfrac{\partial^2 f}{\partial x_3^2} \end{vmatrix} = \begin{vmatrix} 4 & 2 \\ 2 & 4 \end{vmatrix} = 12,$$

we have identified a minimum of f, namely,

$$x_1 = \tfrac{1}{2}(b_1 + b_2) - x_2 - x_3 = \tfrac{1}{2}(b_1 + b_2) - \tfrac{2}{6}(b_1 + b_2) = \tfrac{1}{6}(b_1 + b_2)$$
$$x_2 = \tfrac{1}{6}(b_1 + b_2)$$
$$x_3 = \tfrac{1}{6}(b_1 + b_2).$$

The vector of shortest length is $\mathbf{x}^+ = \dfrac{1}{6}\begin{bmatrix} b_1 + b_2 \\ b_1 + b_2 \\ b_1 + b_2 \end{bmatrix}$. ∎

6.4.4 EXAMPLE

Let $A = \begin{bmatrix} 1 & 1 & 2 \\ -4 & 1 & -3 \\ 3 & -2 & 1 \end{bmatrix}$. The column space of A is the plane with equation $x + y +$

$z = 0$ and the matrix that projects vectors onto this plane is $P = \dfrac{1}{3}\begin{bmatrix} 2 & -1 & -1 \\ -1 & 2 & -1 \\ -1 & -1 & 2 \end{bmatrix}$.

With $\mathbf{x} = \begin{bmatrix} x_1 \\ x_2 \\ x_3 \end{bmatrix}$ and $\mathbf{b} = \begin{bmatrix} b_1 \\ b_2 \\ b_3 \end{bmatrix}$, the system $A\mathbf{x} = P\mathbf{b}$ is

$$\begin{bmatrix} 1 & 1 & 2 \\ -4 & 1 & -3 \\ 3 & -2 & 1 \end{bmatrix}\begin{bmatrix} x_1 \\ x_2 \\ x_3 \end{bmatrix} = \begin{bmatrix} \frac{2}{3}b_1 - \frac{1}{3}b_2 - \frac{1}{3}b_3 \\ -\frac{1}{3}b_1 + \frac{2}{3}b_2 - \frac{1}{3}b_3 \\ -\frac{1}{3}b_1 - \frac{1}{3}b_2 + \frac{2}{3}b_3 \end{bmatrix}.$$

Applying Gaussian elimination to the augmented matrix,

$$\begin{bmatrix} 1 & 1 & 2 & \Big| & \frac{2}{3}b_1 - \frac{1}{3}b_2 - \frac{1}{3}b_3 \\ -4 & 1 & -3 & \Big| & -\frac{1}{3}b_1 + \frac{2}{3}b_2 - \frac{1}{3}b_3 \\ 3 & -2 & 1 & \Big| & -\frac{1}{3}b_1 - \frac{1}{3}b_2 + \frac{2}{3}b_3 \end{bmatrix}$$

$$\rightarrow \begin{bmatrix} 1 & 1 & 2 & \Big| & \frac{2}{3}b_1 - \frac{1}{3}b_2 - \frac{1}{3}b_3 \\ 0 & 5 & 5 & \Big| & \frac{7}{3}b_1 - \frac{2}{3}b_2 - \frac{5}{3}b_3 \\ 0 & -5 & -5 & \Big| & -\frac{7}{3}b_1 + \frac{2}{3}b_2 + \frac{5}{3}b_3 \end{bmatrix} \rightarrow \begin{bmatrix} 1 & 1 & 2 & \Big| & \frac{2}{3}b_1 - \frac{1}{3}b_2 - \frac{1}{3}b_3 \\ 0 & 1 & 1 & \Big| & \frac{7}{15}b_1 - \frac{2}{15}b_2 - \frac{1}{3}b_3 \\ 0 & 0 & 0 & \Big| & 0 \end{bmatrix}.$$

Thus x_3 is free,

$$x_2 = \tfrac{7}{15}b_1 - \tfrac{2}{15}b_2 - \tfrac{1}{3}b_3 - x_3$$

$$x_1 = \tfrac{2}{3}b_1 - \tfrac{1}{3}b_2 - \tfrac{1}{3}b_3 - x_2 - 2x_3$$

$$= \tfrac{2}{3}b_1 - \tfrac{1}{3}b_2 - \tfrac{1}{3}b_3 - \tfrac{7}{15}b_1 + \tfrac{2}{15}b_2 + \tfrac{1}{3}b_3 + x_3 - 2x_3$$

$$= \tfrac{1}{5}b_1 - \tfrac{1}{5}b_2 - x_3,$$

and

$$\mathbf{x} = \begin{bmatrix} \frac{1}{5}b_1 - \frac{1}{5}b_2 - x_3 \\ \frac{7}{15}b_1 - \frac{2}{15}b_2 - \frac{1}{3}b_3 - x_3 \\ x_3 \end{bmatrix}.$$

The length of this vector is a function of x_3, and one can use calculus to show that
the minimum length occurs with $x_3 = \tfrac{2}{9}b_1 - \tfrac{1}{9}b_2 - \tfrac{1}{9}b_3$. We leave it to the reader to

check that for this value of x_3, our desired vector is

$$\mathbf{x}^+ = \frac{1}{45} \begin{bmatrix} -b_1 - 4b_2 + 5b_3 \\ 11b_1 - b_2 - 10b_3 \\ 10b_1 - 5b_2 - 5b_3 \end{bmatrix}. \qquad \blacksquare$$

Reading Challenge 3 *Verify that the column space of the matrix A in **6.4.4** is the plane with equation $x + y + z = 0$. (You must do more than observe that the components of each column satisfy this equation!)*

Reading Challenge 4 *Verify that the matrix that projects vectors onto the plane with equation $x + y + z = 0$ is the matrix P given in **6.4.4**.*

As we have seen, computing the length of a vector whose components involve variables and then finding the values of the variables that minimize this length can be rather formidable tasks when done by hand. Fortunately, there is an alternative method for finding the vector $\mathbf{x}^+$ of shortest length.

In Section 6.3, we showed that the null space and row space of a matrix are orthogonal complements. If A is $m \times n$, the row space of A and the null space of A are subspaces of $\mathbf{R}^n$, so every vector in $\mathbf{R}^n$ can be written uniquely as the sum of a vector in the null space and a vector in the row space—see Theorems 6.3.18 and 6.3.25.

Let $\mathbf{x}_0$ be any solution to

$$A\mathbf{x} = P\mathbf{b} \qquad (1)$$

where P is the matrix that projects vectors in $\mathbf{R}^m$ onto the column space of A and write $\mathbf{x}_0 = \mathbf{x}_r + \mathbf{x}_n$ with $\mathbf{x}_r$ in the row space and $\mathbf{x}_n$ in the null space of A. Since $A\mathbf{x}_n = \mathbf{0}$, we have $A\mathbf{x}_0 = A(\mathbf{x}_r + \mathbf{x}_n) = A\mathbf{x}_r$, so $\mathbf{x}_r$ is also a solution to (1). By the Theorem of Pythagoras—Theorem 6.1.5—$\|\mathbf{x}_0\|^2 = \|\mathbf{x}_r\|^2 + \|\mathbf{x}_n\|^2 \geq \|\mathbf{x}_r\|^2$, so $\|\mathbf{x}_r\| \leq \|\mathbf{x}_0\|$. Since we are seeking a solution to (1) of shortest length, it follows that we should look for a solution in the row space of A.

Wonderfully, there is just one such vector! To see why, suppose $\mathbf{x}_r$ and $\mathbf{x}'_r$ are both solutions to $A\mathbf{x} = P\mathbf{b}$ with both these vectors in the row space, then $A\mathbf{x}_r = A\mathbf{x}'_r$, so $A(\mathbf{x}_r - \mathbf{x}'_r) = \mathbf{0}$. Thus $\mathbf{x}_r - \mathbf{x}'_r$ is in the null space of A as well as the row space. The intersection of a subspace and its orthogonal complement is $\mathbf{0}$. So $\mathbf{x}_r - \mathbf{x}'_r = \mathbf{0}$ and $\mathbf{x}_r = \mathbf{x}'_r$.

We summarize.

Theorem 6.4.5 *Suppose A is an $m \times n$ matrix and $\mathbf{b}$ is a vector in $\mathbf{R}^m$. Then the best solution to $A\mathbf{x} = \mathbf{b}$—that is, the shortest vector among those minimizing $\|\mathbf{b} - A\mathbf{x}\|$—is the unique vector $\mathbf{x}^+$ in the row space of A with the property that $A\mathbf{x}^+ = P\mathbf{b}$ is the projection of $\mathbf{b}$ on the column space of A.*

The matrix A^+ is also known as the *Moore–Penrose* inverse.

Since the map $\mathbf{b} \mapsto \mathbf{x}^+$ is a linear transformation, the "best solution" to $A\mathbf{x} = \mathbf{b}$ can be written in the form $\mathbf{x}^+ = A^+\mathbf{b}$ for a matrix A^+ called the *pseudoinverse* of A. (See Exercise 7.)

6.4.6 The "best solution" to $A\mathbf{x} = \mathbf{b}$ is $\mathbf{x}^+ = A^+\mathbf{b}$.

6.4.7 DEFINITION

The *pseudoinverse* of a matrix A is the matrix A^+ satisfying

1. $A^+\mathbf{b}$ is in the row space of A for all $\mathbf{b}$, and
2. $AA^+ = P$ is the projection matrix onto the column space of A.

Equivalently, A^+ is defined by the condition that $A^+\mathbf{b} = \mathbf{x}^+$ is the shortest vector among those minimizing $\|\mathbf{b} - A\mathbf{x}\|$.

6.4.8 Remark If A is $m \times n$, the equation $\mathbf{x}^+ = A^+\mathbf{b}$ means that A^+ is $n \times m$.

6.4.9 EXAMPLE

If $A = \begin{bmatrix} -1 & 0 & 0 \\ 0 & \frac{1}{2} & 0 \\ 0 & 0 & 0 \end{bmatrix}$ is the matrix of **6.4.1**, we found

$$\mathbf{x}^+ = \begin{bmatrix} -b_1 \\ 2b_2 \\ 0 \end{bmatrix} = \begin{bmatrix} -1 & 0 & 0 \\ 0 & 2 & 0 \\ 0 & 0 & 0 \end{bmatrix} \begin{bmatrix} b_1 \\ b_2 \\ b_3 \end{bmatrix},$$

so $A^+ = \begin{bmatrix} -1 & 0 & 0 \\ 0 & 2 & 0 \\ 0 & 0 & 0 \end{bmatrix}$. ∎

6.4.10 EXAMPLE

If $A = \begin{bmatrix} \frac{1}{2} & 0 & 0 & 0 \\ 0 & 5 & 0 & 0 \\ 0 & 0 & 0 & 0 \end{bmatrix}$ is the matrix of **6.4.2**, we had

$$\mathbf{x}^+ = \begin{bmatrix} 2b_1 \\ \frac{1}{5}b_2 \\ 0 \\ 0 \end{bmatrix} = \begin{bmatrix} 2 & 0 & 0 \\ 0 & \frac{1}{5} & 0 \\ 0 & 0 & 0 \\ 0 & 0 & 0 \end{bmatrix} \begin{bmatrix} b_1 \\ b_2 \\ b_3 \end{bmatrix},$$

so $A^+ = \begin{bmatrix} 2 & 0 & 0 \\ 0 & \frac{1}{5} & 0 \\ 0 & 0 & 0 \\ 0 & 0 & 0 \end{bmatrix}$. ∎

6.4.11 Remark Examples 6.4.9 and 6.4.10 illustrate a general fact. If $A = \begin{bmatrix} D & 0 \\ 0 & 0 \end{bmatrix}$ is an $m \times n$ matrix, where $D = \begin{bmatrix} \mu_1 & & \\ & \ddots & \\ & & \mu_r \end{bmatrix}$ is an $r \times r$ matrix with nonzero diagonal entries, the **0**s representing zero matrices of the sizes needed to complete an $m \times n$ matrix, then A^+ is the $n \times m$ matrix

$$A^+ = \begin{bmatrix} D^{-1} & \mathbf{0} \\ \mathbf{0} & \mathbf{0} \end{bmatrix}, \quad \text{with} \quad D^{-1} = \begin{bmatrix} \frac{1}{\mu_1} & & \\ & \ddots & \\ & & \frac{1}{\mu_r} \end{bmatrix},$$

this time the big **0**s denoting blocks of 0s of the sizes needed for an $n \times m$ matrix.

Reading Challenge 5   *Suppose* $A = \begin{bmatrix} -2 & 0 \\ 0 & 3 \\ 0 & 0 \end{bmatrix}$. *What is* A^+?

6.4.12 EXAMPLE   With $A = \begin{bmatrix} 1 & 1 & 2 \\ -4 & 1 & -3 \\ 3 & -2 & 1 \end{bmatrix}$ the matrix of **6.4.4**, we found

$$\mathbf{x}^+ = \frac{1}{45}\begin{bmatrix} -b_1 - 4b_2 + 5b_3 \\ 11b_1 - b_2 - 10b_3 \\ 10b_1 - 5b_2 - 5b_3 \end{bmatrix} = \frac{1}{45}\begin{bmatrix} -1 & -4 & 5 \\ 11 & -1 & -10 \\ 10 & -5 & -5 \end{bmatrix}\begin{bmatrix} b_1 \\ b_2 \\ b_3 \end{bmatrix},$$

so $A^+ = \frac{1}{45}\begin{bmatrix} -1 & -4 & 5 \\ 11 & -1 & -10 \\ 10 & -5 & -5 \end{bmatrix}$.   ■

6.4.13 EXAMPLE   Here, we use the definition of "pseudoinverse" to find A^+ with $A = \begin{bmatrix} 1 & 1 & 1 \\ 1 & 1 & 1 \end{bmatrix}$, the matrix of **6.4.3**. We require $A^+\mathbf{b}$ to be in the row space of A, so $A^+\mathbf{b} = t\begin{bmatrix} 1 \\ 1 \\ 1 \end{bmatrix}$ for some scalar t. We also want $A(A^+\mathbf{b}) = t\begin{bmatrix} 3 \\ 3 \end{bmatrix}$ to be the projection of $P\mathbf{b}$ of $\mathbf{b}$ on the column space of A. This is the matrix $P = \frac{1}{2}\begin{bmatrix} 1 & 1 \\ 1 & 1 \end{bmatrix}$, as before. The condition $A(A^+\mathbf{b}) = P\mathbf{b}$ says

$$t\begin{bmatrix} 3 \\ 3 \end{bmatrix} = \frac{1}{2}\begin{bmatrix} b_1 + b_2 \\ b_1 + b_2 \end{bmatrix}.$$

Thus $3t = \frac{1}{2}(b_1 + b_2)$ and $t = \frac{1}{6}(b_1 + b_2)$. We obtain

$$A^+\mathbf{b} = \frac{1}{6}(b_1 + b_2)\begin{bmatrix} 1 \\ 1 \\ 1 \end{bmatrix} = \begin{bmatrix} \frac{1}{6}(b_1 + b_2) \\ \frac{1}{6}(b_1 + b_2) \\ \frac{1}{6}(b_1 + b_2) \end{bmatrix} = \frac{1}{6}\begin{bmatrix} 1 & 1 \\ 1 & 1 \\ 1 & 1 \end{bmatrix}\begin{bmatrix} b_1 \\ b_2 \end{bmatrix}$$

so that $A^+ = \frac{1}{6}\begin{bmatrix} 1 & 1 \\ 1 & 1 \\ 1 & 1 \end{bmatrix}$, as before. This approach, without calculus, seems easier. ■

■ Special Classes of Matrices

Certain classes of matrices have predictable pseudoinverses that do not have to be determined from the definition as was necessary in **6.4.13**.

1. A is invertible.

Suppose first that A is an invertible $n \times n$ matrix. In this case, the column space of A is $\mathbf{R}^n$ and the matrix that projects vectors in $\mathbf{R}^n$ onto $\mathbf{R}^n$ is I, the identity matrix. The pseudoinverse A^+ of A has the property that AA^+ is projection onto the column space of A. Here, we have $AA^+ = I$, so, since A is invertible, $A^+ = A^{-1}$. The pseudoinverse is supposed to provide the "best solution" to $A\mathbf{x} = \mathbf{b}$ in all cases. Here, this is certainly the case.

> **6.4.14** If A is invertible, $A^+ = A^{-1}$.

2. A is a projection matrix.

Suppose A is an $n \times n$ projection matrix. Then $A^2 = A$ and $A^T = A$, by Theorem 6.1.17. In this case, the matrix that projects vectors onto the column space of A is A itself, so the second property of A^+ given in 6.4.7 says $AA^+ = A$. Therefore, $A(A^+ - A) = \mathbf{0}$, since $A^2 = A$. This implies $A(A^+\mathbf{b} - A\mathbf{b}) = \mathbf{0}$ for any vector $\mathbf{b}$. So $A^+\mathbf{b} - A\mathbf{b}$ is in the null space of A for any $\mathbf{b}$. But this vector is also in the row space of A because $A^+\mathbf{b}$ is in the row space and so is $A\mathbf{b} = A^T\mathbf{b}$. Since the row space and the null space of A are orthogonal complements (so their intersection is $\mathbf{0}$), we obtain $A^+\mathbf{b} - A\mathbf{b} = \mathbf{0}$ for all $\mathbf{b}$, so $A^+ = A$.

> **6.4.15** If A is a projection matrix, $A^+ = A$.

3. A has linearly independent columns.

Suppose the $m \times n$ matrix A has linearly independent columns. The results of Section 6.1 show that the matrix that projects vectors onto the column space of A is $P = A(A^T A)^{-1} A^T$. Thus $AA^+ = A(A^T A)^{-1} A^T$. Multiplying on the left by A^T gives

$$A^T A A^+ = A^T A (A^T A)^{-1} A^T = A^T$$

(since the product of $A^T A$ and its inverse is I). Since $A^T A$ is invertible, we obtain $A^+ = (A^T A)^{-1} A^T$.

> **6.4.16** If A has linearly independent columns, then $A^+ = (A^T A)^{-1} A^T$.

4. A has linearly independent rows.

Suppose the $m \times n$ matrix A has linearly independent rows. By Theorem 4.4.7, the columns of A span $\mathbf{R}^m$. Thus the column space of A is $\mathbf{R}^m$ and the projection P onto the column space is I. Also, rank $A = m$. By Theorem 4.3.40, rank $AA^T = $ rank A, so the $m \times m$ matrix AA^T is invertible. Now the argument is very similar to that used in the projection matrix case. We have $I = AA^+ = AA^T(AA^T)^{-1}$, hence $A[(A^+ - A^T(AA^T)^{-1})\mathbf{b}] = \mathbf{0}$ for any vector $\mathbf{b}$, so $(A^+ - A^T(AA^T)^{-1})\mathbf{b}$ is in the null space of A, for any $\mathbf{b}$. It is also in the row space of A, so it is $\mathbf{0}$. We obtain $A^+ - A^T(AA^T)^{-1} = \mathbf{0}$.

> **6.4.17** If A has linearly independent rows, then $A^+ = A^T(AA^T)^{-1}$.

6.4.18 Remark There are, of course, matrices not of the types we have considered here. To find the pseudoinverse in these cases, one can always resort to the definition. There are examples in the exercises. In particular, see Exercise 3. (Another way to find the pseudoinverse is described in Section 7.5.)

Answers to Reading Challenges

1. If the $m \times n$ matrix A has linearly independent columns, it has column rank and hence rank n. Thus rank $A^TA = $ rank $A = n$ too. Since A^TA is $n \times n$, it is invertible.

2. If $A\mathbf{x}_1 = A\mathbf{x}_2$, then $A(\mathbf{x}_1 - \mathbf{x}_2) = \mathbf{0}$. The columns of A are linearly independent, so $\mathbf{x}_1 - \mathbf{x}_2 = \mathbf{0}$. Thus $\mathbf{x}_1 = \mathbf{x}_2$.

3. The components of each column satisfy $x + y + z = 0$, so the column space is contained in the plane with this equation. The first two columns of A are linearly independent, so the column space has dimension at least 2. The plane has dimension 2, so the column space must be the plane.

4. First find two linearly independent vectors in the plane. We could use two columns of the matrix, but $\begin{bmatrix} 1 \\ -1 \\ 0 \end{bmatrix}$ and $\begin{bmatrix} 1 \\ 0 \\ -1 \end{bmatrix}$ are simpler to handle. Let $B = \begin{bmatrix} 1 & 1 \\ -1 & 0 \\ 0 & -1 \end{bmatrix}$ be the matrix with these vectors as columns. Then $P = B(B^TB)^{-1}B^T$ is the desired projection matrix—see 6.1.11. We have $B^TB = \begin{bmatrix} 1 & -1 & 0 \\ 1 & 0 & -1 \end{bmatrix} \begin{bmatrix} 1 & 1 \\ -1 & 0 \\ 0 & -1 \end{bmatrix} =$

$\begin{bmatrix} 2 & 1 \\ 1 & 2 \end{bmatrix}$, $(B^TB)^{-1} = \frac{1}{3}\begin{bmatrix} 2 & -1 \\ -1 & 2 \end{bmatrix}$, and

$$P = B(B^TB)^{-1}B^T = \frac{1}{3}\begin{bmatrix} 1 & 1 \\ -1 & 0 \\ 0 & -1 \end{bmatrix}\begin{bmatrix} 2 & -1 \\ -1 & 2 \end{bmatrix}\begin{bmatrix} 1 & -1 & 0 \\ 1 & 0 & -1 \end{bmatrix}$$

$$= \frac{1}{3}\begin{bmatrix} 2 & -1 & -1 \\ -1 & 2 & -1 \\ -1 & -1 & 2 \end{bmatrix}.$$

5. $A^+ = \begin{bmatrix} -\frac{1}{2} & 0 & 0 \\ 0 & \frac{1}{3} & 0 \end{bmatrix}$

True/False Questions

Decide, with as little calculation as possible, whether each of the following statements is true or false and explain your answer whenever you say "false." (Answers can be found in the back of the book.)

1. If A is an $m \times n$ matrix, the pseudoinverse A^+ of A is $n \times m$.

2. If A is an $m \times n$ matrix with pseudoinverse A^+, then AA^+ is the projection matrix onto the column space of A.

3. The pseudoinverse of $A = \begin{bmatrix} 5 & 0 & 0 \\ 0 & 7 & 0 \end{bmatrix}$ is $\begin{bmatrix} -5 & 0 & 0 \\ 0 & -7 & 0 \end{bmatrix}$.

4. If $A = \begin{bmatrix} 2 & 0 & 0 \\ 0 & -3 & 0 \\ 0 & 0 & 0 \\ 0 & 0 & 0 \end{bmatrix}$, then $(A^+)^+ = A$.

5. If $A = \begin{bmatrix} 1 & 0 \\ -2 & 1 \end{bmatrix}$, then $A^+ = \begin{bmatrix} 1 & 0 \\ 2 & 1 \end{bmatrix}$.

6. The pseudoinverse of $\begin{bmatrix} \cos\theta & \sin\theta \\ -\sin\theta & \cos\theta \end{bmatrix}$ is $\begin{bmatrix} \cos(-\theta) & \sin(-\theta) \\ -\sin(-\theta) & \cos(-\theta) \end{bmatrix}$.

7. The pseudoinverse of a matrix A with linearly independent rows is $A^+ = (A^T A)^{-1} A^T$.

8. The pseudoinverse of the matrix $A = \begin{bmatrix} 1 & 2 & 3 & 4 & 5 \\ -2 & 0 & 1 & 1 & 7 \end{bmatrix}$ can be calculated from the formula $A^+ = A^T (AA^T)^{-1}$.

9. The pseudoinverse of a projection matrix is its inverse.

Exercises

*Solutions to exercises marked [BB] can be found in the **B**ack of the **B**ook.*

1. [BB] Let $A = \begin{bmatrix} 1 & 1 & 1 & 1 \\ 0 & 1 & -1 & 0 \end{bmatrix}$.

   (a) The matrix A has a right inverse. Why?

   (b) Why must AA^T be invertible?

   (c) Find the pseudoinverse of A.

2. Let $A = \begin{bmatrix} 1 & 0 & 0 \\ 1 & 1 & 1 \\ 1 & -1 & 1 \\ 1 & 2 & 4 \end{bmatrix}$.

   (a) The matrix A has a left inverse. Why?

   (b) Why must $A^T A$ be invertible?

   (c) Find the pseudoinverse of A.

3. [BB] Let $A = \begin{bmatrix} 1 & 1 & 2 \\ -4 & 1 & -3 \\ 3 & -2 & 1 \end{bmatrix}$ be the matrix of Example **6.4.4**. Use the definition of *pseudoinverse* (6.4.7) to find A^+. (The calculations will be simpler if you first note that the row space of A is spanned by
   $$\mathbf{u} = \begin{bmatrix} 1 \\ 0 \\ 1 \end{bmatrix} \text{ and } \mathbf{v} = \begin{bmatrix} 0 \\ 1 \\ 1 \end{bmatrix}. \text{ Why is this the case?)}$$

4. Find the pseudoinverse of A and the "best solution" to $A\mathbf{x} = \mathbf{b}$ in each of the following cases.

   (a) [BB] $A = \begin{bmatrix} 1 & -1 & 0 \end{bmatrix}$; $\mathbf{b} = [10]$

   (b) $A = \begin{bmatrix} 1 \\ -2 \\ 0 \\ 2 \end{bmatrix}$; $\mathbf{b} = \begin{bmatrix} 1 \\ 1 \\ 1 \\ 1 \end{bmatrix}$

(c) $A = \begin{bmatrix} -1 & 2 & 3 & 1 \\ 1 & 0 & 1 & 1 \end{bmatrix}$; $\mathbf{b} = \begin{bmatrix} -1 \\ 7 \end{bmatrix}$

(d) [BB] $A = \begin{bmatrix} 1 & -1 \\ 1 & -1 \\ 1 & -1 \end{bmatrix}$; $\mathbf{b} = \begin{bmatrix} 2 \\ 0 \\ 1 \end{bmatrix}$

(e) $A = \begin{bmatrix} 1 & 2 \\ 0 & 1 \\ -1 & 0 \end{bmatrix}$; $\mathbf{b} = \begin{bmatrix} 4 \\ -1 \\ 2 \end{bmatrix}$

(f) $A = \begin{bmatrix} 1 & 2 \\ 2 & 4 \\ 3 & 6 \end{bmatrix}$; $\mathbf{b} = \begin{bmatrix} 1 \\ 1 \\ 3 \end{bmatrix}$

(g) [BB] $A = \begin{bmatrix} 1 & -1 & 0 \\ 0 & 0 & 1 \\ 1 & -1 & 0 \end{bmatrix}$; $\mathbf{b} = \begin{bmatrix} -1 \\ 0 \\ 1 \end{bmatrix}$

(h) $A = \begin{bmatrix} 1 & 0 & 1 & 0 \\ 0 & 0 & 1 & 1 \\ 1 & 0 & 1 & 1 \end{bmatrix}$; $\mathbf{b} = \begin{bmatrix} -5 \\ 1 \\ 2 \end{bmatrix}$.

5. (a) [BB] If a matrix A has linearly independent columns, its pseudoinverse is a left inverse of A. Why?

(b) If a matrix A has linearly independent rows, its pseudoinverse is a right inverse of A. Why?

6. Let μ_1, μ_2, μ_3 be nonzero scalars and let
$$A = \begin{bmatrix} \mu_1 & 0 & 0 & 0 & 0 \\ 0 & \mu_2 & 0 & 0 & 0 \\ 0 & 0 & \mu_3 & 0 & 0 \\ 0 & 0 & 0 & 0 & 0 \end{bmatrix}.$$

Show that $A^+ = \begin{bmatrix} \frac{1}{\mu_1} & 0 & 0 & 0 \\ 0 & \frac{1}{\mu_2} & 0 & 0 \\ 0 & 0 & \frac{1}{\mu_3} & 0 \\ 0 & 0 & 0 & 0 \\ 0 & 0 & 0 & 0 \end{bmatrix}$, thus justifying Remark 6.4.11.

7. [BB] Let A be an $m \times n$ matrix and let $\mathbf{b}$ be a vector in $\mathbf{R}^m$. Show that the vector $\mathbf{x}^+$ defined in Theorem 6.4.5 can be written in the form $\mathbf{x}^+ = B\mathbf{b}$ for some matrix B. (Hint: Show that the map $\mathbf{b} \to \mathbf{x}^+$ is a linear transformation $\mathbf{R}^m \to \mathbf{R}^n$ and apply Theorem 5.1.13.)

8. If $A = \begin{bmatrix} D & \mathbf{0} \\ \mathbf{0} & \mathbf{0} \end{bmatrix}$ as in 6.4.11, show that $(A^+)^+ = A$.

■ Critical Reading

9. [BB] Why is the pseudoinverse of a permutation matrix its transpose?

10. Find a simple formula for the pseudoinverse of a matrix A whose columns are orthonormal.

11. Suppose A is an $m \times n$ matrix with the property that $A\mathbf{x} = \mathbf{b}$ has at least one solution for any vector $\mathbf{b}$ in $\mathbf{R}^m$. Find a formula for A^+.

12. Find a formula for the pseudoinverse of an $m \times n$ matrix of rank n.

CHAPTER KEY WORDS AND IDEAS: Here are some technical words and phrases that were used in this chapter. Do you know the meaning of each? If you're not sure, check the glossary or index at the back of the book.

best solution in the sense of least squares
direct sum of subspaces
Gram–Schmidt algorithm
Householder matrix
intersection of subspaces
length of a vector
norm of a vector
orthogonal complement
orthogonal matrix
orthogonal set of vectors
orthogonal subspaces

orthonormal vectors
permutation matrix
projection matrix
projection of a vector
   on a subspace
pseudoinverse
Pythagoras' Theorem
rotation matrix
sum of subspaces
symmetric matrix

Review Exercises for Chapter 6

1. Find the matrix P that projects vectors onto each of the following spaces.

   (a) the subspace of $\mathbf{R}^3$ spanned by $\mathbf{v} = \begin{bmatrix} -2 \\ 7 \\ 1 \end{bmatrix}$;

   (b) the plane spanned by the orthogonal vectors
   $$\mathbf{e} = \begin{bmatrix} 1 \\ -2 \\ 1 \end{bmatrix} \text{ and } \mathbf{f} = \begin{bmatrix} -3 \\ 1 \\ 5 \end{bmatrix};$$

   (c) the plane in $\mathbf{R}^3$ with equation $x + 2y = 0$;

   (d) the subspace U of $\mathbf{R}^4$ consisting of vectors of the form
   $$\begin{bmatrix} -t - s \\ 2t + s \\ s - t \\ s + 2t \end{bmatrix}.$$

2. Let $A = \begin{bmatrix} 3 & 1 \\ 2 & 0 \\ -1 & 1 \end{bmatrix}$ and $\mathbf{b} = \begin{bmatrix} -1 \\ 7 \\ 2 \end{bmatrix}$.

   (a) The column space of A is a plane. Find the equation of this plane.

   (b) The system $A\mathbf{x} = \begin{bmatrix} 2 \\ 1 \\ 0 \end{bmatrix}$ has a solution. Without any calculation, explain why.

   (c) Find the matrix that projects vectors onto the column space of A.

   (d) Find the best solution in the sense of least squares to the system
   $$\begin{aligned} 3x_1 + x_2 &= -1 \\ 2x_1 &= 7 \\ -x_1 + x_2 &= 2. \end{aligned}$$

   (e) What does *best solution in the sense of least squares* mean?

3. (a) Find the matrix P that projects vectors in $\mathbf{R}^3$ onto the plane π with equation $2x - y + z = 0$.

   (b) Find the vector in π closest to $\begin{bmatrix} -3 \\ 1 \\ 1 \end{bmatrix}$.

4. Can a projection matrix have nonzero determinant? Explain.

5. Let $\mathbf{u}_1 = \begin{bmatrix} 1 \\ 2 \\ 3 \end{bmatrix}$, $\mathbf{u}_2 = \begin{bmatrix} 1 \\ 0 \\ -1 \end{bmatrix}$, $\mathbf{u}_3 = \begin{bmatrix} -1 \\ -2 \\ 5 \end{bmatrix}$ and
   $$\mathbf{u}_4 = \begin{bmatrix} 2 \\ 1 \\ -1 \end{bmatrix}.$$
   If the Gram–Schmidt algorithm were applied to these vectors, it would fail. Why?

6. Apply the Gram–Schmidt algorithm to $\mathbf{u}_1 = \begin{bmatrix} 1 \\ 0 \\ 1 \end{bmatrix}$,
   $$\mathbf{u}_2 = \begin{bmatrix} 1 \\ 1 \\ 1 \end{bmatrix}, \mathbf{u}_3 = \begin{bmatrix} -1 \\ 0 \\ 1 \end{bmatrix}.$$

7. Let U be the subspace of $\mathbf{R}^4$ spanned by $\mathbf{u}_1 = \begin{bmatrix} 1 \\ -1 \\ 2 \\ -3 \end{bmatrix}$
   and $\mathbf{u}_2 = \begin{bmatrix} 0 \\ 1 \\ -1 \\ 0 \end{bmatrix}$. Find an orthogonal basis for U and an orthogonal basis for $U^\perp$.

8. (a) Apply the Gram–Schmidt algorithm to
   $$\mathbf{u}_1 = \begin{bmatrix} 1 \\ 0 \\ -1 \\ 0 \end{bmatrix}, \mathbf{u}_2 = \begin{bmatrix} 1 \\ 0 \\ 1 \\ 1 \end{bmatrix}, \mathbf{u}_3 = \begin{bmatrix} 1 \\ -1 \\ 0 \\ 0 \end{bmatrix}.$$

   (b) Find a QR factorization of
   $$A = \begin{bmatrix} 1 & 1 & 1 \\ 0 & 0 & -1 \\ -1 & 1 & 0 \\ 0 & 1 & 0 \end{bmatrix}.$$

   (c) Use the result of Exercise 8(b) to find A^{-1}.

9. Let $A = \begin{bmatrix} -1 & 1 \\ 0 & 1 \\ 1 & -2 \end{bmatrix}$ and $B = \begin{bmatrix} 1 & -1 & 2 \\ 3 & 0 & 1 \end{bmatrix}$.

   One of these matrices has a QR factorization and the other does not. Which does not? Explain. Find a QR factorization of the other.

10. Let $A = QR$ be a QR factorization of an $m \times n$ matrix A with linearly independent columns.

    (a) Show that the columns of R are linearly independent.

    (b) Assuming Q is $m \times n$, why is R invertible?

11. Write $\mathbf{v} = \mathbf{u} + \mathbf{w}$ with $\mathbf{u}$ in U and $\mathbf{w}$ in $U^{\perp}$ in each of the following situations.

(a) $V = \mathbf{R}^2$, $\mathbf{v} = \begin{bmatrix} x \\ y \end{bmatrix}$, $U = \{\begin{bmatrix} -2t \\ 3t \end{bmatrix} \mid t \in \mathbf{R}\}$

(b) $V = \mathbf{R}^3$, $\mathbf{v} = \begin{bmatrix} x \\ y \\ z \end{bmatrix}$, U is the span of $\begin{bmatrix} 1 \\ -3 \\ 2 \end{bmatrix}$

(c) $V = \mathbf{R}^3$, $\mathbf{v} = \begin{bmatrix} x \\ y \\ z \end{bmatrix}$, $U = \{\begin{bmatrix} x \\ y \\ z \end{bmatrix} \mid x + y - 3z = 0\}$

(d) $V = \mathbf{R}^4$, $\mathbf{v} = \begin{bmatrix} 1 \\ 2 \\ -3 \\ -4 \end{bmatrix}$,

$U = \{\begin{bmatrix} -t+r \\ 2t+s+r \\ t-s+r \\ t-s+r \end{bmatrix} \mid t, s, r \in \mathbf{R}\}$.

12. Let A be an $m \times n$ matrix with linearly independent rows. Formulate a factorization theorem for A and prove that your statement is correct.

13. Find an orthogonal basis for the subspace of $\mathbf{R}^4$ defined by $x_1 + x_2 + x_3 + x_4 = 0$.

14. Let $U = \mathrm{sp}\{\mathbf{u}_1, \mathbf{u}_2, \mathbf{u}_3\}$, where $\mathbf{u}_1 = \begin{bmatrix} -1 \\ 0 \\ 1 \\ 0 \end{bmatrix}$, $\mathbf{u}_2 = \begin{bmatrix} 1 \\ 0 \\ 1 \\ 1 \end{bmatrix}$,

$\mathbf{u}_3 = \begin{bmatrix} 1 \\ 0 \\ 0 \\ -1 \end{bmatrix}$.

(a) Use the Gram–Schmidt algorithm to find an orthogonal basis for W.

(b) Find the vector in U closest to $\begin{bmatrix} 1 \\ 1 \\ 1 \\ 0 \end{bmatrix}$.

15. (a) Apply the Gram–Schmidt algorithm to the vectors

$\mathbf{a}_1 = \begin{bmatrix} -1 \\ 0 \\ 1 \\ 1 \end{bmatrix}$, $\mathbf{a}_2 = \begin{bmatrix} 0 \\ 1 \\ 1 \\ 1 \end{bmatrix}$.

(b) Find the matrix that projects vectors in $\mathbf{R}^4$ onto the span of $\mathbf{a}_1$ and $\mathbf{a}_2$.

(c) Find a QR factorization of the matrix

$A = \begin{bmatrix} -1 & 0 \\ 0 & 1 \\ 1 & 1 \\ 1 & 1 \end{bmatrix}$ whose columns are the vectors

of part (a).

(d) Explain how a QR factorization helps find the best solution (in the sense of least squares) to $A\mathbf{x} = \mathbf{b}$ and apply this method to find the best solution to

$$\begin{aligned} -x & & = 1 \\ & y & = 2 \\ x + y & = 2 \\ x + y & = 1. \end{aligned}$$

16. Let A be a 7×3 matrix and B is 3×7. Did you answer Review Exercise 20(c) of Chapter 4? Were you able to explain why $AB = I$ is impossible? What about the possibility that $BA = I$? Explain.

17. Find $U^{\perp}$ in each of the following cases.

(a) $V = \mathbf{R}^3$, U is the span of $\mathbf{u}_1 = \begin{bmatrix} 2 \\ 2 \\ -4 \end{bmatrix}$ and

$\mathbf{u}_2 = \begin{bmatrix} -2 \\ -1 \\ -3 \end{bmatrix}$

(b) $V = \mathbf{R}^4$, $U = \{\begin{bmatrix} x - 2y \\ -x \\ -2y \\ x + 4y \end{bmatrix} \mid x, y \in \mathbf{R}\}$.

18. Let A be a matrix with pseudoinverse A^{+}.

(a) In what sense is $\mathbf{x} = A^{+}\mathbf{b}$ the "best solution" to $A\mathbf{x} = \mathbf{b}$?

(b) Find the pseudoinverse of $A = \begin{bmatrix} -3 \\ 4 \end{bmatrix}$.

19. Let $A = \begin{bmatrix} 1 & 1 & 1 \\ 1 & 2 & 3 \\ 1 & 3 & 6 \end{bmatrix}$.

(a) Are the columns of A linearly independent or linearly dependent?

(b) Are the rows of A linearly independent or linearly dependent?

(c) Find the pseudoinverse of A.

20. Let A be an $m \times n$ matrix and B an $n \times p$ matrix.

    (a) Prove that nullity $A +$ nullity $B \geq$ nullity AB.
[Hint: Let $\mathbf{x}_1, \ldots, \mathbf{x}_k$ be a basis for null sp B and let $B\mathbf{y}_1, \ldots, B\mathbf{y}_\ell$ be a basis for null sp $A \cap$ range B. The vectors $\mathbf{x}_1, \ldots, \mathbf{x}_k, \mathbf{y}_1, \ldots, \mathbf{y}_\ell$ all lie in null sp AB, so it is sufficient to show that these are linearly independent. Explain and complete the proof.]

    (b) Restate the inequality in part (a) as an assertion about rank A, rank B, and rank AB.

21. Let $T: \mathbf{R}^3 \to \mathbf{R}^3$ be a linear transformation such that

$$T\left(\begin{bmatrix} 1 \\ 2 \\ 0 \end{bmatrix}\right) = \begin{bmatrix} 1 \\ 1 \\ 1 \end{bmatrix}, \ T\left(\begin{bmatrix} 1 \\ 0 \\ 2 \end{bmatrix}\right) = \begin{bmatrix} -1 \\ -1 \\ 1 \end{bmatrix} \text{ and}$$

$$T\left(\begin{bmatrix} 1 \\ 1 \\ 2 \end{bmatrix}\right) = \begin{bmatrix} 0 \\ 0 \\ 2 \end{bmatrix}.$$

Find ker $T \cap$ range T and ker $T +$ range T.

7

The Spectral Theorem

7.1   Complex Numbers and Vectors

This section, and the next two sections as well, can be omitted by those wishing to proceed as quickly as possible to a treatment of the diagonalization of real symmetric matrices, a topic discussed without using complex numbers and vectors in Section 7.4.

The negative integers were invented so that we could solve equations such as $x + 7 = 0$, at a time when the natural numbers $1, 2, 3, \ldots$ were the only "numbers" known. The rational numbers (common fractions) were introduced so that equations such as $5x - 7 = 0$ would have solutions. The real numbers include the rationals, but also solutions to equations like $x^2 - 2 = 0$ and numbers like π and e, which are not solutions to any polynomial equations (with integer coefficients). The real numbers are much more numerous than the rationals, but there are still not enough to provide solutions to equations like $x^2 + 1 = 0$. It is a remarkable fact that adding to the real numbers just the solution to this single equation stops the process. If we let i be a number satisfying $i^2 + 1 = 0$, then every polynomial with real coefficients has a root of the form $a + bi$ with a and b real. Such a number is called *complex* and the amazing fact we have just stated—every polynomial with real coefficients has a complex root—is known as the "Fundamental Theorem of Algebra."

A *complex number* is a number of the form $a + bi$, where a and b are real numbers. The *real part* of $a + bi$ is a; the *imaginary part* is b. The set of all complex numbers is denoted **C**.

For example, $2 - 3i$, $\frac{1}{2} + \frac{17}{5}i$, $-16 = -16 + 0i$, and $\pi = \pi + 0i$ are complex numbers, the last two examples serving to illustrate the fact that any real number is also a complex number (with imaginary part 0): $\mathbf{R} \subseteq \mathbf{C}$.

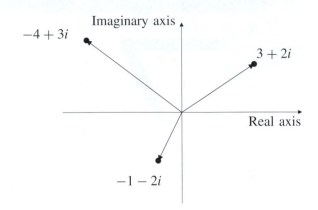

Figure 7.1 The complex plane

Keeping in mind that $i^2 = -1$, complex numbers can be added and multiplied in the more or less obvious way:

$$(a + bi) + (c + di) = (a + c) + (b + d)i,$$

$$(a + bi)(c + di) = (ac - bd) + (ad + bc)i.$$

7.1.1 EXAMPLES

- $(2 - 3i) + (4 + i) = 6 - 2i$;
- $4 + (7 - i) = 11 - i$;
- $i(4 + 5i) = -5 + 4i$;
- $(4 + 3i)(2 - i) = 8 - 4i + 6i - 3i^2 = 11 + 2i$. ■

■ The Complex Plane

The familiar Cartesian plane can be used to picture complex numbers by associating the point (a, b) with the complex number $a + bi$. See Figure 7.1. It is common to picture $a + bi$ as the vector $\begin{bmatrix} a \\ b \end{bmatrix}$ too. Results from Section 1.1 then show that addition of complex numbers can be viewed geometrically as vector addition according to the parallelogram rule: the sum of $a + bi$ and $c + di$ is the diagonal of the parallelogram with sides the vectors $\begin{bmatrix} a \\ b \end{bmatrix}$ and $\begin{bmatrix} c \\ d \end{bmatrix}$—see Figure 1.5. When the Cartesian plane is used to picture complex numbers, it is called the *complex plane*. The x-axis is called the *real axis* and the y-axis is the *imaginary axis*.

■ Complex Conjugation

7.1.2 DEFINITION

The *conjugate* of the complex number $a + bi$ is the complex number $a - bi$. We denote the conjugate of $a + bi$ by $\overline{a + bi}$.

If $z = a + bi$, then $\bar{z} = a - bi$.

7.1.3 EXAMPLE

$\overline{1 + 2i} = 1 - 2i$; $\overline{2 - 3i} = 2 + 3i$; $\overline{-5} = -5$. ■

If $z = a = a + 0i$ is real, then $\bar{z} = a - 0i = a = z$. Conversely, if $z = a + bi$ and $\bar{z} = z$, then $a - bi = a + bi$, so $2bi = 0$ and $b = 0$.

> **7.1.4** A complex number z is real if and only if $\bar{z} = z$.

Suppose we multiply the complex number $z = a + bi$ by its conjugate. We have

$$z\bar{z} = (a + bi)(a - bi) = a^2 - b^2 i^2 = a^2 + b^2.$$

Representing z as the vector $\mathbf{v} = \begin{bmatrix} a \\ b \end{bmatrix}$, the product $z\bar{z}$ is just $\|\mathbf{v}\|^2$, the square of the length of $\mathbf{v}$.

7.1.5 DEFINITION

The *modulus* of the complex number $z = a + bi$ is $\sqrt{a^2 + b^2}$ and is denoted $|z|$.

> If $z = a + bi$, then $|z| = \sqrt{a^2 + b^2}$. Thus $|z|^2 = a^2 + b^2 = z\bar{z}$.

7.1.6 EXAMPLE

If $z = 2 - 3i$, $|z|^2 = z\bar{z} = (2 - 3i)(2 + 3i) = 2^2 + 3^2 = 13$. Thus $|z| = \sqrt{13}$. ∎

The modulus of z is also called its *absolute value* because, when $z = a + 0i$ is real, $|z| = \sqrt{a^2} = |a|$ is the absolute value of the real number z in the usual sense.

Proposition 7.1.7

If z and w are complex numbers, then

1. $\bar{\bar{z}} = z$,
2. $\overline{zw} = \bar{z}\,\bar{w}$, *and*
3. $|zw| = |z||w|$.

Proof 1. Let $z = a + bi$. Then $\bar{z} = a - bi$, so $\bar{\bar{z}} = \overline{a - bi} = a + bi = z$.

2. If $z = a + bi$ and $w = c + di$, then

$$zw = (a + bi)(c + di) = (ac - bd) + (ad + bc)i, \text{ so } \overline{zw} = (ac - bd) - (ad + bc)i.$$

On the other hand, $\bar{z} = a - bi$, $\bar{w} = c - di$, and $(bi)(di) = bdi^2 = -bd$, so

$$\bar{z}\,\bar{w} = (a - bi)(c - di) = (ac - bd) - (ad + bc)i = (ac - bd) - (ad + bc)i.$$

3. We have $|zw|^2 = zw\overline{zw} = zw\bar{z}\,\bar{w} = z\bar{z}w\bar{w} = |z|^2|w|^2$. Taking square roots gives $|zw| = |z||w|$. ∎

7.1.8 EXAMPLE

Let $z = 2 + 3i$ and $w = 4 - 7i$. Then $zw = 29 - 2i$, so $\overline{zw} = 29 + 2i$. On the other hand, $\bar{z}\,\bar{w} = (2 - 3i)(4 + 7i) = 29 + 2i$. Thus $\overline{zw} = \bar{z}\,\bar{w}$. ∎

■ Division of Complex Numbers

We learn early in life that real numbers can be added and multiplied. As we have seen, complex numbers can be added and multiplied as well. What about inverses? What about division? Any nonzero real number b has a *multiplicative inverse* $\frac{1}{b}$, which is a real number satisfying $b\frac{1}{b} = 1$. If a and b are real numbers and $b \neq 0$, there is a real number $\frac{a}{b}$ with the property that $b\frac{a}{b} = a$.

If $a + bi$ is a nonzero complex number, we can write $\frac{1}{a+bi}$ with the expectation that $(a+bi)\frac{1}{a+bi} = 1$, but is $\frac{1}{a+bi}$ also a complex number? Remember that complex numbers are numbers of the form $x + yi$. Here the notion of complex conjugation comes to the rescue. If $z = a + bi$, then $z\bar{z} = a^2 + b^2$ is a **real number**, which is zero if and only if $a = b = 0$, that is, if and only if $z = 0$. Thus, if $z \neq 0$, $\frac{1}{z\bar{z}}$ is a real number and we have

$$\frac{1}{z} = \frac{1}{z\bar{z}}\bar{z} = \frac{1}{a^2+b^2}\bar{z} = \frac{a}{a^2+b^2} - \frac{b}{a^2+b^2}i.$$

Thus $\frac{1}{z}$ is indeed a complex number, with real part $\frac{a}{a^2+b^2}$ and imaginary part $\frac{-b}{a^2+b^2}$.

> **7.1.9** If z is a nonzero complex number, then z has a complex multiplicative inverse, namely, $\frac{1}{z\bar{z}}\bar{z}$.

7.1.10 EXAMPLES

- $\dfrac{1}{2+3i} = \dfrac{1}{(2+3i)(2-3i)}(2-3i) = \dfrac{1}{13}(2-3i) = \dfrac{2}{13} - \dfrac{3}{13}i$;

- $\dfrac{1}{2-i} = \dfrac{1}{5}(2+i) = \dfrac{2}{5} + \dfrac{1}{5}i$ since $(2+i)(2-i) = 2^2 + 1^2 = 5$. ■

With multiplicative inverses in hand, we now have a way to **divide** one complex number by another nonzero complex number. Just as $\frac{a}{b} = a\frac{1}{b}$, so $\frac{w}{z} = w\frac{1}{z} = w\frac{\bar{z}}{z\bar{z}}$:

> **7.1.11** $\dfrac{w}{z} = \dfrac{w}{z\bar{z}}\bar{z}$.

7.1.12 EXAMPLE

$\dfrac{1+2i}{3-i} = \dfrac{1+2i}{(3-i)(3+i)}(3+i) = \dfrac{1}{10}(1+2i)(3+i) = \dfrac{1}{10}(1+7i) = \dfrac{1}{10} + \dfrac{7}{10}i.$ ■

■ The Polar Form of a Complex Number

Just as addition of complex numbers has a nice geometrical interpretation (it's like the addition of vectors), so does multiplication. To describe this, we note that if the point (a, b) in the plane is at distance r from the origin, then

$$\begin{aligned} a &= r\cos\theta \\ b &= r\sin\theta, \end{aligned} \tag{1}$$

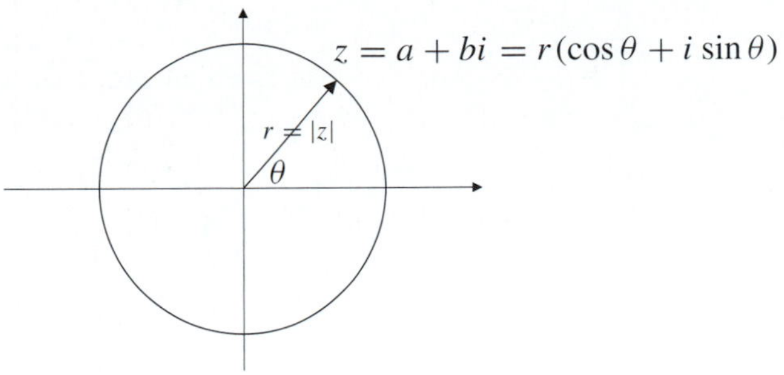

Figure 7.2

where θ is the angle between the real axis and the ray from $(0, 0)$ to (a, b). As usual, we agree that the counterclockwise direction is the positive direction for measuring angles. Thus $a + bi = r\cos\theta + (r\sin\theta)i = r(\cos\theta + i\sin\theta)$. See Figure 7.2. With the agreement that $0 \le \theta < 2\pi$, the representation $z = r(\cos\theta + i\sin\theta)$ is called the *polar form* of the complex number z. By comparison, we call the representation $z = a + bi$ its *standard form*.

7.1.13
$$z = a + bi \qquad \textbf{standard form}$$
$$z = r(\cos\theta + i\sin\theta), \ 0 \le \theta < 2\pi \qquad \textbf{polar form}$$

When $z = r(\cos\theta + i\sin\theta)$ is written in polar form, the number r is just $|z|$, the modulus of z. The angle θ is called its *argument*. Equations (1) express the real and imaginary parts of z in terms of its modulus and its argument. Conversely, the equations

$$|z| = \sqrt{a^2 + b^2}$$
$$\theta = \arccos\frac{a}{\sqrt{a^2 + b^2}} = \arcsin\frac{b}{\sqrt{a^2 + b^2}} \qquad (2)$$

express the modulus and argument in terms of the real and imaginary parts of z. In fact, knowing either $\arccos\theta$ or $\arcsin\theta$ as well as the quadrant where z lies is sufficient to identify θ uniquely in the interval $0 \le \theta < 2\pi$.

7.1.14 EXAMPLE

If $z = 2 + 2i$, then $r = |z| = \sqrt{2^2 + 2^2} = \sqrt{8}$ and $\cos\theta = \frac{2}{2\sqrt{2}} = \frac{1}{\sqrt{2}}$. Since z is in the first quadrant, $\theta = \frac{\pi}{4}$. Thus, in polar form, $z = 2s\sqrt{2}\left(\cos\frac{\pi}{4} + i\sin\frac{\pi}{4}\right)$. ■

7.1.15 EXAMPLE

If $z = -1 + i\sqrt{3}$, then $r = |z| = \sqrt{1 + 3} = \sqrt{4} = 2$ and $\cos\theta = -\frac{1}{2}$. Since θ is in quadrant II, $\theta = \frac{2\pi}{3}$.

The polar form of z is $z = 2\sqrt{2}\left(\cos\frac{2\pi}{s3} + i\sin\frac{2\pi}{3}\right)$. ■

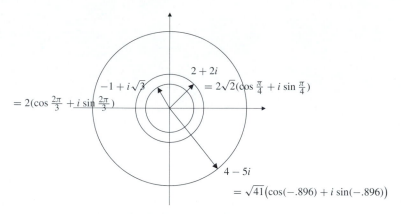

Figure 7.3

7.1.16 EXAMPLE   If $z = 4 - 5i$, then $r = \sqrt{16 + 25} = \sqrt{41}$ and $\cos\theta = \frac{4}{\sqrt{41}}$. A calculator gives $\arccos \frac{4}{\sqrt{41}} \approx .896$ rads $\approx 51°$. Since θ is in quadrant IV, $\theta = 2\pi - .896 \approx 5.387$ and the polar form of z is $\sqrt{41}(\cos 5.387 + i \sin 5.387)$. ∎

Now let

$$z = r(\cos\alpha + i \sin\alpha)$$
$$w = s(\cos\beta + i \sin\beta)$$

be complex numbers in polar form and observe that

$$zw = rs[(\cos\alpha + i \sin\alpha)(\cos\beta + i \sin\beta)]$$
$$= rs[(\cos\alpha \cos\beta - \sin\alpha \sin\beta) + i(\cos\alpha \sin\beta + \sin\alpha \cos\beta)].$$

The terms on the right remind us of the so-called addition rules for sines and cosines:

$$\cos(\alpha + \beta) = \cos\alpha \cos\beta - \sin\alpha \sin\beta$$

and

$$\sin(\alpha + \beta) = \sin\alpha \cos\beta + \cos\alpha \sin\beta.$$

Thus $zw = rs[\cos(\alpha + \beta) + i \sin(\alpha + \beta)]$.

7.1.17 To multiply complex numbers, multiply moduli and add arguments:
$$zw = rs[\cos(\alpha + \beta) + i \sin(\alpha + \beta)].$$

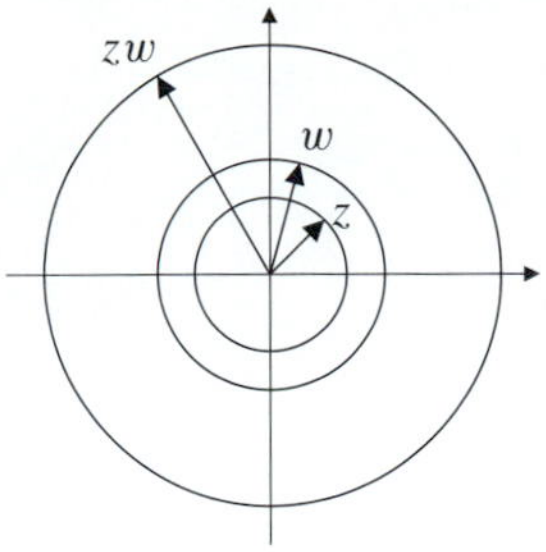

Figure 7.4 To multiply complex numbers, multiply moduli and add arguments.

See Figure 7.4. Similarly,

7.1.18 To divide complex numbers, divide moduli and subtract arguments:
$$\frac{z}{w} = \frac{r}{s}\Big[\cos(\alpha - \beta) + i \sin(\alpha - \beta)\Big].$$

7.1.19 EXAMPLE   Suppose $z = 6\big(\cos\frac{\pi}{3} + i \sin\frac{\pi}{3}\big)$ and $w = 3\big(\cos\frac{\pi}{4} + i \sin\frac{\pi}{4}\big)$. Then

$$zw = 18\Big[\cos\Big(\frac{\pi}{3} + \frac{\pi}{4}\Big) + i \sin\Big(\frac{\pi}{3} + \frac{\pi}{4}\Big)\Big] = 18\Big[\cos\frac{7\pi}{12} + i \sin\frac{7\pi}{12}\Big];$$

$$\frac{z}{w} = 2\Big[\cos\Big(\frac{\pi}{3} - \frac{\pi}{4}\Big) + i \sin\Big(\frac{\pi}{3} - \frac{\pi}{4}\Big)\Big] = 2\Big[\cos\frac{\pi}{12} + i \sin\frac{\pi}{12}\Big]. \qquad \blacksquare$$

The powers z^n of a complex number z are products, of course. Thus, if $z = r(\cos\theta + i \sin\theta)$, multiplying moduli and adding arguments gives

$$z^2 = r^2(\cos 2\theta + i \sin 2\theta)$$

$$z^3 = zz^2 = r^3(\cos 3\theta + i \sin 3\theta)$$

and, in general,

$$z^n = r^n(\cos n\theta + i \sin n\theta).$$

This last statement is known as *De Moivre's Formula*.

Abraham de Moivre (1667–1754), geometer and probabilist, is best known for this formula.

7.1.20   If $z = r(\cos\theta + i \sin\theta)$, then   **De Moivre's Formula**
$$z^n = r^n(\cos n\theta + i \sin n\theta).$$

7.1.21 PROBLEM   Find z^{17} if $z = -1 + i\sqrt{3}$.

Solution.   As in **7.1.15**, the modulus of z is $r = 2$ and the argument is $\theta = \frac{2\pi}{3}$. The polar form of z is $z = 2\big(\cos\frac{2\pi}{3} + i \sin\frac{2\pi}{3}\big)$. By De Moivre's formula,

$$z^{17} = 2^{17}\left(\cos\frac{34\pi}{3} + i\sin\frac{34\pi}{3}\right)$$

$$= 2^{17}\left(\cos\frac{4\pi}{3} + i\sin\frac{4\pi}{3}\right)$$

$$= 2^{17}\left(-\frac{1}{2} - i\frac{\sqrt{3}}{2}\right) = -2^{16} - (2^{16}\sqrt{3})i.$$

Obviously, polar form is the natural form for multiplication of complex numbers. Who would contemplate computing $(-1 + i\sqrt{3})^{17}$ as the notation suggests,

$$(-1 + \sqrt{3}i)^{17} = \underbrace{(-1 + \sqrt{3}i)(-1 + \sqrt{3}i)\cdots(-1 + \sqrt{3}i)}_{17 \text{ factors}} \ ?$$

On the other hand, standard form is the obvious way to add complex numbers. Who would convert to polar form if asked to add $-1 + i\sqrt{3}$ and $2 - 2i$?

> **7.1.22** To add or subtract complex numbers, use standard form.
> To multiply or divide complex numbers, use polar form.

7.1.23 Remark In a calculus course, the student has probably been introduced to the base e of the natural logarithms and perhaps seen the famous *Taylor Expansion*

$$e^x = 1 + x + \frac{x^2}{2} + \frac{x^3}{6} + \frac{x^4}{24} + \cdots,$$

which implies such approximations as

$$e^1 = 1 + 1 + \frac{1}{2} + \frac{1}{6} + \frac{1}{24} + \cdots \approx 2.708$$

(in fact, $e \approx 2.718281828459\ldots$) and

$$\sqrt{e} = e^{1/2} = 1 + \frac{1}{2} + \frac{\frac{1}{4}}{2} + \frac{\frac{1}{8}}{6} + +\frac{\frac{1}{16}}{24} + \cdots \approx 1.648.$$

In the theory of complex variables, the complex valued function e^{ix} is defined by $e^{ix} = \cos x + i\sin x$. With this definition, many of the definitions and results of this section become "obvious." For example, if $z = r(\cos\alpha + i\sin\alpha)$ and $w = s(\cos\beta + i\sin\beta)$, then $z = re^{i\alpha}$ and $w = se^{i\beta}$, so

$$zw = re^{i\alpha}se^{i\beta} = rse^{i(\alpha+\beta)}$$

(see 7.1.17), while

$$\frac{z}{w} = \frac{re^{i\alpha}}{se^{i\beta}} = \frac{r}{x}e^{i\alpha}e^{-i\beta} = \frac{r}{s}e^{i(\alpha-\beta)}$$

(see 7.1.18). De Moivre's formula 7.1.20 for the nth power of z becomes simply

$$z^n = (re^{i\theta})^n = r^n e^{in\theta} = r^n(\cos n\theta + i \sin n\theta).$$

■ Roots of Polynomials

We are now ready to discuss the Fundamental Theorem of Algebra, which was mentioned briefly at the start of this section. This theorem was proven by Karl Friedrich Gauss (1777–1855) in 1797 when he was just 20 years of age!

Theorem 7.1.24   **(Fundamental Theorem of Algebra).** *Every polynomial with real coefficients has a root in the complex numbers.*

Let r_1 be a (complex) root of the polynomial $f(x)$. Then we can factor $f(x)$

$$f(x) = (x - r_1)q_1(x),$$

where $q_1(x)$ is a polynomial of degree one less than the degree of $f(x)$. The Fundamental Theorem of Algebra says that $q_1(x)$ also has a complex root, so $q_1(x) = (x - r_2)q_2(x)$ and

$$f(x) = (x - r_1)(x - r_2)q_2(x),$$

with $q_2(x)$ a polynomial of degree two less than the degree of $f(x)$. Continuing in this way, we eventually have

$$f(x) = (x - r_1)(x - r_2) \cdots (x - r_n)q(x),$$

where $q(x)$ has degree 0; that is, $q(x) = c$ is a constant. Thus the Fundamental Theorem of Algebra implies that a polynomial $f(x)$ of degree n can be expressed

$$f(x) = c(x - r_1)(x - r_2) \cdots (x - r_n)$$

as the product of a constant and n polynomial factors, each of degree one. Each r_i is a root of $f(x)$, that is, $f(r_i) = 0$, so $f(x)$ has n roots, $r_1, r_2, \ldots, r_n$, provided we count *multiplicities* (double roots twice, triple roots three times, and so on).

7.1.25 EXAMPLES
- $x^2 - 3x - 4 = (x + 1)(x - 4)$ is a polynomial of degree two; it has two roots;
- the polynomial $x^4 - 1 = (x^2 - 1)(x^2 + 1) = (x - 1)(x + 1)(x - i)(x + i)$ has degree four and four roots;
- the polynomial $x^2 - 6x + 9 = (x - 3)^2$ has degree two and also two roots if we agree to count the double root 3 twice. We say that 3 is a root of *multiplicity* two. ■

The polynomial $f(z) = z^n - a$ has n roots called the *nth roots of a*, but what are they? To find out, write a and z in polar form:

$$a = r(\cos\theta + i\sin\theta)$$

$$z = s(\cos\alpha + i\sin\alpha).$$

If $z^n - a = 0$, then $z^n = a$, so

$$s^n(\cos n\alpha + i\sin n\alpha) = r(\cos\theta + i\sin\theta).$$

Equating moduli and arguments, we have $s^n = r$ and $n\alpha = \theta + 2k\pi$ for some integer k; that is,

$$s = \sqrt[n]{r} \quad \text{and} \quad \alpha = \frac{\theta + 2k\pi}{n}.$$

7.1.26 The n solutions to $z^n = r(\cos\alpha + i\sin\alpha)$ are

$$z = \sqrt[n]{r}\left(\cos\frac{\theta + 2k\pi}{n} + i\sin\frac{\theta + 2k\pi}{n}\right), \qquad k = 0, 1, 2, \ldots, n-1.$$

These solutions correspond to points distributed evenly around the circumference of the circle with center $(0, 0)$ and radius r, as shown in Figure 7.5.

7.1.27 PROBLEM   Find all complex numbers z such that $z^4 = -64$.

Solution.   In polar form, the complex number $-64 = 64(\cos\pi + i\sin\pi)$. The solutions to $z^4 = 64(\cos\pi + i\sin\pi)$ are

$$\sqrt[4]{64}\left(\cos\frac{\pi + 2k\pi}{4} + i\sin\frac{\pi + 2k\pi}{4}\right), \quad k = 0, 1, 2, 3.$$

Since $(2^6)^{1/4} = 2^{3/2} = 2\sqrt{2}$, these solutions are

Figure 7.5   The n solutions to $z^n = r$.

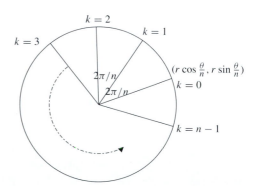

$$z_1 = 2\sqrt{2}\Big(\cos\frac{\pi}{4} + i\sin\frac{\pi}{4}\Big) = 2\sqrt{2}\Big(\frac{\sqrt{2}}{2} + i\frac{\sqrt{2}}{2}\Big) = 2(1+i) \qquad (k=0)$$

$$z_2 = 2\sqrt{2}\Big(\cos\frac{3\pi}{4} + i\sin\frac{3\pi}{4}\Big) = 2\sqrt{2}\Big(-\frac{\sqrt{2}}{2} + i\frac{\sqrt{2}}{2}\Big) = 2(-1+i) \qquad (k=1)$$

$$z_3 = 2\sqrt{2}\Big(\cos\frac{5\pi}{4} + i\sin\frac{5\pi}{4}\Big) = 2(-1-i) \qquad (k=2)$$

$$z_4 = 2\sqrt{2}\Big(\cos\frac{7\pi}{4} + i\sin\frac{7\pi}{4}\Big) = 2(1-i). \qquad (k=3)$$

7.1.28 PROBLEM   Find all complex numbers z such that $z^3 = -1 + i\sqrt{3}$.

Solution.   In polar form, $a = -1 + i\sqrt{3} = 2\Big(\cos\frac{2\pi}{3} + i\sin\frac{2\pi}{3}\Big)$. The solutions of $z^3 = a$ are $\sqrt[3]{2}\Big(\cos\frac{\frac{2\pi}{3}+2k\pi}{3} + i\sin\frac{\frac{2\pi}{3}+2k\pi}{3}\Big)$, $k = 0, 1, 2$; that is,

$$z_1 = \sqrt[3]{2}\Big(\cos\frac{2\pi}{9} + i\sin\frac{2\pi}{9}\Big) \approx .965 + .810i \qquad (k=0)$$

$$z_2 = \sqrt[3]{2}\Big(\cos\frac{8\pi}{9} + i\sin\frac{8\pi}{9}\Big) \approx -1.184 + .431i \qquad (k=1)$$

$$z_3 = \sqrt[3]{2}\Big(\cos\frac{14\pi}{9} + i\sin\frac{14\pi}{9}\Big) \approx .289 - 1.241i. \qquad (k=2)$$

7.1.29 Remark As these problems show, the solutions to $z^n = a$, that is, the nth roots of the complex number a, lie on a circle of radius $\sqrt[n]{|a|}$) and are equally spaced around the circle, successive arguments differing by $\frac{2\pi}{n}$. In particular, taking $a = 1$, the n *roots of unity* are spaced equally around the unit circle.

7.1.30 PROBLEM   Solve $z^2 = 1$.

Solution.   We wish to solve $z^2 = \cos 0 + i\sin 0$. The solutions are $\cos\frac{0+2k\pi}{2} + i\sin\frac{0+2k\pi}{2}$, $k = 0, 1$; that is,

$$z_1 = \cos 0 + i\sin 0 = 1 \qquad (k=0)$$

$$z_2 = \cos\pi + i\sin\pi = -1. \qquad (k=1)$$

There wasn't anything new here!

7.1.31 PROBLEM   Solve $z^3 = 1$.

Solution. We wish to solve $z^3 = \cos 0 + i \sin 0$. The solutions are $\cos \frac{0+2k\pi}{3} + i \sin \frac{0+2k\pi}{3}$, $k = 0, 1, 2$; that is,

$$z_1 = \cos 0 + i \sin 0 = 1 \qquad\qquad (k = 0)$$

$$z_2 = \cos \frac{2\pi}{3} + i \sin \frac{2\pi}{3} = -\frac{1}{2} + i \sin \frac{\sqrt{3}}{2} \qquad (k = 1)$$

$$z_3 = \cos \frac{4\pi}{3} + i \sin \frac{4\pi}{3} = -\frac{1}{2} - i \sin \frac{\sqrt{3}}{2}. \qquad (k = 2)$$

■ Complex Vector Spaces

Until this point, the word "scalar" has been synonymous with "real number." In fact, it could equally well mean "complex number." In the definition of vector space given in Definition 4.2.1, the word "scalar" could have meant "complex number" just as easily. Furthermore, all our theorems and proofs about *real* vector spaces—that is, vector spaces where the scalars are real numbers—carry over verbatim to *complex* vector spaces, where the scalars are complex numbers. The primary differences between real and complex vector spaces concern the dot product.

Just as $\mathbf{R}^n$ is our basic and most important example of a real vector space, so

$$\mathbf{C}^n = \left\{ \begin{bmatrix} z_1 \\ z_2 \\ \vdots \\ z_n \end{bmatrix} \mid z_1, z_2, \ldots, z_n \in \mathbf{C} \right\}$$

is the key example of a complex vector space. But how should we define the dot product in $\mathbf{C}^n$?

Let $\mathbf{v}$ be the vector $\begin{bmatrix} 1 \\ i \end{bmatrix}$ in $\mathbf{C}^2$. Defining dot product as in $\mathbf{R}^n$, we get $\|\mathbf{v}\|^2 = \mathbf{v} \cdot \mathbf{v} = 1^2 + i^2 = 0$, which is surely not very desirable—a nonzero vector should have nonzero length!

7.1.32 DEFINITIONS   The *complex dot product* of vectors $\mathbf{z} = \begin{bmatrix} z_1 \\ \vdots \\ z_n \end{bmatrix}$ and $\mathbf{w} = \begin{bmatrix} w_1 \\ \vdots \\ w_n \end{bmatrix}$ in $\mathbf{C}^n$ is

$$\mathbf{z} \cdot \mathbf{w} = \bar{z}_1 w_1 + \bar{z}_2 w_2 + \cdots + \bar{z}_n w_n$$

and the *norm* or *length* of $\mathbf{z}$ is $\|\mathbf{z}\| = \sqrt{\mathbf{z} \cdot \mathbf{z}}$.

7.1.33 EXAMPLES   If $\mathbf{z} = \begin{bmatrix} 1 \\ i \end{bmatrix}$ and $\mathbf{w} = \begin{bmatrix} 2+i \\ 3-i \end{bmatrix}$ are vectors in $\mathbf{C}^2$,

- $\mathbf{z} \cdot \mathbf{w} = 1(2+i) - i(3-i) = 2+i - 3i - 1 = 1 - 2i$;
- $\mathbf{w} \cdot \mathbf{z} = (2-i)(1) + (3+i)(i) = 2 - i + 3i - 1 = 1 + 2i$;

- $\|\mathbf{z}\|^2 = \mathbf{z} \cdot \mathbf{z} = 1(1) - i(i) = 1 - i^2 = 2;$
- $\|\mathbf{w}\|^2 = \mathbf{w} \cdot \mathbf{w} = (2 - i)(2 + i) + (3 + i)(3 - i) = 4 + 1 + 9 + 1 = 15.$ ■

Notice in particular that with our modified definition of dot product, the length of $\mathbf{z} = \begin{bmatrix} 1 \\ i \end{bmatrix}$ is now $\sqrt{2}$, and no longer 0. This is no accident.

7.1.34 PROBLEM   Show that with the definitions of dot product and norm given in 7.1.32, nonzero vectors in $\mathbf{C}^n$ have nonzero length.

Solution.   Let $\mathbf{z} = \begin{bmatrix} z_1 \\ z_2 \\ \vdots \\ v_n \end{bmatrix}$ be a vector in $\mathbf{C}^n$. We want to show that if $\mathbf{z} \neq 0$, then $\|\mathbf{z}\| \neq 0$. We have

$$\|\mathbf{z}\|^2 = \mathbf{z} \cdot \mathbf{z} = \overline{z_1}z_1 + \overline{z_2}z_2 + \cdots + \overline{z_n}z_n = |z_1|^2 + |z_2|^2 + \cdots + |z_n|^2. \tag{3}$$

For any complex number $z = a + bi$, $|z|^2 = a^2 + b^2$ is the sum of squares of two real numbers, so $z = 0$ if and only if $|z| = 0$. Equation (3) shows that $\|\mathbf{z}\|^2$ is the sum of squares of real numbers, so $\|\mathbf{z}\|^2 \geq 0$, and 0 is possible only if each number being squared is 0; that is, if and only if each $z_i = 0$. This occurs if and only if $\mathbf{z}$ is the zero vector, so, if $\mathbf{z} \neq \mathbf{0}$, then $\|\mathbf{z}\| \neq 0$.

■ Concluding Remarks

1. The complex dot product of two vectors $\mathbf{x} = \begin{bmatrix} x_1 \\ \vdots \\ x_n \end{bmatrix}$ and $\mathbf{y} = \begin{bmatrix} y_1 \\ \vdots \\ y_n \end{bmatrix}$ in $\mathbf{R}^n$—that is, vectors whose entries are all **real**—is just the usual dot product in $\mathbf{R}^n$ because $\mathbf{x} \cdot \mathbf{y} = \overline{x_1}y_1 + \cdots + \overline{x_n}y_n = x_1y_1 + \cdots x_ny_n$. For example, if $\mathbf{x} = \begin{bmatrix} -2 \\ 4 \end{bmatrix}$ and $\mathbf{y} = \begin{bmatrix} 7 \\ -3 \end{bmatrix}$, then $\mathbf{x} \cdot \mathbf{y} = -14 - 12 = -26$ whether one interprets "$\cdot$" as the complex dot product or the usual (real) one.

2. The complex dot product has all the properties of the dot product in $\mathbf{R}^n$ listed in Theorem 1.2.19, with two exceptions.

   i. As illustrated in **7.1.33**,

   $$\mathbf{w} \cdot \mathbf{z} = \overline{\mathbf{z} \cdot \mathbf{w}}. \tag{4}$$

   ii. While $\mathbf{z} \cdot (c\mathbf{w}) = c(\mathbf{z} \cdot \mathbf{w})$ as one might expect—see Exercise 12—if c is a factor of the first argument, it comes outside as $\overline{c}$, the complex conjugate of c:

   $$(c\mathbf{z}) \cdot \mathbf{w} = \overline{c}(\mathbf{z} \cdot \mathbf{w}). \tag{5}$$

To see this, let $\mathbf{z} = \begin{bmatrix} z_1 \\ \vdots \\ z_n \end{bmatrix}$ and $\mathbf{w} = \begin{bmatrix} w_1 \\ \vdots \\ w_n \end{bmatrix}$. Then $c\mathbf{z} = \begin{bmatrix} cz_1 \\ \vdots \\ cz_n \end{bmatrix}$, so

$$(c\mathbf{z}) \cdot \mathbf{w} = \overline{cz_1}w_1 + \overline{cz_2}w_2 + \cdots + \overline{cz_n}w_n$$
$$= (\overline{c}\,\overline{z_1})w_1 + (\overline{c}\,\overline{z_2})w_2 + \cdots + (\overline{c}\,\overline{z_n})w_n$$
$$= \overline{c}(\overline{z_1}w_1 + \overline{z_2}w_2 + \cdots + \overline{z_n}w_n) = \overline{c}(\mathbf{z} \cdot \mathbf{w}).$$

True/False Questions

Decide, with as little calculation as possible, whether each of the following statements is true or false and explain your answer whenever you say "false." (Answers can be found in the back of the book.)

1. The integer 7 is a rational number.

2. The solutions to $x^2 - 2 = 0$ are irrational numbers.

3. The Fundamental Theorem of Algebra says that every polynomial with real coefficients has a root in the complex numbers.

4. If $z = 7$, $\overline{7} = -7$.

5. If z is a complex number, $|z| = z\overline{z}$.

6. $|3 + 4i| = 5$.

7. If $z = a + bi \neq 0$ with a and b real, then there exist real numbers c and d such that $\dfrac{1}{z} = c + di$.

8. $1 + i = 2(\cos \frac{\pi}{4} + i \sin \frac{\pi}{4})$.

9. If $z = 2(\cos \frac{\pi}{3} + i \sin \frac{\pi}{3})$ and $w = 3(\cos \frac{\pi}{4} + i \sin \frac{\pi}{4})$, then $zw = 12(\cos \frac{7\pi}{12} + i \sin \frac{7\pi}{12})$.

10. De Moivre's Formula says that if $z = r(\cos \theta + i \sin \theta)$ and n is a positive integer, then $z^n = r^n(\cos n\theta + i \sin n\theta)$.

11. -1 has five distinct fifth (complex) roots.

12. If $z = \begin{bmatrix} 1 \\ i \end{bmatrix}$ and $w = \begin{bmatrix} -1 \\ 2i \end{bmatrix}$, then $z \cdot w = -3$.

Exercises

*Solutions to exercises marked [BB] can be found in the **B**ack of the **B**ook.*

1. Write each of the following complex numbers in polar form: $1 + \sqrt{3}i$ [BB], $-7i$, $5 - 5i$.

2. Express each of the following complex numbers in standard form.

   (a) [BB] $\dfrac{3 - 4i}{1 + i}$ (b) $\dfrac{1 + 2i}{1 - 4i}$ (c) [BB] $\dfrac{-2 - 2i}{-3 - 3i}$

   (d) $\dfrac{3 + 3i}{2 - i}$ (e) $\dfrac{-i}{4 + 2i}$ (f) $\dfrac{-2 + 3i}{1 - i}$.

3. [BB] Let $z = 2 - 2i$. Express z, $\overline{z}$, z^2, and $1/z$ in standard and polar forms.

4. Answer Exercise 3 with $z = \sqrt{3} - i$.

5. [BB] Let $z = 1 - 2i$ and $w = 3 + i$. Express zw, $\frac{z}{w}$, $z^2 - w^2$, and $\frac{z+w}{z-w}$ in standard form.

6. Answer Exercise 5 with $z = 3 + 3i$ and $w = 2 - 4i$.

7. [BB] Let $z = i$ and $w = 2 + i$, and express $\frac{z\overline{w}+z^2}{w\overline{z}+w^2}$ as a complex number in standard form.

8. Let z and w be complex numbers. Show that the complex conjugate of $z + w$ is $\overline{z} + \overline{w}$.

9. [BB] Show that the real and imaginary parts of a complex number z are, respectively, $\frac{1}{2}(z + \overline{z})$ and $\frac{1}{2i}(z - \overline{z})$.

10. Let $\mathbf{z} = \begin{bmatrix} z_1 \\ z_2 \\ \vdots \\ z_n \end{bmatrix}$ be a vector in $\mathbf{C}^n$. Define the norm or length of $\mathbf{z}$ by $\|\mathbf{z}\|^2 = \mathbf{z} \cdot \mathbf{z}$, "$\cdot$" denoting the complex dot product. Show that $\|\mathbf{z}\|$ is a real number.

11. [BB] Let c be a complex number and $\mathbf{z} = \begin{bmatrix} z_1 \\ z_2 \\ \vdots \\ z_n \end{bmatrix}$ a vector in $\mathbf{C}^n$. Show that $\|c\mathbf{z}\| = |c| \|\mathbf{z}\|$.

12. (a) [BB] Use equations (4) and (5) to show that $\mathbf{z} \cdot (c\mathbf{w}) = c(\mathbf{z} \cdot \mathbf{w})$ for any complex number c and any vectors $\mathbf{z}$ and $\mathbf{w}$ in $\mathbf{C}^n$.

    (b) Let c denote a complex number and $\mathbf{z}, \mathbf{w}$ vectors in $\mathbf{C}^n$. Find a complex number q such that $\mathbf{z} \cdot (c\mathbf{w}) = (q\mathbf{z}) \cdot \mathbf{w}$. (Assume $\mathbf{z} \cdot \mathbf{w} \neq 0$.)

13. [BB] Prove that a complex number z is real if and only if $\overline{z} = z$.

14. (a) Compute $(2 + i)(3 + i)$.

    (b) Convert $2 + i$ and $3 + i$ to polar form and compute the product in (a) again (giving your answer in polar form).

    (c) Recall that $\arctan\theta$ means the angle whose tangent is θ. Use the result of (b) to determine the value of $\arctan\frac{1}{2} + \arctan\frac{1}{3}$.

15. [BB] Suppose $\mathbf{z} = \begin{bmatrix} z_1 \\ \vdots \\ z_n \end{bmatrix}$ is a vector in $\mathbf{C}_n$ and $\overline{\mathbf{z}}$ is the vector $\begin{bmatrix} \overline{z_1} \\ \vdots \\ \overline{z_n} \end{bmatrix}$ whose components are the complex conjugates of those of $\mathbf{z}$. Show that $\mathbf{z}$ and $\overline{\mathbf{z}}$ have the same length.

16. Given nonzero vectors $\mathbf{u}$ and $\mathbf{v}$ in $\mathbf{C}^n$, the *projection of* $\mathbf{u}$ *on* $\mathbf{v}$ is the vector $\mathbf{p} = c\mathbf{v}$ with the property that $\mathbf{u} - \mathbf{p}$ is orthogonal to $\mathbf{v}$.

    (a) [BB] Find a formula for the projection of $\mathbf{u}$ on $\mathbf{v}$.

    Let $\mathbf{u} = \begin{bmatrix} 1 \\ -i \\ 0 \end{bmatrix}$ and $\mathbf{v} = \begin{bmatrix} 3 + 2i \\ -i \\ 2 \end{bmatrix}$ be vectors in $\mathbf{C}^2$.

    (b) Find the projection of $\mathbf{u}$ on $\mathbf{v}$ [BB] and $\mathbf{v}$ on $\mathbf{u}$.

    (c) Find two orthogonal vectors that span the same subspace as $\mathbf{u}$ and $\mathbf{v}$.

■ **Critical Reading**

17. Find a condition that is necessary and sufficient for the product of two complex number numbers to be real.

7.2 **Complex Matrices**

Hermitian matrices are named after the French mathematician Charles Hermite (1822–1901).

If A is any $m \times n$ matrix, the *transpose* of A is the $n \times m$ matrix obtained from A by interchanging respective rows and columns. A matrix A is *symmetric* if $A^T = A$. For complex matrices, the analogue of "transpose" is *conjugate transpose* and the analogue of "symmetric" is *Hermitian*.

7.2.1 DEFINITION The *conjugate transpose* of a complex matrix A, denoted A^*, is the transpose of the matrix whose entries are the conjugates of those of A or, equivalently, the matrix obtained from A^T by taking the conjugates of its entries.

$$A^* = \overline{A}^T = \overline{A^T}.$$

Matrix A is *Hermitian* if $A^* = A$.

7.2.2 EXAMPLES If $A = \begin{bmatrix} 1+i & 3-2i \\ 2-4i & i \\ -5 & 5+7i \end{bmatrix}$, then $A^* = \begin{bmatrix} 1-i & 2+4i & -5 \\ 3+2i & -i & 5-7i \end{bmatrix}$. The 2×2

matrix $A = \begin{bmatrix} -3 & 4-i \\ 4+i & 7 \end{bmatrix}$ is Hermitian. So is $\begin{bmatrix} 1 & 2 & 3 \\ 2 & 4 & 5 \\ 3 & 5 & 6 \end{bmatrix}$. ∎

7.2.3 Remark The last matrix here illustrates an important point. If A is real, then $A^* = A^T$. Thus a real matrix is Hermitian if and only if it is symmetric, so all statements about Hermitian matrices also apply to real symmetric matrices.

Since $(A^T)^T = A$ and $(AB)^T = B^T A^T$ for any matrices A and B (see the last part of Section 2.1), it follows that if A and B have complex entries

7.2.4 $(AB)^* = B^* A^*$ and $(A^*)^* = A$.

We have made great use of the fact that a dot product in $\mathbf{R}^n$ can be viewed as the product of a row matrix and a column matrix: if $\mathbf{x}$ and $\mathbf{y}$ are vectors in $\mathbf{R}^n$, then $\mathbf{x} \cdot \mathbf{y} = \mathbf{x}^T \mathbf{y}$. A similar sort of thing happens in $\mathbf{C}^n$.

Suppose $\mathbf{z} = \begin{bmatrix} z_1 \\ \vdots \\ z_n \end{bmatrix}$ and $\mathbf{w} = \begin{bmatrix} w_1 \\ \vdots \\ w_n \end{bmatrix}$ are vectors in $\mathbf{C}^n$. Then

$$\mathbf{z} \cdot \mathbf{w} = \overline{z}_1 w_1 + \overline{z}_2 w_2 + \cdots + \overline{z}_n w_n = \begin{bmatrix} \overline{z}_1 & \overline{z}_2 & \dots & \overline{z}_n \end{bmatrix} \begin{bmatrix} w_1 \\ w_2 \\ \vdots \\ w_n \end{bmatrix} = \mathbf{z}^* \mathbf{w}.$$

In summary,

7.2.5 $\mathbf{z} \cdot \mathbf{w} = \mathbf{z}^* \mathbf{w}$.

7.2.6 Remark The decision to define $\mathbf{z} \cdot \mathbf{w}$ as we did for complex vectors $\mathbf{z}$ and $\mathbf{w}$—see 7.1.32—was a matter of taste. We wanted to obtain the formula $\mathbf{z} \cdot \mathbf{w} = \mathbf{z}^* \mathbf{w}$ because of its analogy with $\mathbf{u} \cdot \mathbf{v} = \mathbf{u}^T \mathbf{v}$ for real vectors. In other textbooks, you may well see

$$\mathbf{z} \cdot \mathbf{w} = z_1 \overline{w}_1 + z_2 \overline{w}_2 + \cdots + z_n \overline{w}_n,$$

which involves the complex conjugates of the components of $\mathbf{w}$ rather than $\mathbf{z}$.

Reading Challenge 1 *Let* $\mathbf{z} = \begin{bmatrix} z_1 \\ z_2 \\ \vdots \\ z_n \end{bmatrix}$ *and* $\mathbf{w} = \begin{bmatrix} w_1 \\ w_2 \\ \vdots \\ w_n \end{bmatrix}$ *be two complex vectors. The author defined*

$$\mathbf{z} \cdot \mathbf{w} = \overline{z}_1 w_1 + \overline{z}_2 w_2 + \cdots + \overline{z}_n w_n,$$

but the student remembered

$$\mathbf{z} \odot \mathbf{w} = z_1 \overline{w}_1 + z_2 \overline{w}_2 + \cdots + z_n \overline{w}_n.$$

Do the student and author agree, or is there a difference?

Suppose A is an $n \times n$ matrix and λ is a scalar. Recall that a vector $\mathbf{z}$ is an *eigenvector* of A corresponding to the *eigenvalue* λ if $\mathbf{z} \neq \mathbf{0}$ and $A\mathbf{z} = \lambda \mathbf{z}$. In Theorem 3.4.14, we saw that eigenvectors that correspond to different eigenvalues of a matrix are linearly independent. Now we shall see that in the case of real symmetric (or, more generally, Hermitian) matrices, eigenvectors corresponding to different eigenvalues are even orthogonal.

Theorem 7.2.7 **(Properties of Hermitian matrices).** *Let A be a Hermitian matrix. Then the following statements are all true.*

1. $\mathbf{z} \cdot A\mathbf{z}$ *is real for any* $\mathbf{z} \in \mathbf{C}^n$;

2. *All eigenvalues of A are real;*

3. *Eigenvectors corresponding to different eigenvalues are orthogonal.*

As a special case, if A is a real symmetric matrix, then the eigenvalues of A are real and eigenvectors corresponding to different eigenvalues are orthogonal.

Proof The final sentence follows from the statements (2), (3), and Remark 7.2.3: a real symmetric matrix is Hermitian.

1. It is sufficient to prove that $\mathbf{z} \cdot A\mathbf{z}$ equals its conjugate. See 7.1.4. Now $\mathbf{z} \cdot A\mathbf{z} = \mathbf{z}^* A\mathbf{z}$ is a scalar, so it equals its transpose. Using this fact, and $A^* = A$, we have

$$\overline{\mathbf{z} \cdot A\mathbf{z}} = \overline{\mathbf{z}^* A\mathbf{z}} = \overline{(\mathbf{z}^* A\mathbf{z})^T} = (\mathbf{z}^* A\mathbf{z})^* = \mathbf{z}^* A^* (\mathbf{z}^*)^* = \mathbf{z}^* A\mathbf{z} = \mathbf{z} \cdot A\mathbf{z}.$$

2. Suppose λ is an eigenvalue of A; thus $A\mathbf{z} = \lambda \mathbf{z}$ for some scalar λ. Now

$$\mathbf{z} \cdot A\mathbf{z} = \mathbf{z} \cdot \lambda \mathbf{z} = \lambda(\mathbf{z} \cdot \mathbf{z}) = \lambda \|\mathbf{z}\|^2.$$

By part 1, $\mathbf{z} \cdot A\mathbf{z}$ is real. Also, $\|\mathbf{z}\|^2$ is real. Since $\lambda = \frac{\mathbf{z} \cdot A\mathbf{z}}{\|\mathbf{z}\|^2}$ is the quotient of real numbers, it is also real.

3. Let λ and μ be eigenvalues of A corresponding, respectively, to eigenvectors $\mathbf{z}$ and $\mathbf{w}$, and suppose that $\lambda \neq \mu$. We compute $\mathbf{z} \cdot A\mathbf{w}$ two ways. On the one hand,

$$\mathbf{z} \cdot A\mathbf{w} = \mathbf{z} \cdot (\mu\mathbf{w}) = \mu(\mathbf{z} \cdot \mathbf{w}),$$

whereas, on the other hand,

$$\mathbf{z} \cdot A\mathbf{w} = \mathbf{z}^* A\mathbf{w} = \mathbf{z}^* A^*\mathbf{w} = (A\mathbf{z})^*\mathbf{w} = (A\mathbf{z}) \cdot \mathbf{w} = (\lambda\mathbf{z}) \cdot \mathbf{w} = \overline{\lambda}(\mathbf{z} \cdot \mathbf{w}).$$

By part 2, $\overline{\lambda} = \lambda$, so $\mu(\mathbf{z} \cdot \mathbf{w}) = \lambda(\mathbf{z} \cdot \mathbf{w})$ and $(\mu - \lambda)(\mathbf{z} \cdot \mathbf{w}) = 0$. Since $\mu - \lambda \neq 0$, we obtain $\mathbf{z} \cdot \mathbf{w} = 0$ as desired. ∎

A real $n \times n$ matrix may have no real eigenvalues or it may have some that are real and others that are not. The matrix $A = \begin{bmatrix} 0 & -1 \\ 1 & 0 \end{bmatrix}$ has no real eigenvalues since its characteristic polynomial is

$$\begin{vmatrix} -\lambda & -1 \\ 1 & -\lambda \end{vmatrix} = \lambda^2 + 1.$$

The matrix $A = \begin{bmatrix} 1 & 0 & 0 \\ 0 & 0 & -1 \\ 0 & 1 & 0 \end{bmatrix}$ has one real eigenvalue and two eigenvalues that are not real since its characteristic polynomial is

$$\begin{vmatrix} 1-\lambda & 0 & 0 \\ 0 & -\lambda & -1 \\ 0 & 1 & -\lambda \end{vmatrix} = (1-\lambda)(\lambda^2 + 1).$$

It is remarkable that if a matrix is real and symmetric, its eigenvalues must all be real.

7.2.8 EXAMPLE

The matrix $A = \begin{bmatrix} 1 & -2 \\ -2 & -2 \end{bmatrix}$ is real symmetric, so, by Theorem 7.2.7, its eigenvalues must be real. We investigate.

The characteristic polynomial of A is

$$\det(A - \lambda I) = \begin{vmatrix} 1-\lambda & -2 \\ -2 & -2-\lambda \end{vmatrix} = \lambda^2 + \lambda - 6 = (\lambda + 3)(\lambda - 2).$$

The eigenvalues $\lambda = -3, 2$ are real. Since they are different, corresponding eigenvectors should be orthogonal. Let $\mathbf{x} = \begin{bmatrix} x_1 \\ x_2 \end{bmatrix}$ be an eigenvector for $\lambda = -3$. Gaussian elimination applied to $A - \lambda I$, with $\lambda = -3$, gives

$$\begin{bmatrix} 4 & -2 \\ -2 & 1 \end{bmatrix} \rightarrow \begin{bmatrix} 1 & -\frac{1}{2} \\ 0 & 0 \end{bmatrix},$$

so $x_2 = t$ is free, $x_1 = \frac{1}{2}t$, and $\mathbf{x} = \begin{bmatrix} \frac{1}{2}t \\ t \end{bmatrix} = t\begin{bmatrix} \frac{1}{2} \\ 1 \end{bmatrix}$. If $\mathbf{x} = \begin{bmatrix} x_1 \\ x_2 \end{bmatrix}$ is an eigenvector for $\lambda = 2$, Gaussian elimination applied to $A - \lambda I$, with $\lambda = 2$, gives

$$\begin{bmatrix} -1 & -2 \\ -2 & -4 \end{bmatrix} \rightarrow \begin{bmatrix} 1 & 2 \\ 0 & 0 \end{bmatrix},$$

so $x_2 = t$ is free, $x_1 = -2t$, and $\mathbf{x} = \begin{bmatrix} -2t \\ t \end{bmatrix} = t\begin{bmatrix} -2 \\ 1 \end{bmatrix}$. The eigenspaces are orthogonal, as we were assured, with bases $\begin{bmatrix} 1 \\ 2 \end{bmatrix}$ and $\begin{bmatrix} -2 \\ 1 \end{bmatrix}$. ∎

7.2.9 EXAMPLE

We find the eigenvalues and corresponding eigenvectors of the Hermitian matrix $A = \begin{bmatrix} 1 & 1+i \\ 1-i & 2 \end{bmatrix}$. The determinant of $A - \lambda I$ is

$$\begin{vmatrix} 1-\lambda & 1+i \\ 1-i & 2-\lambda \end{vmatrix} = (1-\lambda)(2-\lambda) - (1+i)(1-i)$$

$$= (2 - 3\lambda + \lambda^2) - (1+1) = \lambda^2 - 3\lambda = \lambda(\lambda - 3),$$

so the eigenvalues of A are $\lambda = 0, 3$ (both real). To find the eigenvectors for $\lambda = 0$, we solve the homogeneous system $(A - \lambda I)\mathbf{z} = \mathbf{0}$ with $\lambda = 0$:

$$\begin{bmatrix} 1 & 1+i \\ 1-i & 2 \end{bmatrix} \rightarrow \begin{bmatrix} 1 & 1+i \\ 0 & 0 \end{bmatrix},$$

so if $\mathbf{z} = \begin{bmatrix} z_1 \\ z_2 \end{bmatrix}$, we have $z_2 = t$ free and $z_1 = -(1+i)z_2 = -(1+i)t$. Thus $\mathbf{z} = \begin{bmatrix} -(1+i)t \\ t \end{bmatrix} = t\begin{bmatrix} -(1+i) \\ 1 \end{bmatrix}$. To find the eigenvectors for $\lambda = 3$, we solve the homogeneous system $(A - \lambda I)\mathbf{z} = \mathbf{0}$ with $\lambda = 3$:

$$\begin{bmatrix} -2 & 1+i \\ 1-i & -1 \end{bmatrix} \rightarrow \begin{bmatrix} 1 & -\frac{1+i}{2} \\ 1-i & -1 \end{bmatrix} \rightarrow \begin{bmatrix} 1 & -\frac{1+i}{2} \\ 0 & 0 \end{bmatrix},$$

so with $\mathbf{z} = \begin{bmatrix} z_1 \\ z_2 \end{bmatrix}$, we again have $z_2 = t$ free and $z_1 = \frac{1+i}{2}z_2 = \frac{1+i}{2}t$. Thus $\mathbf{z} = \begin{bmatrix} \frac{1+i}{2}t \\ t \end{bmatrix} = t\begin{bmatrix} \frac{1+i}{2} \\ 1 \end{bmatrix}$. Note that any eigenvector for $\lambda = 0$ is orthogonal to any eigenvector for $\lambda = 3$ (as asserted in Theorem 7.2.7) because

$$\begin{bmatrix} -(1+i) \\ 1 \end{bmatrix} \cdot \begin{bmatrix} \frac{1+i}{2} \\ 1 \end{bmatrix} = -(1-i)\frac{1+i}{2} + 1(1) = -\frac{(1-i)(1+i)}{2} + 1 = 0.$$

[Don't forget the definition (7.1.32) of dot product in the complex case!] ∎

■ Unitary Matrices

Remember that *orthonormal vectors* are orthogonal vectors of length 1 and a real square matrix is *orthogonal* if it has orthonormal columns. The analogue of this concept for complex matrices is "unitary matrix."

7.2.10 DEFINITION

A matrix is *unitary* if it is square and its columns are orthonormal.

7.2.11 EXAMPLE

$U = \begin{bmatrix} \frac{1}{\sqrt{2}} & \frac{1+i}{2} \\ \frac{i}{\sqrt{2}} & \frac{1-i}{2} \end{bmatrix}$ has columns the vectors $\mathbf{z}_1 = \begin{bmatrix} \frac{1}{\sqrt{2}} \\ \frac{i}{\sqrt{2}} \end{bmatrix}$ and $\mathbf{z}_2 = \begin{bmatrix} \frac{1+i}{2} \\ \frac{1-i}{2} \end{bmatrix}$. We have

$$\|\mathbf{z}_1\|^2 = \frac{1}{\sqrt{2}}\frac{1}{\sqrt{2}} + \frac{-i}{\sqrt{2}}\frac{i}{\sqrt{2}} = \frac{1}{2} + \frac{1}{2} = 1$$

and

$$\|\mathbf{z}_2\|^2 = \frac{1-i}{2}\frac{1+i}{2} + \frac{1+i}{2}\frac{1-i}{2} = \frac{2}{4} + \frac{2}{4} = 1,$$

so $\mathbf{z}_1$ and $\mathbf{z}_2$ are unit vectors. Also,

$$\mathbf{z}_1 \cdot \mathbf{z}_2 = \frac{1}{\sqrt{2}}\frac{1+i}{2} + \frac{-i}{\sqrt{2}}\frac{1-i}{2} = \frac{1+i}{2\sqrt{2}} + \frac{-i+i^2}{2\sqrt{2}} = 0,$$

so these vectors are also orthogonal. Thus $\mathbf{z}_1, \mathbf{z}_2$ are orthonormal vectors and the matrix U is unitary. ∎

Since the complex dot product of two vectors in $\mathbf{R}^n$ is the usual dot product, two vectors in $\mathbf{R}^n$ are orthogonal with respect to the complex dot product if and only if they are orthogonal in $\mathbf{R}^n$ with the usual dot product there. In particular, a unitary matrix whose entries are all real numbers is an orthogonal matrix.

7.2.12 EXAMPLE

The matrices $\begin{bmatrix} 0 & 1 \\ 1 & 0 \end{bmatrix}$ and $Q = \begin{bmatrix} \frac{1}{\sqrt{2}} & -\frac{1}{\sqrt{2}} \\ \frac{1}{\sqrt{2}} & \frac{1}{\sqrt{2}} \end{bmatrix}$ are real and unitary, hence orthogonal. As noted in Section 6.2, the observation

$$Q^T Q = \begin{bmatrix} \frac{1}{\sqrt{2}} & \frac{1}{\sqrt{2}} \\ -\frac{1}{\sqrt{2}} & \frac{1}{\sqrt{2}} \end{bmatrix}\begin{bmatrix} \frac{1}{\sqrt{2}} & -\frac{1}{\sqrt{2}} \\ \frac{1}{\sqrt{2}} & \frac{1}{\sqrt{2}} \end{bmatrix} = \begin{bmatrix} 1 & 0 \\ 0 & 1 \end{bmatrix} = I$$

shows that the columns are orthonormal. See 6.2.8. ∎

Since the (i, j) entry of $A^T A$ is the dot product of columns i and j of the matrix A, unitary matrices can be characterized in much the same way as orthogonal matrices.

7.2.13 A square matrix U is unitary if and only if $U^*U = I$, that is, if and only if $U^{-1} = U^*$.

Just as with orthogonal matrices, $U^*U = I$ with U square implies $UU^* = I$. Thus the **rows** of a unitary matrix are also orthonormal.

7.2.14 EXAMPLE   With $U = \begin{bmatrix} \frac{1}{\sqrt{2}} & \frac{1+i}{2} \\ \frac{i}{\sqrt{2}} & \frac{1-i}{2} \end{bmatrix}$ the matrix of **7.2.11**,

$$U^*U = \begin{bmatrix} \frac{1}{\sqrt{2}} & -\frac{i}{\sqrt{2}} \\ \frac{1-i}{\sqrt{2}} & \frac{1+i}{2} \end{bmatrix} \begin{bmatrix} \frac{1}{\sqrt{2}} & \frac{1+i}{2} \\ \frac{i}{\sqrt{2}} & \frac{1-i}{2} \end{bmatrix} = I.$$

Thus U is unitary, $U^{-1} = U^*$, and the rows of U are orthogonal. ■

Theorem 7.2.15   **(Properties of Unitary Matrices).** *Let U be a unitary matrix.*

1. *U preserves dot products and lengths: $U\mathbf{z} \cdot U\mathbf{w} = \mathbf{z} \cdot \mathbf{w}$ and $\|U\mathbf{z}\| = \|\mathbf{z}\|$ for any vectors $\mathbf{z}$ and $\mathbf{w}$ in $\mathbf{C}^n$.*

2. *The eigenvalues of U have modulus 1: $|\lambda| = 1$ for any eigenvalue λ.*

3. *Eigenvectors corresponding to different eigenvalues are orthogonal.*

Proof   1. Since $U^*U = I$, we have $U\mathbf{z} \cdot U\mathbf{w} = (U\mathbf{z})^*(U\mathbf{w}) = \mathbf{z}^*U^*U\mathbf{w} = \mathbf{z}^*\mathbf{w} = \mathbf{z} \cdot \mathbf{w}$, so U preserves the dot product. It follows that U also preserves length since $\|U\mathbf{z}\|^2 = U\mathbf{z} \cdot U\mathbf{z} = \mathbf{z} \cdot \mathbf{z}$.

2. Let λ be an eigenvalue of U and $\mathbf{z}$ a corresponding eigenvector. By part 1, $U\mathbf{z} \cdot U\mathbf{z} = \mathbf{z} \cdot \mathbf{z}$, so $(\lambda\mathbf{z}) \cdot (\lambda\mathbf{z}) = \mathbf{z} \cdot \mathbf{z}$. Thus $\overline{\lambda}\lambda(\mathbf{z} \cdot \mathbf{z}) = \mathbf{z} \cdot \mathbf{z}$. Since $\mathbf{z} \cdot \mathbf{z} = \|\mathbf{z}\|^2 \neq 0$ (by definition eigenvectors are not $\mathbf{0}$), we get $\overline{\lambda}\lambda = 1$, so $|\lambda|^2 = 1$ and $|\lambda| = 1$.

3. This is very similar to part 2. Let λ and μ be different eigenvalues of U corresponding to the eigenvectors $\mathbf{z}$ and $\mathbf{w}$, respectively. Then

$$\mathbf{z} \cdot \mathbf{w} = U\mathbf{z} \cdot U\mathbf{w} = (\lambda\mathbf{z}) \cdot (\mu\mathbf{w}) = (\overline{\lambda}\mu)(\mathbf{z} \cdot \mathbf{w}),$$

so $(\overline{\lambda}\mu - 1)(\mathbf{z} \cdot \mathbf{w}) = 0$ and we can conclude $\mathbf{z} \cdot \mathbf{w} = 0$, as desired, if we can show that $\overline{\lambda}\mu \neq 1$. Now, if $\overline{\lambda}\mu = 1$, multiplying by λ gives $\lambda\overline{\lambda}\mu = \lambda$, so $|\lambda|^2\mu = \lambda$ (because $|\lambda| = 1$) and $\mu = \lambda$. This is not true, so we are done. ■

Answers to Reading Challenges

1. The two versions of dot product are not the same. Since $\overline{\overline{z}w} = z\overline{w}$, the author's and student's versions are complex conjugates: $\mathbf{z} \odot \mathbf{w} = \overline{\mathbf{z} \cdot \mathbf{w}}$.

True/False Questions

Decide, with as little calculation as possible, whether each of the following statements is true or false and explain your answer whenever you say "false." (Answers can be found in the back of the book.)

1. If $A = \begin{bmatrix} 1 & i \\ 1-i & 2i \end{bmatrix}$, $A^* = \begin{bmatrix} 1 & 1+i \\ -i & 2i \end{bmatrix}$.

2. The diagonal entries of a Hermitian matrix are real.

3. All the eigenvalues $\begin{bmatrix} 1 & 1-i & 3-2i \\ 1+i & -5 & -i \\ 3+2i & i & 0 \end{bmatrix}$ are real.

4. A real orthogonal matrix is unitary.

5. Matrix $U = \begin{bmatrix} \frac{1}{\sqrt{3}} & -\frac{i}{\sqrt{2}} \\ \frac{i}{\sqrt{3}} & 0 \\ \frac{1}{\sqrt{3}} & \frac{i}{\sqrt{2}} \end{bmatrix}$ is unitary.

6. If U is a unitary matrix and $\mathbf{z} \cdot \mathbf{w} = 0$, then $U\mathbf{z} \cdot U\mathbf{w} = 0$.

Exercises

*Solutions to exercises marked [BB] can be found in the **B**ack of the **B**ook.*

1. [BB] Let $A = \begin{bmatrix} 1-i & 3i \\ -3 & -i \end{bmatrix}$. Find $\det A$ and $\det(4i\,A)$.

2. Find $A + B$, $A - B$, $2A - B$, AB, and BA (where possible) in each of the following cases.

   (a) [BB] $A = \begin{bmatrix} 1+i & 1 \\ 2-2i & -3i \end{bmatrix}$,
   $B = \begin{bmatrix} 1-i & 3i \\ -3 & -i \end{bmatrix}$

   (b) $A = \begin{bmatrix} 3+i & i \\ 4+i & 0 \end{bmatrix}$, $B = \begin{bmatrix} -1 & -3-3i \\ -i & -1 \end{bmatrix}$

   (c) [BB] $A = \begin{bmatrix} i & 2+i \\ 2-2i & -5 \end{bmatrix}$,
   $B = \begin{bmatrix} i & 2-i \\ 1+3i & 0 \\ -3 & -i \end{bmatrix}$

   (d) $A = \begin{bmatrix} -i & 3+3i \\ -2+3i & -7 \end{bmatrix}$, $B = \begin{bmatrix} 1 \\ i \end{bmatrix}$

   (e) $A = \begin{bmatrix} 1+i & 2i & -1 \\ 3-i & i & 1-i \\ 1 & 1-2i & -i \end{bmatrix}$,
   $B = \begin{bmatrix} 1+i & 2-i \\ -i & 4 \\ 1+2i & 3-2i \end{bmatrix}$

   (f) $A = \begin{bmatrix} -1+2i & 2 & 3 \\ 3-i & 0 & -6+2i \\ i & -i & i \end{bmatrix}$,
   $B = \begin{bmatrix} 1 & -6-6i & -3i \\ i & 2 & 6 \\ 4+4i & 2i & 3+i \end{bmatrix}$.

3. [BB] One of the matrices below is unitary, another is Hermitian, and two others are conjugate transposes. Which is which?

$$A = \begin{bmatrix} 1+i & 1-i \\ -1+i & -1-i \end{bmatrix}, \quad B = \frac{1}{2}\begin{bmatrix} 1+i & 1-i \\ 1-i & 1+i \end{bmatrix},$$

$$C = \frac{1}{\sqrt{2}}\begin{bmatrix} 1-i & -1-i \\ 1+i & -1+i \end{bmatrix}, \quad D = \frac{1}{\sqrt{2}}\begin{bmatrix} 1 & i \\ i & -1 \end{bmatrix},$$

$$E = \frac{1}{\sqrt{2}}\begin{bmatrix} 1 & i \\ -i & 1 \end{bmatrix}.$$

4. Fill in the missing entries in the table below, which is intended to show the analogues in $\mathbf{C}^n$ of ideas about $\mathbf{R}^n$.

$\mathbf{R}^n$	$\mathbf{C}^n$
$\mathbf{x} \cdot \mathbf{y} = x_1 y_1 + x_2 y_2 + \cdots$	$\mathbf{z} \cdot \mathbf{w} =?$
transpose A^T	?
?	$\mathbf{z} \cdot \mathbf{w} = \mathbf{z}^* \mathbf{w}$
symmetric: $A^T = A$	?
?	unitary $U^* = U^{-1}$
orthogonally diagonalizable $Q^T A Q = D$	?
?	$\mathbf{z}^* \mathbf{w}$

5. Let A be an $m \times n$ matrix and B an $n \times p$ complex matrix.

   (a) [BB] Prove that $\overline{AB} = \overline{A}\,\overline{B}$.

   (b) Prove that $(AB)^* = B^* A^*$ and $(A^*)^* = A$.

6. A matrix is *skew-symmetric* if $A^T = -A$.

(a) [BB] Show that $\mathbf{x}^T A\mathbf{x} = \mathbf{0}$ for any real skew-symmetric $n \times n$ matrix A and any vector $\mathbf{x}$ in $\mathbf{R}^n$. [Hint: $\mathbf{x}^T A\mathbf{x}$ is 1×1.]

(b) In 6(a), where is the fact that $\mathbf{x}$ is real used?

(c) Prove that the only real eigenvalue of a real skew-symmetric matrix is 0.

7. Let A and B be $n \times n$ complex matrices. Determine whether each of the following statements is true or false and justify your answer.

(a) [BB] If A and B are unitary, so is AB.

(b) If A and B have determinant 1, so does AB.

(c) [BB] If A and B are symmetric, so is AB.

(d) If A and B are Hermitian, so is AB.

(e) If A and B each commute with their conjugate transposes, so does AB.

8. For each of the following, give an example or explain why no example exists:

(a) [BB] a Hermitian matrix that is not symmetric;

(b) a symmetric matrix that is not Hermitian;

(c) a 3×3 matrix, all entries real, that has no real eigenvalues.

9. [BB] Let A be an $n \times n$ matrix that commutes with its conjugate transpose and let $\mathbf{z}$ be a vector in $\mathbf{C}^n$. Prove that $\|A\mathbf{z}\| = \|A^*\mathbf{z}\|$.

10. Let A be a complex matrix.

(a) [BB] Show that the orthogonal complement of the range of A is the null space of A^*: symbolically, $(\text{range } A)^\perp = A^*$.

(b) Use (a) to show that the null space of A and the column space of A^* are orthogonal complements.

11. [BB] Prove by elementary methods that the eigenvalues of a real 2×2 symmetric matrix are real.

■ Critical Reading

7.2.16 DEFINITION

If V is a Euclidean space (real or complex), U is a subspace of V, and A is a matrix, we say that U is *invariant* under A if AU is contained in U, symbolically, $AU \subseteq U$.

12. [BB] Let $\mathbf{v}$ be an eigenvector of a real symmetric $n \times n$ matrix A and let U be the one-dimensional subspace $\mathbf{Rv}$. Show that $U^\perp$ is invariant under multiplication by A.

13. Let A be a complex $n \times n$ matrix and let U be a subspace of $\mathbf{C}^n$ that is invariant under multiplication by A. Show that $U^\perp$ is invariant under A^*

14. (a) If A and B are matrices that commute and U is an eigenspace of A, show that U is invariant under B.

(b) If U is a subspace of V invariant under A, then $U^\perp$ is also invariant under A. [Hint: Exercise 13.]

(c) Show that any 2×2 matrix over the complex numbers is similar to a diagonal matrix. [Hint: The matrix A has an eigenvector $\mathbf{z}_1$. Why? Let U be the one-dimensional subspace of $V = \mathbf{C}^2$ spanned by $\mathbf{z}_1$ and write $V = U \oplus U^\perp$. Let $\mathbf{z}_2$ be a basis for $U^\perp$. The vector $\mathbf{z}_2$ is also an eigenvector of A. Why? Now let P be the matrix whose columns are $\mathbf{z}_1$ and $\mathbf{z}_2$.]

(d) The matrix P satisfying $P^{-1}AP = D$, that exists by part (c), can be chosen unitary. Why? [This exercise outlines a proof of the "Spectral Theorem" for 2×2 matrices and can be generalized to arbitrary matrices by a technique known as "mathematical induction." See Theorems 7.3.8 and 7.3.16.]

7.3 Unitary Diagonalization

In this section, we prove that every real symmetric matrix can be diagonalized; that is, given a real symmetric matrix A, there exists an invertible matrix P such that $P^{-1}AP$ is a diagonal matrix. In fact, we shall see that P can even be chosen to be an orthogonal matrix. This basic and extremely useful theorem of linear algebra, illustrated in the exercises of Section 3.4 (Exercise 17), comes to us as a corollary of a more general theorem about Hermitian matrices.

Since this section depends so very heavily on the notions of eigenvalue, eigenvector, and diagonalizability, the reader should be very comfortable with the material in Sections 3.3 and 3.4 before continuing.

Unless we state the contrary specifically, we assume in what follows that all matrices have entries that are complex numbers. (This includes the possibility that all entries are real.)

■ A Theorem of Schur

We introduced the concept of similarity in Section 3.4. Square matrices A and B are *similar* if there exists an invertible matrix P such that $P^{-1}AP = B$. An $n \times n$ matrix A is *diagonalizable* if it is similar to a diagonal matrix; equivalently, A has n linearly independent eigenvectors—see Theorem 3.4.9. Here is the best we can say in general.

Theorem 7.3.1

Issai Schur (1875–1941), who lived most of his life in Germany, is considered one of the greatest mathematicians and is especially famous for his work in group theory and number theory.

(Schur). *Any $n \times n$ (complex) matrix is similar to a triangular matrix; in fact, there exists a unitary matrix U such that $U^{-1}AU = T$ is an upper triangular matrix.*

Proof We provide a proof for the case that A is a 3×3 matrix with entries in $\mathbf{C}^3$ (that may be all real). By the Fundamental Theorem of Algebra, the characteristic polynomial of A has a root, so A has an eigenvalue λ_1. Let $\mathbf{z}_1$ be a corresponding eigenvector of length 1. Now $\mathbf{z}_1$ is part of basis for $\mathbf{C}^3$ (any linearly independent set of vectors can be extended to a basis). Applying the Gram–Schmidt Algorithm to this basis—see Section 6.2—we obtain an orthonormal basis $\mathbf{z}_1$, $\mathbf{u}$, $\mathbf{v}$ for $\mathbf{C}^3$. Form the matrix $U_1 = \begin{bmatrix} \mathbf{z}_1 & \mathbf{u} & \mathbf{v} \\ \downarrow & \downarrow & \downarrow \end{bmatrix}$ whose columns are these vectors. Then

$$AU_1 = A \begin{bmatrix} \mathbf{z}_1 & \mathbf{u} & \mathbf{v} \\ \downarrow & \downarrow & \downarrow \end{bmatrix} = \begin{bmatrix} A\mathbf{z}_1 & A\mathbf{u} & A\mathbf{v} \\ \downarrow & \downarrow & \downarrow \end{bmatrix} = \begin{bmatrix} \lambda_1\mathbf{z}_1 & A\mathbf{u} & A\mathbf{v} \\ \downarrow & \downarrow & \downarrow \end{bmatrix}$$

because $A\mathbf{z}_1 = \lambda_1\mathbf{z}_1$ ($\mathbf{z}_1$ is an eigenvector of A corresponding to the eigenvalue λ_1). Now U_1 is invertible, so $AU_1 = U_1X$ for some matrix X ($= U_1^{-1}AU_1$). What does X look like? In particular, what does the first column of X look like?

Let this first column be $\begin{bmatrix} a \\ b \\ c \end{bmatrix}$. The first column of U_1X is then $U_1\begin{bmatrix} a \\ b \\ c \end{bmatrix}$, which is a linear combination of the columns of U_1 with coefficients a, b, c:

$$U_1\begin{bmatrix} a \\ b \\ c \end{bmatrix} = a\mathbf{z}_1 + b\mathbf{u} + c\mathbf{v}.$$

Since $U_1X = AU_1$, the first column of U_1X is the first column of AU_1, which is $\lambda_1\mathbf{z}_1$. This gives

$$a\mathbf{z}_1 + b\mathbf{u} + c\mathbf{v} = \lambda_1\mathbf{z}_1,$$

so the scalars a, b, c are $\lambda_1, 0, 0$, respectively, and

$$X = \begin{bmatrix} \lambda_1 & \star & \star \\ 0 & & \\ & & A_2 \\ 0 & & \end{bmatrix}.$$

The first column of X is $\begin{bmatrix} \lambda_1 \\ 0 \\ 0 \end{bmatrix}$, as we have just seen. We don't know anything about the second or third columns, but we shall be interested in the 2×2 matrix in the lower right corner, which we have labeled A_2. The "$\star$"s are unknown entries.

To summarize so far, and remembering that X was defined by $AU_1 = U_1X$, we have obtained $AU_1 = U_1X$, so

$$U_1^{-1}AU_1 = X = \begin{bmatrix} \lambda_1 & \star & \star \\ 0 & & \\ 0 & & A_2 \end{bmatrix}. \tag{1}$$

Let λ_2 be an eigenvalue of A_2 and, just as in equation (1), find a unitary 2×2 matrix P so that

$$P^{-1}A_2P = \begin{bmatrix} \lambda_2 & \star \\ 0 & \lambda_3 \end{bmatrix}.$$

Let $U_2 = \begin{bmatrix} 1 & 0 & 0 \\ 0 & & \\ 0 & & P \end{bmatrix}$. Then U_2 is invertible, $U_2^{-1} = \begin{bmatrix} 1 & 0 & 0 \\ 0 & & \\ 0 & & P^{-1} \end{bmatrix}$ and

$$U_2^{-1}(U_1^{-1}AU_1)U_2 = \begin{bmatrix} 1 & 0 & 0 \\ 0 & & \\ 0 & & P^{-1} \end{bmatrix} \begin{bmatrix} \lambda_1 & \star & \star \\ 0 & & \\ 0 & & A_2 \end{bmatrix} \begin{bmatrix} 1 & 0 & 0 \\ 0 & & \\ 0 & & P \end{bmatrix}$$

$$= \begin{bmatrix} \lambda_1 & \star & \star \\ 0 & & \\ 0 & & P^{-1}A_2P \end{bmatrix}.$$

Remember that the "$\star$"s indicate entries that we don't know (and about which we usually don't care). What is important is that the first column of this last matrix is $\begin{bmatrix} \lambda_1 \\ 0 \\ 0 \end{bmatrix}$ and that the 2×2 matrix $P^{-1}A_2P = \begin{bmatrix} \lambda_2 & \star \\ 0 & \lambda_3 \end{bmatrix}$ occurs in the lower right corner. Thus

$$U_2^{-1}(U_1^{-1}AU_1)U_2 = \begin{bmatrix} \lambda_1 & \star & \star \\ 0 & \lambda_2 & \star \\ 0 & 0 & \lambda_3 \end{bmatrix} = T$$

is upper triangular, and this is $U^{-1}AU$ with $U = U_1U_2$. This finishes the proof because U is unitary: $U^* = (U_1U_2)^* = U_2^*U_1^* = U_2^{-1}U_1^{-1} = (U_1U_2)^{-1} = U^{-1}$. ∎

Reading Challenge 1 *Why is the matrix P (and hence the matrix U_2) in the proof of Schur's Theorem invertible?*

7.3.2 EXAMPLE   We illustrate the proof of Schur's Theorem with reference to the matrix

$$A = \begin{bmatrix} 1 & 0 & -2 \\ 0 & -1 & 2 \\ 1 & 0 & 0 \end{bmatrix}.$$

Since the second column of A is $A\mathbf{e}_2 = A \begin{bmatrix} 0 \\ 1 \\ 0 \end{bmatrix} = \begin{bmatrix} 0 \\ -1 \\ 0 \end{bmatrix} = -\mathbf{e}_2$, we see quickly that $\mathbf{z}_1 = \mathbf{e}_2$ is an eigenvector of A. This is easily extended to an orthogonal set of three vectors, which gives rise to a unitary matrix $U_1 = \begin{bmatrix} 0 & 1 & 0 \\ 1 & 0 & 0 \\ 0 & 0 & 1 \end{bmatrix}$ whose first column is $\mathbf{z}_1$. The reader can check that

$$U_1^{-1}AU_1 = \begin{bmatrix} -1 & 0 & 2 \\ 0 & 1 & -2 \\ 0 & 1 & 0 \end{bmatrix} = \begin{bmatrix} \lambda_1 & 0 & 2 \\ 0 & & \\ 0 & & A_2 \end{bmatrix}$$

with $A_2 = \begin{bmatrix} 1 & -2 \\ 1 & 0 \end{bmatrix}$. The characteristic polynomial of A_2 is

$$\begin{vmatrix} 1 - \lambda & -2 \\ 1 & -\lambda \end{vmatrix} = \lambda^2 - \lambda + 2,$$

so A_2 has eigenvalues $\lambda = \frac{1 \pm \sqrt{7}i}{2}$. If $\mathbf{x} = \begin{bmatrix} x_1 \\ x_2 \end{bmatrix}$ is an eigenvector for $\lambda = \frac{1 + \sqrt{7}i}{2}$, then $(A_2 - \lambda I)\mathbf{x} = \mathbf{0}$. Gaussian elimination gives

$$\begin{bmatrix} \frac{1 - \sqrt{7}i}{2} & -2 \\ 1 & -\frac{1 + \sqrt{7}i}{2} \end{bmatrix} \rightarrow \begin{bmatrix} 1 & -\frac{1 + \sqrt{7}i}{2} \\ 0 & 0 \end{bmatrix},$$

so that $x_2 = t$ is free, $x_1 = \frac{1 + \sqrt{7}i}{2}t$, and $\mathbf{x} = t \begin{bmatrix} \frac{1 + \sqrt{7}i}{2} \\ 1 \end{bmatrix}$. Let $\mathbf{z}_2 = \begin{bmatrix} \frac{1 + \sqrt{7}i}{2} \\ 1 \end{bmatrix}$. By inspection, $\mathbf{w} = \begin{bmatrix} \frac{1 + \sqrt{7}i}{2} \\ -2 \end{bmatrix}$ is orthogonal to $\mathbf{z}_2$, so, dividing by their lengths and inserting into the columns of a matrix, we get a unitary matrix

$$P = \begin{bmatrix} \frac{\mathbf{z}_2}{\sqrt{3}} & \frac{\mathbf{w}}{\sqrt{6}} \\ \downarrow & \downarrow \end{bmatrix} = \begin{bmatrix} \frac{1 + \sqrt{7}i}{2\sqrt{3}} & \frac{1 + \sqrt{7}i}{2\sqrt{6}} \\ \frac{1}{\sqrt{3}} & -\frac{2}{\sqrt{6}} \end{bmatrix}.$$

Let $U_2 = \begin{bmatrix} 1 & 0 & 0 \\ 0 & & \\ 0 & & P \end{bmatrix} = \begin{bmatrix} 1 & 0 & 0 \\ 0 & \frac{1 + \sqrt{7}i}{2\sqrt{3}} & \frac{1 + \sqrt{7}i}{2\sqrt{6}} \\ 0 & \frac{1}{\sqrt{3}} & -\frac{2}{\sqrt{6}} \end{bmatrix}$. Then

$$U_2^{-1}(U_1^{-1}AU_1)U_2 = U_2^{-1}\begin{bmatrix} -1 & 0 & 2 \\ 0 & 1 & -2 \\ 0 & 1 & 0 \end{bmatrix}U_2$$

$$= \begin{bmatrix} -1 & \frac{2\sqrt{3}}{3} & -\frac{2\sqrt{6}}{3} \\ 0 & \frac{1+\sqrt{7}i}{2} & \frac{3\sqrt{2}-\sqrt{14}i}{4} \\ 0 & 0 & \frac{1-\sqrt{7}i}{2} \end{bmatrix} = T$$

is upper triangular, and this is $U^{-1}AU$ with

$$U = U_1U_2 = \begin{bmatrix} 0 & \frac{1+\sqrt{7}i}{2\sqrt{3}} & \frac{1+\sqrt{7}i}{2\sqrt{6}} \\ 1 & 0 & 0 \\ 0 & \frac{\sqrt{3}}{3} & -\frac{\sqrt{6}}{3} \end{bmatrix}. \qquad ■$$

Schur's Theorem leads to what has always struck the author as one of the most beautiful (and surprising) theorems of linear algebra: Every matrix satisfies its characteristic polynomial!

We demonstrate first with an example.

7.3.3 EXAMPLE   Let $A = \begin{bmatrix} -1 & 2 \\ 3 & 5 \end{bmatrix}$. The characteristic polynomial of A is

$$\begin{vmatrix} -1-\lambda & 2 \\ 3 & 5-\lambda \end{vmatrix} = -5 - 4\lambda + \lambda^2 - 6 = \lambda^2 - 4\lambda - 11.$$

Evaluating this polynomial at the matrix A gives

$$A^2 - 4A - 11I = \begin{bmatrix} 7 & 8 \\ 12 & 31 \end{bmatrix} - 4\begin{bmatrix} -1 & 2 \\ 3 & 5 \end{bmatrix} - 11\begin{bmatrix} 1 & 0 \\ 0 & 1 \end{bmatrix} = \begin{bmatrix} 0 & 0 \\ 0 & 0 \end{bmatrix}. \quad ■$$

As a prelude to the theorem we have mentioned—Theorem 7.3.5—let us try to understand why a 3×3 triangular matrix should satisfy its characteristic polynomial.

7.3.4   So let $T = \begin{bmatrix} \lambda_1 & a & b \\ 0 & \lambda_2 & c \\ 0 & 0 & \lambda_3 \end{bmatrix}$ be any upper triangular 3×3 matrix. The determinant of a triangular matrix is the product of its diagonal entries, so the characteristic polynomial of T is

$$\begin{vmatrix} \lambda_1-\lambda & a & b \\ 0 & \lambda_2-\lambda & c \\ 0 & 0 & \lambda_3-\lambda \end{vmatrix} = (\lambda_1-\lambda)(\lambda_2-\lambda)(\lambda_3-\lambda).$$

Our goal is to discover why this polynomial, evaluated at T, should be 0, that is, why T should satisfy the equation

$$(\lambda_1 I - T)(\lambda_2 I - T)(\lambda_3 I - T) = \mathbf{0}.$$

The factors of the left side are the matrices

$$\lambda_1 I - T = \begin{bmatrix} 0 & -a & -b \\ 0 & \lambda_1 - \lambda_2 & -c \\ 0 & 0 & \lambda_1 - \lambda_3 \end{bmatrix} = A_1$$

$$\lambda_2 I - T = \begin{bmatrix} \lambda_2 - \lambda_1 & -a & -b \\ 0 & 0 & -c \\ 0 & 0 & \lambda_2 - \lambda_3 \end{bmatrix} = A_2$$

$$\lambda_3 I - T = \begin{bmatrix} \lambda_3 - \lambda_1 & -a & -b \\ 0 & \lambda_3 - \lambda_2 & -c \\ 0 & 0 & 0 \end{bmatrix} = A_3.$$

Remember how matrix multiplication works:

$$A \begin{bmatrix} \mathbf{b}_1 & \mathbf{b}_2 & \mathbf{b}_3 \\ \downarrow & \downarrow & \downarrow \end{bmatrix} = \begin{bmatrix} A\mathbf{b}_1 & A\mathbf{b}_2 & A\mathbf{b}_3 \\ \downarrow & \downarrow & \downarrow \end{bmatrix}.$$

Remember too that each vector $A\mathbf{b}_i$ is a linear combination of the columns of A, with coefficients the components of $\mathbf{b}_i$. Consider then the product $A_1 A_2$.

The first column is the product of A_1 and the first column of A_2:

$$\begin{bmatrix} 0 & -a & -b \\ 0 & \lambda_1 - \lambda_2 & -c \\ 0 & 0 & \lambda_1 - \lambda_3 \end{bmatrix} \begin{bmatrix} \lambda_2 - \lambda_1 \\ 0 \\ 0 \end{bmatrix} = (\lambda_2 - \lambda_1) \begin{bmatrix} 0 \\ 0 \\ 0 \end{bmatrix} = \begin{bmatrix} 0 \\ 0 \\ 0 \end{bmatrix}.$$

The second column is the product of A_1 and the second column of A_2:

$$\begin{bmatrix} 0 & -a & -b \\ 0 & \lambda_1 - \lambda_2 & -c \\ 0 & 0 & \lambda_1 - \lambda_3 \end{bmatrix} \begin{bmatrix} -a \\ 0 \\ 0 \end{bmatrix} = -a \begin{bmatrix} 0 \\ 0 \\ 0 \end{bmatrix} = \begin{bmatrix} 0 \\ 0 \\ 0 \end{bmatrix}.$$

The first two columns of $A_1 A_2$ are $\mathbf{0}$, so $A_1 A_2 = \begin{bmatrix} 0 & 0 & \mathbf{x} \\ 0 & 0 & \downarrow \\ 0 & 0 & \end{bmatrix}$ for some column $\mathbf{x}$.

Now consider the product of $A_1 A_2$ and A_3. The first column is

$$\begin{bmatrix} 0 & 0 & \mathbf{x} \\ 0 & 0 & \downarrow \\ 0 & 0 & \end{bmatrix} \begin{bmatrix} \lambda_3 - \lambda_1 \\ 0 \\ 0 \end{bmatrix} = (\lambda_3 - \lambda_1) \begin{bmatrix} 0 \\ 0 \\ 0 \end{bmatrix} = \begin{bmatrix} 0 \\ 0 \\ 0 \end{bmatrix}.$$

The second column is

$$
\begin{bmatrix} 0 & 0 & \mathbf{x} \\ 0 & 0 & \downarrow \\ 0 & 0 & \end{bmatrix} \begin{bmatrix} -a \\ \lambda_3 - \lambda_2 \\ 0 \end{bmatrix} = -a \begin{bmatrix} 0 \\ 0 \\ 0 \end{bmatrix} + (\lambda_3 - \lambda_2) \begin{bmatrix} 0 \\ 0 \\ 0 \end{bmatrix} + 0\mathbf{x} = \begin{bmatrix} 0 \\ 0 \\ 0 \end{bmatrix}.
$$

The third column is

$$
\begin{bmatrix} 0 & 0 & \mathbf{x} \\ 0 & 0 & \downarrow \\ 0 & 0 & \end{bmatrix} \begin{bmatrix} -b \\ -c \\ 0 \end{bmatrix} = -b \begin{bmatrix} 0 \\ 0 \\ 0 \end{bmatrix} + -c \begin{bmatrix} 0 \\ 0 \\ 0 \end{bmatrix} + 0\mathbf{x} = \begin{bmatrix} 0 \\ 0 \\ 0 \end{bmatrix}.
$$

Each column of $(A_1 A_2) A_3$ is $\mathbf{0}$. So $A_1 A_2 A_3 = \mathbf{0}$: The triangular matrix T satisfies its characteristic polynomial.

Having examined the special case of a triangular matrix, we state the result for arbitrary (square) matrices.

Theorem 7.3.5

(Cayley–Hamilton). *Any (square) matrix satisfies its characteristic polynomial.*

The Cayley–Hamilton Theorem is named after Arthur Cayley (1821–1896) and Sir William Rowan Hamilton (1805–1865).

Proof First, we remark that the reasoning applied to a 3×3 triangular matrix in **7.3.4** generalizes to triangular matrices of any size. Specifically, if T is upper triangular with $\lambda_1, \lambda_2, \ldots, \lambda_n$ on the diagonal, then

- $\lambda_1 I - T$ has first column $\mathbf{0}$,

- $(\lambda_1 I - T)(\lambda_2 I - T)$ has first two columns $\mathbf{0}$, and, in general,

- $(\lambda_1 I - T) \cdots (\lambda_k I - T)$ has first k columns $\mathbf{0}$.

Thus all n columns of $(\lambda_1 I - T) \cdots (\lambda_n I - T)$, which is the characteristic polynomial of T evaluated at T, are 0. So this polynomial is 0 and the theorem we are trying to prove is true for T.

Now let A be any $n \times n$ matrix. Schur's Theorem says there is a unitary matrix U with

$$
U^{-1} A U = \begin{bmatrix} \lambda_1 & & & \star \\ & \lambda_2 & & \\ & & \ddots & \\ \mathbf{0} & & & \lambda_n \end{bmatrix} = T
$$

upper triangular. Since A is similar to T, the characteristic polynomial of A is the characteristic polynomial of T (see Reading Challenge 2), which is

$$
\det(T - \lambda I) = (\lambda_1 - \lambda)(\lambda_2 - \lambda) \cdots (\lambda_n - \lambda),
$$

so we have only to show that

$$
(\lambda_1 I - A)(\lambda_2 I - A) \cdots (\lambda_n I - A) = 0.
$$

Since $U^{-1}AU = T$, we have $A = UTU^{-1}$, so $\lambda_1 I - A = \lambda_1 I - UTU^{-1} = U(\lambda_1 I - T)U^{-1}$.

Similarly, each factor $\lambda_i I - A = U(\lambda_i I - T)U^{-1}$, so that

$$(\lambda_1 I - A)(\lambda_2 I - A) \cdots (\lambda_n I - A)$$

$$= U(\lambda_1 I - T)U^{-1}U(\lambda_2 I - T)U^{-1} \cdots U(\lambda_n I - T)U^{-1}$$

$$= U(\lambda_1 I - T)(\lambda_2 I - T) \cdots (\lambda_n I - T)U^{-1},$$

which is $\mathbf{0}$ because $(\lambda_1 I - T)(\lambda_2 I - T) \cdots (\lambda_n I - T)$, the factor between U and U^{-1}, is the characteristic polynomial of the triangular matrix T evaluated at T, which we have already seen is $\mathbf{0}$.   ■

Reading Challenge 2   *Let A and T be matrices and suppose A is similar to T. Show that A and T have the same characteristic polynomial.*

■ Hermitian Matrices Are Unitarily Diagonizable

The subject of diagonalizability was introduced in Section 3.4. To say that a matrix A is *diagonalizable* is to say that A is similar to a diagonal matrix; that is, there is an equation of the form $P^{-1}AP = D$ where P is some invertible matrix and D is a diagonal matrix. When this occurs, the equation $AP = PD$ shows that the diagonal entries of D are eigenvalues of A and the columns of P are eigenvectors corresponding in order to the order of the diagonal entries of D—see 3.4.1. Theorem 3.4.14 stated that an $n \times n$ real matrix A is diagonalizable if and only if A has n linearly independent eigenvectors, hence if and only if $\mathbf{R}^n$ has a basis of eigenvectors of A. Arguments similar to those presented there would show that an $n \times n$ complex matrix is diagonalizable if and only if $\mathbf{C}^n$ has a basis of eigenvectors.

Theorem 7.3.6   *An $n \times n$ matrix A with real (or complex) entries is diagonalizable if and only if $\mathbf{R}^n$ (respectively, $\mathbf{C}^n$) has a basis of eigenvectors of A.*

In this section, we show that there is always a basis of eigenvectors if A is Hermitian (or real symmetric). Thus Hermitian matrices are diagonalizable. In fact, we show that if A is a Hermitian matrix, then there is a **unitary** matrix U such that $U^{-1}AU = D$ is diagonal. Thus the eigenvectors of A can be chosen orthogonal, so $\mathbf{C}^n$ has an orthogonal basis of eigenvectors of A.

7.3.7 DEFINITION   A matrix A is *unitarily diagonalizable* if there exists a unitary matrix U such that $U^{-1}AU$ is diagonal.

The name "Spectral Theorem," by which the following theorem is generally known, is derived from the word "spectrum," which is a term some people use for the set of eigenvalues of a matrix.

Theorem 7.3.8   (**Spectral Theorem**). *Any Hermitian matrix is unitarily diagonalizable; in fact, if A is Hermitian, there exists a unitary matrix U such that $U^{-1}AU = D$ is a real diagonal matrix.*

Proof   Suppose A is an $n \times n$ Hermitian matrix. Schur's Theorem gives a unitary matrix U and an upper triangular matrix T such that $U^{-1}AU = T$. Since $U^* = U^{-1}$, we have $U^*AU = T$. Now A is Hermitian, so $A = A^*$. Thus $T^* = U^*A^*(U^*)^* = U^*AU = T$. Now T is upper triangular, so T^* is lower triangular. The only way an upper triangular matrix and a lower triangular matrix can be equal is if the matrix in question is a diagonal matrix. Thus $T^* = T$ implies that T is diagonal. Now $T^* = \overline{T^T} = \overline{T}$, since the diagonal matrix T is its own transpose. This gives $T = \overline{T}$, which shows that the diagonal entries of T are real. ■

7.3.9 EXAMPLE   Let $A = \begin{bmatrix} 2 & 1-i \\ 1+i & 3 \end{bmatrix}$. Since A is Hermitian, the Spectral Theorem says that A is unitarily diagonalizable. Now the characteristic polynomial of A is

$$\begin{vmatrix} 2-\lambda & 1-i \\ 1+i & 3-\lambda \end{vmatrix} = 6 - 5\lambda + \lambda^2 - 2 = \lambda^2 - 5\lambda + 4 = (\lambda - 1)(\lambda - 4),$$

so the eigenvalues of A are $\lambda = 1$ and $\lambda = 4$. If $\mathbf{z} = \begin{bmatrix} z_1 \\ z_2 \end{bmatrix}$ is an eigenvector for $\lambda = 1$, then $(A - \lambda I)\mathbf{z} = \mathbf{0}$. Using Gaussian elimination,

$$\begin{bmatrix} 1 & 1-i \\ 1+i & 2 \end{bmatrix} \rightarrow \begin{bmatrix} 1 & 1-i \\ 0 & 0 \end{bmatrix},$$

so $z_2 = t$ is free, $z_1 = -(1-i)t$, and $\mathbf{z} = \begin{bmatrix} -(1-i)t \\ t \end{bmatrix}$. The eigenspace has basis $\mathbf{f}_1 = \begin{bmatrix} -1+i \\ 1 \end{bmatrix}$. If $\mathbf{z} = \begin{bmatrix} z_1 \\ z_2 \end{bmatrix}$ is an eigenvector for $\lambda = 4$, then $(A - \lambda I)\mathbf{z} = \mathbf{0}$. Using Gaussian elimination,

$$\begin{bmatrix} -2 & 1-i \\ 1+i & -1 \end{bmatrix} \rightarrow \begin{bmatrix} 1 & -\frac{1-i}{2} \\ 0 & 0 \end{bmatrix},$$

so $z_2 = t$ is free, $z_1 = \frac{1-i}{2}t$, and $\mathbf{z} = \begin{bmatrix} \frac{1-i}{2}t \\ t \end{bmatrix}$. The eigenspace has basis $\mathbf{f}_2 = \begin{bmatrix} 1-i \\ 2 \end{bmatrix}$. Since $\mathbf{f}_1$ and $\mathbf{f}_2$ are eigenvectors of a Hermitian matrix corresponding to different eigenvalues, they must be orthogonal (see Theorem 7.2.7). We verify:

$$\mathbf{f}_1 \cdot \mathbf{f}_2 = (-1-i)(1-i) + 1(2) = -(1+i)(1-i) + 2 = -2 + 2 = 0.$$

The square of the length of $\mathbf{f}_1$ is $(-1-i)(-1+i) + 1(1) = -(1+i)(-1+i) + 1 = 2 + 1 = 3$, so $\mathbf{u}_1 = \begin{bmatrix} \frac{-1+i}{\sqrt{3}} \\ \frac{1}{\sqrt{3}} \end{bmatrix}$ is a unit eigenvector for $\lambda = 1$. The square of the length of $\mathbf{f}_2$ is $(1+i)(1-i) + 2(2) = 2 + 4 = 6$, so $\mathbf{u}_2 = \begin{bmatrix} \frac{1-i}{\sqrt{6}} \\ \frac{2}{\sqrt{6}} \end{bmatrix}$ is a unit eigenvector for

$\lambda = 4$. The matrix

$$U = \begin{bmatrix} \mathbf{u}_1 & \mathbf{u}_2 \\ \downarrow & \downarrow \end{bmatrix} = \begin{bmatrix} -\frac{1+i}{\sqrt{3}} & \frac{1-i}{\sqrt{6}} \\ \frac{1}{\sqrt{3}} & \frac{2}{\sqrt{6}} \end{bmatrix}$$

is unitary and U^*AU is the real diagonal matrix whose diagonal entries are the eigenvalues corresponding to $\mathbf{u}_1, \mathbf{u}_2$, respectively: $D = \begin{bmatrix} 1 & 0 \\ 0 & 4 \end{bmatrix}$. ■

■ Real Symmetric Matrices Are Orthogonally Diagonalizable

7.3.10 DEFINITION

A real $n \times n$ matrix A is *orthogonally diagonalizable* if there exists a (real) orthogonal matrix Q such that $Q^{-1}AQ = D$ is a (real) diagonal matrix.

When $Q^{-1}AQ = D$, the columns of Q are eigenvectors for A. Thus, if an $n \times n$ matrix A is orthogonally diagonalizable, we have not just n linearly independent eigenvectors, but n orthogonal eigenvectors. An $n \times n$ matrix A is orthogonally diagonalizable if and only if $\mathbf{R}^n$ has a basis of **orthogonal** eigenvectors for A.

7.3.11 EXAMPLE

In Example **7.2.8**, we saw that the matrix $A = \begin{bmatrix} 1 & -2 \\ -2 & -2 \end{bmatrix}$ has two orthogonal eigenvectors, $\mathbf{f}_1 = \begin{bmatrix} 1 \\ 2 \end{bmatrix}$ and $\mathbf{f}_2 = \begin{bmatrix} -2 \\ 1 \end{bmatrix}$. Since $\dim \mathbf{R}^2 = 2$, these vectors comprise a basis, so A is orthogonally diagonalizable. Replacing $\mathbf{f}_1$ with $\mathbf{q}_1 = \frac{1}{\|\mathbf{f}_1\|}\mathbf{f}_1$ and $\mathbf{f}_2$ with $\mathbf{q}_2 = \frac{1}{\|\mathbf{f}_2\|}\mathbf{f}_2$, we have an orthogonal matrix

$$Q = \begin{bmatrix} \mathbf{q}_1 & \mathbf{q}_2 \\ \downarrow & \downarrow \end{bmatrix} = \begin{bmatrix} \frac{1}{\sqrt{5}} & -\frac{2}{\sqrt{5}} \\ \frac{2}{\sqrt{5}} & \frac{1}{\sqrt{5}} \end{bmatrix}$$

whose columns are eigenvectors corresponding to -3 and 2, respectively. Thus $Q^{-1}AQ = Q^TAQ = \begin{bmatrix} -3 & 0 \\ 0 & 2 \end{bmatrix}$. ■

Corollary 7.3.12

(**Principal Axes Theorem**). *A real symmetric matrix is orthogonally diagonalizable.*

Proof Suppose A is a real symmetric matrix. By the Spectral Theorem, there is a unitary matrix U such that $U^{-1}AU = D$ is real diagonal. The columns of U are eigenvectors of A. Since A is real and the eigenvalues are real, and the eigenspace corresponding to λ is obtained by solving $(A - \lambda I)\mathbf{x} = \mathbf{0}$, all eigenvectors are real. Thus U is real, in which case the fact that $U^T = \overline{U^T} = U^* = U^{-1}$ says U is orthogonal. ∎

7.3.13 EXAMPLE

Let A be the real symmetric matrix $A = \begin{bmatrix} 1 & -2 & 0 \\ -2 & 0 & 2 \\ 0 & 2 & -1 \end{bmatrix}$. Expanding $\det(A - \lambda I)$ by cofactors along the first row, the characteristic polynomial of this matrix is

$$\begin{vmatrix} 1-\lambda & -2 & 0 \\ -2 & -\lambda & 2 \\ 0 & 2 & -1-\lambda \end{vmatrix} = (1-\lambda) \begin{vmatrix} -\lambda & 2 \\ 2 & -1-\lambda \end{vmatrix} + 2 \begin{vmatrix} -2 & 2 \\ 0 & -1-\lambda \end{vmatrix}$$

$$= (1-\lambda)(\lambda + \lambda^2 - 4) + 2(2 + 2\lambda)$$

$$= -\lambda^3 + 9\lambda = -\lambda(\lambda^2 - 9).$$

Thus the eigenvalues for A are $\lambda = 0$ and $\lambda = \pm 3$. Since these are different, eigenvectors corresponding to $0, 3, -3$, respectively, must be orthogonal, so A is orthogonally similar to one of the six diagonal matrices with diagonal entries $0, 3, -3$, for instance, $D = \begin{bmatrix} 0 & 0 & 0 \\ 0 & 3 & 0 \\ 0 & 0 & -3 \end{bmatrix}$. To find the orthogonal matrix Q with $Q^{-1}AQ = D$, we find eigenvectors corresponding to $\lambda = 0$, $\lambda = 3$, and $\lambda = -3$, respectively, and *normalize*, that is, divide each by its length to ensure length 1. If $\mathbf{x} = \begin{bmatrix} x_1 \\ x_2 \\ x_3 \end{bmatrix}$ is an eigenvector for $\lambda = 0$, then $(A - \lambda I)\mathbf{x} = \mathbf{0}$ with $\lambda = 0$. Using Gaussian elimination,

$$\begin{bmatrix} 1 & -2 & 0 \\ -2 & 0 & 2 \\ 0 & 2 & -1 \end{bmatrix} \rightarrow \begin{bmatrix} 1 & -2 & 0 \\ 0 & -4 & 2 \\ 0 & 2 & -1 \end{bmatrix} \rightarrow \begin{bmatrix} 1 & -2 & 0 \\ 0 & 1 & -\frac{1}{2} \\ 0 & 0 & 0 \end{bmatrix},$$

so that $x_3 = t$ is free, $x_2 = \frac{1}{2}x_3 = \frac{1}{2}t$, $x_1 = 2x_2 = t$, and $\mathbf{x} = \begin{bmatrix} t \\ \frac{1}{2}t \\ t \end{bmatrix}$. The eigenspace is one-dimensional with basis $\mathbf{f}_1 = \begin{bmatrix} 2 \\ 1 \\ 2 \end{bmatrix}$. If $\mathbf{x} = \begin{bmatrix} x_1 \\ x_2 \\ x_3 \end{bmatrix}$ is an eigenvector for $\lambda = 3$, then $(A - \lambda I)\mathbf{x} = \mathbf{0}$ with $\lambda = 3$. Using Gaussian elimination,

$$\begin{bmatrix} -2 & -2 & 0 \\ -2 & -3 & 2 \\ 0 & 2 & -4 \end{bmatrix} \rightarrow \begin{bmatrix} 1 & 1 & 0 \\ 0 & -1 & 2 \\ 0 & 1 & -2 \end{bmatrix} \rightarrow \begin{bmatrix} 1 & 1 & 0 \\ 0 & 1 & -2 \\ 0 & 0 & 0 \end{bmatrix},$$

so that $x_3 = t$ is free, $x_2 = 2x_3 = 2t$, $x_1 = -x_2 = -2t$, and $\mathbf{x} = \begin{bmatrix} -2t \\ 2t \\ t \end{bmatrix}$. The eigenspace is one-dimensional with basis $\mathbf{f}_2 = \begin{bmatrix} -2 \\ 2 \\ 1 \end{bmatrix}$. If $\mathbf{x} = \begin{bmatrix} x_1 \\ x_2 \\ x_3 \end{bmatrix}$ is an eigenvector for $\lambda = -3$, then $(A - \lambda I)\mathbf{x} = \mathbf{0}$ with $\lambda = -3$. Using Gaussian elimination,

$$
\begin{bmatrix} 4 & -2 & 0 \\ -2 & 3 & 2 \\ 0 & 2 & 2 \end{bmatrix} \rightarrow \begin{bmatrix} 1 & -\frac{1}{2} & 0 \\ 0 & 2 & 2 \\ 0 & 1 & 1 \end{bmatrix} \rightarrow \begin{bmatrix} 1 & -\frac{1}{2} & 0 \\ 0 & 1 & 1 \\ 0 & 0 & 0 \end{bmatrix},
$$

so that $x_3 = t$ is free, $x_2 = -x_3 = -t$, $x_1 = \frac{1}{2}x_2 = -\frac{1}{2}t$, and $\mathbf{x} = \begin{bmatrix} -\frac{1}{2}t \\ -t \\ t \end{bmatrix}$. The

eigenspace is one-dimensional with basis $\mathbf{f}_3 = \begin{bmatrix} -1 \\ -2 \\ 2 \end{bmatrix}$. The eigenvectors $\mathbf{f}_1$, $\mathbf{f}_2$, and $\mathbf{f}_3$

are orthogonal. Dividing each by its length, we obtain

$$
\mathbf{q}_1 = \begin{bmatrix} \frac{2}{3} \\ \frac{1}{3} \\ -\frac{2}{3} \end{bmatrix}, \quad \mathbf{q}_2 = \begin{bmatrix} -\frac{2}{3} \\ \frac{2}{3} \\ \frac{1}{3} \end{bmatrix}, \quad \mathbf{q}_3 = \begin{bmatrix} -\frac{1}{3} \\ -\frac{2}{3} \\ \frac{2}{3} \end{bmatrix}.
$$

With

$$
Q = \begin{bmatrix} \mathbf{q}_1 & \mathbf{q}_2 & \mathbf{q}_3 \\ \downarrow & \downarrow & \downarrow \end{bmatrix} = \begin{bmatrix} \frac{2}{3} & -\frac{2}{3} & -\frac{1}{3} \\ \frac{1}{3} & \frac{2}{3} & -\frac{2}{3} \\ \frac{2}{3} & \frac{1}{3} & \frac{2}{3} \end{bmatrix},
$$

we have $Q^{-1}AQ = Q^T AQ = D = \begin{bmatrix} 0 & 0 & 0 \\ 0 & 3 & 0 \\ 0 & 0 & -3 \end{bmatrix}$. ∎

7.3.14 EXAMPLE   Consider the real symmetric matrix $A = \begin{bmatrix} 0 & 2 & -1 \\ 2 & 3 & -2 \\ -1 & -2 & 0 \end{bmatrix}$.

Expanding $\det(A - \lambda I)$ by cofactors along the first row, the characteristic polynomial of this matrix is

$$
\begin{vmatrix} -\lambda & 2 & -1 \\ 2 & 3-\lambda & -2 \\ -1 & -2 & -\lambda \end{vmatrix} = -\lambda \begin{vmatrix} 3-\lambda & -2 \\ -2 & -\lambda \end{vmatrix} - 2 \begin{vmatrix} 2 & -2 \\ -1 & -\lambda \end{vmatrix} - 1 \begin{vmatrix} 2 & 3-\lambda \\ -1 & -2 \end{vmatrix}
$$

$$
= -\lambda(-3\lambda + \lambda^2 - 4) - 2(-2\lambda - 2) - (-4 + 3 - \lambda)
$$

$$
= -\lambda^3 + 3\lambda^2 + 9\lambda + 5 = -(\lambda + 1)^2(\lambda - 5).
$$

Thus the eigenvalues for A are $\lambda = 5$ and $\lambda = -1$. If $\mathbf{x} = \begin{bmatrix} x_1 \\ x_2 \\ x_3 \end{bmatrix}$ is an eigenvector

for $\lambda = 5$, then $(A - \lambda I)\mathbf{x} = \mathbf{0}$ with $\lambda = 5$. Using Gaussian elimination,

$$\begin{bmatrix} -5 & 2 & -1 \\ 2 & -2 & -2 \\ -1 & -2 & -5 \end{bmatrix} \rightarrow \begin{bmatrix} 1 & 2 & 5 \\ 0 & 12 & 24 \\ 0 & -6 & -12 \end{bmatrix} \rightarrow \begin{bmatrix} 1 & 2 & 5 \\ 0 & 1 & 2 \\ 0 & 0 & 0 \end{bmatrix},$$

so that $x_3 = t$ is free, $x_2 = -2x_3 = -2t$, $x_1 = -2x_2 - 5x_3 = 4t - 5t = -t$, and

$$\mathbf{x} = \begin{bmatrix} -t \\ -2t \\ t \end{bmatrix} = t\mathbf{f}_1, \text{ with } \mathbf{f}_1 = \begin{bmatrix} -1 \\ -2 \\ 1 \end{bmatrix}. \text{ If } \mathbf{x} = \begin{bmatrix} x_1 \\ x_2 \\ x_3 \end{bmatrix} \text{ is an eigenvector for } \lambda = -1, \text{ then}$$

$(A - \lambda I)\mathbf{x} = \mathbf{0}$ with $\lambda = -1$. Using Gaussian elimination,

$$\begin{bmatrix} 1 & 2 & -1 \\ 2 & 4 & -2 \\ -1 & -2 & 1 \end{bmatrix} \rightarrow \begin{bmatrix} 1 & 2 & -1 \\ 0 & 0 & 0 \\ 0 & 0 & 0 \end{bmatrix},$$

so that $x_3 = t$ and $x_2 = s$ are free, $x_1 = -2x_2 + x_3 = -2s + t$, and $\mathbf{x} = \begin{bmatrix} -2s+t \\ s \\ t \end{bmatrix} =$

$t\mathbf{u} + s\mathbf{v}$, with $\mathbf{u} = \begin{bmatrix} 1 \\ 0 \\ 1 \end{bmatrix}$ and $\mathbf{v} = \begin{bmatrix} -2 \\ 1 \\ 0 \end{bmatrix}$. Each of these vectors is orthogonal to

$\mathbf{f}_1$, as must be the case—eigenvectors corresponding to different eigenvalues of a symmetric matrix are orthogonal. On the other hand, $\mathbf{u}$ and $\mathbf{v}$ are not orthogonal, but this is just an inconvenience, not a genuine obstacle. Applying the Gram–Schmidt Algorithm in its simplest form, we can produce orthogonal eigenvalues by taking $\mathbf{f}_2 = \mathbf{u}$ and

$$\mathbf{f}_3 = \mathbf{v} - \text{proj}_{\mathbf{u}} \mathbf{v} = \begin{bmatrix} -2 \\ 1 \\ 0 \end{bmatrix} - \frac{-2}{2} \begin{bmatrix} 1 \\ 0 \\ 1 \end{bmatrix} = \begin{bmatrix} -2 \\ 1 \\ 0 \end{bmatrix} + \begin{bmatrix} 1 \\ 0 \\ 1 \end{bmatrix} = \begin{bmatrix} -1 \\ 1 \\ 1 \end{bmatrix}.$$

The set $\{\mathbf{f}_1, \mathbf{f}_2, \mathbf{f}_3\}$ is an orthogonal basis of eigenvectors. We turn this into an orthonormal basis by dividing each vector by its length. We obtain

$$\mathbf{q}_1 = \begin{bmatrix} \frac{1}{\sqrt{6}} \\ \frac{2}{\sqrt{6}} \\ -\frac{1}{\sqrt{6}} \end{bmatrix}, \quad \mathbf{q}_2 = \begin{bmatrix} \frac{1}{\sqrt{2}} \\ 0 \\ \frac{1}{\sqrt{2}} \end{bmatrix}, \quad \mathbf{q}_3 = \begin{bmatrix} -\frac{1}{\sqrt{3}} \\ \frac{1}{\sqrt{3}} \\ \frac{1}{\sqrt{3}} \end{bmatrix}.$$

The matrix

$$Q = \begin{bmatrix} \mathbf{q}_1 & \mathbf{q}_2 & \mathbf{q}_3 \\ \downarrow & \downarrow & \downarrow \end{bmatrix} = \begin{bmatrix} -\frac{1}{\sqrt{6}} & \frac{1}{\sqrt{2}} & -\frac{1}{\sqrt{3}} \\ -\frac{2}{\sqrt{6}} & 0 & \frac{1}{\sqrt{3}} \\ \frac{1}{\sqrt{6}} & \frac{1}{\sqrt{2}} & \frac{1}{\sqrt{3}} \end{bmatrix}$$

is orthogonal and $Q^{-1}AQ = D = \begin{bmatrix} 5 & 0 & 0 \\ 0 & -1 & 0 \\ 0 & 0 & -1 \end{bmatrix}$ is the diagonal matrix whose diagonal entries are eigenvalues for $\mathbf{q}_1, \mathbf{q}_2,$ and $\mathbf{q}_3$, respectively.

■ Unitarily Diagonalizable If and Only If Normal

7.3.15 The converse of the Spectral Theorem is the statement that a unitarily diagonalizable matrix is Hermitian, which is not true. The unitary matrix $V = \begin{bmatrix} 0 & -1 \\ 1 & 0 \end{bmatrix}$ is not Hermitian. The eigenvalues of V are $\pm i$. Corresponding eigenvectors are $\begin{bmatrix} 1 \\ -i \end{bmatrix}$ and $\begin{bmatrix} 1 \\ i \end{bmatrix}$. These vectors are orthogonal and of length $\sqrt{2}$. Dividing by their lengths and using these vectors as columns gives a unitary matrix

$$U = \begin{bmatrix} \frac{1}{\sqrt{2}} & \frac{1}{\sqrt{2}} \\ -\frac{i}{\sqrt{2}} & \frac{i}{\sqrt{2}} \end{bmatrix}.$$

It is straightforward to check that $U^*VU = D = \begin{bmatrix} i & 0 \\ 0 & -i \end{bmatrix}$ is diagonal. Thus V is unitarily diagonalizable, but V is not Hermitian.

Thus, while every Hermitian matrix is unitarily diagonalizable, not every unitarily diagonalizable matrix is Hermitian. It would be satisfying to have a theorem characterizing unitarily diagonalizable matrices, that is, a theorem that says

"A matrix is unitarily diagonalizable if and only if ... "

We conclude this section with just such a theorem, one proof of which was suggested in Exercise 14 of Section 7.2.

Theorem 7.3.16 *A matrix A can be unitarily diagonalized if and only if it commutes with its conjugate transpose, that is, if and only if $AA^* = A^*A$.*

Proof ($\Longrightarrow$) Assume A can be unitarily diagonalized. Then there exists a unitary matrix U and a diagonal matrix D such that $U^{-1}AU = D$. Thus $A = UDU^{-1}$. Since U is unitary, $U^{-1} = U^*$, so $A = UDU^*$. Thus $A^* = (U^*)^*D^*U^* = U\overline{D}U^*$ because D is diagonal (and hence equal to its transpose). Since $U^*U = I$, $AA^* = UDU^*U\overline{D}U^* = UD\overline{D}U^*$, whereas $A^*A = U\overline{D}U^*UDU^* = U\overline{D}DU^*$. The desired result $AA^* = A^*A$ then follows because D and $\overline{D}$ commute: $D\overline{D} = \overline{D}D$. To see why, let $D = \begin{bmatrix} \lambda_1 & & 0 \\ & \ddots & \\ 0 & & \lambda_n \end{bmatrix}$. Then $\overline{D} = \begin{bmatrix} \overline{\lambda_1} & & 0 \\ & \ddots & \\ 0 & & \overline{\lambda_n} \end{bmatrix}$ and

$$D\overline{D} = \overline{D}D = \begin{bmatrix} \lambda_1\overline{\lambda_1} & & 0 \\ & \ddots & \\ 0 & & \lambda_n\overline{\lambda_n} \end{bmatrix}.$$

($\Longleftarrow$) Conversely, suppose $AA^* = A^*A$. By Schur's Theorem, there exists a unitary matrix U such that $U^{-1}AU = T$ is triangular. We claim $TT^* = T^*T$. Why?

Since $U^{-1} = U^*$, $T = U^*AU$ and $T^* = U^*A^*(U^*)^* = U^*A^*U$. Since $U^*U = I$, $TT^* = U^*AUU^*A^*U = U^*AA^*U$ and, similarly, $T^*T = U^*A^*AU$. Since $AA^* = A^*A$, we obtain $TT^* = T^*T$, as claimed.

Now, for any vector $\mathbf{z}$, $\|T\mathbf{z}\|^2 = T\mathbf{z} \cdot T\mathbf{z} = (T\mathbf{z})^*(T\mathbf{z}) = \mathbf{z}^*T^*T\mathbf{z}$ and, similarly, $\|T^*\mathbf{z}\|^2 = \mathbf{z}^*T^{**}T^*\mathbf{z} = \mathbf{z}^*TT^*\mathbf{z}$. Since $TT^* = T^*T$, we have $\|T\mathbf{z}\|^2 = \|T^*\mathbf{z}\|^2$. In particular,

$$\|T\mathbf{e}_i\|^2 = \left\|T^*\mathbf{e}_i\right\|^2 \tag{2}$$

for each of the standard basis vectors $\mathbf{e}_1, \mathbf{e}_2, \dots, \mathbf{e}_n$. Suppose

$$T = \begin{bmatrix} \lambda_1 & \star & \cdots & \star \\ 0 & \lambda_2 & & \star \\ \vdots & & \ddots & \star \\ 0 & 0 & \cdots & \lambda_n \end{bmatrix}$$

is upper triangular (the argument is similar if T is lower triangular) and look at equation (2) with $i = 1$.

The vector $T\mathbf{e}_1$ is the first column of T, the square of whose length is $|\lambda_1|^2$.

The vector $T^*\mathbf{e}_1$ is a vector of the form $\begin{bmatrix} \overline{\lambda_1} \\ c_1 \\ \vdots \\ c_{n-1} \end{bmatrix}$, the square of whose length is

$|\lambda_1|^2 + |c_1|^2 + \cdots + |c_{n-1}|^2$. Thus $|\lambda_1|^2 = |\lambda_1|^2 + |c_1|^2 + \cdots + |c_{n-1}|^2$. It follows that $c_1 = c_2 = \cdots = c_{n-1} = 0$.

Continuing to examine the consequences of (2) in the cases $i = 2, 3, \dots, n$ shows that T is diagonal. ∎

Corollary 7.3.17 *Any unitary matrix is unitarily diagonalizable.*

Proof If U is a unitary matrix, $UU^* = U^*U = I$. ∎

A matrix A satisfying $AA^* = A^*A$ is called *normal*. Theorem 7.3.16 is consistent with Theorem 7.3.8 since any Hermitian matrix is normal. There are, however, normal matrices that are not Hermitian, for example, the unitary matrix V of 7.3.15.

Answers to Reading Challenges

1. P is invertible because it is unitary and U_2 is invertible because its product with $\begin{bmatrix} 1 & 0 & 0 \\ 0 & & \\ & P^{-1} & \\ 0 & & \end{bmatrix}$ is the identity.

2. We are given that $T = P^{-1}AP$ for some invertible matrix P. Using basic properties of the determinant, the characteristic polynomial of T is

$$\det(P^{-1}AP - \lambda I) = \det P^{-1}(A - \lambda I)P$$

$$= \frac{1}{\det P}\det(A - \lambda I)\det P = \det(A - \lambda I),$$

which is the characteristic polynomial of A.

True/False Questions

Decide, with as little calculation as possible, whether each of the following statements is true or false and explain your answer whenever you say "false." (Answers can be found in the back of the book.)

1. Matrices $A = \begin{bmatrix} 1 & 0 \\ 0 & 1 \end{bmatrix}$ and $B = \begin{bmatrix} 3 & 0 \\ 0 & 3 \end{bmatrix}$ are similar.

2. The matrix $A = \begin{bmatrix} 1 & 1 \\ 0 & 1 \end{bmatrix}$ satisfies $A^2 = 2A - I$.

3. If A is a unitary matrix with eigenvalue 0, then A is invertible.

4. If D is a diagonal matrix and Q is a real orthogonal matrix, then $Q^T DQ$ is also diagonal.

5. If A is a real symmetric matrix, there is an orthogonal matrix Q such that $Q^T AQ$ is a diagonal matrix.

6. A Hermitian matrix is normal.

7. A normal matrix is unitary.

Exercises

*Solutions to exercises marked [BB] can be found in the **B**ack of the **B**ook.*

1. For each of the following real symmetric matrices A, find an orthogonal matrix Q and a real diagonal matrix such that $Q^T AQ = D$.

(a) [BB] $A = \begin{bmatrix} 2 & 3 \\ 3 & -6 \end{bmatrix}$

(b) $A = \begin{bmatrix} 17 & 24 \\ 24 & 3 \end{bmatrix}$

(c) [BB] $A = \begin{bmatrix} -2 & 1 & 1 \\ 1 & -2 & 1 \\ 1 & 1 & -2 \end{bmatrix}$

(d) $A = \begin{bmatrix} 1 & -3 & 2 \\ -3 & 1 & -2 \\ 2 & -2 & 2 \end{bmatrix}$

(e) [BB] $A = \begin{bmatrix} 5 & 2 & 2 \\ 2 & 2 & -4 \\ 2 & -4 & 2 \end{bmatrix}$

(f) $A = \begin{bmatrix} 5 & 4 & -4 \\ 4 & 5 & 4 \\ -4 & 4 & 5 \end{bmatrix}$

(g) $A = \begin{bmatrix} 0 & 2 & 1 & 1 \\ 2 & 0 & -1 & -1 \\ 1 & -1 & 0 & -2 \\ 1 & -1 & -2 & 0 \end{bmatrix}$

(h) $A = \begin{bmatrix} 3 & -1 & -1 & 1 \\ -1 & 3 & 1 & -1 \\ -1 & 1 & 3 & -1 \\ 1 & -1 & -1 & 3 \end{bmatrix}$.

(The characteristic polynomial of this matrix is $\lambda^4 - 12\lambda^3 + 48\lambda^2 - 80\lambda + 48$.)

2. For each of the symmetric matrices given here,

i. find an orthogonal matrix Q such that $Q^T A Q$ is diagonal;

ii. use the result of part i to find A^5.

(a) [BB] $A = \begin{bmatrix} 3 & 0 & -1 \\ 0 & -1 & 0 \\ -1 & 0 & 3 \end{bmatrix}$

(b) $A = \begin{bmatrix} 3 & 0 & 2 \\ 0 & 1 & 0 \\ 2 & 0 & 3 \end{bmatrix}$

(c) $A = \begin{bmatrix} -1 & -4 & 4 & 0 \\ -4 & -5 & 0 & -4 \\ 4 & 0 & -5 & -4 \\ 0 & -4 & -4 & -1 \end{bmatrix}$

(d) $A = \begin{bmatrix} 1 & -1 & -9 & 1 \\ -1 & 1 & 9 & -1 \\ -9 & 9 & -15 & -9 \\ 1 & -1 & -9 & 1 \end{bmatrix}$.

3. For each of the following Hermitian matrices A, find a unitary matrix U and a diagonal matrix D such that $U^*AU = D$.

(a) [BB] $A = \begin{bmatrix} 1 & -i \\ i & 1 \end{bmatrix}$

(b) $A = \begin{bmatrix} 3 & 1+i \\ 1-i & 2 \end{bmatrix}$

(c) $A = \begin{bmatrix} 1 & 1+i \\ 1-i & 3 \end{bmatrix}$

(d) $A = \begin{bmatrix} 1 & i & 0 \\ -i & 0 & 1 \\ 0 & 1 & 1 \end{bmatrix}$.

4. Determine whether A is diagonalizable over **R** and over **C** in each of the following cases. Justify your answers.

(a) [BB] $A = \begin{bmatrix} 2 & -3 \\ 1 & 1 \end{bmatrix}$

(b) $A = \begin{bmatrix} 0 & -1 & 1 \\ 1 & 1 & 0 \\ -1 & 0 & 1 \end{bmatrix}$

(c) $A = \begin{bmatrix} 3 & 1 & -1 \\ 2 & 2 & -1 \\ 2 & 2 & 0 \end{bmatrix}$

(d) $A = \begin{bmatrix} 1 & 1 & 1 \\ -1 & 3 & -1 \\ -1 & 2 & 0 \end{bmatrix}$.

5. [BB] Let A be a real skew-symmetric matrix, that is, a real matrix satisfying $A^T = -A$. Prove that there exists a real orthogonal matrix Q and a matrix B such that $A = QBQ^T$ and B^2 is diagonal. [Hint: $A^T A$ is real symmetric.]

6. A matrix K is *skew-Hermitian* if $K^* = -K$. Prove that a skew-Hermitian matrix is unitarily diagonalizable.

7. For each of the following matrices, find a unitary matrix U and an upper triangular matrix T such that $U^*AU = T$.

(a) [BB] $A = \begin{bmatrix} 2 & 2 \\ -1 & 0 \end{bmatrix}$ **(b)** $A = \begin{bmatrix} 6 & 9 \\ -1 & 12 \end{bmatrix}$

(c) $A = \begin{bmatrix} 1 & -1 & 1 \\ 1 & -1 & 1 \\ 1 & -1 & 1 \end{bmatrix}$

(d) $A = \begin{bmatrix} 0 & 1 & 0 \\ 1 & 0 & -3 \\ 0 & 1 & 0 \end{bmatrix}$.

8. Prove that the determinant of a Hermitian matrix is a real number.

9. [BB] Show that the eigenvalues of a 2×2 (real) symmetric matrix are always real numbers.

10. We concluded this section with the remark that there exist normal matrices that are not Hermitian.

(a) Show that for any real numbers a and b,
$A = \begin{bmatrix} a & b \\ -b & a \end{bmatrix}$ is normal, but not symmetric.

(b) Show that if A is a real 2×2 normal matrix that is not symmetric, then A has the form of the matrix in (a).

11. Determine, with as few calculations as possible, whether each of the following matrices is diagonalizable over **C**.

(a) [BB] $A = \begin{bmatrix} 1 & 0 & 1 \\ 0 & 1 & -1 \\ 1 & 0 & 0 \end{bmatrix}$

(b) $A = \begin{bmatrix} 1 & 1 & 0 \\ 0 & 1 & 1 \\ 0 & 0 & -1 \end{bmatrix}$

(c) $A = \begin{bmatrix} \frac{1}{\sqrt{2}} & \frac{1}{\sqrt{2}} & 0 \\ 0 & 0 & 1 \\ -\frac{1}{\sqrt{2}} & \frac{1}{\sqrt{2}} & 0 \end{bmatrix}$

(d) $A = \begin{bmatrix} i & i & 0 \\ -i & i & i \\ 0 & -i & i \end{bmatrix}$.

12. Let $A = \begin{bmatrix} 1 & 0 & -i \\ 0 & 1 & 1 \\ i & -1 & 2 \end{bmatrix}$.

   (a) Is A normal?

   (b) Is A unitarily diagonalizable?

   (c) Is A similar to a diagonal matrix?

13. [BB] Prove that any Hermitian matrix is normal.

14. Prove that a matrix A is normal if and only if $A + A^*$ commutes with $A - A^*$. [Remember that matrices X and Y *commute* if and only if $XY = YX$.]

15. [BB] Suppose A is a normal $n \times n$ (complex) matrix. Prove that null sp $A =$ null sp A^*. [Hint: Let $\mathbf{z} \in \mathbf{C}^n$ and compute $(A^*\mathbf{z}) \cdot (A^*\mathbf{z})$.]

16. Suppose A is a normal $n \times n$ (complex) matrix. Prove that col sp $A =$ col sp A^* (equivalently, A and A^* have the same range). [Hint: Exercise 10(b) of Section 7.2 and Exercise 15.]

17. [BB] Let A be a (real) $n \times n$ matrix with the property that $\mathbf{x}^T A \mathbf{x} > 0$ for all vectors $\mathbf{x}$.

   (a) Let λ be a real eigenvalue of A. Show that $\lambda > 0$.

   (b) If A is symmetric, show that A has a square root; that is, there exists a matrix B such that $A = B^2$.

   (c) If A is symmetric, show that A can be factored in the form $A = RR^T$ for some matrix R.

18. Suppose A is a symmetric $n \times n$ matrix and $\mathbf{q}_1, \mathbf{q}_2, \dots, \mathbf{q}_n$ are orthonormal eigenvectors corresponding to eigenvalues $\lambda_1, \lambda_2, \dots, \lambda_n$, respectively.

   (a) Prove that $A = \lambda_1 \mathbf{q}_1 \mathbf{q}_1^T + \lambda_2 \mathbf{q}_2 \mathbf{q}_2^T + \cdots + \lambda_n \mathbf{q}_n \mathbf{q}_n^T$. [Hint: Call the matrix on the right B. It is sufficient to show that $A\mathbf{x} = B\mathbf{x}$ for any $\mathbf{x}$. Why? For this, note how to rewrite an expression of the form $(\mathbf{v}\mathbf{v}^T)\mathbf{u}$—see equation (5) in Section 6.1.]

   (b) Any real symmetric matrix is a linear combination of projection matrices. Why?

[The decomposition of A given in part (a) is its *spectral decomposition*.]

■ Critical Reading

19. [BB] Think of a way to show that any square complex matrix is unitarily similar to a **lower** triangular matrix. (Use Schur's Theorem but do not modify its proof.)

20. Suppose the only eigenvalue of a real symmetric matrix A is 3. What can you say about A?

21. [BB] Suppose A and B are real symmetric matrices. Why are all the roots of $\det(A + B - \lambda I)$ real?

22. If U is a unitary matrix and every eigenvalue of U is either $+i$ or $-i$, find U^2 and justify your answer.

23. (a) [BB] Let A be a real symmetric matrix with nonnegative eigenvalues. Show that A can be written in the form $A = RR^T$ for some matrix R. (This is a partial converse of the elementary fact that RR^T is always symmetric.) [Hint: A diagonal matrix with real nonnegative diagonal entries has a square root.]

   (b) Let A and B be real symmetric matrices and suppose the eigenvalues of B are all positive. Show that the roots of the polynomial $\det(A - \lambda B)$ are real. [Hint: First explain why $\det(A - \lambda B) = 0$ gives a nonzero $\mathbf{x}$ with $A\mathbf{x} = \lambda B\mathbf{x}$. Then use the result of 23(a) to factor $B = RR^T$ with R invertible.]

24. We noted that the converse of the Spectral Theorem is false. On the other hand, the converse of the Principal Axes Theorem is true. Prove this: If a real matrix is orthogonally diagonalizable, then A is symmetric.

25. (a) What are the possible values for the eigenvalues of a projection matrix P?

   (b) Does every projection matrix have all possible eigenvalues?
   Explain your answers.

26. Let $A = \begin{bmatrix} 3 & 3 & -4 \\ -4 & -3 & 5 \\ 2 & 4 & 0 \end{bmatrix}$.

   (a) Find the characteristic polynomial of A.

   (b) Evaluate $A^6 + 4A^5 - 5A^4 - 50A^3 - 76A^2 - 10A + 50I$.

| 7.4 | **The Orthogonal Diagonalization of Real Symmetric Matrices** |

This section offers an alternative approach to the diagonalization of real symmetric matrices for readers who wish to avoid the arguments involving complex numbers that were used in previous sections. This section is completely redundant and should be considered just light reading for people familiar with the material in Sections 7.1, 7.2, and 7.3.

Since this section depends heavily on the notions of eigenvalue, eigenvector, and diagonalizability, the reader should be totally comfortable with the material in Sections 3.3 and 3.4 before continuing.

We know what it means for a matrix A to be *diagonalizable*: there exists an invertible matrix P and a diagonal matrix D such that $P^{-1}AP = D$. As the name suggests, a matrix is *orthogonally diagonalizable* if the matrix P can be chosen orthogonal. When such is the case, we usually denote the orthogonal *diagonalizing* matrix by Q, instead of P.

7.4.1 DEFINITION   A matrix A is *orthogonally diagonalizable* if there exist an orthogonal matrix Q and a diagonal matrix D such that $Q^{-1}AQ = D$, equivalently, such that $Q^T AQ = D$.

Reading Challenge 1   *Why "equivalently" in this definition?*

Suppose A is an $n \times n$ matrix. Recall from Section 3.4 that the equation $P^{-1}AP = D$, that implies $AP = PD$, says that the columns of P are eigenvectors of A and the diagonal entries of D are eigenvalues in the order that corresponds to the order of the columns of A; that is, the $(1, 1)$ entry of D is an eigenvalue and the first column of P is a corresponding eigenvector, and so on. Since an $n \times n$ matrix is invertible if and only if its columns are a basis for $\mathbf{R}^n$, invertibility of P implies that $\mathbf{R}^n$ has a basis of eigenvectors of A. The converse is also true.

Theorem 7.4.2   *An $n \times n$ matrix A is diagonalizable if and only if $\mathbf{R}^n$ has a basis consisting of eigenvectors of A.*

If A is **orthogonally** diagonalizable, say $Q^{-1}AQ = D$ with Q orthogonal, then the columns of Q, which are eigenvectors of A, are orthogonal (in fact, orthonormal). So,

Corollary 7.4.3   *An $n \times n$ matrix A is orthogonally diagonalizable if and only if $\mathbf{R}^n$ has an orthogonal basis of eigenvectors of A.*

7.4.4 EXAMPLE   Let $A = \begin{bmatrix} 1 & -2 \\ -2 & -2 \end{bmatrix}$. The characteristic polynomial of A is

$$\det(A - \lambda I) = \begin{vmatrix} 1 - \lambda & -2 \\ -2 & -2 - \lambda \end{vmatrix} = \lambda^2 + \lambda - 6 = (\lambda + 3)(\lambda - 2).$$

The eigenvalues are $\lambda = -3$ and $\lambda = 2$. Let $\mathbf{x} = \begin{bmatrix} x_1 \\ x_2 \end{bmatrix}$ be an eigenvector for $\lambda = -3$. Gaussian elimination applied to $A - \lambda I$, with $\lambda = -3$, gives

$$\begin{bmatrix} 4 & -2 \\ -2 & 1 \end{bmatrix} \to \begin{bmatrix} 1 & -\frac{1}{2} \\ 0 & 0 \end{bmatrix},$$

so $x_2 = t$ is free, $x_1 = \frac{1}{2}t$, and $\mathbf{x} = \begin{bmatrix} \frac{1}{2}t \\ t \end{bmatrix} = t \begin{bmatrix} \frac{1}{2} \\ 1 \end{bmatrix}$. If $\mathbf{x} = \begin{bmatrix} x_1 \\ x_2 \end{bmatrix}$ is an eigenvector for $\lambda = 2$, Gaussian elimination applied to $A - \lambda I$, with $\lambda = 2$, gives

$$\begin{bmatrix} -1 & -2 \\ -2 & -4 \end{bmatrix} \to \begin{bmatrix} 1 & 2 \\ 0 & 0 \end{bmatrix},$$

so $x_2 = t$ is free, $x_1 = -2t$, and $\mathbf{x} = \begin{bmatrix} -2t \\ t \end{bmatrix} = t \begin{bmatrix} -2 \\ 1 \end{bmatrix}$. The vectors $\mathbf{f}_1 = \begin{bmatrix} 1 \\ 2 \end{bmatrix}$ and $\mathbf{f}_2 = \begin{bmatrix} -2 \\ 1 \end{bmatrix}$ are orthogonal eigenvectors and, since they are linearly independent, provide a basis for $\mathbf{R}^2$. Thus A is orthogonally diagonalizable. Normalizing $\mathbf{f}_1$ and $\mathbf{f}_2$, that is, dividing each by its length so as to produce unit eigenvectors $\mathbf{q}_1 = \begin{bmatrix} \frac{1}{\sqrt{5}} \\ \frac{2}{\sqrt{5}} \end{bmatrix}$ and $\mathbf{q}_2 = \begin{bmatrix} -\frac{2}{\sqrt{5}} \\ \frac{1}{\sqrt{5}} \end{bmatrix}$, we have an orthogonal matrix

$$Q = \begin{bmatrix} \mathbf{q}_1 & \mathbf{q}_2 \\ \downarrow & \downarrow \end{bmatrix} = \begin{bmatrix} \frac{1}{\sqrt{5}} & -\frac{2}{\sqrt{5}} \\ \frac{2}{\sqrt{5}} & \frac{1}{\sqrt{5}} \end{bmatrix}$$

and $Q^{-1}AQ = D = \begin{bmatrix} -3 & 0 \\ 0 & 2 \end{bmatrix}$, the diagonal matrix whose diagonal entries are eigenvalues in order corresponding to the columns of Q: the first diagonal entry is an eigenvalue for $\mathbf{q}_1$ and the second is an eigenvalue for $\mathbf{q}_2$. ∎

■ Properties of Symmetric Matrices

Symmetric matrices are quite remarkable objects. The characteristic polynomial of $A = \begin{bmatrix} 0 & -1 \\ 1 & 0 \end{bmatrix}$ (which is not symmetric) is $\begin{vmatrix} -\lambda & -1 \\ 1 & -\lambda \end{vmatrix} = \lambda^2 + 1$, which has no real roots, so A has no real eigenvalues. The characteristic polynomial of $A = \begin{bmatrix} 1 & 0 & -1 \\ 0 & 1 & -1 \\ 1 & 2 & 3 \end{bmatrix}$ (also not symmetric) is $(1 - \lambda)(\lambda^2 - 4\lambda + 6)$, which has just one real root, $\lambda = 1$.

Reading Challenge 2 *Why?*

On the other hand, **the characteristic polynomial of a symmetric matrix always factors completely over the real numbers, hence the eigenvalues of a** symmetric matrix are always real. This is easy to show for 2×2 matrices—see Exercise 4—and not hard in general, though the proof involves complex matrices and is omitted here (see Theorem 7.2.7).

Remember that eigenvectors corresponding to different eigenvalues are always linearly independent (see Theorem 3.4.14). For eigenvectors of symmetric matrices, even more is true.

Theorem 7.4.5 *If A is a real symmetric matrix, then eigenvectors that correspond to different eigenvalues are orthogonal.*

Proof   Let $\mathbf{x}$ and $\mathbf{y}$ be eigenvectors for A with corresponding eigenvalues λ and μ, and suppose $\lambda \neq \mu$. We compute $\mathbf{x}^T A\mathbf{y}$ in two different ways. On the one hand, since $A\mathbf{y} = \mu\mathbf{y}$, we have $\mathbf{x}^T A\mathbf{y} = \mu\mathbf{x}^T\mathbf{y} = \mu(\mathbf{x} \cdot \mathbf{y})$. On the other hand, since $A = A^T$,

$$\mathbf{x}^T A\mathbf{y} = \mathbf{x}^T A^T\mathbf{y} = (A\mathbf{x})^T\mathbf{y} = (\lambda\mathbf{x})^T\mathbf{y} = \lambda(\mathbf{x}^T\mathbf{y}) = \lambda(\mathbf{x} \cdot \mathbf{y}).$$

Thus $\lambda(\mathbf{x} \cdot \mathbf{y}) = \mu(\mathbf{x} \cdot \mathbf{y})$, so $(\lambda - \mu)(\mathbf{x} \cdot \mathbf{y}) = 0$ and, since $\lambda - \mu \neq 0$, $\mathbf{x} \cdot \mathbf{y} = 0$. This says $\mathbf{x}$ and $\mathbf{y}$ are orthogonal, as claimed. ∎

The symmetric 2×2 matrix A in **7.4.4** had two different eigenvalues, -3 and 2. Theorem 7.4.5 shows that it is not accidental that corresponding eigenvalues are orthogonal and hence that A is orthogonally diagonalizable. In general, if a symmetric $n \times n$ matrix has n different eigenvalues, corresponding eigenvectors are orthogonal, and since there are n of them, they form an orthogonal basis for $\mathbf{R}^n$. The following corollary is immediate.

Corollary 7.4.6 *A symmetric $n \times n$ matrix with n different eigenvalues is orthogonally diagonalizable.*

7.4.7 EXAMPLE   We leave it to the reader to show that the characteristic polynomial of the symmetric 3×3 matrix $A = \begin{bmatrix} 1 & -2 & 0 \\ -2 & 0 & 2 \\ 0 & 2 & -1 \end{bmatrix}$ is $-\lambda^3 + 9\lambda = -\lambda(\lambda^2 - 9)$. Thus A has eigenvalues $0, 3, -3$ and, since these are different, it is orthogonally diagonalizable, being similar (via an orthogonal matrix) to one of the six diagonal matrices with diagonal entries $0, 3, -3$, for instance, $D = \begin{bmatrix} 0 & 0 & 0 \\ 0 & 3 & 0 \\ 0 & 0 & -3 \end{bmatrix}$. If we want to find an orthogonal matrix Q with $Q^{-1}AQ = D$, we must find eigenvectors corresponding to $\lambda = 0$, $\lambda = 3$, and $\lambda = -3$, respectively, and *normalize*, that is, divide each by its length to ensure length 1. If $\mathbf{x} = \begin{bmatrix} x_1 \\ x_2 \\ x_3 \end{bmatrix}$ is an eigenvector for $\lambda = 0$, then $(A - \lambda I)\mathbf{x} = \mathbf{0}$ with $\lambda = 0$. Using Gaussian elimination,

$$\begin{bmatrix} 1 & -2 & 0 \\ -2 & 0 & 2 \\ 0 & 2 & -1 \end{bmatrix} \rightarrow \begin{bmatrix} 1 & -2 & 0 \\ 0 & -4 & 2 \\ 0 & 2 & -1 \end{bmatrix} \rightarrow \begin{bmatrix} 1 & -2 & 0 \\ 0 & 1 & -\frac{1}{2} \\ 0 & 0 & 0 \end{bmatrix},$$

so that $x_3 = t$ is free, $x_2 = \frac{1}{2}x_3 = \frac{1}{2}t$, $x_1 = 2x_2 = t$, and $\mathbf{x} = \begin{bmatrix} t \\ \frac{1}{2}t \\ t \end{bmatrix}$. The eigenspace

is one-dimensional with basis $\mathbf{f}_1 = \begin{bmatrix} 2 \\ 1 \\ 2 \end{bmatrix}$. If $\mathbf{x} = \begin{bmatrix} x_1 \\ x_2 \\ x_3 \end{bmatrix}$ is an eigenvector for $\lambda = 3$,

then $(A - \lambda I)\mathbf{x} = \mathbf{0}$ with $\lambda = 3$. Using Gaussian elimination,

$$\begin{bmatrix} -2 & -2 & 0 \\ -2 & -3 & 2 \\ 0 & 2 & -4 \end{bmatrix} \rightarrow \begin{bmatrix} 1 & 1 & 0 \\ 0 & -1 & 2 \\ 0 & 1 & -2 \end{bmatrix} \rightarrow \begin{bmatrix} 1 & 1 & 0 \\ 0 & 1 & -2 \\ 0 & 0 & 0 \end{bmatrix},$$

so that $x_3 = t$ is free, $x_2 = 2x_3 = 2t$, $x_1 = -x_2 = -2t$, and $\mathbf{x} = \begin{bmatrix} -2t \\ 2t \\ t \end{bmatrix}$. The

eigenspace is one-dimensional with basis $\mathbf{f}_2 = \begin{bmatrix} -2 \\ 2 \\ 1 \end{bmatrix}$. If $\mathbf{x} = \begin{bmatrix} x_1 \\ x_2 \\ x_3 \end{bmatrix}$ is an eigenvector

for $\lambda = -3$, then $(A - \lambda I)\mathbf{x} = \mathbf{0}$ with $\lambda = -3$. Using Gaussian elimination,

$$\begin{bmatrix} 4 & -2 & 0 \\ -2 & 3 & 2 \\ 0 & 2 & 2 \end{bmatrix} \rightarrow \begin{bmatrix} 1 & -\frac{1}{2} & 0 \\ 0 & 2 & 2 \\ 0 & 1 & 1 \end{bmatrix} \rightarrow \begin{bmatrix} 1 & -\frac{1}{2} & 0 \\ 0 & 1 & 1 \\ 0 & 0 & 0 \end{bmatrix},$$

so that $x_3 = t$ is free, $x_2 = -x_3 = -t$, $x_1 = \frac{1}{2}x_2 = -\frac{1}{2}t$, and $\mathbf{x} = \begin{bmatrix} -\frac{1}{2}t \\ -t \\ t \end{bmatrix}$. The

eigenspace is one-dimensional with basis $\mathbf{f}_3 = \begin{bmatrix} -1 \\ -2 \\ 2 \end{bmatrix}$. The eigenvectors $\mathbf{f}_1$, $\mathbf{f}_2$, and $\mathbf{f}_3$
are orthogonal (in agreement with Theorem 7.4.5). Dividing each by its length, we
obtain

$$\mathbf{q}_1 = \begin{bmatrix} \frac{2}{3} \\ \frac{1}{3} \\ -\frac{2}{3} \end{bmatrix}, \quad \mathbf{q}_2 = \begin{bmatrix} -\frac{2}{3} \\ \frac{2}{3} \\ \frac{1}{3} \end{bmatrix}, \quad \mathbf{q}_3 = \begin{bmatrix} -\frac{1}{3} \\ -\frac{2}{3} \\ \frac{2}{3} \end{bmatrix}.$$

With

$$Q = \begin{bmatrix} \mathbf{q}_1 & \mathbf{q}_2 & \mathbf{q}_3 \\ \downarrow & \downarrow & \downarrow \end{bmatrix} = \begin{bmatrix} \frac{2}{3} & -\frac{2}{3} & -\frac{1}{3} \\ \frac{1}{3} & \frac{2}{3} & -\frac{2}{3} \\ \frac{2}{3} & \frac{1}{3} & \frac{2}{3} \end{bmatrix},$$

we have $Q^{-1}AQ = Q^T AQ = D = \begin{bmatrix} 0 & 0 & 0 \\ 0 & 3 & 0 \\ 0 & 0 & -3 \end{bmatrix}$. ■

Theorem 7.4.5 says that a symmetric $n \times n$ matrix with n different eigenvalues is orthogonally diagonalizable. This statement, however, is far from the whole truth. For example, it does not explain why the 3×3 symmetric matrix $A = \begin{bmatrix} 0 & 2 & -1 \\ 2 & 3 & -2 \\ -1 & -2 & 0 \end{bmatrix}$, which has just two eigenvalues, $\lambda = 5$ and $\lambda = -1$, is also orthogonally diagonalizable. The proof of the "Principal Axes Theorem," given in Corollary 7.3.12, is beyond the scope of this section.

Theorem 7.4.8   **(Principal Axes Theorem).** *A real symmetric matrix is orthogonally diagonalizable.*

7.4.9 EXAMPLE   Consider the real symmetric matrix $A = \begin{bmatrix} 0 & 2 & -1 \\ 2 & 3 & -2 \\ -1 & -2 & 0 \end{bmatrix}$ introduced above. We let the reader verify that the eigenvalues for A are $\lambda = 5$ and $\lambda = -1$, as asserted previously. If $\mathbf{x} = \begin{bmatrix} x_1 \\ x_2 \\ x_3 \end{bmatrix}$ is an eigenvector for $\lambda = 5$, then $(A - \lambda I)\mathbf{x} = \mathbf{0}$ with $\lambda = 5$. Using Gaussian elimination,

$$\begin{bmatrix} -5 & 2 & -1 \\ 2 & -2 & -2 \\ -1 & -2 & -5 \end{bmatrix} \to \begin{bmatrix} 1 & 2 & 5 \\ 0 & 12 & 24 \\ 0 & -6 & -12 \end{bmatrix} \to \begin{bmatrix} 1 & 2 & 5 \\ 0 & 1 & 2 \\ 0 & 0 & 0 \end{bmatrix},$$

so that $x_3 = t$ is free, $x_2 = -2x_3 = -2t$, $x_1 = -2x_2 - 5x_3 = 4t - 5t = -t$, and $\mathbf{x} = \begin{bmatrix} -t \\ -2t \\ t \end{bmatrix} = t\mathbf{f}_1$, with $\mathbf{f}_1 = \begin{bmatrix} -1 \\ -2 \\ 1 \end{bmatrix}$. If $\mathbf{x} = \begin{bmatrix} x_1 \\ x_2 \\ x_3 \end{bmatrix}$ is an eigenvector for $\lambda = -1$, then $(A - \lambda I)\mathbf{x} = \mathbf{0}$ with $\lambda = -1$. Using Gaussian elimination,

$$\begin{bmatrix} 1 & 2 & -1 \\ 2 & 4 & -2 \\ -1 & -2 & 1 \end{bmatrix} \to \begin{bmatrix} 1 & 2 & -1 \\ 0 & 0 & 0 \\ 0 & 0 & 0 \end{bmatrix},$$

so that $x_3 = t$ and $x_2 = s$ are free, $x_1 = -2x_2 + x_3 = -2s + t$, and $\mathbf{x} = \begin{bmatrix} -2s + t \\ s \\ t \end{bmatrix} = t\mathbf{u} + s\mathbf{v}$, with $\mathbf{u} = \begin{bmatrix} 1 \\ 0 \\ 1 \end{bmatrix}$ and $\mathbf{v} = \begin{bmatrix} -2 \\ 1 \\ 0 \end{bmatrix}$. Each of these vectors is orthogonal to $\mathbf{f}_1$, as must be the case—eigenvectors corresponding to different eigenvalues of a symmetric matrix are orthogonal. On the other hand, $\mathbf{u}$ and $\mathbf{v}$ are not orthogonal, but this is just an inconvenience, not a genuine obstacle. Applying the Gram–Schmidt Algorithm in its simplest form, we can produce orthogonal eigenvalues by taking $\mathbf{f}_2 = \mathbf{u}$ and

$$\mathbf{f}_3 = \mathbf{v} - \text{proj}_\mathbf{u}\,\mathbf{v} = \begin{bmatrix} -2 \\ 1 \\ 0 \end{bmatrix} - \frac{-2}{2}\begin{bmatrix} 1 \\ 0 \\ 1 \end{bmatrix} = \begin{bmatrix} -2 \\ 1 \\ 0 \end{bmatrix} + \begin{bmatrix} 1 \\ 0 \\ 1 \end{bmatrix} = \begin{bmatrix} -1 \\ 1 \\ 1 \end{bmatrix}.$$

The set $\{\mathbf{f}_1, \mathbf{f}_2, \mathbf{f}_3\}$ is an orthogonal basis of eigenvectors. We turn this into an orthonormal basis by dividing each vector by its length. We obtain

$$\mathbf{q}_1 = \begin{bmatrix} \frac{1}{\sqrt{6}} \\ \frac{2}{\sqrt{6}} \\ -\frac{1}{\sqrt{6}} \end{bmatrix}, \quad \mathbf{q}_2 = \begin{bmatrix} \frac{1}{\sqrt{2}} \\ 0 \\ \frac{1}{\sqrt{2}} \end{bmatrix}, \quad \mathbf{q}_3 = \begin{bmatrix} -\frac{1}{\sqrt{3}} \\ \frac{1}{\sqrt{3}} \\ \frac{1}{\sqrt{3}} \end{bmatrix}.$$

The matrix

$$Q = \begin{bmatrix} \mathbf{q}_1 & \mathbf{q}_2 & \mathbf{q}_3 \\ \downarrow & \downarrow & \downarrow \end{bmatrix} = \begin{bmatrix} -\frac{1}{\sqrt{6}} & \frac{1}{\sqrt{2}} & -\frac{1}{\sqrt{3}} \\ -\frac{2}{\sqrt{6}} & 0 & \frac{1}{\sqrt{3}} \\ \frac{1}{\sqrt{6}} & \frac{1}{\sqrt{2}} & \frac{1}{\sqrt{3}} \end{bmatrix}$$

is orthogonal, and $Q^{-1}AQ = D = \begin{bmatrix} 5 & 0 & 0 \\ 0 & -1 & 0 \\ 0 & 0 & -1 \end{bmatrix}$ is the diagonal matrix whose entries are eigenvalues for $\mathbf{q}_1$, $\mathbf{q}_2$, and $\mathbf{q}_3$, respectively. ∎

7.4.10 Remark Why the name "Principal Axes Theorem"? A set of n orthonormal vectors in $\mathbf{R}^n$, such as the standard basis vectors $\mathbf{e}_1, \dots, \mathbf{e}_n$, determine a set of so-called "principal axes," a principal axis being the span of one of the given vectors. For example, the principal axes associated with the vectors $\mathbf{i}$, $\mathbf{j}$, and $\mathbf{k}$ in $\mathbf{R}^3$ are the x-, the y-, and the z-axes, respectively.

A symmetric $n \times n$ matrix A has n orthogonal eigenvectors; thus $\mathbf{R}^n$ has an orthonormal basis of eigenvectors and hence a new set of "principal axes," each one the span of an eigenvector.

Answers to Reading Challenges

1. If Q is orthogonal, $Q^{-1} = Q^T$, the transpose of Q.

2. The roots of $\lambda^2 - 4\lambda + 6$, namely, $\lambda = \frac{4 \pm \sqrt{(-4)^2 - 4(6)}}{2} = 2 \pm \sqrt{-2}$, are not real numbers.

True/False Questions

Decide, with as little calculation as possible, whether each of the following statements is true or false and explain your answer whenever you say "false." (Answers can be found in the back of the book.)

1. If A is a matrix, P is an invertible matrix, and D is a diagonal matrix with $P^{-1}AP = D$, then the columns of P are eigenvectors of A.

2. If A is a matrix, Q is an orthogonal matrix, and D is a diagonal matrix with $Q^{-1}AQ = D$, then $Q^T AQ = D$.

3. If D is a diagonal matrix and Q is a real orthogonal matrix, then $Q^T D Q$ is also diagonal.

4. If the eigenvalues of a matrix A are all real, then A is symmetric.

5. If a 3×3 matrix A has three orthogonal eigenvectors, then A is symmetric.

6. A real symmetric matrix can be orthogonally diagonalized.

Exercises

*Solutions to exercises marked [BB] can be found in the **B**ack of the **B**ook.*

1. For each of the following real symmetric matrices A, find an orthogonal matrix Q and a real diagonal matrix such that $Q^T A Q = D$.

   (a) [BB] $A = \begin{bmatrix} 2 & 3 \\ 3 & -6 \end{bmatrix}$ (b) $A = \begin{bmatrix} 17 & 24 \\ 24 & 3 \end{bmatrix}$

   (c) [BB] $A = \begin{bmatrix} -2 & 1 & 1 \\ 1 & -2 & 1 \\ 1 & 1 & -2 \end{bmatrix}$

   (d) $A = \begin{bmatrix} 1 & -3 & 2 \\ -3 & 1 & -2 \\ 2 & -2 & 2 \end{bmatrix}$

   (e) $A = \begin{bmatrix} 5 & 2 & 2 \\ 2 & 2 & -4 \\ 2 & -4 & 2 \end{bmatrix}$

   (f) $A = \begin{bmatrix} 5 & 4 & -4 \\ 4 & 5 & 4 \\ -4 & 4 & 5 \end{bmatrix}$

   (g) $A = \begin{bmatrix} 0 & 2 & 1 & 1 \\ 2 & 0 & -1 & -1 \\ 1 & -1 & 0 & -2 \\ 1 & -1 & -2 & 0 \end{bmatrix}$

   (h) $A = \begin{bmatrix} 3 & -1 & -1 & 1 \\ -1 & 3 & 1 & -1 \\ -1 & 1 & 3 & -1 \\ 1 & -1 & -1 & 3 \end{bmatrix}$.

   (The characteristic polynomial of this matrix is $\lambda^4 - 12\lambda^3 + 48\lambda^2 - 80\lambda + 48$.)

2. For each of the real symmetric matrices given below,

   i. find an orthogonal matrix Q such that $Q^T A Q$ is diagonal;

   ii. use the result of part i to find A^5.

   (a) [BB] $A = \begin{bmatrix} 1 & 2 \\ 2 & 1 \end{bmatrix}$

   (b) $A = \begin{bmatrix} 3 & 0 & -1 \\ 0 & -1 & 0 \\ -1 & 0 & 3 \end{bmatrix}$

   (c) $A = \begin{bmatrix} 3 & 0 & 2 \\ 0 & 1 & 0 \\ 2 & 0 & 3 \end{bmatrix}$

   (d) $A = \begin{bmatrix} -1 & -4 & 4 & 0 \\ -4 & -5 & 0 & -4 \\ 4 & 0 & -5 & -4 \\ 0 & -4 & -4 & -1 \end{bmatrix}$.

3. [BB] Establish the converse of the Principal Axes Theorem: If a real matrix is orthogonally diagonalizable, then A is symmetric.

4. Show that the eigenvalues of a 2×2 (real) symmetric matrix are always real numbers.

5. [BB] If A is a real symmetric $n \times n$ matrix with nonnegative eigenvalues, show that $\mathbf{x}^T A \mathbf{x} \geq 0$ for all vectors $\mathbf{x} \in \mathbf{R}^n$.

6. Let A be a (real) $n \times n$ matrix with the property that $\mathbf{x}^T A \mathbf{x} \geq 0$ for all vectors $\mathbf{x}$.

   (a) Let λ be a real eigenvalue of A. Show that $\lambda \geq 0$.

   (b) If A is symmetric, show that A has a square root; that is, there exists a matrix B such that $A = B^2$.

   (c) If A is symmetric, show that A can be factored in the form $A = R R^T$ for some matrix R.

■ Critical Reading

7. [BB] Suppose A and B are real symmetric matrices. Why are all the roots of $\det(A + B - \lambda I)$ real?

8. Suppose A is a skew-symmetric matrix. Let $\mathbf{x}$ and $\mathbf{y}$ be eigenvectors of A corresponding to eigenvalues λ and μ, respectively. Show that either $\mathbf{x}$ and $\mathbf{y}$ are orthogonal or $\lambda = -\mu$.

9. [BB] Let A be a real skew-symmetric matrix. Prove that there exists a real orthogonal matrix Q and a matrix B such that $A = QBQ^T$ and B^2 is diagonal. [Hint: A^TA is real symmetric.]

10. (a) What are the possible values for the eigenvalues of a projection matrix P?

    (b) Does every projection matrix have all possible eigenvalues?
    Explain your answers.

11. Let $\mathbf{v}$ be an eigenvector of a real symmetric $n \times n$ matrix A and let U be the one-dimensional subspace $\mathbf{Rv}$. Show that $U^\perp$ is invariant under multiplication by A. (See the glossary for the definition of "invariant subspace.")

12. (a) [BB] Let A be real symmetric matrix A with nonnegative eigenvalues. Show that A can be written in the form $A = RR^T$ for some matrix R.

    (This is a partial converse of the elementary fact that RR^T is always symmetric.) [Hint: A diagonal matrix with real nonnegative diagonal entries has a square root.]

    (b) Let A and B be real symmetric matrices and suppose the eigenvalues of B are all positive. Show that the roots of the polynomial $\det(A - \lambda B)$ are real. [Hint: First explain why $\det(A - \lambda B) = 0$ gives a nonzero $\mathbf{x}$ with $A\mathbf{x} = \lambda B\mathbf{x}$. Then use the result of 12(a) to factor $B = RR^T$ with R invertible.]

13. Let A be a real symmetric matrix with nonnegative eigenvalues. It is a fact that $\mathbf{x}^T A\mathbf{x} \geq 0$ for all vectors $\mathbf{x}$ (of compatible size). (See Exercise 5.) Suppose A and B are real symmetric matrices of the same size with nonnegative eigenvalues. Show that all eigenvalues of $A + B$ are nonnegative. [Hint: Exercise 6.]

7.5 The Singular Value Decomposition

Suppose A is a real $m \times n$ matrix. The $n \times n$ matrix A^TA is symmetric, so by the Principal Axes Theorem—Corollary 7.3.12—$\mathbf{R}^n$ has a basis $\mathbf{x}_1, \ldots, \mathbf{x}_n$ of orthonormal eigenvectors of A^TA. If $\lambda_1, \ldots, \lambda_n$ are corresponding eigenvalues, then $A^TA\mathbf{x}_i = \lambda_i\mathbf{x}_i$ for $i = 1, 2, \ldots, n$.

Now $\|\mathbf{x}_i\|^2 = \mathbf{x}_i \cdot \mathbf{x}_i = 1$ for each i, so

$$(A^TA\mathbf{x}_i) \cdot \mathbf{x}_i = (\lambda_i\mathbf{x}_i) \cdot \mathbf{x}_i = \lambda_i(\mathbf{x}_i \cdot \mathbf{x}_i) = \lambda_i,$$

while, on the other hand,

$$(A^TA\mathbf{x}_i) \cdot \mathbf{x}_i \overset{\downarrow}{=} (A^TA\mathbf{x}_i)^T\mathbf{x}_i$$

$$= \mathbf{x}_i^T A^TA\mathbf{x}_i = (A\mathbf{x}_i)^T(A\mathbf{x}_i) = (A\mathbf{x}_i) \cdot (A\mathbf{x}_i) = \|A\mathbf{x}_i\|^2.$$

Reading Challenge 1   *Explain the first equality here, the one marked with the arrow.*

Thus

$$\lambda_i = \|A\mathbf{x}_i\|^2 \quad \text{for } i = 1, 2, \ldots, n, \tag{1}$$

which implies that $\lambda_i \geq 0$ for all i. Now rearrange the λ_i, if necessary, so that $\lambda_1 > 0, \lambda_2 > 0, \ldots, \lambda_r > 0$, and $\lambda_{r+1} = \lambda_{r+2} = \cdots = \lambda_n = 0$. Define

$$\mu_i = \sqrt{\lambda_i} \quad \text{and} \quad \mathbf{y}_i = \frac{1}{\mu_i} A\mathbf{x}_i \quad \text{for } i = 1, 2, \ldots, r.$$

The $\mathbf{y}_i$ are vectors in $\mathbf{R}^m$, each of length 1, since

$$\|\mathbf{y}_i\| = \frac{1}{\mu_i}\|A\mathbf{x}_i\| = \frac{1}{\mu_i}\sqrt{\lambda_i} = 1.$$

Also, for $1 \le i, j \le r$ and $i \ne j$, we have

$$\mathbf{y}_i \cdot \mathbf{y}_j = \frac{1}{\mu_i \mu_j} A\mathbf{x}_i \cdot A\mathbf{x}_j$$

$$= \frac{1}{\mu_i \mu_j}\mathbf{x}_i^T A^T A\mathbf{x}_j = \frac{1}{\mu_i \mu_j}\mathbf{x}_i^T \lambda_j \mathbf{x}_j \overset{\Downarrow}{=} \frac{1}{\mu_i \mu_j}\lambda_j (\mathbf{x}_i \cdot \mathbf{x}_j) = 0. \qquad (2)$$

Reading Challenge 2   *Explain the last equality here, the one marked* $\Downarrow$.

Thus $\{\mathbf{y}_1, \dots, \mathbf{y}_r\}$ is an orthonormal set of vectors that can be extended to an orthonormal basis $\mathbf{y}_1, \dots, \mathbf{y}_m$ of $\mathbf{R}^m$ (for example, by the Gram–Schmidt Algorithm). Set

$$Q_1 = \begin{bmatrix} \mathbf{y}_1 & \cdots & \mathbf{y}_m \\ \downarrow & & \downarrow \end{bmatrix} \quad \text{and} \quad Q_2 = \begin{bmatrix} \mathbf{x}_1 & \cdots & \mathbf{x}_n \\ \downarrow & & \downarrow \end{bmatrix}.$$

Then Q_1 is an orthogonal $m \times m$ matrix, Q_2 is an orthogonal $n \times n$ matrix, and

$$\Sigma = Q_1^T A Q_2 = \begin{bmatrix} \mathbf{y}_1^T & \rightarrow \\ & \vdots \\ \mathbf{y}_m^T & \rightarrow \end{bmatrix}\begin{bmatrix} A\mathbf{x}_1 & \cdots & A\mathbf{x}_n \\ \downarrow & & \downarrow \end{bmatrix}$$

is an $m \times n$ matrix, the (i, j) entry of which is $\mathbf{y}_i^T (A\mathbf{x}_j)$.

If $j \le r$, then $\mathbf{y}_j = \dfrac{1}{\mu_j}A\mathbf{x}_j$, so

$$\mathbf{y}_i^T (A\mathbf{x}_i) = \mathbf{y}_i^T (\mu_j \mathbf{y}_j) = \mu_j \mathbf{y}_i \cdot \mathbf{y}_j = \begin{cases} \mu_j & \text{if } i = j \\ 0 & \text{if } i \ne j \end{cases}$$

because $\mathbf{y}_1, \dots, \mathbf{y}_m$ are orthonormal.

If $j > r$, $\|A\mathbf{x}_j\|^2 = \lambda_j = 0$, so $A\mathbf{x}_j = \mathbf{0}$ and $\mathbf{y}_i^T (A\mathbf{x}_j) = \mathbf{0}$. Thus

$$\Sigma = \begin{bmatrix} \mu_1 & 0 & \cdots & 0 & 0 & & 0 \\ 0 & \mu_2 & & & \vdots & & \vdots \\ \vdots & & \ddots & & & & \\ 0 & & & \mu_r & & & \\ 0 & & & & 0 & & \\ \vdots & & & & & \ddots & \\ 0 & & \cdots & & & & 0 \end{bmatrix}.$$

Since $Q_1^T = Q^{-1}$ and $Q_2^T = Q^{-1}$, the equation $\Sigma = Q_1^T A Q_2$ implies $A = Q_1 \Sigma Q_2^{-1} = Q_1 \Sigma Q_2^T$.

All this establishes the following theorem.

Theorem 7.5.1   *Any $m \times n$ matrix A can be factored $A = Q_1 \Sigma Q_2^T$, where Q_1 is an orthogonal $m \times m$ matrix, Q_2 is an orthogonal $n \times n$ matrix, and*

$$\Sigma = \begin{bmatrix} D & \mathbf{0} \\ \mathbf{0} & \mathbf{0} \end{bmatrix}$$

is a diagonal $m \times n$ matrix with $D = \begin{bmatrix} \mu_1 & & \\ & \ddots & \\ & & \mu_r \end{bmatrix}$, a diagonal $r \times r$ matrix with positive diagonal entries μ_i. The boldface $\mathbf{0}$s in the representation of Σ denote zero matrices of the sizes needed to make Σ an $m \times n$ matrix.

7.5.2 Remark   A factorization of an $m \times n$ matrix A in the form $A = Q_1 \Sigma Q_2^T$, as in Theorem 7.5.1, is called a *singular value decomposition* of A. The nonzero diagonal entries of Σ—$\mu_1, \dots, \mu_r$—are called the *singular values* of A. The orthogonal matrices Q_1 and Q_2 are not unique, but the singular values of A are uniquely determined as the square roots of the nonzero eigenvalues of the matrix $A^T A$. (See Exercise 8.)

7.5.3 EXAMPLE   Let $A = \begin{bmatrix} -2 \\ 1 \\ 2 \end{bmatrix}$. Then $A^T A = \begin{bmatrix} -2 & 1 & 2 \end{bmatrix} \begin{bmatrix} -2 \\ 1 \\ 2 \end{bmatrix} = [9]$. The vector $\mathbf{x}_1 = [1]$ is a basis for $\mathbf{R}^1$ and it's a unit eigenvector for $A^T A$ corresponding to eigenvalue $\lambda_1 = 9$. Set $\mu_1 = \sqrt{\lambda_1} = 3$ and $\mathbf{y}_1 = \frac{1}{\mu_1} A \mathbf{x}_1 = \frac{1}{3} \begin{bmatrix} -2 \\ 1 \\ 2 \end{bmatrix}$.

Extend $\mathbf{y}_1$ to an orthonormal basis of $\mathbf{R}^3$, for example,

$$\mathbf{y}_1 = \frac{1}{3} \begin{bmatrix} -2 \\ 1 \\ 2 \end{bmatrix}, \quad \mathbf{y}_2 = \frac{1}{\sqrt{5}} \begin{bmatrix} 0 \\ 2 \\ -1 \end{bmatrix}, \quad \mathbf{y}_3 = \frac{1}{\sqrt{45}} \begin{bmatrix} 5 \\ 2 \\ 4 \end{bmatrix}.$$

Set

$$Q_1 = \begin{bmatrix} -\frac{2}{3} & 0 & \frac{5}{\sqrt{45}} \\ \frac{1}{3} & \frac{2}{\sqrt{5}} & -\frac{1}{\sqrt{5}} \\ \frac{2}{3} & -\frac{1}{\sqrt{5}} & \frac{4}{\sqrt{45}} \end{bmatrix}, \quad Q_2 = [1], \quad \text{and} \quad \Sigma = \begin{bmatrix} 3 \\ 0 \\ 0 \end{bmatrix},$$

Then $A = Q_1 \Sigma Q_2^T$ is a singular value decomposition. The lone singular value of A is 3.   ∎

7.5.4 EXAMPLE

Suppose $A = \begin{bmatrix} 1 & 1 \\ 1 & 0 \\ 0 & 1 \end{bmatrix}$. Then $A^T A = \begin{bmatrix} 2 & 1 \\ 1 & 2 \end{bmatrix}$ and

$$\mathbf{x}_1 = \frac{1}{\sqrt{2}} \begin{bmatrix} 1 \\ -1 \end{bmatrix}, \quad \mathbf{x}_2 = \frac{1}{\sqrt{2}} \begin{bmatrix} 1 \\ 1 \end{bmatrix}$$

provide an orthonormal basis of eigenvectors of $A^T A$ with corresponding eigenvalues $\lambda_1 = 1$, $\lambda_2 = 3$.

Let $\mu_1 = \sqrt{\lambda_1} = 1$, $\mu_2 = \sqrt{\lambda_2} = \sqrt{3}$, and $\Sigma = \begin{bmatrix} 1 & 0 \\ 0 & \sqrt{3} \\ 0 & 0 \end{bmatrix}$, and set

$$\mathbf{y}_1 = \frac{1}{\mu_1} A\mathbf{x}_1 = \frac{1}{\sqrt{2}} \begin{bmatrix} 0 \\ 1 \\ -1 \end{bmatrix} \quad \text{and} \quad \mathbf{y}_2 = \frac{1}{\mu_2} A\mathbf{x}_2 = \frac{1}{\sqrt{6}} \begin{bmatrix} 2 \\ 1 \\ 1 \end{bmatrix}.$$

Together with $\mathbf{y}_3 = \frac{1}{\sqrt{3}} \begin{bmatrix} -1 \\ 1 \\ 1 \end{bmatrix}$, we obtain an orthonormal basis $\mathbf{y}_1, \mathbf{y}_2, \mathbf{y}_3$ for $\mathbf{R}^3$.

Letting

$$Q_1 = \begin{bmatrix} 0 & \frac{2}{\sqrt{6}} & -\frac{1}{\sqrt{3}} \\ \frac{1}{\sqrt{2}} & \frac{1}{\sqrt{6}} & \frac{1}{\sqrt{3}} \\ -\frac{1}{\sqrt{2}} & \frac{1}{\sqrt{6}} & \frac{1}{\sqrt{3}} \end{bmatrix} \quad \text{and} \quad Q_2 = \begin{bmatrix} \frac{1}{\sqrt{2}} & \frac{1}{\sqrt{2}} \\ -\frac{1}{\sqrt{2}} & \frac{1}{\sqrt{2}} \end{bmatrix},$$

we have

$$\begin{bmatrix} 1 & 1 \\ 1 & 0 \\ 0 & 1 \end{bmatrix} = \begin{bmatrix} 0 & \frac{2}{\sqrt{6}} & -\frac{1}{\sqrt{3}} \\ \frac{1}{\sqrt{2}} & \frac{1}{\sqrt{6}} & \frac{1}{\sqrt{3}} \\ -\frac{1}{\sqrt{2}} & \frac{1}{\sqrt{6}} & \frac{1}{\sqrt{3}} \end{bmatrix} \begin{bmatrix} 1 & 0 \\ 0 & \sqrt{3} \\ 0 & 0 \end{bmatrix} \begin{bmatrix} \frac{1}{\sqrt{2}} & -\frac{1}{\sqrt{2}} \\ \frac{1}{\sqrt{2}} & \frac{1}{\sqrt{2}} \end{bmatrix}$$

$$A \quad = \quad\quad\quad Q_1 \quad\quad\quad\quad \Sigma \quad\quad\quad Q_2^T,$$

which is a singular value decomposition of A. The singular values of A are 1 and $\sqrt{3}$. ∎

■ The Pseudoinverse Again

A singular value decomposition for a matrix A gives us a nice formula for the pseudoinverse A^+, which we first met in Section 6.4:

7.5.5 If $A = Q_1 \Sigma Q_2^T$, then $A^+ = Q_2 \Sigma^+ Q_1^T$.

The formula is easily remembered. Since Q_1 and Q_2 are orthogonal, $A = Q_1 \Sigma Q_2^{-1}$, so if A and Σ were invertible, we would have $A^{-1} = Q_2 \Sigma^{-1} Q_1^{-1} = Q_2 \Sigma^{-1} Q_1^T$. In general, without any assumptions about invertibility, one simply replaces the inverses of A and Σ by their pseudoinverses. But why is 7.5.5 true?

Remember that $\mathbf{x}^+ = A^+\mathbf{b}$ is the vector of shortest length among those vectors $\mathbf{x}$ for which $\|\mathbf{b} - A\mathbf{x}\|$ is a minimum. Given $A = Q_1 \Sigma Q_2^T$ and recalling 6.2.9— multiplication by the orthogonal matrix Q_1^T preserves length—we have

$$\|\mathbf{b} - A\mathbf{x}\| = \left\| \mathbf{b} - Q_1 \Sigma Q_2^T \mathbf{x} \right\| = \left\| Q_1^T \mathbf{b} - Q_1^T (Q_1 \Sigma Q_2^T \mathbf{x}) \right\| = \left\| Q_1^T \mathbf{b} - \Sigma Q_2^T \mathbf{x} \right\|.$$

Thus

$$\|\mathbf{b} - A\mathbf{x}\| = \left\| Q_1^T \mathbf{b} - \Sigma \mathbf{y} \right\| \tag{3}$$

with $\mathbf{y} = Q_2^T \mathbf{x}$. Note that $\|\mathbf{y}\| = \|\mathbf{x}\|$. Thus the shortest $\mathbf{x}$ minimizing the left side of (3) is the shortest $\mathbf{y}$ minimizing the right side. This shortest $\mathbf{y}$ is just $\mathbf{y}^+ = \Sigma^+ (Q_1^T \mathbf{b})$. Thus $\mathbf{x}^+ = Q_2 \mathbf{y}^+ = Q_2 \Sigma^+ Q_1^T \mathbf{b}$, so $A^+ = Q_2 \Sigma^+ Q_1^T$, as asserted.

Formula 7.5.5 makes it quite easy to discover a number of properties of the pseudoinverse. The proof of the following theorem makes use of the fact that if Σ is

the $m \times n$ matrix $\Sigma = \begin{bmatrix} \mu_1 & & & & \\ & \ddots & & & \mathbf{0} \\ & & \mu_r & & \\ & \mathbf{0} & & \mathbf{0} \end{bmatrix}$, with $\mu_1, \ldots, \mu_r$ all nonzero, then Σ^+ is the

$n \times m$ matrix $\Sigma^+ = \begin{bmatrix} \frac{1}{\mu_1} & & & & \\ & \ddots & & & \mathbf{0} \\ & & \frac{1}{\mu_r} & & \\ & \mathbf{0} & & \mathbf{0} \end{bmatrix}$. (As always, the boldface $\mathbf{0}$s indicate zero matri-

ces of the size needed to complete a matrix of the right size.) See paragraph 6.4.11. In particular, $(\Sigma^+)^+ = \Sigma$.

Theorem 7.5.6   (**Properties of the Pseudoinverse**). *The pseudoinverse A^+ of an $m \times n$ matrix A has the following properties:*

1. *A^+ is $n \times m$;*

2. *$A^{++} = A$;*

3. *rank A^+ = rank A;*

4. *col sp A^+ = row sp A;*

5. *row sp A^+ = col sp A;*

6. *A^+A is the projection matrix onto the row space of A.*

Proof   1. We have $A = Q_1 \Sigma Q_2^T$ with Q_1 an orthogonal $m \times m$ matrix, Σ an $m \times n$ matrix, and Q_2 an orthogonal $n \times n$ matrix. Thus $A^+ = Q_2 \Sigma^+ Q_1^T$ is $n \times m$.

2. Let $A = Q_1 \Sigma Q_2^T$. Then $A^+ = Q_2 \Sigma^+ Q_1^T$, so $(A^+)^+ = Q_1 (\Sigma^+)^+ Q_2^T = Q_1 \Sigma Q_2^T = A$ since $(\Sigma^+)^+ = \Sigma$.

3. For this part, we first remember that multiplication by an invertible matrix preserves rank—see Theorem 4.4.3 and Exercise 10 in Section 4.4. Thus rank $A = $ rank $Q_1 \Sigma Q_2^T = $ rank Σ and, similarly, rank $A^+ = $ rank Σ^+. Since the rank of a diagonal matrix with r nonzero entries is r, we have rank $\Sigma = r$.

If Σ is the $m \times n$ matrix $\Sigma = \begin{bmatrix} \mu_1 & & & & \\ & \ddots & & & \mathbf{0} \\ & & \mu_r & & \\ & \mathbf{0} & & \mathbf{0} \end{bmatrix}$, with $\mu_1 \neq 0, \dots \mu_r \neq 0$, then

Σ^+ is the $n \times m$ matrix $\Sigma^+ = \begin{bmatrix} \frac{1}{\mu_1} & & & & \\ & \ddots & & & \mathbf{0} \\ & & \frac{1}{\mu_r} & & \\ & \mathbf{0} & & \mathbf{0} \end{bmatrix}$, so Σ^+ also has rank r. Therefore,

rank $A = $ rank $\Sigma = $ rank $\Sigma^+ = $ rank A^+.

4. A vector in the column space of A^+ has the form $A^+\mathbf{b}$, so it's in the row space of A. (This is part of the definition of the pseudoinverse—see Definition 6.4.7.) Thus col sp $A^+ \subseteq$ row sp A. Since the dimensions of these two spaces are equal by part 3, these spaces are the same.

5. Applying part 4 to the matrix A^+ gives row sp $A^+ = $ col sp$(A^+)^+ = $ col sp A.

6. The definition of the pseudoinverse specifies that $A(A^+\mathbf{b})$ is the projection of $\mathbf{b}$ on the column space of A. Applying this fact to the matrix A^+ tells us that $A^+(A^+)^+\mathbf{b} = A^+A\mathbf{b}$ is the projection $\mathbf{b}$ on the column space of A^+, which is the row space of A. ∎

Answers to Reading Challenges

1. For any vectors $\mathbf{u}$ and $\mathbf{v}$ (in the same Euclidean space), we know that $\mathbf{u} \cdot \mathbf{v} = \mathbf{u}^T \mathbf{v}$.

2. The vectors $\mathbf{x}_1, \dots, \mathbf{x}_n$ are orthogonal.

True/False Questions

Decide, with as little calculation as possible, whether each of the following statements is true or false and explain your answer whenever you say "false." (Answers can be found in the back of the book.)

1. If $c = \|A\mathbf{x}\|$ for some vector $\mathbf{x}$ and matrix A, then $c \geq 0$.

2. If $\mathbf{u}, \mathbf{v}, \mathbf{w}$ are orthonormal vectors in $\mathbf{R}^4$, there is a vector $\mathbf{x}$ of length 1 orthogonal to each of $\mathbf{u}, \mathbf{v}, \mathbf{w}$.

3. If A is a matrix and Q_1 and Q_2 are orthogonal matrices with $\Sigma = Q_1^T A Q_2$, then $A = Q_1 \Sigma Q_2^{-1}$.

4. The singular values of a matrix A are the eigenvalues of $A^T A$.

5. If A and B are matrices and B is invertible, then rank $BA = $ rank A.

6. If $A = Q_1 \Sigma Q_2^T$ is a singular value decomposition, then $A^+ = Q_1^+ \Sigma^+ Q_2$.

Exercises

*Solutions to exercises marked [BB] can be found in the **B**ack of the **B**ook.*

1. Find a singular value decomposition of each of the following matrices and use this to find the pseudoinverse of each matrix as well.

   (a) [BB] $A = \begin{bmatrix} 1 & -1 \\ 0 & 0 \\ 1 & -1 \end{bmatrix}$

   (b) $A = \begin{bmatrix} 1 & -1 \\ -1 & 1 \\ 2 & -2 \end{bmatrix}$

   (c) [BB] $A = \begin{bmatrix} 1 & 2 & 3 \end{bmatrix}$

   (d) $A = \begin{bmatrix} 1 & 1 & -1 \\ 1 & 3 & 1 \end{bmatrix}$

   (e) $A = \begin{bmatrix} 1 & -1 & 0 \\ 0 & 0 & 1 \\ 1 & -1 & 0 \end{bmatrix}$

   (f) $A = \begin{bmatrix} 0 & -1 & 3 \\ 0 & 2 & 0 \\ 0 & 0 & 2 \end{bmatrix}$.

2. For a square matrix A, prove that $\det A$ is either the product of the singular values of A or the negative of this product.

3. [BB] Let A be an $m \times n$ matrix. Show that $(A^+)^T = (A^T)^+$.

4. Explain how to obtain a singular value decomposition of A^{-1} from a singular value decomposition of an invertible matrix A.

5. [BB] Let A be an $m \times n$ matrix with all singular values positive. What is the rank of A? Explain.

6. Let A be a matrix.

   (a) [BB] Prove that $AA^+A = A$.

   (b) Use part (a) to prove that $A^+ = A^+AA^+$.

   (c) Use part (b) to prove that AA^+ is a projection matrix.

7. [BB] Suppose A is a symmetric matrix with positive eigenvalues. Prove that any eigenvalue of A is a singular value of A.

8. Let $A = Q_1 \Sigma Q_2^T$ denote a singular value decomposition of an $m \times n$ matrix A.

   (a) Show that the matrices $A^T A$ and $\Sigma^T \Sigma$ are similar.

   (b) Explain why $A^T A$ and $\Sigma^T \Sigma$ have the same eigenvalues.

   (c) Explain why the nonzero entries of Σ are uniquely determined by A.

9. Let A be any $m \times n$ matrix. Show that A and A^T have the same singular values.

10. Let $A = Q_1 \Sigma Q_2^T$ be a singular value decomposition of the $m \times n$ matrix A.

    (a) [BB] Show that the columns of Q_2 are eigenvectors of $A^T A$.

    (b) Show that the columns of Q_1 are eigenvectors of AA^T.

CHAPTER KEY WORDS AND IDEAS: Here are some technical words and phrases that were used in this chapter. Do you know the meaning of each? If you're not sure, check the glossary or index at the back of the book.

absolute value (of a complex number)
argument (of a complex number)
Cayley–Hamilton theorem
commute
orthonormal vectors
complex conjugate
complex number
conjugate transpose of a matrix

norm of a complex vector
orthogonal matrix
orthogonally diagonalizable
orthonormal vectors
parallelogram rule (for vector addition)
polar form
principal axes
pseudoinverse

De Moivre's formula
diagonalizable
dot product (complex)
eigenvalue
eigenvector
Fundamental Theorem of Algebra
Hermitian
imaginary axis
imaginary part (of a complex number)
length of a complex vector
modulus (of a complex number)
multiplicity of a root
normal matrix
normalize

real axis
real part (of a complex number)
root of unity
Schur's Lemma
similar matrices
singular value decomposition
singular values
Spectral Theorem
spectrum
standard form (of a complex number)
transpose of a matrix
unitarily diagonalizable
unitary matrix

Review Exercises for Chapter 7

1. Suppose the only eigenvalue of a real symmetric matrix A is -7. What can you say about A?

2. Let A and B be Hermitian matrices.

  (a) Suppose A and B *commute*, that is, $AB = BA$. Show that AB is Hermitian.

  (b) Establish the converse of part (a): If AB is Hermitian, then A and B commute.

3. Determine whether the following statement is true or false and explain your answer.

  A symmetric matrix (whose entries are not necessarily real) is diagonalizable.

  [Hint: $A = \begin{bmatrix} 3 & i \\ i & 1 \end{bmatrix}$.]

4. The characteristic polynomial of
$$A = \begin{bmatrix} 9 & -2 & 6 \\ -2 & 9 & 6 \\ 6 & 6 & 7 \end{bmatrix}$$ is $-(\lambda - 11)^2(\lambda + 11)$. There exists an orthogonal matrix Q such that $Q^T A Q$ is diagonal with entries ± 11. Why? Find Q such that
$$Q^T A Q = \begin{bmatrix} 11 & 0 & 0 \\ 0 & -11 & 0 \\ 0 & 0 & 11 \end{bmatrix}.$$

5. Let $A = \begin{bmatrix} 1 & 0 & -1 \\ 0 & 0 & 0 \\ -1 & 0 & 1 \end{bmatrix}$.

  (a) This matrix is orthogonally diagonalizable. Why?

(b) Find an orthogonal matrix Q and a diagonal matrix D such that $Q^{-1} A Q = D$.

6. Let $A = \begin{bmatrix} -7 & 6 & 3 \\ 3 & -4 & 3 \\ 6 & 12 & -4 \end{bmatrix}$.

  (a) Show that $\lambda = -10$ is an eigenvalue of A. Find the corresponding eigenspace.

  (b) Given that $\begin{bmatrix} 1 \\ 1 \\ 2 \end{bmatrix}$ is an eigenvector of A corresponding to $\lambda = 5$, find a matrix P and a diagonal matrix D such that $P^{-1} A P = D$.

  (c) What is the characteristic polynomial of A? Explain.

  (d) Let $Q = \begin{bmatrix} 2 & -3 & -2 \\ 0 & -3 & 1 \\ -2 & -6 & 0 \end{bmatrix}$. Find $Q^{-1} A Q$ without calculation. Explain your reasoning.

7. Find the eigenvalues and corresponding eigenvectors of $A = \begin{bmatrix} 2 & 3 - 3i \\ 3 + 3i & 5 \end{bmatrix}$. Verify that eigenvectors corresponding to different eigenvalues are orthogonal. Why should this be the case?

8. Prove that a matrix that is unitarily similar to a real diagonal matrix is Hermitian.

9. **(a)** Show that if A and Q are orthogonal matrices, so is AQ.

(b) If Q is an orthogonal matrix and $\mathbf{x}$ is a vector, show that $\|Q\mathbf{x}\| = \|\mathbf{x}\|$.

(c) For any $\mathbf{x} \in \mathbf{R}^n$ and any scalar t, show that $\|t\mathbf{x}\| = |t|\,\|\mathbf{x}\|$.

(d) If λ is an eigenvalue of an orthogonal matrix, then $\lambda = \pm 1$.

10. Let A be a real symmetric matrix. Show that the following three conditions are equivalent:

  i. A is orthogonal;

  ii. If λ is an eigenvalue of A, then $\lambda = \pm 1$;

  iii. $A^2 = I$.

11. Answer true or false and justify your choice.

(a) $A = \begin{bmatrix} 1 & i \\ -i & 1+i \end{bmatrix}$ is Hermitian.

(b) If U^*AU is a diagonal matrix and U is unitary, then A is normal.

(c) If $\mathbf{x}$ and $\mathbf{y}$ are different eigenvectors of a normal matrix, then $\mathbf{x}$ is orthogonal to $\mathbf{y}$.

(d) All eigenvalues of a Hermitian matrix are real.

(e) If U is a unitary matrix, then $U - 2I$ is invertible.

(f) For any $n \times n$ matrix A, the eigenvalues of $A + A^*$ are all real.

12. Let $A = \begin{bmatrix} 1 & -1 & 1 \\ 1 & -1 & 1 \\ 1 & -1 & 1 \end{bmatrix}$. True or false: A is similar to a diagonal matrix with at least one 0 on the diagonal.

13. Find a singular value decomposition of $A = \begin{bmatrix} 1 & 0 & -1 \\ 0 & -1 & 2 \end{bmatrix}$ and use this to find the pseudoinverse of A as well.

14. Let A^+ denote the pseudoinverse of a matrix A. Prove that A^+A is symmetric. [Hint: Singular value decomposition.]

15. Suppose A and B are $n \times n$ matrices that have the same set of n linearly independent eigenvectors. Prove that $AB = BA$.

8

Applications

8.1 Data Fitting

In this section, we return to the central idea of Section 6.1—how to find the "best solution" to a system that does not have a solution—and discuss a common application. Given a linear system $A\mathbf{x} = \mathbf{b}$ that does not have a solution, we have found that, if A has linearly independent columns, the "best solution in the sense of least squares" is

$$\mathbf{x}^+ = (A^T A)^{-1} A^T \mathbf{b}.$$

Suppose we want to put the "best" possible line through the three points

$$(-2, 4), \ (-1, -1), \ (4, -3). \tag{1}$$

See Figure 8.1. We'll make clear our concept of "best" in this context a little later.

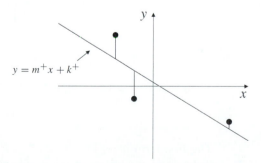

Figure 8.1

The line has an equation of the form $y = mx + k$. If the points in question did in fact lie on a line, their coordinates would have to satisfy $y = mx + k$. So, we would have

$$4 = -2m + k$$

$$-1 = -m + k$$

$$-3 = 4m + k,$$

which is $A\mathbf{x} = \mathbf{b}$, with

$$A = \begin{bmatrix} 1 & -2 \\ 1 & -1 \\ 1 & 4 \end{bmatrix}, \quad \mathbf{x} = \begin{bmatrix} k \\ m \end{bmatrix}, \quad \mathbf{b} = \begin{bmatrix} 4 \\ -1 \\ -3 \end{bmatrix}.$$

The system has no solution, of course; the points don't lie on a line. On the other hand, we can find the "best solution in the sense of least squares." We have

$$A^T A = \begin{bmatrix} 3 & 1 \\ 1 & 21 \end{bmatrix} \quad \text{and} \quad (A^T A)^{-1} = \frac{1}{62} \begin{bmatrix} 21 & -1 \\ -1 & 3 \end{bmatrix},$$

so the best solution to $A\mathbf{x} = \mathbf{b}$ is

$$\mathbf{x}^+ = \begin{bmatrix} k^+ \\ m^+ \end{bmatrix} = (A^T A)^{-1} A^T \mathbf{b}$$

$$= \frac{1}{62} \begin{bmatrix} 21 & -1 \\ -1 & 3 \end{bmatrix} \begin{bmatrix} 1 & 1 & 1 \\ -2 & -1 & 4 \end{bmatrix} \begin{bmatrix} 45 \\ -1 \\ -3 \end{bmatrix} = \begin{bmatrix} \frac{19}{62} \\ -\frac{57}{62} \end{bmatrix}.$$

Thus $k^+ = \frac{19}{62}$, $m^+ = -\frac{57}{62}$, and our "best" line (the one shown in Figure 8.1) is the one with equation $y = -\frac{57}{62}x + \frac{19}{62}$.

Now let us examine the sense in which this line is "best." Remember that $\mathbf{x}^+$ is the $\mathbf{x}$ that minimizes $\|\mathbf{b} - A\mathbf{x}\|$. Here

$$\mathbf{b} - A\mathbf{x}^+ = \begin{bmatrix} 4 \\ -1 \\ -3 \end{bmatrix} - \begin{bmatrix} -2m^+ + k^+ \\ -m^+ + k^+ \\ 4m^+ + k^+ \end{bmatrix},$$

so

$$\|\mathbf{b} - A\mathbf{x}^+\|^2 = [4 - (-2m^+ + k^+)]^2$$

$$+ [-1 - (-m^+ + k^+)]^2 + [-3 - (4m^+ + k^+)]^2.$$

The first term here, $[4 - (-2m^+ + k^+)]^2$, is the square of the *deviation* or vertical distance of the left-most point, where $x = -2$, from the line: 4 is the height of the point, $-2m^+ + k^+$ is the height of the line.

The second term, $[-1 - (-m^+ + k^+)]^2$, is the square of the deviation of the middle point, where $x = -1$, from the line. The third term, $[-3 - (4m^+ + k^+)]^2$, is the square of the deviation of the right-most point from the line. Our line is "best" in the sense that the sum of the squares of the deviations of the points from the line (these are the vertical distances from points to line shown in Figure 8.1) is a minimum. For any other line, this sum of squares of deviations is more.

We consider a more complicated situation. The points

$$(-3, -.5), \ (-1, -.5), \ (1, .9), \ (2, 1.1), \ (3, 3.3) \tag{2}$$

may have arisen in some experiment and there may be reason to think that they should lie on a line. See Figure 8.2. We proceed as before.

If the points were to lie on the line with equation $y = mx + k$, we would have

$$-.5 = -3m + k$$
$$-.5 = -m + k$$
$$.9 = m + k$$
$$1.1 = 2m + k$$
$$3.3 = 3m + k,$$

which is $A\mathbf{x} = \mathbf{b}$, with

$$A = \begin{bmatrix} 1 & -3 \\ 1 & -1 \\ 1 & 1 \\ 1 & 2 \\ 1 & 3 \end{bmatrix}, \quad \mathbf{x} = \begin{bmatrix} k \\ m \end{bmatrix}, \quad \text{and} \quad \mathbf{b} = \begin{bmatrix} -.5 \\ -.5 \\ .9 \\ 1.1 \\ 3.3 \end{bmatrix}.$$

We have $A^T A = \begin{bmatrix} 5 & 2 \\ 2 & 24 \end{bmatrix}$ and $(A^T A)^{-1} = \frac{1}{116} \begin{bmatrix} 24 & -2 \\ -2 & 5 \end{bmatrix}$, so the best solution to $A\mathbf{x} = \mathbf{b}$ is

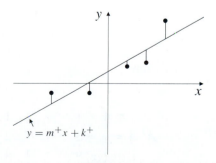

Figure 8.2

$$\mathbf{x}^{+} = \begin{bmatrix} k^{+} \\ m^{+} \end{bmatrix} = (A^{T}A)^{-1}A^{T}\mathbf{b}$$

$$= \frac{1}{116}\begin{bmatrix} 24 & -2 \\ -2 & 5 \end{bmatrix}\begin{bmatrix} 1 & 1 & 1 & 1 & 1 \\ -3 & -1 & 1 & 2 & 3 \end{bmatrix}\begin{bmatrix} -.5 \\ -.5 \\ .9 \\ 1.1 \\ 3.3 \end{bmatrix} = \begin{bmatrix} .631 \\ .572 \end{bmatrix}.$$

Thus $k^{+} = .631$, $m^{+} = .572$, and our "best" line (the one shown in Figure 8.2) is the one with equation $y = .572x + .631$. As before, this line is "best" in the sense that the sum of the squares of the deviations of the points from the line is a minimum.

◼ Other Curves

Perhaps we believe that our points really lie on a parabola, say the one with equation $y = rx^{2} + sx + t$. Then each (x, y) in the list (2) must satisfy this equation. We have

$$-.5 = r(-3)^{2} - 3s + t$$

$$-.5 = r(-1)^{2} - s + t$$

$$.9 = r + s + t$$

$$1.1 = r(2^{2}) + 2s + t$$

$$3.3 = r(3^{2}) + 3s + t,$$

which is $A\mathbf{x} = \mathbf{b}$, with

$$A = \begin{bmatrix} 1 & -3 & 9 \\ 1 & -1 & 1 \\ 1 & 1 & 1 \\ 1 & 2 & 4 \\ 1 & 3 & 9 \end{bmatrix}, \quad \mathbf{x} = \begin{bmatrix} t \\ s \\ r \end{bmatrix}, \quad \mathbf{b} = \begin{bmatrix} -.5 \\ -.5 \\ .9 \\ 1.1 \\ 3.3 \end{bmatrix}.$$

This time

$$A^{T}A = \begin{bmatrix} 1 & 1 & 1 & 1 & 1 \\ -3 & -1 & 1 & 2 & 3 \\ 9 & 1 & 1 & 4 & 9 \end{bmatrix}\begin{bmatrix} 1 & -3 & 9 \\ 1 & -1 & 1 \\ 1 & 1 & 1 \\ 1 & 2 & 4 \\ 1 & 3 & 9 \end{bmatrix} = \begin{bmatrix} 5 & 2 & 24 \\ 2 & 24 & 8 \\ 24 & 8 & 180 \end{bmatrix}$$

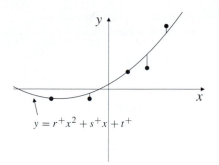

$$y = r^+x^2 + s^+|x| + t^+$$

Figure 8.3

and

$$(A^T A)^{-1} = \frac{1}{1876} \begin{bmatrix} 1064 & -42 & -140 \\ -42 & 81 & 2 \\ -140 & 2 & 29 \end{bmatrix}.$$

The best solution in the sense of least squares is

$$\mathbf{x}^+ = \begin{bmatrix} t^+ \\ s^+ \\ r^+ \end{bmatrix} = (A^T A)^{-1} A^T \mathbf{b}$$

$$= \frac{1}{1876} \begin{bmatrix} 1064 & -42 & -140 \\ -42 & 81 & 2 \\ -140 & 2 & 29 \end{bmatrix} \begin{bmatrix} 1 & 1 & 1 & 1 & 1 \\ -3 & -1 & 1 & 2 & 3 \\ 9 & 1 & 1 & 4 & 9 \end{bmatrix} \begin{bmatrix} -.5 \\ -.5 \\ .9 \\ 1.1 \\ 3.3 \end{bmatrix}$$

$$= \begin{bmatrix} .246 \\ .569 \\ .109 \end{bmatrix}.$$

So $t^+ = .246$, $s^+ = .569$, $r^+ = .109$ and our "best" parabola has the equation $y = .109x^2 + .569x + .246$. This is the one shown in Figure 8.3. This is the parabola for which the sums of the squares of the deviations of the points from the parabola is a minimum.

Exercises

Solutions to exercises marked [BB] can be found in the Back of the Book.

1. Find the equation of the line that best fits each of the following sets of points in the sense of least squares.

(a) [BB] $(0, 0), (1, 1), (3, 12)$

(b) $(3, 2), (-2, -3), (2, 4)$

(c) [BB] $(-1, 1), (0, 0), (1, 1), (2, 0), (3, -8)$

(d) $(-2, -2), (-1, -2), (0, -1), (2, 0)$

(e) $(2, 3), (1, 4), (1, -2), (-3, 1), (-1, 1)$

(f) $(4, -3), (2, 1), (1, -4), (-2, -2), (6, 3), (-1, -2)$.

2. Find the equation of the parabola that best fits each of the following sets of points in the sense of least squares.

   (a) [BB] $(0, 0), (1, 1), (-1, 2), (2, 2)$

   (b) $(-1, -1), (0, 4), (1, 2), (2, 3)$

   (c) $(-2, 2), (-1, 1), (0, 0), (1, 1), (2, 2)$

   (d) $(1, -2), (2, 1), (-1, 1), (2, -2)$.

3. Find the cubic polynomial $f(x) = px^3 + qx^2 + rx + s$ that best fits the given points in the sense of least squares.

   (a) [BB] $(-2, 0), (-1, -\frac{1}{2}), (0, 1), (1, \frac{1}{2}), (2, -1)$

   (b) $(-2, 0), (-1, -2), (0, 0), (1, 2), (2, -5)$

   (c) $(-1, -3), (0, -\frac{1}{2}), (1, 0), (2, 1), (3, 2)$.

4. The "best" quadratic polynomial through three given points, no two of which lie on a vertical line in the (x, y) plane, is, in fact, a perfect fit. Why?

5. Find a function of the form $f(x) = ax + b2^x$ that best fits the given points.

   (a) [BB] $(-1, 1), (0, 0), (2, -1)$

   (b) $(-2, -3), (-1, 0), (1, 4)$

   (c) $(0, 0), (1, -1), (2, 2), (3, -3)$.

6. According to the coach of a diving team, a person's score z on her third dive is a function $z = rx + sy + t$ of her scores x, y on her first two dives. Assuming this is the case,

   (a) find a formula for z based on the scores shown to the right;

   (b) estimate Brenda's score on her third dive, given that she scored 5 and 7, respectively, on her first two dives.

	x	y	z
Gail	6	7	7
Michelle	8	6	8
Wendy	6	8	8
Melanie	7	4	9
Linda	4	6	7

7. Find a curve with equation of the form $y = a \sin(x) + b \sin(2x)$ that best fits the given sets of points.

   (a) [BB] $(-\frac{\pi}{3}, -4), (\frac{\pi}{6}, 2), (\frac{5\pi}{6}, -1)$

   (b) $(-\frac{7\pi}{6}, 0), (-\frac{\pi}{6}, 1), (\frac{\pi}{2}, 5)$

   (c) $(-\frac{5\pi}{3}, 1), (-\frac{2\pi}{3}, -7), (\frac{\pi}{6}, 0), (\frac{7\pi}{6}, -7)$.

8. Suppose we want to find the line that best fits n pairs of points

$$(x_1, y_1), (x_2, y_2), \dots, (x_n, y_n)$$

in the sense of least squares.

   (a) Suppose the best line has equation $y = m^+ x + k^+$. Show that the constants m^+ and k^+ are the solutions to a system of the form $C \begin{bmatrix} k^+ \\ m^+ \end{bmatrix} = \mathbf{b}$ where $C = \begin{bmatrix} n & \sum x_i \\ \sum x_i & \sum x_i^2 \end{bmatrix}$ and $\mathbf{b} = \begin{bmatrix} \sum y_i \\ \sum x_i y_i \end{bmatrix}$.

   (b) Use part (a) to find the line that best fits the points $(1, 2), (2, 3), (3, 3), (4, 4), (5, 4)$.

   (c) Show that the system in (a) has a solution for any $\mathbf{b}$ if and only if the points $(x_1, y_1), (x_2, y_2), \dots, (x_n, y_n)$ do not lie on a vertical line.

8.2 Linear Recurrence Relations

A sequence of real numbers $a_0, a_1, a_2, \dots$ is defined by

$$a_0 = 1, \quad a_1 = 2 \quad \text{and, for } n \geq 1, \quad a_{n+1} = -a_n + 2a_{n-1}. \tag{1}$$

The third equation here,

$$a_{n+1} = -a_n + 2a_{n-1}, \tag{2}$$

tells us how to find the terms after a_0 and a_1. To get a_2, we set $n = 1$ in (2) and discover $a_2 = -a_1 + 2a_0 = 0$. We get a_3 by setting $n = 2$ in (2), and so on. Thus

$$
\begin{aligned}
a_3 &= -a_2 + 2a_1 = -0 + 2(2) = 4 \\
a_4 &= -a_3 + 2a_2 = -4 + 2(0) = -4 \\
a_5 &= -a_4 + 2a_3 = 4 + 2(4) = 12.
\end{aligned}
$$

A sequence defined as in (1) is said to be defined *recursively* because each term, after the first two, depends on previous terms. The definition suggests that we can't compute a_{20}, for example, without first computing $a_0, a_1, a_2, \ldots, a_{19}$. This is wrong, however. As we shall soon see, linear algebra enables us to find an explicit formula for a_n.

For $n \geq 1$, we have

$$
\begin{bmatrix} a_{n+1} \\ a_n \end{bmatrix} = \begin{bmatrix} -1 & 2 \\ 1 & 0 \end{bmatrix} \begin{bmatrix} a_n \\ a_{n-1} \end{bmatrix},
$$

which is $\mathbf{v}_n = A\mathbf{v}_{n-1}$, with $\mathbf{v}_n = \begin{bmatrix} a_{n+1} \\ a_n \end{bmatrix}$ and $A = \begin{bmatrix} -1 & 2 \\ 1 & 0 \end{bmatrix}$. Replacing n by $n - 1$ in $\mathbf{v}_n = A\mathbf{v}_{n-1}$ gives $\mathbf{v}_{n-1} = A\mathbf{v}_{n-2}$. Replacing n by $n - 2$ gives $\mathbf{v}_{n-2} = A\mathbf{v}_{n-3}$. We obtain

$$
\begin{aligned}
\mathbf{v}_n &= A\mathbf{v}_{n-1} = A(A\mathbf{v}_{n-2}) \\
&= A^2\mathbf{v}_{n-2} = A^2(A\mathbf{v}_{n-3}) \\
&= A^3\mathbf{v}_{n-3} = \cdots = A^n\mathbf{v}_0 = A^n \begin{bmatrix} a_1 \\ a_0 \end{bmatrix} = A^n \begin{bmatrix} 2 \\ 1 \end{bmatrix}.
\end{aligned}
$$

Our job then is to find the matrix A^n, a rather unpleasant thought: $A^2 = AA$, $A^3 = AAA$, $A^4 = AAAA, \ldots$ The task is much simpler, however, if we can diagonalize!

The characteristic polynomial of A is

$$
\det(A - \lambda I) = \begin{vmatrix} -1 - \lambda & 2 \\ 1 & -\lambda \end{vmatrix} = \lambda^2 + \lambda - 2 = (\lambda + 2)(\lambda - 1).
$$

Since the 2×2 matrix A has two distinct eigenvalues, it is diagonalizable: for some matrix P, we know that $P^{-1}AP$ is the diagonal matrix $D = \begin{bmatrix} -2 & 0 \\ 0 & 1 \end{bmatrix}$, whose diagonal entries are the eigenvalues of A. The matrix P has first column an eigenvector for -2 and second column an eigenvector for 1—see Section 3.4, especially 3.4.5.

To find the eigenspace for $\lambda = -2$, we solve the homogeneous system $(A - \lambda I)\mathbf{x} = \mathbf{0}$ for $\mathbf{x} = \begin{bmatrix} x_1 \\ x_2 \end{bmatrix}$ with $\lambda = -2$. Gaussian elimination gives

$$
\begin{bmatrix} 1 & 2 \\ 1 & 2 \end{bmatrix} \rightarrow \begin{bmatrix} 1 & 2 \\ 0 & 0 \end{bmatrix},
$$

so $x_2 = t$ is free and $x_1 = -2x_2 = -2t$, giving $\mathbf{x} = \begin{bmatrix} -2t \\ t \end{bmatrix} = t \begin{bmatrix} -2 \\ 1 \end{bmatrix}$.

To find the eigenspace for $\lambda = 1$, we solve the homogeneous system $(A - \lambda I)\mathbf{x} = \mathbf{0}$ for $\mathbf{x} = \begin{bmatrix} x_1 \\ x_2 \end{bmatrix}$ with $\lambda = 1$. Gaussian elimination gives

$$\begin{bmatrix} -2 & 2 \\ 1 & -1 \end{bmatrix} \rightarrow \begin{bmatrix} 1 & -1 \\ 0 & 0 \end{bmatrix},$$

so $x_2 = t$ is free and $x_1 = x_2 = t$, giving $\mathbf{x} = \begin{bmatrix} t \\ t \end{bmatrix} = t \begin{bmatrix} 1 \\ 1 \end{bmatrix}$.

Our matrix $P = \begin{bmatrix} -2 & 1 \\ 1 & 1 \end{bmatrix}$ is the one whose columns are the eigenvectors $\begin{bmatrix} -2 \\ 1 \end{bmatrix}$ and $\begin{bmatrix} 1 \\ 1 \end{bmatrix}$. Thus $P^{-1} = \frac{1}{3} \begin{bmatrix} -1 & 1 \\ 1 & 2 \end{bmatrix}$ and $P^{-1}AP = D$ implies $A = PDP^{-1}$.

Now notice that $A^2 = AA = PDP^{-1}PDP^{-1} = PD^2P^{-1}$ (because $PP^{-1} = I$), $A^3 = A^2A = PD^2P^{-1}PDP^{-1} = PD^3P^{-1}$, $A^4 = PD^4P^{-1}$ and, in general,

$$A^n = PD^nP^{-1} = \frac{1}{3} \begin{bmatrix} -2 & 1 \\ 1 & 1 \end{bmatrix} \begin{bmatrix} (-2)^n & 0 \\ 0 & 1 \end{bmatrix} \begin{bmatrix} -1 & 1 \\ 1 & 2 \end{bmatrix}$$

$$= \frac{1}{3} \begin{bmatrix} -2 & 1 \\ 1 & 1 \end{bmatrix} \begin{bmatrix} -(-2)^n & (-2)^n \\ 1 & 2 \end{bmatrix}$$

$$= \frac{1}{3} \begin{bmatrix} -(-2)^{n+1} + 1 & (-2)^{n+1} + 2 \\ -(-2)^n + 1 & (-2)^n + 2 \end{bmatrix}.$$

Thus

$$\mathbf{v}_n = \begin{bmatrix} a_{n+1} \\ a_n \end{bmatrix} = A^n \begin{bmatrix} 2 \\ 1 \end{bmatrix} = \frac{1}{3} \begin{bmatrix} -(-2)^{n+1} + 1 & (-2)^{n+1} + 2 \\ -(-2)^n + 1 & (-2)^n + 2 \end{bmatrix} \begin{bmatrix} 2 \\ 1 \end{bmatrix}$$

$$= \frac{1}{3} \begin{bmatrix} (-2)^{n+2} + 2 + (-2)^{n+1} + 2 \\ (-2)^{n+1} + 2 + (-2)^n + 2 \end{bmatrix}.$$

The second component of this vector is

$$a_n = \tfrac{1}{3}[(-2)(-2)^n + 2 + (-2)^n + 2] = \tfrac{1}{3}[4 - (-2)^n] = \tfrac{4}{3} - \tfrac{1}{3}(-2)^n,$$

the sought-after formula. Check that

$$a_2 = \tfrac{4}{3} - \tfrac{4}{3} = 0$$

$$a_3 = \tfrac{4}{3} + \tfrac{8}{3} = 4$$

$$a_4 = \tfrac{4}{3} - \tfrac{16}{3} = -4$$

$$a_5 = \tfrac{4}{3} + \tfrac{32}{3} = 12,$$

in agreement with what we found earlier.

■ The Fibonacci Sequence

Leonardo Fibonacci ($\approx$1180–1228), one of the brightest mathematicians of the Middle Ages, is credited with posing a problem about rabbits that goes more or less like this.

> Suppose that newborn rabbits start producing offspring by the end of their second month of life, after which they produce a pair a month, one male and one female. Starting with one pair of rabbits on January 1, and assuming that rabbits never die, how many pairs will be alive at the end of the year (and available for New Year's dinner)?

At the beginning, on January 1, only the initial pair is alive, and the same is true at the start of February. By March 1, however, the initial pair of rabbits has produced its first pair of offspring, so now two pairs of rabbits are alive (and well?). On April 1, the initial pair has produced one more pair and the second pair hasn't started reproducing yet, so three pairs are alive. On May 1, there are five pairs of rabbits. Can you see why?

At the beginning of the month, the number of pairs alive is the number alive at the start of the previous month plus the number alive two months ago (since this last group have produced offspring, one pair per pair). The numbers of pairs of rabbits alive at the start of each month are the terms of the famous *Fibonacci sequence* $1, 1, 2, 3, 5, 8, \ldots$, each number after the first two being the sum of the previous two.

The Fibonacci sequence is best defined recursively like this:

$$f_0 = 1, \quad f_1 = 1 \quad \text{and, for } n \geq 1, \quad f_{n+1} = f_n + f_{n-1}.$$

This can be written

$$\begin{bmatrix} f_{n+1} \\ f_n \end{bmatrix} = \begin{bmatrix} 1 & 1 \\ 1 & 0 \end{bmatrix} \begin{bmatrix} f_n \\ f_{n-1} \end{bmatrix},$$

which is $\mathbf{v}_n = A\mathbf{v}_{n-1}$, with $\mathbf{v}_n = \begin{bmatrix} f_{n+1} \\ f_n \end{bmatrix}$ and $A = \begin{bmatrix} 1 & 1 \\ 1 & 0 \end{bmatrix}$. As before, this implies $\mathbf{v}_n = A^n \mathbf{v}_0$ with $\mathbf{v}_0 = \begin{bmatrix} 1 \\ 1 \end{bmatrix}$ and our job is to find A^n. The characteristic polynomial of A is

$$\det(A - \lambda I) = \begin{vmatrix} 1 - \lambda & 1 \\ 1 & -\lambda \end{vmatrix} = \lambda^2 - \lambda - 1,$$

whose roots are $\lambda_1 = \frac{1+\sqrt{5}}{2}$ and $\lambda_2 = \frac{1-\sqrt{5}}{2}$. Since the eigenvalues are distinct, A is diagonalizable: we have $P^{-1}AP = D$, $D = \begin{bmatrix} \lambda_1 & 0 \\ 0 & \lambda_2 \end{bmatrix}$ and P the matrix whose columns are eigenvectors corresponding to λ_1 and λ_2, respectively.

It will pay dividends to work now in some generality. Let λ be either λ_1 or λ_2 and note that $\lambda^2 - \lambda - 1 = 0$ implies $\lambda(\lambda - 1) = 1$. Hence,

$$\lambda - 1 = \frac{1}{\lambda}, \tag{3}$$

something that will prove quite useful!

To find the eigenspace for λ, we solve the homogeneous system $(A - \lambda I)\mathbf{x} = \mathbf{0}$. Gaussian elimination proceeds

$$\begin{bmatrix} 1 - \lambda & 1 \\ 1 & -\lambda \end{bmatrix} \rightarrow \begin{bmatrix} 1 & \frac{1}{1-\lambda} \\ 1 & -\lambda \end{bmatrix} = \begin{bmatrix} 1 & -\lambda \\ 1 & -\lambda \end{bmatrix} \rightarrow \begin{bmatrix} 1 & -\lambda \\ 0 & 0 \end{bmatrix}.$$

With $\mathbf{x} = \begin{bmatrix} x_1 \\ x_2 \end{bmatrix}$, $x_2 = t$ is free and $x_1 = \lambda x_2 = \lambda t$, giving $\mathbf{x} = \begin{bmatrix} \lambda t \\ t \end{bmatrix} = t \begin{bmatrix} \lambda \\ 1 \end{bmatrix}$.

The matrix P is

$$P = \begin{bmatrix} \lambda_1 & \lambda_2 \\ 1 & 1 \end{bmatrix}$$

(its columns are eigenvectors for λ_1 and λ_2, respectively). The determinant of this matrix is $\det P = \lambda_1 - \lambda_2 = \sqrt{5}$ and its inverse is

$$P^{-1} = \frac{1}{\sqrt{5}} \begin{bmatrix} 1 & -\lambda_2 \\ -1 & \lambda_1 \end{bmatrix}.$$

Now

$$\begin{bmatrix} f_{n+1} \\ f_n \end{bmatrix} = \mathbf{v}_n = A^n \mathbf{v}_0 = PD^n P^{-1} \mathbf{v}_0 = PD^n P^{-1} \begin{bmatrix} 1 \\ 1 \end{bmatrix} = \frac{1}{\sqrt{5}} PD^n \begin{bmatrix} 1 - \lambda_2 \\ -1 + \lambda_1 \end{bmatrix}.$$

Since $PD^n = \begin{bmatrix} \lambda_1 & \lambda_2 \\ 1 & 1 \end{bmatrix} \begin{bmatrix} \lambda_1^n & 0 \\ 0 & \lambda_2^n \end{bmatrix} = \begin{bmatrix} \lambda_1^{n+1} & \lambda_2^{n+1} \\ \lambda_1^n & \lambda_2^n \end{bmatrix}$, we have

$$\begin{bmatrix} f_{n+1} \\ f_n \end{bmatrix} = \frac{1}{\sqrt{5}} \begin{bmatrix} \lambda_1^{n+1} & \lambda_2^{n+1} \\ \lambda_1^n & \lambda_2^n \end{bmatrix} \begin{bmatrix} 1 - \lambda_2 \\ -1 + \lambda_1 \end{bmatrix}$$

$$= \frac{1}{\sqrt{5}} \left[(1 - \lambda_2) \begin{bmatrix} \lambda_1^{n+1} \\ \lambda_1^n \end{bmatrix} + (-1 + \lambda_1) \begin{bmatrix} \lambda_2^{n+1} \\ \lambda_2^n \end{bmatrix} \right].$$

(Never forget 2.1.33!)

Since $\lambda - 1 = \dfrac{1}{\lambda}$ for each λ, by (3), the second component of this vector is

$$f_n = \frac{1}{\sqrt{5}} \left[(1 - \lambda_2)\lambda_1^n + (-1 + \lambda_1)\lambda_2^n \right] = \frac{1}{\sqrt{5}} \left(-\frac{\lambda_1^n}{\lambda_2} + \frac{\lambda_2^n}{\lambda_1} \right)$$

$$= \frac{1}{\sqrt{5}} \left(\frac{-\lambda_1^{n+1} + \lambda_2^{n+1}}{\lambda_1 \lambda_2} \right).$$

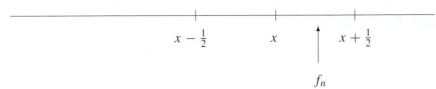

Figure 8.4

Now $\lambda_1\lambda_2 = \dfrac{1+\sqrt{5}}{2}\dfrac{1-\sqrt{5}}{2} = \dfrac{1-5}{4} = -1$, so

$$\frac{-\lambda_1^{n+1}+\lambda_2^{n+1}}{\lambda_1\lambda_2} = \lambda_1^{n+1}-\lambda_2^{n+1}$$

and

$$f_n = \frac{1}{\sqrt{5}}\lambda_1^{n+1} - \frac{1}{\sqrt{5}}\lambda_2^{n+1} = \frac{1}{\sqrt{5}}\left(\frac{1+\sqrt{5}}{2}\right)^{n+1} - \frac{1}{\sqrt{5}}\left(\frac{1-\sqrt{5}}{2}\right)^{n+1}. \qquad (4)$$

It is almost incredible that this formula should give integer values (see Reading Challenge 1), but keeping this in mind, we can get a better formula.

Since $\left|\frac{1-\sqrt{5}}{2}\right| < 1$,

$$\left|\frac{1}{\sqrt{5}}\left(\frac{1-\sqrt{5}}{2}\right)^{n+1}\right| < \frac{1}{\sqrt{5}} < \frac{1}{2}.$$

From (4),

$$\left|f_n - \frac{1}{\sqrt{5}}\left(\frac{1+\sqrt{5}}{2}\right)^{n+1}\right| = \left|\frac{1}{\sqrt{5}}\left(\frac{1-\sqrt{5}}{2}\right)^{n+1}\right| < \frac{1}{2},$$

which says that the integer f_n is at a distance less than $\frac{1}{2}$ from $x = \frac{1}{\sqrt{5}}\left(\frac{1+\sqrt{5}}{2}\right)^{n+1}$. Since there is at most one such integer, we discover that f_n is the integer closest to $\frac{1}{\sqrt{5}}\left(\frac{1+\sqrt{5}}{2}\right)^{n+1}$. For example, since $\frac{1}{\sqrt{5}}\left(\frac{1+\sqrt{5}}{2}\right)^{13} \approx 232.999$, we know that $f_{12} = 233$. (New Year's dinner looks promising.)

Reading Challenge 1   *Why does the expression on the right of formula in (4) give integer values?*

Answers to Reading Challenges

1. The expression gives the values of the Fibonacci sequence, which are all integers.

Exercises

*Solutions to exercises marked [BB] can be found in the **B**ack of the **B**ook.*

1. [BB] Consider the sequence of numbers defined recursively by

$$a_0 = 0, a_1 = 1 \text{ and, for } n \geq 1, a_{n+1} = a_n + 2a_{n-1}.$$

(a) List the first eight terms of this sequence.

(b) Find a general formula for a_n.

2. Repeat Exercise 1 for the sequence defined recursively by $a_{n+1} = -4a_n + 5a_{n-1}$, $n \geq 1$. Assume $a_0 = 1$ and $a_1 = -1$.

3. Repeat Exercise 1 for the sequence defined recursively by $a_{n+1} = 2a_n + 3a_{n-1}$, $n \geq 1$. Assume $a_0 = -3$ and $a_1 = 4$.

4. For each of the following recursively defined sequences, find a formula for a_n.

(a) [BB] $a_0 = -4$, $a_1 = 0$ and, for $n \geq 1$,
$a_{n+1} = 3a_n - 2a_{n-1}$

(b) $a_0 = -1$, $a_1 = 5$ and, for $n \geq 1$,
$a_{n+1} = 4a_n - 3a_{n-1}$

(c) $a_1 = 1$, $a_2 = 2$ and, for $n \geq 1$,
$a_{n+1} = 3a_n + 4a_{n-1}$

(d) $a_1 = -8$, $a_2 = 1$ and, for $n \geq 1$,
$a_{n+1} = a_n + 12a_{n-1}$.

5. Let A be a (real) $n \times n$ matrix.

(a) [BB] The matrix $A^T A$ is diagonalizable over **R**. Why?

(b) Show that the eigenvalues $\lambda_1, \lambda_2, \ldots, \lambda_n$ of $A^T A$ are all nonnegative.

(c) Let $\mathbf{v}_1, \mathbf{v}_2, \ldots$ be eigenvectors of $A^T A$ satisfying $\mathbf{v}_{k+1} = A\mathbf{v}_k$. If all the eigenvalues of $A^T A$ are less than 1, show that $\|\mathbf{v}_1\|^2 \geq \|\mathbf{v}_2\|^2 \geq \|\mathbf{v}_3\|^2 \geq \cdots$.

6. Let $f_0, f_1, f_2, \ldots$ be the terms of the Fibonacci sequence.
Find the values of $\frac{1}{\sqrt{5}}\left(\frac{1+\sqrt{5}}{2}\right)^{n+1}$ for $n = 0, 1, \ldots, 9$ (to three decimal place accuracy) and compare with the values of $f_0, f_1, \ldots, f_9$. What do you observe?

8.3 Markov Processes

Markov matrices are named after the Russian probabilist Andrei Markov (1856–1922).

Let $\mathbf{v}_1, \mathbf{v}_2, \mathbf{v}_3, \ldots$ be a sequence of vectors. In this section, we continue our study of the equation $\mathbf{v}_{n+1} = A\mathbf{v}_n$, but with a very special class of matrices.

8.3.1 DEFINITION

A *Markov matrix* is a matrix $A = [a_{ij}]$, each a_{ij} satisfying $0 \leq a_{ij} \leq 1$, and the entries in each column summing to 1.

8.3.2 EXAMPLES

The matrix $\begin{bmatrix} \frac{1}{2} & \frac{1}{4} \\ \frac{1}{2} & \frac{3}{4} \end{bmatrix}$ is Markov, as is $\begin{bmatrix} 0 & \frac{2}{3} \\ 1 & \frac{1}{3} \end{bmatrix}$. The matrix $\begin{bmatrix} \frac{1}{2} & \frac{3}{4} \\ \frac{1}{2} & \frac{1}{3} \end{bmatrix}$ is not Markov. ■

Markov matrices have some remarkable properties. In order to describe perhaps the most important of these, we first discuss matrices each of whose rows or each of whose columns sums to 0.

To be specific, suppose $A = \begin{bmatrix} \mathbf{a}_1 & \mathbf{a}_2 & \cdots & \mathbf{a}_n \\ \downarrow & \downarrow & & \downarrow \end{bmatrix}$ is an $n \times n$ matrix each of whose

rows sums to 0. Let $\mathbf{v} = \begin{bmatrix} 1 \\ 1 \\ \vdots \\ 1 \end{bmatrix}$. Since $A\mathbf{v}$ is a linear combination of the columns of A

with coefficients the components of **v** (by 2.1.33), we have

$$A\mathbf{v} = 1\mathbf{a}_1 + 1\mathbf{a}_2 + \cdots + 1\mathbf{a}_n = \mathbf{a}_1 + \mathbf{a}_2 + \cdots + \mathbf{a}_n.$$

Each component of this vector is the sum of the components of the vectors $\mathbf{a}_1, \ldots, \mathbf{a}_n$. These are the sums of the rows of A, and we assumed each of these was 0. Thus $A\mathbf{v} = \mathbf{0}$; in particular, A is not invertible and $\det A = 0$.

Now suppose that A is a matrix each of whose **columns** sums to 0. Then each of the rows of A^T sums to 0, so $0 = \det A^T = \det A$. Whether the rows or the columns of a matrix sum to 0, its determinant must be 0.

Let A be a Markov matrix. Since each column of $A - I$ sums to 0, $\det(A - I) = 0$. This equation says that 1 is an eigenvalue of A.

Suppose $\mathbf{x} = \begin{bmatrix} x_1 \\ x_2 \\ \vdots \\ x_n \end{bmatrix}$ and $A\mathbf{x} = \mathbf{y} = \begin{bmatrix} y_1 \\ y_2 \\ \vdots \\ y_n \end{bmatrix}$. If $A = \begin{bmatrix} \mathbf{a}_1 & \mathbf{a}_2 & \cdots & \mathbf{a}_n \\ \downarrow & \downarrow & & \downarrow \end{bmatrix}$, then

$$\begin{bmatrix} y_1 \\ \vdots \\ y_n \end{bmatrix} = x_1\mathbf{a}_1 + x_2\mathbf{a}_2 + \cdots + x_n\mathbf{a}_n.$$

Summing the components of the vector on each side of this equation gives

$$y_1 + y_2 + \cdots + y_n = x_1 + x_2 + \cdots + x_n$$

since the sum of the components of each $\mathbf{a}_i$ is 1. So

$$A\mathbf{x} = \mathbf{y} \text{ implies } \sum x_i = \sum y_i; \tag{1}$$

that is, the sum of the components of $\mathbf{x}$ is the sum of the components of $\mathbf{y}$.

Suppose we have a sequence of vectors $\mathbf{v}_0, \mathbf{v}_1, \mathbf{v}_2, \ldots$ with $\mathbf{v}_1 = A\mathbf{v}_0$, $\mathbf{v}_2 = A\mathbf{v}_1$, $\mathbf{v}_3 = A\mathbf{v}_2$, and so on, and suppose that the sum of the components of $\mathbf{v}_0$ is c. The foregoing shows that the sum of the components of each $\mathbf{v}_k$ is also c. If A is also diagonalizable, if $\mathbf{x}_1, \mathbf{x}_2, \ldots, \mathbf{x}_n$ are linearly independent eigenvectors corresponding to eigenvalues $\lambda_1, \lambda_2, \ldots, \lambda_n$, and if

$$P = \begin{bmatrix} \mathbf{x}_1 & \mathbf{x}_2 & \cdots & \mathbf{x}_n \\ \downarrow & \downarrow & & \downarrow \end{bmatrix},$$

then we know that

$$P^{-1}AP = D = \begin{bmatrix} \lambda_1 & 0 & \cdots & 0 \\ 0 & \lambda_2 & & 0 \\ \vdots & & \ddots & \\ 0 & & & \lambda_n \end{bmatrix}.$$

So $A = PDP^{-1}$ and $A^k = PD^kP^{-1}$ for any positive integer k.

Reading Challenge 1   *Explain why* $A^k = PD^kP^{-1}$.

Reading Challenge 2   *Explain why* $\mathbf{v}_k = A^k\mathbf{v}_0$. *[This fact will be critical in the next line of our reasoning.]*

Thus

$$\mathbf{v}_k = A^k\mathbf{v}_0 = PD^kP^{-1}\mathbf{v}_0$$

$$= \begin{bmatrix} \mathbf{x}_1 & \mathbf{x}_2 & \cdots & \mathbf{x}_n \\ \downarrow & \downarrow & & \downarrow \\ & & & \end{bmatrix} \begin{bmatrix} \lambda_1^k & & & \\ & \lambda_2^k & & \\ & & \ddots & \\ & & & \lambda_n^k \end{bmatrix} \underbrace{P^{-1}\mathbf{v}_0}_{\mathbf{t}}$$

$$= \begin{bmatrix} \lambda_1^k\mathbf{x}_1 & \lambda_2^k\mathbf{x}_2 & \cdots & \lambda_n^k\mathbf{x}_n \\ \downarrow & \downarrow & & \downarrow \\ & & & \end{bmatrix} \begin{bmatrix} t_1 \\ t_2 \\ \vdots \\ t_n \end{bmatrix}$$

$$= t_1\lambda_1^k\mathbf{x}_1 + t_2\lambda_2^k\mathbf{x}_2 + \cdots + t_n\lambda_n^k\mathbf{x}_n,$$

where, as indicated, $\mathbf{t} = \begin{bmatrix} t_1 \\ \vdots \\ t_n \end{bmatrix}$ is the vector $P^{-1}\mathbf{v}_0$. [We have used 2.1.33 several times in these calculations!] Look at what we have shown:

$$\mathbf{v}_k = t_1\lambda_1^k\mathbf{x}_1 + t_2\lambda_2^k\mathbf{x}_2 + \cdots + t_n\lambda_n^k\mathbf{x}_n. \tag{2}$$

This describes what happens to the vectors $\mathbf{v}_k$ in the long run, as k increases.

Suppose, for example, that $\lambda_1 = 1$ and all other λ_i satisfy $|\lambda_i| < 1$. Then $\lambda_i^k \to 0$ as $k \to \infty$ for all $i > 1$ and

$$\mathbf{v}_k \to t_1\mathbf{x}_1 = \mathbf{v}_\infty \tag{3}$$

for some vector that we naturally label $\mathbf{v}_\infty$. As a multiple of $\mathbf{x}_1$, the "limit vector" $t_1\mathbf{x}_1 = \mathbf{v}_\infty$ is also an eigenvector for $\lambda_1 = 1$. This could have been predicted.

Letting $k \to \infty$ in the equation $\mathbf{v}_{k+1} = A\mathbf{v}_k$ gives $\mathbf{v}_\infty = A\mathbf{v}_\infty$. For this reason, the vector $\mathbf{v}_\infty$ is known as the *steady state*. When k is large, so that $\mathbf{v}_k$ is close to $\mathbf{v}_\infty$, $A\mathbf{v}_k \approx \mathbf{v}_k$.

Even more is true. Recall that the sum of the components of each $\mathbf{v}_k$ is c. Equation (3) says that the same is true of $t_1\mathbf{x}_1$. If the components of $\mathbf{x}_1$ have sum $r > 0$, the sum of the components of $t\mathbf{x}_1$ is t_1r, so $t_1r = c$ and $t_1 = c/r$ depends only on the eigenvector $\mathbf{x}_1$ corresponding to 1 and the fact that $\mathbf{v}_0$ has sum of components equal to c. In particular, we get the same limit vector $\mathbf{v}_\infty = t_1\mathbf{x}_1$ starting with **any** $\mathbf{v}_0$ whose components sum to c.

As an application, letting $\mathbf{e}_1, \ldots, \mathbf{e}_n$ denote the standard basis vectors of $\mathbf{R}^n$ (so $c = 1$), all the sequences $A^k\mathbf{e}_1, A^k\mathbf{e}_2, \ldots, A^k\mathbf{e}_n$ converge to the same limit vector $\mathbf{v}_\infty$. If the matrix A^k converges to a matrix B, then $A^k\mathbf{e}_1$ converges to $B\mathbf{e}_1$, which is the first column of B. The sequence $A^k\mathbf{e}_2$ converges to $B\mathbf{e}_2$, the second column of B. The sequence $A^k\mathbf{e}_n$ converges to $B\mathbf{e}_n$, the last column of B. All columns of B are the same,

$$B = \begin{bmatrix} \mathbf{v}_\infty & \mathbf{v}_\infty & \cdots & \mathbf{v}_\infty \\ \downarrow & \downarrow & & \downarrow \\ & & & \end{bmatrix},$$

every column $\mathbf{v}_\infty$ being an eigenvector of A corresponding to $\lambda = 1$ whose components sum to 1. This identifies the limit vector $\mathbf{v}_\infty$ uniquely if the eigenspace of A corresponding to $\lambda = 1$ is one-dimensional and the components of the basis vector do not sum to 0: There is only one vector in this sort of eigenspace whose components sum to 1.

Reading Challenge 3 *If an eigenspace of A is one-dimensional with basis $\mathbf{x}$, and if the components of $\mathbf{x}$ do not sum to 0, find the unique vector in this eigenspace whose components sum to 1.*

8.3.3 EXAMPLE Let A be the Markov matrix $A = \begin{bmatrix} \frac{1}{2} & \frac{1}{3} & \frac{1}{6} \\ \frac{1}{4} & \frac{1}{3} & \frac{1}{3} \\ \frac{1}{4} & \frac{1}{3} & \frac{1}{2} \end{bmatrix}$. Some powers of A are, approximately,

$$A^2 = \begin{bmatrix} .375 & .333 & .278 \\ .292 & .306 & .319 \\ .333 & .361 & .403 \end{bmatrix}, \quad A^3 = \begin{bmatrix} .340 & .329 & .313 \\ .302 & .306 & .310 \\ .358 & .366 & .377 \end{bmatrix},$$

$$A^4 = \begin{bmatrix} .330 & .327 & .323 \\ .304 & .305 & .307 \\ .365 & .367 & .370 \end{bmatrix}, \quad A^{10} = \begin{bmatrix} .327 & .327 & .327 \\ .306 & .306 & .306 \\ .367 & .367 & .367 \end{bmatrix}.$$

The powers of A appear to converge to the limit matrix

$$B = \begin{bmatrix} .327 & .327 & .327 \\ .306 & .306 & .306 \\ .367 & .367 & .367 \end{bmatrix},$$

a matrix with identical columns, each of which we have shown is an eigenvector of A corresponding to $\lambda = 1$ and whose components sum to 1. Let us verify.

The characteristic polynomial of A is

$$\det(A - \lambda I) = -\lambda^3 + \tfrac{4}{3}\lambda^2 - \tfrac{25}{72}\lambda + \tfrac{1}{72} = -\tfrac{1}{72}(\lambda - 1)(72\lambda^2 - 24\lambda + 1).$$

The roots of $72\lambda^2 - 24\lambda + 1$ are

$$\lambda = \frac{24 \pm 12\sqrt{2}}{144} = \frac{1}{6} \pm \frac{\sqrt{2}}{12}$$

so the eigenvalues of A are 1 and two numbers less than 1 in absolute value. The eigenvalues are distinct, so A is diagonalizable. As we have shown, the powers A^k of A converge to a matrix B with identical columns.

Now the eigenspace corresponding to $\lambda = 1$ is one-dimensional: Gaussian elimination applied to $A - I$ proceeds

$$\begin{bmatrix} -\frac{1}{2} & \frac{1}{3} & \frac{1}{6} \\ \frac{1}{4} & -\frac{2}{3} & \frac{1}{3} \\ \frac{1}{4} & \frac{1}{3} & -\frac{1}{2} \end{bmatrix} \rightarrow \begin{bmatrix} 1 & -\frac{2}{3} & -\frac{1}{3} \\ \frac{1}{4} & -\frac{2}{3} & \frac{1}{3} \\ 0 & 1 & -\frac{5}{6} \end{bmatrix} \rightarrow \begin{bmatrix} 1 & -\frac{2}{3} & -\frac{1}{3} \\ 0 & 1 & -\frac{5}{6} \\ 0 & 0 & 0 \end{bmatrix}.$$

So if $\mathbf{x} = \begin{bmatrix} x_1 \\ x_2 \\ x_3 \end{bmatrix}$ is an eigenvector, $x_3 = t$ is free, $x_2 = \frac{5}{6}x_3 = \frac{5}{6}t$, and $x_1 = \frac{2}{3}x_2 + \frac{1}{3}x_3 = \frac{8}{9}t$. The eigenspace corresponding to $\lambda = 1$ consists of vectors of the

form $\begin{bmatrix} \frac{8}{9}t \\ \frac{5}{6}t \\ t \end{bmatrix}$. There is only one vector in this eigenspace whose components sum to

1. To find it, we solve

$$\tfrac{8}{9}t + \tfrac{5}{6}t + t = 1,$$

obtaining $t = \frac{18}{49}$ and $\mathbf{x} = \frac{1}{49}\begin{bmatrix} 16 \\ 15 \\ 18 \end{bmatrix} \approx \begin{bmatrix} .326 \\ .306 \\ .367 \end{bmatrix}$, in close agreement with the

columns of B found earlier.   ∎

8.3.4 EXAMPLE   Suppose that each year one tenth of the population of Canada that resides outside Newfoundland moves into the province, while three tenths of those inside move out. Assume there is no other population movement. Let x_k denote the size of the Canadian population outside Newfoundland after k years and y_k the size of the Newfoundland population after k years. Then

$$x_{k+1} = .9x_k + .3y_k$$
$$y_{k+1} = .1x_k + .7y_k.$$

Letting $\mathbf{v}_k = \begin{bmatrix} x_k \\ y_k \end{bmatrix}$, we have

$$\mathbf{v}_{k+1} = \begin{bmatrix} x_{k+1} \\ y_{k+1} \end{bmatrix} = \begin{bmatrix} .9 & .3 \\ .1 & .7 \end{bmatrix}\begin{bmatrix} x_k \\ y_k \end{bmatrix} = A\mathbf{v}_k,$$

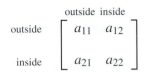

outside inside

	outside	inside
outside	a_{11}	a_{12}
inside	a_{21}	a_{22}

Figure 8.5

with $A = \begin{bmatrix} .9 & .3 \\ .1 & .7 \end{bmatrix}$. It is helpful to think of the entries of the Markov matrix A as probabilities, as suggested in Figure 8.5. The $(1, 1)$ entry, for instance, is the probability that someone living outside Newfoundland is still outside after one year; a_{12} is the probability that someone currently living in Newfoundland is outside after one year, and so on. More generally, the entries of A^k give the probabilities that people outside or inside the province are outside or inside after k years.

The eigenvalues of A are $\lambda_1 = 1$ and $\lambda_2 = 0.6$. The eigenspace corresponding to λ_1 consists of vectors of the form $\begin{bmatrix} 3t \\ t \end{bmatrix}$. The unique such vector whose components sum to 1 is $\frac{1}{4}\begin{bmatrix} 3 \\ 1 \end{bmatrix}$, so the matrices A^k converge to $B = \frac{1}{4}\begin{bmatrix} 3 & 3 \\ 1 & 1 \end{bmatrix}$ and the vectors $\mathbf{v}_k = A^k\mathbf{v}_0$ converge to

$$B\mathbf{v}_0 = \frac{1}{4}\begin{bmatrix} 3 & 3 \\ 1 & 1 \end{bmatrix}\begin{bmatrix} x_0 \\ y_0 \end{bmatrix} = \begin{bmatrix} \frac{3}{4}(x_0 + y_0) \\ \frac{1}{4}(x_0 + y_0) \end{bmatrix}.$$

Eventually $\frac{3}{4}(x_0 + y_0)$, that is, three quarters of the original total population (x_0 outside, y_0 inside) resides outside Newfoundland, while one quarter of the original population will reside inside. ■

Answers to Reading Challenges

1. From $A = PDP^{-1}$, so $A^2 = AA = PDP^{-1}PDP^{-1} = PD^2P^{-1}$ (because $PP^{-1} = I$), $A^3 = A^2A = PD^2P^{-1}PDP^{-1} = PD^3P^{-1}$, $A^4 = PD^4P^{-1}$, and so on.

2. We have $\mathbf{v}_1 = A\mathbf{v}_0$, $\mathbf{v}_2 = A\mathbf{v}_1 = A(A\mathbf{v}_0) = A^2\mathbf{v}_0$, $\mathbf{v}_3 = A\mathbf{v}_2 = A(A^2\mathbf{v}_0) = A^3\mathbf{v}_0$, and so on.

3. Any vector in the eigenspace is of the form $t\mathbf{x}$ for some scalar t. If the sum of the components of $\mathbf{x}$ is r, the sum of the components of $t\mathbf{x}$ is tr. There is just one t giving $tr = 1$; namely, $t = \frac{1}{r}$.

Exercises

Solutions to exercises marked [BB] can be found in the Back of the Book.

1. [BB] Let $A = \begin{bmatrix} \frac{1}{3} & \frac{1}{4} \\ \frac{2}{3} & \frac{3}{4} \end{bmatrix}$.

   (a) Find a matrix P such that $P^{-1}AP$ is diagonal.

   (b) Let $n \geq 1$ be an arbitrary integer. Find A^n.

   (c) Use the result of part (b) to determine the matrix B to which A converges as n gets large.

   (d) Could B have been determined without computing A^n? Explain.

2. Let $A = \begin{bmatrix} \frac{1}{2} & \frac{1}{4} & \frac{1}{4} \\ \frac{1}{4} & \frac{1}{2} & \frac{1}{4} \\ \frac{1}{4} & \frac{1}{4} & \frac{1}{2} \end{bmatrix}$. The powers of A converge to a unique matrix B. Why? Find B.

3. [BB] After the first day of lectures one semester, one tenth of Dr. G's linear algebra students transferred to Dr. L's section and one fifth moved to Dr. P's section. After his first class, poor Dr. L lost one half his students to Dr. G and one fourth to Dr. P. Meanwhile, Dr. P found that three tenths had moved to Dr. G and one tenth to Dr. L. If this movement occurs lecture after lecture, determine the eventual relative proportions of students in each section.

4. Let $A = \begin{bmatrix} -1 & 2 & 2 \\ 2 & 2 & 2 \\ -3 & -6 & -6 \end{bmatrix}$. The eigenvalues of A are $0, -2, -3$.

   (a) Find the eigenspaces of A.

   (b) Find A^4.

   (c) Solve $v_{k+1} = A v_k$ with $v_0 = \begin{bmatrix} 1 \\ 0 \\ 1 \end{bmatrix}$.

5. Find the limit vector $\mathbf{v}_\infty$ given $\mathbf{v}_{k+1} = A\mathbf{v}_k$, $\mathbf{v}_0 = \begin{bmatrix} 1 \\ 2 \\ 3 \end{bmatrix}$ and $A = \begin{bmatrix} \frac{2}{3} & \frac{2}{3} & \frac{1}{3} \\ 0 & 0 & \frac{1}{3} \\ \frac{1}{3} & \frac{1}{3} & \frac{1}{3} \end{bmatrix}$.

6. [BB] Each year the populations of British Columbia, Newfoundland, and Ontario migrate as follows:

   - one quarter of those in British Columbia and one quarter of those in Newfoundland move to Ontario;

   - one sixth of those in Ontario move to British Columbia and one third of those in Ontario move to Newfoundland.

   If the total initial population of these three provinces is 10 million, what is the eventual long-term distribution of the population. Does the answer depend on the initial distribution? Explain.

7. Each year, 20% of the mathematics majors at a certain university become physics majors, 10% of the chemistry majors turn to mathematics and 10% to physics, while 10% of physics majors become mathematics majors. Assuming each of these subjects has 100 students at the outset, how will these 300 eventually be distributed? (Life is so sweet at this university that students choose to stay indefinitely.)

8. Three companies A, B, and C simultaneously introduce new laptop computers. Every year, A loses 30% of its market to B, B loses 40% of its market to C, and C loses 20% of its market to A. Find the percentage share of the market each company will eventually have.

■ **Critical Reading**

9. If A is a Markov matrix and $\mathbf{x}$ is an eigenvector of A corresponding to $\lambda \neq 1$, then the components of $\mathbf{x}$ sum to 0. Why?

10. Let $A = \begin{bmatrix} .2 & .1 \\ .8 & .9 \end{bmatrix}$. Without using paper and pencil, explain why there exists a nonzero vector $\mathbf{x}$ satisfying $A\mathbf{x} = \mathbf{x}$.

8.4 Quadratic Forms and Conic Sections

8.4.1 DEFINITION A *quadratic form* in n variables $x_1, x_2, \ldots, x_n$ is a linear combination of the products $x_i x_j$ of pairs of variables, that is, a linear combination of squares $x_1^2, x_2^2, \ldots, x_n^2$ and *cross terms* $x_1 x_2, x_1 x_3, \ldots, x_1 x_n, x_2 x_3, \ldots, x_2 x_n, \ldots, x_{n-1} x_n$.

8.4.2 EXAMPLES
   - $q = x^2 - y^2 + 4xy$ and $q = x^2 + 3y^2 - 2xy$ are quadratic forms in x and y;
   - $q = -4x_1^2 + x_2^2 + 4x_3^2 + 6x_1 x_3$ is a quadratic form in x_1, x_2 and x_3;
   - the most general quadratic form in x_1, x_2, x_3 is

$$a_1 x_1^2 + a_2 x_2^2 + a_3 x_3^2 + a_{12} x_1 x_2 + a_{13} x_1 x_3 + a_{23} x_2 x_3. \qquad ■$$

   In the study of quadratic forms, our starting point is the observation that any quadratic form can be written in the form $\mathbf{x}^T A \mathbf{x}$, where A is a symmetric matrix and $\mathbf{x} = \begin{bmatrix} x_1 \\ \vdots \\ x_n \end{bmatrix}$. The matrix A is obtained by putting the coefficient of x_i^2 in the ith diagonal position and splitting equally the coefficient of $x_i x_j$ between the (i, j) and (j, i) positions; that is, putting one half the coefficient in each position.

8.4.3 EXAMPLE   Suppose $q = x_1^2 - x_2^2 + 4x_1x_2$. The coefficients of x_1^2 and x_2^2 are 1 and -1, respectively, so we put these into the first two diagonal positions of a matrix A. The coefficient of x_1x_2 is 4, which we split equally between the $(1, 2)$ and $(2, 1)$ positions, putting a 2 in each place.

We obtain $A = \begin{bmatrix} 1 & 2 \\ 2 & -1 \end{bmatrix}$ and, with $\mathbf{x} = \begin{bmatrix} x_1 \\ x_2 \end{bmatrix}$, the reader should check that $\mathbf{x}^T A \mathbf{x} = q = x_1^2 - x_2^2 + 4x_1x_2$:

$$\begin{bmatrix} x_1 & x_2 \end{bmatrix} \begin{bmatrix} 1 & 2 \\ 2 & -1 \end{bmatrix} \begin{bmatrix} x_1 \\ x_2 \end{bmatrix} = x_1^2 - x_2^2 + 4x_1x_2. \qquad \blacksquare$$

Here are some other examples.

8.4.4 EXAMPLES
- $-4x_1^2 + x_2^2 + 4x_3^2 + 6x_1x_3 - 10x_2x_3 = \begin{bmatrix} x_1 & x_2 & x_3 \end{bmatrix} \begin{bmatrix} -4 & 0 & 3 \\ 0 & 1 & -5 \\ 3 & -5 & 4 \end{bmatrix} \begin{bmatrix} x_1 \\ x_2 \\ x_3 \end{bmatrix}$

- $4x^2 - 7xy + y^2 = \begin{bmatrix} x & y \end{bmatrix} \begin{bmatrix} 4 & -\frac{7}{2} \\ -\frac{7}{2} & 1 \end{bmatrix} \begin{bmatrix} x \\ y \end{bmatrix}$

Pay special attention to the last two examples.

- $2x_1^2 - 3x_2^2 + 7x_3^2 = \begin{bmatrix} x_1 & x_2 & x_3 \end{bmatrix} \begin{bmatrix} 2 & 0 & 0 \\ 0 & -3 & 0 \\ 0 & 0 & 7 \end{bmatrix} \begin{bmatrix} x_1 \\ x_2 \\ x_3 \end{bmatrix}$

- $y_1^2 - 2y_2^2 + 3y_3^2 - 4y_4^2 = \begin{bmatrix} y_1 & y_2 & y_3 & y_4 \end{bmatrix} \begin{bmatrix} 1 & 0 & 0 & 0 \\ 0 & -2 & 0 & 0 \\ 0 & 0 & 3 & 0 \\ 0 & 0 & 0 & -4 \end{bmatrix} \begin{bmatrix} y_1 \\ y_2 \\ y_3 \\ y_4 \end{bmatrix}. \qquad \blacksquare$

When a quadratic form $q = \lambda_1 x_1^2 + \lambda_2 x_2^2 + \cdots + \lambda_n x_n^2$ has no cross terms, the matrix A giving $q = \mathbf{x}^T A \mathbf{x}$ is a diagonal matrix, with diagonal entries the coefficients of $x_1^2, x_2^2, \ldots, x_n^2$:

$$\lambda_1 x_1^2 + \lambda_2 x_2^2 + \cdots + \lambda_n x_n^2$$

$$= \begin{bmatrix} x_1 & x_2 & \cdots & x_n \end{bmatrix} \begin{bmatrix} \lambda_1 & 0 & \cdots & 0 \\ 0 & \lambda_2 & & 0 \\ \vdots & & \ddots & \vdots \\ 0 & 0 & \cdots & \lambda_n \end{bmatrix} \begin{bmatrix} x_1 \\ x_2 \\ \vdots \\ x_n \end{bmatrix}. \qquad (1)$$

In general, the matrix A in $\mathbf{x}^T A \mathbf{x}$ is not diagonal, but it is symmetric. The Principal Axes Theorem says that A can be orthogonally diagonalized: There exists an orthogonal matrix Q such that $Q^T A Q = D$ is a real diagonal matrix. Set $\mathbf{y} = Q^T \mathbf{x}$ $(= Q^{-1}\mathbf{x})$. Then $\mathbf{x} = Q\mathbf{y}$ and $\mathbf{x}^T A \mathbf{x} = \mathbf{y}^T Q^T A Q \mathbf{y} = \mathbf{y}^T D \mathbf{y}$. If the diagonal entries of D are $\lambda_1, \lambda_2, \ldots, \lambda_n$, the form $q = \mathbf{x}^T A \mathbf{x}$ becomes $q = \mathbf{y}^T D \mathbf{y} = \lambda_1 y_1^2 + \lambda_2 y_2^2 + \cdots + \lambda_n y_n^2$.

8.4.5 EXAMPLES

- The quadratic form $q = x^2 - y^2 + 4xy$ is $\mathbf{x}^T A \mathbf{x}$ with $A = \begin{bmatrix} 1 & -2 \\ -2 & -1 \end{bmatrix}$. The characteristic polynomial of A is

$$\det(A - \lambda I) = \begin{vmatrix} 1 - \lambda & -2 \\ -2 & -1 - \lambda \end{vmatrix} = \lambda^2 - 5.$$

Thus A has eigenvalues $\lambda_1 = \sqrt{5}$, $\lambda_2 = -\sqrt{5}$ and $q = \mathbf{x}^T A \mathbf{x} = \mathbf{y}^T D \mathbf{y} = \sqrt{5} y_1^2 - \sqrt{5} y_2^2$ with $\mathbf{y} = Q^T \mathbf{x}$ for a certain matrix Q.

- The quadratic form $q = -4x_1^2 + x_2^2 + 4x_3^2 + 6x_1 x_3$ is $\mathbf{x}^T A \mathbf{x}$ with $A = \begin{bmatrix} -4 & 0 & 3 \\ 0 & 1 & 0 \\ 3 & 0 & 4 \end{bmatrix}$.

The characteristic polynomial of A is

$$\begin{vmatrix} -4 - \lambda & 0 & 3 \\ 0 & 1 - \lambda & 0 \\ 3 & 0 & 4 - \lambda \end{vmatrix} = \lambda^3 - \lambda^2 - 25\lambda + 25 = (\lambda - 1)(\lambda + 5)(\lambda - 5).$$

Matrix A has eigenvalues $1, 5, -5$, so $q = y_1^2 + 5y_2^2 - 5y_3^2$, where $\mathbf{y} = Q^T \mathbf{x}$ for a certain matrix Q. ∎

■ Conic Sections

Remember that the *coordinates* of a vector $\mathbf{x}$ relative to a basis $\{\mathbf{v}_1, \mathbf{v}_2, \ldots, \mathbf{v}_n\}$ are the scalars $c_1, c_2, \ldots, c_n$ required as coefficients when $\mathbf{x}$ is written as a linear combination of $\mathbf{v}_1, \ldots, \mathbf{v}_n$. For example, the coordinates of $\begin{bmatrix} -3 \\ 4 \end{bmatrix}$ relative to the standard basis of $\mathbf{R}^2$ are -3 and 4 because

$$\begin{bmatrix} -3 \\ 4 \end{bmatrix} = -3 \begin{bmatrix} 1 \\ 0 \end{bmatrix} + 4 \begin{bmatrix} 0 \\ 1 \end{bmatrix} = -3\mathbf{e}_1 + 4\mathbf{e}_2.$$

Let's examine the equation $\mathbf{y} = Q^T \mathbf{x}$ in more detail with $\mathbf{y} = \begin{bmatrix} y_1 \\ y_2 \end{bmatrix}$ and $Q = \begin{bmatrix} \mathbf{q}_1 & \mathbf{q}_2 \\ \downarrow & \downarrow \end{bmatrix}$. Since $Q^T = Q^{-1}$,

$$\mathbf{x} = Q\mathbf{y} = y_1 \mathbf{q}_1 + y_2 \mathbf{q}_2, \tag{2}$$

so $\mathbf{x}$ has coordinates y_1 and y_2 relative to the basis $\{\mathbf{q}_1, \mathbf{q}_2\}$. See Figure 8.6.

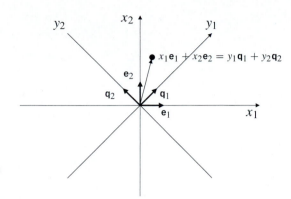

Figure 8.6 The coordinates of (x_1, x_2) relative to the orthonormal basis $\{\mathbf{q}_1, \mathbf{q}_2\}$ are (y_1, y_2).

The set $\mathcal{Q} = \{\mathbf{q}_1, \mathbf{q}_2\}$ is an orthonormal basis for $\mathbf{R}^2$ just like the standard basis $\{\mathbf{e}_1, \mathbf{e}_2\}$. A graph in $\mathbf{R}^2$ has an equation relative to the standard basis and another relative to $\mathcal{Q}$—same graph, different equations.

For example, suppose $\mathbf{q}_1 = \begin{bmatrix} \frac{1}{\sqrt{2}} \\ \frac{1}{\sqrt{2}} \end{bmatrix}$ and $\mathbf{q}_2 = \begin{bmatrix} -\frac{1}{\sqrt{2}} \\ \frac{1}{\sqrt{2}} \end{bmatrix}$. The relationship between the coordinates x_1, x_2 of a vector relative to the standard basis and its coordinates y_1, y_2 relative to $\mathcal{Q}$ is expressed by $\mathbf{x} = Q\mathbf{y}$:

$$\begin{bmatrix} x_1 \\ x_2 \end{bmatrix} = \begin{bmatrix} \frac{1}{\sqrt{2}} & -\frac{1}{\sqrt{2}} \\ \frac{1}{\sqrt{2}} & \frac{1}{\sqrt{2}} \end{bmatrix} \begin{bmatrix} y_1 \\ y_2 \end{bmatrix},$$

so

$$x_1 = \frac{1}{\sqrt{2}} y_1 - \frac{1}{\sqrt{2}} y_2$$

and

$$x_2 = \frac{1}{\sqrt{2}} y_1 + \frac{1}{\sqrt{2}} y_2.$$

The ellipse with equation $\frac{x_1^2}{16} + \frac{x_2^2}{9} = 1$ relative to the standard basis is shown in Figure 8.7. Relative to the basis $\mathcal{Q}$, it has equation

$$\frac{1}{16}\left(\frac{1}{\sqrt{2}} y_1 - \frac{1}{\sqrt{2}} y_2\right)^2 + \frac{1}{9}\left(\frac{1}{\sqrt{2}} y_1 + \frac{1}{\sqrt{2}} y_2\right)^2 = 1,$$

that is, $25y_1^2 + 25y_2^2 + 14y_1 y_2 = 288$. This example is only for the purposes of illustration; if we had to choose between the equations $\frac{x_1^2}{16} + \frac{x_2^2}{9} = 1$ and $25y_1^2 + 25y_2^2 + 14y_1 y_2 = 288$, most of us would choose the former, without the cross terms.

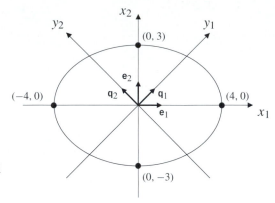

Figure 8.7 The ellipse with equation $\frac{x_1^2}{16} + \frac{x_2^2}{9} = 1$ relative to the standard basis.

On the other hand, given a quadratic form with nonzero cross terms, it would be nice to find a new set of "principal" axes relative to which the cross terms disappear.

8.4.6 PROBLEM

Identify the graph of $5x_1^2 - 4x_1x_2 + 8x_2^2 = 36$.

Solution. The given equation is $q = 36$, where $q = 5x_1^2 - 4x_1x_2 + 8x_2^2$. We have $q = \mathbf{x}^T A \mathbf{x}$ where

$$\mathbf{x} = \begin{bmatrix} x_1 \\ x_2 \end{bmatrix} \quad \text{and} \quad A = \begin{bmatrix} 5 & -2 \\ -2 & 8 \end{bmatrix}.$$

The eigenvalues of A are $\lambda_1 = 4$ and $\lambda_2 = 9$, so that, relative to a certain new orthonormal basis $\mathcal{Q}$, our equation is $4y_1^2 + 9y_2^2 = 36$. The graph is an ellipse, in standard position relative to the basis $\mathcal{Q} = \{\mathbf{q}_1, \mathbf{q}_2\}$. See Figure 8.8. The vectors

$$\mathbf{q}_1 = \frac{1}{\sqrt{5}} \begin{bmatrix} 2 \\ 1 \end{bmatrix} \quad \text{and} \quad \mathbf{q}_2 = \frac{1}{\sqrt{5}} \begin{bmatrix} -1 \\ 2 \end{bmatrix}$$

are eigenvectors of A corresponding to $\lambda_1 = 4$ and $\lambda_2 = 9$, respectively.

Figure 8.8 The ellipse with equation $4y_1^2 + 9y_2^2 = 1$.

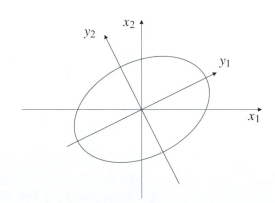

Exercises

*Solutions to exercises marked [BB] can be found in the **B**ack of the **B**ook.*

1. Write each of the given quadratic forms as $\mathbf{x}^T A \mathbf{x}$ for some symmetric matrix A.

   (a) [BB] $q = 2x^2 - 3xy + y^2$

   (b) $q = x_1^2 - 2x_2^2 + 3x_3^2 - x_1 x_3$

   (c) $q = x_2^2 + 4x_3^2 + x_1 x_2 + 4x_1 x_3 - 7x_2 x_3$

   (d) $q = 7y_1^2 - 3y_2^2 + 5y_3^2 - 8y_1 y_3 + 14y_2 y_3 - y_1 y_2$.

2. Let $\mathbf{q}_1 = \frac{1}{5}\begin{bmatrix} 3 \\ -4 \end{bmatrix}$ and $\mathbf{q}_2 = \frac{1}{5}\begin{bmatrix} 4 \\ 3 \end{bmatrix}$. Let (x_1, x_2) be the coordinates of the point P in the Euclidean plane relative to the standard basis and (y_1, y_2) the coordinates of P relative to the orthogonal basis $\mathcal{Q} = \{\mathbf{q}_1, \mathbf{q}_2\}$.

   (a) Write the equations that relate x_1, x_2 to y_1, y_2.

   (b) Find the equations of each of the following quadratic forms in terms of y_1 and y_2.

       i. [BB] $q = 25x_1^2 - 50x_2^2$

       ii. $q = x_1^2 + x_2^2$

       iii. $q = x_1 x_2$

       iv. $q = 2x_1^2 + 7x_2^2$.

3. Rewrite the given equations so that relative to some orthonormal basis the cross terms disappear. (You are not being asked to find the new basis.) If possible, name the graph.

   (a) [BB] $6x^2 - 4xy + 3y^2 = 1$

   (b) $3x^2 + y^2 + 3z^2 + 2xz = 1$

   (c) $5y^2 + 4z^2 + 4xy + 8xz + 12yz = 3$

   (d) $4x^2 + 4y^2 - 8z^2 - 10xy - 4yz = 1$.

4. [BB] Find an orthonormal basis of $\mathbf{R}^3$ relative to which the quadratic form $q = 3x_1^2 + 4x_2^2 + x_3^2 - 4x_2 x_3$ has no cross terms. Give the expression for q relative to this new basis too.

5. Repeat Exercise 4 for each of the following quadratic forms.

   (a) $q = x_1^2 + x_2^2 + 5x_3^2 - 2x_1 x_3 + 4x_2 x_3$

   (b) $q = 2x_1^2 + 2x_3^2 - 2x_1 x_2 - 6x_1 x_3 - 2x_2 x_3$

   (c) $q = 2x_1^2 - 13x_2^2 - 10x_3^2 + 20x_1 x_2 - 12x_1 x_3 + 24x_2 x_3$

   (d) $q = 2x_1 + x_2^2 + 2x_3^2 + 2x_1 x_2 + 2x_2 x_3$.

6. Show that the graph (in the plane) of the equation $ax^2 + bxy + cy^2 = 1$ is an ellipse if $b^2 - 4ac < 0$ and a hyperbola if $b^2 - 4ac > 0$. What is the graph if $b^2 - 4ac = 0$? [Remark: An equation of the given type may not have a graph at all. Consider, for example, $-x^2 - y^2 = 1$. In this exercise, assume that the values of a, b, and c are such that such a "vacuous" case does not occur.]

8.5 Graphs

A *graph* is a pair $(\mathcal{V}, \mathcal{E})$ of sets, $\mathcal{V}$ not empty and each element of $\mathcal{E}$ a set $\{u, v\}$ of two distinct elements of $\mathcal{V}$. The elements of $\mathcal{V}$ are called *vertices* and the elements of $\mathcal{E}$ *edges*. An element $\{u, v\}$ of $\mathcal{E}$ is usually denoted just uv. Edge vu is the same as edge uv because the sets $\{u, v\}$ and $\{v, u\}$ are the same.

    One nice thing about graphs is that they can easily be pictured. In Figure 8.9, we have drawn a picture of a graph $\mathcal{G}$ with five vertices and six edges: $\mathcal{V} = \{v_1, v_2, v_3, v_4, v_5\}$ and $\mathcal{E} = \{v_1 v_2, v_1 v_3, v_1 v_5, v_3 v_4, v_4 v_5, v_3 v_5\}$. There are five vertices labeled $v_1, \ldots, v_5$ and we draw a line between vertices v_i and v_j if $v_i v_j$ is an edge.

    Graphs arise in numerous contexts. In an obvious way, they can be used to illustrate a map of towns and connecting roads or a network of computers on the Internet. Less obvious perhaps are applications in biochemistry concerned with the

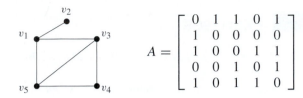

Figure 8.9 A graph and its adjacency matrix.

$$A = \begin{bmatrix} 0 & 1 & 1 & 0 & 1 \\ 1 & 0 & 0 & 0 & 0 \\ 1 & 0 & 0 & 1 & 1 \\ 0 & 0 & 1 & 0 & 1 \\ 1 & 0 & 1 & 1 & 0 \end{bmatrix}$$

recovery of RNA and DNA chains from fragmentary data, to the design of facilities such as hospitals, where it is important that certain rooms be in close proximity to others (an operating room and intensive care unit), and to examination scheduling, where the goal is to minimize conflicts (two exams scheduled for the same time, with one or more students having to write both).

The vertices in Figure 8.9, for instance, might correspond to subjects—math, English, chemistry, physics, biology—with an edge between subjects, meaning that exams in these subjects should not be scheduled at the same time (presumably because there are students taking both courses). The reader might wish to convince herself that three examination periods are required for the problem pictured in the figure.

Why talk about graphs in a linear algebra book? In practice, graphs are often very large (lots of vertices), and it is only feasible to analyze a graph theoretical problem with the assistance of high-speed computers. Since computers are blind, you can't give a computer a picture; instead, it is common to give a computer the *adjacency matrix* of the graph. This is the matrix whose entries consist only of 0s and 1s, a 1 in position (i, j) signifying that $v_i v_j$ is an edge, and a 0 denoting no edge. The adjacency matrix of the graph in Figure 8.9 appears next to its picture.

Reading Challenge 1   *The adjacency matrix of a graph is symmetric. Why?*

If A is the adjacency matrix of a graph $\mathcal{G}$, the powers $A, A^2, A^3, \ldots$ record the number of *walks* between pairs of vertices, where a walk is a sequence of vertices for which there is an edge between each consecutive pair. The *length* of a walk is the number of edges required, which is one less than the number of vertices. In the graph of Figure 8.9, $v_1 v_3$ is a walk of length one from v_1 to v_3, $v_1 v_3 v_5 v_4$ is a walk of length three from v_1 to v_4, and $v_1 v_2 v_1$ is a walk of length two from v_1 to v_1.

The (i, j) entry of the adjacency matrix A is the number of walks of length **one** from v_i to v_j, since a walk of length one exists if and only if there is an edge between v_i and v_j. Interestingly, the (i, j) entry of A^2 is the number of walks of length **two** from v_i to v_j and the (i, j) entry of A^3 is the number of walks of length **three** from v_i to v_j. The general situation is summarized in the next proposition.

Proposition 8.5.1   *If A is the adjacency matrix of a graph with vertices labeled $v_1, v_2, \ldots, v_n$, then the (i, j) entry of A^k is the number of walks of length k from v_i to v_j.*

This proposition can be proved directly using a technique called "mathematical induction," with which we do not presume all our readers to be familiar, so we omit the proof, contenting ourselves with illustrations here and the proof of two specific cases in the exercises. (See Exercise 3.)

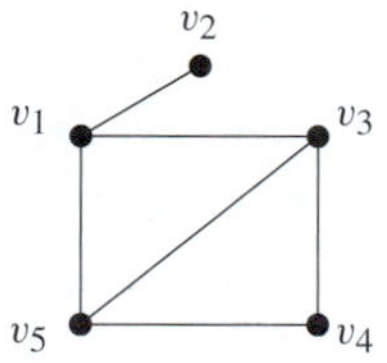

$$A^2 = \begin{bmatrix} 3 & 0 & 1 & 2 & 1 \\ 0 & 1 & 1 & 0 & 1 \\ 1 & 1 & 3 & 1 & 2 \\ 2 & 0 & 1 & 2 & 1 \\ 1 & 1 & 2 & 1 & 3 \end{bmatrix}, \quad A^3 = \begin{bmatrix} 2 & 3 & 6 & 2 & 6 \\ 3 & 0 & 1 & 2 & 1 \\ 6 & 1 & 4 & 5 & 5 \\ 2 & 2 & 5 & 2 & 5 \\ 6 & 1 & 5 & 5 & 4 \end{bmatrix}$$

Figure 8.10 The (i, j) entry of A^k is the number of walks of length k from v_i to v_j.

In Figure 8.10, we reproduce the graph of Figure 8.9 and show the powers A^2 and A^3 of its adjacency matrix A.

The $(1, 1)$ entry of A^2 is 3, corresponding to the fact that there are three walks of length 2 from v_1 to v_1, namely, $v_1 v_2 v_1$, $v_1 v_3 v_1$, and $v_1 v_5 v_1$.

The $(3, 5)$ entry is 2, corresponding to the fact that there are two walks of length 2 from v_3 to v_5, namely, $v_3 v_1 v_5$ and $v_3 v_4 v_5$.

The $(1, 3)$ entry of A^3 is 6, corresponding to the six walks of length 3 from v_1 to v_3:

$$v_1 v_2 v_1 v_3, \quad v_1 v_3 v_1 v_3, \quad v_1 v_5 v_1 v_3, \quad v_1 v_5 v_4 v_3, \quad v_1 v_3 v_4 v_3, \quad \text{and} \quad v_1 v_3 v_5 v_3.$$

Reading Challenge 2 *Explain why the $(2, 4)$ entry of A^3 is 2 and why the $(4, 3)$ entry of A^3 is 5.*

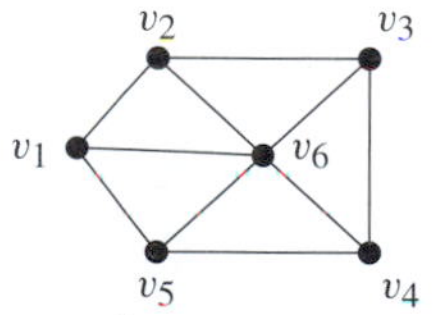

Figure 8.11

"Connected" means just what you'd expect. There's a route from any vertex to any other vertex along a sequence of edges.

Kirchhoff's Theorem is also known as the "Matrix Tree Theorem."

Spanning Trees

A *cycle* in a graph is a walk from a vertex v back to v that passes through different vertices. In the graph shown to the left, $v_2 v_3 v_6 v_2$ and $v_1 v_5 v_4 v_6 v_2 v_1$ are cycles, but $v_1 v_6 v_4 v_3 v_6 v_1$ is not.

A *tree* is a connected graph that contains no cycles. Several trees are illustrated in Figure 8.12.

A *spanning tree* in a connected graph $\mathcal{G}$ is a tree that contains every vertex of $\mathcal{G}$ and all of whose edges are edges of $\mathcal{G}$. A graph and three of its spanning trees are shown in Figure 8.13. A graph and all (eight of) its spanning trees is shown in Figure 8.14.

What do all these interesting pictures have to do with linear algebra? Well, it is of interest to know how many spanning trees a given graph has. Are we sure, for instance, that the graph in Figure 8.14 has just the spanning trees shown? Are there others we have missed?

In 1847, the German physicist Gustav Kirchhoff (1824–1887) gave a remarkable answer.

Theorem 8.5.1 **(Kirchhoff).** *Given a connected graph $\mathcal{G}$, let M be the matrix obtained from the adjacency matrix of $\mathcal{G}$ by changing all 1s to -1s and each diagonal 0 to the degree*

Figure 8.12 These are trees (of the graphical variety).

Figure 8.13 A graph and three of its spanning trees.

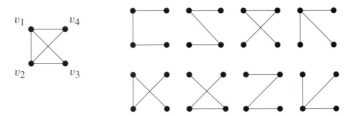

Figure 8.14 A graph and its eight spanning trees.

of the corresponding vertex. Then all the cofactors of M are equal, and this common number is the number of spanning trees in $\mathcal{G}$.

8.5.2 EXAMPLE

The graph in Figure 8.14, labeled as shown, has adjacency matrix $A = \begin{bmatrix} 0 & 1 & 1 & 1 \\ 1 & 0 & 1 & 1 \\ 1 & 1 & 0 & 0 \\ 1 & 1 & 0 & 0 \end{bmatrix}$,

so the matrix specified by Kirchhoff's Theorem is $M = \begin{bmatrix} 3 & -1 & -1 & -1 \\ -1 & 3 & -1 & -1 \\ -1 & -1 & 2 & 0 \\ -1 & -1 & 0 & 2 \end{bmatrix}$.

Expanding by cofactors of the first row, the $(1, 1)$ cofactor of M is

$$\det \begin{bmatrix} 3 & -1 & -1 \\ -1 & 2 & 0 \\ -1 & 0 & 2 \end{bmatrix} = 3 \begin{vmatrix} 2 & 0 \\ 0 & 2 \end{vmatrix} - (-1) \begin{vmatrix} -1 & 0 \\ -1 & 2 \end{vmatrix} + (-1) \begin{vmatrix} -1 & 2 \\ -1 & 0 \end{vmatrix}$$

$$= 3(4) + 1(-2) + (-1)2 = 8.$$

Expanding by cofactors of the third column, the $(2, 3)$ cofactor of M is

$$-\det \begin{bmatrix} 3 & -1 & -1 \\ -1 & -1 & 0 \\ -1 & -1 & 2 \end{bmatrix} = -\left[(-1) \begin{vmatrix} -1 & -1 \\ -1 & -1 \end{vmatrix} + 2 \begin{vmatrix} 3 & -1 \\ -1 & -1 \end{vmatrix} \right]$$

$$= -[(-1)(0) + 2(-3 - 1)] = -(-8) = 8.$$

It is no coincidence that each of these cofactors is the same. This is guaranteed by Kirchhoff's Theorem, which also says that the graph in question has eight spanning trees. ∎

8.5.3 PROBLEM

In Figure 8.13, we showed a graph $\mathcal{G}$ and three of its spanning trees. How many spanning trees does $\mathcal{G}$ have in all?

Solution. Labeling the vertices of $\mathcal{G}$ as shown, the adjacency matrix is

$$A = \begin{bmatrix} 0 & 1 & 0 & 0 & 1 & 1 \\ 1 & 0 & 1 & 0 & 0 & 1 \\ 0 & 1 & 0 & 1 & 0 & 1 \\ 0 & 0 & 1 & 0 & 1 & 1 \\ 1 & 0 & 0 & 1 & 0 & 1 \\ 1 & 1 & 1 & 1 & 1 & 0 \end{bmatrix},$$

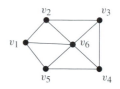

so the matrix specified by Kirchhoff's Theorem is

$$M = \begin{bmatrix} 3 & -1 & 0 & 0 & -1 & -1 \\ -1 & 3 & -1 & 0 & 0 & -1 \\ 0 & -1 & 3 & -1 & 0 & -1 \\ 0 & 0 & -1 & 3 & -1 & -1 \\ -1 & 0 & 0 & -1 & 3 & -1 \\ -1 & -1 & -1 & -1 & -1 & 5 \end{bmatrix}.$$

The $(1, 1)$ cofactor is $\det \begin{bmatrix} 3 & -1 & 0 & 0 & -1 \\ -1 & 3 & -1 & 0 & -1 \\ 0 & -1 & 3 & -1 & -1 \\ 0 & 0 & -1 & 3 & -1 \\ -1 & -1 & -1 & -1 & 5 \end{bmatrix}$, which, using elementary row operations, is

$$-\begin{vmatrix} -1 & 3 & -1 & 0 & -1 \\ 3 & -1 & 0 & 0 & -1 \\ 0 & -1 & 3 & -1 & -1 \\ 0 & 0 & -1 & 3 & -1 \\ -1 & -1 & -1 & -1 & 5 \end{vmatrix} = -\begin{vmatrix} -1 & 3 & -1 & 0 & -1 \\ 0 & 8 & -3 & 0 & -4 \\ 0 & -1 & 3 & -1 & -1 \\ 0 & 0 & -1 & 3 & -1 \\ 0 & -4 & 0 & -1 & 6 \end{vmatrix}$$

$$= +\begin{vmatrix} -1 & 3 & -1 & 0 & -1 \\ 0 & -1 & 3 & -1 & -1 \\ 0 & 8 & -3 & 0 & -4 \\ 0 & 0 & -1 & 3 & -1 \\ 0 & -4 & 0 & -1 & 6 \end{vmatrix} = +\begin{vmatrix} -1 & 3 & -1 & 0 & -1 \\ 0 & -1 & 3 & -1 & -1 \\ 0 & 0 & 21 & -8 & -12 \\ 0 & 0 & -1 & 3 & -1 \\ 0 & 0 & -12 & 3 & 10 \end{vmatrix}$$

$$= -\begin{vmatrix} -1 & 3 & -1 & 0 & -1 \\ 0 & -1 & 3 & -1 & -1 \\ 0 & 0 & -1 & 3 & -1 \\ 0 & 0 & 21 & -8 & -12 \\ 0 & 0 & -12 & 3 & 10 \end{vmatrix} = -\begin{vmatrix} -1 & 3 & -1 & 0 & -1 \\ 0 & -1 & 3 & -1 & -1 \\ 0 & 0 & -1 & 3 & -1 \\ 0 & 0 & 0 & 55 & -33 \\ 0 & 0 & 0 & -33 & 22 \end{vmatrix}$$

$$= -11(11) \begin{vmatrix} -1 & 3 & -1 & 0 & -1 \\ 0 & -1 & 3 & -1 & -1 \\ 0 & 0 & -1 & 3 & -1 \\ 0 & 0 & 0 & 5 & -3 \\ 0 & 0 & 0 & -3 & 2 \end{vmatrix} = -121 \begin{vmatrix} -1 & 3 & -1 & 0 & -1 \\ 0 & -1 & 3 & -1 & -1 \\ 0 & 0 & -1 & 3 & -1 \\ 0 & 0 & 0 & 5 & -3 \\ 0 & 0 & 0 & 0 & \frac{1}{5} \end{vmatrix}$$

$$= -121(-1)(-1)(-1)(5)(\frac{1}{5}) = 121.$$

The graph in question has 121 spanning trees.

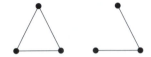

Figure 8.15 The graphs pictured here are not isomorphic.

Isomorphism

The concept of "isomorphism" recurs throughout many fields of mathematics. Objects are isomorphic if they differ only in appearance rather than in some fundamental way. In Figure 8.15, we show the pictures of two graphs that are fundamentally different. The graph pictured at the left has three vertices and three edges. The graph pictured on the right is different. On the other hand, the graphs pictured in Figure 8.16 are the same, technically, *isomorphic*.

8.5.4 DEFINITION Graphs $\mathcal{G}_1$ and $\mathcal{G}_2$ are *isomorphic* if the vertices of $\mathcal{G}_2$ can be labeled using the symbols that label $\mathcal{G}_1$ in such a way that uv is an edge in $\mathcal{G}_1$ if and only if uv is an edge in $\mathcal{G}_2$.

For instance, consider the following relabeling of vertices of the graph $\mathcal{G}_1$, pictured on the left in Figure 8.16.

$$\begin{aligned} v_1 &\rightarrow u_1 \\ v_2 &\rightarrow u_5 \\ v_3 &\rightarrow u_2 \\ v_4 &\rightarrow u_4 \\ v_5 &\rightarrow u_3 \end{aligned}$$

The edges in each graph are the same—v_1v_2, v_1v_3, v_1v_5, v_3v_4, v_3v_5, v_4v_5—so the graphs are isomorphic. Equivalently, the labels of the vertices of $\mathcal{G}_1$ have been

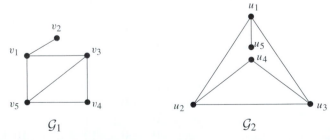

Figure 8.16 Two pictures of the same graph.

reordered so that the adjacency matrices of the two graphs are the same. Before relabeling, $\mathcal{G}_1$ and $\mathcal{G}_2$ had the adjacency matrices A_1, A_2, respectively, shown in (1).

$$A_1 = \begin{bmatrix} 0 & 1 & 1 & 0 & 1 \\ 1 & 0 & 0 & 0 & 0 \\ 1 & 0 & 0 & 1 & 1 \\ 0 & 0 & 1 & 0 & 1 \\ 1 & 0 & 1 & 1 & 0 \end{bmatrix}, \qquad A_2 = \begin{bmatrix} 0 & 1 & 1 & 0 & 1 \\ 1 & 0 & 1 & 1 & 0 \\ 1 & 1 & 0 & 1 & 0 \\ 0 & 1 & 1 & 0 & 0 \\ 1 & 0 & 0 & 0 & 0 \end{bmatrix}. \tag{1}$$

The relabeling of the vertices of $\mathcal{G}_1$ had the effect of rearranging the columns and the rows of A_1 in the order 15243, so that the rearranged matrix became A_2. Rearranging the columns of A_1 is effected by multiplying A_1 on the right by the permutation matrix P whose columns are the standard basis vectors $\mathbf{e}_1, \mathbf{e}_5, \mathbf{e}_2, \mathbf{e}_4, \mathbf{e}_3$ in this order:

$$P = \begin{bmatrix} \mathbf{e}_1 & \mathbf{e}_5 & \mathbf{e}_2 & \mathbf{e}_4 & \mathbf{e}_3 \\ \downarrow & \downarrow & \downarrow & \downarrow & \downarrow \\ & & & & \end{bmatrix} = \begin{bmatrix} 1 & 0 & 0 & 0 & 0 \\ 0 & 0 & 1 & 0 & 0 \\ 0 & 0 & 0 & 0 & 1 \\ 0 & 0 & 0 & 1 & 0 \\ 0 & 1 & 0 & 0 & 0 \end{bmatrix}$$

since

$$A_1 P = \begin{bmatrix} A_1\mathbf{e}_1 & A_1\mathbf{e}_5 & A_1\mathbf{e}_2 & A_1\mathbf{e}_4 & A_1\mathbf{e}_3 \\ \downarrow & \downarrow & \downarrow & \downarrow & \downarrow \\ & & & & \end{bmatrix}$$

and $A_1\mathbf{e}_i$ is column i of A_1.

Rearranging the rows of A_1 in the order 15243 can be accomplished by multiplying A_1 on the left by P^T, the transpose of P. To see why, let $A_1 = \begin{bmatrix} \mathbf{a}_1 & \rightarrow \\ \mathbf{a}_2 & \rightarrow \\ \mathbf{a}_3 & \rightarrow \\ \mathbf{a}_4 & \rightarrow \\ \mathbf{a}_5 & \rightarrow \end{bmatrix}$

have rows $\mathbf{a}_1, \mathbf{a}_2, \dots, \mathbf{a}_5$ as shown. The first column of

$$A_1^T P = A_1^T \begin{bmatrix} \mathbf{e}_1 & \mathbf{e}_5 & \mathbf{e}_2 & \mathbf{e}_4 & \mathbf{e}_3 \\ \downarrow & \downarrow & \downarrow & \downarrow & \downarrow \\ & & & & \end{bmatrix}$$

is $A_1^T \mathbf{e}_1$, which is the first column of A_1^T, namely, $\mathbf{a}_1$. The second column of $A_1^T P$ is $A_1^T \mathbf{e}_5$, which is the fifth column of A_1^T, that is, $\mathbf{a}_5$. We obtain

$$P^T A_1 = (A_1^T P)^T = \begin{bmatrix} \mathbf{a}_1 & \rightarrow \\ \mathbf{a}_5 & \rightarrow \\ \mathbf{a}_2 & \rightarrow \\ \mathbf{a}_4 & \rightarrow \\ \mathbf{a}_3 & \rightarrow \end{bmatrix},$$

so $A_2 = P^T A_1 P$. The following theorem should be apparent.

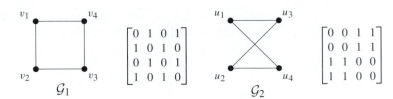

Figure 8.17   The pictures of two isomorphic graphs and their adjacency matrices.

Theorem 8.5.6   *Let graphs G_1 and G_2 have adjacency matrices A_1 and A_2, respectively. Then G_1 is isomorphic to G_2 if and only if there exists a permutation matrix P with the property that $P^T A_1 P = A_2$.*

8.5.7 EXAMPLE   The pictures of two isomorphic graphs, along with their adjacency matrices, are shown in Figure 8.17. If we relabel the vertices of G_1,

$$
\begin{aligned}
v_1 &\rightarrow u_1 \\
v_2 &\rightarrow u_4 \\
v_3 &\rightarrow u_2 \\
v_4 &\rightarrow u_3,
\end{aligned}
$$

and form the permutation matrix $P = \begin{bmatrix} 1 & 0 & 0 & 0 \\ 0 & 0 & 0 & 1 \\ 0 & 1 & 0 & 0 \\ 0 & 0 & 1 & 0 \end{bmatrix}$ whose columns are the standard basis vectors in $\mathbf{R}^4$ in the order 1423, then $P^T A_1 P = A_2$.   ◼

8.5.8 PROBLEM   Let $A_1 = \begin{bmatrix} 0 & 1 & 0 & 1 & 0 \\ 1 & 0 & 1 & 0 & 1 \\ 0 & 1 & 0 & 1 & 0 \\ 1 & 0 & 1 & 0 & 1 \\ 0 & 1 & 0 & 1 & 0 \end{bmatrix}$ and $A_2 = \begin{bmatrix} 0 & 1 & 0 & 1 & 1 \\ 1 & 0 & 1 & 0 & 0 \\ 0 & 1 & 0 & 1 & 0 \\ 1 & 0 & 1 & 0 & 1 \\ 1 & 0 & 0 & 1 & 0 \end{bmatrix}$. Is there a permutation matrix P such that $P^T A_1 P = A_2$?

Solution.   The matrices are the adjacency matrices of the graphs pictured in Figure 8.18. The graphs are not isomorphic since, for example, G_2 contains a *triangle*—three vertices u_1, u_4, u_5, each of which is joined by an edge—whereas G_1 has no triangles. By Theorem 8.5.6, no permutation matrix exists.

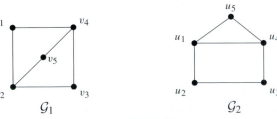

Figure 8.18

Answers to Reading Challenges

1. If there is an edge between v_i and v_j, there is an edge between v_j and v_i. So $a_{ij} = 1$ if and only if $a_{ji} = 1$.

2. There are two walks of length 3 from v_2 to v_4: $v_2v_1v_5v_4$ and $v_2v_1v_3v_4$. There are five walks of length 3 from v_4 to v_3: $v_4v_3v_4v_3$, $v_4v_5v_4v_3$, $v_4v_3v_1v_3$, $v_4v_3v_5v_3$, and $v_4v_5v_1v_3$.

■ Exercises

Solutions to exercises marked [BB] can be found in the Back of the Book.

1. [BB] If A is the adjacency matrix of a graph, the (i, i) entry of A^2 is the number of edges that meet at vertex v_i. Why?

2. Let A be the adjacency matrix of the graph shown in Figure 8.9. Find the $(1, 2)$ [BB], $(2, 5)$ and $(5, 4)$ $(3, 3)$ entries of A^4 without calculating this matrix. Explain your answers.

3. Let A be the adjacency matrix of a graph with vertices $v_1, v_2, \ldots$.

    (a) [BB] Explain why the (i, j) entry of A^2 is the number of walks of length 2 from v_i to v_j.

    (b) Use the result of (a) to explain why the (i, j) entry of A^3 is the number of walks of length 3 from v_i to v_j.

4. The "Petersen" graph shown below is of interest in graph theory for many reasons.

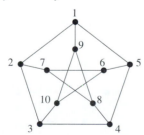

    (a) What is the adjacency matrix of the Petersen graph, as shown?

    (b) Show that the next graph is isomorphic to the Petersen graph by labeling its vertices so that its adjacency matrix is the matrix of part (a).

5. [BB] Let $\mathcal{T}$ be a tree and let u and v be vertices of $\mathcal{T}$ that are not joined by an edge. Adding edge uv to $\mathcal{T}$ must produce a cycle. Why?

6. There is a procedure for obtaining all the spanning trees of a graph, which we describe with reference to the graph in Figure 8.11. First delete edges one at a time until the graph that remains is a spanning tree. In particular, delete edges v_6v_2, v_6v_3, v_6v_4, v_6v_5, and v_3v_4. This should produce the first spanning tree shown in Figure 8.13. Now replace edge v_3v_4.

    (a) [BB] Draw the current graph.

    (b) Your graph should contain exactly one cycle. What is this cycle? How many edges does it contain?

    (c) Delete edges of the cycle one at a time. In each case, you should be left with a spanning tree. Draw pictures of all such spanning trees.

7. Use Kirchhoff's Theorem to determine the number of spanning trees in each of the graphs shown.

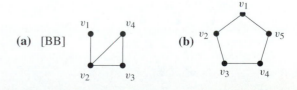

(c)

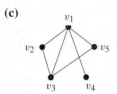

8. The *complete graph* on n vertices is the graph with n vertices and all possible edges. It is denoted $\mathcal{K}_n$.

(a) Draw pictures of $\mathcal{K}_2$, $\mathcal{K}_3$, $\mathcal{K}_4$ [BB], and $\mathcal{K}_4$.

(b) How many spanning trees does $\mathcal{K}_n$ have?

9. Let A be the adjacency matrix of the graph shown in Figure 8.14 and let M be the matrix specified by Kirchhoff's Theorem. Find the $(1, 4)$, the $(3, 3)$, and the $(4, 3)$ cofactors of M and comment on your findings.

10. Find the adjacency matrices A_1, A_2 of each pair of graphs shown below. If there is a permutation matrix P such that $P^T A_1 P = A_2$, write P and explain how you found it.

(a) [BB] $\mathcal{G}_1$: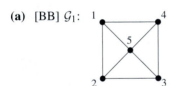

$\mathcal{G}_2$: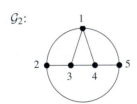

(b) $\mathcal{G}_1$: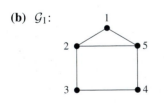

$\mathcal{G}_2$: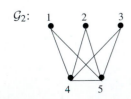

(c) [BB] $\mathcal{G}_1$: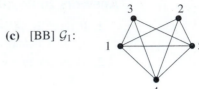

$\mathcal{G}_2$: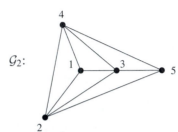

(d) $\mathcal{G}_1$: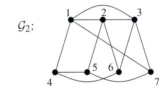

$\mathcal{G}_2$: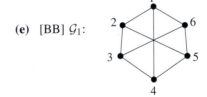

(e) [BB] $\mathcal{G}_1$: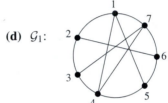

$\mathcal{G}_2$: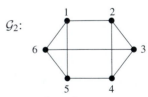

(f) $\mathcal{G}_1$: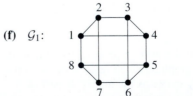

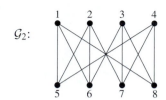

$\mathcal{G}_2$:

$$A_2 = \begin{bmatrix} 0 & 0 & 1 & 1 & 0 \\ 0 & 0 & 1 & 1 & 0 \\ 1 & 1 & 0 & 1 & 1 \\ 1 & 1 & 1 & 0 & 0 \\ 0 & 0 & 1 & 0 & 0 \end{bmatrix}$$

11. For each pair of matrices A_1, A_2 shown below, decide whether there exists a permutation matrix P with $A_2 = P^T A_1 P$. Explain your answers.

(a) [BB] $A_1 = \begin{bmatrix} 0 & 1 & 0 & 1 \\ 1 & 0 & 1 & 1 \\ 0 & 1 & 0 & 1 \\ 1 & 1 & 1 & 0 \end{bmatrix}$

$$A_2 = \begin{bmatrix} 0 & 1 & 1 & 1 \\ 1 & 0 & 1 & 1 \\ 1 & 1 & 0 & 1 \\ 1 & 1 & 1 & 0 \end{bmatrix}$$

(b) $A_1 = \begin{bmatrix} 0 & 1 & 0 & 0 & 1 \\ 1 & 0 & 1 & 0 & 0 \\ 0 & 1 & 0 & 1 & 0 \\ 0 & 0 & 1 & 0 & 1 \\ 1 & 0 & 1 & 1 & 0 \end{bmatrix}$

$$A_2 = \begin{bmatrix} 0 & 1 & 0 & 0 & 1 \\ 1 & 0 & 1 & 0 & 1 \\ 0 & 1 & 0 & 1 & 0 \\ 0 & 0 & 1 & 0 & 1 \\ 1 & 1 & 0 & 1 & 0 \end{bmatrix}$$

(c) [BB] $A_1 = \begin{bmatrix} 0 & 0 & 0 & 0 & 1 \\ 0 & 0 & 1 & 0 & 1 \\ 0 & 1 & 0 & 1 & 1 \\ 0 & 0 & 1 & 0 & 1 \\ 1 & 1 & 1 & 1 & 0 \end{bmatrix}$

(d) $A_1 = \begin{bmatrix} 0 & 1 & 0 & 1 & 0 & 1 \\ 1 & 0 & 1 & 1 & 0 & 0 \\ 0 & 1 & 0 & 1 & 1 & 0 \\ 1 & 0 & 1 & 0 & 1 & 0 \\ 0 & 0 & 1 & 1 & 0 & 1 \\ 1 & 1 & 0 & 0 & 1 & 0 \end{bmatrix}$

$$A_2 = \begin{bmatrix} 0 & 0 & 0 & 1 & 1 & 1 \\ 0 & 0 & 0 & 1 & 1 & 1 \\ 0 & 0 & 0 & 1 & 1 & 1 \\ 1 & 1 & 1 & 0 & 0 & 0 \\ 1 & 1 & 1 & 0 & 0 & 0 \\ 1 & 1 & 1 & 0 & 0 & 0 \end{bmatrix}$$

(e) $A_1 = \begin{bmatrix} 0 & 1 & 1 & 1 & 1 & 1 \\ 1 & 0 & 1 & 0 & 0 & 0 \\ 1 & 1 & 0 & 1 & 0 & 0 \\ 1 & 0 & 1 & 0 & 1 & 0 \\ 1 & 0 & 0 & 1 & 0 & 1 \\ 1 & 0 & 0 & 0 & 1 & 0 \end{bmatrix}$

$$A_2 = \begin{bmatrix} 0 & 1 & 1 & 0 & 0 & 1 \\ 1 & 0 & 1 & 1 & 0 & 0 \\ 1 & 1 & 0 & 1 & 1 & 1 \\ 0 & 1 & 1 & 0 & 1 & 0 \\ 0 & 0 & 1 & 1 & 0 & 0 \\ 1 & 0 & 1 & 0 & 0 & 0 \end{bmatrix}.$$

CHAPTER KEY WORDS AND IDEAS: Here are some technical words and phrases that were used in this chapter. Do you know the meaning of each? If you're not sure, check the glossary or index at the back of the book.

best solution in sense of least squares	permutation matrix
coordinates of a vector	quadratic form
Fibonacci sequence	recursively defined sequence
isomorphic graphs	sequence
Markov matrix	steady state

Appendix: Show and Prove

Gillian came to my office one day seeking help with a homework assignment. She circled the word "show" in one problem and said, "You know, I guess I don't know what that little word means." Words like **show** and **prove** in a question imply that some understanding and writing will be required to answer it, not just a number that can be easily checked with a friend over the phone. In this appendix, we attempt to make "show and prove" problems less difficult than some people think they are.

The author has heard "proof" defined as "a convincing communication that answers why." Proofs are **communications**. When you prove something, you aren't done until you can explain it to others. To give a proof, you have to convince the other person that something is so. This is usually done with a sequence of sentences, **each one following clearly and logically from the previous**, which explain **why** the desired result is true.

Most mathematical theorems and most show and prove problems are statements of the form "if $\mathcal{A}$, then $\mathcal{B}$," where $\mathcal{A}$ and $\mathcal{B}$ are other statements. "If a matrix A is invertible, then its determinant is not 0" is one example. There are other ways to say the same thing:

- Suppose A is an invertible matrix. Then its determinant is different from 0.
- The determinant of an invertible matrix is not 0.

In any show and prove problem, it's a good first step to determine what $\mathcal{A}$ and $\mathcal{B}$ are. In our example,

$\mathcal{A}$ is "A is an invertible matrix" and $\mathcal{B}$ is "The determinant of A is not 0."

The statement $\mathcal{A}$ (which could also be a collection of several statements) is called the *hypothesis*; $\mathcal{B}$ is called the *conclusion*.

Reading Challenge 1   *A student is asked to prove the following:*

"A homogeneous system of equations always has a solution."

State this in the form "If $\mathcal{A}$, then $\mathcal{B}$." What is the hypothesis? What is the conclusion?

To prove "If $\mathcal{A}$, then $\mathcal{B}$" requires writing down a sequence of statements, **each of which follows logically from the preceding**, starting with $\mathcal{A}$ (usually) and finishing with $\mathcal{B}$.

A-1 PROBLEM

If A is an invertible matrix, then $\det A \neq 0$.

Solution.

1. Let A be an invertible matrix.
2. Then $AB = I$, the identity matrix, for some matrix B. (This comes from the definition of "invertible.")
3. Then $\det AB = \det I = 1$.
4. Thus $(\det A)(\det B) = 1$ (because $\det AB = (\det A)(\det B)$).
5. Thus $\det A$ cannot be 0; otherwise, its product with $\det B$ would be 0.

In this proof, and elsewhere in this appendix, we number the statements in proofs so that it is easy to refer to them. While we are not recommending that you number the statements in your proofs, we do suggest that you try writing proofs in single statements, one under the other (rather than in a paragraph), making sure that every statement you write down is a logical consequence of previous statements. In our sample proof, we gave a reason for every statement except 3. We assumed the reader would know that the determinant of the identity matrix is 1. If you find it difficult to know when to justify something and when justification is not necessary, remember the rule: "If in doubt, justify."

A-2 PROBLEM

Suppose A is an $m \times n$ matrix such that the matrix $A^T A$ is invertible. Let $\mathbf{b}$ be any vector in $\mathbf{R}^n$. Show that the linear system $A\mathbf{x} = \mathbf{b}$ has at most one solution. It's pretty clear how we should begin.

Solution.

1. Suppose $A\mathbf{x} = \mathbf{b}$ has two solutions.
2. Let these solutions be $\mathbf{x}_1$ and $\mathbf{x}_2$.
3. Thus $A\mathbf{x}_1 = \mathbf{b}$ and $A\mathbf{x}_2 = \mathbf{b}$.
4. Therefore $A\mathbf{x}_1 = A\mathbf{x}_2$.

We interrupt our proof to make a couple of comments. The first few lines were "obvious," we suggest, what the author calls "follow your nose" steps. But what should come next? So far, we haven't used the most important part of the hypothesis, that $A^T A$ is invertible. Looking at line 4, it should now be clear how to continue.

5. Multiplying by A^T gives $A^T A \mathbf{x}_1 = A^T A \mathbf{x}_2$.
6. So $(A^T A)^{-1} A^T A \mathbf{x}_1 = (A^T A)^{-1} A^T A \mathbf{x}_2$.
7. So $I \mathbf{x}_1 = I \mathbf{x}_2$, since $(A^T A)^{-1}(A^T A) = I$ is the identity matrix.
8. So $\mathbf{x}_1 = \mathbf{x}_2$, as desired.

When Jennifer enrols for a German course, she understands that she'll have to memorize a lot of vocabulary. It is very difficult to translate a sentence if you do not know how to translate the words! So it is with mathematics. If you don't know what the words mean, you aren't going to be able to translate a mathematical sentence into language you understand and you certainly will not be able to solve show or prove problems.

> **A-3** Before attempting to show or prove something, make sure you know the meaning of every word in the statement.

A-4 PROBLEM

Show that $\begin{bmatrix} 7 \\ 7 \end{bmatrix}$ is a linear combination of $\begin{bmatrix} 2 \\ 3 \end{bmatrix}$ and $\begin{bmatrix} -1 \\ 2 \end{bmatrix}$.

This problem is straightforward if you know what it means for one vector to be a **linear combination** of two others. Vector $\mathbf{u}$ is a linear combination of vectors $\mathbf{v}$ and $\mathbf{w}$ if there are scalars a and b so that $\mathbf{u} = a\mathbf{v} + b\mathbf{w}$. Here then, we are asked to show that there are scalars a and b so that

$$\begin{bmatrix} 7 \\ 7 \end{bmatrix} = a \begin{bmatrix} 2 \\ 3 \end{bmatrix} + b \begin{bmatrix} -1 \\ 2 \end{bmatrix}.$$

This vector equation is equivalent to the system

$$2a - b = 7$$
$$3a + 2b = 7,$$

and we find quickly that $a = 3$, $b = -1$. Now we check that $\begin{bmatrix} 7 \\ 7 \end{bmatrix} = 3 \begin{bmatrix} 2 \\ 3 \end{bmatrix} - \begin{bmatrix} -1 \\ 2 \end{bmatrix}$, which shows that $\begin{bmatrix} 7 \\ 7 \end{bmatrix}$ is indeed a linear combination of $\begin{bmatrix} 2 \\ 3 \end{bmatrix}$ and $\begin{bmatrix} -1 \\ 2 \end{bmatrix}$.

A-5 PROBLEM

Suppose $\mathbf{u}$, $\mathbf{v}$, and $\mathbf{w}$ are vectors and $\mathbf{u}$ is a scalar multiple of $\mathbf{v}$. Show that $\mathbf{u}$ is a linear combination of $\mathbf{v}$ and $\mathbf{w}$.

Solution.

1. Since $\mathbf{u}$ is a scalar multiple of $\mathbf{v}$, there is some scalar c so that $\mathbf{u} = c\mathbf{v}$.
2. Thus $\mathbf{u} = c\mathbf{v} + 0\mathbf{w}$. This is a linear combination of $\mathbf{v}$ and $\mathbf{w}$.

This solution required only two lines, but each line required knowledge of linear algebra vocabulary. First, we used the definition of "scalar multiple" to interpret the hypothesis "$\mathbf{u}$ is a scalar multiple of $\mathbf{v}$." Then, after a moment's thought and remembering the definition of "linear combination," adding the term "$0\mathbf{w}$" gave the desired conclusion.

If X is a matrix, show that the product XX^T of X and its transpose is symmetric.

The hypothesis is "X is a matrix." Not much! The desired conclusion is "XX^T is symmetric." We should know what "symmetric" means. We check the index. "Symmetric matrix" is defined on p. 147 (and also in the glossary). A matrix A is *symmetric* if and only if $A^T = A$. Here, the matrix A is XX^T. This is the matrix for which we must show that $A^T = A$.

Solution.

1. The transpose of XX^T is $(XX^T)^T = (X^T)^T X^T$ because the transpose of AB is $B^T A^T$.

2. $(X^T)^T = X$.

3. Therefore $(XX^T)^T = XX^T$ as desired.

Suppose A and B are matrices for which the product AB is defined. Show that the null space of B is contained in the null space of AB.

If you are not sure what "null space" means, refer to the index (or glossary). The null space of a matrix A is the set of all vectors $\mathbf{x}$ for which $A\mathbf{x} = \mathbf{0}$.

To show that one set (the null space of B) is contained in another (the null space of AB), we have to take something in the first set and show that it is in the other. The problem can be rephrased "If $\mathbf{x}$ is in the null space of B, then $\mathbf{x}$ is in the null space of AB." There is only one way to begin the solution.

1. Let $\mathbf{x}$ be in the null space of B.

2. Therefore $B\mathbf{x} = \mathbf{0}$.

The second line is more or less obvious. Let's show the reader we know what it means for something to belong to the null space of B! But now what to do? If you are unsure, you can always try moving to the conclusion of the proof and working upward. The last line must be "Therefore the null space of B is contained in the null space of AB," and this would follow if the previous line read "So $\mathbf{x}$ is in the null space of AB." In turn, this statement would follow from a previous statement saying "$(AB)\mathbf{x} = \mathbf{0}$." (We assure the reader again that we know the meaning of "null space.") So we have the last three lines of our proof.

$n-2.$ $(AB)\mathbf{x} = \mathbf{0}$.

$n-1.$ So $\mathbf{x}$ is in the null space of AB.

$n.$ Therefore the null space of B is contained in the null space of AB.

Altogether, what do we have?

1. Let $\mathbf{x}$ be in the null space of B.

2. Therefore $B\mathbf{x} = \mathbf{0}$.

$\vdots$

$n-2.$ $(AB)\mathbf{x} = \mathbf{0}$.

$n-1.$ So $\mathbf{x}$ is in the null space of AB.

$n.$ Therefore the null space of B is contained in the null space of AB.

It remains only to fill in the missing lines. In this case, however, there are no missing lines! Line $n-2$ follows from line 2 upon multiplication by A, using the fact that matrix multiplication is associative; $A(B\mathbf{x}) = (AB)\mathbf{x}$. Our proof is complete.

In an attempt to prove "if A, then B," an all too common error is for a student to assume at the very beginning that the desired statement B is true and, after a few steps, having reached a statement that is true, to cry "Eureka" and claim to have confirmed that B is true.

Here is a favorite example of the author's that shows the fallacy of this sort of reasoning. It makes use of the complex number i which, as the reader may know, satisfies $i^2 = -1$.

1. $1 = -1$. This is clearly false, but let's see what we can do with it.
2. $1 = i^2$, since $i^2 = -1$.
3. $\dfrac{1}{i} = \dfrac{i}{1}$, dividing by i.
4. $\dfrac{\sqrt{1}}{\sqrt{-1}} = \dfrac{\sqrt{-1}}{\sqrt{1}}$, since $1 = \sqrt{1}$ and $i = \sqrt{-1}$.
5. $\sqrt{\dfrac{1}{-1}} = \sqrt{\dfrac{-1}{1}}$.
6. $\sqrt{-1} = \sqrt{-1}$.
7. $-1 = -1$, squaring both sides.

Since the last statement is correct, so was the first. I don't think so!

Let $\mathbf{u}$ and $\mathbf{v}$ be vectors. The Cauchy–Schwarz inequality is the statement $|\mathbf{u} \cdot \mathbf{v}| \le \|\mathbf{u}\| \, \|\mathbf{v}\|$ while the triangle inequality asserts $\|\mathbf{u} + \mathbf{v}\| \le \|\mathbf{u}\| + \|\mathbf{v}\|$.

To answer a homework exercise that read

Use the Cauchy–Schwarz inequality to prove the triangle inequality,

the author once saw this "proof:"

Solution. $\|\mathbf{u} + \mathbf{v}\| \le \|\mathbf{u}\| + \|\mathbf{v}\|$.
$\|\mathbf{u} + \mathbf{v}\|^2 \le (\|\mathbf{u}\| + \|\mathbf{v}\|)^2$.
$\|\mathbf{u} + \mathbf{v}\|^2 \le \|\mathbf{u}\|^2 + 2\|\mathbf{u}\| \, \|\mathbf{v}\| + \|\mathbf{v}\|^2$.
$(\mathbf{u} + \mathbf{v}) \cdot (\mathbf{u} + \mathbf{v}) \le \mathbf{u} \cdot \mathbf{u} + 2\|\mathbf{u}\| \, \|\mathbf{v}\| + \mathbf{v} \cdot \mathbf{v}$.
$\mathbf{u} \cdot \mathbf{u} + 2\mathbf{u} \cdot \mathbf{v} + \mathbf{v} \cdot \mathbf{v} \le \mathbf{u} \cdot \mathbf{u} + 2\|\mathbf{u}\| \, \|\mathbf{v}\| + \mathbf{v} \cdot \mathbf{v}$.
Since $\mathbf{u} \cdot \mathbf{v} \le \|\mathbf{u}\| \, \|\mathbf{v}\|$ (Cauchy–Schwarz), $\|\mathbf{u} + \mathbf{v}\| \le \|\mathbf{u}\| + \|\mathbf{v}\|$. Eureka!

The last line, the desired conclusion, is hardly surprising since it is with this information that the proof began! To the author, the disappointing part of this alleged "proof" is that it contains the ingredients of a completely correct proof, that would go like this:

Solution.

$$\|\mathbf{u} + \mathbf{v}\|^2 = (\mathbf{u} + \mathbf{v}) \cdot (\mathbf{u} + \mathbf{v}) \qquad \text{since } \|\mathbf{w}\|^2 = \mathbf{w} \cdot \mathbf{w} \text{ for any vector } \mathbf{w}$$

$$= \mathbf{u} \cdot \mathbf{u} + 2\mathbf{u} \cdot \mathbf{v} + \mathbf{v} \cdot \mathbf{v} \qquad \text{expanding}$$

$$\le \mathbf{u} \cdot \mathbf{u} + 2|\mathbf{u} \cdot \mathbf{v}| + \mathbf{v} \cdot \mathbf{v} \qquad \text{since } a \le |a| \text{ for any number } a$$

$$\le \mathbf{u} \cdot \mathbf{u} + 2\,\|\mathbf{u}\|\,\|\mathbf{v}\| + \mathbf{v} \cdot \mathbf{v} \quad \text{using the Cauchy–Schwarz inequality}$$

$$= \|\mathbf{u}\|^2 + 2\,\|\mathbf{u}\|\,\|\mathbf{v}\| + \|\mathbf{v}\|^2$$

$$= (\|\mathbf{u}\| + \|\mathbf{v}\|)^2,$$

and the result now follows by taking the square root of each side.

Some time ago, the author put this question on his linear algebra exam.

> Suppose A and B are matrices and $BA = \mathbf{0}$. Show that the column space of A is contained in the null space of B.

Here is how one student answered:

1. Let y be in the column space of A.
2. Therefore $A\mathbf{x} = \mathbf{y}$ for some $\mathbf{x}$.
3. Let $\mathbf{x}$ be in the null space of B.

   $\vdots$

n. Therefore A is in the null space of B.

My student got off to a great start. We want to show that the column space of A is contained in the null space of B, so we should start with something in the column space of A and show that we know what this means. The first two lines are great, but from where did statement 3 come? It sure does not follow from anything preceding. And the last line is **not** what we wanted to show! The last three lines had to be these:

$n - 2$. $B\mathbf{y} = \mathbf{0}$.
$n - 1$. So y is in the null space of B.
   n. Therefore the column space of A is contained in the null space of B.

Here is a correct solution.

Solution.

1. Let y be in the column space of A.
2. Therefore $A\mathbf{x} = \mathbf{y}$ for some $\mathbf{x}$.
3. Multiplying by B gives $B\mathbf{y} = B(A\mathbf{x}) = (BA)\mathbf{x}$.
4. $B\mathbf{y} = \mathbf{0}$ because $BA = \mathbf{0}$.
5. So y is in the null space of B.
6. Therefore the column space of A is contained in the null space of B.

The first two lines were automatic, as were the last three. We had only to insert line 3 to close the gap between beginning and end.

Answers to Reading Challenges

1. "If a system of equations is homogeneous, then it has a solution." The hypothesis is that a system of equations is homogeneous; the conclusion is that it has a solution.

Solutions to True/False Questions and Selected Exercises

Section 1.1—True/False

1. False. $\overrightarrow{AB} = \begin{bmatrix} -4 \\ 3 \end{bmatrix}$ (the coordinates of B less those of A).

2. False. The same or opposite direction.

3. False in general. True if and only if the vectors are not parallel.

4. True. This is 1.1.20.

5. False. A linear combination of $\begin{bmatrix} 1 \\ 0 \\ 1 \end{bmatrix}$ and $\begin{bmatrix} 2 \\ 0 \\ -1 \end{bmatrix}$ is a vector of the form $a\begin{bmatrix} 1 \\ 0 \\ 1 \end{bmatrix} + b\begin{bmatrix} 2 \\ 0 \\ -1 \end{bmatrix} = \begin{bmatrix} a+2b \\ 0 \\ a-b \end{bmatrix}$. Since $\begin{bmatrix} 1 \\ 2 \\ 3 \end{bmatrix}$ does not have second component 0, it is not such a linear combination.

6. True. $\begin{bmatrix} 0 \\ 0 \end{bmatrix} = 0\begin{bmatrix} 1 \\ 2 \end{bmatrix}$.

7. True. $4\mathbf{u} - 9\mathbf{v} = 2(2\mathbf{u}) + (-\frac{9}{7})(7\mathbf{v})$.

8. False: End of $\mathbf{v}$ to end of $\mathbf{u}$.

9. True.

10. True. $\mathbf{0} = 0\mathbf{u} + 0\mathbf{v} + 0\mathbf{w}$.

11. True. Given $\mathbf{u} = \mathbf{v} + \mathbf{w}$, we have $\mathbf{w} = \mathbf{u} - \mathbf{v}$.

12. False. If $\mathbf{u} = \begin{bmatrix} 2 \\ 2 \end{bmatrix}$, $\mathbf{v} = \begin{bmatrix} 1 \\ 1 \end{bmatrix}$ and $\mathbf{w} = \begin{bmatrix} 1 \\ 0 \end{bmatrix}$, then $\mathbf{u} = 2\mathbf{v} + 0\mathbf{w}$ is a linear combination of $\mathbf{v}$ and $\mathbf{w}$, but $\mathbf{w}$ is not a linear combination of $\mathbf{u}$ and $\mathbf{v}$.

13. True. The vectors are not parallel: Neither vector is a multiple of the other.

Exercises 1.1

1. (a) $\overrightarrow{AB} = \begin{bmatrix} 3 \\ 3 \end{bmatrix}$

2. (a) Let $B = (x, y)$. Then $\overrightarrow{AB} = \begin{bmatrix} x-1 \\ y-4 \end{bmatrix} = \begin{bmatrix} -1 \\ 2 \end{bmatrix}$, so $x - 1 = -1$, $y - 4 = 2$, and $B = (0, 6)$.

3. $\mathbf{x} = -4\mathbf{u} = 0\mathbf{v} = \frac{1}{4}\mathbf{w}$.

5. **(a)** $4\begin{bmatrix}2\\-3\end{bmatrix}+2\begin{bmatrix}3\\1\end{bmatrix}=\begin{bmatrix}14\\-10\end{bmatrix}$ **(c)** $3\begin{bmatrix}2\\1\\3\end{bmatrix}-2\begin{bmatrix}1\\0\\-5\end{bmatrix}-4\begin{bmatrix}0\\-1\\2\end{bmatrix}=\begin{bmatrix}4\\7\\11\end{bmatrix}$

6. **(a)** Since $2\mathbf{u}-\mathbf{v}=\begin{bmatrix}2a-1\\-10-6+b\end{bmatrix}=\begin{bmatrix}2a-1\\-16+b\end{bmatrix}$, we must have $2a-1=3$ and $-16+b=1$, so $a=2$ and $b=17$.

7. $\mathbf{x}=\mathbf{y}+\mathbf{u}$, so $2(\mathbf{y}+\mathbf{u})+3\mathbf{y}=\mathbf{v}$. This is $5\mathbf{y}+2\mathbf{u}=\mathbf{v}$, so $5\mathbf{y}=\mathbf{v}-2\mathbf{u}$ and $\mathbf{y}=\frac{1}{5}(\mathbf{v}-2\mathbf{u})=-\frac{2}{5}\mathbf{u}+\frac{1}{5}\mathbf{v}$. So
$\mathbf{x}=\mathbf{y}+\mathbf{u}=\frac{3}{5}\mathbf{u}+\frac{1}{5}\mathbf{v}$.

8. **(a)**

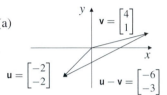

9. 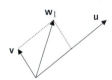 It appears that $\mathbf{w}_1=\frac{1}{2}\mathbf{u}+\mathbf{v}$.

10. **(a)** We wish to find a and b such that $\begin{bmatrix}7\\7\end{bmatrix}=a\begin{bmatrix}2\\3\end{bmatrix}+b\begin{bmatrix}-1\\2\end{bmatrix}$.

Thus we should have $\begin{array}{l}2a-b=7\\3a+2b=7\end{array}$.

The first equation gives $b=2a-7$, so substituting in the second, $3a+2(2a-7)=7$, $7a-14=7$, $a=3$, $b=-1$.

11. **(a)** The typical linear combination of $\begin{bmatrix}-2\\-2\end{bmatrix}$ and $\begin{bmatrix}3\\3\end{bmatrix}$ is a vector of the form $a\begin{bmatrix}-2\\-2\end{bmatrix}+b\begin{bmatrix}3\\3\end{bmatrix}=\begin{bmatrix}-2a+3b\\-2a+3b\end{bmatrix}$. This is a vector
both of whose components are equal, so it can never be $\begin{bmatrix}1\\2\end{bmatrix}$. The answer is "no".

12. **(a)** There are many ways to express $\begin{bmatrix}0\\0\end{bmatrix}$ as a linear combination of $\begin{bmatrix}1\\4\end{bmatrix}$ and $\begin{bmatrix}-2\\-8\end{bmatrix}$. Here are some of these.

$$\begin{bmatrix}0\\0\end{bmatrix}=0\begin{bmatrix}1\\4\end{bmatrix}+0\begin{bmatrix}-2\\-8\end{bmatrix}=2\begin{bmatrix}1\\4\end{bmatrix}+1\begin{bmatrix}-2\\-8\end{bmatrix}=4\begin{bmatrix}1\\4\end{bmatrix}+2\begin{bmatrix}-2\\-8\end{bmatrix}=20\begin{bmatrix}1\\4\end{bmatrix}+10\begin{bmatrix}-2\\-8\end{bmatrix}.$$

13. **(a)** A linear combination of $2\mathbf{u}$ and $-3\mathbf{v}$ is a vector of the form $a(2\mathbf{u})+b(-3\mathbf{v})$. This is the same as $(2a)\mathbf{u}+(-3b)\mathbf{v}$, and hence also a linear combination of $\mathbf{u}$ and $\mathbf{v}$.

14. **(a)** $\mathbf{u}=\begin{bmatrix}2\\2\end{bmatrix}$; $\mathbf{v}=\begin{bmatrix}2\\-1\end{bmatrix}$.

(b) **i.** $\overrightarrow{OD}=\begin{bmatrix}4\\1\end{bmatrix}=\mathbf{u}+\mathbf{v}$.

15. We wish to show that $\overrightarrow{DE} = \frac{1}{2}\overrightarrow{BC}$. Now
$\overrightarrow{AD} = \frac{1}{2}\overrightarrow{AB}$ and $\overrightarrow{AE} = \frac{1}{2}\overrightarrow{AC}$, so $\overrightarrow{DE} =$
$\overrightarrow{DA} + \overrightarrow{AE} = -\frac{1}{2}\overrightarrow{AB} + \frac{1}{2}\overrightarrow{AC} = \frac{1}{2}(\overrightarrow{AC} -$
$\overrightarrow{AB}) = \frac{1}{2}\overrightarrow{BC}$.

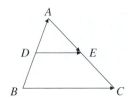

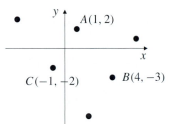

17. The picture at the right shows the position of the points. The answer is not unique. There are three possibilities for $D(x, y)$.

If $\overrightarrow{CA} = \overrightarrow{DB}$, then $\begin{bmatrix} 2 \\ 4 \end{bmatrix} = \begin{bmatrix} 4 - x \\ -3 - y \end{bmatrix}$ and D is $(2, -7)$.

If $\overrightarrow{DA} = \overrightarrow{CB}$, then $\begin{bmatrix} 1 - x \\ 2 - y \end{bmatrix} = \begin{bmatrix} 5 \\ -1 \end{bmatrix}$ and D is $(-4, 3)$.

If $\overrightarrow{AD} = \overrightarrow{CB}$, then $\begin{bmatrix} x - 1 \\ y - 2 \end{bmatrix} = \begin{bmatrix} 5 \\ -1 \end{bmatrix}$ and D is $(6, 1)$.

21. **1.** Closure under addition: Let $\mathbf{u} = \begin{bmatrix} x \\ y \end{bmatrix}$ and $\mathbf{v} = \begin{bmatrix} z \\ w \end{bmatrix}$ be vectors. Then $\mathbf{u} + \mathbf{v} = \begin{bmatrix} x + z \\ y + w \end{bmatrix}$ is a vector.

2. Commutativity of addition: Let $\mathbf{u} = \begin{bmatrix} x \\ y \end{bmatrix}$ and $\mathbf{v} = \begin{bmatrix} z \\ w \end{bmatrix}$ be vectors. Then $\mathbf{u} + \mathbf{v} = \begin{bmatrix} x + z \\ y + w \end{bmatrix}$ and $\mathbf{v} + \mathbf{u} = \begin{bmatrix} z + x \\ w + y \end{bmatrix}$. Since addition of real numbers is a commutative operation, $\mathbf{v} + \mathbf{u} = \begin{bmatrix} x + z \\ y + w \end{bmatrix} = \mathbf{u} + \mathbf{v}$.

5. Negatives: If $\mathbf{u} = \begin{bmatrix} x \\ y \end{bmatrix}$, then $-\mathbf{u} = \begin{bmatrix} -x \\ -y \end{bmatrix}$, so $\mathbf{u} + (-\mathbf{u}) = \begin{bmatrix} x \\ y \end{bmatrix} + \begin{bmatrix} -x \\ -y \end{bmatrix} = \begin{bmatrix} x - x \\ y - y \end{bmatrix} = \begin{bmatrix} 0 \\ 0 \end{bmatrix} = \mathbf{0}$. Similarly, $(-\mathbf{u}) + \mathbf{u} = \mathbf{0}$.

22. A linear combination of $\begin{bmatrix} 1 \\ \frac{3}{2} \\ 0 \end{bmatrix}$ and $\begin{bmatrix} 0 \\ 3 \\ 6 \end{bmatrix}$ is a vector of the form $a \begin{bmatrix} 1 \\ \frac{3}{2} \\ 0 \end{bmatrix} + b \begin{bmatrix} 0 \\ 3 \\ 6 \end{bmatrix}$. This is $\frac{1}{2} \begin{bmatrix} 2 \\ 3 \\ 0 \end{bmatrix} + 3b \begin{bmatrix} 0 \\ 1 \\ 2 \end{bmatrix}$, which is a linear combination of $\begin{bmatrix} 2 \\ 3 \\ 0 \end{bmatrix}$ and $\begin{bmatrix} 0 \\ 1 \\ 2 \end{bmatrix}$.

24. **(a)** Three points X, Y and Z are collinear if and only if the vector $\overrightarrow{XY}$ is a scalar multiple of the vector $\overrightarrow{XZ}$ (or of $\overrightarrow{YZ}$). (There are other similar responses that are equally correct.)

27. **(a)** We are given that $\mathbf{u}$ and $\mathbf{v}$ are parallel. Thus $\mathbf{v} = c\mathbf{u}$ or $\mathbf{u} = c\mathbf{v}$ for some scalar c. Suppose $\mathbf{v} = c\mathbf{u}$. This can be rewritten $1\mathbf{v} - c\mathbf{u} = \mathbf{0}$. This expresses $\mathbf{0}$ as a linear combination of $\mathbf{u}$ and $\mathbf{v}$ in a nontrivial way since at least one coefficient, the coefficient of $\mathbf{v}$, is not zero. The case $\mathbf{u} = c\mathbf{v}$ is similar.

Section 1.2—True/False

1. True. The length is $\sqrt{(-1)^2 + 2^2 + 2^2}$.

2. False. $|c|$ times the length of $\mathbf{v}$.

3. True. The dot product of these vectors is 0.

4. True. This is 1.2.7.

5. False. True if and only if the vectors are unit vectors. In general, $\cos\theta = \frac{\mathbf{u}\cdot\mathbf{v}}{\|\mathbf{u}\|\|\mathbf{v}\|}$.

6. True. Applying the Arithmetic Mean/Geometric Mean inequality with $a = \pi$ and $b = \pi^2$ gives $\sqrt{\pi\pi^2} \le \frac{\pi+\pi^2}{2}$.

7. True. The Cauchy–Schwarz inequality says $|\mathbf{u}\cdot\mathbf{v}| \le \|\mathbf{u}\|\,\|\mathbf{v}\|$. Since $\mathbf{u}\cdot\mathbf{v} \le |\mathbf{u}\cdot\mathbf{v}|$, the given statement also is true.

8. True. $\mathbf{u}\cdot\mathbf{u} = \|\mathbf{u}\|^2$ is the sum of the squares of the components of $\mathbf{u}$. If this is zero, each number being squared is zero, so each component of $\mathbf{u}$ is 0. This means $\mathbf{u} = \mathbf{0}$.

9. False. If $\mathbf{u}$ and $\mathbf{v}$ are unit vectors, $\mathbf{u}\cdot\mathbf{v}$ is the cosine of the angle between them, and this cannot be more than 1.

Exercises 1.2

1. $\|\mathbf{u}\| = \sqrt{0^2 + 3^2 + 4^2} = 5$, so $\frac{1}{5}\mathbf{u} = \frac{1}{5}\begin{bmatrix} 0 \\ 3 \\ 4 \end{bmatrix} = \begin{bmatrix} 0 \\ \frac{3}{5} \\ \frac{4}{5} \end{bmatrix}$ is a unit vector in the direction of $\mathbf{u}$. Since $\|\mathbf{v}\| = \sqrt{1^2 + 2^2 + 2^2} = 3$,

   $\frac{1}{3}\mathbf{v}$ is a unit vector in the direction of $\mathbf{v}$ and $-\frac{2}{3}\mathbf{v} = -\frac{2}{3}\begin{bmatrix} 1 \\ 2 \\ 2 \end{bmatrix} = \begin{bmatrix} -\frac{2}{3} \\ -\frac{4}{3} \\ -\frac{4}{3} \end{bmatrix}$ is a vector of length 2 in the direction opposite to $\mathbf{v}$.

3. (a) Since $\mathbf{u}\cdot\mathbf{v} = 0$, the vectors are orthogonal: $\theta = \frac{\pi}{2}$.

   (b) $\mathbf{u}\cdot\mathbf{v} = -3+4 = 1$, $\|\mathbf{u}\| = \sqrt{3^2+4^2} = 5$, $\|\mathbf{v}\| = \sqrt{1^2+1^2} = \sqrt{2}$, so $\cos\theta = \frac{\mathbf{u}\cdot\mathbf{v}}{\|\mathbf{u}\|\|\mathbf{v}\|} = \frac{1}{5\sqrt{2}} \approx .1414$.
   $\theta \approx \arccos(.1414) \approx 1.43$ rads $\approx 82°$.

4. (a) $\mathbf{u}\cdot\mathbf{v} = \|\mathbf{u}\|\,\|\mathbf{v}\|\cos\theta = 3\frac{\sqrt{2}}{2} = \frac{3\sqrt{2}}{2}$.

5. (a) Let θ be the angle between $\mathbf{u}$ and $\mathbf{v}$. If $\mathbf{u}\cdot\mathbf{v} = -7$, $\|\mathbf{u}\| = 3$, and $\|\mathbf{v}\| = 2$, then $\cos\theta = \frac{\mathbf{u}\cdot\mathbf{v}}{\|\mathbf{u}\|\|\mathbf{v}\|} = \frac{-7}{6}$. This is impossible since $-1 \le \cos\theta \le +1$ for any θ.

7. (a) $\|\mathbf{u}\| = \sqrt{3^2 + 4^2 + 0^2} = 5$; $\mathbf{u} + \mathbf{v} = \begin{bmatrix} 5 \\ 5 \\ 2 \end{bmatrix}$, so $\|\mathbf{u}+\mathbf{v}\| = \sqrt{5^2 + 5^2 + 2^2} = \sqrt{54}$; $\left\|\frac{\mathbf{w}}{\|\mathbf{w}\|}\right\| = 1$ (this is true for any $\mathbf{w} \ne \mathbf{0}$).

   (b) The answer is $\frac{1}{\|\mathbf{u}\|}\mathbf{u} = \frac{1}{5}\mathbf{u} = \begin{bmatrix} \frac{3}{5} \\ \frac{4}{5} \\ 0 \end{bmatrix}$.

   (c) First we find a vector of norm 1 in the direction of $\mathbf{v}$; this vector is $\frac{1}{\|\mathbf{v}\|}\mathbf{v} = \frac{1}{3}\mathbf{v}$. A vector of norm 4 in this direction

   is $\frac{4}{3}\mathbf{v}$, so a vector of norm 4 in the opposite direction is $-\frac{4}{3}\mathbf{v} = \begin{bmatrix} -\frac{8}{3} \\ -\frac{4}{3} \\ -\frac{8}{3} \end{bmatrix}$.

9. (a) We are given $\mathbf{u}\cdot\mathbf{u} = 3^2 = 9$ and $\mathbf{v}\cdot\mathbf{v} = 5^2 = 25$. So $(\mathbf{u}-\mathbf{v})\cdot(2\mathbf{u}-3\mathbf{v}) = 2\mathbf{u}\cdot\mathbf{u} - 5\mathbf{u}\cdot\mathbf{v} + 3\mathbf{v}\cdot\mathbf{v} = 2(9) - 5(8) + 3(25) = 53$.

   (f) $\|\mathbf{u}+\mathbf{v}\|^2 = (\mathbf{u}+\mathbf{v})\cdot(\mathbf{u}+\mathbf{v}) = \mathbf{u}\cdot\mathbf{u} + 2\mathbf{u}\cdot\mathbf{v} + \mathbf{v}\cdot\mathbf{v} = 9 + 2(8) + 25 = 50$.

11. (a) $\mathbf{u} + k\mathbf{v} = \begin{bmatrix} 1+k \\ 2-k \end{bmatrix}$, so

$$\|\mathbf{u} + k\mathbf{v}\| = \sqrt{(1+k)^2 + (2-k)^2} = \sqrt{k^2 + 2k + 1 + 4 - 4k + k^2} = \sqrt{2k^2 - 2k + 5}.$$

We want $2k^2 - 2k + 5 = 3^2$, so $2k^2 - 2k - 4 = 0$, $k^2 - k - 2 = 0$, $(k+1)(k-2) = 0$. Thus $k = -1, 2$.

13. $(\mathbf{u} + k\mathbf{v}) \cdot \mathbf{u} = \mathbf{u} \cdot \mathbf{u} + k\mathbf{v} \cdot \mathbf{u} = \mathbf{u} \cdot \mathbf{u}$ since $\mathbf{v} \cdot \mathbf{u} = 0$. Since $\mathbf{u} \neq \mathbf{0}$, we know that $\mathbf{u} \cdot \mathbf{u} \neq 0$, so $(\mathbf{u} + k\mathbf{v}) \cdot \mathbf{u} \neq 0$.

15. There are two possible answers, as the figure
to the right shows. Since $\overrightarrow{AB} = \begin{bmatrix} 3 \\ 3 \end{bmatrix} = 3\begin{bmatrix} 1 \\ 1 \end{bmatrix}$

and we wish $\overrightarrow{AB} \cdot \overrightarrow{AC} = 0$, we must have
$\overrightarrow{AC} = t\begin{bmatrix} -1 \\ 1 \end{bmatrix}$ for some t. Since
$\left\| \overrightarrow{AC} \right\| = |t|\sqrt{2}$, we want $|t|\sqrt{2} = 2$, so
$|t| = \sqrt{2}$.

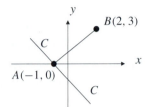

If $t = \sqrt{2}$ and $C = (x, y)$, then $\overrightarrow{AC} = \begin{bmatrix} x+1 \\ y \end{bmatrix} = \sqrt{2}\begin{bmatrix} -1 \\ 1 \end{bmatrix}$, giving $x + 1 = -\sqrt{2}$, $y = \sqrt{2}$, and $C = (-1 - \sqrt{2}, \sqrt{2})$. If
$t = -\sqrt{2}$ and $C = (x, y)$, then $\overrightarrow{AC} = \begin{bmatrix} x+1 \\ y \end{bmatrix} = -\sqrt{2}\begin{bmatrix} -1 \\ 1 \end{bmatrix}$, giving $x + 1 = \sqrt{2}$, $y = -\sqrt{2}$, and $C = (-1 + \sqrt{2}, -\sqrt{2})$.

17. The picture at the right shows the
approximate position of the points. There is
only one possible location for D. Let D have
coordinates (x, y).

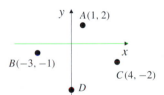

Since $\overrightarrow{CA} = \begin{bmatrix} -3 \\ 4 \end{bmatrix} = \overrightarrow{DB}$, we need $\begin{bmatrix} -3 \\ 4 \end{bmatrix} = \begin{bmatrix} -3-x \\ -1-y \end{bmatrix}$, $-3 - x = -3$ and $4 = -1 - y$. Thus $x = 0$, $y = -5$, and D is $(0, -5)$.

19. (a) $\mathbf{u}$ is orthogonal to $\mathbf{v}$ since $\mathbf{u} \cdot \mathbf{v} = 12 - 8 - 4 = 0$.
$\mathbf{u}$ is orthogonal to $\mathbf{w}$ since $\mathbf{u} \cdot \mathbf{w} = -6 + 20 - 14 = 0$.

20. 1. Commutativity: Let $\mathbf{u} = \begin{bmatrix} x \\ y \end{bmatrix}$ and $\mathbf{v} = \begin{bmatrix} z \\ w \end{bmatrix}$ be vectors. Then $\mathbf{u} \cdot \mathbf{v} = xz + yw$ and $\mathbf{v} \cdot \mathbf{u} = zx + wy = xz + yw$, so
$\mathbf{u} \cdot \mathbf{v} = \mathbf{v} \cdot \mathbf{u}$.

5. For any vector $\mathbf{v} = \begin{bmatrix} x \\ y \end{bmatrix}$, we have $\mathbf{v} \cdot \mathbf{v} = x^2 + y^2$, the sum of squares of real numbers. Since $a^2 \geq 0$ for any real
number a, $x^2 + y^2 \geq 0$. Moreover, $x^2 + y^2$ is positive if either $x \neq 0$ or $y \neq 0$; that is, $x^2 + y^2 = 0$ if and only if
$x = y = 0$. This says precisely that $\|\mathbf{v}\| = 0$ if and only if $\mathbf{v} = \mathbf{0}$.

22. (a) We have $\mathbf{u} \cdot \mathbf{v} = -6 + 5 = -1$, $\|\mathbf{u}\| = \sqrt{5}$ and $\|\mathbf{v}\| = \sqrt{34}$. The Cauchy–Schwarz inequality says that $1 \leq \sqrt{170}$.

24. Let $\mathbf{u} = \begin{bmatrix} a \\ b \end{bmatrix}$ and $\mathbf{v} = \begin{bmatrix} \cos\theta \\ \sin\theta \end{bmatrix}$. Then $\mathbf{u} \cdot \mathbf{v} = a\cos\theta + b\sin\theta$, $\|\mathbf{u}\|^2 = a^2 + b^2$ and $\|\mathbf{v}\|^2 = \cos^2\theta + \sin^2\theta = 1$. Squaring the
Cauchy–Schwarz inequality gives $(\mathbf{u} \cdot \mathbf{v})^2 \leq \|\mathbf{u}\|^2 \|\mathbf{v}\|^2$, which is the given inequality.

25. (a) Let $\mathbf{u} = \begin{bmatrix} \sqrt{3}a \\ b \end{bmatrix}$ and $\mathbf{v} = \begin{bmatrix} \sqrt{3}c \\ d \end{bmatrix}$. We have $\mathbf{u} \cdot \mathbf{v} = 3ac + bd$, $\|\mathbf{u}\| = \sqrt{3a^2 + b^2}$, and $\|\mathbf{v}\| = \sqrt{3c^2 + d^2}$, so the result
follows upon squaring the Cauchy–Schwarz inequality: $(\mathbf{u} \cdot \mathbf{v})^2 \leq \|\mathbf{u}\|^2 \|\mathbf{v}\|^2$.

27. If $\mathbf{u}$ and $\mathbf{v}$ had the same length, the vector $\mathbf{u} + \mathbf{v}$ represented by the diagonal of the parallelogram with sides $\mathbf{u}$ and $\mathbf{v}$ would do.

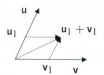

Since we have no information about lengths, we replace $\mathbf{u}$ and $\mathbf{v}$ by unit vectors $\mathbf{u}_1, \mathbf{u}_2$ with the same directions. So a suitable vector is $\mathbf{u}_1 + \mathbf{v}_1 = \frac{1}{\|\mathbf{u}\|}\mathbf{u} + \frac{1}{\|\mathbf{v}\|}\mathbf{v}$.

30. (a) The diagonals are $\overrightarrow{AC} = \mathbf{u} + \mathbf{v}$ and $\overrightarrow{DB} = \mathbf{u} - \mathbf{v}$.

Section 1.3—True/False

1. False. A line (in the xy-plane) is the *graph* of such an equation.

2. False. $\mathbf{u} \times \mathbf{v}$ is a vector and the absolute value of a vector is meaningless.

3. False. $\mathbf{u} \cdot \mathbf{v}$ is a number, and the length of a number is meaningless.

4. False. Removing the absolute value signs on the left, we obtain the correct formula $\mathbf{u} \cdot \mathbf{v} = \|\mathbf{u}\| \, \|\mathbf{v}\| \cos\theta$.

5. True. Since $\|\mathbf{u} \times \mathbf{v}\| = \|\mathbf{u}\| \, \|\mathbf{v}\| \sin\theta$, the given statement says $\sin\theta = 1$, so $\theta = \frac{\pi}{2}$ and the vectors are orthogonal.

6. False. The norm of a vector cannot be negative.

7. True. The coordinates of the point satisfy the equation.

8. False. The line has direction normal to the plane, so the line is perpendicular to the plane and certainly hits it!

9. True. The normal vectors $\begin{bmatrix} -2 \\ 3 \\ -1 \end{bmatrix}$ and $\begin{bmatrix} 4 \\ -6 \\ 2 \end{bmatrix}$ are parallel.

10. False. The line contains the point $(-1, 1, 2)$ but the plane does not.

11. False. $\begin{vmatrix} 2 & -3 \\ 5 & 4 \end{vmatrix} = 8 - (-15) = 23.$

12. False. If $\mathbf{u} \times \mathbf{v} = \mathbf{0}$, then $\mathbf{u}$ and $\mathbf{v}$ are parallel.

13. True. $\mathbf{i} \times (\mathbf{i} \times \mathbf{j}) = \mathbf{i} \times \mathbf{k} = -\mathbf{j}$.

14. True. If the dot product is 0, the direction and normal are perpendicular.

Exercises 1.3

1. (a) $\begin{bmatrix} x \\ y \\ z \end{bmatrix}$ is in the plane if and only if $z = -3x + 2y$. So the plane consists of all vectors of the form

$$\begin{bmatrix} x \\ y \\ -3x + 2y \end{bmatrix} = x \begin{bmatrix} 1 \\ 0 \\ -3 \end{bmatrix} + y \begin{bmatrix} 0 \\ 1 \\ 2 \end{bmatrix}. \text{ This says that every vector in the plane is a linear combination of } \begin{bmatrix} 1 \\ 0 \\ -3 \end{bmatrix} \text{ and } \begin{bmatrix} 0 \\ 1 \\ 2 \end{bmatrix}.$$

2. (a) The plane consists of all vectors of the form $a \begin{bmatrix} 1 \\ 0 \\ 4 \end{bmatrix} + b \begin{bmatrix} -1 \\ 2 \\ 0 \end{bmatrix} = \begin{bmatrix} a - b \\ 2b \\ 4a \end{bmatrix}$. Setting $x = a - b$, $y = 2b$, and $z = 4a$, we have $a = \frac{1}{4}z$ and $b = \frac{1}{2}y$, so $x = a - b = \frac{1}{4}z - \frac{1}{2}y$. The equation is $4x + 2y - z = 0$.

3. (a) The answer is an equation of the form $4x + 2y - z = d$. Substituting $x = 1$, $y = 2$, $z = 3$, we get $d = 4 + 4 - 3 = 5$, so $4x + 2y - z = 5$ works.

4. (a) $\mathbf{u} \times \mathbf{v} = \begin{vmatrix} \mathbf{i} & \mathbf{j} & \mathbf{k} \\ 3 & 0 & 2 \\ -2 & 4 & 1 \end{vmatrix} = \begin{vmatrix} 0 & 2 \\ 4 & 1 \end{vmatrix} \mathbf{i} - \begin{vmatrix} 3 & 2 \\ -2 & 1 \end{vmatrix} \mathbf{j} + \begin{vmatrix} 3 & 0 \\ -2 & 4 \end{vmatrix} \mathbf{k} = -8\mathbf{i} - 7\mathbf{j} + 12\mathbf{k} = \begin{bmatrix} -8 \\ -7 \\ 12 \end{bmatrix}.$

$$\mathbf{v} \times \mathbf{u} = \begin{vmatrix} \mathbf{i} & \mathbf{j} & \mathbf{k} \\ -2 & 4 & 1 \\ 3 & 0 & 2 \end{vmatrix} = \begin{vmatrix} 4 & 1 \\ 0 & 2 \end{vmatrix} \mathbf{i} - \begin{vmatrix} -2 & 1 \\ 3 & 2 \end{vmatrix} \mathbf{j} + \begin{vmatrix} -2 & 4 \\ 3 & 0 \end{vmatrix} \mathbf{k} = 8\mathbf{i} - (-7)\mathbf{j} - 12\mathbf{k} = \begin{bmatrix} 8 \\ 7 \\ -12 \end{bmatrix} = -(\mathbf{u} \times \mathbf{v}).$$

5. (a) $\mathbf{u} \times \mathbf{v} = \begin{vmatrix} \mathbf{i} & \mathbf{j} & \mathbf{k} \\ 1 & -2 & 1 \\ 3 & 1 & -2 \end{vmatrix} = \begin{vmatrix} -2 & 1 \\ 1 & -2 \end{vmatrix} \mathbf{i} - \begin{vmatrix} 1 & 1 \\ 3 & -2 \end{vmatrix} \mathbf{j} + \begin{vmatrix} 1 & -2 \\ 3 & 1 \end{vmatrix} \mathbf{k} = 3\mathbf{i} - (-5)\mathbf{j} + 7\mathbf{k} = \begin{bmatrix} 3 \\ 5 \\ 7 \end{bmatrix}.$ To verify that

this is indeed orthogonal to $\mathbf{u}$ and to $\mathbf{v}$, we compute $(\mathbf{u} \times \mathbf{v}) \cdot \mathbf{u} = \begin{bmatrix} 3 \\ 5 \\ 7 \end{bmatrix} \cdot \begin{bmatrix} 1 \\ -2 \\ 1 \end{bmatrix} = 3 - 10 + 7 = 0$ and

$$(\mathbf{u} \times \mathbf{v}) \cdot \mathbf{v} = \begin{bmatrix} 3 \\ 5 \\ 7 \end{bmatrix} \cdot \begin{bmatrix} 3 \\ 1 \\ -2 \end{bmatrix} = 9 + 5 - 14 = 0.$$

Finally, we have $\mathbf{v} \times \mathbf{u} = \begin{vmatrix} \mathbf{i} & \mathbf{j} & \mathbf{k} \\ 3 & 1 & -2 \\ 1 & -2 & 1 \end{vmatrix} = \begin{vmatrix} 1 & -2 \\ -2 & 1 \end{vmatrix} \mathbf{i} - \begin{vmatrix} 3 & -2 \\ 1 & 1 \end{vmatrix} \mathbf{j} + \begin{vmatrix} 3 & 1 \\ 1 & -2 \end{vmatrix} \mathbf{k}$

$$= -3\mathbf{i} - 5\mathbf{j} + (-7)\mathbf{k} = \begin{bmatrix} -3 \\ -5 \\ -7 \end{bmatrix} = -(\mathbf{u} \times \mathbf{v}).$$

7. One vector perpendicular to both $\mathbf{u}$ and $\mathbf{v}$ is

$$\mathbf{u} \times \mathbf{v} = \begin{vmatrix} \mathbf{i} & \mathbf{j} & \mathbf{k} \\ 1 & 2 & 0 \\ 0 & 0 & 3 \end{vmatrix} = \begin{vmatrix} 2 & 0 \\ 0 & 3 \end{vmatrix} \mathbf{i} - \begin{vmatrix} 1 & 0 \\ 0 & 3 \end{vmatrix} \mathbf{j} + \begin{vmatrix} 1 & 2 \\ 0 & 0 \end{vmatrix} \mathbf{k} = 6\mathbf{i} - 3\mathbf{j} + 0\mathbf{k} = \begin{bmatrix} 6 \\ -3 \\ 0 \end{bmatrix}.$$

Any multiple of this vector, for instance $\begin{bmatrix} 2 \\ -1 \\ 0 \end{bmatrix}$, is another.

9. Call the given vectors $\mathbf{u} = \begin{bmatrix} 1 \\ 3 \\ -1 \end{bmatrix}$ and $\mathbf{v} = \begin{bmatrix} 5 \\ 0 \\ 1 \end{bmatrix}$. One vector perpendicular to both $\mathbf{u}$ and $\mathbf{v}$ is

$$\mathbf{u} \times \mathbf{v} = \begin{vmatrix} \mathbf{i} & \mathbf{j} & \mathbf{k} \\ 1 & 3 & -1 \\ 5 & 0 & 1 \end{vmatrix} = \begin{vmatrix} 3 & -1 \\ 0 & 1 \end{vmatrix} \mathbf{i} - \begin{vmatrix} 1 & -1 \\ 5 & 1 \end{vmatrix} \mathbf{j} + \begin{vmatrix} 1 & 3 \\ 5 & 0 \end{vmatrix} \mathbf{k}$$

$$= 3\mathbf{i} - 6\mathbf{j} - 15\mathbf{k} = \begin{bmatrix} 3 \\ -6 \\ -15 \end{bmatrix}.$$

Any multiple of this is also perpendicular to both, for instance, $\mathbf{n} = \begin{bmatrix} 1 \\ -2 \\ -5 \end{bmatrix}$, a vector of length $\sqrt{1 + 4 + 25} = \sqrt{30}$.

Now $\frac{1}{\sqrt{30}}\mathbf{n}$ has length one and $\frac{5}{\sqrt{30}}\mathbf{n}$ has length five. One answer is $\frac{5}{\sqrt{30}} \begin{bmatrix} 1 \\ -2 \\ -5 \end{bmatrix}$. The only other possibility is the negative

of this vector.

10. (a) i. $\mathbf{u} \times \mathbf{v} = \begin{vmatrix} \mathbf{i} & \mathbf{j} & \mathbf{k} \\ 1 & 2 & 3 \\ 0 & -1 & 1 \end{vmatrix} = \begin{vmatrix} 2 & 3 \\ -1 & 1 \end{vmatrix} \mathbf{i} - \begin{vmatrix} 1 & 3 \\ 0 & 1 \end{vmatrix} \mathbf{j} + \begin{vmatrix} 1 & 2 \\ 0 & -1 \end{vmatrix} \mathbf{k} = 5\mathbf{i} - 1\mathbf{j} - 1\mathbf{k} = \begin{bmatrix} 5 \\ -1 \\ -1 \end{bmatrix}.$

Thus $\|\mathbf{u} \times \mathbf{v}\| = \sqrt{25 + 1 + 1} = \sqrt{27}$.

Also $\|\mathbf{u}\| = \sqrt{1 + 4 + 9} = \sqrt{14}$, $\|\mathbf{v}\| = \sqrt{0 + 1 + 1} = \sqrt{2}$ and $\mathbf{u} \cdot \mathbf{v} = 0 - 2 + 3 = 1$. The cosine of the angle θ between $\mathbf{u}$ and $\mathbf{v}$ is $\cos \theta = \frac{\mathbf{u} \cdot \mathbf{v}}{\|\mathbf{u}\| \|\mathbf{v}\|} = \frac{1}{\sqrt{28}}$. So $\sin^2 \theta = 1 - \cos^2 \theta = \frac{27}{28}$. Assuming $0 \le \theta \le \pi$, $\sin \theta \ge 0$, so

$\sin \theta = \sqrt{\frac{27}{28}}$. Thus $\|\mathbf{u}\| \|\mathbf{v}\| \sin \theta = \sqrt{14}\sqrt{2}\sqrt{\frac{27}{28}} = \sqrt{27} = \|\mathbf{u} \times \mathbf{v}\|$.

11. (a) $\overrightarrow{AB} = \begin{bmatrix} 1 \\ -2 \\ 2 \end{bmatrix}$ and $\overrightarrow{AC} = \begin{bmatrix} -2 \\ 1 \\ -7 \end{bmatrix}$. A normal vector is $\overrightarrow{AB} \times \overrightarrow{AC} = \begin{vmatrix} \mathbf{i} & \mathbf{j} & \mathbf{k} \\ 1 & -2 & 2 \\ -2 & 1 & -7 \end{vmatrix} = 12\mathbf{i} + 3\mathbf{j} - 3\mathbf{k} = \begin{bmatrix} 12 \\ 3 \\ -3 \end{bmatrix}$. We

take $\mathbf{n} = \begin{bmatrix} 4 \\ 1 \\ -1 \end{bmatrix}$. The equation of the plane is $4x + y - z = d$. Since the coordinates of A satisfy the equation, we

have $8 + 1 - 3 = d$, so $d = 6$ and the equation is $4x + y - z = 6$.

12. (a) Setting $t = 0$ and then $t = 1$, we see that $A(1, 0, 2)$ and $B(-3, 3, 3)$ are also on the plane. The plane contains the

arrows from P to A and from P to B, hence it is parallel to the vectors $\begin{bmatrix} -2 \\ -1 \\ 3 \end{bmatrix}$ and $\begin{bmatrix} -6 \\ 2 \\ 4 \end{bmatrix}$. So a normal vector is

$\mathbf{n} = \begin{vmatrix} \mathbf{i} & \mathbf{j} & \mathbf{k} \\ -2 & -1 & 3 \\ -6 & 2 & 4 \end{vmatrix} = \begin{bmatrix} -10 \\ -10 \\ -10 \end{bmatrix}$. The vector $\begin{bmatrix} 1 \\ 1 \\ 1 \end{bmatrix}$ is equally good. The equation of the plane is $x + y + z = d$.

Substituting the coordinates of P gives $d = 3$, so we get $x + y + z = 3$.

13. (a) The direction of the line is $\begin{bmatrix} -1 \\ 1 \end{bmatrix}$, so the slope is -1 and the equation is of the form $y = -x + b$. The point $(1, 2)$

lies on the line, so $2 = -1 + b$, $b = 3$. The line has equation $x + y = 3$.

14. (a) The line has slope 2, so $\begin{bmatrix} 1 \\ 2 \end{bmatrix}$ is a direction vector. The line contains $(0, -3)$, so the vector equation is

$\begin{bmatrix} x \\ y \end{bmatrix} = \begin{bmatrix} 0 \\ -3 \end{bmatrix} + t \begin{bmatrix} 1 \\ 2 \end{bmatrix}$.

15. (a) We need triples (x, y, z) that satisfy both equations. Setting $x = 0$ gives $y + z = 5$, $-y + z = 1$, so $z = 3$ and $y = 2$. This gives the point $A(0, 2, 3)$. Setting $y = 0$ gives $2x + z = 5$, $x + z = 1$, so $x = 4$ and $z = -3$. This gives the point $B(4, 0, -3)$. Setting $z = 0$ gives $2x + y = 5$, $x - y = 1$, so $x = 2$ and $y = 1$. This gives the point $C(2, 1, 0)$. Many other points are possible, of course.

(b) Solution 1. The line has direction $\overrightarrow{AB} = \begin{bmatrix} 4 \\ -2 \\ -6 \end{bmatrix}$ and hence also $\begin{bmatrix} 2 \\ -1 \\ -3 \end{bmatrix}$. A vector equation is $\begin{bmatrix} x \\ y \\ z \end{bmatrix} = \begin{bmatrix} 0 \\ 2 \\ 3 \end{bmatrix} + t \begin{bmatrix} 2 \\ -1 \\ -3 \end{bmatrix}$.

Solution 2. The line has direction perpendicular to the normal vector of each plane, so the cross product of the

normal vectors gives a direction vector. This cross product is $\begin{vmatrix} \mathbf{i} & \mathbf{j} & \mathbf{k} \\ 2 & 1 & 1 \\ 1 & -1 & 1 \end{vmatrix} = 2\mathbf{i} - \mathbf{j} - 3\mathbf{k} = \begin{bmatrix} 2 \\ -1 \\ -3 \end{bmatrix}$. As in Solution 1,

we still need to find a point on the line; $A(0, 2, 3)$ will do. The equation is $\begin{bmatrix} x \\ y \\ z \end{bmatrix} = \begin{bmatrix} 0 \\ 2 \\ 3 \end{bmatrix} + t \begin{bmatrix} 2 \\ -1 \\ -3 \end{bmatrix}$, as before.

[Solutions to this question may look different from the one we have obtained, of course. A correct answer must be

the equation of a line whose direction vector is a multiple of $\begin{bmatrix} 2 \\ -1 \\ -3 \end{bmatrix}$ and passes through a point whose coordinates

satisfy the equation of each plane.]

17. If (x, y, z) is on the line, then, for some t, $x = 1 + 2t$, $y = -2 + 5t$, $z = 3 - t$. If such a point is on the plane, then $4 = x - 3y + 2z = (1 + 2t) - 3(-2 + 5t) + 2(3 - t) = 13 - 15t$, so $15t = 9$, $t = 3/5$ and $(x, y, z) = (11/5, 1, 12/5)$.

18. (a) We are given that ℓ_1 has equation $\begin{bmatrix} x \\ y \\ z \end{bmatrix} = \begin{bmatrix} 2 \\ -3 \\ -4 \end{bmatrix} + t \begin{bmatrix} 1 \\ 2 \\ 3 \end{bmatrix}$. The line ℓ_2 has equation $\begin{bmatrix} x \\ y \\ z \end{bmatrix} = \begin{bmatrix} 2 \\ -2 \\ 5 \end{bmatrix} + t \begin{bmatrix} 2 \\ 3 \\ -3 \end{bmatrix}$. If they

intersect, at the point (x, y, z), then there exist t and s such that

$$
\begin{array}{ccc}
x = 2 + t & & x = 2 + 2s \\
y = -3 + 2t & \text{and} & y = -2 + 3s \\
z = -4 + 3t & & z = 5 - 3s
\end{array}
$$

We find that $s = 1$, $t = 2$, so the lines intersect at $(4, 1, 2)$.

19. (a) $\begin{bmatrix} 2 \\ -3 \\ 5 \end{bmatrix}$ is a normal and hence perpendicular to π.

20. (a) Since $\overrightarrow{AB} = \begin{bmatrix} 1 \\ 1 \\ 0 \end{bmatrix} = \overrightarrow{CD}$, $ABCD$ is a parallelogram. Since

$$
\overrightarrow{AB} \times \overrightarrow{AC} = = \begin{bmatrix} \mathbf{i} & \mathbf{j} & \mathbf{k} \\ 1 & 1 & 0 \\ 1 & -4 & 4 \end{bmatrix} = 4\mathbf{i} - 4\mathbf{j} - 5\mathbf{k} = \begin{bmatrix} 4 \\ -4 \\ -5 \end{bmatrix},
$$

the area is $\left\| \overrightarrow{AB} \times \overrightarrow{AC} \right\| = \sqrt{57}$.

(b) The area of the triangle is $\frac{1}{2}\sqrt{57}$, one-half the area of the parallelogram.

21. Two sides of the triangle are $\overrightarrow{AB} = \begin{bmatrix} 6 \\ 1 \end{bmatrix}$ and $\overrightarrow{AC} = \begin{bmatrix} 7 \\ -6 \end{bmatrix}$. Think of these as lying in 3-space. The triangle has area one half the area of the parallelogram with sides $\mathbf{u} = \begin{bmatrix} 6 \\ 1 \\ 0 \end{bmatrix}$ and $\mathbf{v} = \begin{bmatrix} 7 \\ -6 \\ 0 \end{bmatrix}$. This is one half the length of

$$
\mathbf{u} \times \mathbf{v} = \begin{bmatrix} \mathbf{i} & \mathbf{j} & \mathbf{k} \\ 6 & 1 & 0 \\ 7 & -6 & 0 \end{bmatrix} = 0\mathbf{i} - 0\mathbf{j} - 43\mathbf{k} = \begin{bmatrix} 0 \\ 0 \\ -43 \end{bmatrix}.
$$

The area is $\frac{1}{2}\sqrt{(-43)^2} = \frac{43}{2}$.

22. (a) Since $ad - bc = 0$, we have $ad = bc$.

If $a \neq 0$ and $d \neq 0$, then $b \neq 0$, so $\dfrac{c}{a} = \dfrac{d}{b} = k$. This says $c = ka$ and $d = kb$, so $\begin{bmatrix} c \\ d \end{bmatrix} = k \begin{bmatrix} a \\ b \end{bmatrix}$.

Suppose $a = 0$ and $d \neq 0$. Then $b = 0$ or $c = 0$. In the case $b = 0$, then $\begin{bmatrix} a \\ b \end{bmatrix} = \begin{bmatrix} 0 \\ 0 \end{bmatrix} = 0 \begin{bmatrix} c \\ d \end{bmatrix}$, while if $c = 0$ (and

$b \neq 0$), then $\begin{bmatrix} c \\ d \end{bmatrix} = \begin{bmatrix} 0 \\ d \end{bmatrix} = \dfrac{d}{b} \begin{bmatrix} 0 \\ b \end{bmatrix}$.

Suppose $a \neq 0$ and $d = 0$. Again, $b = 0$ or $c = 0$. In the case $b = 0$, then $\begin{bmatrix} c \\ d \end{bmatrix} = \begin{bmatrix} c \\ 0 \end{bmatrix} = \dfrac{c}{a} \begin{bmatrix} a \\ 0 \end{bmatrix} = [c]a\begin{bmatrix} a \\ b \end{bmatrix}$, while if

$c = 0$ (and $b \neq 0$), then $\begin{bmatrix} c \\ d \end{bmatrix} = \begin{bmatrix} 0 \\ 0 \end{bmatrix} = 0 \begin{bmatrix} a \\ b \end{bmatrix}$.

Finally, if $a = d = 0$, then $b = 0$ or $c = 0$. If $b = 0$, then $\begin{bmatrix} a \\ b \end{bmatrix} = \begin{bmatrix} 0 \\ 0 \end{bmatrix} = 0\begin{bmatrix} c \\ d \end{bmatrix}$, while if $c = 0$ (and $b \neq 0$), then

$$\begin{bmatrix} c \\ d \end{bmatrix} = \begin{bmatrix} 0 \\ d \end{bmatrix} = \frac{d}{b}\begin{bmatrix} 0 \\ b \end{bmatrix} = \frac{d}{b}\begin{bmatrix} a \\ b \end{bmatrix}.$$

In every case, one of the two given vectors is a multiple of the other.

23. Since the lines are perpendicular, the direction vectors must have dot product 0:

$$\begin{bmatrix} -1 \\ 2 \\ 1 \end{bmatrix} \cdot \begin{bmatrix} 1 \\ -1 \\ b \end{bmatrix} = 0 \text{ gives } -1 - 2 + b = 0, \text{ so } b = 3. \text{ Since the lines intersect, there are values of } t \text{ and } s \text{ such that}$$

$$2 - t = 3 + s$$

$$-1 + 2t = 1 - s$$

$$3 + t = a + bs = a + 3s.$$

Adding the first two equations gives $1 + t = 4$, so $t = 3$ and $3 + s = 2 - t = -1$, so $s = -4$. Substituting $t = 3$, $s = -4$ in the third equation gives $6 = a - 12$, so $a = 18$.

25. Solution 1. The point $A(x, y, z)$ is on the plane if and only if $\left\| \overrightarrow{PA} \right\| = \left\| \overrightarrow{QA} \right\|$; that is, if and only if

$$\sqrt{(x-2)^2 + (y+1)^2 + (z-3)^2} = \sqrt{(x-1)^2 + (y-1)^2 + (z+1)^2};$$

that is, if and only if

$$x^2 - 4x + 4 + y^2 + 2y + 1 + z^2 - 6z + 9$$

$$= x^2 - 2x + 1 + y^2 - 2y + 1 + z^2 + 2z + 1.$$

We obtain $-2x + 4y - 8z + 11 = 0$.

Solution 2. Imagine a slingshot. When you put a stone in the middle and pull it back, the length of the rubber band is the same from the stone to either side of the handle and when shot, the stone moves right down the plane in question.

The line joining the points is perpendicular to the plane. So we can take $\mathbf{n} = \overrightarrow{PQ} = \begin{bmatrix} -1 \\ 2 \\ -4 \end{bmatrix}$ and have $-x + 2y - 4z = d$.

The midpoint on the line joining P and Q is on the plane; this is the point $(\frac{1}{2}(2+1), \frac{1}{2}(-1+1), \frac{1}{2}(3-1))$, that is, $(\frac{3}{2}, 0, 1)$. Thus $d = -\frac{3}{2} - 4 = -\frac{11}{2}$. The equation is $-x + 2y - 4z = -\frac{11}{2}$, or $-2x + 4y - 8z + 11 = 0$, as before.

Section 1.4—True/False

1. False. The vectors $\mathbf{e}$ and $\mathbf{f}$ must be *orthogonal*. See Theorem 1.4.13.

2. False. The projection of a vector $\mathbf{u}$ on a nonzero vector $\mathbf{v}$ is a multiple of $\mathbf{v}$. The $\mathbf{u}$ at the end should be $\mathbf{v}$.

3. False. If $\mathbf{u}$ and $\mathbf{v}$ are orthogonal, the projection of $\mathbf{u}$ on $\mathbf{v}$ is the zero vector.

4. True.

5. True. The point $(-2, 3, 1)$ is on the plane.

6. False. The line is perpendicular to the plane since its direction is a normal to the plane.

Exercises 1.4

1. We have $\text{proj}_\mathbf{v}\,\mathbf{u} = c\mathbf{v}$, with $c = \frac{\mathbf{u}\cdot\mathbf{v}}{\mathbf{v}\cdot\mathbf{v}}$. If this vector has direction opposite $\mathbf{v}$, the scalar c is negative. Since the denominator of c is positive, the numerator $\mathbf{u}\cdot\mathbf{v}$ is negative. Since $\mathbf{u}\cdot\mathbf{v} = \|\mathbf{u}\|\,\|\mathbf{v}\|\cos\theta$, we must have $\cos\theta < 0$. Thus the angle θ between $\mathbf{u}$ and $\mathbf{v}$ is *obtuse*: $\frac{\pi}{2} < \theta < \pi$.

2. (a) The plane spanned by vectors is the set of all linear combinations of the vectors. Since $\mathbf{0} = 0\mathbf{u} + 0\mathbf{v}$, the zero vector is always in the plane spanned by vectors.

3. (a) The projection of $\mathbf{u}$ on $\mathbf{v}$ is $\text{proj}_\mathbf{v}\,\mathbf{u} = \dfrac{\mathbf{u}\cdot\mathbf{v}}{\mathbf{v}\cdot\mathbf{v}}\,\mathbf{v} = \dfrac{3+2-1}{9+1+1}\begin{bmatrix}3\\1\\1\end{bmatrix} = \frac{4}{11}\begin{bmatrix}3\\1\\1\end{bmatrix}$.

   The projection of $\mathbf{v}$ on $\mathbf{u}$ is $\text{proj}_\mathbf{u}\,\mathbf{v} = \dfrac{\mathbf{v}\cdot\mathbf{u}}{\mathbf{u}\cdot\mathbf{u}}\,\mathbf{u} = \dfrac{3+2-1}{1+4+1}\begin{bmatrix}1\\2\\-1\end{bmatrix} = \frac{2}{3}\begin{bmatrix}1\\2\\-1\end{bmatrix}$.

   (c) The projection of $\mathbf{u}$ on $\mathbf{v}$ is $\text{proj}_\mathbf{v}\,\mathbf{u} = \dfrac{\mathbf{u}\cdot\mathbf{v}}{\mathbf{v}\cdot\mathbf{v}}\,\mathbf{v} = -\frac{5}{29}\begin{bmatrix}3\\4\\-2\end{bmatrix}$.

   The projection of $\mathbf{v}$ on $\mathbf{u}$ is $\text{proj}_\mathbf{u}\,\mathbf{v} = \dfrac{\mathbf{v}\cdot\mathbf{u}}{\mathbf{u}\cdot\mathbf{u}}\,\mathbf{u} = -\frac{5}{2}\begin{bmatrix}-1\\0\\1\end{bmatrix}$.

4. (a) With reference to the diagram in the text for Problem 1.4.6, we let Q be any point on the line, say $Q(1,2,3)$. Let $\mathbf{w} = \overrightarrow{QP} = \begin{bmatrix}-2\\0\\-2\end{bmatrix}$. The line has direction $\mathbf{d} = \begin{bmatrix}1\\1\\1\end{bmatrix}$ and the distance we want is the length of $\mathbf{w} - \mathbf{p}$, where

   $\mathbf{p} = \text{proj}_\mathbf{d}\,\mathbf{w}$ is the projection of $\mathbf{w}$ on $\mathbf{d}$. We have $\mathbf{p} = \text{proj}_\mathbf{d}\,\mathbf{w} = \dfrac{\mathbf{w}\cdot\mathbf{d}}{\mathbf{d}\cdot\mathbf{d}}\,\mathbf{d} = -\frac{4}{3}\begin{bmatrix}1\\1\\1\end{bmatrix}$,

   so $\mathbf{w} - \mathbf{p} = \begin{bmatrix}-2\\0\\-2\end{bmatrix} + \frac{4}{3}\begin{bmatrix}1\\1\\1\end{bmatrix} = \begin{bmatrix}-\frac{2}{3}\\\frac{4}{3}\\-\frac{2}{3}\end{bmatrix} = \frac{2}{3}\begin{bmatrix}-1\\2\\-1\end{bmatrix}$ and the required distance is $\frac{2}{3}\left\|\begin{bmatrix}-1\\2\\-1\end{bmatrix}\right\| = \frac{2}{3}\sqrt{6}$.

5. (a) Find any point in the plane, say $Q(0,0,1)$. Then the distance we want is the length of the projection of $\overrightarrow{PQ}$ onto a normal $\mathbf{n}$ to the plane. We have $\overrightarrow{PQ} = \begin{bmatrix}-2\\-3\\1\end{bmatrix}$ and $\mathbf{n} = \begin{bmatrix}5\\-1\\1\end{bmatrix}$. So $\text{proj}_\mathbf{n}\,\overrightarrow{PQ} = -\frac{6}{27}\begin{bmatrix}5\\-1\\1\end{bmatrix} = -\frac{2}{9}\begin{bmatrix}5\\-1\\1\end{bmatrix}$. This has length $\frac{2}{9}\sqrt{27} = \frac{2}{3}\sqrt{3}$.

   To find the point $A(x,y,z)$ of the plane closest to P, note that $\overrightarrow{PA}$ is the projection of $\overrightarrow{PQ}$ on $\mathbf{n}$, that is,

   $\overrightarrow{PA} = -\frac{2}{9}\begin{bmatrix}5\\-1\\1\end{bmatrix} = \begin{bmatrix}-\frac{10}{9}\\\frac{2}{9}\\-\frac{2}{9}\end{bmatrix}$. Since $\overrightarrow{PA} = \begin{bmatrix}x-2\\y-3\\z\end{bmatrix}$, we get $A = (\frac{8}{9}, \frac{29}{9}, -\frac{2}{9})$.

7. (a) Let $\mathbf{p} = \text{proj}_\mathbf{v}\,\mathbf{u}$. Then $\mathbf{u} - \mathbf{p}$ is orthogonal to $\mathbf{v}$ and in the plane spanned by $\mathbf{u}$ and $\mathbf{v}$, so we can use these two vectors. We have $\mathbf{p} = \text{proj}_\mathbf{v}\,\mathbf{u} = \dfrac{\mathbf{u}\cdot\mathbf{v}}{\mathbf{v}\cdot\mathbf{v}}\,\mathbf{v} = \frac{2}{5}\begin{bmatrix}2\\0\\1\end{bmatrix}$ and $\mathbf{u} - \mathbf{p} = \begin{bmatrix}-\frac{9}{5}\\1\\\frac{18}{5}\end{bmatrix}$. In applications, we should probably multiply by 5, obtaining $\begin{bmatrix}-9\\5\\18\end{bmatrix}$ and $\begin{bmatrix}2\\0\\1\end{bmatrix}$ as a desired pair of orthogonal vectors.

(d) We proceed as in Problem 1.4.5. First we find two vectors in the plane, for example, $\mathbf{u} = \begin{bmatrix} 3 \\ 2 \\ 0 \end{bmatrix}$ and $\mathbf{v} = \begin{bmatrix} 2 \\ 0 \\ -1 \end{bmatrix}$. Let

$\mathbf{p} = \text{proj}_\mathbf{v}\, \mathbf{u}$. Then $\mathbf{u} - \mathbf{p}$ is orthogonal to $\mathbf{v}$ and in the plane spanned by $\mathbf{u}$ and $\mathbf{v}$, which is precisely the given plane,

so we can use these two vectors. We have $\mathbf{p} = \text{proj}_\mathbf{v}\, \mathbf{u} = \frac{\mathbf{u}\cdot\mathbf{v}}{\mathbf{v}\cdot\mathbf{v}}\mathbf{v} = \frac{6}{5}\begin{bmatrix} 2 \\ 0 \\ -1 \end{bmatrix}$ and $\mathbf{u} - \mathbf{p} = \begin{bmatrix} \frac{3}{5} \\ 2 \\ \frac{6}{5} \end{bmatrix}$. In applications, we

should probably multiply by 5, obtaining $\begin{bmatrix} 3 \\ 10 \\ 6 \end{bmatrix}$ and $\begin{bmatrix} 2 \\ 0 \\ -1 \end{bmatrix}$ as a desired pair of orthogonal vectors. Many answers

are possible: any pair of orthogonal vectors whose components satisfy $2x - 3y + 4z = 0$ is a correct answer.

8. (a) We find two nonparallel vectors $\mathbf{u}$ and $\mathbf{v}$ in the plane, find the projection $\mathbf{p}$ of $\mathbf{u}$ on $\mathbf{v}$, and take $\mathbf{v}$ and $\mathbf{u} - \mathbf{p}$ as our

vectors. Let $\mathbf{u} = \begin{bmatrix} 1 \\ 0 \\ -2 \end{bmatrix}$ and $\mathbf{v} = \begin{bmatrix} 0 \\ 1 \\ 1 \end{bmatrix}$. Then

$$\mathbf{p} = \text{proj}_\mathbf{v}\, \mathbf{u} = \frac{\mathbf{u}\cdot\mathbf{v}}{\mathbf{v}\cdot\mathbf{v}}\mathbf{v} = \frac{-2}{2}\begin{bmatrix} 0 \\ 1 \\ 1 \end{bmatrix} = \begin{bmatrix} 0 \\ -1 \\ -1 \end{bmatrix}$$

and $\mathbf{u} - \mathbf{p} = \begin{bmatrix} 1 \\ 1 \\ -1 \end{bmatrix}$, so $\begin{bmatrix} 0 \\ 1 \\ 1 \end{bmatrix}$ and $\begin{bmatrix} 1 \\ 1 \\ -1 \end{bmatrix}$ are suitable orthogonal vectors. Many answers are possible, of course.

Correct answers consist of two orthogonal vectors whose components satisfy $2x - y + z = 0$.

(b) In part (a), we learned that $\mathbf{e} = \begin{bmatrix} 1 \\ 1 \\ -1 \end{bmatrix}$ and $\mathbf{f} = \begin{bmatrix} 0 \\ 1 \\ 1 \end{bmatrix}$ are orthogonal vectors spanning π, so we just write the answer:

$$\text{proj}_\pi\, \mathbf{w} = \frac{\mathbf{w}\cdot\mathbf{e}}{\mathbf{e}\cdot\mathbf{e}}\mathbf{e} + \frac{\mathbf{w}\cdot\mathbf{f}}{\mathbf{f}\cdot\mathbf{f}}\mathbf{f} = \frac{1}{3}\begin{bmatrix} 1 \\ 1 \\ -1 \end{bmatrix} + \frac{2}{2}\begin{bmatrix} 0 \\ 1 \\ 1 \end{bmatrix} = \begin{bmatrix} \frac{1}{3} \\ \frac{4}{3} \\ \frac{2}{3} \end{bmatrix} = \frac{1}{3}\begin{bmatrix} 1 \\ 4 \\ 2 \end{bmatrix}.$$

9. (a) First we find two orthogonal vectors in the plane: $\mathbf{u} = \begin{bmatrix} -1 \\ 2 \\ 0 \end{bmatrix}$ and $\mathbf{v} = \begin{bmatrix} 1 \\ 0 \\ 2 \end{bmatrix}$ come quickly to mind. These are not

orthogonal but, as illustrated by Problem 1.4.5, we can get orthogonal vectors $\mathbf{e}, \mathbf{f}$ by taking $\mathbf{e} = \mathbf{v} = \begin{bmatrix} 1 \\ 0 \\ 2 \end{bmatrix}$ and

$\mathbf{f} = \mathbf{u} - \mathbf{p}$, with $\mathbf{p}$ the projection of $\mathbf{u}$ on $\mathbf{v}$. We have

$$\mathbf{p} = \text{proj}_\mathbf{v}\, \mathbf{u} = \frac{\mathbf{u}\cdot\mathbf{v}}{\mathbf{v}\cdot\mathbf{v}}\mathbf{v} = -\frac{1}{5}\begin{bmatrix} 1 \\ 0 \\ 2 \end{bmatrix} = \begin{bmatrix} -\frac{1}{5} \\ 0 \\ -\frac{2}{5} \end{bmatrix}$$

and

$$\mathbf{f} = \mathbf{u} - \mathbf{p} = \begin{bmatrix} -1 \\ 2 \\ 0 \end{bmatrix} - \begin{bmatrix} -\frac{1}{5} \\ 0 \\ -\frac{2}{5} \end{bmatrix} = \begin{bmatrix} -\frac{4}{5} \\ 2 \\ \frac{2}{5} \end{bmatrix} = \frac{2}{5}\begin{bmatrix} -2 \\ 5 \\ 1 \end{bmatrix},$$

a vector that is indeed perpendicular to **e**. (Did you check?) Since the purpose is simply to find orthogonal vectors that span π, we can make life much simpler by replacing **f** by a new $\mathbf{f} = \begin{bmatrix} -2 \\ 5 \\ 1 \end{bmatrix}$. Now the projection of $\mathbf{w} = \begin{bmatrix} 3 \\ 1 \\ 1 \end{bmatrix}$ on

the plane is $\text{proj}_\pi \mathbf{w} = \frac{\mathbf{w \cdot e}}{\mathbf{e \cdot e}} \mathbf{e} + \frac{\mathbf{w \cdot f}}{\mathbf{f \cdot f}} \mathbf{f} = \frac{5}{5}\mathbf{e} + \frac{0}{30}\mathbf{f} = \mathbf{e}$. The projection of **w** on π is $\mathbf{e} = \begin{bmatrix} 1 \\ 0 \\ 2 \end{bmatrix}$.

(b) In general, we take as before $\mathbf{e} = \begin{bmatrix} 1 \\ 0 \\ 2 \end{bmatrix}$ and $\mathbf{f} = \begin{bmatrix} -2 \\ 5 \\ 1 \end{bmatrix}$ as orthogonal vectors that span π. With $\mathbf{w} = \begin{bmatrix} x \\ y \\ z \end{bmatrix}$, we have

$$\text{proj}_\pi \mathbf{w} = \frac{\mathbf{w \cdot e}}{\mathbf{e \cdot e}} \mathbf{e} + \frac{\mathbf{w \cdot f}}{\mathbf{f \cdot f}} \mathbf{f} = \frac{x + 2z}{5} \mathbf{e} + \frac{-2x + 5y + z}{30} \mathbf{f}$$

$$= \frac{x + 2z}{5} \begin{bmatrix} 1 \\ 0 \\ 2 \end{bmatrix} + \frac{-2x + 5y + z}{30} \begin{bmatrix} -2 \\ 5 \\ 1 \end{bmatrix}$$

$$= \begin{bmatrix} \frac{x-y+z}{3} \\ \frac{-2x+5y+z}{6} \\ \frac{2x+y+5z}{6} \end{bmatrix} = \frac{1}{6} \begin{bmatrix} 2x - 2y + 2z \\ -2x + 5y + z \\ 2x + y + 5z \end{bmatrix}.$$

10. (a) Two orthogonal vectors in π are $\mathbf{e} = \begin{bmatrix} 1 \\ 1 \\ 1 \end{bmatrix}$ and $\mathbf{f} = \begin{bmatrix} 1 \\ -1 \\ 0 \end{bmatrix}$, so

$$\text{proj}_\pi \mathbf{w} = \frac{\mathbf{w \cdot e}}{\mathbf{e \cdot e}} \mathbf{e} + \frac{\mathbf{w \cdot f}}{\mathbf{f \cdot f}} \mathbf{f}$$

$$= \frac{1}{3} \begin{bmatrix} 1 \\ 1 \\ 1 \end{bmatrix} - \frac{5}{2} \begin{bmatrix} 1 \\ -1 \\ 0 \end{bmatrix} = \begin{bmatrix} -\frac{13}{6} \\ \frac{17}{6} \\ \frac{1}{3} \end{bmatrix} = \frac{1}{6} \begin{bmatrix} -13 \\ 17 \\ 2 \end{bmatrix}.$$

12. (a) This is exactly as on p. 49 of the text. Write $\mathbf{w} = a\mathbf{e} + b\mathbf{f}$. Then $\mathbf{w \cdot e} = a\mathbf{e \cdot e}$, so $a = \frac{\mathbf{w \cdot e}}{\mathbf{e \cdot e}}$. Similarly $b = \frac{\mathbf{w \cdot f}}{\mathbf{f \cdot f}}$, so $\mathbf{w} = \frac{\mathbf{w \cdot e}}{\mathbf{e \cdot e}} \mathbf{e} + \frac{\mathbf{w \cdot f}}{\mathbf{f \cdot f}} \mathbf{f}$.

14. (a) The lines have directions $\mathbf{d}_1 = \begin{bmatrix} -2 \\ 1 \\ 2 \end{bmatrix}$ and $\mathbf{d}_2 = \begin{bmatrix} 6 \\ -3 \\ -6 \end{bmatrix}$. Since $\mathbf{d}_2$ is a scalar multiple of $\mathbf{d}_1$, the lines are parallel.

(b) The distance between parallel lines is the distance from any point on one line to the other line. We take $Q = (0, 1, 4)$ as a point on the first line and $P = (4, -2, 2)$ on the second line and proceed as in Problem 1.4.6.

Let $\mathbf{w} = \overrightarrow{PQ} = \begin{bmatrix} -4 \\ 3 \\ 2 \end{bmatrix}$. The second line has direction $\mathbf{d} = \begin{bmatrix} 2 \\ -1 \\ -2 \end{bmatrix}$ and the distance we want is the length of $\mathbf{w} - \mathbf{p}$, where $\mathbf{p} = \text{proj}_\mathbf{d}\, \mathbf{w}$ is the projection of $\mathbf{w}$ on $\mathbf{d}$: $\mathbf{p} = \text{proj}_\mathbf{d}\, \mathbf{w} = \frac{\mathbf{w} \cdot \mathbf{d}}{\mathbf{d} \cdot \mathbf{d}}\, \mathbf{d} = -\frac{15}{9} \begin{bmatrix} 2 \\ -1 \\ -2 \end{bmatrix}$,

$\mathbf{w} - \mathbf{p} = \begin{bmatrix} -4 \\ 3 \\ 2 \end{bmatrix} + \frac{5}{3} \begin{bmatrix} 2 \\ -1 \\ -2 \end{bmatrix} = \begin{bmatrix} -\frac{2}{3} \\ \frac{4}{3} \\ -\frac{4}{3} \end{bmatrix} = \frac{2}{3} \begin{bmatrix} -1 \\ 2 \\ -2 \end{bmatrix}$, so the required distance is $\frac{2}{3} \left\| \begin{bmatrix} -1 \\ 2 \\ -2 \end{bmatrix} \right\| = \frac{2}{3}(3) = 2.$

17. **(b) i.** The converse of Theorem 1.4.12 says that if a vector $\mathbf{w}$ is orthogonal to every vector in the plane π spanned by $\mathbf{u}$ and $\mathbf{v}$, then $\mathbf{w}$ is orthogonal to $\mathbf{u}$ and to $\mathbf{v}$.

18. **(a)** The picture certainly makes Don's theory appear plausible and the following argument establishes that it is indeed correct. Don takes for the projection the vector $\mathbf{p} = \mathbf{w} - \text{proj}_\mathbf{n}\, \mathbf{w}$.

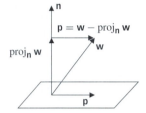

By definition of "projection on a plane," we must show that $\mathbf{w} - \mathbf{p}$ is orthogonal to every vector in π. This is immediately clear since $\mathbf{w} - \mathbf{p} = \mathbf{w} - (\mathbf{w} - \text{proj}_\mathbf{n}\, \mathbf{w}) = \text{proj}_\mathbf{n}\, \mathbf{w}$ is a normal to the plane.

(b) A normal to π is $\mathbf{n} = \begin{bmatrix} 3 \\ -2 \\ 1 \end{bmatrix}$. Thus $\text{proj}_\mathbf{n}\, \mathbf{w} = \frac{\mathbf{w} \cdot \mathbf{n}}{\mathbf{n} \cdot \mathbf{n}}\, \mathbf{n} = \frac{2}{14} \begin{bmatrix} 3 \\ -2 \\ 1 \end{bmatrix} = \frac{1}{7} \begin{bmatrix} 3 \\ -2 \\ 1 \end{bmatrix}$ and

$\mathbf{p} = \mathbf{w} - \text{proj}_\mathbf{n}\, \mathbf{w} = \begin{bmatrix} 1 \\ 1 \\ 1 \end{bmatrix} - \frac{1}{7} \begin{bmatrix} 3 \\ -2 \\ 1 \end{bmatrix} = \frac{1}{7} \begin{bmatrix} 4 \\ 9 \\ 6 \end{bmatrix}.$

Section 1.5—True/False

1. True. These vectors, also labeled $\mathbf{e}_1$, $\mathbf{e}_2$, $\mathbf{e}_3$, are the standard basis for $\mathbf{R}^3$. See 1.5.8.

2. True. Since every vector in $\mathbf{R}^3$ is a linear combination of $\mathbf{i}$, $\mathbf{j}$, $\mathbf{k}$, the same is true of any set containing $\mathbf{i}$, $\mathbf{j}$, and $\mathbf{k}$. If $\mathbf{x} = \begin{bmatrix} x_1 \\ x_2 \\ x_3 \end{bmatrix}$, then $\mathbf{x} = x_1\mathbf{e}_1 + x_2\mathbf{e}_2 + x_3\mathbf{e}_3 + 0\mathbf{u}$.

3. True. The dot product of these vectors is 0.

4. False. $\frac{1}{\sqrt{15}} \begin{bmatrix} 1 \\ 2 \\ 0 \\ 1 \\ 3 \end{bmatrix}$ is a unit vector.

5. False. Any linear combination of $\mathbf{u}$ and $\mathbf{v}$ must have third component 0.

6. False. The statement $0\mathbf{u} + 0\mathbf{v} = \mathbf{0}$ is true for any vectors $\mathbf{u}$ and $\mathbf{v}$ and has nothing to do with linear independence.

7. False. Saying "$a\mathbf{u} + b\mathbf{v} = \mathbf{0}$ where a and b are 0" is the same as saying $0\mathbf{u} + 0\mathbf{v} = \mathbf{0}$, which is true for any vectors $\mathbf{u}$ and $\mathbf{v}$ and has nothing to do with linear independence.

8. True. $0\mathbf{u} + 0\mathbf{v} + 58\mathbf{0}$ is a nontrivial linear combination of $\mathbf{u}$, $\mathbf{v}$, and $\mathbf{0}$ that equals the zero vector.

Exercises 1.5

1. (a) This is not defined.

    (b) $\|\mathbf{y}\|^2 = 0^2 + (-1)^2 + 1^2 + 2^2 = 6$, so $\|\mathbf{y}\| = \sqrt{6}$.

2. (a) $\begin{bmatrix} a \\ -1 \\ 1 \\ b \\ 2 \end{bmatrix} + \begin{bmatrix} b \\ 0 \\ 6 \\ -2 \end{bmatrix} = \begin{bmatrix} a+b \\ -1 \\ 7 \\ b+1 \\ 3 \end{bmatrix}$, so we need $a + b = 0$ and $b + 1 = 0$. So $b = -1$, $a = 1$.

3. (a) Since $3\mathbf{u} - 2\mathbf{v} = \begin{bmatrix} -3 - 2c \\ 3a - 2 \\ 3b - 2 \\ 6 - 2a \\ 8 \end{bmatrix}$, we must have $-3 - 2c = -1$, $3a - 2 = 1$, $3b - 2 = -2$, and $6 - 2a = 4$. We obtain

    $a = 1$, $b = 0$, $c = -1$.

4. (a) Since $\|\mathbf{u}\| = \sqrt{3}$, the answer is $\dfrac{\mathbf{u}}{\|\mathbf{u}\|} = \dfrac{1}{\sqrt{3}} \begin{bmatrix} 1 \\ 0 \\ -1 \\ 1 \end{bmatrix}$.

5. (a) Since $\|c\mathbf{u}\| = |c|\,\|\mathbf{u}\| = |c|\sqrt{6}$, we want $|c| = \frac{8}{\sqrt{6}}$, so $c = \pm\frac{4}{3}\sqrt{6}$.

6. (a) $\mathbf{u} \cdot \mathbf{v} = -2$, $\|\mathbf{u}\| = \sqrt{6}$, $\|\mathbf{v}\| = \sqrt{2}$, so $\cos\theta = \dfrac{\mathbf{u}\cdot\mathbf{v}}{\|\mathbf{u}\|\|\mathbf{v}\|} = \dfrac{-2}{\sqrt{12}} \approx -0.577$, and $\theta \approx \arccos(0.577) \approx 2.2$ rads $\approx 125°$.

7. $\|2\mathbf{u} - \mathbf{v}\|^2 = (2\mathbf{u} - \mathbf{v}) \cdot (2\mathbf{u} - \mathbf{v}) = 4\mathbf{u}\cdot\mathbf{u} - 4\mathbf{u}\cdot\mathbf{v} + \mathbf{v}\cdot\mathbf{v} = 4\|\mathbf{u}\|^2 - 4\mathbf{u}\cdot\mathbf{v} + \|\mathbf{v}\|^2$. We are given that
    $0 = \mathbf{u}\cdot(\mathbf{u} + \mathbf{v}) = \mathbf{u}\cdot\mathbf{u} + \mathbf{u}\cdot\mathbf{v} = \|\mathbf{u}\|^2 + \mathbf{u}\cdot\mathbf{v}$, so $\mathbf{u}\cdot\mathbf{v} = -6$. Thus $\|2\mathbf{u} - \mathbf{v}\|^2 = 4(6) - 4(-6) + 1 = 49$ and $\|2\mathbf{u} - \mathbf{v}\| = 7$.

10. (a) We have $\mathbf{u}\cdot\mathbf{v} = -3$, $\|\mathbf{u}\| = \sqrt{5}$, $\|\mathbf{v}\| = \sqrt{10}$. The Cauchy–Schwarz inequality is the statement $3 \le \sqrt{5}\sqrt{10}$. Note

    that $\sqrt{5}\sqrt{10} \approx 7.1$. Since $\mathbf{u} + \mathbf{v} = \begin{bmatrix} -2 \\ 1 \\ -2 \end{bmatrix}$, $\|\mathbf{u} + \mathbf{v}\| = \sqrt{9} = 3$. The triangle inequality says $3 \le \sqrt{5} + \sqrt{10}$. Note that

    $\sqrt{5} + \sqrt{10} \approx 5.4$.

11. (a) Since $\mathbf{v}_4 = -2\mathbf{v}_1$, the vectors are linearly dependent: $2\mathbf{v}_1 + 0\mathbf{v}_2 + 0\mathbf{v}_3 + \mathbf{v}_4 = \mathbf{0}$.

Section 2.1—True/False

1. False. Take A to be 2×3 and B to be 3×2. Then AB is 2×2 and BA is 3×3, both of which are square.

2. True.

3. False. For example, let $A = \begin{bmatrix} 1 & 0 \end{bmatrix}$ and $B = \begin{bmatrix} 0 \\ 1 \end{bmatrix}$.

4. False. For example, let $A = \begin{bmatrix} 1 & 0 \end{bmatrix}$, $B = \begin{bmatrix} 0 \\ 1 \end{bmatrix}$, and $X = \begin{bmatrix} 1 & 1 \\ 1 & 1 \end{bmatrix}$.

5. False. The $(2, 1)$ entry is 2, the entry in row two, column 1.

6. True. Since $\mathbf{b} = \mathbf{e}_3$, the third standard basis vector, $A\mathbf{b}$ is the third column of A.

7. True. This is the very important fact 2.1.33.

8. False. This says matrix multiplication is *associative*.

9. True. The hypothesis implies, in particular, that $A\mathbf{e}_i = \mathbf{0}$, where $\mathbf{e}_1, \mathbf{e}_2, \ldots$ denote the standard basis. But $A\mathbf{e}_i$ is column i of A, so each column of A is $\mathbf{0}$. Thus $A = \mathbf{0}$.

10. False. $A(2\mathbf{x}_0) = 2A\mathbf{x}_0 = 2\mathbf{b} \ne \mathbf{b}$.

Exercises 2.1

1. (a) Matrices are equal if and only if corresponding columns and hence corresponding entries are equal. Thus

$$
\begin{aligned}
2x - 3y &= 8 \\
-y &= 2 \\
x - y &= 3 \\
x + y &= -1 \\
-x + y &= -3 \\
x + 2y &= -3.
\end{aligned}
$$

We find that $x = 1$, $y = -2$.

2. (a) Equating $(2, 1)$ entries gives $a + x = -3$. Equating $(1, 2)$ entries gives $-2(a + x) = 2$, so $a + x = -1$. No x, y, a, b exist.

3. (a) Since $2A - B = \begin{bmatrix} 4 & 2x \\ 2y & 2 \end{bmatrix} - \begin{bmatrix} z & -4 \\ -1 & 5 \end{bmatrix} = \begin{bmatrix} 4-z & 2x+4 \\ 2y+1 & -3 \end{bmatrix}$, we must have $4 - z = 0$, $2x + 4 = 7$, and $2y + 1 = 2$, so $x = \frac{3}{2}$, $y = \frac{1}{2}$, and $z = 4$.

(b) Since $AB = \begin{bmatrix} 2z - x & -8 + 5x \\ yz - 1 & -4y + 5 \end{bmatrix}$, we need $2z - x = 0$, $-8 + 5x = 7$, $yz - 1 = 2$, and $-4y + 5 = -3$. The second equation says $5x = 15$, so $x = 3$. Then the first equation gives $z = \frac{1}{2}x = \frac{3}{2}$. The third equation says $yz = 3$, so $y = \frac{2}{3}(3) = 2$. Since $y = 2$ satisfies the last equation, we have a solution.

5. (a) $\begin{bmatrix} 1 & 2 & 3 \\ 1 & 2 & 3 \end{bmatrix}$.

7. (a) $A + B = \begin{bmatrix} 4 & -2 & 5 \\ -4 & 0 & 6 \end{bmatrix}$; $A - B = \begin{bmatrix} 0 & 4 & -3 \\ 2 & -2 & 2 \end{bmatrix}$; $2A - B = \begin{bmatrix} 2 & 5 & -2 \\ 1 & -3 & 6 \end{bmatrix}$.

8. (a) AB is not defined; $BA = \begin{bmatrix} -27 & 26 & 16 \\ -34 & 52 & -2 \end{bmatrix}$.

9. (a) The $(1, 1)$ and $(2, 2)$ entries of AB are, respectively, $ax + by + cz$ and $du + ev + fw$.

11. (a) $3\begin{bmatrix} -1 \\ 0 \end{bmatrix} + 4\begin{bmatrix} 2 \\ 1 \end{bmatrix} - \begin{bmatrix} 1 \\ 8 \end{bmatrix} = \begin{bmatrix} -1 & 2 & 1 \\ 0 & 1 & 8 \end{bmatrix}\begin{bmatrix} 3 \\ 4 \\ -1 \end{bmatrix}$.

12. (a) $\begin{bmatrix} 2 & -1 \\ 3 & 5 \\ -1 & 1 \\ 2 & -5 \end{bmatrix}\begin{bmatrix} x \\ y \end{bmatrix} = \begin{bmatrix} 10 \\ 7 \\ -3 \\ 1 \end{bmatrix}$.

13. Substituting $x = 1$, $y = 4$ gives $a + b + c = 4$. Substituting $x = 2$, $y = 8$ gives $4a + 2b + c = 8$. These two equations correspond to the single matrix equation $\begin{bmatrix} 1 & 1 & 1 \\ 4 & 2 & 1 \end{bmatrix}\begin{bmatrix} a \\ b \\ c \end{bmatrix} = \begin{bmatrix} 4 \\ 8 \end{bmatrix}$.

15. A linear combination of the columns of a matrix A is $A\mathbf{x}$ for some vector $\mathbf{x}$.

16. $\mathbf{b} = -4\begin{bmatrix} 1 \\ 4 \\ 7 \end{bmatrix} + \begin{bmatrix} 2 \\ 5 \\ 8 \end{bmatrix} + 7\begin{bmatrix} 3 \\ 6 \\ 9 \end{bmatrix}$.

18. We seek a and b so that $\begin{bmatrix} 2 \\ 3 \end{bmatrix} = a\begin{bmatrix} 1 \\ 0 \end{bmatrix} + b\begin{bmatrix} 1 \\ 1 \end{bmatrix}$. We need $2 = a + b$ and $3 = b$, so $a = -1$. Thus $\begin{bmatrix} 2 \\ 3 \end{bmatrix} = (-1)\begin{bmatrix} 1 \\ 0 \end{bmatrix} + 3\begin{bmatrix} 1 \\ 1 \end{bmatrix}$.

21. (a) $\mathbf{x} = 20\begin{bmatrix}1\\0\\0\end{bmatrix} + 35\begin{bmatrix}0\\1\\0\end{bmatrix} + 47\begin{bmatrix}0\\0\\1\end{bmatrix}$.

22. $AX = \begin{bmatrix} A\mathbf{x}_1 & A\mathbf{x}_2 & A\mathbf{x}_3 \\ \downarrow & \downarrow & \downarrow \end{bmatrix} = \begin{bmatrix} 1 & 0 & 0 \\ 0 & 1 & 0 \\ 0 & 0 & 1 \end{bmatrix}$.

24. This is an application of the result of Problem 2.1.37, but we repeat the argument. The first column of AD is A times the first column of D. By the important fact expressed in 2.1.33, this is

$$-7 \times \text{first column of } A + 0 \times \text{second column of } A + 0 \times \text{third column of } A,$$

which is $\begin{bmatrix}-7\\-7\\-7\end{bmatrix}$. The second and third columns of AD are obtained by similar means, giving $AD = \begin{bmatrix} -7 & 16 & 15 \\ -7 & 16 & 15 \\ -7 & 16 & 15 \end{bmatrix}$.

28. $AB = \begin{bmatrix} 8 & 5 & 0 \\ 16 & -5 & 20 \end{bmatrix}$, $AC = \begin{bmatrix} -3 & -2 & 10 \\ -1 & 6 & 20 \end{bmatrix}$, $AB + AC = \begin{bmatrix} 5 & 3 & 10 \\ 15 & 1 & 40 \end{bmatrix}$, $B + C = \begin{bmatrix} 1 & -1 & 4 \\ 3 & 1 & 7 \end{bmatrix}$,

$A(B + C) = \begin{bmatrix} 5 & 3 & 10 \\ 15 & 1 & 40 \end{bmatrix}$.

31. (a) "No." For example, as suggested by Exercise 37, if $A = \begin{bmatrix} 0 & -1 \\ 0 & -5 \end{bmatrix}$ and $B = \begin{bmatrix} 5 & -1 \\ 0 & 0 \end{bmatrix}$, then $AB = \mathbf{0}$.

33. Let A be $m \times n$. Since $A\mathbf{x}$ exists, $\mathbf{x}$ is $n \times 1$ and $A\mathbf{x}$ is $m \times 1$. Thus $\mathbf{x} + A\mathbf{x}$ is the sum of a vector in $\mathbf{R}^n$ and a vector in $\mathbf{R}^m$, so $m = n$. The matrix A is square.

36. We use the fact that $A\mathbf{e}_i$ is column i of A. (See 2.1.21.) Since $A\mathbf{x} = B\mathbf{x}$ for all $\mathbf{x}$, then certainly $A\mathbf{e}_1 = B\mathbf{e}_1$, so the first columns of A and B are the same. Similarly, $A\mathbf{e}_i = B\mathbf{e}_i$ implies that the ith columns of A and B are equal, for every i. Thus $A = B$ by 2.1.7.

Section 2.2—True/False

1. False. If C is invertible, multiplying $AC = BC$ on the right by C^{-1} gives $A = B$, but not necessarily otherwise.

2. True.

3. False.

4. False, unless $AB = BA$ (which is seldom the case).

5. True. See 2.2.12.

6. False. $X = BA^{-1}$. (In general, this is different from $A^{-1}B$ unless A^{-1} and B commute, which is unlikely.)

7. True. AB is 2×5.

8. False. If A is 2×3 and B is 3×2, then $A^T B^T$ is defined and 3×3, and $B^T A^T$ is defined and 2×2.

Exercises 2.2

1. (a) $AB = BA = I_3$, so these matrices are inverses.

4. (a) $AB = \begin{bmatrix} 1 & -7 & 1 \\ 2 & -9 & 1 \end{bmatrix}\begin{bmatrix} -1 & 3 \\ 0 & 1 \\ 2 & 4 \end{bmatrix} = \begin{bmatrix} 1 & 0 \\ 0 & 1 \end{bmatrix}$; $\qquad BA = \begin{bmatrix} -1 & 3 \\ 0 & 1 \\ 2 & 4 \end{bmatrix}\begin{bmatrix} 1 & -7 & 1 \\ 2 & -9 & 1 \end{bmatrix} = \begin{bmatrix} 5 & -20 & 2 \\ 2 & -9 & 1 \\ 10 & -50 & 6 \end{bmatrix}$.

(b) A is not invertible since $BA \neq I$. Only square matrices can be invertible.

7. $A^T = \begin{bmatrix} 3 & -1 \\ 1 & 2 \end{bmatrix}$, $\quad B^T = \begin{bmatrix} 2 & 4 \\ 0 & -3 \end{bmatrix}$, $\quad AB = \begin{bmatrix} 10 & -3 \\ 6 & -6 \end{bmatrix}$, $\quad BA = \begin{bmatrix} 6 & 2 \\ 15 & -2 \end{bmatrix}$,

$(AB)^T = \begin{bmatrix} 10 & 6 \\ -3 & -6 \end{bmatrix}$, $\quad (BA)^T = \begin{bmatrix} 6 & 15 \\ 2 & -2 \end{bmatrix}$, $\quad A^T B^T = \begin{bmatrix} 6 & 15 \\ 2 & -2 \end{bmatrix}$,

$B^T A^T = \begin{bmatrix} 10 & 6 \\ -3 & -6 \end{bmatrix}$.

9. We are asked to show that $(A^T)^{-1}$, which is the inverse of A^T, is the matrix $X = (A^{-1})^T$. By 2.2.7, it suffices to prove that the product of A^T and X (in either order) is the identity matrix. Using the fact that $(BC)^T = C^T B^T$, we have $A^T X = A^T (A^{-1})^T = (A^{-1}A)^T = I^T = I$.

13. Since $AB = BA$, we have $(AB)^T = (BA)^T = A^T B^T$, as desired.

14. (a) Multiplying $AC = BC$ on the right by C^{-1} gives $ACC^{-1} = BCC^{-1}$, so $AI = BI$ and $A = B$.

15. (a) This follows immediately from 2.1.18: $\mathbf{u} \cdot \mathbf{v} = \mathbf{u}^T \mathbf{v}$ for any vectors $\mathbf{u}$ and $\mathbf{v}$.

18. (a) Multiplying $AXB = A + B$ on the left by A^{-1} gives $XB = I + A^{-1}B$ and multiplying this on the right by B^{-1} gives $X = B^{-1} + A^{-1}$.

Section 2.3—True/False

1. False. The variable x_3 does not appear to the first power.

2. True. The second system is the same as the first, except that the third equation has been replaced by its sum with -2 times the first.

3. False. A system of linear equations has no solution, one solution, or infinitely many solutions.

4. True.

5. True. The pivot columns of U are those containing the pivots that are, by definition, the pivot columns of A.

6. False. The last row corresponds to the equation $0 = 1$, so there are no solutions.

7. False. Free variables correspond to the columns *without* pivots.

8. True. See Problems 2.3.15 and 2.3.16.

9. False. Dividing the rows of the matrix by these numbers, respectively, gives the matrix $\begin{bmatrix} 1 & 0 & \frac{1}{2} \\ 0 & 1 & \frac{3}{7} \\ 0 & 0 & 1 \end{bmatrix}$, which is *not* in row echelon form.

10. False. A pivot cannot be 0.

Exercises 2.3

1. (a) This is not in row echelon form because the leading nonzero entries do not step to the right as you read down the matrix.

2. (a) $\begin{bmatrix} 1 & -1 & -2 \\ 2 & -3 & -5 \\ -1 & 4 & 5 \end{bmatrix} \rightarrow \begin{bmatrix} 1 & -1 & -2 \\ 0 & -1 & -1 \\ 0 & 3 & 3 \end{bmatrix} \rightarrow \begin{bmatrix} 1 & -1 & -2 \\ 0 & -1 & -1 \\ 0 & 0 & 0 \end{bmatrix}$.

The pivots are 1 and -1. The pivot columns are columns one and two.

(c)
$$
\begin{bmatrix}
1 & 4 & 5 & 2 \\
3 & 13 & 20 & 8 \\
-2 & -10 & -16 & -4 \\
1 & 10 & 38 & 28
\end{bmatrix}
\rightarrow
\begin{bmatrix}
1 & 4 & 5 & 2 \\
0 & 1 & 5 & 2 \\
0 & -2 & -6 & 0 \\
0 & 6 & 33 & 26
\end{bmatrix}
\rightarrow
\begin{bmatrix}
1 & 4 & 5 & 2 \\
0 & 1 & 5 & 2 \\
0 & 0 & 4 & 4 \\
0 & 0 & 3 & 14
\end{bmatrix}
\rightarrow
\begin{bmatrix}
1 & 4 & 5 & 2 \\
0 & 1 & 5 & 2 \\
0 & 0 & 4 & 4 \\
0 & 0 & 0 & 11
\end{bmatrix}.
$$

The pivots are 1, 1, 4, and 11. Every column is a pivot column.

3. (a)
$$
\begin{bmatrix}
3 & -1 & 5 & 3 & 6 & -3 \\
7 & -7 & 1 & 5 & 4 & 2 \\
-4 & 6 & 4 & -2 & 4 & 10 \\
16 & -10 & 16 & 14 & 22 & -10 \\
-13 & 9 & -11 & -11 & -16 & 12
\end{bmatrix}
\rightarrow
\begin{bmatrix}
-4 & 6 & 4 & -2 & 4 & 10 \\
7 & -7 & 1 & 5 & 4 & 2 \\
3 & -1 & 5 & 3 & 6 & -3 \\
16 & -10 & 16 & 14 & 22 & -10 \\
-13 & 9 & -11 & -11 & -16 & 12
\end{bmatrix}
$$

$$
\rightarrow
\begin{bmatrix}
2 & -3 & -2 & 1 & -2 & -5 \\
7 & -7 & 1 & 5 & 4 & 2 \\
3 & -1 & 5 & 3 & 6 & -3 \\
8 & -5 & 8 & 7 & 11 & -5 \\
-13 & 9 & -11 & -11 & -16 & 12
\end{bmatrix}
\rightarrow
\begin{bmatrix}
2 & -3 & -2 & 1 & -2 & -5 \\
0 & \frac{7}{2} & 8 & \frac{3}{2} & 11 & \frac{39}{2} \\
0 & \frac{7}{2} & 8 & \frac{3}{2} & 9 & \frac{9}{2} \\
0 & 7 & 16 & 3 & 19 & 15 \\
0 & -\frac{21}{2} & -24 & -\frac{9}{2} & -29 & -\frac{41}{2}
\end{bmatrix}
$$

$$
\rightarrow
\begin{bmatrix}
2 & -3 & -2 & 1 & -2 & -5 \\
0 & 7 & 16 & 3 & 22 & 39 \\
0 & 7 & 16 & 3 & 18 & 9 \\
0 & 7 & 16 & 3 & 19 & 15 \\
0 & -21 & -48 & -9 & -58 & -41
\end{bmatrix}
\rightarrow
\begin{bmatrix}
2 & -3 & -2 & 1 & -2 & -5 \\
0 & 7 & 16 & 3 & 22 & 39 \\
0 & 0 & 0 & 0 & -4 & -30 \\
0 & 0 & 0 & 0 & -3 & -24 \\
0 & 0 & 0 & 0 & 8 & 76
\end{bmatrix}
\rightarrow
\begin{bmatrix}
2 & -3 & -2 & 1 & -2 & -5 \\
0 & 7 & 16 & 3 & 22 & 39 \\
0 & 0 & 0 & 0 & 1 & 8 \\
0 & 0 & 0 & 0 & -4 & -30 \\
0 & 0 & 0 & 0 & 8 & 76
\end{bmatrix}
$$

$$
\rightarrow
\begin{bmatrix}
2 & -3 & -2 & 1 & -2 & -5 \\
0 & 7 & 16 & 3 & 22 & 39 \\
0 & 0 & 0 & 0 & 1 & 8 \\
0 & 0 & 0 & 0 & 0 & 2 \\
0 & 0 & 0 & 0 & 0 & 12
\end{bmatrix}
\rightarrow
\begin{bmatrix}
2 & -3 & -2 & 1 & -2 & -5 \\
0 & 7 & 16 & 3 & 22 & 39 \\
0 & 0 & 0 & 0 & 1 & 8 \\
0 & 0 & 0 & 0 & 0 & 2 \\
0 & 0 & 0 & 0 & 0 & 0
\end{bmatrix}.
$$

The pivot columns are columns one, two, five, and six.

4. (a) $[A|\mathbf{b}] =
\begin{bmatrix}
2 & 2 & 2 & 2 \\
4 & 6 & 6 & 4 \\
6 & 6 & 10 & 2
\end{bmatrix}
\rightarrow
\begin{bmatrix}
1 & 1 & 1 & 1 \\
4 & 6 & 6 & 4 \\
6 & 6 & 10 & 2
\end{bmatrix}
\rightarrow
\begin{bmatrix}
1 & 1 & 1 & 1 \\
0 & 2 & 2 & 0 \\
0 & 0 & 4 & -4
\end{bmatrix}
\rightarrow
\begin{bmatrix}
1 & 1 & 1 & 1 \\
0 & 1 & 1 & 0 \\
0 & 0 & 1 & -1
\end{bmatrix}.$

Thus $x_3 = -1$, $x_2 + x_3 = 0$, so $x_2 = -x_3 = 1$ and $x_1 + x_2 + x_3 = 1$, so $x_1 = 1 - x_2 - x_3 = 1$. The solution is

$\mathbf{x} = \begin{bmatrix} 1 \\ 1 \\ -1 \end{bmatrix}.$

5. (a) $\begin{bmatrix} ① & -2 & 3 & -1 & | & 5 \end{bmatrix}$. The free variables are x_2, x_3, and x_4, corresponding to the columns that are not pivot columns.

(b) Let $x_2 = t$, $x_3 = s$, and $x_4 = r$. Then $x_1 - 2x_2 + 3x_3 - x_4 = 5$, so $x_1 = 5 + 2x_2 - 3x_3 + x_4 = 5 + 2t - 3s + r$. The

solution is $\begin{bmatrix} x_1 \\ x_2 \\ x_3 \\ x_4 \end{bmatrix} = \begin{bmatrix} 5 + 2t - 3s + r \\ t \\ s \\ r \end{bmatrix} = \begin{bmatrix} 5 \\ 0 \\ 0 \\ 0 \end{bmatrix} + t\begin{bmatrix} 2 \\ 1 \\ 0 \\ 0 \end{bmatrix} + s\begin{bmatrix} -3 \\ 0 \\ 1 \\ 0 \end{bmatrix} + r\begin{bmatrix} 1 \\ 0 \\ 0 \\ 1 \end{bmatrix}.$

8. (a) $x_3 = t$ is free; $x_4 = \frac{1}{3}$, $x_2 + x_3 + 2x_4 = 3$, so $x_2 = \frac{7}{3} - t$; $x_1 + 3x_4 = 2$, so $x_1 = 1$.

Thus $\mathbf{x} = \begin{bmatrix} 1 \\ \frac{7}{3} - t \\ t \\ \frac{1}{3} \end{bmatrix} = \begin{bmatrix} 1 \\ \frac{7}{3} \\ 0 \\ \frac{1}{3} \end{bmatrix} + t\begin{bmatrix} 0 \\ -1 \\ 1 \\ 0 \end{bmatrix}.$ There are infinitely many solutions.

9. (a) $[A|\mathbf{b}] = \begin{bmatrix} 2 & -1 & 2 & | & -4 \\ 3 & 2 & 0 & | & 1 \\ 1 & 3 & -6 & | & 5 \end{bmatrix} \rightarrow \begin{bmatrix} 1 & 3 & -6 & | & 5 \\ 0 & -7 & 14 & | & -14 \\ 0 & -7 & 18 & | & -14 \end{bmatrix} \rightarrow \begin{bmatrix} 1 & 3 & -6 & | & 5 \\ 0 & 1 & -2 & | & 2 \\ 0 & 0 & 4 & | & 0 \end{bmatrix}.$

So $z = 0$; $y - 2z = 2$, so $y = 2$; $x + 3y - 6z = 5$, so $x = -1$. The solution is $\begin{bmatrix} -1 \\ 2 \\ 0 \end{bmatrix}$.

(c) $[A|\mathbf{b}] = \begin{bmatrix} 1 & -1 & 2 & | & 4 \end{bmatrix}$. This is row echelon form. The free variables are $y = t$ and $z = s$, so $x = 4 + y - 2z = 4 + t - 2s$. In vector form the solution is

$$\begin{bmatrix} x \\ y \\ z \end{bmatrix} = \begin{bmatrix} 4 + t - 2s \\ t \\ s \end{bmatrix} = \begin{bmatrix} 4 \\ 0 \\ 0 \end{bmatrix} + t \begin{bmatrix} 1 \\ 1 \\ 0 \end{bmatrix} + s \begin{bmatrix} -2 \\ 0 \\ 1 \end{bmatrix}.$$

(e) $[A|\mathbf{b}] = \begin{bmatrix} 1 & 1 & 7 & 2 \\ 2 & -4 & 14 & -1 \\ 5 & 11 & -7 & 8 \\ 2 & 5 & -4 & -3 \end{bmatrix} \rightarrow \begin{bmatrix} 1 & 1 & 7 & 2 \\ 0 & -6 & 0 & -5 \\ 0 & 6 & -42 & -2 \\ 0 & 3 & -18 & -7 \end{bmatrix} \rightarrow \begin{bmatrix} 1 & 1 & 7 & 2 \\ 0 & 1 & 0 & \frac{5}{6} \\ 0 & 6 & -42 & -2 \\ 0 & 3 & -18 & -7 \end{bmatrix} \rightarrow \begin{bmatrix} 1 & 1 & 7 & 2 \\ 0 & 1 & 0 & \frac{5}{6} \\ 0 & 0 & -42 & -7 \\ 0 & 0 & -18 & -\frac{19}{2} \end{bmatrix}.$

The last two rows imply, respectively, that $z = \frac{1}{6}$ and that $z = \frac{19}{36}$. This cannot be. The system has no solution.

(m) $[A|\mathbf{b}] = \begin{bmatrix} -6 & 8 & -5 & -5 & | & -11 \\ -6 & 7 & -10 & -8 & | & -9 \\ -8 & 10 & -10 & -9 & | & -13 \end{bmatrix} \rightarrow \begin{bmatrix} 0 & 1 & 5 & 3 & | & -2 \\ -6 & 7 & -10 & -8 & | & -9 \\ -8 & 10 & -10 & -9 & | & -13 \end{bmatrix}$

$\rightarrow \begin{bmatrix} 0 & 1 & 5 & 3 & | & -2 \\ 2 & -3 & 0 & 1 & | & 4 \\ -8 & 10 & -10 & -9 & | & -13 \end{bmatrix} \rightarrow \begin{bmatrix} 2 & -3 & 0 & 1 & | & 4 \\ 0 & 1 & 5 & 3 & | & -2 \\ -8 & 10 & -10 & -9 & | & -13 \end{bmatrix}$

$\rightarrow \begin{bmatrix} 1 & -\frac{3}{2} & 0 & \frac{1}{2} & | & 2 \\ 0 & 1 & 5 & 3 & | & -2 \\ -8 & 10 & -10 & -9 & | & -13 \end{bmatrix} \rightarrow \begin{bmatrix} 1 & -\frac{3}{2} & 0 & \frac{1}{2} & | & 2 \\ 0 & 1 & 5 & 3 & | & -2 \\ 0 & -2 & -10 & -5 & | & 3 \end{bmatrix} \rightarrow \begin{bmatrix} 1 & -\frac{3}{2} & 0 & \frac{1}{2} & | & 2 \\ 0 & 1 & 5 & 3 & | & -2 \\ 0 & 0 & 0 & 1 & | & -1 \end{bmatrix}.$

The variable $x_3 = t$ is free. Back substitution gives

$x_4 = -1$,

$x_2 + 5x_3 + 3x_4 = -2$, so

$\quad x_2 = -2 - 5x_3 - 3x_4 = -2 - 5t + 3 = 1 - 5t$, and then

$x_1 - \frac{3}{2}x_2 + \frac{1}{2}x_4 = 2$, so

$\quad x_1 = 2 + \frac{3}{2}x_2 - \frac{1}{2}x_4 = 2 + \frac{3}{2}(1 - 5t) - \frac{1}{2}(-1)$

$\quad\quad = 2 + \frac{3}{2} - \frac{15}{2}t + \frac{1}{2} = 4 - \frac{15}{2}t.$

The solution is $\begin{bmatrix} 4 - \frac{15}{2}t \\ 1 - 5t \\ t \\ -1 \end{bmatrix} = \begin{bmatrix} 4 \\ 1 \\ 0 \\ -1 \end{bmatrix} + t \begin{bmatrix} -\frac{15}{2} \\ -5 \\ 1 \\ 0 \end{bmatrix}.$

10. The question asks if there are scalars a and b such that $\begin{bmatrix} 2 \\ -11 \\ -3 \end{bmatrix} = a \begin{bmatrix} 0 \\ -1 \\ 5 \end{bmatrix} + b \begin{bmatrix} -1 \\ 4 \\ 9 \end{bmatrix}$. This is $\begin{bmatrix} 0 & -1 \\ -1 & 4 \\ 5 & 9 \end{bmatrix} \begin{bmatrix} a \\ b \end{bmatrix} = \begin{bmatrix} 2 \\ -11 \\ -3 \end{bmatrix}$.

Gaussian elimination proceeds

$$\begin{bmatrix} 0 & -1 & | & 2 \\ -1 & 4 & | & -11 \\ 5 & 9 & | & -3 \end{bmatrix} \rightarrow \begin{bmatrix} 1 & -4 & | & 11 \\ 0 & -1 & | & 2 \\ 0 & 29 & | & -58 \end{bmatrix} \rightarrow \begin{bmatrix} 1 & -4 & | & 11 \\ 0 & 1 & | & -2 \\ 0 & 0 & | & 0 \end{bmatrix}.$$

There is a unique solution: $b = -2$, $a = 3$. The given vector is indeed a linear combination of the other two.

$$\begin{bmatrix} 2 \\ -11 \\ -3 \end{bmatrix} = 3 \begin{bmatrix} 0 \\ -1 \\ 5 \end{bmatrix} - 2 \begin{bmatrix} -1 \\ 4 \\ 9 \end{bmatrix}.$$

12. We attempt to solve $A\mathbf{x} = \mathbf{b}$, with $\mathbf{x} = \begin{bmatrix} x_1 \\ x_2 \\ x_3 \end{bmatrix}$ and $\mathbf{b} = \begin{bmatrix} 1 \\ 6 \\ -4 \end{bmatrix}$. Gaussian elimination proceeds

$$[A|\mathbf{b}] = \begin{bmatrix} 2 & 3 & 4 & | & 1 \\ 4 & 7 & 5 & | & 6 \\ 6 & -1 & 9 & | & -4 \end{bmatrix} \rightarrow \begin{bmatrix} 2 & 3 & 4 & | & 1 \\ 0 & 1 & -3 & | & 4 \\ 0 & -10 & -3 & | & -7 \end{bmatrix} \rightarrow \begin{bmatrix} 2 & 3 & 4 & | & 1 \\ 0 & 1 & -3 & | & 4 \\ 0 & 0 & -33 & | & 33 \end{bmatrix} \rightarrow \begin{bmatrix} 2 & 3 & 4 & | & 1 \\ 0 & 1 & -3 & | & 4 \\ 0 & 0 & 1 & | & -1 \end{bmatrix},$$

so $x_3 = -1$, $x_2 = 4 + 3x_3 = 1$, $2x_1 = 1 - 3x_2 - 4x_3 = 2$, and $x_1 = 1$. Thus $A \begin{bmatrix} 1 \\ 1 \\ -1 \end{bmatrix} = \begin{bmatrix} 1 \\ 6 \\ -4 \end{bmatrix}$. This says that

$$\begin{bmatrix} 1 \\ 6 \\ -4 \end{bmatrix} = 1 \begin{bmatrix} 2 \\ 4 \\ 6 \end{bmatrix} + 1 \begin{bmatrix} 3 \\ 7 \\ -1 \end{bmatrix} - \begin{bmatrix} 4 \\ 5 \\ 9 \end{bmatrix}.$$

15. The question is whether every $\mathbf{b} = \begin{bmatrix} b_1 \\ b_2 \\ b_3 \end{bmatrix}$ in $\mathbf{R}^3$ is a linear combination of the given vectors. The vector equation $c_1\mathbf{v}_1 + c_2\mathbf{v}_2 + c_3\mathbf{v}_3 + c_4\mathbf{v}_4 = \mathbf{b}$ gives the system of equations

$$\begin{aligned} 2c_1 + c_2 - c_3 \quad\quad &= b_1 \\ c_2 + c_3 \quad + \quad 2c_4 &= b_2 \\ 2c_1 + c_2 - c_3 \quad\quad &= b_3. \end{aligned}$$

Gaussian elimination proceeds

$$\begin{bmatrix} 2 & 1 & -1 & 0 & | & b_1 \\ 0 & 1 & 1 & 2 & | & b_2 \\ 2 & 1 & -1 & 0 & | & b_3 \end{bmatrix} \rightarrow \begin{bmatrix} 1 & \frac{1}{2} & -\frac{1}{2} & 0 & | & \frac{1}{2}b_1 \\ 0 & 1 & 1 & 2 & | & b_2 \\ 0 & 0 & 0 & 0 & | & b_3 - b_1 \end{bmatrix}.$$

This system has a solution if and only if $b_3 - b_1 = 0$. Not every vector in $\mathbf{R}^3$ is a linear combination of the given vectors since not all vectors in $\mathbf{R}^3$ satisfy this condition. In fact, the given vectors span the plane with equation $x - z = 0$.

17. (a) Points on the line are solutions of $\begin{aligned} x \quad\quad + 2z &= 5 \\ 2x + y \quad\quad &= 2. \end{aligned}$

By Gaussian elimination, $\begin{bmatrix} 1 & 0 & 2 & | & 5 \\ 2 & 1 & 0 & | & 2 \end{bmatrix} \rightarrow \begin{bmatrix} 1 & 0 & 2 & | & 5 \\ 0 & 1 & -4 & | & -8 \end{bmatrix}$,

so $z = t$ is free, $y - 4z = -8$, so $y = 4t - 8$ and $x + 2z = 5$, so $x = -2t + 5$. Points on the line are obtained by varying t. For example, with $t = 0$, we see that $P(5, -8, 0)$ is on the line; and with $t = 1$, we see that $Q(3, -4, 1)$ is also on the line.

(b) The vector $\overrightarrow{PQ} = \begin{bmatrix} -2 \\ 4 \\ 1 \end{bmatrix}$ has the direction of the line.

(c) One possibility is $\begin{bmatrix} x \\ y \\ z \end{bmatrix} = \begin{bmatrix} 5 \\ -8 \\ 0 \end{bmatrix} + t \begin{bmatrix} -2 \\ 4 \\ 1 \end{bmatrix}$.

19. (a)
$$3 = a - 2b + 4c$$
$$-11 = a$$
$$24 = a + 5b + 25c$$

(b) $a = -11$, $b = -3$, $c = 2$; thus $p(x) = -11 - 3x + 2x^2$.

25. Gaussian elimination begins $\begin{bmatrix} 5 & 2 & | & a \\ -15 & -6 & | & b \end{bmatrix} \rightarrow \begin{bmatrix} 5 & 2 & | & a \\ 0 & 0 & | & b + 3a \end{bmatrix}$.

(a) If $b + 3a \neq 0$, the second equation is $0 = b + 3a \neq 0$, so there is no solution.

(b) Under no circumstances does the system have a unique solution since if $b + 3a = 0$, y is free and there are infinitely many solutions.

(c) If $b + 3a = 0$, there are infinitely many solutions.

Row echelon form is $\begin{bmatrix} 1 & \frac{2}{5} & | & \frac{a}{5} \\ 0 & 0 & | & 0 \end{bmatrix}$, so $y = t$ is free and $x = \frac{a}{5} - \frac{2}{5}y = \frac{a}{5} - \frac{2}{5}t$.

29. (a) $\begin{bmatrix} 1 & -2 & | & a \\ -5 & 3 & | & b \\ 3 & 1 & | & c \end{bmatrix} \rightarrow \begin{bmatrix} 1 & -2 & | & a \\ 0 & -7 & | & b + 5a \\ 0 & 7 & | & c - 3a \end{bmatrix} \rightarrow \begin{bmatrix} 1 & -2 & | & a \\ 0 & -7 & | & b + 5a \\ 0 & 0 & | & (c - 3a) + (b + 5a) \end{bmatrix}$.

The system has a solution if and only if $(c - 3a) + (b + 5a) = 0$; that is, if and only if $2a + b + c = 0$.

(b) If $2a + b + c = 0$, row echelon form is $\begin{bmatrix} 1 & -2 & | & a \\ 0 & -7 & | & b + 5a \\ 0 & 0 & | & 0 \end{bmatrix}$, which implies a unique solution: $y = -\frac{1}{7}(b + 5a)$, $x = a + 2y$. Infinitely many solutions is not a possibility.

Section 2.4—True/False

1. True.

2. False. Assuming $\mathbf{x} = \begin{bmatrix} x_1 \\ x_2 \\ x_3 \\ x_4 \end{bmatrix}$, variable x_4 is free, so the system has infinitely many solutions.

3. True. We are given that $A\mathbf{u} = \mathbf{b}$ and $A\mathbf{v} = \mathbf{b}$, so $A(\mathbf{u} - \mathbf{v}) = A\mathbf{u} - A\mathbf{v} = \mathbf{b} - \mathbf{b} = \mathbf{0}$.

4. True. Since $A\mathbf{x}$ is a linear combination of the columns of A, this is the definition of linear dependence.

Exercises 2.4

1. (a) This system is homogeneous. The elementary row operations will not change the right column of 0s, so we apply Gaussian elimination to the matrix of coefficients (not the **augmented** matrix of coefficients), simply remembering that there is a final column of 0s.

$$\begin{bmatrix} 1 & 1 & 1 \\ 1 & 3 & 0 \\ 2 & -1 & -1 \end{bmatrix} \rightarrow \begin{bmatrix} 1 & 1 & 1 \\ 0 & 2 & -1 \\ 0 & -3 & -3 \end{bmatrix} \rightarrow \begin{bmatrix} 1 & 1 & 1 \\ 0 & -3 & -3 \\ 0 & 2 & -1 \end{bmatrix} \rightarrow \begin{bmatrix} 1 & 1 & 1 \\ 0 & 1 & 1 \\ 0 & 2 & -1 \end{bmatrix} \rightarrow \begin{bmatrix} 1 & 1 & 1 \\ 0 & 1 & 1 \\ 0 & 0 & -3 \end{bmatrix}.$$

Remembering that the constants are all 0, the last equation reads $-3x_3 = 0$, so $x_3 = 0$, then $x_2 + x_3 = 0$ gives $x_2 = 0$ too and $x_1 + x_2 + x_3 = 0$ gives $x_1 = 0$. The solution is the zero vector, $\begin{bmatrix} 0 \\ 0 \\ 0 \end{bmatrix}$.

2. (a) We seek a, b, c, not all 0, such that

$$a\begin{bmatrix} 1 \\ 0 \\ 1 \end{bmatrix} + b\begin{bmatrix} -1 \\ 1 \\ 0 \end{bmatrix} + c\begin{bmatrix} 1 \\ 1 \\ 2 \end{bmatrix} = \begin{bmatrix} 0 \\ 0 \\ 0 \end{bmatrix}. \tag{*}$$

This leads to the system of equations

$$\begin{aligned} a - b + c &= 0 \\ b + c &= 0 \\ a \quad\;\; + 2c &= 0. \end{aligned}$$

The solutions are $a = -2t$, $b = -t$, $c = t$. Thus, and for example, $a = -2$, $b = -1$, $c = 1$ is a solution to (*).

(b) The result of (a) shows that $A\mathbf{x} = \mathbf{0}$, where $\mathbf{x} = \begin{bmatrix} -2 \\ -1 \\ 1 \end{bmatrix}$. If A were invertible, we would have $A^{-1}A\mathbf{x} = A^{-1}\mathbf{0}$; that is, $\mathbf{x} = \mathbf{0}$, which is not true. Thus A is not invertible.

3. (a) $c_1\mathbf{v}_1 + c_2\mathbf{v}_2 + c_3\mathbf{v}_3 + c_4\mathbf{v}_4 = \mathbf{0}$ is $A\mathbf{c} = \mathbf{0}$ with $A = \begin{bmatrix} 1 & 3 & 2 & 1 \\ 1 & 3 & 5 & 2 \\ 2 & 3 & 10 & 3 \\ 1 & 1 & 4 & 1 \end{bmatrix}$ and $\mathbf{c} = \begin{bmatrix} c_1 \\ c_2 \\ c_3 \\ c_4 \end{bmatrix}$. Gaussian elimination proceeds

$$A \to \begin{bmatrix} 1 & 3 & 2 & 1 \\ 0 & 0 & 3 & 1 \\ 0 & -3 & 6 & 1 \\ 0 & -2 & 2 & 0 \end{bmatrix} \to \begin{bmatrix} 1 & 3 & 2 & 1 \\ 0 & 1 & -1 & 0 \\ 0 & 0 & 3 & 1 \\ 0 & 0 & 3 & 1 \end{bmatrix} \to \begin{bmatrix} 1 & 3 & 2 & 1 \\ 0 & 1 & -1 & 0 \\ 0 & 0 & 1 & \frac{1}{3} \\ 0 & 0 & 0 & 0 \end{bmatrix}.$$

Since c_4 is a free variable, this homogeneous system has nontrivial solutions. The vectors are linearly dependent. For example, $2\mathbf{v}_1 - \mathbf{v}_2 - \mathbf{v}_3 + 3\mathbf{v}_4 = \mathbf{0}$.

4. The principle to follow when the given system involves letters is always to proceed as if the letters were numbers. So apply Gaussian elimination to the matrix of coefficients.

$$\begin{bmatrix} 1 & 1 & 0 \\ 0 & 1 & 1 \\ 1 & 0 & -1 \\ a & b & c \end{bmatrix} \to \begin{bmatrix} 1 & 1 & 0 \\ 0 & 1 & 1 \\ 0 & -1 & -1 \\ 0 & b-a & c \end{bmatrix} \to \begin{bmatrix} 1 & 1 & 0 \\ 0 & 1 & 1 \\ 0 & 0 & 0 \\ 0 & 0 & c-(b-a) \end{bmatrix}$$

(a) If $c - b + a \neq 0$, row echelon form is $\begin{bmatrix} 1 & 1 & 0 \\ 0 & 1 & 1 \\ 0 & 0 & 1 \\ 0 & 0 & 0 \end{bmatrix}$ and there is a unique solution: $x = y = z = 0$.

(b) It is not possible for this system to have no solution. Since it is homogeneous, there is always at least the solution, $x = y = z = 0$.

(c) If $c - b + a = 0$, row echelon form is $\begin{bmatrix} 1 & 1 & 0 \\ 0 & 1 & 1 \\ 0 & 0 & 0 \\ 0 & 0 & 0 \end{bmatrix}$. In this case, z is a free variable and there are infinitely many solutions.

6. (a) The given system is $A\mathbf{x} = \mathbf{b}$ where $A = \begin{bmatrix} 1 & 5 & 7 \\ 0 & 0 & 9 \end{bmatrix}$, $\mathbf{x} = \begin{bmatrix} x_1 \\ x_2 \\ x_3 \end{bmatrix}$, and $\mathbf{b} = \begin{bmatrix} -2 \\ 3 \end{bmatrix}$.

The augmented matrix is $\begin{bmatrix} 1 & 5 & 7 & -2 \\ 0 & 0 & 9 & 3 \end{bmatrix}$. Row echelon form is $U = \begin{bmatrix} 1 & 5 & 7 & -2 \\ 0 & 0 & 1 & \frac{1}{3} \end{bmatrix}$. Variable $x_2 = t$ is free. Back substitution yields $x_3 = \frac{1}{3}$ and $x_1 = -2 - 5x_2 - 7x_3 = -\frac{13}{3} - 5t$, so the solution is

$$\mathbf{x} = \begin{bmatrix} -\frac{13}{3} - 5t \\ t \\ \frac{1}{3} \end{bmatrix} = \begin{bmatrix} -\frac{13}{3} \\ 0 \\ \frac{1}{3} \end{bmatrix} + t \begin{bmatrix} -5 \\ 1 \\ 0 \end{bmatrix},$$

which is of the from $\mathbf{x}_p + \mathbf{x}_h$ with $\mathbf{x}_p = \begin{bmatrix} -\frac{13}{3} \\ 0 \\ \frac{1}{3} \end{bmatrix}$ and $\mathbf{x}_h = t \begin{bmatrix} -5 \\ 1 \\ 0 \end{bmatrix}$.

Section 2.5—True/False

1. False. If an elementary row operation applied to the identity matrix puts a 2 in the $(2, 2)$ position, the operation is "multiply row two by 2". This operation leaves a 0, not a 1, in the $(2, 3)$ position.

2. True. E is elementary, corresponding to the elementary row operation which subtracts twice row three from row two. The inverse of E adds twice row three to row two.

3. True. The elementary matrix E is obtained from the 3×3 identity matrix I by an elementary row operation that not does not change rows two or three of I.

4. False. They are the elements in positions $(1, 1)$, $(2, 2)$, and $(3, 3)$—three 0s.

5. False. For example, $\begin{bmatrix} 0 & 1 \\ 1 & 0 \end{bmatrix}$ is not lower triangular.

6. True. This is Theorem 2.5.15.

7. False. First solve $L\mathbf{y} = \mathbf{b}$ for $\mathbf{y}$ and then $U\mathbf{x} = \mathbf{y}$ for $\mathbf{x}$.

8. True. Since P is a permutation matrix, $P^{-1} = P^T = P$.

Exercises 2.5

1. (a) The second row of EA is the sum of the second row and four times the third row of A; all other rows of EA are the same as those of A.

(b) E is called an elementary matrix.

(c) Any elementary matrix is invertible, its inverse being that elementary matrix that "undoes" E. Here,
$$E^{-1} = \begin{bmatrix} 1 & 0 & 0 \\ 0 & 1 & -4 \\ 0 & 0 & 1 \end{bmatrix}.$$

2. (a) $\begin{bmatrix} 1 & 0 & 0 \\ 0 & 1 & 0 \\ 0 & -1 & 1 \end{bmatrix}$.

3. (a) EA was formed by the operation $R3 \rightarrow R3 - 6R1$, so $E = \begin{bmatrix} 1 & 0 & 0 \\ 0 & 1 & 0 \\ -6 & 0 & 1 \end{bmatrix}$.

6. (a) $E = \begin{bmatrix} 1 & 0 & 0 \\ 4 & 1 & 0 \\ 0 & 0 & 1 \end{bmatrix}$, $F = \begin{bmatrix} 1 & 0 & 0 \\ 0 & 1 & 0 \\ 0 & -3 & 1 \end{bmatrix}$. Since EA is A with row two replaced by row two plus four times row one,

$EF = \begin{bmatrix} 1 & 0 & 0 \\ 4 & 1 & 0 \\ 0 & -3 & 1 \end{bmatrix}$. Since FA is A with row three replaced by row three minus three times row two,

$FE = \begin{bmatrix} 1 & 0 & 0 \\ 4 & 1 & 0 \\ -12 & -3 & 1 \end{bmatrix}$.

(b) Since E^{-1} undoes E, $E^{-1} = \begin{bmatrix} 1 & 0 & 0 \\ -4 & 1 & 0 \\ 0 & 0 & 1 \end{bmatrix}$. Since F^{-1} undoes F, $F^{-1} = \begin{bmatrix} 1 & 0 & 0 \\ 0 & 1 & 0 \\ 0 & 3 & 1 \end{bmatrix}$. Since

$(EF)^{-1} = F^{-1}E^{-1}$ and $F^{-1}A$ is A with row three replaced by row three plus three times row two,

$(EF)^{-1} = \begin{bmatrix} 1 & 0 & 0 \\ -4 & 1 & 0 \\ -12 & 3 & 1 \end{bmatrix}$. Since $(FE)^{-1} = E^{-1}F^{-1}$ and $E^{-1}A$ is A with row two replaced by row two minus

four times row one, $(FE)^{-1} = \begin{bmatrix} 1 & 0 & 0 \\ -4 & 1 & 0 \\ 0 & 3 & 1 \end{bmatrix}$.

8. $\begin{bmatrix} 2 & 5 & 1 \\ 4 & x & 1 \\ 0 & 1 & -1 \end{bmatrix} \begin{bmatrix} 1 & \frac{5}{2} & \frac{1}{2} \\ 4 & x & 1 \\ 0 & 1 & -1 \end{bmatrix} \rightarrow \begin{bmatrix} 1 & \frac{5}{2} & \frac{1}{2} \\ 0 & x-10 & -1 \\ 0 & 1 & -1 \end{bmatrix}$, so if $x = 10$, it will be necessary to exchange rows two and three.

10. (a) A is elementary, so its inverse is the elementary matrix that "undoes" A: $A^{-1} = \begin{bmatrix} 1 & -3 \\ 0 & 1 \end{bmatrix}$.

(c) A is elementary, so its inverse is the elementary matrix that "undoes" A: $A^{-1} = \begin{bmatrix} 1 & 0 & 0 \\ 0 & 1 & 2 \\ 0 & 0 & 1 \end{bmatrix}$.

11. (a) The given matrix is a permutation matrix P, so $P^{-1} = P^T = \begin{bmatrix} 0 & 1 \\ 1 & 0 \end{bmatrix}$.

13. (a) Let $M = \begin{bmatrix} 1 & 0 \\ x & 1 \end{bmatrix}$. Then $LM = \begin{bmatrix} 1 & 0 \\ a+x & 1 \end{bmatrix}$, so we should let $x = -a$. The inverse of L is $\begin{bmatrix} 1 & 0 \\ -a & 1 \end{bmatrix}$.

14. (a) The first row of EA is the same as the first row of A. The second row of EA is (Row 2 of A) − (Row 1 of A); the third row of EA is (Row 3 of A) + 2 (Row 1 of A).

The first row of DA is 4 (Row 1 of A); the second row of DA is the same as the second row of A, and the third row of DA is −(Row 3 of A).

The rows of PA, in order, are the third, first, and second rows of A.

(b) Since P is a permutation matrix, $P^{-1} = P^T = \begin{bmatrix} 0 & 1 & 0 \\ 0 & 0 & 1 \\ 1 & 0 & 0 \end{bmatrix}$.

18. Ian is right. This matrix has no LU factorization. Lynn's first step corresponds to multiplication by $E = \begin{bmatrix} 1 & 1 \\ 0 & 1 \end{bmatrix}$, which is not lower triangular. A lower triangular matrix can be obtained only if the row operations used in the transformation from A to U only change entries on or below the main diagonal.

19. If E is elementary, remember that EA is A transformed in exactly the way that the identity matrix was transformed to make E. This is the basis for our factorizations (without calculation).

(a) $L = E_1 E_2 E_3$ with $E_1 = \begin{bmatrix} 1 & 0 & 0 \\ -2 & 1 & 0 \\ 0 & 0 & 1 \end{bmatrix}$, $E_2 = \begin{bmatrix} 1 & 0 & 0 \\ 0 & 1 & 0 \\ 3 & 0 & 1 \end{bmatrix}$, and $E_3 = \begin{bmatrix} 1 & 0 & 0 \\ 0 & 1 & 0 \\ 0 & 5 & 1 \end{bmatrix}$.

$L^{-1} = E_3^{-1} E_2^{-1} E_1^{-1} = \begin{bmatrix} 1 & 0 & 0 \\ 0 & 1 & 0 \\ 0 & -5 & 1 \end{bmatrix} \begin{bmatrix} 1 & 0 & 0 \\ 0 & 1 & 0 \\ -3 & 0 & 1 \end{bmatrix} \begin{bmatrix} 1 & 0 & 0 \\ 2 & 1 & 0 \\ 0 & 0 & 1 \end{bmatrix} = \begin{bmatrix} 1 & 0 & 0 \\ 2 & 1 & 0 \\ -13 & -5 & 1 \end{bmatrix}$.

22. (a) $A \to \begin{bmatrix} 2 & 1 & 0 \\ 0 & 4 & 2 \\ 0 & 0 & 5 \end{bmatrix} = U$ with $E = \begin{bmatrix} 1 & 0 & 0 \\ 0 & 1 & 0 \\ -3 & 0 & 1 \end{bmatrix}$.

(b) From (a), remembering the single multiplier, we get immediately $L = \begin{bmatrix} 1 & 0 & 0 \\ 0 & 1 & 0 \\ 3 & 0 & 1 \end{bmatrix}$.

(c) $A\mathbf{v} = -2 \begin{bmatrix} 2 \\ 0 \\ 6 \end{bmatrix} + 3 \begin{bmatrix} 1 \\ 4 \\ 3 \end{bmatrix} + 5 \begin{bmatrix} 0 \\ 2 \\ 5 \end{bmatrix}$.

25. (a) Here is a row reduction to upper triangular U.

$\begin{bmatrix} -3 & 3 & 6 \\ 2 & 5 & 10 \\ 0 & 1 & 4 \end{bmatrix} \to \begin{bmatrix} 1 & -1 & -2 \\ 2 & 5 & 10 \\ 0 & 1 & 4 \end{bmatrix} \to \begin{bmatrix} 1 & -1 & -2 \\ 0 & 7 & 14 \\ 0 & 1 & 4 \end{bmatrix}$

$\to \begin{bmatrix} 1 & -1 & -2 \\ 0 & 1 & 2 \\ 0 & 1 & 4 \end{bmatrix} \to \begin{bmatrix} 1 & -1 & -2 \\ 0 & 1 & 2 \\ 0 & 0 & 2 \end{bmatrix} = U.$

The corresponding elementary matrices are

$E_1 = \begin{bmatrix} -\frac{1}{3} & 0 & 0 \\ 0 & 1 & 0 \\ 0 & 0 & 1 \end{bmatrix}$, $E_2 = \begin{bmatrix} 1 & 0 & 0 \\ -2 & 1 & 0 \\ 0 & 0 & 1 \end{bmatrix}$, $E_3 = \begin{bmatrix} 1 & 0 & 0 \\ 0 & \frac{1}{7} & 0 \\ 0 & 0 & 1 \end{bmatrix}$, $E_4 = \begin{bmatrix} 1 & 0 & 0 \\ 0 & 1 & 0 \\ 0 & -1 & 1 \end{bmatrix}$.

We have $E_4 E_3 E_2 E_1 A = U$, so $A = LU$ with $L = (E_4 E_3 E_2 E_1)^{-1}$

$= E_1^{-1} E_2^{-1} E_3^{-1} E_4^{-1}$

$= \begin{bmatrix} -3 & 0 & 0 \\ 0 & 1 & 0 \\ 0 & 0 & 1 \end{bmatrix} \begin{bmatrix} 1 & 0 & 0 \\ 2 & 1 & 0 \\ 0 & 0 & 1 \end{bmatrix} \begin{bmatrix} 1 & 0 & 0 \\ 0 & 7 & 0 \\ 0 & 0 & 1 \end{bmatrix} \begin{bmatrix} 1 & 0 & 0 \\ 0 & 1 & 0 \\ 0 & 1 & 1 \end{bmatrix} = \begin{bmatrix} -3 & 0 & 0 \\ 2 & 7 & 0 \\ 0 & 1 & 1 \end{bmatrix}$.

(b) Here is a row reduction of A to upper triangular U that uses only the third elementary row operation.

$\begin{bmatrix} -3 & 3 & 6 \\ 2 & 5 & 10 \\ 0 & 1 & 4 \end{bmatrix} \xrightarrow{R2 \to R2 - (-\frac{2}{3})R1} \begin{bmatrix} -3 & 3 & 6 \\ 0 & 7 & 14 \\ 0 & 1 & 4 \end{bmatrix} \xrightarrow{R3 \to R3 - \frac{1}{7}(R2)} \begin{bmatrix} -3 & 3 & 6 \\ 0 & 7 & 14 \\ 0 & 0 & 2 \end{bmatrix} = U'$. We write $L = \begin{bmatrix} 1 & 0 & 0 \\ -\frac{2}{3} & 1 & 0 \\ 0 & \frac{1}{7} & 1 \end{bmatrix}$.

27. (a) We bring A to upper triangular form using only the third elementary row operation. This can be done in one step:

$$A = \begin{bmatrix} 2 & 4 \\ -1 & 6 \end{bmatrix} \to \begin{bmatrix} 2 & 4 \\ 0 & 8 \end{bmatrix} = U'$$

with $L = \begin{bmatrix} 1 & 0 \\ -\frac{1}{2} & 1 \end{bmatrix}$ and $A = LU$.

(c) $A \to \begin{bmatrix} 5 & 0 & 15 \\ 0 & 1 & 2 \\ 0 & -3 & -4 \\ 0 & 2 & 5 \end{bmatrix} \to \begin{bmatrix} 5 & 0 & 15 \\ 0 & 1 & 2 \\ 0 & 0 & 2 \\ 0 & 0 & 1 \end{bmatrix} \to \begin{bmatrix} 5 & 0 & 15 \\ 0 & 1 & 2 \\ 0 & 0 & 2 \\ 0 & 0 & 0 \end{bmatrix} = U$. Keeping track of the multipliers, $L = \begin{bmatrix} 1 & 0 & 0 \\ 0 & 1 & 0 \\ \frac{4}{5} & -3 & 1 \end{bmatrix}$.

28. (a) First we solve $L\begin{bmatrix} y_1 \\ y_2 \end{bmatrix} = \begin{bmatrix} -2 \\ 9 \end{bmatrix}$. Using forward substitution, we have $2y_1 = -2$, so $y_1 = -1$, and then $6y_1 + 5y_2 = 9$, so $y_2 = 3$. Thus $\begin{bmatrix} c_1 \\ c_2 \end{bmatrix} = \begin{bmatrix} -1 \\ 3 \end{bmatrix}$. Now we solve $U\begin{bmatrix} x \\ y \end{bmatrix} = \begin{bmatrix} -1 \\ 3 \end{bmatrix}$. Using back substitution, we obtain $y = 3$; then, since $x + \frac{1}{2}y = -1$, we get $x = -\frac{5}{2}$. Thus $\begin{bmatrix} x \\ y \end{bmatrix} = \begin{bmatrix} -\frac{5}{2} \\ 3 \end{bmatrix}$.

(b) We have just shown that $A\begin{bmatrix} -\frac{5}{2} \\ 3 \end{bmatrix} = \begin{bmatrix} -2 \\ 9 \end{bmatrix}$, thus $\begin{bmatrix} -2 \\ 9 \end{bmatrix} = -\frac{5}{2}\begin{bmatrix} 1 \\ 6 \end{bmatrix} + 3\begin{bmatrix} 1 \\ 8 \end{bmatrix}$.

31. We have $A = LU$ with $L = \begin{bmatrix} -2 & 0 \\ 1 & 1 \end{bmatrix}$ and $U = \begin{bmatrix} 1 & 3 \\ 0 & 12 \end{bmatrix}$.

First, we solve $L\mathbf{y} = \mathbf{b}$ for $\mathbf{y} = \begin{bmatrix} y_1 \\ y_2 \end{bmatrix}$. This is the system $\begin{aligned} -2y_1 &= -3 \\ y_1 + y_2 &= 4. \end{aligned}$

By forward substitution, we obtain $y_1 = \frac{3}{2}$ and $y_2 = 4 - y_1 = \frac{5}{2}$, so $\mathbf{y} = \begin{bmatrix} \frac{3}{2} \\ \frac{5}{2} \end{bmatrix}$. Now we solve $U\mathbf{x} = \mathbf{y}$ by back substitution. We have

$$\begin{aligned} x_1 + 3x_2 &= \tfrac{3}{2} \\ 12x_2 &= \tfrac{5}{2} \end{aligned}$$

so $x_2 = \frac{5}{24}$ and $x_1 = \frac{3}{2} - 3x_2 = \frac{3}{2} - \frac{5}{8} = \frac{7}{8}$. Our solution is $\mathbf{x} = \begin{bmatrix} \frac{7}{8} \\ \frac{5}{24} \end{bmatrix}$.

34. (a) $\begin{bmatrix} 1 & -1 & 0 & | & 1 \\ 3 & 0 & 2 & | & 0 \\ -1 & 0 & -1 & | & 0 \end{bmatrix} \to \begin{bmatrix} 1 & -1 & 0 & | & 1 \\ 0 & 3 & 2 & | & -3 \\ 0 & -1 & -1 & | & 1 \end{bmatrix} \to \begin{bmatrix} 1 & -1 & 0 & | & 1 \\ 0 & 3 & 2 & | & -3 \\ 0 & 0 & -\frac{1}{3} & | & 0 \end{bmatrix}$.

Letting $\mathbf{x} = \begin{bmatrix} x \\ y \\ z \end{bmatrix}$, the equations corresponding to this last (upper triangular) system are

$$\begin{aligned} x - y &= 1 \\ 3y + 2z &= -3 \\ -\tfrac{1}{3}z &= 0. \end{aligned}$$

So $z = 0$. Then $3y = -3$ so $y = -1$. Then $x = 1 + y = 0$, so $\mathbf{x} = \begin{bmatrix} 0 \\ -1 \\ 0 \end{bmatrix}$.

(b) From the preceding elimination, we see that $U = \begin{bmatrix} 1 & -1 & 0 \\ 0 & 3 & 2 \\ 0 & 0 & -\frac{1}{3} \end{bmatrix}$.

Keeping track of the multipliers, $L = \begin{bmatrix} 1 & 0 & 0 \\ 3 & 1 & 0 \\ -1 & -\frac{1}{3} & 1 \end{bmatrix}$.

(c) The matrices L and U are invertible since each is square triangular with all diagonal entries nonzero. Thus the product $A = LU$ of L and U is invertible.

(d) The first column of A^{-1} is the vector $\mathbf{x}$ that satisfies $A\mathbf{x} = \begin{bmatrix} 1 \\ 0 \\ 0 \end{bmatrix}$.

We found this in (a); $\mathbf{x} = \begin{bmatrix} 0 \\ -1 \\ 0 \end{bmatrix}$.

37. (a)   **i.** The system is $A\mathbf{x} = \mathbf{b}$ with $A = \begin{bmatrix} 2 & 2 \\ 4 & 9 \end{bmatrix}$, $\mathbf{x} = \begin{bmatrix} x_1 \\ x_2 \end{bmatrix}$ and $\mathbf{b} = \begin{bmatrix} 8 \\ 21 \end{bmatrix}$.

ii. Row reduction gives $A \to \begin{bmatrix} 2 & 2 \\ 0 & 5 \end{bmatrix} = U$ so $A = LU$ with $L = \begin{bmatrix} 1 & 0 \\ 2 & 1 \end{bmatrix}$.

iii. We must solve $L(U\mathbf{x}) = \mathbf{b}$, so let $U\mathbf{x} = \mathbf{y} = \begin{bmatrix} y_1 \\ y_2 \end{bmatrix}$ and first solve $L\mathbf{y} = \mathbf{b}$. The corresponding equations are

$$\begin{aligned} y_1 &= 8 \\ 2y_1 + y_2 &= 21 \end{aligned}$$

so forward substitution gives $y_1 = 8$ and $y_2 = 21 - 2y_1 = 5$. Now we solve $U\mathbf{x} = \mathbf{y}$. The corresponding equations are

$$\begin{aligned} 2x_1 + 2x_2 &= 8 \\ 5x_2 &= 5 \end{aligned}$$

and back substitution gives $x_2 = 1$ and $2x_1 = 8 - 2x_2 = 6$, giving $x_1 = 3$. Our solution is $\mathbf{x} = \begin{bmatrix} 3 \\ 1 \end{bmatrix}$.

38. (a)   **i.** $A \to \begin{bmatrix} 1 & 1 & 1 \\ 0 & 1 & 2 \\ 0 & 2 & 1 \end{bmatrix} \to \begin{bmatrix} 1 & 1 & 1 \\ 0 & 1 & 2 \\ 0 & 0 & -3 \end{bmatrix} = U$ and, keeping track of multipliers, $L = \begin{bmatrix} 1 & 0 & 0 \\ 1 & 1 & 0 \\ 0 & 2 & 1 \end{bmatrix}$.

ii. $A\mathbf{x} = \mathbf{b}$ is $L(U\mathbf{x}) = \mathbf{b}$. Let $U\mathbf{x} = \mathbf{y}$; then $L\mathbf{y} = \mathbf{b}$. Solving for $\mathbf{y} = \begin{bmatrix} y_1 \\ y_2 \\ y_3 \end{bmatrix}$, we find $y_1 = -1$, $y_1 + y_2 = 0$, so

$y_2 = -y_1 = 1$ and $2y_2 + y_3 = 1$ so $y_3 = 1 - 2y_2 = -1$. Thus $\mathbf{y} = \begin{bmatrix} -1 \\ 1 \\ -1 \end{bmatrix}$.

Now we solve $U\mathbf{x} = \mathbf{y}$. We have $x_3 = \frac{1}{3}$, $x_2 + 2x_3 = 1$, so $x_2 = \frac{1}{3}$ and $x_1 + x_2 + x_3 = 1$, so

$x_1 = -1 - x_2 - x_3 = -\frac{5}{3}$. Thus $\mathbf{x} = \begin{bmatrix} -\frac{5}{3} \\ \frac{1}{3} \\ \frac{1}{3} \end{bmatrix}$.

40. (a) $\begin{bmatrix} 2 & 1 \\ 6 & 8 \end{bmatrix} \to \begin{bmatrix} 2 & 1 \\ 0 & 5 \end{bmatrix}$, so $\begin{bmatrix} 2 & 1 \\ 6 & 8 \end{bmatrix} = \begin{bmatrix} 1 & 0 \\ 3 & 1 \end{bmatrix} \begin{bmatrix} 2 & 1 \\ 0 & 5 \end{bmatrix}$

41. (a) Let $P' = \begin{bmatrix} 0 & 1 \\ 1 & 0 \end{bmatrix}$. Then $P'A = \begin{bmatrix} 1 & 2 & 1 \\ 0 & 3 & 1 \end{bmatrix} = U$. So $P'A = LU$ with $L = I$, the 2×2 identity matrix. Thus

$A = PLU$, with $P = (P')^{-1} = P^T = \begin{bmatrix} 0 & 1 \\ 1 & 0 \end{bmatrix}$.

Section 2.6—True/False

1. False.

2. True.

3. True. $\begin{bmatrix} 0 & 0 \\ 0 & 0 \end{bmatrix} = \begin{bmatrix} 1 & 0 \\ 0 & 1 \end{bmatrix} \begin{bmatrix} 0 & 0 \\ 0 & 0 \end{bmatrix} \begin{bmatrix} 1 & 0 \\ 0 & 1 \end{bmatrix}$.

4. False. $\begin{bmatrix} 0 & 0 \\ 0 & 1 \end{bmatrix} = \begin{bmatrix} 1 & 0 \\ a & 1 \end{bmatrix} \begin{bmatrix} 0 & 0 \\ 0 & 1 \end{bmatrix} \begin{bmatrix} 1 & b \\ 0 & 1 \end{bmatrix}$, for any a and b, as shown in the text.

5. False. $\begin{bmatrix} 0 & 1 \\ 1 & 0 \end{bmatrix} \begin{bmatrix} 2 & 0 \\ 0 & 1 \end{bmatrix} = \begin{bmatrix} 0 & 1 \\ 2 & 0 \end{bmatrix}$.

6. True. We know that $(A^{-1})^T = (A^T)^{-1}$ (see Theorem 2.2.19, part 4), so $(A^{-1})^T = A^{-1}$.

Exercises 2.6

1. (a) Call the given matrix A. We reduce A to upper triangular using only the third elementary row operation and keep track of the multipliers.

$$\begin{bmatrix} -2 & 1 & 6 \\ 4 & 0 & 8 \\ 6 & 3 & -10 \end{bmatrix} \to \begin{bmatrix} -2 & 1 & 6 \\ 0 & 2 & 20 \\ 0 & 6 & 8 \end{bmatrix} \to \begin{bmatrix} -2 & 1 & 6 \\ 0 & 2 & 20 \\ 0 & 0 & -52 \end{bmatrix} = U' \text{ with } L = \begin{bmatrix} 1 & 0 & 0 \\ -2 & 1 & 0 \\ -3 & 3 & 1 \end{bmatrix}.$$

We factor $U' = DU$ with $D = \begin{bmatrix} -2 & 0 & 0 \\ 0 & 2 & 0 \\ 0 & 0 & -52 \end{bmatrix}$ and $U = \begin{bmatrix} 1 & -\frac{1}{2} & 3 \\ 0 & 1 & 10 \\ 0 & 0 & 1 \end{bmatrix}$.

2. (a) We have $A = LU'$ with $L = \begin{bmatrix} 1 & 0 & 0 \\ -\frac{2}{3} & 1 & 0 \\ 0 & \frac{1}{7} & 1 \end{bmatrix}$ and $U' = \begin{bmatrix} -3 & 3 & 6 \\ 0 & 7 & 14 \\ 0 & 0 & 2 \end{bmatrix}$. Factoring $U' = DU$ with $D = \begin{bmatrix} -3 & 0 & 0 \\ 0 & 7 & 0 \\ 0 & 0 & 2 \end{bmatrix}$

and $U = \begin{bmatrix} 1 & -1 & -2 \\ 0 & 1 & 2 \\ 0 & 0 & 1 \end{bmatrix}$ gives $A = LDU$.

3. (a) We have $A = LU'$ with $L = \begin{bmatrix} 1 & 0 \\ -\frac{1}{2} & 1 \end{bmatrix}$ and $U' = \begin{bmatrix} 2 & 4 \\ 0 & 8 \end{bmatrix}$ and $U' = DU$ with $D = \begin{bmatrix} 2 & 0 \\ 0 & 8 \end{bmatrix}$ and $U = \begin{bmatrix} 1 & 2 \\ 0 & 1 \end{bmatrix}$.

(c) We have $A = LU'$ with $L = \begin{bmatrix} 1 & 0 & 0 & 0 \\ 0 & 1 & 0 & 0 \\ \frac{4}{5} & -3 & 1 & 0 \\ -\frac{2}{5} & 2 & \frac{1}{2} & 1 \end{bmatrix}$

and $U' = \begin{bmatrix} 5 & 0 & 15 \\ 0 & 1 & 2 \\ 0 & 0 & 2 \\ 0 & 0 & 1 \end{bmatrix} = DU$ with $D = \begin{bmatrix} 5 & 0 & 0 & 0 \\ 0 & 1 & 0 & 0 \\ 0 & 0 & 2 & 0 \\ 0 & 0 & 0 & 0 \end{bmatrix}$ and $U = \begin{bmatrix} 1 & 0 & 3 \\ 0 & 1 & 2 \\ 0 & 0 & 1 \\ 0 & 0 & 0 \end{bmatrix}$.

5. (a) A matrix X is symmetric if $X^T = X$. We are given that $A^T = A$ and $B^T = B$. Since $(AB)^T = B^T A^T = BA$ which is not, in general, the same as AB, AB is not symmetric.

7. $(A + B)^T = A^T + B^T = -A - B = -(A + B)$. Since the transpose of $A + B$ is its negative, $A + B$ is skew-symmetric.

9. (a) We transform A to an upper triangular matrix using only the third elementary row operation:

$$\begin{bmatrix} 2 & 5 \\ 5 & 9 \end{bmatrix} \rightarrow \begin{bmatrix} 2 & 5 \\ 0 & -\frac{7}{2} \end{bmatrix} = U' = DU$$

with $D = \begin{bmatrix} 2 & 0 \\ 0 & -\frac{7}{2} \end{bmatrix}$ and $U = \begin{bmatrix} 1 & \frac{5}{2} \\ 0 & 1 \end{bmatrix} = L^T$. So $L = \begin{bmatrix} 1 & 0 \\ \frac{5}{2} & 1 \end{bmatrix}$.

Section 2.7—True/False

1. True.

2. False. Both $\begin{bmatrix} -1 & 0 & 2 \\ 0 & 2 & 14 \end{bmatrix}$ and $\begin{bmatrix} 1 & 0 & -2 \\ 0 & 1 & 7 \end{bmatrix}$ are row echelon forms of $A = \begin{bmatrix} -1 & 0 & 2 \\ 4 & 2 & 6 \end{bmatrix}$.

3. True. $(AB)^{-1} = B^{-1}A^{-1}$.

4. False. Let A be any invertible matrix and let $B = -A$.

5. True. Since E and F are elementary, they are invertible, and the product of invertible matrices is invertible.

6. False. $(ABC)^{-1} = C^{-1}B^{-1}A^{-1}$. In general, there is no reason why $(ABC)(A^{-1}B^{-1}C^{-1})$ should be I.

7. False. This holds only if A and B are square. (See Problem 2.7.2.)

Exercises 2.7

1. (a) A is square and $AB = I$, so A and B are inverses.

2. (a)
$$\begin{bmatrix} 1 & -1 & 2 \\ -2 & 0 & 1 \\ -3 & 3 & -5 \\ 2 & -1 & 4 \end{bmatrix} \rightarrow \begin{bmatrix} 1 & -1 & 2 \\ 0 & -2 & 5 \\ 0 & 0 & 1 \\ 0 & 1 & 0 \end{bmatrix} \rightarrow \begin{bmatrix} 1 & -1 & 2 \\ 0 & 1 & 0 \\ 0 & 0 & 1 \\ 0 & -2 & 5 \end{bmatrix} \rightarrow \begin{bmatrix} 1 & 0 & 2 \\ 0 & 1 & 0 \\ 0 & 0 & 1 \\ 0 & 0 & 5 \end{bmatrix} \rightarrow \begin{bmatrix} 1 & 0 & 0 \\ 0 & 1 & 0 \\ 0 & 0 & 1 \\ 0 & 0 & 0 \end{bmatrix}.$$
This is reduced row echelon form.

3. (a) $[A|I] = \begin{bmatrix} 3 & 1 & | & 1 & 0 \\ -6 & -3 & | & 0 & 1 \end{bmatrix} \rightarrow \begin{bmatrix} 1 & \frac{1}{3} & | & \frac{1}{3} & 0 \\ 0 & -1 & | & 2 & 1 \end{bmatrix} \rightarrow \begin{bmatrix} 1 & 0 & | & 1 & \frac{1}{3} \\ 0 & 1 & | & -2 & -1 \end{bmatrix}$. The inverse is $\begin{bmatrix} 1 & \frac{1}{3} \\ -2 & -1 \end{bmatrix}$.

(e) $[A|I] = \begin{bmatrix} 2 & 4 & 2 & | & 1 & 0 & 0 \\ 1 & 2 & 3 & | & 0 & 1 & 0 \\ 3 & 2 & 1 & | & 0 & 0 & 1 \end{bmatrix} \rightarrow \begin{bmatrix} 1 & 2 & 3 & | & 0 & 1 & 0 \\ 2 & 4 & 2 & | & 1 & 0 & 0 \\ 3 & 2 & 1 & | & 0 & 0 & 1 \end{bmatrix} \rightarrow \begin{bmatrix} 1 & 2 & 3 & | & 0 & 1 & 0 \\ 0 & 0 & -4 & | & 1 & -2 & 0 \\ 0 & -4 & -8 & | & 0 & -3 & 1 \end{bmatrix}$

$$\rightarrow \begin{bmatrix} 1 & 2 & 3 & | & 0 & 1 & 0 \\ 0 & 1 & 2 & | & 0 & \frac{3}{4} & -\frac{1}{4} \\ 0 & 0 & 1 & | & -\frac{1}{4} & \frac{1}{2} & 0 \end{bmatrix} \rightarrow \begin{bmatrix} 1 & 0 & -1 & | & 0 & -\frac{1}{2} & \frac{1}{2} \\ 0 & 1 & 2 & | & 0 & \frac{3}{4} & -\frac{1}{4} \\ 0 & 0 & 1 & | & -\frac{1}{4} & \frac{1}{2} & 0 \end{bmatrix} \rightarrow \begin{bmatrix} 1 & 0 & 0 & | & -\frac{1}{4} & 0 & \frac{1}{2} \\ 0 & 1 & 0 & | & \frac{1}{2} & -\frac{1}{4} & -\frac{1}{4} \\ 0 & 0 & 1 & | & -\frac{1}{4} & \frac{1}{2} & 0 \end{bmatrix}.$$

The inverse is $\begin{bmatrix} -\frac{1}{4} & 0 & \frac{1}{2} \\ \frac{1}{2} & -\frac{1}{4} & -\frac{1}{4} \\ -\frac{1}{4} & \frac{1}{2} & 0 \end{bmatrix}$.

5. i. The system is $A\mathbf{x} = \mathbf{b}$ with $A = \begin{bmatrix} 1 & -2 & 2 \\ 2 & 1 & 1 \\ 1 & 0 & 1 \end{bmatrix}$, $\mathbf{x} = \begin{bmatrix} x_1 \\ x_2 \\ x_3 \end{bmatrix}$, and $\mathbf{b} = \begin{bmatrix} 3 \\ 0 \\ -2 \end{bmatrix}$.

ii.   We have $[A|I] = \begin{bmatrix} 1 & -2 & 2 & | & 1 & 0 & 0 \\ 2 & 1 & 1 & | & 0 & 1 & 0 \\ 1 & 0 & 1 & | & 0 & 0 & 1 \end{bmatrix} \rightarrow \begin{bmatrix} 1 & -2 & 2 & | & 1 & 0 & 0 \\ 0 & 5 & -3 & | & -2 & 1 & 0 \\ 0 & 2 & -1 & | & -1 & 0 & 1 \end{bmatrix} \rightarrow \begin{bmatrix} 1 & -2 & 2 & | & 1 & 0 & 0 \\ 0 & 1 & -\frac{3}{5} & | & -\frac{2}{5} & \frac{1}{5} & 0 \\ 0 & 2 & -1 & | & -1 & 0 & 1 \end{bmatrix}$

$\rightarrow \begin{bmatrix} 1 & 0 & \frac{4}{5} & | & \frac{1}{5} & \frac{2}{5} & 0 \\ 0 & 1 & -\frac{3}{5} & | & -\frac{2}{5} & \frac{1}{5} & 0 \\ 0 & 0 & \frac{1}{5} & | & -\frac{1}{5} & -\frac{2}{5} & 1 \end{bmatrix} \rightarrow \begin{bmatrix} 1 & 0 & \frac{4}{5} & | & \frac{1}{5} & \frac{2}{5} & 0 \\ 0 & 1 & -\frac{3}{5} & | & -\frac{2}{5} & \frac{1}{5} & 0 \\ 0 & 0 & 1 & | & -1 & -2 & 5 \end{bmatrix} \rightarrow \begin{bmatrix} 1 & 0 & 0 & | & 1 & 2 & -4 \\ 0 & 1 & 0 & | & -1 & -1 & 3 \\ 0 & 0 & 1 & | & -1 & -2 & 5 \end{bmatrix},$

so $A^{-1} = \begin{bmatrix} 1 & 2 & -4 \\ -1 & -1 & 3 \\ -1 & -2 & 5 \end{bmatrix}$ and $\mathbf{x} = A^{-1}\mathbf{b} = \begin{bmatrix} 1 & 2 & -4 \\ -1 & -1 & 3 \\ -1 & -2 & 5 \end{bmatrix}\begin{bmatrix} 3 \\ 0 \\ -2 \end{bmatrix} = \begin{bmatrix} 11 \\ -9 \\ -13 \end{bmatrix}.$

iii.   The given vectors are the columns of A, so $A\mathbf{x}$ is a linear combination of these vectors. (Recall 2.1.33.) We have

just shown that $A\begin{bmatrix} 11 \\ -9 \\ -13 \end{bmatrix} = \begin{bmatrix} 3 \\ 0 \\ -2 \end{bmatrix}$, so $\begin{bmatrix} 3 \\ 0 \\ -2 \end{bmatrix} = 11\begin{bmatrix} 1 \\ 2 \\ 1 \end{bmatrix} - 9\begin{bmatrix} -2 \\ 1 \\ 0 \end{bmatrix} - 13\begin{bmatrix} 2 \\ 1 \\ 1 \end{bmatrix}.$

8. $A = \begin{bmatrix} 1 & 2 \\ 3 & 0 \end{bmatrix}\begin{bmatrix} 5 & 1 \\ -1 & 1 \end{bmatrix}^{-1} = \begin{bmatrix} 1 & 2 \\ 3 & 0 \end{bmatrix}\frac{1}{6}\begin{bmatrix} 1 & -1 \\ 1 & 5 \end{bmatrix} = \frac{1}{6}\begin{bmatrix} 3 & 9 \\ 3 & -3 \end{bmatrix} = \frac{1}{2}\begin{bmatrix} 1 & 3 \\ 1 & -1 \end{bmatrix}.$

11. (a) $\begin{bmatrix} 0 & -1 \\ 2 & 1 \end{bmatrix} \rightarrow \begin{bmatrix} 2 & 1 \\ 0 & -1 \end{bmatrix} = E_1 A \rightarrow \begin{bmatrix} 1 & \frac{1}{2} \\ 0 & -1 \end{bmatrix} = E_2 E_1 A \rightarrow \begin{bmatrix} 1 & \frac{1}{2} \\ 0 & 1 \end{bmatrix} = E_3 E_2 E_1 A \rightarrow \begin{bmatrix} 1 & 0 \\ 0 & 1 \end{bmatrix} = I =$

$E_4 E_3 E_2 E_1 A$, where $E_1 = \begin{bmatrix} 0 & 1 \\ 1 & 0 \end{bmatrix}$, $E_2 = \begin{bmatrix} \frac{1}{2} & 0 \\ 0 & 1 \end{bmatrix}$, $E_3 = \begin{bmatrix} 1 & 0 \\ 0 & -1 \end{bmatrix}$ and $E_4 = \begin{bmatrix} 1 & -\frac{1}{2} \\ 0 & 1 \end{bmatrix}$. Thus

$A = (E_4 E_3 E_2 E_1)^{-1} = E_1^{-1} E_2^{-1} E_3^{-1} E_4^{-1} = \begin{bmatrix} 0 & 1 \\ 1 & 0 \end{bmatrix}\begin{bmatrix} 2 & 0 \\ 0 & 1 \end{bmatrix}\begin{bmatrix} 1 & 0 \\ 0 & -1 \end{bmatrix}\begin{bmatrix} 1 & \frac{1}{2} \\ 0 & 1 \end{bmatrix}.$

12. (a)   You compute XY or YX. If $XY = I$ or $YX = I$, then X and Y are inverses.

14.   No it is not. Consider, for instance, $\begin{bmatrix} 1 & 0 \\ 0 & 1 \end{bmatrix} = \begin{bmatrix} 1 & 1 \\ 0 & 1 \end{bmatrix}\begin{bmatrix} 1 & -1 \\ 0 & 1 \end{bmatrix}.$

Section 3.1—True/False

1.   True. Strike out first row and second column of the matrix. The determinant of what is left is 2, which is the $(1, 2)$ minor. The $(1, 2)$ cofactor is $(-1)^{1+2}2 = -2$.

2.   True. The scalar is $\det A$.

3.   False. $-2I = \begin{bmatrix} -2 & 0 \\ 0 & -2 \end{bmatrix}$, so the determinant is the product of the diagonal entries, which is $+4$.

4.   True. The determinant of a triangular matrix is the product of its diagonal entries. If no diagonal entry is 0, the determinant is not 0, so the matrix is invertible.

5.   False. Let $A = \begin{bmatrix} 2 & 1 \\ 4 & 3 \end{bmatrix}$ and $B = -A$. Then $\det A = \det B = 2$, but $\det(A + B) = 0$.

6.   False. For example, consider $A = \begin{bmatrix} 1 & 0 & 0 \\ 0 & 0 & 0 \\ 0 & 0 & 0 \end{bmatrix}.$

7.   True. Since A is triangular, $\det A$ is the product of the diagonal entries.

8.   True. Since $\det A \neq 0$, A is invertible, so the equation $A\mathbf{x} = \mathbf{b}$ can be solved for any $\mathbf{b}$ (the solution is $\mathbf{x} = A^{-1}\mathbf{b}$). Now remember that $A\mathbf{x}$ is a linear combination of the columns of A.

Exercises 3.1

1. Not possible. Only a square matrix has a determinant.

3. (a) **i.** $M = \begin{bmatrix} 9 & -7 \\ -4 & 2 \end{bmatrix}$, $C = \begin{bmatrix} 9 & 7 \\ 4 & 2 \end{bmatrix}$, $AC^T = \begin{bmatrix} 2 & -4 \\ -7 & 9 \end{bmatrix}\begin{bmatrix} 9 & 4 \\ 7 & 2 \end{bmatrix} = \begin{bmatrix} -10 & 0 \\ 0 & -10 \end{bmatrix}$

        ii. $\det A = -10$

        iii. $A^{-1} = -\frac{1}{10}\begin{bmatrix} 9 & 4 \\ 7 & 2 \end{bmatrix}$

(c) **i.** $M = \begin{bmatrix} -26 & -12 & 4 \\ -13 & -6 & 2 \\ 13 & 6 & -2 \end{bmatrix}$, $C = \begin{bmatrix} -26 & 12 & 4 \\ 13 & -6 & -2 \\ 13 & -6 & -2 \end{bmatrix}$, $AC^T = \begin{bmatrix} 2 & 3 & 4 \\ 0 & -1 & 3 \\ 4 & 7 & 5 \end{bmatrix}\begin{bmatrix} -26 & 13 & 13 \\ 12 & -6 & -6 \\ 4 & -2 & -2 \end{bmatrix} = \begin{bmatrix} 0 & 0 & 0 \\ 0 & 0 & 0 \\ 0 & 0 & 0 \end{bmatrix}$.

        ii. Since $AC^T = (\det A)I$, we must have $\det A = 0$ in this case.

        iii. A is not invertible since $AC^T = \mathbf{0}$ but $C^T \neq \mathbf{0}$. See **3.1.13**.

4. (a) A matrix is not invertible if and only if its determinant is 0. The determinant of the given matrix is $x^2 + 3x + 2 = 0$; so the matrix is singular if and only if $x = -1$ or $x = -2$.

(c) Expanding by cofactors across the second row, we see that the matrix has determinant $-3\begin{vmatrix} 1 & x \\ 4 & 7 \end{vmatrix} = -3(7 - 4x)$, so the matrix is singular if and only if $x = \frac{7}{4}$.

5. (a) Expanding by cofactors of the third row gives

$$\det A = 2\begin{vmatrix} -1 & 2 \\ 1 & 1 \end{vmatrix} + \begin{vmatrix} 1 & 2 \\ 3 & 1 \end{vmatrix} + 3\begin{vmatrix} 1 & -1 \\ 3 & 1 \end{vmatrix} = 2(-3) + (-5) + 3(4) = 1.$$

7. (a) $c_{13} = \begin{vmatrix} 2 & 1 \\ 1 & 2 \end{vmatrix} = 3$, $c_{21} = -\begin{vmatrix} 0 & 1 \\ 2 & 1 \end{vmatrix} = -(-2) = 2$, $c_{32} = -\begin{vmatrix} 1 & 1 \\ 2 & 1 \end{vmatrix} = -(-1) = 1$.

(b) Expanding by cofactors of the first row, $\det A = 1(-1) + 0(-1) + 1(3) = 2$.

(c) A is invertible since $\det A \neq 0$.

(d) $A^{-1} = \frac{1}{\det A}C^T = \frac{1}{2}\begin{bmatrix} -1 & 2 & -1 \\ -1 & 0 & 1 \\ 3 & -2 & 1 \end{bmatrix}$.

9. (a) Let $A = \begin{bmatrix} -1 & 2 & x \\ 0 & 3 & y \\ 2 & -2 & z \end{bmatrix}$. The $(1, 1)$ cofactor is $19 = 3z + 2y$. The $(2, 1)$ cofactor is $-14 = -(2z + 2x)$, the $(3, 1)$ cofactor is $-2 = 2y - 3x$ and the $(2, 2)$ cofactor is $-11 = -z - 2x$. We get $x = 4$, $y = 5$, $z = 3$.

Now let $C = \begin{bmatrix} 19 & 10 & r \\ -14 & -11 & s \\ -2 & 5 & t \end{bmatrix}$. Then $r = -6$, $s = -(2 - 4) = 2$ and $t = -3$.

(b) We compute

$$AC^T = \begin{bmatrix} -1 & 2 & 4 \\ 0 & 3 & 5 \\ 2 & -2 & 3 \end{bmatrix}\begin{bmatrix} 19 & -14 & -2 \\ 10 & -11 & 5 \\ -6 & 2 & -3 \end{bmatrix} = \begin{bmatrix} -23 & 0 & 0 \\ 0 & -23 & 0 \\ 0 & 0 & -23 \end{bmatrix} = -23I$$

and conclude that $A^{-1} = -\frac{1}{23}\begin{bmatrix} 19 & -14 & -2 \\ 10 & -11 & 5 \\ -6 & 2 & -3 \end{bmatrix}$.

11. Let $A = \begin{bmatrix} a & b \\ c & d \end{bmatrix}$. The cofactor matrix is $C = \begin{bmatrix} d & -c \\ -b & a \end{bmatrix}$, so $d = -1$, $-c = 7$, $-b = 2$, $a = 4$. Thus $A = \begin{bmatrix} 4 & -2 \\ -7 & -1 \end{bmatrix}$.

13. (a) Since $C = \begin{bmatrix} 3 & 1 \\ -2 & 1 \end{bmatrix}$, $C^T = \begin{bmatrix} 3 & -2 \\ 1 & 1 \end{bmatrix}$, so $A^{-1} = \frac{1}{\det A} C^T = \frac{1}{5} \begin{bmatrix} 3 & -2 \\ 1 & 1 \end{bmatrix}$.

Thus $A = (A^{-1})^{-1} = \begin{bmatrix} \frac{3}{5} & -\frac{2}{5} \\ \frac{1}{5} & \frac{1}{5} \end{bmatrix}^{-1} = \begin{bmatrix} 1 & 2 \\ -1 & 3 \end{bmatrix}$.

Section 3.2—True/False

1. False in general. If A is $n \times n$, $\det(-A) = (-1)^n \det A$. This is $-\det A$ if and only if n is odd.

2. True. "Singular" means "not invertible."

3. True. "Not singular" means "invertible," and if A is invertible, so is A^{-1}, with $(A^{-1})^{-1} = A$.

4. True. This is probably the most important use of the determinant. It is a test for invertibility.

5. True. $AC^T = (\det A)I = 5I$, so $\det(AC^T) = (\det A)(\det C^T) = 125$ and $\det C^T = \frac{125}{\det A} = 25$.

6. False. Interchanging two rows, for example, changes the sign of the determinant.

7. False. The matrix $A = \begin{bmatrix} 1 & 2 \\ 1 & 2 \end{bmatrix}$ has determinant 0.

8. False. The determinant is **not** a linear function, although it is linear **in each row**.

9. True. $\det AB = (\det A)(\det B) = (\det B)(\det A) = \det BA$.

Exercises 3.2

1. $\det(-3A) = (-3)^5 \det A = -243 \det A = 4$, so $\det A = -\frac{4}{243}$, $\det B^{-1} = \frac{1}{\det B}$, so $\det B = \frac{1}{\det B^{-1}} = \frac{1}{2}$,
$\det AB = \det A \det B = -\frac{4}{243} \cdot \frac{1}{2} = -\frac{2}{243}$.

6. The determinant of a triangular matrix is the product of its diagonal entries, so $\det A = 60$. Also, $\det A^{-1} = \frac{1}{\det A} = \frac{1}{60}$ and $\det A^2 = (\det A)^2 = 3600$.

7. (a) Expanding by cofactors of the first column gives

$$-1 \begin{vmatrix} 1 & 1 & 3 \\ 1 & 1 & 2 \\ 3 & -1 & 2 \end{vmatrix} - 2 \begin{vmatrix} -1 & 1 & 0 \\ 1 & 1 & 2 \\ 3 & -1 & 2 \end{vmatrix} - \begin{vmatrix} -1 & 1 & 0 \\ 1 & 1 & 3 \\ 1 & 1 & 2 \end{vmatrix} = -1(-4) - 2(0) - (2) = 2.$$

(b) $\det A = \begin{vmatrix} -1 & -1 & 1 & 0 \\ 2 & 1 & 1 & 3 \\ 0 & 1 & 1 & 2 \\ 1 & 3 & -1 & 2 \end{vmatrix} = \begin{vmatrix} -1 & -1 & 1 & 0 \\ 0 & -1 & 3 & 3 \\ 0 & 1 & 1 & 2 \\ 0 & 2 & 0 & 2 \end{vmatrix} = \begin{vmatrix} -1 & -1 & 1 & 0 \\ 0 & -1 & 3 & 3 \\ 0 & 0 & 4 & 5 \\ 0 & 0 & 6 & 8 \end{vmatrix}$

$= 4 \begin{vmatrix} -1 & -1 & 1 & 0 \\ 0 & -1 & 3 & 3 \\ 0 & 0 & 1 & \frac{5}{4} \\ 0 & 0 & 6 & 8 \end{vmatrix} = 4 \begin{vmatrix} -1 & -1 & 1 & 0 \\ 0 & -1 & 3 & 3 \\ 0 & 0 & 1 & \frac{5}{4} \\ 0 & 0 & 0 & \frac{1}{2} \end{vmatrix} = 4(\frac{1}{2}) = 2.$

9. (a)

$$\begin{vmatrix} -2 & 1 & 2 \\ 1 & 3 & 6 \\ -4 & 5 & 9 \end{vmatrix} = - \begin{vmatrix} 1 & 3 & 6 \\ -2 & 1 & 2 \\ -4 & 5 & 9 \end{vmatrix} \quad \text{interchanging rows one and two}$$

$$= - \begin{vmatrix} 1 & 3 & 6 \\ 0 & 7 & 14 \\ 0 & 17 & 33 \end{vmatrix} \quad \begin{array}{l}\text{using the third elementary row operation to put 0s in the first column} \\ \text{under the leading 1}\end{array}$$

$$= -7 \begin{vmatrix} 1 & 3 & 6 \\ 0 & 1 & 2 \\ 0 & 17 & 33 \end{vmatrix} \quad \text{factoring 7 from row two}$$

$$= -7 \begin{vmatrix} 1 & 3 & 6 \\ 0 & 1 & 2 \\ 0 & 0 & -1 \end{vmatrix} \quad \text{using the third elementary row operation to put a 0 in the } (2,3) \text{ position}$$

$$= (-7)(-1) = 7 \quad \begin{array}{l}\text{since the determinant of a triangular matrix is the product of its diagonal} \\ \text{entries.}\end{array}$$

(d)

$$\begin{vmatrix} -3 & 0 & 1 & 1 \\ 3 & 1 & 2 & 2 \\ -6 & -2 & -4 & 2 \\ 1 & -1 & 0 & -1 \end{vmatrix} = - \begin{vmatrix} 1 & -1 & 0 & -1 \\ 3 & 1 & 2 & 2 \\ -6 & -2 & -4 & 2 \\ -3 & 0 & 1 & 1 \end{vmatrix} \quad \text{interchanging rows one and four}$$

$$= - \begin{vmatrix} 1 & -1 & 0 & -1 \\ 0 & 4 & 2 & 5 \\ 0 & -8 & -4 & -4 \\ 0 & -3 & 1 & -2 \end{vmatrix} \quad \begin{array}{l}\text{using the third elementary row operation to put 0s in the first column under} \\ \text{the leading 1}\end{array}$$

$$= \begin{vmatrix} 1 & -1 & 0 & -1 \\ 0 & -8 & -4 & -4 \\ 0 & 4 & 2 & 5 \\ 0 & -3 & 1 & -2 \end{vmatrix} \quad \text{interchanging rows two and three}$$

$$= -8 \begin{vmatrix} 1 & -1 & 0 & -1 \\ 0 & 1 & \frac{1}{2} & \frac{1}{2} \\ 0 & 4 & 2 & 5 \\ 0 & -3 & 1 & -2 \end{vmatrix} \quad \text{factoring } -8 \text{ from row two}$$

$$= -8 \begin{vmatrix} 1 & -1 & 0 & -1 \\ 0 & 1 & \frac{1}{2} & \frac{1}{2} \\ 0 & 0 & 0 & 3 \\ 0 & 0 & \frac{5}{2} & -\frac{1}{2} \end{vmatrix} \quad \begin{array}{l}\text{using the third elementary row operation to put 0s under the leading 1 in} \\ \text{column two}\end{array}$$

$$= 8 \begin{vmatrix} 1 & -1 & 0 & -1 \\ 0 & 1 & \frac{1}{2} & \frac{1}{2} \\ 0 & 0 & \frac{5}{2} & -\frac{1}{2} \\ 0 & 0 & 0 & 3 \end{vmatrix} \quad \text{interchanging rows three and four}$$

$$= 8(\tfrac{5}{2})(3) = 60 \quad \text{since the determinant of a triangular matrix is the product of the diagonal entries.}$$

11. $(\det P)(\det A) = \det PA = (\det L)(\det U) = 1(-35) = -35$ and $\det P = \pm 1$, so $\det A = \pm 35$.

14. It is sufficient to show that $\det A = 0$. Let the columns of A be $a_1\mathbf{u} + b_1\mathbf{v}, a_2\mathbf{u} + b_2\mathbf{v}, a_3\mathbf{u} + b_3\mathbf{v}$ for scalars $a_1, b_1, a_2, b_2, a_3, b_3$. Write $\det(\mathbf{w}_1, \mathbf{w}_2, \mathbf{w}_3)$ for the determinant of a matrix whose columns are the vectors $\mathbf{w}_1, \mathbf{w}_2, \mathbf{w}_3$. Then

$$\det A = \det(a_1\mathbf{u} + b_1\mathbf{v}, a_2\mathbf{u} + b_2\mathbf{v}, a_3\mathbf{u} + b_3\mathbf{v})$$

$$= \det(a_1\mathbf{u}, a_2\mathbf{u} + b_2\mathbf{v}, a_3\mathbf{u} + b_3\mathbf{v}) + \det(b_1\mathbf{v}, a_2\mathbf{u} + b_2\mathbf{v}, a_3\mathbf{u} + b_3\mathbf{v})$$

by linearity of the determinant in the first column. Using linearity of the determinant in columns two and three, eventually we obtain

$$\det A = \det(a_1\mathbf{u}, a_2\mathbf{u}, a_3\mathbf{u}) + \det(a_1\mathbf{u}, a_2\mathbf{u}, b_3\mathbf{v}) + \det(a_1\mathbf{u}, b_2\mathbf{v}, a_3\mathbf{u}) + \det(a_1\mathbf{u}, b_2\mathbf{v}, b_3\mathbf{v})$$

$$+ \det(b_1\mathbf{v}, a_2\mathbf{u}, a_3\mathbf{u}) + \det(b_1\mathbf{v}, a_2\mathbf{u}, b_3\mathbf{v}) + \det(b_1\mathbf{v}, b_2\mathbf{v}, a_3\mathbf{u}) + \det(b_1\mathbf{v}, b_2\mathbf{v}, b_3\mathbf{v})$$

and each of these eight determinants is 0 because in each case, either one column is 0 or one column is a multiple of another. In the second determinant—$\det(a_1\mathbf{u}, a_2\mathbf{u}, b_3\mathbf{v})$—for example, if $a_1 \neq 0$, the second column is $\dfrac{a_2}{a_1}$ times the first.

15. **(a)** $\begin{vmatrix} 6 & -3 \\ 1 & -4 \end{vmatrix} = -24 + 3 = -18$

(c) We expand by cofactors along the first row.

$$\begin{vmatrix} 5 & 3 & 8 \\ -4 & 1 & 4 \\ -2 & 3 & 6 \end{vmatrix} = 5\begin{vmatrix} 1 & 4 \\ 3 & 6 \end{vmatrix} - 3\begin{vmatrix} -4 & 4 \\ -2 & 6 \end{vmatrix} + 8\begin{vmatrix} -4 & 1 \\ -2 & 3 \end{vmatrix} = 5(-6) - 3(-16) + 8(-10) = -62.$$

(e) $\begin{vmatrix} -1 & 2 & 3 & 4 \\ 3 & -9 & 2 & 1 \\ 0 & -5 & 7 & 6 \\ 2 & -4 & -6 & -8 \end{vmatrix} = 0$ because the last row is a scalar multiple of the first.

16. $\begin{vmatrix} a & -g & 2d \\ b & -h & 2e \\ c & -i & 2f \end{vmatrix} = -2\begin{vmatrix} a & g & d \\ b & h & e \\ c & i & f \end{vmatrix}$ factoring -1 from the second column and 2 from the third column

$$= -2\begin{vmatrix} a & b & c \\ g & h & i \\ d & e & f \end{vmatrix}$$ since the determinant of a matrix is the determinant of its transpose

$$= +2\begin{vmatrix} a & b & c \\ d & e & f \\ g & h & i \end{vmatrix} = 2(-3) = 6,$$ interchanging rows two and three.

19. Suppose $A = LDU$. Each of L and U is triangular with 1s on the diagonal, so $\det L = \det U = 1$, the product of the diagonal entries in each case. Thus $\det A = \det L \det D \det U = \det D$.

Using just the third elementary row operation to reduce $A = \begin{bmatrix} 2 & -1 & 4 & 1 \\ 1 & 1 & -10 & -2 \\ 4 & 0 & -7 & 6 \\ 6 & -3 & 0 & 1 \end{bmatrix}$ to an upper triangular matrix, we have

$$\begin{bmatrix} 2 & -1 & 4 & 1 \\ 1 & 1 & -10 & -2 \\ 4 & 0 & -7 & 6 \\ 6 & -3 & 0 & 1 \end{bmatrix} \rightarrow \begin{bmatrix} 2 & -1 & 4 & 1 \\ 0 & \frac{3}{2} & -12 & -\frac{5}{2} \\ 0 & 2 & -15 & 4 \\ 0 & 0 & -12 & -2 \end{bmatrix} \rightarrow \begin{bmatrix} 2 & -1 & 4 & 1 \\ 0 & \frac{3}{2} & -12 & -\frac{5}{2} \\ 0 & 0 & 1 & \frac{22}{3} \\ 0 & 0 & -12 & -2 \end{bmatrix}$$

$$\rightarrow \begin{bmatrix} 2 & -1 & 4 & 1 \\ 0 & \frac{3}{2} & -12 & -\frac{5}{2} \\ 0 & 0 & 1 & \frac{22}{3} \\ 0 & 0 & 0 & 86 \end{bmatrix} = U' = DU$$

with $D = \begin{bmatrix} 2 & 0 & 0 & 0 \\ 0 & \frac{3}{2} & 0 & 0 \\ 0 & 0 & 1 & 0 \\ 0 & 0 & 0 & 86 \end{bmatrix}$, so $\det A = \det D = 2(\frac{3}{2})(1)(86) = 258$.

21. We have $\frac{1}{2}(I - A)A = I$, so $(I - A)A = 2I$. Taking the determinant of each side gives $\det(I - A) \det A = 2^n$. If the product of two numbers is a power of 2, each number itself is a power of 2, so the result follows.

Section 3.3—True/False

1. False. An eigenvector is, by definition, not zero, so the eigenvectors of A are the **nonzero** solutions to $(A - \lambda I)\mathbf{x} = \mathbf{0}$.

2. True.

3. True.

4. False. See our "Final Thought," on p. 200.

5. False. The eigenvalues of the triangular matrix $A = \begin{bmatrix} 0 & 1 \\ 0 & 0 \end{bmatrix}$ are all 0, but $A \neq \mathbf{0}$.

6. False. The only eigenvalue of $\begin{bmatrix} 1 & 1 \\ 0 & 1 \end{bmatrix}$ is 1.

Exercises 3.3

1. $A\mathbf{v} = \begin{bmatrix} 8 \\ 16 \\ 0 \end{bmatrix} = 4\mathbf{v}$. Thus $\mathbf{v}$ is an eigenvector of A corresponding to the eigenvalue 4.

2. (a) $A\mathbf{v}_1 = 5\mathbf{v}_1$, so $\mathbf{v}_1$ is an eigenvector corresponding to $\lambda = 5$.

$A\mathbf{v}_2 = 0\mathbf{v}_2$, so $\mathbf{v}_2$ is an eigenvector corresponding to $\lambda = 0$.

$A\mathbf{v}_3 = \begin{bmatrix} 3 \\ 6 \end{bmatrix}$ is not $\lambda\mathbf{v}_3$ for any λ, so $\mathbf{v}_3$ is not an eigenvector.

$A\mathbf{v}_4 = \begin{bmatrix} -3 \\ -6 \end{bmatrix}$ is not $\lambda\mathbf{v}_4$ for any λ, so $\mathbf{v}_4$ is not an eigenvector.

$\mathbf{v}_5$ is not an eigenvector since an eigenvector is a **nonzero** vector.

3. (a) It is easy to compute and factor $\det(A - \lambda I)$ when A is 2×2, so this is the most straightforward way the question for this particular A. The matrix $A - \lambda I = \begin{bmatrix} 1-\lambda & 2 \\ 2 & 4-\lambda \end{bmatrix}$ has determinant

$(1-\lambda)(4-\lambda) - 4 = \lambda^2 - 5\lambda = \lambda(\lambda - 5)$. Thus $\lambda = 0$ and $\lambda = 5$ are eigenvalues, while -1, 1, and 3 are not.

(d) When A is larger than 2×2, it is not so easy to find $\det(A - \lambda I)$ and its roots, so we answer the question in this instance by computing $\det(A - \lambda I)$ for each given value of λ and determining whether or not the matrix has 0 determinant. The matrix $A - \lambda I = \begin{bmatrix} 5-\lambda & -7 & 7 \\ 4 & -3-\lambda & 4 \\ 4 & -1 & 2-\lambda \end{bmatrix}$. When $\lambda = 1$,

$A - \lambda I = \begin{bmatrix} 4 & -7 & 7 \\ 4 & -4 & 4 \\ 4 & -1 & 1 \end{bmatrix} \to \begin{bmatrix} 1 & -1 & 1 \\ 4 & -7 & 7 \\ 4 & -1 & 1 \end{bmatrix} \to \begin{bmatrix} 1 & -1 & 1 \\ 0 & -3 & 3 \\ 0 & 3 & -3 \end{bmatrix}$ has determinant 0, so $\lambda = 1$ is an eigenvalue.

When $\lambda = 2$, $A - \lambda I = \begin{bmatrix} 3 & -7 & 7 \\ 4 & -5 & 4 \\ 4 & -1 & 0 \end{bmatrix} \to \begin{bmatrix} 3 & -7 & 7 \\ 4 & -5 & 4 \\ 0 & 4 & -4 \end{bmatrix} \to \begin{bmatrix} 12 & -28 & 28 \\ 12 & -15 & 12 \\ 0 & 1 & -1 \end{bmatrix} \to \begin{bmatrix} 12 & -28 & 28 \\ 0 & 13 & -16 \\ 0 & 1 & -1 \end{bmatrix}$ has nonzero determinant, so $\lambda = 2$ is not an eigenvalue.

When $\lambda = 4$, $A - \lambda I = \begin{bmatrix} 1 & -7 & 7 \\ 4 & -7 & 4 \\ 4 & -1 & -2 \end{bmatrix} \to \begin{bmatrix} 1 & -7 & 7 \\ 0 & 21 & -24 \\ 0 & 27 & -30 \end{bmatrix}$ has nonzero determinant, so $\lambda = 4$ is not an eigenvalue.

When $\lambda = 5$, $A - \lambda I = \begin{bmatrix} 0 & -7 & 7 \\ 4 & -8 & 4 \\ 4 & -1 & -3 \end{bmatrix} \to \begin{bmatrix} 1 & -2 & 1 \\ 0 & -1 & 1 \\ 0 & 7 & -7 \end{bmatrix}$ has determinant 0, so $\lambda = 5$ is an eigenvalue.

When $\lambda = 6$, $A - \lambda I = \begin{bmatrix} -1 & -7 & 7 \\ 4 & -9 & 4 \\ 4 & -1 & -4 \end{bmatrix} \to \begin{bmatrix} -1 & -7 & 7 \\ 0 & -37 & 31 \\ 0 & -29 & 24 \end{bmatrix}$ has nonzero determinant, so $\lambda = 6$ is not an eigenvalue.

4. (a) The characteristic polynomial of A is $\det(A - \lambda I) = \begin{vmatrix} 1-\lambda & 2 \\ 3 & 2-\lambda \end{vmatrix} = (1-\lambda)(2-\lambda) - 6 = \lambda^2 - 3\lambda - 4$.

(b) Since $\lambda^2 - 3\lambda - 4 = (\lambda - 4)(\lambda + 1) = 0$ for $\lambda = 4$ and $\lambda = -1$, the eigenvalues are 4 and -1.

(c) $A \begin{bmatrix} 2 \\ 3 \end{bmatrix} = \begin{bmatrix} 8 \\ 12 \end{bmatrix} = 4 \begin{bmatrix} 2 \\ 3 \end{bmatrix}$, so $\begin{bmatrix} 2 \\ 3 \end{bmatrix}$ is an eigenvector corresponding to $\lambda = 4$.

6. (a) Call the given matrix A. The characteristic polynomial of A is

$\begin{vmatrix} 5-\lambda & 8 \\ 4 & 1-\lambda \end{vmatrix} = (5-\lambda)(1-\lambda) - 32 = \lambda^2 - 6\lambda - 27 = (\lambda - 9)(\lambda + 3)$, so $\lambda = 9$ and $\lambda = -3$

are the eigenvalues of A. To find the eigenspace corresponding to $\lambda = 9$, we must solve the homogeneous system $(A - \lambda I)\mathbf{x} = \mathbf{0}$ with $\lambda = 9$. We have

$$A - \lambda I = \begin{bmatrix} -4 & 8 \\ 4 & -8 \end{bmatrix} \to \begin{bmatrix} 1 & -2 \\ 0 & 0 \end{bmatrix}.$$

The variable $x_2 = t$ is free and $x_1 = 2x_2 = 2t$. The eigenspace corresponding to $\lambda = 9$ is the set of vectors of the form of $\mathbf{x} = \begin{bmatrix} x_1 \\ x_2 \end{bmatrix} = \begin{bmatrix} 2t \\ t \end{bmatrix} = t \begin{bmatrix} 2 \\ 1 \end{bmatrix}$.

To find the eigenspace corresponding to $\lambda = -3$, we must solve the homogeneous system $(A - \lambda I)\mathbf{x} = \mathbf{0}$ with $\lambda = -3$. We have

$$A - \lambda I = \begin{bmatrix} 8 & 8 \\ 4 & 4 \end{bmatrix} \rightarrow \begin{bmatrix} 1 & 1 \\ 0 & 0 \end{bmatrix}.$$

Again, $x_2 = t$ is free, but $x_1 = -x_2 = -t$. The eigenspace corresponding to $\lambda = -3$ is the set of vectors of the form of $\mathbf{x} = \begin{bmatrix} x_1 \\ x_2 \end{bmatrix} = \begin{bmatrix} -t \\ t \end{bmatrix} = t\begin{bmatrix} -1 \\ 1 \end{bmatrix}$.

(d) The characteristic polynomial of $A = \begin{bmatrix} 1 & -2 & 3 \\ 2 & 6 & -6 \\ 1 & 2 & -1 \end{bmatrix}$ is

$$\det(A - \lambda I) = \begin{vmatrix} 1-\lambda & -2 & 3 \\ 2 & 6-\lambda & -6 \\ 1 & 2 & -1-\lambda \end{vmatrix} = -\lambda^3 + 6\lambda^2 - 12\lambda + 8 = -(\lambda - 2)^3.$$

The only eigenvalue of A is $\lambda = 2$. To find the corresponding eigenspace, we solve the homogeneous system $(A - \lambda I)\mathbf{x} = \mathbf{0}$ for $\mathbf{x} = \begin{bmatrix} x_1 \\ x_2 \\ x_3 \end{bmatrix}$ with $\lambda = 2$. We have

$$A - \lambda I = \begin{bmatrix} -1 & -2 & 3 \\ 2 & 4 & -6 \\ 1 & 2 & -3 \end{bmatrix} \rightarrow \begin{bmatrix} -1 & -2 & 3 \\ 0 & 0 & 0 \\ 0 & 0 & 0 \end{bmatrix}.$$

The solutions are $x_3 = t$, $x_2 = s$, $x_1 = -2s + 3t$. The eigenspace consists of vectors of the form
$$\begin{bmatrix} -2s + 3t \\ s \\ t \end{bmatrix} = s\begin{bmatrix} -2 \\ 1 \\ 0 \end{bmatrix} + t\begin{bmatrix} 3 \\ 0 \\ 1 \end{bmatrix}.$$

7. (a) Since $A\mathbf{v} = \begin{bmatrix} 0 & 1 \\ 1 & 0 \end{bmatrix}\begin{bmatrix} x \\ y \end{bmatrix} = \begin{bmatrix} y \\ x \end{bmatrix}$, Q is (y, x).

(b) The vector $\overrightarrow{PQ} = \begin{bmatrix} y - x \\ x - y \end{bmatrix}$. The dot product of $\overrightarrow{PQ}$ and $\begin{bmatrix} 1 \\ 1 \end{bmatrix}$ (which is the direction of ℓ) is $y - x + x - y = 0$, so PQ and ℓ are perpendicular. Since the midpoint of PQ, which is $(\frac{x+y}{2}, \frac{x+y}{2})$, is on ℓ, ℓ is the right bisector of PQ.

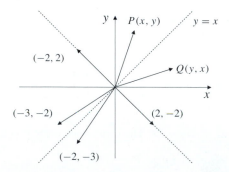

(c) Multiplication by A is reflection in the line with equation $y = x$.

(d) Reflection in a line fixes the line; in fact, it fixes every vector on the line: $A \begin{bmatrix} x \\ x \end{bmatrix} = \begin{bmatrix} 0 & 1 \\ 1 & 0 \end{bmatrix} \begin{bmatrix} x \\ x \end{bmatrix} = \begin{bmatrix} x \\ x \end{bmatrix}$.

Thus every vector on this line is an eigenvector corresponding to $\lambda = 1$. This reflection also fixes the line with equation $y = -x$ because every vector on this line is moved to its negative (which is still on the line):

$$A \begin{bmatrix} x \\ -x \end{bmatrix} = \begin{bmatrix} 0 & 1 \\ 1 & 0 \end{bmatrix} \begin{bmatrix} x \\ -x \end{bmatrix} = \begin{bmatrix} -x \\ x \end{bmatrix} = - \begin{bmatrix} x \\ -x \end{bmatrix}.$$

Every vector on the line with equation $y = -x$ is an eigenvector corresponding to $\lambda = -1$. The matrix A has two eigenspaces, the lines with equations $y = x$ and $y = -x$.

8. (a) Since multiplication by A is reflection in a line, any vector on the line will be fixed by A, so we seek solutions to $A\mathbf{x} = \mathbf{x}$:

$$A\mathbf{x} = \begin{bmatrix} -1 & 0 \\ 0 & 1 \end{bmatrix} \begin{bmatrix} x \\ y \end{bmatrix} = \begin{bmatrix} -x \\ y \end{bmatrix} = \begin{bmatrix} x \\ y \end{bmatrix} \text{ gives } x = 0,$$

so our matrix reflects vectors in the y-axis. Any nonzero vector on the y-axis is an eigenvector corresponding to $\lambda = 1$ and any nonzero vector on the x-axis is an eigenvector corresponding to $\lambda = -1$ since these vectors are mapped to their negatives.

10. We have $A(a\mathbf{u} + b\mathbf{v}) = aA\mathbf{u} + bA\mathbf{v} = a\lambda\mathbf{u} + b\lambda\mathbf{v} = \lambda(a\mathbf{u} + b\mathbf{v})$. This shows that $a\mathbf{u} + b\mathbf{v}$ is an eigenvector corresponding to λ (provided it is not $\mathbf{0}$).

12. Let $\mathbf{x} = \begin{bmatrix} 1 \\ 1 \\ \vdots \\ 1 \end{bmatrix}$ and observe that $A\mathbf{x} = 17\mathbf{x}$.

14. Let λ be an eigenvalue of A and let $\mathbf{x} \neq \mathbf{0}$ be a corresponding eigenvector. Then $A\mathbf{x} = \lambda\mathbf{x}$. Multiplying on the left by P gives $PA\mathbf{x} = \lambda P\mathbf{x}$. Since $PA = BP$, we get $BP\mathbf{x} = \lambda P\mathbf{x}$, that is, $B(P\mathbf{x}) = \lambda(P\mathbf{x})$. Since P is invertible, $P\mathbf{x} \neq \mathbf{0}$. So λ is an eigenvalue of B with corresponding eigenvector $P\mathbf{x}$.

16. The characteristic polynomial of A is the determinant of $A - \lambda I$. Since $\det X^T = \det X$, the determinant of $A - \lambda I$ is the determinant of $(A - \lambda I)^T = A^T - (\lambda I)^T = A^T - \lambda I$ (because the transpose of the diagonal matrix λI is λI). Thus $\det(A - \lambda I) = \det(A^T - \lambda I)$, which is the desired result.

18. (a) The answer is yes. Since $A\mathbf{v} = \lambda\mathbf{v}$, $(5A)\mathbf{v} = 5\lambda\mathbf{v} = (5\lambda)\mathbf{v}$, so $\mathbf{v}$ is an eigenvector of $5A$ with eigenvalue 5λ.

Section 3.4—True/False

1. True. The characteristic polynomial is the determinant of $\begin{bmatrix} 1 - \lambda & 2 & 3 \\ 0 & 4 - \lambda & 5 \\ 0 & 0 & 6 - \lambda \end{bmatrix}$. This matrix is triangular, so its

determinant is the product of its diagonal entries.

2. True. The eigenvalues of the diagonal matrix $A^{10} = \begin{bmatrix} 1 & 0 \\ 0 & 2^{10} \end{bmatrix}$ are its diagonal entries.

3. False. Similar matrices have the same eigenvalues: The given matrices have eigenvalues $2, -1$, and $2, 1$, respectively.

4. True. Any diagonal matrix is diagonalizable: Take : $P = I$.

5. The identity matrix $\begin{bmatrix} 1 & 0 \\ 0 & 1 \end{bmatrix}$ is diagonalizable with 1 as its only eigenvalue.

6. True. This is part of Theorem 3.4.14.

7. False. The 3×3 matrix in Example 3.4.12 is diagonalizable, yet it has just two different eigenvalues.

8. False. The symmetric matrix $A = \begin{bmatrix} 0 & 1 \\ 1 & 0 \end{bmatrix}$ has eigenvalues ± 1.

Exercises 3.4

1. **(a)** A similar to I means $A = P^{-1}IP$ for some invertible matrix P. But $P^{-1}IP = P^{-1}P = I$, so $A = I$.

   (b) No, by 1(a). The given matrix is not I.

   (c) The characteristic polynomial of A is $(\lambda - 1)^2$, so $\lambda = 1$ is the only eigenvalue of A. If A were diagonalizable, it would be similar to I, hence equal to I by Exercise 1(a). Since this is not the case, A is not diagonalizable.

3. We have equations of the type $B = P^{-1}AP$ and $C = Q^{-1}BQ$ for invertible matrices P and Q. Thus $C = Q^{-1}P^{-1}APQ = R^{-1}AR$, with R the invertible matrix PQ. Thus A is similar to C.

4. The characteristic polynomial of A is

$$\begin{vmatrix} 5 - \lambda & 3 \\ 2 & 4 - \lambda \end{vmatrix} = 20 - 9\lambda + \lambda^2 - 6 = \lambda^2 - 9\lambda + 14 = (\lambda - 2)(\lambda - 7).$$

Since the 2×2 matrix A has two distinct eigenvalues, it is diagonalizable by Theorem 3.4.14 and similar to either of the two matrices whose diagonal entries are eigenvalues: $\begin{bmatrix} 2 & 0 \\ 0 & 7 \end{bmatrix}$ and $\begin{bmatrix} 7 & 0 \\ 0 & 2 \end{bmatrix}$.

6. **(a)** $\det(A - \lambda I) = \begin{vmatrix} -\lambda & 1 \\ 0 & -\lambda \end{vmatrix} = \lambda^2$ so $\lambda = 0$ is the only eigenvalue. Corresponding eigenvectors are obtained by

   solving $(A - \lambda I)\mathbf{x} = \mathbf{0}$ for $\mathbf{x} = \begin{bmatrix} x_1 \\ x_2 \end{bmatrix}$ with $\lambda = 0$. Since A is already in row echelon form, we have $x_1 = t$ is free

   and $x_2 = 0$, so $\mathbf{x} = t \begin{bmatrix} 1 \\ 0 \end{bmatrix}$.

   (b) If $P^{-1}AP = D$, the columns of P are eigenvectors. The only possibility for P here is a matrix of the form $\begin{bmatrix} t & s \\ 0 & 0 \end{bmatrix}$ and such a matrix is not invertible.

8. **(a)** Since $A - 2I = \begin{bmatrix} -1 & 2 \\ 2 & -4 \end{bmatrix}$ is not invertible, we can find nonzero vectors $\mathbf{x}$ with $(A - 2I)\mathbf{x} = \mathbf{0}$ using Gaussian

   elimination: $\begin{bmatrix} -1 & 2 \\ 2 & -4 \end{bmatrix} \to \begin{bmatrix} 1 & -2 \\ 0 & 0 \end{bmatrix}$, so if $\mathbf{x} = \begin{bmatrix} x_1 \\ x_2 \end{bmatrix}$, $x_2 = t$ is free and $x_1 = 2x_2 = 2t$. Thus $\mathbf{x} = t \begin{bmatrix} 2 \\ 1 \end{bmatrix}$. The

   eigenspace corresponding to $\lambda = 2$ is the set of scalar multiples of $\begin{bmatrix} 2 \\ 1 \end{bmatrix}$.

   (b) The characteristic polynomial of A is $(\lambda - 2)(\lambda + 3) = \lambda^2 + \lambda - 6$.

   (c) A is diagonalizable because A is 2×2 and A has two different eigenvalues.

   (d) The columns of P should be eigenvectors corresponding to -3 and 2, respectively. So $P = \begin{bmatrix} -1 & 2 \\ 1 & 2 \end{bmatrix}$.

   (e) The columns of Q are eigenvectors of A corresponding to eigenvalues -3 and 2, respectively. So $Q^{-1}AQ$ is the diagonal matrix with these numbers as the diagonal entries, in the given order: $D = \begin{bmatrix} -3 & 0 \\ 0 & 2 \end{bmatrix}$.

9. (a) The characteristic polynomial of A is $\begin{vmatrix} 4-\lambda & 2 & 2 \\ -5 & -3-\lambda & -2 \\ 5 & 5 & 4-\lambda \end{vmatrix}$

$$= (4-\lambda)[(-3-\lambda)(4-\lambda)+10] + 5[2(4-\lambda)-10] + 5[-4+2(3+\lambda)]$$

$$= -\lambda^3 + 5\lambda^2 - 2\lambda - 8 = -(\lambda-4)(\lambda+1)(\lambda-2).$$

(b) Since the 3×3 matrix A has three different eigenvalues, 4, -1, and 2, A is diagonalizable by Theorem 3.4.14.

(c) The desired matrix P is a matrix whose columns are eigenvectors corresponding to 4, -1 and 2, **in that order**. To find the eigenspace for $\lambda = 4$, we solve $(A - \lambda I)\mathbf{x} = \mathbf{0}$ with $\lambda = 4$. Gaussian elimination proceeds

$$\begin{bmatrix} 0 & 2 & 2 \\ -5 & -7 & -2 \\ 5 & 5 & 0 \end{bmatrix} \rightarrow \begin{bmatrix} 1 & 1 & 0 \\ 0 & -2 & -2 \\ 0 & 2 & 2 \end{bmatrix} \rightarrow \begin{bmatrix} 1 & 1 & 0 \\ 0 & 1 & 1 \\ 0 & 0 & 0 \end{bmatrix}.$$ With $\mathbf{x} = \begin{bmatrix} x_1 \\ x_2 \\ x_3 \end{bmatrix}$, we have $x_3 = t$ free,

$x_2 = -x_3 = -t$, and $x_1 = -x_2 = t$, so $\mathbf{x} = t \begin{bmatrix} 1 \\ -1 \\ 1 \end{bmatrix}$. To find the eigenspace for $\lambda = -1$, we solve $(A - \lambda I)\mathbf{x} = \mathbf{0}$

with $\lambda = -1$. Gaussian elimination proceeds $\begin{bmatrix} 5 & 2 & 2 \\ -5 & -2 & -2 \\ 5 & 5 & 5 \end{bmatrix} \rightarrow \begin{bmatrix} 1 & 1 & 1 \\ 0 & 3 & 3 \\ 0 & 3 & 3 \end{bmatrix} \rightarrow \begin{bmatrix} 1 & 1 & 1 \\ 0 & 1 & 1 \\ 0 & 0 & 0 \end{bmatrix}$. With $\mathbf{x} = \begin{bmatrix} x_1 \\ x_2 \\ x_3 \end{bmatrix}$, we have

$x_3 = t$ free, $x_2 = -x_3 = -t$, and $x_1 = -x_2 - x_3 = 0$, so $\mathbf{x} = t \begin{bmatrix} 0 \\ -1 \\ 1 \end{bmatrix}$. To find the eigenspace for $\lambda = 2$, we solve

$(A - \lambda I)\mathbf{x} = \mathbf{0}$ with $\lambda = 2$. Gaussian elimination proceeds $\begin{bmatrix} 2 & 2 & 2 \\ -5 & -5 & -2 \\ 5 & 5 & 2 \end{bmatrix} \rightarrow \begin{bmatrix} 1 & 1 & 1 \\ -5 & -5 & -2 \\ 0 & 0 & 0 \end{bmatrix} \rightarrow \begin{bmatrix} 1 & 1 & 1 \\ 0 & 0 & 1 \\ 0 & 0 & 0 \end{bmatrix}$.

With $\mathbf{x} = \begin{bmatrix} x_1 \\ x_2 \\ x_3 \end{bmatrix}$, we have $x_2 = t$ free, $x_3 = 0$, and $x_1 = -x_2 - x_3 = -t$, so $\mathbf{x} = t \begin{bmatrix} -1 \\ 1 \\ 0 \end{bmatrix}$. We obtain

$$P = \begin{bmatrix} 1 & 0 & -1 \\ -1 & -1 & 1 \\ 1 & 1 & 0 \end{bmatrix}.$$

12. The characteristic polynomial of A is

$$\det(A - \lambda I) = \begin{vmatrix} 2-\lambda & -16 & -2 \\ 0 & 5-\lambda & 0 \\ 2 & -8 & -3-\lambda \end{vmatrix} = (5-\lambda) \begin{vmatrix} 2-\lambda & -2 \\ 2 & -3-\lambda \end{vmatrix}$$

$$= (5-\lambda)(\lambda^2 + \lambda - 2) = (5-\lambda)(\lambda+2)(\lambda-1).$$

The eigenvalues of A $(1, -2, 5)$ are distinct, so A is diagonalizable. One of six diagonal matrices to which A is similar

is $\begin{bmatrix} 1 & 0 & 0 \\ 0 & -2 & 0 \\ 0 & 0 & 5 \end{bmatrix}$.

14. (a) The characteristic polynomial of A is $\begin{vmatrix} 1-\lambda & 0 \\ 2 & 3-\lambda \end{vmatrix} = (1-\lambda)(3-\lambda)$, so A has two distinct eigenvalues and

hence is diagonalizable. For $\lambda = 1$, we find that $\mathbf{x} = \begin{bmatrix} 1 \\ -1 \end{bmatrix}$ is an eigenvector and, for $\lambda = 3$, $\mathbf{x} = \begin{bmatrix} 0 \\ 1 \end{bmatrix}$. For

$P = \begin{bmatrix} 1 & 0 \\ -1 & 1 \end{bmatrix}$ we have $P^{-1}AP = D = \begin{bmatrix} 1 & 0 \\ 0 & 3 \end{bmatrix}$.

(d) The characteristic polynomial of A is

$$\begin{vmatrix} -1-\lambda & 3 & 0 \\ 0 & 2-\lambda & 0 \\ 2 & 1 & -1-\lambda \end{vmatrix} = (2-\lambda)\begin{vmatrix} -1-\lambda & 0 \\ 2 & -1-\lambda \end{vmatrix} = (2-\lambda)(\lambda+1)^2.$$

There are two eigenvalues, $\lambda = -1$ and $\lambda = 2$. For $\lambda = 2$, the eigenspace is spanned by $\begin{bmatrix} 1 \\ 1 \\ 1 \end{bmatrix}$. For $\lambda = -1$, the

eigenspace is spanned by $\begin{bmatrix} 0 \\ 0 \\ 1 \end{bmatrix}$. There are just two linearly independent eigenvectors. The matrix is not

diagonalizable.

(h) The characteristic polynomial of A is

$$\begin{vmatrix} 2-\lambda & 1 & 1 \\ 0 & 1-\lambda & 0 \\ 1 & -1 & 2-\lambda \end{vmatrix} = (1-\lambda)\begin{vmatrix} 2-\lambda & 1 \\ 1 & 2-\lambda \end{vmatrix} = (1-\lambda)(\lambda^2 - 4\lambda + 3)$$

$$= (1-\lambda)(\lambda-1)(\lambda-3),$$

so A has eigenvalues $\lambda = 1$ and $\lambda = 3$. The eigenvectors for $\lambda = 1$ are found by solving $(A - \lambda I)\mathbf{x} = \mathbf{0}$ with $\lambda = 1$. Gaussian elimination proceeds

$$\begin{bmatrix} 1 & 1 & 1 \\ 0 & 0 & 0 \\ 1 & -1 & 1 \end{bmatrix} \to \begin{bmatrix} 1 & 1 & 1 \\ 0 & -2 & 0 \\ 0 & 0 & 0 \end{bmatrix},$$

so $x_3 = t$ is free, $x_2 = 0$, and $x_1 = -x_2 - x_3 = -t$. Eigenvectors are of the form $\begin{bmatrix} x_1 \\ x_2 \\ x_3 \end{bmatrix} = \begin{bmatrix} -t \\ 0 \\ t \end{bmatrix} = t\begin{bmatrix} -1 \\ 0 \\ 1 \end{bmatrix}$. To find

the eigenvectors for $\lambda = 3$, we solve $(A - \lambda I)\mathbf{x} = \mathbf{0}$ with $\lambda = 3$. This time, Gaussian elimination proceeds

$$\begin{bmatrix} -1 & 1 & 1 \\ 0 & -2 & 0 \\ 1 & -1 & -1 \end{bmatrix} \to \begin{bmatrix} 1 & -1 & -1 \\ 0 & 1 & 0 \\ 0 & 0 & 0 \end{bmatrix},$$ so $x_3 = t$ is free, $x_2 = 0$, and $x_1 = x_2 + x_3 = t$. Eigenvectors are of the form

$\begin{bmatrix} x_1 \\ x_2 \\ x_3 \end{bmatrix} = \begin{bmatrix} t \\ 0 \\ t \end{bmatrix} = t\begin{bmatrix} 1 \\ 0 \\ 1 \end{bmatrix}$. Since there are only two linearly independent eigenvectors, the matrix is not diagonalizable.

15. (a) The eigenvalues of A are 1 and 2 with corresponding eigenvectors, respectively, $\begin{bmatrix} 1 \\ 1 \end{bmatrix}$ and $\begin{bmatrix} 2 \\ 1 \end{bmatrix}$. Let $P = \begin{bmatrix} 1 & 2 \\ 1 & 1 \end{bmatrix}$.

Then $P^{-1}AP = D = \begin{bmatrix} 1 & 0 \\ 0 & 2 \end{bmatrix}$.

(b) Following the hint, we note that $D = D_1^2$ where $D_1 = \begin{bmatrix} 1 & 0 \\ 0 & \sqrt{2} \end{bmatrix}$. Now $P^{-1}AP = D_1^2$, so

$A = PD_1^2P^{-1} = (PD_1P^{-1})(PD_1P^{-1}) = B^2$, with $B = PD_1P^{-1}$

$$= \begin{bmatrix} 1 & 2 \\ 1 & 1 \end{bmatrix}\begin{bmatrix} 1 & 0 \\ 0 & \sqrt{2} \end{bmatrix}\begin{bmatrix} -1 & 2 \\ 1 & -1 \end{bmatrix} = \begin{bmatrix} -1+2\sqrt{2} & 2-2\sqrt{2} \\ -1+\sqrt{2} & 2-\sqrt{2} \end{bmatrix}.$$

17. The characteristic polynomial of A is

$$\begin{vmatrix} 103 - \lambda & -96 \\ -96 & 47 - \lambda \end{vmatrix} = \lambda^2 - 150\lambda - 4375 = (\lambda - 175)(\lambda + 25).$$

So A has two eigenvalues, $\lambda = 175$ and $\lambda = -25$. The eigenspace for $\lambda = 175$ is the set of solutions to $(A - \lambda I)\mathbf{x} = \mathbf{0}$ with $\lambda = 175$. Gaussian elimination proceeds

$$\begin{bmatrix} -72 & -96 \\ -96 & -128 \end{bmatrix} \rightarrow \begin{bmatrix} 3 & 4 \\ 6 & 8 \end{bmatrix} \rightarrow \begin{bmatrix} 1 & \frac{4}{3} \\ 0 & 0 \end{bmatrix}.$$

With $\mathbf{x} = \begin{bmatrix} x_1 \\ x_2 \end{bmatrix}$, $x_2 = t$ is free and $x_1 = -\frac{4}{3}x_2 = -\frac{4}{3}t$, so the eigenspace consists of all vectors of the form $t \begin{bmatrix} -\frac{4}{3} \\ 1 \end{bmatrix}$, that is, multiples of $\begin{bmatrix} -4 \\ 3 \end{bmatrix}$. The eigenspace for $\lambda = -25$ is the set of solutions to $(A - \lambda I)\mathbf{x} = \mathbf{0}$ with $\lambda = -25$. Gaussian elimination proceeds

$$\begin{bmatrix} 128 & -96 \\ -96 & 72 \end{bmatrix} \rightarrow \begin{bmatrix} 4 & -3 \\ -4 & 3 \end{bmatrix} \rightarrow \begin{bmatrix} 1 & -\frac{3}{4} \\ 0 & 0 \end{bmatrix}.$$

With $\mathbf{x} = \begin{bmatrix} x_1 \\ x_2 \end{bmatrix}$, $x_2 = t$ is free and $x_1 = \frac{3}{4}x_2 = \frac{3}{4}t$, so the eigenspace consists of all vectors of the form $t \begin{bmatrix} \frac{3}{4} \\ 1 \end{bmatrix}$, that is, multiples of $\begin{bmatrix} 3 \\ 4 \end{bmatrix}$. The matrix $P = \begin{bmatrix} -4 & 3 \\ 3 & 4 \end{bmatrix}$ has orthogonal columns and $P^{-1}AP = \begin{bmatrix} 175 & 0 \\ 0 & -25 \end{bmatrix}$.

18. (a) The characteristic polynomial of A is $\begin{vmatrix} 3 - \lambda & -5 \\ -5 & 3 - \lambda \end{vmatrix} = 9 - 6\lambda + \lambda^2 - 25 = \lambda^2 - 6\lambda - 16 = (\lambda - 8)(\lambda + 2)$. The eigenvalues are $\lambda = -2$ and $\lambda = 8$. When $\lambda = -2$, $A - \lambda I = \begin{bmatrix} 5 & -5 \\ -5 & 5 \end{bmatrix}$, the eigenspace is the set of vectors of the form $\begin{bmatrix} t \\ t \end{bmatrix}$, which is spanned by $\begin{bmatrix} 1 \\ 1 \end{bmatrix}$. When $\lambda = 8$, $A - \lambda I = \begin{bmatrix} -5 & -5 \\ -5 & -5 \end{bmatrix}$, the eigenspace is the set of vectors of the form $\begin{bmatrix} -t \\ t \end{bmatrix}$, which is spanned by $\begin{bmatrix} -1 \\ 1 \end{bmatrix}$. The eigenvectors $\begin{bmatrix} 1 \\ 1 \end{bmatrix}$ and $\begin{bmatrix} -1 \\ 1 \end{bmatrix}$ are orthogonal. We can take $P = \begin{bmatrix} 1 & -1 \\ 1 & 1 \end{bmatrix}$, in which case $P^{-1}AP = D$ with $D = \begin{bmatrix} -2 & 0 \\ 0 & 8 \end{bmatrix}$.

19. Similar matrices have the same characteristic polynomial, so the characteristic polynomial of A is the same as that of the given matrix. The given matrix is triangular; its characteristic polynomial is $(-1 - \lambda)(1 - \lambda)(-3 - \lambda)^2$.

23. For some invertible matrix P, we have $P^{-1}AP = D$ where D is diagonal with diagonal entries the eigenvalues of A. Let D_1 be the diagonal matrix whose diagonal entries are the square roots of those of D. Thus $D_1^2 = D$. Let $B = PD_1P^{-1}$. Then $B^2 = (PD_1P^{-1})(PD_1P^{-1}) = PD_1P^{-1}PD_1P^{-1} = PD_1^2P^{-1} = PDP^{-1} = A$.

Section 4.1—True/False

1. False. It defines a function from $\mathbf{R}^\ell \to \mathbf{R}^k$.

2. True. Every vector in the xz-plane is a linear combination of the columns: $\begin{bmatrix} x \\ 0 \\ z \end{bmatrix} = -x \begin{bmatrix} -1 \\ 0 \\ 0 \end{bmatrix} + \frac{z}{2} \begin{bmatrix} 0 \\ 0 \\ 2 \end{bmatrix} + 0 \begin{bmatrix} 4 \\ 0 \\ 5 \end{bmatrix} + 0 \begin{bmatrix} 1 \\ 0 \\ 1 \end{bmatrix}$.

3. True. The matrix A is $m \times n$ with $m = 3 < 4 = n$, so we apply Theorem 4.1.8.

4. True. There are m pivot columns, so at least m columns altogether, so $n \geq m$.

5. True. The only solution of $A\mathbf{x} = \mathbf{0}$ is $\mathbf{x} = \mathbf{0}$.

6. False. For example, if A is an invertible 2×2 matrix with inverse B, rank $A =$ rank $B = 2 =$ rank AB.

7. True. The hypothesis says each of A and B is invertible (with inverses $\frac{1}{c}B$ and $\frac{1}{c}A$, respectively), so rank $A =$ rank $B = n$.

8. True. There is one leading 1 in each pivot column.

9. True. $\mathbf{x} = \frac{1}{2}A\mathbf{x}$ and $A\mathbf{x}$ is a linear combination of the columns of A.

10. False. There are n pivots. Since these occur in different rows, there must be at least n rows, so $m \geq n$.

11. False. The system has **at most one** solution, by Theorem 4.1.24.

12. True. Let the columns of A be $\mathbf{a}_1, \dots, \mathbf{a}_n$ and suppose $c_1\mathbf{a}_1 + \cdots + c_n\mathbf{a}_n = \mathbf{0}$. By 2.1.33, this is $A\mathbf{c} = \mathbf{0}$ with $\mathbf{c} = \begin{bmatrix} c_1 \\ \vdots \\ c_n \end{bmatrix}$.

13. True. By Theorem 2.4.1, any solution is of the form $\mathbf{x}_p + \mathbf{x}_h$ with $\mathbf{x}_h$ a solution to the corresponding homogeneous system. This is the system $A\mathbf{x} = \mathbf{0}$, so $\mathbf{x}$ is in the null space of A.

14. True. The null space of A consists of the vectors $\mathbf{x}$ that satisfy $A\mathbf{x} = \mathbf{0}$. These are exactly the vectors $\mathbf{x}$ satisfying $U\mathbf{x} = \mathbf{0}$. Passing to U is the way we solve systems of equations!

15. False. The matrices $A = \begin{bmatrix} 1 & 1 \\ 1 & 1 \end{bmatrix}$ and $U = \begin{bmatrix} 1 & 1 \\ 0 & 0 \end{bmatrix}$ have different column spaces.

16. True. "Singular" means "not invertible."

17. False. Corollary 4.1.25 says that this statement is equivalent to two others. For a given square matrix A, all three statements are true or all three are false.

Exercises 4.1

1. If $\mathbf{x}$ is in the null space, $A\mathbf{x} = \mathbf{0}$. In order for $A\mathbf{x}$ to be defined, $\mathbf{x}$ must be 9×1. Thus $n = 9$. Since $A\mathbf{x}$ is 7×1 vector, $m = 7$.

3. (a) We must solve the homogeneous system $A\mathbf{x} = \mathbf{0}$ with A the given matrix and $\mathbf{x} = \begin{bmatrix} x_1 \\ x_2 \\ x_3 \\ x_4 \end{bmatrix}$. We have

$$A = \begin{bmatrix} 1 & -2 & 1 & 1 \\ 3 & 0 & 2 & -2 \\ 0 & 4 & -1 & -1 \\ 5 & 0 & 3 & -1 \end{bmatrix} \rightarrow \begin{bmatrix} 1 & -2 & 1 & 1 \\ 0 & 6 & -1 & -5 \\ 0 & 4 & -1 & -1 \\ 0 & 10 & -2 & -6 \end{bmatrix} \rightarrow \begin{bmatrix} 1 & -2 & 1 & 1 \\ 0 & 1 & -\frac{1}{6} & -\frac{5}{6} \\ 0 & 4 & -1 & -1 \\ 0 & 10 & -2 & -6 \end{bmatrix}$$

$$\rightarrow \begin{bmatrix} 1 & -2 & 1 & 1 \\ 0 & 1 & -\frac{1}{6} & -\frac{5}{6} \\ 0 & 0 & -\frac{1}{3} & \frac{7}{3} \\ 0 & 0 & -\frac{1}{3} & \frac{7}{3} \end{bmatrix} \rightarrow \begin{bmatrix} 1 & -2 & 1 & 1 \\ 0 & 1 & -\frac{1}{6} & -\frac{5}{6} \\ 0 & 0 & 1 & -7 \\ 0 & 0 & 0 & 0 \end{bmatrix},$$

so $x_4 = t$ is free, $x_3 = 7x_4 = 7t$, $x_2 = \frac{1}{6}x_3 + \frac{5}{6}x_4 = 2t$, $x_1 = 2x_2 - x_3 - x_4 = -4t$, and $\mathbf{x} = \begin{bmatrix} -4t \\ 2t \\ 7t \\ t \end{bmatrix} = t \begin{bmatrix} -4 \\ 2 \\ 7 \\ 1 \end{bmatrix}$. The

null space consists of all scalar multiples of $\begin{bmatrix} -4 \\ 2 \\ 7 \\ 1 \end{bmatrix}$. Rank $A = 3$ since there are three pivot columns.

4. (a) The question asks whether there exist scalars a and b so that $\begin{bmatrix} 1 \\ 3 \end{bmatrix} = a \begin{bmatrix} 2 \\ 2 \end{bmatrix} + b \begin{bmatrix} 0 \\ 3 \end{bmatrix}$.

This is true, with $a = \frac{1}{2}$, $b = \frac{2}{3}$, so the answer is yes.

(c) The question asks whether there exist scalars a and b so that $\begin{bmatrix} 1 \\ 0 \\ 0 \end{bmatrix} = a \begin{bmatrix} -1 \\ 2 \\ 3 \end{bmatrix} + b \begin{bmatrix} 4 \\ 5 \\ 6 \end{bmatrix}$. For this, we would need

$-a + 4b = 1$, $2a + 5b = 0$, and $3a + 6b = 0$. There are no solutions, so the answer is no. The given vector is not in the column space.

5. (a) We attempt to solve $A\mathbf{x} = \mathbf{b}_1$ for $\mathbf{x} = \begin{bmatrix} x_1 \\ x_2 \\ x_3 \end{bmatrix}$. Gaussian elimination applied to the augmented matrix proceeds

$$\left[\begin{array}{ccc|c} 2 & -1 & 4 & 1 \\ 0 & 2 & 3 & 1 \\ -1 & -5 & 1 & 1 \\ 1 & -3 & 2 & 1 \end{array}\right] \rightarrow \left[\begin{array}{ccc|c} 1 & -3 & 2 & 1 \\ 0 & 2 & 3 & 1 \\ 0 & 5 & 0 & -1 \\ 0 & -8 & 3 & 2 \end{array}\right] \rightarrow \left[\begin{array}{ccc|c} 1 & -3 & 2 & 1 \\ 0 & 1 & \frac{3}{2} & \frac{1}{2} \\ 0 & 5 & 0 & -1 \\ 0 & -8 & 3 & 2 \end{array}\right]$$

$$\rightarrow \left[\begin{array}{ccc|c} 1 & -3 & 2 & 1 \\ 0 & 1 & \frac{3}{2} & \frac{1}{2} \\ 0 & 0 & -\frac{15}{2} & -\frac{7}{2} \\ 0 & 0 & 15 & 6 \end{array}\right] \rightarrow \left[\begin{array}{ccc|c} 1 & -3 & 2 & 1 \\ 0 & 1 & \frac{3}{2} & \frac{1}{2} \\ 0 & 0 & 15 & 7 \\ 0 & 0 & 0 & 1 \end{array}\right]$$. The last row says $0 = 1$, so there is no solution. The

vector $\mathbf{b}_1$ is not in the column space of A.

6. (a) For $A\mathbf{x} = \mathbf{v}$ with $\mathbf{x} = \begin{bmatrix} x_1 \\ x_2 \\ x_3 \\ x_4 \end{bmatrix}$, $\left[\begin{array}{ccccc|c} 3 & 1 & 7 & 2 & 1 & 13 \\ 2 & -4 & 14 & -2 & 0 & -10 \\ 5 & 11 & -7 & 10 & 3 & 59 \\ 2 & 5 & -4 & -3 & 1 & 39 \end{array}\right]$

$$\rightarrow \left[\begin{array}{ccccc|c} 1 & -2 & 7 & -1 & 0 & -5 \\ 3 & 1 & 7 & 2 & 1 & 13 \\ 5 & 11 & -7 & 10 & 3 & 59 \\ 2 & 5 & -4 & -3 & 1 & 39 \end{array}\right] \rightarrow \left[\begin{array}{ccccc|c} 1 & -2 & 7 & -1 & 0 & -5 \\ 0 & 7 & -14 & 5 & 1 & 28 \\ 0 & 21 & -42 & 15 & 3 & 84 \\ 0 & 9 & -18 & -1 & 1 & 49 \end{array}\right]$$

$$\rightarrow \left[\begin{array}{ccccc|c} 1 & -2 & 7 & -1 & 0 & -5 \\ 0 & 1 & -2 & \frac{5}{7} & \frac{1}{7} & 4 \\ 0 & 0 & 0 & 0 & 0 & 0 \\ 0 & 0 & 0 & -\frac{52}{7} & -\frac{2}{7} & 13 \end{array}\right] \rightarrow \left[\begin{array}{ccccc|c} 1 & -2 & 7 & -1 & 0 & -5 \\ 0 & 1 & -2 & \frac{5}{7} & \frac{1}{7} & 4 \\ 0 & 0 & 0 & 1 & \frac{1}{26} & -\frac{7}{4} \\ 0 & 0 & 0 & 0 & 0 & 0 \end{array}\right]$$.

So $x_3 = t$ and $x_5 = s$ are free. Back substitution gives $x_4 = -\frac{7}{4} - \frac{1}{26}s$, $x_2 = \frac{21}{4} + 2t - \frac{3}{26}s$, and $x_1 = \frac{15}{4} - 3t - \frac{7}{26}s$. The general solution is

$$\mathbf{x} = \begin{bmatrix} x_1 \\ x_2 \\ x_3 \\ x_4 \\ x_5 \end{bmatrix} = \begin{bmatrix} \frac{15}{4} - 3t - \frac{7}{26}s \\ \frac{21}{4} + 2t - \frac{3}{26}s \\ t \\ -\frac{7}{4} - \frac{1}{26}s \\ s \end{bmatrix} = \begin{bmatrix} \frac{15}{4} \\ \frac{21}{4} \\ 0 \\ -\frac{7}{4} \\ 0 \end{bmatrix} + t \begin{bmatrix} -3 \\ 2 \\ 1 \\ 0 \\ 0 \end{bmatrix} + s \begin{bmatrix} -\frac{7}{26} \\ -\frac{3}{26} \\ 0 \\ -\frac{1}{26} \\ 1 \end{bmatrix}.$$

7. (a) $\mathbf{b} = \begin{bmatrix} b_1 \\ b_2 \\ b_3 \end{bmatrix}$ is in the column space if and only if the system $A\mathbf{x} = \mathbf{b}$ has a solution. We solve this system by Gaussian elimination and have

$$[A|\mathbf{b}] \rightarrow \begin{bmatrix} 2 & 3 & 4 & b_1 \\ 4 & 7 & 5 & b_2 \\ 0 & -1 & 3 & b_3 \end{bmatrix} \rightarrow \begin{bmatrix} 2 & 3 & 4 & b_1 \\ 0 & 1 & -3 & b_2 - 2b_1 \\ 0 & -1 & 3 & b_3 \end{bmatrix} \rightarrow \begin{bmatrix} 2 & 3 & 4 & b_1 \\ 0 & 1 & -3 & b_2 - 2b_1 \\ 0 & 0 & 0 & b_3 + b_2 - 2b_1 \end{bmatrix}.$$

The last equation reads $0 = b_3 + b_2 - 2b_1$. The system has a solution if and only if $-2b_1 + b_2 + b_3 = 0$. The column space is the plane with equation $2x - y - z = 0$.

The null space of A is the set of vectors $\mathbf{x}$ satisfying $A\mathbf{x} = \mathbf{0}$. Previous calculations show that row echelon form of

A is $\begin{bmatrix} 2 & 3 & 4 \\ 0 & 1 & -3 \\ 0 & 0 & 0 \end{bmatrix}$ so, if $\mathbf{x} = \begin{bmatrix} x_1 \\ x_2 \\ x_3 \end{bmatrix}$, $x_3 = t$ is free, $x_2 = 3x_3 = 3t$, and $2x_1 = -3x_2 - 4x_3 = -13t$, so $\mathbf{x} = \begin{bmatrix} -\frac{13}{2}t \\ 3t \\ t \end{bmatrix}$.

The null space consists of multiples of $\begin{bmatrix} -\frac{13}{2} \\ 3 \\ 1 \end{bmatrix}$.

8. (a) $\begin{bmatrix} 1 & 2 \\ 4 & 8 \end{bmatrix} \rightarrow \begin{bmatrix} 1 & 2 \\ 0 & 0 \end{bmatrix}$. There is one pivot column, so the rank is 1.

9. (a) The null space is the set of solutions to $A\mathbf{x} = \mathbf{0}$.

$$\begin{bmatrix} 2 & 3 & 4 \\ 4 & 7 & 5 \\ 0 & -1 & 3 \end{bmatrix} \rightarrow \begin{bmatrix} 1 & \frac{3}{2} & 2 \\ 0 & 1 & -3 \\ 0 & -1 & 3 \end{bmatrix} \rightarrow \begin{bmatrix} 1 & \frac{3}{2} & 2 \\ 0 & 1 & -3 \\ 0 & 0 & 0 \end{bmatrix}.$$

Thus $x_3 = t$ is free, $x_2 - 3x_3 = 0$, so $x_2 = 3x_3 = 3t$ and $x_1 + \frac{3}{2}x_2 + 2x_3 = 0$, so

$x_1 = -\frac{3}{2}x_2 - 2x_3 = -\frac{9}{2}t - 2t = -\frac{13}{2}t$. The null space is all vectors of the form $\begin{bmatrix} -\frac{13}{2}t \\ 3t \\ t \end{bmatrix} = t \begin{bmatrix} -\frac{13}{2} \\ 3 \\ 1 \end{bmatrix}$.

(b) The rank of A is the number of pivot columns. From part (a), we see rank $A = 2$.

(c) $\begin{bmatrix} 2 & 3 & 4 & 1 \\ 4 & 7 & 5 & 6 \\ 0 & -1 & 3 & -4 \end{bmatrix} \rightarrow \begin{bmatrix} 1 & \frac{3}{2} & 2 & \frac{1}{2} \\ 0 & 1 & -3 & 4 \\ 0 & -1 & 3 & -4 \end{bmatrix} \rightarrow \begin{bmatrix} 1 & \frac{3}{2} & 2 & \frac{1}{2} \\ 0 & 1 & -3 & 4 \\ 0 & 0 & 0 & 0 \end{bmatrix}.$

Thus $x_3 = t$ is free, $x_2 - 3x_3 = 4$, so $x_2 = 4 + 3x_3 = 4 + 3t$ and $x_1 + \frac{3}{2}x_2 + 2x_3 = \frac{1}{2}$, so

$x_1 = \frac{1}{2} - \frac{3}{2}x_2 - 2x_3 = \frac{1}{2} - \frac{3}{2}(4 + 3t) - 2t = -\frac{11}{2} - \frac{13}{2}t$. The solution is $\mathbf{x} = \begin{bmatrix} -\frac{11}{2} - \frac{13}{2}t \\ 4 + 3t \\ t \end{bmatrix} = \begin{bmatrix} -\frac{11}{2} \\ 4 \\ 0 \end{bmatrix} t + \begin{bmatrix} -\frac{13}{2} \\ 3 \\ 1 \end{bmatrix}$.

(d) From part (c), we see that one particular solution to $A\mathbf{x} = \mathbf{b}$ is $\begin{bmatrix} -\frac{11}{2} \\ 4 \\ 0 \end{bmatrix}$, so $\mathbf{b} = \begin{bmatrix} 1 \\ 6 \\ -4 \end{bmatrix} = -\frac{11}{2}\begin{bmatrix} 2 \\ 4 \\ 0 \end{bmatrix} + 4\begin{bmatrix} 3 \\ 7 \\ -1 \end{bmatrix}$.

(e) The column space consists of all vectors $\mathbf{b}$ with $A\mathbf{x} = \mathbf{b}$ for some $\mathbf{b} = \begin{bmatrix} b_1 \\ b_2 \\ b_3 \end{bmatrix}$. We try to solve $A\mathbf{x} = \mathbf{b}$:

$$\begin{bmatrix} 2 & 3 & 4 & b_1 \\ 4 & 7 & 5 & b_2 \\ 0 & -1 & 3 & b_3 \end{bmatrix} \rightarrow \begin{bmatrix} 1 & \frac{3}{2} & 2 & \frac{1}{2}b_1 \\ 0 & 1 & -3 & b_2 - 2b_1 \\ 0 & -1 & 3 & b_3 \end{bmatrix} \rightarrow \begin{bmatrix} 1 & \frac{3}{2} & 2 & \frac{1}{2}b_1 \\ 0 & 1 & -3 & b_2 - 2b_1 \\ 0 & 0 & 0 & b_3 + b_2 - 2b_1 \end{bmatrix}.$$

The system has a solution if and only if $b_3 + b_2 - 2b_1 = 0$. The column space is the plane with equation $2x - y - z = 0$.

11. (a) Since $A\mathbf{x} = \mathbf{0}$, $(BA)\mathbf{x} = B(A\mathbf{x}) = B\mathbf{0} = \mathbf{0}$, so $\mathbf{x}$ is in the null space of BA.

12. The given matrix is elementary, hence invertible. For any $\mathbf{b}$ in $\mathbf{R}^3$, $\mathbf{b} = A(A^{-1}\mathbf{b})$, so $\mathbf{b}$ is in the column space of A: col sp $A = \mathbf{R}^3$. The null space of A is $\mathbf{0}$ since $A\mathbf{x} = \mathbf{0}$ implies $\mathbf{x} = A^{-1}\mathbf{0} = \mathbf{0}$.

13. A row echelon form is $\begin{bmatrix} 1 & 2 & 3 & 4 \\ 0 & 0 & 0 & 0 \\ 0 & 0 & 0 & 0 \end{bmatrix}$, thus $A = \begin{bmatrix} 1 & 2 & 3 & 4 \\ 2 & 4 & 6 & 8 \\ 3 & 6 & 9 & 12 \end{bmatrix}$.

15. (a) The null space is the set of $\mathbf{x} = \begin{bmatrix} x_1 \\ x_2 \\ x_3 \\ x_4 \end{bmatrix}$ satisfying the homogeneous system $A\mathbf{x} = \mathbf{0}$.

$$\begin{bmatrix} 1 & 2 & 3 & 2 \\ -1 & -2 & -2 & 1 \\ 2 & 4 & 8 & 12 \end{bmatrix} \rightarrow \begin{bmatrix} 1 & 2 & 3 & 2 \\ 0 & 0 & 1 & 3 \\ 0 & 0 & 2 & 8 \end{bmatrix} \rightarrow \begin{bmatrix} 1 & 2 & 0 & -7 \\ 0 & 0 & 1 & 3 \\ 0 & 0 & 0 & 2 \end{bmatrix} \rightarrow \begin{bmatrix} 1 & 2 & 0 & 0 \\ 0 & 0 & 1 & 0 \\ 0 & 0 & 0 & 1 \end{bmatrix}.$$

We have $x_2 = t$ is free, $x_4 = 0$, $x_3 = 0$, $x_1 = -2t$. The null space is the set of vectors of the form $\begin{bmatrix} -2t \\ t \\ 0 \\ 0 \end{bmatrix} = t\begin{bmatrix} -2 \\ 1 \\ 0 \\ 0 \end{bmatrix}$.

(b) To solve $A\mathbf{x} = \mathbf{b} = \begin{bmatrix} -1 \\ 2 \\ 4 \end{bmatrix}$, we apply Gaussian elimination to the augmented matrix:

$$\begin{bmatrix} 1 & 2 & 3 & 2 & -1 \\ -1 & -2 & -2 & 1 & 2 \\ 2 & 4 & 8 & 12 & 4 \end{bmatrix} \rightarrow \begin{bmatrix} 1 & 2 & 3 & 2 & -1 \\ 0 & 0 & 1 & 3 & 1 \\ 0 & 0 & 2 & 8 & 6 \end{bmatrix} \rightarrow \begin{bmatrix} 1 & 2 & 3 & 2 & -1 \\ 0 & 0 & 1 & 3 & 1 \\ 0 & 0 & 0 & 2 & 4 \end{bmatrix} \rightarrow \begin{bmatrix} 1 & 2 & 3 & 2 & -1 \\ 0 & 0 & 1 & 3 & 1 \\ 0 & 0 & 0 & 1 & 2 \end{bmatrix}.$$

We have $x_2 = t$ free; $x_4 = 2$, $x_3 + 3x_4 = 1$, so $x_3 = -5$; $x_1 + 2x_2 + 3x_3 + 2x_4 = -1$, so $x_1 = -2t + 10$. Thus

$$\mathbf{x} = \begin{bmatrix} -2t + 10 \\ t \\ -5 \\ 2 \end{bmatrix} = \begin{bmatrix} 10 \\ 0 \\ -5 \\ 2 \end{bmatrix} + t\begin{bmatrix} -2 \\ 1 \\ 0 \\ 0 \end{bmatrix}.$$

(c) The rank of A is the number of pivot columns: rank $A = 3$.

(d) A vector $\mathbf{b} = \begin{bmatrix} b_1 \\ b_2 \\ b_3 \\ b_4 \end{bmatrix}$ is in the column space of A if and only if $A\mathbf{x} = \mathbf{b}$ has a solution. Applying Gaussian

elimination to the augmented matrix proceeds $\begin{bmatrix} 1 & 2 & 3 & 2 & b_1 \\ -1 & -2 & -2 & 1 & b_2 \\ 2 & 4 & 8 & 12 & b_3 \end{bmatrix} \rightarrow \begin{bmatrix} 1 & 2 & 3 & 2 & b_1 \\ 0 & 0 & 1 & 3 & b_2 + b_1 \\ 0 & 0 & 2 & 8 & b_3 - 2b_1 \end{bmatrix}$

$\rightarrow \begin{bmatrix} 1 & 2 & 3 & 2 & b_1 \\ 0 & 0 & 1 & 3 & b_2 + b_1 \\ 0 & 0 & 0 & 2 & b_3 - 2b_1 - 2(b_2 + b_1) \end{bmatrix} \rightarrow \begin{bmatrix} 1 & 2 & 3 & 2 & b_1 \\ 0 & 0 & 1 & 3 & b_2 + b_1 \\ 0 & 0 & 0 & 1 & \frac{1}{2}(-4b_1 - 2b_2 + b_3) \end{bmatrix}$.

This system has a solution (in fact, infinitely many) for any $\mathbf{b}$. The column space is $\mathbf{R}^3$.

18. Let the first two columns of A be $\mathbf{i}$ and $\mathbf{j}$. This ensures that the column space of A contains $\mathbf{i}$ and $\mathbf{j}$. If the third column is also $\mathbf{i}$, for example, then the column space is not changed, and so certainly does not contain $\mathbf{k}$, which is not a linear combination of $\mathbf{i}$ and $\mathbf{j}$. One matrix that does the job is $\begin{bmatrix} 1 & 0 & 1 \\ 0 & 1 & 0 \\ 0 & 0 & 0 \end{bmatrix}$.

19. **(a)** We are given that $(AB)\mathbf{x} = \mathbf{0}$. Thus $A(B\mathbf{x}) = \mathbf{0}$, so $B\mathbf{x}$ is in the null space of A.

21. The column space is the set of linear combinations of the three columns. The first two columns (for example) are not parallel, so any vector is a linear combination of them and hence of all three columns.

23. **(a)** $\begin{bmatrix} 2 & 1 & 5 & 4 \\ 1 & -1 & 1 & -1 \\ 1 & 0 & 2 & 1 \end{bmatrix} \rightarrow \begin{bmatrix} 1 & -1 & 1 & -1 \\ 0 & 3 & 3 & 6 \\ 0 & 1 & 1 & 2 \end{bmatrix} \rightarrow \begin{bmatrix} 1 & -1 & 1 & -1 \\ 0 & 1 & 1 & 2 \\ 0 & 0 & 0 & 0 \end{bmatrix}$. There are two pivots, so rank $A = 2$.

(b) $\begin{bmatrix} 2 & 1 & 5 & 4 & c^2 + 2 \\ 1 & -1 & 1 & -1 & 4c + 5 \\ 1 & 0 & 2 & 1 & c + 3 \end{bmatrix} \rightarrow \begin{bmatrix} 1 & -1 & 1 & -1 & 4c + 5 \\ 0 & 3 & 3 & 6 & c^2 + 2 - 2(4c + 5) \\ 0 & 1 & 1 & 2 & c + 3 - (4c + 5) \end{bmatrix}$

$\rightarrow \begin{bmatrix} 1 & -1 & 1 & -1 & 4c + 5 \\ 0 & 1 & 1 & 2 & -3c - 2 \\ 0 & 0 & 0 & 0 & c^2 - 8c - 8 - 3(-3c - 2) \end{bmatrix}$.

The system has a solution if and only if the equation corresponding to the last row here reads $0 = 0$; that is, if and only if $c^2 + c - 2 = 0$, that is, if and only if $c = 1$ or $c = -2$. If $c = 1$, row echelon form is

$\begin{bmatrix} 1 & -1 & 1 & -1 & 9 \\ 0 & 1 & 1 & 2 & -5 \\ 0 & 0 & 0 & 0 & 0 \end{bmatrix}$, $x_3 = t$ and $x_4 = s$ are free variables, $x_2 = -5 - x_3 - 2x_4 = -5 - t - 2s$ and

$x_1 = 9 + x_2 - x_3 + x_4 = 4 - 2t - s$. Thus $\mathbf{x} = \begin{bmatrix} 4 - 2t - s \\ -5 - t - 2s \\ t \\ s \end{bmatrix} = \begin{bmatrix} 4 \\ -5 \\ 0 \\ 0 \end{bmatrix} + t\begin{bmatrix} -2 \\ -1 \\ 1 \\ 0 \end{bmatrix} + s\begin{bmatrix} -1 \\ -2 \\ 0 \\ 1 \end{bmatrix}$. If $c = -2$, row echelon

form is $\begin{bmatrix} 1 & -1 & 1 & -1 & -3 \\ 0 & 1 & 1 & 2 & 4 \\ 0 & 0 & 0 & 0 & 0 \end{bmatrix}$, $x_3 = t$ and $x_4 = s$ are free variables, $x_2 = 4 - x_3 - 2x_4 = 4 - t - 2s$ and

$x_1 = -3 + x_2 - x_3 + x_4 = 4 - 2t - s$.

Thus $\mathbf{x} = \begin{bmatrix} -3 - 2t - s \\ 4 - t - 2s \\ t \\ s \end{bmatrix} = \begin{bmatrix} -3 \\ 4 \\ 0 \\ 0 \end{bmatrix} + t\begin{bmatrix} -2 \\ -1 \\ 1 \\ 0 \end{bmatrix} + s\begin{bmatrix} -1 \\ -2 \\ 0 \\ 1 \end{bmatrix}$.

24. If $\mathbf{b}$ is in col sp AB, then $\mathbf{b} = (AB)\mathbf{x}$ for some vector $\mathbf{x}$. Since $\mathbf{b} = A(B\mathbf{x})$, $\mathbf{b}$ is in col sp A. Thus col sp $AB \subseteq$ col sp A. On the other hand, if $\mathbf{b}$ is in col sp A, then $\mathbf{b} = A\mathbf{x}$ for some vector $\mathbf{x}$, so $\mathbf{b} = (ABA)\mathbf{x} = AB(A\mathbf{x})$ is in col sp AB. Thus col sp $A \subseteq$ col sp AB, so these subspaces are the same.

27. The eigenspace of A corresponding to λ is the collection of all vectors $\mathbf{x}$ that satisfy $A\mathbf{x} = \lambda\mathbf{x}$. This equation is equivalent to $(A - \lambda I)\mathbf{x} = \mathbf{0}$, the solution to which is the null space of $A - \lambda I$.

Section 4.2—True/False

1. False. A subspace must contain the zero vector, and this line does not.
2. False. Any two **nonparallel** vectors; neither is a scalar multiple of the other.
3. True. Any scalar multiple of a vector with equal components has equal components.
4. True. $\mathbf{0} = 0\mathbf{u} + 0\mathbf{v}$ is a linear combination of $\mathbf{u}$ and $\mathbf{v}$.
5. True. The column space of the given matrix is $\mathbf{R}^2$.
6. True. The second matrix is a row echelon form of the first.
7. False. The column space of the second matrix is the xy-plane, but the column space of the first matrix contains vectors with nonzero third component.
8. False. Let $\mathbf{u} = \mathbf{0}$, let $\mathbf{v}$ be any nonzero vector and let $\mathbf{w}$ any nonzero vector that is not a scalar multiple of $\mathbf{v}$.
9. True. The zero vector is never part of a linearly independent set.
10. False. The zero vector is in any subspace.
11. False, although the author has seen this statement proposed by many students over a long period of time. The statement is utterly trivial. **Of course** $c_1\mathbf{v}_1 + c_2\mathbf{v}_2 + \cdots + c_n\mathbf{v}_n = \mathbf{0}$ if you tell me all the coefficients are 0. Read the definition of linear independence again. See the glossary or check p. 245.
12. False. Let $\mathbf{u} = \begin{bmatrix} 1 \\ 1 \end{bmatrix}$ and $\mathbf{v} = \begin{bmatrix} 0 \\ 0 \end{bmatrix}$. Then $\{\mathbf{u}, \mathbf{v}\}$ is a linearly dependent set, but $\mathbf{u}$ is not a multiple of $\mathbf{v}$.
13. True. See 4.2.33.
14. False. If the vectors are linearly dependent, there is such an equation with **not all** the scalars $c_i = 0$. "Not all" does not mean "none."

Exercises 4.2

1. **(a)** The question is whether there exist scalars c_1, c_2, c_3 such that $\mathbf{v} = c_1\mathbf{v}_1 + c_2\mathbf{v}_2 + c_3\mathbf{v}_3$; that is,

$$\begin{bmatrix} 1 \\ 3 \\ 2 \end{bmatrix} = c_1\begin{bmatrix} 1 \\ 1 \\ 1 \end{bmatrix} + c_2\begin{bmatrix} 1 \\ 1 \\ 0 \end{bmatrix} + c_3\begin{bmatrix} 1 \\ 0 \\ 2 \end{bmatrix}.$$ This is $A\mathbf{c} = \mathbf{v}$ with $A = \begin{bmatrix} 0 & 1 & 1 \\ 1 & 1 & 0 \\ 1 & 0 & 2 \end{bmatrix}$ and $\mathbf{c} = \begin{bmatrix} c_1 \\ c_2 \\ c_3 \end{bmatrix}$. Gaussian elimination on the augmented matrix is

$$\begin{bmatrix} 0 & 1 & 1 & | & 1 \\ 1 & 1 & 0 & | & 3 \\ 1 & 0 & 2 & | & 2 \end{bmatrix} \rightarrow \begin{bmatrix} 1 & 1 & 0 & | & 3 \\ 0 & 1 & 1 & | & 1 \\ 0 & -1 & 2 & | & -1 \end{bmatrix} \rightarrow \begin{bmatrix} 1 & 1 & 0 & | & 3 \\ 0 & 1 & 1 & | & 1 \\ 0 & 0 & 3 & | & 0 \end{bmatrix}$$

We get $c_3 = 0$, $c_2 = 1$, $c_1 = 3 - c_2 = 2$. Thus $\mathbf{v} = 2\mathbf{v}_1 + \mathbf{v}_2 + 0\mathbf{v}_3$ is a linear combination of $\mathbf{v}_1, \mathbf{v}_2, \mathbf{v}_3$.

2. **(a)** The question is whether there are scalars a and b so that $\mathbf{v} = a\mathbf{v}_1 + b\mathbf{v}_2$. This is $A\mathbf{x} = \mathbf{v}$ with $A = \begin{bmatrix} 1 & 2 \\ -1 & 1 \\ 2 & 1 \\ 1 & 0 \end{bmatrix}$ and $\mathbf{x} = \begin{bmatrix} a \\ b \end{bmatrix}$. Gaussian elimination on the augmented matrix gives

$$\begin{bmatrix} 1 & 2 & | & 1 \\ -1 & 1 & | & 3 \\ 2 & 1 & | & -1 \\ 1 & 0 & | & 1 \end{bmatrix} \rightarrow \begin{bmatrix} 1 & 2 & | & 1 \\ 0 & 3 & | & 4 \\ 0 & -3 & | & -3 \\ 0 & -2 & | & -1 \end{bmatrix} \rightarrow \begin{bmatrix} 1 & 2 & | & 1 \\ 0 & 1 & | & 1 \\ 0 & 0 & | & 1 \\ 0 & 0 & | & 1 \end{bmatrix}.$$

The system has no solution. Vector $\mathbf{v}$ is **not** in the span of the others.

3. (a) These vectors do not span $\mathbf{R}^2$. Each one is a multiple of $\begin{bmatrix} 1 \\ -2 \end{bmatrix}$. Any linear combination of them will also be a multiple of $\begin{bmatrix} 1 \\ -2 \end{bmatrix}$.

5. One approach is to put the given vectors into the rows of a matrix and to investigate the row space using elementary row operations. We have

$$\begin{bmatrix} 1 & 0 & 0 & 1 \\ 1 & 0 & 0 & -1 \\ 0 & 1 & 1 & 0 \\ 0 & 1 & -1 & 0 \end{bmatrix} \rightarrow \begin{bmatrix} 1 & 0 & 0 & 1 \\ 0 & 0 & 0 & -2 \\ 0 & 1 & 1 & 0 \\ 0 & 1 & -1 & 0 \end{bmatrix} \rightarrow \begin{bmatrix} 1 & 0 & 0 & 1 \\ 0 & 1 & 1 & 0 \\ 0 & 0 & -2 & 0 \\ 0 & 0 & 0 & -2 \end{bmatrix}$$

$$\rightarrow \begin{bmatrix} 1 & 0 & 0 & 1 \\ 0 & 1 & 1 & 0 \\ 0 & 0 & 1 & 0 \\ 0 & 0 & 0 & 1 \end{bmatrix} \rightarrow \begin{bmatrix} 1 & 0 & 0 & 0 \\ 0 & 1 & 0 & 0 \\ 0 & 0 & 1 & 0 \\ 0 & 0 & 0 & 1 \end{bmatrix}.$$

The row space of the last matrix is $\mathbf{R}^4$ since its rows are the standard basis vectors for $\mathbf{R}^4$. Since the elementary row operations do not change the row space, the rows of the original matrix span $\mathbf{R}^4$. Thus the given vectors span $\mathbf{R}^4$.

6. (a) Closure under addition. If $\begin{bmatrix} x \\ y_1 \\ 2x_1 - 3y_1 \end{bmatrix}$ and $\begin{bmatrix} x_2 \\ y_2 \\ 2x_2 - 3y_2 \end{bmatrix}$ are both in U, then so is their sum because

$$\begin{bmatrix} x_1 \\ y_1 \\ 2x_1 - 3y_1 \end{bmatrix} + \begin{bmatrix} x_2 \\ y_2 \\ 2x_2 - 3y_2 \end{bmatrix} = \begin{bmatrix} x_1 + x_2 \\ y_1 + y_2 \\ 2x_1 - 3y_1 + 2x_2 - 3y_2 \end{bmatrix} = \begin{bmatrix} x_1 + x_2 \\ y_1 + y_2 \\ 2(x_1 + x_2) - 3(y_1 + y_2) \end{bmatrix},$$

which is of the form $\begin{bmatrix} X \\ Y \\ 2X - 3Y \end{bmatrix}$.

Closure under scalar multiplication. If $\begin{bmatrix} x \\ y \\ 2x - 3y \end{bmatrix}$ is in U and $c \in \mathbf{R}$, then $c\begin{bmatrix} x \\ y \\ 2x - 3y \end{bmatrix} = \begin{bmatrix} cx \\ cy \\ 2cx - 3cy \end{bmatrix}$ is in U since this vector is of the form $\begin{bmatrix} X \\ Y \\ 2X - 3Y \end{bmatrix}$.

7. (a) U is not a subspace since it is not closed under addition:

for example, $\begin{bmatrix} 0 \\ -1 \\ 0 \end{bmatrix}$ $(a = c = 0)$ and $\begin{bmatrix} 1 \\ 0 \\ 1 \end{bmatrix}$ $(a = c = 1)$ are in U but their sum $\begin{bmatrix} 1 \\ -1 \\ 1 \end{bmatrix}$ is not because $\begin{bmatrix} 1 \\ -1 \\ 1 \end{bmatrix} = \begin{bmatrix} a \\ a - 1 \\ c \end{bmatrix}$

implies $a = 1$ (comparing first coordinates) and $a - 1 = -1$ (comparing second coordinates) and there is no such a. (There are other possible answers. For instance, U is not closed under scalar multiplication. Also, U does not contain the zero vector.)

9. $\mathbf{u}_1 - \mathbf{u}_2 = \mathbf{u}_1 + (-1)\mathbf{u}_2$. Since U is closed under scalar multiplication, the vector $(-1)\mathbf{u}_2$ is in U. Since U is closed under addition, the sum of $\mathbf{u}_1$ and $(-1)\mathbf{u}_2$ is in U.

10. (a) The question is whether there are scalars c_1, c_2, c_3, not all 0, such that $c_1\mathbf{v}_1 + c_2\mathbf{v}_2 + c_3\mathbf{v}_3 = \mathbf{0}$. This is $A\mathbf{c} = \mathbf{0}$,

with $A = \begin{bmatrix} 1 & 1 & 2 \\ 0 & 2 & 1 \\ 1 & -1 & -4 \end{bmatrix}$ and $\mathbf{c} = \begin{bmatrix} c_1 \\ c_2 \\ c_3 \end{bmatrix}$. Gaussian elimination gives

$$\begin{bmatrix} 1 & 1 & 2 \\ 0 & 2 & 1 \\ 0 & -2 & -6 \end{bmatrix} \to \begin{bmatrix} 1 & 1 & 2 \\ 0 & 2 & 1 \\ 0 & 0 & -5 \end{bmatrix}.$$

The only solution is $c_1 = c_2 = c_3 = 0$. The vectors are linearly independent.

11. $1\mathbf{v}_1 + 0\mathbf{v}_2 + \cdots + 0\mathbf{v}_n$ is a nontrivial linear combination equal to the zero vector.

14. (a)   **i.** We use Gaussian elimination to move A to row echelon form.

$$\begin{bmatrix} -1 & 1 & 0 & 2 \\ 3 & 1 & 1 & 10 \\ 2 & 4 & -1 & 0 \end{bmatrix} \to \begin{bmatrix} 1 & -1 & 0 & -2 \\ 0 & 4 & 1 & 16 \\ 0 & 6 & -1 & 4 \end{bmatrix} \to \begin{bmatrix} 1 & -1 & 0 & -2 \\ 0 & 1 & \frac{1}{4} & 4 \\ 0 & 0 & -\frac{5}{2} & -20 \end{bmatrix}.$$ If $\mathbf{x} = \begin{bmatrix} x_1 \\ x_2 \\ x_3 \\ x_4 \end{bmatrix}$ is in the null space of A,

$x_4 = t$ is free,
$-\frac{5}{2}x_3 = 20x_4 = 20t$, so $x_3 = -8t$,
$x_2 = -\frac{1}{4}x_3 - 4x_4 = -2t$, and
$x_1 = x_2 + 2x_4 = 0$,

so $\mathbf{x} = \begin{bmatrix} 0 \\ -2t \\ -8t \\ t \end{bmatrix} = t \begin{bmatrix} 0 \\ -2 \\ -8 \\ 1 \end{bmatrix}$. The vector $\begin{bmatrix} 0 \\ -2 \\ -8 \\ 1 \end{bmatrix}$ spans the null space.

ii. The nonzero vectors in a row echelon form of A form are linearly independent and span the row space of A:
$\begin{bmatrix} 1 & -1 & 0 & -2 \end{bmatrix}$, $\begin{bmatrix} 0 & 1 & \frac{1}{4} & 4 \end{bmatrix}$ and $\begin{bmatrix} 0 & 0 & -\frac{5}{2} & -20 \end{bmatrix}$.

iii. We use Gaussian elimination to move A^T to row echelon form.

$$\begin{bmatrix} -1 & 3 & 2 \\ 1 & 1 & 4 \\ 0 & 1 & 1 \\ 2 & 10 & 0 \end{bmatrix} \to \begin{bmatrix} 1 & -3 & -2 \\ 0 & 4 & 6 \\ 0 & 1 & 1 \\ 0 & 16 & 4 \end{bmatrix} \to \begin{bmatrix} 1 & -3 & -2 \\ 0 & 1 & 1 \\ 0 & 0 & 2 \\ 0 & 0 & -12 \end{bmatrix} \to \begin{bmatrix} 1 & -3 & -2 \\ 0 & 1 & 1 \\ 0 & 0 & 1 \\ 0 & 0 & 0 \end{bmatrix}.$$

The nonzero rows of the matrix form a linearly independent set of vectors whose transposes span the column space of A: $\begin{bmatrix} 1 \\ -3 \\ -2 \end{bmatrix}$, $\begin{bmatrix} 0 \\ 1 \\ 1 \end{bmatrix}$ and $\begin{bmatrix} 0 \\ 0 \\ 1 \end{bmatrix}$.

16. As suggested, we imitate the proof of Theorem 3.4.14. First multiply (†) by A and use the fact that $A\mathbf{x}_1 = \lambda_1\mathbf{x}_1$, $A\mathbf{x}_2 = \lambda_2\mathbf{x}_2$, and so on. We obtain

$$c_1\lambda_1\mathbf{x}_1 + c_2\lambda_2\mathbf{x}_2 + \cdots + c_\ell\lambda_\ell\mathbf{x}_\ell = \mathbf{0}. \tag{*}$$

Now we multiply (†) by λ_1 and obtain

$$c_1\lambda_1\mathbf{x}_1 + c_2\lambda_1\mathbf{x}_2 + \cdots + c_\ell\lambda_1\mathbf{x}_\ell = \mathbf{0}. \tag{**}$$

The first terms of equations (*) and (**) are the same, so subtracting (**) from (*), we obtain

$$c_2(\lambda_2 - \lambda_1)\mathbf{x}_2 + c_3(\lambda_3 - \lambda_1)\mathbf{x}_3 + \cdots + c_\ell(\lambda_\ell - \lambda_1)\mathbf{x}_\ell = \mathbf{0}.$$

This is an equation just like (†) but with one less term. Also, each coefficient is nonzero since $c_i \neq 0$ and $\lambda_i \neq \lambda_1$ (the λs were all different). We have contradicted the fact that (†) was the shortest dependence relation with all coefficients nonzero and the result follows.

17. Suppose $a\mathbf{u} + b\mathbf{v} + c\mathbf{w} = \mathbf{0}$. We must show that $a = b = c = 0$. First we note that $c = 0$, for otherwise, dividing by c would show that $\mathbf{w}$ is a linear combination (and in the span) of $\mathbf{u}$ and $\mathbf{v}$. So $a\mathbf{u} + b\mathbf{v} = \mathbf{0}$. This says $b = 0$; otherwise, $\mathbf{v}$ is a scalar multiple of $\mathbf{u}$. Now we have $a\mathbf{u} = \mathbf{0}$ and, since $\mathbf{u} \neq \mathbf{0}$, $a = 0$.

18. If U is contained in W, then $U \cup W = W$ is a subspace. Similarly, if W is contained in U, then $U \cup W = U$ is a subspace. Suppose, conversely, that $U \cup W$ is a subspace and, following the hint, suppose that neither U nor W is contained in the other. Then there is a vector u that is in U but not in W, and a vector w that is in W but not in U. Consider the vector $u + w$ that, by assumption, is in the subspace $U \cup W$. If $u + w = u_1$ is in U, then $w = u_1 - u$ is in U since U is a subspace, contradicting the fact that w was a vector not in U. So $u + w$ is not in U and, similarly, it is not in W. This contradicts the fact that $U \cup W$ is a subspace, so our assumption that neither U nor W was contained in the other must be wrong.

19. (a) Suppose the columns of U are linearly independent. If some linear combination of the columns of A is $\mathbf{0}$, we have $A\mathbf{x} = \mathbf{0}$ for some $\mathbf{x}$. Thus $EA\mathbf{x} = \mathbf{0}$, so $U\mathbf{x} = \mathbf{0}$. Since the columns of U are linearly independent, $\mathbf{x} = \mathbf{0}$, so the columns of A are linearly independent.

    (b) Suppose the columns of A span $\mathbf{R}^m$ and let $\mathbf{y}$ be a vector in $\mathbf{R}^m$. Then $E^{-1}\mathbf{y} = A\mathbf{x}$ for some $\mathbf{x}$, so $\mathbf{y} = EA\mathbf{x} = U\mathbf{x}$. This says $\mathbf{y}$ is a linear combination of the columns of U, so these span $\mathbf{R}^m$ too.

Section 4.3—True/False

1. False. Any spanning set of $\mathbf{R}^4$ needs at least four vectors.

2. True. The given vectors are linearly independent since neither is a multiple of the other. There are two of them in a vector space of dimension 2, so they form a basis. In particular, they span.

3. False. A linearly independent set in $\mathbf{R}^3$ contains at most three vectors.

4. False. **At most** n vectors.

5. True.

6. False. A finite dimensional vector space has a finite **basis**, but an infinite number of vectors: if $\mathbf{v}$ is in the vector space, so are all the vectors $c\mathbf{v}$, for any real number c.

7. True. The matrix has rank 3, so its column space is $\mathbf{R}^3$ and the three given vectors are the standard basis for $\mathbf{R}^3$.

8. True. The nonzero rows of row echelon form constitute a basis.

9. True. Once we have row echelon form for a matrix A, we can identify the pivot columns, and these constitute a basis for the column space of A.

10. False. The first two columns of each matrix give bases for the respective column spaces, but these spaces are different. The vectors $\begin{bmatrix} 1 \\ 5 \\ 9 \end{bmatrix}$ and $\begin{bmatrix} 2 \\ 6 \\ 10 \end{bmatrix}$ give a basis for the column space of A.

11. True. There are n columns, each being a vector in $\mathbf{R}^m$. In $\mathbf{R}^m$, any $n > m$ vectors are linearly dependent.

12. True. rank $AA^T = $ rank $A \leq 3$.

13. True. The matrix has row rank $=$ column rank $= 3$.

14. True. n vectors span $\mathbf{R}^n$ if and only if they are linearly independent.

Exercises 4.3

1. (a) $\dim \mathbf{R}^4 = 4$ and we are given four vectors. These form a basis if and only if they are linearly independent, so we investigate the system $A\mathbf{x} = \mathbf{0}$, where A is the matrix whose columns are the given vectors. Gaussian elimination proceeds

$$
\begin{bmatrix} 3 & 0 & 0 & 1 \\ 6 & -1 & -8 & 0 \\ 3 & -1 & -12 & -1 \\ -6 & 0 & -4 & 2 \end{bmatrix} \rightarrow
\begin{bmatrix} 3 & 0 & 0 & 1 \\ 0 & -1 & -8 & -2 \\ 0 & -1 & -12 & -2 \\ 0 & 0 & -4 & 4 \end{bmatrix} \rightarrow
\begin{bmatrix} 3 & 0 & 0 & 1 \\ 0 & -1 & -8 & -2 \\ 0 & 0 & -4 & 0 \\ 0 & 0 & -4 & 4 \end{bmatrix} \rightarrow
\begin{bmatrix} 3 & 0 & 0 & 1 \\ 0 & -1 & -8 & -2 \\ 0 & 0 & -4 & 0 \\ 0 & 0 & 0 & 4 \end{bmatrix}.
$$

The only solution to $A\mathbf{x} = \mathbf{0}$ is $\mathbf{x} = \mathbf{0}$, so the vectors form a basis.

(c) Two vectors in a vector space of dimension three cannot form a basis.

2. We have $A = \begin{bmatrix} -1 & 0 & 2 & -4 & 4 \\ 0 & 1 & 1 & -4 & 4 \\ 1 & 1 & 1 & -6 & -2 \\ 1 & 0 & -1 & 1 & -5 \end{bmatrix}$. Gaussian elimination proceeds

$$
\begin{bmatrix} -1 & 0 & 2 & -4 & 4 \\ 0 & 1 & 1 & -4 & 4 \\ 1 & 1 & 1 & -6 & -2 \\ 1 & 0 & -1 & 1 & -5 \end{bmatrix} \rightarrow
\begin{bmatrix} 1 & 0 & -2 & 4 & -4 \\ 0 & 1 & 1 & -4 & 4 \\ 0 & 1 & 3 & -10 & 2 \\ 0 & 0 & 1 & -3 & -1 \end{bmatrix}
$$

$$
\rightarrow
\begin{bmatrix} 1 & 0 & -2 & 4 & -4 \\ 0 & 1 & 1 & -4 & 4 \\ 0 & 0 & 2 & -6 & -2 \\ 0 & 0 & 1 & -3 & -1 \end{bmatrix} \rightarrow
\begin{bmatrix} 1 & 0 & -2 & 4 & -4 \\ 0 & 1 & 1 & -4 & 4 \\ 0 & 0 & 1 & -3 & -1 \\ 0 & 0 & 0 & 0 & 0 \end{bmatrix}.
$$

The span of the given vectors is the column space of A which, by **4.3.35**, has basis the pivot columns. These are columns one, two, and three that we found in **4.3.32** by other means.

4. (a) Let $\mathbf{v}_2$ be any vector that is not a scalar multiple of $\mathbf{v}_1$. Then $\mathbf{v}_1$ and $\mathbf{v}_2$ will be linearly independent and hence a basis of $\mathbf{R}^2$. For example, $\mathbf{v}_2 = \mathbf{e}_1 = \begin{bmatrix} 1 \\ 0 \end{bmatrix}$ is one possibility.

5. (a) Since S contains the standard basis vectors for $\mathbf{R}^3$, $V = \mathbf{R}^3$ and the most obvious basis for V contained in S is

$$
\mathbf{e}_1 = \begin{bmatrix} 1 \\ 0 \\ 0 \end{bmatrix}, \mathbf{e}_2 = \begin{bmatrix} 0 \\ 1 \\ 0 \end{bmatrix}, \mathbf{e}_3 = \begin{bmatrix} 0 \\ 0 \\ 1 \end{bmatrix}.
$$

6. (a) Apply Gaussian elimination to the matrix A whose rows are the vectors that span V.

$$
A = \begin{bmatrix} 1 & -1 & 3 & 2 \\ 2 & 1 & 1 & 3 \\ 1 & 5 & -7 & 0 \\ 4 & -1 & 7 & 7 \end{bmatrix} \rightarrow
\begin{bmatrix} 1 & -1 & 3 & 2 \\ 0 & 3 & -5 & -1 \\ 0 & 6 & -10 & -2 \\ 0 & 3 & -5 & -1 \end{bmatrix} \rightarrow
\begin{bmatrix} 1 & -1 & 3 & 2 \\ 0 & 3 & -5 & -1 \\ 0 & 0 & 0 & 0 \\ 0 & 0 & 0 & 0 \end{bmatrix} = U.
$$

The nonzero rows of U are linearly independent and span the same subspace as the rows of A, so they are a basis for V. A basis is $\left\{ \begin{bmatrix} 1 \\ -1 \\ 3 \\ 2 \end{bmatrix}, \begin{bmatrix} 0 \\ 3 \\ -5 \\ -1 \end{bmatrix} \right\}$: $\dim V = 2$.

(b) Notice that none of the given vectors is a multiple of any other. It follows that any pair of vectors is linearly independent, and hence a basis for V.

9. W is the null space of $A = \begin{bmatrix} 1 & 2 & 3 & 5 \\ 1 & 0 & 1 & 3 \\ 2 & 3 & 5 & 9 \end{bmatrix}$. Using Gaussian elimination,

$$\begin{bmatrix} 1 & 2 & 3 & 5 \\ 1 & 0 & 1 & 3 \\ 2 & 3 & 5 & 9 \end{bmatrix} \rightarrow \begin{bmatrix} 1 & 2 & 3 & 5 \\ 0 & -2 & -2 & -2 \\ 0 & -1 & -1 & -1 \end{bmatrix} \rightarrow \begin{bmatrix} 1 & 2 & 3 & 5 \\ 0 & 1 & 1 & 1 \\ 0 & 0 & 0 & 0 \end{bmatrix}$$

so $x_3 = t$ and $x_4 = s$ are free, $x_2 = -t - s$, $x_1 = -2(-t - s) - 3t - 5s = -t - 3s$, and

$$\mathbf{x} = \begin{bmatrix} -t - 3s \\ -t - s \\ t \\ s \end{bmatrix} = t \begin{bmatrix} -1 \\ -1 \\ 1 \\ 0 \end{bmatrix} + s \begin{bmatrix} -3 \\ -1 \\ 0 \\ 1 \end{bmatrix}.$$ The vectors $\begin{bmatrix} -1 \\ -1 \\ 1 \\ 0 \end{bmatrix}$ and $\begin{bmatrix} -3 \\ -1 \\ 0 \\ 1 \end{bmatrix}$ are a basis for W: dim $W = 2$.

11. We have to show that a maximal linearly independent set S of vectors in V also spans V. This is essentially a restatement of Lemma 4.3.21. For any $\mathbf{v} \in V$, $S \cup \{\mathbf{v}\}$ is linearly dependent, so $\mathbf{v}$ must lie in the span of S. Thus S spans V.

13. (a) Any basis for U consists of linearly independent vectors in V. By Corollary 4.3.22, these can be extended to a basis for V, so they are part of a basis for V.

15. (a) Gaussian elimination proceeds

$$\begin{bmatrix} 1 & -1 & 0 & 2 & 3 & 1 \\ 3 & -2 & 4 & 4 & 11 & 6 \\ 5 & -4 & 4 & 9 & 12 & 9 \\ 2 & -1 & 4 & 2 & 8 & 5 \end{bmatrix} \rightarrow \begin{bmatrix} 1 & -1 & 0 & 2 & 3 & 1 \\ 0 & 1 & 4 & -2 & 2 & 3 \\ 0 & 1 & 4 & -1 & -3 & 4 \\ 0 & 1 & 4 & -2 & 2 & 3 \end{bmatrix} \rightarrow \begin{bmatrix} 1 & -1 & 0 & 2 & 3 & 1 \\ 0 & 1 & 4 & -2 & 2 & 3 \\ 0 & 0 & 0 & 1 & -5 & 1 \\ 0 & 0 & 0 & 0 & 0 & 0 \end{bmatrix} = U.$$

(b) The nonzero rows of U are linearly independent and span row sp U, so the row space of U has basis $\begin{bmatrix} 1 & -1 & 0 & 2 & 3 & 1 \end{bmatrix}$, $\begin{bmatrix} 0 & 1 & 4 & -2 & 2 & 3 \end{bmatrix}$, and $\begin{bmatrix} 0 & 0 & 0 & 1 & -5 & 1 \end{bmatrix}$.

Since row sp A = row sp U, these same vectors form a basis for row sp A.

(c) The row rank of U is dim row sp $U = 3$. Since U and A have the same row spaces, this is also the row rank of A.

(d) The pivot columns of U are columns one, two, and four: $\begin{bmatrix} 1 \\ 0 \\ 0 \\ 0 \end{bmatrix}$, $\begin{bmatrix} -1 \\ 1 \\ 0 \\ 0 \end{bmatrix}$, and $\begin{bmatrix} -2 \\ 2 \\ 1 \\ 0 \end{bmatrix}$.

$a \begin{bmatrix} 1 \\ 0 \\ 0 \\ 0 \end{bmatrix} + b \begin{bmatrix} -1 \\ 1 \\ 0 \\ 0 \end{bmatrix} + c \begin{bmatrix} -2 \\ 2 \\ 1 \\ 0 \end{bmatrix} = \begin{bmatrix} 0 \\ 0 \\ 0 \\ 0 \end{bmatrix}$ implies $\begin{bmatrix} a - b - 2c \\ b + 2c \\ c \\ 0 \end{bmatrix} = \begin{bmatrix} 0 \\ 0 \\ 0 \\ 0 \end{bmatrix}$ and back substitution gives $c = b = a = 0$. So these columns are linearly independent.

(e) We must show that $\mathbf{a}_1 = \begin{bmatrix} 1 \\ 3 \\ 5 \\ 2 \end{bmatrix}$, $\mathbf{a}_3 = \begin{bmatrix} -1 \\ -2 \\ -4 \\ -1 \end{bmatrix}$, and $\mathbf{a}_5 = \begin{bmatrix} 2 \\ 4 \\ 9 \\ 2 \end{bmatrix}$ are linearly independent.

$a \begin{bmatrix} 1 \\ 3 \\ 5 \\ 2 \end{bmatrix} + b \begin{bmatrix} -1 \\ -2 \\ -4 \\ -1 \end{bmatrix} + c \begin{bmatrix} 2 \\ 4 \\ 9 \\ 2 \end{bmatrix} = \begin{bmatrix} 0 \\ 0 \\ 0 \\ 0 \end{bmatrix}$ implies $\begin{array}{l} a - b + 2c = 0 \\ 3a - 2b + 4c = 0 \\ 5a - 4b + 9c = 0 \\ 2a - b + 2c = 0 \end{array}$

and

$$\begin{bmatrix} 1 & -1 & 2 \\ 3 & -2 & 4 \\ 5 & -4 & 9 \\ 2 & -1 & 2 \end{bmatrix} \rightarrow \begin{bmatrix} 1 & -1 & 2 \\ 0 & 1 & -2 \\ 0 & 1 & -1 \\ 0 & 1 & -2 \end{bmatrix} \rightarrow \begin{bmatrix} 1 & -1 & 2 \\ 0 & 1 & -2 \\ 0 & 0 & 1 \\ 0 & 0 & 0 \end{bmatrix}$$

gives $c = b = a = 0$, so $\mathbf{a}_1, \mathbf{a}_3, \mathbf{a}_4$ are linearly independent.

(f) The "other" columns are $\mathbf{a}_2 = \begin{bmatrix} -1 \\ -2 \\ -4 \\ -1 \end{bmatrix}$, $\mathbf{a}_5 = \begin{bmatrix} 3 \\ 11 \\ 12 \\ 8 \end{bmatrix}$, and $\mathbf{a}_6 = \begin{bmatrix} 1 \\ 6 \\ 9 \\ 5 \end{bmatrix}$.

We have $\mathbf{a}_2 = \frac{1}{4}\mathbf{a}_3 - \mathbf{a}_1$, $\mathbf{a}_5 = 13\mathbf{a}_1 - 2\mathbf{a}_3 - 5\mathbf{a}_4$, and $\mathbf{a}_6 = -\mathbf{a}_1 + \frac{5}{4}\mathbf{a}_3 + \mathbf{a}_4$.

(g) Since columns $\mathbf{a}_1$, $\mathbf{a}_3$, and $\mathbf{a}_4$ form a basis for the column space, the column space has dimension 3. The column rank of A is 3, the same as the row rank.

(h) The rank of A is its row rank; the rank of A^T is its column rank. These numbers are the same for any matrix.

(i) The nullity of A is $n - \text{rank } A = 6 - 3 = 3$.

17. (a) Gaussian elimination on A proceeds

$$A = \begin{bmatrix} -1 & 0 & -1 & 1 \\ 0 & 1 & -1 & 2 \\ 1 & 1 & 0 & 1 \\ 2 & -1 & 3 & 1 \\ 0 & -1 & 1 & 0 \end{bmatrix} \rightarrow \begin{bmatrix} 1 & 0 & 1 & -1 \\ 0 & 1 & -1 & 2 \\ 0 & 1 & -1 & 2 \\ 0 & -1 & 1 & 3 \\ 0 & -1 & 1 & 0 \end{bmatrix} \rightarrow \begin{bmatrix} 1 & 0 & 1 & -1 \\ 0 & 1 & -1 & 2 \\ 0 & 0 & 0 & 0 \\ 0 & 0 & 0 & 5 \\ 0 & 0 & 0 & 2 \end{bmatrix} \rightarrow \begin{bmatrix} 1 & 0 & 1 & -1 \\ 0 & 1 & -1 & 2 \\ 0 & 0 & 0 & 1 \\ 0 & 0 & 0 & 0 \\ 0 & 0 & 0 & 0 \end{bmatrix} = U.$$

The nonzero rows of U are a basis for the row space:

$$\begin{bmatrix} 1 & 0 & 1 & -1 \end{bmatrix}, \quad \begin{bmatrix} 0 & 1 & -1 & 2 \end{bmatrix}, \quad \text{and} \quad \begin{bmatrix} 0 & 0 & 0 & 1 \end{bmatrix}.$$

The dimension of the row space is 3.

(b) The pivot columns are columns one, two, and four. So columns one, two, and four of A provide a basis for the column space of A:

$$\begin{bmatrix} -1 \\ 0 \\ 1 \\ 2 \\ 0 \end{bmatrix}, \quad \begin{bmatrix} 0 \\ 1 \\ 1 \\ -1 \\ -1 \end{bmatrix}, \quad \text{and} \quad \begin{bmatrix} 1 \\ 2 \\ 1 \\ 1 \\ 0 \end{bmatrix}.$$

The dimension of the column space is 3.

(c) The null space of A is the set of $\mathbf{x} = \begin{bmatrix} x_1 \\ x_2 \\ x_3 \\ x_4 \end{bmatrix}$ that satisfy $A\mathbf{x} = \mathbf{0}$. Using the result of part (a), we see that $x_3 = t$ is

free, $x_4 = 0$, $x_2 = x_3 - 2x_4 = t$, and $x_1 = -x_3 + x_4 = -t$, so $\mathbf{x} = t \begin{bmatrix} -1 \\ 1 \\ 1 \\ 0 \end{bmatrix}$. A basis is $\begin{bmatrix} -1 \\ 1 \\ 1 \\ 0 \end{bmatrix}$. The nullity of A,

which is the dimension of the null space, is 1.

20. (a) B has (column) rank n, so it is invertible by Theorem 4.1.24.

21. (a) First reduce A to row echelon form:

$$A = \begin{bmatrix} 1 & 1 \\ 1 & -1 \\ 2 & 0 \\ 1 & 2 \end{bmatrix} \rightarrow \begin{bmatrix} 1 & 1 \\ 0 & -2 \\ 0 & -2 \\ 0 & 1 \end{bmatrix} \rightarrow \begin{bmatrix} 1 & 1 \\ 0 & 1 \\ 0 & 0 \\ 0 & 0 \end{bmatrix}.$$

    i. If $\mathbf{x} = \begin{bmatrix} x_1 \\ x_2 \end{bmatrix}$ is in the null space, $x_2 = x_1 = 0$, $\mathbf{x} = \mathbf{0}$. The null space is $\{\mathbf{0}\}$ and nullity $A = 0$.

    ii. The row space has basis the nonzero rows of row echelon form: $\begin{bmatrix} 1 & 1 \end{bmatrix}$ and $\begin{bmatrix} 0 & 1 \end{bmatrix}$. The row rank of A is 2.

    iii. The pivot columns of A comprise a basis for the column space, here both columns: $\begin{bmatrix} 1 \\ 1 \\ 2 \\ 1 \end{bmatrix}$ and $\begin{bmatrix} 1 \\ -1 \\ 0 \\ 2 \end{bmatrix}$. The

column rank is 2.

    iv. $n = 2$, the number of columns, and $2 = 2 + 0 = \text{rank } A + \text{nullity } A$.

(d) First reduce A to row echelon form.

$$\begin{bmatrix} 1 & 0 & 1 & 1 \\ 1 & 1 & 2 & 0 \\ 0 & 1 & 1 & -1 \end{bmatrix} \rightarrow \begin{bmatrix} 1 & 0 & 1 & 1 \\ 0 & 1 & 1 & -1 \\ 0 & 1 & 1 & -1 \end{bmatrix} \rightarrow \begin{bmatrix} 1 & 0 & 1 & 1 \\ 0 & 1 & 1 & -1 \\ 0 & 0 & 0 & 0 \end{bmatrix}.$$

    i. If $\mathbf{x} = \begin{bmatrix} x_1 \\ x_2 \\ x_3 \\ x_4 \end{bmatrix}$ is in the null space, $x_3 = t$ and $x_4 = s$ are free, $x_2 = -x_3 + x_4 = -t + s$,

$$x_1 = -x_3 - x_4 = -t - s, \text{ and } \mathbf{x} = \begin{bmatrix} -t - s \\ -t + s \\ t \\ s \end{bmatrix} = t \begin{bmatrix} -1 \\ -1 \\ 1 \\ 0 \end{bmatrix} + s \begin{bmatrix} -1 \\ 1 \\ 0 \\ 1 \end{bmatrix}.$$

The vectors $\begin{bmatrix} -1 \\ -1 \\ 1 \\ 0 \end{bmatrix}$ and $\begin{bmatrix} -1 \\ 1 \\ 0 \\ 1 \end{bmatrix}$ are a basis for the null space: nullity $A = 2$.

    ii. The nonzero rows of row echelon form comprise a basis for the row space: $\begin{bmatrix} 1 & 0 & 1 & 1 \end{bmatrix}$ and $\begin{bmatrix} 0 & 1 & 1 & -1 \end{bmatrix}$. The row rank is 2.

    iii. The pivot columns are columns one and two, and these comprise a basis for the column space: $\begin{bmatrix} 1 \\ 1 \\ 0 \end{bmatrix}$ and $\begin{bmatrix} 0 \\ 1 \\ 1 \end{bmatrix}$.

The column rank is 2.

    iv. $n = 4$, the number of columns of A, and $4 = 2 + 2 = \text{rank } A + \text{nullity } A$.

25. Two vectors can span a vector space of dimension at most two. So $n \leq 2$.

28. (a) Take any two vectors, neither of which is a multiple of the other (so these are linearly independent). These vectors will span a plane. For a third vector, take any linear combination of the first two.

(b) I would cry. This cannot be done. Three vectors in a plane must be linearly dependent.

30. We are given that the row rank of this matrix is 1, so the column rank is 1. Let $\mathbf{v}$ be any nonzero column of A. Then $\mathbf{v}$ is a basis for the column space, so every column is a multiple of $\mathbf{v}$.

Section 4.4—True/False

1. False. A vector space of positive dimension has infinitely many subspaces; for example, any line through **0** is a subspace.

2. False. An $m \times n$ matrix has rank at most the smaller of m and n.

3. True. rank $A = 2$, rank $B = 3$ and rank $AB \leq$ rank A.

4. False. Take $A = \mathbf{0}$.

5. False. Take $A = \mathbf{0}$ and B any singular matrix.

6. True. In part 3 of Theorem 4.1.24, we even showed how to construct a right inverse.

7. True.

8. True.

9. False. The system has a unique solution **if a solution exists**.

10. True. The rows of A are linearly independent.

11. False. The columns of A are linearly dependent (three vectors in $\mathbf{R}^2$).

12. False in general. The hypothesis says A has a right inverse, $\frac{1}{c}B$, so rank $A = m$, and that B has a left inverse, $\frac{1}{c}A$, so rank $B = n$.

Exercises 4.4

1. The rank of this matrix is 2, which is neither the number of columns nor the number of rows.

3. **(a)** The rank of A is 2, the number of rows, so A has a right inverse.

   (b) If A has a right inverse and also a left inverse, then A is square, by Corollary 4.4.12. Thus the given matrix does not have an inverse on both sides.

   (c) To find the general right inverse of A, we use the method demonstrated in **4.4.8**, namely, we solve the equations
   $$A \begin{bmatrix} x_1 \\ y_1 \\ z_1 \end{bmatrix} = \begin{bmatrix} 1 \\ 0 \end{bmatrix}, \; A \begin{bmatrix} x_2 \\ y_2 \\ z_2 \end{bmatrix} = \begin{bmatrix} 0 \\ 1 \end{bmatrix} \text{ simultaneously. The calculation}$$

   $$\left[\begin{array}{ccc|cc} -1 & 2 & 1 & 1 & 0 \\ 3 & 1 & 0 & 0 & 1 \end{array} \right] \rightarrow \left[\begin{array}{ccc|cc} 1 & -2 & -1 & -1 & 0 \\ 0 & 7 & 3 & 3 & 1 \end{array} \right] \rightarrow \left[\begin{array}{ccc|cc} 1 & 0 & -\frac{1}{7} & -\frac{1}{7} & \frac{2}{7} \\ 0 & 1 & \frac{3}{7} & \frac{3}{7} & \frac{1}{7} \end{array} \right]$$

   shows us that $z_1 = t$ free, and then

   $$y_1 = \tfrac{3}{7} - \tfrac{3}{7}z_1 = \tfrac{3}{7} - \tfrac{3}{7}t$$

   $$x_1 = -\tfrac{1}{7} + \tfrac{1}{7}z_1 = -\tfrac{1}{7} + \tfrac{1}{7}t$$

   so that $\begin{bmatrix} x_1 \\ y_1 \\ z_1 \end{bmatrix} = \begin{bmatrix} -\frac{1}{7} + \frac{1}{7}t \\ \frac{3}{7} - \frac{3}{7}t \\ t \end{bmatrix}$. Also, $z_2 = s$ is free,

   $$y_2 = \tfrac{1}{7} - \tfrac{3}{7}z_2 = \tfrac{1}{7} - \tfrac{3}{7}s$$

   $$x_2 = \tfrac{2}{7} + \tfrac{1}{7}z_2 = \tfrac{2}{7} + \tfrac{1}{7}s$$

   and $\begin{bmatrix} x_2 \\ y_2 \\ z_2 \end{bmatrix} = \begin{bmatrix} \frac{2}{7} + \frac{1}{7}s \\ \frac{1}{7} - \frac{3}{7}s \\ s \end{bmatrix}$. The general right inverse for A is $B = \begin{bmatrix} -\frac{1}{7} + \frac{1}{7}t & \frac{2}{7} + \frac{1}{7}s \\ \frac{3}{7} - \frac{3}{7}t & \frac{1}{7} - \frac{3}{7}s \\ t & s \end{bmatrix}$.

7. (a) The rows of A are linearly independent, so A has a right inverse.

(b) The calculation $\begin{bmatrix} 1 & 0 & 3 & 4 & | & 1 & 0 \\ -1 & 1 & 4 & 0 & | & 0 & 1 \end{bmatrix} \rightarrow \begin{bmatrix} 1 & 0 & 3 & 4 & | & 1 & 0 \\ 0 & 1 & 7 & 4 & | & 1 & 1 \end{bmatrix}$ leads to the solution of $A \begin{bmatrix} x_1 \\ y_1 \\ z_1 \\ w_1 \end{bmatrix} = \begin{bmatrix} 1 \\ 0 \end{bmatrix}$ and

$A \begin{bmatrix} x_2 \\ y_2 \\ z_2 \\ w_2 \end{bmatrix} = \begin{bmatrix} 0 \\ 1 \end{bmatrix}$ simultaneously. We have $z_1 = t$ and $w_1 = s$ free, and then

$$y_1 = 1 - 7z_1 - 4w_1 = 1 - 7t - 4s$$

$$x_1 = 1 - 3z_1 - 4w_1 = 1 - 3t - 4s$$

so that $\begin{bmatrix} x_1 \\ y_1 \\ z_1 \\ w_1 \end{bmatrix} = \begin{bmatrix} 1 - 3t - 4s \\ 1 - 7t - 4s \\ t \\ s \end{bmatrix}$. Also, $z_2 = q$ and $w_2 = r$ are free,

$$y_2 = 1 - 7z_2 - 4w_2 = 1 - 7q - 4r$$

$$x_2 = -3z_2 - 4w_2 = -3q - 4r$$

and $\begin{bmatrix} x_2 \\ y_2 \\ z_2 \\ w+2 \end{bmatrix} = \begin{bmatrix} -3q - 4r \\ 1 - 7q - 4r \\ q \\ r \end{bmatrix}$. The general right inverse for A is $B = \begin{bmatrix} 1 - 3t - 4s & -3q - 4r \\ 1 - 7t - 4s & 1 - 7q - 4r \\ t & q \\ s & r \end{bmatrix}$.

(c) Setting $t = s = q = r0$, gives $B = \begin{bmatrix} 1 & 0 \\ 1 & 1 \\ 0 & 0 \\ 0 & 0 \end{bmatrix}$ as one right inverse of A.

10. (a) By Theorem 4.4.3, rank $AB \le$ rank A. Using the very same fact, rank $A =$ rank$(AB)B^{-1} \le$ rank AB, so rank $AB =$ rank A.

11. (a) There exist matrices C and D such that $CA = I$ and $DB = I$. Then $(DC)(AB) = D(CA)B = DIB = DB = I$, so AB has left inverse DC.

13. (a) The answer is "yes." Since the $n \times n$ matrix $A^T A$ is invertible, its rank is n by Corollary 4.4.4, so the rank of A is n by Theorem 4.3.40 and the columns of A are linearly independent by Theorem 4.4.9.

14. (a) Suppose $AB = I$ and B is unique. If $A\mathbf{x} = \mathbf{0}$, then $A(B + \mathbf{x}) = \mathbf{0}$, so $B + \mathbf{x}$ is also a right inverse. Thus $B + \mathbf{x} = B$ and $\mathbf{x} = \mathbf{0}$. Recalling that $A\mathbf{x}$ is a linear combination of the columns of A, it follows that the columns of A are linearly independent, so A has a left inverse. So A is square and invertible by Corollary 4.4.12.

15. Since the rows of U are linearly independent, U has a right inverse.

Section 5.1—True/False

1. False. $\sqrt{x + y} \ne \sqrt{x} + \sqrt{y}$.

2. False. $T(\mathbf{0}) \ne \mathbf{0}$.

3. True. $T(\mathbf{u} + \mathbf{v}) = T(\mathbf{u}) + T(\mathbf{v})$.

4. False. The person who says this has forgotten that "closed" applies to subspaces. A linear transformation is a function that *preserves* addition and scalar multiplication.

5. True. The kernel of a linear transformation is the set of vectors that T maps to **0**.

6. True. A linear transformation is one-to-one if and only if its kernel is **0**.

7. True. Since $\dim \ker T + \dim \operatorname{range} T = 3$, $\dim \operatorname{range} T \leq 3$. Since $\dim \operatorname{range} T = 4$ is impossible, range T cannot be $\mathbf{R}^4$, so T cannot be onto.

8. True. Since $\dim \ker T + \dim \operatorname{range} T = \dim V$, $\ker T = \{0\}$ if and only if $\dim \operatorname{range} T = \dim V$, that is, if and only if range $T = V$.

Exercises 5.1

1. (a) Since $T(\mathbf{0}) = \begin{bmatrix} 0 \\ 1 \\ 0 \end{bmatrix} \neq \mathbf{0}$, T is not a linear transformation.

(b) Let $\mathbf{x} = \begin{bmatrix} x_1 \\ x_2 \\ x_3 \end{bmatrix}$ and $\mathbf{y} = \begin{bmatrix} y_1 \\ y_2 \\ y_3 \end{bmatrix}$ be vectors in $\mathbf{R}^3$. Then $\mathbf{x} + \mathbf{y} = \begin{bmatrix} x_1 + y_1 \\ x_2 + y_2 \\ x_3 + y_3 \end{bmatrix}$, so

$$T(\mathbf{x} + \mathbf{y}) = \begin{bmatrix} (x_1 + y_1) + (x_2 + y_2) \\ (x_2 + y_2) + (x_3 + y_3) \end{bmatrix} = \begin{bmatrix} x_1 + x_2 + y_1 + y_2 \\ x_2 + x_3 + y_2 + y_3 \end{bmatrix} = \begin{bmatrix} x_1 + x_2 \\ x_2 + x_3 \end{bmatrix} + \begin{bmatrix} y_1 + y_2 \\ y_2 + y_3 \end{bmatrix} = T(\mathbf{x}) + T(\mathbf{y}).$$

Also, for any scalar c, $c\mathbf{x} = c \begin{bmatrix} x_1 \\ x_2 \\ x_3 \end{bmatrix} = \begin{bmatrix} cx_1 \\ cx_2 \\ cx_3 \end{bmatrix}$, so

$$T(c\mathbf{x}) = \begin{bmatrix} cx_1 + cx_2 \\ cx_2 + cx_3 \end{bmatrix} = c \begin{bmatrix} x_1 + x_2 \\ x_2 + x_3 \end{bmatrix} = cT(\mathbf{x}).$$

Thus T is a linear transformation.

3. Since T preserves linear combinations,

$$T\left(\begin{bmatrix} x_1 \\ x_2 \\ x_3 \end{bmatrix}\right) = x_1 T\left(\begin{bmatrix} 1 \\ 0 \\ 0 \end{bmatrix}\right) + x_2 T\left(\begin{bmatrix} 0 \\ 1 \\ 0 \end{bmatrix}\right) + x_3 T\left(\begin{bmatrix} 0 \\ 0 \\ 1 \end{bmatrix}\right) = x_1 \begin{bmatrix} 3 \\ 1 \end{bmatrix} + x_2 \begin{bmatrix} 1 \\ -2 \end{bmatrix} + x_3 \begin{bmatrix} -1 \\ 0 \end{bmatrix} = \begin{bmatrix} 3x_1 + x_2 - x_3 \\ x_1 - 2x_2 \end{bmatrix}.$$

5. Solving the equation

$$\begin{bmatrix} x \\ y \\ z \end{bmatrix} = a \begin{bmatrix} 1 \\ 0 \\ 1 \end{bmatrix} + b \begin{bmatrix} -1 \\ 2 \\ 1 \end{bmatrix} + c \begin{bmatrix} 0 \\ 2 \\ 3 \end{bmatrix} = \begin{bmatrix} 1 & -1 & 0 \\ 0 & 2 & 2 \\ 1 & 1 & 3 \end{bmatrix} \begin{bmatrix} a \\ b \\ c \end{bmatrix}$$

gives $\begin{bmatrix} a \\ b \\ c \end{bmatrix} = \begin{bmatrix} 2x + \frac{3}{2}y - z \\ x + \frac{3}{2}y - z \\ -x - y + z \end{bmatrix}$. Since T preserves linear combinations,

$$T\left(\begin{bmatrix} x \\ y \\ z \end{bmatrix}\right) = T\left((2x + \frac{3}{2}y - z)\begin{bmatrix} 1 \\ 0 \\ 1 \end{bmatrix} + (x + \frac{3}{2}y - z)\begin{bmatrix} -1 \\ 2 \\ 1 \end{bmatrix} + (-x - y + z)\begin{bmatrix} 0 \\ 2 \\ 3 \end{bmatrix}\right)$$

$$= (2x + \frac{3}{2}y - z)T\left(\begin{bmatrix} 1 \\ 0 \\ 1 \end{bmatrix}\right) + (x + \frac{3}{2}y - z)T\left(\begin{bmatrix} -1 \\ 2 \\ 1 \end{bmatrix}\right) + (-x - y + z)T\left(\begin{bmatrix} 0 \\ 2 \\ 3 \end{bmatrix}\right)$$

$$= (2x + \frac{3}{2}y - z)\begin{bmatrix} -2 \\ 3 \end{bmatrix} + (x + \frac{3}{2}y - z)\begin{bmatrix} 1 \\ 1 \end{bmatrix} + (-x - y + z)\begin{bmatrix} -2 \\ 5 \end{bmatrix} = \begin{bmatrix} -x + \frac{1}{2}y - z \\ 2x + y + z \end{bmatrix}.$$

7. (a) The answer is no. Let $\mathbf{v} = \begin{bmatrix} 1 \\ 0 \\ -2 \end{bmatrix}$. We are given that $T(\mathbf{v}) = \begin{bmatrix} 1 \\ 2 \\ 3 \end{bmatrix}$. If T were linear, $T(2\mathbf{v}) = 2T(\mathbf{v}) = \begin{bmatrix} 2 \\ 4 \\ 6 \end{bmatrix}$; that is,

$$T\left(\begin{bmatrix} 2 \\ 0 \\ -4 \end{bmatrix}\right) = \begin{bmatrix} 2 \\ 4 \\ 6 \end{bmatrix}.$$

8. (a) Since $T(\mathbf{e}_1) = T\left(\begin{bmatrix} 1 \\ 0 \end{bmatrix}\right) = \begin{bmatrix} 3 \\ 1 \\ -1 \end{bmatrix}$ and $T(\mathbf{e}_2) = T\left(\begin{bmatrix} 0 \\ 1 \end{bmatrix}\right) = \begin{bmatrix} 0 \\ -1 \\ 5 \end{bmatrix}$, $A = \begin{bmatrix} 3 & 0 \\ 1 & -1 \\ -1 & 5 \end{bmatrix}$.

9. (a) The columns of A are $T(\mathbf{e}_1)$ and $T(\mathbf{e}_2)$. Writing $\mathbf{e}_1 = \begin{bmatrix} 1 \\ 0 \end{bmatrix} = a\begin{bmatrix} -3 \\ -2 \end{bmatrix} + b\begin{bmatrix} 2 \\ 1 \end{bmatrix} = \begin{bmatrix} -3 & 2 \\ -2 & 1 \end{bmatrix}\begin{bmatrix} a \\ b \end{bmatrix}$ and solving, we find

$a = 1$, $b = 2$. Thus $\mathbf{e}_1 = \begin{bmatrix} -3 \\ -2 \end{bmatrix} + 2\begin{bmatrix} 2 \\ 1 \end{bmatrix}$. Since T is linear,

$$T(\mathbf{e}_1) = T\left(\begin{bmatrix} -3 \\ -2 \end{bmatrix}\right) + 2T\left(\begin{bmatrix} 2 \\ 1 \end{bmatrix}\right) = \begin{bmatrix} 3 \\ 4 \end{bmatrix} + 2\begin{bmatrix} 2 \\ 5 \end{bmatrix} = \begin{bmatrix} 7 \\ 14 \end{bmatrix}.$$

Writing $\mathbf{e}_2 = \begin{bmatrix} 0 \\ 1 \end{bmatrix} = a\begin{bmatrix} -3 \\ -2 \end{bmatrix} + b\begin{bmatrix} 2 \\ 1 \end{bmatrix} = \begin{bmatrix} -3 & 2 \\ -2 & 1 \end{bmatrix}\begin{bmatrix} a \\ b \end{bmatrix}$ and solving, we find $a = -2$, $b = -3$. Thus

$\mathbf{e}_2 = -2\begin{bmatrix} -3 \\ -2 \end{bmatrix} - 3\begin{bmatrix} 2 \\ 1 \end{bmatrix}$. Since T is linear,

$$T(\mathbf{e}_2) = -2T\left(\begin{bmatrix} -3 \\ -2 \end{bmatrix}\right) - 3T\left(\begin{bmatrix} 2 \\ 1 \end{bmatrix}\right) = -2\begin{bmatrix} 3 \\ 4 \end{bmatrix} - 3\begin{bmatrix} 2 \\ 5 \end{bmatrix} = \begin{bmatrix} -12 \\ -23 \end{bmatrix}.$$

Thus $A = \begin{bmatrix} 7 & -12 \\ 14 & -23 \end{bmatrix}$.

11. (a) Since $T(\mathbf{e}_1) = \begin{bmatrix} -1 \\ 1 \\ 1 \end{bmatrix}$, $T(\mathbf{e}_2) = \begin{bmatrix} 1 \\ 2 \\ 0 \end{bmatrix}$, $T(\mathbf{e}_3) = \begin{bmatrix} 1 \\ 5 \\ 1 \end{bmatrix}$ and $T(\mathbf{e}_4) = \begin{bmatrix} -4 \\ 1 \\ 3 \end{bmatrix}$, $A = \begin{bmatrix} -1 & 1 & 1 & -4 \\ 1 & 2 & 5 & 1 \\ 1 & 0 & 1 & 3 \end{bmatrix}$.

(b) Since $T\begin{bmatrix} 1 \\ -1 \\ 0 \\ 0 \end{bmatrix} = \begin{bmatrix} -2 \\ -1 \\ 1 \end{bmatrix} \neq \begin{bmatrix} 0 \\ 0 \\ 0 \end{bmatrix}$, this vector is not in ker T.

(c) A vector **b** is in range T if $\mathbf{b} = A\mathbf{x}$ has a solution. Using Gaussian elimination on the augmented matrix,

$$\left[\begin{array}{rrrr|r} -1 & 1 & 1 & -4 & -1 \\ 1 & 2 & 5 & 1 & 7 \\ 1 & 0 & 1 & 3 & 3 \end{array}\right] \rightarrow \left[\begin{array}{rrrr|r} 1 & -1 & -1 & 4 & 1 \\ 0 & 3 & 6 & -3 & 6 \\ 0 & 1 & 2 & -1 & 2 \end{array}\right] \rightarrow \left[\begin{array}{rrrr|r} 1 & -1 & -1 & 4 & 1 \\ 0 & 1 & 2 & -1 & 2 \\ 0 & 0 & 0 & 0 & 0 \end{array}\right].$$

There are many solutions. The given vector is in the range of T.

12. (a) i. The matrix A has columns the vectors $T(\mathbf{e}_1) = \begin{bmatrix} 1 \\ 0 \\ 2 \end{bmatrix}$, $T(\mathbf{e}_2) = \begin{bmatrix} -2 \\ 1 \\ -3 \end{bmatrix}$ and $T(\mathbf{e}_3) = \begin{bmatrix} 0 \\ 3 \\ 3 \end{bmatrix}$. Thus

$$A = \begin{bmatrix} 1 & -2 & 0 \\ 0 & 1 & 3 \\ 2 & -3 & 3 \end{bmatrix}.$$

ii. The kernel of T is the null space of A, so we solve $A\begin{bmatrix} x \\ y \\ z \end{bmatrix} = \mathbf{0}$. Gaussian elimination proceeds

$$\begin{bmatrix} 1 & -2 & 0 \\ 0 & 1 & 3 \\ 2 & -3 & 3 \end{bmatrix} \rightarrow \begin{bmatrix} 1 & -2 & 0 \\ 0 & 1 & 3 \\ 0 & 1 & 3 \end{bmatrix} \rightarrow \begin{bmatrix} 1 & -2 & 0 \\ 0 & 1 & 3 \\ 0 & 0 & 0 \end{bmatrix}.$$

Thus $z = t$ is free, $y = -3t$, $x = -6t$, and $\mathbf{x} = \begin{bmatrix} -6t \\ -3t \\ t \end{bmatrix}$. The kernel has basis $\begin{bmatrix} -6 \\ -3 \\ 1 \end{bmatrix}$. The pivot columns of A

are columns one and two. Thus a basis for range T is columns one and two of A; that is, $\begin{bmatrix} 1 \\ 0 \\ 2 \end{bmatrix}$ and $\begin{bmatrix} -2 \\ 1 \\ -3 \end{bmatrix}$. The range has dimension 2.

iii. $\dim \ker T + \dim \operatorname{range} T = 1 + 2 = 3 = \dim V$, $V = \mathbf{R}^3$.

13. (a) Suppose $c_1 T(\mathbf{v}_1) + c_2 T(\mathbf{v}_2) + \cdots + c_n T(\mathbf{v}_n) = \mathbf{0}$. Since T preserves linear combinations, $T(c_1\mathbf{v}_1 + c_2\mathbf{v}_2 + \cdots + c_n\mathbf{v}_n) = \mathbf{0}$; thus $c_1\mathbf{v}_1 + c_2\mathbf{v}_2 + \cdots + c_n\mathbf{v}_n$ is in the kernel of T. Since T is one-to-one, its kernel is $\mathbf{0}$, so $c_1\mathbf{v}_1 + c_2\mathbf{v}_2 + \cdots + c_n\mathbf{v}_n = \mathbf{0}$. Since $\mathbf{v}_1, \ldots, \mathbf{v}_n$ are linearly independent, all the coefficients c_i are 0, as required.

14. (a) True. If $\mathbf{w} \in \operatorname{range} ST$, then $\mathbf{w} = ST(\mathbf{u}) = S(T(\mathbf{u}))$ for some $\mathbf{u}$, so $\mathbf{w} \in \operatorname{range}(S)$.

(c) True. By (a), $\operatorname{range}(ST)$ is a subspace of $\operatorname{range}(S)$, so $\dim(\operatorname{range}(ST)) \leq \dim(\operatorname{range}(S))$.

15. T is one-to-one if and only if $\ker T = \mathbf{0}$. Since $\ker T = \operatorname{null} \operatorname{sp} A$, this occurs if and only if the columns of A are linearly independent, that is, if and only if A has a left inverse.

17. For any vector $\mathbf{v}$, $A\mathbf{v}$ is $\mathbf{v}$ rotated through θ. Since $B(A\mathbf{v})$ is $A\mathbf{v}$ rotated through $-\theta$, $BA\mathbf{v} = \mathbf{v}$, so $BA = I$ is the identity matrix and $B = A^{-1}$.

20. Given $\mathbf{v}$ in V, we have $\mathbf{v} = c_1\mathbf{v}_1 + \cdots + c_k\mathbf{v}_k$ for scalars $c_1, c_2, \ldots, c_k$ because $\mathbf{v}_1, \ldots, \mathbf{v}_k$ span V. Since T preserves linear combinations,

$$T(\mathbf{v}) = T(c_1\mathbf{v}_1 + \cdots + c_k\mathbf{v}_k) = c_1 T(\mathbf{v}_1) + \cdots + c_k T(\mathbf{v}_k) = c_1\mathbf{w}_1 + \cdots c_k\mathbf{w}_k.$$

Section 5.2—True/False

1. False. For any x, $(f \circ g)(x) = f(g(x)) = f(x) = 1$.

2. False. For any x, $(g \circ f)(x) = g(f(x)) = g(1) = 1$.

3. False. If $f, g : \mathbf{R} \to \mathbf{R}$ are defined by $f(x) = 2$ and $g(x) = \dfrac{1}{x}$ for all x, then $(f \circ g)(x) = 2$, while $(g \circ f)(x) = \dfrac{1}{2}$.

4. True. See Theorem 5.2.1.

5. False. The matrix of T is 3×2.

6. True.

7. False. If T is invertible, then $n = m$, but not conversely. The function $T : \mathbf{R}^2 \to \mathbf{R}^2$ defined by $T(\mathbf{x}) = \mathbf{0}$ for all $\mathbf{x}$ is certainly not invertible.

Exercises 5.2

1. **(a)** **i.** $A = \begin{bmatrix} 1 & 1 & -1 \\ 0 & 1 & -1 \\ 1 & 0 & 1 \\ 0 & 1 & 0 \end{bmatrix}$ is the matrix whose columns are $S\left(\begin{bmatrix} 1 \\ 0 \\ 0 \end{bmatrix}\right)$, $S\left(\begin{bmatrix} 0 \\ 1 \\ 0 \end{bmatrix}\right)$, $S\left(\begin{bmatrix} 0 \\ 0 \\ 1 \end{bmatrix}\right)$. $B = \begin{bmatrix} 2 & 1 \\ 1 & -3 \\ 4 & 0 \end{bmatrix}$ is the matrix whose columns are $T\left(\begin{bmatrix} 1 \\ 0 \end{bmatrix}\right)$, $T\left(\begin{bmatrix} 0 \\ 1 \end{bmatrix}\right)$.

   ii. $ST\left(\begin{bmatrix} x \\ y \end{bmatrix}\right) = S\left(\begin{bmatrix} 2x + y \\ x - 3y \\ 4x \end{bmatrix}\right) = S\left(\begin{bmatrix} (2x + y) + (x - 3y) - 4x \\ (x - 3y) - 4x \\ (2x + y) + 4x \\ x - 3y \end{bmatrix}\right) = \begin{bmatrix} -x - 2y \\ -3x - 3y \\ 6x + y \\ x - 3y \end{bmatrix}$.

   iii. The matrix for ST has columns the vectors $ST\left(\begin{bmatrix} 1 \\ 0 \end{bmatrix}\right)$, $ST\left(\begin{bmatrix} 0 \\ 1 \end{bmatrix}\right)$. This is $\begin{bmatrix} -1 & -2 \\ -3 & -3 \\ 6 & 1 \\ 1 & -3 \end{bmatrix}$, which is AB.

3. **(a)** $ST\left(\begin{bmatrix} x \\ y \end{bmatrix}\right) = S\left(T\left(\begin{bmatrix} x \\ y \end{bmatrix}\right)\right) = S\left(\begin{bmatrix} y \\ x \end{bmatrix}\right) = \begin{bmatrix} 2(-y) - 3(x + 7y) \\ -y + 5(x + 7y) \end{bmatrix} = \begin{bmatrix} y - x \\ -y + 2x \end{bmatrix}$;

   $TS\left(\begin{bmatrix} x \\ y \end{bmatrix}\right) = T\left(S\left(\begin{bmatrix} x \\ y \end{bmatrix}\right)\right) = T\left(\begin{bmatrix} x - y \\ -x + 2y \end{bmatrix}\right) = \begin{bmatrix} -x + 2y \\ x - y \end{bmatrix}$.

   (b) $A = \begin{bmatrix} 1 & -1 \\ -1 & 2 \end{bmatrix}$, $B = \begin{bmatrix} 0 & 1 \\ 1 & 0 \end{bmatrix}$, $C = \begin{bmatrix} -1 & 1 \\ 2 & -1 \end{bmatrix}$, $D = \begin{bmatrix} -1 & 2 \\ 1 & -1 \end{bmatrix}$.

   (c) $AB = C$, $BA = D$.

5. The matrix of T is $\begin{bmatrix} 1 & -2 \\ 4 & 0 \end{bmatrix}$ and the matrix of ST is $\begin{bmatrix} -5 & -4 & -1 \\ 12 & 0 & -4 \end{bmatrix}$.

   If A is the matrix of T, then $\begin{bmatrix} 1 & -2 \\ 4 & 0 \end{bmatrix} A = \begin{bmatrix} -5 & -4 & -1 \\ 12 & 0 & -4 \end{bmatrix}$, so

   $$A = \begin{bmatrix} 1 & -2 \\ 4 & 0 \end{bmatrix}^{-1} \begin{bmatrix} -5 & -4 & -1 \\ 12 & 0 & -4 \end{bmatrix} = \frac{1}{8} \begin{bmatrix} 0 & 2 \\ -4 & 1 \end{bmatrix} \begin{bmatrix} -5 & -4 & -1 \\ 12 & 0 & -4 \end{bmatrix} = \begin{bmatrix} 3 & 0 & -1 \\ 4 & 2 & 0 \end{bmatrix}.$$

   It follows that $S\left(\begin{bmatrix} x \\ y \\ z \end{bmatrix}\right) = \begin{bmatrix} 3 & 0 & -1 \\ 4 & 2 & 0 \end{bmatrix} \begin{bmatrix} x \\ y \\ z \end{bmatrix} = \begin{bmatrix} 3x - z \\ 4x + 2y \end{bmatrix}$.

7. (a) $(ST)\left(\begin{bmatrix}x\\y\end{bmatrix}\right) = S\left(T\left(\begin{bmatrix}x\\y\end{bmatrix}\right)\right) = S\left(\begin{bmatrix}x+y\\x-y\end{bmatrix}\right) = \begin{bmatrix}x-y\\x+y\end{bmatrix}; \ (TU)\left(\begin{bmatrix}x\\y\end{bmatrix}\right) = T\left(U\left(\begin{bmatrix}x\\y\end{bmatrix}\right)\right) = T\left(\begin{bmatrix}2x\\3x\end{bmatrix}\right) = \begin{bmatrix}5x\\-x\end{bmatrix}.$

(b) $[(ST)U]\left(\begin{bmatrix}x\\y\end{bmatrix}\right) = ST\left(U\left(\begin{bmatrix}x\\y\end{bmatrix}\right)\right) = ST\left(\begin{bmatrix}2x\\3x\end{bmatrix}\right) = \begin{bmatrix}-x\\5x\end{bmatrix}; \ [S(TU)]\left(\begin{bmatrix}x\\y\end{bmatrix}\right) = S\left(TU\left(\begin{bmatrix}x\\y\end{bmatrix}\right)\right) = S\left(\begin{bmatrix}5x\\-x\end{bmatrix}\right) = \begin{bmatrix}-x\\5x\end{bmatrix}.$

11. Let T and S be the linear transformations $\mathbf{R}^2 \to \mathbf{R}^2$ that are rotation through angles α and $-\beta$, respectively. Then $T = T_A$ is left multiplication by $A = \begin{bmatrix}\cos\alpha & -\sin\alpha\\\sin\alpha & \cos\alpha\end{bmatrix}$ and $S = S_B$ is left multiplication by

$B = \begin{bmatrix}\cos(-\beta) & -\sin(-\beta)\\\sin(-\beta) & \cos(-\beta)\end{bmatrix} = \begin{bmatrix}\cos\beta & \sin\beta\\-\sin\beta & \cos\beta\end{bmatrix}$. [Remember that $\sin(-\theta) = -\sin\theta \ \cos(-\theta) = \cos\theta$.] The composition

ST is rotation through $\alpha - \beta$, which has matrix $C = \begin{bmatrix}\cos(\alpha-\beta) & -\sin(\alpha-\beta)\\\sin(\alpha-\beta) & \cos(\alpha-\beta)\end{bmatrix}$. On the other hand, the matrix of ST

is BA, so $BA = C$ This implies

$$\begin{bmatrix}\cos\beta & \sin\beta\\-\sin\beta & \cos\beta\end{bmatrix}\begin{bmatrix}\cos\alpha & -\sin\alpha\\\sin\alpha & \cos\alpha\end{bmatrix} = \begin{bmatrix}\cos(\alpha-\beta) & -\sin(\alpha-\beta)\\\sin(\alpha-\beta) & \cos(\alpha-\beta)\end{bmatrix}$$

and hence

$$\begin{bmatrix}\cos\beta\cos\alpha + \sin\beta\sin\alpha & -\cos\beta\sin\alpha + \sin\beta\cos\alpha\\-\sin\beta\cos\alpha + \cos\beta\sin\alpha & \sin\beta\sin\alpha + \cos\beta\cos\alpha\end{bmatrix} = \begin{bmatrix}\cos(\alpha-\beta) & -\sin(\alpha-\beta)\\\sin(\alpha-\beta) & \cos(\alpha-\beta)\end{bmatrix}.$$

Equating corresponding entries, we obtain the desired formulas.

12. (a) Let $\mathbf{w}_1$ and $\mathbf{w}_2$ be vectors in W and let $\mathbf{v}_1, \mathbf{v}_2$ be the (unique) vectors in V satisfying $T(\mathbf{v}_1) = \mathbf{w}_1$ and $T(\mathbf{v}_2) = \mathbf{w}_2$. Then $S(\mathbf{w}_1) = \mathbf{v}_1$ and $S(\mathbf{w}_2) = \mathbf{v}_2$. Since T is linear, $T(\mathbf{v}_1 + \mathbf{v}_2) = T(\mathbf{v}_1) + T(\mathbf{v}_2) = \mathbf{w}_1 + \mathbf{w}_2$ so, by the way S is defined, $S(\mathbf{w}_1 + \mathbf{w}_2) = \mathbf{v}_1 + \mathbf{v}_2 = S(\mathbf{w}_1) + S(\mathbf{w}_2)$.

Section 5.3—True/False

1. True.

2. False. Let $\mathbf{v}_1 = \begin{bmatrix}0\\1\end{bmatrix}$ and $\mathbf{v}_2 = \begin{bmatrix}1\\0\end{bmatrix}$. Since $\begin{bmatrix}-2\\3\end{bmatrix} = 3\mathbf{v}_1 - 2\mathbf{v}_2$, the coordinate vector is $\begin{bmatrix}3\\-2\end{bmatrix}$.

3. True. $\mathbf{v}_1 = 1\mathbf{v}_1 + 0\mathbf{v}_2$.

4. True. $\begin{bmatrix}3\\-4\end{bmatrix} = -1\mathbf{v}_1 + 1\mathbf{v}_2$.

5. False. Any invertible matrix is the matrix of the identity transformation, relative to a suitable basis. See Proposition 5.3.11.

Exercises 5.3

1. The coordinates of $\mathbf{x}$ are the scalars a and b that satisfy $\mathbf{x} = a\mathbf{v}_1 + b\mathbf{v}_2$. This is the matrix equation $\begin{bmatrix}2 & -3\\5 & 1\end{bmatrix}\begin{bmatrix}a\\b\end{bmatrix} = \begin{bmatrix}x_1\\x_2\end{bmatrix}$. Gaussian elimination on the augmented matrix proceeds

$$\left[\begin{array}{cc|c}2 & -3 & x_1\\5 & 1 & x_2\end{array}\right] \to \left[\begin{array}{cc|c}1 & -\frac{3}{2} & \frac{1}{2}x_1\\0 & \frac{17}{2} & x_2 - \frac{5}{2}x_1\end{array}\right] \to \left[\begin{array}{cc|c}1 & -\frac{3}{2} & \frac{1}{2}x_1\\0 & 1 & \frac{2}{17}x_2 - \frac{5}{17}x_1\end{array}\right].$$

We obtain $b = -\frac{5}{17}x_1 + \frac{2}{17}x_2$, $a = \frac{1}{17}x_1 + \frac{3}{17}x_2$.

3. The coordinate vector of $\mathbf{x}$ relative to V is the vector $\mathbf{x}_V = \begin{bmatrix} a \\ b \\ c \end{bmatrix}$ where a, b, c are the scalars satisfying

$\mathbf{x} = a\mathbf{v}_1 + b\mathbf{v}_2 + c\mathbf{v}_3$. This is the matrix equation $\begin{bmatrix} 2 & -1 & 0 \\ 3 & -2 & 1 \\ -1 & 4 & -1 \end{bmatrix} \begin{bmatrix} a \\ b \\ c \end{bmatrix} = \begin{bmatrix} x_1 \\ x_2 \\ x_3 \end{bmatrix}$. Gaussian elimination on the augmented

matrix proceeds

$$\begin{bmatrix} 2 & -1 & 0 & x_1 \\ 3 & -2 & 1 & x_2 \\ -1 & 4 & -1 & x_3 \end{bmatrix} \rightarrow \begin{bmatrix} 1 & -\frac{1}{2} & 0 & \frac{1}{2}x_1 \\ 0 & -\frac{1}{2} & 1 & x_2 - \frac{3}{2}x_1 \\ 0 & \frac{7}{2} & -1 & \frac{1}{2}x_2 + x_3 \end{bmatrix} \rightarrow \begin{bmatrix} 1 & -\frac{1}{2} & 0 & \frac{1}{2}x_1 \\ 0 & 1 & -2 & 3x_1 - 2x_2 \\ 0 & 0 & 6 & -10x_1 + 7x_2 + x_3 \end{bmatrix}.$$

We obtain $c = \frac{1}{6}(-10x_1 + 7x_2 + x_3)$, $b = \frac{1}{3}(-x_1 + x_2 + x_3)$, and $a = \frac{1}{6}(2x_1 + x_2 + x_3)$, so $\mathbf{x}_V = \frac{1}{6} \begin{bmatrix} 2x_1 + x_2 + x_3 \\ -2x_1 + 2x_2 + 2x_3 \\ -10x_1 + 7x_2 + x_3 \end{bmatrix}$.

4. (a) Let $\mathbf{x} = \begin{bmatrix} x_1 \\ x_2 \end{bmatrix}$. We want $T(\mathbf{x})$ and know that $[T(\mathbf{x})]_{\mathcal{E}} = A\mathbf{x}_{\mathcal{E}}$. Since $\mathbf{x}_{\mathcal{E}} = \mathbf{x}$, $A\mathbf{x}_{\mathcal{E}} = A\mathbf{x} = \begin{bmatrix} x_1 + 2x_2 \\ 3x_1 + 4x_2 \end{bmatrix}$. This is

$[T(\mathbf{x})]_{\mathcal{E}} = T(\mathbf{x})$; that is, $T\left(\begin{bmatrix} x_1 \\ x_2 \end{bmatrix}\right) = \begin{bmatrix} x_1 + 2x_2 \\ 3x_1 + 4x_2 \end{bmatrix}$.

(b) Let $\mathbf{x} = \begin{bmatrix} x_1 \\ x_2 \end{bmatrix}$. We seek $T(\mathbf{x})$ and know that $[T(\mathbf{x})]_{\mathcal{E}} = A\mathbf{x}_V$. Since $\begin{bmatrix} x_1 \\ x_2 \end{bmatrix} = (x_1 - x_2)\mathbf{v}_1 + x_2\mathbf{v}_2$, we have

$\mathbf{x}_V = \begin{bmatrix} x_1 - x_2 \\ x_2 \end{bmatrix}$. Thus $A\mathbf{x}_V = \begin{bmatrix} 1 & 2 \\ 3 & 4 \end{bmatrix} \begin{bmatrix} x_1 - x_2 \\ x_2 \end{bmatrix} = \begin{bmatrix} x_1 - x_2 + 2x_2 \\ 3(x_1 - x_2) + 4x_2 \end{bmatrix} = \begin{bmatrix} x_1 + x_2 \\ 3x_1 + x_2 \end{bmatrix}$. This is $[T(\mathbf{x})]_{\mathcal{E}} = T(\mathbf{x})$; that is,

$T\left(\begin{bmatrix} x_1 \\ x_2 \end{bmatrix}\right) = \begin{bmatrix} x_1 + x_2 \\ 3x_1 + x_2 \end{bmatrix}$.

(c) Let $\mathbf{x} = \begin{bmatrix} x_1 \\ x_2 \end{bmatrix}$. We seek $T(\mathbf{x})$ and know that $[T(\mathbf{x})]_V = A\mathbf{x}_{\mathcal{E}} = \begin{bmatrix} 1 & 2 \\ 3 & 4 \end{bmatrix} \begin{bmatrix} x_1 \\ x_2 \end{bmatrix} = \begin{bmatrix} x_1 + 2x_2 \\ 3x_1 + 4x_2 \end{bmatrix}$. This is $[T(\mathbf{x})]_V$, so

$T(\mathbf{x}) = (x_1 + 2x_2)\mathbf{v}_1 + (3x_1 + 4x_2)\mathbf{v}_2 = \begin{bmatrix} 4x_1 + 6x_2 \\ 3x_1 + 4x_2 \end{bmatrix}$.

(d) Let $\mathbf{x} = \begin{bmatrix} x_1 \\ x_2 \end{bmatrix}$. We seek $T(\mathbf{x})$ and know that $[T(\mathbf{x})]_V = A\mathbf{x}_V$. As before, $\mathbf{x}_V = \begin{bmatrix} x_1 - x_2 \\ x_2 \end{bmatrix}$, so

$A\mathbf{x}_V = \begin{bmatrix} 1 & 2 \\ 3 & 4 \end{bmatrix} \begin{bmatrix} x_1 - x_2 \\ x_2 \end{bmatrix} = \begin{bmatrix} x_1 + x_2 \\ 3x_1 + x_2 \end{bmatrix}$. This is $[T(\mathbf{x})]_V$, so $T(\mathbf{x}) = (x_1 + x_2)\mathbf{v}_1 + (3x_1 + x_2)\mathbf{v}_2 = \begin{bmatrix} 4x_1 + 2x_2 \\ 3x_1 + x_2 \end{bmatrix}$.

7. (a) Let $\mathbf{x} = \begin{bmatrix} x_1 \\ x_2 \end{bmatrix}$. We want $T(\mathbf{x})$ and know that $[T(\mathbf{x})]_{\mathcal{F}} = A\mathbf{x}_{\mathcal{E}}$. Since $\mathbf{x}_{\mathcal{E}} = \mathbf{x}$, $A\mathbf{x}_{\mathcal{E}} = A\mathbf{x} = \begin{bmatrix} -x_2 \\ 2x_1 + x_2 \\ -4x_1 + x_2 \end{bmatrix}$. This is

$[T(\mathbf{x})]_{\mathcal{F}} = T(\mathbf{x})$; that is, $T\left(\begin{bmatrix} x_1 \\ x_2 \end{bmatrix}\right) = \begin{bmatrix} -x_2 \\ 2x_1 + x_2 \\ -4x_1 + x_2 \end{bmatrix}$.

(b) Let $\mathbf{x} = \begin{bmatrix} x_1 \\ x_2 \end{bmatrix}$. We seek $T(\mathbf{x})$ and know that $[T(\mathbf{x})]_W = A\mathbf{x}_{\mathcal{E}} = A\mathbf{x} == \begin{bmatrix} -x_2 \\ 2x_1 + x_2 \\ -4x_1 + x_2 \end{bmatrix}$. This is $[T(\mathbf{x})]_W$, so

$T(\mathbf{x}) = -x_2\mathbf{w}_1 + (2x_1 + x_2)\mathbf{w}_2 + (-4x_1 + x_2)\mathbf{w}_3 = \begin{bmatrix} -4x_1 \\ -2x_1 + 2x_2 \\ -2x_1 - 2x_2 \end{bmatrix}$.

(c) Let $\mathbf{x} = \begin{bmatrix} x_1 \\ x_2 \end{bmatrix}$. We seek $T(\mathbf{x})$ and know that $[T(\mathbf{x})]_{\mathcal{F}} = A\mathbf{x}_V$. Since $\begin{bmatrix} x_1 \\ x_2 \end{bmatrix} = (x_1 - x_2)\mathbf{v}_1 + x_2\mathbf{v}_2$, we have

$\mathbf{x}_V = \begin{bmatrix} x_1 - x_2 \\ x_2 \end{bmatrix}$, so $A\mathbf{x}_V = \begin{bmatrix} 0 & -1 \\ 2 & 1 \\ -4 & 1 \end{bmatrix} \begin{bmatrix} x_1 - x_2 \\ x_2 \end{bmatrix} = \begin{bmatrix} -x_2 \\ 2x_1 - x_2 \\ -4x_1 + 5x_2 \end{bmatrix}$. This is $[T(\mathbf{x})]_{\mathcal{F}} = T(\mathbf{x}) = \begin{bmatrix} -x_2 \\ 2x_1 - x_2 \\ -4x_1 + 5x_2 \end{bmatrix}$.

(d) Let $\mathbf{x} = \begin{bmatrix} x_1 \\ x_2 \end{bmatrix}$. We seek $T(\mathbf{x})$ and know that $[T(\mathbf{x})]_W = A\mathbf{x}_V$. As before, $\mathbf{x}_V = \begin{bmatrix} x_1 - x_2 \\ x_2 \end{bmatrix}$ and $A\mathbf{x}_V = \begin{bmatrix} -x_2 \\ 2x_1 - x_2 \\ -4x_1 + 5x_2 \end{bmatrix}$.

This is $[T(\mathbf{x})]_W$, so $T(\mathbf{x}) = (-x_2)\mathbf{w}_1 + (2x_1 - x_2)\mathbf{w}_2 + (-4x_1 + 5x_2)\mathbf{w}_3 = \begin{bmatrix} -4x_1 + 4x_2 \\ -2x_1 + 4x_2 \\ -2x_1 \end{bmatrix}$.

9. (a) The columns of $A = M_{\mathcal{E}\leftarrow\mathcal{E}}(T)$ are the coordinate vectors of $T(\mathbf{e}_1)$ and $T(\mathbf{e}_2)$ relative to $\mathcal{E}$. We have

$$T(\mathbf{e}_1) = T\left(\begin{bmatrix} 1 \\ 0 \end{bmatrix}\right) = \begin{bmatrix} 1 \\ 2 \end{bmatrix} \quad \text{and} \quad T(\mathbf{e}_2) = T\left(\begin{bmatrix} 0 \\ 1 \end{bmatrix}\right) = \begin{bmatrix} 4 \\ 7 \end{bmatrix},$$

so $[T(\mathbf{e}_1)]_{\mathcal{E}} = \begin{bmatrix} 1 \\ 2 \end{bmatrix}$, $[T(\mathbf{e}_2)]_{\mathcal{E}} = \begin{bmatrix} 4 \\ 7 \end{bmatrix}$ and $M_{\mathcal{E}\leftarrow\mathcal{E}}(T) = \begin{bmatrix} 1 & 4 \\ 2 & 7 \end{bmatrix}$.

(b) The columns of $A = M_{\mathcal{E}\leftarrow\mathcal{V}}(T)$ are the coordinate vectors of $T(\mathbf{v}_1)$ and $T(\mathbf{v}_2)$ relative to $\mathcal{E}$. We find

$$T(\mathbf{v}_1) = T\left(\begin{bmatrix} 1 \\ -1 \end{bmatrix}\right) = \begin{bmatrix} -3 \\ -5 \end{bmatrix} \quad \text{and} \quad T(\mathbf{v}_2) = T\left(\begin{bmatrix} -2 \\ 1 \end{bmatrix}\right) = \begin{bmatrix} 2 \\ 3 \end{bmatrix}$$

so $[T(\mathbf{v}_1)]_{\mathcal{E}} = \begin{bmatrix} -3 \\ -5 \end{bmatrix}$, $[T(\mathbf{v}_2)]_{\mathcal{E}} = \begin{bmatrix} 2 \\ 3 \end{bmatrix}$ and $M_{\mathcal{V}\leftarrow\mathcal{E}}(T) = \begin{bmatrix} -3 & 2 \\ -5 & 3 \end{bmatrix}$.

(c) The columns of $A = M_{\mathcal{V}\leftarrow\mathcal{E}}(T)$ are the coordinate vectors of $T(\mathbf{e}_1)$ and $T(\mathbf{e}_2)$ relative to $\mathcal{V}$. As before, $T(\mathbf{e}_1) = \begin{bmatrix} 1 \\ 2 \end{bmatrix}$ and $T(\mathbf{e}_2) = \begin{bmatrix} 4 \\ 7 \end{bmatrix}$. Since

$$\begin{bmatrix} 1 \\ 2 \end{bmatrix} = -5\mathbf{v}_1 - 3\mathbf{v}_2 \quad \text{and} \quad \begin{bmatrix} 4 \\ 7 \end{bmatrix} = -18\mathbf{v}_1 - 11\mathbf{v}_2$$

we have $[T(\mathbf{e}_1)]_{\mathcal{V}} = \begin{bmatrix} -5 \\ -3 \end{bmatrix}$, $[T(\mathbf{e}_2)]_{\mathcal{V}} = \begin{bmatrix} -18 \\ -11 \end{bmatrix}$ and $M_{\mathcal{E}\leftarrow\mathcal{V}}(T) = \begin{bmatrix} -5 & -18 \\ -3 & -11 \end{bmatrix}$.

(d) The columns of $M_{\mathcal{V}\leftarrow\mathcal{V}}(T)$ are the coordinate vectors of $T(\mathbf{v}_1)$ and $T(\mathbf{v}_2)$ relative to $\mathcal{V}$. As before, $T(\mathbf{v}_1) = \begin{bmatrix} -3 \\ -5 \end{bmatrix}$ and $T(\mathbf{v}_2) = \begin{bmatrix} 2 \\ 3 \end{bmatrix}$. Since

$$\begin{bmatrix} -3 \\ -5 \end{bmatrix} = 13\mathbf{v}_1 + 8\mathbf{v}_2 \quad \text{and} \quad \begin{bmatrix} 2 \\ 3 \end{bmatrix} = -8\mathbf{v}_1 - 5\mathbf{v}_2$$

we have $[T(\mathbf{v}_1)]_{\mathcal{V}} = \begin{bmatrix} 13 \\ 8 \end{bmatrix}$, $[T(\mathbf{v}_2)]_{\mathcal{V}} = \begin{bmatrix} -8 \\ -5 \end{bmatrix}$ and $M_{\mathcal{E}\leftarrow\mathcal{V}}(T) = \begin{bmatrix} 13 & -8 \\ 8 & -5 \end{bmatrix}$.

11. As in the proof of the lemma, call the linear transformation T, so that $T(\mathbf{x}) = \mathbf{x}_V$ for $\mathbf{x}$ in V, and let $\mathcal{V} = \{\mathbf{v}_1, \ldots, \mathbf{v}_n\}$. To prove T is one-to-one, it suffices to show that $\ker T = \{\mathbf{0}\}$. Thus suppose $\mathbf{x} = x_1\mathbf{v}_1 + x_2\mathbf{v}_2 + \cdots + x_n\mathbf{v}_n$ is in the kernel of T. Then $T(\mathbf{x}) = \mathbf{0}$, so $\begin{bmatrix} x_1 \\ x_2 \\ \vdots \\ x_n \end{bmatrix} = \mathbf{0}$. Thus $x_1 = x_2 = \cdots = x_n = 0$, so $\mathbf{x} = \mathbf{0}$.

13. (a) The columns of $M_{\mathcal{F}\leftarrow\mathcal{E}}(T)$ are the coordinates of $T(\mathbf{e}_1)$, $T(\mathbf{e}_2)$, and $T(\mathbf{e}_3)$ relative to $\mathcal{F}$. Since

$$T(\mathbf{e}_1) = \begin{bmatrix} 1 \\ 1 \\ 5 \\ 3 \end{bmatrix} = [T(\mathbf{e}_1)]_{\mathcal{F}}, \ T(\mathbf{e}_2) = \begin{bmatrix} -1 \\ 0 \\ -2 \\ -1 \end{bmatrix} = [T(\mathbf{e}_2)]_{\mathcal{F}}, \text{ and } T(\mathbf{e}_3) = \begin{bmatrix} 2 \\ 0 \\ 4 \\ 2 \end{bmatrix} = [T(\mathbf{e}_3)]_{\mathcal{F}}, \text{ we have}$$

$$M_{\mathcal{F}\leftarrow\mathcal{E}}(T) = \begin{bmatrix} 1 & -1 & 2 \\ 1 & 0 & 0 \\ 5 & -2 & 4 \\ 3 & -1 & 2 \end{bmatrix}.$$

(b) The kernel of T is the null space of $A = M_{\mathcal{E}\leftarrow\mathcal{E}}(T)$. Using Gaussian elimination,

$$\begin{bmatrix} 1 & -1 & 2 \\ 1 & 0 & 0 \\ 5 & -2 & 4 \\ 3 & -1 & 2 \end{bmatrix} \rightarrow \begin{bmatrix} 1 & -1 & 2 \\ 0 & 1 & -2 \\ 0 & 3 & -6 \\ 0 & 2 & -4 \end{bmatrix} \rightarrow \begin{bmatrix} 1 & 0 & 0 \\ 0 & 1 & -2 \\ 0 & 0 & 0 \\ 0 & 0 & 0 \end{bmatrix},$$

so, if $\mathbf{x} = \begin{bmatrix} x \\ y \\ z \end{bmatrix}$ is in the null space, $z = t$ is free, $y = 2t$, and $x = 0$: the null space is the set of vectors of the form

$\begin{bmatrix} 0 \\ 2t \\ t \end{bmatrix}$. The pivot columns of A are columns one and two, so these two columns comprise a basis for T; that is, the

vectors $\begin{bmatrix} 1 \\ 1 \\ 5 \\ 3 \end{bmatrix}$ and $\begin{bmatrix} -1 \\ 0 \\ -2 \\ -1 \end{bmatrix}$.

We have $\dim \ker T = 1$ and $\dim \text{range } T = 2$ and $\dim \ker T + \dim \text{range } T = 3$ as asserted by Theorem 5.1.24.

15. (a) Since $T(\mathbf{e}_1) = \begin{bmatrix} 1 \\ 1 \end{bmatrix} = [T(\mathbf{e}_1)]_{\mathcal{E}}$ and $T(\mathbf{e}_2) = \begin{bmatrix} 3 \\ 4 \end{bmatrix} = [T(\mathbf{e}_2)]_{\mathcal{E}}$, we have $A = M_{\mathcal{E}\leftarrow\mathcal{E}}(T) = \begin{bmatrix} 1 & 3 \\ 1 & 4 \end{bmatrix}$.

(b) $A^{-1} = \begin{bmatrix} 4 & -3 \\ -1 & 1 \end{bmatrix}$.

(c) By Theorem 5.3.14, $A^{-1} = M_{\mathcal{E}\leftarrow\mathcal{E}}^{-1}(T) = M_{\mathcal{E}\leftarrow\mathcal{E}}(T^{-1})$. Thus $T^{-1}\left(\begin{bmatrix} x \\ y \end{bmatrix}\right) = A^{-1}\begin{bmatrix} x \\ y \end{bmatrix} = \begin{bmatrix} 4x - 3y \\ -x + y \end{bmatrix}$.

18. The vectors in $\mathcal{V}$ span V, so there are scalars $x_1, x_2, \dots, x_n$ so that $\mathbf{x} = x_1\mathbf{v}_1 + x_2\mathbf{v}_2 + \cdots + x_n\mathbf{v}_n$. If $\mathbf{x}$ could also be written $\mathbf{x} = x_1'\mathbf{v}_1 + x_2'\mathbf{v}_2 + \cdots + x_n'\mathbf{v}_n$, we would have

$$x_1\mathbf{v}_1 + x_2\mathbf{v}_2 + \cdots + x_n\mathbf{v}_n = x_1'\mathbf{v}_1 + x_2'\mathbf{v}_2 + \cdots + x_n'\mathbf{v}_n$$

and hence $(x_1 - x_1')\mathbf{v}_1 + (x_2 - x_2')\mathbf{v}_2 + \cdots + (x_n - x_n')\mathbf{v}_n = \mathbf{0}$. The vectors in $\mathcal{V}$ are linearly independent, so all coefficients here are 0; that is, $x_1 = x_1', x_2 = x_2', \dots, x_n = x_n'$. The coefficients $x_1, x_2, \dots, x_n$ are unique.

Section 5.4—True/False

1. True. Let $\mathbf{x} = a\mathbf{v}_1 + b\mathbf{v}_2$. Then $\mathbf{x}_{\mathcal{V}} = \begin{bmatrix} a \\ b \end{bmatrix}$ and the equation $P_{\mathcal{W}\leftarrow\mathcal{V}}\begin{bmatrix} a \\ b \end{bmatrix} = \begin{bmatrix} c \\ d \end{bmatrix}$ is $P_{\mathcal{W}\leftarrow\mathcal{V}}\mathbf{x}_{\mathcal{V}} = \begin{bmatrix} c \\ d \end{bmatrix}$, which says that the

coordinate vector of $\mathbf{x}$ relative to $\mathcal{W}$ is $\begin{bmatrix} c \\ d \end{bmatrix}$, that is, $\mathbf{x} = c\mathbf{w}_1 + d\mathbf{w}_2$.

2. True. This is Definition 5.4.1.

3. True.

4. False. The correct matrix is the inverse of the one given.

5. True, since A and B are similar and $\det P^{-1}AP = \det A$.

Exercises 5.4

1. (a) We can simply write $P = P_{\mathcal{E}\leftarrow\mathcal{V}}$ since the columns of this matrix are the coordinates of $\mathbf{v}_1$, $\mathbf{v}_2$ and $\mathbf{v}_3$ relative to $\mathcal{E}$:

$$P = \begin{bmatrix} 1 & 0 & 3 \\ 1 & 0 & 0 \\ 0 & 1 & 0 \end{bmatrix}. \text{ Thus } P_{\mathcal{V}\leftarrow\mathcal{E}} = P^{-1} = \begin{bmatrix} 0 & 1 & 0 \\ 0 & 0 & 1 \\ \frac{1}{3} & -\frac{1}{3} & 0 \end{bmatrix}.$$

(b) The coordinates of $\mathbf{v} = \begin{bmatrix} 2 \\ 1 \\ -1 \end{bmatrix}$ relative to $\mathcal{V}$ are the components of $P_{\mathcal{V}\leftarrow\mathcal{E}}\begin{bmatrix} 2 \\ 1 \\ -1 \end{bmatrix} = \begin{bmatrix} 1 \\ 2 \\ -\frac{2}{3} \end{bmatrix}$.

4. If $\mathbf{x}_\mathcal{V} = \begin{bmatrix} 2 \\ -2 \\ -3 \\ 1 \end{bmatrix}$, then $\mathbf{x} = 2\mathbf{v}_1 - 2\mathbf{v}_2 - 3\mathbf{v}_3 + \mathbf{v}_4 = 2\begin{bmatrix} 1 \\ 0 \\ 0 \\ 0 \end{bmatrix} - 2\begin{bmatrix} 1 \\ 0 \\ 0 \\ 1 \end{bmatrix} - 3\begin{bmatrix} 0 \\ 1 \\ 1 \\ 0 \end{bmatrix} + \begin{bmatrix} 1 \\ 1 \\ 0 \\ 1 \end{bmatrix} = \begin{bmatrix} -2 \\ -3 \\ -3 \\ 1 \end{bmatrix}$. If $\mathbf{y}_\mathcal{V} = \begin{bmatrix} -2 \\ -3 \\ 10 \\ -9 \end{bmatrix}$, then

$$\mathbf{y} = -2\mathbf{v}_1 - 3\mathbf{v}_2 + 10\mathbf{v}_3 - 9\mathbf{v}_4 = -2\begin{bmatrix} 1 \\ 0 \\ 0 \\ 0 \end{bmatrix} - 3\begin{bmatrix} 1 \\ 0 \\ 0 \\ 1 \end{bmatrix} + 10\begin{bmatrix} 0 \\ 1 \\ 1 \\ 0 \end{bmatrix} - 9\begin{bmatrix} 1 \\ 1 \\ 0 \\ 1 \end{bmatrix} = \begin{bmatrix} -14 \\ 1 \\ 10 \\ -12 \end{bmatrix}.$$

We are asked to check that $T(\mathbf{x}) = \mathbf{y}$. This is the case since $A = M_{\mathcal{E}\leftarrow\mathcal{E}}(T)$ implies that $T(\mathbf{x}) = A\mathbf{x} = \begin{bmatrix} -14 \\ 1 \\ 10 \\ -12 \end{bmatrix}$.

5. (a) Since $A\begin{bmatrix} 1 \\ 0 \end{bmatrix} = \begin{bmatrix} -109 \\ -24 \end{bmatrix}$, $A\begin{bmatrix} -2 \\ 1 \end{bmatrix} = \begin{bmatrix} 712 \\ 157 \end{bmatrix}$, and $A\begin{bmatrix} 3 \\ 3 \end{bmatrix} = \begin{bmatrix} 1155 \\ 255 \end{bmatrix}$, we have $T(\mathbf{u}) = \begin{bmatrix} -109 \\ -24 \end{bmatrix}$, $T(\mathbf{v}) = \begin{bmatrix} 712 \\ 157 \end{bmatrix}$, and $T(\mathbf{w}) = \begin{bmatrix} 1155 \\ 255 \end{bmatrix}$.

(b) For any $\mathbf{x} \in \mathbf{R}^2$, the coordinate vector $T(\mathbf{x})_\mathcal{V}$ of $T(\mathbf{x})$ relative to $\mathcal{V}$ is $A\begin{bmatrix} a \\ b \end{bmatrix}$, where a and b are the coordinates of $\mathbf{x}$ relative to $\mathcal{E}$. Since $A\begin{bmatrix} 1 \\ 0 \end{bmatrix} = \begin{bmatrix} -109 \\ -24 \end{bmatrix}$, the coordinates of $T(\mathbf{u})$ relative to $\mathcal{V}$ are -109 and -24; so

$$T(\mathbf{u}) = -109\begin{bmatrix} -1 \\ 1 \end{bmatrix} - 24\begin{bmatrix} 4 \\ -5 \end{bmatrix} = \begin{bmatrix} 13 \\ 11 \end{bmatrix}.$$

Similarly, $T(\mathbf{v}) = 712\begin{bmatrix} -1 \\ 1 \end{bmatrix} + 157\begin{bmatrix} 4 \\ -5 \end{bmatrix} = \begin{bmatrix} -84 \\ -73 \end{bmatrix}$ and

$$T(\mathbf{u}) = 1155\begin{bmatrix} -1 \\ 1 \end{bmatrix} + 255\begin{bmatrix} 4 \\ -5 \end{bmatrix} = \begin{bmatrix} -135 \\ -120 \end{bmatrix}.$$

7. The matrix $B = M_{\mathcal{W}\leftarrow\mathcal{W}}(T) = P_{\mathcal{W}\leftarrow\mathcal{V}}M_{\mathcal{V}\leftarrow\mathcal{V}}P_{\mathcal{V}\leftarrow\mathcal{W}}P_{\mathcal{W}\leftarrow\mathcal{V}}AP_{\mathcal{V}\leftarrow\mathcal{W}} = P^{-1}AP$, where $P = P_{\mathcal{V}\leftarrow\mathcal{W}}$. The first column of P is the coordinate vector of $\mathbf{w}_1$ relative to $\mathcal{V}$. Solving

$$\begin{bmatrix} 2 \\ -3 \end{bmatrix} = a\begin{bmatrix} 0 \\ 1 \end{bmatrix} + b\begin{bmatrix} 2 \\ 4 \end{bmatrix} = \begin{bmatrix} 0 & 2 \\ 1 & 4 \end{bmatrix}\begin{bmatrix} a \\ b \end{bmatrix}$$

gives $\begin{bmatrix} a \\ b \end{bmatrix} = -\frac{1}{2} \begin{bmatrix} 4 & -2 \\ -1 & 0 \end{bmatrix} \begin{bmatrix} 2 \\ -3 \end{bmatrix} = -\frac{1}{2} \begin{bmatrix} 14 \\ -2 \end{bmatrix} = \begin{bmatrix} -7 \\ 1 \end{bmatrix}$. The second column of P is the coordinate vector of $\mathbf{w}_2$ relative to $\mathcal{V}$. Solving

$$\begin{bmatrix} -2 \\ 5 \end{bmatrix} = a \begin{bmatrix} 0 \\ 1 \end{bmatrix} + b \begin{bmatrix} 2 \\ 4 \end{bmatrix} = \begin{bmatrix} 0 & 2 \\ 1 & 4 \end{bmatrix} \begin{bmatrix} a \\ b \end{bmatrix}$$

gives $\begin{bmatrix} a \\ b \end{bmatrix} = -\frac{1}{2} \begin{bmatrix} 4 & -2 \\ -1 & 0 \end{bmatrix} \begin{bmatrix} -2 \\ 5 \end{bmatrix} = -\frac{1}{2} \begin{bmatrix} -18 \\ 2 \end{bmatrix} = \begin{bmatrix} 9 \\ -1 \end{bmatrix}$. So $P = \begin{bmatrix} -7 & 9 \\ 1 & -1 \end{bmatrix}$ and

$B = P^{-1}AP = -\frac{1}{2} \begin{bmatrix} -1 & -9 \\ -1 & -7 \end{bmatrix} \begin{bmatrix} 3 & 2 \\ 2 & 1 \end{bmatrix} \begin{bmatrix} -7 & 9 \\ 1 & -1 \end{bmatrix} = \begin{bmatrix} -68 & 59 \\ -55 & 72 \end{bmatrix}$.

8. (a) Let $\mathbf{v}_1 = \begin{bmatrix} 1 \\ 0 \\ 1 \end{bmatrix}$, $\mathbf{v}_2 = \begin{bmatrix} 0 \\ 1 \\ 1 \end{bmatrix}$, $\mathbf{v}_3 = \begin{bmatrix} 1 \\ 1 \\ 0 \end{bmatrix}$, so that $\mathcal{V} = \{\mathbf{v}_1, \mathbf{v}_2, \mathbf{v}_3\}$. We seek $B = M_{\mathcal{V} \leftarrow \mathcal{V}}(T)$. Let $\mathcal{E}$ denote the standard

basis of $\mathbf{R}^3$. Since $T(\mathbf{e}_1) = \begin{bmatrix} 2 \\ -2 \\ 1 \end{bmatrix}$, $T(\mathbf{e}_2) = \begin{bmatrix} 3 \\ 1 \\ -1 \end{bmatrix}$, $T(\mathbf{e}_3) = \begin{bmatrix} 0 \\ 1 \\ -1 \end{bmatrix}$, the matrix of T relative to $\mathcal{E}$ of $\mathbf{R}^3$ is

$A = M_{\mathcal{E} \leftarrow \mathcal{E}}(T) = \begin{bmatrix} 2 & 3 & 0 \\ -2 & 1 & 1 \\ 1 & -1 & -1 \end{bmatrix}$.

The diagram to the right shows that $B = P^{-1}AP$ where $P = P_{\mathcal{E} \leftarrow \mathcal{V}}$ is the indicated change of coordinates matrix. Thus

$P = \begin{bmatrix} 1 & 0 & 1 \\ 0 & 1 & 1 \\ 1 & 1 & 0 \end{bmatrix}$, so

$P^{-1} = \frac{1}{2} \begin{bmatrix} 1 & -1 & 1 \\ -1 & 1 & 1 \\ 1 & 1 & -1 \end{bmatrix}$ and $B = \frac{1}{2} \begin{bmatrix} 3 & -1 & 6 \\ -3 & -3 & -6 \\ 1 & 2 & 4 \end{bmatrix}$.

$$\begin{array}{ccc} \mathbf{R}^3_{\mathcal{E}} & \xrightarrow{\;\;T\;\;} & \mathbf{R}^3_{\mathcal{E}} \\ & A & \\ \mathrm{id} \uparrow P & & P^{-1} \downarrow \mathrm{id} \\ \mathbf{R}^3_{\mathcal{V}} & \xrightarrow{\;\;B\;\;} & \mathbf{R}^3_{\mathcal{V}} \\ & T & \end{array}$$

9. (a) $\mathbf{u} = 1\mathbf{v}_1 + 2\mathbf{v}_2 + 3\mathbf{v}_3 = \begin{bmatrix} -8 \\ 4 \\ 16 \end{bmatrix}$; $\mathbf{v} = 0\mathbf{v}_1 + 1\mathbf{v}_2 + 0\mathbf{v}_3 = \begin{bmatrix} 0 \\ 1 \\ 1 \end{bmatrix}$; $\mathbf{w} = -1\mathbf{v}_1 + 0\mathbf{v}_2 + 1\mathbf{v}_3 = \begin{bmatrix} -4 \\ -2 \\ 7 \end{bmatrix}$.

11. (a) We can simply write the change of coordinates matrix $\mathcal{V}$ to $\mathcal{E}$ since the columns of this matrix are just the coordinates of $\mathbf{v}_1$, $\mathbf{v}_2$, and $\mathbf{v}_3$ relative to $\mathcal{E}$:

$P_{\mathcal{E} \leftarrow \mathcal{V}} = \begin{bmatrix} 1 & 1 & -1 \\ 1 & -1 & 0 \\ 1 & 0 & 1 \end{bmatrix}$.

Section 6.1—True/False

1. False. The length of $\begin{bmatrix} 1 \\ 1 \\ 1 \\ 1 \end{bmatrix}$ is $\sqrt{4} = 2$.

2. True. $\mathbf{x} \cdot \mathbf{x}$ is a sum of squares of real numbers.

3. False. $\mathbf{u} \cdot \mathbf{v} = \mathbf{u}^T \mathbf{v}$.

4. False. The columns of A must be linearly independent.

5. False. It's the vector $\mathbf{x}$ that makes $\|\mathbf{b} - A\mathbf{x}\|$ as small as possible.

6. True.

7. True. Multiply $P^2 = P$ on both sides by P^{-1}.

8. True. This is just Reading Challenge 8 in the case that $\mathbf{v} \cdot \mathbf{v} = 1$.

9. False. The vectors $\mathbf{u}$ and $\mathbf{v}$ must be orthogonal.

10. True. The vector $\mathbf{w}$ clearly has the property that it is in $\mathbf{u}$ and $\mathbf{w} - \mathbf{w}$ is orthogonal to U.

Exercises 6.1

1. In general, A (and hence A^T) are not square matrices, so the step $(AA^T)^{-1} = (A^T)^{-1}A^{-1}$ is invalid.

3. The given equation $\mathbf{y} \cdot (A^T\mathbf{b} - A^T A\mathbf{x}) = 0$ implies $\mathbf{y}^T(A^T\mathbf{b} - A^T A\mathbf{x}) = 0$, so $(\mathbf{y}^T A^T)(\mathbf{b} - A\mathbf{x}) = 0$, hence $(A\mathbf{y})^T(\mathbf{b} - A\mathbf{x}) = 0$ and $(A\mathbf{y}) \cdot (\mathbf{b} - A\mathbf{x}) = 0$. Since this happens for all $A\mathbf{y}$ (and these vectors form exactly the column space of A), the vector $\mathbf{b} - A\mathbf{x}$ is indeed perpendicular to the column space of A.

5. (a) The system is $A\mathbf{x} = \mathbf{b}$ with $A = \begin{bmatrix} 2 & -1 \\ 1 & -1 \\ 2 & 1 \\ 1 & 3 \end{bmatrix}$, $\mathbf{x} = \begin{bmatrix} x_1 \\ x_2 \end{bmatrix}$ and $\mathbf{b} = \begin{bmatrix} 1 \\ 0 \\ -1 \\ 3 \end{bmatrix}$. We have $A^T A = \begin{bmatrix} 10 & 2 \\ 2 & 12 \end{bmatrix}$.

The best solution is $\mathbf{x}^+ = (A^T A)^{-1} A^T \mathbf{b} = \frac{1}{58} \begin{bmatrix} 6 & -11 \\ -1 & 5 \end{bmatrix} \begin{bmatrix} 2 & 1 & 2 & 1 \\ -1 & -1 & 1 & 3 \end{bmatrix} \begin{bmatrix} 1 \\ 0 \\ -1 \\ 3 \end{bmatrix} = \frac{1}{58} \begin{bmatrix} 11 \\ 16 \end{bmatrix}$.

(c) The system is $A\mathbf{x} = \mathbf{b}$ with $A = \begin{bmatrix} 1 & 0 & -2 \\ 1 & 1 & -1 \\ 0 & -1 & -1 \\ 2 & 3 & 1 \end{bmatrix}$, $\mathbf{x} = \begin{bmatrix} x_1 \\ x_2 \\ x_3 \end{bmatrix}$ and $\mathbf{b} = \begin{bmatrix} 0 \\ 2 \\ 3 \\ -4 \end{bmatrix}$. We have $A^T A = \begin{bmatrix} 6 & 7 & -1 \\ 7 & 11 & 3 \\ -1 & 3 & 7 \end{bmatrix}$.

The best solution is

$$\mathbf{x}^+ = (A^T A)^{-1} A^T \mathbf{b} = \frac{1}{12} \begin{bmatrix} 68 & -39 & 32 \\ -52 & 41 & -25 \\ 32 & -25 & 17 \end{bmatrix} \begin{bmatrix} 1 & 1 & 0 & 2 \\ 0 & 1 & -1 & 3 \\ -2 & -1 & -1 & 1 \end{bmatrix} \begin{bmatrix} 0 \\ 2 \\ 3 \\ -4 \end{bmatrix} = \frac{1}{3} \begin{bmatrix} -5 \\ 1 \\ -5 \end{bmatrix}.$$

6. (a) Since $P^2 \neq P$, P is not a projection matrix.

(d) Since $P^2 = P = P^T$, P is a projection matrix.

7. (a) Since the columns of A are linearly independent, we seek the matrix $P = A(A^T A)^{-1} A^T$.

We have $A^T A = \begin{bmatrix} 1 & 2 & -2 \\ 1 & 1 & 1 \end{bmatrix} \begin{bmatrix} 1 & 1 \\ 2 & 1 \\ -2 & 1 \end{bmatrix} = \begin{bmatrix} 9 & 1 \\ 1 & 3 \end{bmatrix}$ and $(A^T A)^{-1} = \frac{1}{26} \begin{bmatrix} 3 & -1 \\ -1 & 9 \end{bmatrix}$, so

$$P = \frac{1}{26} \begin{bmatrix} 1 & 1 \\ 2 & 1 \\ -2 & 1 \end{bmatrix} \begin{bmatrix} 3 & -1 \\ -1 & 9 \end{bmatrix} \begin{bmatrix} 1 & 2 & -2 \\ 1 & 1 & 1 \end{bmatrix} = \frac{1}{13} \begin{bmatrix} 5 & 6 & 2 \\ 6 & 17 & -3 \\ 4 & -3 & 25 \end{bmatrix}.$$

(d) The columns of A are not linearly independent, so $A^T A$ is not invertible. Our job, however, is to project vectors

onto the column space of A, which has basis the first three columns of A: $\mathbf{f}_1 = \begin{bmatrix} 1 \\ 1 \\ -1 \\ -1 \end{bmatrix}$, $\mathbf{f}_2 = \begin{bmatrix} 0 \\ 1 \\ 0 \\ 1 \end{bmatrix}$, $\mathbf{f}_3 = \begin{bmatrix} 1 \\ 1 \\ 3 \\ -1 \end{bmatrix}$. These vectors are orthogonal, so the projection matrix is

$$P = \frac{\mathbf{f}_1 \mathbf{f}_1^T}{\mathbf{f}_1 \cdot \mathbf{f}_1} + \frac{\mathbf{f}_2 \mathbf{f}_2^T}{\mathbf{f}_2 \cdot \mathbf{f}_2} + \frac{\mathbf{f}_3 \mathbf{f}_3^T}{\mathbf{f}_3 \cdot \mathbf{f}_3} = \frac{1}{6} \begin{bmatrix} 2 & 2 & 0 & -2 \\ 2 & 5 & 0 & 1 \\ 0 & 0 & 1 & 0 \\ -2 & 1 & 0 & 5 \end{bmatrix}.$$

10. Let $A = c_1 P_1 + c_2 P_2 + \cdots + c_n P_n$ be a linear combination of projection matrices $P_1, P_2, \ldots, P_n$. Since $P_1^T = P_1$, $P_2^T = P_2$, and so on, $A^T = c_1 P_1^T + c_2 P_2^T + \cdots + c_n P_n^T = c_1 P_1 + c_2 P_2 + \cdots P_n = A$. Thus A is symmetric.

13. If $P = AA^T$, then $P^2 = AA^T AA^T = AIA^T = P$, because $A^T A = I$, and $P^T = (A^T)^T A^T = AA^T = P$.

14. (a) **i.** We have $A^T A = \begin{bmatrix} 3 & -3 \\ -3 & 15 \end{bmatrix}$ and $(A^T A)^{-1} = \frac{1}{12} \begin{bmatrix} 5 & 1 \\ 1 & 1 \end{bmatrix}$.

The least squares solution to $A\mathbf{x} = \mathbf{b} = \begin{bmatrix} 1 \\ 2 \\ -1 \\ 3 \end{bmatrix}$ is

$$\mathbf{x}^+ = (A^T A)^{-1} A^T \mathbf{b} = \frac{1}{12} \begin{bmatrix} 5 & 1 \\ 1 & 1 \end{bmatrix} \begin{bmatrix} 1 & 0 & 1 & 1 \\ -1 & 2 & 1 & -3 \end{bmatrix} \begin{bmatrix} 1 \\ 2 \\ -1 \\ 3 \end{bmatrix} = \frac{1}{3} \begin{bmatrix} 2 \\ -1 \end{bmatrix}.$$

ii. $A\mathbf{x}^+ = \frac{1}{3} \begin{bmatrix} 3 \\ -2 \\ 1 \\ 5 \end{bmatrix}$ is the vector in the column space of A that minimizes $\|\mathbf{b} - A\mathbf{x}\|$; equivalently, the vector in the column space of A that is closest to $\mathbf{b}$.

iii. $P = A(A^T A)^{-1} A^T = \frac{1}{12} \begin{bmatrix} 1 & -1 \\ 0 & 2 \\ 1 & 1 \\ 1 & -3 \end{bmatrix} \begin{bmatrix} 5 & 1 \\ 1 & 1 \end{bmatrix} \begin{bmatrix} 1 & 0 & 1 & 1 \\ -1 & 2 & 1 & -3 \end{bmatrix} = \frac{1}{3} \begin{bmatrix} 1 & 0 & 1 & 1 \\ 0 & 1 & 1 & -1 \\ 1 & 1 & 2 & 0 \\ 1 & -1 & 0 & 2 \end{bmatrix}.$

iv. The best solution to $A\mathbf{x} = \mathbf{b}$ is the vector $\mathbf{x}^+$ that satisfies $A\mathbf{x}^+ = P\mathbf{b} = \frac{1}{3} \begin{bmatrix} 1 & 0 & 1 & 1 \\ 0 & 1 & 1 & -1 \\ 1 & 1 & 2 & 0 \\ 1 & -1 & 0 & 2 \end{bmatrix} \begin{bmatrix} 1 \\ 2 \\ -1 \\ 3 \end{bmatrix} = \begin{bmatrix} 1 \\ -\frac{2}{3} \\ \frac{1}{3} \\ \frac{5}{3} \end{bmatrix}.$

Gaussian elimination on the augmented matrix $[A|P\mathbf{b}]$ proceeds

$$\begin{bmatrix} 1 & -1 & | & 1 \\ 0 & 2 & | & -\frac{2}{3} \\ 1 & 1 & | & \frac{1}{3} \\ 1 & -3 & | & \frac{5}{3} \end{bmatrix} \rightarrow \begin{bmatrix} 1 & -1 & | & 1 \\ 0 & 2 & | & -\frac{2}{3} \\ 0 & 2 & | & -\frac{2}{3} \\ 0 & -2 & | & \frac{2}{3} \end{bmatrix} \rightarrow \begin{bmatrix} 1 & -1 & | & 1 \\ 0 & 1 & | & -\frac{1}{3} \\ 0 & 0 & | & 0 \\ 0 & 0 & | & 0 \end{bmatrix},$$

so $x_2^+ = -\frac{1}{3}$, $x_1^+ = 1 + x_2^+ = \frac{2}{3}$ and $\mathbf{x}^+ = \begin{bmatrix} \frac{2}{3} \\ -\frac{1}{3} \end{bmatrix}$ in agreement with our answer to part i.

(c)   **i.** We have $A^T A = \begin{bmatrix} 7 & 2 & 1 \\ 2 & 2 & 0 \\ 1 & 0 & 3 \end{bmatrix}$ and $(A^T A)^{-1} = \frac{1}{14}\begin{bmatrix} 3 & -3 & -1 \\ -3 & 10 & 1 \\ -1 & 1 & 5 \end{bmatrix}$. The least squares solution to

$$A\mathbf{x} = \mathbf{b} = \begin{bmatrix} -1 \\ 0 \\ 0 \\ 3 \end{bmatrix} \text{ is}$$

$$\mathbf{x}^+ = (A^T A)^{-1} A^T \mathbf{b} = \frac{1}{14}\begin{bmatrix} 3 & -3 & -1 \\ -3 & 10 & 1 \\ -1 & 1 & 5 \end{bmatrix}\begin{bmatrix} 1 & 1 & 2 & 1 \\ 0 & 1 & 0 & 1 \\ 1 & 1 & 0 & -1 \end{bmatrix}\begin{bmatrix} -1 \\ 0 \\ 0 \\ 3 \end{bmatrix} = \frac{1}{14}\begin{bmatrix} 1 \\ 20 \\ -19 \end{bmatrix}.$$

ii. $A\mathbf{x}^+ = \frac{1}{7}\begin{bmatrix} -9 \\ 1 \\ 1 \\ 20 \end{bmatrix}$ is the vector in the column space of A that minimizes $\|\mathbf{b} - A\mathbf{x}\|$; equivalently, the vector in the

column space of A that is closest to $\mathbf{b}$.

iii. $P = A(A^T A)^{-1} A^T = \frac{1}{14}\begin{bmatrix} 1 & 0 & 1 \\ 1 & 1 & 1 \\ 2 & 0 & 0 \\ 1 & 1 & -1 \end{bmatrix}\begin{bmatrix} 3 & -3 & -1 \\ -3 & 10 & 1 \\ -1 & 1 & 5 \end{bmatrix}\begin{bmatrix} 1 & 1 & 2 & 1 \\ 0 & 1 & 0 & 1 \\ 1 & 1 & 0 & -1 \end{bmatrix} = \frac{1}{7}\begin{bmatrix} 3 & 2 & 2 & -2 \\ 2 & 6 & -1 & 1 \\ 2 & -1 & 6 & 1 \\ -2 & 1 & 1 & 6 \end{bmatrix}.$

iv. The best solution to $A\mathbf{x} = \mathbf{b}$ is the vector $\mathbf{x}^+$ that satisfies

$$A\mathbf{x}^+ = P\mathbf{b} = \frac{1}{7}\begin{bmatrix} 3 & 2 & 2 & -2 \\ 2 & 6 & -1 & 1 \\ 2 & -1 & 6 & 1 \\ -2 & 1 & 1 & 6 \end{bmatrix}\begin{bmatrix} -1 \\ 0 \\ 0 \\ 3 \end{bmatrix} = \begin{bmatrix} -\frac{9}{7} \\ \frac{1}{7} \\ \frac{1}{7} \\ \frac{20}{7} \end{bmatrix}.$$

Gaussian elimination applied to the augmented matrix $[A\,|\,P\mathbf{b}]$ proceeds

$$\begin{bmatrix} 1 & 0 & 1 & -\frac{9}{7} \\ 1 & 1 & 1 & \frac{1}{7} \\ 2 & 0 & 0 & \frac{1}{7} \\ 1 & 1 & -1 & \frac{20}{7} \end{bmatrix} \rightarrow \begin{bmatrix} 1 & 0 & 1 & -\frac{9}{7} \\ 0 & 1 & 0 & \frac{10}{7} \\ 0 & 0 & -2 & \frac{19}{7} \\ 0 & 1 & -2 & \frac{29}{7} \end{bmatrix} \rightarrow \begin{bmatrix} 1 & 0 & 1 & -\frac{9}{7} \\ 0 & 1 & 0 & \frac{10}{7} \\ 0 & 0 & -2 & \frac{19}{7} \\ 0 & 0 & -2 & \frac{19}{7} \end{bmatrix} \rightarrow \begin{bmatrix} 1 & 0 & 1 & -\frac{9}{7} \\ 0 & 1 & 0 & \frac{10}{7} \\ 0 & 0 & 1 & -\frac{19}{14} \\ 0 & 0 & 0 & 0 \end{bmatrix},$$

so $x_3{}^+ = -\frac{19}{14}$, $x_2{}^+ = \frac{10}{7}$, $x_1{}^+ = -\frac{9}{7} - x_3{}^+ = \frac{1}{14}$ and $\mathbf{x}^+ = \begin{bmatrix} \frac{1}{14} \\ \frac{10}{7} \\ -\frac{19}{14} \end{bmatrix}$, in agreement with our answer to i.

15. Method 1: We form the matrix $A = \begin{bmatrix} 1 & 1 \\ 0 & 1 \\ 1 & 0 \end{bmatrix}$ whose columns are the given vectors. We wish to project vectors onto the

column space of A. We have $A^T A = \begin{bmatrix} 2 & 1 \\ 1 & 2 \end{bmatrix}$ and $(A^T A)^{-1} = \frac{1}{3}\begin{bmatrix} 2 & -1 \\ -1 & 2 \end{bmatrix}$. The desired projection matrix is

$$P = A(A^T A)^{-1} A^T = \frac{1}{3}\begin{bmatrix} 1 & 1 \\ 0 & 1 \\ 1 & 0 \end{bmatrix}\begin{bmatrix} 2 & -1 \\ -1 & 2 \end{bmatrix}\begin{bmatrix} 1 & 0 & 1 \\ 1 & 1 & 0 \end{bmatrix} = \frac{1}{3}\begin{bmatrix} 2 & 1 & 1 \\ 1 & 2 & -1 \\ 1 & -1 & 2 \end{bmatrix}.$$

Method 2: We replace the given vectors by two **orthogonal** vectors that span the same plane, taking, for example,

$$\mathbf{e} = \begin{bmatrix} 1 \\ 0 \\ 1 \end{bmatrix} \text{ and } \mathbf{f} = \begin{bmatrix} 1 \\ 2 \\ -1 \end{bmatrix}. \text{ Then } P = \frac{\mathbf{e}\mathbf{e}^T}{\mathbf{e}\cdot\mathbf{e}} + \frac{\mathbf{f}\mathbf{f}^T}{\mathbf{f}\cdot\mathbf{f}} = \frac{1}{2}\begin{bmatrix} 1 & 0 & 1 \\ 0 & 0 & 0 \\ 1 & 0 & 1 \end{bmatrix} + \frac{1}{6}\begin{bmatrix} 1 & 2 & -1 \\ 2 & 4 & -2 \\ -1 & -2 & 1 \end{bmatrix} = \frac{1}{3}\begin{bmatrix} 2 & 1 & 1 \\ 1 & 2 & -1 \\ 1 & -1 & 2 \end{bmatrix} \text{ as before.}$$

17. A vector is in U if and only if it is of the form $s\begin{bmatrix} 2 \\ 1 \\ 0 \\ -5 \end{bmatrix} + t\begin{bmatrix} -1 \\ 0 \\ 1 \\ 2 \end{bmatrix}$. Thus U is the column space of the matrix

$$A = \begin{bmatrix} 2 & -1 \\ 1 & 0 \\ 0 & 1 \\ -5 & 2 \end{bmatrix}. \text{ We have } A^T A = \begin{bmatrix} 30 & -12 \\ -12 & 6 \end{bmatrix} \text{ and } (A^T A)^{-1} = \frac{1}{6}\begin{bmatrix} 1 & 2 \\ 2 & 5 \end{bmatrix}, \text{ so the matrix we seek is}$$

$$P = A(A^T A)^{-1} A^T = \frac{1}{6}\begin{bmatrix} 2 & -1 \\ 1 & 0 \\ 0 & 1 \\ -5 & 2 \end{bmatrix}\begin{bmatrix} 1 & 2 \\ 2 & 5 \end{bmatrix}\begin{bmatrix} 2 & 1 & 0 & -5 \\ -1 & 0 & 1 & 2 \end{bmatrix} = \frac{1}{6}\begin{bmatrix} 1 & 0 & -1 & -2 \\ 0 & 1 & 2 & -1 \\ -1 & 2 & 5 & 0 \\ -2 & -1 & 0 & 5 \end{bmatrix}.$$

18. (a) We solve the "system" $x_1 - x_2 + x_4 = 0$, which is $A\mathbf{x} = \mathbf{0}$ with $A = \begin{bmatrix} 1 & -1 & 0 & 1 \end{bmatrix}$. This is in row echelon form. The variables $x_2 = t$, $x_3 = s$, and $x_4 = r$ are free, while $x_1 = x_2 - x_4 = t - r$. A vector in U is of the form

$$\mathbf{x} = \begin{bmatrix} x_1 \\ x_2 \\ x_3 \\ x_4 \end{bmatrix} = \begin{bmatrix} t-r \\ t \\ s \\ r \end{bmatrix} = t\begin{bmatrix} 1 \\ 1 \\ 0 \\ 0 \end{bmatrix} + s\begin{bmatrix} 0 \\ 0 \\ 1 \\ 0 \end{bmatrix} + r\begin{bmatrix} -1 \\ 0 \\ 0 \\ 1 \end{bmatrix}. \text{ Thus } U \text{ is spanned by the vectors } \mathbf{u}_1 = \begin{bmatrix} 1 \\ 1 \\ 0 \\ 0 \end{bmatrix}, \mathbf{u}_2 = \begin{bmatrix} 0 \\ 0 \\ 1 \\ 0 \end{bmatrix} \text{ and}$$

$$\mathbf{u}_3 = \begin{bmatrix} -1 \\ 0 \\ 0 \\ 1 \end{bmatrix}.$$

19. (a) The projection matrix is $P = \frac{\mathbf{v}\mathbf{v}^T}{\mathbf{v}\cdot\mathbf{v}} = \frac{1}{35}\begin{bmatrix} 1 \\ -3 \\ 5 \end{bmatrix}\begin{bmatrix} 1 & -3 & 5 \end{bmatrix} = \frac{1}{35}\begin{bmatrix} 1 & -3 & 5 \\ -3 & 9 & -15 \\ 5 & -15 & 25 \end{bmatrix}.$

(d) The projection matrix is

$$P = \frac{\mathbf{e}\mathbf{e}^T}{\mathbf{e}\cdot\mathbf{e}} + \frac{\mathbf{f}\mathbf{f}^T}{\mathbf{f}\cdot\mathbf{f}} = \frac{1}{6}\mathbf{e}\mathbf{e}^T + \frac{1}{11}\mathbf{f}\mathbf{f}^T = \frac{1}{6}\begin{bmatrix} 1 & -2 & -1 \\ -2 & 4 & 1 \\ -1 & 2 & 1 \end{bmatrix} + \frac{1}{11}\begin{bmatrix} 9 & 3 & 3 \\ 3 & 1 & 1 \\ 3 & 1 & 1 \end{bmatrix} = \frac{1}{66}\begin{bmatrix} 65 & -4 & 7 \\ -4 & 50 & 28 \\ 7 & 28 & 17 \end{bmatrix}.$$

(f) **Method 1:** We find two linearly independent vectors that span the plane, choosing $\begin{bmatrix} 0 \\ 5 \\ 1 \end{bmatrix}$ and $\begin{bmatrix} 1 \\ 2 \\ 0 \end{bmatrix}$ and let

$A = \begin{bmatrix} 0 & 1 \\ 5 & 2 \\ 1 & 0 \end{bmatrix}$ be the matrix with these vectors as columns. We want the matrix P that projects vectors onto the column space of A.

We have $A^T A = \begin{bmatrix} 26 & 10 \\ 10 & 5 \end{bmatrix}$, $(A^T A)^{-1} = \frac{1}{30}\begin{bmatrix} 5 & -10 \\ -10 & 26 \end{bmatrix}$, so $P = A(A^T A)^{-1} A^T = \frac{1}{30}\begin{bmatrix} 26 & 2 & -10 \\ 2 & 29 & 5 \\ -10 & 5 & 5 \end{bmatrix}.$

Method 2: We find two **orthogonal** vectors that span the plane, for example, $\mathbf{e} = \begin{bmatrix} 1 \\ 2 \\ 0 \end{bmatrix}$ and $\mathbf{f} = \begin{bmatrix} 2 \\ -1 \\ -1 \end{bmatrix}$. Then

$$P = \frac{\mathbf{e}\mathbf{e}^T}{\mathbf{e}\cdot\mathbf{e}} + \frac{\mathbf{f}\mathbf{f}^T}{\mathbf{f}\cdot\mathbf{f}} = \tfrac{1}{5}\begin{bmatrix} 1 & 2 & 0 \\ 2 & 4 & 0 \\ 0 & 0 & 0 \end{bmatrix} + \tfrac{1}{6}\begin{bmatrix} 4 & -2 & -2 \\ -2 & 1 & 1 \\ -2 & 1 & 1 \end{bmatrix} = \tfrac{1}{30}\begin{bmatrix} 26 & 2 & -10 \\ 2 & 29 & 5 \\ -10 & 5 & 5 \end{bmatrix} \text{ as before.}$$

21. (a) We have $A^T A = \begin{bmatrix} 14 & -28 \\ -28 & 56 \end{bmatrix}$, a matrix that is not invertible. The methods of this section do not apply directly.

The object, however, is to project vectors onto the column space of A, which is the span of $\mathbf{v} = \begin{bmatrix} 1 \\ 2 \\ 3 \end{bmatrix}$.

The answer is $P = \frac{\mathbf{v}\mathbf{v}^T}{\mathbf{v}\cdot\mathbf{v}} = \tfrac{1}{14}\begin{bmatrix} 1 \\ 2 \\ 3 \end{bmatrix}\begin{bmatrix} 1 & 2 & 3 \end{bmatrix} = \tfrac{1}{14}\begin{bmatrix} 1 & 2 & 3 \\ 2 & 4 & 6 \\ 3 & 6 & 9 \end{bmatrix}$.

22. (a) Let $B = A^T A$. We are asked to prove that the inverse of B^T is $(B^{-1})^T$. Since these matrices are square, it is sufficient to show that the product of these matrices is the identity matrix. Remembering that $X^T Y^T = (XY)^T$, we simply compute $B^T (B^{-1})^T = (B^{-1}B)^T = I^T = I$.

(b) $P^T = [A(A^T A)^{-1}A^T]^T = A[(A^T A)^{-1}]^T A^T = A[(A^T A)^T]^{-1}A^T = A[(A^T A)^{-1}]A^T = P$;

$P^2 = [A(A^T A)^{-1}A^T][A(A^T A)^{-1}A^T] = A[(A^T A)^{-1}A^T A][(A^T A)^{-1}A^T] = AI(A^T A)^{-1}A^T = P.$

23. (a) The (i, j) entry of $Q^T Q$ is the product of row i of Q^T, which is $\mathbf{q}_i^T$, and column j of Q, which is $\mathbf{q}_j$. Since $\mathbf{q}_i^T \mathbf{q}_j = \mathbf{q}_i \mathbf{q}_j = 1$ if $i = j$ and otherwise 0, we have $Q^T Q = I$. Now this part is just Exercise 13.

(b) It is sufficient to show that $Q^T Q\mathbf{x} = (\mathbf{q}_1^T\mathbf{q}_1 + \mathbf{q}_2^T\mathbf{q}_2 + \cdots + \mathbf{q}_n^T\mathbf{q}_n)\mathbf{x}$ for any vector $\mathbf{x}$. This follows from

$$Q Q^T\mathbf{x} = Q\begin{bmatrix} \mathbf{q}_1^T & \rightarrow \\ \mathbf{q}_2^T & \rightarrow \\ \vdots \\ \mathbf{q}_n^T & \rightarrow \end{bmatrix}\mathbf{x} = \begin{bmatrix} \mathbf{q}_1 & \mathbf{q}_2 & \cdots & \mathbf{q}_n \\ \downarrow & \downarrow & & \downarrow \end{bmatrix}\begin{bmatrix} \mathbf{q}_1 \cdot \mathbf{x} \\ \mathbf{q}_2 \cdot \mathbf{x} \\ \vdots \\ \mathbf{q}_n \cdot \mathbf{x} \end{bmatrix}$$

$$= (\mathbf{q}_1 \cdot \mathbf{x})\mathbf{q}_1 + (\mathbf{q}_2 \cdot \mathbf{x})\mathbf{q}_2 + \cdots + (\mathbf{q}_n \cdot \mathbf{x})\mathbf{q}_n$$

$$= \mathbf{q}_1(\mathbf{q}_1 \cdot \mathbf{x}) + \mathbf{q}_2(\mathbf{q}_2 \cdot \mathbf{x}) + \cdots + \mathbf{q}_n(\mathbf{q}_n \cdot \mathbf{x}) \qquad \text{using the hint}$$

$$= \mathbf{q}_1\mathbf{q}_1^T\mathbf{x} + \mathbf{q}_2\mathbf{q}_2^T\mathbf{x} + \cdots + \mathbf{q}_n\mathbf{q}_n^T\mathbf{x} = (\mathbf{q}_1\mathbf{q}_1^T + \mathbf{q}_2\mathbf{q}_2^T + \cdots + \mathbf{q}_n\mathbf{q}_n^T)\mathbf{x}.$$

25. Every vector perpendicular to the xy-plane goes to the zero vector. So vectors of the form $\begin{bmatrix} 0 \\ 0 \\ z \end{bmatrix}$ are eigenvectors corresponding to the eigenvalue $\lambda = 0$. Every vector in the xy-plane is fixed, so vectors of the form $\begin{bmatrix} x \\ y \\ 0 \end{bmatrix}$ are eigenvectors corresponding to $\lambda = 1$.

27. The column space of A is a subspace of $\mathbf{R}^3$ that contains $\mathbf{e}_1$ and $\mathbf{e}_3$ (columns one and three). Since the other columns are linear combinations of these two, these two are a basis for the column space. The column space is the xz-plane. The matrix that projects vectors in $\mathbf{R}^3$ onto this space is the matrix P that sends $\begin{bmatrix} x \\ y \\ z \end{bmatrix}$ to $\begin{bmatrix} x \\ 0 \\ z \end{bmatrix}$: $P = \begin{bmatrix} 1 & 0 & 0 \\ 0 & 0 & 0 \\ 0 & 0 & 1 \end{bmatrix}$.

Section 6.2—True/False

1. True.

2. False. The given vectors are not linearly independent.

3. True. An orthonormal basis is an orthogonal basis. On the other hand, an orthogonal basis can be converted to an orthonormal basis by dividing each vector by its length.

4. True. A permutation matrix is square with orthonormal columns since its columns are the standard basis vectors $e_1, e_2, \ldots, e_n$ in some order.

5. False. The columns are not of length 1.

6. False. This matrix has orthonormal columns but it is not square.

7. True. The given matrix is rotation through θ; its inverse is rotation through $-\theta$.

8. True. Orthonormal vectors are linearly independent and any linearly independent set of vectors can always be extended to a basis.

9. True.

10. False. Since $Q^T Q = I$, $\det Q^2 = 1$, so $\det Q = \pm 1$.

11. False. $A = \begin{bmatrix} 1 & 0 \\ 1 & 1 \end{bmatrix}$ has determinant 1 but A is not orthogonal.

12. True.

13. False. The $(1, 1)$ and $(2, 2)$ entries are not equal.

Exercises 6.2

1. **(a)** The Gram–Schmidt algorithm begins

$$f_1 = u_1 = \begin{bmatrix} 0 \\ 1 \\ 3 \end{bmatrix}$$

$$f_2 = u_2 - \frac{u_2 \cdot f_1}{f_1 \cdot f_1} f_1 = u_2 - \frac{3}{10} f_1 = \begin{bmatrix} -1 \\ 0 \\ 1 \end{bmatrix} - \frac{3}{10} \begin{bmatrix} 0 \\ 1 \\ 3 \end{bmatrix} = \begin{bmatrix} -1 \\ -\frac{3}{10} \\ \frac{1}{10} \end{bmatrix}.$$

In applications, we would probably replace by f_2 by $10f_3$ giving $f_1 = \begin{bmatrix} 0 \\ 1 \\ 3 \end{bmatrix}$ and $f_2 = \begin{bmatrix} -10 \\ -3 \\ 1 \end{bmatrix}$ as our final orthogonal set.

3. **(a)** The columns of A are the vectors $a_1 = \begin{bmatrix} 1 \\ 1 \end{bmatrix}$ and $a_2 = \begin{bmatrix} 4 \\ 0 \end{bmatrix}$. We apply the Gram–Schmidt algorithm to these.

$$f_1 = a_1 = \begin{bmatrix} 1 \\ 1 \end{bmatrix}$$

$$f_2 = a_2 - \frac{a_2 \cdot f_1}{f_1 \cdot f_1} f_1 = a_2 - \frac{4}{2} f_1 = \begin{bmatrix} 4 \\ 0 \end{bmatrix} - 2 \begin{bmatrix} 1 \\ 1 \end{bmatrix} = \begin{bmatrix} 2 \\ -2 \end{bmatrix}.$$

To assist further calculations, it will help to replace $\mathbf{f}_2$ by $\frac{1}{2}\mathbf{f}_2$, obtaining a new $\mathbf{f}_2 = \begin{bmatrix} 1 \\ -1 \end{bmatrix}$. Dividing $\mathbf{f}_1$ and $\mathbf{f}_2$ by their lengths, we obtain the orthonormal vectors

$$\mathbf{q}_1 = \frac{1}{\|\mathbf{f}_1\|}\mathbf{f}_1 = \frac{1}{\sqrt{2}}\mathbf{f}_1 = \begin{bmatrix} \frac{1}{\sqrt{2}} \\ \frac{1}{\sqrt{2}} \end{bmatrix}$$

$$\text{and } \mathbf{q}_2 = \frac{1}{\|\mathbf{f}_2\|}\mathbf{f}_2 = \frac{1}{\sqrt{2}}\mathbf{f}_2 = \begin{bmatrix} \frac{1}{\sqrt{2}} \\ -\frac{1}{\sqrt{2}} \end{bmatrix}.$$

Now

$$\mathbf{f}_1 = \mathbf{a}_1 \qquad \text{so that} \qquad \mathbf{a}_1 = \mathbf{f}_1 = \sqrt{2}\mathbf{q}_1$$

$$\mathbf{f}_2 = \frac{1}{2}(\mathbf{a}_2 - 2\mathbf{f}_1) \qquad \text{so that} \qquad \mathbf{a}_2 = 2\mathbf{f}_1 + 2\mathbf{f}_2 = 2\sqrt{2}\mathbf{q}_1 + 2\sqrt{2}\mathbf{q}_2.$$

Thus $A = \begin{bmatrix} \mathbf{a}_1 & \mathbf{a}_2 \\ \downarrow & \downarrow \end{bmatrix} = \begin{bmatrix} \mathbf{q}_1 & \mathbf{q}_2 \\ \downarrow & \downarrow \end{bmatrix}\begin{bmatrix} \sqrt{2} & 2\sqrt{2} \\ 0 & 2\sqrt{2} \end{bmatrix} = QR$, with $Q = \begin{bmatrix} \frac{1}{\sqrt{2}} & \frac{1}{\sqrt{2}} \\ \frac{1}{\sqrt{2}} & -\frac{1}{\sqrt{2}} \end{bmatrix}$ and $R = \begin{bmatrix} \sqrt{2} & 2\sqrt{2} \\ 0 & 2\sqrt{2} \end{bmatrix}$.

(e) We apply the Gram–Schmidt algorithm to the columns $\mathbf{a}_1 = \begin{bmatrix} 1 \\ 0 \\ 1 \end{bmatrix}$, $\mathbf{a}_2 = \begin{bmatrix} -1 \\ 1 \\ 1 \end{bmatrix}$ and $\mathbf{a}_3 = \begin{bmatrix} 2 \\ 1 \\ 1 \end{bmatrix}$ of A. The algorithm begins

$$\mathbf{f}_1 = \mathbf{a}_1 = \begin{bmatrix} 1 \\ 0 \\ 1 \end{bmatrix}$$

$$\mathbf{f}_2 = \mathbf{a}_2 - \frac{\mathbf{a}_2 \cdot \mathbf{f}_1}{\mathbf{f}_1 \cdot \mathbf{f}_1}\mathbf{f}_1 = \mathbf{a}_2 - \frac{0}{\mathbf{f}_1 \cdot \mathbf{f}_1}\mathbf{f}_1 = \begin{bmatrix} -1 \\ 1 \\ 1 \end{bmatrix}$$

$$\mathbf{f}_3 = \mathbf{a}_3 - \frac{\mathbf{a}_3 \cdot \mathbf{f}_1}{\mathbf{f}_1 \cdot \mathbf{f}_1}\mathbf{f}_1 - \frac{\mathbf{a}_3 \cdot \mathbf{f}_2}{\mathbf{f}_2 \cdot \mathbf{f}_2}\mathbf{f}_2 = \mathbf{a}_3 - \frac{3}{2}\mathbf{f}_1 - \frac{0}{\mathbf{f}_2 \cdot \mathbf{f}_2}\mathbf{f}_2$$

$$= \begin{bmatrix} 2 \\ 1 \\ 1 \end{bmatrix} - \frac{3}{2}\begin{bmatrix} 1 \\ 0 \\ 1 \end{bmatrix} = \begin{bmatrix} \frac{1}{2} \\ 1 \\ -\frac{1}{2} \end{bmatrix},$$

which we replace by $\mathbf{f}_3 = \begin{bmatrix} 1 \\ 2 \\ -1 \end{bmatrix}$, for convenience. Dividing each of these by their length, we obtain the orthonormal vectors

$$\mathbf{q}_1 = \frac{1}{\|\mathbf{f}_1\|}\mathbf{f}_1 = \frac{1}{\sqrt{2}}\mathbf{f}_1 = \begin{bmatrix} \frac{1}{\sqrt{2}} \\ 0 \\ \frac{1}{\sqrt{2}} \end{bmatrix}, \quad \mathbf{q}_2 = \frac{1}{\|\mathbf{f}_2\|}\mathbf{f}_2 = \frac{1}{\sqrt{3}}\mathbf{f}_2 = \begin{bmatrix} -\frac{1}{\sqrt{3}} \\ \frac{1}{\sqrt{3}} \\ \frac{1}{\sqrt{3}} \end{bmatrix}$$

$$\mathbf{q}_3 = \frac{1}{\|\mathbf{f}_3\|}\mathbf{f}_3 = \frac{1}{\sqrt{6}}\mathbf{f}_3 = \begin{bmatrix} \frac{1}{\sqrt{6}} \\ \frac{2}{\sqrt{6}} \\ -\frac{1}{\sqrt{6}} \end{bmatrix}.$$

Now

$$\mathbf{f}_1 = \mathbf{a}_1 \qquad\qquad \text{so that} \qquad\qquad \mathbf{a}_1 = \mathbf{f}_1 = \sqrt{2}\mathbf{q}_1$$

$$\mathbf{f}_2 = \mathbf{a}_2 \qquad\qquad \text{so that} \qquad\qquad \mathbf{a}_2 = \mathbf{f}_2 = \sqrt{3}\mathbf{q}_2$$

$$\mathbf{f}_3 = 2\left(\mathbf{a}_3 - \frac{3}{2}\mathbf{f}_1\right) \qquad \text{so that} \qquad \mathbf{a}_3 = \frac{3}{2}\mathbf{f}_1 + \frac{1}{2}\mathbf{f}_3 = \frac{3\sqrt{2}}{2}\mathbf{q}_1 + \frac{\sqrt{6}}{2}\mathbf{q}_3.$$

Finally,

$$A = \begin{bmatrix} \mathbf{a}_1 & \mathbf{a}_2 & \mathbf{a}_3 \\ \downarrow & \downarrow & \downarrow \end{bmatrix} = \begin{bmatrix} \mathbf{q}_1 & \mathbf{q}_2 & \mathbf{q}_3 \\ \downarrow & \downarrow & \downarrow \end{bmatrix} \begin{bmatrix} \sqrt{2} & 0 & \frac{3\sqrt{2}}{2} \\ 0 & \sqrt{3} & 0 \\ 0 & 0 & \frac{\sqrt{6}}{2} \end{bmatrix} = QR$$

with $Q = \begin{bmatrix} \frac{1}{\sqrt{2}} & -\frac{1}{\sqrt{3}} & \frac{1}{\sqrt{6}} \\ 0 & \frac{1}{\sqrt{3}} & \frac{2}{\sqrt{6}} \\ \frac{1}{\sqrt{2}} & \frac{1}{\sqrt{3}} & -\frac{1}{\sqrt{6}} \end{bmatrix}$ and $R = \begin{bmatrix} \sqrt{2} & 0 & \frac{3\sqrt{2}}{2} \\ 0 & \sqrt{3} & 0 \\ 0 & 0 & \frac{\sqrt{6}}{2} \end{bmatrix}$.

4. (a) We have $A = QR$ with $Q = \begin{bmatrix} \frac{1}{\sqrt{2}} & \frac{1}{\sqrt{2}} \\ \frac{1}{\sqrt{2}} & -\frac{1}{\sqrt{2}} \end{bmatrix}$ and $R = \begin{bmatrix} \sqrt{2} & 2\sqrt{2} \\ 0 & 2\sqrt{2} \end{bmatrix}$. So $A^{-1} = R^{-1}Q^{-1} = R^{-1}Q^T$

$$= \begin{bmatrix} \frac{\sqrt{2}}{2} & -\frac{\sqrt{2}}{2} \\ 0 & \frac{\sqrt{2}}{4} \end{bmatrix} \begin{bmatrix} \frac{1}{\sqrt{2}} & \frac{1}{\sqrt{2}} \\ \frac{1}{\sqrt{2}} & -\frac{1}{\sqrt{2}} \end{bmatrix} = \begin{bmatrix} 0 & 1 \\ \frac{1}{4} & -\frac{1}{4} \end{bmatrix}$$

5. (a) The columns of A are the vectors $\mathbf{a}_1 = \begin{bmatrix} 1 \\ 2 \\ 2 \end{bmatrix}$ and $\mathbf{a}_2 = \begin{bmatrix} 1 \\ 3 \\ 1 \end{bmatrix}$. We apply the Gram–Schmidt algorithm to these.

$$\mathbf{f}_1 = \mathbf{a}_1 = \begin{bmatrix} 1 \\ 2 \\ 2 \end{bmatrix}$$

$$\mathbf{f}_2 = \mathbf{a}_2 - \frac{\mathbf{a}_2 \cdot \mathbf{f}_1}{\mathbf{f}_1 \cdot \mathbf{f}_1}\mathbf{f}_1 = \mathbf{a}_2 - \frac{9}{9}\mathbf{f}_1 = \begin{bmatrix} 1 \\ 3 \\ 1 \end{bmatrix} - \begin{bmatrix} 1 \\ 2 \\ 2 \end{bmatrix} = \begin{bmatrix} 0 \\ 1 \\ -1 \end{bmatrix}.$$

Dividing $\mathbf{f}_1$ and $\mathbf{f}_2$ by their lengths, we obtain the orthonormal vectors

$$\mathbf{q}_1 = \frac{1}{\|\mathbf{f}_1\|}\mathbf{f}_1 = \frac{1}{3}\mathbf{f}_1 = \begin{bmatrix} \frac{1}{3} \\ \frac{2}{3} \\ \frac{2}{3} \end{bmatrix}, \qquad \mathbf{q}_2 = \frac{1}{\|\mathbf{f}_2\|}\mathbf{f}_2 = \frac{1}{\sqrt{2}}\mathbf{f}_2 = \begin{bmatrix} 0 \\ \frac{1}{\sqrt{2}} \\ -\frac{1}{\sqrt{2}} \end{bmatrix}.$$

Now

$$\mathbf{f}_1 = \mathbf{a}_1 \qquad\qquad \text{so that} \qquad\qquad \mathbf{a}_1 = \mathbf{f}_1 = 3\mathbf{q}_1$$

$$\mathbf{f}_2 = \mathbf{a}_2 - \mathbf{f}_1 \qquad\qquad \text{so that} \qquad\qquad \mathbf{a}_2 = \mathbf{f}_1 + \mathbf{f}_2 = 3\mathbf{q}_1 + \sqrt{2}\mathbf{q}_2.$$

Thus $A = \begin{bmatrix} \mathbf{a}_1 & \mathbf{a}_2 \\ \downarrow & \downarrow \end{bmatrix} = \begin{bmatrix} \mathbf{q}_1 & \mathbf{q}_2 \\ \downarrow & \downarrow \end{bmatrix} \begin{bmatrix} 3 & 3 \\ 0 & \sqrt{2} \end{bmatrix} = QR$, with $Q = \begin{bmatrix} \frac{1}{3} & 0 \\ \frac{2}{3} & \frac{1}{\sqrt{2}} \\ \frac{2}{3} & -\frac{1}{\sqrt{2}} \end{bmatrix}$ and $R = \begin{bmatrix} 3 & 3 \\ 0 & \sqrt{2} \end{bmatrix}$.

(b) The given system is $A\begin{bmatrix} x \\ y \end{bmatrix} = \mathbf{b}$, with $A = \begin{bmatrix} 1 & 1 \\ 2 & 3 \\ 2 & 1 \end{bmatrix}$ and $\mathbf{b} = \begin{bmatrix} 1 \\ -1 \\ 1 \end{bmatrix}$. From part (a), we know that $A = QR$ with

$$Q = \begin{bmatrix} \frac{1}{3} & 0 \\ \frac{2}{3} & \frac{1}{\sqrt{2}} \\ \frac{2}{3} & -\frac{1}{\sqrt{2}} \end{bmatrix} \quad \text{and} \quad R = \begin{bmatrix} 3 & 3 \\ 0 & \sqrt{2} \end{bmatrix}.$$

The desired vector $\mathbf{x}^+$ is the solution to

$$R\mathbf{x} = Q^T\mathbf{b} = \begin{bmatrix} \frac{1}{3} & \frac{2}{3} & \frac{2}{3} \\ 0 & \frac{1}{\sqrt{2}} & -\frac{1}{\sqrt{2}} \end{bmatrix} \begin{bmatrix} 1 \\ -1 \\ 1 \end{bmatrix} = \begin{bmatrix} \frac{1}{3} \\ -\frac{2}{\sqrt{2}} \end{bmatrix}.$$

The augmented matrix is $\begin{bmatrix} 3 & 3 & \Big| & \frac{1}{3} \\ 0 & \sqrt{2} & \Big| & -\frac{2}{\sqrt{2}} \end{bmatrix}$. Back substitution gives $x_2 = -1$ and $x_1 = \frac{1}{9} - x_2 = \frac{10}{9}$.

The best solution is $\mathbf{x}^+ = \begin{bmatrix} \frac{10}{9} \\ -1 \end{bmatrix}$.

6. (a) The system is $A\mathbf{x} = \mathbf{b}$ with $A = \begin{bmatrix} 1 & -1 \\ 2 & 1 \\ 1 & 3 \end{bmatrix}$ and $\mathbf{b} = \begin{bmatrix} 1 \\ -2 \\ -3 \end{bmatrix}$. Applying the Gram–Schmidt algorithm to the columns $\mathbf{a}_1$ and $\mathbf{a}_2$ of A, we obtain

$$\mathbf{f}_1 = \mathbf{a}_1 = \begin{bmatrix} 1 \\ 2 \\ 1 \end{bmatrix}$$

$$\mathbf{f}_2 = \mathbf{a}_2 - \frac{\mathbf{a}_2 \cdot \mathbf{f}_1}{\mathbf{f}_1 \cdot \mathbf{f}_1}\mathbf{f}_1 = \mathbf{a}_2 - \frac{4}{6}\mathbf{f}_1 = \begin{bmatrix} -1 \\ 1 \\ 3 \end{bmatrix} - \frac{2}{3}\begin{bmatrix} 1 \\ 2 \\ 1 \end{bmatrix} = \begin{bmatrix} -\frac{5}{3} \\ -\frac{1}{3} \\ \frac{7}{3} \end{bmatrix},$$

which we replace with $f_2 = \begin{bmatrix} -5 \\ -1 \\ 7 \end{bmatrix}$. Dividing f_1 and f_2 by their lengths, we obtain the orthonormal vectors

$$q_1 = \frac{1}{\|f_1\|}f_1 = \frac{1}{\sqrt{6}}f_1 = \begin{bmatrix} \frac{1}{\sqrt{6}} \\ \frac{2}{\sqrt{6}} \\ \frac{1}{\sqrt{6}} \end{bmatrix}, q_2 = \frac{1}{\|f_2\|}f_2 = \frac{1}{\sqrt{75}}f_2 = \begin{bmatrix} -\frac{5}{\sqrt{75}} \\ -\frac{1}{\sqrt{75}} \\ \frac{7}{\sqrt{75}} \end{bmatrix}$$

Now

$$f_1 = a_1 \qquad\qquad \text{so that} \qquad\qquad a_1 = f_1 = \sqrt{6}q_1$$

$$f_2 = 3\left(a_2 - \frac{2}{3}a_1\right) \qquad \text{so that} \qquad a_2 = \frac{2}{3}f_1 + \frac{1}{3}f_2 = \frac{2\sqrt{6}}{3}q_1 + \frac{\sqrt{75}}{3}q_2.$$

Thus $A = \begin{bmatrix} a_1 & a_2 \\ \downarrow & \downarrow \end{bmatrix} = \begin{bmatrix} q_1 & q_2 \\ \downarrow & \downarrow \end{bmatrix}\begin{bmatrix} \sqrt{6} & \frac{2\sqrt{6}}{3} \\ 0 & \frac{\sqrt{75}}{3} \end{bmatrix} = QR,$

with $Q = \begin{bmatrix} \frac{1}{\sqrt{6}} & -\frac{5}{\sqrt{75}} \\ \frac{2}{\sqrt{6}} & -\frac{1}{\sqrt{75}} \\ \frac{1}{\sqrt{6}} & \frac{7}{\sqrt{75}} \end{bmatrix}$ and $R = \begin{bmatrix} \sqrt{6} & \frac{2\sqrt{6}}{3} \\ 0 & \frac{\sqrt{75}}{3} \end{bmatrix}$. The desired best vector x^+ is the solution to

$$R\mathbf{x} = Q^T\mathbf{b} = \begin{bmatrix} \frac{1}{\sqrt{6}} & \frac{2}{\sqrt{6}} & \frac{1}{\sqrt{6}} \\ -\frac{5}{\sqrt{75}} & -\frac{1}{\sqrt{75}} & \frac{7}{\sqrt{75}} \end{bmatrix}\begin{bmatrix} 1 \\ -1 \\ 1 \end{bmatrix} = \begin{bmatrix} -\frac{6}{\sqrt{6}} \\ \frac{24}{\sqrt{75}} \end{bmatrix}.$$

The augmented matrix of coefficients is $\begin{bmatrix} \sqrt{6} & \frac{2\sqrt{6}}{3} & \Big| & -\frac{6}{\sqrt{6}} \\ 0 & \frac{\sqrt{75}}{3} & \Big| & -\frac{24}{\sqrt{75}} \end{bmatrix}.$

Back substitution gives $x_2 = -\frac{24}{25}$ and $x_1 = -1 - \frac{2}{3}x_2 = -\frac{9}{25}$. The best solution is $\mathbf{x}^+ = \frac{1}{25}\begin{bmatrix} -9 \\ -24 \end{bmatrix}.$

7. QR factorizations require linearly independent columns. Any set of more than m vectors in $\mathbf{R}^m$ is linearly dependent so, if $m < n$, the n columns (which are vectors in $\mathbf{R}^m$) would be linearly dependent and there would be no QR factorization.

9. Let $Q = (I + K)(I - K)^{-1}$. It suffices to show that $Q^T Q = I$. Recalling that $(XY)^T = Y^T X^T$ for matrices X and Y and noting that $(X^{-1})^T = (X^T)^{-1}$ if X is invertible, we have

$$Q^T Q = [(I - K)^T]^{-1}(I + K)^T(I + K)(I - K)^{-1}.$$

Now $(I - K)^T = I^T - K^T = I + K$ and $(I + K)^T = I^T + K^T = I - K$. Also, the matrices $I + K$ and $I - K$ commute. Thus

$$Q^T Q = (I + K)^{-1}(I - K)(I + K)(I - K)^{-1} = (I + K)^{-1}(I + K)(I - K)(I - K)^{-1} = I.$$

So Q is orthogonal, as required.

11. (a) Since Q is orthogonal, $Q^T = Q^{-1}$ and since Q is symmetric, $Q^T = Q$. Thus $Q = Q^{-1}$, so $Q^2 = I$.

13. (a) It suffices to show that $(Q_1 Q_2)^T (Q_1 Q_2) = I$. Since $Q_1^T Q_1 = I$ and $Q_2^T Q_2 = I$, we have
$(Q_1 Q_2)^T (Q_1 Q_2) = Q_2^T Q_1^T Q_1 Q_2 = Q_2^T Q_2 = I$.

15. (a) The subspace spanned by $u_1, u_2, \ldots, u_k$ has dimension less than k, so it cannot contain k orthogonal (and hence linearly independent) vectors.

   (b) Since $u_1, u_2, \ldots, u_k$ are linearly dependent, some u_t is a linear combination of the previous vectors in the list. Let U be the span of these vectors; that is, $U = \text{sp}\{u_1, \ldots, u_{t-1}\}$. Since u_t is in U, the projection of u_t on U is u_t and the next vector $f_t = u_t - \text{proj}_U u_t$ will be 0.

17. Let $x_1, \ldots, x_n$ be any basis and apply the Gram–Schmidt algorithm.

19. Solution 1. We know that $x = c_1 v_1 + c_2 v_2 + \cdots + c_n v_n$ is a linear combination of $v_1, \ldots, v_n$. Linearity of the dot product gives

$$x \cdot x = x \cdot (c_1 v_1 + c_2 v_2 + \cdots + c_n v_n) = c_1(x \cdot v_1) + c_2(x \cdot v_2) + \cdots + c_n(x \cdot v_n)$$

$$= c_1(0) + c_2(0) + \cdots + c_n(0) = 0,$$

so $\|x\|^2 = 0$ and $x = 0$.

Solution 2. Let v be any vector in V. Then $v = c_1 v_1 + c_2 v_2 + \cdots + c_n v_n$ for scalars $c_1, c_2, \ldots, c_n$. It follows that

$$v \cdot x = c_1(v_1 \cdot x) + c_2(v_2 \cdot x) + \cdots + c_n(v_n \cdot x) = c_1(0) + c_2(0) + \cdots + c_n(0) = 0.$$

Thus $v \cdot x = 0$ for any v, in particular for $v = x$. The proof concludes as with Solution 1.

21. (a) We have $y = Hx - x = (I - 2ww^T)x - x = Ix - 2ww^T x - x = -2w(w \cdot x) = -(2w \cdot x)w$.

   (b) Using the result of (a), we have $x + \frac{1}{2}y = x - (w \cdot x)w$. Thus

$$[x + \frac{1}{2}y] \cdot w = x \cdot w - (w \cdot x)(w \cdot w) = 0$$

because $w \cdot w = \|w\|^2 = 1$.

Section 6.3—True/False

1. True. The parallelogram law for vector addition shows that any vector in $\mathbf{R}^2$ is the sum of a multiple of $\begin{bmatrix} 1 \\ 1 \end{bmatrix}$ and $\begin{bmatrix} 1 \\ 2 \end{bmatrix}$.

2. True. A two-dimensional subspace of $\mathbf{R}^3$ is a plane and the intersection of two planes is a line.

3. True. Since $\dim(U + W) \leq 6$, the formula $\dim(U + W) + \dim(U \cap W) = \dim U + \dim W = 6$ implies $\dim(U \cap W) \neq 0$.

4. True. Either statement is equivalent to the assertion that $u \cdot w = 0$ for any u in U and any w in W.

5. False. These planes have nonzero intersection (the y-axis).

6. False. Let $V = \mathbf{R}^3$, let U be the x-axis and W the y-axis.

7. True.

8. False. In $\mathbf{R}^3$, the orthogonal complement of the x-axis is the xy-plane.

9. True. In Theorem 6.3.18, we showed that $R^\perp = \text{null sp } A$, so $(\text{null sp } A)^\perp = (R^\perp)^\perp = R$.

10. True. $W = U^\perp$.

11. True. $Ax = 0$ says x is in the null space of A, which is the orthogonal complement of the row space of A. Since $A^T = A$, the row space of A is U, its column space.

Exercises 6.3

1. (a) We follow the approach of **6.3.4**. Vectors in U have the form $x\begin{bmatrix} 1 \\ 1 \\ 2 \\ 0 \end{bmatrix} + y\begin{bmatrix} -1 \\ 1 \\ -1 \\ 1 \end{bmatrix}$. Thus U has basis $\mathbf{u}_1 = \begin{bmatrix} 1 \\ 1 \\ 2 \\ 0 \end{bmatrix}$,

$\mathbf{u}_2 = \begin{bmatrix} -1 \\ 1 \\ -1 \\ 1 \end{bmatrix}$. These vectors, together with the given basis $\mathbf{w}_1 = \begin{bmatrix} -1 \\ 5 \\ 0 \\ 3 \end{bmatrix}$, $\mathbf{w}_2 = \begin{bmatrix} 0 \\ 0 \\ 1 \\ 0 \end{bmatrix}$ for W, span $U + W$. We put the

basis vectors for U and those for W into the columns of a matrix A and apply the Gaussian elimination process.

$$A = \begin{bmatrix} 1 & -1 & -1 & 0 \\ 1 & 1 & 5 & 0 \\ 2 & -1 & 0 & 1 \\ 0 & 1 & 3 & 0 \end{bmatrix} \rightarrow \begin{bmatrix} 1 & -1 & -1 & 0 \\ 0 & 2 & 6 & 0 \\ 0 & 1 & 2 & 1 \\ 0 & 1 & 3 & 0 \end{bmatrix} \rightarrow \begin{bmatrix} 1 & -1 & -1 & 0 \\ 0 & 1 & 3 & 0 \\ 0 & 0 & 1 & -1 \\ 0 & 0 & 0 & 0 \end{bmatrix}.$$ There are pivots in columns one, two, and

three, so the first three columns of A are a basis for the column space of A, which is $U + W$. A basis for $U + W$ is $\mathbf{u}_1, \mathbf{u}_2, \mathbf{w}_1$.

To find $U \cap W$, we find the null space of A. The Gaussian elimination above shows that the null space consists of

vectors of the form $\begin{bmatrix} -2t \\ -3t \\ t \\ t \end{bmatrix}$, so $-2t\mathbf{u}_1 - 3t\mathbf{u}_2 + t\mathbf{w}_1 + t\mathbf{w}_2 = \mathbf{0}$ and vectors in $U \cap W$ look like

$t(2\mathbf{u}_1 + 3Bu_2) = t(\mathbf{w}_1 + \mathbf{w}_2)$. So a basis for $U \cap W$ is $\mathbf{w}_1 + \mathbf{w}_2 = \begin{bmatrix} 1 \\ -5 \\ -1 \\ -3 \end{bmatrix}$. Finally, since $\dim U = \dim W = 2$, we

have $\dim(U + W) + \dim(U \cap W) = 3 + 1 = 2 + 2 = 4$.

(c) Following the approach of Example 6.3.4, to find $U + W$, we put the vectors $\mathbf{u}_i$ and $\mathbf{w}_i$ into the columns of a matrix A and apply the Gaussian elimination process.

$$A = \begin{bmatrix} -2 & 3 & 6 & 1 & 3 \\ 0 & 5 & 3 & 2 & 1 \\ 2 & 2 & 3 & -5 & 10 \\ 4 & 4 & -6 & 2 & -4 \\ 1 & 1 & -1 & 0 & 0 \end{bmatrix} \rightarrow \begin{bmatrix} 1 & 1 & -1 & 0 & 0 \\ 0 & 5 & 3 & 2 & 1 \\ 0 & 0 & 5 & -5 & 10 \\ 0 & 0 & -2 & 2 & -4 \\ 0 & 5 & 4 & 1 & 3 \end{bmatrix} \rightarrow \begin{bmatrix} 1 & 1 & -1 & 0 & 0 \\ 0 & 5 & 3 & 2 & 1 \\ 0 & 0 & 1 & -1 & 2 \\ 0 & 0 & 1 & -1 & 2 \\ 0 & 0 & 1 & -1 & 25 \end{bmatrix} \rightarrow \begin{bmatrix} 1 & 1 & -1 & 0 & 0 \\ 0 & 5 & 3 & 2 & 1 \\ 0 & 0 & 1 & -1 & 2 \\ 0 & 0 & 0 & 0 & 0 \\ 0 & 0 & 0 & 0 & 0 \end{bmatrix}.$$

The pivots appear in columns one, two, and three, so the first three columns of A are a basis for the column space of A, which is $U + W$. A basis for $U + W$ is $\mathbf{u}_1, \mathbf{u}_2, \mathbf{u}_3$. This implies that $U + W = U$, so W is contained in U and $U \cap W = W$ has basis $\mathbf{w}_1, \mathbf{w}_2$.

2. (a) U is a line; $U^\perp$ is the plane through the origin with this normal, that is, the plane with equation $2x - 3y - 7z = 0$.

(c) $U^\perp$ is the one-dimensional subspace consisting of multiples of the normal $\begin{bmatrix} 1 \\ -2 \\ 1 \end{bmatrix}$ to the plane.

(e) Let $A = \begin{bmatrix} 1 & 1 & -3 \\ 1 & 2 & -1 \end{bmatrix}$ be the matrix whose rows are the (transposes of the) given vectors. Then U is the row space

of A and $U^\perp$ is the null space of A. Row reduction $A \rightarrow \begin{bmatrix} 1 & 1 & -3 \\ 0 & 1 & 2 \end{bmatrix}$ shows that if $\begin{bmatrix} x \\ y \\ z \end{bmatrix}$ is in null sp A, then $z = t$

is free, $y = -2t$, and $x = -y + 3z = 5t$. Thus $U^\perp$ is the set of multiples of $\begin{bmatrix} 5 \\ -2 \\ 1 \end{bmatrix}$.

3. (a) Let $f_1 = \begin{bmatrix} 1 \\ 1 \\ 1 \\ 1 \end{bmatrix}$, $f_2 = \begin{bmatrix} 1 \\ 1 \\ -1 \\ -1 \end{bmatrix}$ and $f_3 = \begin{bmatrix} 1 \\ -1 \\ 1 \\ -1 \end{bmatrix}$. The desired vector u is the projection of v on U:

$$\text{proj}_U v = \frac{v \cdot f_1}{f_1 \cdot f_1} f_1 + \frac{v \cdot f_2}{f_2 \cdot f_2} f_2 + \frac{v \cdot f_3}{f_3 \cdot f_3} f_3 = \frac{9}{4} f_1 - \frac{5}{4} f_2 - \frac{3}{4} f_3$$

$$= \frac{9}{4} \begin{bmatrix} 1 \\ 1 \\ 1 \\ 1 \end{bmatrix} - \frac{5}{4} \begin{bmatrix} 1 \\ 1 \\ -1 \\ -1 \end{bmatrix} - \frac{3}{4} \begin{bmatrix} 1 \\ -1 \\ 1 \\ -1 \end{bmatrix} = \frac{1}{4} \begin{bmatrix} 1 \\ 7 \\ 11 \\ 17 \end{bmatrix}.$$

Then $w = v - u = \frac{1}{4} \begin{bmatrix} 1 \\ -1 \\ -1 \\ 1 \end{bmatrix}$ which, we note, is indeed orthogonal to U.

4. (a) The subspace U has basis $x = \begin{bmatrix} 1 \\ 1 \end{bmatrix}$, so $u = \text{proj}_U v = \text{proj}_x v = \frac{x+y}{2} \begin{bmatrix} 1 \\ 1 \end{bmatrix} = \begin{bmatrix} \frac{x+y}{2} \\ \frac{x+y}{2} \end{bmatrix}$ and $w = v - u = \begin{bmatrix} \frac{x-y}{2} \\ \frac{-x+y}{2} \end{bmatrix}$.

(c) Let $x = \begin{bmatrix} -2 \\ 1 \\ 0 \end{bmatrix}$ We want $v = u + v$ with u in U and w in $U^\perp$. The vector u is $\text{proj}_U v = \frac{v \cdot x}{x \cdot x} x = \frac{-2x + y}{5} \begin{bmatrix} -2 \\ 1 \\ 0 \end{bmatrix}$.

Thus $u = \frac{1}{5} \begin{bmatrix} 4x - 2y \\ -2x + y \\ 0 \end{bmatrix}$ and $w = \frac{1}{5} \begin{bmatrix} x + 2y \\ 2x + 4y \\ z \end{bmatrix}$.

5. (a) Let $u = \begin{bmatrix} -2 \\ 5 \\ 1 \end{bmatrix}$. The desired matrix is $\frac{uu^T}{u \cdot u} = \frac{1}{30} \begin{bmatrix} 4 & -10 & -2 \\ -10 & 25 & 5 \\ -2 & 5 & 1 \end{bmatrix}$.

(b) The orthogonal complement of U is the plane (through the origin) with normal $\begin{bmatrix} -2 \\ 5 \\ 1 \end{bmatrix}$, the direction of the line.

This is the plane with equation $-2x + 5y + z = 0$.

(c) We wish to project vectors in $\mathbf{R}^3$ onto the plane with equation $-2x + 5y + z = 0$. An orthogonal basis for this plane is $\{f_1, f_2\}$, where $f_1 = \begin{bmatrix} 1 \\ 0 \\ 2 \end{bmatrix}$, $f_2 = \begin{bmatrix} 2 \\ 1 \\ -1 \end{bmatrix}$, so the projection matrix is

$$\frac{f_1 f_1^T}{f_1 \cdot f_1} + \frac{f_2 f_2^T}{f_2 \cdot f_2} = \frac{1}{5} \begin{bmatrix} 1 & 0 & 2 \\ 0 & 0 & 0 \\ 2 & 0 & 4 \end{bmatrix} + \frac{1}{6} \begin{bmatrix} 4 & 2 & -2 \\ 2 & 1 & -1 \\ -2 & -1 & 1 \end{bmatrix} = \frac{1}{30} \begin{bmatrix} 26 & 10 & 2 \\ 10 & 5 & -5 \\ 2 & -5 & 29 \end{bmatrix}.$$

(d) The vector we seek is the projection of $\begin{bmatrix} -1 \\ 2 \\ 3 \end{bmatrix}$ on the plane. With P the matrix of part (c), the answer is

$$P \begin{bmatrix} -1 \\ 2 \\ 3 \end{bmatrix} = \frac{1}{2} \begin{bmatrix} 0 \\ -1 \\ 5 \end{bmatrix}.$$

7. This exercise is just a test of understanding of Theorem 6.3.25. We know that $x = \text{proj}_U v$ is in U and that $y = v - \text{proj}_U v$ is in $U^\perp$. Thus $v = x + y$ is one way to write v as the sum of a vector in U and another in $U^\perp$. Part 2 of Theorem 6.3.25 says this is the only way.

8. Since U and W are subspaces, each contains $\mathbf{0}$. Thus $\mathbf{0} = \mathbf{0} + \mathbf{0} \in U + W$. So $U + W$ is not empty. Suppose $\mathbf{v}_1 = \mathbf{u}_1 + \mathbf{w}_1$ and $\mathbf{v}_2 = \mathbf{u}_2 + \mathbf{w}_2$, $\mathbf{u}_1, \mathbf{u}_2 \in U$, $\mathbf{w}_1, \mathbf{w}_2 \in W$ are each in $U + W$. Since U is a subspace $\mathbf{u}_1 + \mathbf{u}_2 \in U$ and, since W is a subspace, $\mathbf{w}_1 + \mathbf{w}_2 \in W$. Thus $\mathbf{v}_1 + \mathbf{v}_2 = (\mathbf{u}_1 + \mathbf{u}_2) + (\mathbf{w}_1 + \mathbf{w}_2) \in U + W$, showing that $U + W$ is closed under addition. Let $\mathbf{v} = \mathbf{u} + \mathbf{w}$, $\mathbf{u} \in U$, $\mathbf{w} \in W$ belong to $U + W$ and let c be a scalar. Since U is a subspace, $c\mathbf{u} \in U$ and, since W is a subspace, $c\mathbf{w} \in W$. Thus $c\mathbf{v} = (c\mathbf{u}) + (c\mathbf{w}) \in U + W$, showing that $U + W$ is also closed under scalar multiplication. Thus $U + W$ is a subspace.

9. Certainly if $\mathbf{x} \in U^\perp$, then each $\mathbf{x} \cdot \mathbf{u}_i = 0$ since $\mathbf{u}_i \in U$. Conversely, if $\mathbf{x} \cdot \mathbf{u}_i = 0$ for all i and $\mathbf{u}$ is any vector in U, then $u = a_1\mathbf{u}_1 + \cdots + a_k\mathbf{u}_k$ for some scalars $a_1, \ldots, a_k$. So $\mathbf{x} \cdot \mathbf{u} = a_1\mathbf{x} \cdot \mathbf{u}_1 + a_2\mathbf{x} \cdot \mathbf{u}_k + \cdots + a_k\mathbf{x} \cdot \mathbf{u}_k = 0$, which shows that $\mathbf{x} \in U^\perp$.

11. **(a)** A vector in U is of the form $s\begin{bmatrix} 2 \\ 1 \\ 0 \\ -5 \end{bmatrix} + t\begin{bmatrix} -1 \\ 0 \\ 1 \\ 2 \end{bmatrix}$. Thus U has basis $\mathbf{u}_1 = \begin{bmatrix} 2 \\ 1 \\ 0 \\ -5 \end{bmatrix}$ and $\mathbf{u}_2 = \begin{bmatrix} -1 \\ 0 \\ 1 \\ 2 \end{bmatrix}$. Let

$A = \begin{bmatrix} 2 & 1 & 0 & -5 \\ -1 & 0 & 1 & 2 \end{bmatrix}$ be the basis whose rows are $\mathbf{u}_1^T$ and $\mathbf{u}_2^T$. By Theorem 6.3.18, $U^\perp$ is the null space of A. By

Gaussian elimination, $A \to \begin{bmatrix} 1 & 0 & -1 & -2 \\ 0 & 1 & 2 & -1 \end{bmatrix}$, so, if $\mathbf{x} = \begin{bmatrix} x_1 \\ x_2 \\ x_3 \\ x_4 \end{bmatrix}$ is in the null space, $x_3 = t$ and $x_4 = s$ are free,

$x_2 = -2x_3 + x_4 = -2t + s$, $x_1 = x_3 + 2x_4 = t + 2s$, so $\mathbf{x} = \begin{bmatrix} t + 2s \\ -2t + s \\ t \\ s \end{bmatrix} = t\begin{bmatrix} 1 \\ -2 \\ 1 \\ 0 \end{bmatrix} + s\begin{bmatrix} 2 \\ 1 \\ 0 \\ 1 \end{bmatrix}$. The orthogonal

complement of U has basis $\mathbf{v}_1 = \begin{bmatrix} 1 \\ -2 \\ 1 \\ 0 \end{bmatrix}$ and $\mathbf{v}_2 = \begin{bmatrix} 2 \\ 1 \\ 0 \\ 1 \end{bmatrix}$.

(b) The vector in U closest to $\mathbf{w} = \begin{bmatrix} -1 \\ 0 \\ 0 \\ 1 \end{bmatrix}$ is $\text{proj}_U \mathbf{w}$. To find this projection, we require an orthogonal basis for U.

Take $\mathbf{f}_1 = \mathbf{u}_1$ and

$$\mathbf{f}_2 = \mathbf{u}_2 - \text{proj}_{\mathbf{u}_1} \mathbf{u}_2 = \begin{bmatrix} -1 \\ 0 \\ 1 \\ 2 \end{bmatrix} - \frac{-12}{30}\begin{bmatrix} 2 \\ 1 \\ 0 \\ -5 \end{bmatrix} = \begin{bmatrix} -1 \\ 0 \\ 1 \\ 2 \end{bmatrix} + \frac{2}{5}\begin{bmatrix} 2 \\ 1 \\ 0 \\ -5 \end{bmatrix} = \begin{bmatrix} -\frac{1}{5} \\ \frac{2}{5} \\ 1 \\ 0 \end{bmatrix}$$

which, for convenience, we replace by $\mathbf{f}_2 = \begin{bmatrix} -1 \\ 2 \\ 5 \\ 0 \end{bmatrix}$. Then

$$\text{proj}_U \mathbf{w} = \frac{\mathbf{w} \cdot \mathbf{f}_1}{\mathbf{f}_1 \cdot \mathbf{f}_1}\mathbf{f}_1 + \frac{\mathbf{w} \cdot \mathbf{f}_2}{\mathbf{f}_2 \cdot \mathbf{f}_2}\mathbf{f}_2 = \frac{-7}{30}\begin{bmatrix} 2 \\ 1 \\ 0 \\ -5 \end{bmatrix} + \frac{1}{30}\begin{bmatrix} -1 \\ 2 \\ 5 \\ 0 \end{bmatrix} = \begin{bmatrix} -\frac{1}{2} \\ -\frac{1}{6} \\ \frac{1}{6} \\ \frac{7}{6} \end{bmatrix}.$$

The vector in $U^\perp$ closest to $\mathbf{w}$ is $\text{proj}_{U^\perp} \mathbf{w}$. Since $\mathbf{v}_1$ and $\mathbf{v}_2$ are orthogonal, this is

$$\text{proj}_{U^\perp} \mathbf{w} = \frac{\mathbf{w} \cdot \mathbf{v}_1}{\mathbf{v}_1 \cdot \mathbf{v}_1} \mathbf{v}_1 + \frac{\mathbf{w} \cdot \mathbf{v}_2}{\mathbf{v}_2 \cdot \mathbf{v}_2} \mathbf{v}_2 = \frac{-1}{6} \begin{bmatrix} 1 \\ -2 \\ 1 \\ 0 \end{bmatrix} + \frac{-1}{6} \begin{bmatrix} 2 \\ 1 \\ 0 \\ 1 \end{bmatrix} = \begin{bmatrix} -\frac{1}{2} \\ \frac{1}{6} \\ -\frac{1}{6} \\ -\frac{1}{6} \end{bmatrix}.$$

13. **(a)** U is given as the null space of $A = \begin{bmatrix} 1 & 0 & -1 & 2 \\ 2 & 3 & 1 & 1 \end{bmatrix}$. Using Gaussian elimination,

$$\begin{bmatrix} 1 & 0 & -1 & 2 \\ 2 & 3 & 1 & 1 \end{bmatrix} \rightarrow \begin{bmatrix} 1 & 0 & -1 & 2 \\ 0 & 3 & 3 & -3 \end{bmatrix} \rightarrow \begin{bmatrix} 1 & 0 & -1 & 2 \\ 0 & 1 & 1 & -1 \end{bmatrix}.$$

Thus $x_3 = t$ and $x_4 = s$ are free, $x_2 = -x_3 + x_4 = -t + s$, $x_1 = x_3 - 2x_4 = t - 2s$, and a vector $\mathbf{x}$ in the null

space is of the form $\begin{bmatrix} t - 2s \\ -t + s \\ t \\ s \end{bmatrix} = t \begin{bmatrix} 1 \\ -1 \\ 1 \\ 0 \end{bmatrix} + s \begin{bmatrix} -2 \\ 1 \\ 0 \\ 1 \end{bmatrix}$.

A basis is $\mathbf{u}_1 = \begin{bmatrix} 1 \\ -1 \\ 1 \\ 0 \end{bmatrix}$, $\mathbf{u}_2 = \begin{bmatrix} -2 \\ 1 \\ 0 \\ 1 \end{bmatrix}$ and an orthogonal basis is $\mathbf{f}_1 = \mathbf{u}_1$ and

$$\mathbf{f}_2 = \mathbf{u}_2 - \frac{-3}{3} \mathbf{f}_1 = \begin{bmatrix} -2 \\ 1 \\ 0 \\ 1 \end{bmatrix} + \begin{bmatrix} 1 \\ -1 \\ 1 \\ 0 \end{bmatrix} = \begin{bmatrix} -1 \\ 0 \\ 1 \\ 1 \end{bmatrix}.$$

(b) The orthogonal complement of U is the orthogonal complement of the null space of A. This is the row space of A.

So $U^\perp$ is spanned by $\begin{bmatrix} 1 \\ 0 \\ -1 \\ 2 \end{bmatrix}$ and $\begin{bmatrix} 2 \\ 3 \\ 1 \\ 1 \end{bmatrix}$.

(c) $P = \frac{1}{3} \mathbf{u}_1 \mathbf{u}_1^T + \frac{1}{3} \mathbf{u}_2 \mathbf{u}_2^T$

$$= \frac{1}{3} \begin{bmatrix} 1 & -1 & 1 & 0 \\ -1 & 1 & -1 & 0 \\ 1 & -1 & 1 & 0 \\ 0 & 0 & 0 & 0 \end{bmatrix} + \frac{1}{3} \begin{bmatrix} 1 & 0 & -1 & -1 \\ 0 & 0 & 0 & 0 \\ -1 & 0 & 1 & 1 \\ -1 & 0 & 1 & 1 \end{bmatrix} = \frac{1}{3} \begin{bmatrix} 2 & -1 & 0 & -1 \\ -1 & 1 & -1 & 0 \\ 0 & -1 & 2 & 1 \\ -1 & 0 & 1 & 1 \end{bmatrix}.$$

(d) The vector closest to $\begin{bmatrix} 1 \\ 2 \\ -1 \\ 1 \end{bmatrix}$ is $P \begin{bmatrix} 1 \\ 2 \\ -1 \\ 1 \end{bmatrix} = \frac{1}{3} \begin{bmatrix} 1 \\ 1 \\ 0 \\ 1 \end{bmatrix}$.

14. **(b)** i. $\Longrightarrow$ ii. Assume U and W are orthogonal. Let $\mathbf{u}$ be a vector in U. We wish to show that $\mathbf{u}$ is in $W^\perp$, that is, that $\mathbf{u} \cdot \mathbf{w} = \mathbf{0}$ for any $\mathbf{w}$ in W. But this is true because U and W are orthogonal.

ii. $\Longrightarrow$ iii. Assume $U \subseteq W^\perp$. By part (a), $(W^\perp)^\perp \subseteq U$. Since $(W^\perp)^\perp = W$, this part is complete.

iii. $\Longrightarrow$ i. Assume $W \subseteq U^\perp$. We wish to that U and W are orthogonal, that is, that $\mathbf{u} \cdot \mathbf{w} = \mathbf{0}$ for any $\mathbf{u}$ in U and any $\mathbf{w}$ in W. This is true because $\mathbf{w}$ is in $U^\perp$.

16. (a) First, let $\mathbf{x} = \begin{bmatrix} x_1 \\ \vdots \\ x_k \\ y_1 \\ \vdots \\ y_\ell \end{bmatrix}$ and $\mathbf{x}' = \begin{bmatrix} x'_1 \\ \vdots \\ x'_k \\ y'_1 \\ \vdots \\ y'_\ell \end{bmatrix}$. Then $\mathbf{x} + \mathbf{x}' = \begin{bmatrix} x_1 + x'_1 \\ \vdots \\ x_k + x'_k \\ y_1 + y'_1 \\ \vdots \\ y_\ell + y'_\ell \end{bmatrix}$, so

$$T(\mathbf{x} + \mathbf{x}') = (x_1 + x'_1)\mathbf{u}_1 + \cdots + (x_k + x'_k)\mathbf{u}_k = (x_1\mathbf{u}_1 + \cdots + x_k\mathbf{u}_k) + (x'_1\mathbf{u}_1 + \cdots + x'_k\mathbf{u}_k) = T(\mathbf{x}) + T(\mathbf{x}').$$

Second, let $\mathbf{x} = \begin{bmatrix} x_1 \\ \vdots \\ x_k \\ y_1 \\ \vdots \\ y_\ell \end{bmatrix}$ and let c be a scalar. Then $c\mathbf{x} = \begin{bmatrix} cx_1 \\ \vdots \\ cx_k \\ cy_1 \\ \vdots \\ cy_\ell \end{bmatrix}$, so $T(c\mathbf{x}) = cx_1\mathbf{u}_1 + \cdots + cx_k\mathbf{u}_k$

$= c(x_1\mathbf{u}_1 + \cdots + x_k\mathbf{u}_k) = cT(\mathbf{x})$.

17. $U + W = V$ if and only if $\dim(U + W) = \dim V$. In this case, since $\dim(U + W) = \dim U + \dim W$, $U + W = V$ if and only if $\dim U + \dim W = \dim V$.

19. Since A has linearly independent columns, rank $A = $ rank $A^T A = n$, so col sp $A^T A = \mathbf{R}^n$ and $U^\perp = \{\mathbf{0}\}$.

21. A vector $\mathbf{x}$ is an eigenvector corresponding to $\lambda = 1$ if and only if $P\mathbf{x} = \mathbf{x}$ and this occurs if and only if $\mathbf{x}$ is in the column space of P, that is, the subspace onto which P projects. The eigenspace corresponding to $\lambda = 1$ is the column space of P. Call this U. A vector $\mathbf{y}$ is an eigenvector corresponding to $\lambda = 0$ if and only if $P\mathbf{y} = \mathbf{0}$. Thus the eigenspace corresponding to $\lambda = 0$ is the null space of P, which is the orthogonal complement of the row space. Since $P^T = P$, the row space and column space of P are the same. So the eigenspace of 0 is the null space, which is the orthogonal complement of U.

Section 6.4—True/False

1. True.

2. True. $P\mathbf{b} = A\mathbf{x}^+ = AA^+\mathbf{b}$.

3. False. $A^+ = \begin{bmatrix} \frac{1}{5} & 0 & 0 \\ 0 & \frac{1}{7} & 0 \end{bmatrix}$.

4. True. $A^+ = \begin{bmatrix} \frac{1}{2} & 0 & 0 \\ 0 & -\frac{1}{3} & 0 \\ 0 & 0 & 0 \\ 0 & 0 & 0 \end{bmatrix}$, by 6.4.11, and the same observation gives $(A^+)^+ = A$.

5. True. A is an elementary matrix, so it is invertible and $A^+ = A^{-1}$.

6. True. The given matrix is rotation through θ, so its pseudoinverse is its inverse, which is rotation through $-\theta$.

7. False. The given formula for the pseudoinverse is that for a matrix A with independent **columns**.

8. True. Matrix A has linearly independent rows.

9. False. Most projection matrices are not invertible. If P is a projection matrix, $P^+ = P$.

Exercises 6.4

1. (a) The matrix has row rank 2, which is the number of rows.

(b) The 2×2 matrix AA^T has linearly independent rows.

(c) $AA^T = \begin{bmatrix} 4 & 0 \\ 0 & 2 \end{bmatrix}$, so $A^+ = A^T(AA^T)^{-1} = \begin{bmatrix} \frac{1}{4} & 0 \\ \frac{1}{4} & \frac{1}{2} \\ \frac{1}{4} & -\frac{1}{2} \\ \frac{1}{4} & 0 \end{bmatrix}$.

3. Gaussian elimination on A yields

$$A \to \begin{bmatrix} 1 & 1 & 2 \\ 0 & 5 & 5 \\ 0 & -5 & -5 \end{bmatrix} \to \begin{bmatrix} 1 & 1 & 2 \\ 0 & 1 & 1 \\ 0 & 0 & 0 \end{bmatrix} \to \begin{bmatrix} 1 & 0 & 1 \\ 0 & 1 & 1 \\ 0 & 0 & 0 \end{bmatrix}.$$

Since row operations do not affect the row space, row sp A is spanned by $\mathbf{u} = \begin{bmatrix} 1 \\ 0 \\ 1 \end{bmatrix}$ and $\mathbf{v} = \begin{bmatrix} 0 \\ 1 \\ 1 \end{bmatrix}$ verifying the parenthetical comment. We seek a vector $\mathbf{x}^+ = A^+\mathbf{b}$ in the row space of A, hence a vector of the form

$$\mathbf{x}^+ = s\mathbf{u} + t\mathbf{v} = \begin{bmatrix} s \\ t \\ s+t \end{bmatrix}$$

such that $A\mathbf{x}^+ = P\mathbf{b}$, where P is the matrix that projects vectors in $\mathbf{R}^3$ onto the column space of A. As in **6.4.4**, we have $P = \frac{1}{3} \begin{bmatrix} 2 & -1 & -1 \\ -1 & 2 & -1 \\ -1 & -1 & 2 \end{bmatrix}$, so $P\mathbf{b} = \frac{1}{3} \begin{bmatrix} 2b_1 - b_2 - b_3 \\ -b_1 + 2b_2 - b_3 \\ -b_1 - b_2 + 2b_3 \end{bmatrix}$ for $\mathbf{b} = \begin{bmatrix} b_1 \\ b_2 \\ b_3 \end{bmatrix}$. Since

$$A\mathbf{x}^+ = \begin{bmatrix} 1 & 1 & 2 \\ -4 & 1 & -3 \\ 3 & -2 & 1 \end{bmatrix} \begin{bmatrix} s \\ t \\ s+t \end{bmatrix} = \begin{bmatrix} 3s + 3t \\ -7s - 2t \\ 4s - t \end{bmatrix}$$

the system $A\mathbf{x}^+ = P\mathbf{b}$ is

$$3s + 3t = \tfrac{1}{3}(2b_1 - b_2 - b_3)$$
$$-7s - 2t = \tfrac{1}{3}(-b_1 + 2b_2 - b_3)$$
$$4s - t = \tfrac{1}{3}(-b_1 - b_2 + 2b_3)$$

Gaussian elimination on the augmented matrix yields

$$\left[\begin{array}{cc|c} 3 & 3 & \tfrac{1}{3}(2b_1 - b_2 - b_3) \\ -7 & -2 & \tfrac{1}{3}(-b_1 + 2b_2 - b_3) \\ 4 & -1 & \tfrac{1}{3}(-b_1 - b_2 + 2b_3) \end{array}\right] \to \left[\begin{array}{cc|c} 1 & 1 & \tfrac{1}{9}(2b_1 - b_2 - b_3) \\ 0 & 5 & \tfrac{1}{9}(11b_1 - b_2 - 10b_3) \\ 0 & 0 & 0 \end{array}\right],$$

so $t = \frac{1}{45}(11b_1 - b_2 - 10b_3)$, $s = \frac{1}{9}(2b_1 - b_2 - b_3) - t = \frac{1}{45}(-b_1 - 4b_2 + 5b_3)$ and

$$\mathbf{x}^+ = A^+\mathbf{b} = s\mathbf{u} + t\mathbf{v} = \frac{1}{45} \begin{bmatrix} -b_1 - 4b_2 + 5b_3 \\ 11b_1 - b_2 - 10b_3 \\ 10b_1 - 5b_2 - 5b_3 \end{bmatrix} \text{ exactly as before.}$$

4. (a) Since A has linearly independent rows, $A^+ = A^T(AA^T)^{-1}$. We have $AA^T = \begin{bmatrix} 1 & -1 & 0 \end{bmatrix} \begin{bmatrix} 1 \\ -1 \\ 0 \end{bmatrix} = [2]$, so

$A^+ = \frac{1}{2} \begin{bmatrix} 1 \\ -1 \\ 0 \end{bmatrix}$. The best solution to $A\mathbf{x} = \mathbf{b}$ is $\mathbf{x}^+ = A^+\mathbf{b} = \frac{1}{2} \begin{bmatrix} 1 \\ -1 \\ 0 \end{bmatrix} [10] = \begin{bmatrix} 5 \\ -5 \\ 0 \end{bmatrix}$.

(d) Let $\mathbf{b} = \begin{bmatrix} b_1 \\ b_2 \\ b_3 \end{bmatrix}$. We require $A^+\mathbf{b}$ to be in the row space of A, so $A^+\mathbf{b} = t\begin{bmatrix} 1 \\ -1 \end{bmatrix}$ and $A(A^+\mathbf{b}) = t\begin{bmatrix} 2 \\ 2 \\ 2 \end{bmatrix}$. The column

space of A is spanned by $\mathbf{u} = \begin{bmatrix} 1 \\ 1 \\ 1 \end{bmatrix}$, so the matrix P that projects $\mathbf{b}$ onto the column space of A is

$P = \dfrac{\mathbf{u}\mathbf{u}^T}{\|\mathbf{u}\|^2} = \dfrac{1}{3}\begin{bmatrix} 1 \\ 1 \\ 1 \end{bmatrix}\begin{bmatrix} 1 & 1 & 1 \end{bmatrix} = \dfrac{1}{3}\begin{bmatrix} 1 & 1 & 1 \\ 1 & 1 & 1 \\ 1 & 1 & 1 \end{bmatrix}$. The condition $A(A^+\mathbf{b}) = P\mathbf{b}$ says

$$t\begin{bmatrix} 2 \\ 2 \\ 2 \end{bmatrix} = \frac{1}{3}\begin{bmatrix} 1 & 1 & 1 \\ 1 & 1 & 1 \\ 1 & 1 & 1 \end{bmatrix}\begin{bmatrix} b_1 \\ b_2 \\ b_3 \end{bmatrix} = \frac{1}{3}\begin{bmatrix} b_1 + b_2 + b_3 \\ b_1 + b_2 + b_3 \\ b_1 + b_2 + b_3 \end{bmatrix}$$

so $2t = \frac{1}{3}(b_1 + b_2 + b_3)$, $t = \frac{1}{6}(b_1 + b_2 + b_3)$ and

$$A^+\begin{bmatrix} b_1 \\ b_2 \\ b_3 \end{bmatrix} = t\begin{bmatrix} 1 \\ -1 \end{bmatrix} = \frac{1}{6}(b_1 + b_2 + b_3)\begin{bmatrix} 1 \\ -1 \end{bmatrix}$$

gives $A^+ = \dfrac{1}{6}\begin{bmatrix} 1 & 1 & 1 \\ -1 & -1 & -1 \end{bmatrix}$. The best solution to $A\mathbf{x} = \mathbf{b}$ is $\mathbf{x}^+ = A^+\mathbf{b} = \dfrac{1}{6}\begin{bmatrix} 1 & 1 & 1 \\ -1 & -1 & -1 \end{bmatrix}\begin{bmatrix} 2 \\ 0 \\ 1 \end{bmatrix} = \dfrac{1}{2}\begin{bmatrix} 1 \\ -1 \end{bmatrix}$.

(g) The column space of A is spanned by $\mathbf{u} = \begin{bmatrix} 1 \\ 0 \\ 1 \end{bmatrix}$ and $\mathbf{v} = \begin{bmatrix} 0 \\ 1 \\ 0 \end{bmatrix}$. Since these vectors are orthogonal, the matrix that

projects vectors onto this space is $P = \dfrac{\mathbf{u}\mathbf{u}^T}{\|\mathbf{u}\|^2} + \dfrac{\mathbf{v}\mathbf{v}^T}{\|\mathbf{v}\|^2} = \dfrac{1}{2}\begin{bmatrix} 1 \\ 0 \\ 1 \end{bmatrix}\begin{bmatrix} 1 & 0 & 1 \end{bmatrix} + \begin{bmatrix} 1 \\ 0 \\ 1 \end{bmatrix}\begin{bmatrix} 1 & 0 & 1 \end{bmatrix} = \dfrac{1}{2}\begin{bmatrix} 1 & 0 & 1 \\ 0 & 2 & 0 \\ 1 & 0 & 1 \end{bmatrix}$. For

$\mathbf{b} = \begin{bmatrix} b_1 \\ b_2 \\ b_3 \end{bmatrix}$, we must have $A^+\mathbf{b} = \begin{bmatrix} t \\ -t \\ s \end{bmatrix}$ for some t and s since $A^+\mathbf{b}$ is in the row space of A, hence $t\begin{bmatrix} 1 \\ -1 \\ 0 \end{bmatrix} + s\begin{bmatrix} 0 \\ 0 \\ 1 \end{bmatrix}$.
The equation $AA^+\mathbf{b} = P\mathbf{b}$ gives

$$\begin{bmatrix} 1 & -1 & 0 \\ 0 & 0 & 1 \\ 1 & -1 & 0 \end{bmatrix}\begin{bmatrix} t \\ -t \\ s \end{bmatrix} = \begin{bmatrix} \frac{1}{2}b_1 + \frac{1}{2}b_3 \\ b_2 \\ \frac{1}{2}b_1 + \frac{1}{2}b_3 \end{bmatrix},$$

that is, $\begin{bmatrix} 2t \\ s \\ 2t \end{bmatrix} = \begin{bmatrix} \frac{1}{2}b_1 + \frac{1}{2}b_3 \\ b_2 \\ \frac{1}{2}b_1 + \frac{1}{2}b_3 \end{bmatrix}$. Thus $t = \frac{1}{4}(b_1 + b_3)$, $s = b_2$ and $A^+\begin{bmatrix} b_1 \\ b_2 \\ b_3 \end{bmatrix} = \begin{bmatrix} \frac{1}{4}(b_1 + b_3) \\ -\frac{1}{4}(b_1 + b_3) \\ b_2 \end{bmatrix}$ means

$A^+ = \dfrac{1}{4}\begin{bmatrix} 1 & 0 & 1 \\ -1 & 0 & 1 \\ 0 & 4 & 0 \end{bmatrix}$. The best solution to $A\mathbf{x} = \mathbf{b}$ is $\mathbf{x}^+ = A^+\mathbf{b} = \dfrac{1}{4}\begin{bmatrix} 1 & 0 & 1 \\ -1 & 0 & 1 \\ 0 & 4 & 0 \end{bmatrix}\begin{bmatrix} -1 \\ 0 \\ 1 \end{bmatrix} = \dfrac{1}{2}\begin{bmatrix} 0 \\ 1 \\ 0 \end{bmatrix}$.

5. (a) If A has linearly independent columns, $A^+ = (A^T A)^{-1} A^T$, so $A^+ A = (A^T A)^{-1} (A^T A) = I$.

7. The vector $\mathbf{x}^+$ is the unique vector in the row space of A with the property that $A\mathbf{x}^+ = P\mathbf{b}$ is the projection of $\mathbf{b}$ on the column space of A. Let $T : \mathbf{R}^m \to \mathbf{R}^n$ be defined by $T(\mathbf{b}) = \mathbf{x}^+$. First we show that T preserves vector addition. Given vectors $\mathbf{b}_1$ and $\mathbf{b}_2$, let $\mathbf{x}_1^+$ and $\mathbf{x}_2^+$ be the unique vectors in the row space of A with the property that $A\mathbf{x}_1^+ = P\mathbf{b}_1$ and $A\mathbf{x}_2^+ = P\mathbf{b}_2$. To show that $T(\mathbf{b}_1 + \mathbf{b}_2) = T(\mathbf{b}_1) + T(\mathbf{b}_2)$, we must show that the unique vector $\mathbf{x}^+$ in the row space of A with the property that $A\mathbf{x}^+ = P(\mathbf{b}_1 + \mathbf{b}_2)$ is $\mathbf{x}^+ = \mathbf{x}_1^+ + \mathbf{x}_2^+$. Now $A(\mathbf{x}_1^+ + \mathbf{x}_2^+) = A\mathbf{x}_1^+ + A\mathbf{x}_2^+ = P\mathbf{b}_1 + P\mathbf{b}_2 = P(\mathbf{b}_1 + \mathbf{b}_2)$. So $\mathbf{x} = \mathbf{x}_1^+ + \mathbf{x}_2^+$ satisfies $A\mathbf{x} = P(\mathbf{b}_1 + \mathbf{b}_2)$. Since this vector is also in the row space of A (since the row space is closed under vector addition), we have what we want. To see that T preserves scalar multiplication, let $\mathbf{b}$ be a vector in $\mathbf{R}^m$ and let c be a scalar. Let $\mathbf{x}^+ = T(\mathbf{b})$ be the unique vector in the row space of A satisfying $A\mathbf{x}^+ = P\mathbf{b}$. To show $T(c\mathbf{b}) = cT(\mathbf{b})$, we must show that the unique vector $\mathbf{x}$ in the row space of A that satisfies $A\mathbf{x} = P(c\mathbf{b})$ is $\mathbf{x} = c\mathbf{x}^+$. This follows because $c\mathbf{x}^+$ is in the row space of A (the row space is closed under scalar multiplication) and $A(c\mathbf{x}^+) = cA\mathbf{x}^+ = cP\mathbf{b} = P(c\mathbf{b})$.

9. A permutation matrix is invertible, so its pseudoinverse is its inverse, and the inverse of a permutation matrix is its transpose.

Section 7.1—True/False

1. True. $7 = \frac{7}{1}$ is the quotient of integers.

2. True. $\sqrt{2}$ cannot be written as the quotient of integers.

3. True.

4. False. Every real number is its own complex conjugate.

5. False. $z\bar{z} = |z|^2$.

6. True.

7. True. The inverse of nonzero complex number is also a complex number.

8. False. $1 + i = \sqrt{2}(\cos\frac{\pi}{4} + i\sin\frac{\pi}{4})$.

9. True. Multiply moduli and add arguments.

10. True.

11. True. The equation $z^5 + 1 = 0$ has five roots.

12. False. $z \cdot w = 1(-1) + \bar{i}2i = -1 - 2i^2 = +1$.

Exercises 7.1

1. $1 + \sqrt{3}i = 2(\cos\frac{\pi}{3} + i\sin\frac{\pi}{3})$.

2. (a) $\dfrac{3 - 4i}{1 + i} = \dfrac{3 - 4i}{1 + i}\dfrac{1 - i}{1 - i} = \dfrac{-7 - 7i}{2} = -\dfrac{7}{2} - \dfrac{7}{2}i$.

   (c) $\dfrac{-2 - 2i}{-3 - 3i} = \dfrac{2}{3}\dfrac{1 + i}{1 + i} = \dfrac{2}{3}$

3. We have $|z| = \sqrt{4 + 4} = 2\sqrt{2}$ and $\theta = -\frac{\pi}{4}$, so $z = 2 - 2i = 2[\cos(-\frac{\pi}{4}) + i\sin(-\frac{\pi}{4})]$.
   Also, $\bar{z} = 2 + 2i = 2\sqrt{2}\cos\frac{\pi}{4} + i\sin\frac{\pi}{4})$, $z^2 = -8i = 8(\cos-\frac{\pi}{2} + i\sin-\frac{\pi}{2})$, and
   $\dfrac{1}{z} = \dfrac{\bar{z}}{z\bar{z}} = \frac{1}{8}(2 + 2i) = \frac{1}{4} + \frac{1}{4}i = \dfrac{1}{2\sqrt{2}}(\cos\frac{\pi}{4} + i\sin\frac{\pi}{4})$.

5. $zw = 1(3 + i) - 2i(3 + i) = 3 + i - 6i + 2 = 5 - 5i$; $\dfrac{z}{w} = \dfrac{z\bar{w}}{w\bar{w}} = \frac{1}{10}(1 - 7i)$;
   $z^2 - w^2 = (-3 - 4i) - (8 + 6i) = -11 - 10i$; $\dfrac{z + w}{z - w} = \dfrac{4 - i}{-2 - 3i} = \dfrac{(4 - i)(-2 + 3i)}{(-2 - 3i)(-2 + 3i)} = \frac{1}{13}(-5 + 14i)$.

7. $\dfrac{z\overline{w}+z^2}{w\overline{z}+w^2} = \dfrac{z(\overline{w}+z)}{w(\overline{z}+w)} = \dfrac{i(2-i+i)}{(2+i)(-i+2+i)} = \dfrac{2i}{2(2+i)} = \dfrac{i}{2+i} = \dfrac{i(2-i)}{(2+i)(2-i)} = \dfrac{2i+1}{5} = \frac{1}{5} + \frac{2}{5}i.$

9. Write $z = a + bi$. Then $\overline{z} = a - bi$, so $z + \overline{z} = 2a$ and the real part of z is $a = \frac{1}{2}(z+\overline{z})$. Also, $z - \overline{z} = 2bi$, so the imaginary part of z is $b = \frac{1}{2i}(z - \overline{z})$.

11. We have $c\mathbf{z} = \begin{bmatrix} cz_1 \\ cz_2 \\ \vdots \\ cz_n \end{bmatrix}$, so that

$$\|c\mathbf{z}\|^2 = (c\mathbf{z}) \cdot (c\mathbf{z}) = \overline{cz_1}cz_1 + \overline{cz_2}cz_2 + \cdots + \overline{cz_n}cz_n$$

$$= \overline{c}\,\overline{z_1}cz_1 + \overline{c}\,\overline{z_2}cz_2 + \cdots + \overline{c}\,\overline{z_n}cz_n$$

$$= \overline{c}c\overline{z_1}z_1 + \overline{c}c\overline{z_2}z_2 + \cdots + \overline{c}c\,\overline{z_n}z_n$$

$$= |c|^2|z_1|^2 + |c|^2|z_2|^2 + \cdots + |c|^2|z_n|^2$$

$$= |c|^2(|z_1|^2 + |z_2|^2 + \cdots + |z_n|^2) = |c|^2\,\|\mathbf{z}\|^2.$$

The result follows upon taking square roots.

12. (a) $\begin{aligned}[t] \mathbf{z} \cdot (c\mathbf{w}) &= \overline{(c\mathbf{w}) \cdot \mathbf{z}} && \text{by (4)} \\ &= \overline{\overline{c}(\mathbf{w} \cdot \mathbf{z})} && \text{by (5)} \\ &= \overline{\overline{c}}\,\overline{\mathbf{w} \cdot \mathbf{z}} && \text{by Proposition 7.1.7} \\ &= c\,\overline{\mathbf{w} \cdot \mathbf{z}} && \text{by Proposition 7.1.7} \\ &= c(\mathbf{z} \cdot \mathbf{w}) && \text{using (4) again.} \end{aligned}$

13. This was shown in the text. Let $z = a + bi$. If z is real, then $b = 0$ and $\overline{z} = \overline{a + 0i} = a - 0i = a = z$. Conversely, if $\overline{z} = z$, then $a - bi = a + bi$, so $2bi = 0$, so $b = 0$ and $z = a$ is real.

15. If $\mathbf{w} = \begin{bmatrix} w_1 \\ \vdots \\ w_n \end{bmatrix}$, we know that $\|\mathbf{w}\|^2 = \overline{w}_1 w_1 + \cdots + \overline{w}_n w_n$. Since $\overline{\mathbf{z}} = \begin{bmatrix} \overline{z_1} \\ \vdots \\ \overline{z_n} \end{bmatrix}$, $\|\overline{\mathbf{z}}\|^2 = \overline{\overline{z}_1}\overline{z}_1 + \cdots + \overline{\overline{z}_n}\overline{z}_n$. Since $\overline{\overline{z}_i} = z_i$, we have $\|\overline{\mathbf{z}}\|^2 = z_1\overline{z}_1 + \cdots + z_n\overline{z}_n$, and since z_i and $\overline{z}_i$ *commute* (that is, $z_i\overline{z}_i = \overline{z}_i z_i$), we have $\|\overline{\mathbf{z}}\|^2 = \overline{z}_1 z_1 + \cdots + \overline{z}_n z_n$, which is precisely $\|\mathbf{z}\|^2$.

16. (a) We want $(\mathbf{u} - c\mathbf{v}) \cdot \mathbf{v} = 0$; thus $\mathbf{u} \cdot \mathbf{v} - \overline{c}(\mathbf{v} \cdot \mathbf{v}) = 0$ and $\overline{c} = \frac{\mathbf{u} \cdot \mathbf{v}}{\mathbf{v} \cdot \mathbf{v}}$. Thus $c = \frac{\overline{\mathbf{u} \cdot \mathbf{v}}}{\mathbf{v} \cdot \mathbf{v}} = \frac{\mathbf{v} \cdot \mathbf{u}}{\mathbf{v} \cdot \mathbf{v}}$ since $\overline{\mathbf{u} \cdot \mathbf{v}} = \mathbf{v} \cdot \mathbf{u}$. So the projection of $\mathbf{u}$ on $\mathbf{v}$ is the vector $\mathbf{p} = \frac{\mathbf{v} \cdot \mathbf{u}}{\mathbf{v} \cdot \mathbf{v}}\,\mathbf{v}$.

(b) Note first that $\mathbf{v} \cdot \mathbf{u} = (3 - 2i)1 + i(-i) = 4 - 2i$, $\mathbf{v} \cdot \mathbf{v} = (3 - 2i)(3 + 2i) + (-i)(i) + 2(2) = 9 + 4 + 1 = 18$ and $\mathbf{u} \cdot \mathbf{u} = 1(1) + i(-i) = 2$.

The projection of $\mathbf{u}$ on $\mathbf{v}$ is $\frac{4-2i}{18}\mathbf{v} = \frac{2-i}{9}\begin{bmatrix} 3 + 2i \\ -i \\ 2 \end{bmatrix} = \frac{1}{9}\begin{bmatrix} 8 + i \\ -1 - 2i \\ 4 - 2i \end{bmatrix}.$

Checking our answer, we have $\mathbf{u} - \mathbf{p} = \frac{1}{9}\begin{bmatrix} 1 - i \\ 1 - 7i \\ -4 + 2i \end{bmatrix}$ and

$\mathbf{v} \cdot (\mathbf{u} - \mathbf{p}) = \frac{1}{9}\Big((3 - 2i)(1 - i) + i(1 - 7i) + 2(-4 + 2i)\Big) = 0.$

Section 7.2—True/False

1. False. $A^* = \begin{bmatrix} 1 & 1+i \\ -i & -2i \end{bmatrix}$.

2. True.

3. True. The matrix is Hermitian.

4. True.

5. False. The matrix is not square.

6. True. Unitary matrices preserve the dot product.

Exercises 7.2

1. $\det A = \begin{vmatrix} 1-i & 3i \\ -3 & -i \end{vmatrix} = (1-i)(-i) - (3i)(-3) = -i - 1 + 9i = -1 + 8i$.

   $\det(4i\,A) = (4i)^2 \det A = -16(-1+8i) = 16 - 128i$.

2. (a) $A + B = \begin{bmatrix} 2 & 1+3i \\ -1-2i & -4i \end{bmatrix}$; $A - B = \begin{bmatrix} 2i & 1-3i \\ 5-2i & -2i \end{bmatrix}$; $2A - B = \begin{bmatrix} 1+3i & 2-3i \\ 7-4i & -5i \end{bmatrix}$;

   $AB = \begin{bmatrix} -1 & -3+2i \\ 5i & 3+6i \end{bmatrix}$, $BA = \begin{bmatrix} 8+6i & 10-i \\ -5-5i & -6 \end{bmatrix}$.

   (c) $BA = \begin{bmatrix} 1-6i & -11+7i \\ -3+i & -1+7i \\ -2-5i & -6+2i \end{bmatrix}$; $A+B, A-B, 2A-B, AB$ are not defined.

3. A and C are conjugate transposes; B is unitary; E is Hermitian.

5. (a) The (i, j) entry of AB is a sum of the form $z_1 w_1 + z_2 w_2 + \cdots z_n w_n$. So the (i, j) entry of $\overline{AB}$ is $\overline{z_1 w_1} + \overline{z_2 w_2} + \cdots + \overline{z_n w_n}$. Since $\overline{zw} = \overline{z}\,\overline{w}$ for any complex numbers, this entry is $\overline{z_1}\,\overline{w_1} + \overline{z_2}\,\overline{w_2} + \cdots + \overline{z_n}\,\overline{w_n}$ and this is precisely the i, j entry of $\overline{A}\,\overline{B}$.

6. (a) The hint gives this away! Since $\mathbf{x}^T A\mathbf{x}$ is 1×1, this matrix equals its transpose. Thus $\mathbf{x}^T A\mathbf{x} = (\mathbf{x}^T A\mathbf{x})^T = \mathbf{x}^T A^T \mathbf{x} = -\mathbf{x}^T A\mathbf{x}$, so $2\mathbf{x}^T A\mathbf{x} = 0$, so $\mathbf{x}^T A\mathbf{x} = 0$.

7. (a) True. If $A^* = A^{-1}$ and $B^* = B^{-1}$, then $(AB)^* = B^* A^* = B^{-1} A^{-1} = (AB)^{-1}$.

   (c) False. Consider $A = \begin{bmatrix} 2 & 1 \\ 1 & 2 \end{bmatrix}$ and $B = \begin{bmatrix} 1 & -3 \\ -3 & 10 \end{bmatrix}$. These matrices are symmetric but their product,

   $AB = \begin{bmatrix} -1 & 4 \\ -5 & 17 \end{bmatrix}$, is not.

8. (a) $\begin{bmatrix} 0 & i \\ -i & 0 \end{bmatrix}$

9. $\|A\mathbf{z}\|^2 = (A\mathbf{z}) \cdot (A\mathbf{z}) = (A\mathbf{z})^*(A\mathbf{z}) = \mathbf{z}^* A^* A\mathbf{z} = \mathbf{z}^* A A^* \mathbf{z} = (A^*\mathbf{z})^*(A^*\mathbf{z}) = \|A\mathbf{z}\|^2$.

10. (a) A typical vector in range A is of the form $A\mathbf{w}$ for some $\mathbf{w}$. Thus A vector $\mathbf{z}$ is in $(\text{range } A)^{\perp}$ if and only if $A\mathbf{w} \cdot \mathbf{z} = 0$ for all $\mathbf{w}$, and this occurs if and only if $(A\mathbf{w})^*\mathbf{z} = 0$ for all $\mathbf{w}$; that is, if and only if $\mathbf{w}^* A^*\mathbf{z} = 0$ for all $\mathbf{w}$. Since $\mathbf{w}^* A^*\mathbf{z} = \mathbf{w} \cdot A^*\mathbf{z}$ and this is 0 for all $\mathbf{w}$ if and only if $A^*\mathbf{z} = \mathbf{0}$, we have the result.

11. Let $A = \begin{bmatrix} a & b \\ b & c \end{bmatrix}$ have real entries. The characteristic polynomial of A is

    $\begin{vmatrix} a-\lambda & b \\ b & c-\lambda \end{vmatrix} = ac - (a+c)\lambda + \lambda^2 - b^2\lambda^2 - (a+c)\lambda + (ac - b^2)$.

12. Let $\mathbf{x} \in U^{\perp}$. We have to show that $A\mathbf{x}$ is also in $U^{\perp}$, that is, that $A\mathbf{x}$ is orthogonal to any vector in U. So take a vector $\mathbf{u} \in U$ (and remember that $\mathbf{u}$ is also an eigenvector of A, corresponding to a scalar λ, say). We have

$$A\mathbf{x} \cdot \mathbf{u} = (A\mathbf{x})^T \mathbf{u} = \mathbf{x}^T A^T \mathbf{u} = \mathbf{x}^T A\mathbf{u} = \mathbf{x}^T \lambda \mathbf{u} = \lambda(\mathbf{x}^T \mathbf{u}) = \lambda(\mathbf{x} \cdot \mathbf{u}) = 0,$$

as desired.

Section 7.3—True/False

1. False. $A = I$. If X is similar to I, then $X = P^{-1}IP = I$.

2. False. The characteristic polynomial of A is $(1 - \lambda)^2 = 1 - 2\lambda + \lambda^2$, and this equation is satisfied by A (by the Cayley–Hamilton theorem).

3. True. We have $U^*AU = D$ for some unitary matrix U and diagonal matrix D. The diagonal entries of D are the eigenvalues of A, so U is invertible. Thus $A = UDU^*$ is also invertible.

4. False. Let A be any real symmetric matrix that is not diagonal. Then there exists a diagonal matrix and an orthogonal matrix Q_1 such that $Q_1^{-1}AQ_1 = D$. Then $A = Q_1 D Q_1^{-1}$. Letting $Q = Q_1^{-1}$, $Q^{-1}DQ = Q^T DQ$ is not diagonal.

5. True. This is the Principal Axes Theorem.

6. True. If H is Hermitian, $HH^* = HH^* = H^2$.

7. False. $A = \begin{bmatrix} 1 & i \\ -i & 1 \end{bmatrix}$ is Hermitian, hence normal, but it is not unitary because its columns do not have modulus 1 (neither are its columns orthogonal).

Exercises 7.3

1. (a) The characteristic polynomial of A is $\begin{vmatrix} 2 - \lambda & 3 \\ 3 & -6 - \lambda \end{vmatrix} = \lambda^2 + 4\lambda - 21 = (\lambda + 7)(\lambda - 3)$. The eigenvalues $\lambda = 3, -7$ are real. If $\mathbf{x} = \begin{bmatrix} x_1 \\ x_2 \end{bmatrix}$ is an eigenvector for $\lambda = 3$, Gaussian elimination applied to $A - \lambda I$, with $\lambda = 3$, proceeds

$$\begin{bmatrix} -1 & 3 \\ 3 & -9 \end{bmatrix} \rightarrow \begin{bmatrix} 1 & -3 \\ 0 & 0 \end{bmatrix},$$

so $x_2 = t$ is free, $x_1 = 3t$ and $\mathbf{x} = \begin{bmatrix} 3t \\ t \end{bmatrix} = t\begin{bmatrix} 3 \\ 1 \end{bmatrix}$. If $\mathbf{x} = \begin{bmatrix} x_1 \\ x_2 \end{bmatrix}$ is an eigenvector for $\lambda = -7$, Gaussian elimination applied to $A - \lambda I$, with $\lambda = -7$, proceeds

$$\begin{bmatrix} 9 & 3 \\ 3 & 1 \end{bmatrix} \rightarrow \begin{bmatrix} 1 & 3 \\ 0 & 0 \end{bmatrix},$$

so $x_2 = t$ is free, $x_1 = -\frac{1}{3}t$ and $\mathbf{x} = \begin{bmatrix} -\frac{1}{3}t \\ t \end{bmatrix} = t\begin{bmatrix} -\frac{1}{3} \\ 1 \end{bmatrix}$. The eigenspaces are orthogonal with bases $\mathbf{f}_1 = \begin{bmatrix} 3 \\ 1 \end{bmatrix}$ and $\mathbf{f}_2 = \begin{bmatrix} -1 \\ 3 \end{bmatrix}$. Let $Q = \begin{bmatrix} \frac{3}{\sqrt{10}} & -\frac{1}{\sqrt{10}} \\ \frac{1}{\sqrt{10}} & \frac{3}{\sqrt{10}} \end{bmatrix}$ be the matrix whose columns are the orthonormal eigenvectors $\mathbf{q}_1 = \frac{1}{\|\mathbf{f}_1\|}\mathbf{f}_1$ and $\mathbf{q}_2 = \frac{1}{\|\mathbf{f}_2\|}\mathbf{f}_2$. Then $Q^T A Q = D = \begin{bmatrix} 3 & 0 \\ 0 & -7 \end{bmatrix}$.

(c) The characteristic polynomial of A is

$$\begin{bmatrix} -2-\lambda & 1 & 1 \\ 1 & -2-\lambda & 1 \\ 1 & 1 & -2-\lambda \end{bmatrix} = (-2-\lambda)(4+4\lambda+\lambda^2-1)-(-2-\lambda-1)+(1+2+\lambda)$$

$$= -(\lambda+2)(\lambda^2+4\lambda+3)+\lambda+3+\lambda+3$$

$$= -(\lambda+2)(\lambda+1)(\lambda+3)+2(\lambda+3)$$

$$= -(\lambda+3)((\lambda^2+3\lambda+2-2) = -\lambda(\lambda+3)^2$$

(expanding by cofactors along the first row). So the eigenvalues of A are $\lambda=0,-3$. If $\mathbf{x} = \begin{bmatrix} x_1 \\ x_2 \\ x_3 \end{bmatrix}$ is an eigenvector for $\lambda=0$, then $(A-\lambda I)\mathbf{x} = \mathbf{0}$. Gaussian elimination proceeds

$$\begin{bmatrix} -2 & 1 & 1 \\ 1 & -2 & 1 \\ 1 & 1 & -2 \end{bmatrix} \to \begin{bmatrix} 1 & 1 & -2 \\ 0 & 3 & -3 \\ 0 & -3 & 3 \end{bmatrix} \to \begin{bmatrix} 1 & 1 & -2 \\ 0 & 1 & -1 \\ 0 & 0 & 0 \end{bmatrix},$$

so $x_3 = t$ is free, $x_2 = x_3 = t$, $x_1 = -x_2+2x_3 = -t+2t = t$ and $\mathbf{x} = \begin{bmatrix} t \\ t \\ t \end{bmatrix} = t\mathbf{f}_1$, with $\mathbf{f}_1 = \begin{bmatrix} 1 \\ 1 \\ 1 \end{bmatrix}$. If $\mathbf{x} = \begin{bmatrix} x_1 \\ x_2 \\ x_3 \end{bmatrix}$ is an

eigenvector for $\lambda=-3$, then $(A-\lambda I)\mathbf{x} = \mathbf{0}$. Gaussian elimination proceeds

$$\begin{bmatrix} 1 & 1 & 1 \\ 1 & 1 & 1 \\ 1 & 1 & 1 \end{bmatrix} \to \begin{bmatrix} 1 & 1 & 1 \\ 0 & 0 & 0 \\ 0 & 0 & 0 \end{bmatrix},$$

so $x_3 = t$ and $x_2 = s$ are free, $x_1 = -x_2-x_3 = -t-s$ and $\mathbf{x} = \begin{bmatrix} -t-s \\ s \\ t \end{bmatrix} = t\mathbf{u}+s\mathbf{v}$, with $\mathbf{u} = \begin{bmatrix} -1 \\ 0 \\ 1 \end{bmatrix}$ and

$\mathbf{v} = \begin{bmatrix} -1 \\ 1 \\ 0 \end{bmatrix}$. These are orthogonal to $\mathbf{f}_1$ (as they must be), but not to each other. So we replace with $\mathbf{f}_2 = \mathbf{u}$ and

$\mathbf{f}_3 = \mathbf{v} - \text{proj}_{\mathbf{u}}\mathbf{v} = \begin{bmatrix} -1 \\ 1 \\ 0 \end{bmatrix} - \frac{1}{2}\begin{bmatrix} -1 \\ 0 \\ 1 \end{bmatrix} = \begin{bmatrix} -\frac{1}{2} \\ 1 \\ -\frac{1}{2} \end{bmatrix}$, which we double, giving $\mathbf{f}_3 = -\begin{bmatrix} -1 \\ 2 \\ -1 \end{bmatrix}$. Now $\mathbf{f}_1, \mathbf{f}_2, \mathbf{f}_3$ are orthogonal.

Dividing each by its length gives orthonormal vectors, which we use as the columns of an orthogonal matrix

$$Q = \begin{bmatrix} \frac{1}{\sqrt{3}} & -\frac{1}{\sqrt{2}} & -\frac{1}{\sqrt{6}} \\ \frac{1}{\sqrt{3}} & 0 & \frac{2}{\sqrt{6}} \\ \frac{1}{\sqrt{3}} & \frac{1}{\sqrt{2}} & -\frac{1}{\sqrt{6}} \end{bmatrix}. \text{ Then } Q^{-1}AQ = D = \begin{bmatrix} 0 & 0 & 0 \\ 0 & -3 & 0 \\ 0 & 0 & -3 \end{bmatrix}.$$

(e) The characteristic polynomial of A is

$$\begin{bmatrix} 5-\lambda & 2 & 2 \\ 2 & 2-\lambda & -4 \\ 2 & -4 & 2-\lambda \end{bmatrix} = (5-\lambda)(4-4\lambda+\lambda^2-16)-2(4-2\lambda+8)+2(-8-4+2\lambda)$$

$$= (5-\lambda)(\lambda^2-4\lambda-12)-24+4\lambda-24+4\lambda$$

$$= (5-\lambda)(\lambda-6)(\lambda+2)+8(\lambda-6)$$

$$= (\lambda-6)(-\lambda^2+3\lambda+10+8) = -(\lambda-6)^2(\lambda-3)$$

(expanding by cofactors along the first row). So the eigenvalues of A are $\lambda = -3, 6$. If $\mathbf{x} = \begin{bmatrix} x_1 \\ x_2 \\ x_3 \end{bmatrix}$ is an eigenvector for $\lambda = -3$, then $(A - \lambda I)\mathbf{x} = \mathbf{0}$. Gaussian elimination proceeds

$$\begin{bmatrix} 8 & 2 & 2 \\ 2 & 5 & -4 \\ 2 & -4 & 5 \end{bmatrix} \to \begin{bmatrix} 1 & \frac{1}{4} & \frac{1}{4} \\ 0 & \frac{9}{2} & -\frac{9}{2} \\ 0 & -9 & 9 \end{bmatrix} \to \begin{bmatrix} 1 & \frac{1}{4} & \frac{1}{4} \\ 0 & 1 & -1 \\ 0 & 0 & 0 \end{bmatrix},$$

so $x_3 = t$ is free, $x_2 = x_3 = t$, $x_1 = -\frac{1}{4}x_2 - \frac{1}{4}x_3 = -\frac{1}{2}$, and $\mathbf{x} = \begin{bmatrix} -\frac{1}{2}t \\ t \\ t \end{bmatrix} = t\mathbf{f}_1$, with $\mathbf{f}_1 = \begin{bmatrix} -1 \\ 2 \\ 2 \end{bmatrix}$. If $\mathbf{x} = \begin{bmatrix} x_1 \\ x_2 \\ x_3 \end{bmatrix}$ is an

eigenvector for $\lambda = 6$, then $(A - \lambda I)\mathbf{x} = \mathbf{0}$. Gaussian elimination proceeds

$$\begin{bmatrix} -1 & 2 & 2 \\ 2 & -4 & -4 \\ 2 & -4 & -4 \end{bmatrix} \to \begin{bmatrix} 1 & -2 & -2 \\ 0 & 0 & 0 \\ 0 & 0 & 0 \end{bmatrix},$$

so $x_3 = t$ and $x_2 = s$ are free, $x_1 = 2x_2 + 2x_3 = 2t + 2s$, and $\mathbf{x} = \begin{bmatrix} 2t+2s \\ s \\ t \end{bmatrix} = t\mathbf{u} + s\mathbf{v}$, with $\mathbf{u} = \begin{bmatrix} 2 \\ 0 \\ 1 \end{bmatrix}$ and $\mathbf{v} = \begin{bmatrix} 2 \\ 1 \\ 0 \end{bmatrix}$.

These are orthogonal to $\mathbf{f}_1$ (as they must be), but not to each other. So we replace with $\mathbf{f}_2 = \mathbf{u}$ and

$$\mathbf{f}_3 = \mathbf{v} - \text{proj}_{\mathbf{u}}\,\mathbf{v} = \begin{bmatrix} 2 \\ 1 \\ 0 \end{bmatrix} - \frac{4}{5}\begin{bmatrix} 2 \\ 0 \\ 1 \end{bmatrix} = \begin{bmatrix} \frac{2}{5} \\ 1 \\ -\frac{4}{5} \end{bmatrix},$$ which we multiply by 5, giving $\mathbf{f}_3 = \begin{bmatrix} 2 \\ 5 \\ -4 \end{bmatrix}$. Now $\mathbf{f}_1, \mathbf{f}_2, \mathbf{f}_3$ are

orthogonal. Dividing each by its length gives orthonormal vectors that we use as the columns of an orthogonal

matrix $Q = \begin{bmatrix} \frac{1}{3} & \frac{2}{\sqrt{5}} & \frac{2}{3\sqrt{5}} \\ \frac{2}{3} & 0 & \frac{4}{3\sqrt{5}} \\ \frac{2}{3} & \frac{1}{\sqrt{5}} & -\frac{4}{3\sqrt{5}} \end{bmatrix}$. Then $Q^{-1}AQ = D = \begin{bmatrix} -3 & 0 & 0 \\ 0 & 6 & 0 \\ 0 & 0 & 6 \end{bmatrix}$.

2. (a) **i.** The characteristic polynomial of A is

$$\begin{bmatrix} 3-\lambda & 0 & -1 \\ 0 & -1-\lambda & 0 \\ -1 & 0 & 3-\lambda \end{bmatrix} = -(-1-\lambda)(9-6\lambda+\lambda^2-1)$$

$$= -(\lambda+1)(\lambda^2-6\lambda+8) = -(\lambda+1)(\lambda-2)(\lambda-4)$$

(expanding by cofactors along the second row). So the eigenvalues of A are $\lambda = -1, 2, 4$. If $\mathbf{x} = \begin{bmatrix} x_1 \\ x_2 \\ x_3 \end{bmatrix}$ is an

eigenvector for $\lambda = -1$, then $(A - \lambda I)\mathbf{x} = \mathbf{0}$, so

$$\begin{bmatrix} 4 & 0 & -1 \\ 0 & 0 & 0 \\ -1 & 0 & 4 \end{bmatrix} \rightarrow \begin{bmatrix} 1 & 0 & -4 \\ 0 & 0 & 11 \\ 0 & 0 & 0 \end{bmatrix} \rightarrow \begin{bmatrix} 1 & 0 & -4 \\ 0 & 0 & 1 \\ 0 & 0 & 0 \end{bmatrix},$$

so $x_2 = t$ is free, $x_3 = 0$, $x_1 = 4x_3 = 0$, and $\mathbf{x} = \begin{bmatrix} 0 \\ t \\ 0 \end{bmatrix} = t\mathbf{f}_1$, with $\mathbf{f}_1 = \begin{bmatrix} 0 \\ 1 \\ 0 \end{bmatrix}$. If $\mathbf{x} = \begin{bmatrix} x_1 \\ x_2 \\ x_3 \end{bmatrix}$ is an eigenvector for

$\lambda = 2$, then $(A - \lambda I)\mathbf{x} = \mathbf{0}$. Using Gaussian elimination

$$\begin{bmatrix} 1 & 0 & -1 \\ 0 & -3 & 0 \\ -1 & 0 & 1 \end{bmatrix} \rightarrow \begin{bmatrix} 1 & 0 & -1 \\ 0 & 1 & 0 \\ 0 & 0 & 0 \end{bmatrix},$$

so $x_3 = t$ is free, $x_2 = 0$, $x_1 = x_3 = t$, and $\mathbf{x} = \begin{bmatrix} t \\ 0 \\ t \end{bmatrix} = t\mathbf{f}_2$, with $\mathbf{f}_2 = \begin{bmatrix} 1 \\ 0 \\ 1 \end{bmatrix}$. If $\mathbf{x} = \begin{bmatrix} x_1 \\ x_2 \\ x_3 \end{bmatrix}$ is an eigenvector for

$\lambda = 4$, then $(A - \lambda I)\mathbf{x} = \mathbf{0}$. Using Gaussian elimination

$$\begin{bmatrix} -1 & 0 & -1 \\ 0 & -5 & 0 \\ -1 & 0 & -1 \end{bmatrix} \rightarrow \begin{bmatrix} 1 & 0 & 1 \\ 0 & 1 & 0 \\ 0 & 0 & 0 \end{bmatrix},$$

so $x_3 = t$ is free, $x_2 = 0$, $x_1 = -x_3 = -t$, and $\mathbf{x} = \begin{bmatrix} -t \\ 0 \\ t \end{bmatrix} = t\mathbf{f}_3$, with $\mathbf{f}_3 = \begin{bmatrix} -1 \\ 0 \\ 1 \end{bmatrix}$. Notice that $\mathbf{f}_1, \mathbf{f}_2, \mathbf{f}_3$ are

orthogonal as they must be (they are eigenvectors of a symmetric matrix corresponding to different eigenvalues). Dividing each by its length gives an orthonormal basis of eigenvectors. Using these as the

columns of a matrix gives an orthogonal matrix $Q = \begin{bmatrix} 0 & \frac{1}{\sqrt{2}} & -\frac{1}{\sqrt{2}} \\ 1 & 0 & 0 \\ 0 & \frac{1}{\sqrt{2}} & \frac{1}{\sqrt{2}} \end{bmatrix}$ and $Q^{-1}AQ = D = \begin{bmatrix} -1 & 0 & 0 \\ 0 & 2 & 0 \\ 0 & 0 & 4 \end{bmatrix}$.

ii. Since $Q^{-1}AQ = D$, $A = QDQ^{-1}$ and $A^5 = QD^5Q^{-1} = QD^5Q^T$

$$= \begin{bmatrix} 0 & \frac{1}{\sqrt{2}} & -\frac{1}{\sqrt{2}} \\ 1 & 0 & 0 \\ 0 & \frac{1}{\sqrt{2}} & \frac{1}{\sqrt{2}} \end{bmatrix} \begin{bmatrix} -1 & 0 & 0 \\ 0 & 32 & 0 \\ 0 & 0 & 1024 \end{bmatrix} \begin{bmatrix} 0 & 1 & 0 \\ \frac{1}{\sqrt{2}} & 0 & \frac{1}{\sqrt{2}} \\ -\frac{1}{\sqrt{2}} & 0 & \frac{1}{\sqrt{2}} \end{bmatrix} = \begin{bmatrix} 528 & 0 & -496 \\ 0 & -1 & 0 \\ -496 & 0 & 528 \end{bmatrix}.$$

3. (a) The characteristic polynomial of A is

$$\begin{vmatrix} 1 - \lambda & -i \\ i & 1 - \lambda \end{vmatrix} = 1 - 2\lambda + \lambda^2 - 1 = \lambda(\lambda - 2),$$

so the eigenvalues of A are $\lambda = 0, 2$. If $\mathbf{x} = \begin{bmatrix} x_1 \\ x_2 \end{bmatrix}$ is an eigenvector for $\lambda = 0$, then $(A - \lambda I)\mathbf{x} = \mathbf{0}$. Using Gaussian elimination,

$$\begin{bmatrix} 1 & -i \\ i & 1 \end{bmatrix} \to \begin{bmatrix} 1 & -i \\ 0 & 0 \end{bmatrix},$$

so $x_2 = t$ is free, $x_1 = it$, and $\mathbf{x} = \begin{bmatrix} it \\ t \end{bmatrix} = t\mathbf{f}_1$, where $\mathbf{f}_1 = \begin{bmatrix} i \\ 1 \end{bmatrix}$. If $\mathbf{x} = \begin{bmatrix} x_1 \\ x_2 \end{bmatrix}$ is an eigenvector for $\lambda = 4$, then $(A - \lambda I)\mathbf{x} = \mathbf{0}$. Using Gaussian elimination,

$$\begin{bmatrix} -1 & -i \\ i & -1 \end{bmatrix} \to \begin{bmatrix} 1 & i \\ 0 & 0 \end{bmatrix},$$

so $x_2 = t$ is free, $x_1 = -it$, and $\mathbf{x} = \begin{bmatrix} -it \\ t \end{bmatrix} = t\mathbf{f}_2$, where $\mathbf{f}_2 = \begin{bmatrix} -i \\ 1 \end{bmatrix}$. Dividing each of $\mathbf{f}_1, \mathbf{f}_2$ by its length and using these unit vectors as the columns of U, we have $U = \begin{bmatrix} \frac{i}{\sqrt{2}} & -\frac{i}{\sqrt{2}} \\ \frac{1}{\sqrt{2}} & \frac{1}{\sqrt{2}} \end{bmatrix}$ and $U^*AU = D = \begin{bmatrix} 0 & 0 \\ 0 & 2 \end{bmatrix}$.

4. **(a)** $\det(A - \lambda I) = (2 - \lambda)(1 - \lambda) + 3 = \lambda^2 - 3\lambda + 5$ has no real roots, so A is not diagonalizable over **R**. Over **C**, however, the characteristic polynomial has the distinct roots $\lambda = \frac{3 \pm i\sqrt{11}}{2}$, so A is diagonalizable.

5. Since $A^T A$ is a real symmetric matrix, there exists an orthogonal matrix Q such that $Q^T A^T A Q = D$ is diagonal. Thus $-Q^T A^2 Q = D$. Since $Q^T = Q^{-1}$, $Q^T A^2 Q = Q^{-1} A^2 Q = Q^{-1} A Q Q^{-1} A Q = (Q^{-1} A Q)^2 = (Q^T A Q)^2$. Let $B = Q^T A Q$. Then $B^2 = A^T A^2 A = -D$ is diagonal and $A = QBQ^{-1} = QBQ^T$.

7. **(a)** We mimic the proof of Schur's Theorem.

The characteristic polynomial of A is $\begin{vmatrix} 2-\lambda & 2 \\ -1 & -\lambda \end{vmatrix} = \lambda^2 - 2\lambda + 2$, so the eigenvalues of A are $\lambda = 1 \pm i$. If $\mathbf{x} = \begin{bmatrix} x_1 \\ x_2 \end{bmatrix}$ is an eigenvector for $\lambda = 1 + i$, $(A - \lambda I)\mathbf{x} = \mathbf{0}$. Gaussian elimination proceeds

$$\begin{bmatrix} 1-i & 2 \\ -1 & -1-i \end{bmatrix} \to \begin{bmatrix} 1 & 1+i \\ 0 & 0 \end{bmatrix},$$

so $x_2 = t$ is free, $x_1 = -(1+i)t$, and $\mathbf{x} = t\mathbf{z}_1$, with $\mathbf{z}_1 = \begin{bmatrix} -(1+i) \\ 1 \end{bmatrix}$. By inspection, $\mathbf{z}_2 = \begin{bmatrix} 1+i \\ 2 \end{bmatrix}$ is orthogonal to $\mathbf{z}_1$, so dividing each of these vectors by its length and using as the columns of a matrix, we obtain a unitary matrix $U_1 = \begin{bmatrix} -\frac{1+i}{\sqrt{3}} & \frac{1+i}{\sqrt{6}} \\ \frac{1}{\sqrt{3}} & \frac{2}{\sqrt{6}} \end{bmatrix}$ and $U_1^{-1}AU_1 = U^*AU_1 = \begin{bmatrix} 1+i & -\frac{3}{\sqrt{2}} + \frac{i}{\sqrt{2}} \\ 0 & 1-i \end{bmatrix} = T$ is upper triangular.

9. Let $A = \begin{bmatrix} a & b \\ b & c \end{bmatrix}$ be a 2×2 symmetric matrix with real entries. The characteristic polynomial of A is

$$\begin{vmatrix} a-\lambda & b \\ b & c-\lambda \end{vmatrix} = (a-\lambda)(c-\lambda) - b^2 = \lambda^2 - (a+c)\lambda + (ac - b^2).$$

The roots of this polynomial are

$$\lambda = \frac{(a+c) \pm \sqrt{(a+c)^2 - 4(ac - b^2)}}{2} = \frac{(a+c) \pm \sqrt{(a-c)^2 + 4b^2}}{2}.$$

Since the quantity under the square root sign is the sum of two squares, it is nonnegative, so the square root is a real number and so is λ.

11. (a) The characteristic polynomial of this 3×3 matrix is $-(\lambda - 1)(\lambda^2 - \lambda - 1)$ has three different roots. So the matrix is diagonalizable over **C**.

13. Let A be Hermitian. Then $A^* = A$, so $A^*A = A^2 = AA^*$.

15. Remember that $\mathbf{z} \cdot \mathbf{w} = \mathbf{z}^*\mathbf{w}$ for complex vectors $\mathbf{z}$ and $\mathbf{w}$. Thus, following the hint, we have

$$(A^*\mathbf{z}) \cdot (A^*\mathbf{z}) = (A^*\mathbf{z})^* A^*\mathbf{z} = \mathbf{z}^*(A^*)^* A^*\mathbf{z} = \mathbf{z}^* AA^*\mathbf{z} = \mathbf{z}^* A^* A\mathbf{z}$$

using $AA^* = A^*A$. Thus $(A^*\mathbf{z}) \cdot (A^*\mathbf{z}) = (A\mathbf{z}) \cdot (A\mathbf{z})$. It follows that $(A^*\mathbf{z}) \cdot (A^*\mathbf{z}) = 0$ if and only if $(A\mathbf{z}) \cdot (A\mathbf{z}) = 0$. Now $\mathbf{z} \in \text{null sp } A$ if and only if $A\mathbf{z} = \mathbf{0}$, and this occurs if and only if $\|A\mathbf{z}\|^2 = (A\mathbf{z}) \cdot (A\mathbf{z}) = 0$. Similarly, $\mathbf{z} \in \text{null sp } A^*$ if and only if $(A^*\mathbf{z}) \cdot (A^*\mathbf{z}) = 0$. So we have the desired result.

17. (a) Let $\mathbf{x}$ be an eigenvector corresponding to λ. Then $\mathbf{x}^T A\mathbf{x} = \mathbf{x}^T(\lambda\mathbf{x}) = \lambda \|\mathbf{x}\|^2$. Since $\mathbf{x} \neq \mathbf{0}$, $\|\mathbf{x}\| \neq 0$, so this number is positive. Now the fact that $\lambda \|\mathbf{x}\|^2 > 0$ implies $\lambda > 0$.

(b) There exists an orthogonal matrix Q such that $Q^{-1}AQ = D$ is a diagonal matrix. The diagonal entries of D are the eigenvalues of A, so these are positive, hence each is the square of a real number. If

$$D = \begin{bmatrix} \lambda_1 & 0 & \cdots & 0 \\ 0 & \lambda_2 & \cdots & 0 \\ \vdots & \vdots & \ddots & \vdots \\ 0 & 0 & \cdots & \lambda_n \end{bmatrix} = \begin{bmatrix} \mu_1^2 & 0 & \cdots & 0 \\ 0 & \mu_2^2 & \cdots & 0 \\ \vdots & \vdots & \ddots & \vdots \\ 0 & 0 & \cdots & \mu_n^2 \end{bmatrix}, \text{ then } D = E^2 \text{ with } E = \begin{bmatrix} \mu_1 & 0 & \cdots & 0 \\ 0 & \mu_2 & \cdots & 0 \\ \vdots & \vdots & \ddots & \vdots \\ 0 & 0 & \cdots & \mu_n \end{bmatrix}, \text{ so}$$

$A = QDQ^{-1} = QE^2Q^{-1} = (QEQ^{-1})(QEQ^{-1}) = B^2$, with $B = QEQ^{-1}$.

(c) There exists an orthogonal matrix Q such that $Q^{-1}AQ = D$ is a diagonal matrix with positive entries on the diagonal. As in (b), $D = E^2$ for some matrix E, so $A = QDQ^{-1} = QE^2Q^T = (QE)(EQ^T)$. This is RR^T with $R = QE$ since $R^T = (QE)^T = E^T Q^T = EQ^T$ (because E is diagonal, implying $E^T = E$).

19. Given a square matrix A, apply Schur's Theorem to A^*. Thus we have a unitary matrix U and an upper triangular matrix T such that $U^{-1}A^*U = T$. So $U^*A^*U = T$. Applying the conjugate transpose, $U^*AU = T^*$, and T^* is lower triangular.

21. The given polynomial is the characteristic polynomial of $A + B$. Since A and B are symmetric, so is $A + B$: $(A + B)^T = A^T + B^T = A + B$. Thus the characteristic polynomial of $A + B$ has real roots.

23. (a) There exists an orthogonal matrix Q and a diagonal matrix D with nonnegative diagonal entries such that $Q^T AQ = D$. Following the hint, we note that D has a square root; that is, $D = D_1^2$ for some diagonal matrix D_1. In fact, if the diagonal entries of D are $\lambda_1, \lambda_2, \ldots, \lambda_n$, let D_1 be the matrix whose diagonal entries are $\sqrt{\lambda_1}, \sqrt{\lambda_2}, \ldots, \sqrt{\lambda_n}$. We have $A = QDQ^T = QD_1^2Q^T = (QD_1)(D_1Q^T)$. Let $R = QD_1$. Then $R^T = D_1^T Q^T = D_1 Q^T$ (because D_1 is diagonal). So $A = RR^T$ as required.

Section 7.4—True/False

1. True.

2. True. Since Q is orthogonal, $Q^{-1} = Q^T$.

3. False. Let A be any real symmetric matrix that is not diagonal. Then there exists a diagonal matrix and an orthogonal matrix Q_1 such that $Q_1^{-1}AQ_1 = D$. Then $A = Q_1DQ_1^{-1}$. Letting $Q = Q_1^{-1}$, $Q^{-1}DQ = Q^TDQ$ is not diagonal.

4. False. The eigenvalues of the upper triangular matrix $A = \begin{bmatrix} 1 & 2 \\ 0 & 3 \end{bmatrix}$ are real (its diagonal entries), but A is not symmetric.

5. True. The hypothesis implies that $Q^TAQ = D$ is diagonal, where Q is the orthogonal matrix whose columns are the given vectors divided by their lengths. So $A = QDQ^T$ is symmetric. [Check this!]

6. True. This is one of the most useful theorems in linear algebra.

Exercises 7.4

1. (a) The characteristic polynomial of A is $\begin{vmatrix} 2 - \lambda & 3 \\ 3 & -6 - \lambda \end{vmatrix} = \lambda^2 + 4\lambda - 21 = (\lambda + 7)(\lambda - 3)$. The eigenvalues $\lambda = 3, -7$ are real. If $\mathbf{x} = \begin{bmatrix} x_1 \\ x_2 \end{bmatrix}$ is an eigenvector for $\lambda = 3$, Gaussian elimination applied to $A - \lambda I$, with $\lambda = 3$, gives

$$\begin{bmatrix} -1 & 3 \\ 3 & -9 \end{bmatrix} \rightarrow \begin{bmatrix} 1 & -3 \\ 0 & 0 \end{bmatrix},$$

so $x_2 = t$ is free, $x_1 = 3t$, and $\mathbf{x} = \begin{bmatrix} 3t \\ t \end{bmatrix} = t\begin{bmatrix} 3 \\ 1 \end{bmatrix}$. If $\mathbf{x} = \begin{bmatrix} x_1 \\ x_2 \end{bmatrix}$ is an eigenvector for $\lambda = -7$, Gaussian elimination applied to $A - \lambda I$, with $\lambda = -7$, proceeds

$$\begin{bmatrix} 9 & 3 \\ 3 & 1 \end{bmatrix} \rightarrow \begin{bmatrix} 1 & 3 \\ 0 & 0 \end{bmatrix},$$

so $x_2 = t$ is free, $x_1 = -\frac{1}{3}t$, and $\mathbf{x} = \begin{bmatrix} -\frac{1}{3}t \\ t \end{bmatrix} = t\begin{bmatrix} -\frac{1}{3} \\ 1 \end{bmatrix}$. The eigenspaces are orthogonal with bases $\mathbf{f}_1 = \begin{bmatrix} 3 \\ 1 \end{bmatrix}$ and $\mathbf{f}_2 = \begin{bmatrix} -1 \\ 3 \end{bmatrix}$. Let $Q = \begin{bmatrix} \frac{3}{\sqrt{10}} & -\frac{1}{\sqrt{10}} \\ \frac{1}{\sqrt{10}} & \frac{3}{\sqrt{10}} \end{bmatrix}$ be the matrix whose columns are the orthonormal eigenvectors $\mathbf{q}_1 = \frac{1}{\|\mathbf{f}_1\|}$ and $\mathbf{q}_2 = \frac{1}{\|\mathbf{f}_2\|}$. Then $Q^TAQ = D = \begin{bmatrix} 3 & 0 \\ 0 & -7 \end{bmatrix}$.

(c) The characteristic polynomial of A is

$$\begin{vmatrix} -2 - \lambda & 1 & 1 \\ 1 & -2 - \lambda & 1 \\ 1 & 1 & -2 - \lambda \end{vmatrix} = (-2 - \lambda)(4 + 4\lambda + \lambda^2 - 1) - (-2 - \lambda - 1) + (1 + 2 + \lambda)$$

$$= -(\lambda + 2)(\lambda^2 + 4\lambda + 3) + \lambda + 3 + \lambda + 3$$

$$= -(\lambda + 2)(\lambda + 1)(\lambda + 3) + 2(\lambda + 3)$$

$$= -(\lambda + 3)((\lambda^2 + 3\lambda + 2 - 2) = -\lambda(\lambda + 3)^2$$

(expanding by cofactors along the first row). So the eigenvalues of A are $\lambda = 0, -3$. If $\mathbf{x} = \begin{bmatrix} x_1 \\ x_2 \\ x_3 \end{bmatrix}$ is an eigenvector

for $\lambda = 0$, then $(A - \lambda I)\mathbf{x} = \mathbf{0}$. Gaussian elimination proceeds

$$\begin{bmatrix} -2 & 1 & 1 \\ 1 & -2 & 1 \\ 1 & 1 & -2 \end{bmatrix} \rightarrow \begin{bmatrix} 1 & 1 & -2 \\ 0 & 3 & -3 \\ 0 & -3 & 3 \end{bmatrix} \rightarrow \begin{bmatrix} 1 & 1 & -2 \\ 0 & 1 & -1 \\ 0 & 0 & 0 \end{bmatrix},$$

so $x_3 = t$ is free, $x_2 = x_3 = t$, $x_1 = -x_2 + 2x_3 = -t + 2t = t$, and $\mathbf{x} = \begin{bmatrix} t \\ t \\ t \end{bmatrix} = t\mathbf{f}_1$, with $\mathbf{f}_1 = \begin{bmatrix} 1 \\ 1 \\ 1 \end{bmatrix}$. If $\mathbf{x} = \begin{bmatrix} x_1 \\ x_2 \\ x_3 \end{bmatrix}$ is

an eigenvector for $\lambda = -3$, then $(A - \lambda I)\mathbf{x} = \mathbf{0}$. Gaussian elimination proceeds

$$\begin{bmatrix} 1 & 1 & 1 \\ 1 & 1 & 1 \\ 1 & 1 & 1 \end{bmatrix} \rightarrow \begin{bmatrix} 1 & 1 & 1 \\ 0 & 0 & 0 \\ 0 & 0 & 0 \end{bmatrix},$$

so $x_3 = t$ and $x_2 = s$ are free, $x_1 = -x_2 - x_3 = -t - s$, and $\mathbf{x} = \begin{bmatrix} -t - s \\ s \\ t \end{bmatrix} = t\mathbf{u} + s\mathbf{v}$, with $\mathbf{u} = \begin{bmatrix} -1 \\ 0 \\ 1 \end{bmatrix}$ and

$\mathbf{v} = \begin{bmatrix} -1 \\ 1 \\ 0 \end{bmatrix}$. These are orthogonal to $\mathbf{f}_1$ (as they must be), but not to each other. So we replace with $\mathbf{f}_2 = \mathbf{u}$ and

$\mathbf{f}_3 = \mathbf{v} - \text{proj}_\mathbf{u}\,\mathbf{v} = \begin{bmatrix} -1 \\ 1 \\ 0 \end{bmatrix} - \frac{1}{2}\begin{bmatrix} -1 \\ 0 \\ 1 \end{bmatrix} = \begin{bmatrix} -\frac{1}{2} \\ 1 \\ -\frac{1}{2} \end{bmatrix}$ that we double, giving $\mathbf{f}_3 = -\begin{bmatrix} -1 \\ 2 \\ -1 \end{bmatrix}$. Now $\mathbf{f}_1, \mathbf{f}_2, \mathbf{f}_3$ are orthogonal.

Dividing each by its length gives orthonormal vectors that we use as the columns of an orthogonal matrix

$$Q = \begin{bmatrix} \frac{1}{\sqrt{3}} & -\frac{1}{\sqrt{2}} & -\frac{1}{\sqrt{6}} \\ \frac{1}{\sqrt{3}} & 0 & \frac{2}{\sqrt{6}} \\ \frac{1}{\sqrt{3}} & \frac{1}{\sqrt{2}} & -\frac{1}{\sqrt{6}} \end{bmatrix}. \text{ Then } Q^{-1}AQ = D = \begin{bmatrix} 0 & 0 & 0 \\ 0 & -3 & 0 \\ 0 & 0 & -3 \end{bmatrix}.$$

2. (a)   i. The characteristic polynomial of A is $\begin{vmatrix} 1 - \lambda & 2 \\ 2 & 1 - \lambda \end{vmatrix} = \lambda^2 - 2\lambda - 3 = (\lambda + 1)(\lambda - 3)$, so the eigenvalues of A

are $\lambda = -1, 3$. If $\mathbf{x} = \begin{bmatrix} x_1 \\ x_2 \end{bmatrix}$ is an eigenvector for $\lambda = -1$, then $(A - \lambda I)\mathbf{x} = \mathbf{0}$. Row reduction proceeds

$$\begin{bmatrix} 2 & 2 \\ 2 & 2 \end{bmatrix} \rightarrow \begin{bmatrix} 1 & 1 \\ 0 & 0 \end{bmatrix},$$

so $x_2 = t$ is free, $x_1 = -t$, and $\mathbf{x} = \begin{bmatrix} -t \\ t \end{bmatrix} = t\mathbf{f}_1$, with $\mathbf{f}_1 = \begin{bmatrix} -1 \\ 1 \end{bmatrix}$. If $\mathbf{x} = \begin{bmatrix} x_1 \\ x_2 \end{bmatrix}$ is an eigenvector for $\lambda = 3$, then $(A - \lambda I)\mathbf{x} = \mathbf{0}$. Row reduction proceeds

$$\begin{bmatrix} -2 & 2 \\ 2 & -2 \end{bmatrix} \rightarrow \begin{bmatrix} 1 & -1 \\ 0 & 0 \end{bmatrix},$$

so $x_2 = t$ is free, $x_1 = t$, and $\mathbf{x} = \begin{bmatrix} t \\ t \end{bmatrix} = t\mathbf{f}_2$, with $\mathbf{f}_2 = \begin{bmatrix} 1 \\ 1 \end{bmatrix}$. Notice that $\mathbf{f}_1$ and $\mathbf{f}_2$ are orthogonal as they must be (they are eigenvectors of a symmetric matrix corresponding to different eigenvalues). Dividing each by its length gives an orthonormal basis of eigenvectors. Using these as the columns of a matrix gives an orthogonal

matrix $Q = \begin{bmatrix} -\frac{1}{\sqrt{2}} & \frac{1}{\sqrt{2}} \\ \frac{1}{\sqrt{2}} & \frac{1}{\sqrt{2}} \end{bmatrix}$ and $Q^{-1}AQ = D = \begin{bmatrix} -1 & 0 \\ 0 & 3 \end{bmatrix}$.

ii. Since $Q^{-1}AQ = D$, $A = QDQ^{-1}$, and

$$A^5 = QD^5Q^{-1} = QD^5Q^T$$

$$= \begin{bmatrix} -\frac{1}{\sqrt{2}} & \frac{1}{\sqrt{2}} \\ \frac{1}{\sqrt{2}} & \frac{1}{\sqrt{2}} \end{bmatrix} \begin{bmatrix} -1 & 0 \\ 0 & 3 \end{bmatrix} \begin{bmatrix} -\frac{1}{\sqrt{2}} & \frac{1}{\sqrt{2}} \\ \frac{1}{\sqrt{2}} & \frac{1}{\sqrt{2}} \end{bmatrix} = \begin{bmatrix} 121 & 122 \\ 122 & 121 \end{bmatrix}.$$

3. Suppose $Q^{-1}AQ = D$ is diagonal for some orthogonal matrix Q. Thus $A = QDQ^{-1}$. Since $Q^T = Q^{-1}$, $A = QDQ^T$, so $A^T = (Q^T)^T D^T Q^T = QDQ^T = A$, since the diagonal matrix D is its own transpose.

5. There exists an orthogonal matrix Q and a diagonal matrix D with nonnegative diagonal entries $\lambda_1, \lambda_2, \ldots, \lambda_n$ such

that $Q^T AQ = D$. Thus $A = QDQ^T$ and for any $\mathbf{x}$, $\mathbf{x}^T A\mathbf{x} = \mathbf{x}^T QDQ^T\mathbf{x}$. Set $\mathbf{y} = Q^T\mathbf{x} = \begin{bmatrix} y_1 \\ \vdots \\ y_n \end{bmatrix}$. Then $\mathbf{x}^T Q = \mathbf{y}^T$ and

$\mathbf{x}^T A\mathbf{x} = \mathbf{y}^T D\mathbf{y} = \lambda_1 y_1^2 + \lambda_2 y_2^2 + \cdots + \lambda_n y_n^2$ and this is nonnegative because each $\lambda_i \geq 0$ and each $y_i^2 \geq 0$.

7. The given polynomial is the characteristic polynomial of $A + B$. Since A and B are symmetric, so is $A + B$: $(A + B)^T = A^T + B^T = A + B$. Thus the characteristic polynomial of $A + B$ has real roots.

9. Since $A^T A$ is a real symmetric matrix, there exists an orthogonal matrix Q such that $Q^T A^T AQ = D$ is diagonal. Thus $-Q^T A^2 Q = D$. Since $Q^T = Q^{-1}$, $Q^T A^2 Q = Q^{-1}A^2 Q = Q^{-1}AQQ^{-1}AQ = (Q^{-1}AQ)^2 = (Q^T AQ)^2$. Let $B = Q^T AQ$. Then $B^2 = A^T A^2 A = -D$ is diagonal and $A = QBQ^{-1} = QBQ^T$.

12. (a) There exists an orthogonal matrix Q and a diagonal matrix D with nonnegative diagonal entries such that $Q^T AQ = D$. Following the hint, we note that D has a square root; that is, $D = D_1^2$ for some diagonal matrix D_1. In fact, if the diagonal entries of D are $\lambda_1, \lambda_2, \ldots, \lambda_n$, let D_1 be the matrix whose diagonal entries are $\sqrt{\lambda_1}, \sqrt{\lambda_2}, \ldots, \sqrt{\lambda_n}$. We have $A = QDQ^T = QD_1^2 Q^T = (QD_1)(D_1 Q^T)$. Let $R = QD_1$. Then $R^T = D_1^T Q^T = D_1 Q^T$ (because D_1 is diagonal). So $A = RR^T$ as required.

Section 7.5—True/False

1. True. The norm of a vector is a nonnegative real number.

2. True. Extend $\{\mathbf{u}, \mathbf{v}, \mathbf{w}\}$ to a basis $\{\mathbf{u}, \mathbf{v}, \mathbf{w}, \mathbf{t}\}$ of $\mathbf{R}^4$ and apply the Gram–Schmidt Algorithm to these four vectors.

3. True, since $Q_1^T = Q_1^{-1}$.

4. False. The singular values of A are the **square roots** of the **nonzero** eigenvalues of $A^T A$.

5. True. See Exercise 10 in Section 4.4.

6. False. $A^+ = Q_2\Sigma^+ Q_1^T$.

Exercises 7.5

1. (a) We have $A^T A = \begin{bmatrix} 2 & -2 \\ -2 & 2 \end{bmatrix}$. The vectors $\mathbf{x}_1 = \begin{bmatrix} \frac{1}{\sqrt{2}} \\ -\frac{1}{\sqrt{2}} \end{bmatrix}$ and $\mathbf{x}_2 = \begin{bmatrix} \frac{1}{\sqrt{2}} \\ \frac{1}{\sqrt{2}} \end{bmatrix}$ comprise an orthonormal basis of

eigenvectors of $A^T A$ with corresponding eigenvalues $\lambda_1 = 4$, $\lambda_2 = 0$. Let $\mu_1 = \sqrt{\lambda_1} = 2$ and $\Sigma = \begin{bmatrix} 2 & 0 \\ 0 & 0 \\ 0 & 0 \end{bmatrix}$. Let

$$\mathbf{y}_1 = \frac{1}{\mu_1} A\mathbf{x}_1 = \frac{1}{2} \begin{bmatrix} \frac{2}{\sqrt{2}} \\ 0 \\ \frac{2}{\sqrt{2}} \end{bmatrix} = \begin{bmatrix} \frac{1}{\sqrt{2}} \\ 0 \\ \frac{1}{\sqrt{2}} \end{bmatrix}$$

and extend to an orthonormal basis $\mathbf{y}_1 = \begin{bmatrix} \frac{1}{\sqrt{2}} \\ 0 \\ \frac{1}{\sqrt{2}} \end{bmatrix}$, $\mathbf{y}_2 = \begin{bmatrix} \frac{1}{\sqrt{2}} \\ 0 \\ -\frac{1}{\sqrt{2}} \end{bmatrix}$, $\mathbf{y}_3 = \begin{bmatrix} 0 \\ 1 \\ 0 \end{bmatrix}$ of $\mathbf{R}^3$. Letting $Q_1 = \begin{bmatrix} \frac{1}{\sqrt{2}} & \frac{1}{\sqrt{2}} & 0 \\ 0 & 0 & 1 \\ \frac{1}{\sqrt{2}} & -\frac{1}{\sqrt{2}} & 0 \end{bmatrix}$

and $Q_2 = \begin{bmatrix} \frac{1}{\sqrt{2}} & \frac{1}{\sqrt{2}} \\ -\frac{1}{\sqrt{2}} & \frac{1}{\sqrt{2}} \end{bmatrix}$, a singular value decomposition of A is $A = Q_1 \Sigma Q_2^T$:

$$\begin{bmatrix} 1 & 1 \\ 1 & 0 \\ 0 & 1 \end{bmatrix} = \begin{bmatrix} 0 & \frac{2}{\sqrt{6}} & -\frac{1}{\sqrt{3}} \\ \frac{1}{\sqrt{2}} & \frac{1}{\sqrt{6}} & \frac{1}{\sqrt{3}} \\ -\frac{1}{\sqrt{2}} & \frac{1}{\sqrt{6}} & \frac{1}{\sqrt{3}} \end{bmatrix} \begin{bmatrix} 2 & 0 \\ 0 & 0 \\ 0 & 0 \end{bmatrix} \begin{bmatrix} \frac{1}{\sqrt{2}} & -\frac{1}{\sqrt{2}} \\ \frac{1}{\sqrt{2}} & \frac{1}{\sqrt{2}} \end{bmatrix}.$$

and the pseudoinverse of A is

$$A^+ = Q_2 \Sigma^+ Q_1^T = \begin{bmatrix} \frac{1}{\sqrt{2}} & \frac{1}{\sqrt{2}} \\ -\frac{1}{\sqrt{2}} & \frac{1}{\sqrt{2}} \end{bmatrix} \begin{bmatrix} \frac{1}{2} & 0 & 0 \\ 0 & 0 & 0 \end{bmatrix} \begin{bmatrix} \frac{1}{\sqrt{2}} & 0 & \frac{1}{\sqrt{2}} \\ \frac{1}{\sqrt{2}} & 0 & -\frac{1}{\sqrt{2}} \\ 0 & 1 & 0 \end{bmatrix} = \begin{bmatrix} \frac{1}{4} & 0 & \frac{1}{4} \\ -\frac{1}{4} & 0 & -\frac{1}{4} \end{bmatrix}.$$

(c) We have $A^T A = \begin{bmatrix} 1 & 2 & 3 \\ 2 & 4 & 6 \\ 3 & 6 & 9 \end{bmatrix}$. The vectors $\mathbf{x}_1 = \begin{bmatrix} \frac{1}{\sqrt{14}} \\ \frac{2}{\sqrt{14}} \\ \frac{3}{\sqrt{14}} \end{bmatrix}$, $\mathbf{x}_2 = \begin{bmatrix} -\frac{2}{\sqrt{5}} \\ \frac{1}{\sqrt{5}} \\ 0 \end{bmatrix}$ and $\mathbf{x}_3 = \begin{bmatrix} -\frac{3}{\sqrt{70}} \\ -\frac{6}{\sqrt{70}} \\ \frac{5}{\sqrt{70}} \end{bmatrix}$ comprise an orthonormal

basis of eigenvectors of $A^T A$ with corresponding eigenvalues $\lambda_1 = 14$, $\lambda_2 = 0$ and $\lambda_3 = 0$. Let $\mu_1 = \sqrt{\lambda_1} = \sqrt{14}$

and $\Sigma = \begin{bmatrix} \sqrt{14} & 0 & 0 \end{bmatrix}$. The vector $\mathbf{y}_1 = \frac{1}{\mu_1} A\mathbf{x}_1 = \frac{1}{\sqrt{14}} \begin{bmatrix} 1 & 2 & 3 \end{bmatrix} \begin{bmatrix} \frac{1}{\sqrt{14}} \\ \frac{2}{\sqrt{14}} \\ \frac{3}{\sqrt{14}} \end{bmatrix} = [1]$ is an orthonormal basis of $\mathbf{R}$.

Letting $Q_1 = [1]$ and $Q_2 = \begin{bmatrix} \frac{1}{\sqrt{14}} & -\frac{2}{\sqrt{5}} & -\frac{3}{\sqrt{70}} \\ \frac{2}{\sqrt{14}} & \frac{1}{\sqrt{5}} & -\frac{6}{\sqrt{70}} \\ \frac{3}{\sqrt{14}} & 0 & \frac{5}{\sqrt{70}} \end{bmatrix}$, a singular value decomposition of A is $A = Q_1 \Sigma Q_2^T$:

$$\begin{bmatrix} 1 & 2 & 3 \end{bmatrix} = [1] \begin{bmatrix} \sqrt{14} & 0 & 0 \end{bmatrix} \begin{bmatrix} \frac{1}{\sqrt{14}} & \frac{2}{\sqrt{14}} & \frac{3}{\sqrt{14}} \\ -\frac{2}{\sqrt{5}} & \frac{1}{\sqrt{5}} & 0 \\ -\frac{3}{\sqrt{70}} & -\frac{6}{\sqrt{70}} & \frac{5}{\sqrt{70}} \end{bmatrix}$$

and the pseudoinverse of A is

$$A^+ = Q_2 \Sigma^+ Q_1^T = \begin{bmatrix} \frac{1}{\sqrt{14}} & -\frac{2}{\sqrt{5}} & -\frac{3}{\sqrt{70}} \\ \frac{2}{\sqrt{14}} & \frac{1}{\sqrt{5}} & -\frac{6}{\sqrt{70}} \\ \frac{3}{\sqrt{14}} & 0 & \frac{5}{\sqrt{70}} \end{bmatrix} \begin{bmatrix} \frac{1}{\sqrt{14}} \\ 0 \\ 0 \end{bmatrix} [1] = \frac{1}{14} \begin{bmatrix} 1 \\ 2 \\ 3 \end{bmatrix}.$$

3. Let $A = Q_1 \Sigma Q_2^T$ be a singular value decomposition of A. Then $A^+ = Q_2 \Sigma^+ Q_1^T$, by 7.5.5, so $(A^+)^T = Q_1 (\Sigma^+)^T Q_2^T$. Now $A^T = (Q_1 \Sigma Q_2^T)^T = Q_2 \Sigma^T Q_1^T$ is a singular value decomposition of A^T, so 7.5.5 says that $(A^T)^+ = Q_1 (\Sigma^T)^+ Q_2^T$ and the result will follow if $(\Sigma^+)^T = (\Sigma^T)^+$. This is certainly the case since, with Σ the

$m \times n$ matrix $\Sigma = \begin{bmatrix} \mu_1 & & & & \\ & \ddots & & & \mathbf{0} \\ & & \mu_r & & \\ & \mathbf{0} & & & \mathbf{0} \end{bmatrix}$, we know that Σ^+ is the $n \times m$ matrix $\Sigma^+ = \begin{bmatrix} \frac{1}{\mu_1} & & & & \\ & \ddots & & & \mathbf{0} \\ & & \frac{1}{\mu_r} & & \\ & \mathbf{0} & & & \mathbf{0} \end{bmatrix}$. So

$(\Sigma^+)^T = (\Sigma^T)^+$ is an $m \times n$ matrix that looks like Σ^+.

5. Let $A = Q_1 \Sigma Q_2^T$ be a singular value decomposition of A. The matrix Σ is $m \times n$ and its rank is the number of nonzero diagonal entries. By assumption, rank $\Sigma = n$. Since multiplication of a matrix by an invertible matrix preserves rank, rank $A = n$ too. [See Section 4.4 and also Exercise 10 of that section.]

6. (a) Writing $A = Q_1 \Sigma Q_2^T$, we have $A^+ = Q_2 \Sigma^+ Q_1^T$, so

$$AA^+ A = Q_1 \Sigma Q_2^T Q_2 \Sigma^+ Q_1^T = Q_1 \Sigma \Sigma^+ \Sigma Q_2^T = Q_1 \Sigma Q_2^T = A,$$

since $Q_1^T Q_1 = I$ by orthogonality and $\Sigma \Sigma^+ \Sigma = \Sigma$.

7. Let $\lambda > 0$ be an eigenvalue of A and $\mathbf{x}$ a corresponding eigenvector. Thus $A\mathbf{x} = \lambda \mathbf{x}$ and, since $A^T = A$, we have $A^T A\mathbf{x} = AA\mathbf{x} = A(\lambda \mathbf{x}) = \lambda(A\mathbf{x}) = \lambda^2 \mathbf{x}$. Thus $\lambda^2 > 0$ is an eigenvector of $A^T A$, and $\mu = \sqrt{\lambda^2} = \lambda$ is a singular value of A.

10. (a) From $A = Q_1 \Sigma Q_2^T$, we have $A^T = Q_2 \Sigma^T Q_1^T$, so $A^T A = Q_2 \Sigma^T Q_1^T Q_1 \Sigma Q_2^T = Q_2 \Sigma^T \Sigma Q_2^{-1}$ because Q_1, Q_2 orthogonal implies $Q_1^T = Q_1^{-1}$ and $Q_2^T = Q_2^{-1}$. Hence $A^T A Q_2 = Q_2 \Sigma^T \Sigma$. Now denote $A^T A$ by B and note that the matrix $\Sigma^T \Sigma$ is a diagonal matrix D. We have an equation of the from $BP = PD$, with $P = Q_2$. The result now follows exactly as shown at the start of Section 3.4. We repeat the argument that appears there. Let

$$P = \begin{bmatrix} \mathbf{x}_1 & \mathbf{x}_2 & \cdots & \mathbf{x}_n \\ \downarrow & \downarrow & & \downarrow \end{bmatrix} \text{ and } D = \begin{bmatrix} \lambda_1 & 0 & \cdots & 0 \\ 0 & \lambda_2 & & 0 \\ \vdots & & & \vdots \\ 0 & 0 & \cdots & \lambda_n \end{bmatrix}.$$

Then $BP = PD$ says

$$\begin{bmatrix} B\mathbf{x}_1 & B\mathbf{x}_2 & \cdots & B\mathbf{x}_n \\ \downarrow & \downarrow & & \downarrow \end{bmatrix} = \begin{bmatrix} \lambda_1\mathbf{x}_1 & \lambda_2\mathbf{x}_2 & \cdots & \lambda_n\mathbf{x}_n \\ \downarrow & \downarrow & & \downarrow \end{bmatrix},$$

which says $B\mathbf{x}_i = \lambda_i \mathbf{x}_i$ as claimed.

Exercises 8.1

1. (a) Let the line have equation $y = mx + k$. We want the "best solution" to

$$\begin{aligned} 0 &= & k \\ 1 &= & m + k \\ 12 &= & 3m + k. \end{aligned}$$

This is $A\mathbf{x} = \mathbf{b}$ with $A = \begin{bmatrix} 1 & 0 \\ 1 & 1 \\ 1 & 3 \end{bmatrix}$, $\mathbf{x} = \begin{bmatrix} k \\ m \end{bmatrix}$ and $\mathbf{b} = \begin{bmatrix} 0 \\ 1 \\ 12 \end{bmatrix}$. We have $A^T A = \begin{bmatrix} 3 & 4 \\ 4 & 10 \end{bmatrix}$ and the best solution to $A\mathbf{x} = \mathbf{b}$ is

$$\mathbf{x}^+ = \begin{bmatrix} k^+ \\ m^+ \end{bmatrix} = (A^T A)^{-1} A^T \mathbf{b} = \frac{1}{14} \begin{bmatrix} 10 & -4 \\ -4 & 3 \end{bmatrix} \begin{bmatrix} 1 & 1 & 1 \\ 0 & 1 & 3 \end{bmatrix} \begin{bmatrix} 0 \\ 1 \\ 12 \end{bmatrix} = \frac{1}{14} \begin{bmatrix} -18 \\ 59 \end{bmatrix}.$$

Thus $k^+ = -\frac{9}{7}$, $m^+ = \frac{59}{14}$ and our "best" line has equation $y = \frac{59}{14}x - \frac{9}{7}$.

(c) Let the line have equation $y = mx + k$. We want the "best solution" to

$$\begin{aligned} 1 &= -1m + k \\ 0 &= & k \\ 1 &= & m + k \\ 0 &= & 2m + k \\ -8 &= & 3m + k. \end{aligned}$$

This is, $A\mathbf{x} = \mathbf{b}$, with $A = \begin{bmatrix} 1 & -1 \\ 1 & 0 \\ 1 & 1 \\ 1 & 2 \\ 1 & 3 \end{bmatrix}$, $\mathbf{x} = \begin{bmatrix} k \\ m \end{bmatrix}$ and $\mathbf{b} = \begin{bmatrix} 1 \\ 0 \\ 1 \\ 0 \\ -8 \end{bmatrix}$. We have $A^T A = \begin{bmatrix} 5 & 5 \\ 5 & 15 \end{bmatrix}$ and the best solution to

$A\mathbf{x} = \mathbf{b}$ is

$$\mathbf{x}^+ = \begin{bmatrix} k^+ \\ m^+ \end{bmatrix} = (A^T A)^{-1} A^T \mathbf{b} = \frac{1}{10} \begin{bmatrix} 3 & -1 \\ -1 & 1 \end{bmatrix} \begin{bmatrix} 1 & 1 & 1 & 1 & 1 \\ -1 & 0 & 1 & 2 & 3 \end{bmatrix} \begin{bmatrix} 1 \\ 0 \\ 1 \\ 0 \\ -8 \end{bmatrix} = \frac{1}{5} \begin{bmatrix} 3 \\ -9 \end{bmatrix}.$$

Thus $k^+ = 3$, $m^+ = -9$, and our "best" line has equation $y = -\frac{9}{5}x + \frac{3}{5}$.

2. (a) Let the parabola have equation $y = rx^2 + sx + t$. We want to solve

$$\begin{aligned} t &= 0 \\ r + s + t &= 1 \\ r - s + t &= 2 \\ 4r + 2s + t &= 2. \end{aligned}$$

This is $A\mathbf{x} = \mathbf{b}$ with $A = \begin{bmatrix} 1 & 0 & 0 \\ 1 & 1 & 1 \\ 1 & -1 & 1 \\ 1 & 2 & 4 \end{bmatrix}$, $\mathbf{x} = \begin{bmatrix} t \\ s \\ r \end{bmatrix}$, $\mathbf{b} = \begin{bmatrix} 0 \\ 1 \\ 2 \\ 2 \end{bmatrix}$. We have $A^T A = \begin{bmatrix} 4 & 2 & 6 \\ 2 & 6 & 8 \\ 6 & 8 & 18 \end{bmatrix}$. The best solution is

$$\mathbf{x}^+ = (A^T A)^{-1} A^T \mathbf{b} = \frac{1}{20} \begin{bmatrix} 11 & 3 & -5 \\ 3 & 9 & -5 \\ -5 & -5 & 5 \end{bmatrix} \begin{bmatrix} 1 & 1 & 1 & 1 \\ 0 & 1 & -1 & 2 \\ 0 & 1 & 1 & 4 \end{bmatrix} \begin{bmatrix} 0 \\ 1 \\ 2 \\ 2 \end{bmatrix} = \frac{1}{20} \begin{bmatrix} 9 \\ -13 \\ 15 \end{bmatrix}$$

and the best parabola has equation $y = \frac{3}{4}x^2 - \frac{13}{20}x + \frac{9}{20}$.

3. (a) We want to solve

$$\begin{aligned} (-2)^3 p + (-2)^2 q - 2r + s &= 0 \\ (-1)^3 p + (-1)^2 q - 1r + s &= -\tfrac{1}{2} \\ 0^3 p + 0^2 q + 0r + s &= 1 \\ 1^3 p + 1^2 q + 1r + s &= \tfrac{1}{2} \\ 2^3 p + 2^2 q + 2r + s &= -1 \end{aligned}$$

This is $A\mathbf{x} = \mathbf{b}$ with $A = \begin{bmatrix} 1 & -2 & 4 & -8 \\ 1 & -1 & 1 & -1 \\ 1 & 0 & 0 & 0 \\ 1 & 1 & 1 & 1 \\ 1 & 2 & 4 & 8 \end{bmatrix}$, $\mathbf{x} = \begin{bmatrix} s \\ r \\ q \\ p \end{bmatrix}$, $\mathbf{b} = \begin{bmatrix} 0 \\ -\tfrac{1}{2} \\ 1 \\ \tfrac{1}{2} \\ -1 \end{bmatrix}$. We have $A^T A = \begin{bmatrix} 5 & 0 & 10 & 0 \\ 0 & 10 & 0 & 34 \\ 10 & 0 & 34 & 0 \\ 0 & 34 & 0 & 130 \end{bmatrix}$.

The best solution is

$$\mathbf{x}^+ = (A^T A)^{-1} A^T \mathbf{b} = \begin{bmatrix} \frac{17}{35} & 0 & -\frac{1}{7} & 0 \\ 0 & \frac{65}{72} & 0 & -\frac{17}{72} \\ -\frac{1}{7} & 0 & \frac{1}{14} & 0 \\ 0 & -\frac{17}{72} & 0 & \frac{5}{72} \end{bmatrix} \begin{bmatrix} 1 & 1 & 1 & 1 & 1 \\ -2 & -1 & 0 & 1 & 2 \\ 4 & 1 & 0 & 1 & 4 \\ -8 & -1 & 0 & 1 & 8 \end{bmatrix} \begin{bmatrix} 0 \\ -\frac{1}{2} \\ 1 \\ \frac{1}{2} \\ -1 \end{bmatrix} = \begin{bmatrix} \frac{4}{7} \\ \frac{3}{4} \\ -\frac{2}{7} \\ -\frac{1}{4} \end{bmatrix},$$

so the best cubic is $f(x) = -\frac{1}{4}x^3 - \frac{2}{7}x^2 + \frac{3}{4}x + \frac{4}{7}$.

5. (a) Ideally, we should have $f(-1) = 1$, $f(0) = 0$ and $f(2) = 1$. Thus we want the "best solution," in the sense of least squares, to the system

$$-a + \tfrac{1}{2}b = 1$$
$$b = 0$$
$$2a + 4b = -1.$$

This is $A\mathbf{x} = \mathbf{b}$ with $\mathbf{x} = \begin{bmatrix} a \\ b \end{bmatrix}$, $\mathbf{b} = \begin{bmatrix} 1 \\ 0 \\ -1 \end{bmatrix}$ and $A = \begin{bmatrix} -1 & \frac{1}{2} \\ 0 & 1 \\ 2 & 4 \end{bmatrix}$. We have $A^T A = \begin{bmatrix} 5 & \frac{15}{2} \\ \frac{15}{2} & \frac{69}{4} \end{bmatrix}$. The best solution is

$$\mathbf{x}^+ = (A^T A)^{-1} A^T \mathbf{b} = \begin{bmatrix} \frac{23}{40} & -\frac{1}{4} \\ -\frac{1}{4} & \frac{1}{6} \end{bmatrix} \begin{bmatrix} -1 & 0 & 2 \\ \frac{1}{2} & 1 & 4 \end{bmatrix} \begin{bmatrix} 1 \\ 0 \\ -1 \end{bmatrix} = \begin{bmatrix} -\frac{17}{20} \\ \frac{1}{6} \end{bmatrix}.$$

so the best function is $f(x) = -\frac{17}{20}x + \frac{1}{6}2^x \approx 0.85x + 0.17(2^x)$.

7. (a) We want the "best solution," in the sense of least squares, to the system

$$\tfrac{\sqrt{3}}{2}a - \tfrac{\sqrt{3}}{2}b = -4$$
$$\tfrac{1}{2}a + \tfrac{\sqrt{3}}{2}b = 2$$
$$\tfrac{1}{2}a - \tfrac{\sqrt{3}}{2}b = -1.$$

This is $A\mathbf{x} = \mathbf{b}$ with $\mathbf{x} = \begin{bmatrix} a \\ b \end{bmatrix}$, $\mathbf{b} = \begin{bmatrix} -4 \\ 2 \\ -1 \end{bmatrix}$ and $A = \begin{bmatrix} -\frac{\sqrt{3}}{2} & -\frac{\sqrt{3}}{2} \\ \frac{1}{2} & \frac{\sqrt{3}}{2} \\ \frac{1}{2} & -\frac{\sqrt{3}}{2} \end{bmatrix}$. We have $A^T A = \begin{bmatrix} \frac{5}{4} & \frac{3}{4} \\ \frac{3}{4} & \frac{9}{4} \end{bmatrix}$. The best solution is

$$\mathbf{x}^+ = (A^T A)^{-1} A^T \mathbf{b} = \begin{bmatrix} 1 & -\frac{1}{3} \\ -\frac{1}{3} & \frac{5}{9} \end{bmatrix} \begin{bmatrix} -\frac{\sqrt{3}}{2} & \frac{1}{2} & \frac{1}{2} \\ -\frac{\sqrt{3}}{2} & \frac{\sqrt{3}}{2} & -\frac{\sqrt{3}}{2} \end{bmatrix} \begin{bmatrix} -4 \\ 2 \\ -1 \end{bmatrix} = \begin{bmatrix} \frac{7\sqrt{3}+9}{6} \\ \frac{13\sqrt{3}-9}{18} \end{bmatrix}.$$

so the best curve has equation $y = \frac{7\sqrt{3}+9}{6} \sin x + \frac{13\sqrt{3}-9}{18} \sin 2x \approx 3.52 \sin x + 0.75 \sin 2x$.

Exercises 8.2

1. (a) The first eight terms are $a_0, a_1, \ldots, a_7$:

$$a_0 = 0$$

$$a_1 = 1$$

$$a_2 = a_1 + 2a_0 = 1 \qquad\qquad \text{setting } n = 1 \text{ in } a_{n+1} = a_n + 2a_{n-1}$$

$$a_3 = a_2 + 2a_1 = 3 \qquad\qquad \text{setting } n = 2$$

$$a_4 = a_3 + 2a_2 = 5 \qquad\qquad \text{setting } n = 3$$

$$a_5 = a_4 + 2a_3 = 11 \qquad\qquad \text{setting } n = 4$$

$$a_6 = a_5 + 2a_4 = 21 \qquad\qquad \text{setting } n = 5$$

$$a_7 = a_6 + 2a_5 = 43 \qquad\qquad \text{setting } n = 6$$

(b) Let $\mathbf{v}_n = \begin{bmatrix} a_{n+1} \\ a_n \end{bmatrix}$. Then $\mathbf{v}_n = A^n \mathbf{v}_0$ where $A = \begin{bmatrix} 1 & 2 \\ 1 & 0 \end{bmatrix}$. The eigenvalues of A are -1 and 2 with corresponding

eigenvectors, respectively, $\begin{bmatrix} -1 \\ 1 \end{bmatrix}$ and $\begin{bmatrix} 2 \\ 1 \end{bmatrix}$. Let $P = \begin{bmatrix} -1 & 2 \\ 1 & 1 \end{bmatrix}$. Then $P^{-1}AP = D = \begin{bmatrix} -1 & 0 \\ 0 & 2 \end{bmatrix}$ and

$\mathbf{v}_n = A^n \mathbf{v}_0 = (PDP^{-1})^n \mathbf{v}_0 = PD^n P^{-1} \begin{bmatrix} 1 \\ 0 \end{bmatrix}$. Now $P^{-1} \begin{bmatrix} 1 \\ 0 \end{bmatrix} = \frac{1}{3} \begin{bmatrix} -1 \\ 1 \end{bmatrix}$ and

$PD^n = P \begin{bmatrix} (-1)^n & 0 \\ 0 & 2^n \end{bmatrix} = \begin{bmatrix} (-1)^{n+1} & 2^{n+1} \\ (-1)^n & 2^n \end{bmatrix}$, so $\mathbf{v}_n = \frac{1}{3} \left[-\begin{bmatrix} (-1)^{n+1} \\ (-1)^n \end{bmatrix} + \begin{bmatrix} 2^{n+1} \\ 2^n \end{bmatrix} \right]$ and the second component of this

vector is $a_n = \frac{1}{3}[(-1)^{n+1} + 2^n]$.

4. (a) Let $\mathbf{v}_n = \begin{bmatrix} a_{n+1} \\ a_n \end{bmatrix}$. Then $\mathbf{v}_n = A^n \mathbf{v}_0$ where $A = \begin{bmatrix} 3 & -2 \\ 1 & 0 \end{bmatrix}$. The eigenvalues of A are 1 and 2 with corresponding

eigenvectors, respectively, $\begin{bmatrix} 1 \\ 1 \end{bmatrix}$ and $\begin{bmatrix} 2 \\ 1 \end{bmatrix}$. Let $P = \begin{bmatrix} 1 & 2 \\ 1 & 1 \end{bmatrix}$. Then $P^{-1}AP = D = \begin{bmatrix} 1 & 0 \\ 0 & 2 \end{bmatrix}$ and

$\mathbf{v}_n = A^n \mathbf{v}_0 = (PDP^{-1})^n \mathbf{v}_0 = PD^n P^{-1} \begin{bmatrix} 0 \\ -4 \end{bmatrix}$. Now $P^{-1} \begin{bmatrix} 0 \\ -4 \end{bmatrix} = \begin{bmatrix} -1 & 2 \\ 1 & -1 \end{bmatrix} \begin{bmatrix} 0 \\ -4 \end{bmatrix} = \begin{bmatrix} -8 \\ 4 \end{bmatrix}$ and

$PD^n = P \begin{bmatrix} 1 & 0 \\ 0 & 2^n \end{bmatrix} = \begin{bmatrix} 1 & 2^{n+1} \\ 1 & 2^n \end{bmatrix}$. Thus $\mathbf{v}_n = \begin{bmatrix} 1 & 2^{n+1} \\ 1 & 2^n \end{bmatrix} \begin{bmatrix} -8 \\ 4 \end{bmatrix} = -8 \begin{bmatrix} 1 \\ 1 \end{bmatrix} + 4 \begin{bmatrix} 2^{n+1} \\ 2^n \end{bmatrix}$ and the second component of this

vector is $a_n = -8 + 4(2^n)$.

5. (a) $A^T A$ is symmetric, hence diagonalizable by the Principal Axes Theorem.

Exercises 8.3

1. (a) The characteristic polynomial of A is $\begin{vmatrix} \frac{1}{3} - \lambda & \frac{1}{4} \\ \frac{2}{3} & \frac{3}{4} - \lambda \end{vmatrix} = (\frac{1}{3} - \lambda)(\frac{3}{4} - \lambda) - \frac{1}{6} = \frac{1}{12} - \frac{13}{12}\lambda + \lambda^2 = (\lambda - 1)(\lambda - \frac{1}{12})$,

so the eigenvalues of A are $\lambda = 1, \frac{1}{12}$. If $\mathbf{x} = \begin{bmatrix} x_1 \\ x_2 \end{bmatrix}$ is an eigenvector for $\lambda = 1$, then $\mathbf{x}$ is a solution to the

homogeneous system $(A - I)\mathbf{x} = \mathbf{0}$. We find $\mathbf{x} = t \begin{bmatrix} 3 \\ 8 \end{bmatrix}$. If $\mathbf{x} = \begin{bmatrix} x_1 \\ x_2 \end{bmatrix}$ is an eigenvector for $\lambda = \frac{1}{12}$, then $\mathbf{x}$ is a solution

to the homogeneous system $(A - \frac{1}{12}I)\mathbf{x} = \mathbf{0}$. We find $\mathbf{x} = t \begin{bmatrix} 1 \\ -1 \end{bmatrix}$. With $P = \begin{bmatrix} 3 & 1 \\ 8 & 1 \end{bmatrix}$, we have $P^{-1}AP = \begin{bmatrix} 1 & 0 \\ 0 & \frac{1}{12} \end{bmatrix}$.

(b) Since $P^{-1}AP = D$, $A = PDP^{-1}$ and

$$A^n = PD^nP^{-1} = \frac{1}{11}\begin{bmatrix} 3 & 1 \\ 8 & -1 \end{bmatrix}\begin{bmatrix} 1 & 0 \\ 0 & (\frac{1}{12})^n \end{bmatrix}\begin{bmatrix} 1 & 1 \\ 8 & -3 \end{bmatrix}$$

$$= \begin{bmatrix} \frac{3}{11} + \frac{8}{11}(\frac{1}{12})^n & \frac{3}{11} - \frac{3}{11}(\frac{1}{12})^n \\ \frac{8}{11} - \frac{8}{11}(\frac{1}{12})^n & \frac{8}{11} + \frac{3}{11}(\frac{1}{12})^n \end{bmatrix}.$$

(c) As $n \to \infty$, $(\frac{1}{12})^n \to 0$, so $A \to B = \begin{bmatrix} \frac{3}{11} & \frac{3}{11} \\ \frac{8}{11} & \frac{8}{11} \end{bmatrix}$.

(d) Since A is a Markov matrix with all eigenvalues but one less than 1 in absolute value and the eigenspace for $\lambda = 1$ one-dimensional, A will converge to a matrix each of whose columns is an eigenvector corresponding to $\lambda = 1$ whose entries sum to 1; that is, each column is $\begin{bmatrix} 3t \\ 8t \end{bmatrix}$ with $3t + 8t = 1$. So $t = \frac{1}{11}$ and $B = \begin{bmatrix} \frac{3}{11} & \frac{3}{11} \\ \frac{8}{11} & \frac{8}{11} \end{bmatrix}$, as before.

3. Let $\mathbf{v}_0 = \begin{bmatrix} x_0 \\ y_0 \\ z_0 \end{bmatrix}$ be the vector giving the initial numbers of students in the classes of Drs. G, L, and P, respectively, and, in general, $\mathbf{v}_k$ the vector giving the numbers in each section after k lectures. Then $\mathbf{v}_{k+1} = A\mathbf{v}_k$ where

$$A = \begin{bmatrix} \frac{7}{10} & \frac{1}{2} & \frac{3}{10} \\ \frac{1}{10} & \frac{1}{4} & \frac{1}{10} \\ \frac{1}{5} & \frac{1}{4} & \frac{3}{5} \end{bmatrix}.$$

This is a Markov matrix. Its characteristic polynomial is

$$-\lambda^3 + \frac{31}{20}\lambda^2 - \frac{61}{100}\lambda + \frac{3}{50} = -\frac{1}{100}(\lambda - 1)(20\lambda - 3)(5\lambda - 2).$$

Since there are three distinct eigenvalues, A is diagonalizable. Since the eigenvalues satisfy $|\lambda| \le 1$ and all but one satisfy $|\lambda| < 1$, the powers of A converge to a matrix with identical columns. The eigenspace for $\lambda = 1$ is determined by solving $(A - I)\mathbf{x} = \mathbf{0}$. Gaussian elimination proceeds

$$A - I = \begin{bmatrix} -\frac{3}{10} & \frac{1}{2} & \frac{3}{10} \\ \frac{1}{10} & -\frac{3}{4} & \frac{1}{10} \\ \frac{1}{5} & \frac{1}{4} & -\frac{2}{5} \end{bmatrix} \to \begin{bmatrix} 1 & -\frac{3}{5} & -1 \\ \frac{1}{10} & -\frac{3}{4} & \frac{1}{10} \\ \frac{1}{5} & \frac{1}{4} & -\frac{2}{5} \end{bmatrix} \to \begin{bmatrix} 1 & -\frac{3}{5} & -1 \\ 0 & -\frac{7}{12} & \frac{1}{5} \\ 0 & \frac{7}{12} & -\frac{1}{5} \end{bmatrix} \to \begin{bmatrix} 1 & -\frac{3}{5} & -1 \\ 0 & 1 & -\frac{12}{35} \\ 0 & 0 & 0 \end{bmatrix}.$$

The eigenspace for $\lambda = 1$ is one-dimensional and, if $\mathbf{x} = \begin{bmatrix} x_1 \\ x_2 \\ x_3 \end{bmatrix}$ is an eigenvector, $x_3 = t$ is free, $x_2 = \frac{12}{35}t$ and $x_1 = \frac{3}{5}x_2 + x_3 = \frac{11}{7}t$. There is a unique such eigenvector whose components sum to 1, obtained by solving $\frac{11}{7}t + \frac{12}{35}t + t = 1$. We obtain $t = \frac{35}{102}$ and $\mathbf{x} = \frac{1}{102}\begin{bmatrix} 55 \\ 12 \\ 35 \end{bmatrix}$. The powers of A converge to the limit matrix that has this vector in every column. The eventual distribution of students is $\frac{55}{102} \approx .539$ for Dr. G, $\frac{12}{102} \approx .118$ for Dr. L, and $\frac{35}{102} \approx .343$ for Dr. P.

6. Let $\mathbf{v}_0 = \begin{bmatrix} x_0 \\ y_0 \\ z_0 \end{bmatrix}$ be the vector giving the initial populations of British Columbia, Ontario, and Newfoundland, respectively. In general, let $\mathbf{v}_k$ be the vector giving the numbers in each province after k years. Then $\mathbf{v}_{k+1} = A\mathbf{v}_k$ where

$$A = \begin{bmatrix} \frac{3}{4} & \frac{1}{6} & 0 \\ \frac{1}{4} & \frac{1}{2} & \frac{1}{4} \\ 0 & \frac{1}{3} & \frac{3}{4} \end{bmatrix}.$$

This is a Markov matrix. Its characteristic polynomial is

$$-\lambda^3 + 2\lambda^2 - \frac{19}{16}\lambda + \frac{3}{16} = -\frac{1}{16}(\lambda - 1)(4\lambda - 3)(4\lambda - 1).$$

Since there are three distinct eigenvalues, A is diagonalizable. Since the eigenvalues satisfy $|\lambda| \le 1$ and all but one satisfy $|\lambda| < 1$, the powers of A converge to a matrix with identical columns. The eigenspace for $\lambda = 1$ is determined by solving $(A - I)\mathbf{x} = \mathbf{0}$. Gaussian elimination proceeds

$$A - I = \begin{bmatrix} -\frac{1}{4} & \frac{1}{6} & 0 \\ \frac{1}{4} & -\frac{1}{2} & \frac{1}{4} \\ 0 & \frac{1}{3} & -\frac{1}{4} \end{bmatrix} \rightarrow \begin{bmatrix} 1 & -\frac{2}{3} & 0 \\ 0 & -\frac{1}{3} & \frac{1}{4} \\ 0 & \frac{1}{3} & -\frac{1}{4} \end{bmatrix} \rightarrow \begin{bmatrix} 1 & -\frac{2}{3} & 0 \\ 0 & 1 & -\frac{3}{4} \\ 0 & 0 & 0 \end{bmatrix}.$$

The eigenspace for $\lambda = 1$ is one-dimensional and, if $\mathbf{x} = \begin{bmatrix} x_1 \\ x_2 \\ x_3 \end{bmatrix}$ is an eigenvector, $x_3 = t$ is free, $x_2 = \frac{3}{4}t$,

$x_1 = \frac{2}{3}x_2 = \frac{1}{2}t$ and $\mathbf{x} = \begin{bmatrix} \frac{1}{2}t \\ \frac{3}{4}t \\ t \end{bmatrix}$. There is a unique such eigenvector whose components sum to 1, obtained by solving

$\frac{1}{2}t + \frac{3}{4}t + t = 1$. We obtain $t = \frac{4}{9}$ and $\mathbf{x} = \begin{bmatrix} \frac{2}{9} \\ \frac{1}{3} \\ \frac{4}{9} \end{bmatrix}$. The powers of A converge to the limit matrix that has this vector in every column. The eventual distribution is $\frac{2}{9} \approx 22\%$ of the initial total population in British Columbia, 33% in Ontario, and $\frac{4}{9} \approx 44\%$ in Newfoundland; that is, 2.2 million in British Columbia, 3.3 million in Ontario and 4.4 million in Newfoundland. This is independent of the initial distribution.

Exercises 8.4

1. (a) $q = \mathbf{x}^T A\mathbf{x}$ with $\mathbf{x} = \begin{bmatrix} x \\ y \end{bmatrix}$ and $A = \begin{bmatrix} 2 & -\frac{3}{2} \\ -\frac{3}{2} & 1 \end{bmatrix}$.

2. (b) i. $q = 25(\frac{3}{5}y_1 + \frac{4}{5}y_2)^2 - 50(-\frac{4}{5}y_1 + \frac{3}{5}y_2)^2$
$= -23y_1^2 + 72y_1y_2 - 2y_2^2$

3. (a) We have $q = 6x^2 - 4xy + 3y^2 = \mathbf{x}^T A\mathbf{x}$, where $\mathbf{x} = \begin{bmatrix} x \\ y \end{bmatrix}$ and $A = \begin{bmatrix} 3 & -2 \\ -2 & 3 \end{bmatrix}$.

The eigenvalues of A are $\lambda = 2, 7$ so that relative to a certain new orthonormal basis, our equation is $2x^2 + 7y^2 = 1$. The graph is an ellipse.

4. Our form is $q = \mathbf{x}^T A \mathbf{x}$ where $\mathbf{x} = \begin{bmatrix} x_1 \\ x_2 \\ x_3 \end{bmatrix}$ and $A = \begin{bmatrix} 3 & 0 & 0 \\ 0 & 4 & -2 \\ 0 & -2 & 1 \end{bmatrix}$.

We find that A has three distinct eigenvalues, $\lambda_1 = 0$, $\lambda_2 = 3$ and $\lambda_3 = 5$. Thus $q = 3y_2^2 + 5y_3^2$ relative to an orthonormal basis $\{\mathbf{q}_1, \mathbf{q}_2, \mathbf{q}_3\}$ of eigenvectors corresponding to $0, 3, 5$, respectively. We find

$$\mathbf{q}_1 = \frac{1}{\sqrt{5}} \begin{bmatrix} 0 \\ 1 \\ 2 \end{bmatrix}, \quad \mathbf{q}_2 = \begin{bmatrix} 1 \\ 0 \\ 0 \end{bmatrix}, \quad \mathbf{q}_3 = \frac{1}{\sqrt{5}} \begin{bmatrix} 0 \\ -2 \\ 1 \end{bmatrix}.$$

Exercises 8.5

1. The (i, i) entry of A^2 is the number of walks of length two from v_i to v_i. Such a walk is of the form $v_i v_j v_i$ and this is possible if and only if $v_i v_j$ is an edge.

2. The $(1, 2)$ entry of A^4 is the number of walks of length 4 from v_1 to v_2. This number is 2 since the only such walks are $v_1 v_5 v_3 v_1 v_2$ and $v_1 v_3 v_5 v_1 v_2$.

3. (a) The (i, j) entry of A^2 is the dot product of row i of A with column j of A. This is the number of positions k for which a_{ik} and a_{kj} are both 1. This is the number of vertices v_k for which $v_i v_k$ and $v_k v_j$ are edges and this is the number of walks of length 2 from v_i to v_j since any such walk is of the form $v_i v_k v_j$.

5. $\mathcal{T}$ is connected, so you can get from u to v within the tree. Follow this route in the larger graph and return to u along the new edge. This gives a cycle.

6. (a)

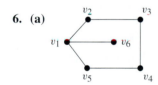

7. (a) The adjacency matrix relative to the given labeling is $A = \begin{bmatrix} 0 & 1 & 0 & 0 \\ 1 & 0 & 1 & 1 \\ 0 & 1 & 0 & 1 \\ 0 & 1 & 1 & 0 \end{bmatrix}$. The matrix defined in Kirchhoff's theorem

is $M = \begin{bmatrix} 1 & -1 & 0 & 0 \\ -1 & 3 & -1 & -1 \\ 0 & -1 & 2 & -1 \\ 0 & -1 & -1 & 2 \end{bmatrix}$. There are three spanning trees, 3 being the value of any cofactor of M.

8. (a) $\mathcal{K}_4$:

10. (a) The adjacency matrices are $A_1 = \begin{bmatrix} 0 & 1 & 0 & 1 & 1 \\ 1 & 0 & 1 & 0 & 1 \\ 0 & 1 & 0 & 1 & 1 \\ 1 & 0 & 1 & 0 & 1 \\ 1 & 1 & 1 & 0 & 0 \end{bmatrix}$ and $A_2 = \begin{bmatrix} 0 & 1 & 1 & 1 & 0 \\ 1 & 0 & 1 & 0 & 1 \\ 1 & 1 & 0 & 1 & 0 \\ 1 & 0 & 1 & 0 & 1 \\ 1 & 1 & 0 & 1 & 0 \end{bmatrix}$, respectively. The graphs are

isomorphic as we see by relabeling the vertices of $\mathcal{G}_1$.

Letting $P = \begin{bmatrix} 0 & 1 & 0 & 0 & 0 \\ 0 & 0 & 1 & 0 & 0 \\ 0 & 0 & 0 & 1 & 0 \\ 0 & 0 & 0 & 0 & 1 \\ 1 & 0 & 0 & 0 & 0 \end{bmatrix}$ be the permutation matrix that has the rows of the identity in written in the order 23451, we have $A_2 = P^T A_1 P$.

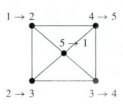

(c) The adjacency matrices are $A_1 = \begin{bmatrix} 0 & 1 & 1 & 1 & 1 \\ 1 & 0 & 0 & 1 & 1 \\ 1 & 0 & 0 & 1 & 1 \\ 1 & 1 & 1 & 0 & 1 \\ 1 & 1 & 1 & 1 & 0 \end{bmatrix}$ and $A_2 = \begin{bmatrix} 0 & 1 & 1 & 1 & 0 \\ 1 & 0 & 1 & 1 & 1 \\ 1 & 1 & 0 & 1 & 1 \\ 1 & 1 & 1 & 0 & 1 \\ 0 & 1 & 1 & 1 & 0 \end{bmatrix}$, respectively. The graphs are isomorphic as we see by relabeling the vertices of $\mathcal{G}_1$.

Letting $P = \begin{bmatrix} 0 & 1 & 0 & 0 & 0 \\ 0 & 0 & 0 & 0 & 1 \\ 1 & 0 & 0 & 0 & 0 \\ 0 & 0 & 1 & 0 & 0 \\ 0 & 0 & 0 & 1 & 0 \end{bmatrix}$ be the permutation matrix that has the rows of the identity in written in the order 25134, we have $A_2 = P^T A_1 P$.

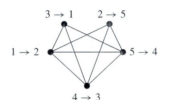

(e) The adjacency matrices are $A_1 = \begin{bmatrix} 0 & 1 & 0 & 1 & 0 & 1 \\ 1 & 0 & 1 & 0 & 1 & 0 \\ 0 & 1 & 0 & 1 & 0 & 1 \\ 1 & 0 & 1 & 0 & 1 & 0 \\ 0 & 1 & 0 & 1 & 0 & 1 \\ 1 & 0 & 1 & 0 & 1 & 0 \end{bmatrix}$ and $A_2 = \begin{bmatrix} 0 & 1 & 0 & 0 & 1 & 1 \\ 1 & 0 & 1 & 1 & 0 & 0 \\ 0 & 1 & 0 & 1 & 0 & 1 \\ 0 & 1 & 1 & 0 & 1 & 0 \\ 1 & 0 & 0 & 1 & 0 & 1 \\ 1 & 0 & 1 & 0 & 0 & 1 \end{bmatrix}$, respectively. No such

permutation matrix P exists because the graphs are not isomorphic: graph $\mathcal{G}_2$ has several *triangles*—156, 234, for instance—but graph $\mathcal{G}_1$ contains no such triangles.

11. **(a)** The given matrices are the adjacency matrices of the graphs $\mathcal{G}_1, \mathcal{G}_2$ shown to the right. The graphs are not isomorphic ($\mathcal{G}_2$ has more edges than $\mathcal{G}_1$), so no such permutation matrix exists.

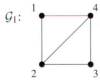

$\mathcal{G}_1$: $\mathcal{G}_2$;

(c) The given matrices are two adjacency matrices of the graph shown on the right, as the labelings show. We have $A_2 = P^T A_1 P$ with
$P = \begin{bmatrix} 0 & 0 & 0 & 0 & 1 \\ 1 & 0 & 0 & 0 & 0 \\ 0 & 0 & 0 & 1 & 0 \\ 0 & 1 & 0 & 0 & 0 \\ 0 & 0 & 1 & 0 & 0 \end{bmatrix}$.

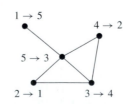

Glossary

If you do not know what the words mean, it is impossible to read anything with understanding. This fact, which is completely obvious to students of German or Russian, is often lost on mathematics students, but it is just as applicable. What follows is a vocabulary list of all the technical terms discussed in this book together with the page where each was first introduced. In most cases, each definition is followed by an example.

Angle between vectors: If $\mathbf{u}$ and $\mathbf{v}$ are nonzero vectors in $\mathbf{R}^n$, the angle between $\mathbf{u}$ and $\mathbf{v}$ is that angle θ, $0 \leq \theta \leq \pi$, whose cosine satisfies $\cos\theta = \frac{\mathbf{u}\cdot\mathbf{v}}{\|\mathbf{u}\|\|\mathbf{v}\|}$. For example, if $\mathbf{u} = \begin{bmatrix} 1 \\ 2 \\ 1 \\ 0 \\ 3 \end{bmatrix}$ and $\mathbf{v} = \begin{bmatrix} 1 \\ -1 \\ 1 \\ 1 \\ 1 \end{bmatrix}$, then $\cos\theta = \frac{\mathbf{u}\cdot\mathbf{v}}{\|\mathbf{u}\|\|\mathbf{v}\|} = \frac{3}{\sqrt{15}\sqrt{5}} = \frac{\sqrt{3}}{5} \approx .346$, so $\theta \approx 1.217$ rads $\approx 70°$. *p. 60*

Basis: A basis of a vector space V is a linearly independent set of vectors that span V. For example, the standard basis vectors $\mathbf{e}_1, \mathbf{e}_2, \ldots, \mathbf{e}_n$ in $\mathbf{R}^n$ form a basis for $\mathbf{R}^n$, and the polynomials $1, x, x^2, \ldots$ form a basis for the vector space of all polynomials in x with real coefficients. *p. 255*

Change of Coordinates Matrix: The change of coordinates matrix from V to V', where V and V' are bases for a vector space V, is the matrix of the identity linear transformation relative to these bases. It is denoted $P_{V'\leftarrow V}$. Thus $P_{V'\leftarrow V} = M_{V'\leftarrow V}(\text{id})$. For example, let $\mathcal{E} = \{\mathbf{e}_1, \mathbf{e}_2\}$ be the standard basis of $\mathbf{R}^2$ and let $V = \{\mathbf{v}_1, \mathbf{v}_2\}$ be the basis with $\mathbf{v}_1 = \begin{bmatrix} -1 \\ 2 \end{bmatrix}$ and $\mathbf{v}_2 = \begin{bmatrix} 2 \\ -3 \end{bmatrix}$. The change of coordinates matrix V to $\mathcal{E}$ is $P = P_{\mathcal{E}\leftarrow V} = M_{\mathcal{E}\leftarrow V}(\text{id}) = \begin{bmatrix} -1 & 2 \\ 2 & -3 \end{bmatrix}$. The change of coordinates matrix $\mathcal{E}$ to V is most easily determined by the fact that $P_{V\leftarrow\mathcal{E}} = P_{\mathcal{E}\leftarrow V}^{-1}$, so here, $P_{V\leftarrow\mathcal{E}} = \begin{bmatrix} -1 & 2 \\ 2 & -3 \end{bmatrix}^{-1} = \begin{bmatrix} 3 & 2 \\ 2 & 1 \end{bmatrix}$. *p. 321*

Closed under: See *Subspace*.

Closest: See *Distance*.

Cofactor: See *Minor*.

Column space: The column space of a matrix A is the set of all linear combinations of the columns of A. It is denoted col sp A. For example, the column space of $A = \begin{bmatrix} 1 & 0 \\ 0 & 1 \end{bmatrix}$ is $\mathbf{R}^2$, the entire Euclidean plane, while the column space of $A = \begin{bmatrix} 1 & 3 & 7 \\ 2 & 6 & 14 \end{bmatrix}$ is the set of scalar multiples of $\begin{bmatrix} 1 \\ 2 \end{bmatrix}$. Since a linear combination of the columns of A is a vector of the form $A\mathbf{x}$, if you are given that $\mathbf{b}$ is in the column space of A, you know that $\mathbf{b} = A\mathbf{x}$ for some vector $\mathbf{x}$. *p. 221*

Complex conjugate: The conjugate of the complex number $a + bi$ is the complex number $a - bi$, denoted $\overline{a + bi}$. For example, $\overline{1 + 2i} = 1 - 2i$, $\overline{-3i} = +3i$ and $\overline{-5} = -5$. *p. 402*

Component (of a vector): See *Vector*.

Conjugate transpose: See *Transpose*.

Coordinate vector: The coordinate vector of a vector $\mathbf{x}$ relative to a basis $V = \{\mathbf{v}_1, \mathbf{v}_2, \ldots, \mathbf{v}_n\}$ is the vector $\mathbf{x}_V = \begin{bmatrix} x_1 \\ x_2 \\ \vdots \\ x_n \end{bmatrix}$ whose components are the unique numbers satisfying $\mathbf{x} = x_1\mathbf{v}_1 + x_2\mathbf{v}_2 + \cdots + x_n\mathbf{v}_n$. For example, if $\mathbf{x} = \begin{bmatrix} x_1 \\ x_2 \end{bmatrix}$ is a vector in $\mathbf{R}^2$, and $\mathcal{E} = \{\mathbf{e}_1, \mathbf{e}_2\}$ is the standard basis, then $x_\varepsilon = \begin{bmatrix} x_1 \\ x_2 \end{bmatrix}$ because $x = x_1 e_1 + x_2 e_2$.

If $V = \{v_1, v_2\}$ with $v_1 = \begin{bmatrix} -1 \\ 2 \end{bmatrix}$ and $v_2 = \begin{bmatrix} 2 \\ -3 \end{bmatrix}$, then

$x_V = \begin{bmatrix} 3x_1 + 2x_2 \\ 2x_1 + x_2 \end{bmatrix}$ because $x = (3x_1 + 2x_2)v_1 + (2x_1 + x_2)v_2$.

p. 308

Cross product: The cross product of vectors $u = \begin{bmatrix} u_1 \\ u_2 \\ u_3 \end{bmatrix}$

and $v = \begin{bmatrix} v_1 \\ v_2 \\ v_3 \end{bmatrix}$ is a vector whose calculation we remember

by the expression

$$u \times v = \begin{vmatrix} i & j & k \\ u_1 & u_2 & u_3 \\ v_1 & v_2 & v_3 \end{vmatrix}.$$

Thus

$$u \times v = \begin{vmatrix} u_2 & u_3 \\ v_2 & v_3 \end{vmatrix} i - \begin{vmatrix} u_1 & u_3 \\ v_1 & v_3 \end{vmatrix} j + \begin{vmatrix} u_1 & u_2 \\ v_1 & v_2 \end{vmatrix} k.$$

For example, let $u = \begin{bmatrix} 1 \\ 3 \\ 2 \end{bmatrix}$ and $v = \begin{bmatrix} 0 \\ 2 \\ -1 \end{bmatrix}$. We have

$$u \times v = \begin{vmatrix} i & j & k \\ 1 & 3 & 2 \\ 0 & -2 & 1 \end{vmatrix} = \begin{vmatrix} 3 & 2 \\ -2 & 1 \end{vmatrix} i - \begin{vmatrix} 1 & 2 \\ 0 & 1 \end{vmatrix}$$

$$\times j + \begin{vmatrix} 1 & 3 \\ 0 & -2 \end{vmatrix} k$$

$$= 7i - 1j + (-2)k = \begin{bmatrix} 7 \\ -1 \\ -2 \end{bmatrix}.$$

p. 31

Determinant: The determinant of the 2×2 matrix $A = \begin{bmatrix} a & b \\ c & d \end{bmatrix}$ is the number $ad - bc$. It is denoted det A or $\begin{vmatrix} a & b \\ c & d \end{vmatrix}$. Thus

$$\det A = \begin{vmatrix} a & b \\ c & d \end{vmatrix} = ad - bc.$$

For $n > 2$, the determinant of an $n \times n$ matrix A is defined by the equation $AC^T = (\det A)I$, where C denotes the matrix of cofactors of A. *pp. 30, 170*

Diagonal: See *Main diagonal.*

Diagonalizable: A matrix A is diagonalizable if it is similar to a diagonal matrix; that is, if there exists an invertible matrix P and a diagonal matrix D such that $P^{-1}AP = D$. For example, $A = \begin{bmatrix} -11 & 18 \\ -6 & 10 \end{bmatrix}$ is diagonalizable because $P^{-1}AP = D$, with $P = \begin{bmatrix} 2 & 3 \\ 1 & 2 \end{bmatrix}$ and $D = \begin{bmatrix} -2 & 0 \\ 0 & 1 \end{bmatrix}$. A real $n \times n$ matrix A is orthogonally diagonalizable if there exists a (real) orthogonal matrix Q such that $Q^{-1}AQ = D$ is a (real) diagonal matrix. For example, let $A = \begin{bmatrix} 1 & -2 \\ -2 & -2 \end{bmatrix}$. The matrix $Q = \begin{bmatrix} \frac{1}{\sqrt{5}} & -\frac{2}{\sqrt{5}} \\ \frac{2}{\sqrt{5}} & \frac{1}{\sqrt{5}} \end{bmatrix}$ is orthogonal and $Q^{-1}AQ = Q^TAQ = \begin{bmatrix} -3 & 0 \\ 0 & 2 \end{bmatrix} = D$. A complex matrix A is unitarily diagonalizable if there exists a unitary matrix U such that $U^{-1}AU$ is diagonal. For example, let $A = \begin{bmatrix} 2 & 1-i \\ 1+i & 3 \end{bmatrix}$. The matrix $U = \begin{bmatrix} \frac{-1+i}{\sqrt{3}} & \frac{1-i}{\sqrt{6}} \\ \frac{1}{\sqrt{3}} & \frac{2}{\sqrt{6}} \end{bmatrix}$ is unitary and $U^{-1}AU = \begin{bmatrix} 1 & 0 \\ 0 & 4 \end{bmatrix}$. *pp. 205, 432, 430.*

Distance: The distance between vectors u and v is $\|u - v\|$. If v is in some Euclidean space V and U is a subspace of V, a vector u_0 in U is *closest* to v if $\|u_0 - v\| \leq \|u - v\|$ for all u in U. *p. 336*

Dot product: The dot product of n-dimensional vectors $x = \begin{bmatrix} x_1 \\ x_2 \\ \vdots \\ x_n \end{bmatrix}$ and $y = \begin{bmatrix} y_1 \\ y_2 \\ \vdots \\ y_n \end{bmatrix}$ is the number $x \cdot y = x_1 y_1 + x_2 y_2 + \cdots + x_n y_n$. For example, with $x = \begin{bmatrix} -1 \\ 2 \end{bmatrix}$ and $y = \begin{bmatrix} 4 \\ -3 \end{bmatrix}$, we have $x \cdot y = -1(4) + 2(-3) = -10$.

If $x = \begin{bmatrix} -1 \\ 0 \\ 2 \end{bmatrix}$ and $y = \begin{bmatrix} 1 \\ 2 \\ 3 \end{bmatrix}$, then $x \cdot y = -1(1) + 0(2) +$

$2(3) = 5$. The complex dot product of vectors $\mathbf{z} = \begin{bmatrix} z_1 \\ \vdots \\ z_n \end{bmatrix}$

and $\mathbf{w} = \begin{bmatrix} w_1 \\ \vdots \\ w_n \end{bmatrix}$ in $\mathbf{C}^n$ is $\mathbf{z} \cdot \mathbf{w} = \bar{z}_1 w_1 + \bar{z}_2 w_2 + \cdots + \bar{z}_n w_n$

For example, if $\mathbf{z} = \begin{bmatrix} 1 \\ i \end{bmatrix}$ and $\mathbf{w} = \begin{bmatrix} 2+i \\ 3-i \end{bmatrix}$ are vectors in

$\mathbf{C}^2$, then $\mathbf{z} \cdot \mathbf{w} = 1(2+i) - i(3-i) = 2+i-3i-1 = 1-2i$ and $\mathbf{w} \cdot \mathbf{z} = (2-i)(1) + (3+i)(i) = 2-i+3i-1 = 1+2i$.
pp. 20, 57, 412

Eigenspace; Eigenvalue; Eigenvector: An eigenvalue of a (square) matrix is a real number λ with the property that $A\mathbf{x} = \lambda\mathbf{x}$ for some nonzero vector $\mathbf{x}$. The vector $\mathbf{x}$ is called an eigenvector of A corresponding to λ. The set of all solutions to $A\mathbf{x} = \lambda\mathbf{x}$ is called the eigenspace of A corresponding to λ. For example, for $A = \begin{bmatrix} 1 & 3 \\ 2 & -4 \end{bmatrix}$, we have

$$A\begin{bmatrix} 3 \\ 1 \end{bmatrix} = \begin{bmatrix} 1 & 3 \\ 2 & -4 \end{bmatrix}\begin{bmatrix} 3 \\ 1 \end{bmatrix} = \begin{bmatrix} 6 \\ 2 \end{bmatrix} = 2\mathbf{x},$$

so $\begin{bmatrix} 3 \\ 1 \end{bmatrix}$ is an eigenvector of A corresponding to the eigenvalue $\lambda = 2$. The eigenspace corresponding to $\lambda = 2$ is the set of multiples of $\begin{bmatrix} 3 \\ 1 \end{bmatrix}$. You find this by solving the homogeneous system $(A - 2I)\mathbf{x} = \mathbf{0}$, which is a useful way to rewrite $A\mathbf{x} = 2\mathbf{x}$.
p. 193

Elementary matrix: An elementary matrix is a (square) matrix obtained from the identity matrix by a single elementary row operation. For example, $E = \begin{bmatrix} 0 & 1 & 0 \\ 1 & 0 & 0 \\ 0 & 0 & 1 \end{bmatrix}$ is an elementary matrix, obtained from the 3×3 identity matrix I by interchanging rows one and two. The matrix $E = \begin{bmatrix} 1 & 0 & 0 \\ 0 & 3 & 0 \\ 0 & 0 & 1 \end{bmatrix}$ is elementary; it is obtained from I by multiplying row two by 3.
p. 125

Elementary row operations: The three elementary row operations on a matrix are

i. the interchange of two rows,

ii. the multiplication of a row by a nonzero scalar, and

iii. replacing a row by that row minus a multiple of another.
p. 101

Euclidean n-space: Euclidean n-space is the set of all n-dimensional vectors. It is denoted $\mathbf{R}^n$. Euclidean 2-space, $\mathbf{R}^2 = \{\begin{bmatrix} x \\ y \end{bmatrix} \mid x, y \in \mathbf{R}\}$ is more commonly called the Euclidean plane and Euclidean 3-space $\mathbf{R}^3 = \{\begin{bmatrix} x \\ y \\ z \end{bmatrix} \mid x, y, z \in \mathbf{R}\}$ is often called simply 3-*space*.
p. 54

Finite dimension: A vector space V is finite dimensional (has finite dimension) if $V = \{\mathbf{0}\}$ or if V has a basis of n vectors for some $n \geq 1$. A vector space that is not finite dimensional is infinite dimensional. The dimension of a finite dimensional vector space V is the number of elements in any basis of V. For example, the set of all polynomials over $\mathbf{R}$ in a variable x, with usual addition and scalar multiplication, is an infinite dimensional vector space, while $\mathbf{R}^n$ has finite dimension n. Planes through the origin in $\mathbf{R}^3$ have dimension 2, while lines through the origin have dimension 1.
p. 256

Hermitian: A complex matrix A is Hermitian if it equals its conjugate transpose: $A = A^*$. For example, the matrix $A = \begin{bmatrix} -3 & 4-i \\ 4+i & 7 \end{bmatrix}$ is Hermitian, as is $\begin{bmatrix} 1 & 2 & 3 \\ 2 & 4 & 5 \\ 3 & 5 & 6 \end{bmatrix}$.
p. 415

Homogeneous system: A system of linear equations is homogeneous if it has the form $A\mathbf{x} = \mathbf{0}$, that is, the vector of constants to the right of $=$ is the zero vector. For example, the system

$$\begin{aligned} 3x_1 + 2x_2 - x_3 &= 0 \\ 2x_1 - 5x_2 + 7x_3 &= 0, \end{aligned}$$

which is $A\mathbf{x} = \mathbf{0}$ with $A = \begin{bmatrix} 3 & 2 & -1 \\ 2 & -5 & 7 \end{bmatrix}$ and $\mathbf{x} = \begin{bmatrix} x_1 \\ x_2 \end{bmatrix}$, is homogeneous.
p. 117

Identity matrix: An identity matrix is a square matrix with 1s on the diagonal and 0s everywhere else. For example, $\begin{bmatrix} 1 & 0 \\ 0 & 1 \end{bmatrix}$ and $\begin{bmatrix} 1 & 0 & 0 \\ 0 & 1 & 0 \\ 0 & 0 & 1 \end{bmatrix}$ are identity matrices, the first being the 2×2 identity matrix and the second the 3×3 identity matrix. *p. 81*

Inconsistent: See *Linear equation/Linear system.*

Intersection of subspaces: The intersection of subspaces U and W of $\mathbf{R}^n$ is the set $U \cap W = \{\mathbf{x} \in \mathbf{R}^n \mid \mathbf{x} \in U \text{ and } \mathbf{x} \in W\}$ of vectors that are in both U and in W. For example, if $U = \{\begin{bmatrix} c \\ 2c \end{bmatrix} \in \mathbf{R}^2$ and $W = \{\begin{bmatrix} 0 \\ c \end{bmatrix} \in \mathbf{R}^2\}$, $U \cap W = \{\mathbf{0}\}$ since if $\mathbf{v} = \begin{bmatrix} x \\ y \end{bmatrix}$ is in both U and W, then $\mathbf{v} = \begin{bmatrix} x \\ y \end{bmatrix} = \begin{bmatrix} c \\ 2c \end{bmatrix} = \begin{bmatrix} 0 \\ d \end{bmatrix}$ for certain scalars c and d. So $x = c = 0$ and $y = 2c = d = 0$, implying $\mathbf{v} = \mathbf{0}$. *p. 371*

Invariant subspace: A subspace U of a vector space is invariant under multiplication by a matrix A if AU is contained in U, symbolically, $AU \subseteq U$. For example, the xy-plane, U, is invariant under multiplication by
$$A = \begin{bmatrix} 2 & -3 & 1 \\ -1 & 5 & 1 \\ 0 & 0 & 8 \end{bmatrix} \text{ because } A \begin{bmatrix} x \\ y \\ 0 \end{bmatrix} = \begin{bmatrix} 2x - 3y \\ -x + 5y \\ 0 \end{bmatrix} \text{ is in}$$
U for any $\begin{bmatrix} x \\ y \\ 0 \end{bmatrix}$ in U. *p. 423*

Inverse: See *Invertible linear transformation; Invertible matrix; Right and/or Left inverse.*

Invertible linear transformation: A linear transformation $T: V \to W$ is invertible if and only if there exists a linear transformation $S: W \to V$ such that $TS = \text{id}_W$ and $ST = \text{id}_V$. The linear transformation S is called the inverse of T. For example, if $T, S: \mathbf{R}^2 \to \mathbf{R}^2$ are defined by $T\left(\begin{bmatrix} x \\ y \end{bmatrix}\right) = \begin{bmatrix} x + y \\ x - y \end{bmatrix}$ and $S\left(\begin{bmatrix} x \\ y \end{bmatrix}\right) = \begin{bmatrix} \frac{1}{2}(x + y) \\ \frac{1}{2}(x - y) \end{bmatrix}$, then $TS = ST = \text{id}$, so T is invertible with inverse S (and S is invertible with inverse T). *p. 304*

Invertible matrix: A square matrix A is invertible (or has an inverse) if there is another matrix B such that $AB = I$ and $BA = I$. The matrix B is called the inverse of A and we write $B = A^{-1}$. For example, the matrix $A = \begin{bmatrix} 1 & -2 \\ 2 & -3 \end{bmatrix}$ has an inverse, namely, $B = $

$\begin{bmatrix} -3 & 2 \\ -2 & 1 \end{bmatrix}$, since

$$AB = \begin{bmatrix} 1 & -2 \\ 2 & -3 \end{bmatrix} \begin{bmatrix} -3 & 2 \\ -2 & 1 \end{bmatrix} = \begin{bmatrix} 1 & 0 \\ 0 & 1 \end{bmatrix} = I$$

and

$$BA = \begin{bmatrix} -3 & 2 \\ -2 & 1 \end{bmatrix} \begin{bmatrix} 1 & -2 \\ 2 & -3 \end{bmatrix} = \begin{bmatrix} 1 & 0 \\ 0 & 1 \end{bmatrix} = I.$$

From theory, we know that if $AB = I$ with A and B square, then BA must also be I so, in point of fact, we had only to compute one of the above two products to be sure that A and B are invertible. The matrix $A = \begin{bmatrix} 1 & 2 \\ 2 & 4 \end{bmatrix}$ is not invertible because, if we let $B = \begin{bmatrix} x & y \\ z & w \end{bmatrix}$, then $AB = \begin{bmatrix} x + 2z & y + 2w \\ 2x + 4z & 2y + 4w \end{bmatrix}$. Since the second row of AB is twice the first, AB can never be I. *p. 89*

Kernel: The kernel of a linear transformation $T: V \to W$ is the set $\ker T = \{\mathbf{v} \in V \mid T(\mathbf{v}) = \mathbf{0}\}$. For example, if $T: \mathbf{R}^2 \to \mathbf{R}^3$ is the linear transformation defined by $T\left(\begin{bmatrix} x \\ y \end{bmatrix}\right) = \begin{bmatrix} x - y \\ 2x + y \\ -x + 4y \end{bmatrix}$, then $\mathbf{x} = \begin{bmatrix} x \\ y \end{bmatrix}$ is in $\ker T$ if and only if

$$\begin{aligned} x + y &= 0 \\ 2x + y &= 0 \\ -x + 4y &= 0 \end{aligned}.$$

We obtain $x = y = 0$, so $\mathbf{x} = \mathbf{0}$ and $\ker T = \{\mathbf{0}\}$. If $T: \mathbf{R}^4 \to \mathbf{R}^3$ is left multiplication by

$$A = \begin{bmatrix} 1 & 1 & -3 & 5 \\ 3 & -2 & 1 & 0 \\ 4 & -2 & 0 & 2 \end{bmatrix},$$

$\mathbf{x} = \begin{bmatrix} x_1 \\ x_2 \\ x_3 \\ x_4 \end{bmatrix} \in \ker T$ if and only if $A\mathbf{x} = \mathbf{0}$. Solving this system gives so $x_3 = t$ and $x_4 = s$ free, $x_2 = 2t - 3s$ and $x_1 = t - 2s$, so $\ker T$ is the two-dimensional subspace of $\mathbf{R}^4$ $\ker T = \{\begin{bmatrix} t - 2s \\ 2t - 3s \\ t \\ s \end{bmatrix} \mid t, s \in \mathbf{R}\}$ *p. 294*

LDU factorization: An LDU factorization of a matrix A is a representation of $A = LDU$ as the product of a lower triangular matrix L with 1s on the diagonal, an upper triangular matrix U with 1s on the diagonal and a diagonal matrix D. For example,

$$\begin{bmatrix} -1 & 2 & 0 \\ 2 & 4 & 6 \\ -3 & 2 & 4 \end{bmatrix} = \begin{bmatrix} 1 & 0 & 0 \\ -2 & 1 & 0 \\ 3 & -\frac{1}{2} & 1 \end{bmatrix}$$

$$\times \begin{bmatrix} -1 & 0 & 0 \\ 0 & 8 & 0 \\ 0 & 0 & 7 \end{bmatrix} \begin{bmatrix} 1 & -2 & 0 \\ 0 & 1 & \frac{3}{4} \\ 0 & 0 & 1 \end{bmatrix}$$

is an LDU factorization of $A = \begin{bmatrix} -1 & 2 & 0 \\ 2 & 4 & 6 \\ -3 & 2 & 4 \end{bmatrix}$.

p. 145

Left inverse: A left inverse of an $m \times n$ matrix A is an $n \times m$ matrix C such that $CA = I_n$, the $n \times n$ identity matrix. For example, any matrix C of the form $C = \begin{bmatrix} 1-5t & t & t \\ -2-5s & 1+s & s \end{bmatrix}$ is a left inverse of $A = \begin{bmatrix} 1 & 0 \\ 2 & 1 \\ 3 & -1 \end{bmatrix}$ since $CA = I_2$, the 2×2 identity matrix. *p. 275*

Length: See *Norm.*

Linear combination: A linear combination of k vectors $\mathbf{u}_1, \mathbf{u}_2, \ldots, \mathbf{u}_k$ is a vector of the form $c_1\mathbf{u}_1 + c_2\mathbf{u}_2 + \cdots + c_k\mathbf{u}_k$, where $c_1, \ldots, c_k$ are scalars. For example, $\begin{bmatrix} -5 \\ 9 \end{bmatrix}$ is a linear combination of $\mathbf{u} = \begin{bmatrix} -2 \\ 3 \end{bmatrix}$ and $\mathbf{v} = \begin{bmatrix} -1 \\ 1 \end{bmatrix}$ since $\begin{bmatrix} -5 \\ 9 \end{bmatrix} = 4\begin{bmatrix} -2 \\ 3 \end{bmatrix} - 3\begin{bmatrix} -1 \\ 1 \end{bmatrix} = 4\mathbf{u} - 3\mathbf{v}$ and $\begin{bmatrix} 2 \\ -6 \end{bmatrix}$ is a linear combination of $\mathbf{u}_1 = \begin{bmatrix} -2 \\ 3 \end{bmatrix}$, $\mathbf{u}_2 = \begin{bmatrix} 6 \\ -5 \end{bmatrix}$ and $\mathbf{u}_3 = \begin{bmatrix} 4 \\ 5 \end{bmatrix}$ since $\begin{bmatrix} 2 \\ -6 \end{bmatrix} = 3\begin{bmatrix} -2 \\ 3 \end{bmatrix} + 2\begin{bmatrix} 6 \\ -5 \end{bmatrix} - \begin{bmatrix} 4 \\ 5 \end{bmatrix} = 3\mathbf{u}_1 + 2\mathbf{u}_2 + (-1)\mathbf{u}_3$. A linear combination of matrices $A_1, A_2, \ldots, A_k$ (all of the same size) is a matrix of the form $c_1 A_1 + c_2 A_2 + \cdots + c_k A_k$, where $c_1, c_2, \ldots, c_k$ are scalars. *pp. 7, 81*

Linear equation/Linear system: A linear equation is an equation of the form

$$a_1 x_1 + a_2 x_2 + \cdots + a_n x_n = b,$$

where $a_1, a_2, \ldots, a_n$ and b are real numbers and $x_1, x_2, \ldots, x_n$ are variables. A set of one or more linear equations is called a linear system. To solve a linear system means to find values of the variables that make each equation true. A system that has no solution is called inconsistent. *p. 98*

Linear transformation: A linear transformation is a function $T: V \to W$ from a vector space V to another vector space W that satisfies

  i. $T(\mathbf{u} + \mathbf{v}) = T(\mathbf{u}) + T(\mathbf{v})$

  ii. $T(c\mathbf{u}) = cT(\mathbf{u})$

for all vectors $\mathbf{u}, \mathbf{v}$ in V and all scalars c. We describe these two properties by saying that T "preserves" addition and scalar multiplication. The most important example of a linear transformation in this book is left multiplication by a matrix, which is defined as follows. Let A be an $m \times n$ matrix. Then left multiplication by A is the function $T: \mathbf{R}^n \to \mathbf{R}^m$ defined by $T(x) = Ax$. The function $T: \mathbf{R}^1 \to \mathbf{R}^1$ defined by $T(x) = 2x$ is a linear transformation because $T(x + y) = 2(x + y) = 2x + 2y = T(x) + T(y)$ and $T(cx) = 2(cx) = c(2x) = cT(x)$ for all vectors x, y in $\mathbf{R}^1 = \mathbf{R}$ and all scalars c. The function $\sin: \mathbf{R} \to \mathbf{R}$ does not define a linear transformation because the sine function satisfies neither of the required properties. *p. 285*

Linearly dependent: See *Linearly independent.*

Linearly independent: A set of vectors $\mathbf{x}_1, \mathbf{x}_2, \ldots, \mathbf{x}_n$ is linearly independent if the only linear combination of them that equals the zero vector is the trivial one, where all the coefficients are 0:

$$c_1 \mathbf{x}_1 + c_2 \mathbf{x}_2 + \cdots + c_n \mathbf{x}_n = \mathbf{0} \quad \text{implies}$$

$$c_1 = c_2 = \cdots = c_n = 0.$$

Vectors that are not linearly independent are linearly dependent. For example, $\mathbf{x}_1 = \begin{bmatrix} 1 \\ 2 \end{bmatrix}$ and $\mathbf{x}_2 = \begin{bmatrix} -1 \\ 0 \end{bmatrix}$ are linearly independent (solve the equation $c_1\mathbf{x}_1 + c_2\mathbf{x}_2 = \mathbf{0}$ and obtain $c_1 = c_2 = 0$) but $\mathbf{u}_1 = \begin{bmatrix} 1 \\ 2 \end{bmatrix}$ and $\mathbf{u}_2 = \begin{bmatrix} 2 \\ 4 \end{bmatrix}$ are linearly dependent because $2\mathbf{u}_1 - \mathbf{u}_2 = \mathbf{0}$. *pp. 61, 245*

LU Factorization: An LU factorization of a matrix A is the representation of $A = LU$ as the product of a lower triangular matrix L and an upper triangular matrix U. For example, $\begin{bmatrix} 1 & 2 \\ 4 & 6 \end{bmatrix} = \begin{bmatrix} 1 & 0 \\ 4 & 1 \end{bmatrix}\begin{bmatrix} 1 & 2 \\ 0 & -2 \end{bmatrix}$ is an LU factorization of $A = \begin{bmatrix} 1 & 2 \\ 4 & 6 \end{bmatrix}$. *p. 128*

Main diagonal: The main diagonal of an $m \times n$ matrix $A = [a_{ij}]$ is the list of elements $a_{11}, a_{22}, a_{33}, \dots$. For instance, the main diagonal of $\begin{bmatrix} 1 & 2 & 3 \\ 4 & 5 & 6 \\ 7 & 8 & 9 \end{bmatrix}$ is $1, 5, 9$ and the main diagonal of $\begin{bmatrix} 7 & 9 & 3 \\ 6 & 5 & 4 \end{bmatrix}$ is $7, 5$. A matrix is diagonal if and only if its only nonzero entries lie on the main diagonal. For example, the matrix $\begin{bmatrix} 1 & 0 & 0 \\ 0 & -7 & 0 \\ 0 & 0 & 2 \end{bmatrix}$ is diagonal. So is $\begin{bmatrix} -2 & 0 & 0 \\ 0 & 7 & 0 \end{bmatrix}$. *p. 128*

Markov matrix: A Markov matrix is a matrix $A = [a_{ij}]$ all of whose entries satisfy $0 \le a_{ij} \le 1$ and each of whose columns sums to 1. For example, $\begin{bmatrix} \frac{1}{2} & \frac{1}{4} \\ \frac{1}{2} & \frac{3}{4} \end{bmatrix}$ is Markov and so is $\begin{bmatrix} 0 & \frac{2}{3} \\ 1 & \frac{1}{3} \end{bmatrix}$. *p. 468*

Matrix: A matrix is a rectangular array of numbers, enclosed in square brackets. If there are m rows and n columns in the array, the matrix is called $m \times n$ (read "m by n") and said to have *size* $m \times n$. For example,
$$A = \begin{bmatrix} 2 & 1 & -1 \\ 3 & -2 & 6 \\ 1 & 0 & -5 \end{bmatrix}$$ is a 3×3 ("three by three")matrix, while $B = \begin{bmatrix} 1 & 2 & 3 \\ 4 & 5 & 6 \end{bmatrix}$ is 2×3 and $\mathbf{x} = \begin{bmatrix} x_1 \\ x_2 \\ x_3 \end{bmatrix}$ is 3×1. The numbers in the matrix are its entries. We always use a capital letter to denote a matrix and the corresponding lower case letter, with subscripts, for its entries. Thus if $A = [a_{ij}]$, the $(2, 3)$ entry is a_{23}. If $A = \begin{bmatrix} 2 & 1 & 3 \\ 3 & 2 & 8 \\ 9 & 0 & 1 \end{bmatrix}$, then $a_{32} = 0$. If $A = [2i - j]$ is a 3×4 matrix, the $(1, 3)$ entry of A is $2(1) - 3 = -1$. *p. 70*

Matrix of a linear transformation: The matrix of a linear transformation $T: V \to W$ with respect to bases $V = \{\mathbf{v}_1, \mathbf{v}_2, \dots, \mathbf{v}_n\}$ of V and W for W is the matrix $M_{W \leftarrow V}(T)$ whose columns are the coordinate vectors of $T(\mathbf{v}_1), T(\mathbf{v}_2), \dots, T(\mathbf{v}_n)$ relative to W. For example, let $V = W = \mathbf{R}^2$ and let $T: V \to V$ be the linear transformation defined by
$$T\left(\begin{bmatrix} x_1 \\ x_2 \end{bmatrix}\right) = \begin{bmatrix} x_2 \\ x_1 \end{bmatrix}.$$

Let $\mathcal{E} = \{\mathbf{e}_1, \mathbf{e}_2\}$ be the standard basis for V and let
$$V = \{\mathbf{v}_1, \mathbf{v}_2\}, \quad \text{where} \quad \mathbf{v}_1 = \begin{bmatrix} -1 \\ 2 \end{bmatrix} \text{ and } \mathbf{v}_2 = \begin{bmatrix} 2 \\ -3 \end{bmatrix}.$$

Then V is also a basis for V. Since
$$T(\mathbf{e}_1) = T\left(\begin{bmatrix} 1 \\ 0 \end{bmatrix}\right) = \begin{bmatrix} 0 \\ 1 \end{bmatrix} \text{ and } T(\mathbf{e}_2) = T\left(\begin{bmatrix} 0 \\ 1 \end{bmatrix}\right) = \begin{bmatrix} 1 \\ 0 \end{bmatrix},$$
the matrix of T relative to $\mathcal{E}$ and $\mathcal{E}$ is $M_{\mathcal{E} \leftarrow \mathcal{E}}(T) = \begin{bmatrix} 0 & 1 \\ 1 & 0 \end{bmatrix}$. Since
$$T\left(\begin{bmatrix} -1 \\ 2 \end{bmatrix}\right) = \begin{bmatrix} 2 \\ -1 \end{bmatrix} = 2\mathbf{e}_1 + (-1)\mathbf{e}_2$$
so that $[T(\mathbf{v}_1)]_{\mathcal{E}} = \begin{bmatrix} 2 \\ -1 \end{bmatrix}$ and
$$T\left(\begin{bmatrix} 2 \\ -3 \end{bmatrix}\right) = \begin{bmatrix} -3 \\ 2 \end{bmatrix} = -3\mathbf{e}_1 + 2\mathbf{e}_2$$
so that $[T(\mathbf{v}_2)]_{\mathcal{E}} = \begin{bmatrix} -3 \\ 2 \end{bmatrix}$, the matrix of T with respect to $\mathcal{E}$ and V is $M_{\mathcal{E} \leftarrow V}(T) = \begin{bmatrix} 2 & -3 \\ -1 & 2 \end{bmatrix}$. *p. 312*

Minor: If A is a square matrix, the (i, j) minor, denoted m_{ij}, is the determinant of the $(n - 1) \times (n - 1)$ matrix obtained from A by deleting row i and column j. The (i, j) cofactor of A is $(-1)^{i+j}m_{ij}$ and is denoted c_{ij}. For example, let $A = \begin{bmatrix} 1 & 2 & 3 \\ 0 & -2 & 2 \\ 3 & 7 & -4 \end{bmatrix}$. The $(1, 1)$ minor is the determinant of the matrix $\begin{bmatrix} -2 & 2 \\ 7 & -4 \end{bmatrix}$, which is what remains when we remove row one and column one of A. So $m_{11} = -2(-4) - 2(7) = -6$ and $c_{11} = (-1)^{1+1}m_{11} = (-1)^2(-6) = -6$. The $(-1)^{i+j}$ in the definition of cofactor has the affect of either leaving the minor alone or changing its sign according to the pattern $\begin{bmatrix} + & - & + & - & \cdots \\ - & + & - & + & \cdots \\ \vdots & \vdots & & & \end{bmatrix}$. Thus the $(2, 3)$ minor of A is the determinant of $\begin{bmatrix} 1 & 2 \\ 3 & 7 \end{bmatrix}$, which is what remains after removing row two and column three of A, $m_{23} = 1$, while $c_{23} = -1$ since there is a $-$ in the $(2, 3)$ position of the pattern. *p. 168*

Modulus of a complex number: The modulus of the complex number $z = a + bi$ is $\sqrt{a^2 + b^2}$. It is denoted $|z|$. For example, if $z = 2 - 3i$, $|z|^2 = z\bar{z} = (2 - 3i)(2 + 3i) = 2^2 + 3^2 = 13$, so $|z| = \sqrt{13}$. *p. 403*

Negative: The negative of a matrix A is the matrix $(-1)A$, which we denote $-A$ and call "minus A": $-A = (-1)A$. For example, if $A = \begin{bmatrix} -1 & 3 \\ -2 & 4 \end{bmatrix}$, then $-A = (-1)\begin{bmatrix} -1 & 3 \\ -2 & 4 \end{bmatrix} = \begin{bmatrix} 1 & -3 \\ 2 & -4 \end{bmatrix}$. *p. 74*

Norm: The norm or length of the n-dimensional vector $\mathbf{x} = \begin{bmatrix} x_1 \\ x_2 \\ \vdots \\ x_n \end{bmatrix}$ is the number

$$\|\mathbf{x}\| = \sqrt{\mathbf{x} \cdot \mathbf{x}} = \sqrt{x_1^2 + x_2^2 + \cdots + x_n^2}.$$

For example, if $\mathbf{x} = \begin{bmatrix} 2 \\ 1 \\ 2 \end{bmatrix}$, then $\|\mathbf{x}\| = \sqrt{4 + 1 + 4} = 3$. The rule is the same for complex vectors (but you must use the complex dot product). For instance, if $\mathbf{z} = \begin{bmatrix} 1 \\ i \end{bmatrix}$ and $\mathbf{w} = \begin{bmatrix} 2 + i \\ 3 - i \end{bmatrix}$, $\|\mathbf{z}\|^2 = \mathbf{z} \cdot \mathbf{z} = 1(1) - i(i) = 1 - i^2 = 2$, so $\|\mathbf{z}\| = \sqrt{2}$, and $\|\mathbf{w}\|^2 = \mathbf{w} \cdot \mathbf{w} = (2 - i)(2 + i) + (3 + i)(3 - i) = 4 + 1 + 9 + 1 = 15$, so $\|\mathbf{w}\| = \sqrt{15}$. *pp. 16, 57, 412*

Null space: The null space of a matrix A is the set of all vectors $\mathbf{x}$ that satisfy $A\mathbf{x} = \mathbf{0}$. It is denoted null sp A. For example, the null space of $A = \begin{bmatrix} 1 & 0 \\ 0 & 1 \end{bmatrix}$ is just the zero vector, while the null space of $A = \begin{bmatrix} 1 & 0 & 2 \\ 0 & 1 & 3 \end{bmatrix}$ is

the set of vectors of the form $\begin{bmatrix} -2t \\ -3t \\ t \end{bmatrix} = t\begin{bmatrix} -2 \\ -3 \\ 1 \end{bmatrix}$. [Solve the homogeneous system $A\mathbf{x} = \mathbf{0}$!] *p. 218*

Nullity: The nullity of a matrix A is the dimension of its null space. For example, the nullity of $A = \begin{bmatrix} 1 & -3 & -2 & 4 \\ 2 & 0 & 2 & 2 \\ 0 & 4 & 4 & -4 \end{bmatrix}$ is 2. (Solve $A\mathbf{x} = \mathbf{0}$ and discover that the two vectors $\begin{bmatrix} -1 \\ -1 \\ 1 \\ 0 \end{bmatrix}$ and $\begin{bmatrix} -1 \\ 1 \\ 0 \\ 1 \end{bmatrix}$ form a basis for the null space.) *p. 258*

One-to-one: A linear transformation $T: V \to W$ is one-to-one if and only if $T(\mathbf{v}_1) = T(\mathbf{v}_2)$ implies $\mathbf{v}_1 = \mathbf{v}_2$. The linear transformation $T: \mathbf{R}^2 \to \mathbf{R}^3$ defined by $T\left(\begin{bmatrix} x \\ y \end{bmatrix}\right) = \begin{bmatrix} x + y \\ x - y \\ 2x \end{bmatrix}$ is one-to-one since, if $T\left(\begin{bmatrix} x_1 \\ y_1 \end{bmatrix}\right) = T\left(\begin{bmatrix} x_2 \\ y_2 \end{bmatrix}\right)$, then $\begin{bmatrix} x_1 + y_1 \\ x_1 - y_1 \\ 2x_1 \end{bmatrix} = \begin{bmatrix} x_2 + y_2 \\ x_2 - y_2 \\ 2x_2 \end{bmatrix}$ and it is easy to see that $x_1 = x_2$, $y_1 = y_2$; that is, $\begin{bmatrix} x_1 \\ y_1 \end{bmatrix} = \begin{bmatrix} x_2 \\ y_2 \end{bmatrix}$. This, however, is not the best way to show that a linear transformation is one-to-one: A linear transformation T is one-to-one if and only if its kernel is $\{\mathbf{0}\}$. *p. 296*

Onto: A linear transformation $T: V \to W$ is onto if and only if its range is W, that is, if and only if every $\mathbf{w}$ in W is of the form $\mathbf{w} = T(\mathbf{v})$ for some $\mathbf{v}$ in V. For example, the linear transformation $T: \mathbf{R}^2 \to \mathbf{R}^2$ defined by $T\left(\begin{bmatrix} x \\ y \end{bmatrix}\right) = \begin{bmatrix} x + y \\ x - y \end{bmatrix}$ is onto because any $\mathbf{w} = \begin{bmatrix} w_1 \\ w_2 \end{bmatrix}$ is $T(\mathbf{v})$ with $\mathbf{v} = \begin{bmatrix} \frac{1}{2}(w_1 + w_2) \\ \frac{1}{2}(w_1 - w_2) \end{bmatrix}$. The linear transformation $T: \mathbf{R}^2 \to \mathbf{R}^2$ defined by $T\left(\begin{bmatrix} x \\ y \end{bmatrix}\right) = \begin{bmatrix} x \\ 2x \end{bmatrix}$ is not onto because, for instance, $\begin{bmatrix} 1 \\ 3 \end{bmatrix}$ is not $T(\mathbf{v})$ for any $\mathbf{v}$: only vectors with second coordinate double the first are of the form $T(\mathbf{v})$. *p. 297*

Opposite direction: Vectors $\mathbf{u}$ and $\mathbf{v}$ have opposite direction if $\mathbf{u} = c\mathbf{v}$ for some scalar $c < 0$. For example, $\mathbf{u} = \begin{bmatrix} 2 \\ -3 \end{bmatrix}$ and $\mathbf{v} = \begin{bmatrix} -3 \\ \frac{9}{2} \end{bmatrix}$ have opposite direction because $\mathbf{u} = -\frac{2}{3}\mathbf{v}$ with the scalar $-\frac{2}{3} < 0$. *p. 4*

Orthogonal: A set $\{f_1, f_2, \ldots, f_k\}$ of vectors is orthogonal if and only if each pair of vectors in this set is orthogonal: $f_i \cdot f_j = 0$ if $i \neq j$. (When this holds, we sometimes say that the vectors are pairwise orthogonal.) If, in addition, each f_i has norm 1, then the set is called orthonormal. For example, any subset of the set of standard basis vectors $e_1, e_2, \ldots, e_n$ is an orthonormal set in $\mathbf{R}^n$, say, $\left\{ \begin{bmatrix} 0 \\ 1 \\ 0 \end{bmatrix}, \begin{bmatrix} 0 \\ 0 \\ 1 \end{bmatrix} \right\}$ in $\mathbf{R}^3$. The vectors $u = \begin{bmatrix} 3 \\ 2 \\ -1 \\ 4 \end{bmatrix}$ and $v = \begin{bmatrix} 1 \\ -1 \\ 1 \\ 0 \end{bmatrix}$ are orthogonal vectors in $\mathbf{R}^4$. The four vectors $f_1 = \begin{bmatrix} 1 \\ 1 \\ 1 \\ -1 \end{bmatrix}$, $f_2 = \begin{bmatrix} 1 \\ 0 \\ 1 \\ 2 \end{bmatrix}$, $f_3 = \begin{bmatrix} -1 \\ 0 \\ 1 \\ 0 \end{bmatrix}$, $f_4 = \begin{bmatrix} -1 \\ 3 \\ -1 \\ 1 \end{bmatrix}$ form an orthogonal set in $\mathbf{R}^4$.

pp. 21, 60, 353

Orthogonal complement: Let U be a subspace of a vector space V. The orthogonal complement of U, denoted $U^\perp$ (and read "U perp"), is the set of all vectors in V that are orthogonal to all vectors of U:

$$U^\perp = \{v \in V \mid v \cdot u = 0 \text{ for all } u \text{ in } U\}.$$

For example, if $V = \mathbf{R}^3$ and $U = \left\{ \begin{bmatrix} 0 \\ 0 \\ z \end{bmatrix} \mid z \in \mathbf{R} \right\}$ is the z-axis, then $U^\perp$ is the set of vectors of the form $\begin{bmatrix} x \\ y \\ 0 \end{bmatrix}$, which is the entire xy-plane. Subspaces U and W are orthogonal complements if each is the orthogonal complement of the other; for example, the null space and the row space of a matrix are orthogonal complements. *p. 378*

Orthogonal matrix: An orthogonal matrix is a square matrix with orthonormal columns. For example, the rotation matrix $Q = \begin{bmatrix} \cos\theta & -\sin\theta \\ \sin\theta & \cos\theta \end{bmatrix}$ is orthogonal for any angle θ. Any permutation matrix is orthogonal because its columns are just the standard basis vectors $e_1, e_2, \ldots, e_n$ in some order. One such matrix is $P = \begin{bmatrix} 0 & 0 & 1 \\ 1 & 0 & 0 \\ 0 & 1 & 0 \end{bmatrix}$. *p. 363*

Orthogonal subspaces: Subspaces U and W of $\mathbf{R}^n$ are orthogonal if $u \cdot w = 0$ for any vector u in U and any vector w in W. For example, in $\mathbf{R}^3$, the x-axis and the z-axis are orthogonal subspaces since if $u = \begin{bmatrix} x \\ 0 \\ 0 \end{bmatrix}$ is on the x-axis and $w = \begin{bmatrix} 0 \\ 0 \\ z \end{bmatrix}$ is on the z-axis, then $u \cdot w = 0$. Let

$$U = \left\{ \begin{bmatrix} u_1 \\ u_2 \\ u_3 \\ u_4 \end{bmatrix} \in \mathbf{R}^4 \mid 2u_1 - u_2 + u_3 - u_4 = 0 \right\} \text{ and}$$

$$W = \left\{ c \begin{bmatrix} 2 \\ -1 \\ 1 \\ -1 \end{bmatrix} \mid c \in \mathbf{R} \right\}.$$

Then U and W are orthogonal since if $u = \begin{bmatrix} u_1 \\ u_2 \\ u_3 \\ u_4 \end{bmatrix} \in U$ and $w = c \begin{bmatrix} 2 \\ -1 \\ 1 \\ -1 \end{bmatrix} \in W$, then $u \cdot w = c(2u_1 - u_2 + u_3 - u_4) = 0$. *p. 376*

Orthogonally diagonalizable: See *Diagonalizable*.

Orthonormal: See *Orthogonal*.

Pairwise orthogonal: See *Orthogonal*.

Parallel: Vectors u and v are parallel if one is a scalar multiple of the other, that is, if $u = cv$ or $v = cu$ for some scalar c. For example, $u = \begin{bmatrix} -2 \\ 1 \end{bmatrix}$ and $v = \begin{bmatrix} 6 \\ -3 \end{bmatrix}$ are parallel because $v = -3u$. Any vector u is parallel to the zero vector because $u = 15(0)$. *p. 3*

Pivot: A pivot of a row echelon matrix is the first nonzero entry in a nonzero row and a pivot column is a column containing a pivot. For example, the pivots of $\begin{bmatrix} 0 & -3 & 1 & 2 & 1 \\ 0 & 0 & 2 & 0 & -3 \\ 0 & 0 & 0 & 0 & 8 \end{bmatrix}$ are -3, 2, and 8 and the pivot columns are two, three, and five. If U is a row echelon form of a matrix A, the pivot columns of A are the pivot columns of U. If U was obtained without multiplying any row by a scalar, then the pivots of A are the pivots of U. For example,

$$A = \begin{bmatrix} -2 & 3 & 4 \\ 1 & 0 & 1 \\ 4 & -5 & 6 \end{bmatrix}$$

$$\rightarrow \begin{bmatrix} 1 & 0 & 1 \\ 0 & 3 & 6 \\ 0 & -5 & 2 \end{bmatrix} \rightarrow \begin{bmatrix} 1 & 0 & 1 \\ 0 & 3 & 6 \\ 0 & 0 & 12 \end{bmatrix} = U$$

so the pivots of A are 1, 3, and 12, and the pivot columns of A are columns one, two, and three.

Projection: The projection of a vector $\mathbf{u}$ on a (nonzero) vector $\mathbf{v}$ is the vector $\mathbf{p} = \text{proj}_\mathbf{v}\,\mathbf{u}$ parallel to $\mathbf{v}$ with the property that $\mathbf{u} - \mathbf{p}$ is orthogonal to $\mathbf{v}$. One can show that $\text{proj}_\mathbf{v}\,\mathbf{u} = \frac{\mathbf{u}\cdot\mathbf{v}}{\mathbf{v}\cdot\mathbf{v}}\,\mathbf{v}$. For example, if $\mathbf{u} = \begin{bmatrix} 1 \\ -3 \\ 4 \end{bmatrix}$ and

$\mathbf{v} = \begin{bmatrix} -3 \\ 1 \\ 2 \end{bmatrix}$, $\text{proj}_\mathbf{v}\,\mathbf{u} = \frac{1}{7}\mathbf{v} = \begin{bmatrix} -\frac{3}{7} \\ \frac{1}{7} \\ \frac{2}{7} \end{bmatrix}$.

The projection of a vector $\mathbf{w}$ onto a plane π is the vector $\mathbf{p} = \text{proj}_\pi\,\mathbf{w}$ in π with the property that $\mathbf{w} - \mathbf{p}$ is orthogonal to every vector in π. If the plane π is spanned by orthogonal vectors $\mathbf{e}$ and $\mathbf{f}$, one can show that $\text{proj}_\pi\,\mathbf{w} = \frac{\mathbf{w}\cdot\mathbf{e}}{\mathbf{e}\cdot\mathbf{e}}\,\mathbf{e} + \frac{\mathbf{w}\cdot\mathbf{f}}{\mathbf{f}\cdot\mathbf{f}}\,\mathbf{f}$. For example, the plane π with equation $3x - 2y + z = 0$ is spanned by the orthogonal vectors $\mathbf{e} = \begin{bmatrix} 5 \\ 6 \\ -3 \end{bmatrix}$ and $\mathbf{f} = \begin{bmatrix} 0 \\ 1 \\ 2 \end{bmatrix}$. If $\mathbf{w} = \begin{bmatrix} 5 \\ -3 \\ -7 \end{bmatrix}$,

$$\text{proj}_\pi\,\mathbf{w} = \frac{\mathbf{w}\cdot\mathbf{e}}{\mathbf{e}\cdot\mathbf{e}}\,\mathbf{e} + \frac{\mathbf{w}\cdot\mathbf{f}}{\mathbf{f}\cdot\mathbf{f}}\,\mathbf{f}$$

$$= \frac{28}{70}\begin{bmatrix} 5 \\ 6 \\ -3 \end{bmatrix} - \frac{17}{5}\begin{bmatrix} 0 \\ 1 \\ 2 \end{bmatrix}$$

$$= \frac{2}{5}\begin{bmatrix} 5 \\ 6 \\ -3 \end{bmatrix} - \frac{17}{5}\begin{bmatrix} 0 \\ 1 \\ 2 \end{bmatrix}$$

$$= \begin{bmatrix} 2 \\ -1 \\ -8 \end{bmatrix}.$$

In general, the projection of a vector $\mathbf{w}$ on a subspace U is the vector $\mathbf{p} = \text{proj}_U\,\mathbf{w}$ with the property that $\mathbf{w} - \mathbf{p}$ is orthogonal to U. If $\mathbf{f}_1, \mathbf{f}_2, \dots, \mathbf{f}_n$ is an orthogonal basis for U, then

$$\text{proj}_U\,\mathbf{w} = \frac{\mathbf{w}\cdot\mathbf{f}_1}{\mathbf{f}_1\cdot\mathbf{f}_1}\,\mathbf{f}_1 + \frac{\mathbf{w}\cdot\mathbf{f}_2}{\mathbf{f}_2\cdot\mathbf{f}_2}\,\mathbf{f}_2 + \cdots + \frac{\mathbf{w}\cdot\mathbf{f}_n}{\mathbf{f}_n\cdot\mathbf{f}_n}\,\mathbf{f}_n.$$

pp. 45, 48, 346

Projection matrix: A (necessarily square) matrix P is a projection matrix if and only if $\mathbf{b} - P\mathbf{b}$ is orthogonal to the column space of P for all vectors $\mathbf{b}$. For example, $P = \frac{1}{3}\begin{bmatrix} 2 & -1 & 1 \\ -1 & 2 & 1 \\ 1 & 1 & 2 \end{bmatrix}$ is a projection matrix. (This is perhaps most easily verified by noticing that P is symmetric and checking that $P^2 = P$. These two properties characterize projection matrices.) *p. 340*

Pseudoinverse: The pseudoinverse of a matrix A is the matrix A^+ satisfying

- $A^+\mathbf{b}$ is in the row space of A for all $\mathbf{b}$, and
- $A(A^+\mathbf{b}) = P\mathbf{b}$ is the projection of $\mathbf{b}$ on the column space of A.

Equivalently, A^+ is defined by the condition that $A^+\mathbf{b} = \mathbf{x}^+$ is the shortest vector amongst those minimizing $\|\mathbf{b} - A\mathbf{x}\|$.

For example, if $A = \begin{bmatrix} -1 & 0 & 0 \\ 0 & \frac{1}{2} & 0 \\ 0 & 0 & 0 \end{bmatrix}$, then $\mathbf{x}^+ = \begin{bmatrix} -b_1 \\ 2b_2 \\ 0 \end{bmatrix} =$

$\begin{bmatrix} -1 & 0 & 0 \\ 0 & 2 & 0 \\ 0 & 0 & 0 \end{bmatrix}\begin{bmatrix} b_1 \\ b_2 \\ b_3 \end{bmatrix}$, so $A^+ = \begin{bmatrix} -1 & 0 & 0 \\ 0 & 2 & 0 \\ 0 & 0 & 0 \end{bmatrix}$. *p. 392*

Quadratic form: A quadratic form in n variables $x_1, x_2, \dots, x_n$ is a linear combination of terms of the form $x_i x_j$; that is, a linear combination of $x_1^2, x_2^2, \dots, x_n^2$ and the cross terms such as $x_1 x_2$, $x_2 x_5$, and $x_4 x_9$. For example, $q = x^2 - y^2 + 4xy$ and

$q = 3y^2 - 2xy$ are quadratic forms in x and y, while $q = -4x_1^2 + x_2^2 + 4x_3^2 + 6x_1x_3$ is a quadratic form in x_1, x_2, and x_3. *p. 474*

Range: The range of a linear transformation $T : V \to W$ is the set $\{\mathbf{w} \in W \mid \mathbf{w} = T(\mathbf{v}) \text{ for some } \mathbf{v} \in V\}$.

For example, if $T : \mathbf{R}^2 \to \mathbf{R}^3$ is defined by $T\left(\begin{bmatrix} x \\ y \end{bmatrix}\right) = \begin{bmatrix} x - y \\ 2x + y \\ -x + 4y \end{bmatrix}$, then $\mathbf{b} = \begin{bmatrix} b_1 \\ b_2 \\ b_3 \end{bmatrix}$ is in range T if and only if $\mathbf{b} = T(\mathbf{x})$ for some $\mathbf{x} = \begin{bmatrix} x \\ y \end{bmatrix}$. This is the case if and only if there exist x and y so that

$$\begin{aligned} x - \ y &= b_1 \\ 2x + \ y &= b_2 \\ -x + 4y &= b_3 \end{aligned}$$

Solving this system, we find that $\mathbf{b} \in$ range T if and only if $(b_1 + b_2) - (b_2 - 2b_1) = 0$, that is, if and only if $3b_1 - b_2 + b_3 = 0$. Thus range $T = \{\begin{bmatrix} x \\ y \\ z \end{bmatrix} \in \mathbf{R}^3 \mid 3x - y + z = 0\}$. *p. 294*

Rank: The rank of a matrix is the number of pivot columns. For example, the rank of $A = \begin{bmatrix} 4 & 0 & 2 \\ 0 & 1 & 3 \end{bmatrix}$ is 2, while the rank of $A = \begin{bmatrix} 1 & 2 & 3 \\ 2 & 4 & 6 \end{bmatrix}$ is 1. *p. 224*

Reduced row echelon form: See *Row echelon form.*

Right inverse: A right inverse for an $m \times n$ matrix A is an $n \times m$ matrix B such that $AB = I_m$, the $m \times m$ identity matrix. For example, any matrix B of the form $B = \begin{bmatrix} t - \frac{5}{3} & s + \frac{2}{3} \\ \frac{4}{3} - 2t & -\frac{1}{3} - 2s \\ t & s \end{bmatrix}$ is a right inverse of $A = \begin{bmatrix} 1 & 2 & 3 \\ 4 & 5 & 6 \end{bmatrix}$ since $AB = I_2$. *p. 275*

Row echelon form: A matrix is in row echelon form if and only if it is upper triangular with the following properties:

i. all rows consisting entirely of 0s are at the bottom,

ii. the leading nonzero entries of the nonzero rows step from left to right as you read down the matrix, and

iii. every entry below a leading nonzero entry is a 0.

For example, $\begin{bmatrix} 7 & 1 & 3 & 4 & 5 \\ 0 & 0 & 5 & 3 & 1 \\ 0 & 0 & 0 & 0 & 1 \end{bmatrix}$ and $\begin{bmatrix} 0 & 0 & 1 & 1 & 5 \\ 0 & 0 & 0 & 0 & 1 \\ 0 & 0 & 0 & 0 & 0 \end{bmatrix}$ are in row echelon form, while $\begin{bmatrix} 1 & 0 & 3 & 2 \\ 0 & -1 & 2 & 4 \\ 0 & 1 & 0 & 1 \end{bmatrix}$ and $\begin{bmatrix} 1 & 0 & 3 & 2 \\ 0 & 0 & 1 & 4 \\ 0 & 1 & 0 & 2 \end{bmatrix}$ are not. A matrix is in reduced row echelon form if it is in row echelon form, each pivot (that is, each nonzero leading entry in a nonzero row) is a 1, and all other entries in a column containing a leading 1 are 0. For example, $\begin{bmatrix} 1 & 0 \\ 0 & 1 \end{bmatrix}$, $\begin{bmatrix} 1 & 0 & -2 \\ 0 & 1 & 1 \end{bmatrix}$ and $\begin{bmatrix} 1 & 0 & 0 & 0 \\ 0 & 0 & 1 & 0 \\ 0 & 0 & 0 & 1 \end{bmatrix}$ are all in reduced row echelon form. *pp. 106, 154*

Same direction: Vectors $\mathbf{u}$ and $\mathbf{v}$ have the same direction if $\mathbf{u} = c\mathbf{v}$ with $c > 0$. For example, $\mathbf{u} = \begin{bmatrix} -2 \\ 4 \\ 6 \end{bmatrix}$ and $\mathbf{v} = \begin{bmatrix} -1 \\ 2 \\ 3 \end{bmatrix}$ have the same direction because $\mathbf{v} = \frac{1}{2}\mathbf{u}$ with the scalar $\frac{1}{2} > 0$. *p. 4*

Similar: Matrices A and B are similar if there is an invertible matrix P such that $B = P^{-1}AP$. For example, $A = \begin{bmatrix} 1 & 2 \\ 3 & 4 \end{bmatrix}$ and $B = \begin{bmatrix} -55 & -97 \\ 34 & 60 \end{bmatrix}$ are similar since

$$B = \begin{bmatrix} -55 & -97 \\ 34 & 60 \end{bmatrix} = \begin{bmatrix} 2 & -5 \\ -1 & 3 \end{bmatrix} \begin{bmatrix} 1 & 2 \\ 3 & 4 \end{bmatrix}$$
$$\times \begin{bmatrix} 3 & 5 \\ 1 & 2 \end{bmatrix} = P^{-1}AP$$

with $P = \begin{bmatrix} 3 & 5 \\ 1 & 2 \end{bmatrix}$. *p. 204*

Singular matrix: A matrix is singular if it is not invertible, a condition equivalent to having determinant 0. For example, $A = \begin{bmatrix} 2 & 3 \\ 4 & 6 \end{bmatrix}$ is singular, because it has determinant 0. *p. 183*

Singular values: The singular values of a matrix A are the positive diagonal entries $\mu_1, \ldots, \mu_r$ of $D = \begin{bmatrix} \mu_1 & & \\ & \ddots & \\ & & \mu_r \end{bmatrix}$ when A has been factored $A = Q_1 \Sigma Q_2^T$

with Q_1 and Q_2 orthogonal matrices and $\Sigma = \begin{bmatrix} D & \mathbf{0} \\ \mathbf{0} & \mathbf{0} \end{bmatrix}$.

For example, if $A = \begin{bmatrix} 1 & 1 \\ 1 & 0 \\ 0 & 1 \end{bmatrix}$, $Q_1 = \begin{bmatrix} 0 & \frac{2}{\sqrt{6}} & -\frac{1}{\sqrt{3}} \\ \frac{1}{\sqrt{2}} & \frac{1}{\sqrt{6}} & \frac{1}{\sqrt{3}} \\ -\frac{1}{\sqrt{2}} & \frac{1}{\sqrt{6}} & \frac{1}{\sqrt{3}} \end{bmatrix}$,

and $Q_2 = \begin{bmatrix} \frac{1}{\sqrt{2}} & \frac{1}{\sqrt{2}} \\ -\frac{1}{\sqrt{2}} & \frac{1}{\sqrt{2}} \end{bmatrix}$, then Q_1 and Q_2 are orthogo-

nal matrices and $A = Q_1 \Sigma Q_2^T$ with $\Sigma = \begin{bmatrix} 1 & 0 \\ 0 & \sqrt{3} \\ 0 & 0 \end{bmatrix}$, so

the singular values of A are 1 and $\sqrt{3}$. *p. 450*

Size: See *Matrix*.

Span, spanned by: The span of vectors $\mathbf{u}_1, \mathbf{u}_2, \ldots, \mathbf{u}_n$ is the set of all linear combinations of them. It is denoted $\mathrm{sp}\{\mathbf{u}_1, \ldots, \mathbf{u}_n\}$. We say that the space $\mathrm{sp}\{\mathbf{u}_1, \ldots, \mathbf{u}_n\}$ is spanned by the vectors $\mathbf{u}_1, \ldots, \mathbf{u}_n$. For example, the span of $\mathbf{i}$ and $\mathbf{j}$ in $\mathbf{R}^2$ is all $\mathbf{R}^2$ since any vector $\mathbf{u} = \begin{bmatrix} a \\ b \end{bmatrix}$ in $\mathbf{R}^2$ is $a\mathbf{i} + b\mathbf{j}$. We say that $\mathbf{R}^2$ is spanned by $\mathbf{i}$ and $\mathbf{j}$. On the other hand, the span of $\mathbf{u} = \begin{bmatrix} -1 \\ 2 \end{bmatrix}$ and $\mathbf{v} = \begin{bmatrix} -3 \\ 6 \end{bmatrix}$ is the set of scalar multiples of $\mathbf{u}$ since $a\mathbf{u} + b\mathbf{v} = \begin{bmatrix} -a - 3b \\ 2a + 6b \end{bmatrix} = (a + 3b)\begin{bmatrix} -1 \\ 2 \end{bmatrix}$. The span of two nonparallel vectors is a plane. For example, the span of $\begin{bmatrix} 1 \\ 2 \end{bmatrix}$ and $\begin{bmatrix} -3 \\ 4 \end{bmatrix}$ is the xy-plane. Any plane through the origin in $\mathbf{R}^3$ is spanned by any two non-parallel vectors it contains. For example, the plane with equation $2x - y + 3z = 0$ is spanned by $\mathbf{u} = \begin{bmatrix} 1 \\ 2 \\ 0 \end{bmatrix}$ and $\mathbf{v} = \begin{bmatrix} 0 \\ 3 \\ 1 \end{bmatrix}$. *pp. 48, 240*

Standard basis vectors: The standard basis vectors in $\mathbf{R}^n$ are the n-dimensional vectors $\mathbf{e}_1, \mathbf{e}_2, \ldots, \mathbf{e}_n$ where $\mathbf{e}_i$ has ith component 1 and all other components 0; that is,

$$\mathbf{e}_1 = \begin{bmatrix} 1 \\ 0 \\ 0 \\ \vdots \\ 0 \end{bmatrix}, \quad \mathbf{e}_2 = \begin{bmatrix} 0 \\ 1 \\ 0 \\ \vdots \\ 0 \end{bmatrix}, \quad \mathbf{e}_3 = \begin{bmatrix} 0 \\ 0 \\ 1 \\ \vdots \\ 0 \end{bmatrix}, \quad \ldots, \quad \mathbf{e}_n = \begin{bmatrix} 0 \\ 0 \\ 0 \\ \vdots \\ 1 \end{bmatrix}.$$

In $\mathbf{R}^2$, the standard basis vectors $\mathbf{e}_1 = \begin{bmatrix} 1 \\ 0 \end{bmatrix}$ and $\mathbf{e}_2 = \begin{bmatrix} 0 \\ 1 \end{bmatrix}$ are also denoted $\mathbf{i}$ and $\mathbf{j}$, respectively, and, in $\mathbf{R}^3$, the standard basis vectors $\mathbf{e}_1 = \begin{bmatrix} 1 \\ 0 \\ 0 \end{bmatrix}$, $\mathbf{e}_2 = \begin{bmatrix} 0 \\ 1 \\ 0 \end{bmatrix}$ and $\mathbf{e}_3 = \begin{bmatrix} 0 \\ 0 \\ 1 \end{bmatrix}$ are also denoted $\mathbf{i}$, $\mathbf{j}$, and $\mathbf{k}$, respectively.

pp. 8, 56

Subspace: A subspace of a vector space V is a subset U of V that is itself a vector space under the operations of addition and scalar multiplication given for V. It is very useful to note that a nonempty subset U is a subspace if and only if it is

- closed under addition: If $\mathbf{u}$ and $\mathbf{v}$ are in U, so is $\mathbf{u} + \mathbf{v}$, and

- closed under scalar multiplication: If $\mathbf{u}$ is in U and c is a scalar, then $c\mathbf{u}$ is also in U.

For example, the set of all vectors of the form $\begin{bmatrix} x \\ y \\ 0 \end{bmatrix}$ is a subspace of $\mathbf{R}^3$ as is the plane with equation $x + y + 2z = 0$. The set of all polynomials in x with real coefficients and constant term 0 is a subspace of the vector space of all polynomials in x. *p. 235*

Sum of subspaces: The sum of subspaces U and W of $\mathbf{R}^n$ is the set $U + W = \{\mathbf{u} + \mathbf{w} \mid u \in U, w \in W\}$ of vectors in $\mathbf{R}^n$ that can be written as the sum of a vector in U and a vector in W. For example, if $U = \{\begin{bmatrix} c \\ 2c \end{bmatrix} \in \mathbf{R}^2\}$ and $W = \{\begin{bmatrix} 0 \\ c \end{bmatrix} \in \mathbf{R}^2\}$, then $U + W = \mathbf{R}^2$ since every $\mathbf{v} = \begin{bmatrix} x \\ y \end{bmatrix} \in \mathbf{R}^2$ can be written $\begin{bmatrix} x \\ y \end{bmatrix} = \begin{bmatrix} x \\ 2x \end{bmatrix} + \begin{bmatrix} 0 \\ y - 2x \end{bmatrix}$ as the sum of a vector in U and a vector in W. A sum $U + W$ is direct, and we write $U \oplus W$, if every vector in $U + W$ can be written in just one way in the form $u + w$. For example, the representation of $\mathbf{R}^2 = U + W$ just given is direct, since if $\mathbf{u}_1 = \begin{bmatrix} x_1 \\ 2x_1 \end{bmatrix}$, $\mathbf{w}_1 = \begin{bmatrix} 0 \\ y_1 - 2x_1 \end{bmatrix}$, $\mathbf{u}_2 = \begin{bmatrix} x_1 \\ 2x_1 \end{bmatrix}$, $\mathbf{w}_2 = \begin{bmatrix} 0 \\ y_1 - 2x_1 \end{bmatrix}$, and $\mathbf{u}_1 + \mathbf{w}_1 = \mathbf{u}_2 + \mathbf{w}_2$, then (it is easily seen that) $x_1 = x_2$ and $y_1 = y_2$, so

$u_1 = u_2$ and $w_1 = w_2$. On the other hand, let

$$U = \left\{ \begin{bmatrix} x \\ y \\ z \end{bmatrix} \mid x + y - z = 0 \right\} \text{ and}$$

$$W = \left\{ \begin{bmatrix} x \\ y \\ z \end{bmatrix} \mid 2x - 3y + z = 0 \right\}.$$

Then $\mathbf{R}^3 = U + W$ since any $\begin{bmatrix} x \\ y \\ z \end{bmatrix}$ in $\mathbf{R}^3$ can be written

$$\begin{bmatrix} x \\ y \\ z \end{bmatrix} = \begin{bmatrix} -y + \frac{1}{3}z \\ -x \\ -x - y + \frac{1}{3}z \end{bmatrix} + \begin{bmatrix} x + y - \frac{1}{3}z \\ x + y \\ x + y + \frac{2}{3}z \end{bmatrix}$$

as the sum of a vector in U and a vector in W. The sum is not direct, however, since (for example), $\begin{bmatrix} 3 \\ 0 \\ 0 \end{bmatrix}$ can be written in two ways as the sum of vectors in U and in W, namely, as $\begin{bmatrix} 0 \\ -3 \\ -3 \end{bmatrix} + \begin{bmatrix} 3 \\ 3 \\ 3 \end{bmatrix}$ and $\begin{bmatrix} 2 \\ 0 \\ 2 \end{bmatrix} + \begin{bmatrix} 1 \\ 0 \\ -2 \end{bmatrix}$. *p. 371*

Symmetric: A matrix A is symmetric if it equals its transpose: $A = A^T$. For example, $\begin{bmatrix} 1 & 2 \\ 2 & 4 \end{bmatrix}$ is a symmetric matrix, as is $\begin{bmatrix} 1 & 2 & 3 \\ 2 & 7 & -8 \\ 3 & -8 & 5 \end{bmatrix}$. *p. 147*

Transpose: The transpose of an $m \times n$ **real** matrix A is the $n \times m$ matrix whose rows are the columns of A in the same order. The transpose of A is denoted A^T. For example, if $A = \begin{bmatrix} 1 & 2 & 3 \\ 4 & 5 & 6 \\ 7 & 8 & 9 \end{bmatrix}$, then $A^T = \begin{bmatrix} 1 & 4 & 7 \\ 2 & 5 & 8 \\ 3 & 6 & 9 \end{bmatrix}$ and if $A = \begin{bmatrix} -1 & 2 \\ 3 & 4 \\ 0 & -5 \end{bmatrix}$, then $A^T = \begin{bmatrix} -1 & 3 & 0 \\ 2 & 4 & -5 \end{bmatrix}$. The conjugate transpose of a complex matrix A, denoted A^*, is the transpose of the matrix whose entries are the conjugates of those of A: $A^* = \overline{A}^T = \overline{A^T}$. For example, if $A = \begin{bmatrix} 1+i & 3-2i \\ 2-4i & i \\ -5 & 5+7i \end{bmatrix}$, then $A^* = \begin{bmatrix} 1-i & 2+4i & -5 \\ 3+2i & -i & 5-7i \end{bmatrix}$. *pp. 71, 415*

Triangular: A matrix is upper triangular if all entries below the main diagonal are 0 and lower triangular if all entries above the main diagonal are 0. For example, $U = \begin{bmatrix} 1 & -1 & 2 \\ 0 & 1 & 1 \end{bmatrix}$ is upper triangular, while $L = \begin{bmatrix} -1 & 0 \\ 2 & 3 \end{bmatrix}$ is lower triangular. A matrix is triangular if it is either upper triangular or lower triangular. *p. 128*

Unit vector: A unit vector is a vector of norm 1. For example, the vector $\begin{bmatrix} \frac{1}{\sqrt{2}} \\ \frac{1}{\sqrt{2}} \end{bmatrix}$ is a unit two-dimensional vector and $\begin{bmatrix} \frac{1}{\sqrt{6}} \\ -\frac{2}{\sqrt{6}} \\ \frac{1}{\sqrt{6}} \end{bmatrix}$ is a unit three-dimensional vector. *p.17*

Unitarily diagonalizable: See *Diagonalizable*.

Unitary matrix: A (complex) matrix is unitary if it is square and its columns are orthonormal. For example, $U = \begin{bmatrix} \frac{1}{\sqrt{2}} & \frac{1+i}{2} \\ \frac{i}{\sqrt{2}} & \frac{1-i}{2} \end{bmatrix}$ is unitary. *p. 420*

Vector: An n-dimensional vector ($n \geq 1$) is a column of n numbers enclosed in brackets. The numbers are called the components of the vector. For example, $\begin{bmatrix} 1 \\ 2 \end{bmatrix}$ is a two-dimensional vector with components 1 and 2, and $\begin{bmatrix} -1 \\ 2 \\ 2 \\ 5 \end{bmatrix}$ is a four-dimensional vector with components -1, 2, 2, and 5. *pp. 1, 9, 54*

Vector space: A vector space is a nonempty set V of objects called vectors together with two operations called addition and scalar multiplication. For any two vectors $\mathbf{u}$ and $\mathbf{v}$ in V, there is another vector $\mathbf{u} + \mathbf{v}$ in V called the sum of $\mathbf{u}$ and $\mathbf{v}$. For any vector $\mathbf{u}$ and any scalar c, there is a vector $c\mathbf{u}$ called a scalar multiple of $\mathbf{v}$. Addition and scalar multiplication have the following properties:

1. (Closure under addition) If $\mathbf{u}$ and $\mathbf{v}$ are in V, so is $\mathbf{u} + \mathbf{v}$.

2. (Commutativity of addition) $\mathbf{u} + \mathbf{v} = \mathbf{v} + \mathbf{u}$ for all $\mathbf{u}, \mathbf{v} \in V$.

3. (Associativity of addition) $\mathbf{u} + (\mathbf{v} + \mathbf{w}) = (\mathbf{u} + \mathbf{v}) + \mathbf{w}$ for all $\mathbf{u}, \mathbf{v}, \mathbf{w} \in V$.

4. (Zero) There is an element **0** in V with the property that $\mathbf{0} + \mathbf{v} = \mathbf{v}$ for any $\mathbf{v} \in V$.

5. (Negatives) For each $\mathbf{v} \in V$, there is another vector $-\mathbf{v}$ (called the negative of $\mathbf{v}$) with the property that $\mathbf{v} + (-\mathbf{v}) = \mathbf{0}$.

6. (Closure under scalar multiplication) If $\mathbf{v} \in V$ and c is a scalar, then $c\mathbf{v}$ is in V.

7. (Scalar associativity) $c(d\mathbf{v}) = (cd)\mathbf{v}$ for all $\mathbf{v} \in V$ and all scalars c, d.

8. (One) $1\mathbf{v} = \mathbf{v}$ for any $\mathbf{v} \in V$.

9. (Distributivity)

   - $c(\mathbf{v} + \mathbf{w}) = c\mathbf{v} + c\mathbf{w}$ for all $\mathbf{v}, \mathbf{w} \in V$ and any scalar c.
   - $(c+d)\mathbf{v} = c\mathbf{v} + d\mathbf{v}$ for all $\mathbf{v} \in V$ and all scalars c, d.

The most important vector space in this book is $\mathbf{R}^n$, Euclidean n-space, with addition and scalar multiplication defined componentwise. The set of all polynomials in x with real coefficients is an example of an infinite dimensional vector space. *p. 233*

Index